Topological Recursion and its Influence in Analysis, Geometry, and Topology

Proceedings of Symposia in
PURE MATHEMATICS

Volume 100

Topological Recursion and its Influence in Analysis, Geometry, and Topology

2016 AMS von Neumann Symposium
Topological Recursion and its Influence
in Analysis, Geometry, and Topology
July 4–8, 2016
Charlotte, North Carolina

Chiu-Chu Melissa Liu
Motohico Mulase
Editors

AMS
AMERICAN
MATHEMATICAL
SOCIETY
Providence, Rhode Island

2010 *Mathematics Subject Classification.* Primary 14H10, 14H15, 14H60, 14H60, 14J33, 14N10, 14N35, 53D45, 81T45.

The AMS von Neumann Symposium was supported by a grant from the National Science Foundation, NSF DMS-1619760.

Library of Congress Cataloging-in-Publication Data

Names: AMS von Neumann Symposium (2016: Charlotte, North Carolina) | Liu, Chiu-Chu Melissa, 1974– editor. | Mulase, Motohico, 1954– editor.
Title: Topological recursion and its influence in analysis, geometry, and topology: 2016 AMS von Neumann Symposium, July 4–8, 2016, Charlotte, North Carolina / Chiu-Chu Melissa Liu, Motohico Mulase, editors.
Description: Providence, Rhode Island: American Mathematical Society, [2018] | Series: Proceedings of symposia in pure mathematics; volume 100 | Includes bibliographical references.
Identifiers: LCCN 2018033790 | ISBN 9781470435417 (alk. paper)
Subjects: LCSH: Topology–Congresses.
Classification: LCC QA611.A1 A47 2018 | DDC 516.3/5–dc23
LC record available at https://lccn.loc.gov/2018033790
DOI: http://dx.doi.org/10.1090/pspum/100

♾ The paper used in this book is acid-free and falls within the guidelines
established to ensure permanence and durability.
Visit the AMS home page at `https://www.ams.org/`

10 9 8 7 6 5 4 3 2 1   23 22 21 20 19 18

Contents

Preface

This volume is intended for presenting a snapshot of rapid and rich development in an emerging research field, *topological recursion*. As we review below, topological recursion was conceived in 2003–2004, appearing completely independently in two disparate contexts: One is random matrix theory/matrix models, where the recursion structure was discovered in calculating multi-resolvent correlation functions of random matrices [**3, 13, 34**]. The other is in geometry. Working on her Harvard thesis, Mirzakhani encountered the same recursive structure in calculating the Weil-Petersson volume of moduli spaces of bordered hyperbolic surfaces. Her thesis of 2004 was later published as [**71, 72**]. Although the shape of formulas of Eynard [**34**] and Mirzakhani [**71**] are similar, they are not exactly the same. Surprisingly, they become *identical* after the Laplace transform.

Once the mathematics behind the scene was understood, topological recursion began to play an unexpected role in connecting a wide area of mathematics and theoretical physics through its universal recursive structures. The exact same formulas are found in such topics, not only as random matrix theory/matrix models and the Weil-Petersson volume of moduli spaces as mentioned above but also in topological quantum field theory and cohomological field theory, enumeration of various Hurwitz numbers, counting Grothendieck's dessins d'enfants and more general graphs drawn on surfaces, intersection numbers of cohomology classes on the moduli space $\overline{\mathcal{M}}_{g,n}$ of stable curves, Gromov-Witten theory, A-polynomials and polynomial invariants of hyperbolic knots and algebro-geometric invariants of algebraic knots, WKB analysis of classical ordinary differential equations, Painlevé equations, moduli spaces of Higgs bundles and character varieties of surface groups, and *opers*.

1. The 2016 AMS von Neumann Symposium

Several experts of topological recursion convened at the 2016 American Mathematical Society von Neumann Symposium, *The topological recursion and its influence in analysis, geometry, and topology*, held in Charlotte, North Carolina, during the week of July 4–8, 2016. The conference was organized by Bertrand Eynard, Chiu-Chu Melissa Liu, and Motohico Mulase serving as chair of the organizing committee. The meeting was planned right at the time when many discoveries and crucial theorems were established, and at the same time, numerous new mysteries were arising.

The symposium featured four mini-courses to illuminate the state of the art of topological recursion, from its origin to the most significant achievements and to a glimpse of future developments. The speakers and titles are as follows:

- Jørgen Ellegaard Andersen: *Geometric quantisation of moduli spaces and topological recursion.*
- Petr Dunin-Barkowski: *Topological recursion and Givental's formalism: spectral curves for Gromov-Witten theories.*
- Bertrand Eynard: *An introduction to topological recursion.*
- Bohan Fang: *Topological recursion and mirror symmetry.*

We also invited many key players of the field to present the most recent developments as plenary lecturers:

- Gaëtan Borot: *2d CFTs, Verlinde formula and indices via topological recursion.*
- Tom Bridgeland: *Quadratic differentials and wall-crossing.*
- Leonid Chekhov: *Topological recursion for classical ensembles and cohomological field theories.*
- Alessandro Chiodo: *Generalized Borcea-Voisin mirror duality in any dimension.*
- Laura Fredrickson: *From the Hitchin component to opers.*
- Kohei Iwaki: *Exact WKB analysis, Painlevé equations and the Stokes phenomenon.*
- Felix Janda: *Frobenius manifolds near the discriminant.*
- Rinat Kashaev: *Invariants of finite cyclic covers of knot complements from Teichmüller TQFT.*
- Paul Norbury: *Primary invariants of Frobenius manifolds and periods on Riemann surfaces.*
- Nicolas Orantin: *Topological recursion, twisted periods and almost Frobenius manifolds.*
- Masa-Hiko Saito: *An explicit geometry of moduli spaces of Higgs bundles and singular connections on a smooth curve and differential equations of Painlevé types.*
- Yan Soibelman: (1) *Holomorphic Floer theory and Riemann-Hilbert correspondence.* (2) *Airy structures and topological recursion.*
- Piotr Sułkowski: *Matrix models, quantum curves, and conformal field theory.*
- Ravi Vakil: *Topological recursions in the Grothendieck ring of varieties.*

Several contributed short talks were also given:

- Bojko Bakalov: *Vertex operators in Gromov-Witten theory.*
- Vincent Bouchard: *An elliptic quantum curve.*
- Andrea Brini: *Quantum invariants at large N and the topological recursion.*
- Qingtao Chen: *Congruent relations, cyclotomic expansions and volume conjectures for various quantum invariants.*
- João N. Esteves: *Hopf algebras and topological recursion.*
- Taro Kimura: *Double quantization of Seiberg-Witten curve and W-algebras.*
- Dmitry Korotkin: *Periods of meromorphic quadratic differentials and Goldman bracket.*
- Karol Kozlowski: *Partition function of the* sinh*-model with varying interactions.*
- Danilo Leo Lewanski: *Ramifications of Hurwitz theory, KP integrability and quantum curves.*

- Olivier Marchal: *A lonely runner problem, asymptotics of Toeplitz determinants and topological recursion.*
- Ran Tessler: *Intersection theory on moduli of bordered Riemann surfaces, and related integrable systems.*
- Zhengyu Zong: *Some applications of the Remodeling Conjecture.*

After the conclusion of the symposium, the invited speakers and participants, including those originally invited but could not participate due to unanticipated circumstances, were invited to submit their papers, which are related to the theme of the conference for publication in this volume. The submitted papers were then sent to referees for peer review. We are pleased to present the papers recommended for publication in this volume.

2. What is topological recursion?

Since Eynard's excellent survey of topological recursion based on his invited address at the ICM-2014 in Seoul is available [**38**], we refer to it for a discoverer's own description of the subject. In particular, his paper presents how it was encountered in random matrix theory and how it is appearing in various problems in geometry, and also cites many original contributions to topological recursion by researchers in a wide variety of comminities. Therefore, in this preface we restrict ourselves to pure mathematical side of the story relevant to the papers collected in this volume, and we refer readers who wish to know other sides of the story, such as mathematical and theoretical physics points of view, to [**38, 40**].

Sir George Biddel Airy [**2**] devised a function

$$Ai(x) := \frac{1}{2\pi}\int_{-\infty}^{\infty} \exp\left(ipx + i\frac{p^3}{3}\right) dp,$$

which he called the *rainbow integral*, to explain the diffraction pattern of the rainbow. Since the diffraction phenomena are a quantum mechanical process, his method based on classical analysis does not explain the physics of rainbows. Phenomenologically, still the location of the local maxima and minima of the Airy function on the $x < 0$ side tells us where we see the sequence of arches of the rainbow, and their brightness. The asymptotic formula

$$Ai(x) \sim \frac{1}{\sqrt{\pi}x^{\frac{1}{4}}} \exp\left(-\frac{2}{3}x^{\frac{3}{2}}\right), \qquad |\mathrm{Arg}(x)| < \frac{\pi}{3} \tag{2.1}$$

tells us that the Airy function exponentially decays on the right side $x > 0$. Therefore, underneath the brightest arch there are no rainbows.

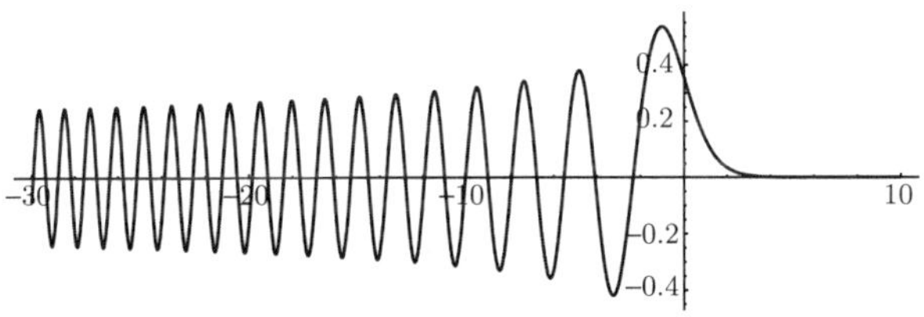

FIGURE 2.1. The Airy function

The above asymptotic formula (2.1) has a refined expansion in terms of the intersection numbers of $\overline{\mathcal{M}}_{g,n}$:

$$(2.2)\quad Ai(x) = \exp\left(-\frac{2}{3}x^{\frac{3}{2}} - \frac{1}{4}\log x - \frac{1}{2}\log\pi\right)$$

$$\cdot\exp\left(\sum_{2g-2+n>0} \frac{(-1)^n}{n!}\left(\frac{1}{2}\right)^{2g-2+n} x^{-\frac{(6g-6+3n)}{2}} \sum_{\substack{d_1+\cdots+d_n\\=3g-3+n}} \langle\tau_{d_1}\cdots\tau_{d_n}\rangle_{g,n}\prod_{i=1}^{n}|2d_i-1|!!\right)$$

$$= \exp\left(-\frac{2}{3}x^{\frac{3}{2}} - \frac{1}{4}\log x - \frac{1}{2}\log\pi - \frac{5}{48}x^{-\frac{3}{2}} + \frac{5}{64}x^{-3} + \cdots\right),$$

where the coefficients

$$(2.3)\qquad \langle\tau_{d_1}\cdots\tau_{d_n}\rangle_{g,n} = \int_{\overline{\mathcal{M}}_{g,n}} \psi_1^{d_1}\cdots\psi_n^{d_n}$$

are the cotangent class intersection numbers on the moduli space $\overline{\mathcal{M}}_{g,n}$ of stable curves of genus g with n non-singular marked points. Let us introduce an $\hbar$-dependence by defining

$$Ai(x,\hbar) := Ai\left(\frac{x}{\hbar^{2/3}}\right).$$

Then this new function satisfies a stationary *Schrödinger equation*

$$(2.4)\qquad \left(\left(\hbar\frac{d}{dx}\right)^2 - x\right)Ai(x,\hbar) = 0.$$

This equation is referred to as a *quantum curve* in many articles in this volume. The *semi-classical limit* of this quantum curve is a quadratic equation, called a *spectral curve*:

$$(2.5)\qquad y^2 - x = 0.$$

Topological recursion is a powerful induction process to calculate every single intersection number $\langle\tau_{d_1}\cdots\tau_{d_n}\rangle_{g,n}$ from the spectral curve (2.5). For this particular case, the recursion is the same as the Virasoro constraint condition of these intersection numbers predicted by Dijkgraaf, Verlinde, and Verlinde [**20**]. We also note that the differential operator of (2.4) at $\hbar = 1$ is the initial value of the Lax operator which contains information of all cotangent class intersection numbers. That this Lax operator satisfies the KdV equation was conjectured by Witten [**87**] and proved by Kontsevich [**64**].

3. The origin

Topological recursion, as we call today, was first devised by Chekhov, Eynard, and Orantin in 2006 [**13**] as a *universal* analytic formula to calculate the asymptotic expansions of the *multi-resolvent correlation functions* of 2-matrix models. The derivation of this formula is very simple, which is the idea of Virasoro constraint condition [**20**] mentioned above. A definite integral, such as a convergent Hermitian matrix integral, is invariant under a change of integration parameters. From an infinitesimal change of variables, one obtains a series of equations as a result of invariance of the integral. This technique is known as *loop equations* in physics. In random matrix theory, the main concern has been the asymptotic behavior of correlation functions as the size of the matrix N tends to infinity. Identities coming from loop equations produce an infinite series of equations in terms of power series in $1/N$ for an arbitrary N as a variable. The degree of powers of $1/N$ in the series expansion takes the form $2g-2+n$, where n is the number of resolvents in the n-point correlation function, and g is the genus parameter of the loop expansion. The formula provides an effective method for calculating correlation functions of the resolvent of random matrices. The new observation here is that the coefficients of the $1/N$-expansion can be calculated in a closed formula in terms of complex analysis on the Riemann surface, which appears as the maximum domain of holomorphy of the resolvent. This Riemann surface is commonly called the spectral curve of the starting matrix integral.

A year later in 2007, Eynard and Orantin [**40**] changed the point of view 180°. They propose the induction formula as an axiom to compute a series of *invariants* from a given algebraic equation in 2-variables, such as (2.5), that determines the spectral curve of the theory. It was noticed that many matrix models appearing in topological string theory can be solved, in a perturbative way, by topological recursion [**19**, **83**].

QUESTION 3.1. *What does the abstract topological recursion calculate when it is not related to matrix models? What is the geometric meaning of the invariants?*

In geometry, topological recursion has a very different origin. It first appeared in 2004 as an integral recursion formula discovered by Mirzakhani [**71**, **72**] in her work on the Weil-Petersson volume of the moduli space of bordered hyperbolic surfaces. Eynard [**35**] noticed that the *Laplace transform* of the Mirzakhani recursion is an example of topological recursion of [**40**]. The spectral curve for this case is the sine curve that was discovered by Mulase and Safnuk [**74**] as the intertwining operator between two Virasoro representations, one for the intersection numbers (2.3) of [**20**], and the other for the Weil-Petersson volumes. Mirzakhani recursion is based on pair-of-pants decompositions of bordered topological surfaces. Her formula for the volume of moduli spaces satisfies an integral recursion equation representing removal of a pair-of-pants in all possible ways. Its Laplace transform, as a complex analytic equation, leads to a topological recursion formulated on the sine curve.

The Laplace transform turns out to be the key to understanding the mirror symmetry behind the scene. In 2009, Eynard, Mulase, Safnuk, and Zhang [**39**, **77**] solved a conjecture for simple Hurwitz numbers predicted by the physicists Bouchard and Mariño [**11**]. The key discovery of [**39**] is that the conjectural topological recursion for Hurwitz numbers is a consequence of the Laplace transform of a combinatorial equation known as the *cut-and-join equation* of simple Hurwitz numbers [**55**, **56**, **84**] (see also [**46**]).

The Laplace transform appearing in these examples is then identified as mirror symmetry, connecting the A-model side of enumerative geometry with a B-model of holomorphic geometry. Evidence for the importance of topological recursion is seen in the work of [**77**]. It is shown that both Witten's conjecture [**87**] and the λ_g conjecture (first proved in [**42**]) are straightforward consequences of the formula established in [**77**], which is the differential equation obtained via the Laplace transform of the cut-and-join equation. The same method is used for proving the λ_g-conjecture for orbifold Hurwitz numbers in [**69**].

Around the same time, Norbury [**79**, **80**] showed that the lattice point counting problem of the combinatorial moduli space $\mathcal{M}_{g,n}^{\text{comb}} = \mathcal{M}_{g,n} \times \mathbb{R}_+^n$ can be solved via topological recursion. Following his lead, numerous examples of topological recursion were found in various graph enumeration problems [**12, 22, 27, 73, 82**]. The example coming out of Catalan numbers [**27**] is particularly simple and elegant. In all of these examples, topological recursion is a consequence of the Laplace transform of edge-contraction operations appearing in graph enumeration. In 2013, Dumitrescu and Mulase noticed that this operation itself is exactly the same as the multiplication and comultiplication operations in a Frobenius algebra, and hence it is equivalent to a two-dimensional topological quantum field theory (2D TQFT) [**25**]. A 2D TQFT is the degree 0 part of a cohomological field theory (CohFT) of Kontsevich and Manin [**65**]. Therefore, it suggests a relation between topological recursion and CohFT.

Double Hurwitz numbers exhibits a rather different nature compared to simple Hurwitz numbers. A topological recursion for a restricted case of orbifold Hurwitz numbers is established in [**8, 21**]. In the paper by Do and Karev in this volume, a conjecture regarding double Hurwitz numbers in a more general context is proposed with abundant evidence.

4. Topological recursion and semi-simple cohomological field theories

The genus zero Gromov-Witten theory of a projective manifold (or, more generally, a compact symplectic manifold) defines a Frobenius manifold, to which the associated 2D TQFT is the degree zero part of the CohFT defined by the Gromov-Witten theory of the projective manifold. The genus zero Gromov-Witten theory of a target with generically semi-simple quantum cohomology defines a semi-simple Frobenius manifold. For any semi-simple Frobenius manifold, Givental [**54**] conjectured a formula of higher genus potentials in terms of functions on the semi-simple Frobenius manifold and the Kontsevich-Witten τ-function (which is equal to the all-genus total descendant Gromov-Witten potential of a point); the formula implies the Virasoro constraint condition. This is known as Givental's higher genus reconstruction conjecture, and has been proved in full generality by Teleman [**86**].

The topological recursion determines higher genus invariants $\omega_{g,n}$ recursively from genus zero invariants $\omega_{0,1}$ and $\omega_{0,2}$. It is natural to speculate a relation between the topological recursion and Givental's formula for higher genus invariants. Dunin-Barkowski, Orantin, Shadrin, and Spitz showed that any semi-simple CohFT, and in particular Gromov-Witten theory of a target with semi-simple quantum cohomology, can be reproduced by a *local* version of topological recursion [**31**]. This volume contains a thorough treatment of the equivalence between topological recursion and Givental's formula based on Dunin-Barkowski's mini-course on the results in [**30, 31**]. Conversely, in their contribution to this volume, Dunin-Barkowski, Norbury, Oranin, Popolitov, and Shadrin show that the topological recursion on a *global* compact spectral curve computes invariants of the CohFT associated with Dubrovin's Hurwitz Frobenius manifold.

5. Topological recursion and toric mirror symmetry

As an application of their main result, the authors of [**31**] proved the Norbury-Scott conjecture [**82**], which provides a precise correspondence between stationary Gromov-Witten invariants of the complex projective line $\mathbb{P}^1$ and invariants obtained by applying topological recursion to the spectral curve $x = z + z^{-1}$, $y = \log z$. This proof is also explained in Dunin-Barkowski's contribution to this volume. It is interesting to observe that the holomorphic Morse function $W : \mathbb{C}^* \to \mathbb{C}$ given by $z \mapsto z + z^{-1}$ is the superpotential of the Landau-Ginzburg mirror of $\mathbb{P}^1$ at the stationary point. Therefore, the Norbury-Scott conjecture can be viewed as a version of all-genus mirror symmetry. Similar correspondence is expected to hold for higher dimensional compact toric manifolds such as complex projective spaces. An equivariant version of the Norbury-Scott conjecture, which can be interpreted as an equivariant all-genus mirror symmetry for $\mathbb{P}^1$, is formulated and proved by Fang, Liu, and Zong in [**47**] using the main result in [**31**] and a formula for the total descendant potential of equivariant Gromov-Witten theory of toric manifolds derived by Givental using torus localization [**54**]. From this result, the Bouchard-Mariño conjecture (cf. Section 3 above) and the Norbury-Scott conjecture can be recovered by taking the large radius limit and the non-equivariant limit, respectively.

Topological recursion also arises in higher genus mirror symmetry for toric Calabi-Yau 3-folds. Mirror symmetry relates the A-model on a Calabi-Yau manifold X, defined in terms of the symplectic structure on X, to the B-model on its mirror Calabi-Yau manifold $\check{X}$, defined in terms of the complex structure on $\check{X}$. Gromov-Witten theory can be viewed as a mathematical theory of A-model topological string. Gromov-Witten invariants are defined for any projective manifolds at all genera; for toric Calabi-Yau 3-folds, which are non-compact, we consider equivariant Gromov-Witten invariants defined by torus action. On the B-model side, genus zero (resp., genus one) B-model topological string is defined in terms of period integrals (resp., Ray-Singer analytic torsion), but a mathematical theory of genus $g \geq 2$ B-model on a general Calabi-Yau manifold has not been established. In 2007-08, Mariño [**70**], and then in collaboration with Bouchard, Klemm, and Pasquetti [**9, 10**], proposed a mathematically defined, *remodeled* all-genus B-model on the mirror of a symplectic toric Calabi-Yau 3-manifold/3-orbifold X, in both closed and open string sectors, based on topological recursion on the mirror curve of X. (Indeed, the original topological recursion [**40**], defined for spectral curves in $\mathbb{C}^2$, needs to be suitably modified since the mirror curve is an affine curve in $(\mathbb{C}^*)^2$ instead of $\mathbb{C}^2$.) The *remodeling conjecture* of Bouchard, Klemm, Mariño, and Pasquetti provides a precise correspondence between closed and open Gromov-Witten invariants of a symplectic toric 3-manifold/3-orbifold X and the invariants $\omega_{g,n}$ obtained by applying topological recursion to the mirror curve of X. Here, open Gromov-Witten invariants (corresponding to $\omega_{g,n}$ with $n > 0$) count parametrized holomorphic curves in X with boundaries in Aganagic-Vafa Lagrangian branes, and may also be interpreted as certain relative Gromov-Witten invariants.

In 2009, Lin Chen proved the remodeling conjecture for $\mathbb{C}^3$ in the open string sector based on the method of [**39**]. Because of the tragic death of the author (in December 2009) shortly after the completion of the paper (in October 2009), Chen's paper has never been submitted to a journal. We publish Chen's paper in this volume based on strong recommendation by the referee. (An alternative proof of the same result can be found in [**88**].) The closed string sector of the remodeling conjecture for $\mathbb{C}^3$ was proved independently by Bouchard, Catuneanu, Marchal, and Sułkowski [**7**], and by Zhu [**91**].

In 2011, Eynard derived graph sum formulas for the unique solution to the topological recursion on an arbitrary spectral curve [**36, 37**], which are key ingredients in the identification of topological recursion and Givental formalism due to

[**31**] (cf. Section 4 above), and also in the work on the remodeling conjecture for general smooth symplectic toric Calabi-Yau 3-folds by Eynard and Orantin [**41**] of 2012.

Fang, Liu, and Zong proved the remodeling conjecture for affine toric Calabi-Yau 3-orbifolds $[\mathbb{C}^3/G]$ in the open string sector [**44, 45**], and finally for general symplectic toric Calabi-Yau 3-orbifolds (including smooth symplectic toric Calabi-Yau 3-folds) in both open and closed string sectors in [**47**]. Key ingredients of the proof include orbifold version of Givental's formula for higher genus equivariant Gromov-Witten invariants [**92**] (specialized to toric Calabi-Yau 3-orbifolds) and the graph sum formula for invariants defined by topological recursion [**31, 36**] (applied to mirror curves). This volume contains an expository article by Fang and Zong on this proof, which is based on the mini-course by Fang and the contributed talk of Zong at the symposium.

6. A knot theory twist and quantum curves

In topology, it has been known since 1994 [**14**] that the $SL(2,\mathbb{C})$-character variety of the fundamental group of the complement of a knot in S^3 determines a plane algebraic curve defined over $\mathbb{Z}$, and that its defining equation, called the *A-polynomial*, is an invariant of the knot. The plane $\mathbb{C}^2$ here is an extension of $\mathbb{C}^* \times \mathbb{C}^*$, which is the $SL(2,\mathbb{C})$-character variety of the fundamental group of the torus T^2 that appears as the boundary of a tubular neighborhood of the knot complement in S^3. By choosing the meridian and longitude of the torus as a basis for $H_1(T^2,\mathbb{Z}) = \pi_1(T^2)$, the plane acquires a canonical coordinate (x, y), and every A-polynomial is a $\mathbb{Z}$-coefficient polynomial in these variables. Based on an earlier work [**50**], Garoufalidis [**52**] conjectured in 2004 that the A-polynomial of a hyperbolic knot should be *quantized* into a *difference operator*, which should then characterize the *colored Jones polynomial* of the knot as a solution to this difference equation. This conjecture is known as the AJ-conjecture, since the quantization of the A-polynomial annihilates the colored Jones polynomial. Around the same time, physicist Gukov [**57**] arrived at the same conjecture independently, and noticed that this was closely related to the notion of *quantum curve* of Aganagic, Dijkgraaf, Klemm, Mariño, and Vafa [**1**]. Physical theory of quantum curves has been further developed by Dijkgraaf, Hollands, Sułkowski, and Vafa [**17, 18, 59**].

Then in 2012, Gukov and Sułkowski [**58**] discovered a conjectural perturbative mechanism to construct the quantization of an A-polynomial so that the AJ-conjecture holds, using a topological recursion on the character variety as its spectral curve. A possible relation between the hyperbolic volume conjecture of Kashaev [**63**] and topological recursion was first pointed out in [**16**]. The mechanism of Gukov and Sułkowski [**58**] is the same as the exact WKB analysis of a Schrödinger equation, and their discovery is a conjectural closed formula for each term of the WKB expansion in terms of the integral of the differential forms determined by topological recursion. An algebraic K-theory condition, that the Steinberg symbol $\{x, y\} \in K_2(F)$ of the K_2 of the function field F of the spectral curve being a 2-torsion element [**14**], is speculated to be the condition for the unique quantizability of the spectral curve in [**58**]. (See also [**6**].)

These developments from physics and topology around the idea of quantum curves energized the topological recursion community. Many rigorous examples of quantum curves have been constructed by Bouchard, Do, Dunin-Barkowski, X. Liu,

Manescu, Mulase, Norbury, Popolitov, Shadrin, Spitz, Sułkowski, Zhou, and others [**8, 22, 29, 75, 76, 81, 89, 90**]. The conjectural WKB-type analysis was first proved for two cases, the Catalan numbers and simple Hurwitz numbers, in [**76**]. Surprisingly, a number of these examples exhibit a deep connection to representation theory of symmetric groups and fermionic operator calculus. In these examples, if the spectral curve is *trigonometric*, i.e., if it is contained in $\mathbb{C}^* \times \mathbb{C}^*$, then the quantum curves are not derived from topological recursion. It is also surprising that there seems to be the unique, or preferred, quantization in each case. The mechanism of this uniqueness is not fully understood as of now. At least the torsion condition of the Steinberg symbol holds for all these examples.

Relations between exact WKB analysis, counting problems, Painlevé equations, and topological recursion has been actively investigated in recent papers [**15, 60, 61, 67**].

Toric mirror symmetry discussed above has a direct relation to torus knots. The topological recursion appearing in the context of mirror to the open Gromov-Witten invariants of the resolved conifold provides all-genera invariants of torus knots [**49**].

7. Relation to the Hitchin theory of Higgs bundles, and quantum curves as opers

The formulation of the original topological recursion is restricted to a plane algebraic (or an analytic) curve. The recursion formula, which is a sequence of residue calculations, is considered as a local operation, and the significance of a global nature of the complete spectral curve has not been fully developed in theory until recently. We note that for many cases of enumerative problems, topological recursion is established by calculating residues explicitly via global contour deformation [**8, 27, 39, 77**]. Thus the global geometry is indeed reflected in topological recursion.

In 2014, Dumitrescu and Mulase [**24**] observed that the notion of the spectral curve of the topological recursion due to Eynard and Orantin [**40**], which comes from spectral curves in random matrix theory, is exactly the same as the notion of the Hitchin spectral curves, and thus a topological recursion can be formulated for *any* Hitchin spectral curve in the cotangent bundle T^*C of a base curve C. The original context is the case of $C = \mathbb{C}$. This unexpected relation between the Hitchin theory of Higgs bundles and topological recursion opened a door to a new uncharted territory (see, for example, [**5**]).

In this geometric framework of spectral curves of topological recursion, the formalism can be further generalized for *singular* spectral curves, as discussed in the paper by Dumitrescu and Mulase in this volume. The Airy function example (2.5) corresponds to a singular spectral curve in the Hirzebruch surface, which comes from a rank 2 vector bundle $\mathcal{O}_{\mathbb{P}^1}(-1) \oplus \mathcal{O}_{\mathbb{P}^1}(1)$ over $\mathbb{P}^1$ and a Higgs field $\phi = \begin{bmatrix} & x(dx)^2 \\ 1 & \end{bmatrix}$. Although (2.5) is a non-singular curve in $\mathbb{P}^2$, we need to consider it in the compactification of the cotangent bundle $T^*\mathbb{P}^1$, which has a quintic cusp singularity at the infinity. This singularity reflects in the essential singularity of the entire solution $Ai(x)$ of the differential equation (2.4) at ∞. Still the relationship between geometry of Higgs bundles and the quantum invariants that the generalized topological recursion should calculate, such as (2.2), are not fully explored.

After this relation to Hitchin theory is established, the notion of a quantum curve corresponding to a given spectral curve becomes clear in terms of the mathematical language of *opers* [**23, 26**]. A spectral curve Σ embedded in the cotangent bundle T^*C of a base curve C is called *rational*, contrasting to the trigonometric case $\Sigma \subset \mathbb{C}^* \times \mathbb{C}^*$ and the *elliptic* case, which is a divisor in a $K3$ surface. For the rational case, if we identify quantum curves being opers corresponding to Hitchin spectral curves, then the unique quantization mechanism is well understood in terms of Kostant's three-dimensional subgroups (TDS) [**66**]. Since topological recursion does not refer to the TDS condition in general, it usually does not produce opers. It is unknown at this moment how to implement the TDS condition into topological recursion associated with Hitchin spectral curves for an arbitrary algebraic group so that we can directly identify the two quantization procedures: one through opers and the other through topological recursion.

While [**24**] is the first attempt of establishing an algebro-geometric framework for topological recursion by identifying its spectral curves with Hitchin spectral curves, the geometry of the recursive structure of topological recursion itself is first mathematically identified as CohFT in [**31**]. In more concrete examples, a formula based transformation from a Kontsevich-Manin CohFT [**65**] to the multilinear differential forms on the spectral curve defined by topological recursion has been rigorously obtained. Andersen, Borot, and Orantin report in this volume a different perspective on the relation between CohFT and topological recursion than those found in [**31**]. The authors describe a systematic way of constructing vector bundles on $\overline{\mathcal{M}}_{g,n}$ through modular functors and then define a CohFT using their characteristic classes. They then prove that in certain cases these CohFT directly give rise to a solution to topological recursion. Important examples in their paper not covered in [**31**] and above mentioned work include the Verlinde formula [**85**].

A striking byproduct of this line of research is a new proof [**28, 68**] of the ELSV-type formulae for various Hurwitz numbers, originally due to Ekedahl, Lando, Shapiro, and Vainshtein [**33**], and its generalizations such as the one by Johnson, Pandharipande, and Tseng [**62**] for orbifold Hurwitz numbers. Lewanski provides a concise article for this volume reviewing recent developments in this direction generalizing the ELSV formula to several cases.

8. New developments

We now briefly mention other papers of this volume that are not cited in the above sections.

Many geometric works on topological recursion, as described above, are trying to identify the recursion itself as a mechanism of quantization, such as CohFT. A totally new point of view is proposed by Kontsevich and Soibelman in this volume. The authors propose a different formulation, which include a much larger context of validity, for topological recursion. They put together topological recursion, quantum curves, and WKB analysis from the perspective of (holomorphic) symplectic geometry. The recursion formula of [**40**] does not exhibit manifest symmetry of the n-point function. For enumerative geometry problems, these n-point functions are obtained by the Laplace transform of symmetric functions, such as Hurwitz numbers and the Weil-Petersson volume of moduli spaces of bordered hyperbolic surfaces. The fundamental point of departure of Kontsevich-Soibelman is the manifest symmetry of the quantities to be calculated. They treat topological recursion as a quantization mechanism from the point of view of *deformation quantization*. Then they produce families of cyclic $\mathcal{D}$-modules, which are their definition of what others call quantum curves. The influence of this work can be seen, for example, in [**4**]. More concrete relations between the new framework and other computational aspect of topological recursion is scheduled to be discussed elsewhere by the authors.

A quantum curve is a $\mathcal{D}$-module over a compact Riemann surface, and these $\mathcal{D}$-modules with singularities are of great interest. Although this is a classical problem since the time of Painlevé, recently there have been explosive developments in this direction. In a paper contained in this volume, Korotkin proves a striking result that directly relates the standard symplectic structure of $T^*\mathcal{M}_{g,n}$ and the Goldman bracket on the $SL(2,\mathbb{C})$-character variety of n-punctured surface group, using a second order linear differential equation on a Riemann surface with singularities.

Extending the physical investigation of quantum curves developed in [**17, 18, 58**], Ciosmak, Hadasz, Manabe, and Sułkowski present a new algebraic way of viewing quantum curves in this volume. They consider both Virasoro and super Virasoro representations, and they identify quantum curves as singular vectors of Virasoro operators. In their definition, quantum curves can also be constructed via topological recursion. A particularly interesting direction is the supersymmetric generalization. So far the known relation between topological recursion and CohFT is based on commutative Frobenius algebras, corresponding to Gromov-Witten theory of a manifold without odd cohomologies. The super quantum curves of this paper are expected to be related to Gromov-Witten theory where odd degree cohomologies play an essential role.

Brini gives a concise survey of his new results concerning the Gopakumar-Vafa and Ooguri-Vafa correspondences. This physical theory relates Reshetikhin-Turaev-Witten invariants of knots, Gromov-Witten and Donaldson-Thomas invariants of local Calabi-Yau 3-folds, and topological recursion. Brini discusses the most general framework in which this correspondence holds.

Chiodo and Nagel present a cohomological Landau-Ginzburg/Calabi-Yau (LG/CY) correspondence which says the orbifold Chen-Ruan cohomology $H^{p,q}_{\mathrm{CR}}$ of a Calabi-Yau complete intersection in a weighted projective space is isomorphic, as a bi-graded vector space, to the state space of a hybrid Landau-Ginzburg model (which is more general than the hybrid Landau-Ginzburg model in the terminology of Fan-Jarvis-Ruan [**43**]). This state space isomorphism is the first step toward the LG/CY correspondence.

The article by Esteves explores intriguing algebraic structures lying at the heart of topological recursion. By using the Loday-Ronco Hopf algebra of planar rooted trees, he constructs a representation of the space of genus zero correlation functions that enjoys a Hopf algebra structure compatible with the residue formula appearing in topological recursion.

Kimura gives a review of remarkable relations between W-algebras, quiver gauge theories, qq-characters, and integrable systems. An important aspect of this work is the construction of W-algebras out of quivers, as well as an elliptic deformation of this construction. Since it has been speculated that quantum curves would appear in the construction of W-algebra representations in the work of Nakajima [**78**], we expect to see a new discovery in this direction of research.

At the symposium, we had a talk on opers by Fredrickson solving a conjecture of Gaiotto [**51**]. We refer to her paper [**23**] on this development. We also had a talk by Iwaki on Painlevé equations and topological recursion. On this important contribution, we refer to his papers [**60**, **61**].

Chiu-Chu Melissa Liu
Motohico Mulase

Bibliography

[1] Mina Aganagic, Robbert Dijkgraaf, Albrecht Klemm, Marcos Mariño, and Cumrun Vafa, *Topological strings and integrable hierarchies*, Comm. Math. Phys. **261** (2006), no. 2, 451–516, DOI 10.1007/s00220-005-1448-9. MR2191887

[2] G.B. Airy, *On the intensity of light in the neighbourhood of a caustic*, Transactions of the Cambridge Philosophical Society **6**, 379–402 (1838).

[3] A. Alexandrov, A. Mironov, and A. Morozov, *Solving Virasoro constraints in matrix models*, Fortschr. Phys. **53** (2005), no. 5-6, 512–521, DOI 10.1002/prop.200410212. MR2145334

[4] J.E. Andersen, G. Borot, and N. Orantin, *Geometric recursion*, arXiv:1711.04729v2 [math.GT] (2017).

[5] D. Baraglia and Z. Huang, *Special Kähler geometry of the Hitchin system and topological recursion*, arXiv:1707.04975v2 [math.DG] (2017).

[6] Gaëtan Borot and Bertrand Eynard, *All order asymptotics of hyperbolic knot invariants from non-perturbative topological recursion of A-polynomials*, Quantum Topol. **6** (2015), no. 1, 39–138, DOI 10.4171/QT/60. MR3335006

[7] Vincent Bouchard, Andrei Catuneanu, Olivier Marchal, and Piotr Sułkowski, *The remodeling conjecture and the Faber-Pandharipande formula*, Lett. Math. Phys. **103** (2013), no. 1, 59–77, DOI 10.1007/s11005-012-0588-z. MR3004817

[8] Vincent Bouchard, Daniel Hernández Serrano, Xiaojun Liu, and Motohico Mulase, *Mirror symmetry for orbifold Hurwitz numbers*, J. Differential Geom. **98** (2014), no. 3, 375–423. MR3263522

[9] Vincent Bouchard, Albrecht Klemm, Marcos Mariño, and Sara Pasquetti, *Remodeling the B-model*, Comm. Math. Phys. **287** (2009), no. 1, 117–178, DOI 10.1007/s00220-008-0620-4. MR2480744

[10] Vincent Bouchard, Albrecht Klemm, Marcos Mariño, and Sara Pasquetti, *Topological open strings on orbifolds*, Comm. Math. Phys. **296** (2010), no. 3, 589–623, DOI 10.1007/s00220-010-1020-0. MR2628817
78, 263–283 (2008).

[11] Vincent Bouchard and Marcos Mariño, *Hurwitz numbers, matrix models and enumerative geometry*, From Hodge theory to integrability and TQFT tt*-geometry, Proc. Sympos. Pure Math., vol. 78, Amer. Math. Soc., Providence, RI, 2008, pp. 263–283, DOI 10.1090/pspum/078/2483754. MR2483754

[12] Kevin M. Chapman, Motohico Mulase, and Brad Safnuk, *The Kontsevich constants for the volume of the moduli of curves and topological recursion*, Commun. Number Theory Phys. **5** (2011), no. 3, 643–698, DOI 10.4310/CNTP.2011.v5.n3.a3. MR2864463

[13] Leonid Chekhov, Bertrand Eynard, and Nicolas Orantin, *Free energy topological expansion for the 2-matrix model*, J. High Energy Phys. **12** (2006), 053, 31, DOI 10.1088/1126-6708/2006/12/053. MR2276699
118, 47–84 (1994).

[14] D. Cooper, M. Culler, H. Gillet, D. D. Long, and P. B. Shalen, *Plane curves associated to character varieties of 3-manifolds*, Invent. Math. **118** (1994), no. 1, 47–84, DOI 10.1007/BF01231526. MR1288467

[15] M. Cutimanco, P. Labelle, and V. Shramchenko, *Topological recursion of Eynard-Orantin and the harmonic oscillator*, arXiv:1701.08913v2 [math-ph] (2017).

[16] Robbert Dijkgraaf, Hiroyuki Fuji, and Masahide Manabe, *The volume conjecture, perturbative knot invariants, and recursion relations for topological strings*, Nuclear Phys. B **849** (2011), no. 1, 166–211, DOI 10.1016/j.nuclphysb.2011.03.014. MR2795276

[17] Robbert Dijkgraaf, Lotte Hollands, and Piotr Sułkowski, *Quantum curves and $\mathcal{D}$-modules*, J. High Energy Phys. **11** (2009), 047, 59, DOI 10.1088/1126-6708/2009/11/047. MR2628886

[18] Robbert Dijkgraaf, Lotte Hollands, Piotr Sułkowski, and Cumrun Vafa, *Supersymmetric gauge theories, intersecting branes and free fermions*, J. High Energy Phys. **2** (2008), 106, 57, DOI 10.1088/1126-6708/2008/02/106. MR2385939

[19] R. Dijkgraaf and C. Vafa, *Two Dimensional Kodaira-Spencer Theory and Three Dimensional Chern-Simons Gravity*, arXiv:0711.1932 [hep-th] (2007).

[20] Robbert Dijkgraaf, Herman Verlinde, and Erik Verlinde, *Loop equations and Virasoro constraints in nonperturbative two-dimensional quantum gravity*, Nuclear Phys. B **348** (1991), no. 3, 435–456, DOI 10.1016/0550-3213(91)90199-8. MR1083914

[21] Norman Do, Oliver Leigh, and Paul Norbury, *Orbifold Hurwitz numbers and Eynard-Orantin invariants*, Math. Res. Lett. **23** (2016), no. 5, 1281–1327, DOI 10.4310/MRL.2016.v23.n5.a3. MR3601067

[22] Norman Do and David Manescu, *Quantum curves for the enumeration of ribbon graphs and hypermaps*, Commun. Number Theory Phys. **8** (2014), no. 4, 677–701, DOI 10.4310/CNTP.2014.v8.n4.a2. MR3318387

[23] O. Dumitrescu, L. Fredrickson, G. Kydonakis, R. Mazzeo, M. Mulase, and A. Neitzke, *From the Hitchin section to opers through nonabelian Hodge*, Journal of Differential Geometry, to appear.

[24] Olivia Dumitrescu and Motohico Mulase, *Quantum curves for Hitchin fibrations and the Eynard-Orantin theory*, Lett. Math. Phys. **104** (2014), no. 6, 635–671, DOI 10.1007/s11005-014-0679-0. MR3200933

[25] Olivia Dumitrescu and Motohico Mulase, *Edge contraction on dual ribbon graphs and 2D TQFT*, J. Algebra **494** (2018), 1–27, DOI 10.1016/j.jalgebra.2017.09.027. MR3723168

[26] O. Dumitrescu and M. Mulase, *Interplay between opers, quantum curves, WKB analysis, and Higgs bundles*, arXiv:1702.00511 (2017).

[27] Olivia Dumitrescu, Motohico Mulase, Brad Safnuk, and Adam Sorkin, *The spectral curve of the Eynard-Orantin recursion via the Laplace transform*, Algebraic and geometric aspects of integrable systems and random matrices, Contemp. Math., vol. 593, Amer. Math. Soc., Providence, RI, 2013, pp. 263–315, DOI 10.1090/conm/593/11867. MR3087960

[28] P. Dunin-Barkowski, M. Kazarian, N. Orantin, S. Shadrin, and L. Spitz, *Polynomiality of Hurwitz numbers, Bouchard-Mariño conjecture, and a new proof of the ELSV formula*, Adv. Math. **279** (2015), 67–103, DOI 10.1016/j.aim.2015.03.016. MR3345179

[29] Petr Dunin-Barkowski, Motohico Mulase, Paul Norbury, Alexander Popolitov, and Sergey Shadrin, *Quantum spectral curve for the Gromov-Witten theory of the complex projective line*, J. Reine Angew. Math. **726** (2017), 267–289, DOI 10.1515/crelle-2014-0097. MR3641659

[30] P. Dunin-Barkowski, P. Norbury, N. Orantin, A. Popolitov, and S. Shadrin, *Dubrovin's superpotential as a global spectral curve*, Journal of the Institute of Mathematics of Jussieu, https://doi.org/10.1017/S147474801700007X (2017).

[31] P. Dunin-Barkowski, N. Orantin, S. Shadrin, and L. Spitz, *Identification of the Givental formula with the spectral curve topological recursion procedure*, Comm. Math. Phys. **328** (2014), no. 2, 669–700, DOI 10.1007/s00220-014-1887-2. MR3199996

[32] Petr Dunin-Barkowski, Sergey Shadrin, and Loek Spitz, *Givental graphs and inversion symmetry*, Lett. Math. Phys. **103** (2013), no. 5, 533–557, DOI 10.1007/s11005-013-0606-9. MR3041758

[33] Torsten Ekedahl, Sergei Lando, Michael Shapiro, and Alek Vainshtein, *Hurwitz numbers and intersections on moduli spaces of curves*, Invent. Math. **146** (2001), no. 2, 297–327, DOI 10.1007/s002220100164. MR1864018

[34] Bertrand Eynard, *Topological expansion for the 1-Hermitian matrix model correlation functions*, J. High Energy Phys. **11** (2004), 031, 35 pp. (2005), DOI 10.1088/1126-6708/2004/11/031. MR2118807

[35] Bertrand Eynard, *Recursion between Mumford volumes of moduli spaces*, Ann. Henri Poincaré **12** (2011), no. 8, 1431–1447, DOI 10.1007/s00023-011-0113-4. MR2855174

[36] B. Eynard, *Intersection numbers of spectral curves*, arXiv:1104.0176 (2011).

[37] B. Eynard, *Invariants of spectral curves and intersection theory of moduli spaces of complex curves*, Commun. Number Theory Phys. **8** (2014), no. 3, 541–588, DOI 10.4310/CNTP.2014.v8.n3.a4. MR3282995

[38] Bertrand Eynard, *An overview of the topological recursion*, Proceedings of the International Congress of Mathematicians—Seoul 2014. Vol. III, Kyung Moon Sa, Seoul, 2014, pp. 1063–1085. MR3729064

[39] Bertrand Eynard, Motohico Mulase, and Bradley Safnuk, *The Laplace transform of the cut-and-join equation and the Bouchard-Mariño conjecture on Hurwitz numbers*, Publ. Res. Inst. Math. Sci. **47** (2011), no. 2, 629–670, DOI 10.2977/PRIMS/47. MR2849645

[40] B. Eynard and N. Orantin, *Invariants of algebraic curves and topological expansion*, Commun. Number Theory Phys. **1** (2007), no. 2, 347–452, DOI 10.4310/CNTP.2007.v1.n2.a4. MR2346575

[41] B. Eynard and N. Orantin, *Computation of open Gromov-Witten invariants for toric Calabi-Yau 3-folds by topological recursion, a proof of the BKMP conjecture*, Comm. Math. Phys. **337** (2015), no. 2, 483–567, DOI 10.1007/s00220-015-2361-5. MR3339157

[42] C. Faber and R. Pandharipande, *Hodge integrals, partition matrices, and the λ_g conjecture*, Ann. of Math. (2) **157** (2003), no. 1, 97–124, DOI 10.4007/annals.2003.157.97. MR1954265

[43] Huijun Fan, Tyler J. Jarvis, and Yongbin Ruan, *A mathematical theory of the gauged linear sigma model*, Geom. Topol. **22** (2018), no. 1, 235–303, DOI 10.2140/gt.2018.22.235. MR3720344

[44] Bohan Fang, Chiu-Chu Melissa Liu, and Zhengyu Zong, *Equivariant Gromov-Witten theory of affine smooth toric Deligne-Mumford stacks*, Int. Math. Res. Not. IMRN **7** (2016), 2127–2144, DOI 10.1093/imrn/rnv201. MR3509949

[45] Bohan Fang, Chiu-Chu Melissa Liu, and Zhengyu Zong, *All genus mirror symmetry for toric Calabi-Yau 3-orbifolds*, String-Math 2014, Proc. Sympos. Pure Math., vol. 93, Amer. Math. Soc., Providence, RI, 2016, pp. 1–19. MR3524231

[46] Bohan Fang, Chiu-Chu Melissa Liu, and Zhengyu Zong, *The Eynard-Orantin recursion and equivariant mirror symmetry for the projective line*, Geom. Topol. **21** (2017), no. 4, 2049–2092, DOI 10.2140/gt.2017.21.2049. MR3654104

[47] Bohan Fang, Chiu-Chu Melissa Liu, and Zhengyu Zong, *All genus mirror symmetry for toric Calabi-Yau 3-orbifolds*, String-Math 2014, Proc. Sympos. Pure Math., vol. 93, Amer. Math. Soc., Providence, RI, 2016, pp. 1–19. MR3524231

[48] Bohan Fang, Chiu-Chu Melissa Liu, and Zhengyu Zong, *The SYZ mirror symmetry and the BKMP remodeling conjecture*, Adv. Theor. Math. Phys. **20** (2016), no. 1, 165–192, DOI 10.4310/ATMP.2016.v20.n1.a3. MR3531992

[49] B. Fang and Z. Zong, *Topological recursion for the conifold transition of a torus knot*, arXiv:1607.01208v2 [math.AG] (2016).
Soc. **354**, 735–747 (2002).

[50] Charles Frohman, Răzvan Gelca, and Walter Lofaro, *The A-polynomial from the noncommutative viewpoint*, Trans. Amer. Math. Soc. **354** (2002), no. 2, 735–747, DOI 10.1090/S0002-9947-01-02889-6. MR1862565

[51] D. Gaiotto, *Opers and TBA*, arXiv:1403.6137 [hep-th] (2014).

[52] Stavros Garoufalidis, *On the characteristic and deformation varieties of a knot*, Proceedings of the Casson Fest, Geom. Topol. Monogr., vol. 7, Geom. Topol. Publ., Coventry, 2004, pp. 291–309, DOI 10.2140/gtm.2004.7.291. MR2172488

[53] Stavros Garoufalidis and Thang T. Q. Lê, *The colored Jones function is q-holonomic*, Geom. Topol. **9** (2005), 1253–1293, DOI 10.2140/gt.2005.9.1253. MR2174266

[54] Alexander B. Givental, *Semisimple Frobenius structures at higher genus*, Internat. Math. Res. Notices **23** (2001), 1265–1286, DOI 10.1155/S1073792801000605. MR1866444

[55] I. P. Goulden, *A differential operator for symmetric functions and the combinatorics of multiplying transpositions*, Trans. Amer. Math. Soc. **344** (1994), no. 1, 421–440, DOI 10.2307/2154724. MR1249468

[56] I. P. Goulden and D. M. Jackson, *Transitive factorisations into transpositions and holomorphic mappings on the sphere*, Proc. Amer. Math. Soc. **125** (1997), no. 1, 51–60, DOI 10.1090/S0002-9939-97-03880-X. MR1396978

[57] Sergei Gukov, *Three-dimensional quantum gravity, Chern-Simons theory, and the A-polynomial*, Comm. Math. Phys. **255** (2005), no. 3, 577–627, DOI 10.1007/s00220-005-1312-y. MR2134725

[58] Sergei Gukov and Piotr Sułkowski, *A-polynomial, B-model, and quantization*, J. High Energy Phys. **2** (2012), 070, front matter+56. MR2996110

[59] L. Hollands, *Topological strings and quantum curves*, Ph.D. Thesis, University of Amsterdam arXiv:0911.3413 [hep-th] (2009).

[60] K. Iwaki, O. Marchal, and A. Saenz, *Painlevé equations, topological type property and reconstruction by the topological recursion*, J. Geom. Phys. **124** (2018), 16–54, DOI 10.1016/j.geomphys.2017.10.009. MR3754496

[61] Kohei Iwaki and Axel Saenz, *Quantum curve and the first Painlevé equation*, SIGMA Symmetry Integrability Geom. Methods Appl. **12** (2016), Paper No. 011, 24 pages, DOI 10.3842/SIGMA.2016.011. MR3453982

[62] P. Johnson, R. Pandharipande, and H.-H. Tseng, *Abelian Hurwitz-Hodge integrals*, Michigan Math. J. **60** (2011), no. 1, 171–198, DOI 10.1307/mmj/1301586310. MR2785870

[63] R. M. Kashaev, *The hyperbolic volume of knots from the quantum dilogarithm*, Lett. Math. Phys. **39** (1997), no. 3, 269–275, DOI 10.1023/A:1007364912784. MR1434238

[64] Maxim Kontsevich, *Intersection theory on the moduli space of curves and the matrix Airy function*, Comm. Math. Phys. **147** (1992), no. 1, 1–23. MR1171758

[65] M. Kontsevich and Yu. Manin, *Gromov-Witten classes, quantum cohomology, and enumerative geometry*, Comm. Math. Phys. **164** (1994), no. 3, 525–562. MR1291244

[66] Bertram Kostant, *The principal three-dimensional subgroup and the Betti numbers of a complex simple Lie group*, Amer. J. Math. **81** (1959), 973–1032, DOI 10.2307/2372999. MR0114875

[67] K.G. Krishna, P. Labelle, and V. Shramchenko, *Feynman diagrams, ribbon graphs, and topological recursion of Eynard-Orantin*, arXiv:1802.01773v2 [hep-th] (2018).

[68] Danilo Lewanski, Alexandr Popolitov, Sergey Shadrin, and Dimitri Zvonkine, *Chiodo formulas for the r-th roots and topological recursion*, Lett. Math. Phys. **107** (2017), no. 5, 901–919, DOI 10.1007/s11005-016-0928-5. MR3633029

[69] Wei Luo and Shengmao Zhu, *Hurwitz-Hodge integral identities from the cut-and-join equation*, Ann. Comb. **19** (2015), no. 4, 749–763, DOI 10.1007/s00026-015-0282-9. MR3415010

[70] Marcos Mariño, *Open string amplitudes and large order behavior in topological string theory*, J. High Energy Phys. **3** (2008), 060, 34, DOI 10.1088/1126-6708/2008/03/060. MR2391060

[71] Maryam Mirzakhani, *Simple geodesics and Weil-Petersson volumes of moduli spaces of bordered Riemann surfaces*, Invent. Math. **167** (2007), no. 1, 179–222, DOI 10.1007/s00222-006-0013-2. MR2264808

[72] Maryam Mirzakhani, *Weil-Petersson volumes and intersection theory on the moduli space of curves*, J. Amer. Math. Soc. **20** (2007), no. 1, 1–23, DOI 10.1090/S0894-0347-06-00526-1. MR2257394

[73] Motohico Mulase and Michael Penkava, *Topological recursion for the Poincaré polynomial of the combinatorial moduli space of curves*, Adv. Math. **230** (2012), no. 3, 1322–1339, DOI 10.1016/j.aim.2012.03.027. MR2921181

[74] Motohico Mulase and Brad Safnuk, *Mirzakhani's recursion relations, Virasoro constraints and the KdV hierarchy*, Indian J. Math. **50** (2008), no. 1, 189–218. MR2379144

[75] M. Mulase, S. Shadrin, and L. Spitz, *The spectral curve and the Schrödinger equation of double Hurwitz numbers and higher spin structures*, Commun. Number Theory Phys. **7** (2013), no. 1, 125–143, DOI 10.4310/CNTP.2013.v7.n1.a4. MR3108774

[76] Motohico Mulase and Piotr Sułkowski, *Spectral curves and the Schrödinger equations for the Eynard-Orantin recursion*, Adv. Theor. Math. Phys. **19** (2015), no. 5, 955–1015, DOI 10.4310/ATMP.2015.v19.n5.a2. MR3487649

[77] Motohico Mulase and Naizhen Zhang, *Polynomial recursion formula for linear Hodge integrals*, Commun. Number Theory Phys. **4** (2010), no. 2, 267–293, DOI 10.4310/CNTP.2010.v4.n2.a1. MR2725053

[78] Hiraku Nakajima, *More lectures on Hilbert schemes of points on surfaces*, Development of moduli theory—Kyoto 2013, Adv. Stud. Pure Math., vol. 69, Math. Soc. Japan, [Tokyo], 2016, pp. 173–205. MR3586508

[79] Paul Norbury, *Counting lattice points in the moduli space of curves*, Math. Res. Lett. **17** (2010), no. 3, 467–481, DOI 10.4310/MRL.2010.v17.n3.a7. MR2653682

[80] Paul Norbury, *String and dilaton equations for counting lattice points in the moduli space of curves*, Trans. Amer. Math. Soc. **365** (2013), no. 4, 1687–1709, DOI 10.1090/S0002-9947-2012-05559-0. MR3009643

[81] Paul Norbury, *Quantum curves and topological recursion*, String-Math 2014, Proc. Sympos. Pure Math., vol. 93, Amer. Math. Soc., Providence, RI, 2016, pp. 41–65. MR3524233

[82] Paul Norbury and Nick Scott, *Gromov-Witten invariants of* $\mathbb{P}^1$ *and Eynard-Orantin invariants*, Geom. Topol. **18** (2014), no. 4, 1865–1910, DOI 10.2140/gt.2014.18.1865. MR3268770

[83] Hirosi Ooguri, Piotr Sułkowski, and Masahito Yamazaki, *Wall crossing as seen by matrix models*, Comm. Math. Phys. **307** (2011), no. 2, 429–462, DOI 10.1007/s00220-011-1330-x. MR2837121

[84] R. Vakil, *Enumerative Geometry of Curves via Degeneration Methods*, Ph.D. thesis, Harvard University (1997), 149 pp.

[85] Erik Verlinde, *Fusion rules and modular transformations in 2D conformal field theory*, Nuclear Phys. B **300** (1988), no. 3, 360–376, DOI 10.1016/0550-3213(88)90603-7. MR954762

[86] Constantin Teleman, *The structure of 2D semi-simple field theories*, Invent. Math. **188** (2012), no. 3, 525–588, DOI 10.1007/s00222-011-0352-5. MR2917177

[87] Edward Witten, *Two-dimensional gravity and intersection theory on moduli space*, Surveys in differential geometry (Cambridge, MA, 1990), Lehigh Univ., Bethlehem, PA, 1991, pp. 243–310. MR1144529

[88] J. Zhou, *Local Mirror Symmetry for One-Legged Topological Vertex*, arXiv:0910.4320 [math.AG] (2009).

[89] J. Zhou, *Intersection numbers on Deligne-Mumford moduli spaces and quantum Airy curve*, arXiv:1206.5896 [math.AG] (2012).

[90] J. Zhou, *Quantum mirror curves for* $\mathbb{C}^3$ *and the resolved conifold*, arXiv:1207.0598v1 [math.AG] (2012).

[91] Shengmao Zhu, *On a proof of the Bouchard-Sulkowski conjecture*, Math. Res. Lett. **22** (2015), no. 2, 633–643, DOI 10.4310/MRL.2015.v22.n2.a14. MR3342249

[92] Zhengyu Zong, *Equivariant Gromov-Witten Theory of GKM Orbifolds*, ProQuest LLC, Ann Arbor, MI, 2015. Thesis (Ph.D.)–Columbia University. MR3312922

Proceedings of Symposia in Pure Mathematics
Volume **100**, 2018
https://doi.org/10.1090/pspum/100/01772

Modular functors, cohomological field theories, and topological recursion

Jørgen Ellegaard Andersen, Gaëtan Borot, and Nicolas Orantin

ABSTRACT. Given a topological modular functor $\mathcal{V}$ in the sense of Walker, we construct vector bundles $[\mathcal{Z}_{\vec{\lambda}}]_{g,n}$ over $\overline{\mathcal{M}}_{g,n}$, whose Chern characters define semi-simple cohomological field theories. This construction depends on a determination of the logarithm of the eigenvalues of the Dehn twist and central element actions. We show that the intersection indices of the Chern character with the ψ-classes in $\overline{\mathcal{M}}_{g,n}$ is computed by the topological recursion of Eynard and Orantin, for a local spectral curve that we describe. In particular, we show how the Verlinde formula for the ranks $[D_{\vec{\lambda}}]_{g,n} = \mathrm{rank}\,[\mathcal{V}_{\vec{\lambda}}]_{g,n}$ is recovered from the topological recursion. We analyze the consequences of our result on two examples: modular functors associated to a finite group G (for which $[D_{\vec{\lambda}}]_{g,n}$ enumerates certain G-principle bundles over a genus g surface with n boundary conditions specified by $\vec{\lambda}$), and the modular functor obtained from Wess-Zumino-Witten conformal field theory associated to a simple, simply-connected Lie group G (for which $[\mathcal{Z}_{\vec{\lambda}}]_{g,n}$ is the Verlinde bundle).

1. Introduction

Background. The pioneering works of Atiyah, Segal and Witten [**9, 84, 96**] turned $2d$ conformal field theories (CFT) [**15**] into an effective machinery to design interesting 3-manifold invariants, known under the name of "quantum invariants". More thoroughly, it allowed the construction of $3d$ topological quantum field theories (TQFT), where the $2d$ CFT is thought of as living on the boundary of 3-manifolds. The axioms of a $3d$ TQFT were worked out in details by Reshetikhin and Turaev, and they further constructed the first and most important example from the quantum group $U_q(\mathfrak{sl}_2)$ at q = a root of unity [**80, 81, 89**].

There exist several variants of axiomatizations that embody the concept of a CFT. This article deals with one of these axiomatizations for the topological part of a CFT, called a "modular functor". It was proposed by Segal in the context of rational conformal field theory [**84**] covering the holomorphic part of CFT, and developed in Walker's notes [**94**] in the purely topological context. Namely, we consider functors from the category of marked surfaces – with projective tangent vectors and labels in a finite set Λ at punctures, and a Lagrangian subspace of the

2010 *Mathematics Subject Classification.* Primary 51Pxx, 81T40.

The first author is supported in part by the center of excellence grant "Center for Quantum Geometry of Moduli Spaces (QGM)" from the Danish National Research Foundation (DNRF95).

The work of the second author is supported by the Max-Planck Gesellschaft.

first homology –, to the category of finite dimensional complex vector spaces. In particular, such a functor determines representations of a central extension of the mapping class groups. The main property required to be a modular functor is that the vector spaces attached to a surface enjoy a factorization property when the surface is pinched. The full definitions are given in Section 2. From the data of a modular functor in this sense, [**52, 63**] show how to obtain a $(2+1)$-dimensional TQFT.

The main source of modular functors are modular tensor categories (MTC) [**8, 12, 89**]. At present, it seems that all known examples of modular functors come from a modular tensor category, but it is not known whether all modular functors are of this kind. Among examples of MTC, we find some categories of representations of quantum groups [**89**], and categories of representations of vertex operator algebras (VOA) [**43, 56**].

The Wess-Zumino-Witten models form a well-studied class of examples of this type. The MTC here arises from representations of a VOA constructed from affine Kač-Moody algebras $\widehat{\mathfrak{g}}$ [**62**]. It gives rise to Hilbert spaces of a TQFT, which come as vector bundles (the so-called Verlinde bundles) over a family of complex curves with coordinates, equipped with a projectively flat connection [**88**], which coincides with Hitchin's connection from the point of view of geometric quantization [**66**]. The choice of coordinates can actually be bypassed [**67, 87, 88**] and the Verlinde bundles exist as bundles $\mathcal{V}_{g,n}^{\mathrm{WZW}}$ over the moduli space of curves, which extend nicely to the Deligne-Mumford compactification. The explicit construction of a modular functor from this perspective – also called the CFT approach – was described in [**3, 4**]. There is another approach, based on a category of representations of a quantum group $U_q(\mathfrak{g}_{\mathbb{C}})$ at certain roots of unity. It leads to the Witten-Reshetikhin-Turaev TQFT, constructed in [**18, 80, 81**] for $\mathfrak{g}_{\mathbb{C}} = \mathfrak{sl}_2$, and in [**90**] for any simple Lie algebra of type ABCD. This theory for $\mathfrak{sl}_2$ was also constructed using skein theory by Blanchet, Habegger, Masbaum and Vogel in [**17, 18**] and by Blanchet in [**16**] for $\mathfrak{sl}_N$. As anticipated by Witten, the CFT approach and the quantum group approach should give equivalent TQFTs. For instance, the equivalence of the modular functors was established by the first author of this paper and Ueno in [**5, 7**] for $\mathfrak{g} = \mathfrak{su}_N$.

The rank of the Verlinde bundle is already a non-trivial invariant, which is computed by the famous Verlinde formula [**14, 40, 74, 93**]. Marian et al. [**70**] lately showed that the Chern character of $\mathcal{V}_{g,n}^{\mathrm{WZW}}$ defines a semi-simple cohomological field theory (CohFT). It can be characterized in terms of its R-matrix thanks to the classification results of Givental and Teleman [**48, 86**]: from the R-matrix, one can build the exponential of a second-order differential operator, which acts on a product of several copies of the Witten-Kontsevich generating series of ψ-classes (the matrix Airy function/KdV tau function of [**64**]), and returns the generating series of the intersection of the Chern character at hand with an arbitrary product of ψ-classes. We call these invariants the "CohFT correlation functions". When we introduce a variable t in the Chern character keeping track of the cohomological degree, and send $t \to 0$, these correlation functions return the rank of the bundle, which can be thought of as the "$2d$ TQFT correlation functions".

Contribution of the article. In the present article, we generalize the results of [**70**] to any modular functor – hence not relying on the peculiarities of the Wess-Zumino-Witten models. For a given modular functor, we construct a trivial bundle

$[\tilde{\mathcal{Z}}^0_{\vec{\lambda}}]_{g,n}$ over Teichmüller space, which, after twisting by suitable line bundles, descends to a bundle $[\mathcal{Z}_{\vec{\lambda}}]_{g,n}$ over $\mathcal{M}_{g,n}$ (Theorem 2.5). We can use Chern-Weil theory to compute the Chern character of $[\mathcal{Z}_{\vec{\lambda}}]_{g,n}$ (Proposition 2.6) in terms of ψ-classes and the first Chern class of the Hodge bundle – their coefficients are related respectively to the central charge c and the log of Dehn twist eigenvalues r_λ (aka conformal weights). Besides, we show that our bundle extends to the boundary of $\overline{\mathcal{M}}_{g,n}$ (Theorem 2.9). All together, this constitutes our first main result. Since $\overline{\mathcal{M}}_{g,n}$ is an orbifold, the Chern character of our bundle must have rational coefficients, hence a new (geometric) proof of Vafa's theorem [**92**] stating that c and r_λ must be rational. Our proof actually shows this for any modular functor, including non-unitary cases.

Since our bundle enjoys factorization implied by the axioms of a modular functor, we can conclude that $\mathrm{Ch}_t([\mathcal{Z}_{\vec{\lambda}}]_{g,n})$ defines a semi-simple cohomological field theory on a Frobenius algebra $\mathcal{A}$ whose underlying vector space is the space that the modular functor assigns to a torus (Theorem 3.2). Because the twists can depend on log-determinations for the central charge and the conformal weights, and because we have the variable $t \in \mathbb{C}$ (introduced to keep track of each cohomological degree separately), we actually produce a 1-parameter family of CohFTs. The existence of an S-matrix that diagonalizes the product in $\mathcal{A}$ ensures the semi-simplicity of these theories, and we compute the R-matrix of these CohFT in terms of the S-matrix (Proposition 3.4).

Then, from the general result of [**33**], we know that the correlation functions of these CohFTs is computed by the topological recursion of [**38**] for a local spectral curve. We describe explicitly this local spectral curve and the relevant initial data $(\omega_{0,1}, \omega_{0,2})$. It depends solely on the $\mathrm{SL}(2,\mathbb{Z})$ representation provided by the genus 1 part of the modular functor (Proposition 4.5). If the variable t is set to 0, we recover Verlinde's formula for the rank of our bundle as a special case of the topological recursion (Proposition 4.4); in general, we obtain that the $\omega'_{g,n}$s of the topological recursion for this spectral curve are expressed in terms of integrals of the Chern character and ψ-classes (Equation 4.25). This formula is our second main result.

We illustrate our findings on two classes of Wess-Zumino-Witten models. In Section 5, we address the modular functors associated to a finite group G [**28, 30, 42**]. They are also called "orbifold holomorphic models". In the "untwisted case", their Frobenius algebra contains simultaneously the fusion rules of the representation ring of G and (albeit undirectly) the decomposition of the product of conjugacy classes. The dimensions of the TQFT vector spaces count certain G-principle bundles over the surface in question. We therefore find – in a rather trivial way – a topological recursion for these numbers, where the induction concerns the Euler characteristics of the base. In the "untwisted case" we obtain a degree 0 CohFT, which only remembers the dimension of the vector spaces, but in the twisted case, the CohFT is in general non-trivial (see Lemma 5.1). In Section 6, we examine the Wess-Zumino-Witten models based on a compact Lie group G at level ℓ, for which $[\mathcal{Z}_{\vec{\lambda}}]_{g,n}$ is the Verlinde bundle studied in [**70**]. Remarkably, for $SU(N)_\ell$, we find that $\omega_{0,2}$ for the local spectral curve is expressible in terms of a suitable truncation of the generating series of double Hurwitz numbers (number of branched coverings over the Riemann sphere). This poses the question of the combinatorial interpretation of the correlation function of these CohFTs in the context of enumeration of coverings over a surface of arbitrary topology.

To summarize, from a physical point of view, we have associated to a modular functor a CohFT that should encode Gromov-Witten theory of a target space X. Having a modular functor means that the worldsheet (a surface of genus g with n boundaries) roughly speaking carries a CFT. As we comment in Section 7, the local spectral curve used in Section 3 for the topological recursion, should describe the vicinity of an isolated singularities in a Landau-Ginzburg model $\tilde{X}$. As of now, the description of the global geometry of X and $\tilde{X}$ is unclear to us.

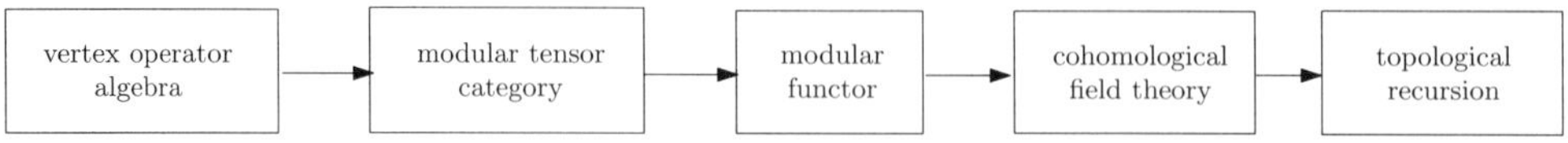

FIGURE 1. $A \to B$ means that B (or a set of quantities fulfilling the axioms of B) can be obtained from the data of A ; these are not equivalences, i.e. in general A cannot be fully recovered from B, and neither all B come from an A. A wealth of definitions sometimes fall under the same names. To avoid ambiguities: VOA are as in [**57**] ; the notion of modular functor we adopt is described in Section 2.3, following the approach of Walker [**94**] ; the notion of CohFT is described in Section 3.1, simplifying the axiomatization given by Kontsevich and Manin [**65**] ; the topological recursion is described in Section 4.1, following Eynard and the third author [**39**].

2. Construction of vector bundles from a modular functor

We introduce the category of marked surfaces, their automorphism groups, and review the axioms of a modular functor. The target category is that of finite dimensional vector spaces over the field of complex numbers. The motivation to introduce marked surfaces is explained e.g. in [**94**]: in quantum field theory, if one works with a naive category of surfaces, the states have a phase ambiguity, so that the target category would rather be that of projective vector spaces. The marking allows us to resolve phase ambiguities and work with vector spaces.

We then explain, in Section 2.6, how to obtain from any given modular functor, a family of vector bundles over the moduli space of curves. We compute its Chern character in terms of the basic data of the modular functor (Theorem 2.6). A delicate but essential point for Section 3 is to prove that the bundles extend to the Deligne-Mumford compactification of the moduli space. This is achieved in Section 2.7, with a detour via a bordification of the Teichmüller space.

2.1. The category of marked surfaces. Let us start by fixing notations. By a closed surface we mean a smooth real 2-dimensional, compact manifold. For an oriented closed surface Σ of genus g we have the non-degenerate skew-symmetric intersection pairing

$$(\cdot,\cdot)\,:\, H_1(\Sigma,\mathbb{Z}) \times H_1(\Sigma,\mathbb{Z}) \to \mathbb{Z}.$$

Suppose Σ is connected. In this case a Lagrangian $L \subseteq H_1(\Sigma,\mathbb{Z})$ is by definition a sublattice which is maximally isotropic with respect to the intersection pairing. A $\mathbb{Z}$-basis $(\vec{\alpha},\vec{\beta}) = (\alpha_1,\ldots,\alpha_g,\beta_1,\ldots\beta_g)$ for $H_1(\Sigma,\mathbb{Z})$ is called a symplectic basis if

$$(\alpha_i,\beta_j) = \delta_{ij},\quad (\alpha_i,\alpha_j) = (\beta_i,\beta_j) = 0,$$

for all $i, j \in [\![1, g]\!]$. If Σ is not connected, then $H_1(\Sigma, \mathbb{Z}) = \bigoplus_i H_1(\Sigma_i, \mathbb{Z})$, where Σ_i are the connected components of Σ. In this context, by definition in this paper, a Lagrangian is a sublattice of the form $L = \bigoplus_i L_i$, where $L_i \subset H_1(\Sigma_i, \mathbb{Z})$ is Lagrangian. Likewise a symplectic basis for $H_1(\Sigma, \mathbb{Z})$ is a $\mathbb{Z}$-basis of the form $((\vec{\alpha}^i, \vec{\beta}^i))$, where $(\vec{\alpha}^i, \vec{\beta}^i)$ is a symplectic basis for $H_1(\Sigma_i, \mathbb{Z})$.

For any real vector space V, we define $\mathcal{P}V = (V - \{0\})/\mathbb{R}_+$.

DEFINITION 2.1. A *pointed surface* (Σ, P) is an oriented closed surface Σ with a finite set of points $P \subset \Sigma$.

DEFINITION 2.2. A *morphism of pointed surfaces* $f : (\Sigma_1, P_1) \to (\Sigma_2, P_2)$ is an isotopy class of orientation preserving diffeomorphisms which map P_1 to P_2. The group of automorphisms of a pointed surface (Σ, P) is the mapping class group, and it will be denoted $\Gamma_{\Sigma,P}$. It consists of the isotopy classes of orientation preserving diffeomorphisms of Σ which are the identity on P. The *framed mapping class group* of a pointed surface is denoted $\tilde{\Gamma}_{\Sigma,P}$ and it consists of isotopy classes of orientation preserving diffeomorphisms which are the identity on P as well as on the tangent spaces at each $p \in P$.

We stress that for the (framed) mapping class group, the isotopies allowed in the equivalence relation must also be the identity on P (and on the tangent spaces at P). We clearly have the following Lemma.

LEMMA 2.1. *There is a short exact sequence*

$$1 \longrightarrow \mathbb{Z}^P \longrightarrow \tilde{\Gamma}_{\Sigma,P} \longrightarrow \Gamma_{\Sigma,P} \to 1,$$

where the generator of the factor of $\mathbb{Z}$ corresponding to $p \in P$ is given by the Dehn twist δ_p in the boundary of an embedded disk in $\Sigma - (P - \{p\})$ and centred in p.

DEFINITION 2.3. A *marked surface* $\mathbf{\Sigma} = (\Sigma, P, V, L)$ is the data of an oriented closed surface Σ with a finite subset $P \subset \Sigma$ of points, projective tangent vectors $V \in \prod_{p \in P} \mathcal{P}T_p\Sigma$ and a Lagrangian subspace $L \subseteq H_1(\Sigma, \mathbb{Z})$.

DEFINITION 2.4. A *morphism* $\mathbf{f} : \mathbf{\Sigma}_1 \to \mathbf{\Sigma}_2$ of marked surfaces $\mathbf{\Sigma}_i = (\Sigma_i, P_i, V_i, L_i)$ is an isotopy class of orientation preserving diffeomorphisms $f : \Sigma_1 \to \Sigma_2$ that map (P_1, V_1) to (P_2, V_2) together with an integer s. Hence we write $\mathbf{f} = (f, s)$.

Let σ be Wall's signature cocycle for triples of Lagrangian subspaces of $H_1(\Sigma, \mathbb{R})$ [**95**].

DEFINITION 2.5. Let $\mathbf{f}_1 = (f_1, s_1) : \mathbf{\Sigma}_1 \to \mathbf{\Sigma}_2$ and $\mathbf{f}_2 = (f_2, s_2) : \mathbf{\Sigma}_2 \to \mathbf{\Sigma}_3$ be morphisms of marked surfaces $\mathbf{\Sigma}_i = (\Sigma_i, P_i, V_i, L_i)$. Then, the *composition* of $\mathbf{f}_1$ and $\mathbf{f}_2$ is

$$\mathbf{f}_2\mathbf{f}_1 = (f_2 f_1, s_2 + s_1 - \sigma((f_2 f_1)_* L_1^{\mathbb{R}}, f_{2*} L_2^{\mathbb{R}}, L_3^{\mathbb{R}})).$$

where $L_i^{\mathbb{R}} = L_i \otimes \mathbb{R}$.

With the objects being marked surfaces and the morphisms and their composition being defined as above, we have constructed the category of marked surfaces.

The mapping class group $\Gamma_{\mathbf{\Sigma}}$ of a marked surface $\mathbf{\Sigma} = (\Sigma, P, V, L)$ is the group of automorphisms of $\mathbf{\Sigma}$. The group $\Gamma_{\mathbf{\Sigma}}$ is a central extension of the framed mapping class group $\tilde{\Gamma}_{\Sigma,P}$ of the pointed surface (Σ, P)

$$1 \longrightarrow \mathbb{Z} \longrightarrow \Gamma_{\mathbf{\Sigma}} \longrightarrow \tilde{\Gamma}_{\Sigma,P} \longrightarrow 1$$

defined by the 2-cocycle $\gamma : \Gamma_{\mathbf{\Sigma}} \to \mathbb{Z}$, $\gamma(f_2, f_1) = \sigma((f_2 f_1)_* L^{\mathbb{R}}, f_{2*} L^{\mathbb{R}}, L^{\mathbb{R}})$. It is known that this cocycle is equivalent to the cocycle obtained by considering 2-framings on mapping cylinders, see [**10**] and [**2**]. Briefly, the relation is as follows: a 2-framing is determined by the first Pontryagin number p_1; Hirzebruch's formula says that p_1 is three times the signature of the 4-manifold and, by construction [**95**], σ expresses the non-additivity of the signature.

Notice also that for any morphism $(f, s) : \mathbf{\Sigma}_1 \to \mathbf{\Sigma}_2$, we can factor

$$\begin{aligned} (f,s) &= ((\mathrm{Id}, s') : \mathbf{\Sigma}_2 \to \mathbf{\Sigma}_2) \circ (f, s - s') \\ &= (f, s - s') \circ ((\mathrm{Id}, s') : \mathbf{\Sigma}_1 \to \mathbf{\Sigma}_1). \end{aligned}$$

In particular $(\mathrm{Id}, s) : \mathbf{\Sigma} \to \mathbf{\Sigma}$ is $(\mathrm{Id}, 1)^s$.

2.2. Operations on marked surfaces.

DEFINITION 2.6. The operation of *disjoint union of marked surfaces* is

$$(\Sigma_1, P_1, V_1, L_1) \sqcup (\Sigma_2, P_2, V_2, L_2) = (\Sigma_1 \sqcup \Sigma_2, P_1 \sqcup P_2, V_1 \sqcup V_2, L_1 \oplus L_2).$$

Morphisms on disjoint unions are accordingly $(f_1, s_1) \sqcup (f_2, s_2) = (f_1 \sqcup f_2, s_1 + s_2)$.

We see that the disjoint union is an operation on the category of marked surfaces.

DEFINITION 2.7. Let $\mathbf{\Sigma}$ be a marked surface. We denote by $-\mathbf{\Sigma}$ the marked surface obtained from $\mathbf{\Sigma}$ by the *operation of reversal of the orientation.* For a morphism $\mathbf{f} = (f, s) : \mathbf{\Sigma}_1 \to \mathbf{\Sigma}_2$ we let the orientation reversed morphism be given by $-\mathbf{f} = (f, -s) : -\mathbf{\Sigma}_1 \to -\mathbf{\Sigma}_2$.

We also see that orientation reversal is an operation on the category of marked surfaces.

Let us now consider glueing of marked surfaces. Let $(\Sigma, \{p_-, p_+\} \sqcup P, \{v_-, v_+\} \sqcup V, L)$ be a marked surface, where we have selected an ordered pair (p_-, p_+) of marked points with projective tangent vectors v_- at p_- and v_+ at p_+, at which we will perform the glueing. Let $C : \mathcal{P}(T_{p_-}\Sigma) \to \mathcal{P}(T_{p_+}\Sigma)$ be an orientation reversing projective isomorphism such that $C(v_-) = v_+$. Such a C is called a *glueing map* for Σ. Let $\tilde{\Sigma}$ be the oriented surface with boundary obtained from Σ by blowing up p_- and p_+, i.e.

$$\tilde{\Sigma} = (\Sigma - \{p_-, p_+\}) \sqcup \mathcal{P}(T_{p_-}\Sigma) \sqcup \mathcal{P}(T_{p_+}\Sigma),$$

with the natural smooth structure induced from Σ. Let now Σ_C be the oriented closed surface obtained from $\tilde{\Sigma}$ by using C to glue the two boundary components of $\tilde{\Sigma}$ corresponding to $p_\pm$. We call Σ_C the glueing of Σ at the ordered pair $((p_-, v_-), (p_+, v_+))$ with respect to C.

Let now Σ' be the topological space obtained from Σ by identifying p_- and p_+. We then have natural continuous maps $q : \Sigma_C \to \Sigma'$ and $n : \Sigma \to \Sigma'$. On the first homology group n induces an injection and q a surjection, so we can define a Lagrangian $L_C \subseteq H_1(\Sigma_C, \mathbb{Z})$ by $L_C = q_*^{-1}(n_*(L))$. We note that the image of $\mathcal{P}(T_{p_-}\Sigma)$ (with the orientation induced from $\tilde{\Sigma}$) induces naturally a line in $H_1(\Sigma_C, \mathbb{R})$ and as such it is contained in $L_C^{\mathbb{R}}$.

REMARK 2.8. If we have two glueing maps $C_i : \mathcal{P}(T_{p_-}\Sigma) \to \mathcal{P}(T_{p_+}\Sigma)$, $i = 1, 2$, we note that there is a diffeomorphism $f : \Sigma \to \Sigma$ inducing the identity on $(p_-, v_-) \sqcup (p_+, v_+) \sqcup (P, V)$ which is isotopic to the identity among such maps, and such that $(\mathrm{d}f_{p_+})^{-1} \circ C_2 \circ \mathrm{d}f_{p_-} = C_1$. In particular f induces a diffeomorphism

$f : \Sigma_{C_1} \to \Sigma_{C_2}$ compatible with $f : \Sigma \to \Sigma$, which maps L_{C_1} to L_{C_2}. Any two such diffeomorphisms of Σ induce isotopic diffeomorphisms from Σ_1 to Σ_2.

DEFINITION 2.9. Let $\mathbf{\Sigma} = (\Sigma, \{p_-, p_+\} \sqcup P, \{v_-, v_+\} \sqcup V, L)$ be a marked surface. Let

$$C : \mathcal{P}(T_{p_-}\Sigma) \to \mathcal{P}(T_{p_+}\Sigma)$$

be a glueing map and Σ_C the glueing of Σ at the ordered pair $((p_-, v_-), (p_+, v_+))$ with respect to C. Let $L_C \subseteq H_1(\Sigma_C, \mathbb{Z})$ be the Lagrangian subspace constructed above from L. Then the marked surface $\mathbf{\Sigma}_C = (\Sigma_C, P, V, L_C)$ is defined to be the *glueing* of $\mathbf{\Sigma}$ at the ordered pair $((p_-, v_-), (p_+, v_+))$ with respect to C.

We observe that glueing also extends to morphisms of marked surfaces which preserve the ordered pair $((p_-, v_-), (p_+, v_+))$, by using glueing maps which are compatible with the morphism in question.

2.3. The axioms for a modular functor. We now give the axioms for a modular functor. This notion is due to G. Segal and appeared first in [**84**]. We present them here in a topological form, which is closer to Walker [**94**]. We note that similar, but different, axioms for a modular functor are given in [**89**], relying on modular tensor categories. At present, it is not known whether the definition in [**89**] is equivalent to ours.

DEFINITION 2.10. A *label set* Λ is a finite set equipped with an involution $\lambda \mapsto \lambda^\dagger$ and a distinguished element 1 such that $1^\dagger = 1$.

DEFINITION 2.11. Let Λ be a label set. The category of Λ-*labeled marked surfaces* consists of marked surfaces with an element of Λ assigned to each of the marked point. An assignment of elements of Λ to the marked points of $\mathbf{\Sigma}$ is called a labeling of $\mathbf{\Sigma}$ and we denote the labeled marked surface by $(\mathbf{\Sigma}, \vec{\lambda})$, where $\vec{\lambda}$ is the labeling. Morphisms of labeled marked surfaces are required to preserve the labelings.

We define a labeled pointed surface similarly.

REMARK 2.12. The operation of disjoint union clearly extends to labeled marked surfaces. When we extend the operation of orientation reversal to labeled marked surfaces, we also apply the involution † to all the labels.

DEFINITION 2.13. A *modular functor* based on the label set Λ is a functor $\mathcal{V}$ from the category of labeled marked surfaces to the category of finite dimensional complex vector spaces satisfying the axioms MF1 to MF5 below.

MF1. Disjoint union axiom. The operation of disjoint union of labeled marked surfaces is taken to the operation of tensor product, i.e. for any pair of labeled marked surfaces there is an isomorphism

$$\mathcal{V}((\mathbf{\Sigma}_1, \vec{\lambda}_1) \sqcup (\mathbf{\Sigma}_2, \vec{\lambda}_2)) \cong \mathcal{V}(\mathbf{\Sigma}_1, \vec{\lambda}_1) \otimes \mathcal{V}(\mathbf{\Sigma}_2, \vec{\lambda}_2).$$

The identification is associative.

MF2. Glueing axiom. Let $\mathbf{\Sigma}$ and $\mathbf{\Sigma}_C$ be marked surfaces such that $\mathbf{\Sigma}_C$ is obtained from $\mathbf{\Sigma}$ by glueing at an ordered pair of points and projective tangent vectors with respect to a glueing map C. Then there is an isomorphism

$$\mathcal{V}(\mathbf{\Sigma}_C, \vec{\lambda}) \cong \bigoplus_{\mu \in \Lambda} \mathcal{V}(\mathbf{\Sigma}, \mu, \mu^\dagger, \vec{\lambda}),$$

which is associative, compatible with glueing of morphisms, disjoint unions and it is independent of the choice of the glueing map in the obvious way (see Remark 2.8).

MF3. Empty surface axiom. Let $\varnothing$ denote the empty labeled marked surface. Then

$$\dim \mathcal{V}(\varnothing) = 1.$$

MF4. Once punctured sphere axiom. Let $\mathbf{\Sigma} = (S^2, \{p\}, \{v\}, 0)$ be a marked sphere with one marked point. Then

$$\dim \mathcal{V}(\mathbf{\Sigma}, \lambda) = \begin{cases} 1, & \lambda = 1 \\ 0, & \lambda \neq 1. \end{cases}$$

MF5. Twice punctured sphere axiom. Let $\mathbf{\Sigma} = (S^2, \{p_1, p_2\}, \{v_1, v_2\}, \{0\})$ be a marked sphere with two marked points. Then

$$\dim \mathcal{V}(\mathbf{\Sigma}, \lambda, \mu) = \begin{cases} 1, & \lambda = \mu^\dagger \\ 0, & \lambda \neq \mu^\dagger. \end{cases}$$

In addition to the above axioms one may require extra properties, namely:

MF-D. Orientation reversal axiom. The operation of orientation reversal of labeled marked surfaces is compatible with the operation of taking the dual vector space, i.e for any labeled marked surface $(\mathbf{\Sigma}, \vec{\lambda})$ there is a pairing

$$\langle \cdot, \cdot \rangle : \mathcal{V}(\mathbf{\Sigma}, \vec{\lambda}) \otimes \mathcal{V}(-\mathbf{\Sigma}, \vec{\lambda}^\dagger) \longrightarrow \mathbb{C}, \tag{2.1}$$

compatible with disjoint unions, glueings and orientation reversals (in the sense that the induced isomorphisms $\mathcal{V}(\mathbf{\Sigma}, \vec{\lambda}) \cong \mathcal{V}(-\mathbf{\Sigma}, \vec{\lambda}^\dagger)^*$ and $\mathcal{V}(-\mathbf{\Sigma}, \vec{\lambda}^\dagger) \cong \mathcal{V}(\mathbf{\Sigma}, \vec{\lambda})^*$ are adjoints).

MF-U. Unitarity axiom. Every vector space $\mathcal{V}(\mathbf{\Sigma}, \vec{\lambda})$ is furnished with a hermitian inner product

$$(\cdot, \cdot) : \overline{\mathcal{V}(\mathbf{\Sigma}, \vec{\lambda})} \otimes \mathcal{V}(\mathbf{\Sigma}, \vec{\lambda}) \longrightarrow \mathbb{C}$$

so that morphisms induce unitary transformations. The hermitian structure must be compatible with disjoint union and glueing. If we have the orientation reversal property, then compatibility with the unitary structure means that we have commutative diagrams

$$\begin{array}{ccc} \mathcal{V}(\mathbf{\Sigma}, \vec{\lambda}) & \xrightarrow[\cong]{} & \mathcal{V}(-\mathbf{\Sigma}, \vec{\lambda}^\dagger)^* \\ \downarrow \cong & & \cong \downarrow \\ \overline{\mathcal{V}(\mathbf{\Sigma}, \vec{\lambda})^*} & \xrightarrow{\cong} & \overline{\mathcal{V}(-\mathbf{\Sigma}, \vec{\lambda}^\dagger)}, \end{array}$$

where the vertical identifications come from the hermitian structure and the horizontal identifications from the pairing (2.1).

2.4. The Teichmüller space of marked surfaces. Let us first review some basic Teichmüller theory. Let Σ be an oriented closed smooth surface and let P be a finite set of points on Σ. The usual Teichmüller space for a pointed surface (Σ, P) consists of equivalence classes of diffeomorphisms $\varphi : (\Sigma, P) \to (C, Q)$, where C is a Riemann surface and $Q \subset C$ is finite set of points, namely

$$\mathcal{T}_{\Sigma, P} = \{\varphi : (\Sigma, P) \to (C, Q)\} / \sim \tag{2.2}$$

where we declare that two diffeomorphisms $\varphi_i : (\Sigma, P) \to (C_i, Q_i)$ for $i = 1, 2$ are equivalent if there exists a biholomorphic map $\Phi : (C_1, Q_1) \to (C_2, Q_2)$ such that

$\Phi \circ \varphi_1$ is isotopic to φ_2 by diffeomorphisms preserving P. If $P = \emptyset$, this space is simply denoted $\mathcal{T}_\Sigma$. We will also consider the "decorated Teichmüller space" consisting of equivalence classes of diffeomorphisms $\varphi : (\Sigma, P) \to (C, Q)$, where C is a Riemann surface, $Q \subset C$ is finite set of points, and $W \in T_Q C$ non-zero tangent vectors

$$\tilde{\mathcal{T}}_{\Sigma,P} = \{\varphi : (\Sigma, P) \to (C, Q, W)\} / \approx$$

where $\approx$ is now the equivalence relation where we ask that the isotopies preserve $(P, (\mathrm{d}_P\varphi)^{-1}(W))$. We have natural projection maps $\tilde{\pi}_{\Sigma,P} : \tilde{\mathcal{T}}_{\Sigma,P} \to \mathcal{T}_{\Sigma,P}$, $\pi_P : \mathcal{T}_{\Sigma,P} \to T_P\Sigma$ and $\pi_\Sigma : \mathcal{T}_{\Sigma,P} \to \mathcal{T}_\Sigma$.

THEOREM 2.2 (Bers). *There is a natural structure of a finite dimensional complex analytic manifold on the Teichmüller spaces $\mathcal{T}_{\Sigma,P}$ and $\tilde{\mathcal{T}}_{\Sigma,P}$. The mapping class group $\Gamma_{\Sigma,P}$ acts biholomorphically on $\mathcal{T}_{\Sigma,P}$, as does $\tilde{\Gamma}_{\Sigma,P}$ on $\tilde{\mathcal{T}}_{\Sigma,P}$.*

PROPOSITION 2.3. *$\tilde{\mathcal{T}}_{\Sigma,P}$ is a principal $\mathbb{C}^P$-bundle over $\mathcal{T}_{\Sigma,P}$ on which $\tilde{\Gamma}_{\Sigma,P}$ acts, covering the action of $\Gamma_{\Sigma,P}$ on $\mathcal{T}_{\Sigma,P}$. Moreover $\mathcal{T}'_{\Sigma,P} = \tilde{\mathcal{T}}_{\Sigma,P}/\langle \delta_p \mid p \in P\rangle$ is a principal $(\mathbb{C}^*)^P$-bundle over $\mathcal{T}_{\Sigma,P}$, such that the induced projection $\tilde{\pi}'_{\Sigma,P} : \tilde{\mathcal{T}}_{\Sigma,P} \to \mathcal{T}'_{\Sigma,P}$ is the fiberwise universal cover with respect to the projection $\pi'_{\Sigma,P} : \mathcal{T}'_{\Sigma,P} \to \mathcal{T}_{\Sigma,P}$, compatible with the exponential map $e^{2\pi\mathrm{i}\cdot} : \mathbb{C}^P \to (\mathbb{C}^*)^P$ on the structure groups.*

PROOF. For every $\varphi : (\Sigma, P) \to (C, Q, W)$ representing a point in $\mathcal{T}_{\Sigma,P}$, we have the map

$$\pi_P|_{\tilde{\pi}^{-1}_{\Sigma,P}([\varphi])} : \tilde{\pi}^{-1}_{\Sigma,P}([\varphi]) \to T_P\Sigma,$$

induced by assigning $(\mathrm{d}_P\varphi')^{-1}(W)$ to a diffeomorphism $\varphi' : (\Sigma, P) \to (C, Q, W)$ which represents a point $\tilde{\pi}^{-1}_{\Sigma,P}([\varphi])$. By the very definition of $\tilde{\mathcal{T}}_{\Sigma,P}$ this map is independent of the representative φ' of a point in $\tilde{\pi}^{-1}_{\Sigma,P}([\varphi])$. Now pick a $\varphi' : (\Sigma, P) \to (C, Q, W)$ representing a point in $\tilde{\pi}^{-1}_{\Sigma,P}([\varphi])$ and let $V \in T_P\Sigma$ be given by

$$V = (\mathrm{d}_P\varphi')^{-1}(W).$$

Then consider the set $(\pi_P|_{\tilde{\pi}^{-1}([\varphi])})^{-1}(V)$. We claim that the group $\langle \delta_p \mid p \in P\rangle$ acts transitively on this set. To see this, let $\varphi'' : (\Sigma, P) \to (C, Q, W)$ represent another point in $(\pi_P|_{\tilde{\pi}^{-1}_{\Sigma,P}([\varphi])})^{-1}(V)$. Since the two diffeomorphisms φ' and φ'' must represent the same element in $\mathcal{T}_{\Sigma,P}$, there exists Φ such that

$$(\varphi'')^{-1} \circ \Phi \circ \varphi' : (\Sigma, P) \to (\Sigma, P)$$

is isotopic to the identity. But then there exists a diffeomorphism $\psi : (\Sigma, P) \to (\Sigma, P)$ which represents an element in $\langle \delta_p \mid p \in P\rangle$ such that

$$(\varphi'')^{-1} \circ \Phi \circ (\varphi' \circ \psi) : (\Sigma, P) \to (\Sigma, P)$$

is isotopic to the identity within diffeomorphisms of (Σ, P, V). This means that $\varphi' \circ \psi$ represents the same point in $\tilde{\mathcal{T}}_{\Sigma,P}$ as φ'' does.

It now follows that $\tilde{\mathcal{T}}_{\Sigma,P}$ is a principal $\mathbb{C}^P$-bundle over $\mathcal{T}_{\Sigma,P}$ and that $\mathcal{T}'_{\Sigma,P} = \tilde{\mathcal{T}}_{\Sigma,P}/\langle \delta_p \mid p \in P\rangle$ is a principal $(\mathbb{C}^*)^P$-bundle over $\mathcal{T}_{\Sigma,P}$. □

2.5. Construction of line bundles over Teichmüller spaces.

2.5.1. *Fiber products.* For each $p \in P$, we consider the representation $\varrho_p : (\mathbb{C}^*)^P \to \mathrm{Aut}(\mathbb{C})$, which is obtained by projection on the factor corresponding to p. We denote $\tilde{\mathcal{L}}_p$ the line bundle over $\mathcal{T}_{\Sigma,P}$ associated to $\mathcal{T}'_{\Sigma,P}$ and the representation ϱ_p.

Take $\alpha \in \mathbb{C}$. We now show how to construct the α'th power of $\tilde{\mathcal{L}}_p$ over $\mathcal{T}_{\Sigma,P}$. To this end consider the map $\tilde{\varrho}_p : \mathbb{C}^P \to \mathbb{C}$, which is simply the projection onto the factor corresponding to $p \in P$. Now we define

$$\tilde{\mathcal{L}}_p^{\alpha} = \tilde{\mathcal{T}}_{\Sigma,P} \times_{\alpha\tilde{\varrho}_p} \mathbb{C},$$

where the action of $\mathbb{C}^P$ on $\tilde{\mathcal{T}}_{\Sigma,P} \times \mathbb{C}$ is given by

$$w(\varphi, z) = (\varphi w, \mathbf{e}(\alpha\tilde{\varrho}_p(w))z), \qquad \mathbf{e}(t) = \exp(2\mathrm{i}\pi t).$$

We observe that $\tilde{\Gamma}_{\Sigma,P}$ acts on $\tilde{\mathcal{L}}_p^{\alpha}$ covering the action of $\Gamma_{\Sigma,P}$ on $\mathcal{T}_{\Sigma,P}$, and that δ_p acts by multiplication by $e^{2\pi i\alpha}$ while $\delta_{p'}$ acts trivially for $p' \in P - \{p\}$.

2.5.2. *Determinant of the Hodge bundle.* The Hodge bundle is the vector bundle over $\mathcal{T}_{\Sigma}$, whose fiber at the class of $\varphi : \Sigma \to C$ is $H^0(C, K_C)$. There is a natural action of Γ_{Σ} on the Hodge bundle. We denote the determinant line bundle associated to this bundle by $\mathcal{L}_D$. It is isomorphic to the line bundle $\mathcal{V}^{\dagger}_{\mathrm{ab}}(\Sigma)$, which the first author and Ueno constructs over $\mathcal{T}_{\Sigma}$ using a certain abelian CFT, namely the CFT associated to the bc-ghost system [**3**]. We observe that for a marked surface $\mathbf{\Sigma} = (\Sigma, P, V, L)$, the Lagrangian L induces a section s_L of $\mathcal{L}_D^2$ over $\mathcal{T}_{\Sigma}$, given by

$$s_L = (u_1 \wedge \cdots \wedge u_g)^{\otimes 2}$$

where $(u_1, \ldots, u_g)$ is normalized on an integral basis of L. The isomorphism between $\mathcal{L}_D$ and $\mathcal{V}^{\dagger}_{\mathrm{ab}}(\Sigma)$ takes this section of $\mathcal{L}_D^2$ to the preferred section of $(\mathcal{V}^{\dagger}_{\mathrm{ab}}(\Sigma))^2$ over $\mathcal{T}_{\Sigma}$ as described in [**3**]. By [**4**, Theorem 11.3], this section allows us to construct a line bundle $\mathcal{L}_D^{-c/2}$ over $\mathcal{T}_{\Sigma}$ for any $c \in \mathbb{C}$, on which $\Gamma_{\mathbf{\Sigma}}$ acts, such that $(\mathrm{Id}, 1)$ acts by $e^{-\mathrm{i}\pi c/2}$. We denote by $\tilde{\mathcal{L}}_D^{-c/2}$ the pullback to $\mathcal{T}_{\Sigma,P}$ of $\mathcal{L}_D^{-c/2}$.

REMARK 2.14. We recall that the Hodge bundle over $\mathcal{T}_{\Sigma}$ has a natural hermitian structure, which is Γ_{Σ} invariant. Hence it induces a hermitian structure on the holomorphic bundle $\mathcal{L}_D$ which is also Γ_{Σ} invariant. Hence we get a unique unitary Chern connection in $\mathcal{L}_D$ compatible with the holomorphic structure on this line bundle, which by uniqueness is also Γ_{Σ} invariant. By the proof of [**4**, Theorem 11.3] we see that for any $c \in \mathbb{R}$, we get an induced $\Gamma_{\mathbf{\Sigma}}$-invariant unitary connection in $\tilde{\mathcal{L}}_D^{-c/2}$, whose curvature is $-c/2$ times the curvature of the Chern connection in $\tilde{\mathcal{L}}_D$.

The moduli space of (Σ, P) is by definition

$$\mathcal{M}_{\Sigma,P} = \mathcal{T}_{\Sigma,P}/\Gamma_{\Sigma,P} \tag{2.3}$$

When $\chi(\Sigma, P) < 0$, the stabilizers are finite, so there is a natural structure of an orbifold on $\mathcal{M}_{\Sigma,P}$. We see that there is a natural action of $\Gamma_{\Sigma,P}$ on $\tilde{\mathcal{L}}_p$, also acting with finite stabilizers and we define the orbifold line bundle over $\mathcal{M}_{\Sigma,P}$

$$\mathcal{L}_p = \tilde{\mathcal{L}}_p/\Gamma_{\Sigma,P}$$

REMARK 2.15. By picking an orbifold Hermitian structure on the holomorphic bundle $\mathcal{L}_p$ and pulling it back to $\tilde{\mathcal{L}}_p$, we see that the corresponding Chern connection in $\tilde{\mathcal{L}}_p$ is Γ_Σ invariant. Hence, for any non-zero $r_p \in \mathbb{R}$, we get an induced unitary connection in $\tilde{\mathcal{L}}_p^{r_p}$, which is $\Gamma_{\mathbf{\Sigma}}$ invariant, and whose curvature is r_p times that of the Chern connection of $\tilde{\mathcal{L}}_p$

The moduli space $\mathcal{M}_{\Sigma,P}$ also carries the Hodge bundle, simply by pulling the Hodge bundle over $\mathcal{M}_\Sigma$ back to $\mathcal{M}_{\Sigma,P}$. We also denote the determinant bundle of this pull back of the Hodge bundle over $\mathcal{M}_{\Sigma,P}$ by $\mathcal{L}_D$ and it remains of course an orbifold line bundle over $\mathcal{M}_{\Sigma,P}$. We denote $\psi_D = c_1(\mathcal{L}_D)$ and $\psi_p = c_1(\mathcal{L}_p)$ and think of them as rational cohomology classes over $\mathcal{M}_{\Sigma,P}$.

2.6. Modular functor and vector bundles over Teichmüller space. We now fix a modular functor, without assuming the unitarity and orientation reversal axioms.

2.6.1. *Scalars.*

- The modular functor gives a linear isomorphism
$$\mathcal{V}_{\vec{\lambda}}(\mathrm{id},1) : \mathcal{V}_{\vec{\lambda}}(\mathbf{\Sigma}) \to \mathcal{V}_{\vec{\lambda}}(\mathbf{\Sigma})$$
and the axioms imply that it acts as multiplication by a scalar $\tilde{c} \in \mathbb{C}^*$ independent of $\vec{\lambda}$ and the topology of $\mathbf{\Sigma}$. This is proved by factoring the $(\mathrm{Id}_\Sigma, 1)$ along the boundary of an embedded disk Σ_0 in $\Sigma - P$ and writing it as $(\mathrm{Id}_\Sigma, 1) = (\mathrm{Id}_{\Sigma_0}, 1) \cup (\mathrm{Id}_{\Sigma'}, 0)$, where $\Sigma' = \Sigma - \Sigma_0$.
- For the sphere with 2 points, we have
$$\begin{aligned} \mathrm{Aut}(\mathbb{S}^2, \{p_1,p_2\}, \{v_1,v_2\}, L) &= \mathrm{Aut}(\mathbb{S}^2, \{p_1,p_2\}, \{v_1,v_2\}) \times \mathbb{Z} \\ \mathrm{Aut}(\mathbb{S}^2, \{p_1,p_2\}, \{v_1,v_2\}) &= \langle \delta \rangle. \end{aligned}$$
Here δ is the Dehn twist along the equator. The modular functor gives a linear isomorphism
$$\mathcal{V}_\lambda(\delta) : \mathcal{V}_{\lambda,\lambda^\dagger}(\mathbb{S}^2, P, V, 0) \to \mathcal{V}_{\lambda,\lambda^\dagger}(\mathbb{S}^2, P, V, 0).$$
Since $\mathcal{V}_{\lambda,\lambda^\dagger}(\mathbb{S}^2, P, V, 0)$ is one dimensional, this linear map equals multiplication by a scalar $\tilde{r}_\lambda \in \mathbb{C}^*$.

LEMMA 2.4. *The Dehn twists around the punctures have the following properties.*

(i) $\tilde{r}_1 = 1$.

(ii) *For any marked surface $(\mathbf{\Sigma}, \vec{\lambda})$, the Dehn twist around a marked point p with label λ_p acts on $\mathcal{V}_{\vec{\lambda}}(\mathbf{\Sigma})$ by multiplication with $\tilde{r}_{\lambda_p}$.*

(iii) *For any $\lambda \in \Lambda$, $\tilde{r}_\lambda = \tilde{r}_{\lambda^\dagger}$.*

PROOF. For (i) and (ii), we use the factorization axiom. (i) – The Dehn twist is trivial on $\mathbb{S}^2$ and we can factor $\mathbb{S}^2$ along two curves, which are respectively the equator pushed a small amount into the northern (resp. southern) hemisphere. It then follows by considering this factorisation that $\tilde{r}_1 = 1$. (ii) – We factor in the boundary of an embedded disk in $\Sigma - (P - \{p\})$, centred in p and then we do the Dehn twist around p inside the disk. The result only depends on the label λ_p at p, and not on the topology of the remaining surface neither on the labels at other punctures. (iii) – We can choose the points (p_1, p_2) and vectors (v_1, v_2) such that the (orientation preserving) map $z \to -1/z$ takes $(\mathbb{S}^2, \{p_1,p_2\}, \{v_1,v_2\})$ to itself, and takes the equator to the itself. Hence, it commutes with the Dehn twist, and that implies $\tilde{r}_\lambda = \tilde{r}_{\lambda^\dagger}$ for all $\lambda \in \Lambda$. □

2.6.2. *Trivial vector bundles over $\mathcal{T}_{\Sigma,P}$.* To a given Λ-marked surface $(\mathbf{\Sigma},\vec{\lambda})$, we associate the trivial vector bundle

$$\tilde{\mathcal{Z}}^0_{(\mathbf{\Sigma},\vec{\lambda})} := \mathcal{T}_{\Sigma,P} \times \mathcal{V}_{\vec{\lambda}}(\mathbf{\Sigma}).$$

If the modular functor is unitary, this bundle carries a $\Gamma_{\mathbf{\Sigma}}$-invariant unitary structure. In any case, we can equip it with the trivial flat connection, which is of course $\Gamma_{\mathbf{\Sigma}}$-equivariant. Subsequently, the $\Gamma_{\mathbf{\Sigma}}$-equivariant Chern character is trivial. According to the above discussion the Dehn twist δ_p around a point p acts by multiplication by $\tilde{r}_\lambda$. Moreover $(\mathrm{Id},1)$ in $\Gamma_{\mathbf{\Sigma}}$ acts by multiplication by $\tilde{c}$.

2.6.3. *Vector bundle over the moduli space.* Pick up $c, r_\lambda \in \mathbb{C}$ such that[1] $\tilde{c} = \mathbf{e}(c/4)$ and $\tilde{r}_\lambda = \mathbf{e}(r_\lambda)$. We observe that δ_p, $p \in P$ as well as $(\mathrm{Id}_\Sigma, 1)$ act trivially on the vector bundle

$$\tilde{\mathcal{Z}}_{(\mathbf{\Sigma},\vec{\lambda})} = \tilde{\mathcal{Z}}^0_{(\mathbf{\Sigma},\vec{\lambda})} \otimes \tilde{\mathcal{L}}_D^{-c/2} \bigotimes_{p\in P} \tilde{\mathcal{L}}_p^{r_{\lambda_p}}.$$

Therefore we get that

THEOREM 2.5. *The action of $\Gamma_{\mathbf{\Sigma}}$ factors to an action of $\Gamma_{\Sigma,P}$ and hence we can define*

$$\mathcal{Z}_{(\mathbf{\Sigma},\vec{\lambda})} = \tilde{\mathcal{Z}}_{(\mathbf{\Sigma},\vec{\lambda})}/\Gamma_{\Sigma,P}$$

as an orbifold bundle over the moduli space $\mathcal{M}_{\Sigma,P}$.

For a variable t, the weighted Chern character is obtained from the Chern character by weighting the term in cohomological degree $2j$ by the weight t^j.

PROPOSITION 2.6. *For any modular functor $\mathcal{V}$ the weighted Chern character of this vector bundle is*

$$\mathrm{Ch}_t(\mathcal{Z}_{(\mathbf{\Sigma},\vec{\lambda})}) = \dim \mathcal{V}_{\vec{\lambda}}(\mathbf{\Sigma}) \exp\Big\{t\Big(-\frac{c}{2}\Lambda_1 + \sum_{p\in P} r_{\lambda_p}\psi_p\Big)\Big\} \tag{2.4}$$

where Λ_1 is the first Chern class of the Hodge bundle.

PROOF. We consider the tensor product connection of the trivial connection in $\tilde{\mathcal{Z}}^0_{(\mathbf{\Sigma},\vec{\lambda})}$ and then the $\Gamma_{\mathbf{\Sigma}}$-invariant unitary connections constructed in the bundles $\tilde{\mathcal{L}}_D^{-c/2}$ and $\tilde{\mathcal{L}}_p^{r_{\lambda_p}}$ in Remarks 2.14-2.15. By Chern-Weil theory we therefore have that

$$\begin{aligned}
\mathrm{Ch}_t(\mathcal{Z}_{(\mathbf{\Sigma},\vec{\lambda})}) &= \dim \mathcal{V}_{\vec{\lambda}}(\mathbf{\Sigma}) \exp\Big\{t\Big(-\frac{c}{2}c_1(\mathcal{L}_D) + \sum_{p\in P} r_{\lambda_p} c_1(\mathcal{L}_p)\Big)\Big\} \\
&= \dim \mathcal{V}_{\vec{\lambda}}(\mathbf{\Sigma}) \exp\Big\{t\Big(-\frac{c}{2}\Lambda_1 + \sum_{p\in P} r_{\lambda_p}\psi_p\Big)\Big\}.
\end{aligned}$$

□

Since Chern classes of vector bundles over the orbifold $\mathcal{M}_{\Sigma,P}$ are rational, c and r_λ must all be rationals. We thus get an alternative proof and generalisation of Vafa's theorem [**92**], as follows.

COROLLARY 2.7. *For any modular functor, $\tilde{c}$ and $\tilde{r}_\lambda$ for any $\lambda \in \Lambda$ are roots of unity.*

We recall that Vafa's original proof was written for unitary modular functors, based on an arithmetic argument following from relations in the mapping class group. Corollary 2.7 can be improved to show the following corollary.

[1] With these convention, c is the Virasoro central charge, and r_λ the conformal dimension.

COROLLARY 2.8. *For any $\gamma \in \mathrm{Aut}(\mathbf{\Sigma})$ which is the product of Dehn twists in non-intersecting curves, the element $\mathcal{V}_{\vec{\lambda}}(\gamma) \in \mathrm{Aut}(\mathcal{V}_{\vec{\lambda}}(\mathbf{\Sigma}))$ has finite order.*

This follows immediately by factoring along two simple closed curves on either side of the simple closed curve of the Dehn twist. We recall that in general, not all elements in the representations of the mapping class group provided by $\mathcal{V}$ have finite order.

2.7. Extension to the boundary. We prove in this section the following theorem.

THEOREM 2.9. *$\mathcal{Z}_{(\mathbf{\Sigma},\vec{\lambda})}$ extends to an orbifold bundle over the Deligne-Mumford compactification $\overline{\mathcal{M}}_{\Sigma,P}$.*

In order to extend the above constructions to the Deligne-Mumford compactification of the moduli space, we introduce the augmented Teichmüller space and extend all our constructions in a mapping class group equivariant way to the augmented Teichmüller space.

Let (Σ, P) be a pointed surface. We introduce the set of contraction cycles $\mathcal{C}_{(\Sigma,P)}$ on (Σ, P). It consists of isotopy classes of 1-dimensional submanifolds C of $\Sigma - P$, such that connected components of C are non-contractible, nor are any two connected components of C isotopic in $\Sigma - P$, nor are any of the components contractible into any of the points in P. We remark that $C = \varnothing$ is allowed.

The augmented Teichmüller space $\mathcal{T}^a_{\Sigma,P}$ for a pointed surface (Σ, P) consists of equivalence classes of continuous maps $\varphi : (\Sigma, C, P) \to (X, N, Q)$, where $C \in \mathcal{C}_{(\Sigma,P)}$, X is a nodal Riemann surface with nodes N, and $Q \subset X$ is finite set of points of $X - N$, such that $\varphi(C) = N$ and the restricted map $\varphi : (\Sigma - C, P) \to (X - N, Q)$ is a diffeomorphism.

$$\mathcal{T}^a_{\Sigma,P} = \{\varphi : (\Sigma, C, P) \to (X, N, Q)\} / \sim^a$$

where we declare that two continuous maps $\varphi_i : (\Sigma, C, P) \to (X_i, N_i, Q_i)$ for $i = 1, 2$ are $\sim^a$-equivalent if there exists a biholomorphic map $\Phi : (X_1, N_1, Q_1) \to (X_2, N_2, Q_2)$ such that $\Phi \circ \varphi_1$ is isotopic to φ_2 via continuous maps from (Σ, C, P) to (X_2, N_2, Q_2) which are diffeomorphisms from $(\Sigma - C, P)$ to $(X_2 - N_2, Q_2)$. If $P = \varnothing$, this space is simply denoted $\mathcal{T}^a_\Sigma$.

The augmented Teichmüller space $\mathcal{T}^a_{\Sigma,P}$ has the topology uniquely determined by following property. Suppose $\pi : Z \to D$ is a holomorphic map from a complex 2-dimensional manifold Z to the unit disk D in the complex plane, such that $\pi^{-1}(x)$ is a nodal Riemann surface for all $x \in D$. Suppose further that we are given a continuous map $\Phi : \Sigma \times D \to Z$, satisfying the two conditions:

- $\Phi(\Sigma \times \{x\}) = \pi^{-1}(\{x\})$
- if N_x are the nodes of $\pi^{-1}(x)$ then $C_x = \Phi^{-1}(N_x)$ is a submanifold of $\Sigma - P$ such that $[C_x] \in \mathcal{C}_{(\Sigma,P)}$, and the restricted map $\Phi : \Sigma - C_x \to \pi^{-1}(x) - N_x$ is a diffeomorphism for all $x \in D$.

Then, the map from D to $\mathcal{T}^a_{\Sigma,P}$, which sends x to the restricted map $\Phi : (\Sigma, C_x, P) \to (\pi^{-1}(x), N_x, \Phi(P \times \{x\}))$ is continuous.

We observe that the mapping class group $\Gamma_{\Sigma,P}$ acts on $\mathcal{T}^a_{\Sigma,P}$, and the quotient is the Deligne-Mumford compactification $\overline{\mathcal{M}}_{g,n}/\mathfrak{S}_n$ of the moduli space of genus g curves with $n = |P|$ marked unordered points (see section 2.8).

We will also consider the "decorated augmented Teichmüller space" consisting of equivalence classes of continuous maps $\varphi : (\Sigma, C, P) \to (X, N, Q)$, where X is a Riemann surface, $Q \subset X$ is a finite set of points, and $W \in T_Q X$ non-zero tangent vectors.

$$\tilde{\mathcal{T}}^a_{\Sigma,P} = \{\varphi : (\Sigma, C, P) \to (X, N, Q, W)\} / \approx^a$$

where $\approx^a$ is now the equivalence relation where we ask that the isotopies preserve $(P, (\mathrm{d}_P\varphi)^{-1}(W))$. We have the following natural projection maps among the augmented Teichmüller spaces.

$$\tilde{\pi}^a_{\Sigma,P} : \tilde{\mathcal{T}}^a_{\Sigma,P} \to \mathcal{T}^a_{\Sigma,P}, \qquad \pi^a_P : \mathcal{T}^a_{\Sigma,P} \to T_P\Sigma, \quad \text{and} \quad \pi^a_\Sigma : \mathcal{T}^a_{\Sigma,P} \to \mathcal{T}^a_\Sigma.$$

The proof of Proposition 2.3 applies word for word to extend the Proposition to the augmented setting. But then we get that all the constructions of Section 2.6 extend to constructions over augmented Teichmüller spaces and hence also to the Deligne-Mumford compactification $\overline{\mathcal{M}}_{\Sigma,P}$ of the moduli spaces of genus g curves with n marked unordered points.

Concerning the extension of the bundle $\tilde{\mathcal{L}}_D$ and the section s_L of $\tilde{\mathcal{L}}^2_D$ to augmented Teichmüller space, we appeal to the constructions of the first author and Ueno presented in [**3**]. By the constructions of [**3**, Section 5] we see that the bundle $\mathcal{V}^\dagger_{\mathrm{ab}}(\Sigma)$ extends to a holomorphic bundle over augmented Teichmüller space. Moreover the preferred section of $\mathcal{V}^\dagger_{\mathrm{ab}}(\Sigma)$ extends to a nowhere vanishing section of the extension of $\mathcal{V}^\dagger_{\mathrm{ab}}(\Sigma)$ to augmented Teichmüller space as is proved in [**3**, Section 6]. From this we conclude that the bundle $\tilde{\mathcal{L}}^{-c/2}_D$ extends to augmented Teichmüller space and that the action of $\Gamma_{\boldsymbol{\Sigma}}$ also extends.

2.8. Remark on ordering of punctures. In the usual definition of the moduli space $\mathcal{M}_{g,n}$, it is assumed that the marked points are ordered from 1 to n.

If (Σ, P) is a pointed surface such that Σ has genus g and $|P| = n$, in our definition of the Teichmüller space $\mathcal{T}_{\Sigma,P}$ in (2.2), the permutations of the n points are divided out. We therefore have

$$\mathcal{M}_{\Sigma,P} \simeq \mathcal{M}_{g,n}/\mathfrak{S}_n$$

and likewise for the Deligne-Mumford compactifications. Later in the text, we work only with the pull-back to $\overline{\mathcal{M}}_{g,n}$ of the bundle $\mathcal{V}_{\vec{\lambda}}$ that was so far obtained over $\overline{\mathcal{M}}_{\Sigma,P}$. The formula for the Chern character in Theorem 2.9 is the same for the bundle over $\mathcal{M}_{g,n}$.

3. Cohomological field theories

3.1. Generalities.

3.1.1. *Frobenius algebras.* A Frobenius algebra is a finite dimensional complex vector space $\mathcal{A}$, equipped with a symmetric, non-degenerate bilinear form $b : \mathcal{A} \otimes \mathcal{A} \to \mathbb{C}$, and an associative, commutative $\mathbb{C}$-linear morphism $\times : \mathcal{A} \otimes \mathcal{A} \to \mathcal{A}$ such that

$$\forall a_1, a_2, a_3 \in \mathcal{A}, \qquad b(a_1, a_2 \times a_3) = b(a_1 \times a_2, a_3).$$

We require the existence of a unit for the product, denoted $\mathbb{1}$.

The algebra $\mathcal{A}$ is semi-simple if there exists a $\mathbb{C}$-linear basis $(\tilde{\varepsilon}_1, \ldots, \tilde{\varepsilon}_n)$ such that

$$\forall (i,j) \in [\![1,n]\!], \qquad \tilde{\varepsilon}_i \times \tilde{\varepsilon}_j = \delta_{ij}\, \tilde{\varepsilon}_i. \tag{3.1}$$

The unit is then $\mathbb{1} = \sum_i \tilde{\varepsilon}_i$, and (3.1) implies that the bilinear form is diagonal in this basis and reads

$$b(\tilde{\varepsilon}_i, \tilde{\varepsilon}_j) = \frac{\delta_{ij}}{\Delta_i}.$$

We say that $(\tilde{\varepsilon}_i)_i$ is a canonical basis. It is sometimes more convenient to work with the orthonormal basis $\underline{\varepsilon}_i = \Delta_i^{1/2}\tilde{\varepsilon}_i$ which satisfies that

$$\underline{\varepsilon}_i \times \underline{\varepsilon}_j = \delta_{ij}\Delta_i^{1/2}\,\underline{\varepsilon}_i, \qquad b(\underline{\varepsilon}_i, \underline{\varepsilon}_j) = \delta_{ij}.$$

Then b induces a bivector $b^\dagger \in \mathcal{A}\otimes\mathcal{A}$, that will play an important role. In a canonical or an orthonormal basis it reads

$$b^\dagger = \sum_i \underline{\varepsilon}_i \otimes \underline{\varepsilon}_i.$$

3.1.2. *CohFTs.* A cohomological field theory (CohFT) is the data of a finite dimensional complex vector space $\mathcal{A}$ with a symmetric bilinear non-degenerate $b : \mathcal{A}\otimes\mathcal{A} \to \mathbb{C}$ and a sequence $\Omega_{g,n} \in H^\bullet(\overline{\mathcal{M}}_{g,n}) \otimes \mathcal{A}^{\otimes n}$ indexed by integers $g \geq 0$ and $n \geq 0$ such that $2g - 2 + n > 0$, satisfying the axioms given below. Since b gives a canonical identification of $\mathcal{A}$ with its dual $\mathcal{A}^*$, we can equivalently consider $\Omega^*_{g,n} \in H^\bullet(\overline{\mathcal{M}}_{g,n}) \otimes (\mathcal{A}^*)^{\otimes n}$. The axioms are.

- The unit $\mathbb{1} \in \mathcal{A}$ is such that the pairing is given by

$$\forall a_1, a_2 \in \mathcal{A}, \qquad b(a_1, a_2) = \int_{\overline{\mathcal{M}}_{0,3}} \Omega^*_{0,3}(a_1 \otimes a_2 \otimes \mathbb{1}).$$

- $\Omega^*_{g,n}$ is symmetric by simultaneous permutations of the n factors in $(\mathcal{A}^*)^{\otimes n}$ and the n punctures in $\overline{\mathcal{M}}_{g,n}$.
- Pulling back by the glueing map $\pi : \overline{\mathcal{M}}_{g,2+n} \to \overline{\mathcal{M}}_{1+g,n}$, we should have that

$$\pi^*\Omega^*_{1+g,n}(\cdot) = \Omega^*_{g,2+n}(b^\dagger \otimes \cdot).$$

- Pulling back by the glueing map $\pi : \overline{\mathcal{M}}_{g_1,1+n_1} \times \overline{\mathcal{M}}_{g_2,1+n_2} \to \overline{\mathcal{M}}_{g_1+g_2,n_1+n_2}$, we should have that

$$\pi^*\Omega^*_{g_1+g_2,n_1+n_2}(\cdot) = \sum_{i,j}(\Omega^*_{g_1,1+n_1} \otimes \Omega^{g_2,1+n_2})(p_{n_1}(b^\dagger \otimes \cdot))$$

where $p_{n_1} : (\mathcal{A}^*)^{\otimes 2} \otimes (\mathcal{A}^*)^{\otimes n_1} \otimes (\mathcal{A}^*)^{\otimes n_2} \to (\mathcal{A}^*)^{\otimes(1+n_1)} \otimes (\mathcal{A}^*)^{\otimes(1+n_2)}$ puts the second factor of $\mathcal{A}^*$ into the (n_1+2)-th position in the target.
- Pulling back by the forgetful map $\pi : \overline{\mathcal{M}}_{g,1+n} \to \overline{\mathcal{M}}_{g,n}$, we should have that

$$\pi^*\Omega^*_{g,n}(\cdot) = \Omega^*_{g,1+n}(\mathbb{1} \otimes \cdot).$$

The axioms imply that $\mathcal{A}$ is a Frobenius algebra with the product

$$b(a_1 \times a_2, a_3) = \int_{\overline{\mathcal{M}}_{0,3}} \Omega^*_{0,3}(a_1 \otimes a_2 \otimes a_3).$$

Givental [**47**] describes two basic actions on the set of CohFTs over the same Frobenius algebra. We follow the presentation of [**70**].

3.1.3. *Translations.* Let $T(u) \in u^2\mathcal{A}[[u]]$, and consider the forgetful maps $\pi_m : \overline{\mathcal{M}}_{g,m+n} \to \overline{\mathcal{M}}_{g,n}$. One can define a new CohFT by the formula

$$\begin{aligned}&(\hat{T}\Omega^*)_{g,n}(a_1 \otimes \ldots \otimes a_n)\\ &= \sum_{m\geq 0} \frac{1}{m!} \sum_{k_1,\ldots,k_m \geq 2} (\pi_m)_*\Big\{\Omega^*_{g,n+m}(T_{k_1} \otimes \cdots \otimes T_{k_m} \otimes a_1 \otimes \cdots \otimes a_n)\psi_1^{k_1}\ldots\psi_m^{k_m}\Big\}\end{aligned}$$

where $T(u) = \sum_{k\geq 2} T_k u^k$ for $T_m \in \mathcal{A}$. Here $(\pi_m)_*$ is the pushforward in cohomology classes, induced on smooth forms on the smooth compact orbifolds $\overline{\mathcal{M}}_{g,n}$.

3.1.4. *R-matrix actions.* Let $R(u) \in \mathrm{End}(\mathcal{A})[[u]]$ such that $R(0) = \mathrm{id}$ and satisfying the symplectic condition

$$R(u)R^{\dagger}(-u) = \mathrm{id}$$

where $R^{\dagger}(u) \in \mathrm{End}(\mathcal{A})[[u]]$ is the adjoint for the pairing b. One then defines[2]

$$B(u_1,u_2) := \frac{b^{\dagger} - R(u_1)\otimes R(u_2)\cdot b^{\dagger}}{u_1+u_2} \in (\mathcal{A}\otimes\mathcal{A})\,[[u_1,u_2]].$$

The symplectic condition guarantees that B is a formal power series in u_1 and u_2. One can define a new CohFT by the formula

(3.2)
$$(\hat{R}\Omega)_{g,n} := \sum_{\Gamma\in\mathcal{G}_{g,n}} \frac{1}{|\mathrm{Aut}\,\Gamma|}(\pi_\Gamma)_*\bigg(\prod_{l=\mathrm{leaf}} R(\psi_l) \prod_{\substack{e=\mathrm{edge}\\ \{v'_e,v''_e\}}} B(\psi_{v'_e},\psi_{v''_e}) \prod_{v=\mathrm{vertex}} \Omega_{g(v),n(v)}\bigg)$$

The sum is over the set $\mathcal{G}_{g,n}$ of stable graphs of topology (g,n), namely Γ meeting the following requirements.

- vertices v are trivalent, carry an integer label $g(v) \geq 0$ (the genus), and their valency $n(v)$ satisfy $2g(v) - 2 + n(v) > 0$.
- there are n leaves (1-valent vertices), labeled from 1 to n.
- $\mathrm{rank}(H_1(\Gamma,\mathbb{Z})) + \sum_{v=\mathrm{vertex}} g(v) = g$.

In (3.2), the endomorphisms are naturally composed along the graph, and we use the pushforward by the following glueing map along the graph

$$\pi_\Gamma : \prod_v \overline{\mathcal{M}}_{g(v),n(v)} \to \overline{\mathcal{M}}_{g,n}.$$

3.1.5. *Classification of semi-simple CohFT.* Given a CohFT defined by correlators $\Omega_{g,n}$, its restriction to the degree 0 part $\omega_{g,n} \in H^0(\mathcal{M}_{g,n})\otimes\mathcal{A}^{\otimes n}$ is sometimes called (abusively) a topological quantum field theory ($2d$ TQFT). Teleman [**86**], building upon the work of Givental [**47, 48**], has classified semi-simple CohFT whose underlying Frobenius algebra has dimension k.

THEOREM 3.1. [**86**] *Any semi-simple CohFT can be obtained from a degree 0 CohFT by the composition of the action of an R-matrix, and a translation such that*

$$T(u) = u\big(\mathbb{1} - R(u)\cdot\mathbb{1}\big). \tag{3.3}$$

More precisely, if Ω denote the correlators of the CohFT, and $\Omega^{\deg 0}$ the correlators of its underlying $2d$ TQFT, we have $\Omega = \hat{R}\hat{T}\Omega^{\deg 0}$.

[2]Let us remark that this notation differs from the one used in the topological recursion literature. In the topological recursion setup, B refers usually to the so-called Bergman kernel while our B corresponds to its Laplace transform often denoted by $\check{B}$.

This reconstruction is a powerful tool since the correlators $\Omega_{g,n}^{\deg 0}$ of a degree 0 CohFT with canonical basis $(\tilde{\varepsilon}_i)_i$ such that $b(\tilde{\varepsilon}_i, \tilde{\varepsilon}_j) = \delta_{ij}/\Delta_i$ and $\tilde{\varepsilon}_i \times \tilde{\varepsilon}_j = \delta_{ij}\,\tilde{\varepsilon}_i$ are

$$\Omega_{g,n}^{\deg 0} = \sum_{i=1}^{k} \Delta_i^{g-1}\,[\overline{\mathcal{M}}_{g,n}] \otimes \tilde{\varepsilon}_i \otimes \cdots \otimes \tilde{\varepsilon}_i, \tag{3.4}$$

where the Poincaré duality is implicitly used in this formula. The knowledge of the R-matrix is enough for reconstructing the correlators of a semi-simple CohFT.

3.2. Reference vector spaces attached to a modular functor. We shall now describe the CohFT defined by a modular functor, starting by the definition of the reference vector spaces underlying it. We start with a general modular functor $\mathcal{V}$. We shall work with marked surfaces of reference, and choose basis in their corresponding vector spaces.

3.2.1. *Once-punctured sphere.* $\mathbf{\Sigma}_{0,1}$ is the 2-sphere with $P = \{0\}$ and $V = \{v_0\}$ with v_0 pointing to the positive real axis. $\mathcal{V}_1(\mathbf{\Sigma}_{0,1})$ is a line, and we pick up a generator $\tilde{\zeta}[1]$. This induces an isomorphism $\mathcal{V}_1(\mathbf{\Sigma}_{0,1}) \cong \mathcal{V}_1(\mathbf{\Sigma}_{0,1})^{\star}$, and allows us to project the isomorphism from propagation of vacua

$$\mathcal{V}_\lambda(\Sigma, P, V, L) \cong \mathcal{V}_{\lambda,1}(\Sigma, P \sqcup \{p'\}, V \sqcup \{v'\}, L) \otimes \mathcal{V}_1(\mathbf{\Sigma}_{0,1})$$

to an isomorphism

$$\mathcal{V}_\lambda(\Sigma, P, V, L) \cong \mathcal{V}_{\lambda,1}(\Sigma, P \sqcup \{p'\}, V \sqcup \{v'\}, L).$$

Later on, this will be used systematically.

3.2.2. *Twice-punctured sphere.* $\mathbf{\Sigma}_{0,2}$ is the 2-sphere with $P = \{0, \infty\}$ and $V = \{v_0, v_\infty\}$ with v_∞ pointing to the negative real axis. $\mathcal{V}_{\lambda,\lambda^\dagger}(\mathbf{\Sigma}_{0,2})$ is a line. The property of propagation of vacua provides an isomorphism

$$\mathcal{V}_{1,1}(\mathbf{\Sigma}_{0,2}) \cong \mathcal{V}_1(\mathbf{\Sigma}_{0,1}) \otimes \mathcal{V}_{1,1}(\mathbf{\Sigma}_{0,2}) \tag{3.5}$$

and the property of factorization provides isomorphisms

$$\mathcal{V}_{\lambda,\lambda^\dagger}(\mathbf{\Sigma}_{0,2}) \cong \mathcal{V}_{\lambda,\lambda^\dagger}(\mathbf{\Sigma}_{0,2}) \otimes \mathcal{V}_{\lambda^\dagger,\lambda}(\mathbf{\Sigma}_{0,2}) = \mathcal{V}_{\lambda,\lambda^\dagger}(\mathbf{\Sigma}_{0,2})^{\otimes 2}. \tag{3.6}$$

They determine for each $\lambda \in \Lambda$ a unique generator $\zeta[\lambda]$ of $\mathcal{V}_{\lambda,\lambda^\dagger}(\mathbf{\Sigma}_{0,2})$ such that $\zeta[1] = \zeta[1] \otimes \tilde{\zeta}[1]$ using (3.5), and $\zeta[\lambda] = \zeta[\lambda] \otimes \zeta[\lambda]$ using (3.6).

3.2.3. *Thrice-punctured sphere.* $\mathbf{\Sigma}_{0,3}$ is the 2-sphere with $P = \{0, 1, \infty\}$ and $V = \{v_0, v_1, v_\infty\}$ with v_1 pointing in real direction to ∞. We denote the dimension of $\mathcal{V}_{\lambda,\mu,\nu}(\mathbf{\Sigma}_{0,3})$ by

$$N_{\lambda\mu\nu} = \dim\,\mathcal{V}_{\lambda,\mu,\nu}(\mathbf{\Sigma}_{0,3}).$$

Since the homomorphism from the mapping class group of $\mathbf{\Sigma}_{0,3}$ to the permutation group of the marked points is surjective, this symbol is invariant under permutation of (λ, μ, ν). Propagation of vacua gives an isomorphism

$$\mathcal{V}_{\lambda,\lambda^\dagger}(\mathbf{\Sigma}_{0,2}) \cong \mathcal{V}_{\lambda,\lambda^\dagger,1}(\mathbf{\Sigma}_{0,3}) \otimes \mathcal{V}_1(\mathbf{\Sigma}_{0,1}). \tag{3.7}$$

We fix a basis $\zeta_i[\lambda\mu\nu]$ of the space $\mathcal{V}_{\lambda,\mu,\nu}(\mathbf{\Sigma}_{0,3})$, indexed by $i = 1, \ldots, N_{\lambda\mu\nu}$. Without loss of generality, we can require that, under (3.7), we have

$$\zeta[\lambda] = \zeta_1[\lambda\lambda^\dagger 1] \otimes \tilde{\zeta}[1].$$

3.2.4. *Torus.* $\mathbf{\Sigma}_1$ is the torus $\mathbb{C}/(\mathbb{Z} \oplus \mathrm{i}\mathbb{Z})$. We denote α – resp. β – the oriented simple closed curve based at o and following the positive real axis – resp. the positive imaginary axis.

3.3. The Frobenius algebra of a modular functor.

3.3.1. *As a vector space.* If γ is a oriented simple closed curve on $\mathbf{\Sigma}_1$, we obtain a marked surface $\mathbf{\Sigma}_1^{(\gamma)}$ by considering the Lagrangian spanned by the homology class of γ. In this paragraph, we shall define a structure of Frobenius algebra on the vector space of a torus. Let

$$\mathcal{A} := \mathcal{V}_1(\mathbf{\Sigma}_1^{(\alpha)}). \tag{3.8}$$

By applying a diffeomorphism that takes (α, β) to $(\beta, -\alpha)$, we also have a natural isomorphism

$$\mathcal{A} \cong \mathcal{V}_1(\mathbf{\Sigma}_1^{(\beta)}). \tag{3.9}$$

By propagation of vacua, then factorization along α and application of a suitable diffeomorphism, we obtain from (3.8) an isomorphism

$$\mathcal{A} \cong \bigoplus_{\lambda \in \Lambda} \mathcal{V}_{\lambda,\lambda^\dagger}(\mathbf{\Sigma}_{0,2}).$$

Our previous choices of generators in the right-hand side carries to a basis e_λ of $\mathcal{A}$. Similarly, the factorization along β from (3.9) gives another basis ε_λ of $\mathcal{A}$. The change of basis is called the "S-matrix". It is the linear map $S : \mathcal{A} \to \mathcal{A}$ defined by

$$S(e_\lambda) := \varepsilon_\lambda = \sum_{\mu \in \Lambda} S_{\lambda\mu} e_\mu.$$

3.3.2. *Pairing.* We define a pairing on $\mathcal{A}$ by the following formula.

$$b(e_\lambda, e_\mu) := \delta_{\lambda\mu^\dagger} \tag{3.10}$$

3.3.3. *Involutions.* We define the charge conjugation, which is the involutive linear map $C : \mathcal{A} \to \mathcal{A}$ such that $C(e_\lambda) = e_{\lambda^\dagger}$. If $O : \mathcal{A} \to \mathcal{A}$ is a linear map, we denote $O^\top$ its adjoint with respect to the scalar product in which e_λ is an orthonormal basis, and $O^\dagger$ its adjoint for the bilinear product b. We have

$$C = C^\top = C^\dagger = C^{-1}.$$

We may sometimes confuse the operator O with its matrix $(O_{\lambda,\mu})_{\lambda,\mu \in \Lambda}$ in the basis $(e_\lambda)_{\lambda \in \Lambda}$. In terms of matrices, $O^\top$ is the transpose, while $M^\dagger = CO^\top C$.

3.3.4. *Curve operators and S-matrix.* For any marked surface (Σ, P) and a oriented simple closed curve γ on $\Sigma \setminus P$ and label $\lambda \in \Lambda$, following [**5**], we introduce curve operators, that we denote $\mathcal{C}[\gamma; \lambda]$. We consider in particular the curve operators $(\mathcal{C}[\beta; \lambda])_{\lambda \in \Lambda}$ acting on $\mathcal{A}$, as multiplication by e_λ. In the basis related to α they have the following expression

$$\mathcal{C}[\beta; \lambda](e_\mu) = \sum_{\nu \in \Lambda} N_{\lambda\mu\nu^\dagger}\, e_\nu.$$

In the basis related to β, they are simultaneously diagonalized

$$\mathcal{C}[\beta; \lambda](\varepsilon_\mu) = \mathfrak{c}_\mu[\lambda]\, \varepsilon_\mu. \tag{3.11}$$

From these two facts, the eigenvalue can easily be computed. Indeed, we compute from the definitions that we have

$$\mathcal{C}[\beta; \lambda][e_1] = \sum_{\mu \in \Lambda} N_{\lambda 1 \nu^\dagger}\, e_\nu = e_\lambda = \sum_{\mu \in \Lambda} (S^{-1})_{\lambda\mu}\, \varepsilon_\mu,$$

while, if we first go to the ε-basis, we get

$$\mathcal{C}[\beta; \lambda][e_1] = \sum_{\mu \in \Lambda} (S^{-1})_{1\mu}\, \mathfrak{c}_\mu[\lambda]\, \varepsilon_\mu.$$

The comparison gives that $(S^{-1})_{1\mu}$ is non-zero and the eigenvalue reads

$$\mathfrak{c}_{\mu}[\lambda] = \frac{(S^{-1})_{\lambda\mu}}{(S^{-1})_{1\mu}}. \tag{3.12}$$

One then deduces the following standard formula.

$$N_{\lambda\mu^{\dagger}\nu} = \sum_{\tau\in\Lambda} \frac{(S^{-1})_{\lambda\tau} S_{\tau\mu} (S^{-1})_{\nu\tau}}{(S^{-1})_{1\tau}}. \tag{3.13}$$

This formula does not depend on the normalization of the basis diagonalizing the curve operator action, i.e. it is invariant under rescalings $S_{\tau\mu} \to a_{\tau} S_{\tau\mu}$ with $(a_{\tau})_{\tau\in\Lambda} \in [\mathbb{C}^*]^{\Lambda}$.

3.3.5. *Relation between S and C.* Setting $\mu = 1$ in (3.13) yields $N_{\lambda 1\nu} = \delta_{\lambda\nu^{\dagger}}$, which gives the following relation.

$$C_{\lambda\mu} := \delta_{\lambda\mu^{\dagger}} = \sum_{\tau\in\Lambda} \frac{S_{\tau 1}}{(S^{-1})_{1\tau}} (S^{-1})_{\lambda\tau} (S^{-1})_{\mu\tau}. \tag{3.14}$$

In terms of the rescaled[3] S-matrix, it can be rewritten

$$\underline{S}_{\lambda\mu} := \sqrt{\frac{(S^{-1})_{1\lambda}}{S_{\lambda 1}}}\, S_{\lambda\mu}, \qquad C = \underline{S}^{-1}(\underline{S}^{-1})^{\top}, \tag{3.15}$$

or equivalently

$$\underline{S}^{\top}\underline{S} = C. \tag{3.16}$$

Whenever possible, we prefer to avoid the occurrence of $\dagger$ indices, so we will use (3.16) to convert it in entries of the inverse $\underline{S}$-matrix.

3.3.6. *Symmetric formula.* This relation between S and C allows us to write down the action of the curve operator in a more symmetric form avoiding the $\dagger$ indices.

$$N_{\lambda\mu\nu} = \sum_{\tau\in\Lambda} \frac{(S^{-1})_{\lambda\tau}(S^{-1})_{\mu\tau}(S^{-1})_{\nu\tau}}{[(S^{-1})_{1\tau}]^2} S_{\tau 1}.$$

Once again, one can express this symmetric formula in the orthonormal basis to get a more natural form

$$N_{\lambda\mu\nu} = \sum_{\tau\in\Lambda} \frac{(\underline{S}^{-1})_{\lambda\tau}(\underline{S}^{-1})_{\mu\tau}(\underline{S}^{-1})_{\nu\tau}}{(\underline{S}^{-1})_{1\tau}}. \tag{3.17}$$

The rescaling from S to $\underline{S}$ also does not affect the formula for the eigenvalues of the curve operators

$$\mathfrak{c}_{\mu}[\lambda] = \frac{(\underline{S}^{-1})_{\lambda\mu}}{(\underline{S}^{-1})_{1\mu}}. \tag{3.18}$$

---

[3]It follows from the definition that b is non-degenerate, so $S_{\lambda 1}/(S^{-1})_{1\lambda} = b(\varepsilon_{\lambda},\varepsilon_{\lambda})$ computed in (3.20) below cannot be 0, i.e. $S_{\lambda 1} \neq 0$. Then, $\underline{S}_{\lambda\mu}$ depends on the arbitrary choice of a sign for the squareroot, which does not affect any of the Verlinde formula, since an even number of $\underline{S}^{-1}$ factors appear.

3.3.7. *Extra relations.* If **MF-D** is satisfied or if the modular functor comes from a modular tensor category [**89**, page 97-98], then $S_{\lambda\mu} = S^{-1}_{\mu\lambda^\dagger}$. In particular, we have $(S^{-1})_{1\lambda} = S_{\lambda 1}$, therefore $\underline{S} = S$ and the basis ε_λ is already orthonormal. If **MF-U** is satisfied, the matrix S is unitary. In particular, $S^{-1}_{1\lambda} = S^*_{\lambda 1}$. These properties are justified in Appendix A. In this text, we study modular functors where neither **MF-D** **MF-U** is assumed and in the following computations, we do not use duality nor unitarity properties of the S-matrix.

3.3.8. *As a Frobenius algebra.* We define a product on $\mathcal{A}$ by the following formula

$$e_\lambda \times e_\mu := \mathcal{C}[\beta;\lambda](e_\mu). \tag{3.19}$$

A direct check from the previous formulas shows that $\mathcal{A}$ is now a Frobenius algebra, with unit $e_1 = \mathbb{1}$. We also find respectively from (3.13) and (3.14) that one has

$$\varepsilon_\lambda \times \varepsilon_\mu = \delta_{\lambda\mu}\,\frac{\varepsilon_\lambda}{(S^{-1})_{1\lambda}} \qquad \text{and} \qquad b(\varepsilon_\lambda, \varepsilon_\mu) = \frac{S_{\lambda 1}}{(S^{-1})_{1\lambda}}\,\delta_{\lambda\mu}.$$

The canonical basis $(\tilde{\varepsilon}_\lambda)_\lambda$ is obtained by a rescaling of the ε_λ as follows

$$\tilde{\varepsilon}_\lambda = (S^{-1})_{1\lambda}\,\varepsilon_\lambda.$$

This basis satisfies

$$\tilde{\varepsilon}_\lambda \times \tilde{\varepsilon}_\mu = \delta_{\lambda\mu}\,\tilde{\varepsilon}_\lambda \qquad \text{and} \qquad b(\tilde{\varepsilon}_\lambda, \tilde{\varepsilon}_\mu) = (S^{-1})_{1\lambda} S_{\lambda 1}\,\delta_{\lambda,\mu}. \tag{3.20}$$

A third interesting normalization gives the orthonormal basis defined by

$$\underline{\varepsilon}_\lambda = \frac{\tilde{\varepsilon}_\lambda}{\sqrt{(S^{-1})_{1\lambda} S_{\lambda 1}}} = \sqrt{\frac{(S^{-1})_{1\lambda}}{S_{\lambda 1}}}\,\varepsilon_\lambda := \sum_{\mu\in\Lambda} \underline{S}_{\lambda\mu} e_\mu \tag{3.21}$$

which satisfies

$$\underline{\varepsilon}_\lambda \times \underline{\varepsilon}_\mu = \delta_{\lambda\mu}\frac{\underline{\varepsilon}_\lambda}{\sqrt{(S^{-1})_{1\lambda} S_{\lambda 1}}} \qquad \text{and} \qquad b(\underline{\varepsilon}_\lambda, \underline{\varepsilon}_\mu) = \delta_{\lambda\mu}.$$

The unit expressed in the various basis is

$$\mathbb{1} = e_1 = \sum_{\lambda\in\Lambda} \tilde{\varepsilon}_\lambda = \sum_{\lambda\in\Lambda} \sqrt{(S^{-1})_{1\lambda} S_{\lambda 1}}\,\underline{\varepsilon}_\lambda = \sum_{\lambda\in\Lambda} (S^{-1})_{1\lambda}\varepsilon_\lambda.$$

The norm of the canonical basis will appear in subsequent computations and is read from (3.20) as

$$\Delta_\lambda := \frac{1}{b(\tilde{\varepsilon}_\lambda, \tilde{\varepsilon}_\lambda)} = \frac{1}{(S^{-1})_{1\lambda} S_{\lambda 1}} = \frac{1}{\big[(\underline{S}^{-1})_{1\lambda}\big]^2}.$$

3.4. CohFTs associated to a modular functor. We shall build, for each log-determination of the central charge and Dehn twist eigenvalues,

$$\tilde{r}_\lambda = \exp(2\mathrm{i}\pi r_\lambda), \qquad \tilde{c} = \exp(\mathrm{i}\pi c/2), \qquad r_\lambda, c \in \mathbb{Q} \quad \text{such that } r_\lambda = r_{\lambda^\dagger}, \tag{3.22}$$

a 1-parameter family of CohFTs based on the Frobenius algebra $\mathcal{A}$ described in the previous paragraph. The parameter here is denoted $t \in \mathbb{C}$. We rely on the result of Theorem 2.7: We have defined for each n-tuple $\vec{\lambda}$ of labels, a complex vector bundle $[\mathcal{Z}_{\vec{\lambda}}]_{g,n} \to \overline{\mathcal{M}}_{g,n}$. Then, we simply take its Chern character.

$$\Omega_{g,n} = \sum_{\vec{\lambda}\in\Lambda^n} \mathrm{Ch}_t([\mathcal{Z}_{\vec{\lambda}}]_{g,n})\, e_{\lambda_1} \otimes \cdots \otimes e_{\lambda_n}.$$

THEOREM 3.2. *For any $t \in \mathbb{C}$ and choice of log-determinations (3.22), $\Omega_{g,n}$ is a semi-simple CohFT.*

PROOF. Our bundle has been defined over the boundary of the moduli space, and the axioms of a CohFT for the Chern character immediately follow from the factorization properties of the underlying bundles. □

To describe explicitly this CohFT, we need to find the operator which transports it to a degree 0 CohFT. The strategy is the same as in [**70**] which was written in the example where $\mathcal{V}_{g,n}$ is the Verlinde bundle. The task of identifying the R-matrix is facilitated by the following result, whose proof relies on Teleman's classification of semi simple CohFTs [**86**].

LEMMA 3.3. [**70**] *In a semi-simple CohFT, the restriction of $\Omega_{g,n}$ to the smooth locus $\mathcal{M}_{g,n}$ completely determines the R-matrix.*

For any modular functor we already have computed in Proposition 2.6 the Chern character of $[\mathcal{Z}_{\vec{\lambda}}]_{g,n}$ on the smooth locus $\mathcal{M}_{g,n}$.

$$\Omega^*_{g,n}(e_{\lambda_1}\otimes\cdots\otimes e_{\lambda_n}) = [D_{\vec{\lambda}}]_{g,n}\exp\Big\{t\Big(-\frac{c}{2}\Lambda_1+\sum_{i=1}^n r_{\lambda_i}\psi_i\Big)\Big\},\qquad [D_{\vec{\lambda}}]_{g,n} := \dim\mathcal{V}_{\vec{\lambda}}(\mathbf{\Sigma}_{g,n}).$$

Here, Λ_1 is the first Chern class of the Hodge bundle, not to be confused with the label set Λ and $\mathbf{\Sigma}_{g,n}$ is any marked surface of genus g with n points. By the Verlinde formula, $[D_{\vec{\lambda}}]_{g,n}$ only depends on g, n and $\vec{\lambda}$.

PROPOSITION 3.4. *Assume $r_1 = 0$. We have the following diagonal R-matrix in the e-basis, which can also be written non-diagonally in the ε-basis*

$$R(u) = \sum_{\lambda\in\Lambda} e^{ut(r_\lambda+c/24)}\,\mathrm{id}_{e_\lambda}. \tag{3.23}$$

The corresponding translation is $T(u) = u\big(1-\exp(utc/24)\big)e_1$.

Since u here is not a quantized parameter, the log-determination of r_λ and c do matter in (3.23).

REMARK 3.1. We observe that, up to the scalar $e^{utc/24}$, the R-matrix can be identified with the action of the flow at time $ut/4\mathrm{i}\pi$ generated by an infinitesimal Dehn twist around a puncture, on the space $\mathcal{V}_{\mathrm{tot}}(\mathbf{\Sigma}_{g,n}) = \bigoplus_{\vec{\lambda}\in\Lambda^n}\mathcal{V}_{\vec{\lambda}}(\mathbf{\Sigma}_{g,n})$. This awaits an interpretation in hyperbolic geometry.

PROOF. We first remark that

$$\Omega^*_{g,n}(e_{\lambda_1}\otimes\cdots\otimes e_{\lambda_n})|_{t=0} := [D_{\vec{\lambda}}]_{g,n}\,[\overline{\mathcal{M}}_{g,n}] \tag{3.24}$$

are the correlators of the degree 0 part of the CohFT of the modular functor. Denote $\iota : \mathcal{M}_{g,n}\to\overline{\mathcal{M}}_{g,n}$ the natural inclusion. With the formula $\iota^*(\Lambda_1) = \iota^*(\kappa_1/12)$ [**75**], we obtain that

$$\begin{aligned}\iota^*\Omega^*_{g,n}(e_{\lambda_1}\otimes\cdots\otimes e_{\lambda_n}) &= \iota^*\mathrm{Ch}_t([\mathcal{Z}_{\vec{\lambda}}]_{g,n})\\ &= \exp(-t\,c\,\kappa_1/24)\,\exp\Big(\sum_{i=1}^n t\,r_{\lambda_i}\psi_i\Big).\end{aligned} \tag{3.25}$$

Comparing with Teleman's classification theorem, one gets

$$\Omega_{g,n} = \hat{R}\hat{T}\,\Omega_{g,n}|_{t=0}.$$

We see by definition of the κ-classes that the first factor in (3.25) can only arise by the action of the translation operator, and the second factor arises from the action of the R-matrix given in (3.23). The expression of the translation follows from the fact that we have a CohFT and the general result (3.3). Since $r_1 = 0$ and e_1 is the unit, we find $T(u) = u\big(1 - \exp(utc/24)\big)e_1$. □

3.5. Remark on log-determinations. When the modular functor comes from a category of representations of a vertex operator algebra $\mathfrak{A}$, $\mathcal{A}$ is the character ring of $\mathfrak{A}$. Characters are functions on the upper-half plane $\{\tau \in \mathbb{C},\ \mathrm{Im}\,\tau > 0\}$, and after multiplication by a suitable rational power of $q = e^{2i\pi\tau}$, the characters fit in a vector-valued modular form. The S-matrix is implemented by the transformation $\tau \mapsto -1/\tau$, and the Dehn twist is implemented by $\tau \mapsto \tau + 1$. The modularity of characters then provides canonical log-determinations r_λ and c, since multiplying by another power of q will destroy modularity.

4. Topological recursion

The main result of [**33**] is that the intersection of the classes $\Omega_{g,n}$ of a semi-simple CohFT against ψ-classes in $\overline{\mathcal{M}}_{g,n}$ are computed by the topological recursion of [**38**]. We quickly review this theory, in the context of local spectral curves. We will comment on the setting of global spectral curves in Section 7.

4.1. Local spectral curve and residue formula. A local spectral curve consists in the following set of data.

- $U = \bigsqcup_i U_i$ which is a disjoint union of formal neighbourhoods of complex dimension 1 of points $o_i \in U_i$;
- a branched covering $x : U \to V$ whose ramification divisor is $O = \bigsqcup_i \{o_i\}$. Here V itself is also a disjoint union of formal neighbourhoods of points in $\mathbb{P}^1$.

We assume that x has only simple ramifications. Then, we can choose a coordinate on U_i, denoted ζ_i such that $x(\zeta_i) = \zeta_i^2/2 + x(o_i)$. We make this choice once for all in each U_i. Note that U_i carries a holomorphic involution σ_i, which sends the point in U_i with coordinate ζ_i, to the point in U_i with coordinate $-\zeta_i$. In the cases, where we consider $U_{i_{\ell_1}} \times \ldots \times U_{i_{\ell_m}}$, we denote points in this cartesian product $(z_{\ell_1}, \ldots z_{\ell_m})$ and we shall denote by $\zeta_{i_{\ell_i}}(z_{\ell_i})$ their respective coordinates.

Let $\Delta = \{(z,z) \in U^2 \mid z \in U\}$ be the diagonal divisor and K_U be the canonical bundle of U. The initial data of the topological recursion is

$$\omega_{0,1} \in H^0(U, K_U), \qquad \omega_{0,2} \in H^0\big(U^2, K_U^{\boxtimes 2}(2\Delta)\big)^{\mathfrak{S}_2}$$

with the extra condition that $\omega_{0,1}(z) - \omega_{0,1}(\sigma_i(z))$ has at most double zeros at o_i. In coordinates this means that we have

$$\text{(4.1)} \qquad \forall z \in U_i, \qquad \omega_{0,1}(z) = \sum_{d \geq 0} \omega_{0,1}\left[\begin{smallmatrix} i \\ d \end{smallmatrix}\right] \zeta_i^d(z)\,\mathrm{d}\zeta_i(z)$$

$$\forall (z_1, z_2) \qquad \in U_{i_1} \times U_{i_2}, \omega_{0,2}(z_1, z_2) = \frac{\delta_{i_1 i_2}\, t_{i_1}\, \mathrm{d}\zeta_{i_1}(z_1)\,\mathrm{d}\zeta_{i_2}(z_2)}{(\zeta_{i_1}(z_1) - \zeta_{i_2}(z_2))^2}$$
$$+ \sum_{d_1, d_2 \geq 0} \omega_{0,2}\left[\begin{smallmatrix} i_1 & i_2 \\ d_1 & d_2 \end{smallmatrix}\right] \zeta_{i_1}^{d_1}(z_1)\zeta_{i_2}^{d_2}(z_2)\,\mathrm{d}\zeta_{i_1}(z_1)\mathrm{d}\zeta_{i_2}(z_2)$$

with $\omega_{0,1}\left[\begin{smallmatrix} i \\ 0 \end{smallmatrix}\right] \neq 0$ or $\omega_{0,1}\left[\begin{smallmatrix} i \\ 2 \end{smallmatrix}\right] \neq 0$. We usually assume that the coefficients of the double poles are $t_i = 1$ for all i.

The topological recursion provides a sequence of symmetric forms in n variables,

$$\omega_{g,n} \in H^0(U^n, (K_U(*O))^{\boxtimes n})^{\mathfrak{S}_n}, \qquad \text{for } 2g-2+n>0, \tag{4.2}$$

which describe the unique normalized solution of the "abstract loop equations" with initial data $(\omega_{0,1}, \omega_{0,2})$ [**20, 21**]. In (4.2), we take forms that are invariant under the symmetric group $\mathfrak{S}_n$ acting by permutation of the n factors of U. The definition of $\omega_{g,n}$ proceeds by induction on $2g-2+n>0$. We introduce the recursion kernel (4.3)

$$\mathcal{K}_i \in H^0\big(U \times U_i, [K_U \boxtimes K_{U_i}^{-1}(O)](\Delta \sqcup \Delta_\sigma)\big), \qquad \mathcal{K}_i(z_0, z) = \frac{1}{2}\, \frac{\int_{\sigma_i(z)}^{z} \omega_{0,2}(\cdot, z_0)}{\omega_{0,1}(z) - \omega_{0,1}(\sigma_i(z))}.$$

where $\Delta_\sigma = \bigsqcup_i \{(z, \sigma_i(z)) \mid z \in U_i\}$. Denote by $I = \{z_2, \ldots, z_n\}$ a set of $(n-1)$ variables. The topological recursion formula defining the symmetric forms is

$$\begin{aligned}\omega_{g,n}(z_1, I) = \sum_{o_i \in O} \operatorname*{Res}_{z \to o_i} \mathcal{K}_i(z_0, z) \Big\{ &\omega_{g-1,n+1}(z, \sigma_i(z), I) \\ &+ \sum_{\substack{h+h'=g \\ J \sqcup J' = I \\ (h,J) \neq (0,\varnothing),(g,I)}} \omega_{h,1+|J|}(z, J) \otimes \omega_{h',1+|J'|}(\sigma_i(z), J') \Big\}.\end{aligned} \tag{4.4}$$

The induction reduces $\omega_{g,n}$ to an expression involving $2g-2+n$ residues of products of $\mathcal{K}_i$s and $g+n-1$ factors of $\omega_{0,2}$. If the induction is completely unfolded, $\omega_{g,n}$ is a sum over certain trivalent graphs containing a spanning tree, having $2g-2+n$ vertices and $g+n-1$ extra edges in the complement of the tree [**38**]. The topological recursion also define a sequence of numbers $(F_g)_{g \geq 2}$ by

$$F_g = \frac{1}{2-2g} \sum_{o_i \in O} \operatorname*{Res}_{z \to o_i} \left(\oint_{o_i}^{z} \omega_{0,1} \right) \omega_{g,1}(z)$$

REMARK 4.1. Sometimes, the data of the differential form $\omega_{0,1}$ is replaced by the data of a germ of functions y holomorphic at all the o_i such that $\omega_{0,1}(z) = y(z)dx(z)$ for any $z \in U$.

4.2. Relation to cohomological field theories. For any local spectral curve, the $\omega_{g,n}$ can be represented in terms of intersection numbers on $\overline{\mathcal{M}}_{g,n}$ [**36**]. There is a partial converse presented in [**33**], where it is established that the correlators of a semi-simple CohFT are computed by the topological recursion for a local spectral curve prescribed by the corresponding R-matrix (we stress that not all local spectral curves can arise from a CohFT). We now review this correspondence.

4.2.1. *CohFT data.* Let $(\tilde{\varepsilon}_i)_i$ be a canonical basis, and $\underline{\varepsilon}_i = \tilde{\varepsilon}_i / \Delta_i^{1/2}$ the orthonormal basis. We write the formal series expansions

$$T(u) = u(\mathbb{1} - R(u)\cdot \mathbb{1}) = \sum_i \sum_{d \geq 2} T\left[\begin{smallmatrix} i \\ d \end{smallmatrix}\right] u^d \,\underline{\varepsilon}_i \tag{4.5}$$

$$\begin{aligned} B(u_1, u_2) &= \frac{b^\dagger - R(u_1) \otimes R(u_2) \cdot b^\dagger}{u_1 + u_2} \\ &= \sum_{\substack{i_1, i_2 \\ d_1, d_2 \geq 0}} B\left[\begin{smallmatrix} i_1 & i_2 \\ d_1 & d_2 \end{smallmatrix}\right] u_1^{d_1} u_2^{d_2}\, \underline{\varepsilon}_{i_1} \otimes \underline{\varepsilon}_{i_2} \end{aligned} \tag{4.6}$$

of the R-matrix and translation matrix defining uniquely a CohFT with canonical basis $(\tilde{\varepsilon}_i)_i$. We denote by $\Omega_{g,n}$ its correlators and $\Omega_{g,n}^{\deg 0}$ the restriction to their degree 0 part (which only depends on the norms Δ_i).

4.2.2. *Local spectral curve data.* We now define a local spectral curve, in terms of the above CohFT data. Its ramification points are indexed by a canonical basis $(\tilde{\varepsilon}_i)_i$ of the underlying Frobenius manifold, and we set for $z \in U_i$

$$\begin{aligned} x(z) &= x(o_i) + \zeta_i^2(z)/2 \\ \omega_{0,1}^{\mathrm{odd}}(z) &= -\Delta_i^{-1/2}\,\zeta_i^2(z)\mathrm{d}\zeta_i(z) + \sum_{d\geq 2} \frac{T\left[\begin{smallmatrix} i \\ d \end{smallmatrix}\right]}{(2d-1)!!}\,\zeta_i^{2d}(z)\mathrm{d}\zeta_i(z) \end{aligned}$$

and for $(z_1,z_2) \in U_{i_1} \times U_{i_2}$

$$\begin{aligned} \omega_{0,2}^{\mathrm{odd}}(z_1,z_2) &= \frac{\delta_{i_1 i_2}\,\mathrm{d}\zeta_{i_1}(z_1)\,\mathrm{d}\zeta_{i_2}(z_2)}{(\zeta_{i_1}(z_1) - \zeta_{i_2}(z_2))^2} \\ &\quad + \sum_{d_1,d_2\geq 0} \frac{B\left[\begin{smallmatrix} i_1 & i_2 \\ d_1 & d_2 \end{smallmatrix}\right]}{(2d_1-1)!!(2d_2-1)!!}\,\zeta_{i_1}^{2d_1}(z_1)\,\zeta_{i_2}^{2d_2}(z_2)\,\mathrm{d}\zeta_{i_1}(z_1)\mathrm{d}\zeta_{i_2}(z_2). \end{aligned} \tag{4.7}$$

By convention, $(2d-1)!! = 1$ for $d = 0$. Then, let $\omega_{0,1}^{\not{\mathrm{odd}}}(z)$ be an arbitrary holomorphic 1-form which is invariant under the local involution $\zeta_i(z) \to -\zeta_i(z)$ in the patch $z \in U_i$, and an arbitrary bidifferential of the form, for $(z_1,z_2) \in U_{i_1} \times U_{i_2}$,

$$\omega_{0,2}^{\not{\mathrm{odd}}}(z_1,z_2) = \sum_{\substack{d_1,d_2\geq 0 \\ d_1 \text{ or } d_2 \text{ even}}} \omega_{0,2}\left[\begin{smallmatrix} i_1 & i_2 \\ d_1 & d_2 \end{smallmatrix}\right] \zeta_{i_1}^{d_1}(z_1)\zeta_{i_2}^{d_2}(z_2)\,\mathrm{d}\zeta_{i_1}(z_1)\mathrm{d}\zeta_{i_2}(z_2).$$

We then consider the following quantities as initial data for the topological recursion

$$\omega_{0,1} = \omega_{0,1}^{\mathrm{odd}} + \omega_{0,1}^{\not{\mathrm{odd}}}, \qquad \omega_{0,2} = \omega_{0,2}^{\mathrm{odd}} + \omega_{0,2}^{\not{\mathrm{odd}}}. \tag{4.8}$$

To sum up, in the decomposition

$$\begin{aligned} \omega_{0,2}(z_1,z_2) &= \frac{\delta_{i_1 i_2}\mathrm{d}\zeta_{i_1}(z_1)\mathrm{d}\zeta_{i_2}(z_2)}{(\zeta_{i_1}(z_1) - \zeta_{i_2}(z_2))^2} \\ &\quad + \sum_{d_1,d_2\geq 0} \omega_{0,2}\left[\begin{smallmatrix} i_1 & i_2 \\ d_1 & d_2 \end{smallmatrix}\right] \zeta_{i_1}^{d_1}(z_1)\zeta_{i_2}^{d_2}(z_2)\,\mathrm{d}\zeta_{i_1}(z_1)\mathrm{d}\zeta_{i_2}(z_2), \end{aligned}$$

the R-matrix of the CohFT only fixes the following coefficients.

$$\omega_{0,2}\left[\begin{smallmatrix} i_1 & i_2 \\ 2d_1 & 2d_2 \end{smallmatrix}\right] = \frac{B\left[\begin{smallmatrix} i_1 & i_2 \\ d_1 & d_2 \end{smallmatrix}\right]}{(2d_1-1)!!(2d_2-1)!!}$$

All the other coefficients can be arbitrarily chosen. With this $\omega_{0,2}$ at hand, we build a family of meromorphic 1-forms $\Xi_{d,i}(z_0)$ indexed by ramification points and an integer $d \geq 0$

$$\Xi_{i,d}(z_0) := \operatorname*{Res}_{z\to o_i} \frac{(2d+1)!!\,\mathrm{d}\zeta_i(z)}{\zeta_i(z)^{2d+2}} \int_{o_i}^{z} \omega_{0,2}(\cdot, z_0). \tag{4.9}$$

The only singularity of this 1-form is a pole at $z_0 \to o_i$, and we actually have, for $z_0 \in U_{i_0}$,

$$\begin{aligned} &\Xi_{i,d}(z_0) \\ &= \delta_{i i_0} \frac{(2d+1)!!\,\mathrm{d}\zeta_{i_0}(z_0)}{\zeta_{i_0}^{2d+2}(z_0)} + (2d-1)!! \sum_{d_0\geq 0} \omega_{0,2}\left[\begin{smallmatrix} i & i_0 \\ 2d & d_0 \end{smallmatrix}\right] \zeta_{i_0}^{d_0}(z_0)\,\mathrm{d}\zeta_{i_0}(z_0). \end{aligned} \tag{4.10}$$

We stress that the domain of definition of $\Xi_{i,d}$ is the whole U – as for $\omega_{0,2}$ – and not only U_i.

THEOREM 4.1. **[33]** *Consider the n-forms $\omega_{g,n}(z_1,\ldots,z_n)$ defined by*

$$\omega_{g,n}(z_1,\ldots,z_n) = \sum_{\substack{i_1,\ldots,i_n \\ d_1,\ldots,d_n\geq 0}} \int_{\overline{\mathcal{M}}_{g,n}} \Omega^*_{g,n}(\underline{\varepsilon}_{i_1}\otimes\cdots\otimes\underline{\varepsilon}_{i_n}) \prod_{\ell=1}^{n}\psi_\ell^{d_\ell} \prod_{\ell=1}^{n}\Xi_{i_\ell,d_\ell}(z_\ell). \tag{4.11}$$

Then, $\omega_{g,n}$ is computed by the topological recursion (4.4) with initial data (4.8).

Note that, in the right-hand side, the only dependence in the choice of $\omega_{0,2}^{\text{odd}}$ lies in the meromorphic forms $\Xi_{d,i}$ defined in (4.9) and on which the correlators are decomposed. Further, the $\omega_{g,n}$ do not depend on the constant $x(o_i)$ and the 1-form $\omega_{0,1}^{\text{even}}$. Allowing them to be non-zero can simplify expressions, and does matter if one is interested in building a Landau-Ginzburg model for which (4.7)-(4.7) is the local expansion near the ramification divisors. This matter will be discussed in Section 7.

We observe that

$$\int_{\mathbb{R}} \frac{e^{-\zeta^2/2u}}{2(2\pi u)^{1/2}}\,\zeta^{2d}\,\mathrm{d}\zeta = (2d-1)!!\,u^d. \tag{4.12}$$

Therefore, the relation between the initial data (4.5) and the R-matrix of the CohFT is given by a Laplace transform, as it is well-known

$$\begin{aligned}
\mathbb{1}u + T(u) &= \sum_i \bigg(\int_{\gamma_i} \frac{\exp\big[-\frac{x(z)-x(o_i)}{u}\big]}{2(2\pi u)^{1/2}}\,\omega_{0,1}(z)\bigg)\underline{\varepsilon}_i \\
B(u_1,u_2) &= \sum_{i,j}\bigg(\int_{\gamma_i\times\gamma_j} \frac{\exp\big[-\frac{x(z_1)-x(o_i)}{u_1}-\frac{x(z_2)-x(o_j)}{u_2}\big]}{4(2\pi u_1)^{1/2}(2\pi u_2)^{1/2}} \\
&\qquad\cdot\bigg\{\omega_{0,2}(z_1,z_2) - \frac{\delta_{ij}\,\mathrm{d}\zeta_i(z_1)\mathrm{d}\zeta_j(z_2)}{(\zeta_i(z_1)-\zeta_j(z_2))^2}\bigg\}\bigg)\underline{\varepsilon}_i\otimes\underline{\varepsilon}_j.
\end{aligned}$$

Here, γ_i is a steepest descent contour in the spectral curve passing through the ramification point o_i and going to ∞ in the direction $\mathrm{Re}[x(z)/u]\to+\infty$. For the case of local spectral curves we consider here, the meaning of the right-hand side must be precised. We expand the non-exponential part of the integrand as power series when $z\to o_i$, then integrate term by term against $e^{-x(z)/u}$ using (4.12). Each term yields a monomial in u, and thus the right-hand side is a well-defined formal power series in u.

4.2.3. 0-*point correlation functions.* The following lemma did not appear in **[33]**, but is an easy consequence of the content of that paper.

THEOREM 4.2. *For $g\geq 2$, we have*

$$\int_{\overline{\mathcal{M}}_{g,0}} \Omega_{g,0} = F_g := \frac{1}{2-2g}\sum_i \operatorname*{Res}_{z\to o_i}\bigg(\int_{o_i}^{z}\omega_{0,1}\bigg)\omega_{g,1}(z).$$

In order to prove it, we shall use the dilaton equation for the classes produced by the action of the Givental group.

LEMMA 4.3. *Given a semi-simple 2d TQFT* $(\Omega_{g,n}^{\deg 0})_{g,n}$, *a translation* $T(u) \in u^2\mathcal{A}[[u]]$ *and an element* $R(u) \in \mathrm{End}(\mathcal{A})[[u]]$, *the cohomology classes* $\Omega = \hat{T}\hat{R}\Omega^{\deg 0}$ *satisfy*

$$(2g-2+n)\int_{\overline{\mathcal{M}}_{g,n}} \Omega_{g,n}^*(\cdot) = \int_{\overline{\mathcal{M}}_{g,n+1}} \Omega_{g,n+1}^*(\cdot \otimes \mathbb{1})\psi_{n+1} - \sum_k \int_{\overline{\mathcal{M}}_{g,n+1}} \Omega_{g,n+1}^*(\cdot \otimes T_k)\psi_{n+1}^k \tag{4.13}$$

for $2g-2+n>0$.

In particular, this result holds for CohFTs, where R and T are related by $T(u) = u(\mathrm{Id} - R(u))\mathbb{1}$. Theorem 4.2 is then a simple consequence of (4.13) using a residue representation.

PROOF. We prove this formula in three steps. First, let us remark that this statement is true for a vanishing $T(u)$ and $R(u) = \mathrm{Id}$ by the dilaton equation on the moduli space of curves

$$\int_{\overline{\mathcal{M}}_{g,n+1}} \Big[\prod_{i=1}^n \psi_i^{d_i}\Big]\psi_{n+1} = (2g-2+n)\int_{\overline{\mathcal{M}}_{g,n}} \prod_{i=1}^n \psi_i^{d_i}. \tag{4.14}$$

Next, we check the validity of (4.13) for arbitrary $R(u)$ with $T(u) \equiv 0$. By definition, we have that

$$\begin{aligned}\check{\Omega}_{g,n} &:= (\hat{R}\Omega)_{g,n} \\ &= \sum_{\Gamma \in \mathcal{G}_{g,n}} \frac{1}{|\mathrm{Aut}\,\Gamma|}(\pi_\Gamma)_*\Big(\prod_{l=\text{leaf}} R(\psi_l) \prod_{\substack{e=\text{edge}\\ \{v'_e,v''_e\}}} B(\psi_{v'_e},\psi_{v''_e}) \prod_{v=\text{vertex}} \Omega_{g(v),n(v)}^{\deg 0}\Big),\end{aligned}$$

where $G_{g,n}$ denotes the stable graphs of genus g with n legs. We can now compute

$$\begin{aligned}&\int_{\overline{\mathcal{M}}_{g,n+1}} \check{\Omega}_{g,n+1}^*(\cdot \otimes \mathbb{1})\psi_{n+1} \\ &\quad= \sum_{\Gamma \in \mathcal{G}_{g,n+1}} \frac{1}{|\mathrm{Aut}\,\Gamma|}\int_{\overline{\mathcal{M}}_{g,n+1}} \psi_{n+1}\,(\pi_\Gamma)_* \\ &\qquad\cdot\Big(\prod_{l=\text{leaf}\setminus\{n+1\}} R(\psi_l) \prod_{\substack{e=\text{edge}\\ \{v'_e,v''_e\}}} B(\psi_{v'_e},\psi_{v''_e}) \prod_{v=\text{vertex}} \Omega_{g(v),n(v)}^{\deg 0\,*}\Big).\end{aligned}$$

Denoting by v^* the unique vertex of Γ adjacent to the leaf corresponding to the $(n+1)$-th marked point, we can write it as a product of integrals over the moduli spaces associated to each vertex

$$\begin{aligned}
&\int_{\overline{\mathcal{M}}_{g,n+1}} \check{\Omega}^*_{g,n+1}(\cdot \otimes \mathbb{1})\psi_{n+1} \\
&= \sum_{\Gamma\in\mathcal{G}_{g,n+1}} \frac{1}{|\mathrm{Aut}\,\Gamma|} \sum_{\substack{d(h)\geq 0 \ h-\text{half-edge} \\ i(v)\in[\![1,k]\!] \ v=\text{vertex}}} \prod_{\substack{e=\text{edge} \\ \{v'_e,v''_e\}}} B\begin{bmatrix} i(v'_e) & i(v''_e) \\ d(h_1(e)) & d(h_2(e)) \end{bmatrix} \\
&\quad \times \prod_{v\neq v^*} \left(\Delta_{i(v)}^{g(v)-1} \int_{\overline{\mathcal{M}}_{g(v),n(v)}} \prod_{j=1}^{n(v)} \psi_j^{d_j(v)} \right) \\
&\quad \times \Delta_{i(v^*)}^{g(v^*)-1} \int_{\overline{\mathcal{M}}_{g(v^*),n(v^*)}} \left[\prod_{j=1}^{n(v^*)-1} \psi_j^{d_j(v^*)} \right] \psi_{n(v^*)}
\end{aligned}$$

with the following notations. Each vertex v is assigned a label $i(v) \in [\![1,k]\!]$ where k is the dimension of $\mathcal{A}$. Each half-edge h is assigned a nonnegative integer $d(h)$. Each edge e is decomposed into a pair of half edges $(h_1(e), h_2(e))$ such that $h_1(e)$ (resp. $h_2(e)$) is adjacent to v'_e (resp. v''_e). The incident half-edges to a vertex v are labeled (in some arbitrary way) from 1 to $n(v)$ and we denote by $d_j(v)$ the label $d(h)$ where h is the j-th half edge incident to v. By convention, the $d(v^*)$-th half edge is the one from v^* to the $(n+1)$-th leaf.

We can now apply the dilaton equation to the vertex v^* in order to remove the $(n+1)$-th marked point leading to

$$\begin{aligned}
&\int_{\overline{\mathcal{M}}_{g,n+1}} \check{\Omega}^*_{g,n+1}(\cdot \otimes \mathbb{1})\psi_{n+1} \\
&= \sum_{\Gamma\in\mathcal{G}_{g,n+1}} \frac{1}{|\mathrm{Aut}\,\Gamma|} \sum_{\substack{d_1(e),d_2(e) \\ e=\text{edge}}} \sum_{\substack{i(v) \\ v=\text{vertex}}} \prod_{\substack{e=\text{edge} \\ \{v'_e,v''_e\}}} B\begin{bmatrix} i(v'_e) & i(v''_e) \\ d(h_1(e)) & d(h_2(e)) \end{bmatrix} \\
&\quad \times \prod_{v\neq v^*} \left[\Delta_{i(v)}^{g(v)-1} \int_{\overline{\mathcal{M}}_{g(v),n(v)}} \prod_{j=1}^{n(v)} \psi_j^{d_j(v)} \right] \\
&\quad \times (2g(v^*) - 2 + n(v^*) - 1)\, \Delta_{i(v^*)}^{g(v^*)-1} \int_{\overline{\mathcal{M}}_{g(v^*),n(v^*)-1}} \left[\prod_{j=1}^{n(v)-1} \psi_j^{d_j(v)} \right].
\end{aligned}$$

Finally, recall that a graph of genus g with $n+1$ leaves is nothing but a graph a genus g with n leaves and a distinguished vertex – one recovers the first by adding an $(n+1)$-th leaf at the distinguished vertex. So, we can convert the sum into a sum over $\Gamma \in \mathcal{G}_{g,n}$, and since $\sum_v (2g(v) - 2 + n(v)) = 2g - 2 + n$, we obtain

$$\int_{\overline{\mathcal{M}}_{g,n+1}} \check{\Omega}^*_{g,n+1}(\cdot \otimes \mathbb{1})\psi_{n+1} = (2g-2+n) \int_{\overline{\mathcal{M}}_{g,n}} \check{\Omega}^*_{g,n}(\cdot).$$

In the last step of the argument, we study the action of the translation on the dilaton equation. Let $(\Omega^*_{g,n})_{g,n}$ be a set of cohomology classes satisfying $\int_{\overline{\mathcal{M}}_{g,n+1}} \Omega^*_{g,n+1}(\cdot \otimes \mathbb{1})\psi_{n+1} = (2g-2+n) \int_{\overline{\mathcal{M}}_{g,n}} \Omega^*_{g,n}(\cdot)$, and let $T(u) = \sum_{k\geq 2} T_k u^k \in u^2\mathcal{A}[[u]]$ be a formal series. For $g \geq 2$, we have by definition that

$$(T\Omega^*)_{g,1}(\mathbb{1})\psi_1 = \sum_{m\geq 0} \frac{1}{m!} \sum_{k_1,\dots k_m\geq 2} (\pi_m)_* \left\{ \Omega^*_{g,1+m}(\mathbb{1} \otimes T_{k_1} \otimes \cdots \otimes T_{k_m}) \psi_1^{k_1} \dots \psi_m^{k_m} \right\} \psi_1$$

where we denote by $\pi_m : \overline{\mathcal{M}}_{g,1+m} \to \overline{\mathcal{M}}_{g,1}$ the map forgetting the last m marked points. This implies that

$$\int_{\overline{\mathcal{M}}_{g,1}} (T\Omega^*)_{g,1}(\mathbb{1})\psi_1 = \sum_{m\geq 0} \frac{1}{m!} \sum_{k_1,\ldots k_m\geq 2} \int_{\overline{\mathcal{M}}_{g,1+m}} \Omega^*_{g,1+m}(\mathbb{1}\otimes T_{k_1}\otimes\cdots\otimes T_{k_m})\psi_1^{k_1}\ldots\psi_m^{k_m}\psi_1.$$

Using the dilaton equation for the classes $(\Omega_{g,n})_{g,n}$, one can push it down to

$$\int_{\overline{\mathcal{M}}_{g,1}} (T\Omega^*)_{g,1}(\mathbb{1})\psi_1 = \sum_{m\geq 0} \frac{(2g-2+m)}{m!} \sum_{k_1,\ldots k_m\geq 2} \int_{\overline{\mathcal{M}}_{g,m}} \Omega^*_{g,m}(T_{k_1}\otimes\cdots\otimes T_{k_m})\psi_1^{k_1}\ldots\psi_m^{k_m}.$$

This can be decomposed into a sum of two terms which reads

$$\begin{aligned}&\int_{\overline{\mathcal{M}}_{g,1}} (T\Omega^*)_{g,1}(\mathbb{1})\psi_0\\ = \;& (2g-2)\int_{\overline{\mathcal{M}}_{g,0}} (T\Omega^*)_{g,0}(\mathbb{1})\\ &+ \sum_{m\geq 1} \frac{1}{(m-1)!} \sum_{k_1,\ldots k_m\geq 2} \int_{\overline{\mathcal{M}}_{g,1+m-1}} \Omega^*_{g,1+m}(T_{k_1}\otimes\cdots\otimes T_{k_m})\psi_1^{k_1}\ldots\psi_m^{k_m}.\end{aligned}$$

One can see the second term as a contribution to $(\hat{T}\Omega)_{g,1}$ by marking the first point for example and making the change of variable $m \to m-1$. This proves the lemma for $n=0$. The generalization to $n>0$ follows by similar arguments. □

In the next paragraphs, we apply Theorems 4.1-4.2 to the CohFT obtained from a modular functor.

4.3. Topological recursion and Verlinde formula. We shall not reproduce the general proof of Theorem 4.1, but it is easy and nevertheless instructive to derive it directly for the degree 0 part of the theory, i.e. for the $2d$ TQFT associated to a modular functor. In this way, we exhibit the Verlinde formula computing the dimension $[D_{\vec{\lambda}}]_{g,n} = \dim \mathcal{V}_{\vec{\lambda}}(\mathbf{\Sigma}_{g,n})$ of the spaces provided by the modular functor as a special case of topological recursion.

4.3.1. *Verlinde formula.* Let us start with a brief review of the Verlinde formula. In terms of matrix elements in the $(e_\lambda)_\lambda$ basis, it takes the following form

$$[D_{\vec{\lambda}}]_{g,n} = \big[\mathcal{C}[\beta;\lambda_1]\cdots\mathcal{C}[\beta;\lambda_n]\,\mathcal{W}^g\big]_{11} = b(e_1, e_{\lambda_1}\times\cdots\times e_{\lambda_n}\times\mathcal{W}^g(e_1)).$$

In this expression, we used the curve operators introduced in § 3.3.4, and

$$\mathcal{W} = \sum_{\mu\in\Lambda} \mathrm{Tr}\big(\mathcal{C}[\beta;\mu^\dagger]\big)\,\mathcal{C}[\beta;\mu].$$

Equation (4.15) – and its equivalent form below (4.16) – have been first conjectured by Verlinde [**93**]. It was then proved by Moore and Seiberg [**73**] for modular functors coming from modular tensor categories (though the notion had not been coined yet), but the proof actually holds for any modular functor (see also [**89**]). The strategy consists (a) in degenerating $\mathbf{\Sigma}_{g,n}$ in an arbitrary way to a nodal surface whose smooth components all are thrice-punctured spheres, and then use the factorization, thus expressing $[D_\bullet]_{g,n}$ in terms of $[D_\bullet]_{0,3}$; (b) to show that the

S-matrix diagonalizes the multiplication in the Frobenius algebra, which implies (4.16) for $(g,n) = (0,3)$. For modular functors without further assumptions, this is for instance proved in [**84**] and also in [**5**].

We can derive another expression by working in the $\tilde{\varepsilon}$-basis, taking advantage of the fact that it diagonalizes simultaneously the curve operators and exploiting $e_1 = \sum_{\tau\in\Lambda} \tilde{\varepsilon}_\tau$. The eigenvalues of the curve operators are (3.11)

$$\mathcal{C}[\beta;\mu](\tilde{\varepsilon}_\tau) = \frac{(\underline{S}^{-1})_{\mu\tau}}{(\underline{S}^{-1})_{1\tau}}\,\tilde{\varepsilon}_\tau,$$

so that the trace yields

$$\mathrm{Tr}\,\mathcal{C}[\beta;\mu^\dagger] = \sum_{\nu\in\Lambda} \frac{(\underline{S}^{-1})_{\mu^\dagger\nu}}{(\underline{S}^{-1})_{1\nu}} = \sum_{\nu\in\Lambda} \frac{\underline{S}_{\nu\mu}}{(\underline{S}^{-1})_{1\nu}}. \tag{4.15}$$

Then the eigenvalues of $\mathcal{W}$ read, using (3.15),

$$\mathcal{W}(\tilde{\varepsilon}_\tau) = \bigg(\sum_{\mu,\nu\in\Lambda} \frac{\underline{S}_{\nu\mu}}{(\underline{S}^{-1})_{1\nu}}\,\frac{(\underline{S}^{-1})_{\mu\tau}}{(\underline{S}^{-1})_{1\tau}}\bigg)\tilde{\varepsilon}_\tau = \frac{\tilde{\varepsilon}_\tau}{\big[(\underline{S}^{-1})_{1\tau}\big]^2}.$$

Using that $b(\tilde{\varepsilon}_\tau,\tilde{\varepsilon}_{\tau'}) = \big[(\underline{S}^{-1})_{1\tau}\big]^2\delta_{\tau,\tau'}$, we obtain the equivalent – and maybe better known form – of (4.15)

$$[D_{\vec{\lambda}}]_{g,n} = \sum_{\tau\in\Lambda} \frac{(\underline{S}^{-1})_{\lambda_1\tau}\cdots(\underline{S}^{-1})_{\lambda_n\tau}}{\big[(\underline{S}^{-1})_{1\tau}\big]^{2g-2+n}}. \tag{4.16}$$

4.3.2. *Topological recursion for the* 2d *TQFT.* We denote $\omega^{\mathrm{KdV}}_{g,n}$ the correlation functions returned by the topological recursion for the Airy curve defined by

$$U = \mathbb{C},\qquad x(\zeta) = \zeta^2/2,\qquad y(z) = -z,\qquad \omega_{0,2}(\zeta_1,\zeta_2) = \frac{\mathrm{d}\zeta_1\mathrm{d}\zeta_2}{(\zeta_1-\zeta_2)^2}. \tag{4.17}$$

Although it is not needed in the proof of Proposition 4.4 below, we recall the following well-known formula[4].

$$\omega^{\mathrm{KdV}}_{g,n}(\zeta_1,\ldots,\zeta_n) = \sum_{d_1,\ldots,d_n\geq 0} \int_{\overline{\mathcal{M}}_{g,n}} \prod_{i=1}^{n} \psi_i^{d_i}\,\frac{(2d_i+1)!!\,\mathrm{d}\zeta_i}{\zeta_i^{2d_i+2}}. \tag{4.18}$$

PROPOSITION 4.4. *Consider the local spectral curve defined by the data of* $U = \bigsqcup_{\lambda\in\Lambda} U_\lambda$ *and, for* $z\in U_\lambda$ *and* $(z_1,z_2)\in U_{\lambda_1}\times U_{\lambda_2}$ *in the expression of the following data local coordinates.*

$$\left\{\begin{array}{rcl} x(z) & = & x(o_\lambda) + (\zeta_\lambda(z))^2/2 \\ y(z) & = & -(\underline{S}^{-1})_{1\lambda}\,\zeta_\lambda(z) \\ \omega_{0,2}(z_1,z_2) & = & \delta_{\lambda_1\lambda_2}\frac{\mathrm{d}\zeta_{\lambda_1}(z_1)\,\mathrm{d}\zeta_{\lambda_2}(z_2)}{(\zeta_{\lambda_1}(z_1)-\zeta_{\lambda_2}(z_2))^2} \end{array}\right. . \tag{4.19}$$

Then, the correlation functions returned by the topological recursion for $2g-2+n>0$ *are, for* $z_i\in U_{\lambda_i}$,

$$\omega_{g,n}(z_1,\ldots,z_n) = \bigg(\sum_{\vec{\mu}\in\Lambda^n}\prod_{i=1}^{n}\underline{S}_{\lambda_i\mu_i}\,[D_{\vec{\mu}}]_{g,n}\bigg)\,\omega^{\mathrm{KdV}}_{g,n}(\zeta_\lambda(z_1),\ldots,\zeta_\lambda(z_n)). \tag{4.20}$$

---

[4] See [**35**] for a proof based on matrix model techniques. It can also be obtained directly from the Virasoro constraints [**29, 97**] satisfied by the intersection numbers.

Remark that, since $\omega_{0,2}$ is purely diagonal in (4.19), the left-hand side in (4.20) vanishes unless all z_i belongs to the same open set U_{λ_1}. This is consistent with (4.16) which shows that the bracket in the right-hand side of (4.20) vanishes unless $\lambda_i = \lambda$ for all i, and explains the notation used in the KdV factor. If we want to compare with Theorem 4.1, we use (4.18) and the following basis of 1-forms.

$$z \in U_\mu, \qquad \Xi_{d,\lambda}(z) = \delta_{\lambda\mu} \frac{(2d+1)!!}{[\zeta_\lambda(z)]^{2d+2}}.$$

PROOF. In (4.19), we recognize several copies of the Airy curve, except for a rescaling of $y(z) \to (\underline{S}^{-1})_{1\lambda}\, y(z)$ in the patch $z \in U_\lambda$. By the previous remark, we can assume that for $i \in [\![1,n]\!]$, we have all $z_i \in U_\tau$ for some $\tau \in \Lambda$. Since $\omega_{g,n}$ is obtained from the initial data by a sequence of $2g-2+n$ residues at the ramification point $o_\tau \in U_\tau$ involving the kernel (4.3), we have that

$$\omega_{g,n}(z_1,\ldots,z_n) = \frac{\omega_{g,n}^{\mathrm{KdV}}(z_1,\ldots,z_n)}{\left[(\underline{S}^{-1})_{1\tau}\right]^{2g-2+n}}.$$

To conclude, we recognize with (4.16) that

$$\frac{1}{\left[(\underline{S}^{-1})_{1\tau}\right]^{2g-2+n}} = \sum_{\tau\in\Lambda} \prod_{i=1}^{n} \underline{S}_{\tau\mu_i}\, [D_{\vec{\mu}}]_{g,n}.$$

□

The expression of the topological recursion in terms of intersection numbers is naturally written in terms of local coordinates around the ramification points, which are associated to the canonical basis $(\underline{\varepsilon}_\lambda)_\lambda$. To recover the expression in the $(e_\lambda)_\lambda$-basis, one has to make a linear combination which cancels the $\underline{S}$ in (4.20). So, the property that $(\omega_{g,n})_{g,n}$ appearing in (4.11) satisfy the topological recursion is equivalent to the Verlinde formula (4.16) for $[D_\bullet]_{g,n}$.

We can reformulate the proof by saying that, in the sum over trivalent graphs that computes the $\omega_{g,n}$ of the topological recursion, the dependence in the indices $\lambda_i \in \Lambda$ of the weight of a graph is the same for all graphs[5].

4.3.3. *Change of basis.* Denote $z_\lambda(\zeta) \in U_\lambda$ the point such that $\zeta_\lambda(z_\lambda(\zeta)) = \zeta$, and define the following forms

$$\omega_{g,n}^{[\vec{\lambda}]}(\zeta_1,\ldots,\zeta_n) := \omega_{g,n}(z_{\lambda_1}(\zeta_1),\ldots,z_{\lambda_n}(\zeta_n)).$$

Proposition 4.4 can be alternatively written as

$$\sum_{\vec{\lambda}\in\Lambda^n} \prod_{i=1}^{n} (\underline{S}^{-1})_{\mu_i\lambda_i}\, \omega_{g,n}^{[\vec{\lambda}]}(\zeta_1,\ldots,\zeta_n) = [D_{\vec{\mu}}]_{g,n}\, \omega_{g,n}^{\mathrm{KdV}}(\zeta_1,\ldots,\zeta_n). \tag{4.21}$$

In anticipation of Section 7, we remark that if there exists a global spectral curve, it corresponds to expressing the correlation functions in a different basis of coordinates. In the Landau-Ginzburg picture, this amounts to taking a different linear combination of vanishing cycles, while in the GW theory picture, it means taking a different basis of the cohomology of the target space pulled back to the moduli space by the evaluation map.

[5] While finishing this project, we heard that Dumitrescu and Mulase arrived independently in [**32**] to Proposition 4.4.

4.4. Local spectral curves for modular functors. For the CohFTs built from a modular functor, we computed the R-matrix in (3.23). Working in the orthonormal basis $(\underline{\varepsilon}_\lambda)_\lambda$ introduced in (3.21), we deduce an expression for the local spectral curve, but we exploit the freedom to add non-odd parts to the initial data to simplify the result. Define the function $\mathcal{B}(\zeta)$ by

$$\mathcal{B}(\zeta) = \frac{\cosh(\zeta)}{\zeta^2} - \frac{\sinh(\zeta)}{\zeta}, \tag{4.22}$$

and the renormalized Dehn twist by

$$h_\lambda = 2t(r_\lambda + c/24). \tag{4.23}$$

From Lemma 2.4, we deduce that $h_1 = ct/12$ and $h_\lambda = h_{\lambda^\dagger}$.

THEOREM 4.5. *The topological recursion applied to the initial data* ($\omega_{0,1} = y\mathrm{d}x$, $\omega_{0,2}$) *with*

$$\begin{aligned} x(z) &= x(o_\lambda) + [\zeta_\lambda(z)]^2/2 \\ y(z) &= -(\underline{S}^{-1})_{1\lambda}\,\frac{\exp[(ct/12)^{1/2}\zeta_\lambda(z)]}{(ct/12)^{1/2}} \\ \omega_{0,2}(z_1,z_2) &= \sum_{\tau\in\Lambda} \underline{S}_{\lambda_2\tau}\,(\underline{S}^{-1})_{\tau\lambda_1}\,h_\tau\,\mathcal{B}\big[h_\tau^{1/2}(\zeta_{\lambda_1}(z_1)-\zeta_{\lambda_2}(z_2))\big]\,\mathrm{d}\zeta_{\lambda_1}(z_1)\mathrm{d}\zeta_{\lambda_2}(z_2) \end{aligned} \tag{4.24}$$

where $z \in U_\lambda$ and $(z_1,z_2) \in U_{\lambda_1}\times U_{\lambda_2}$, computes the intersection indices of Chern classes of the bundles $[\mathcal{Z}_{\vec{\lambda}}]_{g,n}$, in the sense of Theorem 4.1-4.2 *when decomposed on the following basis of* 1*-forms.*

$$\forall z \in U_\mu, \qquad \Xi_{d,\lambda}(z) = \sum_{\tau\in\Lambda} \underline{S}_{\lambda\tau}(\underline{S}^{-1})_{\tau\mu}\,\frac{\Gamma[2d+2\,;\,h_\tau^{1/2}\zeta_\mu(z)]}{2^d\,d!\,[\zeta_\mu(z)]^{2d+2}}\,\mathrm{d}\zeta_\mu(z)$$

where $\Gamma[a\,;x] := \int_x^\infty \mathrm{d}v\, v^{a-1}e^{-v}$ is the incomplete Gamma function[6].

With the notational remark of Section 4.3.3, we therefore obtain a generalization of (4.21), which is our second main result

$$\boxed{\sum_{\vec{\lambda}\in\Lambda^n}\prod_{i=1}^n(\underline{S}^{-1})_{\mu_i\lambda_i}\,\omega_{g,n}^{[\vec{\lambda}]}(\zeta_1,\ldots,\zeta_n) = \int_{\overline{\mathcal{M}}_{g,n}} \mathrm{Ch}_t(\mathcal{Z}_{\vec{\mu}}(\mathbf{\Sigma}_{g,n}))\prod_{i=1}^n \Xi(\psi_i, h_{\mu_i}^{1/2}\zeta_i)} \tag{4.25}$$

where $\Xi(u,\zeta)$ is the following formal series in u with coefficients meromorphic forms in a neighbourhood of $\zeta = 0$

$$\Xi(u,\zeta) = \sum_{d\geq 0}\frac{\Gamma[2d+2;\zeta]}{2^d d!\,\zeta^{2d+2}}\,u^d.$$

And, for $g \geq 2$, we have as well that

$$\boxed{F_g = \int_{\overline{\mathcal{M}}_{g,0}} \mathrm{Ch}_t(\mathcal{Z}(\mathbf{\Sigma}_{g,0})).}$$

[6]Since $\Gamma[a\,;x] = \Gamma(a) + O(x^{a-1})$ when $x \to 0$, we see that $\Xi_{d,\lambda}(z) = \delta_{\lambda\mu}\,(2d+1)!!\,[\zeta_\mu(z)]^{-(2d+2)}\,\mathrm{d}\zeta_\mu(z) + O(1)$ as it should be.

PROOF. This is pure algebra. The inverse norm of $\underline{\varepsilon}_\lambda$ is introduced in (3.21), and gives $\Delta_\lambda^{-1/2} = \big(\underline{S}^{-1}\big)_{1\lambda}$. Inserting $r_1 = 0$ in (3.23), we obtain $T(u) = u(1 - e^{utc/24})\mathbb{1}$. Then, for $z \in U_\lambda$, we compute with (4.7)-(4.7) that

$$\begin{aligned}
\omega_{0,1}^{\mathrm{odd}}(z) &= (\underline{S}^{-1})_{1\lambda}\bigg(-[\zeta_\lambda(z)]^2 - \sum_{d\geq 1}\frac{(ct/24)^d\,[\zeta_\lambda(z)]^{2d+2}}{d!\,(2d+1)!!}\bigg)\mathrm{d}\zeta_\lambda(z)\\
&= -(\underline{S}^{-1})_{1\lambda}\bigg(\zeta_\lambda(z) + \sum_{d\geq 1}\frac{(ct/12)^d\,[\zeta_\lambda(z)]^{2d+1}}{(2d+1)!}\bigg)\zeta_\lambda(z)\mathrm{d}\zeta_\lambda(z)\\
&= -(\underline{S}^{-1})_{1\lambda}\,\frac{\sinh[(ct/12)^{1/2}\,\zeta_\lambda(z)]}{(ct/12)^{1/2}}\,\mathrm{d}x(z).
\end{aligned} \tag{4.26}$$

By adding a suitable even part, we can choose

$$\omega_{0,1}(z) = -(\underline{S}^{-1})_{1\lambda}\,\frac{\exp[(ct/12)^{1/2}\zeta_\lambda(z)]}{(ct/12)^{1/2}}\,\mathrm{d}\zeta_\lambda(z).$$

We now proceed to the 2-point function. We have $b^\dagger = \sum_{\lambda\in\Lambda} e_\lambda \otimes e_{\lambda^\dagger}$. Thus, we obtain from the definition (4.6) that

$$B(u_1,u_2) = \sum_{\lambda_1,\lambda_2\in\Lambda}\bigg(\sum_{\tau\in\Lambda}\frac{1-\exp\big[(u_1h_\tau + u_2h_{\tau^\dagger})/2\big]}{u_1+u_2}\,(\underline{S}^{-1})_{\tau\lambda_1}\,(\underline{S}^{-1})_{\tau^\dagger\lambda_2}\bigg)\underline{\varepsilon}_{\lambda_1}\otimes\underline{\varepsilon}_{\lambda_2}.$$

There is some simplification because $h_\tau = h_{\tau^\dagger}$, and $(\underline{S}^{-1})_{\tau^\dagger\lambda_2} = \underline{S}_{\lambda_2\tau}$ according to (3.16), so we get that

$$B(u_1,u_2) = \sum_{\lambda_1,\lambda_2\in\Lambda}\bigg(\sum_{\tau\in\Lambda}\frac{1-\exp\big[(u_1+u_2)h_\tau/2\big]}{u_1+u_2}\,\underline{S}_{\lambda_2\tau}\,\underline{S}^{-1}_{\tau\lambda_1}\bigg)\underline{\varepsilon}_{\lambda_1}\otimes\underline{\varepsilon}_{\lambda_2}. \tag{4.27}$$

Let $z_1 \in U_{\lambda_1}$ and $z_2 \in U_{\lambda_2}$. We decompose in power series of (u_1,u_2) and compute with (4.7)-(4.7) for the odd part

$$\begin{aligned}
&\omega_{0,2}^{\mathrm{odd}}(z_1,z_2) - \frac{\delta_{\lambda_1\lambda_2}\,\mathrm{d}\zeta_{\lambda_1}(z_1)\mathrm{d}\zeta_{\lambda_2}(z_2)}{\big(\zeta_{\lambda_1}(z_1)-\zeta_{\lambda_2}(z_2)\big)^2}\\
&= -\sum_{\tau\in\Lambda}\underline{S}_{\lambda_2\tau}\,(\underline{S}^{-1})_{\tau\lambda_1}\sum_{k\geq 0}\frac{(h_\tau/2)^{k+1}}{(k+1)!}\bigg(\sum_{\substack{d_1,d_2\geq 0\\ d_1+d_2=k}}\frac{k!}{d_1!d_2!}\,\frac{[\zeta_{\lambda_1}(z_1)]^{2d_1}\,[\zeta_{\lambda_2}(z_2)]^{2d_2}}{(2d_1-1)!!(2d_2-1)!!}\bigg)\\
&\qquad\cdot\mathrm{d}\zeta_{\lambda_1}(z_1)\mathrm{d}\zeta_{\lambda_2}(z_2)\\
&= -\sum_{\tau\in\Lambda}\underline{S}_{\lambda_2\tau}\,(\underline{S}^{-1})_{\tau\lambda_1}\sum_{k\geq 0}\frac{h_\tau^{k+1}}{4(k+1)\cdot(2k)!}\Big\{\big(\zeta_{\lambda_1}(z_1)+\zeta_{\lambda_2}(z_2)\big)^{2k} + \big(\zeta_{\lambda_1}(z_1)-\zeta_{\lambda_2}(z_2)\big)^{2k}\Big\}\\
&\qquad\cdot\mathrm{d}\zeta_{\lambda_1}(z_1)\mathrm{d}\zeta_{\lambda_2}(z_2)\\
&= -\frac{1}{2}\sum_{\tau\in\Lambda}\underline{S}_{\lambda_2\tau}\,(\underline{S}^{-1})_{\tau\lambda_1}\,h_\tau\Big\{\tilde{\mathcal{B}}[h_\tau^{1/2}(\zeta_{\lambda_1}(z_1)+\zeta_{\lambda_2}(z_2))] + \tilde{\mathcal{B}}[h_\tau^{1/2}(\zeta_{\lambda_1}(z_1)-\zeta_{\lambda_2}(z_2))]\Big\}\\
&\qquad\cdot\mathrm{d}\zeta_{\lambda_1}(z_1)\mathrm{d}\zeta_{\lambda_2}(z_2).
\end{aligned}$$

It is expressed in terms of the renormalized Dehn twist introduced in (4.22) and the following series

$$\tilde{\mathcal{B}}(\zeta) = \sum_{k\geq 0}\frac{\zeta^{2k}}{2(k+1)\cdot(2k)!} = \frac{1-\cosh(\zeta)}{\zeta^2} + \frac{\sinh(\zeta)}{\zeta} = \frac{1}{\zeta^2} - \mathcal{B}(\zeta).$$

Writing $\delta_{\lambda_1\lambda_2} = \sum_{\tau\in\Lambda} \underline{S}_{\lambda_2\tau}(\underline{S}^{-1})_{\tau\lambda_1}$, the double pole can also be incorporated in the sum in the right-hand side. By adding a suitable non-odd part, we can choose a bidifferential with a rather simple expression

(4.28)
$$\omega_{0,2}(z_1,z_2) := \sum_{\tau\in\Lambda} \underline{S}_{\lambda_2\tau}(\underline{S}^{-1})_{\tau\lambda_1}\, h_\tau\, \mathcal{B}\big[h_\tau^{1/2}(\zeta_{\lambda_1}(z_1)-\zeta_{\lambda_2}(z_2))\big]\,\mathrm{d}\zeta_{\lambda_1}(z_1)\mathrm{d}\zeta_{\lambda_2}(z_2).$$

This is the result announced in (4.24). Eventually, we compute from (4.9) the basis of 1-form induced by this choice of $\omega_{0,2}$. Taking advantage of the fact that

$$\mathcal{B}(\zeta) = \frac{\cosh(\zeta)}{\zeta^2} - \frac{\sinh(\zeta)}{\zeta} = -\frac{f'(\zeta)+f'(-\zeta)}{2}, \qquad f(\zeta) = \frac{\exp(\zeta)}{\zeta},$$

we find that, for $z_0 \in U_\mu$,

$$\begin{aligned}
\frac{\Xi_{d,\lambda}(z_0)}{\mathrm{d}\zeta_\mu(z_0)}
&= \operatorname*{Res}_{z\to o_\lambda} \frac{(2d+1)!!}{\zeta_\lambda^{2d+2}(z)} \int_{o_\lambda}^{z} \frac{\omega_{0,2}(\cdot,z_0)}{\mathrm{d}\zeta_\mu(z_0)} \\
&= -\sum_{\tau\in\Lambda} \underline{S}_{\mu\tau}(\underline{S}^{-1})_{\tau\lambda} \operatorname*{Res}_{\zeta\to 0} \frac{\exp[h_\tau^{1/2}(\zeta-\zeta_\mu(z_0))]}{\zeta-\zeta_\mu(z_0)}\,\frac{(2d+1)!!\,\mathrm{d}\zeta}{\zeta^{2d+2}} \\
&= \sum_{\tau\in\Lambda} \underline{S}_{\mu\tau}(\underline{S}^{-1})_{\tau\lambda}\bigg(\frac{(2d+1)!!}{[\zeta_\mu(z_0)]^{2d+2}} - \sum_{m\geq 0}\frac{(2d+1)!!}{(2d+1)!}\,\frac{(-1)^m h_\tau^{d+1+m/2}\,[\zeta_\mu(z_0)]^m}{(2d+2+m)\cdot m!}\bigg) \\
&= \sum_{\tau\in\Lambda} \underline{S}_{\mu\tau}(\underline{S}^{-1})_{\tau\lambda}\bigg(\frac{(2d+1)!!}{[\zeta_\mu(z_0)]^{2d+2}} - \frac{h_\tau^{d+1}}{2^d\,d!}\int_0^1 \mathrm{d}v\, v^{2d+1}\,\exp[-h_\tau^{1/2}\zeta_\mu(z_0)v]\bigg) \\
&= \sum_{\tau\in\Lambda} \underline{S}_{\mu\tau}(\underline{S}^{-1})_{\tau\lambda}\,\frac{\Gamma[2d+2\,;\,h_\tau^{1/2}\zeta_\mu(z_0)]}{2^d\,d!\,[\zeta_\mu(z_0)]^{2d+2}}.
\end{aligned}$$

□

REMARK 4.2. For $t \neq 0$, the dependence in t can easily be absorbed by defining $\tilde{\zeta}_\lambda(z) = t^{1/2}\zeta_\lambda(z)$. More precisely

$$\omega_{g,n}|_t(z_1,\ldots,z_n) = t^{3g-3+3n/2}\Big[\omega_{g,n}(z_1,\ldots,z_n)\Big|_{t=1}\Big]_{\zeta_\lambda(z)\to t^{1/2}\zeta_\lambda(z)}$$

and subsequently

$$F_g|_t = t^{3g-3}\,F_g|_{t=1}$$

The latter is also true for $t=0$, as it is known that $F_g^{\mathrm{KdV}} = 0$ for all $g \geq 2$.

In the remaining of the text, we comment on the local spectral curves obtained for two important families of modular functors, which are both related to quantum Chern-Simons theory in 2+1-dimensions, with a gauge group G which is either finite (Section 5) or simply-connected compact (Section 6). Without entering into the details of the construction, we present the main facts necessary to apply effectively the results of the previous sections. We also discuss in Section 7 the problem of constructing global spectral curves in which the local expansion at the ramification gives (4.24).

5. Example: Modular functors associated to finite groups

The simplest examples of modular functors are provided by quantum Chern-Simons theory with finite gauge groups G in $(2+1)$-dimensions. These TQFT was studied by Dijkgraaf and Witten as a particular case of quantization of Chern-Simons theory with arbitrary compact (maybe non simply connected) gauge group, and its construction depends on a cocycle $[\alpha] \in H^3(BG, \mathrm{U}(1))$ [**30**], where BG is the classifying space for G. When G is a finite group, the path integral producing the TQFT correlation functions over a manifold X is reduced to a finite sum over isomorphism classes of certain G-principal bundles on X. The theory is therefore an attractive playground to have a grasp on TQFTs. The construction of the modular functor was presented in [**42**] shortly after. The central extension of the mapping class group plays no role here, i.e. $\tilde{c} = 1$ and the central charge is always 0. This model is also known as the "holomorphic orbifold model", and fits in the framework of VOA.

We summarize below the untwisted theory $[\alpha] = 0$. The modular functor in this case can also be obtained from the modular tensor category of representations of $D(G)$, the quantum double of the finite group G [**12**]. We have that $\mathcal{A} = \mathrm{Rep}(D(G))$, which is also equipped with the structure of a Frobenius manifold. Then, the Dehn twist eigenvalues are also trivial: $\tilde{d}_\lambda = 1$ for all $\lambda \in \Lambda$. Therefore, the CohFT we produce is rather trivial, like in Section 4.3, and does not remember more than the dimensions of the TQFT vector spaces.

The twisted theory deals with projective representations of G, with cocycle determined by the class $[\alpha]$. We refer to [**28, 30, 42**] for a full presentation. But, we will point out in Section 5.2 that it gives Dehn twist eigenvalues depending in a non-trivial way on the class $[\alpha]$, and therefore leads to CohFTs which do not sit only in degree 0.

5.1. The untwisted theory.

5.1.1. *Frobenius algebra.* If $g \in G$, C_g denotes the conjugacy class of g, and Z_g the centralizer of g in G, i.e. the set of all elements commuting with g. We obviously have that

$$\#C_g = \frac{\#G}{\#Z_g}.$$

The label set Λ consists of ordered pairs (i, ϱ) where C_i is a conjugacy class of G, and $\varrho = \varrho_{i,g_i}$ an (isomorphism class of) irreducible representation of Z_{g_i} for some $g_i \in C_i$. Remark that, for any two representatives $g_i, g_i' \in C_i$, Z_{g_i} and $Z_{g_i'}$ (hence their representations) are canonically isomorphic, by the following formula.

$$\varrho_{i,kgk^{-1}}(h) = \varrho_{i,g}(k^{-1}hk). \tag{5.1}$$

We denote $i^\dagger$ the index of the conjugacy class containing g_i^{-1}, and $\varrho^\dagger$ the dual representation. Then, $(i, \varrho) \mapsto (i^\dagger, \varrho^\dagger)$ endows Λ with an involution. The vector space $\mathcal{A} = \bigoplus_{(i,\varrho)} \mathbb{C}.e_{(i,\varrho)}$ is equipped with a product reflecting the operation of tensor product of representations. If ϱ_i and ϱ_j are representations of Z_{g_i} and Z_{g_j}, one has to consider $\varrho_i \otimes \varrho_j$ as defining a representation of Z_{g_k} for each g_k that can appear as the product of two elements in C_i and C_j.

The algebra $\mathcal{A}$ contains two remarkable subalgebras $\mathcal{A}_{\text{conj}}$ and $\mathcal{A}_{\text{rep}}$ which encode respectively the product induced by the group algebra of G on its center, and the representation theory of G.

The space $\mathcal{A}_{\text{rep}}$ is a subalgebra spanned by the vectors $e_{(1,\varrho)}$, where $i = 1$ is reserved for the conjugacy class of the identity in G, and ϱ is an irreducible representation of $G = Z_1$. The subalgebra $\mathcal{A}_{\text{conj}}$ is spanned by the vectors $e_{(i,1)}$, which indexes conjugacy classes C_i of G, and the centralizer is equipped with the trivial representation, so that $\varrho_{i,1}(g)$ is constant equal to 1. It is isomorphic as an algebra to the center of the group algebra of G, and as such it has a natural structure of commutative associative algebra. The structure constants in the product

$$e_{(i,1)} \underset{\text{conj}}{\times} e_{(j,1)} = \sum_k N_{ijk}\, e_{(k,1)} \tag{5.2}$$

compute the number of factorizations of the identity in G by elements in fixed conjugacy classes. It can be expressed as follows

$$\begin{aligned} N_{i_1 i_2 i_3} &= \frac{\#\{(g_1, g_2, g_3) \in C_{i_1} \times C_{i_2} \times C_{i_3} \quad | \quad g_1 g_2 g_3 = 1\}}{\#C_{i_3}} \\ &= \frac{\#\{p \in \mathrm{Hom}(\pi_1(\mathbf{\Sigma}_{0,3}), G) \quad | \quad p(\mathfrak{l}_j) \in C_{i_j}\ \ j = 1,2,3\}}{\#C_{i_3}}. \end{aligned}$$

Here, $\mathfrak{l}_j$ is a loop around the j-th puncture of $\mathbf{\Sigma}_{0,3}$. The numerator is also the number of G-bundles on a sphere with 3 punctures, in which the monodromy around the j-th point belongs to the conjugacy class C_{i_j}.

5.1.2. *Scalars and S-matrix.* These modular functors have $\tilde{c} = 1$ and $\tilde{r}_{(i,\varrho)} = 1$, but their VOA origin provides canonical log-determinations (see Section 3.5):

$$c = 0, \qquad r_\lambda = 0.$$

For each i labeling a conjugacy class, $g_i \in C_i$, and a representation $\varrho := \varrho_{i,g_i}$ of Z_{g_i}, we denote by $\chi_{i,g_i}(h) = \mathrm{Tr}\, \varrho_{i,g_i}(h)$ its character. We take as convention $\chi_{i,g_i}(h) = 0$ whenever $h \notin Z_{g_i}$. Let Z_i be the centralizer of some (arbitrary) representative of C_i, and $V_{i,\varrho}$ be the vector space on which Z_i acts via the representation ϱ. The dimension of this module is $\dim V_{i,\varrho} = \varrho_{i,g_i}(1)$. We reserve the label $i = 1$ for the conjugacy class of the identity in G, and then in $(1,\varrho)$, ϱ must be a representation of G. Its character will be denoted χ_ϱ, and $V_{1,\varrho} := V_\varrho$ the vector space of the representation.

The S-matrix is computed in [**28**, **42**]. Keeping the notations of Section 3.3, it reads

$$\underline{S}_{(i_1,\varrho_1)(i_2,\varrho_2)} = \frac{1}{\#G} \sum_{\substack{g_j \in C_{i_j} \\ [g_1,g_2]=1}} \chi_{i_1,g_1}(g_2^{-1})\, \chi_{i_2,g_2}(g_1^{-1}). \tag{5.3}$$

We list for bookkeeping some special entries of the S-matrix.

$$
\begin{aligned}
\underline{S}_{(1,1)(1,1)} &= \frac{1}{\#G} \\
\underline{S}_{(i,1)(1,1)} &= \frac{\#C_i}{\#G} \\
\underline{S}_{(1,\varrho)(1,1)} &= \frac{\dim V_\varrho}{\#G} \\
\underline{S}_{(i,1)(1,\varrho)} &= \frac{\#C_i\,\chi_\varrho(C_i)}{\#G} \\
\underline{S}_{(i,\varrho)(1,1)} &= \frac{\#C_i}{\#G}\dim V_{i,\varrho} = \frac{\dim V_{i,\varrho}}{\#Z_i} \\
\underline{S}_{(i,1)(j,1)} &= \frac{\#\mathrm{Commutant}(C_i,C_j)}{\#G} \\
\underline{S}_{(1,\varrho)(1,\tau)} &= \frac{\dim V_\tau\ \dim V_\varrho}{\#G} \\
\underline{S}_{(i,\varrho)(1,\tau)} &= \frac{\dim V_{i,\varrho}}{\#G}\,\#C_i\cdot\chi_\tau(C_{i^\dagger}).
\end{aligned}
\tag{5.4}
$$

The matrix $\underline{S}$ can be thought as an extension of the character table of G which appears in (5.4), so that both indices i and ϱ are treated on the same footing. This matrix is clearly symmetric, and it can be checked by direct computation that $\underline{S}^2 = C$ is the operator sending e_λ to $e_{\lambda^\dagger}$.

5.1.3. *Local spectral curve.* Since $c = 0$ and $r_{(i,\varrho)} = 0$, the CohFT just consists of several (suitably rescaled) copies of the trivial CohFT, and the local spectral curve consists of several copies of the Airy curve (4.17). The only information left from the modular functor is the dimension of the representations, appearing as the rescaling

$$x(z) = [\zeta_{i,\varrho}(z)]^2/2, \qquad y(z) = -\frac{\dim V_{i,\varrho}}{\#G}\,\zeta_{i,\varrho}(z), \qquad z \in U_{i,\varrho}$$

and as we have seen in Section 4.3, this is just enough to compute the dimensions of the TQFT vector spaces.

5.2. The twisted theory. Let us choose a 3-cocycle α representing the class $[\alpha]$, and introduce for any $h \in G$, the following 2-cocycle for Z_h

$$c_h(g_1,g_2) = \frac{\alpha(h,g_1,g_2)\alpha(g_1,g_2,h)}{\alpha(g_1,h,g_2)}. \tag{5.5}$$

Choosing another representative α', we would obtain a 2-cocycle c'_h differing from c_h only by a coboundary. A projective representation of Z_h with cocycle c_h is a map $\varrho_h : Z_h \to \mathrm{GL}(V)$ for a finite dimensional vector space, such that $\varrho_h(1) = 1$ and one has that

$$\forall (g_1,g_2) \in G^2, \qquad \varrho_h(g_1)\varrho_h(g_2) = c_h(g_1,g_2)\,\varrho_h(g_1g_2).$$

In the construction of the modular functor for the twisted theory, $\varrho_{i,g}$ are now the projective representations of Z_{g_i} with cocycle c_{g_i} [**42**]. The S-matrix takes the form

$$\underline{S}_{(i_1,\varrho_1)(i_2,\varrho_2)} = \frac{1}{\#G} \sum_{\substack{g_j \in C_{i_j} \\ [g_1,g_2]=1}} \chi_{i_1,g_1}(g_2^{-1})\,\chi_{i_2,g_2}(g_1^{-1})\,\sigma(g_1|g_2)$$

and the Dehn twist eigenvalues are

$$\tilde{r}_{(i,\varrho)} = \sigma(g_i|g_i)^{-1/2}$$

for some arbitrary $g_i \in C_i$. The function $\sigma(g_1|g_2)$ depends in a non-trivial way on $[\alpha]$, but must satisfy $\sigma(g_1|g_2) = \sigma(g_2|g_1)$ and another condition of cohomological nature [**28**]. The relation between σ and α is explained in full generality in [**41**]. In the simpler case where all 2-cocycles c_h defined in (5.5) are coboundaries, let us take a 1-cocycle β_h such that

$$c_h(g_1,g_2) = \frac{\beta_h(g_1)\beta_h(g_2)}{\beta_h(g_1g_2)}.$$

The consistency conditions impose $\beta_{h^{-1}}(g) = \beta_{g^{-1}}(h) = [\beta_g(h)]^{-1}$, and the expression for σ is then [**28**]

$$\sigma(g_1|g_2) = \beta_{g_1}(g_2)\beta_{g_2}(g_1).$$

As we see, the Dehn twist eigenvalue $\tilde{r}_{(i,\varrho)}$ now depends on i, but still not on the representation ϱ of Z_i. In the computation of $\omega_{0,2}$ from (4.24), one encounters the following terms, for a fixed i,

$$\begin{aligned}(5.6)\qquad & \sum_{\varrho} \underline{S}_{(i_2,\varrho_2)(i,\varrho)}(\underline{S}^{-1})_{(i,\varrho)(i_1,\varrho_1)} \\ &= \sum_{\varrho} \underline{S}_{(i_1^\dagger,\varrho_1^\dagger)(i,\varrho)}\underline{S}_{(i_2,\varrho_2)(i,\varrho)} \\ &= \frac{1}{(\#G)^2} \sum_{\substack{g_1\in C_{i_1},\ g_2\in C_{i_2} \\ h_1,h_2\in C_i \\ [g_j,h_j]=0}} \sigma(g_1^{-1}|h_1)\sigma(g_2|h_2)\,\chi_{i_1,g_1}(g)\,\chi_{i_2,g_2}(h_2^{-1}) \\ &\quad \cdot \bigg(\sum_{\chi} \chi_{i,h_2}(g_2^{-1})\,\chi_{i,h_1}(g_1)\bigg).\end{aligned}$$

We focus on the quantity in the brackets, which is a sum over all projective representations of Z_i. Since there exists $k \in G$ such that $h_1 = kh_2k^{-1}$, we have $\chi_{i,h_2}(g_2^{-1}) = \chi_{i,h_1}(kg_2^{-1}k^{-1})$. The characters of projective representations of Z_i form an orthonormal basis of c_{h_1}-class functions – the proof, as for representations, only relies on Schur's lemma. Therefore, the sum over χ in (5.6) vanishes when g_1 and $kg_2^{-1}k^{-1}$ do not belong to the same conjugacy class in Z_i. A fortiori, to have a non-zero result in (5.6), we need g_1 and g_2 to be in the same conjugacy class in G, i.e. $i_1 = i_2$. Therefore, $\omega_{0,2}(z_1,z_2)$ for $z_j \in U_{(i_j,\varrho_j)}$ is proportional to $\delta_{i_1i_2}$, and it follows from the residue formula (4.4) that

LEMMA 5.1. *For $z_j \in U_{(i_j,\varrho_j)}$, $\omega_{g,n}(z_1,\ldots,z_n)$ vanishes unless all i_j are equal for $j \in [\![1,n]\!]$.*

We find that the CohFT does not couple the representation theory of different centralizers, or in the words of [**30**], it does not couple different interaction channels.

However, $\omega_{0,2}$ in a given channel will mix various ϱ's, and thus the CohFT is non-trivial, unlike the untwisted case.

6. Example: WZW model for compact Lie groups

6.1. Short presentation. We briefly review the definition of the Verlinde bundle that arises from the Wess-Zumino-Witten model, and the corresponding modular functor. It is based on the representation theory of a VOA, namely an algebra that incorporates the Virasoro algebra – describing infinitesimal coordinate reparametrizations in a disk neighbourhood of a puncture on a Riemann surface – together with a central extension and extra symmetries coming from a simple complex Lie algebra $\mathfrak{g}$. Since the seminal work of Tsuchiya, Ueno and Yamada [**88**], there is a vast literature on the Verlinde bundles, that readers with different backgrounds will appreciate differently; we can suggest the nice article of Beauville [**14**] giving a proof of the Verlinde formula for the non-exceptional $\mathfrak{g}$, the book of Ueno [**91**], and the review of Looijenga [**67**]. A complete proof that these theories gives modular functors was given by Andersen and Ueno in [**3**, **4**].

6.1.1. *Affine Kač-Moody algebras.* We refer to [**62**] for the detailed theory. Let $\mathfrak{h}$ be the Cartan subalgebra, Δ_+ the set of positive roots, θ the highest root – which is the highest weight for the adjoint representation – and $\varrho = \frac{1}{2}\sum_{\alpha>0}\alpha$ the Weyl vector. We normalize the Killing form $\langle\cdot,\cdot\rangle$ such that $\langle\theta,\theta\rangle = 2$. It induces an isomorphism between $\mathfrak{h}$ and $\mathfrak{h}^*$, which we use systematically to identify $\mathfrak{h}$ and $\mathfrak{h}^*$. In particular, we use the notation ϱ for the Weyl vector in $\mathfrak{h}$ or the corresponding element in $\mathfrak{h}^*$. Let $(e_i)_{i=1}^{\dim\mathfrak{h}}$ be an orthonormal basis of $\mathfrak{h}$ for the Killing form. The quadratic Casimir is the element of the universal enveloping algebra of $\mathfrak{g}$ defined by $\mathcal{Q} = \frac{1}{2}\sum_{i=1}^{\dim\mathfrak{h}} e_i \otimes e_i$. It acts as a scalar on the highest weight $\mathfrak{g}$-module V_λ, more precisely, one has that

$$\mathcal{Q}|_{V_\lambda} = \frac{1}{2}\langle\lambda,\lambda+2\varrho\rangle\,\mathrm{id}_{V_\lambda}.$$

This scalar on the adjoint is denoted $h^\vee(\mathfrak{g})$, and called the dual Coxeter number of $\mathfrak{g}$, namely $h^\vee(\mathfrak{g}) = 1 + \langle\theta,\varrho\rangle$. Let $\mathfrak{g}_\mathbb{R}$ be the compact real form of $\mathfrak{g}$, and $\mathfrak{h}_\mathbb{R}$ its Cartan subalgebra. The weight lattice in $\mathfrak{h}^*$ is denoted L_W, it consists of all $\lambda \in \mathfrak{h}_\mathbb{R}^*$ such that

$$\forall\alpha\in\Delta_+,\qquad 2\,\frac{\lambda(\alpha)}{\langle\alpha,\alpha\rangle}\in\mathbb{Z}.$$

For a fixed $\ell\geq 1$, we introduce the set of highest weights at level ℓ defined by

$$\mathcal{P}_\ell = \big\{\lambda\in L_W \mid 0\leq\lambda(\theta)\leq\ell\big\}. \tag{6.1}$$

Consider the untwisted affine Lie algebra

$$\widehat{\mathfrak{g}} = \mathbb{C}.\mathrm{c}\oplus\mathfrak{g}\otimes\mathbb{C}((\xi)).$$

It is a central extension of $\mathfrak{g}\otimes\mathbb{C}((\xi))$, with the Lie bracket defined by

$$\forall X_1,X_2\in\mathfrak{g},\quad \forall f_1,f_2\in\mathbb{C}((\xi)),$$
$$[X_1\otimes f_1, X_2\otimes f_2] = \langle X,Y\rangle\big(\operatorname*{Res}_{\xi\to 0} f_1\mathrm{d}f_2\big)\cdot\mathrm{c} + [X_1,X_2]\otimes f_1f_2.$$

We can use the following decomposition

$$\widehat{\mathfrak{g}} = \mathfrak{p}_+\oplus\mathfrak{p}_-,\qquad \mathfrak{p}_+ = \mathfrak{g}_+\oplus\mathfrak{g}\oplus\mathbb{C}.\mathrm{c},\qquad \mathfrak{g}_+ = \xi\cdot\mathfrak{g}[[\xi]].$$

If $\lambda \in \mathcal{P}_\ell$, we promote the $\mathfrak{g}$-module V_λ to a $\mathfrak{p}_+$ module, by declaring that $\mathfrak{g}_+$ annihilates V_λ, and the central element c acts as $\ell \cdot \mathrm{id}_{V_\lambda}$. Then, one introduces the Verma module as

$$M_\lambda = U(\mathfrak{g}) \otimes_{\mathfrak{p}_+} V_\lambda.$$

It is a left $\widehat{\mathfrak{g}}$-module, which is not irreducible. However, it contains a maximal proper submodule defined by

$$J_\lambda = U(\widehat{\mathfrak{p}}_-) \cdot (\theta \otimes \xi^{-1})^{\ell - \langle \theta, \lambda \rangle + 1}.$$

Then, $H_\lambda = M_\lambda / J_\lambda$ becomes irreducible, and it is actually an integrable highest weight module for $\widehat{\mathfrak{g}}$. If P is a finite set, define the multi-variable analogue of $\widehat{\mathfrak{g}}$ by

$$\widehat{\mathfrak{g}}_P = \mathbb{C}.\mathrm{c} \oplus \bigoplus_{p \in P} \mathfrak{g} \otimes \mathbb{C}((\xi_p)).$$

Now, if $\mathfrak{X} = (\Sigma, P; \xi)$ is a Riemann surface Σ equipped with a finite marked set of points P and $\xi = (\xi_p)_{p \in P}$ a set of local coordinates at the points of P, then the Lie algebra

$$\widehat{\mathfrak{g}}(\mathfrak{X}) = \mathfrak{g} \otimes_{\mathbb{C}} H^0(\Sigma, \mathcal{O}_\Sigma(*P))$$

is naturally embedded as a subalgebra of $\widehat{\mathfrak{g}}_P$.

6.1.2. *From the space of vacua to the modular functor.* For any $\mathfrak{X}$ as above, and $\vec{\lambda} : P \to \mathcal{P}_\ell$ a set of labels, we define the space of vacua by

$$\mathcal{V}_{\vec{\lambda}}(\mathfrak{X}) = H_{\vec{\lambda}} / \widehat{\mathfrak{g}}(\mathfrak{X}) H_{\vec{\lambda}}.$$

Then, one can show [**67, 87, 88**] that, if $\mathfrak{X}$ and $\mathfrak{X}'$ differ by a change of coordinates $\xi \to \xi'$, there is a canonical isomorphism between $\mathcal{V}_{\vec{\lambda}}(\mathfrak{X})$ and $\mathcal{V}_{\vec{\lambda}}(\mathfrak{X}')$. This allows us to define $\mathcal{V}_{\vec{\lambda}}(\mathfrak{X})$ as a bundle over the Teichmüller space $\widetilde{\mathcal{T}}_{\Sigma,P}$.

By the work of Tsuchiya, Ueno and Yamada [**88**], this bundle carries a projectively flat connection, and enjoys nice factorization properties over families where the surface is pinched. Exploiting this connection, Andersen and Ueno proved [**3, 4**], that one can make a definition independent of the complex structure: They assign unambiguously a vector space $\mathcal{V}_{\vec{\lambda}}(\mathbf{\Sigma})$ to any marked surface $\mathbf{\Sigma}$, and prove that this assignment defines a modular functor in the sense of Section 2. Moreover, in [**5, 7**] Andersen and Ueno established that, for $\mathfrak{g} = \mathfrak{sl}_N$, this modular functor is isomorphic to the modular functor obtained from the modular tensor category of representations of the quantum group $U_{-q^{1/2}}(\mathfrak{g})$ with

$$q = \exp[2\mathrm{i}\pi/(\ell + h^\vee(\mathfrak{g}))]. \tag{6.2}$$

6.1.3. *More notations.* Let $\mathfrak{W}$ be the Weyl group of $\mathfrak{g}$. We denote $w_0 \in \mathfrak{W}$ its longest element. Its length is $|\Delta_+|$, and it is the unique element that sends positive roots to negative roots. By definition of the Weyl vector, $w_0(\varrho) = -\varrho$.

We consider as label set $\Lambda := \mathcal{P}_\ell$, the set of representations of $\mathfrak{g}_{\mathbb{R}}$ at level ℓ. It is equipped with the involution $\lambda^\dagger = -w_0(\lambda)$. Note that $\lambda^\dagger$ is actually the highest weight of V_λ^*.

If $\beta \in \mathfrak{h}$, q^β stands for the function $\mathfrak{h} \to \mathbb{C}$ defined by $x \mapsto q^{\langle \beta, x \rangle}$ on $\mathfrak{h}$. The character of V_λ is denoted $\mathrm{ch}_\lambda : \mathfrak{h} \to \mathbb{C}$, and is given by the Weyl character formula

$$\mathrm{ch}_\lambda(q^\beta) = \frac{\sum_{w \in \mathfrak{W}} \mathrm{sgn}(w)\, q^{\langle \lambda + \varrho, w(\beta) \rangle}}{\sum_{w \in \mathfrak{W}} \mathrm{sgn}(w)\, q^{\langle \varrho, w(\beta) \rangle}}.$$

The q-dimension is defined as $\dim_q V_\lambda = \mathrm{ch}_\lambda(1)$, and using the Weyl denominator formula, it reads

$$\dim_q \lambda := \prod_{\alpha>0} \frac{[\langle\alpha, \lambda+\varrho\rangle]_q}{[\langle\alpha,\varrho\rangle]_q}, \qquad [x]_q = q^{x/2} - q^{-x/2}.$$

The weight lattice L_W has a sublattice L_θ spanned by the elements $\langle w(\theta), \cdot\rangle$ with $w \in \mathfrak{W}$. The integer

$$D_\ell := \#\big(L_W/(\ell+h^\vee)L_\theta\big)$$

appears as a normalization constant in

$$\sum_{\lambda\in\Lambda} \big|\mathrm{ch}_\lambda(q^\beta)\big|^2 = D_\ell^{-1}\cdot\Big|\prod_{\alpha>0}[\langle\alpha,\beta\rangle]_q\Big|^2.$$

Consider the matrix

$$(6.3)\qquad \begin{aligned}\mathcal{S}_{\lambda\mu} &= D_\ell^{-1/2}\sum_{w\in\mathfrak{W}} \mathrm{sgn}(w)\, q^{\langle\lambda+\varrho, w(\mu+\varrho)\rangle}\\ &= D_\ell^{-1/2}\cdot\prod_{\alpha>0}[\langle\alpha,\varrho\rangle]_q\cdot \dim_q V_\lambda\cdot \mathrm{ch}_\mu(q^{\lambda+\varrho})\end{aligned}$$

with the value $q = \exp[2\mathrm{i}\pi/(\ell+h^\vee)]$. From the first equality we see that $\mathcal{S}$ is symmetric, and since the scalar product is $\mathfrak{W}$-invariant and $\mathrm{sgn}(w_0) = (-1)^{|\Delta_+|}$, we have that

$$C\,\mathcal{S} = (-1)^{|\Delta_+|}\,\mathcal{S}^*,$$

where we recall that C is the matrix of the involution $\dagger$. It is less obvious but also true using orthogonality of characters that $\mathcal{S}$ is unitary which means that

$$\mathcal{S}\mathcal{S}^* = 1.$$

6.1.4. *Scalars and S-matrix.* The modularity of characters of the underlying VOA gives canonical log-determinations for the eigenvalues of the central element and the Dehn twist. These are the conformal weights and the central charge [**59**, **88**]

$$(6.4)\qquad r_\lambda = \frac{\langle\lambda,\lambda+2\varrho\rangle}{2(\ell+h^\vee)}, \qquad c = \frac{\ell\dim\mathfrak{g}}{\ell+h^\vee(\mathfrak{g})}.$$

The formulas manifestly satisfy $r_1 = 0$ and $r_\lambda = r_{\lambda^\dagger}$. With Freudenthal strange formula, we can also write that

$$\frac{c}{24} = \frac{\ell}{h^\vee}\,\frac{\langle\varrho,\varrho\rangle}{2(\ell+h^\vee(\mathfrak{g}))}.$$

The S-matrix appearing in Section 3.3 is given by the Kač-Peterson formula [**59**, Proposition 4.6(d)]

$$(6.5)\qquad \underline{S}_{\lambda\mu} = \mathrm{i}^{|\Delta_+|}\,\mathcal{S}^*_{\lambda\mu} \qquad \text{or equivalently} \qquad \underline{S}^{-1}_{\lambda\mu} = (-\mathrm{i})^{|\Delta_+|}\,\mathcal{S}_{\lambda\mu}.$$

Since the diversity of notations in the literature can be confusing, let us make two checks ensuring that (6.5) in our notations is correct. Firstly, with the properties of $\mathcal{S}$ just pointed out, we can write

$$\underline{S}^{\top}\underline{S} = (-1)^{|\Delta_+|}\,\mathcal{S}^*\mathcal{S}^* = C\,\mathcal{S}\mathcal{S}^* = C$$

as it should be according to (3.16). Secondly, the Verlinde formula (3.17)-(4.16) in our notations does agree, once we insert (6.5) in the Verlinde formula [**14**, Corollary 9.8].

6.1.5. *The case* $\mathfrak{g} = \mathfrak{sl}_N$. Let $\mathbb{C}^N$ be equipped with its canonical orthonormal basis $(e_i)_{i=1}^N$, and e_i^* be the dual basis. The Cartan algebra of $\mathfrak{sl}_N$ can be identified with the hyperplane in $\mathbb{C}^N$ orthogonal to $\sum_{i=1}^N e_i$, and the Killing form is induced from the scalar product on $\mathbb{C}^N$. The positive roots are $e_i - e_j$ for $1 \leq i < j \leq N$, the highest root is $\theta = e_1 - e_N$, and the Weyl vector is

$$\varrho = \frac{1}{2}\sum_{i=1}^N (N+1-2i)\, e_i.$$

Here $\mathfrak{h}^*$ (resp. the weight lattice L_W) is the $\mathbb{C}$-span (resp. $\mathbb{Z}$-span) of $(e_i^*)_{i=1}^N$ modulo the relation $\sum_{i=1}^N e_i^* = 0$. If $\lambda \in \mathfrak{h}^*$, let us denote $|\lambda| = \sum_{i=1}^N \lambda_i$. The Killing form induced on $\mathfrak{h}^*$ is [**45**]

$$\Big\langle \sum_{i=1}^N \lambda_i\, e_i^*\,,\, \sum_{j=1}^N \mu_j\, e_j^* \Big\rangle = \sum_{i=1}^N \lambda_i \mu_i - \frac{|\lambda||\mu|}{N}.$$

The element representing the Weyl vector in $\mathfrak{h}^*$ is

$$\varrho = \frac{1}{2}\sum_{i=1}^N (N+1-2i)\, e_i^* = \sum_{i=1}^{N-1} (N-i) e_i^* \bmod \Big(\sum_{i=1}^N e_i^*\Big).$$

The fundamental weights are

$$1 \leq i \leq N-1, \qquad w_i = \sum_{j=1}^{i} e_j^*$$

and the corresponding highest weight $\mathfrak{sl}_N$-modules are $\bigwedge^i \mathbb{C}^N$. Irreducible highest weight representations of $\mathfrak{sl}_N$ are encoded in an $(N-1)$-tuple $\lambda = (\lambda_1, \ldots, \lambda_{N-1})$ such that $\lambda_1 \geq \cdots \geq \lambda_{N-1} \geq 0$, which corresponds to the highest weight vector

$$\lambda = \sum_{j=1}^N \lambda_j\, e_j^* = \sum_{j=1}^{N-1} (\lambda_j - \lambda_{j+1})\, w_j$$

with the convention $\lambda_N = 0$. The representations at level ℓ are those with $\lambda_1 \leq \ell$. The Weyl group permutes the $(e_i)_{i=1}^N$, and its longest element is the permutation $e_i \mapsto e_{N+1-i}$ for all $i \in [\![1, N]\!]$. Since $e_N^* = -\sum_{i=1}^{N-1} e_i^*$ on the hyperplane $\mathfrak{h}$, we see that the involution takes λ to its "complement" as in Figure 2, i.e. $\lambda_i^\dagger = \lambda_1 - \lambda_{N-i+1}$. The characters ch_λ are the Schur polynomials, and the central charge and Dehn twist eigenvalues are

$$h^\vee(\mathfrak{sl}_N) = N, \qquad c = \frac{\ell(N^2-1)}{\ell+N}, \qquad r_\lambda = \frac{1}{2(\ell+N)}\bigg(\sum_{i=1}^{N-1} \lambda_i(\lambda_i - 2i + 1) + N|\lambda| - \frac{|\lambda|^2}{N}\bigg).$$

Since the Weyl group acts transitively on the set of roots, L_θ is just the root lattice

$$L_\theta = \Big\{ \sum_{i=1}^N a_i e_i^*,\ \ \sum_{j=1}^N a_j = 0 \Big\} \Big/ \Big\{ \sum_{i=1}^N e_i^* = 0 \Big\}$$

and it has index N in the weight lattice. Therefore, the normalization integer reads

$$D_\ell = N\,(\ell+N)^{N-1}.$$

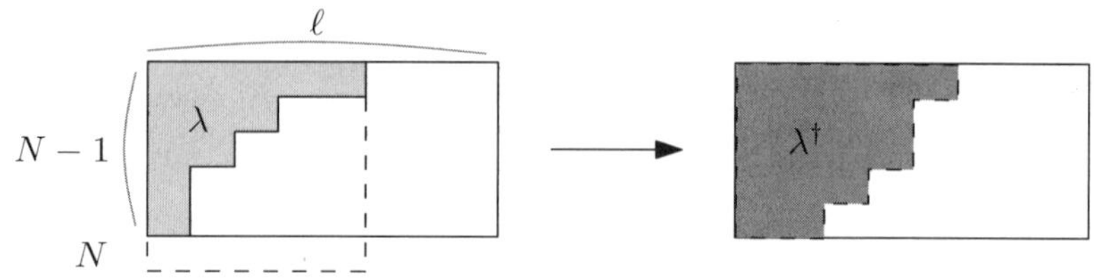

FIGURE 2. The involution $\lambda^\dagger = -w_0(\lambda)$ in terms of Young tableaux.

In the case $N = 2$ for level $\ell \geq 0$, Λ is the set of integers from 0 to ℓ, the involution is the identity, and $D_\ell = 2(\ell+2)$. The scalars and S-matrix read

$$\forall \lambda, \mu \in [\![0,\ell]\!] \qquad c = \frac{3\ell}{\ell+2}, \qquad r_\lambda = \frac{\lambda(\lambda+2)}{4(\ell+2)},$$

$$\underline{S}^{-1}_{\lambda\mu} = \sqrt{\frac{2}{\ell+2}}\,\sin\Big[\frac{\pi(\lambda+1)(\mu+1)}{\ell+2}\Big].$$

6.2. CohFT and double Hurwitz numbers. The Chern character of the Verlinde bundle has already been studied in [**69**, **87**], and [**70**] showed that it defines a CohFT – this was in fact the example motivating our work. In this particular case, our work only completes [**70**] by remarking that the CohFT correlation functions are computed by the topological recursion with the local spectral curve of Theorem 4.5. It would be interesting to know the meaning of this CohFT from an enumerative geometry perspective. For $\mathfrak{g} = \mathfrak{sl}_N$, we show that $\omega_{0,2}$ given in (4.24) is related in some way to double Hurwitz numbers.

6.2.1. *Hurwitz numbers.* Let us first review the definition of Hurwitz numbers. For any finite group G, we consider the number of G-principal bundles over a surface of genus g and n punctures, whose monodromy around the j-th punctures belongs to the conjugacy class C_{i_j} of G. It is computed by the Frobenius formula [**99**]

$$\frac{\#\{R \in \mathrm{Hom}(\pi_1(\Sigma_{g,n}),G)\mid R(\mathfrak{l}_j)\in C_j\}}{\#G} = \sum_\nu \Big[\frac{\chi_\nu(1)}{\#G}\Big]^{2-2g}\prod_{j=1}^n f_\nu(C_j) \tag{6.6}$$

where the sum ranges over irreducible representations of G, and

$$f_\nu(C) := \frac{\#C\,\chi_\nu(C)}{\chi_\nu(1)}.$$

When $G = \mathfrak{S}_d$, this is the number of (possibly disconnected) branched coverings of degree d over a surface of genus g. Conjugacy classes of $\mathfrak{S}_d$ are labeled by partitions μ of d: The parts of μ are the lengths of the cycles of a representative of C_μ. For a collection of partitions $\mu_1,\ldots,\mu_n \vdash d$, the numbers

$$H_g^d(\vec{\mu}) = \frac{\#\{R \in \mathrm{Hom}(\pi_1(\Sigma_{g,n}),\mathfrak{S}_d)\mid R(\mathfrak{l}_j)\in C_{\mu_j}\}}{d!}$$

are called "Hurwitz numbers" of genus g. Though it is a good starting point, the formula (6.6) involving the character table of $\mathfrak{S}_d$ is not the end of the story.

Let $C_{(2)}$ be the conjugacy class of a transposition. One often would like to count branched coverings with arbitrary number b of simple ramification points.

The corresponding Hurwitz numbers are

$$H_g(\vec{\mu}\,|\,T) := \sum_{b\geq 0} \frac{T^b}{b!}\, H_g^d(\vec{\mu}, \underbrace{C_{(2)},\ldots,C_{(2)}}_{b \text{ times}}).$$

If we keep n points with arbitrary ramifications $\vec{\mu}$, this is the generating series of n-tuples Hurwitz numbers in genus[7] g. For instance, the generating series of double Hurwitz numbers in genus 0 and degree d is, according to (6.6),

$$H_0^d(\mu_1,\mu_2\,|\,T) = \frac{1}{d!^2}\sum_{\nu\vdash d} \#C_{\mu_1}\#C_{\mu_2}\,\chi_\nu(C_{\mu_1})\chi_\nu(C_{\mu_2})\,e^{T f_\nu(C_{(2)})}. \tag{6.7}$$

The generating series of genus 0 simple [**19, 23, 37, 49, 50**] and double [**25, 53, 54, 58, 78**] Hurwitz numbers have been intensively studied from the point of view of combinatorics, mirror symmetry, and integrable systems. The generating series of Hurwitz numbers in genus 1 is somewhat simpler because the coupling in (6.6) disappears, and there is a nice theory relating them to quasimodular forms [**27, 60**]. The realm of genus $g \geq 2$ seems uncharted.

6.2.2. *Rewriting of the 2-point function.* We recall two elementary facts on symmetric functions [**68**]. By Schur-Weyl duality, for a partition with $|\lambda| = d$ boxes, the Schur polynomials decompose on the power sums as

$$\mathrm{ch}_\lambda(x) = \frac{1}{d!}\sum_{\mu\vdash d} \#C_\mu\,\chi_\lambda(C_\mu)\,p_\mu(x).$$

Further, $f_\lambda(C_{(2)}) = \frac{1}{2}\sum_{i=1}^N \lambda_i(\lambda_i - 2i + 1)$ is related to the quadratic Casimir, hence to the Dehn twist eigenvalues by the formula

$$r_\lambda = \frac{1}{\ell+N}\Big[f_\lambda(C_{(2)}) + \frac{N|\lambda|}{2} - \frac{|\lambda|^2}{2N}\Big].$$

Putting together the expression (6.3)-(6.5) of the S-matrix and these two observations, we can compute with (4.27) the 2-point function

$$B(u_1,u_2) = \sum_{\lambda_1,\lambda_2\in\Lambda} \frac{\delta_{\lambda_1\lambda_2} - B_{\lambda_1,\lambda_2}(u_1+u_2)}{u_1+u_2}\,\underline{\varepsilon}_{\lambda_1}\otimes\underline{\varepsilon}_{\lambda_2}.$$

We recall that $B_{\lambda_1,\lambda_2}(u_1+u_2)$ is essentially the Laplace transform of $\omega_{0,2}$ in $U_{\lambda_1}\times U_{\lambda_2}$.

LEMMA 6.1. *With* $T = ut/(\ell+N)$, *we have that*

$$B_{\lambda_1,\lambda_2}(u) = D_\ell^{-1}\Big|\prod_{\alpha>0}[\langle\alpha,\varrho\rangle]_q\Big|^2 \dim_q V_{\lambda_1}\,(\dim_q V_{\lambda_2})^*$$
$$\cdot\Big\{\sum_{d\geq 0} e^{(Nd/2-d^2/2N+c/24)T}\sum_{\mu_1,\mu_2\vdash d} p_{\mu_1}(q^{\lambda_1+\varrho})p_{\mu_2}(q^{-(\lambda_2+\varrho)})\,H_0^{d,[\ell,N]}(\mu_1,\mu_2\,|\,T)\Big\}$$

[7]For connected coverings, g should not be confused with the genus $\tilde{g}$ of the total space, which is given by the Riemann-Hurwitz formula.

$$d(2-2g) = 2-2\tilde{g} - b - \sum_{i=1}^n \{|\mu_i| - \ell(\mu_i)\}.$$

with

$$H_0^{d,[\ell,N]}(\mu_1,\mu_2\,|\,T) = \frac{1}{d!^2} \sum_{\substack{\nu\vdash d \\ \ell(\nu)\leq N-1 \\ \nu_1\leq \ell}} \#C_{\mu_1}\#C_{\mu_2}\,\chi_\nu(C_{\mu_1})\chi_\nu(C_{\mu_2})\,e^{T\,f_\nu(C_{(2)})}.$$

The generating series $H_0^{d,[\ell,N]}(\mu_1,\mu_2\,|\,T)$ only differs from the generating series of double Hurwitz numbers in genus 0 (6.7) by the fact that, instead of using Frobenius formula (6.6), we only sum over partitions $\nu\vdash d$ which belong to Λ, i.e. included in the rectangle of size $\ell\times(N-1)$. For $d\leq\min(\ell,N-1)$, this is not a restriction, namely $H_0^{d,[\ell,N]} = H_0^d$ exactly encodes double Hurwitz numbers.

It would be interesting to devise a combinatorial meaning for $H_0^{d,[\ell,N]}$, maybe by counting coverings with extra geometric constraints, or counting paths between C_{μ_1} and C_{μ_2} in $\mathfrak{S}_d$ with certain properties [**49**, **53**], or to relate it to tau functions [**53**]. It is natural to ask if such a combinatorial interpretation could be extended to higher genus g of the base. Since we already know that we have a CohFT – or equivalently since we have the topological recursion formula – all correlation functions $\omega_{g,n}$ are determined by these $[\ell,N]$-restricted double Hurwitz numbers. As there is already a good knowledge of the Chern character of the Verlinde bundle – including its expression on the boundary for $n=0$ [**69**] – this would provide an ELSV-like formula [**34**] for a combinatorial problem yet to be found.

6.2.3. *Remark.* In [**61**], Karev considered a generating series for the numbers (6.6), and by combinatorial means, he expressed it in terms of an Airy-like integral over the center of the group algebra of G, which can be reexpressed as a product of $1d$ Airy integrals after a suitable change of basis. We shall explain how his results square from the CohFT perspective.

The product on $\mathbb{C}[G]$ and the trace $\sum_{k\in G} c_k\,k\mapsto c_1$ induces the structure of a semi-simple Frobenius algebra on the center $Z(\mathbb{C}[G])$ – which is the $\mathcal{A}_{\rm conj}$ of Section 5.1.1. The conjugacy classes define a basis

$$e_\lambda = \frac{1}{\sqrt{\#C_\lambda}}\sum_{k\in C_\lambda} k$$

while the characters define another basis

$$\underline{\varepsilon}_\lambda = \sum_{k\in G}\frac{\chi_\lambda(k)}{\sqrt{\#G}}\,k$$

and by these notations we mean that the properties of Section 3.1.1 are satisfied, with $\Delta_\lambda^{-1} = (\dim\lambda)^2$. The change of basis matrix is

$$e_\lambda = \sum_\mu \underline{S}^{-1}_{\lambda\mu}\,\underline{\varepsilon}_\mu,\qquad \underline{S}^{-1}_{\lambda\mu} = \sqrt{\frac{\#C_\lambda}{\#G}}\,\chi_\mu(C_\lambda)$$

and up to rescaling, this is a (unitary) submatrix of the S-matrix of the modular functor described in Section 3.1.1. Then, we consider the trivial CohFT on $Z(\mathbb{C}[G])$, and apply a rescaling to define

$$\Omega^*_{g,n}(\underline{\varepsilon}_{\lambda_1},\ldots,\underline{\varepsilon}_{\lambda_n}) = \Big(\frac{\#G}{\dim\lambda}\Big)^{2g-2+n}[\overline{\mathcal{M}}_{g,n}].$$

If we change basis, we obtain, according to Frobenius formula (6.6),

$$\Omega^*_{g,n}\Big(\sum_{k_1\in C_{\lambda_1}}\frac{k_1}{\#G}\otimes\cdots\otimes\sum_{k_n\in C_{\lambda_n}}\frac{k_n}{\#G}\Big) = \frac{\#\{R\in\mathrm{Hom}(\pi_1(\boldsymbol{\Sigma}_{g,n}),G)\mid R(\mathfrak{l}_i)\in C_{\lambda_i}\}}{\#G}\cdot[\overline{\mathcal{M}}_{g,n}].$$

These numbers are encoded in the CohFT partition function

(6.8)

$$Z^{\mathrm{CohFT}} := \exp\bigg(\sum_{n\geq 1}\sum_{g\geq 0}\frac{\hbar^{g-1}}{n!}\sum_{\substack{d_1,\ldots,d_n\geq 0\\ \lambda_1,\ldots,\lambda_n\geq 0}}\bigg\{\int_{\overline{\mathcal{M}}_{g,n}}\Omega^*_{g,n}\Big(\bigotimes_{j=1}^n\varepsilon_{\lambda_j}\Big)\prod_{j=1}^n\psi_j^{d_j}\bigg\}\prod_{j=1}^n t_{d_j,\lambda_j}\bigg).$$

Since we have several copies of the rescaled trivial CohFT, Z^{CohFT} is a product of suitably rescaled matrix Airy function [**64**]. Karev's partition function is (6.8) after specialization of the times to $t_{d,\lambda}=\xi^d p_\lambda$ for a set of formal variables p_λ.

6.3. Bundle comparison and families index formulae. The Wess-Zumino-Witten $\mathrm{SU}(N)_\ell$ theory can also be approached from the perspective of geometric quantization of moduli spaces of flat $\mathrm{SU}(N)$ connections. This allows us to give an interpretation of the CohFT correlations functions via families index formulae – (6.15) below – modulo some identifications (Hypothesis **I** below) which have not been proved yet to the best of our knowledge. As the statement of current vs. unknown results is subtle, we thought it useful to explain it in detail.

For $g\geq 2$, let $\mathcal{U}^{(N)}_{g,0}$ be the moduli space of pairs (X,E) where X is a Riemann surface of genus g and E a semistable holomorphic vector bundle on X of rank N with trivial determinant. By the Narasimhan-Seshadri theorem [**76**], $\mathcal{U}^{(N)}_{g,0}$ is homeomorphic to $\mathcal{M}_{g,0}\times M^{\mathrm{SU}(N)}_{g,0}$, where the last factor is the moduli space of flat $\mathrm{SU}(N)$ connections on a smooth surface of genus g. The fibration $\pi:\mathcal{U}^{(N)}_{g,0}\to\mathcal{M}_{g,0}$ is proper, in particular for each choice of complex structure σ on the smooth surface of genus g, $\pi^{-1}(\sigma)$ is the moduli space of semi-stable bundles of rank N and trivial determinant, which compactifies the smooth part of the fiber consisting of the moduli space of stable rank N bundles with trivial determinant. Caporaso for $N=1$ [**24**], and Pandharipande [**79**] for general N, constructed a compactification $\overline{\mathcal{U}}^{(N)}_{g,0}$ of this moduli space, as a projective variety together with a proper fibration $\pi:\overline{\mathcal{U}}^{(N)}_{g,0}\to\overline{\mathcal{M}}_{g,0}$.

This picture was later generalized to genus g surfaces with $n\geq 1$ punctures. Here, one fixes a sequence of conjugacy classes $\mathbf{c}=(c_1,\ldots,c_n)$ with finite order in $\mathrm{SU}(N)$, and consider, on the one hand the moduli space $M^{\mathrm{SU}(N)}_{g,n}(\mathbf{c})$ of flat $\mathrm{SU}(N)$ flat connections on a smooth surface such that the holonomies of the connection around the i-th puncture belongs to c_i and on the other hand, the moduli space $\mathcal{U}^{(N)}_{g,n}(\mathbf{c})$ of pairs (X,E) where X is a Riemann surface, E is a semistable parabolic holomorphic vector bundle of rank N with trivial determinant with a parabolic structure at the i-th puncture prescribed by c_i. By the Mehta-Seshadri theorem [**72**], $\mathcal{U}^{(N)}_{g,n}(\mathbf{c})$ is homeomorphic to $\mathcal{M}_{g,n}\times M^{\mathrm{SU}(N)}_{g,n}(\mathbf{c})$. In a recent thesis, Schlüter [**85**] constructs a compactification $\overline{\mathcal{U}}^{(N)}_{g,n}(\mathbf{c})$, giving a fibration of projective varieties $\pi:\overline{\mathcal{U}}^{(N)}_{g,n}(\mathbf{c})\to\overline{\mathcal{M}}_{g,n}$.

For the remaining, we fix the level $\ell \geq 1$. For any $\vec{\lambda} \in \Lambda^n$, we consider the sequence of conjugacy classes $\mathbf{c}_{\vec{\lambda}} := ([\exp(\lambda_i/\ell)])_{i=1}^n$, recalling that λ_i is a highest weight at level ℓ. We shall denote

$$M_{g,n,\vec{\lambda}}^{(N)} := M_{g,n}^{\mathrm{SU}(N)}(\mathbf{c}_{\vec{\lambda}}), \qquad \mathcal{U}_{g,n,\vec{\lambda}}^{(N)} := \mathcal{U}_{g,n}^{(N)}(\mathbf{c}_{\vec{\lambda}})$$

6.3.1. *Geometric quantization using Quillen determinant line bundles.* The moduli space $\overline{\mathcal{U}}_{g,n,\vec{\lambda}}^{(N)}$ carries the Quillen determinant line bundle $\mathscr{L}_Q$, and one can consider the ℓ-th power

$$\begin{array}{ccc} \mathscr{L}_Q^{\otimes \ell} & \longrightarrow & \overline{\mathcal{U}}_{g,n,\vec{\lambda}}^{(N)} \\ & & \Big\downarrow \pi \\ & & \overline{\mathcal{M}}_{g,n} \end{array}$$

Then, by push-forward we obtain the coherent sheaf

$$[\mathcal{Z}_{\vec{\lambda}}^{(Q)}]_{g,n} = \pi_!(\mathscr{L}_Q^{\otimes \ell}).$$

As the restriction of $\mathscr{L}_Q^{\otimes \ell}$ to fibers of π over $\mathcal{M}_{g,n}$ have no higher cohomology, the restriction of this sheaf to $\mathcal{M}_{g,n}$ defines a bundle. This bundle by its very definition extends to a sheaf over $\overline{\mathcal{M}}_{g,n}$ (it is expected to remain locally free, but we do not use this). By the version of Grothendieck-Riemann-Roch of [**13**], we have that

$$\mathrm{Ch}\big([\mathcal{Z}_{\vec{\lambda}}^{(Q)}]_{g,n}\big) = \pi_*\big(e^{\ell c_1(\mathscr{L}_Q)} \cap \mathrm{Td}(\mathrm{Ker}\, d\pi)\big). \tag{6.9}$$

If we restrict to the degree 0 part, we recover the index formula for the rank of the bundles

$$\dim\,[\mathcal{Z}_{\vec{\lambda}}^{(Q)}]_{g,n} = e^{\ell \varpi} \cap \mathrm{Td}\big(M_{g,n,\vec{\lambda}}^{(N)}\big), \tag{6.10}$$

where ϖ is the first Chern class of the restriction of $\mathscr{L}_Q$ to a fiber of π. We return to ϖ in the next paragraph, but we can already state that it does not depend on the complex structure, compatible with that the dimension is independent of the complex structure, since the higher cohomology groups of the restriction of $\mathscr{L}_Q$ to the fibers of π vanishes.

6.3.2. *Geometric quantization using the Chern-Simons line bundles and comparison.* The moduli space of flat $\mathrm{SU}(N)$ connections over a surface Σ of genus g minus the n punctures, with prescribed holonomies $\mathbf{c}$ around the punctures, is equipped with the Goldman symplectic form ϖ, and for each $\sigma \in \mathcal{M}_{g,n}$, with a complex structure coming from the Riemann surface Σ_σ, making it a Kähler variety $(M_{g,n}^{\mathrm{SU}(N)}(\mathbf{c}))_\sigma$ with Kähler form ϖ. It further supports the Chern-Simons line bundle $\mathscr{L}_{\mathrm{CS}}$, which supports the Chern-Simons connection whose curvature is $-i\varpi$. The geometric quantization procedure associates to it the bundle $[\mathcal{Z}_{\vec{\lambda}}^{(\mathrm{CS})}]_{g,n} \longrightarrow \mathcal{M}_{g,n}$ whose fiber above σ is $H^0\big((M_{g,n}^{\mathrm{SU}(N)}(\mathbf{c}))_\sigma, \mathscr{L}_{\mathrm{CS}}^k\big)$. For each $\sigma \in \mathcal{M}_{g,n}$, using the isomorphism of Narasimhan-Seshadri, one finds that the Chern-Simons line bundle $\mathscr{L}_{\mathrm{CS}}$ is isomorphic to $\mathscr{L}_Q$, as topological bundles on $M_{g,n}^{\mathrm{SU}(N)}(\mathbf{c})$ and as smooth line bundles over the smooth part of the moduli space. Therefore, $c_1(\mathscr{L}_Q)$ restricted to the smooth locus of $M_{g,n}^{\mathrm{SU}(N)}(\mathbf{c})$ is represented by ϖ. Over $M_{g,n}^{\mathrm{SU}(N)}(\mathbf{c})$, one has the comparison isomorphism [**44**] for $n = 0$ of holomorphic vector bundles

$$\iota^*\mathscr{L}_Q \otimes \pi^*\mathcal{L}_D^{\otimes N} \simeq \mathscr{L}_{\mathrm{CS}}. \tag{6.11}$$

From this we conclude that over $\mathcal{M}_{g,0}$ we have the isomorphism

$$[\mathcal{Z}^{(Q)}]_{g,0} \otimes \mathcal{L}_D^{\otimes N\ell} \simeq [\mathcal{Z}^{(\mathrm{CS})}]_{g,0}. \tag{6.12}$$

It is expected, though currently unproved, that these results also hold for $n > 0$, maybe upon twisting $[\mathcal{Z}_{\vec{\lambda}}^{(Q)}]_{g,n}$ by $\bigotimes_{i=1}^n \mathcal{L}_i^{\ell n_{\lambda_i}}$ for some integers $n_i \in \mathbb{Z}$.

6.3.3. *Comparison with our bundle.* It is clear from the construction of the WZW modular functor in [**4**] that $[\mathcal{Z}_{\vec{\lambda}}^{(\mathrm{CFT})}]_{g,n}$ is isomorphic to the bundle that we obtain in Theorem 2.9 by specializing to the WZW modular functor and for the specific choice (6.4) of log-determinations of twists and central charge. Besides, for $n = 0$ combining the isomorphism (6.12) with the isomorphism from [**69**, Section 5.3], we have the following isomorphism of holomorphic bundles over $\mathcal{M}_{g,0}$

$$[\mathcal{Z}^{(\mathrm{CS})}]_{g,0} \simeq [\mathcal{Z}^{(Q)}]_{g,0} \otimes \mathcal{L}_D^{\otimes N\ell} \simeq [\mathcal{Z}^{(\mathrm{CFT})}]_{g,0}. \tag{6.13}$$

The fact that the first isomorphism hold projectively over $\mathcal{M}_{g,0}$ follows directly from the fact that the Picard group of the moduli space semi-stable bundles of rank n bundles with trivial determinant is a copy of the integers. That the second isomorphisms holds projectively over $\mathcal{M}_{g,0}$ is an older result of Laszlo [**66**], which actually also identifies projective the Hitchin connection [**55**] in the second bundle with the TUY-connection in the last bundle [**88**]. See also [**6**] for the projective identification of the Axelrod, Della-Pietra and Witten [**11**] connection in the first bundle and the Hitchin connection in the second bundle.

Restricting over $\mathcal{M}_{g,0}$, we see that the bundle $[\mathcal{Z}_{\vec{\lambda}}^{(Q)}]_{g,0}$ coincide with our bundle of Theorem 2.9 for a non-standard choice (compare to (6.4)) of log-determination c' of the central charge

$$c' = \frac{\ell(N^2-1)}{\ell+N} + 2N\ell \tag{6.14}$$

It is expected, but to our knowledge currently unproved, that

- the sheaf $[\mathcal{Z}^{(Q)}]_{g,0}$ over $\overline{\mathcal{M}}_{g,0}$ remains locally free over the Deligne-Mumford compactification, i.e. exists as a bundle over $\overline{\mathcal{M}}_{g,0}$.
- the second isomorphism in (6.13) extends to $\overline{\mathcal{M}}_{g,0}$.
- these results extend for $n > 0$, so that there is an isomorphism of bundles over $\overline{\mathcal{M}}_{g,n}$

$$[\mathcal{Z}_{\vec{\lambda}}^{(Q)}]_{g,n} \otimes \mathcal{L}_D^{\otimes N\ell} \otimes \bigotimes_{i=1}^n \mathcal{L}_i^{\ell n_{\lambda_i}} \simeq [\mathcal{Z}_{\vec{\lambda}}^{(\mathrm{CFT})}]_{g,n}$$

for some integers $(n'_\lambda)_{\lambda\in\Lambda}$.

For convenience, we call this expectation Hypothesis **I**. If it were true, it would mean that $[\mathcal{Z}_{\vec{\lambda}}^{(Q)}]_{g,n}$ coincide with the bundle we construct in Theorems 2.5-2.9 for the choice (6.14) of log-determination of the central charge, and for the non-standard choice (compare to (6.4)) of log-determination of the Dehn twist eigenvalues

$$r'_\lambda = \frac{\langle\lambda, \lambda + 2\varrho\rangle}{2(\ell+N)} - n'_\lambda.$$

6.3.4. *Families Index interpretation of the CohFT correlation functions.* If Hypothesis **I** holds, we deduce that the $\omega_{g,n}$ of the CohFT we produce for the specific choice of log-determinations (6.4), have an index interpretation. By homogeneity, it is enough to state it for $t = 1$ and $t = 0$. For $t = 1$, following (6.9) we have for $n \geq 1$

(6.15)
$$\sum_{\vec{\lambda}\in\Lambda^n} \prod_{i=1}^{n} \underline{S}^{-1}_{\mu_i\lambda_i}\omega^{[\vec{\lambda}]}_{g,n}(\zeta_1,\ldots,\zeta_n) = \tilde{\pi}_*\bigg(\prod_{i=1}^{n} \Xi(\psi_i, h_{\mu_i}^{1/2}\zeta_i) \cup \pi_*\big(e^{\ell c_1(\mathscr{L}_Q)} \cap \mathrm{Td}(\mathrm{Ker}\,\mathrm{d}\pi)\big)\bigg)$$

where $\tilde{\pi}$ is the morphism from $\overline{\mathcal{M}}_{g,n}$ to a point. On the left-hand side, we use the notations of § 4.3.3, and on the right-hand side, we recall that

$$h_\mu = r_\mu + \frac{c}{24}, \qquad \Xi(u,\zeta) = \sum_{d\geq 0} \frac{\Gamma[2d+2;\zeta]\,\mathrm{d}\zeta}{2^d d!\,\zeta^{2d+2}}\,u^d.$$

And, for $n = 0$ and $g \geq 2$

$$F_g = \tilde{\pi}_*\big(e^{\ell c_1(\mathscr{L}_Q)} \cap \mathrm{Td}(\mathrm{Ker}\,\mathrm{d}\pi)\big).$$

For $t = 0$, we do not need Hypothesis **I**. As explained in § 4.3.2 the left-hand side computed by topological recursion factors into the Verlinde dimensions and the intersection numbers over $\overline{\mathcal{M}}_{g,n}$

$$\sum_{\vec{\lambda}\in\Lambda^n} \prod_{i=1}^{n} \underline{S}^{-1}_{\mu_i\lambda_i}\omega^{[\vec{\lambda}]}_{g,n}(\zeta_1,\ldots,\zeta_n) = [D_{\vec{\mu}}]_{g,n}\,\omega^{\mathrm{KdV}}_{g,n}(\zeta_1,\ldots,\zeta_n).$$

On the other hand, it is known, see e.g. [**2**], Theorem 8.1, that $[D_{\vec{\mu}}]_{g,n}$ are given by the index formula (6.10). Recalling that $\omega^{\mathrm{KdV}}_{g,n}$ is a generating series of intersection of ψ-classes over $\overline{\mathcal{M}}_{g,n}$, this can be rewritten as follows

$$\sum_{\vec{\lambda}\in\Lambda^n} \prod_{i=1}^{n} \underline{S}^{-1}_{\mu_i\lambda_i}\omega^{[\vec{\lambda}]}_{g,n}(\zeta_1,\ldots,\zeta_n) = \Big(e^{\ell\varpi} \cap \mathrm{Td}\big(M^{(N)}_{g,n,\vec{\lambda}}\big)\Big)\Big(\int_{\overline{\mathcal{M}}_{g,n}} \prod_{i=1}^{n} \Xi(\psi_i,\zeta_i)\Big) \tag{6.16}$$

with

$$\Xi(u,\zeta) = \sum_{d\geq 0} \frac{(2d+1)!!\,\mathrm{d}\zeta}{\zeta^{2d+2}}\,u^d.$$

The factorization between the moduli space of curves, and the moduli space of flat connections over a smooth surface, is nicely displayed in (6.16), while for general t the $\omega_{g,n}$ is interpreted by (6.15) as a families index of the determinant line bundle over the universal moduli space of bundles, seen as a family over $\overline{\mathcal{M}}_{g,n}$.

7. Discussion about global spectral curves

A global spectral curve is by definition a Riemann surface Σ equipped with a branched cover $x : \Sigma \to \mathbb{P}^1$. One recover a local spectral curve by considering the disconnected neighbourhoods of the ramification points of x. The only difference in the axiomatics is that, in the global case, we require that[8]

$$\omega_{0,2} \in H^0(\Sigma^2, K_\Sigma^{\boxtimes 2}(-2\Delta))^{\mathfrak{S}_2}$$

and therefore (4.9) defines meromorphic 1-forms on Σ, *i.e.* $\Xi_{d,i} \in H^0(\Sigma, K_\Sigma((2d+2)\{o_i\}))$.

[8]Although not strictly necessary, one often adds the assumption that $\omega_{0,1} \in H^0(\Sigma', K_{\Sigma'})$ for a dense open subset $\Sigma' \subseteq \Sigma$.

Although the topological recursion is well-defined for local spectral curves as presented in Section 4.1, it was originally defined on global spectral curves and this allows further properties and manipulations [**22**, **38**]. It is thus desirable, whenever we have a local spectral curve, to realize it embed in a global spectral curve. This is not always possible, but given the geometric origin of modular functors and general ideas from mirror symmetry, we may still hope it admits a global description.

7.1. Landau-Ginzburg models and Frobenius manifolds. Among the different realizations of a Frobenius manifold, let us focus on a particular one which gives rise in a natural way to global spectral curves – or more generally, varieties.

In singularity theory, one is interested in the study of Landau-Ginzburg (LG) models. They are defined by a family of germs of functions

$$x_a : \mathbb{C}^d \longrightarrow \mathbb{C},$$

called "potentials" and parameterized by a point a living in a k-dimensional manifold A obtained as a miniversal deformations of a function x_0 with an isolated singularity at $z = 0$ with Milnor number k. The tangent spaces $\mathcal{A}_a := T_a A$ are naturally identified with the Jacobi ring

$$\mathrm{Jac}(x_a) := \frac{\mathbb{C}[z_1, \ldots, z_d]}{\langle \partial_{z_1} x_a, \ldots, \partial_{z_d} x_a \rangle}.$$

The multiplication in this polynomial ring is denoted $\times$, and the unit is 1. For any holomorphic volume form Ω, one can define the residue pairing by

(7.1)

$$\forall (\varphi, \psi) \in \mathrm{Jac}(x_a), \qquad b(\varphi, \psi) := \frac{1}{(2\mathrm{i}\pi)^d} \oint_{|\partial x_a/\partial z_1| = \varepsilon_1} \cdots \oint_{|\partial x_a/\partial z_d| = \varepsilon_d} \frac{\varphi(z)\,\psi(z)\,\Omega(z)}{\prod_{i=1}^d \partial x_a/\partial z_i}$$

for small enough ε_i's. According to Saito [**83**], there exists a volume form – called a primitive – such that (7.1) is a flat metric. It thus provides $\mathrm{Jac}(x_a)$ with a Frobenius structure, promoting $(\mathcal{A}_a)_{a \in A}$ to a Frobenius manifold A. If for a generic $a \in A$, x_a has only isolated Morse singularities, the associated CohFT is semi-simple. We denote by $(t_i)_{i=1}^k$ flat coordinates on an open set $A' \subseteq A$, $(\varphi_i)_{i=1}^k$ the corresponding frame in the tangent bundle, ∇ the Levi-Civita connection, and

$$\varphi_i \times \varphi_j = \sum_{\ell=1}^k N_{ij}^{\ell}(t)\, \varphi_\ell$$

the multiplication in the Jacobi ring. We also denote $(a_i)_{i=1}^k$ canonical coordinates (that we can also assume to be defined on A'), i.e. such that $(\partial_{a_i})_{i=1}^k$ form a canonical basis of the tangent bundle. To connect with the notations of Section 3.1.1, $\partial_{a_i} = \tilde{\varepsilon}_i$ and $\Delta_i = 1/b(\partial_{a_i}, \partial_{a_i})$, and let $\Psi \in \mathrm{End}(T^*A')$ the change of basis from the flat to the canonical coordinates

$$\Psi(\mathrm{d}t_i) = \Delta_i^{-1/2}\, \mathrm{d}a_i.$$

In this example of a Frobenius manifold, the R-matrix can be written in terms of oscillating integrals. Indeed, given a Frobenius manifold A, the axioms of a CohFT ensure that $\nabla_i^{(u)} := \nabla_i + u^{-1}\varphi_i \times$ forms a flat pencil of connections on A parametrized by $u \in \mathbb{C}^*$. This ensures the compatibility of the following PDEs for a section $J(u) = \sum_{i=1}^k J_i(u)\mathrm{d}t_i$ of T^*A'

$$\forall i, j \in [\![1, k]\!], \qquad u\partial_{t_i} J_j(u;t) = \sum_{\ell=1}^k N_{ij}^{\ell}(t)\, J_\ell(u;t).$$

If A' is semi-simple and conformal[9], there is a unique basis of solutions [**31**, **48**] such that, in matrix form one has that

$$J(u;t) = \Psi R(u;t)\exp(\mathbf{a}(t)/u) \tag{7.2}$$

where $\mathbf{a} \in \mathrm{End}(T^*A')$ is defined such that $\mathbf{a}(\mathrm{d}a_i) = a_i\, \mathrm{d}a_i$ and $R(u;t) \in \mathrm{End}(T^*A')[[u]]$ is an operator satisfying a unitarity and homogeneity condition. This R-matrix is the one appearing in Section 3.1.4.

For the Frobenius manifold attached to a Landau-Ginzburg model with potential x_t, one can find a simple integral representation of this solution of the form (in flat coordinates)

$$J_{ij}(u;t) = \frac{1}{(2\mathrm{i}\pi u)^{d/2}} \int_{\Gamma_j} \exp[-x_t(z)/u]\, \varphi_i(z)\, \Omega(z), \tag{7.3}$$

where $(\Gamma_j)_{j=1}^k$ are cycles in $\mathbb{C}^d$ which can be constructed through Morse theory of the function $\mathrm{Re}[x_t/u]$. By the Fubini theorem, this integral can be rewritten as an integral along the path $\mathcal{X}_j = x_t(\Gamma_j) \subset \mathbb{C}$, whose integrand itself is an integral over some vanishing cycle $\gamma_j(X) \subseteq x_t^{-1}\{X\}$ over a point $X \in \mathcal{X}_j$

$$J_{ij}(u;t) = \frac{1}{(2\mathrm{i}\pi u)^{d/2}} \int_{\mathcal{X}_j} e^{-X/u} \int_{\gamma_j(X)} \varphi_i\, \Omega.$$

Comparing with (7.2), this provides us with an integral representation of the R-matrix of the form

$$R_{ij}(u;t) = \frac{\Delta_i^{1/2}}{(2\mathrm{i}\pi u)^{d/2}} \int_{\mathcal{X}_j} e^{-(X-a_j(t))/u} \int_{\gamma_j(X)} \varphi_i\, \Omega$$

and the condition that $R(u=0) = \mathrm{id}$ implies that the paths $\mathcal{X}_j$ start from $a_j(t)$, and extends to $-\infty$.

7.2. $1d$ Landau-Ginzburg models. The realization of a given Frobenius manifold via a LG model, when it exists, has no reason to be unique, and several realizations may have different dimensions. It is particularly interesting, for tractability, if a 1-dimensional LG model can be found. This means having a family of Riemann surfaces Σ together with a family of potentials $x_t : \Sigma \to \mathbb{C}$ such that the local algebra of x_t at the ramification points is isomorphic to the tangent space T_tA of our Frobenius manifold A. In this case, the vanishing cycle $\gamma_j(X)$ is just an ordered pair of points $\{z(X), \sigma_j(z(X))\} \subseteq \Sigma$ above $X \in \mathbb{C}$, related to each other by the (analytic continuation of the) local involution σ_j permuting the two sheets that meet at the ramification point above a_j. In flat coordinates, the R-matrix then reads

$$R_{ij}(u;t) = \frac{\Delta_i^{1/2}}{(2\mathrm{i}\pi u)^{1/2}} \oint_{\mathcal{X}_j} e^{-(X-a_j(t))/u}\, [\varphi_i\, \omega]_{\sigma_j}^{\mathrm{odd}} \tag{7.4}$$

for some 1-form ω on Σ, and thimbles $(\mathcal{X}_j)_{j=1}^k$. Here we have used the notation $[f]_\sigma^{\mathrm{odd}} := f - \sigma^* f$.

[9]Conformal here means there exists a vector field E on A' such that the Lie derivative $\mathfrak{L}_E$ acts by multiplication by a scalar on the metric tensor, on the product tensor, and on the unit vector field.

7.3. Fusion potentials for modular functors. In this paragraph, we give an easy argument for how to obtain a $1d$ LG model for the Frobenius algebra $\mathcal{A}$ of any modular functor (see Section 3.3), i.e. a function x_0 – called the "fusion potential" – whose Jacobi ring is isomorphic to $\mathcal{A}$. Then the Frobenius manifold will be (at least locally) defined by considering miniversal deformations $(x_a)_a$ of the potential. For the Wess-Zumino-Witten models (Section 6), this has been posed as a conjecture by Gepner [**46**], which was later answered by Di Francesco and Zuber [**26**] and further studied by Aharony [**1**]. We borrow the idea of [**26**] to propose a family of fusion potentials for the Frobenius algebra of any modular functor, which hinges on the commutativity of the curve operators $(\mathcal{C}[\beta;\lambda])_{\lambda\in\Lambda}$.

For any vector $f \in \mathbb{C}[\Lambda]$, let us define

$$\mathcal{C}_f[\beta] := \sum_{\lambda\in\Lambda} f_\lambda\, \mathcal{C}[\beta;\lambda].$$

Its characteristic polynomial reads

$$P_f(\eta) := \det(\eta - \mathcal{C}_f[\beta]) = \prod_{\mu\in\Lambda}(\eta - \mathfrak{c}_{\mu,f}), \qquad \mathfrak{c}_{\mu,f} = \sum_{\lambda\in\Lambda} f_\lambda\, \mathfrak{c}_\mu[\lambda]$$

with the eigenvalues $\mathfrak{c}_\mu[\lambda]$ of the curve operators given in (3.18). Let $Q_f(\eta)$ be any polynomial such that $Q_f'(\eta) = P_f(\eta)$.

LEMMA 7.1. *For generic $f \in \mathbb{C}[\Lambda]$, the Jacobi ring of $x_0 : \mathbb{C} \to \mathbb{C}$ defined by $x_0 = Q_f(\eta)$ is isomorphic to the Frobenius algebra $\mathcal{A}$ of the modular functor.*

PROOF. The Jacobi ring is $\mathrm{Jac}(x_0) = \mathbb{C}[\eta]/\langle P_f(\eta)\rangle$. For generic f, the roots $(\mathfrak{c}_{\mu,f})_{\mu\in\Lambda}$ are pairwise distinct. Therefore, we can construct by Lagrange interpolation a unique set of polynomials $(\varphi_{\lambda,f})_{\lambda\in\Lambda}$ such that

$$\forall \mu \in \Lambda, \qquad \varphi_{\lambda,f}(\mathfrak{c}_{\mu,f}) = \mathfrak{c}_\mu[\lambda].$$

This is actually a basis of the Jacobi ring. Indeed, if there is a relation $\sum_{\lambda\in\Lambda} k_\lambda\, P_\lambda(\eta) = 0$, by evaluation at $\mathfrak{c}_{\mu,f}$ and using the expression (3.18) we would have that

$$\forall \mu \in \Lambda, \qquad \sum_{\lambda\in\Lambda} k_\lambda\, (\underline{S}^{-1})_{\lambda\mu} = 0$$

which implies $k_\lambda = 0$ for any $\lambda \in \Lambda$ since $(\underline{S}^{-1})^{\top}$ is invertible. Then, by definition and evaluation at the eigenvalues, their multiplication modulo the annihilating polynomial P_f reads

$$\varphi_\lambda(\eta)\varphi_\mu(\eta) = \sum_{\nu\in\Lambda} N_{\lambda\mu\nu^\dagger}\, \varphi_\nu(\eta) \mod P_f(\eta).$$

Thus, we have build a basis of the Jacobi ring in which we identify the Frobenius algebra of Section 3.3. □

Notice that, knowing $(\mathfrak{c}_{\mu,f})_\mu$, the position of the ramification points appearing in (7.4) is

$$a_\mu := x(o_\mu) = Q_f(\mathfrak{c}_{\mu,f}).$$

Alternatively, one can decide to impose the values $(a_\mu)_{\mu\in\Lambda}$, and ask for the determination of roots $(\eta_\mu)_{\mu\in\Lambda}$ and a polynomial Q of degree $|\Lambda|+1$ such that

$$\forall \mu \in \Lambda, \qquad Q(\eta_\mu) = a_\mu \qquad \text{and} \qquad Q'(\eta_\mu) = 0. \tag{7.5}$$

Remark that $Q(\eta), (\eta_\mu)_\mu$ is a solution iff $Q(\eta) \leftarrow Q(\gamma\eta + \gamma')$ and $\eta_\mu \leftarrow \gamma^{-1}(\eta_\mu - \gamma')$ is a solution. Hence, (7.5) represents $2|\Lambda|$ constraints for the same number of independent unknowns, and can be solved for generic a's, and we can find the corresponding f's by the formula

$$f_\lambda = \sum_{\mu\in\Lambda} \underline{S}_{\mu\lambda}\,\eta_\mu\,(\underline{S}^{-1})_{1\mu}.$$

7.4. Open question: Global spectral curve for modular functors. This defines, at least locally around an (arbitrary, generic) origin a^0, a $1d$ LG model $(x_a)_a$ that we can try to use to describe as a "global curve" Σ_a for the modular functor. However, the explicit construction of a primitive form and a good set of integration cycles is a serious issue in general which we have not been able address so far. To restate (a part of) the problem in an elementary way, it is not easy to find a 1-form $\omega_{0,1}$ on Σ_a whose expansion at the ramification point is equal to (4.24) modulo the even part – at least when $a := a(t)$ is coupled in some way to the single parameter t on which our local curve depend.

In a handful of examples, one can rely on an indirect derivation of a (higher-dimensional) Landau-Ginzburg model through mirror symmetry. For instance, for the $\mathfrak{sl}_N$, level ℓ WZW models recapped in Section 6, Witten [**98**] proved that the Frobenius algebra $\mathcal{A}$ is isomorphic to the quantum cohomology of a Grassmannian. On the other hand, the Landau-Ginzburg model mirror dual to Grassmannians has been built in [**71**], but, in most cases, the issue of finding a primitive form and a good basis of cycles has not been solved yet, except in some very simple examples such as $\mathbb{P}^2$ [**51**]. The complete construction of a global spectral curve thus remains an open problem, even in simple examples. It must be addressed if, for instance, one wishes to take advantage of the topological recursion to study the level $\ell \to \infty$ limit in WZW models.

Appendix A. Extra properties of the S-matrix

In this section we derive the symmetries of the S-matrix in the case the modular functor in question has duality or is unitary.

MF-U. Let us first assume that the modular functor $\mathcal{V}$ is unitary. We recall from Section 3.2 that Σ_1 is an oriented closed surface of genus 1 and α, β two oriented simple closed curves on Σ_1, such that they intersect in exactly one point with intersection number one. Then we have two marked surfaces $\mathbf{\Sigma}_1^{(\alpha)} = (\Sigma_1, \{\alpha\})$ and $\mathbf{\Sigma}_1^{(\beta)} = (\Sigma_1, \{\beta\})$. Both $\mathcal{V}(\mathbf{\Sigma}_1^{(\alpha)})$ and $\mathcal{V}(\mathbf{\Sigma}_1^{(\beta)})$ are equipped with hermitian inner products, which are compatible with the factorization isomorphisms (Section 3.3.1)

$$\mathcal{V}(\mathbf{\Sigma}_1^{(\alpha)}) \cong \bigoplus_{\mu\in\Lambda} \mathcal{V}(\mathbf{\Sigma}_{0,2}, \mu, \mu^\dagger)$$

and

$$\mathcal{V}(\mathbf{\Sigma}_1^{(\beta)}) \cong \bigoplus_{\mu\in\Lambda} \mathcal{V}(\mathbf{\Sigma}_{0,2}, \mu, \mu^\dagger).$$

The basis $\zeta[\mu] \in \mathcal{V}(\mathbf{\Sigma}_{0,2}, \mu, \mu^\dagger)$ induces via these isomorphisms the basis e_μ of $\mathcal{V}(\mathbf{\Sigma}_1^{(\alpha)})$ and ε_μ of $\mathcal{V}(\mathbf{\Sigma}_1^{(\beta)})$. The unique orientation preserving diffeomorphism S that sends (α, β) to $(\beta, -\alpha)$ sends the e-basis to the ε-basis

$$\varepsilon_\mu = \sum_{\lambda\in\Lambda} S_{\mu\lambda}\, e_\lambda\,.$$

Now we simply compute

$$\begin{aligned} \delta_{\lambda\mu} &= \big(\zeta[\lambda], \zeta[\mu]\big) \\ &= \big(e_\lambda, e_\mu\big) \\ &= \big(\mathcal{V}(S)(e_\lambda), \mathcal{V}(S)(e_\mu)\big) \\ &= \sum_{\varrho, \tilde{\varrho} \in \Lambda} S^*_{\lambda\varrho} S_{\mu\tilde{\varrho}} \big(\varepsilon_\varrho, \varepsilon_{\tilde{\varrho}}\big) \\ &= \sum_{\varrho \in \Lambda} S^*_{\lambda\varrho} \overline{S}_{\mu\varrho} . \end{aligned}$$

where * here denotes complex conjugation. Hence we get that the S-matrix is unitary

$$(S^*)^\top S = 1$$

when the modular functor is unitary.

MF-D. Let us now instead assume that the modular functor $\mathcal{V}$ satisfies the orientation reversal axiom. Let now C_α be the orientation reversing self-diffeomorphism of $\mathbf{\Sigma}_1$ which maps (α, β) to $(\alpha, -\beta)$. Let C_β the same as C_α, except that the roles of α and β are exchanged. Then, we have the following commutative diagram, where * denotes the dual operation

$$\begin{array}{ccc} \mathcal{V}(\mathbf{\Sigma}_1^{(\alpha)}) & \xrightarrow{\mathcal{V}(S)} & \mathcal{V}(\mathbf{\Sigma}_1^{(\beta)}) \\ {\scriptstyle \mathcal{V}(C_\alpha)}\big\downarrow & & \big\downarrow{\scriptstyle \mathcal{V}(C_\beta)} \\ \mathcal{V}(-\mathbf{\Sigma}_1^{(\alpha)}) & \xrightarrow{\mathcal{V}(S)} & \mathcal{V}(-\mathbf{\Sigma}_1^{(\beta)}) \\ {\scriptstyle \cong}\big\downarrow & & \big\downarrow{\scriptstyle \cong} \\ \mathcal{V}(\mathbf{\Sigma}_1^{(\alpha)})^\star & \xrightarrow{\mathcal{V}(S^{-1})^\star} & \mathcal{V}(\mathbf{\Sigma}_1^{(\beta)})^\star . \end{array}$$

Let us denote the dual basis of $(e_\lambda)_\lambda$ by $(e^\star_\lambda)_\lambda$. We see, by compatibility between the glueing isomorphism and the orientation reversal isomorphism, that e_λ, under the composition of the two maps in the first column in the above diagram, is taken to $(e^\star_{\lambda^\dagger})_\lambda$ and likewise for β in the last column. But since we have the following easy computation

$$\begin{aligned} \mathcal{V}(S^{-1})^\star(e^\star_{\lambda^\dagger})(\varepsilon_\mu) &= e^\star_{\lambda^\dagger}(\mathcal{V}(S^{-1})(\varepsilon_\mu)) \\ &= e^\star_{\lambda^\dagger}\left(\sum_{\varrho\in\Lambda} S^{-1}_{\mu\varrho} e_\varrho\right) \\ &= S^{-1}_{\mu\lambda^\dagger}, \end{aligned}$$

we see that the orientation reversal axiom implies that

$$S_{\lambda\mu} = S^{-1}_{\mu\lambda^\dagger} .$$

Acknowledgments

The authors are grateful to the organizers of the AIM workshop "Quantum curves, Hitchin systems, and the Eynard-Orantin theory" in October 2014, where this work was initiated and to AIM for their hospitality during this meeting. We thank Bertrand Eynard, Nikolai Reshetikhin, Chris Schommer-Pries and Dimitri

Zvonkine for discussions. GB would like to thank the QGM at Aarhus University, Berkeley, Caltech and EPFL Mathematics Departments, and the Caltech Theoretical Physics Group for hospitality, while this work was conducted. NO would like to thank the QGM at Aarhus University and the Max-Planck Institute for their hospitality at different stages of this work.

References

[1] Ofer Aharony, *Generalized fusion potentials*, Phys. Lett. B **306** (1993), no. 3-4, 276–282, DOI 10.1016/0370-2693(93)90079-W. MR1223057

[2] Jørgen Ellegaard Andersen, *The Witten-Reshetikhin-Turaev invariants of finite order mapping tori I*, J. Reine Angew. Math. **681** (2013), 1–38, DOI 10.1515/crelle-2012-0033. MR3181488

[3] Jørgen Ellegaard Andersen and Kenji Ueno, *Abelian conformal field theory and determinant bundles*, Internat. J. Math. **18** (2007), no. 8, 919–993, DOI 10.1142/S0129167X07004369. MR2339577

[4] Jørgen Ellegaard Andersen and Kenji Ueno, *Geometric construction of modular functors from conformal field theory*, J. Knot Theory Ramifications **16** (2007), no. 2, 127–202, DOI 10.1142/S0218216507005233. MR2306213

[5] Jørgen Ellegaard Andersen and Kenji Ueno, *Modular functors are determined by their genus zero data*, Quantum Topol. **3** (2012), no. 3-4, 255–291. MR2928086

[6] Jørgen Ellegaard Andersen, *Hitchin's connection, Toeplitz operators, and symmetry invariant deformation quantization*, Quantum Topol. **3** (2012), no. 3-4, 293–325. MR2928087

[7] Jørgen Ellegaard Andersen and Kenji Ueno, *Construction of the Witten-Reshetikhin-Turaev TQFT from conformal field theory*, Invent. Math. **201** (2015), no. 2, 519–559, DOI 10.1007/s00222-014-0555-7. MR3370620

[8] Jørgen Ellegaard Andersen and William Elbæk Petersen, *Construction of modular functors from modular tensor categories*, Quantum geometry of moduli spaces with applications to TQFT and RNA folding, Trav. Math., vol. 25, Fac. Sci. Technol. Commun. Univ. Luxemb., Luxembourg, 2017, pp. 147–211. MR3700062

[9] Michael Atiyah, *Topological quantum field theories*, Inst. Hautes Études Sci. Publ. Math. **68** (1988), 175–186 (1989). MR1001453

[10] Michael Atiyah, *On framings of 3-manifolds*, Topology **29** (1990), no. 1, 1–7, DOI 10.1016/0040-9383(90)90021-B. MR1046621

[11] Scott Axelrod, Steve Della Pietra, and Edward Witten, *Geometric quantization of Chern-Simons gauge theory*, J. Differential Geom. **33** (1991), no. 3, 787–902. MR1100212

[12] Bojko Bakalov and Alexander Kirillov Jr., *Lectures on tensor categories and modular functors*, University Lecture Series, vol. 21, American Mathematical Society, Providence, RI, 2001. MR1797619

[13] Paul Baum, William Fulton, and Robert MacPherson, *Riemann-Roch and topological K theory for singular varieties*, Acta Math. **143** (1979), no. 3-4, 155–192, DOI 10.1007/BF02392091. MR549773

[14] Arnaud Beauville, *Conformal blocks, fusion rules and the Verlinde formula*, Proceedings of the Hirzebruch 65 Conference on Algebraic Geometry (Ramat Gan, 1993), Israel Math. Conf. Proc., vol. 9, Bar-Ilan Univ., Ramat Gan, 1996, pp. 75–96. MR1360497

[15] A. A. Belavin, A. M. Polyakov, and A. B. Zamolodchikov, *Infinite conformal symmetry in two-dimensional quantum field theory*, Nuclear Phys. B **241** (1984), no. 2, 333–380, DOI 10.1016/0550-3213(84)90052-X. MR757857

[16] Christian Blanchet, *Hecke algebras, modular categories and 3-manifolds quantum invariants*, Topology **39** (2000), no. 1, 193–223, DOI 10.1016/S0040-9383(98)00066-4. MR1710999

[17] C. Blanchet, N. Habegger, G. Masbaum, and P. Vogel, *Three-manifold invariants derived from the Kauffman bracket*, Topology **31** (1992), no. 4, 685–699, DOI 10.1016/0040-9383(92)90002-Y. MR1191373

[18] C. Blanchet, N. Habegger, G. Masbaum, and P. Vogel, *Topological quantum field theories derived from the Kauffman bracket*, Topology **34** (1995), no. 4, 883–927, DOI 10.1016/0040-9383(94)00051-4. MR1362791

[19] Gaëtan Borot, Bertrand Eynard, Motohico Mulase, and Brad Safnuk, *A matrix model for simple Hurwitz numbers, and topological recursion*, J. Geom. Phys. **61** (2011), no. 2, 522–540, DOI 10.1016/j.geomphys.2010.10.017. MR2746135

[20] Gaëtan Borot, Bertrand Eynard, and Nicolas Orantin, *Abstract loop equations, topological recursion and new applications*, Commun. Number Theory Phys. **9** (2015), no. 1, 51–187, DOI 10.4310/CNTP.2015.v9.n1.a2. MR3339853

[21] Gaëtan Borot and Sergey Shadrin, *Blobbed topological recursion: properties and applications*, Math. Proc. Cambridge Philos. Soc. **162** (2017), no. 1, 39–87, DOI 10.1017/S0305004116000323. MR3581899

[22] Vincent Bouchard and Bertrand Eynard, *Think globally, compute locally*, J. High Energy Phys. **2** (2013), 143, front matter + 34. MR3046532

[23] Vincent Bouchard and Marcos Mariño, *Hurwitz numbers, matrix models and enumerative geometry*, From Hodge theory to integrability and TQFT tt*-geometry, Proc. Sympos. Pure Math., vol. 78, Amer. Math. Soc., Providence, RI, 2008, pp. 263–283, DOI 10.1090/pspum/078/2483754. MR2483754

[24] Lucia Caporaso, *A compactification of the universal Picard variety over the moduli space of stable curves*, J. Amer. Math. Soc. **7** (1994), no. 3, 589–660, DOI 10.2307/2152786. MR1254134

[25] Renzo Cavalieri, Paul Johnson, and Hannah Markwig, *Chamber structure for double Hurwitz numbers* (English, with English and French summaries), 22nd International Conference on Formal Power Series and Algebraic Combinatorics (FPSAC 2010), Discrete Math. Theor. Comput. Sci. Proc., AN, Assoc. Discrete Math. Theor. Comput. Sci., Nancy, 2010, pp. 227–238. MR2673838

[26] P. Di Francesco and J.-B. Zuber, *Fusion potentials. I*, J. Phys. A **26** (1993), no. 6, 1441–1454. MR1212013

[27] Robbert Dijkgraaf, *Mirror symmetry and elliptic curves*, The moduli space of curves (Texel Island, 1994), Progr. Math., vol. 129, Birkhäuser Boston, Boston, MA, 1995, pp. 149–163, DOI 10.1007/978-1-4612-4264-2_5. MR1363055

[28] Robbert Dijkgraaf, Cumrun Vafa, Erik Verlinde, and Herman Verlinde, *The operator algebra of orbifold models*, Comm. Math. Phys. **123** (1989), no. 3, 485–526. MR1003430

[29] Robbert Dijkgraaf, Herman Verlinde, and Erik Verlinde, *Loop equations and Virasoro constraints in nonperturbative two-dimensional quantum gravity*, Nuclear Phys. B **348** (1991), no. 3, 435–456, DOI 10.1016/0550-3213(91)90199-8. MR1083914

[30] Robbert Dijkgraaf and Edward Witten, *Topological gauge theories and group cohomology*, Comm. Math. Phys. **129** (1990), no. 2, 393–429. MR1048699

[31] Boris Dubrovin, *Geometry of 2D topological field theories*, Integrable systems and quantum groups (Montecatini Terme, 1993), Lecture Notes in Math., vol. 1620, Springer, Berlin, 1996, pp. 120–348, DOI 10.1007/BFb0094793. MR1397274

[32] Olivia Dumitrescu and Motohico Mulase, *Edge contraction on dual ribbon graphs and 2D TQFT*, J. Algebra **494** (2018), 1–27, DOI 10.1016/j.jalgebra.2017.09.027. MR3723168

[33] P. Dunin-Barkowski, N. Orantin, S. Shadrin, and L. Spitz, *Identification of the Givental formula with the spectral curve topological recursion procedure*, Comm. Math. Phys. **328** (2014), no. 2, 669–700, DOI 10.1007/s00220-014-1887-2. MR3199996

[34] Torsten Ekedahl, Sergei Lando, Michael Shapiro, and Alek Vainshtein, *Hurwitz numbers and intersections on moduli spaces of curves*, Invent. Math. **146** (2001), no. 2, 297–327, DOI 10.1007/s002220100164. MR1864018

[35] Bertrand Eynard, *Recursion between Mumford volumes of moduli spaces*, Ann. Henri Poincaré **12** (2011), no. 8, 1431–1447, DOI 10.1007/s00023-011-0113-4. MR2855174

[36] ———, *Invariants of spectral curves and intersection theory of moduli spaces of complex curves*, Commun. Number Theory and Physics **8** 3 (2014) ; math-ph/1110.2949.

[37] Bertrand Eynard, Motohico Mulase, and Bradley Safnuk, *The Laplace transform of the cut-and-join equation and the Bouchard-Mariño conjecture on Hurwitz numbers*, Publ. Res. Inst. Math. Sci. **47** (2011), no. 2, 629–670, DOI 10.2977/PRIMS/47. MR2849645

[38] B. Eynard and N. Orantin, *Invariants of algebraic curves and topological expansion*, Commun. Number Theory Phys. **1** (2007), no. 2, 347–452, DOI 10.4310/CNTP.2007.v1.n2.a4. MR2346575

[39] ———, *Topological recursion in random matrices and enumerative geometry*, J. Phys. A: Math. Theor. **42** 29 (2009) ; math-ph/0811.3531.

[40] Gerd Faltings, *A proof for the Verlinde formula*, J. Algebraic Geom. **3** (1994), no. 2, 347–374. MR1257326

[41] Daniel S. Freed, *Higher algebraic structures and quantization*, Comm. Math. Phys. **159** (1994), no. 2, 343–398. MR1256993

[42] Daniel S. Freed and Frank Quinn, *Chern-Simons theory with finite gauge group*, Comm. Math. Phys. **156** (1993), no. 3, 435–472. MR1240583

[43] Igor B. Frenkel, Yi-Zhi Huang, and James Lepowsky, *On axiomatic approaches to vertex operator algebras and modules*, Mem. Amer. Math. Soc. **104** (1993), no. 494, viii+64, DOI 10.1090/memo/0494. MR1142494

[44] Hajime Fujita, *On the functoriality of the Chern-Simons line bundle and the determinant line bundle*, Commun. Contemp. Math. **8** (2006), no. 6, 715–735, DOI 10.1142/S0219199706002271. MR2274939

[45] William Fulton and Joe Harris, *Representation theory*, Graduate Texts in Mathematics, vol. 129, Springer-Verlag, New York, 1991. A first course; Readings in Mathematics. MR1153249

[46] Doron Gepner, *Fusion rings and geometry*, Comm. Math. Phys. **141** (1991), no. 2, 381–411. MR1133272

[47] Alexander B. Givental, *Gromov-Witten invariants and quantization of quadratic Hamiltonians* (English, with English and Russian summaries), Mosc. Math. J. **1** (2001), no. 4, 551–568, 645. Dedicated to the memory of I. G. Petrovskii on the occasion of his 100th anniversary. MR1901075

[48] Alexander B. Givental, *Semisimple Frobenius structures at higher genus*, Internat. Math. Res. Notices **23** (2001), 1265–1286, DOI 10.1155/S1073792801000605. MR1866444

[49] I. P. Goulden and D. M. Jackson, *Transitive factorisations into transpositions and holomorphic mappings on the sphere*, Proc. Amer. Math. Soc. **125** (1997), no. 1, 51–60, DOI 10.1090/S0002-9939-97-03880-X. MR1396978

[50] I. P. Goulden, D. M. Jackson, and R. Vakil, *The Gromov-Witten potential of a point, Hurwitz numbers, and Hodge integrals*, Proc. London Math. Soc. (3) **83** (2001), no. 3, 563–581, DOI 10.1112/plms/83.3.563. MR1851082

[51] Mark Gross, *Tropical geometry and mirror symmetry*, CBMS Regional Conference Series in Mathematics, vol. 114, Published for the Conference Board of the Mathematical Sciences, Washington, DC; by the American Mathematical Society, Providence, RI, 2011. MR2722115

[52] Jakob Grove, *Constructing TQFTs from modular functors*, J. Knot Theory Ramifications **10** (2001), no. 8, 1085–1131, DOI 10.1142/S0218216501001347. MR1871220

[53] Mathieu Guay-Paquet and J. Harnad, *Generating functions for weighted Hurwitz numbers*, J. Math. Phys. **58** (2017), no. 8, 083503, 28, DOI 10.1063/1.4996574. MR3683833

[54] J. Harnad and A. Yu. Orlov, *Hypergeometric τ-functions, Hurwitz numbers and enumeration of paths*, Comm. Math. Phys. **338** (2015), no. 1, 267–284, DOI 10.1007/s00220-015-2329-5. MR3345377

[55] N. J. Hitchin, *Flat connections and geometric quantization*, Comm. Math. Phys. **131** (1990), no. 2, 347–380. MR1065677

[56] Yi-Zhi Huang, *Rigidity and modularity of vertex tensor categories*, Commun. Contemp. Math. **10** (2008), no. suppl. 1, 871–911, DOI 10.1142/S0219199708003083. MR2468370

[57] Yi-Zhi Huang, *Vertex operator algebras, the Verlinde conjecture, and modular tensor categories*, Proc. Natl. Acad. Sci. USA **102** (2005), no. 15, 5352–5356, DOI 10.1073/pnas.0409901102. MR2140309

[58] Paul Johnson, *Double Hurwitz numbers via the infinite wedge*, Trans. Amer. Math. Soc. **367** (2015), no. 9, 6415–6440, DOI 10.1090/S0002-9947-2015-06238-2. MR3356942

[59] V. G. Kac and D. H. Peterson, *Affine Lie algebras and Hecke modular forms*, Bull. Amer. Math. Soc. (N.S.) **3** (1980), no. 3, 1057–1061, DOI 10.1090/S0273-0979-1980-14854-5. MR585190

[60] Masanobu Kaneko and Don Zagier, *A generalized Jacobi theta function and quasimodular forms*, The moduli space of curves (Texel Island, 1994), Progr. Math., vol. 129, Birkhäuser Boston, Boston, MA, 1995, pp. 165–172, DOI 10.1007/978-1-4612-4264-2_6. MR1363056

[61] M. Karev, *Covers counting via Feynman calculus* (English, with English and Russian summaries), Zap. Nauchn. Sem. S.-Peterburg. Otdel. Mat. Inst. Steklov. (POMI) **403** (2012), no. Teoriya Predstavleniĭ, Dinamicheskie Sistemy, Kombinatornye Metody. XXI, 58–80, 198; English transl., J. Math. Sci. (N.Y.) **190** (2013), no. 3, 438–450. MR3029580

[62] Victor G. Kac, *Infinite-dimensional Lie algebras*, 3rd ed., Cambridge University Press, Cambridge, 1990. MR1104219

[63] M. Kontsevich, *Rational conformal field theory and invariants of 3-manifolds*, Preprint of Centre de Physique Théorique Marseille (1988), no. CPT-88/p2189.

[64] Maxim Kontsevich, *Intersection theory on the moduli space of curves and the matrix Airy function*, Comm. Math. Phys. **147** (1992), no. 1, 1–23. MR1171758

[65] M. Kontsevich and Yu. Manin, *Gromov-Witten classes, quantum cohomology, and enumerative geometry*, Comm. Math. Phys. **164** (1994), no. 3, 525–562. MR1291244

[66] Yves Laszlo, *Hitchin's and WZW connections are the same*, J. Differential Geom. **49** (1998), no. 3, 547–576. MR1669720

[67] Eduard Looijenga, *From WZW models to modular functors*, Handbook of moduli. Vol. II, Adv. Lect. Math. (ALM), vol. 25, Int. Press, Somerville, MA, 2013, pp. 427–466. MR3184182

[68] I. G. Macdonald, *Symmetric functions and Hall polynomials*, 2nd ed., Oxford Mathematical Monographs, The Clarendon Press, Oxford University Press, New York, 1995. With contributions by A. Zelevinsky; Oxford Science Publications. MR1354144

[69] A. Marian, D. Oprea, and R. Pandharipande, *The first Chern class of the Verlinde bundle* (2013); math.AG/1308.4425.

[70] Alina Marian, Dragos Oprea, Rahul Pandharipande, Aaron Pixton, and Dimitri Zvonkine, *The Chern character of the Verlinde bundle over* $\overline{\mathcal{M}}_{g,n}$, J. Reine Angew. Math. **732** (2017), 147–163, DOI 10.1515/crelle-2015-0003. MR3717090

[71] R. Marsh and K. Rietsch, *The B-model connection and T-equivariant mirror symmetry for Grassmannians* (2013) ; math.AG/1307.1085.

[72] V. B. Mehta and C. S. Seshadri, *Moduli of vector bundles on curves with parabolic structures*, Math. Ann. **248** (1980), no. 3, 205–239, DOI 10.1007/BF01420526. MR575939

[73] Gregory Moore and Nathan Seiberg, *Classical and quantum conformal field theory*, Comm. Math. Phys. **123** (1989), no. 2, 177–254. MR1002038

[74] ———, *Polynomial equations for rational conformal field theories*, Phys. Lett. B **212** 4 (1989).

[75] David Mumford, *Towards an enumerative geometry of the moduli space of curves*, Arithmetic and geometry, Vol. II, Progr. Math., vol. 36, Birkhäuser Boston, Boston, MA, 1983, pp. 271–328. MR717614

[76] M. S. Narasimhan and C. S. Seshadri, *Stable and unitary vector bundles on a compact Riemann surface*, Ann. of Math. (2) **82** (1965), 540–567, DOI 10.2307/1970710. MR0184252

[77] Siu-Hung Ng and Peter Schauenburg, *Congruence subgroups and generalized Frobenius-Schur indicators*, Comm. Math. Phys. **300** (2010), no. 1, 1–46, DOI 10.1007/s00220-010-1096-6. MR2725181

[78] Andrei Okounkov, *Toda equations for Hurwitz numbers*, Math. Res. Lett. **7** (2000), no. 4, 447–453, DOI 10.4310/MRL.2000.v7.n4.a10. MR1783622

[79] Rahul Pandharipande, *A compactification over* $\overline{M}_g$ *of the universal moduli space of slope-semistable vector bundles*, J. Amer. Math. Soc. **9** (1996), no. 2, 425–471, DOI 10.1090/S0894-0347-96-00173-7. MR1308406

[80] N. Yu. Reshetikhin and V. G. Turaev, *Ribbon graphs and their invariants derived from quantum groups*, Comm. Math. Phys. **127** (1990), no. 1, 1–26. MR1036112

[81] N. Reshetikhin and V. G. Turaev, *Invariants of 3-manifolds via link polynomials and quantum groups*, Invent. Math. **103** (1991), no. 3, 547–597, DOI 10.1007/BF01239527. MR1091619

[82] Eric Rowell, Richard Stong, and Zhenghan Wang, *On classification of modular tensor categories*, Comm. Math. Phys. **292** (2009), no. 2, 343–389, DOI 10.1007/s00220-009-0908-z. MR2544735

[83] Kyoji Saito, *Period mapping associated to a primitive form*, Publ. Res. Inst. Math. Sci. **19** (1983), no. 3, 1231–1264, DOI 10.2977/prims/1195182028. MR723468

[84] Graeme Segal, *The definition of conformal field theory*, Topology, geometry and quantum field theory, London Math. Soc. Lecture Note Ser., vol. 308, Cambridge Univ. Press, Cambridge, 2004, pp. 421–577. MR2079383

[85] D. Schlüter, *Universal moduli of parabolic sheaves on stable marked curves*, Oxford PhD thesis, 2011.

[86] Constantin Teleman, *The structure of 2D semi-simple field theories*, Invent. Math. **188** (2012), no. 3, 525–588, DOI 10.1007/s00222-011-0352-5. MR2917177

[87] Yoshifumi Tsuchimoto, *On the coordinate-free description of the conformal blocks*, J. Math. Kyoto Univ. **33** (1993), no. 1, 29–49, DOI 10.1215/kjm/1250519338. MR1203889

[88] Akihiro Tsuchiya, Kenji Ueno, and Yasuhiko Yamada, *Conformal field theory on universal family of stable curves with gauge symmetries*, Integrable systems in quantum field theory and statistical mechanics, Adv. Stud. Pure Math., vol. 19, Academic Press, Boston, MA, 1989, pp. 459–566. MR1048605

[89] V. G. Turaev, *Quantum invariants of knots and 3-manifolds*, De Gruyter Studies in Mathematics, vol. 18, Walter de Gruyter & Co., Berlin, 1994. MR1292673

[90] V. Turaev and H. Wenzl, *Quantum invariants of* 3*-manifolds associated with classical simple Lie algebras*, Internat. J. Math. **4** (1993), no. 2, 323–358, DOI 10.1142/S0129167X93000170. MR1217386

[91] Kenji Ueno, *Conformal field theory with gauge symmetry*, Fields Institute Monographs, vol. 24, American Mathematical Society, Providence, RI; Fields Institute for Research in Mathematical Sciences, Toronto, ON, 2008. MR2433154

[92] Cumrun Vafa, *Toward classification of conformal theories*, Phys. Lett. B **206** (1988), no. 3, 421–426, DOI 10.1016/0370-2693(88)91603-6. MR944264

[93] Erik Verlinde, *Fusion rules and modular transformations in* 2*D conformal field theory*, Nuclear Phys. B **300** (1988), no. 3, 360–376, DOI 10.1016/0550-3213(88)90603-7. MR954762

[94] K. Walker, *On Witten's* 3*-manifold invariants*, (1991), Preliminary version # 2, preprint.

[95] C. T. C. Wall, *Non-additivity of the signature*, Invent. Math. **7** (1969), 269–274, DOI 10.1007/BF01404310. MR0246311

[96] Edward Witten, *Quantum field theory and the Jones polynomial*, Comm. Math. Phys. **121** (1989), no. 3, 351–399. MR990772

[97] Edward Witten, *Two-dimensional gravity and intersection theory on moduli space*, Surveys in differential geometry (Cambridge, MA, 1990), Lehigh Univ., Bethlehem, PA, 1991, pp. 243–310. MR1144529

[98] Edward Witten, *The Verlinde algebra and the cohomology of the Grassmannian*, Geometry, topology, & physics, Conf. Proc. Lecture Notes Geom. Topology, IV, Int. Press, Cambridge, MA, 1995, pp. 357–422. MR1358625

[99] Sergei K. Lando and Alexander K. Zvonkin, *Graphs on surfaces and their applications*, Encyclopaedia of Mathematical Sciences, vol. 141, Springer-Verlag, Berlin, 2004. With an appendix by Don B. Zagier; Low-Dimensional Topology, II. MR2036721

Centre for Quantum Geometry of Moduli Spaces, Department of Mathematics, Ny Munkegade 118, 8000 Aarhus C, Denmark
Email address: andersen@qgm.au.dk

Max Planck Institut für Mathematik, Vivatsgasse 7, 53111 Bonn, Germany
Email address: gborot@mpim-bonn.mpg.de

École Polytechnique Fédérale de Lausanne, Département de Mathématiques, 1015 Lausanne, Switzerland
Email address: nicolas.orantin@epfl.ch

Proceedings of Symposia in Pure Mathematics
Volume **100**, 2018
https://doi.org/10.1090/pspum/100/01759

On the Gopakumar–Ooguri–Vafa correspondence for Clifford–Klein 3-manifolds

Andrea Brini

Abstract. Gopakumar, Ooguri and Vafa famously proposed the existence of a correspondence between a topological gauge theory on one hand–U(N) Chern–Simons theory on the three-sphere–and a topological string theory on the other–the topological A-model on the resolved conifold. On the physics side, this duality provides a concrete instance of the large N gauge/string correspondence where exact computations can be performed in detail; mathematically, it puts forward a triangle of striking relations between quantum invariants (Reshetikhin–Turaev–Witten) of knots and 3-manifolds, curve-counting invariants (Gromov–Witten/Donaldson–Thomas) of local Calabi-Yau 3-folds, and the Eynard–Orantin recursion for a specific class of spectral curves. I here survey recent results on the most general frame of validity of this correspondence and discuss some of its implications.

Contents

1. Introduction

In a series of works [**47**, **81**, **82**], Gopakumar, Ooguri and Vafa postulated and gave strong evidence for an identity between two rather different physical theories: on the one hand, U(N) Chern–Simons theory on S^3 [**94**], and on the other, topologically A-twisted string theory on the resolved conifold, $\mathrm{Tot}(\mathcal{O}_{\mathbb{P}^1}^{\oplus 2}(-1))$. From the physics side, such identification is a concrete example of 't Hooft's idea [**91**] that perturbative gauge theories with classical gauge groups in the $1/N$ expansion should be equivalent to some first-quantised string theory on a given background. The topological nature of both Chern–Simons theory and the A-model topological

2010 *Mathematics Subject Classification.* Primary 57M27; Secondary 14N35, 81T30.

string allow for detailed checks of the correspondence, which might be regarded as a simplified setting for gauge/string dualities for the type II superstring, such as the AdS/CFT correspondence [**70**].

Mathematically, the implications of the Gopakumar–Ooguri–Vafa (GOV) correspondence are perhaps even more noteworthy: the correspondence ties together, in a highly non-trivial way, two different theories of geometric invariants having *a priori* little resemblance to each other. In Witten's landmark paper [**94**] Chern–Simons observables were proposed to give rise to topological invariants of framed 3-manifolds and links therein in light of the Schwarz-type topological invariance of the quantum theory. The relation to rational conformal field theory leads in particular to the skein relations typical of knot invariants such as the HOMFLY-PT and Kaufmann invariants, and more generally, to the Reshetikhin–Turaev invariants arising in the representation theory of quantum groups and modular tensor categories [**85, 86**]. The A-type topological string side, in turn, also enjoys a mathematical definition – of a quite different flavour. The partition function for a given Calabi–Yau target X is a formal generating function of various virtual counts of curves in X, either via stable maps [**64**] or ideal sheaves [**35**]. In particular, the GOV correspondence asserts that the asymptotic expansion of the Reshetikhin–Turaev invariant associated to the quantum group $U_q(\mathrm{sl}_N)$ at large N and fixed q^N equates the formal Gromov–Witten potential of X in the genus expansion. This has a B-model counterpart due to recent developments in higher genus toric mirror symmetry a [**19, 42**], where the same genus expansion can be phrased in terms of the Eynard–Orantin topological recursion [**41**] on the Hori–Iqbal–Vafa mirror curve of X [**58, 59**].

The GOV correspondence has had a profound impact for both communities involved. In Gromov–Witten / Donaldson–Thomas theory, it has laid the foundations of the use of large N dualities to solve the topological string on toric backgrounds [**4, 6, 77**] and to obtain all-genus results that went well beyond the existing localisation computations at the time, as well as some striking results for the intersection theory on moduli spaces of curves [**75**] and, via the relation of Chern–Simons theory to random matrices, an embryo of the remodeling proposal [**19, 71, 73**]. In the other direction, the integral structure of BPS invariants leads to non-trivial constraints for the structure of quantum knot invariants [**66, 68, 74**].

Since the original correspondence of [**47, 81**] was confined to the case where the gauge group is the unitary group $\mathrm{U}(N)$, the base manifold is S^3, and the knot is the trivial knot, a natural question to ask is whether a similar connection could be generalised to other classical gauge groups [**17, 18, 89**], knots other than the unknot [**24, 65, 66**] (see also [**3**] for a significant generalisation, in a rather different setting), as well as categorified/refined invariants of various types [**7, 51**]. A further natural direction would be to seek the broadest generalisation of the correspondence in its original form beyond the basic case of the three-sphere; this would require to provide a description of the string dual of Chern–Simons both in terms of Gromov–Witten theory and of the Eynard–Orantin theory on a specific spectral curve setup. This program was initiated in [**5**] (see also [**25, 52, 53**]) for the case of lens spaces, and what is presumably its widest frame of validity has been recently described in [**15, 27**].

This short survey is meant as an overview of results contained in [**15**, **27**], putting them into context and highlighting their implications for quantum invariants and Calabi–Yau enumerative geometry alike. It is structured as follows: we first give a lightning review of the three-way relation between matrix models, intersection theory on $\overline{\mathcal{M}}_{g,n}$, and the Eynard–Orantin recursion. We then move on to review the uplift of these notions to the GOV correspondence for S^3. Finally, we discuss its generalisation to the setting of Clifford–Klein 3-manifolds and outline the proof of the B-model side of the correspondence for the case of the Lê–Murakami–Ohtsuki invariant [**67**].

Acknowledgments. I would like to thank my collaborators Gaëtan Borot, Bertrand Eynard, Luca Griguolo, Marcos Mariño, Domenico Seminara and Alessandro Tanzini for sharing their insights and educating me on various aspects of the Chern–Simons/topological strings correspondence as well as for most enjoyable collaborations on this topic. This paper is a write-up of material presented at a talk at the 2016 AMS Von Neumann Symposium on *Topological Recursion and its Influence in Analysis, Geometry, and Topology*, July 4–8, 2016, at Charlotte (NC); I wish to express my thanks to Motohico Mulase and Chiu-Chu Melissa Liu for the wonderful scientific environment at the workshop and for giving me the opportunity to present my work. This research was partially supported by a fellowship under the ERC-2015-CoG "GWT".

2. A 0-dimensional aperçu: Matrix models, the topological recursion and enumerative geometry

As a warm-up for the description of the triangle of relations linking 3D TQFT, curve-counting invariants and the topological recursion, let us consider the much more basic setting of 0D QFTs – namely, matrix models. For any $N, r \in \mathbb{Z}^+$, let $\mathrm{U}(N)$ denote the rank-N unitary group, H_N be its Lie algebra of hermitian matrices, and for x in a formal neighbourhood of the origin of $\mathbb{C}^r$ we let $\mathrm{d}\mu_N(x)$ be a family of formally finite Ad-invariant measures, absolutely continuous with respect to the standard gaussian measure $\mathrm{d}\mu_N^{(0)} = \mathrm{d}\mu_N(0)$ on $H_N \simeq \mathbb{R}^{N^2}$. The typical setup here is

$$\begin{aligned}(1)\qquad \mathrm{d}\mu_N(x) &= \mathcal{N}_N \prod_{i=1}^{N} \mathrm{d}M_{ii} \prod_{i<j}^{N} \mathrm{d}\mathfrak{Re}M_{ij}\mathrm{d}\mathfrak{Im}M_{ij}\mathrm{e}^{-\frac{N}{2t}\mathrm{Tr}M^2+\sum_j x_j \mathrm{Tr}M^j} \\ &= \mathcal{N}_N \mathrm{d}\mu_N^{(0)} \mathrm{e}^{\sum_j x_j \mathrm{Tr}M^j}\end{aligned}$$

where $\mathcal{N}_N \in \mathbb{C}^\star$, and the exponential deformation in the last equality should be treated as formal. For fixed t, we will denote by $Z_N(x)$ the formal total mass of $\mathrm{d}\mu_N(x)$,

$$(2)\qquad Z_N(x) = \int_{H_N} \mathrm{d}\mu_N(x) \in \mathbb{C}[[x]],$$

and for any collection of positive integers $k_1, \ldots, k_h$, $h \geq 1$, we write

$$(3)\qquad \mathsf{W}_{k_1,\ldots,k_h;N}(x) \triangleq \partial^{\sum_i k_i}_{x_{k_1},\ldots,x_{k_h}} \log Z_N(x) \in \mathbb{C}[[x]]$$

for the cumulants of the measure. Eqs. (1) and (2) define respectively a *formal hermitian 1-matrix model* and its partition function, and we are particularly interested in their formal behaviour as $N \to \infty$. It is a matter of book-keeping in the application of Wick's theorem to Eqs. (2) and (3) to show that both $\ln Z_N$ and $W_{k_1,\dots,k_m}$ have a formal connected ribbon graph (*fat-graph*) expansion [**20**, **91**]

$$\ln Z_N(x) = \sum_{g\geq 0} N^{2-2g}\mathcal{F}_g(x), \tag{4}$$

$$\mathsf{W}_{k_1,\dots,k_h;N}(x) = \sum_{g\geq 0} N^{2-2g-h}\mathcal{W}_{g,h;k_1,\dots,k_h}(x). \tag{5}$$

The $1/N$ expansion of Eqs. (4) and (5) lends itself to two interpretations.

- The first one stems from the observation that, as the dual of a ribbon graph is a polygonulation of a connected, oriented topological 2-manifold, this $1/N$ expansion can be regarded as a genus expansion in a sum over random polygonulations of a Riemann surface. The $1/N$ expansion gives then an answer to an *enumeration problem*: it is easy to show in particular, when $\mathcal{N}_N = (t/N)^{N(N+1)/2}(\pi/N)^{-N^2/2}2^{-N}$, that the coefficients of the formal expansion in x have an enumerative meaning as a count of topologically inequivalent dual graphs. One consequence is that, as (metric) ribbon graphs can be used as a means to describe simplicial decompositions of the moduli space $\mathcal{M}_{g,n}$ of pointed Riemann surfaces via the Strebel correspondence, the enumerative content of Eq. (4) provides a combinatorial answer to topological questions on this moduli space [**34**, **56**, **83**].
- The second (physical) interpretation comes from a special case of the observation, due to 't Hooft, that the perturbative $1/N$ expansion of a $\mathrm{U}(N)$ gauge theory with only adjoint fields takes the shape of the expansion in the string coupling of a first quantised closed string theory, possibly with probe branes insertions [**91**]; in a sense the observation above that the enumeration problem of fatgraphs is related to intersection-theoretic problems on moduli spaces of curves is the most basic example of this phenomenon. Indeed, 1-matrix models are possibly the most basic avatars of such QFTs, and it is only natural to ask *what is the string dual of the formal 1-matrix model?* This is a sharp, and already non-trivial question in the context of *gauged* matrix models, as can be seen in the following example.

EXAMPLE 2.1. Take the simplest example of a gaussian measure $\mathrm{d}\mu_N(x) = \mathrm{d}\mu_N^{(0)}$, with a normalisation factor given by

$$\mathcal{N}_N \triangleq [\mathrm{Vol}_g(\mathrm{U}(N))]^{-1} \tag{6}$$

where the metric g is the Ad-invariant metric induced by the Killing form on H_N. The gaussian integration is trivial, and yields

$$\int_{H_N} \mathrm{d}\mu_N^{(0)} = \left(\frac{2\pi t}{N}\right)^{N^2/2} \tag{7}$$

The normalisation factor in Eq. (6) is less trivially computed as a product of volumes of spheres [**69**] and it takes the form of a Barnes double gamma function:

$$\mathrm{Vol}_g(\mathrm{U}(N)) = \frac{(2\pi)^{N(N+1)/2}}{G_2(N+1)} \tag{8}$$

where $G_2(x+1) = \Gamma(x)G_2(x)$, $G_2(1) = 1$. Computing the large N asymptotics of $\ln Z_N$ is an exercise in the use of the all-order Stirling formula for the Gamma function; the result is

$$\ln Z_N(t) = \frac{N^2}{2}\left(\log(t) - \frac{3}{2}\right) - \frac{1}{12}\log N + \zeta'(-1) + \sum_{g\geq 2} \frac{B_{2g}}{2g(2g-2)} N^{2-2g}. \tag{9}$$

We point out three features of this example.

i. In this case we can give a full answer to the question of finding a string dual for the matrix model. Indeed, Eq. (9) coincides with the perturbative expansion of the $c=1$ string theory at the self-dual radius [**46**, **49**, **84**], upon identifying the string coupling constant g_s and the cosmological constant μ as $g_s = t/N$, $t = \mathrm{i}\mu$.

ii. The coefficients of N^{2-2g} in the genus expansion of Eq. (9) have a particular geometrical significance of their own: by the Harer–Zagier formula [**55**], they coincide with the Euler characteristic of the moduli space of genus g-curves with no marked points and $g>1$,

$$\chi(\mathcal{M}_{g,0}) = \frac{B_{2g}}{2g(2g-2)} = \frac{(-1)^{g-1}2^{2g-1}(2g-1)(2g-3)!}{2^{2g-1}-1}\int_{\overline{\mathcal{M}}_{g,1}} \psi_1^{2g-2}\lambda_g \tag{10}$$

This is perhaps the simplest setting where an answer to a topological problem on the moduli space of Riemann surfaces (respectively, in the second equality, an intersection theoretic problem on $\overline{\mathcal{M}}_{g,n}$) is encoded into a formal 1-matrix model[1]. This is not an accident, and the rest of this paper will be devoted to non-trivial extensions of this phenomenon.

iii. A specular (in a precise sense) point of view to the previous statement is given by the study of the large N expansion in Eq. (9) using *loop equations* [**20**, **79**]; an infinite set of differential constraints on the cumulants of Eq. (3) (see [**40**, **72**] for reviews). Its upshot in our case is that the planar free energy, $\mathcal{F}_0$, in Eq. (4) can be recovered as an integral transform of Wigner's semi-circle function,

$$\begin{aligned} \mathcal{F}_0(t) &= -t\int_0^{2\sqrt{t}} z^2\rho(z)\mathrm{d}z, \\ \rho(z) &\triangleq \frac{1}{2\pi\mathrm{i}}\sqrt{z^2-4t}. \end{aligned} \tag{11}$$

Equivalently, for $t \in \mathbb{C}^\star$ consider the *spectral curve* $(C_t, \mathrm{d}\lambda_t)$ given by the smooth, genus zero affine plane curve $C_t = \{(z,y) \in \mathbb{C}^2 | y^2 = z^2 - 4t\}$ endowed with the meromorphic differential $\mathrm{d}\lambda_t = ydz$. Then $F_0(t)$ is recovered from the rigid special Kähler geometry relations [**90**]

$$t = \frac{1}{2\pi\mathrm{i}}\oint_A \mathrm{d}\lambda, \qquad \frac{\partial \mathcal{F}_0}{\partial t} = \frac{1}{2}\oint_B \mathrm{d}\lambda, \tag{12}$$

[1] The distinction between *gauged* and *ungauged* matrix models is somewhat in the eye of the beholder: the same result would be obtained via the large N study of the (ungauged) Penner matrix model in a double-scaling limit; see [**34**].

where the integrals over $A \in H_1(C_t, \mathbb{Z})$, $B \in H_1^{\rm BM}(C_t, \mathbb{Z})$ denote respectively a contour integral encircling the cut $[-2\sqrt{t}, 2\sqrt{t}]$ of $y(z)$ in the z plane, and the sum of principal value integrals from $-\infty^-$ to $-2\sqrt{t}$ and from $-2\sqrt{t}$ to $-\infty^+$; here $-\infty^\pm$ refers to the two pre-images of $z = -\infty$ under the branched covering map $z : C_t \to \mathbb{C}$.

By the looks of it, Point ii) above relates the formal $1/N$ expansion of the gauged gaussian matrix model to the computation of a particular class of intersection numbers on the moduli space of curves; in physics language, these are a subset of observables in a simple example of an *A-type* topologically twisted string theory[2]. Point iii) does the same thing (at the leading order) for something superficially rather different: special geometry governs the dependence on complex (vector) moduli of the prepotential in type IIB compactification, and is precisely what is captured by the planar limit of a *B-type* topologically twisted string theory. This can actually be made more precise: it was shown in [**46**] that the $c = 1$ string at the self-dual radius reproduces the genus expansion of the topological A-model on the singular conifold, which is its own self-mirror on the B-side.

2.1. A-model: intersection theory on $\overline{\mathcal{M}}_{g,n}$. The triangle of relations we found in this example between formal matrix models, intersection theory on $\overline{\mathcal{M}}_{g,n}$ and special geometry on a family of curves is part of a more general story. For the general formally deformed measure $\mathrm{d}\mu_N(x)$, we write

$$W_{g,h}(z_1, \ldots, z_h; x) \triangleq \prod_{i=1}^h z_i^{-1} \sum_{k_1,\ldots,k_h} \mathcal{W}_{g,h;k_1,\ldots,k_h}(x) \prod_{i=1}^h z_i^{-k_i} \tag{13}$$

for the generating function of the cumulants of Eq. (4) at the $g^{\rm th}$ order in the $1/N$ expansion of the formal 1-matrix model of Eq. (1). At the leading (planar) order in $1/N$, the first loop equation reduces to

$$W_{0,1}(z) = \frac{1}{2t}P'(z) - \frac{1}{2t}M(z)\sqrt{\sigma(z)}, \tag{14}$$

where $\sigma(z) = (z - b_1(x,t))(z - b_2(x,t))$ and the *moment function* $M(z;x)$ as well as the branchpoints $b_i(x,t)$ are entirely determined by $P(z)$ and t. The second planar loop equation fixes the two-point function to take the form [**9**]

$$W_{0,2}(z_1, z_2) = -\frac{1}{2(z_1 - z_2)^2} + \frac{\sqrt{\sigma(z_1)}}{2(z_1 - z_2)^2\sqrt{\sigma(z_1)}\sqrt{\sigma(z_2)}} - \frac{\sigma'(z_1)}{4(z_1 - z_2)\sqrt{\sigma(z_1)\sigma(z_2)}}. \tag{15}$$

In two remarkable papers [**38, 39**] (see also [**36**]), Eynard established the following

THEOREM 2.1. *There exist explicit generating functions of tautological classes $\Lambda_{g,n}(x) \in \mathbb{C}[[x]] \otimes R^\bullet(\overline{\mathcal{M}}_{g,n})$, $\mathcal{B}_{g,n}(z_1, \ldots, z_n; x) \in \mathbb{C}[[x; z_1, \ldots z_n]] \otimes R^\bullet(\overline{\mathcal{M}}_{g,n})$ such that*

$$\begin{aligned} W_{g,n}(z_1, \ldots, z_n; x) &= \int_{\overline{\mathcal{M}}_{g,n}} \Lambda_{g,n}(x)\mathcal{B}_{g,n}(x, z_1, \ldots, z_n), \\ \mathcal{F}_g(x) &= \int_{\overline{\mathcal{M}}_{g,0}} \Lambda_{g,0}(x). \end{aligned} \tag{16}$$

[2]In particular, these numbers are closely related to the equivariant Gromov–Witten theory of the affine line – a deformation of topological gravity by linear insertions of Chern classes of the Hodge bundle.

In Theorem 2.1, the classes $\Lambda_{g,n}(x)$ and $\mathcal{B}_{g,n}(x, z_1, \ldots, z_n)$ are entirely determined by knowledge of $W_{0,1}$ and $W_{0,2}$ in Eqs. (14) *and* (15).

Theorem 2.1 encodes the solution of an intersection theoretic problem on $\overline{\mathcal{M}}_{g,n}$ into the knowledge of the correlators $W_{g,n}(z_1, \ldots, z_n; x)$ of an associated formal 1-matrix model. In fact, the applicability of Theorem 2.1 goes far beyond the realm of (multi-)matrix models: it applies to any spectral curve setup and higher order correlators defined through the Eynard–Orantin topological recursion as in the following section.

2.2. B-model: the Eynard–Orantin topological recursion. Theorem 2.1 gives a means to compute the $1/N$ expansion of the correlators of the 1-matrix model in terms of intersection numbers on $\overline{\mathcal{M}}_{g,n}$, which in general is hardly a simplification from the computational point of view. An independent way to compute them was found in 2004 by Eynard, and further developed in subsequent works with Chekhov and Orantin, in terms of a recursive solution of the loop equations, as in the following

THEOREM 2.2 ([**29, 37, 41**]). *Let $\overline{C}_t$ be the normalisation of the projective closure of the plane curve given by the zero locus of $y^2 - \sigma(z)$ in $\mathbb{C}^2$ with coordinates (z, y), and write $\mathrm{d}E_w(z) \in \Omega^1(\overline{C}_t \setminus \{w = z\})$ for the logarithmic derivative of the prime form on $\overline{C}_t$, normalised as $\oint_A \mathrm{d}E_w = 0$; here A is an oriented loop winding once counter-clockwise around the segment $[b_1(x,t), b_2(x,t)]$. Then, for $2g - 2 + h > 0$ the following recursions hold:*

$$W_{g,h+1}(z_0, z_1 \ldots, z_h) = \sum_{i=1,2} \operatorname*{Res}_{z=b_i} \frac{\mathrm{d}E_{z_0}(z)}{2M(z)\sqrt{\sigma(z)}} \Big(W_{g-1,h+2}(z, \overline{z}, z_1, \ldots, z_h)$$

$$+ \sum_{l=0}^{g} \sum_{J \subset H}^{\prime} W^{(g-l)}_{|J|+1}(z, z_J) W^{(l)}_{|H|-|J|+1}(\overline{z}, z_{H \setminus J}) \Big), \tag{17}$$

$$\mathcal{F}_g = \sum_{i=1,2} \operatorname*{Res}_{z=b_i} \frac{\mathrm{d}E_{z_0}(z)}{2M(z)\sqrt{\sigma(z)}} \int^z \mathrm{d}z' W_{g,1}(z') \tag{18}$$

Here $I \cup J = \{z_1, \ldots, z_h\}$, $I \cap J = \emptyset$, $\sum'$ denotes omission of the terms $(h, I) = (0, \emptyset)$ and (g, J), and the primitive in Eq. (18) *is independent of its base point.*

Eqs. (17) and (18) together make up the *topological recursion*: they determine recursively all the higher orders of the correlators $W_{g,h}$ starting from $(g, h) = (0, 1), (0, 2)$ in Eqs. (14) and (15), and consequently the all-order expansion of the partition function. Its frame of applicability goes far beyond the formal analysis of single-cut 1-matrix models, encompassing also multi-cut solutions, multi-matrix models, and non-polynomial potentials. The overall picture that emerges can be codified by the diagram of Figure 1.

3. The GOV correspondence for S^3

How much of this story is specific to matrix models? It should be noticed that 't Hooft's original argument in no way depended on the dimensionality of the gauge theory we are doing perturbation theory on, as long as the interacting theory has a weakly coupled limit given by a free Gaussian theory, a sensible UV-regularisation scheme exists, and renormalisation does not lead to contributions to

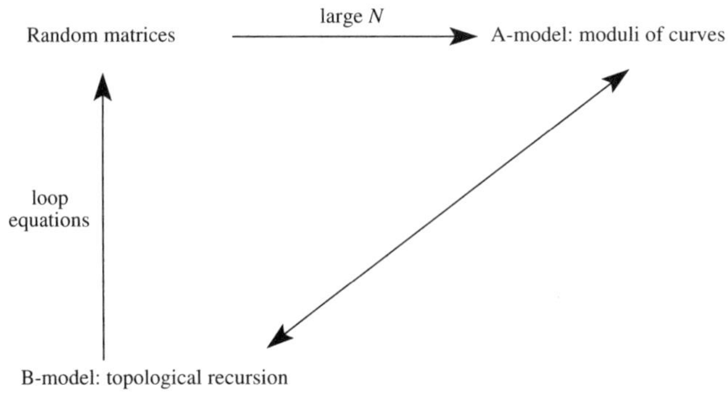

FIGURE 1. Large N duality and mirror symmetry in $d = 0$

the perturbation series which are individually factorially divergent at each order in the topological expansion. A natural, if perhaps bold, question that could be asked is then: "Can the top left corner in Figure 1 be replaced by a higher dimensional (topological) gauge theory?".

Evidence that this is not a mere speculation comes from a non-trivial example, found by Gopakumar, Ooguri and Vafa in [**47**,**81**], where the *entire* setup of Figure 1 carries forth verbatim to higher dimension. Let M be a smooth, closed, oriented real 3-manifold and $\mathcal{K}$ be a link in it. For a smooth $\mathrm{U}(N)$ gauge connection $\mathcal{A}$ on M let

$$\mathrm{CS}[\mathcal{A}] \triangleq \int_M \mathrm{Tr}_\square \left(\mathcal{A} \wedge \mathrm{d}\mathcal{A} + \frac{2}{3} \mathcal{A} \wedge \mathcal{A} \wedge \mathcal{A} \right) \tag{19}$$

be the Chern–Simons functional of $\mathcal{A}$. It was Witten's realisation that the formal functional integrals [**94**]

$$\begin{aligned} Z_N^{\mathrm{CS}}(M,k) &\triangleq \int D[\mathcal{A}] \exp\left[\frac{ik}{4\pi}\mathrm{CS}[\mathcal{A}]\right], \\ W_{N,\rho}^{\mathrm{CS}}(M,\mathcal{K},k) &\triangleq Z_N^{-1} \int D[\mathcal{A}]\, \mathrm{Tr}_\rho\left(\mathrm{Hol}_\mathcal{K}(\mathcal{A})\right) \exp\left[\frac{ik}{4\pi}\mathrm{CS}[\mathcal{A}]\right] \end{aligned} \tag{20}$$

lead to (framed) topological invariants of M and $\mathcal{K}$ in the form of the Reshethikin and Turaev's sl_N-quantum invariants; here $k \in \mathbb{Z}^*$ is the Chern–Simons level, and the coloring ρ is an irreducible representation of $\mathrm{U}(N)$. Their large N expansion can be worked out entirely explicitly in different ways, for example by resorting to the surgery techiniques of [**94**]: for the partition function we find

$$\ln Z_N^{\mathrm{CS}}(M,k) = \sum_{g \geq 0} g_s^{2g-2} \mathcal{F}_g^{\mathrm{CS}}(t) \tag{21}$$

where $g_s = 2\pi\mathrm{i}(k+N)^{-1}$, $t = g_s N$, and

$$\mathcal{F}_g^{\mathrm{CS}}(t) = \frac{|B_{2g}|}{2g(2g-2)!}\mathrm{Li}_{3-2g}(e^{-t}) + \frac{(-1)^g B_{2g} B_{2g-2}}{2g(2g-2)(2g-2)!} \tag{22}$$

for $g \geq 2$, with similar formulas for $g = 0, 1$ [**47**]. If we trust that a higher dimensional analogue of the discussion of the previous section exists, the two questions we are tasked to answer are:

(1) characterise the large N A-model dual of Chern–Simons theory in terms of a precise intersection theoretic problem on $\overline{\mathcal{M}}_{g,n}$;
(2) characterise its mirror, large N B-model dual in terms of the topological recursion on a specific spectral curve setup.

3.1. A-model. The A-model dual found in [**47**] is a sort of q-deformation of the setup of Example 2.1, and it has an explicit presentation in terms of an intersection theory problem on a moduli space of stable *maps*. First off, it was found in [**95**] that the partition function of U(N) Chern–Simons theory on any closed three manifold M is equal to the partition function of the *open* topological A-model on the total space of the cotangent bundle T^*M, with N Lagrangian A-branes wrapping its zero section. The proposal of [**47**] is to relate the latter at large N to the *ordinary, closed* A-model/Gromov–Witten theory on a target space obtained from T^*S^3 via a complex deformation to a normal singular variety and a bimeromorphic resolution of its singularities (the *conifold transition*), as follows. When S^3 has radius one in the canonical metric, we have an obvious isometry $T^*S^3 \simeq \mathbb{R}^3 \times S^3 \simeq \mathrm{SL}(2,\mathbb{C})$ given by the decomposition of a special linear matrix A into radial (positive definite) and polar (unitary) part:

$$A = U\mathrm{e}^H, \qquad U \in \mathrm{SU}(2), \quad H \in \mathcal{H}_0(2,\mathbb{C}). \tag{23}$$

In particular this endows T^*S^3 with a complex structure given by its presentation as a quadric hypersurface $\det A = 1$ in $\mathrm{Mat}(2,\mathbb{C}) \simeq \mathbb{C}^4$.

We now perform the following two operations on this A-model target space:

(1) First we vary the radius of the base unit sphere to give a flat family $\psi : X = \mathrm{GL}(2,\mathbb{C}) \to \mathbb{C}^*$ via the determinant map, whose fiber $X_{[\mu]}$ at a point μ with $\mathrm{Im}\,\mu = 0$ and $\mathrm{Re}\,\mu > 0$ is isomorphic to the cotangent bundle $T^*S^3_{[\mu]}$. Notice that we are doing *nothing here* either from the point of view of Chern–Simons theory or the open topological A-model on the cotangent bundle, as a homeomorphism of the base (resp. a complex deformation of the total space) leave invariant the CS partition function (resp. the A-model partition function).
(2) The second is to add the locus of non-invertible matrices to form:

$$\tilde{\psi} : \mathrm{Mat}(2,\mathbb{C}) \longrightarrow \mathbb{C}\,. \tag{24}$$

The singular fiber $X_{[0]}$ above $\mu = 0$ is the singular quadric $\det A = 0$: this is the singular conifold of Example 2.1. The latter admits a canonical toric minimal resolution

$$\pi : \widehat{X} \longrightarrow X_{[0]}, \qquad \widehat{X} \triangleq \big\{(\rho(A), v) \in X_{[0]} \times \mathbb{P}^1, \quad \rho(A)v = 0\big\}, \tag{25}$$

where π is the projection to the first factor. The point $A = 0$ is singular in $X_{[0]}$, and its fiber is a complex projective line with $[v_1 : v_2]$ as homogeneous coordinates. Using coordinate charts on $\mathbb{P}^1$ exhibits $\widehat{X}$ as the total space of $\mathcal{O}(-1) \oplus \mathcal{O}(-1) \to \mathbb{P}^1$, i.e. the *resolved conifold.*

The proposal of [**47**] is that

$$\mathcal{F}_g^{\mathrm{CS}}(t) = \mathrm{GW}_g^{\widehat{X}}(t) \tag{26}$$

where we have identified the 't Hooft parameter on Chern–Simons theory with the A-model Kähler parameter dual to the curve class of the base $\mathcal{O}(-1)^{\oplus 2}_{\mathbb{P}^1}$. In Eq. (26), $\mathrm{GW}_g^{\widehat{X}}$ denotes the genus-g primary Gromov–Witten potential of $\widehat{X}$,

$$\mathrm{GW}_g^{\widehat{X}}(t) = \sum_{d\geq 0} \mathrm{e}^{td} \int_{[\overline{\mathcal{M}}_{g,0}(\widehat{X},d)]^{\mathrm{vir}}} 1 \tag{27}$$

where $\overline{\mathcal{M}}_g(\widehat{X},d)$ is the stack of degree d stable maps from genus g curves to $\widehat{X}$ and $[\overline{\mathcal{M}}_g(\widehat{X},d)]^{\mathrm{vir}}$ denotes its virtual fundamental cycle.

There are a number of reasons, at various degrees of rigour, to take this proposal seriously:

(1) Firstly, it holds true asymptotically in the $t \to 0$ limit: indeed on the CS side, the partition function is known to reduce in this limit to the gauged gaussian matrix model of Example 2.1 [**84**]; and on the GW side, the $t \to 0$ regime corresponds to the small volume limit (the *singular conifold*). The claim then follows from the conifold/$c = 1$ duality we discussed in Example 2.1;

(2) There are also heuristic physics arguments in favour of the duality. One hinges on viewing Eq. (26) as an open/closed duality in the A-model, the equality can be interpreted as a brane/flux duality in the physical type IIA superstring: in principle, both sides compute superpotential observables in effective $\mathcal{N} = 1$ theories, obtained either as the world-volume theory of wrapped $D6$-branes in T^*S^3 of via turning on internal RR field strengths on the resolved side [**13, 92**]. These string configurations can then be related by lifting them to M-theory [**10**].

(3) A further microscopic derivation of the duality can also be given [**82**] upon relating the open partition function on the deformed conifold and the closed partition function on the resolved conifold by their UV completion in terms of a gauged linear σ-model: here the $t \to 0$ limit leads to a coexistence of Coulomb/Higgs phases on the closed string world-sheet, with the holes of the open theory arising from integrating out the contribution of Coulomb regions.

(4) What is more, the equality of Eq. (26) is a sharp mathematical statement about RTW invariants and GW invariants that can be rigorously proven, or disproven, by an explicit calculation. The l.h.s. can be rigorously shown to lead to the expansion of Eq. (22) from an explicit MacLaurin expansion of the finite sums appearing in the surgery formula for the Reshetikhin–Turaev invariant [**47**]; and on the other hand the r.h.s. is amenable to a direct localisation analysis in Gromov–Witten theory, first performed in [**43, 48**]. The end result is an exact agreement of the two generating functions.

(5) Finally, a natural question is how the story generalises when we incorporate links in Chern–Simons theory. It was first proposed in [**81**] that when $\mathcal{K} = \bigcirc$ is the trivial knot, Eq. (26) should generalise to an equality between the colored HOMFLY invariants of Section 3 and *open* GW potentials for a suitable choice of Lagrangian A-branes L:

$$W^{\mathrm{CS}}_{g,h,\vec{d}}(S^3, t, \bigcirc) = \mathrm{GW}^{\widehat{X},L}_{g,h,\vec{d}}(t) \tag{28}$$

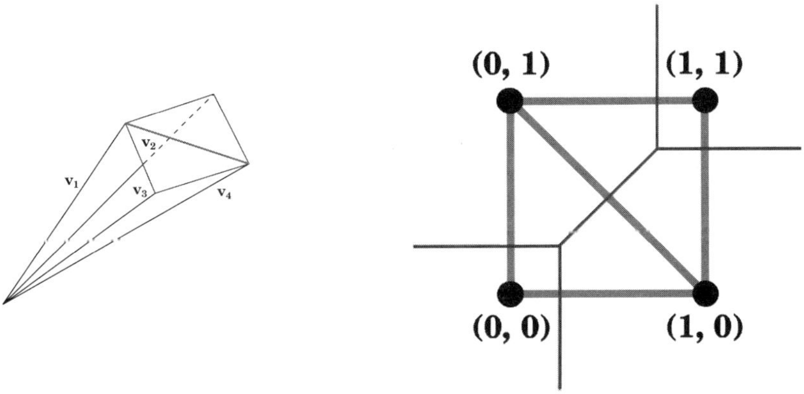

FIGURE 2. The fan (left) of the resolved conifold: its skeleton is given by four rays, labelled v_i in the picture, whose tips lie on an affine hyperplane at unit distance from the origin of $\mathbb{Z}^3$. The intersection of the fan with the hyperplane is shown on the right (the *toric diagram*); superimposed is the pq-web diagram.

where in the l.h.s. we consider a knot invariant obtained as a power sum (instead of Schur functions, as in Section 3) holonomy invariant specified by a vector of integers $d_1, \ldots, d_h$, we take its connected part, and we expand it at large N as in Eq. (5), and in the r.h.s. we take the generating function of open GW invariants with boundary on the fixed locus of a real involution (a real bundle on the equator of the base $\mathbb{P}^1$), with fixed genus g and number of holes h for the source of curves and winding numbers $d_1, \ldots, d_h$ around $S^1 \subset \mathbb{P}^1$. Indeed, the line of reasoning of [**92**] carries through to this setting, and also in this case a localisation definition/computation of open GW invariants can be performed [**63**], confirming the prediction of [**81**].

3.2. B-model. Since the A-model target space is a (non-compact) Calabi–Yau threefold, we would expect a geometric mirror picture in terms of rigid special Kähler geometry of a family of local CY3s, akin to what happened in Example 2.1 for the case of matrix models. Not only is this case, but the analogy with the spectral curve setup of random matrices is even more poignant – the special geometry relations on the local CY3 mirror (which, in this particular case, is just the deformed conifold geometry of the previous section) can be shown to reduce to Seiberg–Witten-type relations on a family of genus zero spectral curves, given by the Hori–Iqbal–Vafa mirror of $\widehat{X}$:

$$t = \frac{1}{2\pi \mathrm{i}} \oint_A \ln y d \ln x, \qquad \frac{\partial \mathrm{GW}_0^{\widehat{X}}}{\partial t} = \frac{1}{2} \oint_B \ln y d \ln x \tag{29}$$

where A, B are homology 1-cycles relative to the principal divisors $x = 0$, $y = 0$ of the plane curve defined by $P_{\widehat{X}}(x, y) = 1 + x + y + \mathrm{e}^t = 0$; here $P_{\widehat{X}}$ is the Newton polynomial of the diagram in Figure 2. What is more, motivated by the study of the chiral boson theory of [**2**], it was first suggested in [**73**], and then postulated in full detail in [**19**], that the higher genus B-model potentials (and, by mirror symmetry, the higher genus open GW invariants of $\widehat{X}$) should be computed by the topological recursion of Eqs. (17) and (18) upon identiying $ydx = W_{0,1}(x)$. This prediction has been subsequently proved in full rigour in [**42**] for all toric Calabi–Yau manifolds that are symplectic quotients[3].

4. The GOV correspondence for Clifford–Klein 3-manifolds

We have seen how the picture of Figure 1 generalises verbatim to the higher dimensional setting of Chern–Simons theory on S^3: the top-right corner has a concrete A-model picture in terms of the Gromov–Witten theory of the resolved conifold, and the bottom-left corner is encoded by the topological recursion on its Hori–Iqbal–Vafa mirror curve. The emerging picture is not only beautiful and unexpected; from a practical point of view, it has deeply affected the relation between quantum invariants, curve-counting invariants, and the topological recursion. Appealing and impactful as it has been, this is however just *one* example – placed in the classical setting of $\mathrm{U}(N)$ theories, encompassing the simplest closed 3-manifold, and with the simplest knot therein. It is then natural to seek a strict generalisation of the GOV correspondence to other[4]

(1) **knots**: in its strictest sense, the GOV correspondence has been shown to carry through to the case of torus knots both in terms of an enumerative theory of open GW invariants on the A-side [**32**] and the topological recursion on a specific spectral curve setup on the B-side [**24**]. For general knots, it would seem that substantially new ideas are needed both on the A- and the B-side of topological string theory [**3**]; the role of the topological recursion in particular is however less clear in this setting [**50**].
(2) **(classical) gauge groups**: $\mathrm{SO}(N)$ and $\mathrm{Sp}(N)$ Chern–Simons theory at large N, which compute in particular the Kauffmann invariant of links, can also fit in the picture of the previous section by considering suitable orientifolds of the resolved conifold [**89**], for which an operative definition of unoriented invariants can be given either by localisation or via the topological vertex [**17, 18**].
(3) **3-manifolds**: this is possibly the boldest generalisation – replace altogether S^3 by an arbitrary closed 3-manifold M. At face value this boils down to solving the problem for arbitrary knots on S^3: the partition function of a link $\mathcal{L}$ in M would then be recovered by the surgery formula from the partition function of a 2-component link $\mathcal{L} \sqcup \mathcal{L}_M$ in S^3, where $\mathcal{L}_M$ is a framed link M can be obtained from (this always exists by Lickorish's theorem). However, both the surgery formula in Reshetikhin–Turaev theory [**54, 94**], and functional localisation in Chern–Simons theory [**12, 14, 61**] lead to an expression of the partition function as a sum over contribution labelled by flat connections (classical vacua) on M:

$$Z_N^{\mathrm{CS}}(M,k) = \sum_{\mathrm{v} \in \mathrm{Hom}(\pi_1(M), \mathrm{U}(N))/\mathrm{U}(N)} Z_{N,\mathrm{v}}^{\mathrm{CS}}(M,k). \tag{30}$$

---

[3] See [**44**] for a recent generalisation of this result to the case of orbifolds.

[4] The following list leaves out one important generalisation, tying together categorification on the knot theory side [**51**], refined BPS counting on the A-model side [**30, 60, 62**], and some version of the β-deformation on the B-model side [**1, 7, 26**].

It was first proposed in [**5**] that the finer invariants given by the *individual summands* $Z^{\rm CS}_{N,\mathsf{v}}$ may also be interpreted as the A/B-model partition function on a background specified by v; notice that this is a more refined object to deal with than the partition function of the link $\mathcal{L}_M$, where all these contributions are summed over. That a dual curve counting theory exists is encouraged by the successful test of this proposal for the case of $L(p,1)$ lens spaces in [**5, 53**]. The case of more general 3-manifolds was considered in [**15, 16, 25, 27**], and we review it below.

4.1. CS theory on Clifford–Klein 3-manifolds. We start by recalling the following

DEFINITION 4.1. A Clifford–Klein 3-manifold (M,g) is a closed oriented smooth 3-manifold M admitting a smooth metric g of everywhere strictly positive Ricci curvature.

Equivalently, by Hamilton's theorem, it is a spherical space form, $M = S^3/\Gamma$ for Γ a freely acting finite isometry group of S^3 w.r.t. its canonical metric; and by Perelman's elliptisation theorem, it follows that these coincide with the orientable 3-manifolds having finite fundamental group. The classification of the possible Γ goes back to Hopf [**57**]: these are central extensions of the left-acting finite subgroups of $\mathrm{SL}_2(\mathbb{C})$, which admit an ADE classification; see [**15**, Appendix A] for more details. We restrict henceforth for simplicity of exposition to the case where the central extension is trivial and Γ is one such $\mathrm{SL}_2(\mathbb{C})$-subgroup; most of our arguments to follow will be unaffected by this.

Now since $|\Gamma| < \infty$, the sum in Eq. (30) truncates at finite N. Denote by $\mathfrak{V}_{\Gamma,N}$ the finite set of gauge-equivalent flat connections and by $\mathfrak{V}_\Gamma = \lim_{N\to\infty}\mathfrak{V}_{\Gamma,N}$ be its direct limit with respect to the composition of morphisms given by the embedding $\mathrm{U}(N) \hookrightarrow \mathrm{U}(N+1)$. Then

$$\mathfrak{V}_\Gamma = \mathbb{N}^{R+1}, \qquad \mathfrak{V}_{\Gamma,N} = \left\{ (N_0,\ldots,N_R) \in \mathbb{N}^{R+1}, \quad N_0 + \sum_{i=1}^{R} D_i N_i = N \right\}. \tag{31}$$

where R is the number of nodes in the simply-laced Dynkin diagram associated to Γ, and D_i are the respective Dynkin indices. When $N \to \infty$, we thus consider a CS vacuum $[\mathcal{A}]_t$ parametrized by $t \triangleq N_i g_s$ for $i \in \{0,\ldots,R\}$, and in particular the rank is encoded in the 't Hooft parameter $t = N g_s$. The resulting partition functions at $N = \infty$ have a standard $1/N$ expansion

$$\ln Z^{\rm CS}_{N,\mathsf{v}(t)}(g_s) = \sum_{g\geq 0} g_s^{2g-2}\mathcal{F}^{\rm CS}_g(t_0,\ldots,t_R) \tag{32}$$

with free energies depending now on $R+1$ 't Hooft parameters.

These manifolds also carry a natural class of knots with them. For S^3, the standard GOV correspondence focused on the unknot – which could be regarded as the fibre knot of the Hopf fibration on S^2. Now Clifford–Klein manifolds are also Seifert-fibred, and in the ADE case they can be regarded as Seifert fibrations over an ADE $\mathbb{P}^1$-orbifold with one (resp. three) exceptional fibres for Dynkin type A (resp. D and E). As for S^3, we will similarly be interested in the $1/N$ expansion of the RTW invariant for the knots $\mathcal{K}_f$ running around these exceptional fibres:

$$W^{\rm CS}_{N,\mathsf{v}(t)}(\mathcal{K}_f, g_s, \vec{d}) = \sum_{g\geq 0} g_s^{2g-2}\mathcal{F}_{g,h}(t_0,\ldots,t_R;\vec{d}) \tag{33}$$

The quest is to find now a Calabi–Yau threefold geometry $\widehat{X}_\Gamma$ with special Lagrangians L_Γ, as well as a spectral curve setup Σ_Γ for each such Γ, such that open/closed GW theory on $(\widehat{X}_\Gamma, L_\Gamma)$ and the topological recursion on Σ_Γ lead to Eqs. (32) and (33). This program was completed in [**15**, **27**], generalising ideas on the lens space case in [**5**].

4.2. A-model from geometric transition. On the A-model the idea is fairly simple: we take seriously the geometric transition argument of Section 3.2 and apply it to the setting at hand. To this aim, notice that the Γ-action on the resolved conifold $\widehat{X}$ only acts on the first factor of Eq. (25), giving a fibrewise action on $p : \widehat{X} \to \mathbb{P}^1$ (the second factor in Eq. (25)); the fibre over a point $z \in \mathbb{P}^1$ is isomorphic to a Gorenstein surface singularity of the same ADE type of Γ. The topological A-string on this target geometry can be studied in several phases, two of which are distinguished:

- in the *orbifold* chamber, we are looking at a theory of twisted stable maps on the Calabi–Yau stack $\widehat{X}^\Gamma_{\rm orb} \triangleq [\widehat{X}/\Gamma]$. This is the maximally singular phase containing the Γ-orbifold of the conifold point, which is the natural point of expansion for the dual Chern–Simons theory;
- in the *large radius* chamber, we take a crepant resolution $\widehat{X}^\Gamma_{\rm res}$ of the singularities of (the coarse moduli space of) $\widehat{X}^\Gamma_{\rm orb}$ obtained by canonically resolving the surface singularity $\mathbb{C}^2/\Gamma$ fibrewise, and we are looking at the ordinary GW theory of $Y^\Gamma_{\rm res}$.

We have four remarks about the resulting target space geometry.

R1: in either case the target supports at least a $(\mathbb{C}^\star)^2$ torus action; this is the product of the 1-torus action that rotates the base $\mathbb{P}^1$ and the fibrewise 1-torus action inherited from the scalar action on the $\mathbb{C}^2$-fibre of $\widehat{X}$. In particular, as for the usual toric case, we will specialise to a resonant 1-subtorus by imposing that its action is trivial on the canonical bundle of the target; this acts with compact fixed 0- and 1- dimensional fixed loci. This in turn allows to define equivariant GW invariants by localisation on either $\widehat{X}^\Gamma_{\rm res}$ or $\widehat{X}^\Gamma_{\rm orb}$.

R2: As we already mentioned, the Γ-action is fibrewise and it covers the trivial action on the base $\mathbb{P}^1$; this circumstance will be important in a moment.

R3: For all Γ, there are natural anti-holomorphic involutions whose fixed loci define Lagrangians for both $\widehat{X}^\Gamma_{\rm res}$ and $\widehat{X}^\Gamma_{\rm orb}$. These generalise the toric Lagrangian branes of [**8**, **22**, **81**], and the residual Calabi–Yau $\mathbb{C}^\star$ action of **R1** above allows to define/compute open GW invariants by localisation. In particular, the methods of [**22**] carry through verbatim to the setting at hand.

R4: The anti-holomorphic involution of [**89**] turns out to commute with the Γ-action, for all ADE types, and an orientifold theory can be defined in the same way.

4.3. B-model from geometric engineering. **R1-R2** above define completely what would be the top-right (A-model) corner of a diagram like Figure 1 for the case at hand. Now notice that in **R1** above, for type D and E the torus action we highlighted does not extend to a full three-torus action on $\widehat{X}^\Gamma$ since Γ is non-abelian. In particular, we cannot resort to the toric mirror symmetry methods of [**58, 59**] to find a spectral curve setup for these cases, unlike for type A. A way out on physical grounds is however pointed at by **R2**: since the action is fibrewise, the topological A-string on these backgrounds is known to geometrically engineer in a suitable limit a 4-dimensional $\mathcal{N}=2$ gauge theory with simply-laced compact gauge group corresponding to the Dynkin type of Γ (and no adjoint hypermultiplets, since $\mathbb{P}^1$ has genus zero) [**62**]; so it is expected in this degenerate limit, which corresponds to a divisor at infinity in the stringy Kähler moduli space of $\widehat{X}^\Gamma$, to have a mirror picture in terms of spectral curves of Seiberg–Witten type [**62, 88**]. As a matter of fact, even away from this limit the A-model is still expected to give rise to a gauge theory with eight supercharges, albeit in one dimension higher – namely $\mathcal{N}=1$ pure super Yang–Mills theory on $\mathbb{R}^4\times S^1$, with the "field-theory limit" of [**62**] corresponding to the fifth-dimensional circle shrinking to zero-size.

Now for type A_N, the five-dimensional theory also enjoys a description in terms of spectral curves of Seiberg–Witten/Hori–Iqbal–Vafa type: these are the spectral curves of the periodic Ruijsenaars system (relativistic Toda chain) with N sites [**80**]; the 4d limit corresponds to the non-relativistic limit. Furthermore, it has long been known that for all ADE types the relevant SW curves should coincide with the spectral curves of the Lie-theoretic generalisation of the non-relativistic Toda chain [**76**] specialised to ADE Lie algebras. Putting all this together, it is then natural to look for a relativistic deformation of [**76**] to supply the spectral curves defining a candidate B-model mirror for the non-toric geometries of the previous section, and these can be computed from the setup of [**45, 93**]. We refer the reader to [**27**, Section 2] for a more detailed review; the upshot is that the spectral curves can be computed in the following two steps:

(1) fix ρ to be a minimal irreducible $\mathcal{G}_\Gamma$-module, where $\mathcal{G}_\Gamma$ is the simply-connected simple Lie group over $\mathbb{C}$ of ADE type specified by Γ, and consider the characteristic polynomial of a group element g in the representation ρ

$$\mathfrak{p}_\rho^\Gamma(g) \triangleq \det_\rho(\mathrm{id}\, x - g) : \mathcal{T}_\Gamma/\mathrm{Weyl}(\mathcal{G}_\Gamma) \to \mathbb{C}[x] \tag{34}$$

Here $\mathcal{T}_\Gamma$ is the Cartan torus of $\mathcal{G}_\Gamma$. We can decompose this on the Weyl character ring upon expanding the determinant in a polynomial with coefficients given by anti-symmetric characters of the representation ρ, and then write the latter as polynomials of the fundamental characters u_i:

$$\mathfrak{p}_\rho^\Gamma(g) = \sum_{n=0}^{\dim\rho} (-1)^{\dim\rho - n}\chi_{\wedge^{\dim\rho-n}\rho}(g)x^n, \qquad \chi_{\wedge^n} \in \mathbb{Z}[u_1,\dots,u_R] \tag{35}$$

(2) Now let u_{k} be the character of the maximally dimensional fundamental representation[5]. Then the spectral curve is defined by the family of non-compact Riemann surfaces given by the polynomial equation

$$\mathfrak{p}_\rho^\Gamma\left(x; u_i + \delta_{i,\mathsf{k}}(y + u_0 y^{-1}\right) = 0 \tag{36}$$

along with a choice of recursion differential given by the Poincaré one form $W_{0,1} = \ln y \mathrm{d}\ln x$, and higher order generating functions obtained by the topological recursion on Eq. (36).

The resulting web of relations is pictured in Figure 3.

[5]In two cases there is an ambiguity which is resolved as follows. For type E_6, pick either of the **27** or $\overline{\mathbf{27}}$. For type A_{2n}, Eq. (36) is modified by shifting $u_i \to u_i + \delta_{i,n}y + \delta_{i,n+1}u_0 y^{-1}$.

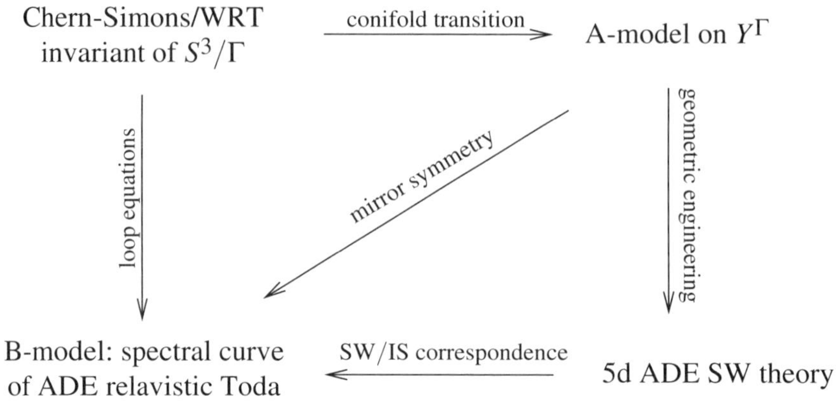

FIGURE 3. The GOV correspondence for S^3/Γ.

4.4. The LMO invariant and the topological recursion. Our proposal fills all the vertices in the diagram of Figure 3: the sl_N-Reshetikhin–Turaev knot invariants of fibre knots in S^3/Γ in the $1/N$ expansion around a fixed flat connection would equate the open Gromov–Witten potentials of $(\widehat{X}^\Gamma, L^\Gamma)$ on one hand, and the Eynard–Orantin invariants of Eq. (36) on the other; a similar statement applies to the RT invariant of S^3/Γ itself, the closed GW potential of $\widehat{X}^\Gamma$, and the topological recursion free energies.

However this is all conjectural for the moment – so how do we prove this? One way to proceed is to resort to the relation of Chern–Simons theory on Seifert fibred spaces (such as the Clifford–Klein manifolds) and matrix models. It was shown by Mariño in [**71**] that Witten's surgery formula for these spaces allows to rewrite the CS partition function as well as the Wilson loops around certain classes of knots as a matrix integral; in particular, the quantum invariants Eqs. (32) and (33) of S^3/Γ and fibre knots therein around the exceptional fibres fall squarely in this category [**12, 14, 71**]. When $\mathsf{v} = 0$ is the trivial flat connection, the resulting matrix model turns out to be a trigonometric deformation of the gauged gaussian matrix model of Example 2.1: it is a canonical ensemble with gaussian 1-body interaction and a sum of q-deformed Vandermonde 2-body interactions, with coefficients determined by the orders of the exceptional fibres of the Seifert fibration [**11, 16, 24, 71**]: this restriction gives rise to the so-called LMO invariant.

This presentation is amenable to a large N analysis via loop equations, akin to that of Point iii) in Example 2.1, leading for all Γ to a singular integral equation to be solved by the input datum of the recursion – the planar disk function $W_{0,1}$. Such an analysis was performed in [**15, 16, 24, 27**]; the strategy, and ensuing results are as follows.

(1) It can readily be shown that these are single-cut matrix models due to the gaussian nature of the 1-body potential, and that the (exponentiated) planar resolvent $W_{0,1}(x)$ is never a log-algebraic function of its argument except when Γ is the trivial group. However, a strategy introduced in [**24**] for torus knots and then employed on a full scale on [**16**] is to considered a symmetrised version of the resolvent, which leads to the same large N eigenvalue density as the original resolvent on the physical cut and may

be such that the sheet transitions given by crossing the cuts close to a finite group.

(2) The previous step can be performed for any Seifert 3-manifold as that is the level of generality to which the ideas of [**71**] apply. It can be shown however that the *only* Seifert spaces for which the group of sheet transitions is finite, and whose planar resolvent thus gives rise to *algebraic* spectral curves, are precisely the Clifford–Klein 3-manifolds: there is no hope to extract an algebraic setup for the large N/B-model curve, even for the restriction to the trivial flat connection, for parabolic and hyperbolic Seifert 3-manifolds. In the elliptic case, instead, $\mathrm{e}^{W_{0,1}}$ is a root of an algebraic equation in $\mathbb{C}^\star \times \mathbb{C}^\star$:

$$P^\Gamma(x, \mathrm{e}^{W_{0,1}}; t) = 0 \tag{37}$$

where t is the 't Hooft parameter $g_s N$. It should be noticed that this symmetrisation is not unique; however the freedom of choice here is in bijection with the freedom of choice of an irreducible $\mathcal{G}_\Gamma$-module on the Toda side: in either case (see [**15, 27**] and [**78**] respectively) it can be shown that these choices do not affect the calculation of the partition function, so we will henceforth drop the subscript ρ from Eq. (36).

(3) For these cases, a computational tour-de-force leads to determine in full detail the spectral curve for type A [**24**], D [**16**], and E_6 [**15**]; substantial information can be extracted for $E_{7,8}$, with a full solution available in all cases in the limit $t \to 0$.

(4) Now, proving the B-model side of the GOV correspondence for the LMO invariant amounts to finding a restriction $u_i(t)$ of the Toda hamiltonians of the ADE relativistic Toda chain such that

$$\mathfrak{p}^\Gamma\left(x; u_i + \delta_{i,\mathsf{k}}(y + u_0 y^{-1}\right)|_{u_i = u_i(t)} = P^\Gamma(x, y; t) \tag{38}$$

A detailed analysis of both spectral curves setup shows[6] that this is indeed the case for all ADE types [**15, 27**].

(5) Finally, having a matrix integral representation for the partition function of Chern–Simons theory on these spaces allows to rigorously derive the topological recursion of Eqs. (17) and (18) for the cumulants of the resulting distribution. On the other hand, on the B-model side, the topological recursion can be either regarded as the definition of the higher genus open/closed topological B-string on a family of curves, or, from a physics standpoint, it could be derived from the chiral boson theory on the spectral curve obtained from dimensional reduction of the BCOV Kodaira–Spencer theory of gravity [**13, 33**]. Since the two theories boil down to the same recursion with the same input datum, we reach the conclusion that they give rise to the same invariants to *all orders in* $1/N$ and arbitrary colourings of the invariants. This yields an all-genus proof of the B-model side of the GOV correspondence for this type of manifolds, restricted to the LMO invariant.

[6]Technically, the polynomials almost never match on the nose, but it can always be shown that upon restriction the Chern–Simons spectral curve arises as a non-trivial reducible component of a degenerate limit of the Toda curve.

5. Conclusions

We have reviewed how a strict generalisation of the GOV correspondence that bears all the ingredients of the simplest original setting of [**47**, **81**], including a geometric A-model theory of open/closed Gromov–Witten invariants and an all-genus B-model theory governed by the topological recursion on a specific spectral curve setup, can be given for the case of Clifford–Klein 3-manifolds – and these alone, according to our remarks in the previous section. This opens severals directions for future research, including an extension to the setting of refined/categorified invariants (particularly on the B-model side), quantum integrability, and the relation to gauge theories. We single out here three more topics in particular on which we hope to report in the near future.

B-model general flat backgrounds. Since most of the analysis of the previous section was restricted to the study of the LMO invariant, it is natural to ask how to extend the GOV strings correspondence to a general Chern–Simons vacuum, thus completing the proof of the B-model version of the GOV correspondence. On the B-model side, the family of relativistic Toda spectral curves was constructed in [**15**] for type $\mathrm{ADE}_{6,7}$, and the missing E_8 case has recently been treated in detail in [**27**]. The more difficult bit here is to provide a complete large N analysis of the matrix model, although this might be possible by suitably rewriting the finite sums expressions of [**14**, **71**] in terms of ordinary eigenvalue models on the real line. Establishing an explicit solution of the loop equations for this matrix model in terms of the topological recursion applied on the corresponding Toda spectral curve would give a full proof of the B-side of the GOV correspondence. This would also shed light on some of the difficulties encountereed in the analysis of arbitrary flat backgrounds to the case of general lens spaces and non-SU(2) abelian quotients of the resolved conifold [**25**].

Non-toric "remodeling-the-B-model". Another line of developement consists of extending the remodeling proposal of [**19**] to the non-toric setting at hand for type D and E. A first step here would be to fully spell out the computation of the disk functions as in [**21**, **22**] for the case at hand, and then derive the topological recursion from the analysis of the descendent theory. A promising route would be to derive the J- and R-calibrations for the quantum cohomology of $\widehat{X}^\Gamma$ from the steepest descent analysis of oscillating integrals of the Toda differential, as in [**23**], and then retrieve the topological recursion from Givental's R-action on the associated cohomological field theory [**36**, **44**]. This would lead to a proof of the remodeling conjecture on an important class of examples, beyond the toric case.

The quantised McKay correspondence. One of the more intriguing consequences of the B-model GOV correspondence is that several limits of the B-model geometries of Section 4.4 would provide a unified construction of spectral curves relevant for both non-toric mirror symmetry and the theory of Frobenius manifolds. In particular, the $u_0 \to 0$ limit, corresponding to the limit where the Kähler volume of the base $\mathbb{P}^1$ in $\widehat{X}^\Gamma$ is sent to infinity, gives rise to a family of 1-dimensional Landau–Ginzburg models for the stack $[\mathbb{C}^2/\Gamma]$ and its crepant resolution: this would finally grant access to a host of explicit computations on the descendent theory, which are likely to be instrumental in the proof of the quantum McKay correspondence in full generality [**28**, **31**, **87**]. This would notably include the higher genus theory, by following the arguments employed for the type A case in [**23**]. Moreover, Dubrovin's almost duality would relate these mirrors to the LG formulation of the Frobenius manifold structure on the orbit spaces of extended affine Weyl groups and ordinary Weyl groups associated to simply-laced root systems upon considering, respectively, the relativistic and non-relativistic limit of the Toda curves of Section 4.4 [**27**]. We plan to further explore this in future work.

Acknowledgments

I would like to thank my collaborators Gaëtan Borot, Bertrand Eynard, Luca Griguolo, Marcos Mariño, Domenico Seminara and Alessandro Tanzini for sharing their insights and educating me on various aspects of the Chern–Simons/topological strings correspondence as well as for most enjoyable collaborations on this topic. This paper is a write-up of material presented at a talk at the 2016 AMS Von Neumann Symposium on *Topological Recursion and its Influence in Analysis, Geometry, and Topology*, July 4–8, 2016, at Charlotte (NC); I wish to express my thanks to Motohico Mulase and Chiu-Chu Melissa Liu for the wonderful scientific environment at the workshop and for giving me the opportunity to present my work. This research was partially supported by a fellowship under the ERC-2015-CoG "GWT".

References

[1] Mina Aganagic, Miranda C. N. Cheng, Robbert Dijkgraaf, Daniel Krefl, and Cumrun Vafa, *Quantum geometry of refined topological strings*, J. High Energy Phys. **11** (2012), 019, front matter + 52. MR3036500

[2] Mina Aganagic, Robbert Dijkgraaf, Albrecht Klemm, Marcos Mariño, and Cumrun Vafa, *Topological strings and integrable hierarchies*, Comm. Math. Phys. **261** (2006), no. 2, 451–516, DOI 10.1007/s00220-005-1448-9. MR2191887

[3] Mina Aganagic, Tobias Ekholm, Lenhard Ng, and Cumrun Vafa, *Topological strings, D-model, and knot contact homology*, Adv. Theor. Math. Phys. **18** (2014), no. 4, 827–956. MR3277674

[4] Mina Aganagic, Albrecht Klemm, Marcos Mariño, and Cumrun Vafa, *The topological vertex*, Comm. Math. Phys. **254** (2005), no. 2, 425–478, DOI 10.1007/s00220-004-1162-z. MR2117633

[5] Mina Aganagic, Albrecht Klemm, Marcos Mariño, and Cumrun Vafa, *Matrix model as a mirror of Chern-Simons theory*, J. High Energy Phys. **2** (2004), 010, 46, DOI 10.1088/1126-6708/2004/02/010. MR2046799

[6] Mina Aganagic, Marcos Mariño, and Cumrun Vafa, *All loop topological string amplitudes from Chern-Simons theory*, Comm. Math. Phys. **247** (2004), no. 2, 467–512, DOI 10.1007/s00220-004-1067-x. MR2063269

[7] Mina Aganagic and Shamil Shakirov, *Knot homology and refined Chern-Simons index*, Comm. Math. Phys. **333** (2015), no. 1, 187–228, DOI 10.1007/s00220-014-2197-4. MR3294947

[8] M. Aganagic and C. Vafa, *Mirror symmetry, D-branes and counting holomorphic discs* (2000), available at `hep-th/0012041`.

[9] Gernot Akemann, *Higher genus correlators for the Hermitian matrix model with multiple cuts*, Nuclear Phys. B **482** (1996), no. 1-2, 403–430, DOI 10.1016/S0550-3213(96)00542-1. MR1424003

[10] Michael Atiyah, Juan Maldacena, and Cumrun Vafa, *An M-theory flop as a large N duality*, J. Math. Phys. **42** (2001), no. 7, 3209–3220, DOI 10.1063/1.1376159. Strings, branes, and M-theory. MR1840340

[11] D. Bar-Natan and R. Lawrence, *A rational surgery formula for the LMO invariant*, Israel J. Math. **140** (2004), 29–60. MR2054838

[12] Chris Beasley and Edward Witten, *Non-abelian localization for Chern-Simons theory*, J. Differential Geom. **70** (2005), no. 2, 183–323. MR2192257

[13] M. Bershadsky, S. Cecotti, H. Ooguri, and C. Vafa, *Kodaira-Spencer theory of gravity and exact results for quantum string amplitudes*, Comm. Math. Phys. **165** (1994), no. 2, 311–427. MR1301851

[14] Matthias Blau and George Thompson, *Chern-Simons theory on Seifert 3-manifolds*, J. High Energy Phys. **9** (2013), 033, front matter+25. MR3144954

[15] G. Borot and A. Brini, *Chern-Simons theory on spherical Seifert manifolds, topological strings and integrable systems*, Adv. Theor. Math. Phys., to appear (2018), available at 1506.06887.

[16] Gaëtan Borot, Bertrand Eynard, and Alexander Weisse, *Root systems, spectral curves, and analysis of a Chern-Simons matrix model for Seifert fibered spaces*, Selecta Math. (N.S.) **23** (2017), no. 2, 915–1025, DOI 10.1007/s00029-016-0266-6. MR3624903

[17] Vincent Bouchard, Bogdan Florea, and Marcos Mariño, *Counting higher genus curves with crosscaps in Calabi-Yau orientifolds*, J. High Energy Phys. **12** (2004), 035, 35 pp. (2005), DOI 10.1088/1126-6708/2004/12/035. MR2130115

[18] Vincent Bouchard, Bogdan Florea, and Marcos Mariño, *Topological open string amplitudes on orientifolds*, J. High Energy Phys. **2** (2005), 002, 35, DOI 10.1088/1126-6708/2005/02/002. MR2140507

[19] Vincent Bouchard, Albrecht Klemm, Marcos Mariño, and Sara Pasquetti, *Remodeling the B-model*, Comm. Math. Phys. **287** (2009), no. 1, 117–178, DOI 10.1007/s00220-008-0620-4. MR2480744

[20] E. Brézin, C. Itzykson, G. Parisi, and J. B. Zuber, *Planar diagrams*, Comm. Math. Phys. **59** (1978), no. 1, 35–51. MR0471676

[21] Andrea Brini, *Open topological strings and integrable hierarchies: remodeling the A-model*, Comm. Math. Phys. **312** (2012), no. 3, 735–780, DOI 10.1007/s00220-012-1489-9. MR2925133

[22] A. Brini and R. Cavalieri, *Open orbifold Gromov-Witten invariants of $[\mathbb{C}^3/\mathbb{Z}_n]$: localization and mirror symmetry*, Selecta Math. (N.S.) **17** (2011), no. 4, 879–933, available at 1007.0934. MR2861610 (2012i:14068)

[23] A. Brini, R. Cavalieri, and D. Ross, *Crepant resolutions and open strings*, J. Reine Angew. Math. in press (2013), https://doi.org/10.1515/crelle{2017{0011, available at 1309.4438.

[24] Andrea Brini, Bertrand Eynard, and Marcos Mariño, *Torus knots and mirror symmetry*, Ann. Henri Poincaré **13** (2012), no. 8, 1873–1910, DOI 10.1007/s00023-012-0171-2. MR2994763

[25] Andrea Brini, Luca Griguolo, Domenico Seminara, and Alessandro Tanzini, *Chern-Simons theory on $L(p,q)$ lens spaces and Gopakumar-Vafa duality*, J. Geom. Phys. **60** (2010), no. 3, 417–429, DOI 10.1016/j.geomphys.2009.11.006. MR2600004

[26] Andrea Brini, Marcos Mariño, and Sébastien Stevan, *The uses of the refined matrix model recursion*, J. Math. Phys. **52** (2011), no. 5, 052305, 24, DOI 10.1063/1.3587063. MR2839073

[27] Andrea Brini, *E_8 spectral curves* (2017), available at 1711.05958.

[28] Jim Bryan and Tom Graber, *The crepant resolution conjecture*, Algebraic geometry—Seattle 2005. Part 1, Proc. Sympos. Pure Math., vol. 80, Amer. Math. Soc., Providence, RI, 2009, pp. 23–42, DOI 10.1090/pspum/080.1/2483931. MR2483931

[29] Leonid Chekhov and Bertrand Eynard, *Hermitian matrix model free energy: Feynman graph technique for all genera*, J. High Energy Phys. **3** (2006), 014, 18, DOI 10.1088/1126-6708/2006/03/014. MR2222762

[30] Jinwon Choi, Sheldon Katz, and Albrecht Klemm, *The refined BPS index from stable pair invariants*, Comm. Math. Phys. **328** (2014), no. 3, 903–954, DOI 10.1007/s00220-014-1978-0. MR3201216

[31] Tom Coates, Hiroshi Iritani, and Hsian-Hua Tseng, *Wall-crossings in toric Gromov-Witten theory. I. Crepant examples*, Geom. Topol. **13** (2009), no. 5, 2675–2744, DOI 10.2140/gt.2009.13.2675. MR2529944

[32] D.-E. Diaconescu, V. Shende, and C. Vafa, *Large N duality, Lagrangian cycles, and algebraic knots*, Comm. Math. Phys. **319** (2013), no. 3, 813–863, DOI 10.1007/s00220-012-1563-3. MR3040377

[33] R. Dijkgraaf and C. Vafa, *Two-dimensional Kodaira–Spencer theory and three-dimensional Chern–Simons gravity* (2007), available at 0711.1932.

[34] Jacques Distler and Cumrun Vafa, *A critical matrix model at $c = 1$*, Modern Phys. Lett. A **6** (1991), no. 3, 259–270, DOI 10.1142/S0217732391000221. MR1087735

[35] S. K. Donaldson and R. P. Thomas, *Gauge theory in higher dimensions*, The geometric universe (Oxford, 1996), Oxford Univ. Press, Oxford, 1998, pp. 31–47. MR1634503

[36] P. Dunin-Barkowski, N. Orantin, S. Shadrin, and L. Spitz, *Identification of the Givental formula with the spectral curve topological recursion procedure*, Comm. Math. Phys. **328** (2014), no. 2, 669–700, DOI 10.1007/s00220-014-1887-2. MR3199996

[37] Bertrand Eynard, *Topological expansion for the 1-Hermitian matrix model correlation functions*, J. High Energy Phys. **11** (2004), 031, 35 pp. (2005), DOI 10.1088/1126-6708/2004/11/031. MR2118807

[38] B. Eynard, *Intersection numbers of spectral curves* (2011), available at `1104.0176`.

[39] B. Eynard, *Invariants of spectral curves and intersection theory of moduli spaces of complex curves*, Commun. Number Theory Phys. **8** (2014), no. 3, 541–588, DOI 10.4310/CNTP.2014.v8.n3.a4. MR3282995

[40] B. Eynard, T. Kimura, and S. Ribault, *Random matrices* (2015), available at `1510.04430`.

[41] B. Eynard and N. Orantin, *Invariants of algebraic curves and topological expansion*, Commun. Number Theory Phys. **1** (2007), no. 2, 347–452, DOI 10.4310/CNTP.2007.v1.n2.a4. MR2346575

[42] B. Eynard and N. Orantin, *Computation of open Gromov-Witten invariants for toric Calabi-Yau 3-folds by topological recursion, a proof of the BKMP conjecture*, Comm. Math. Phys. **337** (2015), no. 2, 483–567, DOI 10.1007/s00220-015-2361-5. MR3339157

[43] C. Faber and R. Pandharipande, *Hodge integrals and Gromov-Witten theory*, Invent. Math. **139** (2000), no. 1, 173–199, DOI 10.1007/s002229900028. MR1728879

[44] B. Fang, C-C. M. Liu, and Z. Zong, *On the Remodeling Conjecture for Toric Calabi-Yau 3-Orbifolds* (2016), available at `1604.07123`.

[45] Vladimir V. Fock and Andrey Marshakov, *Loop groups, clusters, dimers and integrable systems*, Geometry and quantization of moduli spaces, Adv. Courses Math. CRM Barcelona, Birkhäuser/Springer, Cham, 2016, pp. 1–66. MR3675462

[46] Debashis Ghoshal and Cumrun Vafa, *$c = 1$ string as the topological theory of the conifold*, Nuclear Phys. B **453** (1995), no. 1-2, 121–128, DOI 10.1016/0550-3213(95)00408-K. MR1358307

[47] Rajesh Gopakumar and Cumrun Vafa, *On the gauge theory/geometry correspondence*, Adv. Theor. Math. Phys. **3** (1999), no. 5, 1415–1443, DOI 10.4310/ATMP.1999.v3.n5.a5. MR1796682

[48] T. Graber and R. Pandharipande, *Localization of virtual classes*, Invent. Math. **135** (1999), no. 2, 487–518, DOI 10.1007/s002220050293. MR1666787

[49] David J. Gross and Igor Klebanov, *One-dimensional string theory on a circle*, Nuclear Phys. B **344** (1990), no. 3, 475–498, DOI 10.1016/0550-3213(90)90667-3. MR1076752

[50] Jie Gu, Hans Jockers, Albrecht Klemm, and Masoud Soroush, *Knot invariants from topological recursion on augmentation varieties*, Comm. Math. Phys. **336** (2015), no. 2, 987–1051, DOI 10.1007/s00220-014-2238-z. MR3322393

[51] Sergei Gukov, Albert Schwarz, and Cumrun Vafa, *Khovanov-Rozansky homology and topological strings*, Lett. Math. Phys. **74** (2005), no. 1, 53–74, DOI 10.1007/s11005-005-0008-8. MR2193547

[52] Nick Halmagyi, Takuya Okuda, and Vadim Yasnov, *Large-N duality, lens spaces and the Chern-Simons matrix model*, J. High Energy Phys. **4** (2004), 014, 10, DOI 10.1088/1126-6708/2004/04/014. MR2080896

[53] Nick Halmagyi and Vadim Yasnov, *The spectral curve of the lens space matrix model*, J. High Energy Phys. **11** (2009), 104, 21, DOI 10.1088/1126-6708/2009/11/104. MR2628861

[54] Søren Kold Hansen, *Reshetikhin-Turaev invariants of Seifert 3-manifolds and a rational surgery formula*, Algebr. Geom. Topol. **1** (2001), 627–686, DOI 10.2140/agt.2001.1.627. MR1875611

[55] J. Harer and D. Zagier, *The Euler characteristic of the moduli space of curves*, Invent. Math. **85** (1986), no. 3, 457–485. MR848681

[56] J. L. Harer, *The cohomology of the moduli space of curves*, Theory of moduli (Montecatini Terme, 1985), 1988, pp. 138–221. MR963064

[57] H. Hopf, *Zum Clifford-Kleinschen Raumproblem*, Math. Ann. **95** (1926), no. 1, 313–339. MR1512281

[58] K. Hori, A. Iqbal, and C. Vafa, *D-branes and mirror symmetry* (2000), available at `hep-th/0005247`.

[59] K. Hori and C. Vafa, *Mirror symmetry* (2000), available at `hep-th/0002222`.

[60] Amer Iqbal, Can Kozçaz, and Cumrun Vafa, *The refined topological vertex*, J. High Energy Phys. **10** (2009), 069, 58, DOI 10.1088/1126-6708/2009/10/069. MR2607441

[61] Johan Källén, *Cohomological localization of Chern-Simons theory*, J. High Energy Phys. **8** (2011), 008, 32, DOI 10.1007/JHEP08(2011)008. MR2876080
[62] Sheldon Katz, Albrecht Klemm, and Cumrun Vafa, *M-theory, topological strings and spinning black holes*, Adv. Theor. Math. Phys. **3** (1999), no. 5, 1445–1537, DOI 10.4310/ATMP.1999.v3.n5.a6. MR1796683
[63] Sheldon Katz and Chiu-Chu Melissa Liu, *Enumerative geometry of stable maps with Lagrangian boundary conditions and multiple covers of the disc*, The interaction of finite-type and Gromov-Witten invariants (BIRS 2003), Geom. Topol. Monogr., vol. 8, Geom. Topol. Publ., Coventry, 2006, pp. 1–47, DOI 10.2140/gtm.2006.8.1. MR2402818
[64] Maxim Kontsevich, *Enumeration of rational curves via torus actions*, The moduli space of curves (Texel Island, 1994), Progr. Math., vol. 129, Birkhäuser Boston, Boston, MA, 1995, pp. 335–368, DOI 10.1007/978-1-4612-4264-2_12. MR1363062
[65] J. M. F. Labastida and Marcos Mariño, *Polynomial invariants for torus knots and topological strings*, Comm. Math. Phys. **217** (2001), no. 2, 423–449, DOI 10.1007/s002200100374. MR1821231
[66] José M. F. Labastida, Marcos Mariño, and Cumrun Vafa, *Knots, links and branes at large N*, J. High Energy Phys. **11** (2000), Paper 7, 42, DOI 10.1088/1126-6708/2000/11/007. MR1806596
[67] Thang T. Q. Le, Jun Murakami, and Tomotada Ohtsuki, *On a universal perturbative invariant of 3-manifolds*, Topology **37** (1998), no. 3, 539–574, DOI 10.1016/S0040-9383(97)00035-9. MR1604883
[68] Kefeng Liu and Pan Peng, *Proof of the Labastida-Mariño-Ooguri-Vafa conjecture*, J. Differential Geom. **85** (2010), no. 3, 479–525. MR2739811
[69] I. G. Macdonald, *The volume of a compact Lie group*, Invent. Math. **56** (1980), no. 2, 93–95, DOI 10.1007/BF01392542. MR558859
[70] Juan Maldacena, *The large N limit of superconformal field theories and supergravity*, Adv. Theor. Math. Phys. **2** (1998), no. 2, 231–252, DOI 10.4310/ATMP.1998.v2.n2.a1. MR1633016
[71] Marcos Mariño, *Chern-Simons theory, matrix integrals, and perturbative three-manifold invariants*, Comm. Math. Phys. **253** (2005), no. 1, 25–49, DOI 10.1007/s00220-004-1194-4. MR2105636
[72] Marcos Mariño, *Chern-Simons theory, matrix models and topological strings*, International Series of Monographs on Physics, vol. 131, The Clarendon Press, Oxford University Press, Oxford, 2005. MR2177747
[73] Marcos Mariño, *Open string amplitudes and large order behavior in topological string theory*, J. High Energy Phys. **3** (2008), 060, 34, DOI 10.1088/1126-6708/2008/03/060. MR2391060
[74] Marcos Mariño, *String theory and the Kauffman polynomial*, Comm. Math. Phys. **298** (2010), no. 3, 613–643, DOI 10.1007/s00220-010-1088-6. MR2670922
[75] M. Mariño and C. Vafa, *Framed knots at large N*, Contemp. Math. **310** (2002), 185–204, available at `hep-th/0108064`. MR1950947 (2004c:57024)
[76] E. J. Martinec and N. P. Warner, *Integrable systems and supersymmetric gauge theory*, Nucl. Phys. **B459** (1996), 97–112, available at `hep-th/9509161`.
[77] D. Maulik, A. Oblomkov, A. Okounkov, and R. Pandharipande, *Gromov-Witten/Donaldson-Thomas correspondence for toric 3-folds*, Invent. Math. **186** (2011), no. 2, 435–479, DOI 10.1007/s00222-011-0322-y. MR2845622
[78] A. McDaniel and L. Smolinsky, *A Lie-theoretic Galois theory for the spectral curves of an integrable system. II*, Trans. Amer. Math. Soc. **349** (1997), no. 2, 713–746. MR1401779
[79] A. A. Migdal, *Loop Equations and 1/N Expansion*, Phys. Rept. **102** (1983), 199–290.
[80] Nikita Nekrasov, *Five-dimensional gauge theories and relativistic integrable systems*, Nuclear Phys. B **531** (1998), no. 1-3, 323–344, DOI 10.1016/S0550-3213(98)00436-2. MR1657857
[81] H. Ooguri and C. Vafa, *Knot invariants and topological strings*, Nucl. Phys. **B577** (2000), 419–438, available at `hep-th/9912123`.
[82] Hirosi Ooguri and Cumrun Vafa, *Worldsheet derivation of a large N duality*, Nuclear Phys. B **641** (2002), no. 1-2, 3–34, DOI 10.1016/S0550-3213(02)00620-X. MR1928179
[83] R. C. Penner, *Perturbative series and the moduli space of Riemann surfaces*, J. Differential Geom. **27** (1988), no. 1, 35–53. MR918455
[84] Vipul Periwal, *Topological closed-string interpretation of Chern-Simons theory*, Phys. Rev. Lett. **71** (1993), no. 9, 1295–1298, DOI 10.1103/PhysRevLett.71.1295. MR1232813

[85] N. Reshetikhin and V. G. Turaev, *Invariants of 3-manifolds via link polynomials and quantum groups*, Invent. Math. **103** (1991), no. 3, 547–597, DOI 10.1007/BF01239527. MR1091619

[86] N. Yu. Reshetikhin and V. G. Turaev, *Ribbon graphs and their invariants derived from quantum groups*, Comm. Math. Phys. **127** (1990), no. 1, 1–26. MR1036112

[87] Yongbin Ruan, *The cohomology ring of crepant resolutions of orbifolds*, Gromov-Witten theory of spin curves and orbifolds, Contemp. Math., vol. 403, Amer. Math. Soc., Providence, RI, 2006, pp. 117–126, DOI 10.1090/conm/403/07597. MR2234886

[88] N. Seiberg and E. Witten, *Erratum: "Electric-magnetic duality, monopole condensation, and confinement in $N = 2$ supersymmetric Yang-Mills theory"*, Nuclear Phys. B **430** (1994), no. 2, 485–486, DOI 10.1016/0550-3213(94)00449-8. MR1303306

[89] S. Sinha and C. Vafa, *SO and Sp Chern–Simons at large N* (2000), available at `hep-th/0012136`.

[90] Andrew Strominger, *Special geometry*, Comm. Math. Phys. **133** (1990), no. 1, 163–180. MR1071240

[91] G. 't Hooft, *A Planar Diagram Theory for Strong Interactions*, Nucl. Phys. **B72** (1974), 461.

[92] Cumrun Vafa, *Superstrings and topological strings at large N*, J. Math. Phys. **42** (2001), no. 7, 2798–2817, DOI 10.1063/1.1376161. Strings, branes, and M-theory. MR1840317

[93] Harold Williams, *Double Bruhat cells in Kac-Moody groups and integrable systems*, Lett. Math. Phys. **103** (2013), no. 4, 389–419, DOI 10.1007/s11005-012-0604-3. MR3029326

[94] Edward Witten, *Quantum field theory and the Jones polynomial*, Comm. Math. Phys. **121** (1989), no. 3, 351–399. MR990772

[95] E. Witten, *Chern-Simons gauge theory as a string theory*, The Floer memorial volume, Progr. Math., vol. 133, Birkhäuser, Basel, 1995, pp. 637–678. MR1362846

Department of Mathematics, Imperial College London, 180 Queen's Gate, SW7 2AZ, London, United Kingdom

On leave from IMAG, Univ. Montpellier, CNRS, Montpellier, France
Email address: `a.brini@imperial.ac.uk`

Proceedings of Symposia in Pure Mathematics
Volume **100**, 2018
https://doi.org/10.1090/pspum/100/01771

Bouchard-Klemm-Marino-Pasquetti conjecture for $\mathbb{C}^3$

Lin Chen

Lin Chen joined the Simons Center for Geometry and Physics as a Research Assistant Professor in September 2009 after receiving his Ph.D. in Mathematics from UCLA under the supervision of Kefeng Liu. This paper was submitted to arXiv in October 2009. It is the last paper Lin Chen completed before his tragic death in December 2009 at the age of 28.

ABSTRACT. In this paper, we give a proof of the Bouchard-Klemm-Marino-Pasquetti conjecture for a framed vertex, by using the symmetrized Cut-Join Equation developed in a previous paper.

1. Introduction

In their seminal paper [**2**], Bouchard, Klemm, Marino and Pasquetti propose a new approach to compute both the open and closed Gromov-Witten invariants of local Calabi-Yau manifolds, including the mirrors of toric varieties. The approach is based on the matrix models theory of Eynard and Orantin. To each toric Calabi-Yau three-fold, there is an algebraic curve Σ, living inside $\mathbb{C}^* \times \mathbb{C}^*$, associated to the toric diagram. The curve Σ is called the framed mirror curve, with genus equal to the genus of the toric diagram. Bouchard, Klemm, Marino and Pasquetti conjecture that the Gromov-Witten invariants of the toric Calabi-Yau three folds can be computed by applying the recursion of Eynard and Orantin to the framed mirror curve.

The first example of Bouchard-Klemm-Marino-Pasquetti theory is the non-compact toric three fold $\mathbb{C}^3$, the socalled framed vertex, which is the building block of three dimensional toric varieties. In this case, the framed mirror curve is the algebraic curve $x = y^f(1-y)$. The theory of Eynard and Orantin produce a topological recursion relations, when plug in the information of the framed mirror curve. On the other hand, there is an existing topological recursion for $\mathbb{C}^3$, i.e, the cut-join equation proved in [**8**] and its symmetrized version [**3**]. In this paper, we will prove that the Symmetrized Cut-Join Equation obtained in [**3**] implies the topological recursion of Eynard and Orantin, thus prove the Bouchard-Klemm-Marino-Pasquetti conjecture in the $\mathbb{C}^3$ case.

2010 *Mathematics Subject Classification.* Primary 14N35; Secondary 14H10.

Our strategy is strongly motivated by the recent work of Eynard-Mulase-Safnuk on the Bouchard-Marino conjecure. The Bouchard-Marino conjecture on Hurwitz numbers is a topological recursion of Eynard-Orantin type. This conjecture was first proved by Borot, Eynard, Mulase and Safnuk, using very deep results in matrix model theory. Later, Eynard, Mulase and Safnuk give another proof by comparing the conjectural recursion with the symmetrized cut-join equation of Hurwitz numbers, which was discovered by Goulden-Jackson-Vainshtein [**5**]. We now describe their second approach. First, writing the residues in the conjectural recursion as a contour integral, and a residue theorem calculus switch the calculation to the two nearby simple poles. This computation provide an equivalent form of the conjectural formula, which is an identity of polynomials. Then, by pushing forward the Symmetrized Cut-Join Equation of Goulden-Jackson-Vainstein via the projection $\pi : \Sigma' \to \mathbb{C}$ from the mirror curve Σ' associated to the Hurwitz numbers, they are able to show the resulting equation, modulo the principle (singular) part, is precisely the equation obtained in the first step.

In this paper, we will give a proof of the BKMP conjecture for the framed vertex $\mathbb{C}^3$. In fact, the Bouchard-Marino conjecture is a specialization of the BKMP conjecture for $\mathbb{C}^3$, by letting the framing $f \to \infty$. The Hurwitz numbers are then replaced by the Gromov-Witten invariants, which can be written as Hodge integrals involves three λ classes. The generating series of such Hodge integrals satisfies a similar cut-join equation. The corresponding Symmetrized Cut-Join Equation is obtained by the author in a previous paper [**3**]. Following the line of Eynard-Mulase-Safnuk, we prove the BKMP conjecture by switching the residues calculation to the two nearby simple poles in the question, and pushing forward the Symmetrized Cut-Join Equation of [**3**] via the projection from the framed mirror curve.

We will describe some known results in Section 2. The BKMP conjecture for $\mathbb{C}^3$ is stated in Section 3. In Section 4, we compute the residues appeared in the conjectural recursion. Replacing the complex analysis of the functions $\eta_n(v)$, invent by Eynard-Mulase-Safnuk, by formal power series argument in Section 5, we simplify the proof the technical Lemma 5.1, avoid the convergence and analytic continuation difficulty in [**4**]. Finally, we compute the push forward of the Symmetrized Cut-Join Equation in the last two sections, and establish the BKMP conjecture.

I would like to thank Professor Kefeng Liu and Bailin Song for helpful discussion. I also want to thank Professor Mulase for explaining their work to me. This work is supported by Simons Center for Geometry and Physics, Stony Brook University.

2. Marino-Vafa formula and Symmetrized Cut-Join Equation

The celebrated topological vertex theory developed by Li-Liu-Liu-Zhou [**7**] establishes a correspondence between two different types of physics theory, topological string theory and the Chern-Simons theory. More precisely, it gives a closed formula of Gromov-Witten invariants of certain toric Calabi-Yau three folds in terms of link invariants (see [**8**] [**9**] [**7**]).

The simplest example is the so-called Marino-Vafa formula proved by Liu-Liu-Zhou [**8**]. Via virtual localization, one side of the Marino-Vafa formula can be written as a generating series of Hodge integrals involves three lambda classes

$\mathcal{C} = \sum_{g\geqslant 0, n\geqslant 1} \mathcal{C}_n^g \lambda^{2g-2+n}$, where

$$\begin{aligned}
\mathcal{C}_n^g =& \sum_{d\geqslant 1} \sum_{\mu\vdash d, l(\mu)=n} \\
& - \frac{\sqrt{-1}^{d+n}}{|\mathrm{Aut}\mu|}(f(1+f))^{n-1}\prod_{i=1}^{n}\frac{\prod_{a=1}^{\mu_i-1}(\mu_i f+a)}{(\mu_i-1)!}\int_{\overline{\mathcal{M}}_{g,n}}\frac{\Gamma_g(f)}{\prod_{i=1}^{n}(1-\mu_i\psi_i)}\cdot \mathbf{p}_\mu \\
=& -\frac{\sqrt{-1}^n(f(1+f))^{n-1}}{n!}\sum_{\mu_1,\mu_2\cdots\mu_n\geqslant 1}^{+\infty}\sqrt{-1}^{|\mu|}\prod_{i=1}^{n}\frac{\prod_{a=1}^{\mu_i-1}(\mu_i f+a)}{(\mu_i-1)!} \\
& \cdot\sum_{b_1,\cdots,b_n}\int_{\overline{\mathcal{M}}_{g,n}}\Gamma_g(f)\prod_{i=1}^{n}\psi_i^{b_i}\prod_{i=1}^{n}\mu_i^{b_i}\cdot\mathbf{p}_\mu \\
=& -\frac{(f(1+f))^{n-1}}{n!}\sum_{b_1,\cdots,b_n} < \tau_{b_1}\cdots\tau_{b_n}\Gamma_g(f) > \prod_{i=1}^{n}\varphi_{b_i}(\overrightarrow{\mathbf{p}}).
\end{aligned}$$

The formula proved in [**8**] relates the generating series $\mathcal{C}$ to a truncated version of the framing dependent Chern-Simons partition function (generating series of colored HOMFLY polynomials) of unknot. In the above formula, we use the notation

$$\varphi_i(\overrightarrow{\mathbf{p}}) = \sum_{m\geqslant 1}\sqrt{-1}^{m+1}\mathbf{p}_m\frac{\prod_{a=1}^{m-1}(mf+a)}{(m-1)!}m^i = \frac{1}{f}\sum_{m\geqslant 1}\sqrt{-1}^{m+1}\mathbf{p}_m\frac{\prod_{a=0}^{m-1}(mf+a)}{m!}m^i$$

for a formal sum involve infinitely many formal variables $\overrightarrow{\mathbf{p}} = \{\mathbf{p}_1, \mathbf{p}_2, \cdots\}$, and the class

$$\Gamma_g(f) = \Lambda_g^\vee(1)\Lambda_g^\vee(f)\Lambda_g^\vee(-f-1)$$

for

$$\Lambda_g^\vee(u) = u^g - \lambda_1 u^{g-1} + \cdots + (-1)^g\lambda_g.$$

The relation between these Hodge integrals and the Gromov-Witten invariants of Calabi-Yau three-folds can be found in [**6**].

The generating function $\mathcal{C}$ satisfy a Cut-Join Equation:

$$\text{(2.1)}\quad \frac{\partial\mathcal{C}}{\partial f} = \frac{\sqrt{-1}\lambda}{2}\sum_{i,j\geqslant 1}(ij\mathbf{p}_{i+j}\frac{\partial^2\mathcal{C}}{\partial\mathbf{p_i}\partial\mathbf{p_j}} + ij\mathbf{p}_{i+j}\frac{\partial\mathcal{C}}{\partial\mathbf{p}_i}\frac{\partial\mathcal{C}}{\partial\mathbf{p}_j} + (i+j)\mathbf{p}_i\mathbf{p}_j\frac{\partial\mathcal{C}}{\partial\mathbf{p}_{i+j}})$$

In [**3**], the author introduced symmetrization operators

$$\Xi_n : \mathbb{C}[\mathbf{p}_1, \mathbf{p}_2, \cdots] \to \mathbb{C}[x_1, \cdots, x_n]$$

with values in the subring of $\mathbb{C}[x_1, \cdots, x_n]$ consists of symmetric polynomials, by letting

$$\Xi_n\mathbf{p}_\alpha = (\sqrt{-1})^{-n-|\alpha|}\sum_{\sigma\in S_n} x_{\sigma(1)}^{\alpha_1}\cdots x_{\sigma(n)}^{\alpha_n}$$

for $n \geqslant 1$ if $l(\alpha) = n$ with $\alpha = (\alpha_1, \cdots, \alpha_n)$, and 0 otherwise.

After a transcendental change of variables

$$y(x)^f(1-y(x)) = x$$

$$t = \frac{1}{(1+f)y-f},$$

or explicitly

$$\begin{aligned}t &= 1+(\frac{1+f}{f})((1+f)x\frac{d}{dx}(1-y)-(1-y))\\ &= 1+(\frac{1+f}{f})(\frac{f(1-y)}{(1+f)y-f})\\ &= 1+(\frac{1+f}{f})\sum_{n=1}^{+\infty}\frac{\prod_{a=0}^{n-1}(nf+a)}{n!}x^n,\end{aligned}$$

the generating series become

$$H_{g,n}=-(f(1+f))^{n-1}\sum_{b_1,\cdots,b_n}<\tau_{b_1}\cdots\tau_{b_n}\Gamma_g(f)>_g\prod_{i=1}^{n}\phi_{b_i}(t_i).$$

In the above expression,

$$\phi_b(t)=\left(\frac{t(t-1)(ft+1)}{f+1}\frac{d}{dt}\right)^b\left(\frac{t-1}{f+1}\right)$$

for $b\geq 0$ is a polynomial in t of degree $2b+1$, with coefficients in $\mathbb{Q}(f)$. For convenience, we denote

$$\phi_{-1}(t)=-\log(1+\frac{1}{ft})$$

for later use. It is easy to check that

$$\phi_0(t)=\frac{t-1}{f+1}=\left(\frac{t(t-1)(ft+1)}{f+1}\frac{d}{dt}\right)\phi_{-1}(t)$$

compatible with our notations.

Apply the symmetrization operator Ξ_n and change the variables from x_i into t_i, Equation 2.1 transformed into the following Symmetrized Cut-Join Equation proved in [**3**]

$$(\frac{\partial}{\partial f}+\sum_{l=1}^{n}\frac{t_l(t_l-1)}{f+1}\cdot\frac{\partial}{\partial t_l})H_n^g(t_1,\cdots,t_n,f)=T_1+T_2+T_3+T_4,$$

where

$$\begin{aligned}T_1 =& -\frac{1}{2}\sum_{l=1}^{n}t_l(t_l-1)(\frac{t_lf+1}{f+1})\frac{\partial}{\partial t_l}\cdot t_{n+1}(t_{n+1}-1)(\frac{t_{n+1}f+1}{f+1})\frac{\partial}{\partial t_{n+1}}H_{n+1}^{g-1}|_{t_{n+1}=t_l}\\ T_2 =& -\frac{1}{2}\sum_{1\leqslant a\leqslant g-1}\sum_{1\leqslant k\leqslant n}\Theta_{k-1}(t_1(t_1-1)(\frac{t_1f+1}{f+1})\frac{\partial}{\partial t_1}H_k^a(t_1,\cdots,t_k,f))\\ &\cdot(t_1(t_1-1)(\frac{t_1f+1}{f+1})\frac{\partial}{\partial t_1}H_{n-k+1}^{g-a}(t_1,t_{k+1},\cdots,t_n,f))\\ T_3 =& -\sum_{k=3}^{n}\Theta_{k-1}(t_1(t_1-1)(\frac{t_1f+1}{f+1})\frac{\partial}{\partial t_1}H_k^0(t_1,\cdots,t_k,f))(t_1(t_1-1)\\ &\cdot(\frac{t_1f+1}{f+1})\frac{\partial}{\partial t_1}H_{n-k+1}^g(t_1,t_{k+1},\cdots,t_n,f))\\ T_4 =&\Theta_1\frac{t_1^2(t_1-1)(t_2-1)}{t_1-t_2}(\frac{t_1f+1}{f+1})^2\frac{\partial}{\partial t_1}H_{n-1}^g(t_1,t_3,\cdots,t_n,f).\end{aligned}$$

We remark that this is an equality in the polynomial ring $\mathbb{C}(f)[t_1,\cdots,t_n]$. The symbol Θ_k means take the symmetric sum. For more detail about this Symmetrized Cut-Join Equation, we refer to [**3**].

3. The BKMP Conjecture

The Symmetrized Cut-Join Equation is a set of topological recursion of Hodge integrals, i.e, it computes genus g Hodge integrals in terms of integrals with less genus or less marked points.

The BKMP conjecture is another set of topological recursion of Hodge integrals, coming from the conjectural duality between topological string theory and matrix model theory. We now describe the recursions in the BKMP conjecture for $\mathbb{C}^3$. In this case, the Framed Mirror Curve C is given by

$$x = y^f(1-y),$$

with the formal power series inverse function

$$y(x) = 1 - \sum_{n=1}^{\infty} \frac{\prod_{j=0}^{n-2}(nf+j)}{n!} x^n.$$

Denote ζ_n (depends on f) the differential one form

$$\begin{aligned}\zeta_n = d(\phi_n) &= d\left((x\frac{d}{dx})(\frac{t-1}{f+1})\right)\\ &= d\left((\frac{t(t-1)(ft+1)}{f+1})^n(\frac{t-1}{f+1})\right)\\ &= d(\frac{1}{\tau}\sum_{m=1}^{\infty}\frac{\prod_{a=0}^{m-1}(mf+a)}{m!}m^n x^m).\end{aligned}$$

Let $W_g(x_1,\cdots,x_n)$ be the differential n-form defined by the formula

$$\begin{aligned}&W_g(x_1,\cdots,x_n)\\ &\quad = (-1)^{g+n}(f(f+1))^{n-1}\sum_{b_1,\cdots,b_n} < \tau_{b_1}\cdots\tau_{b_n}\Gamma_g(f) > \prod_{i=1}^{n}\zeta_{b_i}(y_i(x_i),f).\end{aligned}$$

For example

$$\begin{aligned}W_0(x) &= \log y(x)\cdot\frac{dx}{x}\\ W_0(x_1,x_2) &= B(y_1,y_2) - \frac{dx_1dx_2}{(x_1-x_2)^2} = \frac{dy_1dy_2}{(y_1-y_2)^2} - \frac{dx_1dx_2}{(x_1-x_2)^2}.\\ W_0(x_1,x_2,x_3) &= -\frac{f^2}{(f+1)}dt_1dt_2dt_3\end{aligned}$$

If we denote $d_i = \frac{\partial}{\partial x_i}dx_i = \frac{\partial}{\partial t_i}dt_i$ the differential with respect to the i-th variable, then the differential n-form

$$W_g(x_1,\cdots,x_n) = (-1)^{g+n-1}d_1\cdots d_n H_g^n$$

becomes an element in the canonical module $\mathbb{C}(f)[t_1,\cdots,t_n]dt_1\cdots dt_n$, if switch to the t_i variables.

The Framed Mirror Curve C has a critical point $(x, y) = (\frac{f^f}{(f+1)^{f+1}}, \frac{f}{f+1})$. Near the critical point, the morphism

$$C \to \mathbb{C}$$

sending (x, y) to x is (locally) a branching double cover. Let q and $\overline{q}$ be two points on the Framed Mirror Curve close to the critical point such that $x(q) = x(\overline{q})$. Define one forms

$$\omega(q) = (\log y(q) - \log y(\overline{q}))\frac{dx(q)}{x(q)}$$

$$dE(q, \overline{q}, y_2) = \frac{dy_2}{2}\left(\frac{1}{y_2 - y(q)} - \frac{1}{y_2 - y(\overline{q})}\right).$$

The kernel function is defined to be the formal quotient

$$K(q, \overline{q}, y_2) = \frac{dE(q, \overline{q}, y_2)}{\omega(q)}.$$

The BKMP conjectural recursion reads:

CONJECTURE 1 (BKMP). *The differential forms $W_g(y, y_1, \cdots, y_n)$ are completely determined by the following topological recursion relation*

$$W_g(y, y_1, \cdots, y_n) = \operatorname{Res}_{y(q)=\frac{f}{f+1}} \frac{dE(q, \overline{q}, y)}{w(q)}[W_{g-1}(y(q), y(\overline{q}), y_1, \cdots, y_n)$$
$$+ \sum_{g_1+g_2=g} \sum_{I \coprod J=H}^{\text{stable}} \sum_{J \subset H} W_{g_1}(y(q), y_I)W_{g_2}(y(\overline{q}), y_J)$$
$$+ \sum_{i=1}^{n}(W_g(y(q), y_{H\backslash\{i\}}) \otimes B(y(\overline{q}), y_i) + B(y(q), y_i) \otimes W_g(y(\overline{q}), y_{H\backslash\{i\}}))],$$

together with the initial conditions

$$W_0(x_1, x_2, x_3) = -\frac{f^2}{(f+1)}dt_1dt_2dt_3$$

$$W_1(x_1) = \frac{1}{24}((1 + f + f^2)\zeta_0 - f(1 + f)\zeta_1).$$

In the rest of this paper, we will prove that the Symmetrized Cut-Join Equation implies the above BKMP conjecture.

4. Residue Calculus

The RHS of the BKMP conjecture involves the following two types of residues. Type I: Comes from W_{g-1} and the stable sums.

$$R_{a,b}(y) = \operatorname{Res}_{y(q)=\frac{f}{f+1}} \frac{dE(q, \overline{q}, y)}{\omega(q)}\zeta_a(y(q))\zeta_b(y(\overline{q})).$$

Type II: Comes from the unstable contribution of integral over the point $\overline{M}_{0,2}$.

$$R_a(y, y_i) = \operatorname{Res}_{y(q)=\frac{f}{f+1}} \frac{dE(q, \overline{q}, y)}{\omega(q)}\left(\zeta_a(y(q))B(y(\overline{q}), y_i) + \zeta_a(y(\overline{q}))B(y(q), y_i)\right).$$

Before going into the computations, we first explain the meaning of the above two residues. Take a small open neighborhood U of $\frac{f^f}{(f+1)^{f+1}}$ in $\mathbb{C}$, such that the projection

$$\pi : C \to \mathbb{C}$$

is two to one on the open set $\pi^{-1}(U \setminus \{\frac{f^f}{(f+1)^{f+1}}\})$. Denote by $s : q \mapsto \overline{q}$ the holomorphic inversion, which is well defined on the open neighborhood $\pi^{-1}U$ of the critical point.

In the formula of Type I case,

$$\zeta_a(y(q)) = \phi'_a(y(q))dx(q)$$
$$\zeta_b(y(\overline{q})) = \phi'_b(y(\overline{q}))s'(x(q))dx(q)$$

where $\phi'_n = \frac{d}{dx}\phi_n$ is the derivative of ϕ_n. The denominator $w(q)$ of the formal quotient $K(q, \overline{q}, y_2)$ canceled one of the $dx(q)$, thus the product $K(q, \overline{q}, y)\zeta_a(y(q))\zeta_b(y(\overline{q}))$ is a two form. After taking the residue, $R_{a,b}(y)$ became a meromorphic one form. Similar description holds for the two form $R_a(y, y_i)$.

LEMMA 4.1. *Change into the variable t, i.e, let $R_{a,b}(y) = P_{a,b}(t)dt$, then $P_{a,b}(t) \in \mathbb{C}(f)[t]$ is a polynomial of t with coefficients in the field $\mathbb{C}(f)$.*

PROOF. Let $y = \frac{f+\frac{1}{t}}{f+1}$ and let $z = \frac{1}{t}$. The critical point of the Framed Mirror Curve is then the point 0 in the z coordinate. Formally, the kernel function $K(q, \overline{q}, y)$ is the product of $\frac{(f+1)dy}{2} \otimes \frac{x(q)}{dx(q)}$ and the function

$$\frac{(\frac{1}{z-z(q)} - \frac{1}{z-z(\overline{q})})}{\log(1+\frac{z(q)}{f}) - \log(1+\frac{z(\overline{q})}{f})}$$
$$= \frac{1}{(z-z(q))(z-z(\overline{q}))} \cdot [\frac{z(q)-z(\overline{q})}{\log(1+\frac{z(q)}{f}) - \log(1+\frac{z(\overline{q})}{f})}]$$

holomorphic around $z(q) = 0$.

We abuse the notation

$$\zeta_a(y(q)) = \zeta_a(t(q)) = \zeta_a(\frac{1}{z(q)}),$$

which is in the module of differentials $\mathbb{C}(f)[t(q)]dt(q)$.

Recall that

$$\frac{x(q)}{dx(q)} = \frac{t(q)(t(q)-1)(ft(q)+1)}{(f+1)dt(q)} = -\frac{(1-z)(f+z)}{z(f+1)dz}.$$

By linearity, we only need to consider the residue

$$\mathrm{Res}_{z(q)=0}\frac{F(z(q))}{(z-z(q))(z-z(\overline{q}))}\cdot\frac{1}{z(q)^a}\frac{dz(q)}{z(q)^5}\otimes dy$$

for integer $a \geq 0$ and functions $F(z)$ holomorphic around $z = 0$. Since $dy = -\frac{dt}{t^2(f+1)}$, we have

$$-\frac{dt}{f+1}\cdot\mathrm{Res}_{z(q)=0}\left(F(z(q))\cdot\sum_{k=0}^{+\infty}\frac{1}{z^k}(\sum_{i=0}^{k}z(q)^i z(\overline{q})^{k-i})\frac{dz(q)}{z(q)^{5+a}}\right)$$

by expanding the denominator. This expression only contains non-negative power of $\frac{1}{z} = t$. Moreover, if $k \geq 5+a$, then the summand is then holomorphic at $z(q) = 0$, and has no residue. This shows that the above residue is a polynomial times dt. A more careful analysis of the parameter f in the above process shows that the coefficients are in fact rational functions in the variable f. □

Under the z coordinate, the projection $\pi : C \to \mathbb{C}$ maps the critical point to $z = 0$. By shrinking the open neighborhood U suitably, we can find a simple closed curve around $z = 0$. The holomorphic inversion s is thus defined in the interior of the area bounded by γ. Let the variable y sufficiently close to the critical point such that it is inside the curve γ.

Let $\zeta_a(y) = F_a(t)dt$ for some polynomial $F_a(t) \in \mathbb{C}(f)[t]$. By compactness, the functions $F_a(y(q))$, $F_b(y(\overline{q}))$, $\frac{1}{\log(y(q))-\log(y(\overline{q}))}$ and $t(q)(t(q)-1)(ft(q)+1)$ are all bounded along the circle γ. We thus have the following estimate of the contour integral:

$$\begin{aligned}
&\frac{1}{2\pi i}\int_\gamma\frac{dE(q,\overline{q},y)}{\omega(q)}\zeta_a(y(q))\zeta_b(y(\overline{q}))\\
&=\frac{1}{2\pi i}\int_\gamma\frac{\frac{1}{2}(\frac{1}{y-y(q)}-\frac{1}{y-y(\overline{q})})}{\log y(q)-\log y(\overline{q})}\cdot(\frac{x(q)}{dx(q)}\otimes dy)\cdot F_a(t(q))dt(q)\cdot F_b(t(\overline{q}))dt(\overline{q})\\
&=\frac{dy}{2\pi i}\int_\gamma\frac{\frac{1}{2}(\frac{1}{y-y(q)}-\frac{1}{y-y(\overline{q})})}{\log y(q)-\log y(\overline{q})}\cdot\frac{t(q)(t(q)-1)(ft(q)+1)}{(f+1)}\\
&\quad\cdot F_a(t(q))\cdot F_b(t(\overline{q}))s'(t(q))dt(q)\\
&\leq M\cdot\frac{dt}{|t^2|}\sim\frac{dt}{|t^2|}.
\end{aligned}$$

By hypothesis, we choose z small enough (sufficiently close to the critical point), such that it is in the interior bounded by γ. Then we choose another simple closed curve γ_∞ around $z = 0$ (i.e, $t = \infty$), such that both z and $s(z)$ are outside γ_∞.

The above estimate together with Lemma 4.1 imply that the above integral is the principle part of the following integral

$$A := \frac{1}{2\pi i}\int_{\gamma\cup-\gamma_\infty}\frac{dE(q,\overline{q},y)}{\omega(q)}\zeta_a(y(q))\zeta_b(y(\overline{q})).$$

There are two poles $z(q) = z$ and $z(q) = s(z)$ between the two circle γ and γ_∞. The integral A thus can be computed by residue theorem:

$$\begin{aligned}
A = &\operatorname{Res}_{z(q)=z,s(z)} \frac{\frac{f+1}{2}(\frac{1}{z-z(q)} - \frac{1}{z-z(\overline{q})})}{\log(1+\frac{z(q)}{f}) - \log(1+\frac{z(\overline{q})}{f})} dy \zeta_a(\frac{1}{z(q)})\zeta_b(\frac{1}{z(\overline{q})}) \cdot \frac{x(q)}{dx(q)} \\
&- \operatorname{Res}_{z(q)=z,s(z)} \frac{\frac{z^2 dt}{2}(\frac{1}{z-z(q)} - \frac{1}{z-z(\overline{q})})}{\log(1+\frac{z(q)}{f}) - \log(1+\frac{z(\overline{q})}{f})} \cdot F_a(\frac{1}{z(q)})F_b(\frac{1}{z(\overline{q})})s'(\frac{1}{z(q)})\frac{dz(q)}{z(q)^2} \\
&\cdot \frac{(1-z(q))(z(q)+f)}{(f+1)z(q)^3} \\
= &\frac{-\frac{1}{2}dt}{\log(1+\frac{1}{ft}) - \log(1+\frac{1}{fs(t)})} F_a(t)F_b(s(t))s'(t) \cdot \frac{t(t-1)(ft+1)}{(f+1)} \\
&+ \frac{-\frac{1}{2}dt}{\log(1+\frac{1}{ft}) - \log(1+\frac{1}{fs(t)})} F_a(s(t))F_b(t) \cdot \frac{s(t)(s(t)-1)(fs(t)+1)}{(f+1)} \\
= &\frac{-\frac{1}{2}}{\log(1+\frac{1}{ft}) - \log(1+\frac{1}{fs(t)})} (\phi_{a+1}(t)d\phi_b(s(t)) + \phi_{a+1}(s(t))d\phi_b(t)) \\
= &\frac{-\frac{1}{2}}{\log(1+\frac{1}{ft}) - \log(1+\frac{1}{fs(t)})} (\phi_{a+1}(t)\phi_{b+1}(s(t)) + \phi_{a+1}(s(t))\phi_{b+1}(t)) \\
&\cdot \frac{(f+1)dt}{t(t-1)(ft+1)}.
\end{aligned}$$

Let $f(t) = \sum_{n\geq -N}^\infty a_n t^n \in \mathbb{C}(f)((t))$ be a Laurent series. Denote by $f(t)_+ = \sum_{n\geq 0}^\infty a_n t^n$ the regular part of $f(t)$. The above formula of A and the estimate of the contour integral along γ give the following proposition.

PROPOSITION 4.2.

$$P_{a,b}(t) = \left(\frac{\frac{1}{2}}{\log(1+\frac{1}{ft}) - \log(1+\frac{1}{fs(t)})} (\phi_{a+1}(t)\phi_{b+1}(s(t)) + \phi_{a+1}(s(t))\phi_{b+1}(t)) \cdot \frac{(f+1)}{t(t-1)(ft+1)} \right)_+$$

Next we consider the Type II residues. A similar argument as in the Type I case leads to the following lemma, the proof of which we left for the reader.

LEMMA 4.3. *Change to the t and t_i variable, $R_a(y, y_i) = P_a(t, t_i)dtdt_i$, then $P_a(t, t_i) \in \mathbb{C}(f)[t, t_i]$ is a polynomial of two variables t and t_i with coefficients in the field $\mathbb{C}(f)$.*

Let the variable t_i take values in the area such that its inverse z_i and $s(z_i)$ are outside the area bounded by γ. Let us now consider the following integral

$$B := \frac{1}{2\pi i}\int_{\gamma\bigsqcup -\gamma_\infty} \frac{dE(q,\overline{q},y)}{w(q)}(\zeta_z(y(q))B(y(\overline{q}),y_i) + \zeta_a(y(\overline{q}))B(y(q),y_i)).$$

We expect the integral along γ is the principle part, while the γ_∞ part gives the residue we compute.

By our hypothesis, there are two simple poles in the area bounded by the contour, so by residue theorem:

$$
\begin{aligned}
B =& \frac{1}{2\pi i}\int_{\gamma-\gamma_\infty} \frac{dE(q,\overline{q},y)}{\omega(q)}\left(\zeta_n(y(q))B(y(\overline{q}),y_i)+\zeta_n(y(\overline{q}))B(y(q),y_i)\right)\\
=& \mathrm{Res}_{z(q)=t^{-1},s(t)^{-1}} \frac{\frac{f+1}{2}\left(\frac{1}{z-z(q)}-\frac{1}{z-z(\overline{q})}\right)}{\log(1+\frac{z(q)}{f})-\log(1+\frac{z(\overline{q})}{f})}dydy_i\cdot\frac{t(q)(t(q)-1)(ft(q)+1)}{(f+1)}\\
&\cdot\left(F_n(t(q))\frac{dy(\overline{q})}{(y(\overline{q})-y_i)^2}+F_n(t(\overline{q}))\frac{s'(t(q))dy(q)}{(y(q)-y_i)^2}\right)\\
=& \frac{-s'(t)dtdt_i}{2(\log(1+\frac{1}{ft})-\log(1+\frac{1}{fs(t)}))}\cdot\frac{t(t-1)(ft+1)}{(f+1)}\cdot\left(\frac{F_n(t)}{(s(t)-t_i)^2}+\frac{F_n(s(t))}{(t-t_i)^2}\right)\\
&+\frac{-dtdt_i}{2(\log(1+\frac{1}{ft})-\log(1+\frac{1}{fs(t)}))}\cdot\frac{s(t)(s(t)-1)(fs(t)+1)}{(f+1)}\\
&\cdot\left(\frac{F_n(s(t))}{(t-t_i)^2}+\frac{F_n(t)}{(s(t)-t_i)^2}\right)\\
=& \frac{-s'(t)dtdt_i}{\log(1+\frac{1}{ft})-\log(1+\frac{1}{fs(t)})}\cdot\frac{t(t-1)(ft+1)}{(f+1)}\cdot\left(\frac{F_n(t)}{(s(t)-t_i)^2}+\frac{F_n(s(t))}{(t-t_i)^2}\right)\\
=& \frac{-dtdt_i}{\log(1+\frac{1}{ft})-\log(1+\frac{1}{fs(t)})}\left(\frac{\phi_{n+1}(t)s'(t)}{(s(t)-t_i)^2}+\frac{\phi_{n+1}(s(t))}{(t-t_i)^2}\right)\\
=& -\frac{\phi_{n+1}(t)B(s(t),t_i)+\phi_{n+1}(s(t))B(t,t_i)}{\log(1+\frac{1}{ft})-\log(1+\frac{1}{fs(t)})}.
\end{aligned}
$$

After a more careful examination of the expression B, by expand it as a formal Laurent series, it is not hard to see that in fact $B \in \mathbb{C}(f)[t_i]((t))$. For a Laurent series

$$f(t,t_i)=\sum_{n\geq -N}^{\infty} a_n(t_i)t^n \in \mathbb{C}(f)[t_i]((t)),$$

we denote by $f(t,t_i)_+$ the truncation $\sum_{n\geq 0}^{\infty} a_n(t_i)t^n \in \mathbb{C}(f)[t_i][[t]]$ of the regular part of $f(t,t_i)$. By a similar argument as we have done for the Type I residue, we have the estimate

$$\left|\frac{1}{2\pi i}\int_{\gamma}\frac{dE(q,\overline{q},y)}{w(q)}(\zeta_a(y(q))B(y(\overline{q}),y_i)+\zeta_a(y(\overline{q}))B(y(q),y_i))\right| \leq M\cdot\frac{dtdt_i}{|t^2|}\sim\frac{dtdt_i}{|t^2|},$$

which has to be the principle part of the Laurent series, according to Lemma 4.3. This finishes the proof the the following

PROPOSITION 4.4.

$$P_n(t,t_i)=\left(\frac{\phi_{n+1}(t)B(s(t),t_i)+\phi_{n+1}(s(t))B(t,t_i)}{\log(1+\frac{1}{ft})-\log(1+\frac{1}{fs(t)})}\right)_+$$

The calculation of this section proves the following theorem:

THEOREM 4.5. *The topological recursion in the BKMP conjecture* 1 *is equivalent to*

$$(f(f+1))^{n-1}\sum_{b_1,\cdots,b_n} <\tau_{b_1}\cdots\tau_{b_n}\Gamma_g(f)>_g \prod_{k=1}^n d\phi_{b_k}(t_k)$$

$$=(f(f+1))^n \sum_{a_1,a_2,b_2,\cdots,b_n} <\tau_{a_1}\tau_{a_2}\prod_{k=2}^n \Gamma_{g-1}(f)>_{g-1}\prod_{k=2}^n d\phi_{b_k}(t_k)\cdot P_{a_1,a_2}(t_1)dt_1$$

$$-(f(f+1))^{n-1}\sum_{g_1+g+2=g, I\coprod J=\{2,\cdots,n\}}^{stable}\sum_{a_1,a_2,b_2,\cdots,b_n} <\tau_{a_1}\prod_{i\in I}\tau_{b_i}\Gamma_{g_1}(f)>_{g_1}$$

$$\cdot <\tau_{a_2}\prod_{j\in J}\tau_{b_j}\Gamma_{g_2}(f)>_{g_2}\prod_{k=2}^n d\phi_{b_k}(t_k)\cdot P_{a_1,a_2}(t_1)dt_1$$

$$-(f(f+1))^{n-2}\sum_{j=2}^n \sum_{b,b_i,i\in\{2,\cdots,n\}\backslash\{j\}} <\tau_b\prod_{k=2,k\neq j}^n \tau_{b_k}>\prod_{k=2,k\neq j}^n d\phi_{b_k}(t_k)\cdot P_b(t_1,t_j).$$

5. Some Formal Analysis

The critical point of the curve $x = y^f(1-f)$ is

$$(x,y) = (\frac{f^f}{(f+1)^{f+1}}, \frac{f}{f+1}).$$

Let

$$x = \frac{f^f}{(f+1)^{f+1}}e^{-\frac{f+1}{f}w} \quad \text{and} \quad w = \frac{1}{2}v^2,$$

then we have

$$v^2 = z^2\left(1+\sum_{k=1}^{+\infty}\frac{z^k}{k+2}(\frac{1-(\frac{-1}{f})^{k+1}}{1-(\frac{-1}{f})})\right)\in\mathbb{Q}(f)[[z]].$$

Let $F(z)$ be the unique formal power series in $\mathbb{Q}(f)[[z]]$ such that $F(0)=1$ and

$$F(z)^2 = \left(1+\sum_{k=1}^{+\infty}\frac{z^k}{k+2}(\frac{1-(\frac{-1}{f})^{k+1}}{1-(\frac{-1}{f})})\right).$$

The coefficients of $F(z)F(z) = 1+\sum_{k=1}^{\infty}a_kz^k$ are determined recursively by the relations

$$a_0 = 1$$

$$\sum_{i=0}^k a_ia_{k-i} = \frac{1}{k+2}\left(\frac{1-(\frac{-1}{f})^{k+1}}{1+\frac{1}{f}}\right) \quad \text{for } k\geq 1.$$

It is obvious that the two possible solutions of v are $\pm zF(z)$. We take v to be the solution $zF(z)$, which admit a formal inverse function

$$z = vG(v) = v+\sum_{k=1}^{\infty}g_kv^{k+1}\in v\mathbb{Q}(f)[[v]].$$

The holomorphic map s can be described as by sending $t = \frac{1}{vG(v)}$ to $s(t) = \frac{1}{-vG(-v)}$, i.e, sending $v \mapsto -v$. It is easy to see that

$$t = \frac{1}{v[1+\sum_{k=1}^{\infty} g_k v^k]} = \frac{1}{v}[1+\sum_{k=1}^{\infty} h_k v^k] \in v^{-1}\mathbb{Q}(f)[[v]].$$

LEMMA 5.1. *Let*

$$\eta_{-1}(v) = -\frac{1}{2}(\log(1+\frac{1}{ft}) - \log(1+\frac{1}{fs(t)})) = \frac{1}{2}(\phi_{-1}(t) - \phi_{-1}(s(t))),$$

and let $\eta_{n+1}(v) = -\frac{f}{f+1}\frac{1}{v}\frac{d}{dv}\eta_n(v)$ *for all* $n \geq -1$, *then we have*

$$\eta_n(v) = \frac{1}{2}(\phi_n(t) - \phi_n(s(t))).$$

Moreover, for each $n \geq -1$

$$\eta_n(v) = \phi_n(t) + F_n(w)$$

for some $F_n(w) \in \mathbb{Q}(f)[[w]]$.

PROOF. By definition, we have

$$-\frac{f}{f+1}\cdot\frac{1}{v}\frac{d}{dv} = -\frac{f}{f+1}\cdot\frac{d}{dw} = x\frac{d}{dx} = \frac{t(t-1)(ft+1)}{(f+1)}\frac{d}{dt}$$

is invariant under the action of $s : t \mapsto s(t)$. Applying this operator n times to the definition of $\eta_{-1}(v)$, we get the first claim.

It is obvious that $\eta_n(v)$ is odd in the variable v, so $\eta_{-1}(v) = \frac{-v}{f}(1+\cdots) \in v\mathbb{Q}(f)[[v]]$ contains only odd power of v. By applying the operator $-\frac{f}{f+1}\frac{1}{v}\frac{d}{dv}$ to $\eta_{-1}(v)$, we get

$$\eta_n(v) = \frac{f^n}{(f+1)^{n+1}}\cdot\frac{(2n-1)!!}{v^{2n+1}}(1+\cdots) \in \frac{1}{v^{2n+1}}\mathbb{Q}(f)[[w]].$$

We have

$$\begin{aligned}\eta_{-1}(v) - \phi_{-1}(t) &= -\frac{1}{2}(\phi_{-1}(t) + \phi_{-1}(s(t))) \\ &= \frac{1}{2}\sum_{n=1}^{+\infty}\frac{(-1)^{n-1}}{nf^n}\left(v^nG(v)^n + (-v)^nG(-v)^n\right),\end{aligned}$$

which is in the formal power series ring $\mathbb{Q}(f)[[v]]$. Moreover, it is even in the variable v, thus in the ring $\mathbb{Q}(f)[[w]]$. We denote it by $F_{-1}(w)$. Since there is no constant term, $F_{-1}(w) \in \mathbb{Q}(f)[[w]]$. This proves the second claim for $n = -1$.

For $n > -1$, apply the operator $-\frac{f}{f+1}\frac{1}{v}\frac{d}{dv}$ to the equation

$$\eta_{-1}(v) - \phi_{-1}(t) = F_{-1}(w)$$

repeatedly n times, we obtain

$$\eta_n(v) = \phi_n(t) + F_n(w)$$

for $F_n(w) = (-\frac{f+1}{f}\frac{d}{dw})^n F_{-1}(w) \in \mathbb{Q}(f)[[w]]$ by induction. □

COROLLARY 5.2.

$$P_{a,b}dt = -\frac{1}{2}\left(\frac{\eta_{a+1}(v)\eta_{b+1}(v)}{\eta_{-1}(v)}(\frac{f+1}{f})vdv|_{v=v(t)}\right)_+$$

PROOF. By Proposition 4.2, we have

$$\frac{\frac{1}{2}}{\log(1+\frac{1}{ft})-\log(1+\frac{1}{fs(t)})}(\phi_{a+1}(t)\phi_{b+1}(s(t))+\phi_{a+1}(s(t))\phi_{b+1}(t))\cdot\frac{(f+1)dt}{t(t-1)(ft+1)}$$
$$=\frac{1}{2\eta_{-1}(v)}\cdot\frac{(f+1)dt}{t(t-1)(ft+1)}\cdot(\frac{\phi_{a+1}(t)-\phi_{a+1}(s(t))}{2}\frac{\phi_{b+1}(t)-\phi_{b+1}(s(t))}{2}$$
$$-\frac{\phi_{a+1}(t)+\phi_{a+1}(s(t))}{2}\frac{\phi_{b+1}(t)+\phi_{b+1}(s(t))}{2})$$
$$=-\frac{(f+1)vdv}{2f\eta_{-1}(v)}(\eta_{a+1}(v)\eta_{b+1}(v)-F_{a+1}(w)F_{b+1}(w)).$$

To complete the proof, only need to notice that

$$\left(\frac{F_{a+1}(w)F_{b+1}(w)vdv}{\eta_{-1}(v)}|_{v\mapsto v(t)}\right)_+=0.$$

□

6. The Left Hand Side

After all the preparations, we are now ready to prove the BKMP conjecture. We will show that the push forward of the Symmetrized Cut-Join Equation via the first variable from the Framed Mirror Curve to $\mathbb{C}$ gives the BKMP conjecture.

In this section, we will deal with the LHS of the Symmetrized Cut-Join Equation. Unlike the Hurwitz number case studied in [**4**], the LHS of our Symmetrized Cut-Join Equation involves taking derivative with respect to the parameter f that need more special treatment.

The LHS of the symmetrized cut-join equation reads

$$\left(\frac{\partial}{\partial f}|_{t_i}+\sum_{l=1}^{n}\frac{t_l(t_l-1)}{f+1}\cdot\frac{\partial}{\partial t_l}\right)H_n^g(t_1,\cdots,t_n,f)$$
$$=\frac{\partial}{\partial f}|_{y_i}H_n^g(t_1(y_1,f),\cdots,t_n(y_n,f),f)$$
$$=\left(\frac{\partial}{\partial f}|_{x_i}+\sum_{l=1}^{n}\log y_l\cdot x_l\frac{\partial}{\partial x_l}\right)H_n^g.$$

If we change the variable

$$x=\frac{f^f}{(f+1)^{f+1}}e^{-(\frac{f+1}{2f})v^2},$$

then

$$\frac{dx}{df}=x\left(\frac{f}{(f+1)^2}+\frac{v^2}{2f^2}+\log(\frac{f}{f+1})\right),$$

and the differential operator above then becomes

$$\frac{\partial}{\partial f}|_{x_i}+\sum_{l=1}^{m}\log y_l\cdot x_l\frac{\partial}{\partial x_l}$$
$$=\frac{\partial}{\partial f}|_{v_i}-\sum_{l=1}^{m}\left(\log(1+\frac{1}{ft_l})-\frac{f}{(f+1)^2}-\frac{v_l^2}{2f^2}\right)\frac{f}{v_l(f+1)}\frac{\partial}{\partial v_l}.$$

Taking the direct image of a function F on the framed curve via the map $\pi : C \to \mathbb{C}$ with respect to the first variable, i.e, $\pi_*(F) = F(v_1) + F(-v_1)$. The term with $\frac{\partial}{\partial f}$ becomes

$$\frac{\partial}{\partial f}|_{v_i}\left(-(f(f+1))^{n-1}\sum_{b_1,\cdots,b_n} < \tau_{b_1}\cdots\tau_{b_n}\Gamma_g(f) >_g (\phi_{b_1}(t_1)+\phi_{b_1}(s(t_1)))\right.$$
$$\left.\cdot\prod_{k=2}^n \phi_{b_k}(t_k)\right)$$
$$=\frac{\partial}{\partial f}|_{v_i}\left(-2(f(f+1))^{n-1}\sum_{b_1,\cdots,b_n} < \tau_{b_1}\cdots\tau_{b_n}\Gamma_g(f) >_g F_{b_1}(w_1)\prod_{k=2}^n \phi_{b_k}(t_k)\right),$$

which we denoted by $\widetilde{A}$ for short. In the summation, the $l \neq 1$ term becomes a sum of

$$\left(\log(1+\frac{1}{ft_l}) - \frac{f}{(f+1)^2} - \frac{v_l^2}{2f^2}\right)\frac{f}{(f+1)v_l}\frac{\partial}{\partial v_l}\left(F_{w_1}(t_1)\prod_{k=2}^n \phi_{b_k}(t_k)\right),$$

which we denote by $A_l(b_1,\cdots,b_n)$. The term with $l = 1$ contributes a sum of

$$\left(\log(1+\frac{1}{ft_1}) - \frac{f}{(f+1)^2} - \frac{v_1^2}{2f^2}\right)\frac{f}{(f+1)v_1}\frac{\partial}{\partial v_1}(\eta_{b_1}(v_1) - F_{b_1}(w_1))\prod_{k=2}^n \phi_{b_k}(t_k)$$
$$+\left(\log(1+\frac{1}{fs(t_1)}) - \frac{f}{(f+1)^2} - \frac{v_1^2}{2f^2}\right)\frac{f}{(f+1)v_1}\frac{\partial}{\partial v_1}(\eta_{b_1}(-v_1) - F_{b_1}(w_1))\prod_{k=2}^n \phi_{b_k}(t_k)$$
$$=\left(\log(1+\frac{1}{ft_1}) - \log(1+\frac{1}{fs(t_1)})\right)\frac{f}{(f+1)v_1}\frac{\partial}{\partial v_1}\eta_{b_1}(v_1)\prod_{k=2}^n \phi_{b_k}(t_k)$$
$$+\left(\log(1+\frac{1}{ft_1}) + \log(1+\frac{1}{fs(t_1)}) - \frac{2f}{(f+1)^2} - \frac{v_1^2}{f^2}\right)F_{b_1+1}(w_1)\prod_{k=2}^n \phi_{b_k}(t_k)$$
$$=2\eta_{-1}(v_1)\eta_{b_1+1}(v_1)\prod_{k=2}^n \phi_{b_k}(t_k) + A_1(b_1,\cdots,b_n),$$

where $A_1(b_1,\cdots,b_n)$ denote the part involves $F_{b_1+1}(w_1)$.

In summary, the push forward of the LHS of the Symmetrized Cut-Join Equation is the following:

$$L :=\widetilde{A} + (f(f+1))^{n-1}\sum_{b_1,\cdots,b_n} < \tau_{b_1}\cdots\tau_{b_n}\Gamma_g(f) >_g$$
$$\cdot\left(2\eta_{-1}(v_1)\eta_{b_1+1}(v_1)\prod_{k=2}^n \phi_{b_k}(t_k) + \sum_{l=1}^n A_l(b_1,\cdots,b_n)\right)$$
$$=2(f(f+1))^{n-1}\sum_{b_1,\cdots,b_n} < \tau_{b_1}\cdots\tau_{b_n}\Gamma_g(f) >_g \eta_{-1}(v_1)\eta_{b_1+1}(v_1)\prod_{k=2}^n \phi_{b_k}(t_k) + \widetilde{B},$$

where we write

$$\widetilde{B} := \widetilde{A} + (f(f+1))^{n-1} \sum_{b_1,\cdots,b_n} < \tau_{b_1} \cdots \tau_{b_n} \Gamma_g(f) >_g \sum_{l=1}^{n} A_l(b_1,\cdots,b_n).$$

for short. The crucial observation is that $\widetilde{A}$ and $A_l(b_1,\cdots,b_n)$ for $l=1,\cdots,n$ are all regular in w_1. This can be check from their definition. Thus their sum $\widetilde{B}$ is also regular in w_1, and we have the important estimate

$$\left(\frac{\widetilde{B}}{2\eta_{-1}(v_1)} \cdot \frac{(f+1)dt}{t(t-1)(ft+1)}\right)_+ = 0.$$

On the other hand, we have

$$\begin{aligned}
W_g =&(-1)^{g+n}(f(f+1))^{n-1} \sum_{b_1,\cdots,b_n} < \tau_{b_1} \cdots \tau_{b_n} \Gamma_g(f) >_g \prod_{k=1}^{n} d\phi_{b_k}(t_k) \\
=& (-1)^{g+n} \frac{(f+1)dt_1}{2\eta_{-1}(v_1)t_1(t_1-1)(ft_1+1)} d_2 \cdots d_n \\
&\cdot \left(L - \widetilde{B} + (f(f+1))^{n-1} \sum_{b_1,\cdots,b_n} < \tau_{b_1} \cdots \tau_{b_n} \Gamma_g(f) >_g F_{b_1+1}(w_1) \prod_{k=2}^{n} \phi_{b_k}(t_k)\right) \\
=& (-1)^{g+n} \left(\frac{(f+1)dt_1}{2\eta_{-1}(v_1)t_1(t_1-1)(ft_1+1)} d_2 \cdots d_n L\right)_+
\end{aligned}$$

This finishes the computation of the LHS of the BKMP conjecture.

7. The Right Hand Side

We now compute the push forward of the RHS of the Symmetrized Cut-Join Equation. Recall that it is the sum of the following four terms:

$$\begin{aligned}
T_1 =& -\frac{1}{2}\sum_{l=1}^{n} t_l(t_l-1)(\frac{t_l f+1}{f+1})\frac{\partial}{\partial t_l} \cdot t_{n+1}(t_{n+1}-1)(\frac{t_{n+1}f+1}{f+1})\frac{\partial}{\partial t_{n+1}} H_{n+1}^{g-1}|_{t_{n+1}=t_l} \\
T_2 =& -\frac{1}{2}\sum_{1\leqslant a\leqslant g-1}\sum_{1\leqslant k\leqslant n} \Theta_{k-1}(t_1(t_1-1)(\frac{t_1 f+1}{f+1})\frac{\partial}{\partial t_1} H_k^a(t_1,\cdots,t_k,f)) \\
&\cdot (t_1(t_1-1)(\frac{t_1 f+1}{f+1})\frac{\partial}{\partial t_1} H_{n-k+1}^{g-a}(t_1,t_{k+1},\cdots,t_n,f)) \\
T_3 =& -\sum_{k=3}^{n} \Theta_{k-1}(t_1(t_1-1)(\frac{t_1 f+1}{f+1})\frac{\partial}{\partial t_1} H_k^0(t_1,\cdots,t_k,f)) \\
&\cdot (t_1(t_1-1)(\frac{t_1 f+1}{f+1})\frac{\partial}{\partial t_1} H_{n-k+1}^{g}(t_1,t_{k+1},\cdots,t_n,f)) \\
T_4 =&\Theta_1 \frac{t_1^2(t_1-1)(t_2-1)}{t_1-t_2}(\frac{t_1 f+1}{f+1})^2 \frac{\partial}{\partial t_1} H_{n-1}^{g}(t_1,t_3,\cdots,t_n,f).
\end{aligned}$$

The push forward of the T_1 part of the RHS of the Symmetrized Cut-Join Equation is the sum of

$$\begin{aligned}
&\frac{1}{2}(f(f+1))^n \sum_{b_1,\cdots,b_{n+1}} < \tau_{b_1}\cdots\tau_{b_{n+1}}\Gamma_{g-1}(f) >_{g-1} \\
&\qquad \cdot (\phi_{b_1+1}(t_1)\phi_{b_{n+1}+1}(t_1) + \phi_{b_1+1}(s(t_1))\phi_{b_{n+1}+1}(s(t_1))) \cdot \prod_{k=2}^{n} \phi_{b_k}(t_k) \\
&= (f(f+1))^n \sum_{b_1,\cdots,b_{n+1}} < \tau_{b_1}\cdots\tau_{b_{n+1}}\Gamma_{g-1}(f) >_{g-1} \\
&\qquad \cdot (\eta_{b_1+1}(v_1)\eta_{b_{n+1}+1}(v_1) + F_{b_1+1}(w_1)F_{b_{n+1}+1}(w_1)) \cdot \prod_{k=2}^{n} \phi_{b_k}(t_k)
\end{aligned}$$

and

$$\begin{aligned}
&-\sum_{l=2}^{m}(f(f+1))^n \sum_{b_1,\cdots,b_{n+1}} < \tau_{b_1}\cdots\tau_{b_{n+1}}\Gamma_{g-1}(f) >_{g-1} F_{b_1}(w_1)(\phi_{b_l+1}(t_l)\phi_{b_{n+1}+1}(t_l)) \\
&\qquad \cdot \prod_{i=2,i\neq l}^{n} \phi_{b_i}(t_i).
\end{aligned}$$

Thus we have

$$\begin{aligned}
&\left(\frac{\pi_* T_1 \cdot (f+1)dt_1}{2\eta_{-1}(v_1)t_1(t_1-1)(ft_1+1)}\right)_+ \\
=&(f(f+1))^n \sum_{b_1,\cdots,b_{n+1}} < \tau_{b_1}\cdots\tau_{b_{n+1}}\Gamma_{g-1}(f) >_{g-1} \prod_{k=2}^{n} \phi_{b_k}(t_k) \\
&\cdot \left(-\frac{(f+1)\eta_{b_1+1}(v_1)\eta_{b_{n+1}+1}(v_1)v_1 dv_1}{2f\eta_{-1}(v_1)}\right)_+ \\
=&(f(f+1))^n \sum_{b_1,\cdots,b_{n+1}} < \tau_{b_1}\cdots\tau_{b_{n+1}}\Gamma_{g-1}(f) >_{g-1} \prod_{k=2}^{n} \phi_{b_k}(t_k) \cdot P_{b_1,b_{n+1}}(t_1)dt_1.
\end{aligned}$$

By completely the same computation, applied to T_2+T_3, we have

$$\begin{aligned}
&\left(\frac{\pi_*(T_2+T_3) \cdot (f+1)dt_1}{2\eta_{-1}(v_1)t_1(t_1-1)(ft_1+1)}\right)_+ \\
=&-(f(f+1))^{n-1} \sum_{g_1+g_2=g} \sum_{I\coprod J=\{2,\cdots,n\}}^{stable} \sum_{a_1,a_2,b_2,\cdots,b_n} < \tau_{a_1}\prod_{i\in I}\tau_{b_i}\Gamma_{g_1}(f) >_{g_1} \\
&\cdot < \tau_{a_2}\prod_{j\in J}\tau_{b_j}\Gamma_{g_2}(f) >_{g_2} P_{a_1,a_2}(t_1)dt_1 \prod_{k=2}^{n} \phi_{b_k}(t_k).
\end{aligned}$$

The T_4 part is much more complicated, and need additional care. First of all, it is equal to

$$-\sum_{1\le i,j\le n}\frac{t_i(ft_i+1)(t_j-1)}{(f+1)(t_i-t_j)}(f(f+1))^{n-2}$$
$$\cdot\sum_{b,b_i,i\in\{1,\cdots,n\}\backslash\{i,j\}}<\tau_b\prod_{k=1,k\neq i,j}^{n}\tau_{b_k}\Gamma_g(f)>_g\phi_{b+1}(t_i)\prod_{k=1,k\neq i,j}^{n}\phi_{b_k}(t_k).$$

The push forward $\pi_*T_4=S_1+S_2+S_3$, for

$$S_1=\sum_{2\le i\neq j\le n}2(f(f+1))^{n-2}\frac{t_i(ft_i+1)(t_j-1)}{(f+1)(t_i-t_j)}$$
$$\cdot\sum_{b,b_i,i\in\{1,\cdots,n\}\backslash\{i,j\}}<\tau_b\prod_{k=1,k\neq i,j}^{n}\tau_{b_k}\Gamma_g(f)>_g F_{b_1}(w_1)\phi_{b+1}(t_i)\prod_{k=2,k\neq i,j}^{n}\phi_{b_k}(t_k)$$
$$S_2=-\sum_{j=2}^{n}(f(f+1))^{n-2}\sum_{b,b_i,i\in\{2,\cdots,n\}\backslash\{j\}}<\tau_b\prod_{k=2,k\neq j}^{n}\tau_{b_k}\Gamma_g(f)>_g$$
$$\cdot\left(\frac{t_1(ft_1+1)(t_j-1)}{(f+1)(t_1-t_j)}\phi_{b+1}(t_1)+\frac{s(t_1)(fs(t_1)+1)(t_j-1)}{(f+1)(s(t_1)-t_j)}\phi_{b+1}(s(t_1))\right)$$
$$\cdot\prod_{k=2,k\neq j}^{n}\phi_{b_k}(t_k)$$
$$S_3=-\sum_{i=2}^{n}(f(f+1))^{n-2}\sum_{b,b_i,i\in\{2,\cdots,n\}\backslash\{i\}}<\tau_b\prod_{k=2,k\neq i}^{n}\tau_{b_k}\Gamma_g(f)>_g$$
$$\cdot\left(\frac{t_i(ft_i+1)(t_1-1)}{(f+1)(t_i-t_1)}+\frac{t_i(ft_i+1)(s(t_1)-1)}{(f+1)(t_i-s(t_1))}\right)\phi_{b+1}(t_i)\prod_{k=2,k\neq i}^{n}\phi_{b_k}(t_k).$$

It is easy to see that

$$\left(\frac{S_1\cdot(f+1)dt_1}{2\eta_{-1}(v_1)t_1(t_1-1)(ft_1+1)}\right)_+=0,$$

since S_1 is holomorphic in the variable w_1. Because of

$$\frac{t_i(ft_i+1)(t_1-1)}{(f+1)(t_i-t_1)}+\frac{t_i(ft_i+1)(s(t_1)-1)}{(f+1)(t_i-s(t_1))}$$
$$=-2\frac{t_i(ft_i+1)}{f+1}-\frac{t_i(t_i-1)(ft_i)}{f+1}\cdot\sum_{k=0}^{\infty}\left(\frac{t_i^k}{t_1^{k+1}}+\frac{t_i^k}{s(t_1)^{k+1}}\right),$$

we also have

$$\left(\frac{S_3\cdot(f+1)dt_1}{2\eta_{-1}(v_1)t_1(t_1-1)(ft_1+1)}\right)_+=0.$$

Finally, let us compute the contribution of S_2. Consider the differentials:

$$
\begin{aligned}
&d_j\left(\frac{(f+1)dt_1}{2\eta_{-1}(v_1)t_1(t_1-1)(ft_1+1)}\cdot\frac{t_1(ft_1+1)(t_j-1)}{(f+1)(t_1-t_j)}\phi_{b+1}(t_1)\right)\\
=&d_j\left(\frac{(t_j-1)dt_1}{2\eta_{-1}(v_1)(t_1-1)(t_1-t_j)}\phi_{b+1}(t_1)\right)\\
=&d_j\left(-\frac{\phi_{b+1}(t_1)dt_1}{2\eta_{-1}(v_1)(t_1-1)}+\frac{\phi_{b+1}(t_1)dt_1}{2\eta_{-1}(v_1)(t_1-t_j)}\right)\\
=&\frac{\phi_{b+1}(t_1)B(t_1,t_j)}{2\eta_{-1}(v_1)},\\
&d_j\left(\frac{(f+1)dt_1}{2\eta_{-1}(v_1)t_1(t_1-1)(ft_1+1)}\cdot\frac{s(t_1)(fs(t_1)+1)(t_j-1)}{(f+1)(s(t_1)-t_j)}\phi_{b+1}(s(t_1))\right)\\
=&d_j\left(\frac{(f+1)s'(t_1)dt_1}{2\eta_{-1}(v_1)s(t_1)(s(t_1)-1)(fs(t_1)+1)}\cdot\frac{s(t_1)(fs(t_1)+1)(t_j-1)}{(f+1)(s(t_1)-t_j)}\phi_{b+1}(s(t_1))\right)\\
=&d_j\left(\frac{(t_j-1)s'(t_1)dt_1}{2\eta_{-1}(v_1)(s(t_1)-1)(s(t_1)-t_j)}\phi_{b+1}(s(t_1))\right)\\
=&d_j\left(-\frac{\phi_{b+1}(s(t_1))s'(t_1)dt_1}{2\eta_{-1}(v_1)(s(t_1)-1)}+\frac{\phi_{b+1}(s(t_1))s'(t_1)dt_1}{2\eta_{-1}(v_1)(s(t_1)-t_j)}\right)\\
=&\frac{\phi_{b+1}(s(t_1))B(s(t_1),t_j)}{2\eta_{-1}(v_1)}.
\end{aligned}
$$

We have the following simplified expression

$$
\begin{aligned}
&d_2\cdots d_n\left(\frac{S_2\cdot(f+1)dt_1}{2\eta_{-1}(v_1)t_1(t_1-1)(ft_1+1)}\right)\\
=&-\sum_{j=2}^{n}(f(f+1))^{n-2}\sum_{b,b_i,i\in\{2,\cdots,n\}\setminus\{j\}}<\tau_b\prod_{k=2,k\neq j}^{n}\tau_{b_k}\Gamma_g(f)>_g\prod_{k=2,k\neq j}^{n}d\phi_{b_k}(t_k)\\
&\cdot\left(\frac{\phi_{b+1}(t_1)B(t_1,t_j)+\phi_{b+1}(s(t_1))B(s(t_1),t_j)}{2\eta_{-1}(v_1)}\right).
\end{aligned}
$$

To obtain the $P_n(t_1,t_j)$ term in the BKMP conjecture, we need to switch one of the t_1 with $s(t_1)$ in the above expression. This can be done by the following estimate:

$$
\left(\frac{B(t_1,t_j)}{2\eta_{-1}(v_1)}(\phi_{b+1}(t_1)+\phi_{b+1}(s(t_1)))\right)_+=\left(-\frac{B(t_1,t_j)F_{b+1}(w_1)}{\eta_{-1}(v_1)}\right)_+=0,
$$

and similarly

$$
\left(\frac{B(s(t_1),t_j)}{2\eta_{-1}(v_1)}(\phi_{b+1}(t_1)+\phi_{b+1}(s(t_1)))\right)_+=\left(-\frac{B(s(t_1),t_j)F_{b+1}(w_1)}{\eta_{-1}(v_1)}\right)_+=0.
$$

Thus we have the relation

$$P_b(t_1,t_j)dt_1dt_j = \left(\frac{\phi_{b+1}(s(t_1))B(t_1,t_j)+\phi_{b+1}(t_1)B(s(t_1),t_j)}{\log(1+\frac{1}{ft})-\log(1+\frac{1}{fs(t)})}\right)_+$$
$$= \left(\frac{\phi_{b+1}(t_1)B(t_1,t_j)+\phi_{b+1}(s(t_1))B(s(t_1),t_j)}{2\eta_{-1}(v_1)}\right)_+$$

and the estimate of π_*T_4 and S_2:

$$d_2\cdots d_n\left(\frac{\pi_*T_4\cdot(f+1)dt_1}{2\eta_{-1}(v_1)t_1(t_1)(ft_1+1)}\right)_+ = d_2\cdots d_n\left(\frac{S_2\cdot(f+1)dt_1}{2\eta_{-1}(v_1)t_1(t_1)(ft_1+1)}\right)_+$$
$$= -(f(f+1))^{n-2}\sum_{j=2}^{n}\sum_{b,b_i,i\in\{2,\cdots,n\}\setminus\{j\}} <\tau_b\prod_{k=1,k\neq i,j}^{n}\tau_{b_k}\Gamma_g(f)>_g$$
$$\cdot P_b(t_1,t_j)\prod_{k=2,k\neq j}^{n}d\phi_{b_k}(t_k).$$

Collect all the pieces, we have finish the computation of the push forward of the RHS of the Symmetrized Cut-Join Equation:

$$d_2\cdots d_n\left(\frac{\pi_*(T_1+T_2+T_3+T_4)\cdot(f+1)dt_1}{2\eta_{-1}(v_1)t_1(t_1)(ft_1+1)}\right)_+$$
$$=(f(f+1))^n\sum_{a_1,a_2,b_2,\cdots,b_n}<\tau_{a_1}\tau_{a_2}\prod_{k=2}^{n}\Gamma_{g-1}(f)>_{g-1}\prod_{k=2}^{n}d\phi_{b_k}(t_k)\cdot P_{a_1,a_2}(t_1)dt_1$$
$$-(f(f+1))^{n-1}\sum_{g_1+g+2=g,I\coprod J=\{2,\cdots,n\}}^{stable}\sum_{a_1,a_2,b_2,\cdots,b_n}<\tau_{a_1}\prod_{i\in I}\tau_{b_i}\Gamma_{g_1}(f)>_{g_1}$$
$$\cdot<\tau_{a_2}\prod_{j\in J}\tau_{b_j}\Gamma_{g_2}(f)>_{g_2}\prod_{k=2}^{n}d\phi_{b_k}(t_k)\cdot P_{a_1,a_2}(t_1)dt_1$$
$$-(f(f+1))^{n-2}\sum_{j=2}^{n}\sum_{b,b_i,i\in\{2,\cdots,n\}\setminus\{j\}}<\tau_b\prod_{k=2,k\neq j}^{n}\tau_{b_k}>\prod_{k=2,k\neq j}^{n}d\phi_{b_k}(t_k)\cdot P_b(t_1,t_j).$$

The recursion in Theorem 4.5 then follows from plugging in the results of this and the previous section into the equality $L=\pi_*(T_1+T_2+T_3+T_4)$.

References

[1] Gaëtan Borot, Bertrand Eynard, Motohico Mulase, and Brad Safnuk, *A matrix model for simple Hurwitz numbers, and topological recursion*, J. Geom. Phys. **61** (2011), no. 2, 522–540, DOI 10.1016/j.geomphys.2010.10.017. MR2746135

[2] Vincent Bouchard, Albrecht Klemm, Marcos Mariño, and Sara Pasquetti, *Remodeling the B-model*, Comm. Math. Phys. **287** (2009), no. 1, 117–178, DOI 10.1007/s00220-008-0620-4. MR2480744

[3] Lin Chen, *Symmetrized cut-join equation of the Mariño-Vafa formula*, Pacific J. Math. **235** (2008), no. 2, 201–212, DOI 10.2140/pjm.2008.235.201. MR2386219

[4] Bertrand Eynard, Motohico Mulase, and Bradley Safnuk, *The Laplace transform of the cut-and-join equation and the Bouchard-Mariño conjecture on Hurwitz numbers*, Publ. Res. Inst. Math. Sci. **47** (2011), no. 2, 629–670, DOI 10.2977/PRIMS/47. MR2849645

[5] I. P. Goulden, D. M. Jackson, and A. Vainshtein, *The number of ramified coverings of the sphere by the torus and surfaces of higher genera*, Ann. Comb. **4** (2000), no. 1, 27–46, DOI 10.1007/PL00001274. MR1763948

[6] Sheldon Katz and Chiu-Chu Melissa Liu, *Enumerative geometry of stable maps with Lagrangian boundary conditions and multiple covers of the disc*, Adv. Theor. Math. Phys. **5** (2001), no. 1, 1–49, DOI 10.4310/ATMP.2001.v5.n1.a1. MR1894336

[7] Jun Li, Chiu-Chu Melissa Liu, Kefeng Liu, and Jian Zhou, *A mathematical theory of the topological vertex*, Geom. Topol. **13** (2009), no. 1, 527–621, DOI 10.2140/gt.2009.13.527. MR2469524

[8] Chiu-Chu Melissa Liu, Kefeng Liu, and Jian Zhou, *A proof of a conjecture of Mariño-Vafa on Hodge integrals*, J. Differential Geom. **65** (2003), no. 2, 289–340. MR2058264

[9] Chiu-Chu Melissa Liu, Kefeng Liu, and Jian Zhou, *A formula of two-partition Hodge integrals*, J. Amer. Math. Soc. **20** (2007), no. 1, 149–184, DOI 10.1090/S0894-0347-06-00541-8. MR2257399

Simons Center for Geometry and Physics, State University of New York, Stony Brook, New York

Proceedings of Symposia in Pure Mathematics
Volume **100**, 2018
https://doi.org/10.1090/pspum/100/01760

The hybrid Landau–Ginzburg models of Calabi–Yau complete intersections

Alessandro Chiodo and Jan Nagel

Abstract. We observe that the state space of Landau–Ginzburg isolated singularities is simply a special case of Chen–Ruan orbifold cohomology relative to the generic fibre of the potential. This leads to the definition of the cohomology of hybrid Landau–Ginzburg models and its identification via an explicit isomorphism to the cohomology of Calabi–Yau complete intersections inside weighted projective spaces. The combinatorial method used in the case of hypersurfaces proven by the first named author in collaboration with Ruan is streamlined and generalised after an orbifold version of the Thom isomorphism and of the Tate twist.

1. Introduction

Landau–Ginzburg models play a central role in mirror symmetry. The simplest examples are the following isolated singularities: weighted-homogeneous polynomials $W\colon \mathbb{C}^n \to \mathbb{C}$ with smooth fibres outside the origin and a single isolated singularity in the special fibre over the origin. They played a key role in the early development of quantum cohomology, giving some of the first examples of Frobenius manifolds and allowing to verify mirror symmetry statements. They have recently acquired a new status with the definition of their quantum invariants by Fan, Jarvis and Ruan [**18**] (FJRW theory) and later by Polishchuk and Vaintrob [**37**] (cohomological field theory of matrix factorisations). These advances make it possible to obtain equivalences between Landau–Ginzburg models and Calabi–Yau models without necessarily passing through mirror symmetry [**10**]. It is now also possible to state mirror symmetry between Landau–Ginzburg models both at the level of cohomological invariants and at the level of quantum cohomology invariants (see [**3**, **18**, **31**]). There are many consequences: for instance the absence of any Calabi–Yau conditions in mirror statements can be effectively used (see for instance [**1**, **18**, **31**]); furthermore, the computational power appears higher: we refer to approaches beyond concavity and in higher genus, see [**11**, **18**, **20**–**23**].

This paper starts from the definition by Fan, Jarvis and Ruan [**18**] of the state space $\mathcal{H}(W)$ of the Landau–Ginzburg model $W\colon [\mathbb{C}^n/\mu_d] \to \mathbb{C}$, where W is a degree-d weighted homogeneous polynomial with isolated singularity and μ_d is the corresponding monodromy action operating on the ith variable with character

2010 *Mathematics Subject Classification.* Primary 14J33.

$w_i \in \mathbb{Z}/d$

$$\mathcal{H}(W) := \bigoplus_{g\in\mu_d} (\mathcal{Q}_{W_g})^{\mu_d}.$$

We refer to W as the potential. Here $\mathcal{Q}_f$ stands for the Jacobian ring, or chiral algebra, attached to a potential f

$$\begin{aligned}\mathcal{Q}_f &= dx_1 \wedge \cdots \wedge dx_n \otimes \mathrm{Jac}(f) \\ &= dx_1 \wedge \cdots \wedge dx_n \otimes \mathbb{C}[x_1, \ldots, x_n]/(\partial_1 f, \ldots, \partial_n f),\end{aligned} \tag{1}$$

and W_g is the restriction of W to the g-fixed space $\mathbb{C}^{n_g}$ in $\mathbb{C}^n$. As pointed out by the authors of [**18**], the crucial ingredient $\mathcal{Q}$ and the orbifolding procedure appear already in the definition of the states of Landau–Ginzburg models in Intriligator–Vafa [**27**] and in Kaufmann's work [**28**],[**29**],[**30**] in the definition of B-model invariants. This state space is bi-graded by applying to each term $(Q_{W_g})^{\mu_d}$ an explicit formula: the somewhat *ad hoc* bigrading

$$(n_g - \deg(f), \deg(f)) + (\mathrm{age}(g), \mathrm{age}(g)) - \left(\textstyle\sum_i \frac{w_i}{d}, \sum_i \frac{w_i}{d}\right), \tag{2}$$

where age is defined additively by mapping the representations of μ_r to rational numbers as follows:

$$\mathrm{age}\colon R\mu_{\mathbf{r}} \to \mathbb{Q}; \qquad \text{for } k \in (\mu_r)^\vee = \mathbb{Z}/r, \quad \mathrm{age}(h) = k/r \quad \text{with } 0 \leq k < r.$$

The cohomological LG/CY correspondence for CICY. In this paper we identify, as a graded vector space, the above $\mathcal{H}(W)$ to an occurrence of relative orbifold Chen–Ruan cohomology. For a Calabi–Yau hypersurface $X_W = V(W)$ we consider

$$H^{p+1,q+1}_{\mathrm{CR}}([\mathbb{C}^n/\mu_d], W^{-1}(t_0); \mathbb{C}),$$

with $t_0 \in \mathbb{C}^\times$. We refer to its as the cohomological Landau–Ginzburg model for any $t_0 \in \mathbb{C}^\times$ and we have a natural (canonical) identification

$$\mathcal{H}^{p,q}(W) = H^{p+1,q+1}_{\mathrm{CR}}([\mathbb{C}^n/\mu_d], W^{-1}(t_0); \mathbb{C}).$$

More generally, note that $[\mathbb{C}^n/\mu_d]$ is $\mathcal{O}(-\mathbf{w}) := \mathcal{O}(-w_1) \oplus \cdots \oplus \mathcal{O}(-w_r)$ on the zero-dimensional weighted projective space $\mathbb{P}(d)$. Then, when $W_1, \ldots, W_r$ define a smooth complete intersection inside the weighted projective stack $\mathbb{P}(\mathbf{w}) := \mathbb{P}(w_1, \ldots, w_n)$, we set

$$\mathcal{H}^{p,q}(W_1, \ldots, W_r) := H^{p+r,q+r}_{\mathrm{CR}}(\mathcal{O}_{\mathbb{P}(\mathbf{d})}(-\mathbf{w}); W^{-1}(t_0); \mathbb{C}), \tag{3}$$

where W may be regarded as a $\mathbb{C}$-valued function, via the Cayley trick $W = \sum_{i=1}^r p_i W_i(\mathbf{x})$ (see for instance [**41**, 5.4]). Here W may be regarded as a function defined over the homogeneous variables of $\mathbb{P}(\mathbf{w})$, on which $\mathbb{C}^\times$ acts with positive weights, and the variables of the fibres of $\mathcal{O}_{\mathbb{P}(\mathbf{d})}(-\mathbf{w})$, on which $\mathbb{C}^\times$ acts with opposite weights. For this reason, we refer to this setup as hybrid.

REMARK 1.1. In the physics literature, it is common to refer to the coordinates of $\mathbb{P}(\mathbf{d})$ as massive or ground coordinates and to the coordinates along the fibres of $\mathcal{O}(-\mathbf{w})$ as massless coordinates, and to refer to $\mathbb{C}$-valued potentials such as W effectively involving both types of variables as *hybrid* (see, for instance, [**41**, 5.4] and [**24**]). The term "hybrid" is used in a more restrictive sense by Fan, Jarvis, and Ruan [**19**]. In their terminology a complete intersection is a "hybrid model" if all of the hypersurfaces have the same degree.

Finally, by a variation of stability condition one gets the counterpart of this hybrid model: the morphism $W\colon \mathcal{O}_{\mathbb{P}(\mathbf{w})}(-\mathbf{d}) \to \mathbb{C}$ which, after the same relative cohomology computation as above, turns out to be isomorphic to the cohomology of the complete Calabi–Yau intersection $V(W_1, \dots, W_r)$ inside $\mathbb{P}(\mathbf{w})$ (see Prop. 3.3 with $G = 1$). The two LG and CY models are related by the following theorem. We refer to Theorem 5.1 and Corollary 5.2 for the complete statement involving group quotients of complete intersections.

THEOREM (cohomological LG/CY correspondence for CICY). *For any smooth complete intersection $X_W = V(W_1, W_2, \dots, W_r)$ of Calabi–Yau type inside the weighted projective space $\mathbb{P}(\mathbf{w})$ we have*

$$\mathcal{H}^{p,q}(W_1, \dots, W_r) \cong H^{p,q}_{\mathrm{CR}}(X_W; \mathbb{C}).$$

REMARK 1.2. A related result by Libgober [**33**] proves the invariance of the elliptic genus in a more general setup which includes complete intersections and several other GIT quotients (we recall that, after specialisation, the elliptic genus gives a combination of Hodge numbers, see *e.g.* [**4**]). It is also worth-while pointing out how, in Libgober's treatment, Calabi–Yau models, Landau–Ginzburg models and their hybrid versions are precisely characterised in geometric terms, see [**33**, Defn. 2.3].

Here, our main focus is how the Landau–Ginzburg/Calabi–Yau correspondence follows from several properties of Chen–Ruan cohomology of independent interest. We illustrate it in the rest of the introduction.

Thom isomorphism in Chen–Ruan cohomology and the Tate twist. The LG/CY statement, and the more general Thm. 5.1 including group actions, relies on a version of the Thom isomorphism in ordinary cohomology

$$H^*(X; \mathbb{C}) = H^*(\mathbb{P}(\mathbf{w}), \mathbb{P}(\mathbf{w}) \setminus X; \mathbb{C})\ (r), \tag{4}$$

where X is a complete intersection of codimension r and we denote by "(r)" the Tate twist making the above isomorphism degree-preserving. Notice that, due to its age-shifted degree, Chen–Ruan orbifold cohomology does not satisfy (4) in the above form. However, by replacing $\mathbb{P}(\mathbf{w})$ with the total space of the vector bundle $\mathcal{O}(-d_1) \oplus \cdots \oplus \mathcal{O}(-d_r)$ we get a Thom isomorphism statement in Chen–Ruan orbifold cohomology (see Prop. 3.3 for the general statement).

PROPOSITION 1.3 (Thom isomorphism in orbifold cohomology). *For any smooth complete intersection $X_W = V(W_1, W_2, \dots, W_r)$ of Calabi–Yau type inside the weighted projective space $\mathbb{P}(\mathbf{w})$ we have a canonical isomorphism*

$$H^{p-r,q-r}_{\mathrm{CR}}(X_W; \mathbb{C}) \cong H^{p,q}_{\mathrm{CR}}\left(\mathcal{O}_{\mathbb{P}(\mathbf{w})}(-\mathbf{d}), W^{-1}(t_0)\right).$$

This is interesting in its own right and explains in a geometric way the *ad hoc* bi-degree shift (2) involving w_i/d: in Calabi–Yau cases it is merely the Tate twist in the Thom isomorphism.[1]

In the same spirit, let us point out how the present description of the state space clarifies in classical geometric terms the dichotomy between narrow and broad sectors, which, following [**18**], depends on the condition that the space of g-fixed

[1]This interpretation of the overall shift-by-total-charge $\sum_i \frac{w_i}{d}$ in (2), may lead to the correct interpretation of the age shift too; it may be worth-while to point out here Denef–Loeser's [**14**] view on the age grading as a representation-theoretic analogue of the weight in Hodge theory, see Miles Reid [**35**, §2].

points is or is not reduced to the origin. This dichotomy does not generalise as such in the complete intersection case. However, in our relative cohomology picture broad states are simply those that lie in the kernel of the natural morphism $H^*_{\mathrm{CR}}(\mathscr{O}(-\mathbf{w}), W^{-1}(t_0)) \to H^*_{\mathrm{CR}}(\mathscr{O}(-\mathbf{w}))$. In this way "broad" is nothing but a generalisation of the classical notion of primitive.

We also feel that the generalisation of Gysin sequences and Thom isomorphisms in Chen–Ruan orbifold cohomology is worth pursuing further in view of a well-behaved setup for an orbifold cohomology theory.

An algebraic model for the quantum theory of hybrid models. We finally notice how, using the Thom isomorphism, the theorem stated above follows from a bi-degree preserving isomorphism

$$H^*_{\mathrm{CR}}(\mathscr{O}_{\mathbb{P}(\mathbf{w})}(-\mathbf{d}); \mathbb{C}) \cong H^*_{\mathrm{CR}}(\mathscr{O}_{\mathbb{P}(\mathbf{d})}(-\mathbf{w}); \mathbb{C}).$$

This isomorphism holds by K-equivalence but can be expressed directly via a combinatorial argument which, with respect to [**8**], is greatly simplified.

By relying on this identification and on the classical descriptions of the primitive cohomology of complete intersections [**15, 34**], we derive an algebraic model of the cohomology of Calabi–Yau complete intersection in terms of the Jacobian ring; see Theorem 4.3. This type of algebraic presentation of cohomology classes has been proven crucial on many occasions in the formulation and proof of mirror symmetry statements.

The cohomology groups $\mathcal{H}(W_1, \dots, W_r)$ developed in this paper, have already been used to provide a quantum correspondence between Calabi–Yau complete intersections inside projective spaces and hybrid Landau–Ginzburg models, see Clader [**12**]. It is worth mentioning that the work of Clader has been later extended by Clader–Ross to a much more general class of complete intersections, [**13**]. Recently, Fan, Jarvis and Ruan provided a generalisation of their theory based on the state space $\mathcal{H}(W_1, \dots, W_r)$, see [**19**]. Clader establishes a correspondence between the quantum D-modules attached to complete intersection of CY type within $\mathbb{P}(\mathbb{C}^n)$ and the corresponding hybrid LG models. Her correspondence depends on the choice of an analytic continuation which should mirror the parallel transport along the Gauß–Manin connection on the B-side. In this way, all isomorphisms arising at the cohomological level from these quantum correspondences should depend on the path chosen for the parallel transport. In this perspective it is interesting how our proof of Theorem 5.1 provides a combinatorial method for writing an isomorphism explicitly. Therefore, our isomorphism could provide a useful tool to study the monodromy operators in the future.

Structure of the paper. We recall the setup in §2, develop the Calabi–Yau side in §3, the Landau–Ginzburg side in §4, and their correspondence in §5.

Acknowledgments. The first named author is grateful to the IMB, Dijon, for the hospitality during the finalisation of this work. We are extremely grateful to Anatoly Libgober for his comments and suggestions. We would like to thank the anonymous referee for his/her careful reading and his/her precise corrections and suggestions.

2. Setup

We consider a complete intersection of $r < n$ hypersurfaces in the weighted projective space $\mathbb{P}(\mathbf{w}) := \mathbb{P}(w_1, \dots, w_n)$. Let $W_1, \dots, W_r$ be quasi-homogeneous polynomials of degree $d_1, \dots, d_r$ in the variables $x_1, \dots, x_n$ of coprime weights $w_1, \dots, w_n$. We assume that X_W is a non-degenerate complete intersection, *i.e.*,

(i) the choice of weights $w_1, \dots, w_n$ is unique;
(ii) $\{W_1 = 0, \dots, W_r = 0\} \subset \mathbb{C}^n$ is smooth outside the origin.

Assumption (ii) holds if and only if the Jacobian matrix $J = (\partial_j W_i(p))_{i,j}$ has rank r for any $p \in \{\mathbf{x} \in \mathbb{C}^n \mid W_1 = \dots = W_r = 0\} \setminus \{\mathbf{0}\}$. Then, inside the Deligne–Mumford stack $\mathbb{P}(\mathbf{w})$, we get a smooth complete intersection

$$X_W = \{W_1 = \ldots = W_r = 0\} \subset \mathbb{P}(\mathbf{w}).$$

We further assume that X_W is of Calabi–Yau type in the following sense: we impose that the canonical sheaf is trivial, in other words

(iii) $\sum_{i=1}^r d_i = \sum_{j=1}^n w_j$.

We consider a group $\Gamma \subset (\mathbb{C}^\times)^n$ of diagonal matrices γ whose diagonal entries are of the form $(\gamma_1, \dots, \gamma_n) = \alpha\overline{\lambda} := (\alpha_1\lambda^{w_1}, \dots, \alpha_n\lambda^{w_n})$ with $W_i(\alpha_1 x_1, \dots, \alpha_n x_n) = W_i(x_1, \dots, x_n)$, all i and $\lambda \in \mathbb{C}^\times$. We assume that Γ contains the torus $\{\overline{\lambda} = (\lambda^{w_1}, \dots, \lambda^{w_n}) \mid \lambda \in \mathbb{C}^\times\}$; this is automatically its identity component Γ^0, and we write G for the group of connected components Γ/Γ^0. By (i), G is finite (see for instance [**18**, Lem. 2.18]). The quotient stack $[U/\Gamma]$, with $U = \{W_1 = W_2 = \dots = W_r = 0\} \setminus \mathbf{0} \subset \mathbb{C}^n$, is a smooth Deligne–Mumford stack and, following Romagny's treatment [**39**, Rem. 2.4] of actions on stacks, it is canonically equivalent to the 2-stack

$$[X_W/G] = [U/\Gamma].$$

We introduce for all $\gamma \in \Gamma$ the following notation

$$\mathbb{C}^n_\gamma := \{\mathbf{x} \in \mathbb{C}^n \setminus \mathbf{0} \mid \gamma \cdot \mathbf{x} = \mathbf{x}\}, \qquad n_\gamma := \dim \mathbb{C}^n_\gamma, \qquad W_{i,\gamma} := W_i|_{\mathbb{C}^n_\gamma}.$$

For any $i = 1, \dots, r$, we can define a character

$$\chi_i \colon \Gamma \to \mathbb{C}^\times$$

satisfying

$$\gamma^* W_i = W_i(\gamma \cdot \mathbf{x}) = \chi_i(\gamma) W_i.$$

The character χ_i maps as $\alpha\overline{\lambda} \mapsto \lambda^{d_i}$. Note that if α and $\beta \in (\mathbb{C}^\times)^n$ preserve all polynomials W_i and satisfy $\alpha\overline{\lambda} = \gamma = \beta\overline{\mu} \in \Gamma$ for some $\lambda, \mu \in \mathbb{C}^\times$, then we have $\lambda^{d_i} = (\gamma^* W_i)/W_i = \mu^{d_i}$ all i.

We keep track of the number of polynomials W_i on which γ acts trivially

$$r_\gamma := \#\{W_{i,\gamma} \mid \chi_i(\gamma) = 1\} \in [0, r] \cap \mathbb{N}.$$

The following lemma describes the two cases where $\gamma^* W_i$ equals W_i or not.

LEMMA 2.1. *We have $W_{i,\gamma} = 0$ if $\chi_i(\gamma) \neq 1$. Otherwise, for $\chi(\gamma) = 1$, we have $\gamma^* W_i = W_i$ and the partial derivatives of $W_{i,\gamma}$ coincide with those of W_i on $\mathbb{C}^n_\gamma$.*

PROOF. For $\chi_i(\gamma) \neq 1$ we have a polynomial W_i of positive degree which is not preserved by γ and whose restriction $W_{i,\gamma}$ to $\mathbb{C}^n_\gamma$ satisfies $W_{i,\gamma} = \gamma^* W_{i,\gamma} = \chi_i(\gamma) W_{i,\gamma}$; therefore $W_{i,\gamma} = 0$.

For $\chi_i(\gamma) = 1$, W_i is of the form $W_{i,\gamma} + f$, where f belongs to the maximal ideal $\mathfrak{m}_\gamma$ spanned by the variables x_j not fixed by γ (*i.e.* $\gamma_j = \alpha_j \lambda^{w_j} \neq 1$). In fact, since the action is diagonal, we have $f \in (\mathfrak{m}_\gamma)^2$, which implies that, after restriction to $\mathbb{C}^n_\gamma$, the partial derivatives $\partial_j W_i$ and $\partial_j W_{i,\gamma}$ coincide for all $j = 1, \dots, n$. $\square$

3. Calabi–Yau side

On the Calabi–Yau side of the correspondence, we consider the Chen–Ruan orbifold cohomology of $[X_W/G] = [U/\Gamma]$. Since the groups acting are Abelian, we can express this explicitly as a direct sum of Γ-invariant parts of cohomology groups

$$H^{p,q}_{\mathrm{CR}}([X_W/G];\mathbb{Q}) = \bigoplus_{\gamma\in\Gamma} H^{p-a_\gamma,q-a_\gamma}([X_{W,\gamma}];\mathbb{Q})^\Gamma,$$

where a_γ is the age of the action of γ on the tangent bundle of X_W at a point of $X_{W,\gamma}$ and $X_{W,\gamma}$ is the quotient stack

$$[\{\mathbf{x} \mid W_{1,\gamma}(\mathbf{x}) = \cdots = W_{r,\gamma}(\mathbf{x}) = 0, \quad \mathbf{x} \in \mathbb{C}^n_\gamma \setminus \mathbf{0}\}/\mathbb{C}^\times] \hookrightarrow \mathbb{P}(\mathbf{w})_\gamma,$$

where $\mathbb{P}(\mathbf{w})_\gamma$ is the weighted projective space spanned by the γ-fixed coordinates x_j. Notice that the sum above is taken over an infinite set; however, $\mathbb{C}^n_\gamma \setminus \mathbf{0}$ is nonempty only for a finite number of elements in Γ.

REMARK 3.1. By Lemma 2.1, on $\mathbb{C}^n_\gamma$, we can ignore the restricted polynomials $W_{i,\gamma}$ with $\chi_i(\gamma) \neq 1$. There are r_γ remaining polynomials

$$W_{i_1,\gamma}, \ldots, W_{i_{r_\gamma},\gamma}$$

invariant with respect to γ; they define the complete intersection $X_{W,\gamma}$ in $\mathbb{P}(\mathbf{w})_\gamma$. We notice that $X_{W,\gamma}$ is smooth since for every point of $\mathbb{C}^n_\gamma \setminus \mathbf{0}$ for which $W_{i_1,\gamma}, \ldots,$ $W_{i_{r_\gamma},\gamma}$ vanish, the Jacobian matrix has r_γ independent rows which, as proven in Lemma 2.1, coincide with the restrictions of the corresponding rows in the Jacobian matrix of X_W. These are independent by *(ii)*.

REMARK 3.2. All stacks considered in this paper are Deligne–Mumford stacks which are global quotients $[Y/H]$. In these cases when we consider ordinary cohomology with coefficients in $\mathbb{Q}$ of the stack $[Y/H]$ and of the coarse scheme Y/H we get the same result via a canonical isomorphism. The groups H considered in this paper are at worse finite extensions of tori operating with finite stabilisers. Sometimes we consider non-faithful actions: $H \mapsto \mathrm{Aut}(Y)$ with nontrivial kernel K. In these cases it is important to notice that, whereas for ordinary cohomology it does not matter if we consider $[Y/H]$ or $[Y/(H/K)]$, when it comes to Chen–Ruan orbifold cohomology these two stacks yield different cohomology groups, as the formula recalled above obviously shows.

As a consequence of these considerations we point out that in the above formula for Chen–Ruan orbifold cohomology $H^{p,q}_{\mathrm{CR}}([X_W/G];\mathbb{Q})$, we can replace on the right hand side $H^*([X_{W,\gamma}];\mathbb{Q})^\Gamma$ by $H^*([X_{W,\gamma}];\mathbb{Q})^G$; indeed the identity component $\Gamma^0 \cong \mathbb{C}^\times$ acts trivially on X_W and $\mathbb{P}(\mathbf{w})$ and on each sector $X_{W,\gamma}$ and $\mathbb{P}(\mathbf{w})_\gamma$.

In order to relate the Calabi–Yau complete intersection X_W to a Landau–Ginzburg model, we rephrase its cohomology in terms of relative cohomology of a vector bundle.

We can define a Γ-action on $\mathbb{C}^{n+r}$ by

$$\gamma \cdot (x_1, \ldots, x_n, p_1, \ldots, p_r) = (\gamma_1 x_1, \ldots, \gamma_n x_n, \chi_1(\gamma)^{-1} p_1, \ldots, \chi_r^{-1}(\gamma) p_r),$$

or, more explicitly, by $\alpha\overline{\lambda} \cdot (\mathbf{x}, \mathbf{p}) = (\alpha_1 \lambda^{w_1} x_1, \ldots, \alpha_n \lambda^{w_n} x_n, \lambda^{-d_1} p_1, \ldots, \lambda^{-d_r} p_r)$. We consider the Γ-invariant $\mathbb{C}$-valued function

$$\begin{aligned} W\colon \mathbb{C}^{n+r} &\to \mathbb{C} \\ (\mathbf{x}, \mathbf{p}) &\mapsto p_1 W_1(\mathbf{x}) + \ldots + p_r W_r(\mathbf{x}). \end{aligned}$$

The fibre $M = \{W = t_0\}$ over any point $t_0 \neq 0$ in $\mathbb{C}$ is smooth and its cohomology does not depend on the choice of $t_0 \in \mathbb{C}^\times$.

We consider

$$U_{\mathrm{CY}} = \mathbb{C}^n \setminus \{0\} \times \mathbb{C}^r$$

and the corresponding vector bundle $[U_{\mathrm{CY}}/\mathbb{C}^\times]$

$$\mathcal{O}_{\mathbf{w}}(-\mathbf{d}) := \bigoplus_{i=1}^{r} \mathcal{O}_{\mathbb{P}(\mathbf{w})}(-d_i),$$

which we identify with the total space over $\mathbb{P}(\mathbf{w})$. If we consider the Γ-action on $\mathbb{C}^{n+r}$, we study the corresponding (total space of the) vector bundle $[\mathcal{O}_{\mathbf{w}}(-\mathbf{d})/G] \to [\mathbb{P}(\mathbf{w})/G]$.

The function W descends to a $\mathbb{C}$-valued function W_{CY} on $[\mathcal{O}_{\mathbf{w}}(-\mathbf{d})/G]$

$$W_{\mathrm{CY}} \colon [\mathcal{O}_{\mathbf{w}}(-\mathbf{d})/G] \to \mathbb{C}.$$

PROPOSITION 3.3. *We have a canonical isomorphism*

$$H^{p-r,q-r}_{\mathrm{CR}}([X_W/G]) \cong H^{p,q}_{\mathrm{CR}}([\mathcal{O}_{\mathbf{w}}(-\mathbf{d})/G], [\mathfrak{F}/G]),$$

where $\mathfrak{F}$ *is the quotient stack* $[M/\mathbb{C}^\times]$ *inside* $\mathcal{O}_{\mathbf{w}}(-\mathbf{d})$.

REMARK 3.4. $\mathfrak{F}$ is the generic fibre of the morphism $\mathcal{O}_{\mathbf{w}}(-\mathbf{d}) \to \mathcal{O} \to \mathbb{C}$.

PROOF. On both sides we have a direct sum over $\gamma \in \Gamma$ involving Γ-invariant classes of $X_{W,\gamma}$ and of the pair $(\mathcal{O}_{\mathbf{w}}(-\mathbf{d})_\gamma, \mathfrak{F}_\gamma)$. We notice that γ acts with age a_γ on the tangent bundle of X_W and with age $a_\gamma + (r - r_\gamma)$ on the tangent bundle of $\mathcal{O}_{\mathbf{w}}(-\mathbf{d})$. Indeed, using the chain of inclusions $X_W \subset \mathbb{P}(\mathbf{w}) \subset \mathcal{O}_{\mathbf{w}}(-\mathbf{d})$ (the last one being obtained by identifying $\mathbb{P}(\mathbf{w})$ with the zero section) one readily shows that the normal bundle of X_W inside $\mathcal{O}_{\mathbf{w}}(-\mathbf{d})$ is $\mathcal{O}_{\mathbf{w}}(\mathbf{d}) \oplus \mathcal{O}_{\mathbf{w}}(-\mathbf{d})$. For any fixed element $\gamma \in \Gamma$, in view of age computations, we may ignore the r_γ summands of $\mathcal{O}_{\mathbf{w}}(-\mathbf{d})$ where γ acts trivially. Since, for a nontrivial character χ, $\mathrm{age}(\chi) + \mathrm{age}(\chi^{-1}) = 1$ we obtain the desired age shift difference $r - r_\gamma$.

Finally we need to check

$$H^{p-r_\gamma,q-r_\gamma}(X_{W,\gamma}) = H^{p,q}(\mathcal{O}_{\mathbf{w}}(-\mathbf{d})_\gamma, \mathfrak{F}_\gamma), \tag{5}$$

which requires the following analysis of $\mathfrak{F}_\gamma$.

LEMMA 3.5. *For any* $\gamma \in \Gamma$, *if* $r_\gamma < n_\gamma$, *then* $X_{W,\gamma}$ *is nondegenerate and we have*

$$H^k(\mathfrak{F}_\gamma) = \begin{cases} H^k(\mathbb{P}(\mathbf{w})_\gamma) & 0 \leq k \leq 2r_\gamma - 2 \\ H^{n_\gamma - r_\gamma - 1}_{\mathrm{pr}}(X_{W,\gamma}) & k = n_\gamma + r_\gamma - 2 \\ 0 & \text{otherwise} \end{cases}$$

If $r_\gamma \geq n_\gamma$ *then* $H^k(\mathfrak{F}_\gamma) \cong H^k(\mathbb{P}(\mathbf{w})_\gamma)$ *for all* k.

PROOF. Set $X = X_{W,\gamma}$ and $\mathfrak{F} = \mathfrak{F}_\gamma$. By abuse of notation we write $\mathbb{P}(\mathbf{w})_\gamma$ as $\mathbb{P}(\mathbf{w})$, n_γ and r_γ as n and r. Consider the projection map $\pi \colon \mathfrak{F} \to \mathbb{P}(\mathbf{w})$. Since we may regard $\mathfrak{F}$ as the variety defined by the equation $\sum_i p_i W_i(x) = 1$, the fiber $\pi^{-1}(x)$ is empty if $x \in X$ and an affine hyperplane isomorphic to $\mathbb{A}^{r-1}$ if $x \notin X$. More precisely, π is a locally trivial fibration over $\mathbb{P}(\mathbf{w})\backslash X$ with fiber $\mathbb{A}^{r-1}$ (trivialize over the open subsets $U_i = \{\mathbf{x} \in \mathbb{P}(\mathbf{w}) | W_i(\mathbf{x}) \neq 0\}$). Hence $H^*(F) \cong H^*(\mathbb{P}(\mathbf{w})\backslash X)$. As X is a $\mathbb{Q}$-homology manifold we have an isomorphism of Hodge structures

$$H^k(\mathbb{P}(\mathbf{w}), \mathbb{P}(\mathbf{w}) \setminus X; \mathbb{Q}) \cong H^{k-2r}(X; \mathbb{Q})(-r).$$

Hence the long exact sequence of relative cohomology for the pair $(\mathbb{P}(\mathbf{w}), \mathbb{P}(\mathbf{w}) \setminus X)$ can be rewritten as

$$H^k(\mathbb{P}(\mathbf{w})) \to H^k(\mathbb{P}(\mathbf{w}) \setminus X) \to H^{k-2r+1}(X) \xrightarrow{i_*} H^{k+1}(\mathbb{P}(\mathbf{w}))$$

where cohomology is with $\mathbb{Q}$-coefficients and the Gysin map i_* is Poincaré dual to the map

$$i_* : H_{2n-k-3}(X) \to H_{2n-k-3}(\mathbb{P}(\mathbf{w})).$$

The result then follows from the Lefschetz hyperplane theorem. □

We now continue the proof of Proposition 3.3 If $r_\gamma < n_\gamma$ and X_W is nondegenerate then

$$H^k(\mathcal{O}_{\mathbf{w}}(-\mathbf{d})_\gamma, \mathfrak{F}_\gamma) \cong H^{k-2r_\gamma}(X_\gamma)(-r_\gamma). \tag{6}$$

because the pullback via $\pi\colon \mathcal{O}_{\mathbf{w}}(-\mathbf{d}) \to \mathbb{P}(\mathbf{w})$ induces an isomorphism

$$H^k(\mathbb{P}(\mathbf{w})_\gamma, \mathbb{P}(\mathbf{w})_\gamma \setminus X_{W,\gamma})) \xrightarrow{\sim} H^k(\mathcal{O}_{\mathbf{w}}(-d)_\gamma, \mathfrak{F}_\gamma)$$

and $H^k(\mathbb{P}(\mathbf{w})_\gamma, \mathbb{P}(\mathbf{w})_\gamma \setminus X_{W,\gamma})) \cong H^{k-2r}(X_{W,\gamma})(-r)$ as we have seen above. If $r_\gamma \geq n_\gamma$ the same results hold, with $X = \varnothing$. □

4. Hybrid Landau–Ginzburg side

On the other side we consider the open set

$$U_{\text{LG}} = \mathbb{C}^n \times (\mathbb{C}^r \setminus \{\mathbf{0}\})$$

and the (total space of the) vector bundle

$$[U_{\text{LG}}/\mathbb{C}^\times] = \bigoplus_{j=1}^n \mathcal{O}_{\mathbb{P}(\mathbf{d})}(-w_i),$$

which we simply denote by $\mathcal{O}_{\mathbf{d}}(-\mathbf{w})$. We notice that the Milnor fiber $M = W^{-1}(t_0)$ for $t_0 \neq 0$ is contained in both U_{CY} and U_{LG}. Its cohomology does not depend on t_0 and, after modding out the $\mathbb{C}^\times$-action, has been completely described in Lemma 3.5.

We recall that Γ is a group of diagonal symmetries acting on $\mathbb{C}^n$; as in the previous section we extend its action to $\mathbb{C}^n \times \mathbb{C}^r$. We now consider the *hybrid LG model* $(\mathcal{O}_{\mathbf{d}}(-\mathbf{w}), W)$ with superpotential

$$W\colon [\mathcal{O}_{\mathbf{d}}(-\mathbf{w})/G] \to \mathbb{C}.$$

As above, the quotient stack $[\mathcal{O}_{\mathbf{d}}(-\mathbf{w})/G]$ may be presented as $[U_{\text{LG}}/\Gamma]$ and may be regarded as a vector bundle over $[\mathbb{P}(\mathbf{d})/G]$.

DEFINITION 4.1. The generalized state space of the hybrid Landau–Ginzburg model $(\mathcal{O}_{\mathbf{d}}(-\mathbf{w}), W)$ is

$$\mathcal{H}^{p,q}_\Gamma(W_1, \dots, W_r) := H^{p+r,q+r}_{\text{CR}}([\mathcal{O}_{\mathbf{d}}(-\mathbf{w})/G], [\mathfrak{F}/G]),$$

where G is the component group Γ/Γ^0.

REMARK 4.2. We point out that the space $\mathcal{H}^{p,q}(W_1, \dots, W_r)$ defined in (3) coincides with the above definition for $\Gamma = \mathbb{C}^\times$.

We can describe in greater detail the relative Chen–Ruan orbifold cohomology

$$H^{p,q}_{\mathrm{CR}}([\mathcal{O}_{\mathbb{P}(\mathbf{d})}(-\mathbf{w})/G],[\mathfrak{F}/G]) = \bigoplus_{g\in G} H^{p-a_g,q-a_g}([\mathcal{O}_{\mathbb{P}(\mathbf{d})}(-\mathbf{w})/G]_g,[\mathfrak{F}/G]_g)$$

or, more explicitly,

$$H^{p,q}_{\mathrm{CR}}([U_{\mathrm{LG}}/\Gamma],[M/\Gamma]) = \bigoplus_{\gamma\in\Gamma} H^{p-a_\gamma,q-a_\gamma}(\mathcal{O}_{\mathbf{d}_\gamma}(-\mathbf{w}_\gamma),\mathfrak{F}_\gamma)^\Gamma, \tag{7}$$

where $\mathbf{d}_\gamma$ and $\mathbf{w}_\gamma$ are the multi-indices obtained by suppressing the entries i and j for which $\gamma_i \neq 1$, and a_γ is the age of the action of γ on $\mathbb{C}^n\times\mathbb{C}^r$. We recall again that $H^*(\mathcal{O}_{\mathbf{d}_\gamma}(-\mathbf{w}_\gamma),\mathfrak{F}_\gamma)^\Gamma$ is just the group of G-invariant classes $H^*(\mathcal{O}_{\mathbf{d}_\gamma}(-\mathbf{w}_\gamma),\mathfrak{F}_\gamma)^G$.

This allows us to give an explicit presentation of FJRW cohomology as a direct sum of two types of terms reflecting the hybrid nature of our Landau–Ginzburg models: a chiral algebra term computing the primitive cohomology via the Jacobian ring, and a term computing the fixed cohomology.

The analogue of the chiral algebra of the Jacobian ring of an isolated singularity is defined as follows. For any smooth complete intersection of $\widetilde{r}$ hypersurfaces $\{W_i = 0\}$ (with $i = 1,\dots,\widetilde{r}$) inside a weighted projective space of $\widetilde{n} > \widetilde{r}$ coordinates we consider the generalised chiral algebra

$$\mathcal{Q}_{\widetilde{W}} = dx_1\wedge\ldots\wedge dx_{\widetilde{n}}\wedge dp_1\wedge\ldots\wedge dp_{\widetilde{r}}\otimes\mathbb{C}[x_1,\dots,x_{\widetilde{n}},p_1,\dots,p_{\widetilde{r}}]/(\partial_{x_j}\widetilde{W},\partial_{p_i}\widetilde{W}),$$

where $\widetilde{W} = \sum_{i=1}^{\widetilde{r}} p_i W_i(\mathbf{x})$. This is the Jacobian ring from [**15**] or [**34**] tensored with the top-degree form. Its $\mathbb{C}^\times$-invariant part is finite dimensional. For $\widetilde{D} = \widetilde{n}-\widetilde{r}-1$, by [**15**, Thm. 7], [**34**, Thm. 2.16], we have

$$(\mathcal{Q}_{\widetilde{W}})^{\mathbb{C}^\times} \cong H^{\widetilde{D}}_{\mathrm{pr}}(X_{W,\gamma})(-r).$$

If we assign bi-degree $(\widetilde{D}-k,k)$ to each term of $\mathcal{Q}_{\widetilde{W}}$ degree k in the variables $p_1,\dots,p_{\widetilde{r}}$ we may regard the above isomorphism as a bi-degree preserving isomorphism. Note that in both papers cited above the result is stated for complete intersections in projective space; the same proof goes through in the case of weighted projective spaces [**15**, Remark 18 (i)].

Finally for any non-negative integer $\widetilde{n}$ and for any $\widetilde{r}\geq\widetilde{n}$ we need to consider the cohomology of $\mathbb{P}(\mathbf{d})$ in degrees $0,\dots,2r-2n-2$. We write the bi-degree preserving identification

$$dt(-\widetilde{n})\otimes\mathbb{C}[t]/(t^{\widetilde{r}-\widetilde{n}}) \cong \bigoplus_{k=0}^{\widetilde{r}-\widetilde{n}-1} H^{2k}(\mathbb{P}(\mathbf{d})).$$

where $dt(-\widetilde{n})\otimes t^k$ has bidegree $(\widetilde{n},\widetilde{n})+(k,k)$.

THEOREM 4.3. *We have*

$$\mathcal{H}^*_\Gamma(W_1,\dots,W_r) = \oplus_{\gamma\in\Gamma}H_\gamma(-a_\gamma+r),$$

where H_γ with its double grading is given by

$$H_\gamma = \begin{cases} (\mathcal{Q}_{W_\gamma})^\Gamma & \text{if } r_\gamma < n_\gamma,\\ dt(-n_\gamma)\otimes\mathbb{C}[t]/(t^{r_\gamma-n_\gamma}) & \text{if } r_\gamma\geq n_\gamma. \end{cases}$$

The proof follows from analysis of the relative cohomology of $(\mathcal{O}_{\mathbb{P}(\mathbf{d}_\gamma)}(-\mathbf{w}_\gamma)_\gamma, F_\gamma)$ for each sector. We assume $G = 1$ for simplicity; as we saw above, nontrivial groups G can be treated easily by the same argument. By abuse of notation we write

$\mathbf{d}, \mathbf{w}$ and F instead of $\mathbf{d}_\gamma, \mathbf{w}_\gamma$ and F_γ. The following lemma provides a complete picture.

LEMMA 4.4. *If $r < n$ and X is a nondegenerate complete intersection, then*

$$H^k(\mathcal{O}_{\mathbf{d}}(-\mathbf{w}), F) = \begin{cases} H^{n-r-1}_{\mathrm{pr}}(X)(-r) & \text{if } k = n+r-2 \\ 0 & \text{if } k \neq n+r-2. \end{cases}$$

If $r \geq n$ then

$$H^k(\mathcal{O}_{\mathbf{d}}(-\mathbf{w}), F) \cong \begin{cases} H^{k-2n}(\mathbb{P}(\mathbf{d}))(-n) & \text{if } 2n \leq k \leq 2r-2 \\ 0 & \text{otherwise.} \end{cases}$$

PROOF. Let $Z_1 \subset \mathcal{O}_{\mathbf{w}}(-\mathbf{d})$ and $Z_2 \subset \mathcal{O}_{\mathbf{d}}(-\mathbf{w})$ be the zero sections. Write

$$\mathcal{O}_{\mathbf{w}}(-\mathbf{d})^\times = \mathcal{O}_{\mathbf{w}}(-\mathbf{d}) \setminus Z_1, \quad \mathcal{O}_{\mathbf{d}}(-\mathbf{w})^\times = \mathcal{O}_{\mathbf{d}}(-\mathbf{w}) \setminus Z_2.$$

The vector bundles $\mathcal{O}_{\mathbf{d}}(-\mathbf{w})$ and $\mathcal{O}_{\mathbf{w}}(-\mathbf{d})$ contain a common open subset

$$[U_{CY} \cap U_{LG}/\mathbb{C}^\times] \cong \mathcal{O}_{\mathbf{w}}(-\mathbf{d})^\times \cong \mathcal{O}_{\mathbf{d}}(-\mathbf{w})^\times.$$

The inclusions $F \subset \mathcal{O}_{\mathbf{d}}(-\mathbf{w})^\times \subset \mathcal{O}_{\mathbf{d}}(-\mathbf{w})$ give an inclusion of pairs

$$(\mathcal{O}_{\mathbf{d}}(-\mathbf{w})^\times, F) \subset (\mathcal{O}_{\mathbf{d}}(-\mathbf{w}), F) \subset (\mathcal{O}_{\mathbf{d}}(-\mathbf{w}), \mathcal{O}_{\mathbf{d}}(-\mathbf{w})^\times).$$

We shall calculate the relative cohomology $H^k(\mathcal{O}_{\mathbf{d}}(-\mathbf{w}), F)$ using the exact sequence of relative cohomology

$$H^k(\mathcal{O}_{\mathbf{d}}(-\mathbf{w}), \mathcal{O}_{\mathbf{d}}(-\mathbf{w})^\times) \to H^k(\mathcal{O}_{\mathbf{d}}(-\mathbf{w}), F) \to H^k(\mathcal{O}_{\mathbf{d}}(-\mathbf{w})^\times, F). \tag{8}$$

We have

$$H^k(\mathcal{O}_{\mathbf{d}}(-\mathbf{w}), \mathcal{O}_{\mathbf{d}}(-\mathbf{w})^\times) = H^k_{Z_2}(\mathcal{O}_{\mathbf{d}}(-\mathbf{w})) \cong H^{k-2n}(Z_2) \cong H^{k-2n}(\mathbb{P}(\mathbf{d})).$$

To calculate $H^k(\mathcal{O}_{\mathbf{d}}(-\mathbf{w})^\times, F)$, we use the isomorphism $H^k(\mathcal{O}_{\mathbf{d}}(-\mathbf{w})^\times, F) \cong H^k(\mathcal{O}_{\mathbf{w}}(-\mathbf{d})^\times, F)$ and the previous results for $\mathcal{O}_{\mathbf{w}}(-\mathbf{d})$. Using the long exact sequence

$$\ldots \to H^k(\mathcal{O}_{\mathbf{w}}(-\mathbf{d}), \mathcal{O}_{\mathbf{w}}(-\mathbf{d})^\times) \to H^k(\mathcal{O}_{\mathbf{w}}(-\mathbf{d}), F) \to H^k(\mathcal{O}_{\mathbf{w}}(-\mathbf{d})^\times, F) \to \ldots$$

coming from the inclusion of pairs

$$(\mathcal{O}_{\mathbf{w}}(-\mathbf{d})^\times, F) \subset (\mathcal{O}_{\mathbf{w}}(-\mathbf{d}), F) \subset (\mathcal{O}_{\mathbf{w}}(-\mathbf{d}), \mathcal{O}_{\mathbf{w}}(-\mathbf{d})^\times),$$

the isomorphisms

$$\begin{aligned} H^k(\mathcal{O}_{\mathbf{w}}(-\mathbf{d}), \mathcal{O}_{\mathbf{w}}(-\mathbf{d})^\times) &\cong H^{k-2r}(\mathbb{P}(\mathbf{w})) \\ H^k(\mathcal{O}_{\mathbf{w}}(-\mathbf{d}), F) &\cong H^{k-2r}(X) \ (\textit{see } (6)) \end{aligned}$$

and the five lemma, we obtain

$$H^k(\mathcal{O}_{\mathbf{w}}(-\mathbf{d})^\times, F) \cong H^{k-2r+1}(\mathbb{P}(\mathbf{w}), X).$$

We now turn to the calculation of $H^k(\mathcal{O}_{\mathbf{d}}(-\mathbf{w}), F)$ in the case $r < n$. If $k \geq 2r$ the exact sequence

$$H^{k-1}(F) \to H^k(\mathcal{O}_{\mathbf{d}}(-\mathbf{w}), F) \to H^k(\mathcal{O}_{\mathbf{d}}(-\mathbf{w}))$$

shows that

$$H^k(\mathcal{O}_{\mathbf{d}}(-\mathbf{w}), F) \cong H^{k-1}(F) = \begin{cases} H^{n-r-1}_{\mathrm{pr}}(X) & \text{if } k = n+r-1 \\ 0 & \text{otherwise.} \end{cases}$$

If $k \leq 2r-1$ the exact sequence (8) shows that $H^k(\mathcal{O}_{\mathbf{d}}(-\mathbf{w}),F)=0$ since the terms $H^{k-2n}(\mathbb{P}(\mathbf{d}))$ and

$$H^k(\mathcal{O}_{\mathbf{d}}(-\mathbf{w})^\times,F)=H^k(\mathcal{O}_{\mathbf{w}}(-\mathbf{d})^\times,F)\cong H^{k-2r+1}(\mathbb{P}(\mathbf{w}),X)$$

vanish (also for $k=2r-1$, since $H^0(\mathbb{P}(\mathbf{w}))\cong H^0(X)$).

Hence in the case $r<n$ we find

$$H^k(\mathcal{O}_{\mathbf{d}}(\ \mathbf{w}),F)=\begin{cases} H^{n-r-1}_{\mathrm{pr}}(X) & \text{if } k=n+r-1\\ 0 & \text{otherwise.}\end{cases}$$

We now consider the case $r\geq n$. The calculations go through as before, with $X=\varnothing$. Using the exact sequence of relative cohomology for the pair $(\mathcal{O}_{\mathbf{d}}(-\mathbf{w}),F)$ we find that $H^k(\mathcal{O}_{\mathbf{d}}(-\mathbf{w}),F)=0$ if $k\geq 2r$.

Next we consider the case $k\leq 2r-2$. The exact sequence (8) then shows that $H^k(\mathcal{O}_{\mathbf{d}}(-\mathbf{w}),F)\cong H^{k-2n}(\mathbb{P}(\mathbf{d}))$, hence

$$H^k(\mathcal{O}_{\mathbf{d}}(-\mathbf{w}),F)=\begin{cases} \mathbb{Q} & \text{if } 2n\leq k\leq 2r-2, k \text{ even}\\ 0 & \text{otherwise.}\end{cases}$$

(Note that if $r=n$ the formula says that $H^k(\mathcal{O}_{\mathbf{d}}(-\mathbf{w}),F)=0$ for all $k\leq 2r-2$.)

For the remaining case $k=2r-1$ we use an Euler characteristic calculation. The exact sequence (8) shows that

$$\sum_i(-1)^i(h^i(\mathcal{O}_{\mathbf{d}}(-\mathbf{w}),\mathcal{O}_{\mathbf{d}}(-\mathbf{w})^\times)-h^i(\mathcal{O}_{\mathbf{d}}(-\mathbf{w}),F)+h^i(\mathcal{O}_{\mathbf{d}}(-\mathbf{w})^\times,F))=0.$$

By the previous calculations we have

$$\begin{aligned}\sum_i(-1)^ih^i(\mathcal{O}_{\mathbf{d}}(-\mathbf{w}),\mathcal{O}_{\mathbf{d}}(-\mathbf{w})^\times) &= \sum_i(-1)^ih^{i-2n}(\mathbb{P}(\mathbf{d}))=r\\ \sum_i(-1)^ih^i(\mathcal{O}_{\mathbf{d}}(-\mathbf{w}),F) &= r-n-h^{2r-1}(\mathcal{O}_{\mathbf{d}}(-\mathbf{w}),F)\\ \sum_i h^i(\mathcal{O}_{\mathbf{d}}(-\mathbf{w})^\times,F) &= \sum_i(-1)^ih^{i-2r+1}(\mathbb{P}(\mathbf{w}))=-n.\end{aligned}$$

Hence we find that $H^{2r-1}(\mathcal{O}_{\mathbf{d}}(-\mathbf{w}),F)=0$.

So in the case $r\geq n$ we obtain

$$H^k(\mathcal{O}_{\mathbf{d}}(-\mathbf{w}),F)=\begin{cases} H^{k-2n}(\mathbb{P}(\mathbf{d})) & \text{if } 2n\leq k\leq 2r-2\\ 0 & \text{otherwise.}\end{cases}$$ □

Proof of Theorem 4.3. The claim follows immediately from the formula (7) and Lemma 4.4. □

5. The correspondence

Theorem 5.1. *Let $X_W=\{W_1=0,\dots,W_r=0\}\subset\mathbb{P}(w_1,\dots,w_n)$ be a non-degenerate Calabi–Yau complete intersection. There exists an explicit bidegree preseving isomorphism*

$$H^*_{\mathrm{CR}}([\mathcal{O}_{\mathbb{P}(\mathbf{d})}(-\mathbf{w})/G])\cong H^*_{\mathrm{CR}}([\mathcal{O}_{\mathbb{P}(\mathbf{w})}(-\mathbf{d})/G]).$$

As an immediate corollary we get the desired isomorphism.

Corollary 5.2. *There exists an explicit bidegree preserving isomorphism*

$$\mathcal{H}^*_\Gamma(W_1,\dots,W_r)\cong H^*_{\mathrm{CR}}([X_W/G]).$$

PROOF OF THEOREM 5.1. Let us refer to the left hand side as the CY side, and to the right hand side as the LG side. By definition, both sides decompose as the direct sum over Γ

$$\mathrm{CY} = \bigoplus_{\gamma\in\Gamma}\bigoplus_{0\leq k<n_\gamma}[\mathrm{CY}^k_\gamma]\mathbb{Q} \qquad \text{and} \qquad \mathrm{LG} = \bigoplus_{\gamma\in\Gamma}\bigoplus_{0\leq k<r_\gamma}[\mathrm{LG}^k_\gamma]\mathbb{Q},$$

where all terms CY^k_γ and LG^k_γ are elements of the form $H^k\cap \mathbf{1}_\gamma$, where H is an hyperplane section and $\mathbf{1}_\gamma$ is the fundamental class of the corresponding sector. The bidegree (p,p) of such elements is $p = k-1+a_\gamma$ by definition.

We shall identify the CY and the LG side for every connected component of Γ. Each component is of the form $g\overline{\lambda}$ for a fixed diagonal symmetry of $W_1,\dots,W_n$. We express g as the diagonal matrix whose $n+r$ entries on the diagonal are $(\exp(2\pi i a_1),\dots,\exp(2\pi i a_n),\mathbf{1})$ with $a_i\in[0,1[$. We write $\lambda = \exp(-2\pi i t)$.

Let $\gamma = g\overline{\lambda}$ with g and λ as above. We denote the fractional part of x by $\langle x\rangle = x-\lfloor x\rfloor$. By definition we have

$$\begin{aligned} a_{g\overline{\lambda}} &= \sum_{j=1}^n\langle -tw_j+a_j\rangle + \sum_{i=1}^r\langle td_i\rangle \\ &= \sum a_j - \sum_{i=1}^r\lfloor d_i t\rfloor - \sum_{j=1}^n\lfloor a_j - w_j t\rfloor \\ &= \sum a_j - \sum_{i=1}^r\lfloor d_i t\rfloor + n + \sum_{j=1}^n\begin{cases}\lfloor w_j t - a_j\rfloor & w_jt-a_j\notin\mathbb{Z}\\ \lfloor w_jt-a_j\rfloor - 1 & w_jt-a_j\in\mathbb{Z}.\end{cases}\end{aligned}$$

The generators of CY and LG arising from Γ can be easily pictured in the following diagram representing on concentric circles the angular coordinates of the elements of $\mu_{w_1}\sqcup\cdots\sqcup\mu_{w_N}$ (black dots) and of $\mu_{d_1}\sqcup\cdots\sqcup\mu_{d_r}$ (white dots).

Let us illustrate how the white and black dots correspond to generators of LG and CY, respectively. Notice that, by construction, a ray $\{\rho\lambda\mid\rho\in\mathbb{R}^+\}$ carries r_λ white dots and n_λ black dots. The white dots correspond to the states $\mathrm{LG}^0_\gamma,\dots,\mathrm{LG}^{r_\gamma-1}_\gamma$ whereas the black dots correspond to the states $\mathrm{CY}^0_\gamma,\dots,\mathrm{CY}^{n_\gamma-1}_\gamma$. The total number of black dots equals the total number of white dots, *i.e.* $D = \sum_j w_j = \sum_i d_i$. We can order all the dots in lexicographic order starting from CY_0; we get $x_1,x_2,\dots,x_{2D}$. Then, by the above formulae, the cohomological Chen–Ruan degree of a state corresponding to x_i is $\sum_j a_j$ plus the following function

$$f(x_i) = \#\{y \text{ black } \mid x_1\leq y<x_i\} - \#\{y\text{ white }\mid x_1\leq y\leq x_i\}.$$

We can identify each term x_i with the integer i and extend this degree function to a piecewise linear real function defined on $]\frac12, 2D+\frac12]$

$$f(x) = \begin{cases} f(x_i)+(x-i) & \text{if } x \text{ black and } x\in]x_i-\frac12,x_i+\frac12],\\ f(x_i)-(x-i) & \text{if } x \text{ white and } x\in]x_i-\frac12,x_i+\frac12].\end{cases}$$

This function is continuous and its values at the boundary of the interval coincide, so it can be regarded as a continuous function on a circle. It is strictly increasing at all black dots and strictly decreasing at all white dots; its relative maxima occur for values of the form $\frac12(x_i+x_{i+1})$ where x_i is black and x_{i+1} is white. Similarly, with inverted colours, for the relative minima. In this way if f attains a given (integer) value at a given number of black dots it must attain the same value at the same

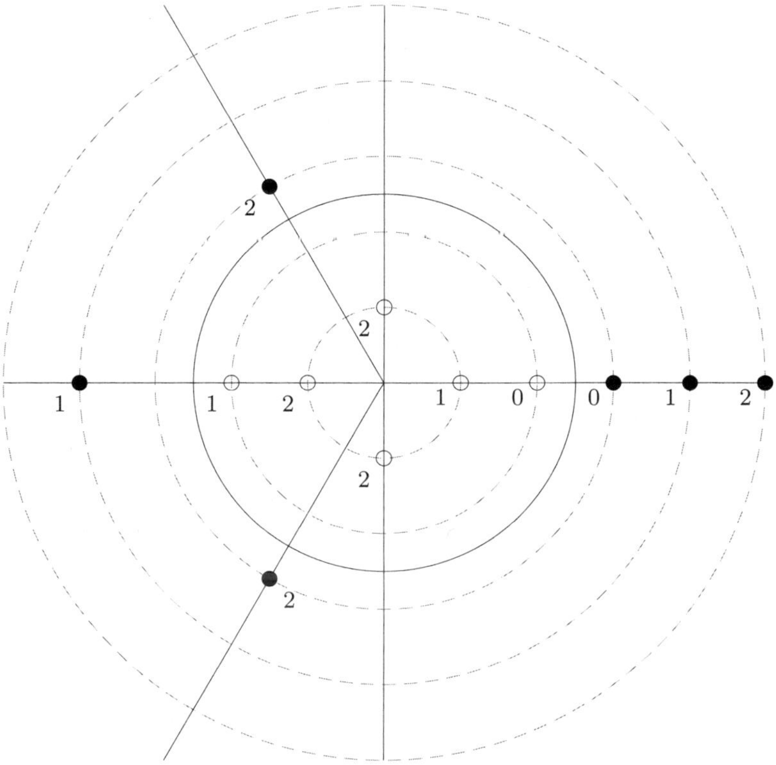

FIGURE 1. Diagram of $\{x_1^2 + x_2 = 0, x_1^4 + x_2^2 + x_3x_1 = 0\} \subseteq \mathbb{P}(1,2,3)$; $(w_1, w_2, w_3, d_1, d_2) = (1,2,3,-2,-4)$.

number of white dots. (We illustrate this in Fig. 1 by writing the value of f next to each dot). This yields the desired bidegree preserving LG/CY isomorphism.

We can use the function f to define an explicit isomorphism, simply by pairing each black dot for which $f = k$ to the following white dot for which $f = k$. This yields an explicit isomorphism. $\square$

References

[1] Pedro Acosta and Mark Shoemaker, *Quantum cohomology of toric blowups and Landau-Ginzburg correspondences*, Algebr. Geom. **5** (2018), no. 2, 239–263. MR3769893

[2] Per Berglund and Tristan Hübsch, *A generalized construction of mirror manifolds*, Nuclear Phys. B **393** (1993), no. 1-2, 377–391, DOI 10.1016/0550-3213(93)90250-S. MR1214325

[3] Lev A. Borisov, *Berglund-Hübsch mirror symmetry via vertex algebras*, Comm. Math. Phys. **320** (2013), no. 1, 73–99, DOI 10.1007/s00220-013-1717-y. MR3046990

[4] Lev A. Borisov and Anatoly Libgober, *Elliptic genera of toric varieties and applications to mirror symmetry*, Invent. Math. **140** (2000), no. 2, 453–485, DOI 10.1007/s002220000058. MR1757003

[5] Lev Borisov and Anatoly Libgober, *Elliptic genera of singular varieties*, Duke Math. J. **116** (2003), no. 2, 319–351, DOI 10.1215/S0012-7094-03-11625-7. MR1953295

[6] Weimin Chen and Yongbin Ruan, *Orbifold Gromov-Witten theory*, Orbifolds in mathematics and physics (Madison, WI, 2001), Contemp. Math., vol. 310, Amer. Math. Soc., Providence, RI, 2002, pp. 25–85, DOI 10.1090/conm/310/05398. MR1950941

[7] Alessandro Chiodo and Yongbin Ruan, *Landau-Ginzburg/Calabi-Yau correspondence for quintic three-folds via symplectic transformations*, Invent. Math. **182** (2010), no. 1, 117–165, DOI 10.1007/s00222-010-0260-0. MR2672282

[8] Alessandro Chiodo and Yongbin Ruan, *LG/CY correspondence: the state space isomorphism*, Adv. Math. **227** (2011), no. 6, 2157–2188, DOI 10.1016/j.aim.2011.04.011. MR2807086

[9] Alessandro Chiodo and Yongbin Ruan, *A global mirror symmetry framework for the Landau-Ginzburg/Calabi-Yau correspondence* (English, with English and French summaries), Ann. Inst. Fourier (Grenoble) **61** (2011), no. 7, 2803–2864. MR3112509

[10] Alessandro Chiodo, Hiroshi Iritani, and Yongbin Ruan, *Landau-Ginzburg/Calabi-Yau correspondence, global mirror symmetry and Orlov equivalence*, Publ. Math. Inst. Hautes Études Sci. **119** (2014), 127–216, DOI 10.1007/s10240-013-0056-z. MR3210178

[11] Alessandro Chiodo, An algorithm computing genus-one invariants of the Landau–Ginzburg model of the quintic Calabi–Yau three-fold (joint work with Yongbin Ruan and Dimitri Zvonkine), Mirror Symmetry, Hodge Theory and Differential Equations, (Oberwolfach, 19-25 April 2015), To appear in the Oberwolfach Reports 2015, available at `http://webusers.imj-prg.fr/~alessandro.chiodo/g1.pdf`

[12] Emily Clader, *Landau-Ginzburg/Calabi-Yau correspondence for the complete intersections* $X_{3,3}$ *and* $X_{2,2,2,2}$, Adv. Math. **307** (2017), 1–52, DOI 10.1016/j.aim.2016.11.010. MR3590512

[13] Emily Clader and Dusty Ross, Sigma Models and Phase Transitions for Complete Intersections, International Mathematical Research Notices (to appear), `https://doi.org/10.1093/imrn/rnx`

[14] Jan Denef and François Loeser, *Motivic integration, quotient singularities and the McKay correspondence*, Compositio Math. **131** (2002), no. 3, 267–290, DOI 10.1023/A:1015565912485. MR1905024

[15] Alexandru Dimca, *Residues and cohomology of complete intersections*, Duke Math. J. **78** (1995), no. 1, 89–100, DOI 10.1215/S0012-7094-95-07805-3. MR1328753

[16] L. Dixon, J. A. Harvey, C. Vafa, and E. Witten, *Strings on orbifolds*, Nuclear Phys. B **261** (1985), no. 4, 678–686, DOI 10.1016/0550-3213(85)90593-0. MR818423

[17] L. Dixon, J. Harvey, C. Vafa, and E. Witten, *Strings on orbifolds. II*, Nuclear Phys. B **274** (1986), no. 2, 285–314, DOI 10.1016/0550-3213(86)90287-7. MR851703

[18] Huijun Fan, Tyler Jarvis, and Yongbin Ruan, *The Witten equation, mirror symmetry, and quantum singularity theory*, Ann. of Math. (2) **178** (2013), no. 1, 1–106, DOI 10.4007/annals.2013.178.1.1. MR3043578

[19] Huijun Fan, Tyler Jarvis, and Yongbin Ruan, *The moduli space in the gauged linear sigma model*, Chin. Ann. Math. Ser. B **38** (2017), no. 4, 913–936, DOI 10.1007/s11401-017-1104-7. MR3673175

[20] Jérémy Guéré, *A Landau-Ginzburg mirror theorem without concavity*, Duke Math. J. **165** (2016), no. 13, 2461–2527, DOI 10.1215/00127094-3477235. MR3546967

[21] Shuai Guo, Dustin Ross, The Genus-One Global Mirror Theorem for the Quintic Threefold, `https://arxiv.org/abs/1703.06955`

[22] Shuai Guo, Felix Janda, Yongbin Ruan, A mirror theorem for genus two Gromov-Witten invariants of quintic threefolds, `https://arxiv.org/abs/1709.07392`

[23] W. He, S. Li, Y. Shen, and R. Webb, Landau–Ginzburg mirror symmetry conjecture, available at `http://arxiv.org/pdf/1503.01757`

[24] Manfred Herbst, Kentaro Hori, and David C. Page, Phases of $N = 2$ theories in $1 + 1$ dimensions with boundary, available at `http://arxiv.org/pdf/0803.2045`

[25] K. Hori and D. Tong, Aspects of Non-Abelian Gauge Dynamics in Two-Dimensional $N = (2,2)$ Theories, JHEP 0705 (2007), available at `http://arxiv.org/pdf/hep-th/0609032`

[26] M.-x. Huang, A. Klemm, and S. Quackenbush, *Topological string theory on compact Calabi-Yau: modularity and boundary conditions*, Homological mirror symmetry, Lecture Notes in Phys., vol. 757, Springer, Berlin, 2009, pp. 45–102. MR2596635

[27] Kenneth Intriligator and Cumrun Vafa, *Landau-Ginzburg orbifolds*, Nuclear Phys. B **339** (1990), no. 1, 95–120, DOI 10.1016/0550-3213(90)90535-L. MR1061738

[28] Ralph M. Kaufmann, *Singularities with symmetries, orbifold Frobenius algebras and mirror symmetry*, Gromov-Witten theory of spin curves and orbifolds, Contemp. Math., vol. 403, Amer. Math. Soc., Providence, RI, 2006, pp. 67–116, DOI 10.1090/conm/403/07596. MR2234885

[29] Ralph M. Kaufmann, *Orbifolding Frobenius algebras*, Internat. J. Math. **14** (2003), no. 6, 573–617, DOI 10.1142/S0129167X03001831. MR1997832

[30] Ralph M. Kaufmann, *Orbifold Frobenius algebras, cobordisms and monodromies*, Orbifolds in mathematics and physics (Madison, WI, 2001), Contemp. Math., vol. 310, Amer. Math. Soc., Providence, RI, 2002, pp. 135–161, DOI 10.1090/conm/310/05402. MR1950945

[31] Marc Krawitz, *FJRW rings and Landau-Ginzburg mirror symmetry*, ProQuest LLC, Ann Arbor, MI, 2010. Thesis (Ph.D.)–University of Michigan. MR2801653

[32] Yuan-Pin Lee, Nathan Priddis, and Mark Shoemaker, *A proof of the Landau-Ginzburg/Calabi-Yau correspondence via the crepant transformation conjecture* (English, with English and French summaries), Ann. Sci. Éc. Norm. Supér. (4) **49** (2016), no. 6, 1403–1443, DOI 10.24033/asens.2312. MR3592361

[33] Anatoly Libgober, *Elliptic genus of phases of $N = 2$ theories*, Comm. Math. Phys. **340** (2015), no. 3, 939–958, DOI 10.1007/s00220-015-2465-y. MR3406638

[34] Jan Nagel, *The Abel-Jacobi map for complete intersections*, Indag. Math. (N.S.) **8** (1997), no. 1, 95–113, DOI 10.1016/S0019-3577(97)83353-8. MR1617814

[35] Miles Reid, *La correspondance de McKay*, Astérisque **276** (2002), 53–72. Séminaire Bourbaki, Vol. 1999/2000. MR1886756

[36] Alexander Polishchuk and Arkady Vaintrob, *Chern characters and Hirzebruch-Riemann-Roch formula for matrix factorizations*, Duke Math. J. **161** (2012), no. 10, 1863–1926, DOI 10.1215/00127094-1645540. MR2954619

[37] Alexander Polishchuk and Arkady Vaintrob, *Matrix factorizations and cohomological field theories*, J. Reine Angew. Math. **714** (2016), 1–122, DOI 10.1515/crelle-2014-0024. MR3491884

[38] Einar Andreas Rødland, *The Pfaffian Calabi-Yau, its mirror, and their link to the Grassmannian $G(2,7)$*, Compositio Math. **122** (2000), no. 2, 135–149, DOI 10.1023/A:1001847914402. MR1775415

[39] Matthieu Romagny, *Group actions on stacks and applications*, Michigan Math. J. **53** (2005), no. 1, 209–236, DOI 10.1307/mmj/1114021093. MR2125542

[40] Joseph Steenbrink, *Intersection form for quasi-homogeneous singularities*, Compositio Math. **34** (1977), no. 2, 211–223. MR0453735

[41] Edward Witten, *Phases of $N = 2$ theories in two dimensions*, Nuclear Phys. B **403** (1993), no. 1-2, 159–222, DOI 10.1016/0550-3213(93)90033-L. MR1232617

[42] Edward Witten, *Algebraic geometry associated with matrix models of two-dimensional gravity*, Topological methods in modern mathematics (Stony Brook, NY, 1991), Publish or Perish, Houston, TX, 1993, pp. 235–269. MR1215968

IMJ-PRG UMR7586, CNRS, UPMC, case 247, 4 Place Jussieu, 75252 Paris cedex 5, France

Email address: alessandro.chiodo@imj-prg.fr

IMB UMR5584, CNRS, Univ. Bourgogne Franche-Comté, 21000 Dijon, France

Email address: johannes.nagel@u-bourgogne.fr

Proceedings of Symposia in Pure Mathematics
Volume **100**, 2018
https://doi.org/10.1090/pspum/100/01766

Singular vector structure of quantum curves

Paweł Ciosmak, Leszek Hadasz, Masahide Manabe, and Piotr Sułkowski

ABSTRACT. We show that quantum curves arise in infinite families and have the structure of singular vectors of a relevant symmetry algebra. We analyze in detail the case of the hermitian one-matrix model with the underlying Virasoro algebra, and the super-eigenvalue model with the underlying super-Virasoro algebra. In the Virasoro case we relate singular vector structure of quantum curves to the topological recursion, and in the super-Virasoro case we introduce the notion of super-quantum curves. We also discuss the double quantum structure of the quantum curves and analyze specific examples of Gaussian and multi-Penner models.

CONTENTS

1. Introduction

Riemann surfaces and plane algebraic curves, which can be defined simply by the equation

$$A(x, y) = 0, \tag{1.1}$$

with $A(x, y)$ being a polynomial in complex variables, play an important role in modern theoretical and mathematical physics. They arise as spectral curves in matrix models, Seiberg-Witten curves, mirror curves in topological string theory, A-polynomials in knot theory, in various problems in enumerative geometry, and in many other contexts. In all these systems there is a natural notion of quantization, and algebraic curves mentioned above encode some information that can be interpreted as classical, which arises in the vanishing limit of the appropriate Planck constant, denoted $\hbar$ or g_s. This suggests that there exist quantum counterparts of

---

2010 *Mathematics Subject Classification.* Primary 81T40; Secondary 81T60, 81R10, 14H81.

This work is supported by the ERC Starting Grant no. 335739 *"Quantum fields and knot homologies"* funded by the European Research Council under the European Union's Seventh Framework Programme.

those curves, referred to simply as quantum curves, which take form of differential (or difference) operators $\widehat{A}(\widehat{x},\widehat{y})$ annihilating appropriate partition functions $\psi(x)$

$$\widehat{A}(\widehat{x},\widehat{y})\psi(x)=0, \tag{1.2}$$

with $\widehat{x}$ and $\widehat{y}$ satisfying the commutation relation

$$[\widehat{y},\widehat{x}]=g_s. \tag{1.3}$$

In the classical, $g_s\to 0$ limit $\widehat{A}(\widehat{x},\widehat{y})$ is expected to reduce to $A(x,y)$. Such quantum curves and partition functions are also referred to, respectively, as Schrödinger operators and wave-functions.

Examples of quantum curves have been found from several seemingly unrelated perspectives. First, they were identified in the system of intersecting branes, which encode Seiberg-Witten theory and provide quantization of Seiberg-Witten curves [**1**, **2**]. Second, it was realized that partition functions of branes in topological string theory satisfy difference equations, which quantize mirror curves [**3**]. Third, it is known that determinant expectation values in matrix models satisfy differential equations, which can be regarded as quantizations of spectral curves [**4**, **5**]. Furthermore, analogous statements were found in knot theory in the context of the AJ conjecture, where colored knot polynomials turn out to satisfy difference equations that quantize A-polynomials [**6**, **7**]. Finally, enumerative geometry provides yet another source of such examples [**8**, **9**].

While the above examples may seem unrelated, it appears that there is a quantization framework that works universally for all of them [**4**], which is based on the topological recursion [**10**]. The topological recursion was originally found as the solution to loop equations in matrix models [**11**–**13**], so it is not that surprising that it provides a quantization of matrix model spectral curves. It is however more surprising that the topological recursion provides a solution to the topological string theory on toric Calabi-Yau threefolds [**14**–**16**] – and so, via geometric engineering, to Seiberg-Witten theory – as well as reconstructs the asymptotic expansion of colored knot polynomials [**17**–**19**]. Based on these observations, a universal quantization procedure based on the topological recursion was proposed for all those and some other systems, with its universality explained by the underlying B-model structure (manifest possibly after some chains of dualities) of all those systems [**4**].

In the systems mentioned above, it is usually claimed that there is a unique quantum curve assigned to a given classical curve. However, as we summarize in this note following [**20**, **21**], it turns out that to a given classical curve one can naturally assign not just a single quantum curve, but an infinite family of quantum curves, which are in one-to-one correspondence with singular vectors (also known as null vectors) of the underlying symmetry algebra. In case of the original formulation of the topological recursion, and in all systems mentioned above (topological strings, Seiberg-Witten theory, knot polynomials, etc.) this is the Virasoro algebra. Quantum curves mentioned above and discussed so far in literature, from the perspective of [**20**], correspond to Virasoro singular vectors at level 2. The relation of (previously considered) quantum curves to singular vectors at level 2 was noticed also in [**5**]. However, in addition to those quantum curves at level 2 we identify an infinite family of quantum curves, which correspond to, and have the structure of, Virasoro singular vectors at higher levels. The first aim of this note is to discuss the structure and properties of such quantum curves, corresponding to all Virasoro singular vectors.

The second aim of this note is to present the relation between quantum curves and underlying symmetry algebra in another example, that of super-Virasoro algebra [**21**]. On one hand, this example nicely illustrates the general idea of the correspondence between quantum curves and symmetries. On the other hand, it leads to the construction of supersymmetric quantum curves, or super-quantum curves for short, which quantize underlying supersymmetric algebraic curves, and are interesting in their own right.

As mentioned above, the quantization procedure for algebraic curves can be very generally formulated by means of the topological recursion, at least in the context of the Virasoro algebra. However, to identify quantum curves corresponding to higher level singular vectors, it is of advantage to consider matrix model formulation of this problem, which is equivalent (at least once both matrix model and corresponding topological recursion formulations exist). In the matrix model formulation the relation to the underlying symmetry algebra is more explicit, and can be identified upon rewriting of the loop equations in the form of Virasoro-like constraints [**22**–**25**]. The approach based on matrix models and loop equations has also other advantages. First, it is not difficult to generalize this approach to the β-deformed case, which makes contact with conformal field theory with arbitrary central charge [**20**, **26**]. Second, it enables to analyze other types of matrix models, with different underlying algebras. We will illustrate this statement in the example of a supersymmetric version of one-matrix model, the so called super-eigenvalue model introduced in [**27**] and further analyzed in [**28**–**32**], characterized by the underlying super-Virasoro algebra mentioned above. The topological recursion formalism is not known in this case, so it is of particular advantage to consider matrix model formulation. On the other hand, in the Virasoro case, having conducted matrix model analysis we reformulate the construction of higher level quantum curves in the language of the topological recursion, which is well understood in this case.

More precisely, our strategy is as follows. To start with, we identify a putative wave-function with some particular matrix integral, that we refer to as the α/β-deformed matrix integral, or α/β-deformed matrix model. It takes form of the β-deformed expectation value of an x-dependent determinantal expression, raised to a power parameterized by a parameter α. We find that the wave-function defined in this way satisfies a differential equation in x only for discrete series of values of α, which turn out to coincide with degenerate momenta in conformal field theory interpretation. In case of the Virasoro algebra these values take form $\alpha = \alpha_{r,s} = -\frac{1}{2}(r-1)\epsilon_1 - \frac{1}{2}(s-1)\epsilon_2$, with $r, s = 1, 2, 3, \ldots$, and ϵ_1, ϵ_2 are simply related to parameters β, g_s (and encode the Omega-background in gauge theory interpretation). Furthermore, these differential equations have the structure of singular vectors of the relevant symmetry algebra, and can be expressed in terms of relevant representation of the symmetry generators.

Our results lead to several other interesting conclusions. First, we find interesting universal expressions for singular vectors, which do not seem to have been known before. For a fixed level, these expressions depend on the parameter α, and setting this value to be equal to the value of the degenerate momentum at the same or any lower level, these expressions reduce to appropriate expressions for singular vectors corresponding to the given degenerate momentum. Second, we note that there are two interesting classical limits of the quantum curves that we consider: the 't Hooft limit of large size of matrices N (or equivalently vanishing g_s), and the classical limit of infinite central charge from the viewpoint of Liouville conformal field theory, which is equivalent to the Nekrasov-Shatashvili limit and corresponds to vanishing or infinite value of β. In this sense the quantum curves that we consider can be interpreted as doubly-quantized objects, similarly (and especially after replacing parameters g_s and β by ϵ_1 and ϵ_2) to considerations in the context of the Langlands correspondence in [**33**]. Third, upon specialization of our analysis to multi-Penner or super-multi-Penner models, corresponding quantum curves and other objects reduce to expressions familiar from conformal field theory studies, such as (supersymmetric) BPZ equations, Ward identities, correlation functions and a representation of (super-)Virasoro operators acting thereon, etc.

We also note that results summarized in this note – at least in the context of the Virasoro algebra and standard β-deformed matrix models – can be rephrased in the language of M-theory, surface operators, as well as refined topological string theory (more precisely, the β-deformed hermitian matrix model can be regarded as a definition of refined topological string theory [**5**, **34**], whose worldsheet formulation has not been found to date). In the M-theory interpretation, one considers a background geometry of the form

$$S^1 \times \text{Taub-NUT} \times X, \tag{1.4}$$

where X is a Calabi-Yau threefold mirror to the geometry $A(x,y) = uv$, with $A(x,y)$ given in (1.1). The Taub-NUT space is parametrized by complex parameters z_1 and z_2, and twisted such that a rotation along the circle S^1 induces a rotation $z_1 \mapsto e^{i\epsilon_1} z_1, z_2 \mapsto e^{i\epsilon_2} z_2$. In addition, in this background one can introduce M5-branes that wrap S^1, a lagrangian subspace of X, and a complex line in the Taub-NUT space parametrized by z_1 or z_2. These M5-branes are referred to as ϵ_1-branes and ϵ_2-branes respectively. The quantum curves and wave-functions labeled by Virasoro degenerate momentum $\alpha = \alpha_{r,s}$ (with $r, s = 1, 2, 3, \ldots$) represent a stack that consists of $(s-1)$ overlapping ϵ_1-branes and $(r-1)$ overlapping ϵ_2-branes. Reducing this system along X leads to analogous interpretation with a stack of $(s-1)$ surface operators of one type and $(r-1)$ surface operators of another type in four or five-dimensional supersymmetric gauge theory, with non-trivial Omega-background parametrized by ϵ_1 and ϵ_2. On the other hand, reducing this system to the internal space and taking the mirror leads to the refined B-model topological string configuration on the background $A(x,y) = uv$, with two types of B-branes, whose wave-function is also labeled by $\alpha = \alpha_{r,s}$. Such systems with a single ϵ_1-brane or a single ϵ_2-brane are discussed in [**5**]. The plan of this note is as follows. In section 2 we present the singular vector structure of quantum curves in the bosonic case, with the underlying Virasoro algebra. In section 3 we discuss super-eigenvalue models and introduce super-quantum curves, which have the structure of singular vectors of the super-Virasoro algebra. Our presentation is based in particular on [**20**, **21**]; however we mostly present main statements and results, and recommend the original papers to those readers who wish to get acquainted with proofs and technical computations. Section 4 includes some comments and closing remarks.

2. Higher level (Virasoro) quantum curves

In this section we determine higher level quantum curves corresponding to singular vectors of the Virasoro algebra, by considering β-deformation of the hermitian one-matrix model; we closely follow [**20**], where more details and proofs are presented.

2.1. α/β-deformed matrix integrals. To start with we introduce an expression that we refer to as α/β-deformed matrix integral, which will play role of the wave-function

(2.1)
$$\widehat{\psi}_\alpha(x) = \frac{e^{-\frac{2\alpha}{\epsilon_1\epsilon_2}V(x)}}{(2\pi)^N N!}\int_{\mathbb{R}^N}\Delta(z)^{2\beta}\Big(\prod_{a=1}^N (x-z_a)^{-\frac{2\alpha}{\epsilon_2}}\Big)e^{-\frac{\sqrt{\beta}}{\hbar}\sum_{a=1}^N V(z_a)}\prod_{a=1}^N dz_a.$$

Ignoring the overall normalization and the term in the bracket under the integral (or simply for $\alpha = 0$), this expression reduces to the eigenvalue representation of the partition function $Z \equiv \widehat{\psi}_{\alpha=0}(x)$ of the standard β-deformed matrix model. By $\Delta(z) = \prod_{1\leq a<b\leq N}(z_a - z_b)$ we denote the Vandermonde determinant, and various parameters that we use are related as follows

$$\epsilon_1 = -\beta^{1/2}g_s, \qquad \epsilon_2 = \beta^{-1/2}g_s, \qquad g_s = 2\hbar, \qquad b^2 = -\beta = \frac{\epsilon_1}{\epsilon_2}. \tag{2.2}$$

The exponential term under the integral is the eigenvalue representation of $e^{-\frac{\sqrt{\beta}}{\hbar}\mathrm{Tr}V(M)}$, with the potential $V = V(x)$ parametrized by times t_n and assumed to take form

$$V(x) = \sum_{n=0}^\infty t_n x^n. \tag{2.3}$$

For $\beta = 1, \frac{1}{2}, 2$ the expression (2.1) is an eigenvalue representation of an integral over respectively hermitian, orthogonal, and symplectic matrices M. The term in the bracket under the integral is the eigenvalue representation of the determinant $\det(x-M)$, raised to a power parametrized by α. To the above model one can also associate the chiral boson field $\phi(x)$, with the background charge $Q = i\big(b+\frac{1}{b}\big) = \frac{\epsilon_1+\epsilon_2}{g_s}$ and central charge $c = 1-6Q^2$

(2.4)
$$\phi(x) = -\frac{g_s N}{\epsilon_2}\log x + \frac{1}{g_s}\sum_{n=0}^\infty t_n x^n - \frac{g_s}{2}\sum_{n=1}^\infty \frac{1}{nx^n}\partial_{t_n} = \frac{1}{g_s}V(x) - \frac{g_s}{\epsilon_2}\sum_{a=1}^N \log(x-z_a),$$

where we identified derivatives with respect to times of the partition function Z (or expectation values of time-independent operators) with unnormalized expectation values (see (2.52) for the definition)

$$\Big\langle \sum_{a=1}^N z_a^n \cdots \Big\rangle = -\frac{\epsilon_2}{2}\partial_{t_n}\Big\langle \cdots \Big\rangle. \tag{2.5}$$

The wave-function (2.1) arises then as the expectation value of the exponent $e^{\frac{2\alpha}{g_s}\phi(x)}$, which in particular gives rise to the prefactor $e^{-\frac{2\alpha}{\epsilon_1\epsilon_2}V(x)}$ (and keeping it simplifies some calculations).

2.2. Higher level quantum curves as Virasoro singular vectors. The main question that we ask is whether a putative wave-function (2.1) satisfies a finite order differential equation in parameter x. In [**20**] we show that this is so only for discrete values of α of the form

$$\alpha = \alpha_{r,s} = -\frac{r-1}{2}\epsilon_1 - \frac{s-1}{2}\epsilon_2, \qquad \text{for } r,s = 1,2,3,\dots \tag{2.6}$$

These values agree with degenerate momenta of the chiral boson in presence of the background charge, and up to a normalization by ig_s can be written in the form $(1-r)b+(1-s)b^{-1}$. Furthermore, we find that for a particular value of $\alpha = \alpha_{r,s}$ the wave-function $\widehat{\psi}_\alpha(x)$ satisfies a differential equation of order $n = rs$ in x, which we write as

$$\widehat{A}_n^\alpha \widehat{\psi}_\alpha(x) = 0. \tag{2.7}$$

We refer to operators $\widehat{A}_n^\alpha$ as (higher level) quantum curves. In general they take form

$$\widehat{A}_n^\alpha = \sum_{p_1+p_2+\ldots+p_k=n} \widehat{c}_{p_1,p_2,\ldots,p_k}(\alpha)\, \widehat{L}_{-p_1}\widehat{L}_{-p_2}\cdots\widehat{L}_{-p_k}, \tag{2.8}$$

where operators $\widehat{L}_{-p}$ form a representation of the Virasoro algebra

$$[L_m, L_n] = (m-n)L_{m+n} + \frac{c}{12}(m^3-m)\delta_{m+n,0}. \tag{2.9}$$

The crucial observation is that (2.8) has the same form as an operator $A_{r,s}$ that acting on a primary state $|\Delta_{r,s}\rangle$ creates a Virasoro singular vector corresponding to a given value $\alpha_{r,s}$. The constants $\widehat{c}_{p_1,p_2,\ldots,p_k}(\alpha)$, after specialization to $\alpha = \alpha_{r,s}$, have the same values as in known expressions for singular vectors. Indeed, in general, Virasoro singular vectors (at level $n = rs$) are labeled by two positive integers r, s (that also label corresponding degenerate momenta (2.6)) and take form $A_{r,s}|\Delta_{r,s}\rangle$, where $|\Delta_{r,s}\rangle$ is the primary state of weight $\Delta_{r,s} = \frac{1-r^2}{4}b^2 + \frac{1-s^2}{4}b^{-2} + \frac{1-rs}{2}$; the operators $A_{r,s}$ for the first few singular vectors take form

(2.10)

$$\begin{aligned}
A_{2,1} &= L_{-1}^2 + b^2 L_{-2}\\
A_{3,1} &= L_{-1}^3 + 2b^2 L_{-1}L_{-2} + 2b^2 L_{-2}L_{-1} + 4b^4 L_{-3}\\
&= L_{-1}^3 + 4b^2 L_{-2}L_{-1} + (2b^2+4b^4)L_{-3}\\
A_{4,1} &= L_{-1}^4 + 10b^2 L_{-1}^2 L_{-2} + 9b^4 L_{-2}^2 + (24b^4 - 10b^2)L_{-1}L_{-3}\\
&\qquad + (36b^6 - 24b^4 + 6b^2)L_{-4}\\
A_{2,2} &= L_{-1}^4 + \left(b^2 - b^{-2}\right)^2 L_{-2}^2 + \frac{3}{2}\left(b+b^{-1}\right)^2 L_{-1}^2 L_{-2} + \frac{3}{2}\left(b-b^{-1}\right)^2 L_{-2}L_{-1}^2 +\\
&\qquad - \left(b^2+b^{-2}\right)L_{-1}L_{-2}L_{-2}
\end{aligned}$$

and operators $A_{s,r}$ are related to $A_{r,s}$ by replacing b by b^{-1}.

It is noteworthy that expressions of the form (2.8) that we find (and which do not seem to have been known before) have general, universal dependence on α, and our work provides an algorithm to determine coefficients $\widehat{c}_{p_1,p_2,\ldots,p_k}(\alpha)$ that encode at once all singular vectors up to a given level; such coefficients determined explicitly up to level 5 are summarized below. This result is interesting, especially that, even though it is known that Virasoro singular vectors can be written in terms of Jack polynomials [**35**], determining their explicit form is still an important problem in conformal field theory (which is solved only in some particular cases, such as $r = 1$ or $s = 1$ [**36**]).

Furthermore, the operators $\widehat{L}_{-p}$ in (2.8) with $p \geq 0$ form the following interesting representation of the Virasoro algebra on a space of functions in x and times t_k

(2.11)
$$\widehat{L}_0 = \Delta_\alpha \equiv \frac{\alpha}{g_s}\Big(\frac{\alpha}{g_s} - Q\Big), \qquad \widehat{L}_{-1} = \partial_x,$$
$$\widehat{L}_{-n} = -\frac{1}{\epsilon_1\epsilon_2(n-2)!}\Big(\partial_x^{n-2}\big(V'(x)^2\big) + (\epsilon_1+\epsilon_2)\partial_x^n V(x) + \partial_x^{n-2}\widehat{f}(x)\Big), \text{ for } n \geq 2,$$

where $\partial_x^n \widehat{f}(x) \equiv [\partial_x, \partial_x^{n-1}\widehat{f}(x)]$ and

$$\widehat{f}(x) = -\epsilon_1\epsilon_2 \sum_{m=0}^{\infty} x^m \partial_{(m)}, \qquad \partial_{(m)} = \sum_{k=m+2}^{\infty} k t_k \frac{\partial}{\partial t_{k-m-2}}. \tag{2.12}$$

The operator $\widehat{f}(x)$ acting on the partition function Z has the same effect as computation of the expectation value of the expression

$$f(x) = 2\epsilon_1 \sum_{a=1}^{N} \frac{V'(x) - V'(z_a)}{x - z_a}. \tag{2.13}$$

The representation (2.11) can be found by interpreting (2.1) in conformal field theory language, as an insertion of an operator $\prod_a (x - z_a)^{-\frac{2\alpha}{\epsilon_2}}$ at a position x, and determining modes of the corresponding energy-momentum tensor. As $\widehat{L}_{-n}$ involves derivatives with respect to times, encoded in the term $\partial_x^{n-2}\widehat{f}(x)$, it follows that operators (2.8) are time-dependent quantum curves, so in general they impose partial differential equations in x and times t_k. However, for certain choices of the potential $V(x)$ – e.g. for Gaussian or multi-Penner potentials – quantum curves become time-independent and impose ordinary differential equations in x for $\widehat{\psi}_\alpha(x)$.

Apart from (2.1), in certain situations it is useful to consider the wave-function normalized by the partition function Z

$$\Psi_\alpha = \frac{\widehat{\psi}_\alpha(x)}{Z}. \tag{2.14}$$

Taking advantage of the relation $Z^{-1}\widehat{L}_{-n}\widehat{\psi}_\alpha(x) = Z^{-1}\widehat{L}_{-n}Z\widehat{\Psi}_\alpha(x) \equiv \widehat{\mathcal{L}}_{-n}\widehat{\Psi}_\alpha(x)$, when acting on Ψ_α, Virasoro generators (2.11) are transformed into

$$\begin{aligned} \widehat{\mathcal{L}}_{-n} &= Z^{-1}\widehat{L}_{-n}Z \\ &= -\frac{1}{\epsilon_1\epsilon_2(n-2)!}\Big(\partial_x^{n-2}\big(V'(x)^2\big) + (\epsilon_1+\epsilon_2)\partial_x^n V(x) \\ &\qquad + \partial_x^{n-2}\widehat{f}(x) + \big[\partial_x^{n-2}\widehat{f}(x), \log Z\big]\Big). \end{aligned} \tag{2.15}$$

In the rest of this section we present explicit expressions for higher level quantum curves, up to level 5. Imposing the condition that $\widehat{\psi}_\alpha(x)$ satisfies a second order differential equation in x, we find that quantum curves at level 2 take form

$$\widehat{A}_2^\alpha \widehat{\psi}_\alpha(x) \equiv \Big(\widehat{L}_{-1}^2 + \frac{4\alpha^2}{\epsilon_1\epsilon_2}\widehat{L}_{-2}\Big)\widehat{\psi}_\alpha(x) = 0, \quad \text{for } \alpha = -\frac{\epsilon_1}{2}, -\frac{\epsilon_2}{2}. \tag{2.16}$$

For the special values $\alpha = \alpha_{2,1} = -\frac{\epsilon_1}{2}$ or $\alpha = \alpha_{1,2} = -\frac{\epsilon_2}{2}$, the above differential operator takes form $\widehat{A}_2^\alpha = \widehat{L}_{-1}^2 + b^{\pm 2}\widehat{L}_{-2}$ respectively, with b^2 defined in (2.2). This indeed agrees with well known expressions for operators that create Virasoro singular vectors at level 2, see (2.10), and analogous calculation was first presented in [**5**]. We stress that the form (2.16) is universal, and reduces to either of the two singular vectors at level 2 upon appropriate choice of α. In terms of the representation (2.11), quantum curves in (2.16) have an explicit form

$$\widehat{A}_2^\alpha = \partial_x^2 - \frac{4\alpha^2}{\epsilon_1^2\epsilon_2^2}V'(x)^2 - \frac{4\alpha^2}{\epsilon_1^2\epsilon_2^2}(\epsilon_1+\epsilon_2)V''(x) - \frac{4\alpha^2}{\epsilon_1^2\epsilon_2^2}\widehat{f}(x). \tag{2.17}$$

Quantum curves at higher levels can be found analogously, by demanding that $\widehat{\psi}_\alpha(x)$ satisfies a differential equation in x of some particular order. At level 3 we find the result

$$\widehat{A}_3^\alpha = \widehat{L}_{-1}\widehat{A}_2^\alpha + \frac{2\alpha^2}{\epsilon_1^2\epsilon_2^2}(2\alpha+\epsilon_1)(2\alpha+\epsilon_2)\widehat{L}_{-3}. \tag{2.18}$$

For $\alpha = \alpha_{3,1} = -\epsilon_1$ or $\alpha = \alpha_{1,3} = -\epsilon_2$ this operator reduces to known expressions (2.10) for singular vectors at level 3; in our context they can be written down more explicitly, using the representation (2.11). Furthermore, substituting values of α corresponding to level 2, i.e. $\alpha_{2,1} = -\frac{\epsilon_1}{2}$ or $\alpha_{1,2} = -\frac{\epsilon_2}{2}$, the second term in (2.18) vanishes and $\widehat{A}_3^\alpha$ reduces to quantum curves $\widehat{A}_2^\alpha$ at level 2. In this sense the expression (2.18) encodes all singular vectors up to and including level 3. Furthermore, quantum curve at level 4 takes form

(2.19)

$$\widehat{A}_4^\alpha = \widehat{L}_{-1}\widehat{A}_3^\alpha + \frac{4\alpha(\alpha+\epsilon_1)(\alpha+\epsilon_2)}{\epsilon_1\epsilon_2(5\alpha+3\epsilon_1+3\epsilon_2)}\widehat{L}_{-2}\widehat{A}_2^\alpha +$$
$$-\frac{2\alpha(2\alpha+\epsilon_1)(2\alpha+\epsilon_2)(\alpha+\epsilon_1)(\alpha+\epsilon_2)}{\epsilon_1^3\epsilon_2^3(5\alpha+3\epsilon_1+3\epsilon_2)}\Big(\epsilon_1\epsilon_2\widehat{L}_{-1}\widehat{L}_{-3} - 2(2\alpha+\epsilon_1)(2\alpha+\epsilon_2)\widehat{L}_{-4}\Big).$$

This expression reduces to known expressions (2.10) for singular vectors at level 4 upon substitution of the relevant values of degenerate momenta $\alpha_{4,1} = -\frac{3\epsilon_1}{2}$, $\alpha_{1,4} = -\frac{3\epsilon_2}{2}$, $\alpha_{2,2} = -\frac{\epsilon_1+\epsilon_2}{2}$. For degenerate momenta corresponding to lower levels, (2.19) factorizes and reduces to expressions for singular vectors at those lower levels.

Taking advantage of the following notation

$$\begin{aligned}
&\gamma_1 = \frac{\alpha^2}{\epsilon_1\epsilon_2}, \qquad \gamma_2 = \frac{(2\alpha+\epsilon_1)(2\alpha+\epsilon_2)}{\epsilon_1\epsilon_2}, \qquad \gamma_3 = \frac{(\alpha+\epsilon_1)(\alpha+\epsilon_2)}{\epsilon_1\epsilon_2},\\
&\gamma_4 = \frac{(2\alpha+3\epsilon_1)(2\alpha+3\epsilon_2)(2\alpha+\epsilon_1+\epsilon_2)}{\epsilon_1\epsilon_2\alpha},\\
&\delta_1 = \frac{\alpha}{5\alpha+3\epsilon_1+3\epsilon_2}, \qquad \delta_2 = \frac{\alpha}{7\alpha+6\epsilon_1+6\epsilon_2},
\end{aligned} \tag{2.20}$$

quantum curves at level 5, and also all lower levels, can be written compactly in the form

$$\begin{aligned}
\widehat{A}_2^\alpha &= \widehat{L}_{-1}^2 + 4\gamma_1\widehat{L}_{-2},\\
\widehat{A}_3^\alpha &= \widehat{L}_{-1}\widehat{A}_2^\alpha + 2\gamma_1\gamma_2\widehat{L}_{-3},\\
\widehat{A}_4^\alpha &= \widehat{L}_{-1}\widehat{A}_3^\alpha + 4\delta_1\gamma_3\widehat{L}_{-2}\widehat{A}_2^\alpha - 2\delta_1\gamma_2\gamma_3\big(\widehat{L}_{-1}\widehat{L}_{-3} - 2\gamma_2\widehat{L}_{-4}\big),\\
\widehat{A}_5^\alpha &= \widehat{L}_{-1}\widehat{A}_4^\alpha + 2\delta_1\delta_2\gamma_4\big(2\widehat{L}_{-2}\widehat{A}_3^\alpha + \gamma_3\widehat{L}_{-3}\widehat{A}_2^\alpha\big)\\
&\quad - 4\delta_2\gamma_2\gamma_3\gamma_4\big(\delta_1\widehat{L}_{-1}\widehat{L}_{-4} - (\gamma_1+3\delta_1)\widehat{L}_{-5}\big).
\end{aligned} \tag{2.21}$$

In particular, setting α to be equal to $\alpha_{5,1} = -2\epsilon_1$ or $\alpha_{1,5} = -2\epsilon_2$, the operator $\widehat{A}_5^\alpha$ reduces to known expressions for singular vectors at level 5, and substituting values of α corresponding to lower levels this operator reduces to the form of singular vectors at those lower levels.

With some effort one can find analogous universal, α-dependent expressions for quantum curves and Virasoro singular vectors at higher levels. Simplifying the algorithm presented in [**20**] that enables to find such higher level quantum curves, and determining their form explicitly at arbitrary level, are important tasks for future work.

2.3. Double quantum structure. The wave-functions and quantum curves that we consider, apart from a discrete value of α, depend on two continuous parameters g_s and β, or equivalently ϵ_1 and ϵ_2 defined in (2.2). It turns out that one can consider two corresponding classical limits, namely 't Hooft large N (or vanishing g_s) limit, and the classical CFT limit of an infinite central charge. It follows that quantum curves that we consider can be regarded as doubly quantized objects, analogously to the discussion in the context of the Langlands correspondence in [**33**]. Let us briefly discuss these two quantization procedures. Note that to obtain a well defined, non-divergent classical limit, one has to subtract the partition function Z from the wave-function; therefore in this section we consider quantum curves and wave-functions $\Psi_\alpha(x)$ in the normalization (2.14).

To start with, we consider the classical 't Hooft limit for quantum curves at level 2, for $\beta = 1$. In this case (2.17), for either value $\alpha = -\epsilon_1/2$ or $-\epsilon_2/2$, reduces to

$$\Big(g_s^2\partial_x^2 - V'(x)^2 - \widehat{f}(x)\Big)\widehat{\psi}_\alpha(x) = 0. \tag{2.22}$$

Dividing this equation by the partition function Z we obtain a differential equation for Ψ_α, and using the factorization of expectation values, in the large N limit we find the equation

$$y^2 - V'(x)^2 - f_{cl}(x) = 0, \tag{2.23}$$

where y is identified with the classical limit of $g_s\partial_x$, while

$$f_{cl}(x) = \lim_{N\to\infty} \frac{\langle f(x)\rangle}{Z}. \tag{2.24}$$

The result (2.23) is precisely the spectral curve of the hermitian one-matrix model (as defined by (2.1) with $\alpha = 0$ and $\beta = 1$). It follows that the quantum curve (2.17) can be regarded as the quantization of the spectral curve.

Similarly one can analyze the classical 't Hooft limit for quantum curves at higher level. As shown in [**20**], in this limit $\widehat{L}_{-p}$ with $p \geq 3$ can be set to zero, and $\widehat{L}_{-1}$ and $\widehat{L}_{-2}$ become commuting, which is apparently the same limit as analyzed in [**37, 38**]. In this limit resulting classical curve equations factorize. In particular, the curves corresponding to $\alpha_{r,1}$, reduce to

$$\begin{aligned} 0 &= \prod_{k=1}^{r/2}\Big(y^2 - \frac{(2k-1)^2}{(r-1)^2}\big(V'(x)^2 + f_{cl}(x)\big)\Big), &&\text{for } r \text{ even} \\ 0 &= y\prod_{k=1}^{(r-1)/2}\Big(y^2 - \frac{4k^2}{(r-1)^2}\big(V'(x)^2 + f_{cl}(x)\big)\Big), &&\text{for } r \text{ odd} \end{aligned} \tag{2.25}$$

where y is identified with the limit of $\frac{\epsilon_2}{r-1}\partial_x$. Each factor in the above expressions represents the spectral curve (2.23), with $(V'(x)^2 + f_{cl}(x))$ term (or simply y) rescaled by a simple factor. In other words, a product of several classical curves is non-trivially resolved in the process of quantization, in a way which is consistent with the structure of singular vectors, and quantum curves at higher levels can be regarded as non-trivial resolutions of multiple copies of the underlying spectral curve.

The second classical limit that we consider is that of infinite central charge, analogous to the classical limit in Liouville theory. This is also equivalent to the Nekrasov-Shatashvili limit, where either ϵ_1 or ϵ_2 is set to zero. For definiteness we choose the case $\epsilon_1 \to 0$, which is equivalent to vanishing b, as follows from the relation $b^2 = \frac{\epsilon_1}{\epsilon_2}$. In terms of b and g_s parameters, in order to keep ϵ_2 constant we need to consider a double scaling limit, with both of these parameters vanishing with a constant ratio. In the limit $\epsilon_1 \to 0$ it is natural to consider quantum curves and wave-functions labeled by

$$\alpha = \alpha_{r+1,1} = -\frac{r}{2}\epsilon_1. \tag{2.26}$$

As mentioned above, we consider wave-functions normalized by the partition function (2.14), which in the $\epsilon_1 \to 0$ limit we denote by $\Psi^{\mathrm{NS}}_\alpha(x)$, and which factorize in this limit as follows

$$\Psi^{\mathrm{NS}}_{-\frac{r}{2}\epsilon_1}(x) \equiv \lim_{\epsilon_1\to 0} \Psi_{-\frac{r}{2}\epsilon_1}(x) = \left(\Psi^{\mathrm{NS}}_{-\frac{1}{2}\epsilon_1}(x)\right)^r. \tag{2.27}$$

These wave-functions are annihilated by operators that arise in the limit of (2.8)

$$\widehat{\mathcal{A}}^{\mathrm{NS}}_{r+1}\Psi^{\mathrm{NS}}_{-\frac{r}{2}\epsilon_1}(x) = 0, \qquad \widehat{\mathcal{A}}^{\mathrm{NS}}_{r+1} = \lim_{\epsilon_1\to 0} \epsilon_2^{r+1} Z^{-1} \widehat{A}^{-\frac{r}{2}\epsilon_1}_{r+1} Z. \tag{2.28}$$

The operators $\widehat{\mathcal{A}}^{\mathrm{NS}}_{r+1}$ have the structure of singular vectors, written in terms of Virasoro generators that arise in the appropriate limit of (2.15)

$$\widehat{\mathcal{L}}^{\mathrm{NS}}_{-n} = \lim_{\epsilon_1\to 0} \epsilon_1\epsilon_2 \widehat{\mathcal{L}}_{-n} = -\frac{1}{(n-2)!}\Big(\partial_x^{n-2}\big(V'(x)^2\big) + \epsilon_2 \partial_x^n V(x) + F^{(0)}_{n-2}(x,\epsilon_2)\Big), \tag{2.29}$$

where $F^{(0)}_0(x,0) = f_{cl}(x)$ given in (2.24), and more generally we defined

$$F^{(0)}(\epsilon_2) = -\lim_{\epsilon_1\to 0} \epsilon_1\epsilon_2 \log Z, \qquad F^{(0)}_k(x,\epsilon_2) = \sum_{n=k}^{\infty} \frac{n!}{(n-k)!} x^{n-k} \partial_{(n)} F^{(0)}(\epsilon_2). \tag{2.30}$$

For example, in the $\epsilon_1 \to 0$ limit the quantum curve equation (2.16) at level 2 reduces to

$$\widehat{\mathcal{A}}^{\mathrm{NS}}_2 \Psi^{\mathrm{NS}}_{-\frac{1}{2}\epsilon_1}(x) = \left(\epsilon_2^2\partial_x^2 + \widehat{\mathcal{L}}^{\mathrm{NS}}_{-2}\right)\Psi^{\mathrm{NS}}_{-\frac{1}{2}\epsilon_1}(x) = 0. \tag{2.31}$$

Now it is useful to define inductively differential operators $\widehat{a}^{r+1}_q$ for $q = 0, 1, \ldots,$ $r+1$

$$\widehat{a}^{r+1}_0 = 1, \quad \widehat{a}^{r+1}_1 = \epsilon_2\partial_x, \quad \widehat{a}^{r+1}_{q+1} = \epsilon_2\partial_x \widehat{a}^{r+1}_q + q(r-q+1)\widehat{\mathcal{L}}^{\mathrm{NS}}_{-2}\widehat{a}^{r+1}_{q-1}. \tag{2.32}$$

By induction one can show that

$$\widehat{a}^{r+1}_{q+1}\Psi^{\mathrm{NS}}_{-\frac{r}{2}\epsilon_1}(x) = r(r-1)(r-2)\cdots(r-q)\left(\Psi^{\mathrm{NS}}_{-\frac{1}{2}\epsilon_1}(x)\right)^{r-q-1}\left(\epsilon_2\partial_x\Psi^{\mathrm{NS}}_{-\frac{1}{2}\epsilon_1}(x)\right)^{q+1}. \tag{2.33}$$

Therefore the wave-function $\Psi^{\rm NS}_{-\frac{r}{2}\epsilon_1}(x)$ satisfies an ordinary differential equation of order $(r+1)$, and the corresponding quantum curve can be identified as $\widehat{\mathcal{A}}^{\rm NS}_{r+1} = \widehat{a}^{r+1}_{r+1}$. In particular

$$\begin{aligned} \widehat{\mathcal{A}}^{\rm NS}_2 &= \epsilon_2^2\partial_x^2 + \widehat{\mathcal{L}}^{\rm NS}_{-2}, \\ \widehat{\mathcal{A}}^{\rm NS}_3 &= \epsilon_2^3\partial_x^3 + 4\epsilon_2\widehat{\mathcal{L}}^{\rm NS}_{-2}\partial_x + 2\epsilon_2\widehat{\mathcal{L}}^{\rm NS}_{-3}, \\ \widehat{\mathcal{A}}^{\rm NS}_4 &= \epsilon_2^4\partial_x^4 + 10\epsilon_2^2\widehat{\mathcal{L}}^{\rm NS}_{-2}\partial_x^2 + 10\epsilon_2^2\widehat{\mathcal{L}}^{\rm NS}_{-3}\partial_x + 9\big(\widehat{\mathcal{L}}^{\rm NS}_{-2}\big)^2 + 6\epsilon_2^2\widehat{\mathcal{L}}^{\rm NS}_{-4}, \\ \widehat{\mathcal{A}}^{\rm NS}_5 &= \epsilon_2^5\partial_x^5 + 20\epsilon_2^3\widehat{\mathcal{L}}^{\rm NS}_{-2}\partial_x^3 + 30\epsilon_2^3\widehat{\mathcal{L}}^{\rm NS}_{-3}\partial_x^2 + 64\epsilon_2\big(\widehat{\mathcal{L}}^{\rm NS}_{-2}\big)^2\partial_x + 36\epsilon_2^3\widehat{\mathcal{L}}^{\rm NS}_{-4}\partial_x \\ &\quad + 64\epsilon_2\widehat{\mathcal{L}}^{\rm NS}_{-2}\widehat{\mathcal{L}}^{\rm NS}_{-3} + 24\epsilon_2^3\widehat{\mathcal{L}}^{\rm NS}_{-5}. \end{aligned} \tag{2.34}$$

Using the Virasoro algebra $\partial_x^n\widehat{\mathcal{L}}^{\rm NS}_{-2} = n!\widehat{\mathcal{L}}^{\rm NS}_{-n-2}$, each $\widehat{\mathcal{A}}^{\rm NS}_{r+1}$ can also be expressed in terms of ∂_x and (derivatives of) $\widehat{\mathcal{L}}^{\rm NS}_{-2}$ only. If we further identify the energy-momentum tensor in classical Liouville theory as $T^{(c)} \equiv \widehat{\mathcal{L}}^{\rm NS}_{-2}$, then the operators $\widehat{\mathcal{A}}^{\rm NS}_{r+1}$ take the same form as operators imposing differential equations for the fields $e^{-r\varphi/2}$ in the classical Liouville theory [**39**]. Furthermore, taking the limit $\epsilon_2 \to 0$, the results we just obtained reduce to those in the 't Hooft limit discussed first.

2.4. Examples: Gaussian and multi-Penner models. In this section we illustrate our general considerations in two examples, of Gaussian and multi-Penner models. One interesting feature is that in these cases quantum curves can be reduced to ordinary (not partial) differential equations. Second, multi-Penner model is particularly interesting, as in this case our formalism reduces essentially to various results familiar from conformal field theory.

We discuss first the Gaussian model, i.e. the model defined by (2.1) with the quadratic potential, which is a specialization of (2.3) to only one non-zero time $t_2 = \frac{1}{2}$

$$V(x) = \frac{1}{2}x^2. \tag{2.35}$$

In this case (2.12) simplifies and its derivatives vanish

$$\widehat{f}(x) = -\epsilon_1\epsilon_2\partial_{t_0}, \qquad \partial_x^k\widehat{f}(x) = 0 \qquad \text{for } k \geq 1, \tag{2.36}$$

while operators (2.11) (when acting on $\widehat{\psi}_\alpha(x)$) take form

$$\widehat{L}_{-1} = \partial_x, \quad \widehat{L}_{-2} = -\frac{x^2+\epsilon_1+\epsilon_2-4\mu+2\alpha}{\epsilon_1\epsilon_2}, \quad \widehat{L}_{-3} = -\frac{2x}{\epsilon_1\epsilon_2}, \quad \widehat{L}_{-4} = -\frac{1}{\epsilon_1\epsilon_2}, \tag{2.37}$$

where $\mu = \beta^{1/2}\hbar N$ is the 't Hooft parameter, and $\widehat{L}_{-n} = 0$ for $n \geq 5$. In this way we obtain an interesting realization of a subalgebra of the Virasoro algebra – truncated at $\widehat{L}_{-5}$ – in terms of differential operators in one variable x. For polynomial potentials of higher degree one can analogously obtain realization of larger Virasoro subalgebras, in terms of differential operators in several variables. The operators (2.37) are building blocks of higher level quantum curves. For example, at level 2, for $\alpha = -\frac{\epsilon_1}{2}, -\frac{\epsilon_2}{2}$ the equation (2.16) takes form

$$\big(\widehat{L}_{-1}^2 + b^{\pm 2}\widehat{L}_{-2}\big)\widehat{\psi}_\alpha(x) = \Big(\partial_x^2 - \frac{4\alpha^2}{\epsilon_1^2\epsilon_2^2}(x^2+\epsilon_1+\epsilon_2-4\mu+2\alpha)\Big)\widehat{\psi}_\alpha(x) = 0, \tag{2.38}$$

where $b^2 = \frac{\epsilon_1}{\epsilon_2} = -\beta$, and a choice of a sign in the exponent of $b^{\pm 2}$ corresponds respectively to the choice of $\alpha = -\frac{\epsilon_1}{2}$ or $\alpha = -\frac{\epsilon_2}{2}$. For $\alpha = -\frac{\epsilon_1}{2}$ or $\alpha = -\frac{\epsilon_2}{2}$, and in the unrefined limit $\beta = 1$, from (2.38) we obtain unrefined quantum curves

$$(g_s^2\partial_x^2 - x^2 + 4\mu \mp g_s)\widehat{\psi}_\alpha(x) = 0, \tag{2.39}$$

and then (a unique) algebraic curve in the classical limit

$$y^2 - x^2 + 4\mu = 0, \tag{2.40}$$

with y representing the classical limit of $g_s\partial_x$. At level 3, the unrefined limit of (2.18) yields

$$\Big(g_s^3\partial_x^3 - 4(x^2 - 4\mu \pm 2g_s)g_s\partial_x + 4g_s x\Big)\widehat{\psi}_\alpha(x) = 0, \tag{2.41}$$

where $\pm$ corresponds respectively to the choice of $\alpha = -\epsilon_1, -\epsilon_2$, and in the classical limit we find an algebraic curve which factorizes as

$$2y\big((2y)^2 - 4x^2 + 16\mu\big) = 0, \tag{2.42}$$

in agreement with (2.25) and the earlier discussion.

The second interesting example is the multi-Penner model, defined by the potential

$$V(x) = \sum_{i=1}^{M} \alpha_i \log(x - x_i). \tag{2.43}$$

As is well known, this matrix model computes, in minimal models or in Liouville theory, correlation functions of $(M+1)$ primary fields with momenta α_i and α_∞, inserted respectively at positions x_i and at infinity on $\mathbb{P}^1$ [**34, 40–42**]. In what follows we show that various other objects that we introduced earlier, e.g. the representation of Virasoro operators $\widehat{L}_n$, higher level quantum curves, etc., reduce to familiar objects in minimal models or in Liouville theory for the potential (2.43), with the dependence on an infinite set of times t_n replaced by the dependence on parameters x_i, i.e. positions of operator insertions. We also recall that the multi-Penner model (2.43) describes a four-dimensional $\mathcal{N} = 2$ $SU(2)^{M-2}$ superconformal linear quiver gauge theory [**34**]; for example for $M = 2$ this model describes a theory of four free hypermultiplets. Via this connection, our results found from the perspective of matrix models can be interpreted in the language of supersymmetric gauge theories and related topics.

Note that for the potential (2.43), the exponential term in the matrix model integrand (2.1) takes form

$$e^{-\frac{\sqrt{\beta}}{\hbar}\sum_{a=1}^N V(z_a)} = \prod_{i=1}^{M}\prod_{a=1}^{N}(z_a - x_i)^{-\frac{2\alpha_i}{\epsilon_2}}. \tag{2.44}$$

Each factor in this expression, corresponding to fixed i, has the same form as the determinant-like insertion that defines the wave-function $\widehat{\psi}_\alpha(x)$. Therefore for the potential (2.43), the wave-function can be interpreted as a correlation function of $(M+2)$ primary fields in the presence of an additional field at infinity, or $(M+1)$ fields once the momentum conservation condition is imposed. This suggests that quantum curve equations in this case should reduce to differential equations for correlation functions of a number of primary fields, which include a distinguished field with a degenerate momentum $\alpha_{r,s}$ inserted at position x. These are nothing but the BPZ equations, and in what follows we show that they indeed arise from our general formalism. In particular, at level 2 we obtain in this way familiar in minimal models or Liouville theory hypergeometric differential equations for four-point correlation functions.

For the potential (2.43) we see that the operator $\widehat{f}(x)$ in (2.12) takes form

$$\widehat{f}(x) = -\epsilon_1\epsilon_2 \sum_{i=1}^{M} \frac{1}{x - x_i}\partial_{x_i}. \tag{2.45}$$

In what follows it is useful to introduce another normalization and consider the wave-function

$$\widetilde{\psi}_\alpha(x) = \widehat{\psi}_\alpha(x) \cdot \prod_{i\neq j}(x_i - x_j)^{-\frac{\alpha_i\alpha_j}{\epsilon_1\epsilon_2}}. \tag{2.46}$$

Then, rewriting (2.45) accordingly, the representation of Virasoro algebra (2.11), when acting on $\widetilde{\psi}_\alpha(x)$, takes form $\widetilde{L}_0 = \Delta_\alpha, \widetilde{L}_{-1} = \partial_x$, and

$$\widetilde{L}_{-n} = \sum_{i=1}^{M}\Big(\frac{(n-1)\Delta_{\alpha_i}}{(x_i - x)^n} - \frac{1}{(x_i - x)^{n-1}}\partial_{x_i}\Big), \quad \text{for } n \geq 2. \tag{2.47}$$

These generators coincide with well-known expressions for Virasoro generators acting on correlation functions in conformal field theory. These generators can be further simplified, taking advantage of $SL(2,\mathbb{C})$ invariance, which in the matrix model formulation gives rise to three additional equations that constrain the wave-function $\widetilde{\psi}_\alpha(x)$. These equations imply that three among M partial derivatives ∂_{x_i} in Virasoro generators $\widetilde{L}_{-n}$ can be expressed in terms ∂_x. In particular, for $M = 2$, with the momentum conservation condition imposed, we find

$$\widetilde{L}_{-2}\widetilde{\psi}_\alpha(x) = \Big[-\sum_{i=1}^{2}\frac{1}{x - x_i}\partial_x + \sum_{i=1}^{2}\frac{\Delta_{\alpha_i}}{(x - x_i)^2} - \frac{\Delta_{\alpha_1} + \Delta_{\alpha_2} + \Delta_\alpha}{(x - x_1)(x - x_2)}\Big]\widetilde{\psi}_\alpha(x), \tag{2.48}$$

while for $M = 3$ we find

$$\begin{aligned}\widetilde{L}_{-2}\widetilde{\psi}_\alpha(x) &\\ = \Big[-\sum_{i=1}^{3}\frac{1}{x - x_i}\partial_x + \sum_{i=1}^{3}\frac{\Delta_{\alpha_i}}{(x - x_i)^2} &+ \sum_{I=1}^{3}\frac{\Delta_{\alpha_I} - \Delta_{\alpha_{I+1}} - \Delta_{\alpha_{I+2}} - \Delta_\alpha}{(x - x_{I+1})(x - x_{I+2})}\Big]\widetilde{\psi}_\alpha(x).\end{aligned} \tag{2.49}$$

For both (2.48) and (2.49), other generators $\widetilde{L}_{-n}$ with $n > 2$ can be similarly obtained from (2.47).

We can now use the representation of Virasoro operators (2.47) to construct quantum curves that annihilate the wave-function $\widetilde{\psi}_\alpha(x)$. Quantum curve equations in this case have the same form as the original BPZ equations in conformal field theory. We stress that these equations make sense only for degenerate values $\alpha = \alpha_{r,s}$ given in (2.6). Including an additional determinant-like insertion that defines the wave-function $\widetilde{\psi}_\alpha(x)$ itself, this wave-function is identified with a correlation function of $(M+1)$ fields with momenta $\alpha = \alpha_{r,s}$ and α_i, $i = 1,\ldots,M$, that are parameters of the potential (2.43).

For $M = 3$ quantum curves can be built from Virasoro operators (2.49) and other relevant $\widetilde{L}_{-n}$. These quantum curves are written in terms of ∂_x and no other derivatives, thus we obtain time-independent equations for the wave-function, which can be interpreted as the four-point function that includes a field with a degenerate momentum. At level 2 we find a hypergeometric differential equation, coinciding with the original BPZ equations at level 2 [43]

$$(\partial_x^2 + b^{\pm 2}\widetilde{L}_{-2})\widetilde{\psi}_\alpha(x) = 0, \tag{2.50}$$

respectively for the choice of momentum $\alpha = \alpha_{2,1}$ and $\alpha = \alpha_{1,2}$, and with $\widetilde{L}_{-2}$ given in (2.49). In this case $\widetilde{\psi}_\alpha(x)$ is identified with the four-point function of fields with momenta α and α_i for $i = 1, 2, 3$ that are parameters of the potential (2.43). One can also analogously write down higher level BPZ equations.

2.5. Quantization and the topological recursion. We also briefly summarize the construction of quantum curves and wave-functions by means of the topological recursion, and in particular β-deformed topological recursion in case $\beta \neq 1$. The main advantage of this formulation is that it works more generally, for a large class of algebraic curves, which are not necessarily spectral curves of some matrix models. More details of this formulation, which are quite technical and involved, are presented in [20].

Recall that for the β-deformed matrix model, the connected h-point differentials

$$W_h(x_1, \ldots, x_h) = \beta^{h/2} \Big\langle \prod_{i=1}^{h} \sum_{a=1}^{N} \frac{dx_i}{x_i - z_a} \Big\rangle^{(c)} \tag{2.51}$$

are defined through the connected part $\langle X \rangle^{(c)}$ of the normalized expectation value

$$\frac{\langle X \rangle}{Z} = \frac{1}{(2\pi)^N N! Z} \int_{\mathbb{R}^N} \Delta(z)^{2\beta} X\, e^{-\frac{\sqrt{\beta}}{\hbar}\sum_{a=1}^{N} V(z_a)} \prod_{a=1}^{N} dz_a. \tag{2.52}$$

In the 't Hooft limit $N \to \infty$, $\hbar \to 0$, $\mu = \beta^{1/2}\hbar N = const$, (2.51) has an asymptotic expansion

$$W_h(x_1, \ldots, x_h) = \sum_{g,\ell=0}^{\infty} \hbar^{2g-2+h+\ell}\gamma^\ell W_\ell^{(g,h)}(x_1, \ldots, x_h), \tag{2.53}$$

where $\gamma = \beta^{1/2} - \beta^{-1/2}$. The so-called multi-resolvents $W_\ell^{(g,h)}$ that arise in this expansion satisfy a set of recurrence equations referred to as the (β-deformed) topological recursion – which are reformulation of the loop equations – with the initial condition specified by the spectral curve [44–46]. Similarly as in the original (unrefined) topological recursion [10], one can regard the recurrence equations as a definition of multi-resolvents assigned to a given algebraic curve, and not necessarily related to (the existence of) a matrix model.

Recall now that the wave-function (2.1) can be regarded as an expectation value of a determinant-like insertion. Taking advantage of the properties of the determinant, the asymptotic expansion of this wave-function can be shown to take form

(2.54)

$$\begin{aligned}\log \frac{\widehat{\psi}_\alpha(x)}{Z} &= -\frac{2\alpha}{\epsilon_1\epsilon_2}V(x) + \sum_{h=1}^{\infty} \frac{1}{h!}\Big(-\frac{2\alpha}{g_s}\Big)^h \int_\infty^x \cdots \int_\infty^x W_h(x_1', \ldots, x_h') \\ &= \sum_{g,\ell=0,h=1}^{\infty} \frac{(-1)^{g+\ell+h-1}}{h!} 2^{2-2g-\ell}\alpha^h(\epsilon_1\epsilon_2)^{g-1}(\epsilon_1+\epsilon_2)^\ell F_\ell^{(g,h)}(x, \ldots, x).\end{aligned}$$

This equation can be also regarded as a definition of the wave-function assigned to a given algebraic curve, even in case there is no underlying matrix model. One can then reconstruct the operator that annihilates this wave-function order by order. In this way, starting from a (classical) algebraic curve, the quantum curve (at appropriate level, determined by the value of α) can be reconstructed.

One important remark concerning the resulting wave-function is that it depends on the choice of the base point of integrals. In (2.54) this point is taken to infinity, and it leads to quantum curves that coincide with those discussed in previous chapters (in case the relevant matrix model exists). Another choice, considered e.g. in [**8**, **9**] is to identify the base point with the conjugate point $\overline{x}$. In [**20**] it is shown that wave-functions corresponding to these two base points are related in some definite way, and this phenomenon is analyzed in more detail in [**47**]. More details and various calculations using the formalism of the (β-deformed) topological recursion are presented in [**20**].

3. Higher level (super-Virasoro) quantum curves

In this section we consider quantum curves which have the structure of singular vectors of super-Virasoro algebra. This algebra underlies the supersymmetric version of a matrix model, also known as super-eigenvalue model, which we therefore analyze. This analysis is quite analogous to what we presented in section 2, and we follow the original presentation in [**21**].

3.1. α/β-deformed super-eigenvalue models. To start with we introduce the β-deformed super-eigenvalue model, which is a β-deformation of the super-eigenvalue model introduced in [**27**] and analyzed in [**27**, **28**, **30**]. This super-eigenvalue model was formulated during the search of a supersymmetric generalization of a hermitian matrix model. Even though a formulation in terms of a proper integral over some supersymmetric ensemble of matrices has not been found, it was shown that the super-eigenvalue model satisfies constraint equations that form $\mathcal{N}=1$ super-Virasoro algebra, analogously to Virasoro constraints for hermitian matrix model. This super-Virasoro algebra is defined by the following (anti)commutation relations

$$\begin{aligned}
&\{G_r, G_s\} = 2L_{r+s} + \frac{c}{3}\big(r^2 - \frac{1}{4}\big)\delta_{r+s,0},\\
&[L_m, G_r] = \big(\frac{m}{2} - r\big)G_{m+r},\\
&[L_m, L_n] = (m-n)L_{m+n} + \frac{c}{12}(m^3 - m)\delta_{m+n,0},
\end{aligned} \tag{3.1}$$

and of our interest is the NS (Neveu-Schwarz) sector, in which the indices of generators G_r take half-integer values, $r \in \mathbb{Z} + \frac{1}{2}$. Generators L_n form the Virasoro subalgebra of the NS superalgebra, and their indices n are integer.

The partition function of the β-deformed super-eigenvalue model that we consider takes form of a formal integral over an even number N of bosonic eigenvalues z_a and fermionic eigenvalues ϑ_a

$$Z = \int \prod_{a=1}^{N} dz_a d\vartheta_a \Delta(z,\vartheta)^{\beta} e^{-\frac{\sqrt{\beta}}{\hbar}\sum_{a=1}^{N} V(z_a,\vartheta_a)}, \tag{3.2}$$

where

$$\Delta(z,\vartheta) = \prod_{1\le a<b\le N} (z_a - z_b - \vartheta_a\vartheta_b), \tag{3.3}$$

and the potential depends on bosonic t_n and fermionic $\xi_{n+1/2}$ times

$$\begin{aligned} V(x,\theta) &= V_B(x) + V_F(x)\theta, \\ V_B(x) &= \sum_{n=0}^{\infty} t_n x^n, \qquad V_F(x) = \sum_{n=0}^{\infty} \xi_{n+1/2} x^n, \end{aligned} \tag{3.4}$$

such that $\{\vartheta_a, \xi_{n+1/2}\} = 0$. Instead of $\hbar$ and β, we also use another pair of parameters (note slight different normalization compared to (2.2))

$$\epsilon_1 = -\beta^{1/2}\hbar, \qquad \epsilon_2 = \beta^{-1/2}\hbar. \tag{3.5}$$

In what follows by $\langle \cdots \rangle$ we denote an unnormalized expectation value, defined as

$$\langle \mathcal{O} \rangle = \int \prod_{a=1}^{N} dz_a d\vartheta_a \Delta(z,\vartheta)^{\beta} e^{-\frac{\sqrt{\beta}}{\hbar}\sum_{a=1}^{N} V(z_a,\vartheta_a)} \mathcal{O}. \tag{3.6}$$

In the analysis of the partition function (3.2) or expectation values of time-independent operators, it is useful to identify their time derivatives with the following expectation values

$$\Big\langle \sum_{a=1}^{N} z_a^n \cdots \Big\rangle = -\frac{\hbar}{\sqrt{\beta}} \partial_{t_n} \Big\langle \cdots \Big\rangle, \quad \Big\langle \sum_{a=1}^{N} z_a^n \vartheta_a \cdots \Big\rangle = -\frac{\hbar}{\sqrt{\beta}} \partial_{\xi_{n+1/2}} \Big\langle \cdots \Big\rangle, \tag{3.7}$$

analogously to (2.5). Similarly as in (2.4), we can associate to the super-eigenvalue model free boson and free fermion fields, and using the above identifications they can be written as

$$\phi(x) = \frac{1}{\hbar} V_B(x) - \sqrt{\beta} \sum_{a=1}^{N} \log(x - z_a), \tag{3.8}$$

$$\psi(x) = \frac{1}{\hbar} V_F(x) - \sqrt{\beta} \sum_{a=1}^{N} \frac{\vartheta_a}{x - z_a}. \tag{3.9}$$

We introduce then a supersymmetric wave-function; it can be defined by means of the above fields (analogously to the non-supersymmetric case, as mentioned below (2.5)), and we also call it as the α/β-deformed super-eigenvalue integral

$$\widehat{\chi}_\alpha(x,\theta) = \Big\langle e^{\frac{\alpha}{\hbar}\left(\phi(x) + \psi(x)\theta\right)} \Big\rangle, \tag{3.10}$$

where θ is a fermionic variable with $\{\theta, \vartheta_a\} = \{\theta, \xi_{n+1/2}\} = 0$. The momentum α is bosonic, and we find that finite order differential equations for $\widehat{\chi}_\alpha(x,\theta)$ arise only for some special values of α. We also decompose the wave-function into bosonic $\widehat{\chi}_{B,\alpha}(x)$ and fermionic $\widehat{\chi}_{F,\alpha}(x)$ components

$$\begin{aligned} \widehat{\chi}_\alpha(x,\theta) &= \widehat{\chi}_{B,\alpha}(x) + \widehat{\chi}_{F,\alpha}(x)\theta, \\ \widehat{\chi}_{B,\alpha}(x) &\equiv \Big\langle e^{\frac{\alpha}{\hbar}\phi(x)} \Big\rangle = \widehat{\chi}_\alpha(x,0), \\ \widehat{\chi}_{F,\alpha}(x) &\equiv \frac{\alpha}{\hbar} \Big\langle \psi(x) e^{\frac{\alpha}{\hbar}\phi(x)} \Big\rangle = -\partial_\theta \widehat{\chi}_\alpha(x,\theta). \end{aligned} \tag{3.11}$$

For the super-eigenvalue model we find a representation of the super-Virasoro algebra analogous to (2.12), which will be used to build higher level quantum curves

(3.12)
$$\begin{aligned}
\widehat{G}_{-1/2} &= \theta\partial_x - \partial_\theta, \qquad\qquad \widehat{L}_{-1} = \partial_x,\\
\widehat{G}_{-n+1/2} &= \frac{1}{\hbar^2(n-2)!}\Big(\partial_x^{n-2}\big(V_B'(x)V_F(x)\big) + Q\hbar\partial_x^{n-1}V_F(x) + \partial_x^{n-2}\widehat{h}(x)\Big), \text{ for } n \geq 2,\\
\widehat{L}_{-n} &= \frac{1}{\hbar^2(n-2)!}\Big(\frac{1}{2}\partial_x^{n-2}\big(V_B'(x)\big)^2 + \frac{1}{2}\partial_x^{n-2}\big(V_F'(x)V_F(x)\big)\\
&\qquad + \frac{1}{2}Q\hbar\partial_x^n V_B(x) + \partial_x^{n-2}\widehat{f}(x)\Big), \quad \text{for } n \geq 2.
\end{aligned}$$

Here we denote

$$\tag{3.13}\begin{aligned}
\widehat{h}(x) &= \widehat{h}_t(x) + \widehat{h}_\xi(x),\\
\widehat{h}_t(x) &\equiv \hbar^2\sum_{n=0}^{\infty}x^n\sum_{k=n+1}^{\infty}\xi_{k+1/2}\partial_{t_{k-n-1}},\\
\widehat{h}_\xi(x) &\equiv \hbar^2\sum_{n=0}^{\infty}x^n\sum_{k=n+2}^{\infty}kt_k\partial_{\xi_{k-n-3/2}},
\end{aligned}$$

and

$$\tag{3.14}\begin{aligned}
\widehat{f}(x) &= \widehat{f}_t(x) + \widehat{f}_\xi(x),\\
\widehat{f}_t(x) &\equiv \hbar^2\sum_{n=0}^{\infty}x^n\sum_{k=n+2}^{\infty}kt_k\partial_{t_{k-n-2}},\\
\widehat{f}_\xi(x) &\equiv \hbar^2\sum_{n=0}^{\infty}x^n\sum_{k=n+2}^{\infty}\Big(k-\frac{n+1}{2}\Big)\xi_{k+1/2}\partial_{\xi_{k-n-3/2}},
\end{aligned}$$

and inductively we define

$$\tag{3.15} \partial_x^n\widehat{h}(x) \equiv \Big[\partial_x, \partial_x^{n-1}\widehat{h}(x)\Big], \qquad \partial_x^n\widehat{f}(x) \equiv \Big[\partial_x, \partial_x^{n-1}\widehat{f}(x)\Big].$$

By direct calculation one can check that the generators in (3.12) indeed satisfy the super-Virasoro algebra, in particular

$$\begin{aligned}
\{\widehat{G}_{-m-1/2}, \widehat{G}_{-n-1/2}\} &= 2\widehat{L}_{-m-n-1},\\
[\widehat{L}_{-m}, \widehat{G}_{-n-1/2}] &= \big(n-(m-1)/2\big)\widehat{G}_{-m-n-1/2}
\end{aligned}$$

and

$$[\widehat{L}_{-m}, \widehat{L}_{-n}] = (n-m)\widehat{L}_{-m-n} \text{ for } n \geq 1.$$

Using (3.12) we also introduce another representation of the super-Virasoro algebra, as acting on bosonic and fermionic components $\widehat{\chi}_{B,\alpha}(x)$ and $\widehat{\chi}_{F,\alpha}(x)$ in (3.11); for $n \geq 2$ we get

(3.16)
$$\begin{aligned}
&\widehat{\mathsf{G}}_{-1/2}\widehat{\chi}_{B,\alpha}(x) = \widehat{\chi}_{F,\alpha}(x), \qquad \widehat{\mathsf{G}}_{-1/2}\widehat{\chi}_{F,\alpha}(x) = \partial_x\widehat{\chi}_{B,\alpha}(x),\\
&\widehat{\mathsf{G}}_{-n+1/2}\widehat{\chi}_{B,\alpha}(x) = \widehat{G}_{-n+1/2}\widehat{\chi}_{B,\alpha}(x), \qquad \widehat{\mathsf{G}}_{-n+1/2}\widehat{\chi}_{F,\alpha}(x) = \widehat{G}_{-n+1/2}\widehat{\chi}_{F,\alpha}(x),\\
&\widehat{\mathsf{L}}_{-1}\widehat{\chi}_{B,\alpha}(x) = \partial_x\widehat{\chi}_{B,\alpha}(x), \qquad \widehat{\mathsf{L}}_{-1}\widehat{\chi}_{F,\alpha}(x) = \partial_x\widehat{\chi}_{F,\alpha}(x),\\
&\widehat{\mathsf{L}}_{-n}\widehat{\chi}_{B,\alpha}(x) = \widehat{L}_{-n}\widehat{\chi}_{B,\alpha}(x), \qquad \widehat{\mathsf{L}}_{-n}\widehat{\chi}_{F,\alpha}(x) = \widehat{L}_{-n}\widehat{\chi}_{F,\alpha}(x).
\end{aligned}$$

From this representation, $\widehat{\mathsf{G}}^2_{-1/2} = \widehat{\mathsf{L}}_{-1}$ on $\widehat{\chi}_{B,\alpha}(x)$, and we find commutation relations

$$\{\widehat{\mathsf{G}}_{-m-1/2}, \widehat{\mathsf{G}}_{-n-1/2}\} = 2\widehat{\mathsf{L}}_{-m-n-1}, \left[\widehat{\mathsf{L}}_{-m}, \widehat{\mathsf{G}}_{-n-1/2}\right] = \big(n-(m-1)/2\big)\widehat{\mathsf{G}}_{-m-n-1/2},$$

as well as $\left[\widehat{\mathsf{L}}_{-m}, \widehat{\mathsf{L}}_{-n}\right] = (n-m)\widehat{\mathsf{L}}_{-m-n}$ for $n \geq 1$. Although commutation relations

$$\{\widehat{\mathsf{G}}_{-1/2}, \widehat{\mathsf{G}}_{-n+1/2}\} = 2\widehat{\mathsf{L}}_{-n}, \quad \left[\widehat{\mathsf{G}}_{-1/2}, \widehat{\mathsf{L}}_{-n}\right] = \frac{n-1}{2}\widehat{\mathsf{G}}_{-n-1/2}, \quad n \geq 1, \tag{3.17}$$

are not obvious at this stage, they are needed to construct super-quantum curves.

Yet another representation of super-Virasoro algebra that we consider is associated to the wave-function normalized by (3.2)

$$\Psi_\alpha(x,\theta) = \frac{\widehat{\chi}_\alpha(x,\theta)}{Z}. \tag{3.18}$$

In this case the operators $\widehat{G}_{n+1/2}$ and $\widehat{L}_n$ in (3.12) are converted to $\widehat{\mathcal{G}}_{-n+1/2} \equiv Z^{-1}\widehat{G}_{-n+1/2}Z$ and $\widehat{\mathcal{L}}_{-n} \equiv Z^{-1}\widehat{L}_{-n}Z$, and for $n \geq 2$ they take form

(3.19)
$$\begin{aligned}
\widehat{\mathcal{G}}_{-n+1/2} &= \frac{1}{\hbar^2(n-2)!}\Big(\partial_x^{n-2}\big(V_B'(x)V_F(x)\big) + Q\hbar\partial_x^{n-1}V_F(x) + \partial_x^{n-2}\widehat{h}(x)\\
&\qquad + \big[\partial_x^{n-2}\widehat{h}(x), \log Z\big]\Big),\\
\widehat{\mathcal{L}}_{-n} &= \frac{1}{\hbar^2(n-2)!}\Big(\frac{1}{2}\partial_x^{n-2}\big(V_B'(x)^2\big) + \frac{1}{2}\partial_x^{n-2}\big(V_F'(x)V_F(x)\big) + \frac{1}{2}Q\hbar\partial_x^n V_B(x)\\
&\qquad + \partial_x^{n-2}\widehat{f}(x) + \big[\partial_x^{n-2}\widehat{f}(x), \log Z\big]\Big).
\end{aligned}$$

3.2. Higher level quantum curves as super-Virasoro singular vectors.

We present now super-quantum curves $\widehat{A}_n^\alpha$, which annihilate wave-functions defined in (3.10)

$$\widehat{A}_n^\alpha\widehat{\chi}_\alpha(x,\theta) = 0. \tag{3.20}$$

These super-quantum curves have the structure of the Neveu-Schwarz (NS) super-Virasoro singular vectors. Recall that NS super-Virasoro singular vectors exist only at levels $n = pq/2$, for integer p and q such that $p-q$ is even, and correspond to degenerate momenta of the form

$$\alpha = \alpha_{p,q} = \frac{(p-1)\beta^{1/2} - (q-1)\beta^{-1/2}}{2}\hbar, \qquad \text{with } p-q \in 2\mathbb{Z} \tag{3.21}$$

which label primary states $|\Delta_{p,q}\rangle$ of weight $\Delta_{p,q} = -\frac{pq-1}{4} + \frac{p^2-1}{8}\beta + \frac{q^2-1}{8}\frac{1}{\beta}$. The first few NS super-Virasoro singular vectors take form

$$\begin{aligned}
\text{(3.22)} \quad &\text{level } \frac{1}{2} : G_{-1/2}|\Delta_{1,1}\rangle, \\
&\text{level } \frac{3}{2} : \left(L_{-1}G_{-1/2} - \beta G_{-3/2}\right)|\Delta_{3,1}\rangle, \\
&\text{level } 2 : \Big(L_{-1}^2 - \frac{1}{2}(\beta^{-1/2} - \beta^{1/2})^2 L_{-2} - G_{-3/2}G_{-1/2}\Big)|\Delta_{2,2}\rangle, \\
&\text{level } \frac{5}{2} : \left(L_{-1}^2 G_{-1/2} + 2\beta(3\beta-1) - 3\beta G_{-3/2}L_{-1} - 2\beta L_{-2}G_{-1/2}\right)|\Delta_{5,1}\rangle.
\end{aligned}$$

Imposing the condition that the wave-function $\widehat{\chi}_\alpha(x,\theta)$ satisfies a differential equation of a finite order in x, we find that it can be satisfied only for special values of α in (3.21), and operators $\widehat{A}_\alpha^n$ encode the structure of NS singular vectors, such as those in (3.22). In the notation $\widehat{A}_n^\alpha$ the superscript α refers to the deformation parameter of the corresponding wave-function, and the subscript n denotes the level of the singular vector encoding the structure of a given super-quantum curve. In particular the super-quantum curve at level 3/2 reduces to super-spectral curve of the super-eigenvalue model in the classical limit.

The operators $\widehat{A}_n^\alpha$ encode the structure of super-Virasoro singular vectors as follows. First, using the decomposition (3.11) and the representation (3.16), we write the wave-function as

$$\widehat{\chi}_\alpha(x,\theta) = (1 - \theta\widehat{\mathsf{G}}_{-1/2})\widehat{\chi}_{B,\alpha}(x), \tag{3.23}$$

so that all information is essentially encoded in its bosonic component $\widehat{\chi}_{B,\alpha}(x)$. Note that

$$\theta\widehat{\chi}_\alpha(x,\theta) = \theta\widehat{\chi}_{B,\alpha}(x), \qquad \widehat{\chi}_{F,\alpha}(x) = -\partial_\theta\widehat{\chi}_\alpha(x,\theta) = \widehat{\mathsf{G}}_{-1/2}\widehat{\chi}_{B,\alpha}(x). \tag{3.24}$$

In consequence one can introduce an operator $\widehat{\mathsf{A}}_n^{(0)}$ which annihilates the bosonic wave-function

$$\widehat{\mathsf{A}}_n^{(0)}\widehat{\chi}_{B,\alpha_{p,q}}(x) = 0. \tag{3.25}$$

As we will show, this is the operator $\widehat{\mathsf{A}}_n^{(0)}$ that takes form of a super-Virasoro singular vector (3.22), written in terms of generators (3.16). The above equation can be also rewritten in terms of an operator $\widehat{A}_n^{(0)}$ – which arises by replacing $\widehat{\mathsf{G}}_{-1/2}$ by $(-\partial_\theta)$ in $\widehat{\mathsf{A}}_n^{(0)}$ – acting on $\widehat{\chi}_\alpha(x,\theta)$

$$\theta\widehat{A}_n^{(0)}\widehat{\chi}_{\alpha_{p,q}}(x,\theta) = \theta\widehat{\mathsf{A}}_n^{(0)}\widehat{\chi}_{B,\alpha_{p,q}}(x) = 0. \tag{3.26}$$

Furthermore, the bosonic wave-function is annihilated also by the following operator $\widehat{\mathsf{A}}_n^{(1)}$ (expressed in terms of generators (3.16))

$$\widehat{\mathsf{A}}_n^{(1)} = \widehat{\mathsf{G}}_{-1/2}\widehat{\mathsf{A}}_n^{(0)}, \tag{3.27}$$

related to the operator acting on $\widehat{\chi}_\alpha(x,\theta)$ as

$$\theta\widehat{A}_n^{(1)}\widehat{\chi}_{\alpha_{p,q}}(x,\theta) = \theta\widehat{\mathsf{A}}_n^{(1)}\widehat{\chi}_{B,\alpha_{p,q}}(x) = 0. \tag{3.28}$$

The original super-quantum curve $\widehat{A}_n^\alpha$ can be reconstructed from $\widehat{A}_n^{(0)}$ and $\widehat{A}_n^{(1)}$ given above

$$\widehat{A}_n^\alpha = \widehat{A}_n^{(0)} - \theta\partial_\theta\widehat{A}_n^{(0)} - \theta\widehat{A}_n^{(1)}, \tag{3.29}$$

and this operator is expressed in terms of generators (3.12). The operator (3.29) can be also transformed into an operator that annihilates the normalized wave-function (3.18)

$$\widehat{\mathcal{A}}_n^{\alpha} = Z^{-1}\widehat{A}_n^{\alpha} Z, \tag{3.30}$$

which is then expressed in terms of generators (3.19).

We present now explicit form of super-quantum curves at first few levels. At level 3/2, for specific values of momenta

$$\alpha = 0, \quad \beta^{1/2}\hbar, \quad \text{or} \quad -\beta^{-1/2}\hbar, \tag{3.31}$$

we find

$$\widehat{A}_{3/2}^{(0)} = -\partial_x\partial_\theta - \frac{\alpha^2}{\hbar^2}\widehat{G}_{-3/2}, \tag{3.32}$$

and as the operators acting on $\widehat{\chi}_B(x)$

$$\begin{aligned} \widehat{\mathsf{A}}_{3/2}^{(0)} &= \widehat{\mathsf{L}}_{-1}\widehat{\mathsf{G}}_{-1/2} - \frac{\alpha^2}{\hbar^2}\widehat{\mathsf{G}}_{-3/2}, \\ \widehat{\mathsf{A}}_{3/2}^{(1)} &= \widehat{\mathsf{G}}_{-1/2}\widehat{\mathsf{A}}_{3/2}^{(0)} = \widehat{\mathsf{L}}_{-1}^2 - \frac{2\alpha^2}{\hbar^2}\widehat{\mathsf{L}}_{-2} + \frac{\alpha^2}{\hbar^2}\widehat{\mathsf{G}}_{-3/2}\widehat{\mathsf{G}}_{-1/2}. \end{aligned} \tag{3.33}$$

We find that $\widehat{\mathsf{A}}_{3/2}^{(0)}$ indeed has the form of singular vectors at level 3/2, see (3.22), for $\alpha = \pm\beta^{\pm 1/2}\hbar$. For the remaining value $\alpha = 0$ in (3.31), $\widehat{\mathsf{A}}_{3/2}^{(0)}$ reduces to the singular vector at level 1/2, see (3.22), which can be identified with the partition function (3.2). We also obtain

$$\widehat{A}_{3/2}^{(1)} = \partial_x^2 - \frac{2\alpha^2}{\hbar^2}\widehat{L}_{-2} - \frac{\alpha^2}{\hbar^2}\widehat{G}_{-3/2}\partial_\theta, \tag{3.34}$$

and from (3.29) we finally find the super-quantum curve equation

$$\widehat{\mathcal{A}}_{3/2}^{\alpha}\widehat{\chi}_\alpha(x,\theta) = 0, \quad \widehat{\mathcal{A}}_{3/2}^{\alpha} = -\partial_x\partial_\theta - \frac{\alpha^2}{\hbar^2}\widehat{G}_{-3/2} - \theta\Big(\partial_x^2 - \frac{2\alpha^2}{\hbar^2}\widehat{L}_{-2}\Big), \tag{3.35}$$

for special values of α in (3.31).

At level 2, for specific values of momenta

$$\alpha = 0, \quad \beta^{1/2}\hbar, \quad -\beta^{-1/2}\hbar, \quad -\frac{Q\hbar}{2} = \frac{\beta^{1/2}-\beta^{-1/2}}{2}\hbar, \tag{3.36}$$

we find

$$\widehat{A}_2^{(0)} = \partial_x^2 - \frac{2\alpha^2}{\hbar^2}\widehat{L}_{-2} - \frac{2\alpha^2 + Q\hbar\alpha - \hbar^2}{\hbar^2}\widehat{G}_{-3/2}\partial_\theta. \tag{3.37}$$

This leads to the following operators acting on $\widehat{\chi}_{B,\alpha}(x)$

$$\begin{aligned} \widehat{\mathsf{A}}_2^{(0)} &= \widehat{\mathsf{L}}_{-1}^2 - \frac{2\alpha^2}{\hbar^2}\widehat{\mathsf{L}}_{-2} + \frac{2\alpha^2 + Q\hbar\alpha - \hbar^2}{\hbar^2}\widehat{\mathsf{G}}_{-3/2}\widehat{\mathsf{G}}_{-1/2} \\ &= \widehat{\mathsf{A}}_{3/2}^{(1)} + \frac{\alpha^2 + Q\hbar\alpha - \hbar^2}{\hbar^2}\widehat{\mathsf{G}}_{-3/2}\widehat{\mathsf{G}}_{-1/2}, \\ \widehat{\mathsf{A}}_2^{(1)} &= \widehat{\mathsf{G}}_{-1/2}\widehat{\mathsf{A}}_2^{(0)} = \widehat{\mathsf{L}}_{-1}^2\widehat{\mathsf{G}}_{-1/2} - \frac{2\alpha^2}{\hbar^2}\widehat{\mathsf{L}}_{-2}\widehat{\mathsf{G}}_{-1/2} - \frac{\alpha^2}{\hbar^2}\widehat{\mathsf{G}}_{-5/2} \\ &\qquad + \frac{2\alpha^2 + Q\hbar\alpha - \hbar^2}{\hbar^2}\big(2\widehat{\mathsf{L}}_{-2}\widehat{\mathsf{G}}_{-1/2} - \widehat{\mathsf{G}}_{-3/2}\widehat{\mathsf{L}}_{-1}\big), \end{aligned} \tag{3.38}$$

and for $\alpha = -\frac{Q\hbar}{2}$ the operator $\widehat{\mathsf{A}}_2^{(0)}$ takes form of singular vector at level 2 (3.22). Furthermore

(3.39)
$$\widehat{A}_2^{(1)} = -\partial_x^2\partial_\theta + \frac{2\alpha^2}{\hbar^2}\widehat{L}_{-2}\partial_\theta - \frac{\alpha^2}{\hbar^2}\widehat{G}_{-5/2} - \frac{2\alpha^2 + Q\hbar\alpha - \hbar^2}{\hbar^2}\big(2\widehat{L}_{-2}\partial_\theta + \widehat{G}_{-3/2}\partial_x\big),$$

and finally from (3.29) we obtain the super-quantum curve equation at level 2

$$\begin{aligned}
&\widehat{A}_2^{\alpha}\widehat{\chi}_\alpha(x,\theta) = 0,\\
&\widehat{A}_2^{\alpha} = \partial_x^2 - \frac{2\alpha^2}{\hbar^2}\widehat{L}_{-2} - \frac{2\alpha^2 + Q\hbar\alpha - \hbar^2}{\hbar^2}\widehat{G}_{-3/2}\partial_\theta + \frac{\alpha^2}{\hbar^2}\theta\widehat{G}_{-5/2}\\
&\qquad + \frac{2\alpha^2 + Q\hbar\alpha - \hbar^2}{\hbar^2}\theta\Big(2\widehat{L}_{-2}\partial_\theta + \widehat{G}_{-3/2}\partial_x\Big),
\end{aligned} \tag{3.40}$$

for the momenta (3.36). Upon substitution those values of α that correspond to lower levels (3.31), this result reduces to quantum curves at lower levels.

At level 5/2, for specific values of momenta

$$\alpha = 0, \quad \beta^{1/2}\hbar, \quad -\beta^{-1/2}\hbar, \quad -\frac{Q\hbar}{2}, \quad 2\beta^{1/2}\hbar, \quad -2\beta^{-1/2}\hbar, \tag{3.41}$$

we find

$$\begin{aligned}
\widehat{A}_{5/2}^{(0)} &= -\partial_x^2\partial_\theta + \frac{2\alpha(\alpha^2 + Q\hbar\alpha - \hbar^2)}{\hbar^2(3\alpha + 2Q\hbar)}\widehat{L}_{-2}\partial_\theta - \frac{\alpha(2\alpha^2 + Q\hbar\alpha + \hbar^2)}{\hbar^2(3\alpha + 2Q\hbar)}\widehat{G}_{-3/2}\partial_x\\
&\quad + \frac{\alpha^2(2\alpha^3 + 3Q\hbar\alpha^2 + (Q^2 - 5)\hbar^2\alpha - 3Q\hbar^3)}{\hbar^4(3\alpha + 2Q\hbar)}\widehat{G}_{-5/2},
\end{aligned} \tag{3.42}$$

and acting on the bosonic component this operator is represented as

(3.43)
$$\widehat{\mathsf{A}}_{5/2}^{(0)} = \widehat{\mathsf{A}}_2^{(1)} + \frac{(\alpha^2 + Q\hbar\alpha - \hbar^2)(2\alpha + Q\hbar)}{\hbar^2(3\alpha + 2Q\hbar)}\Big(\frac{\alpha^2}{\hbar^2}\widehat{\mathsf{G}}_{-5/2} + 2\widehat{\mathsf{G}}_{-3/2}\widehat{\mathsf{L}}_{-1} - 4\widehat{\mathsf{L}}_{-2}\widehat{\mathsf{G}}_{-1/2}\Big).$$

As expected, for $\alpha = \pm 2\beta^{\pm 1/2}\hbar$ it has the same form as a singular vector at level 5/2 (3.22). Finally, determining $\widehat{\mathsf{A}}_{5/2}^{(1)}$ and using (3.29), we obtain the super-quantum curve at level 5/2

(3.44)
$$\begin{aligned}
&\widehat{A}_{5/2}^{\alpha}\widehat{\chi}_\alpha(x,\theta) = 0,\\
&\widehat{A}_{5/2}^{\alpha} = -\partial_x^2\partial_\theta + \frac{2\alpha(\alpha^2 + Q\hbar\alpha - \hbar^2)}{\hbar^2(3\alpha + 2Q\hbar)}\widehat{L}_{-2}\partial_\theta - \frac{\alpha(2\alpha^2 + Q\hbar\alpha + \hbar^2)}{\hbar^2(3\alpha + 2Q\hbar)}\widehat{G}_{-3/2}\partial_x\\
&\quad + \frac{\alpha^2(2\alpha^3 + 3Q\hbar\alpha^2 + (Q^2 - 5)\hbar^2\alpha - 3Q\hbar^3)}{\hbar^4(3\alpha + 2Q\hbar)}\widehat{G}_{-5/2} - \theta\Big(\partial_x^3 - \frac{2\alpha^2}{\hbar^2}\widehat{L}_{-2}\partial_x\\
&\quad + \frac{\alpha(\alpha^2 + Q\hbar\alpha - \hbar^2)}{\hbar^2(3\alpha + 2Q\hbar)}\widehat{G}_{-5/2}\partial_\theta\\
&\quad + \frac{2\alpha^2(2\alpha^3 + 3Q\hbar\alpha^2 + (Q^2 - 5)\hbar^2\alpha - 3Q\hbar^3)}{\hbar^4(3\alpha + 2Q\hbar)}\widehat{L}_{-3}\Big),
\end{aligned}$$

for the momenta (3.41). For the momenta (3.36) at level 2 the operator $\widehat{\mathsf{A}}_{5/2}^{(0)}$ reduces to the operator $\widehat{\mathsf{A}}_2^{(1)}$ in (3.38), and encodes singular vectors up to level 2.

3.3. Double quantum structure. Similarly as in the bosonic case discussed in section 2.3, also two classical limits and correspondingly a double quantum structure can be considered in the supersymmetric case. The first classical limit is the large N 't Hooft limit (of vanishing $\hbar$), and the second limit is that of infinite central charge in the conformal field theory interpretation.

Let us consider 't Hooft limit first, such that $N \to \infty$, $\hbar \to 0$, and $\mu = \beta^{1/2}\hbar N = const.$ In this limit we introduce

$$y_B(x) = \lim_{\substack{N\to\infty \\ \beta=1}} \frac{\hbar}{Z}\langle \partial_x\phi(x)\rangle, \qquad y_F(x) = \lim_{\substack{N\to\infty \\ \beta=1}} \frac{\hbar}{Z}\langle \psi(x)\rangle, \tag{3.45}$$

and then we find

$$\begin{aligned} \hbar\partial_\theta\Psi_\alpha(x,\theta) &\overset{N\to\infty,\ \beta\to1}{\longrightarrow} -\frac{\alpha}{\hbar}y_F(x)\Psi_\alpha(x,\theta), \\ \hbar\partial_x\Psi_\alpha(x,\theta) &\overset{N\to\infty,\ \beta\to1}{\longrightarrow} \frac{\alpha}{\hbar}\big(y_B(x)+y_F'(x)\theta\big)\Psi_\alpha(x,\theta). \end{aligned} \tag{3.46}$$

A redefinition of the super-quantum curve at level 3/2 in (3.35), which annihilates the normalized wave-function (3.18), takes form

$$-\hbar^2\widehat{\mathcal{A}}^{\alpha}_{3/2} = \hbar^2\partial_x\partial_\theta + \alpha^2\widehat{\mathcal{G}}_{-3/2} + \hbar^2\theta\partial_x^2 - 2\theta\alpha^2\widehat{\mathcal{L}}_{-2}, \tag{3.47}$$

for momenta $\alpha = \pm\beta^{\pm1/2}\hbar$. This quantum curve reduces to the supersymmetric algebraic curve, which can be written as

$$\widehat{\mathcal{A}}^{\alpha}_{3/2}\Psi_\alpha(x,\theta) = 0 \overset{N\to\infty,\ \beta\to1}{\longrightarrow} A_F(x,y_B|y_F) = \theta A_B(x,y_B|y_F), \tag{3.48}$$

where

$$\begin{cases} A_F(x,y_B|y_F) \equiv y_B(x)y_F(x) + G(x) = 0, \\ A_B(x,y_B|y_F) \equiv y_B(x)^2 + y_F'(x)y_F(x) + 2L(x) = 0, \end{cases} \tag{3.49}$$

and

$$G(x) = -\lim_{\substack{\hbar\to0 \\ \beta=1}} \hbar^2\widehat{\mathcal{G}}_{-3/2}, \qquad L(x) = -\lim_{\substack{\hbar\to0 \\ \beta=1}} \hbar^2\widehat{\mathcal{L}}_{-2}. \tag{3.50}$$

The supersymmetric algebraic curve defined by (3.48) coincides with the spectral curve of the super-eigenvalue model, which can be determined by the analysis of eigenvalue distribution in the large N limit, and that we also refer to as super-spectral curve [**21**]. It follows that the super-quantum curve at level 3/2 can be regarded as a quantization of the super-spectral curve. Classical curves at higher levels can be obtained as appropriate limits of the quantum curves in the second interesting limit (Nekrasov-Shatashvili limit) listed in (3.61); for completeness, these classical curves at a few higher levels take form

(3.51)
$$\begin{aligned}\widehat{\mathcal{A}}^{\mathrm{cl}}_{5/2} &= -2\Big(3y_B(x)A_F(x,y_B|y_F) + 2y_F(x)A_B(x,y_B|y_F)\Big)\\ &\quad + 2\theta\Big(3y'_F(x)A_F(x,y_B|y_F) + \partial_x A_F(x,y_B|y_F)y_F(x) + 8y_B(x)A_B(x,y_B|y_F)\Big),\\ \widehat{\mathcal{A}}^{\mathrm{cl}}_{7/2} &= -18\Big(\big(3y_B(x)^2 + L(x)\big)A_F(x,y_B|y_F) + 3y_F(x)y_B(x)A_B(x,y_B|y_F)\Big)\\ &\quad + 9\theta\Big(2\big(9y_B(x)^2 + 3y'_F(x)y_F(x) + 2L(x)\big)A_B(x,y_B|y_F)\\ &\quad + \big(15y'_F(x)y_B(x) + 3y_F(x)y'_B(x) + 4\partial_x G(x)\big)A_F(x,y_B|y_F)\\ &\quad + 3G(x)\partial_x A_F(x,y_B|y_F)\Big).\end{aligned}$$

The second limit of interests, that can be identified as the Nekrasov-Shatashvili limit or a classical limit in the super-Liouville theory, in terms of parameters (3.5) arises for $\epsilon_1 \to 0$ with ϵ_2 fixed. We consider normalized wave-functions $\Psi_{\alpha_{2p+1,1}}(x,\theta)$ for the momenta

$$\alpha = \alpha_{2p+1,1} = -p\epsilon_1, \tag{3.52}$$

which in this limit factorize as

$$\Psi^{\mathrm{NS}}_{-p\epsilon_1}(x,\theta) \equiv \lim_{\epsilon_1\to 0}\Psi_{-p\epsilon_1}(x,\theta) = \big(\Psi^{\mathrm{NS}}_{-\epsilon_1}(x,\theta)\big)^p. \tag{3.53}$$

The corresponding super-quantum curves arise at level $p+1/2$, and we rescale them as follows

$$\widehat{\mathcal{A}}^{\mathrm{NS}}_{p+1/2}\Psi^{\mathrm{NS}}_{-p\epsilon_1}(x,\theta) = 0, \qquad \widehat{\mathcal{A}}^{\mathrm{NS}}_{p+1/2} = -\epsilon_2^{p+1}\lim_{\epsilon_1\to 0}\widehat{\mathcal{A}}^{-p\epsilon_1}_{p+1/2}, \tag{3.54}$$

and express in terms of operators

(3.55)
$$\begin{aligned}\widehat{\mathcal{G}}^{\mathrm{NS}}_{-n+1/2} &= \lim_{\epsilon_1\to 0}\epsilon_1\epsilon_2\widehat{\mathcal{G}}_{-n+1/2}\\ &= -\frac{1}{(n-2)!}\Big(\partial_x^{n-2}\big(V'_B(x)V_F(x)\big) + \epsilon_2\partial_x^{n-1}V_F(x) + \partial_x^{n-2}F_F^{(0)}(x,\epsilon_2)\Big),\\ \widehat{\mathcal{L}}^{\mathrm{NS}}_{-n} &= \lim_{\epsilon_1\to 0}\epsilon_1\epsilon_2\widehat{\mathcal{L}}_{-n}\\ &= -\frac{1}{(n-2)!}\Big(\frac{1}{2}\partial_x^{n-2}\big(V'_B(x)^2\big) + \frac{1}{2}\partial_x^{n-2}\big(V'_F(x)V_F(x)\big)\\ &\qquad\qquad + \frac{1}{2}\epsilon_2\partial_x^n V_B(x) + \partial_x^{n-2}F_B^{(0)}(x,\epsilon_2)\Big),\end{aligned}$$

where

(3.56)
$$F_F^{(0)}(x,\epsilon_2) = -\frac{1}{\epsilon_1\epsilon_2}\big[\widehat{h}(x), F^{(0)}(\epsilon_2)\big], \qquad F_B^{(0)}(x,\epsilon_2) = -\frac{1}{\epsilon_1\epsilon_2}\big[\widehat{f}(x), F^{(0)}(\epsilon_2)\big],$$

are defined in terms of the deformed prepotential

$$F^{(0)}(\epsilon_2) = -\lim_{\epsilon_1\to 0}\epsilon_1\epsilon_2\log Z. \tag{3.57}$$

Quantities $\widehat{\mathcal{G}}^{\rm NS}_{-n+1/2}$ and $\widehat{\mathcal{L}}^{\rm NS}_{-n}$ are simply fermionic and bosonic functions of x, such that

$$\partial_x^n \widehat{\mathcal{G}}^{\rm NS}_{-3/2} = n!\widehat{\mathcal{G}}^{\rm NS}_{-n-3/2}, \qquad \partial_x^n \widehat{\mathcal{L}}^{\rm NS}_{-2} = n!\widehat{\mathcal{L}}^{\rm NS}_{-n-2}. \tag{3.58}$$

The super-quantum curve equation (3.35) at level 3/2 in the limit $\epsilon_1 \to 0$ yields

$$\widehat{\mathcal{A}}^{\rm NS}_{3/2}\Psi^{\rm NS}_{-\epsilon_1}(x,\theta) = \left(\epsilon_2^2\partial_x\partial_\theta - \widehat{\mathcal{G}}^{\rm NS}_{-3/2} + \theta\epsilon_2^2\partial_x^2 + 2\theta\widehat{\mathcal{L}}^{\rm NS}_{-2}\right)\Psi^{\rm NS}_{-\epsilon_1}(x,\theta) = 0, \tag{3.59}$$

and equivalently can be written as

$$\theta\left(\epsilon_2^2\partial_x\partial_\theta - \widehat{\mathcal{G}}^{\rm NS}_{-3/2}\right)\Psi^{\rm NS}_{-\epsilon_1}(x,\theta) = \theta\left(\epsilon_2^2\partial_x^2 + \widehat{\mathcal{G}}^{\rm NS}_{-3/2}\partial_\theta + 2\widehat{\mathcal{L}}^{\rm NS}_{-2}\right)\Psi^{\rm NS}_{-\epsilon_1}(x,\theta) = 0. \tag{3.60}$$

Higher level quantum curves in this limit can be determined by generalizing the recurrence relation (2.32) to the supersymmetric case [**21**], which yields the following results

$$\begin{aligned}
\widehat{\mathcal{A}}^{\rm NS}_{5/2} &= \epsilon_2^3\partial_x^2\partial_\theta + 2\epsilon_2\widehat{\mathcal{L}}^{\rm NS}_{-2}\partial_\theta - 3\epsilon_2\widehat{\mathcal{G}}^{\rm NS}_{-3/2}\partial_x - 2\epsilon_2\widehat{\mathcal{G}}^{\rm NS}_{-5/2} \\
&\quad + \theta\Big(\epsilon_2^3\partial_x^3 - \epsilon_2\widehat{\mathcal{G}}^{\rm NS}_{-5/2}\partial_\theta + 8\epsilon_2\widehat{\mathcal{L}}^{\rm NS}_{-2}\partial_x + 4\epsilon_2\widehat{\mathcal{L}}^{\rm NS}_{-3}\Big), \\
\widehat{\mathcal{A}}^{\rm NS}_{7/2} &= \epsilon_2^4\partial_x^3\partial_\theta + 8\epsilon_2^2\widehat{\mathcal{L}}^{\rm NS}_{-2}\partial_x\partial_\theta + 4\epsilon_2^2\widehat{\mathcal{L}}^{\rm NS}_{-3}\partial_\theta - 6\epsilon_2^2\widehat{\mathcal{G}}^{\rm NS}_{-3/2}\partial_x^2 - 8\epsilon_2^2\widehat{\mathcal{G}}^{\rm NS}_{-5/2}\partial_x - 6\epsilon_2^2\widehat{\mathcal{G}}^{\rm NS}_{-7/2} \\
&\quad - 18\widehat{\mathcal{G}}^{\rm NS}_{-3/2}\widehat{\mathcal{L}}^{\rm NS}_{-2} + \theta\Big(\epsilon_2^4\partial_x^4 - 4\epsilon_2^2\widehat{\mathcal{G}}^{\rm NS}_{-5/2}\partial_x\partial_\theta - 4\epsilon_2^2\widehat{\mathcal{G}}^{\rm NS}_{-7/2}\partial_\theta + 20\epsilon_2^2\widehat{\mathcal{L}}^{\rm NS}_{-2}\partial_x^2 \\
&\quad + 20\epsilon_2^2\widehat{\mathcal{L}}^{\rm NS}_{-3}\partial_x + 12\epsilon_2^2\widehat{\mathcal{L}}^{\rm NS}_{-4} - 9\widehat{\mathcal{G}}^{\rm NS}_{-3/2}\widehat{\mathcal{G}}^{\rm NS}_{-5/2} + 36\big(\widehat{\mathcal{L}}^{\rm NS}_{-2}\big)^2\Big), \\
\widehat{\mathcal{A}}^{\rm NS}_{9/2} &= \epsilon_2^5\partial_x^4\partial_\theta + 20\epsilon_2^3\widehat{\mathcal{L}}^{\rm NS}_{-2}\partial_x^2\partial_\theta + 20\epsilon_2^3\widehat{\mathcal{L}}^{\rm NS}_{-3}\partial_x\partial_\theta + 12\epsilon_2^3\widehat{\mathcal{L}}^{\rm NS}_{-4}\partial_\theta - 10\epsilon_2^3\widehat{\mathcal{G}}^{\rm NS}_{-3/2}\partial_x^3 \\
&\quad - 20\epsilon_2^3\widehat{\mathcal{G}}^{\rm NS}_{-5/2}\partial_x^2 - 30\epsilon_2^3\widehat{\mathcal{G}}^{\rm NS}_{-7/2}\partial_x - 24\epsilon_2^3\widehat{\mathcal{G}}^{\rm NS}_{-9/2} - \epsilon_2\widehat{\mathcal{G}}^{\rm NS}_{-3/2}\widehat{\mathcal{G}}^{\rm NS}_{-5/2}\partial_\theta \\
&\quad + 36\epsilon_2\big(\widehat{\mathcal{L}}^{\rm NS}_{-2}\big)^2\partial_\theta - 110\epsilon_2\widehat{\mathcal{G}}^{\rm NS}_{-3/2}\widehat{\mathcal{L}}^{\rm NS}_{-2}\partial_x - 56\epsilon_2\widehat{\mathcal{G}}^{\rm NS}_{-3/2}\widehat{\mathcal{L}}^{\rm NS}_{-3} - 72\epsilon_2\widehat{\mathcal{G}}^{\rm NS}_{-5/2}\widehat{\mathcal{L}}^{\rm NS}_{-2} \\
&\quad + \theta\Big(\epsilon_2^5\partial_x^5 - 10\epsilon_2^3\widehat{\mathcal{G}}^{\rm NS}_{-5/2}\partial_x^2\partial_\theta - 20\epsilon_2^3\widehat{\mathcal{G}}^{\rm NS}_{-7/2}\partial_x\partial_\theta - 18\epsilon_2^3\widehat{\mathcal{G}}^{\rm NS}_{-9/2}\partial_\theta + 40\epsilon_2^3\widehat{\mathcal{L}}^{\rm NS}_{-2}\partial_x^3 \\
&\quad + 60\epsilon_2^3\widehat{\mathcal{L}}^{\rm NS}_{-3}\partial_x^2 + 72\epsilon_2^3\widehat{\mathcal{L}}^{\rm NS}_{-4}\partial_x + 48\epsilon_2^3\widehat{\mathcal{L}}^{\rm NS}_{-5} - 34\epsilon_2\widehat{\mathcal{G}}^{\rm NS}_{-5/2}\widehat{\mathcal{L}}^{\rm NS}_{-2}\partial_\theta - 2\epsilon_2\widehat{\mathcal{G}}^{\rm NS}_{-3/2}\widehat{\mathcal{L}}^{\rm NS}_{-3}\partial_\theta \\
&\quad - 56\epsilon_2\widehat{\mathcal{G}}^{\rm NS}_{-3/2}\widehat{\mathcal{G}}^{\rm NS}_{-5/2}\partial_x - 56\epsilon_2\widehat{\mathcal{G}}^{\rm NS}_{-3/2}\widehat{\mathcal{G}}^{\rm NS}_{-7/2} + 256\epsilon_2\big(\widehat{\mathcal{L}}^{\rm NS}_{-2}\big)^2\partial_x + 256\epsilon_2\widehat{\mathcal{L}}^{\rm NS}_{-2}\widehat{\mathcal{L}}^{\rm NS}_{-3}\Big).
\end{aligned} \tag{3.61}$$

The form of these operators coincides with that of operators that implement classical equations of motion for certain fields in the classical super-Liouville theory [**48**], which is a nice check of our formalism.

3.4. Examples of super-quantum curves. It is also interesting to consider specific potentials in super-eigenvalue models and corresponding super-quantum curves. We consider analogous examples as in the bosonic case in section 2.4, i.e. super-Gaussian and super-multi-Penner models. More details, in particular the analysis of planar one-cut solutions in these examples, are given in [**21**].

To start with we consider a supersymmetric version of the Gaussian model, whose potential has a quadratic bosonic term, and includes bosonic and fermionic linear terms depending on bosonic and fermionic times t and ξ

$$V_t(x,\theta) = V_{B,t}(x) + V_F(x)\theta, \qquad V_{B,t}(x) = tx + \frac{1}{2}x^2, \qquad V_F(x) = \xi x. \tag{3.62}$$

Specializing the super-quantum curve equations (3.35) at level 3/2 to this case we find

$$\widehat{A}_{3/2}\widehat{\chi}_{\alpha_\pm}(x,\theta)=0,\quad \widehat{A}_{3/2}=-\partial_x\partial_\theta-\beta^{\pm 1}\widehat{G}_{-3/2}-\theta\Big(\partial_x^2-2\beta^{\pm 1}\widehat{L}_{-2}\Big), \tag{3.63}$$

where $\alpha_\pm=\pm\beta^{\pm 1/2}\hbar$, and after some algebra we find

$$\begin{aligned}\hbar^2\widehat{G}_{-3/2}\widehat{\chi}_\alpha(x,\theta)&=\Big(\xi(x+t)x+Q\hbar\xi-2\xi(\mu-\alpha)-\frac{\hbar^2}{t}\big(\partial_\xi+\partial_\theta+(\xi-\theta)\partial_x\big)\Big)\widehat{\chi}_\alpha(x,\theta),\\ \hbar^2\widehat{L}_{-2}&=\frac{1}{2}(x+t)^2+\frac{1}{2}Q\hbar-\mu+\alpha.\end{aligned} \tag{3.64}$$

The second example we consider is the super-multi-Penner super-eigenvalue model with the potential

$$\begin{aligned}V(x,\theta)&=\sum_{i=1}^{M}\alpha_i\log(x-x_i+\theta_i\theta)=V_B(x)+V_F(x)\theta,\\ V_B(x)&=\sum_{i=1}^{M}\alpha_i\log(x-x_i),\quad V_F(x)=\sum_{i=1}^{M}\frac{\alpha_i\theta_i}{x-x_i}.\end{aligned} \tag{3.65}$$

It is convenient to rescale the wave-function in (3.10) and introduce the following normalization

$$\widetilde{\chi}_\alpha(x,\theta)=\widehat{\chi}_\alpha(x,\theta)\prod_{i\neq j}(x_i-x_j-\theta_i\theta_j)^{\frac{\alpha_i\alpha_j}{2\hbar^2}}. \tag{3.66}$$

Similarly as in the bosonic case, in the super-multi-Penner model the potential term

$$e^{-\frac{\sqrt{\beta}}{\hbar}\sum_{a=1}^{N}V(z_a,\vartheta_a)} \tag{3.67}$$

takes an analogous form as M determinant-like insertions of primary fields (3.10). It follows that the wave-function (3.66) represents a correlation function in the super-Liouville theory, which involves $(M+2)$ Neveu-Schwarz primary fields inserted on $\mathbb{P}^1$: M of those fields are encoded in the potential, one field is represented by the determinant-like insertion, and one additional field is inserted at $x=\infty\in\mathbb{P}^1$. The primary field at $x=\infty$ can also be removed by the momentum conservation condition, and then the model with the above potential describes super-Liouville theory with $(M+1)$ primary fields. It follows that various objects familiar in super-Liouville theory arise upon the specialization of our formalism to the super-multi-Penner potential. To construct super-quantum curves for the model with the potential (3.65), we note first that fermionic $\widehat{h}(x)$ in (3.13) and the bosonic $\widehat{f}(x)$ in (3.14) operators take form

$$\begin{aligned}\widehat{h}(x)&=\hbar^2\sum_{i=1}^{M}\frac{1}{x-x_i}D_i,\\ \widehat{f}(x)&=\hbar^2\sum_{i=1}^{M}\Big(\frac{1}{x-x_i}\partial_{x_i}+\frac{\theta_i}{2(x-x_i)^2}\partial_{\theta_i}\Big),\end{aligned} \tag{3.68}$$

where $D_i=-\partial_{\theta_i}+\theta_i\partial_{x_i}$. Second, we introduce the super-Virasoro generators represented on the wave-function $\widetilde{\chi}_\alpha(x,\theta)$ in (3.66); taking advantage of (3.12), for

$n \geq 2$ they take form

$$\begin{aligned}\widetilde{G}_{-n+1/2} &= \Big(\prod_{i\neq j}(x_i - x_j - \theta_i\theta_j)^{\frac{\alpha_i\alpha_j}{2\hbar^2}}\Big)\widehat{G}_{-n+1/2}\Big(\prod_{i\neq j}(x_i - x_j - \theta_i\theta_j)^{-\frac{\alpha_i\alpha_j}{2\hbar^2}}\Big) \\ &= \sum_{i=1}^{M}\Big(\frac{2(n-1)\Delta_{\alpha_i}\theta_i}{(x_i - x)^n} - \frac{1}{(x_i - x)^{n-1}}D_i\Big),\end{aligned} \tag{3.69}$$

and

$$\begin{aligned}\widetilde{L}_{-n} &= \Big(\prod_{i\neq j}(x_i - x_j - \theta_i\theta_j)^{\frac{\alpha_i\alpha_j}{2\hbar^2}}\Big)\widehat{L}_{-n}\Big(\prod_{i\neq j}(x_i - x_j - \theta_i\theta_j)^{-\frac{\alpha_i\alpha_j}{2\hbar^2}}\Big) \\ &= \sum_{i=1}^{M}\Big(\frac{(n-1)\Delta_{\alpha_i}}{(x_i - x)^n} - \frac{1}{(x_i - x)^{n-1}}\partial_{x_i} + \frac{(n-1)\theta_i}{2(x_i - x)^n}\partial_{\theta_i}\Big).\end{aligned} \tag{3.70}$$

Super-quantum curves for the super-multi-Penner model can now be constructed using the above representation of super-Virasoro generators in expressions for super-Virasoro singular vectors. For example, super-quantum curve equations (3.35) at level 3/2 take form

$$\widetilde{A}_{3/2}\widetilde{\chi}_{\alpha=\pm\beta^{\pm1/2}\hbar}(x,\theta) = 0, \qquad \widetilde{A}_{3/2} = -\partial_x\partial_\theta - \beta^{\pm1}\widetilde{G}_{-3/2} - \theta\Big(\partial_x^2 - 2\beta^{\pm1}\widetilde{L}_{-2}\Big). \tag{3.71}$$

Moreover, using the $\mathfrak{osp}(1|2)$ invariance of the super-eigenvalue model, it is possible to remove some time-dependence from operators $\widetilde{G}_{-n+1/2}$ and $\widetilde{L}_{-n}$. In particular, for $M = 2$ and imposing momentum conservation condition, time derivatives can be completely removed, and the action of the above operators on $\widetilde{\chi}_\alpha(x,\theta)$ takes for example the following form

$$\widetilde{G}_{-3/2}\widetilde{\chi}_\alpha(x,\theta) = \Big[-\sum_{i=1,2}\frac{1}{x - x_i}D + \sum_{i=1,2}\frac{2\Delta_{\alpha_i}\theta_i}{(x - x_i)^2} - 2\frac{\Delta_\alpha\theta + \Delta_{\alpha_1}\theta_1 + \Delta_{\alpha_2}\theta_2}{(x - x_1)(x - x_2)}\Big]\widetilde{\chi}_\alpha(x,\theta), \tag{3.72}$$

and

$$\begin{aligned}\widetilde{L}_{-2}\widetilde{\chi}_\alpha(x,\theta) = \Big[&- \sum_{i=1,2}\frac{1}{x - x_i}\partial_x + \sum_{i=1,2}\frac{\theta_i D + 2\Delta_{\alpha_i}}{2(x - x_i)^2} - \frac{\theta\partial_\theta + 2\Delta_\alpha + 2\Delta_{\alpha_1} + 2\Delta_{\alpha_2}}{2(x - x_1)(x - x_2)} \\ &+ \frac{1}{(x - x_1)(x - x_2)}\sum_{i=1,2}\frac{\Delta_\alpha\theta_i\theta}{x - x_i} - \frac{\big((x - x_1)\Delta_{\alpha_1} - (x - x_2)\Delta_{\alpha_2}\big)\theta_1\theta_2}{(x - x_1)^2(x - x_2)^2}\Big]\widetilde{\chi}_\alpha(x,\theta).\end{aligned} \tag{3.73}$$

Therefore super-quantum curve equations in this case are time-independent, and take form of ordinary super-differential equations. These equations are directly related to the ordinary differential equation considered by Dotsenko and Fateev [**40**, **49**]. This equation was analyzed in [**50**] and some properties of its solutions, which can be expressed in term of certain two-fold contour integrals, were discussed in [**51**].

We can also consider the classical limit (3.46) in the above example. The super-quantum curves at level 3/2 reduces in this limit to a system of equations

$$
\begin{aligned}
y_B(x)y_F(x) &= \sum_{i=1,2} \frac{\alpha_i^2\theta_i}{(x-x_i)^2} - \frac{\alpha_1^2\theta_1+\alpha_2^2\theta_2}{(x-x_1)(x-x_2)}, \\
y_B(x)^2 + y_F'(x)y_F(x) &= \sum_{i=1,2} \frac{\alpha_i^2}{(x-x_i)^2} - \frac{\alpha_1^2+\alpha_2^2}{(x-x_1)(x-x_2)} + \\
&\quad - \frac{\left((x-x_1)\alpha_1^2-(x-x_2)\alpha_2^2\right)\theta_1\theta_2}{(x-x_1)^2(x-x_2)^2}.
\end{aligned} \tag{3.74}
$$

The same result can be also found from the analysis of the planar solution of this model [**21**].

3.5. Asymptotic expansion. In the super-eigenvalue model one can introduce the following connected differentials

$$
W_{(h,0)}(x_1,\ldots,x_h) = \beta^{h/2}\left\langle \prod_{i=1}^{h}\sum_{a=1}^{N} \frac{dx_i}{x_i-z_a} \right\rangle^{(c)}, \tag{3.75}
$$

$$
W_{(h,1)}(x_1,\ldots,x_h|x,\theta) = \beta^{(h+1)/2}\left\langle \prod_{i=1}^{h}\sum_{a=1}^{N} \frac{dx_i}{x_i-z_a} \cdot \sum_{a=1}^{N} \frac{\vartheta_a\theta}{x-z_a} \right\rangle^{(c)}, \tag{3.76}
$$

where $\langle\cdots\rangle^{(c)}$ denotes the connected part of the normalized expectation value $\langle\cdots\rangle/Z$. From the definition (3.10) we find that (3.18) has the following expansion

(3.77)

$$
\begin{aligned}
\log\Psi_\alpha(x,\theta) &= \frac{\alpha}{\hbar^2}V(x,\theta) \\
&\quad + \sum_{h=1}^{\infty}\frac{1}{h!}\left(-\frac{\alpha}{\hbar}\right)^h \int_\infty^x \cdots \int_\infty^x \Big(W_{(h,0)}(x_1,\ldots,x_h) + W_{(h,1)}(x_1,\ldots,x_h|x,\theta)\Big),
\end{aligned}
$$

analogously to the asymptotic expansion in the bosonic case presented in section 2.5.

Note that in the bosonic case the differentials (2.51) analogous to (3.75) can be reconstructed using the topological recursion, which then gives rise to the quantization procedure sketched in section 2.5. A supersymmetric version of the topological recursion that would compute differentials (3.75) and (3.76) in super-eigenvalue models is not known to date. However, once such a formulation would be established, the expression (3.77) would enable the construction of super-quantum curves beyond the realm of super-eigenvalue models.

4. Summary

In this work we have shown that quantum curves have the structure of singular vectors of appropriate symmetry algebra. We discussed in detail the case of Virasoro algebra and the Neveu-Schwarz sector of the super-Virasoro algebras. From our perspective quantum curves typically considered in the literature to date correspond simply to level 2 of the Virasoro algebra. In the case of Virasoro algebra we also presented the reformulation of the quantization procedure in terms of the topological recursion.

There are many interesting directions in which these results could be generalized. In [**52**] we present more explicit formulation of quantum curves from the perspective of conformal field theory, which in particular enables to determine super-quantum curves corresponding to the Ramond sector. Furthermore, considering more general matrix models with more general underlying algebras (e.g. W-algebras) should lead to more general classes of quantum curves.

An interesting outcome of our work is identification of universal α-dependent expressions, which reduce to singular vectors for values of α corresponding to degenerate momenta. It is an important problem to simplify the algorithm that enables us to determine these expressions, and generalize them to arbitrary level.

Another important challenge is to find the topological recursion for the super-eigenvalue model. Such a recursion would enable quantization of a general class of supersymmetric algebraic curves, not necessarily associated to super-eigenvalue models.

It is also important to understand in more detail relations of our formalism to surface operators and brane systems, sketched briefly in the introduction.

We are convinced that higher level quantum curves presented in this work should play important role also in enumerative geometry, integrable systems, knot theory, and many other systems where the topological recursion has already proved very useful.

Acknowledgments

We are grateful to the American Mathematical Society for the organization of the 2016 AMS von Neumann Symposium *"Topological Recursion and its Influence in Analysis, Geometry, and Topology"*. P.S. is indebted to Bertrand Eynard, Chiu-Chu Melissa Liu, and Motohico Mulase for providing the opportunity to present these results, inspiration and encouragement. We also thank Vincent Bouchard, Zbigniew Jaskólski and Chaiho Rim for discussions.

References

[1] Robbert Dijkgraaf, Lotte Hollands, Piotr Sulkowski, and Cumrun Vafa, *Supersymmetric gauge theories, intersecting branes and free fermions*, J. High Energy Phys. **2** (2008), 106, 57, DOI 10.1088/1126-6708/2008/02/106. MR2385939

[2] Robbert Dijkgraaf, Lotte Hollands, and Piotr Sulkowski, *Quantum curves and $\mathcal{D}$-modules*, J. High Energy Phys. **11** (2009), 047, 59, DOI 10.1088/1126-6708/2009/11/047. MR2628886

[3] M. Aganagic, R. Dijkgraaf, A. Klemm, M. Marino and C. Vafa, *Topological strings and integrable hierarchies*, *Commun. Math. Phys.* **261** (2006) 451–516, [hep-th/0312085].

[4] Sergei Gukov and Piotr Sulkowski, *A-polynomial, B-model, and quantization*, J. High Energy Phys. **2** (2012), 070, front matter+56. MR2996110

[5] Mina Aganagic, Miranda C. N. Cheng, Robbert Dijkgraaf, Daniel Krefl, and Cumrun Vafa, *Quantum geometry of refined topological strings*, J. High Energy Phys. **11** (2012), 019, front matter + 52. MR3036500

[6] Stavros Garoufalidis, *On the characteristic and deformation varieties of a knot*, Proceedings of the Casson Fest, Geom. Topol. Monogr., vol. 7, Geom. Topol. Publ., Coventry, 2004, pp. 291–309, DOI 10.2140/gtm.2004.7.291. MR2172488

[7] Stavros Garoufalidis and Thang T. Q. Lê, *The colored Jones function is q-holonomic*, Geom. Topol. **9** (2005), 1253–1293, DOI 10.2140/gt.2005.9.1253. MR2174266

[8] Paul Norbury, *Quantum curves and topological recursion*, String-Math 2014, Proc. Sympos. Pure Math., vol. 93, Amer. Math. Soc., Providence, RI, 2016, pp. 41–65. MR3524233

[9] O. Dumitrescu and M. Mulase, *Lectures on the topological recursion for Higgs bundles and quantum curves*, 1509.09007.

[10] B. Eynard and N. Orantin, *Invariants of algebraic curves and topological expansion*, *Commun. Num. Theor. Phys.* **1** (2007) 347–452, [math-ph/0702045].

[11] Bertrand Eynard, *Topological expansion for the 1-Hermitian matrix model correlation functions*, J. High Energy Phys. **11** (2004), 031, 35 pp. (2005), DOI 10.1088/1126-6708/2004/11/031. MR2118807

[12] Leonid Chekhov and Bertrand Eynard, *Hermitian matrix model free energy: Feynman graph technique for all genera*, J. High Energy Phys. **3** (2006), 014, 18, DOI 10.1088/1126-6708/2006/03/014. MR2222762

[13] Leonid Chekhov and Bertrand Eynard, *Matrix eigenvalue model: Feynman graph technique for all genera*, J. High Energy Phys. **12** (2006), 026, 29, DOI 10.1088/1126-6708/2006/12/026. MR2276715

[14] Vincent Bouchard, Albrecht Klemm, Marcos Mariño, and Sara Pasquetti, *Remodeling the B-model*, Comm. Math. Phys. **287** (2009), no. 1, 117–178, DOI 10.1007/s00220-008-0620-4. MR2480744

[15] Bertrand Eynard, Amir-Kian Kashani-Poor, and Olivier Marchal, *A matrix model for the topological string I: deriving the matrix model*, Ann. Henri Poincaré **15** (2014), no. 10, 1867–1901, DOI 10.1007/s00023-013-0290-4. MR3257452

[16] Vincent Bouchard and Piotr Sulkowski, *Topological recursion and mirror curves*, Adv. Theor. Math. Phys. **16** (2012), no. 5, 1443–1483. MR3056954

[17] Robbert Dijkgraaf and Hiroyuki Fuji, *The volume conjecture and topological strings*, Fortschr. Phys. **57** (2009), no. 9, 825–856, DOI 10.1002/prop.200900067. MR2567388

[18] Robbert Dijkgraaf, Hiroyuki Fuji, and Masahide Manabe, *The volume conjecture, perturbative knot invariants, and recursion relations for topological strings*, Nuclear Phys. B **849** (2011), no. 1, 166–211, DOI 10.1016/j.nuclphysb.2011.03.014. MR2795276

[19] Gaëtan Borot and Bertrand Eynard, *All order asymptotics of hyperbolic knot invariants from non-perturbative topological recursion of A-polynomials*, Quantum Topol. **6** (2015), no. 1, 39–138, DOI 10.4171/QT/60. MR3335006

[20] M. Manabe and P. Sulkowski, *Quantum curves and conformal field theory*, *Phys. Rev.* **D95** (2017) 126003, [1512.05785].

[21] Paweł Ciosmak, Leszek Hadasz, Masahide Manabe, and Piotr Sulkowski, *Super-quantum curves from super-eigenvalue models*, J. High Energy Phys. **10** (2016), 044, front matter+55. MR3577513

[22] J. Ambjørn, J. Jurkiewicz, and Yu. M. Makeenko, *Multiloop correlators for two-dimensional quantum gravity*, Phys. Lett. B **251** (1990), no. 4, 517–524, DOI 10.1016/0370-2693(90)90790-D. MR1084539

[23] Masafumi Fukuma, Hikaru Kawai, and Ryuichi Nakayama, *Continuum Schwinger-Dyson equations and universal structures in two-dimensional quantum gravity*, Internat. J. Modern Phys. A **6** (1991), no. 8, 1385–1406, DOI 10.1142/S0217751X91000733. MR1093776

[24] H. Itoyama and Y. Matsuo, *$w_{1+\infty}$-type constraints in matrix models at finite N*, Phys. Lett. B **262** (1991), no. 2-3, 233–239, DOI 10.1016/0370-2693(91)91560-I. MR1113624

[25] Robbert Dijkgraaf, Herman Verlinde, and Erik Verlinde, *Loop equations and Virasoro constraints in nonperturbative two-dimensional quantum gravity*, Nuclear Phys. B **348** (1991), no. 3, 435–456, DOI 10.1016/0550-3213(91)90199-8. MR1083914

[26] H. Awata, Y. Matsuo, S. Odake and J. Shiraishi, *Collective field theory, Calogero-Sutherland model and generalized matrix models*, *Phys. Lett.* **B347** (1995) 49–55, [hep-th/9411053].

[27] L. Alvarez-Gaumé, H. Itoyama, J. L. Mañes, and A. Zadra, *Superloop equations and two-dimensional supergravity*, Internat. J. Modern Phys. A **7** (1992), no. 21, 5337–5367, DOI 10.1142/S0217751X92002441. MR1179977

[28] K. Becker and M. Becker, *Nonperturbative solution of the super-Virasoro constraints*, Modern Phys. Lett. A **8** (1993), no. 13, 1205–1214, DOI 10.1142/S0217732393002695. MR1218570

[29] I. N. McArthur, *The partition function for the supersymmetric eigenvalue model*, Modern Phys. Lett. A **8** (1993), no. 35, 3355–3361, DOI 10.1142/S0217732393003780. MR1249853

[30] J. C. Plefka, *Supersymmetric generalizations of matrix models*. PhD thesis, Hannover U., 1996. hep-th/9601041.

[31] Gordon W. Semenoff and Richard J. Szabo, *Fermionic matrix models*, Internat. J. Modern Phys. A **12** (1997), no. 12, 2135–2291, DOI 10.1142/S0217751X97001328. MR1449225

[32] H. Itoyama and H. Kanno, *Supereigenvalue model and Dijkgraaf-Vafa proposal*, Phys. Lett. B **573** (2003), no. 1-4, 227–234, DOI 10.1016/j.physletb.2003.08.052. MR2028259

[33] J. Teschner, *Quantization of the Hitchin moduli spaces, Liouville theory and the geometric Langlands correspondence I*, Adv. Theor. Math. Phys. **15** (2011), no. 2, 471–564. MR2924236

[34] R. Dijkgraaf and C. Vafa, *Toda Theories, Matrix Models, Topological Strings, and N=2 Gauge Systems* (2009), `0909.2453`.

[35] Katsuhisa Mimachi and Yasuhiko Yamada, *Singular vectors of the Virasoro algebra in terms of Jack symmetric polynomials*, Comm. Math. Phys. **174** (1995), no. 2, 447–455. MR1362173

[36] Louis Benoit and Yvan Saint-Aubin, *Degenerate conformal field theories and explicit expressions for some null vectors*, Phys. Lett. B **215** (1988), no. 3, 517–522, DOI 10.1016/0370-2693(88)91352-4. MR972438

[37] B. L. Feĭgin and D. B. Fuchs, *Verma modules over the Virasoro algebra*, Topology (Leningrad, 1982), Lecture Notes in Math., vol. 1060, Springer, Berlin, 1984, pp. 230–245, DOI 10.1007/BFb0099939. MR770243

[38] A. Kent, *Projections of Virasoro singular vectors*, `http://dx.doi.org/10.1016/0370-2693(92)90582-O` *Phys. Lett.* **B278** (1992) 443–448.

[39] Al. Zamolodchikov, *Higher equations of motion in Liouville field theory*, Proceedings of 6th International Workshop on Conformal Field Theory and Integrable Models, Internat. J. Modern Phys. A **19** (2004), no. May, suppl., 510–523, DOI 10.1142/S0217751X04020592. MR2087128

[40] Vl. S. Dotsenko and V. A. Fateev, *Conformal algebra and multipoint correlation functions in 2D statistical models*, Nuclear Phys. B **240** (1984), no. 3, 312–348, DOI 10.1016/0550-3213(84)90269-4. MR762194

[41] A. Mironov, A. Morozov, and Sh. Shakirov, *Conformal blocks as Dotsenko-Fateev integral discriminants*, Internat. J. Modern Phys. A **25** (2010), no. 16, 3173–3207, DOI 10.1142/S0217751X10049141. MR2659902

[42] A. Mironov, A. Morosov, and Sh. Shakirov, *On the 'Dotsenko-Fateev' representation of the toric conformal blocks*, J. Phys. A **44** (2011), no. 8, 085401, 11, DOI 10.1088/1751-8113/44/8/085401. MR2770380

[43] A. A. Belavin, A. M. Polyakov, and A. B. Zamolodchikov, *Infinite conformal symmetry in two-dimensional quantum field theory*, Nuclear Phys. B **241** (1984), no. 2, 333–380, DOI 10.1016/0550-3213(84)90052-X. MR757857

[44] B. Eynard and O. Marchal, *Topological expansion of the Bethe ansatz, and non-commutative algebraic geometry*, J. High Energy Phys. **3** (2009), 094, 52, DOI 10.1088/1126-6708/2009/03/094. MR2495622

[45] B. Eynard and O. Marchal, *Topological expansion of the Bethe ansatz, and non-commutative algebraic geometry*, J. High Energy Phys. **3** (2009), 094, 52, DOI 10.1088/1126-6708/2009/03/094. MR2495622

[46] L. O. Chekhov, B. Eynard, and O. Marchal, *Topological expansion of the β-ensemble model and quantum algebraic geometry in the sectorwise approach*, Theoret. and Math. Phys. **166** (2011), no. 2, 141–185, DOI 10.1007/s11232-011-0012-3. Russian version appears in Teoret. Mat. Fiz. **166** (2011), no. 2, 163–215. MR3165804

[47] Vincent Bouchard and Bertrand Eynard, *Reconstructing WKB from topological recursion* (English, with English and French summaries), J. Éc. polytech. Math. **4** (2017), 845–908, DOI 10.5802/jep.58. MR3694097

[48] A. Belavin and A. Zamolodchikov, *Higher equations of motion in $N = 1$ SUSY Liouville field theory*, *JETP Lett.* **84** (2006) 418–424, [`hep-th/0610316`].

[49] Vl. S. Dotsenko and V. A. Fateev, *Four-point correlation functions and the operator algebra in 2D conformal invariant theories with central charge $C \leq 1$*, Nuclear Phys. B **251** (1985), no. 5-6, 691–734, DOI 10.1016/S0550-3213(85)80004-3. MR789026

[50] V. A. Belavin, *N=1 SUSY Conformal Block Recursive Relations*, `hep-th/0611295`.

[51] A. Belavin, V. Belavin, A. Neveu, and Al. Zamolodchikov, *Bootstrap in supersymmetric Liouville field theory. I. NS sector*, Nuclear Phys. B **784** (2007), no. 3, 202–233, DOI 10.1016/j.nuclphysb.2007.04.018. MR2353413

[52] Paweł Ciosmak, Leszek Hadasz, Zbigniew Jaskólski, Masahide Manabe, and Piotr Sułkowski, *From CFT to Ramond super-quantum curves*, J. High Energy Phys. **5** (2018), 133, DOI 10.1007/jhep05(2018)133. MR3815054

Faculty of Mathematics, Informatics and Mechanics, University of Warsaw, ul. Banacha 2, 02-097 Warsaw, Poland
Email address: p.ciosmak@mimuw.edu.pl

M. Smoluchowski Institute of Physics, Jagiellonian University, ul. Łojasiewicza 11, 30-348 Kraków, Poland
Email address: leszek.hadasz@uj.edu.pl

Faculty of Physics, University of Warsaw, ul. Pasteura 5, 02-093 Warsaw, Poland
Email address: masahidemanabe@gmail.com

Faculty of Physics, University of Warsaw, ul. Pasteura 5, 02-093 Warsaw, Poland; and Walter Burke Institute for Theoretical Physics, California Institute of Technology, Pasadena, California 91125
Email address: psulkows@fuw.edu.pl

Proceedings of Symposia in Pure Mathematics
Volume **100**, 2018
https://doi.org/10.1090/pspum/100/01761

Towards the topological recursion for double Hurwitz numbers

Norman Do and Maksim Karev

ABSTRACT. Single Hurwitz numbers enumerate branched covers of the Riemann sphere with specified genus, prescribed ramification over infinity, and simple branching elsewhere. They exhibit a remarkably rich structure. In particular, they arise as intersection numbers on moduli spaces of curves and are governed by the topological recursion of Chekhov, Eynard and Orantin. Double Hurwitz numbers are defined analogously, but with prescribed ramification over both zero and infinity. Goulden, Jackson and Vakil have conjectured that double Hurwitz numbers also arise as intersection numbers on moduli spaces.

In this paper, we repackage double Hurwitz numbers as enumerations of branched covers weighted by certain monomials and conjecture that they are governed by the topological recursion. Evidence is provided in the form of the associated quantum curve and low genus calculations. We furthermore reduce the conjecture to a weaker one, concerning a certain polynomial structure of double Hurwitz numbers. Via the topological recursion framework, our main conjecture should lead to a direct connection to enumerative geometry, thus shedding light on the aforementioned conjecture of Goulden, Jackson and Vakil.

1. Introduction

1.1. Double Hurwitz numbers. Single Hurwitz numbers — also known as simple Hurwitz numbers — enumerate branched covers of the Riemann sphere with specified genus, prescribed ramification over infinity, and simple branching elsewhere. Although first studied by Hurwitz towards the end of the nineteenth century [**29**], the extent of their remarkably rich structure only became apparent towards the end of the twentieth. The revival of interest in Hurwitz numbers was in part sparked by the empirical observation of Goulden, Jackson and Vainshtein that they possess a certain polynomial structure [**26**]. This was later proved by Ekedahl, Lando, Shapiro and Vainshtein, who showed that single Hurwitz numbers are equal to Hodge integrals over the Deligne–Mumford compactification of the moduli space of curves [**19**]. Their so-called ELSV formula not only makes the polynomial structure of Hurwitz numbers apparent, but connects them to the realms of enumerative geometry and mathematical physics. More recently, work motivated by topological string theory led Bouchard and Mariño to conjecture that

2010 *Mathematics Subject Classification.* Primary 14N10, 05A15, 32G15, 14N35.

The first author was supported by the Australian Research Council grant DE130100650.

single Hurwitz numbers are governed by the topological recursion of Chekhov, Eynard and Orantin [**7**]. This was subsequently proven [**23**] and it has furthermore been demonstrated that the ELSV formula and the Bouchard–Mariño conjecture are in some sense equivalent [**16**].

Double Hurwitz numbers enumerate branched covers of the Riemann sphere with specified genus, prescribed ramification over both zero and infinity, and simple branching elsewhere. Okounkov showed that they arise naturally as coefficients of a certain tau-function of the Toda integrable hierarchy [**34**]. Goulden, Jackson and Vakil demonstrated that double Hurwitz numbers possess a certain piecewise polynomial structure. Moreover, they presented evidence to suggest that double Hurwitz numbers are equal to integrals over moduli spaces of curves equipped with a line bundle [**27**]. However, to this date, a rigorous definition of these moduli spaces and their intersection theory is yet to be determined.

In this paper, we interpret double Hurwitz numbers as weighted enumerations of branched covers in the following way.

Definition 1.1. Fix a positive integer d and weights $s, q_1, q_2, \ldots, q_d \in \mathbb{C}$. Define the *double Hurwitz number* $DH_{g,n}(\mu_1, \ldots, \mu_n)$ to be the weighted count of connected genus g branched covers of the Riemann sphere $f : (\Sigma; p_1, \ldots, p_n) \to (\mathbb{CP}^1; \infty)$ such that

- all branching away from 0 and ∞ is simple and occurs at some number m of fixed points;
- $f^{-1}(\infty) = \mu_1 p_1 + \cdots + \mu_n p_n$; and
- no preimage of 0 has ramification index larger than d.

If such a branched cover has ramification profile $(\lambda_1, \lambda_2, \ldots, \lambda_\ell)$ over 0, then we assign it the weight

$$\frac{q_{\lambda_1} q_{\lambda_2} \cdots q_{\lambda_\ell}}{|\mathrm{Aut}\, f|} \frac{s^m}{m!}.$$

Here, the automorphism group Aut f consists of Riemann surface automorphisms $\phi : \Sigma \to \Sigma$ that preserve the marked points $p_1, \ldots, p_n$ and satisfy $f \circ \phi = f$.

Remark 1.2. We make note of the following initial remarks concerning Definition 1.1.

- The Riemann–Hurwitz formula asserts that the number of simple branch points must satisfy $m = 2g - 2 + n + \ell$, where ℓ is the number of preimages of 0.
- Each double Hurwitz number is a weighted homogeneous polynomial in $q_1, q_2, \ldots, q_d$ with positive rational coefficients.
- The double Hurwitz numbers $H^g_{\boldsymbol{\lambda},\boldsymbol{\mu}}$ defined by Goulden, Jackson and Vakil [**27**] may be recovered as a combinatorial factor multiplied by the coefficient of $q_{\lambda_1} q_{\lambda_2} \cdots q_{\lambda_\ell}$ in the polynomial $DH_{g,n}(\mu_1, \ldots, \mu_n)$, as long as we fix $d \geq \max(\lambda_1, \lambda_2, \ldots, \lambda_\ell)$.
- The parameter s is redundant in the sense that its exponent in each monomial of a double Hurwitz number can be recovered via the Riemann–Hurwitz formula. However, the advantages of retaining it should become apparent in the following.
- The single Hurwitz numbers are recovered by taking $q_1 = 1$ and $q_i = 0$ for $i \neq 1$. More generally, one recovers a-orbifold Hurwitz numbers by taking $q_a = 1$ and $q_i = 0$ for $i \neq a$ [**5, 12**].

See Appendix A for a table of double Hurwitz numbers.

1.2. The main conjecture. The particular way in which we have assembled the double Hurwitz numbers allows us to consider potential analogues of results pertaining to single Hurwitz numbers. As an example, we propose a vast generalisation of the Bouchard–Mariño conjecture, namely that the double Hurwitz numbers defined above are governed by the topological recursion.

The topological recursion of Chekhov, Eynard and Orantin arose from the theory of matrix models and has subsequently found widespread applications to various areas of mathematics and physics [**8**, **21**]. Beyond the realm of matrix models, it is now either known or conjectured to govern the following problems: the enumeration of ribbon graphs and hypermaps [**13**, **15**, **18**]; Hurwitz numbers of various flavours [**5**, **7**, **10**–**12**, **23**]; Gromov–Witten invariants of $\mathbb{CP}^1$ [**17**, **32**]; Gromov–Witten invariants of toric Calabi–Yau threefolds [**6**, **22**, **24**]; and asymptotics of coloured Jones polynomials of knots [**2**, **9**].

Let us first state our main conjecture relating double Hurwitz numbers to the topological recursion before presenting a rigorous treatment of the topological recursion itself.

CONJECTURE 1.3. Let $P(z) = q_1 z + q_2 z^2 + \cdots + q_d z^d$. Topological recursion applied to the rational spectral curve

$$x(z) = z\exp(-sP(z)) \qquad \text{and} \qquad y(z) = P(z) \tag{1.1}$$

produces correlation differentials whose expansions at $x_i = 0$ satisfy

$$\omega_{g,n} = \sum_{\mu_1,\ldots,\mu_n=1}^{\infty} DH_{g,n}(\mu_1,\ldots,\mu_n)\prod_{i=1}^{n}\mu_i x_i^{\mu_i-1}\mathrm{d}x_i, \qquad \text{for } (g,n)\neq(0,2).$$

In general, the topological recursion takes as input the data of a spectral curve and outputs correlation differentials $\omega_{g,n}$ for integers $g \geq 0$ and $n \geq 1$. We now describe explicitly the topological recursion in the context of Conjecture 1.3; for an exposition of the topological recursion in greater generality, the reader may consult the literature [**21**].

First, let $a_1, a_2, \ldots, a_d$ be the branch points of the spectral curve — in other words, the zeroes of

$$\mathrm{d}x(z) = 0 \qquad \Leftrightarrow \qquad szP'(z) - 1 = 0.$$

For the following discussion, we will require the mild assumption that these zeroes are distinct and hence simple.[1] It follows that there exists a local involution $\sigma_i : U_i \to U_i$ defined on a small open neighbourhood of the branch point $z = a_i$ on the spectral curve, such that σ_i is meromorphic and satisfies the equation $x(\sigma_i(z)) = x(z)$, but is not the identity on U_i.

[1]One can also state the conjecture in the case of higher order zeroes by invoking the global topological recursion of Bouchard and Eynard [**4**]. We consider only the generic case in order to streamline the presentation. Indeed, Bouchard and Eynard demonstrate that non-simple spectral curves and their correlation differentials can be obtained in the limit of simple spectral curves.

Define the base cases[2]

$$\omega_{0,1}(z_1) = \frac{y(z_1)\,\mathrm{d}x(z_1)}{x(z_1)} \qquad \text{and} \qquad \omega_{0,2}(z_1, z_2) = \frac{\mathrm{d}z_1 \otimes \mathrm{d}z_2}{(z_1 - z_2)^2}. \tag{1.2}$$

Next, define the recursion kernel

$$K_i(z_1, z) = \frac{\int_o^z \omega_{0,2}(z_1, \cdot)}{\omega_{0,1}(z) - \omega_{0,1}(\sigma_i(z))} = \frac{\mathrm{d}z_1}{z_1 - z} \frac{1}{\omega_{0,1}(z) - \omega_{0,1}(\sigma_i(z))}, \tag{1.3}$$

which exists only on the small open neighbourhood U_i of the branch point $z = a_i$. The topological recursion is not sensitive to the choice of basepoint o on the spectral curve, so we have taken it to be $z = \infty$ for convenience.

For $2g - 2 + n > 0$, let

$$\omega_{g,n}(z_1, \ldots, z_n) = \sum_{i=1}^{d} \operatorname*{Res}_{z=a_i} K_i(z_1, z) \Bigg[\omega_{g-1,n+1}(z, \sigma_i(z), \mathbf{z}_S) + \sum_{\substack{g_1+g_2=g \\ I \sqcup J = S\setminus\{1\}}}^{\circ} \omega_{g_1,|I|+1}(z, \mathbf{z}_I)\, \omega_{g_2,|J|+1}(\sigma_i(z), \mathbf{z}_J) \Bigg].$$

Here, we let $S = \{1, 2, \ldots, n\}$ and let $\mathbf{z}_I = \{z_{i_1}, z_{i_2}, \ldots, z_{i_k}\}$ for $I = \{i_1, i_2, \ldots, i_k\}$. The symbol $\circ$ over the inner summation means that we exclude terms involving $\omega_{0,1}$.

The correlation differential $\omega_{g,n}$ is a multidifferential on the spectral curve $\mathcal{C}$ rather than a differential form. More precisely, $\omega_{g,n}$ is a meromorphic section of the line bundle $\pi_1^*(T^*\mathcal{C}) \otimes \pi_2^*(T^*\mathcal{C}) \otimes \cdots \otimes \pi_n^*(T^*\mathcal{C})$, on the Cartesian product $\mathcal{C}^n$, where $\pi_i : \mathcal{C}^n \to \mathcal{C}$ denotes projection onto the ith factor. For notational convenience, we will subsequently drop the $\otimes$ symbol when writing multidifferentials. Despite the fact that the topological recursion is asymmetric in nature, the resulting correlation differentials are indeed symmetric. Furthermore, for $2g - 2 + n > 0$, it is known that the correlation differential $\omega_{g,n}(z_1, \ldots, z_n)$ has poles only at the branch points $a_1, a_2, \ldots, a_d$ of the spectral scurve.

Finally, to interpret the statement of Conjecture 1.3, one is required to expand the correlation differentials $\omega_{g,n}(z_1, \ldots, z_n)$ at $x_i = 0$. We do this by setting $x_i = x(z_i)$. This notation and the analogous notation $y_i = y(z_i)$ will be used throughout the paper.

1.3. Evidence for the conjecture. There is substantial evidence to support our main conjecture. The cut-and-join recursion, stated below as Proposition 2.1, allows one to compute double Hurwitz numbers recursively. It is natural to present this at the level of generating functions $F_{g,n}$ known as free energies, which store all double Hurwitz numbers of the form $DH_{g,n}(\mu_1, \ldots, \mu_n)$. In particular, we obtain a relation between the free energies, stated below as Corollary 2.8, that superficially resembles the topological recursion. In particular, it expresses $F_{g,n}$ in terms of $F_{g-1,n+1}$ and products $F_{g_1,n_1} \times F_{g_2,n_2}$ that do not involve $F_{0,1}$, where $g_1 + g_2 = g$ and $n_1 + n_2 = n + 1$.

---

[2]The original formulation of the topological recursion usually sets $\omega_{0,1}(z_1) = -y(z_1)\,\mathrm{d}x(z_1)$. The modification here assumes the opposing sign convention and uses $y(z_1)\,\mathrm{d}\log x(z_1)$ instead, which applies in various settings, such as the Bouchard–Mariño conjecture concerning single Hurwitz numbers. In such cases, it is common in the literature to refer to x as a $\mathbb{C}^*$-coordinate, rather than a $\mathbb{C}$-coordinate. Thus, equation (1.1) describes what is known as a $\mathbb{C}^* \times \mathbb{C}$ spectral curve.

In many instances of the topological recursion on a spectral curve, there is an associated quantum curve that underlies it. In short, a quantum curve is a differential operator that annihilates a wave function constructed from the correlation differentials of the topological recursion. One can consider the semi-classical limit of the quantum curve, which is a plane curve that coincides with the original spectral curve in many cases. The quantum curve for double Hurwitz numbers was previously computed by Alexandrov, Lewanski and Shadrin [**1**] and its semi-classical limit is $y = P(x \exp(sy))$, which recovers the spectral curve of equation (1.1). It has been proposed that the existence of a quantum curve can be used to predict the structure of topological recursion and the associated spectral curve [**31**].

Finally, one can attempt to verify the main conjecture via direct computation in low genus. We carry this out in the cases $(g, n) = (0, 1)$, $(0, 2)$, $(0, 3)$ and $(1, 1)$.

1.4. Ramifications and applications. Our main conjecture subsumes some existing results and conjectures in the literature, which can be obtained by specialising the weights $s, q_1, q_2, \ldots$. It is inspired by the Bouchard–Mariño conjecture, which is recovered by considering the case $P(z) = z$. More generally, one can consider $P(z) = z^a$ for a positive integer a and recover the fact that the topological recursion governs orbifold Hurwitz numbers [**5, 12**]. Alexandrov, Lewanski and Shadrin propose a 1-parameter deformation of single Hurwitz numbers, via the enumeration of double Hurwitz numbers, weighted by c to the power of the colength of the ramification profile over 0 [**1**]. Our main conjecture recovers theirs by specialising the weights to $q_i = ic^{i-1}$ and sending d to infinity or, equivalently, by considering the case $P(z) = \frac{z}{(1-cz)^2}$.

Our strategy of packaging double Hurwitz numbers as enumerations of branched covers weighted by monomials allows us to generalise results concerning single Hurwitz numbers to their double counterparts. This approach may also be employed in the context of other problems of a similar nature, such as double montone Hurwitz numbers, though we do not pursue that line of reasoning here [**10, 11**].

Our main conjecture, stated above as Conjecture 1.3, leads to certain previously unidentified structure for double Hurwitz numbers. For instance, it implies that the free energy generating functions for double Hurwitz numbers defined below in equation (2.2) are actually rational, with poles only at the branch points of the spectral curve satisfying a certain symmetry with respect to the local involution. We state this conjecture explicitly below as Conjecture 4.1. At the level of the double Hurwitz numbers themselves, this manifests as a particular polynomial-like structure, which we state explicitly below as Conjecture 4.6. Perhaps the main consequence of this paper is that these three conjectures are all equivalent.

We foresee that the main application of our conjecture is towards the geometry of double Hurwitz numbers. The general theory of topological recursion has been motivated in part by its connections to enumerative geometry. A rather simple statement is that spectral curves are locally modelled on the Airy curve $x(z) = \frac{1}{2}z^2$ and $y(z) = z$ at their branch points. A deeper statement is that correlation differentials of arbitrary spectral curves are also locally modelled on those of the Airy curve. Furthermore, the Airy correlation differentials are known to store psi-class intersection numbers on the moduli space of curves, the main objects of the celebrated Witten–Kontsevich theorem [**21**]. Eynard managed to push this result further and show that the lower order terms of correlation differentials can also be related to intersection numbers on moduli spaces [**20**]. More recently, Dunin-Barkowski, Orantin, Shadrin and Spitz made a connection between topological recursion and the Givental formalism, proving that correlation differentials store ancestor invariants of a cohomological field theory [**17**].

From the previous discussion, a consequence of our main conjecture is a direct relation between double Hurwitz numbers and intersection theory on moduli spaces of curves. Such a connection should shed light on the conjecture of Goulden, Jackson and Vakil, which asserts that double Hurwitz numbers arise as intersection numbers on certain moduli spaces of curves equipped with a line bundle [**27**].

- In Section 2, we state the cut-and-join recursion for double Hurwitz numbers, which previously appeared in the work of Zhu [**35**]. We write these at the level of the generating functions known as free energies. Finally, we discuss the notion of pruned double Hurwitz numbers, which previously appeared in the work of Hahn [**28**], and their relation to the free energies.
- In Section 3, we give evidence to support our main conjecture. First, we present the quantum curve for double Hurwitz numbers, which was first deduced by Alexandrov, Lewanski and Shadrin [**1**], and show that its semi-classical limit does indeed recover the spectral curve of equation (1.1). We then provide low genus evidence by calculating the free energies $F_{0,3}$ and $F_{1,1}$ and demonstrating that they are consistent with our main conjecture.
- In Section 4, we outline a possible proof of our main conjecture. In particular, we show that the conjecture can be reduced to proving that the free energies satisfy so-called linear loop equations. We furthermore show that these constraints are equivalent to a polynomial-like structure for double Hurwitz numbers.

2. Combinatorics of double Hurwitz numbers

2.1. Cut-and-join recursion. A natural way to compute Hurwitz numbers is via the cut-and-join recursion. It was originally formulated by Goulden and Jackson in the case of genus 0 single Hurwitz numbers [**25**], but has much broader applicability. At the level of branched covers, the cut-and-join recursion arises by examining the behaviour of the ramification profile over infinity as one of the simple branch points approaches infinity. Otherwise, one may interpret Hurwitz numbers as an enumeration of transitive factorisations

$$\tau_0\tau_1\tau_2\cdots\tau_m = \rho \tag{2.1}$$

in symmetric groups, by passing to the monodromy representation of a branched cover and appealing to the Riemann existence theorem. Here, τ_0 represents the monodromy over 0, the transpositions $\tau_1, \tau_2, \ldots, \tau_m$ represent the monodromy of the simple branch points, and ρ represents the inverse of the monodromy over ∞. The cut-and-join recursion then arises simply by considering the result of multiplying both sides of equation (2.1) on the right by the transposition τ_m.

The cut-and-join recursion for double Hurwitz numbers appears in the work of Zhu [**35**]. The statement below paraphrases the result using our particular definition of double Hurwitz numbers.

PROPOSITION 2.1 (Cut-and-join recursion). *The double Hurwitz numbers satisfy the equation*

$$\begin{aligned}\left(2g-2+n+\sum_{i=1}^{d} q_i \frac{\partial}{\partial q_i}\right) & DH_{g,n}(\mu_1,\ldots,\mu_n) \\ &= s\sum_{i<j}(\mu_i+\mu_j)\, DH_{g,n-1}(\boldsymbol{\mu}_{S\setminus\{i,j\}},\mu_i+\mu_j) \\ &+ \frac{s}{2}\sum_{i=1}^{n}\sum_{\alpha+\beta=\mu_i}\alpha\beta\Bigg[DH_{g-1,n+1}(\alpha,\beta,\boldsymbol{\mu}_{S\setminus\{i\}}) \\ &+ \sum_{\substack{g_1+g_2=g\\ I\sqcup J=S\setminus\{i\}}} DH_{g_1,|I|+1}(\alpha,\boldsymbol{\mu}_I)\, DH_{g_2,|J|+1}(\beta,\boldsymbol{\mu}_J)\Bigg].\end{aligned}$$

Here, we let $S=\{1,2,\ldots,n\}$ and $\boldsymbol{\mu}_I=\{\mu_{i_1},\mu_{i_2},\ldots,\mu_{i_k}\}$ for $I=\{i_1,i_2,\ldots,i_k\}$. Furthermore, all double Hurwitz numbers can be calculated from this recursion along with the base cases

$$DH_{0,1}(\mu)\big|_{s=0}=\frac{1}{\mu}q_\mu \qquad \textit{and} \qquad DH_{g,n}(\mu_1,\ldots,\mu_n)\big|_{s=0}=0 \quad \textit{for } (g,n)\neq(0,1).$$

Recall the introduction of the parameter s, which records the number of simple branch points, in our definition of the double Hurwitz numbers. One advantage of retaining the parameter is the following slightly more compact form of the cut-and-join recursion [**30**].

COROLLARY 2.2. *The double Hurwitz numbers satisfy the equation*

$$\begin{aligned}\frac{\partial}{\partial s}DH_{g,n}(\mu_1,\ldots,\mu_n) &= \sum_{i<j}(\mu_i+\mu_j)\, DH_{g,n-1}(\boldsymbol{\mu}_{S\setminus\{i,j\}},\mu_i+\mu_j) \\ &+ \frac{1}{2}\sum_{i=1}^{n}\sum_{\alpha+\beta=\mu_i}\alpha\beta\Bigg[DH_{g-1,n+1}(\alpha,\beta,\boldsymbol{\mu}_{S\setminus\{i\}}) \\ &+ \sum_{\substack{g_1+g_2=g\\ I\sqcup J=S\setminus\{i\}}} DH_{g_1,|I|+1}(\alpha,\boldsymbol{\mu}_I)\, DH_{g_2,|J|+1}(\beta,\boldsymbol{\mu}_J)\Bigg].\end{aligned}$$

REMARK 2.3. Observe that the mechanism behind the cut-and-join recursion is in some sense local. In the interpretation of Hurwitz numbers as enumerations of transitive factorisations, the recursion is not sensitive to the permutation τ_0 in equation (2.1). So using the parameter s allows us to express the cut-and-join recursion identically in the case of simple Hurwitz numbers [**25**], orbifold Hurwitz numbers [**5, 12**], as well as double Hurwitz numbers [**35**]. In fact, we remark that the local nature of the cut-and-join recursion implies that it is not even sensitive to the topology of the base curve. Thus, it can also be expressed in the same way for Hurwitz numbers on base curves of higher genus, though with a different set of base cases [**30**].

2.2. Generating functions. It is natural to define the following generating functions for the double Hurwitz numbers. In the context of the topological recursion, such generating functions are commonly referred to as *free energies.*

$$F_{g,n}(x_1, \ldots, x_n) = \sum_{\mu_1,\ldots,\mu_n=1}^{\infty} DH_{g,n}(\mu_1, \ldots, \mu_n) \prod_{i=1}^{n} x_i^{\mu_i} \tag{2.2}$$

We also define the following formal multidifferentials, where d_i denotes the exterior derivative with respect to the ith slot.

$$\begin{aligned}\Omega_{g,n}(x_1, \ldots, x_n) &= \mathrm{d}_1 \cdots \mathrm{d}_n F_{g,n}(x_1, \ldots, x_n) \\ &= \sum_{\mu_1,\ldots,\mu_n=1}^{\infty} DH_{g,n}(\mu_1, \ldots, \mu_n) \prod_{i=1}^{n} \mu_i x_i^{\mu_i-1} \mathrm{d}x_i\end{aligned}$$

Note that our main conjecture implies that when $2g - 2 + n > 0$, these are in fact expansions of rational multidifferentials on the spectral curve of equation (1.1) and coincide with the correlation differentials generated by the topological recursion applied to the spectral curve.

PROPOSITION 2.4. *The free energies satisfy the equation*

$$\begin{aligned}\frac{\partial}{\partial s} F_{g,n}(x_1, \ldots, x_n) &= \sum_{i<j} \frac{x_i x_j}{x_i - x_j} \left[\frac{\partial}{\partial x_i} F_{g,n-1}(\mathbf{x}_{S \setminus \{j\}}) - \frac{\partial}{\partial x_j} F_{g,n-1}(\mathbf{x}_{S \setminus \{i\}}) \right] \\ &+ \frac{1}{2} \sum_{i=1}^{n} \left[u_1 u_2 \frac{\partial^2}{\partial u_1 \, \partial u_2} F_{g-1,n+1}(u_1, u_2, \mathbf{x}_{S \setminus \{i\}}) \right]_{\substack{u_1=x_i \\ u_2=x_i}} \\ &+ \frac{1}{2} \sum_{i=1}^{n} \sum_{\substack{g_1+g_2=g \\ I \sqcup J = S \setminus \{i\}}} \left[x_i \frac{\partial}{\partial x_i} F_{g_1,|I|+1}(x_i, \mathbf{x}_I) \right] \left[x_i \frac{\partial}{\partial x_i} F_{g_2,|J|+1}(x_i, \mathbf{x}_J) \right].\end{aligned}$$

Here, we let $S = \{1, 2, \ldots, n\}$ and let $\mathbf{x}_I = \{x_{i_1}, x_{i_2}, \ldots, x_{i_k}\}$ for $I = \{i_1, i_2, \ldots, i_k\}$. Furthermore, all free energies can be calculated from the initial conditions

$$F_{0,1}(x_1)\big|_{s=0} = \sum_{i=1}^{d} \frac{1}{i} q_i x^i \qquad \textit{and} \qquad F_{g,n}(x_1, \ldots, x_n)\big|_{s=0} = 0 \quad \textit{for } (g, n) \neq (0, 1).$$

PROOF. The strategy is to multiply both sides of the cut-and-join recursion of Corollary 2.2 by $x_1^{\mu_1} \cdots x_n^{\mu_n}$ and to sum over all $\mu_1, \ldots, \mu_n$. The left side of the equation becomes

$$\sum_{\mu_1,\ldots,\mu_n=1}^{\infty} \frac{\partial}{\partial s} DH_{g,n}(\mu_1, \ldots, \mu_n) \prod_{i=1}^{n} x_i^{\mu_i} = \frac{\partial}{\partial s} F_{g,n}(x_1, \ldots, x_n).$$

To calculate the first line on the right side of the equation, we use

$$\begin{aligned}&\sum_{\mu_i,\mu_j=1}^{\infty}(\mu_i+\mu_j)\,DH(\mu_i+\mu_j)\,x_i^{\mu_i}x_j^{\mu_j}\\&=\sum_{\mu=1}^{\infty}\mu\,DH(\mu)\sum_{\alpha+\beta=\mu}x_i^{\alpha}x_j^{\beta}\\&=\sum_{\mu=1}^{\infty}\mu\,DH(\mu)\,x_ix_j\frac{x_i^{\mu-1}-x_j^{\mu-1}}{x_i-x_j}=\frac{x_ix_j}{x_i-x_j}\left[\frac{\partial}{\partial x_i}F(x_i)-\frac{\partial}{\partial x_j}F(x_j)\right].\end{aligned}$$

To calculate the second and third lines on the right side of the equation, we use

$$\begin{aligned}&\sum_{\mu=1}^{\infty}\sum_{\alpha+\beta=\mu}\alpha\beta\,DH(\alpha,\beta)\,x^{\mu}\\&=\sum_{\alpha,\beta=1}^{\infty}\alpha\beta\,DH(\alpha,\beta)\,x^{\alpha+\beta}=\left[u_1u_2\frac{\partial^2}{\partial u_1\partial u_2}F(u_1,u_2)\right]_{\substack{u_1=x\\u_2=x}}\end{aligned}$$

Note that we have dropped extraneous subscripts in the previous two equations. Finally, the initial conditions precisely capture the base cases of Proposition 2.1. □

In this particular form, the cut-and-join recursion can be used to determine the free energies uniquely. For instance, let us apply this to the case $(g,n)=(0,1)$.

It will be useful to consider the change of variables

$$x(z,s')=z\exp\left(-s'P(z)\right)\qquad\text{and}\qquad s(z,s')=s',\tag{2.3}$$

which we will be using throughout the remainder of the paper. Here, we introduce s' in order to specify the system of variables in which we are working, as this distinction turns out to be important later.

PROPOSITION 2.5. *The following equation holds.*

$$x\frac{\partial}{\partial x}F_{0,1}(x)=P(z)$$

PROOF. The change of variables of equation (2.3) yields $x\frac{\partial}{\partial x}=\frac{z}{1-s'zP'(z)}\frac{\partial}{\partial z}$ and $\frac{\partial}{\partial s}=\frac{\partial}{\partial s'}+\frac{zP(z)}{1-s'zP'(z)}\frac{\partial}{\partial z}$. Following Proposition 2.4, the differential equation satisfied by $F_{0,1}(x)$ is

$$\frac{\partial}{\partial s}F_{0,1}(x)=\frac{1}{2}\left[x\frac{\partial}{\partial x}F_{0,1}(x)\right]^2.$$

Apply the operator $x\frac{\partial}{\partial x}$ to both sides and switch to the variables (z,s') to obtain the equation

$$\begin{aligned}&\left(\frac{\partial}{\partial s'}+\frac{zP(z)}{1-s'zP'(z)}\frac{\partial}{\partial z}\right)x\frac{\partial}{\partial x}F_{0,1}(x)\\&=\left(x\frac{\partial}{\partial x}F_{0,1}(x)\right)\frac{z}{1-s'zP'(z)}\frac{\partial}{\partial z}\left(x\frac{\partial}{\partial x}F_{0,1}(x)\right).\end{aligned}$$

We have the initial condition $x\frac{\partial}{\partial x}F_{0,1}(x)\big|_{s'=0}=P\big(\,x(z,s')\big|_{s'=0}\big)=P(z)$. So one can solve the equation perturbatively by introducing an expansion of the form

$$x\frac{\partial}{\partial x}F_{0,1}(x)=P(z)+P_1(z)s'+P_2(z)s'^2+\cdots.$$

Substituting the perturbative expansion into the equation produces a system of equations, allowing one to explicitly determine the coefficients. Solving the system, we conclude that $P_i(z) = 0$ for every positive integer i. □

COROLLARY 2.6. *From the previous proposition, we immediately deduce that* $\Omega_{0,1}(x_1) = \frac{y(z_1)\,\mathrm{d}x(z_1)}{x(z_1)}$. *Due to equation* (1.2), *this verifies the* $(g,n) = (0,1)$ *case of our main conjecture.*

Next, we use the cut-and-recursion in the case $(g,n) = (0,2)$.

PROPOSITION 2.7. *The following equations hold.*

$$\Omega_{0,2}(x_1,x_2) = \frac{\mathrm{d}z_1\,\mathrm{d}z_2}{(z_1-z_2)^2} - \frac{\mathrm{d}x_1\,\mathrm{d}x_2}{(x_1-x_2)^2}$$

$$x_1\frac{\partial}{\partial x_1}F_{0,2}(x_1,x_2) = -\frac{x_2}{x_1-x_2} + \frac{z_2}{(z_1-z_2)(1-s'z_1P'(z_1))}$$

PROOF. The cut-and-join recursion of Proposition 2.4 in the case $(g,n) = (0,2)$ states the following.

$$\frac{\partial}{\partial s}F_{0,2}(x_1,x_2) = \sum_{i=1}^{2}\left[x_i\frac{\partial}{\partial x_i}F_{0,1}(x_i)\right]\left[x_i\frac{\partial}{\partial x_i}F_{0,2}(x_1,x_2)\right] + \frac{x_1x_2}{x_1-x_2}\left[\frac{\partial}{\partial x_1}F_{0,1}(x_1) - \frac{\partial}{\partial x_2}F_{0,1}(x_2)\right].$$

From the change of variables of equation (2.3)

$$x_i = z_i\exp(-s'P(z_i)) \qquad \text{and} \qquad s = s',$$

we have $x_i\frac{\partial}{\partial x_i} = \frac{z_i}{1-s'z_iP'(z_i)}\frac{\partial}{\partial z_i}$ and $\frac{\partial}{\partial s} = \frac{\partial}{\partial s'} + \frac{z_1P(z_1)}{1-s'z_1P'(z_1)}\frac{\partial}{\partial z_1} + \frac{z_2P(z_2)}{1-s'z_2P'(z_2)}\frac{\partial}{\partial z_2}$. If we furthermore substitute $x_i\frac{\partial}{\partial x_i}F_{0,1}(x_i) = P(z_i)$ from Proposition 2.5 and simplify, we obtain

$$\frac{\partial}{\partial s'}F_{0,2}(x_1,x_2) = \frac{x_2P(z_1) - x_1P(z_2)}{x_1-x_2}.$$

Now integrate and use the fact that $F_{0,2}(x_1,x_2)|_{s=s'=0} = 0$ to obtain

$$F_{0,2}(x_1,x_2) = -\log\left(\frac{x_1-x_2}{z_1-z_2}\right) - s'P(z_1) - s'P(z_2).$$

One deduces the first equation of the proposition by applying $\mathrm{d}_1\mathrm{d}_2$ directly to both sides and the second by applying $x_1\frac{\partial}{\partial x_1}$. □

We may now substitute the calculations of $F_{0,1}$ and $F_{0,2}$ from Propositions 2.5 and 2.7, respectively, into the cut-and-join recursion. We obtain the following result, which bears a similar structure to that of the topological recursion.

COROLLARY 2.8. *The free energies satisfy the following for $2g - 2 + n > 0$.*

$$\begin{aligned}
&\left[\frac{\partial}{\partial s} - \sum_{i=1}^{n} P(z_i)x_i\frac{\partial}{\partial x_i}\right] F_{g,n}(x_1,\ldots,x_n)\\
&= \frac{1}{2}\sum_{i=1}^{n}\left[u_1u_2\frac{\partial^2}{\partial u_1\,\partial u_2}F_{g-1,n+1}(u_1,u_2,\mathbf{x}_{S\setminus\{i\}})\right]_{u_1=x_i,u_2=x_i}\\
&+ \frac{1}{2}\sum_{i=1}^{n}\sum_{\substack{g_1+g_2=g\\ I\sqcup J=S\setminus\{i\}}}^{\text{stable}}\left[x_i\frac{\partial}{\partial x_i}F_{g_1,|I|+1}(x_i,\mathbf{x}_I)\right]\left[x_i\frac{\partial}{\partial x_i}F_{g_2,|J|+1}(x_i,\mathbf{x}_J)\right]\\
&+ \sum_{i\neq j}\frac{z_j}{(z_i-z_j)(1-s'z_iP'(z_i))}x_i\frac{\partial}{\partial x_i}F_{g,n-1}(\mathbf{x}_{S\setminus\{j\}})
\end{aligned}$$

The word "stable" over the inner summation on the second line means that we exclude terms involving $F_{0,1}$ or $F_{0,2}$. Equivalently, using the change of variables $x_i = z_i\exp(-s'P(z_i))$ and $s = s'$, we have $x_i\frac{\partial}{\partial x_i} = \frac{z_i}{1-s'z_iP'(z_i)}\frac{\partial}{\partial z_i}$ and $\frac{\partial}{\partial s} = \frac{\partial}{\partial s'} + \sum P(z_i)x_i\frac{\partial}{\partial x_i}$. So the free energies satisfy the following equation for $2g-2+n>0$.

$$\begin{aligned}
&\frac{\partial}{\partial s'}F_{g,n}(x_1,\ldots,x_n)\\
&= \frac{1}{2}\sum_{i=1}^{n}\frac{z_i^2}{(1-s'z_iP'(z_i))^2}\left[\frac{\partial^2}{\partial v_1\,\partial v_2}F_{g-1,n+1}(u_1,u_2,\mathbf{x}_{S\setminus\{i\}})\right]_{\substack{u_1=u_2=x_i\\ v_1=v_2=z_1}}\\
&+ \frac{1}{2}\sum_{i=1}^{n}\sum_{\substack{g_1+g_2=g\\ I\sqcup J=S\setminus\{i\}}}^{\text{stable}}\frac{z_i^2}{(1-s'z_iP'(z_i))^2}\left[\frac{\partial}{\partial z_i}F_{g_1,|I|+1}(x_i,\mathbf{x}_I)\right]\left[\frac{\partial}{\partial z_i}F_{g_2,|J|+1}(x_i,\mathbf{x}_J)\right]\\
&+ \sum_{i\neq j}\frac{z_iz_j}{(z_i-z_j)(1-s'z_iP'(z_i))^2}\frac{\partial}{\partial z_i}F_{g,n-1}(\mathbf{x}_{S\setminus\{j\}})
\end{aligned}$$

2.3. Pruned double Hurwitz numbers. The notion of pruned Hurwitz numbers was introduced by Norbury and the first author [**14**]. The general idea is to interpret Hurwitz numbers as an enumeration of branching graphs and then to restrict the enumeration to those branching graphs without leaves. The upshot is that pruned Hurwitz numbers store the same information as Hurwitz numbers but are in some sense better behaved. Similar to their unpruned counterparts, pruned Hurwitz numbers also satisfy a cut-and-join recursion and exhibit a polynomial structure.

Okounkov and Pandharipande demonstrated how to associate a branching graph to a branched cover $f : (\Sigma; p_1, \ldots, p_n) \to (\mathbb{CP}^1; \infty)$ [**33**]. It is essentially the graph embedded on Σ formed by the preimage of the star graph on $\mathbb{CP}^1$. To construct the star graph on $\mathbb{CP}^1$, fix the branch points at the mth roots of unity and consider *half-edges* connecting a vertex at 0 to the mth roots of unity by line segments. The branching graph is endowed with extra labels — the faces are labelled with the point p_i that they contain, while the half-edges are labelled by the corresponding root of unity. Note that two half-edges meet to create a *full-edge* precisely at the m simple ramification points on Σ. These considerations lead naturally to the following definition.

DEFINITION 2.9. A *branching graph* of type (g, n) is a graph comprising half-edges and full-edges meeting at vertices, embedded on a genus g oriented surface Σ such that the complement consists of n disks labelled from 1 up to n. We furthermore require that

- the half-edges adjacent to a vertex are cyclically labelled
$$1, 2, \ldots, m; 1, 2, \ldots, m; \ldots; 1, 2, \ldots, m;$$
- there are precisely m full-edges, which are labelled $1, 2, \ldots, m$.

Define the *perimeter of a face* and the *degree of a vertex* to be the number of times each label appears on an adjacent half-edge. We consider two branching graphs to be equivalent if there is an orientation-preserving diffeomorphism of their underlying surfaces that maps one branching graph to the other, while preserving all of the labels.

PROPOSITION 2.10. *The double Hurwitz number $DH_{g,n}(\mu_1, \ldots \mu_n)$ is equal to the weighted count of branching graphs of type (g, n) such that the face labelled i has perimeter μ_i. The weight of such a branching graph is*
$$\frac{q_{\lambda_1} q_{\lambda_2} \cdots q_{\lambda_\ell}}{|\mathrm{Aut}\ \Gamma|} \frac{s^m}{m!},$$
where m is the number of full-edges, $\mathrm{Aut}\ \Gamma$ is the automorphism group of the branching graph, and $\lambda_1, \lambda_2, \ldots, \lambda_\ell$ denote the degrees of the vertices.

DEFINITION 2.11. A *leaf* of a branching graph is a vertex that is adjacent to exactly one full-edge, where that full-edge is not a loop. Define the *pruned double Hurwitz number* $PH_{g,n}(\mu_1, \ldots \mu_n)$ to be the weighted count of Proposition 2.10, restricted to those branching graphs without leaves.

A table of pruned double Hurwitz numbers appears in Appendix B.

The pruned viewpoint was applied to double Hurwitz numbers in the recent work of Hahn [**28**]. Hahn derived a cut-and-join recursion for pruned double Hurwitz numbers, using a different notation and normalisation convention to the one we have adopted in this paper. He furthermore demonstrated that they exhibit piecewise polynomial behaviour and derived a pruning correspondence that relates double Hurwitz numbers to their pruned counterparts. We state the pruning correspondence without proof, since it is a restatement of the pruning correspondence in Hahn's paper [**28**, Theorem 3.4]. However, we wish to draw attention to the fact that it is expressed very naturally using our packaging of the double Hurwitz numbers.

PROPOSITION 2.12 (Pruning correspondence, version 1). *For $(g, n) \neq (0, 1)$, we have the following equations.*
$$DH_{g,n}(\mu_1, \ldots, \mu_n) = \sum_{\nu_1, \ldots, \nu_n = 1}^{\mu_1, \ldots, \mu_n} PH_{g,n}(\nu_1, \ldots, \nu_n) \prod_{i=1}^{n} C(\mu_i, \nu_i),$$
$$PH_{g,n}(\nu_1, \ldots, \nu_n) = \sum_{\mu_1, \ldots, \mu_n = 1}^{\nu_1, \ldots, \nu_n} DH_{g,n}(\mu_1, \ldots, \mu_n) \prod_{i=1}^{n} \widehat{C}(\nu_i, \mu_i),$$
where $C(\mu, \nu) = \frac{\nu}{\mu}[z^{\mu-\nu}] \exp(\mu s P(z))$ and $\widehat{C}(\nu, \mu) = \frac{\mu}{\nu}[z^{\nu-\mu}] \frac{1-szP'(z)}{\exp(\mu s P(z))}$.

Note that $C(\mu, \nu)$ is a weighted homogeneous polynomial in $sq_1, sq_2, \ldots$ with positive rational coefficients, while $\widehat{C}(\nu, \mu)$ is a weighted homogeneous polynomial in $sq_1, sq_2, \ldots$ with rational coefficients.

In fact, the pruning correspondence can be even more succinctly stated in the language of the free energies. In particular, the pruned double Hurwitz numbers arise as coefficients in the expansion with respect to the rational parameters $z_1, \ldots, z_n$.

COROLLARY 2.13 (Pruning correspondence, version 2). *For $(g, n) \neq (0, 1)$, we have the following equations.*

$$F_{g,n} = \sum_{\nu_1,\ldots,\nu_n=1}^{\infty} PH_{g,n}(\nu_1, \ldots, \nu_n) \prod_{i=1}^{n} z_i^{\nu_i}$$

$$\Omega_{g,n} = \sum_{\nu_1,\ldots,\nu_n=1}^{\infty} PH_{g,n}(\nu_1, \ldots, \nu_n) \prod_{i=1}^{n} \nu_i z_i^{\nu_i - 1} \mathrm{d}z_i$$

PROOF. We simply compute the coefficients of $\Omega_{g,n}$ in the z_i-expansion.

$$\begin{aligned}
&\operatorname*{Res}_{z_1=0} \cdots \operatorname*{Res}_{z_n=0} \Omega_{g,n} \prod_{i=1}^{n} z_i^{-\nu_i} \\
&= \operatorname*{Res}_{z_1=0} \cdots \operatorname*{Res}_{z_n=0} \sum_{\mu_1,\ldots,\mu_n=1}^{\infty} DH_{g,n}(\mu_1, \ldots, \mu_n) \prod_{i=1}^{n} \mu_i z_i^{-\nu_i} x_i^{\mu_i - 1} \mathrm{d}x_i \\
&= \sum_{\mu_1,\ldots,\mu_n=1}^{\infty} DH_{g,n}(\mu_1, \ldots, \mu_n) \prod_{i=1}^{n} \mu_i [z_i^{\nu_i - 1}] x_i^{\mu_i - 1} \frac{\mathrm{d}x_i}{\mathrm{d}z_i} \\
&= \sum_{\mu_1,\ldots,\mu_n=1}^{\infty} DH_{g,n}(\mu_1, \ldots, \mu_n) \prod_{i=1}^{n} \mu_i [z_i^{\nu_i - \mu_i}] \frac{1 - s z_i P'(z_i)}{\exp(\mu_i s P(z_i))} \\
&= \sum_{\mu_1,\ldots,\mu_n=1}^{\nu_1,\ldots,\nu_n} DH_{g,n}(\mu_1, \ldots, \mu_n) \prod_{i=1}^{n} \nu_i \widehat{C}(\nu_i, \mu_i) = PH_{g,n}(\nu_1, \ldots, \nu_n) \prod_{i=1}^{n} \nu_i
\end{aligned}$$

The desired statement for the multidifferentials $\Omega_{g,n}$ follows immediately, while the desired statement for the free energies $F_{g,n}$ can be obtained by integrating. □

We do not state here the cut-and-join recursion for pruned double Hurwitz numbers. However, we remark that it can be obtained either combinatorially, as performed by Hahn [**28**], or equivalently by expressing Corollary 2.8 at the level of coefficients in the z_i-expansion.

In Section 4, we will reduce our main conjecture to a weaker conjecture concerning the structure of double Hurwitz numbers. We have introduced pruned double Hurwitz numbers since enumeration of pruned objects in general can help to reveal structure of the underlying enumeration [**14**].

3. Evidence for the conjecture

3.1. The quantum curve. It is known that there is a quantum curve underlying many instances of the topological recursion and conversely, the existence of a quantum curve can predict spectral curves and topological recursion [**31**]. We now consider the quantum curve for double Hurwitz numbers and propose it as evidence towards the main conjecture.

Informally, a quantum curve for a spectral curve $P(x, y) = 0$ is a certain operator that can be expressed in terms of a non-commutative polynomial $\widehat{P}(\widehat{x}, \widehat{y})$ in the multiplication operator $\widehat{x} = x$ and the differential operator $\widehat{y} = \hbar x \frac{\partial}{\partial x}$ such that[3]

- the semi-classical limit of $\widehat{P}(\widehat{x}, \widehat{y})$ recovers $P(x, y)$; and
- $\widehat{P}(\widehat{x}, \widehat{y})$ annihilates the wave function.

The semi-classical limit is obtained by sending $\widehat{x}$ and $\widehat{y}$ to the commuting variables x and y, while also sending $\hbar$ to 0. The wave function is a certain generating function constructed from the coefficients of the correlation differentials obtained from topological recursion. We define the wave function only in the case of double Hurwitz numbers and point the reader to the literature for more information on quantum curves [**31**].

In the case of double Hurwitz numbers, the *partition function* and *wave function* are generating functions defined by the following formulas, respectively.

$$Z(p_1, p_2, \ldots; \hbar) = \exp\left[\sum_{g=0}^{\infty}\sum_{n=1}^{\infty}\sum_{\mu_1,\ldots,\mu_n=1}^{\infty} DH_{g,n}(\mu_1, \ldots, \mu_n)\frac{\hbar^{2g-2+n}}{n!}p_{\mu_1}\cdots p_{\mu_n}\right]$$

$$\psi(x, \hbar) = Z(p_1, p_2, \ldots; \hbar)\big|_{p_i = x^i}$$

PROPOSITION 3.1 (Alexandrov–Lewanski–Shadrin [**1**]). *The wave function for the double Hurwitz numbers satisfies the quantum curve equation $\mathcal{Q}\,\psi(x, \hbar) = 0$, where*

$$\mathcal{Q} = \widehat{y} - \sum_{k=1}^{d} q_k \exp\left(\tfrac{1}{2}s\hbar k(k-1)\right)\widehat{x}^k \exp(sk\widehat{y}).$$

One can immediately observe that the semi-classical limit of the operator $\mathcal{Q}$ is simply $y - P(x\exp(sy))$, for which we have the rational parametrisation of equation (1.1). The existence of the quantum curve lends strong support to Conjecture 1.3.

REMARK 3.2. As previously mentioned, the quantum curve may be used to predict the spectral curve for a topological recursion. It is also often the case that the $(0, 1)$ information of a problem may be used to predict the spectral curve for a topological recursion [**15**]. We remark here that the quantum curve may be used to pass directly to the $(0, 1)$ information. In the following, we use the quantum curve to provide another proof of Proposition 2.5, which states that $x\frac{\partial}{\partial x}F_{0,1}(x) = P(z)$.

We begin with the two observations

$$\psi(x, \hbar) = \exp\left(\frac{1}{\hbar}F_{0,1}(x) + O(1)\right) \quad \text{and} \quad \widehat{y}\psi(x, \hbar) = \left(x\frac{\partial}{\partial x}F_{0,1}(x) + O(\hbar)\right)\psi(x, \hbar),$$

which follow directly from the definition of the wave function. Therefore, we have the equation

$$\exp\left(\frac{1}{2}s\hbar k(k-1)\right)\widehat{x}^k \exp(sk\widehat{y})\psi(x, \hbar) = (1 + O(\hbar))x^k\left(e^{skx\frac{\partial}{\partial x}F_{0,1}(x)} + O(\hbar)\right)\psi(x, \hbar).$$

[3]The choice of polarisation — in other words, the operators $\widehat{x}$ and $\widehat{y}$ — relates to the fact that we are dealing with a $\mathbb{C}^* \times \mathbb{C}$ spectral curve. For a $\mathbb{C} \times \mathbb{C}$ spectral curve, it is common to use $\widehat{y} = \hbar\frac{\partial}{\partial x}$. In both cases though, we have the canonical commutation relation $[\widehat{x}, \widehat{y}] = -\hbar$.

The vanishing of $\mathcal{Q}\psi(x,\hbar)$ implies the following equation on $x\frac{\partial}{\partial x}F_{0,1}(x)$.

$$x\frac{\partial}{\partial x}F_{0,1}(x) - \sum_{k=1}^{d} q_k x^k e^{skx\frac{\partial}{\partial x}F_{0,1}(x)} = 0$$

Now use the substitution $x = z\exp\left(-sx\frac{\partial}{\partial x}F_{0,1}(x)\right)$, where z is a new variable and $x\frac{\partial}{\partial x}F_{0,1}(x)$ in the exponent is considered as a function of z. Then the previous equation recovers Proposition 2.5, namely

$$x\frac{\partial}{\partial x}F_{0,1}(x) = \sum_{k=1}^{d} q_k z^k.$$

3.2. Low genus evidence. One can attempt to verify the main conjecture via direct computation in low genus. We include below the calculations in the case $(g,n) = (0,3)$ and $(1,1)$.

Calculation of $F_{0,3}$. The cut-and-join recursion of Corollary 2.8 in the case $(g,n) = (0,3)$ reads

$$\begin{aligned}\frac{\partial}{\partial s'}F_{0,3} = &\sum_{\text{cyclic}} \frac{x_1x_2}{x_1-x_2}\left[\frac{\partial}{\partial x_1}F_{0,2}(x_1,x_3) - \frac{\partial}{\partial x_2}F_{0,2}(x_2,x_3)\right] \\ &+ \sum_{\text{cyclic}}\left[x_1\frac{\partial}{\partial x_1}F_{0,2}(x_1,x_2)\right]\left[x_1\frac{\partial}{\partial x_1}F_{0,2}(x_1,x_3)\right].\end{aligned}$$

Substituting the formula of Proposition 2.7 yields great simplification and we are ultimately left with

$$\frac{\partial}{\partial s'}F_{0,3} = -1 + \sum_{\text{cyclic}} \frac{z_2z_3}{(z_1-z_2)(z_1-z_3)}\frac{1}{(1-s'z_1P'(z_1))^2}.$$

One can integrate this equation, using the initial conditions to deduce the constant of integration, in order to obtain

$$F_{0,3}(x_1,x_2,x_3) = -s' + \sum_{\text{cyclic}} \frac{s'z_2z_3}{(z_1-z_2)(z_1-z_3)(1-s'z_1P'(z_1))}. \tag{3.1}$$

From the previous equation, $F_{0,3}(x_1,x_2,x_3)$ can be expressed as follows.

$$\frac{s'\det\begin{bmatrix} s'z_1P'(z_1) & s'z_2P'(z_2) & s'z_3P'(z_3) \\ z_1(1-s'z_1P'(z_1)) & z_2(1-s'z_2P'(z_2)) & z_3(1-s'z_3P'(z_3)) \\ z_1^2(1-s'z_1P'(z_1)) & z_2^2(1-s'z_2P'(z_2)) & z_3^2(1-s'z_3P'(z_3))\end{bmatrix}}{(1-s'z_1P'(z_1))(1-s'z_2P'(z_2))(1-s'z_3P'(z_3))(z_1-z_2)(z_2-z_3)(z_3-z_1)}$$

Observe that the determinant in the numerator is divisible by $(z_1-z_2)(z_2-z_3)(z_3-z_1)$, so that $F_{0,3}(x_1,x_2,x_3)$ does not have poles along the diagonals $z_i - z_j$. Now consider the expression

$$F_{0,3}(x_1,x_2,x_3)\prod_{i=1}^{3}(1-s'z_iP'(z_i)) = \frac{\text{Pol}(z_1,z_2,z_3,s)}{(z_1-z_2)(z_2-z_3)(z_3-z_1)}.$$

We see that the degree of z_1 in $\text{Pol}(z_1,z_2,z_3,s)$ is $d+2$. However, we have already deduced that $\text{Pol}(z_1,z_2,z_3)$ is divisible by $(z_1-z_2)(z_2-z_3)(z_3-z_1)$. It follows that the entire expression is in fact a polynomial of degree d in z_1. Similarly, one can deduce that the entire expression has degree d in z_2 and z_3. Now as the determinant in the numerator is divisible by $z_1z_2z_3$, there must exist $C_{i_1,i_2,i_3} \in \mathbb{Q}[s, q_1, q_2, \ldots]$ such that

$$F_{0,3}(x_1, x_2, x_3) = \prod_{i=1}^{3} \frac{1}{1 - s'z_iP'(z_i)} \times \sum_{i_1,i_2,i_3=1}^{d} C_{i_1,i_2,i_3} z_1^{i_1} z_2^{i_2} z_3^{i_3}. \tag{3.2}$$

Note in particular that $F_{0,3}(x_1, x_2, x_3)$ is a rational function in z_1, z_2, z_3 with poles only at the branch points of the spectral curve of equation (1.1). We may now check that the topological recursion recovers the correct correlation differential $\omega_{0,3}$ predicted by our main conjecture. We start with the following formula for $\omega_{0,3}$ [**21**].

$$\begin{aligned}\omega_{0,3}(z_1, z_2, z_3) &= -\sum_{i=1}^{d} \operatorname*{Res}_{z=a_i} \frac{\omega_{0,2}(z, z_1)\, \omega_{0,2}(z, z_2)\, \omega_{0,2}(z, z_3)\, x(z)}{\mathrm{d}x(z)\, \mathrm{d}y(z)} \\ &= \mathrm{d}z_1\, \mathrm{d}z_2\, \mathrm{d}z_3 \sum_{i=1}^{d} \frac{-sa_i^3}{(z_1 - a_i)^2\, (z_2 - a_i)^2\, (z_3 - a_i)^2\, (1 + sa_i^2P''(a_i))} \qquad (3.3)\end{aligned}$$

Observe that this equation decomposes $\omega_{0,3}(z_1, z_2, z_3)$ into a sum of its principal parts with respect to the variable z_1 [**12**, Proposition 16]. Recall that the principal part of a meromorphic form may be defined via the formula

$$[\omega(z_1)]_a = \operatorname*{Res}_{z=a} \frac{\mathrm{d}z_1}{z_1 - z}\omega(z). \tag{3.4}$$

Now we use this formula to evaluate the principal part of $\Omega_{0,3}(x_1, x_2, x_3) = \mathrm{d}_1\mathrm{d}_2\mathrm{d}_3F_{0,3}(x_1, x_2, x_3)$ at the branch point a_i.

$$\begin{aligned}\frac{[\Omega_{0,3}(x_1, x_2, x_3)]_{a_i}}{\mathrm{d}z_1\, \mathrm{d}z_2\, \mathrm{d}z_3} &= \frac{1}{\mathrm{d}z_1\, \mathrm{d}z_2\, \mathrm{d}z_3} \operatorname*{Res}_{z=a_i} \frac{\mathrm{d}z_1}{z_1 - z} \mathrm{d}\mathrm{d}_2\mathrm{d}_3F_{0,3}(x, x_2, x_3) \\ &= \frac{\partial}{\partial z_1}\frac{\partial}{\partial z_2}\frac{\partial}{\partial z_3} \operatorname*{Res}_{z=a_i} \frac{\mathrm{d}z}{z_1 - z} F_{0,3}(x, x_2, x_3) \\ &= \frac{\partial}{\partial z_1}\frac{\partial}{\partial z_2}\frac{\partial}{\partial z_3} \operatorname*{Res}_{z=a_i} \frac{\mathrm{d}z}{z_1 - z} \left[\frac{s'z_2z_3}{(z - z_2)(z - z_3)(1 - s'zP'(z))}\right]\end{aligned}$$

The first equality uses the definition of the principal part, the second exchanges the order of taking the residue and the derivative, while the third uses equation (3.1). Note that we have only kept the summand in equation (3.1) that will contribute to the residue. We may now evaluate the residue and derivatives explicitly, noting that the expression has a simple pole at a_i.

$$\begin{aligned}\frac{[\Omega_{0,3}(x_1, x_2, x_3)]_{a_i}}{\mathrm{d}z_1\, \mathrm{d}z_2\, \mathrm{d}z_3} &= \frac{\partial}{\partial z_1}\frac{\partial}{\partial z_2}\frac{\partial}{\partial z_3} \lim_{z\to a_i} \frac{1}{z_1 - z} \left[\frac{s'z_2z_3(z - a_i)}{(z - z_2)(z - z_3)(1 - s'zP'(z))}\right] \\ &= \frac{\partial}{\partial z_1}\frac{\partial}{\partial z_2}\frac{\partial}{\partial z_3} \frac{z_2z_3}{(z_1 - a_i)(z_2 - a_i)(z_3 - a_i)(a_iP''(a_i) + P'(a_i))} \\ &= -\frac{sa_i^3}{(z_1 - a_i)^2(z_2 - a_i)^2(z_3 - a_i)^2(1 + sa_i^2P''(a_i))} \qquad (3.5)\end{aligned}$$

The first equality evaluates the residue via limit, the second is an application of L'Hôpital's rule, while the third evaluates the derivatives explicitly and uses the fact that $1 - sa_iP'(a_i) = 0$.

Comparing equations (3.3) and (3.5), we see that the principal parts of $\omega_{0,3}$ and $\Omega_{0,3}$ match precisely at the branch points. Furthermore, we know that they are both rational multidifferentials with poles only at the branch points, so we may conclude that they are equal. It follows that Conjecture 1.3 holds in the case $(g,n) = (0,3)$.

Calculation of $F_{1,1}$. The cut-and-join recursion of Corollary 2.8 in the case $(g,n) = (1,1)$ reads

$$\left[\frac{\partial}{\partial s} - P(z)x\frac{\partial}{\partial x}\right]F_{1,1}(x) = \frac{1}{2}\left[u_1u_2\frac{\partial^2}{\partial u_1 \partial u_2}F_{0,2}(u_1,u_2)\right]_{\substack{u_1=x\\u_2=x}}.$$

Passing to (z,s') coordinates and substituting the explicit expression for $\frac{\partial^2}{\partial z_1 \partial z_2}F_{0,2}$ from Proposition 2.7, we see that $\frac{\partial}{\partial s'}F_{1,1}(x)$ is equal to

$$\frac{1}{2}\left[\frac{z_1}{1-s'z_1P'(z_1)}\frac{z_2}{1-s'z_2P'(z_2)}\frac{(x(z_1)-x(z_2))^2-(z_1-z_2)^2x'(z_1)x'(z_2)}{(z_1-z_2)^2\,(x(z_1)-x(z_2))^2}\right]_{z_1=z_2=z}.$$

One can evaluate the right side of this equation using L'Hôpital's rule to obtain

$$\frac{\partial}{\partial s'}F_{1,1}(x) = \frac{z^2\left(3x''(z)^2-2x'(z)x'''(z)\right)}{24\left(1-s'zP'(z)\right)^2x'(z)^2}.$$

Now integrate this equation, using the initial conditions to deduce the constant of integration and verify that $F_{1,1}(x)$ is equal to

$$\frac{s'^2z^2\left(3P''+zP'''+3s'P'^2-s'P'^2-s'zP'P''+s'z^2P''^2-s'z^2P'P'''-s'^2zP'^3\right)}{24(1-s'zP')^3}.$$

One can explicitly verify that

$$F_{1,1}(x) = \frac{s'^2}{24}\left(\frac{z}{1-s'zP'(z)}\frac{\partial}{\partial z}\right)^2[zP'(z)] - \frac{s'^2}{24}\left(\frac{z}{1-s'zP'(z)}\frac{\partial}{\partial z}\right)[P(z)]. \tag{3.6}$$

Note that $F_{1,1}(x)$ is a rational function in z with poles only at the branch points of the spectral curve of equation (1.1). Rather than verify that $\Omega_{1,1}(x) = \mathrm{d}F_{1,1}(x)$ coincides with the correlation differential $\omega_{1,1}(z)$, we propose a more general approach in the next section. We will prove that our main conjecture follows if the free energies satisfy so-called linear loop equations.

4. Towards a proof of the conjecture

4.1. The structure of double Hurwitz numbers. There is a certain technique for starting with a combinatorial recursion of cut-and-join type and deducing the topological recursion that works in several known cases. For instance, it was successfully employed in the case of single Hurwitz numbers [**23**], orbifold Hurwitz numbers [**5, 12**] and monotone Hurwitz numbers [**11**]. One further ingredient usually enters into these proofs and that is a polynomial structure theorem for the enumerative problem. However, such a result is not known in the case of double Hurwitz numbers. In this section, we present a conjecture on the structure of double Hurwitz numbers and then proceed to show that it can be used to deduce our main conjecture.

It is known that the correlation differentials produced by the topological recursion satisfy so-called *linear loop equations* [**3**]. These assert that at a branch point a of the spectral curve with associated involution s, the correlation differentials $\omega_{g,n}$ for $2g-2+n>0$ satisfy the following condition: the sum

$$\omega_{g,n}(z_1,z_2,\ldots,z_n) + \omega_{g,n}(\sigma(z_1),z_2,\ldots,z_n)$$

is analytic at $z_1 = a$. The linear loop equations imply structure to the underlying coefficients of the correlation differentials. For example, in the case of single Hurwitz numbers, they are equivalent to the fact that

$$H_{g,n}(\mu_1, \ldots, \mu_n) = \prod_{i=1}^{n} \frac{\mu_i^{\mu_i}}{\mu_i!} P_{g,n}(\mu_1, \ldots, \mu_n),$$

where $P_{g,n}$ is a symmetric polynomial, a fact that follows from the ELSV formula. The linear loop equations are equivalent to an analogous quasi-polynomial structure in the case of orbifold Hurwitz numbers. As previously mentioned, an analogous result is not known in the case of double Hurwitz numbers. However, given our main conjecture, it is natural to posit that the following is true.

CONJECTURE 4.1. For $2g - 2 + n > 0$,

- $F_{g,n}(z_1, \ldots, z_n)$ is a rational function with poles only at $z_i = a_j$ for $i = 1, 2, \ldots, n$ and $j = 1, 2, \ldots, d$; and
- $F_{g,n}(z_1, z_2, \ldots, z_n) + F_{g,n}(\sigma_i(z_1), z_2, \ldots, z_n)$ is analytic at $z_1 = a_i$ for $i = 1, 2, \ldots, d$.

We will subsequently refer to these constraints as *linear loop equations*.

From the general theory of topological recursion, we know that Conjecture 1.3 implies Conjecture 4.1. In fact, we will prove in Theorem 4.8 that the converse is also true. Therefore, the linear loop equations are equivalent to the statement that topological recursion governs the double Hurwitz numbers.

DEFINITION 4.2. Define the $\mathbb{C}(s)$-vector space $V(z)$, whose elements are rational functions $p(z)$ such that

- $p(z)$ has poles only at the branch points $a_1, a_2, \ldots, a_d$; and
- $p(z) + p(\sigma_i(z))$ is analytic at $z = a_i$ for $i = 1, 2, \ldots, d$.

LEMMA 4.3. *A basis for $V(z)$ is formed by $\phi_k^i(z)$ for $i = 1, 2, \ldots, d$ and $k = 0, 1, 2, \ldots$, where we define $\phi_{-1}^i(z) = z^i$ and*

$$\phi_{k+1}^i = \frac{z}{1 - szP'(z)} \frac{\partial}{\partial z} \phi_k^i(z).$$

One can find a short proof of the lemma in the literature [**12**, Lemma 14]. Given the fact that $F_{g,n}(z_1, \ldots, z_n)$ is symmetric in its arguments, it satisfies the linear loop equations if and only if

$$F_{g,n}(z_1, \ldots, z_n) \in V(z_1) \otimes V(z_2) \otimes \cdots \otimes V(z_n).$$

PROPOSITION 4.4. *The linear loop equations hold for $(g, n) = (0, 3)$ and $(1, 1)$.*

PROOF. From equation (3.2), we have

$$F_{0,3}(x_1, x_2, x_3) = \sum_{i_1, i_2, i_3 = 1}^{d} \frac{C_{i_1, i_2, i_3}}{i_1 i_2 i_3} \phi_0^{i_1}(z_1)\, \phi_0^{i_2}(z_2)\, \phi_0^{i_3}(z_3) \in V(z_1) \otimes V(z_2) \otimes V(z_3).$$

From equation (3.6), we have

$$F_{1,1}(x) = \frac{s^2}{24} \sum_{i=1}^{d} \left[i q_i \phi_1^i(z) - q_i \phi_0^i(z) \right] \in V(z). \qquad \square$$

REMARK 4.5. By Lemma 4.3, we know that $x\frac{\partial}{\partial x} = \frac{z}{1-s'zP'(z)}\frac{\partial}{\partial z}$ preserves $V(z)$ in the sense that for any $p(z) \in V(z)$, we also have $x\frac{\partial}{\partial x}p(z) \in V(z)$. Since x and s are commuting variables, we also know that $\frac{\partial}{\partial s} = \frac{\partial}{\partial s'} + \frac{zP(z)}{(1-s'zP'(z)}\frac{\partial}{\partial z}$ preserves $V(z)$. This seemingly trivial remark will prove to be useful later.

The linear loop equations imply structure on the free energies and hence, also on the double Hurwitz numbers themselves. We have $F_{g,n}(z_1, \ldots, z_n) \in V(z_1) \otimes V(z_2) \otimes \cdots \otimes V(z_n)$, supposing that the linear loop equations hold. So there exist $C_{g,n}\binom{i_1,\ldots,i_n}{m_1,\ldots,m_n}$ that depend on $s, q_1, q_2, \ldots$ for $1 \leq i_1, \ldots, i_n \leq d$ and $m_1, \ldots, m_n \geq 0$ such that the following is true.

$$\begin{aligned} F_{g,n}(z_1, \ldots, z_n) &= \sum_{i_1,\ldots,i_n=1}^{d} \sum_{m_1,\ldots,m_n=0}^{\text{finite}} C_{g,n}\binom{i_1,\ldots,i_n}{m_1,\ldots,m_n} \prod_{k=1}^{n} \phi_{m_k}^{i_k}(z_k) \\ &= \sum_{i_1,\ldots,i_n=1}^{d} \sum_{m_1,\ldots,m_n=0}^{\text{finite}} C_{g,n}\binom{i_1,\ldots,i_n}{m_1,\ldots,m_n} \prod_{k=1}^{n} \left(x_k \frac{\partial}{\partial x_k}\right)^{m_k+1} z_k^{i_k} \end{aligned}$$

Write $z^i = \sum_{\mu=0}^{\infty} A_\mu^i x^\mu$ and observe that one can deduce the coefficients A_μ^i via the Lagrange inversion formula. This then leads to the following conjectural expression for the double Hurwitz numbers, which is in fact equivalent to Conjecture 4.1.

CONJECTURE 4.6. For $2g - 2 + n > 0$, there exist $C_{g,n}\binom{i_1,\ldots,i_n}{m_1,\ldots,m_n}$ independent of $\mu_1, \ldots, \mu_n$ such that the double Hurwitz numbers may be expressed as

$$DH_{g,n}(\mu_1, \ldots, \mu_n) = \sum_{i_1,\ldots,i_n=1}^{d} \prod_{k=1}^{n} A_{\mu_k}^{i_k} \sum_{m_1,\ldots,m_n=0}^{\text{finite}} C_{g,n}\binom{i_1,\ldots,i_n}{m_1,\ldots,m_n} \prod_{k=1}^{n} \mu_k^{m_k+1},$$

where

$$A_\mu^i = i \sum_{|\lambda|=\mu-i} \frac{\mu^{\ell(\lambda)-1}}{|\text{Aut } \lambda|} q_{\lambda_1} q_{\lambda_2} \cdots q_{\lambda_{\ell(\lambda)}} s^{\ell(\lambda)}.$$

Here, λ represents an integer partition with $\ell(\lambda)$ parts and Aut λ is the set of permutations of the tuple $(\lambda_1, \lambda_2, \ldots, \lambda_{\ell(\lambda)})$ that leave it invariant.

REMARK 4.7. The previous conjecture reduces to the polynomial structure of single Hurwitz numbers in the case $P(z) = z$ and the quasi-polynomial structure of orbifold Hurwitz numbers in the case $P(z) = z^a$. We remark that the proof of polynomiality for single Hurwitz numbers was originally achieved via the ELSV formula, a rather deep result from algebraic geometry [**19**]. A second proof arose via a thorough analysis of the infinite wedge space expression for single Hurwitz numbers [**16**]. In the context of double Hurwitz numbers, the previously mentioned conjecture of Goulden, Jackson and Vakil may play the role of the ELSV formula, although the underlying algebraic geometry is unclear at present. Thus, one may hope that the infinite wedge space technology may lead to progress the combinatorial structure of double Hurwitz numbers and thence towards the geometry of double Hurwitz numbers.

4.2. From cut-and-join recursion to topological recursion. At the start of the section, we alluded to a certain technique for starting with a combinatorial recursion of cut-and-join type and deducing the topological recursion. We now carry this out in the context of double Hurwitz numbers, using the linear loop equations as an unproven assumption. In other words, the rest of the paper is dedicated to proving the following result.

THEOREM 4.8. *Conjecture* 4.1 *implies Conjecture* 1.3.

PROOF OF THEOREM 4.8. Consider the cut-and-join recursion of Corollary 2.8 for $2g - 2 + n > 1$. We will use the inductive hypothesis that $\Omega_{g',n'} = \omega_{g',n'}$ for $2g' - 2 + n' < 2g - 2 + n$. We will adopt a gentle abuse of notation and write $F_{g,n}(z_1, \ldots, z_n)$ for the rational function obtained by taking $F_{g,n}(x_1, \ldots, x_n)$ and changing coordinates to the z variables. Recall that it is rational due to the assumption of the linear loop equations of Conjecture 4.1.

$$\begin{aligned}
&\frac{\partial}{\partial s'} F_{g,n}(z_1, \ldots, z_n) \\
&= \frac{1}{2} \sum_{i=1}^{n} \frac{z_i^2}{(1 - s' z_i P'(z_i))^2} \left[\frac{\partial^2}{\partial u_1 \, \partial u_2} F_{g-1,n+1}(u_1, u_2, \mathbf{z}_{S \setminus \{i\}}) \right]_{u_1 = z_i, u_2 = z_i} \\
&\quad + \frac{1}{2} \sum_{i=1}^{n} \sum_{\substack{g_1 + g_2 = g \\ I \sqcup J = S \setminus \{i\}}}^{\text{stable}} \frac{z_i^2}{(1 - s' z_i P'(z_i))^2} \left[\frac{\partial}{\partial z_i} F_{g_1, |I|+1}(z_i, \mathbf{z}_I) \right] \left[\frac{\partial}{\partial z_i} F_{g_2, |J|+1}(z_i, \mathbf{z}_J) \right] \\
&\quad + \sum_{i \neq j} \frac{z_i z_j}{(z_i - z_j)(1 - s' z_i P'(z_i))^2} \frac{\partial}{\partial z_i} F_{g,n-1}(\mathbf{z}_{S \setminus \{j\}})
\end{aligned}$$

Pick a branch point a_i of the spectral curve with associated local involution σ_i. The general strategy is to take the principal part of both sides at $z_1 = a_i$. As we have done previously, denote the principal part by the notation $[\,\cdot\,]_{a_i}$. We take the result, apply σ_i to z_1 to both sides of the equation, then add it to the original. We denote this procedure by the notation $\mathrm{Sym}_i[f(z_1)] = f(z_1) + f(\sigma_i(z_1))$. Note that applying $\mathrm{Sym}_i[\,\cdot\,]_{a_i}$ kills all terms that satisfy the linear loop equations with respect to z_1. Thus, we obtain the following equation.

$$\begin{aligned}
&\mathrm{Sym}_i \left[\frac{\partial}{\partial s'} F_{g,n}(z_1, \ldots, z_n) \right]_{a_i} \\
&\quad = \frac{1}{2} \mathrm{Sym}_i \left[\frac{z_1^2}{(1 - s' z_1 P'(z_1))^2} \left[\frac{\partial^2}{\partial u_1 \, \partial u_2} F_{g-1,n+1}(u_1, u_2, \mathbf{z}_{S \setminus \{1\}}) \right]_{u_1 = z_1, u_2 = z_1} \right]_{a_i} \\
&+ \frac{1}{2} \sum_{\substack{g_1 + g_2 = g \\ I \sqcup J = S \setminus \{1\}}}^{\text{stable}} \mathrm{Sym}_i \left[\frac{z_1^2}{(1 - s' z_1 P'(z_1))^2} \left[\frac{\partial}{\partial z_1} F_{g_1, |I|+1}(z_1, \mathbf{z}_I) \right] \left[\frac{\partial}{\partial z_1} F_{g_2, |J|+1}(z_1, \mathbf{z}_J) \right] \right]_{a_i} \\
&+ \sum_{j=2}^{n} \mathrm{Sym}_i \left[\frac{z_1 z_j}{z_1 - z_j} \left[\frac{1}{(1 - z_1 s' P'(z_1))^2} \frac{\partial}{\partial z_1} F_{g,n-1}(\mathbf{z}_{S \setminus \{j\}}) \right.\right. \\
&\qquad \left.\left. - \frac{1}{(1 - z_j s' P'(z_j))^2} \frac{\partial}{\partial z_j} F_{g,n-1}(\mathbf{z}_{S \setminus \{1\}}) \right] \right]_{a_i}
\end{aligned}$$

Now if $f_i(z_1)$ and $g(z_1)$ satisfy the linear loop equation, then $f(z_1) + f(\sigma(z_1))$ and $g(z_1) + g(\sigma(z_1))$ are analytic at a_i, so their product is as well. So we have the following equations.

$$\sum_k [f_k(z_1)g_k(z_1) + f_k(\sigma(z_1))g_k(z_1) + f_k(z_1)g_k(\sigma(z_1)) + f_k(\sigma(z_1))g_k(\sigma(z_1))]_{a_i} = 0$$

$$\sum_k [f_k(z_1)g_k(z_1) + f_k(\sigma(z_1))g_k(\sigma(z_1))]_{a_i} = -\sum_k [f_k(\sigma(z_1))g_k(z_1) + f_k(z_1)g_k(\sigma(z_1))]_{a_i}$$

$$\mathrm{Sym}_i\Big[\sum_k f_k(z_1)g_k(z_1))\Big]_{a_i} = -\mathrm{Sym}_i\Big[\sum_k f_k(\sigma(z_1))g_k(z_1)\Big]_{a_i}$$

Use this on the first and second lines of the right side, keeping in mind that if $f(z)$ satisfies the linear loop equation, then so does $x\frac{\partial}{\partial x} f(z)$. We also rewrite the third line in a more convenient form.

$$\begin{aligned}
&\mathrm{Sym}_i\left[\frac{\partial}{\partial s'} F_{g,n}(z_1, \ldots, z_n)\right]_{a_i} \\
&= -\frac{1}{2}\mathrm{Sym}_i\left[x_1^2\left[\frac{\partial^2}{\partial u_1\,\partial u_2} F_{g-1,n+1}(z(u_1), \sigma(z(u_2)), \mathbf{z}_{S\setminus\{1\}})\right]_{u_1=x_1, u_2=x_1}\right]_{a_i} \\
&\quad -\frac{1}{2}\sum_{\substack{g_1+g_2=g\\ I\sqcup J=S\setminus\{1\}}}^{\text{stable}} \mathrm{Sym}_i\left[x_1^2\left[\frac{\partial}{\partial x_1} F_{g_1,|I|+1}(z_1, \mathbf{z}_I)\right]\sigma\left[\frac{\partial}{\partial x_1} F_{g_2,|J|+1}(z_1, \mathbf{z}_J)\right]\right]_{a_i} \\
&\quad +\sum_{j=2}^{n} \mathrm{Sym}_i\left[\frac{z_j}{z_1 - z_j}\,\frac{x_1\frac{\partial}{\partial x_1} F_{g,n-1}(\mathbf{z}_{S\setminus\{j\}})}{1 - z_1 s' P'(z_1)} - \frac{z_1}{z_1 - z_j}\,\frac{x_j\frac{\partial}{\partial x_j} F_{g,n-1}(\mathbf{z}_{S\setminus\{1\}})}{1 - z_j s' P'(z_j)}\right]_{a_i}
\end{aligned}$$

Now apply the derivative operator $\mathrm{d}_2 \cdots \mathrm{d}_n$ to both sides. The third line transforms in the following way, due to Lemma 4.9, whose statement and proof we postpone in order to declutter the present argument. Note that we do indeed require $\omega_{0,2}$ in the third line, which is not equal to $\Omega_{0,2}$.

$$\begin{aligned}
&\mathrm{d}_2 \cdots \mathrm{d}_n\, \mathrm{Sym}_i\left[\frac{\partial}{\partial s'} F_{g,n}(z_1, \ldots, z_n)\right]_{a_i} \\
&= -\frac{1}{2}\mathrm{Sym}_i\left[\frac{x_1^2}{(\mathrm{d}x_1)^2}\Omega_{g-1,n+1}(z_1, \sigma(z_1), \mathbf{z}_{S\setminus\{1\}})\right]_{a_i} \\
&\quad -\frac{1}{2}\sum_{\substack{g_1+g_2=g\\ I\sqcup J=S\setminus\{1\}}}^{\text{stable}} \mathrm{Sym}_i\left[\frac{x_1^2}{(\mathrm{d}x_1)^2}\Omega_{g_1,|I|+1}(z_1, \mathbf{z}_I)\,\Omega_{g_2,|J|+1}(\sigma(z_1), \mathbf{z}_J)\right]_{a_i} \\
&\quad -\sum_{j=2}^{n} \mathrm{Sym}_i\left[\frac{x_1^2}{(\mathrm{d}x_1)^2}\omega_{0,2}(z_1, z_j)\,\Omega_{g,n-1}(\mathbf{z}_{S\setminus\{j\}})\right]_{a_i}
\end{aligned}$$

Use the inductive hypothesis to replace the occurrences of $\Omega_{g',n'}$ on the right side with $\omega_{g',n'}$. Furthermore, gather together the second and third lines and use the symmetry of the situation to obtain the following.

$$\begin{aligned} &\mathrm{d}_2 \cdots \mathrm{d}_n \, \mathrm{Sym}_i \left[\frac{\partial}{\partial s'} F_{g,n}(z_1, \ldots, z_n) \right]_{a_i} \\ &= - \left[\frac{x_1^2}{(\mathrm{d}x_1)^2} \left[\omega_{g-1,n+1}(z_1, \sigma(z_1), \mathbf{z}_{S \setminus \{1\}}) \right.\right. \\ &\quad \left.\left. + \sum_{\substack{g_1+g_2=g \\ I \sqcup J = S\setminus\{1\}}}^{\circ} \omega_{g_1,|I|+1}(z_1, \mathbf{z}_I) \, \omega_{g_2,|J|+1}(\sigma(z_1), \mathbf{z}_J) \right] \right]_{a_i} \end{aligned}$$

Multiply both sides by $-\frac{\mathrm{d}x_1}{x_1} \frac{1}{y_1 - \sigma(y_1)}$ and take the principal part at a_i again. Since the expression we are multiplying by is analytic at a_i, this just moves the expression inside the principal part.

$$\begin{aligned} &- \left[\frac{\mathrm{d}x_1}{x_1} \frac{1}{(y_1 - \sigma(y_1))} \mathrm{d}_2 \cdots \mathrm{d}_n \, \mathrm{Sym}_i \left[\frac{\partial}{\partial s'} F_{g,n}(z_1, \ldots, z_n) \right] \right]_{a_i} \\ &= \left[\frac{x_1}{(y_1 - \sigma(y_1)) \, \mathrm{d}x_1} \left[\omega_{g-1,n+1}(z_1, \sigma(z_1), \mathbf{z}_{S \setminus \{1\}}) \right.\right. \\ &\quad \left.\left. + \sum_{\substack{g_1+g_2=g \\ I \sqcup J = S\setminus\{1\}}}^{\circ} \omega_{g_1,|I|+1}(z_1, \mathbf{z}_I) \, \omega_{g_2,|J|+1}(\sigma(z_1), \mathbf{z}_J) \right] \right]_{a_i} \end{aligned}$$

Now use equation (3.4) to evaluate the principal part and recall the definition of the recursion kernel from equation (1.3).

$$\begin{aligned} &- \left[\frac{\mathrm{d}x_1}{x_1} \frac{1}{(y_1 - \sigma(y_1))} \mathrm{d}_2 \cdots \mathrm{d}_n \, \mathrm{Sym}_i \left[\frac{\partial}{\partial s'} F_{g,n}(z_1, \ldots, z_n) \right] \right]_{a_i} \\ &= \operatorname*{Res}_{z=a} K(z_1, z) \left[\omega_{g-1,n+1}(z, \sigma(z), \mathbf{z}_{S \setminus \{1\}}) \right. \\ &\quad \left. + \sum_{\substack{g_1+g_2=g \\ I \sqcup J = S\setminus\{1\}}}^{\circ} \omega_{g_1,|I|+1}(z, \mathbf{z}_I) \, \omega_{g_2,|J|+1}(\sigma(z), \mathbf{z}_J) \right] \end{aligned}$$

Finally, sum over all branch points.

$$\begin{aligned} &- \sum_{i=1}^{d} \left[\frac{\mathrm{d}x_1}{x_1} \frac{1}{(y_1 - \sigma_i(y_1))} \mathrm{d}_2 \cdots \mathrm{d}_n \, \mathrm{Sym}_i \left[\frac{\partial}{\partial s'} F_{g,n}(z_1, \ldots, z_n) \right] \right]_{a_i} \\ &= \sum_{i=1}^{d} \operatorname*{Res}_{z=a_i} K_i(z_1, z) \left[\omega_{g-1,n+1}(z, \sigma_i(z), \mathbf{z}_{S \setminus \{1\}}) \right. \\ &\quad \left. + \sum_{\substack{g_1+g_2=g \\ I \sqcup J = S\setminus\{1\}}}^{\circ} \omega_{g_1,|I|+1}(z, \mathbf{z}_I) \, \omega_{g_2,|J|+1}(\sigma_i(z), \mathbf{z}_J) \right] \end{aligned}$$

The left side is simply $\Omega_{g,n}(z_1, \ldots, z_n)$ by Lemma 4.10, whose statement and proof we postpone in order to declutter the present argument. On the other hand, the right side is simply $\omega_{g,n}(z_1, \ldots, z_n)$ by the definition of the topological recursion. This completes the inductive argument.

All that is left to check are the base cases $\Omega_{0,3} = \omega_{0,3}$ and $\Omega_{1,1} = \omega_{1,1}$. The former statement appears in Subsection 3.2. For the latter, we can track through the arguments of the current proof using $(g, n) = (1, 1)$ to find that

$$[\Omega_{1,1}(z_1)]_{a_i} = \left[\frac{1}{y_1 - \sigma_i(y_1)} \frac{x_1}{\mathrm{d}x_1} \omega_{0,2}(z_1, \sigma_i(z_1)]_{a_i}\right]_{a_i}$$

Now sum over all branch points to obtain $\Omega_{1,1}$ on the left side and the expression for $\omega_{1,1}$ given by the topological recursion on the right side. $\square$

The proof of Theorem 4.8 required two lemmas, whose statements and proofs we presently discuss.

LEMMA 4.9. *If $\mathcal{F}(z)$ satisfies the linear loop equations and $\Omega = \frac{\mathrm{d}x_1}{x_1}\mathcal{F}(z)$, then*

$$\begin{aligned}\mathrm{d}_2 \operatorname{Sym}_i \left[\frac{z_2}{z_1 - z_2} \frac{\mathcal{F}(z_1)}{1 - s' z_1 P'(z_1)} - \frac{z_1}{z_1 - z_2} \frac{\mathcal{F}(z_2)}{1 - s' z_2 P'(z_2)}\right]_{a_i} \\ = - \operatorname{Sym}_i \left[\frac{x_1^2}{(\mathrm{d}x_1)^2} \omega_{0,2}(z_1, z_2)\, \Omega(\sigma(z_1))\right]_{a_i}.\end{aligned}$$

PROOF. The proof is by direct calculation.

$$\begin{aligned}
&\mathrm{d}_2 \operatorname{Sym}_i \left[\frac{z_2}{z_1 - z_2} \frac{\mathcal{F}(z_1)}{1 - s' z_1 P'(z_1)} - \frac{z_1}{z_1 - z_2} \frac{\mathcal{F}(z_2)}{1 - s' z_2 P'(z_2)}\right]_{a_i} \\
&= \mathrm{d}_2 \operatorname{Sym}_i \left[\frac{z_1}{z_1 - z_2} \frac{\mathcal{F}(z_1)}{1 - s' z_1 P'(z_1)} - \frac{\mathcal{F}(z_1)}{1 - s' z_1 P'(z_1)} \right. \\
&\quad \left. - \frac{z_2}{z_1 - z_2} \frac{\mathcal{F}(z_2)}{1 - s' z_2 P'(z_2)} - \frac{\mathcal{F}(z_2)}{1 - s' z_2 P'(z_2)}\right]_{a_i} \\
&= \mathrm{d}_2 \operatorname{Sym}_i \left[\frac{z_1}{z_1 - z_2} \frac{\mathcal{F}(z_1)}{1 - s' z_1 P'(z_1)}\right]_{a_i} \\
&= - \mathrm{d}_2 \operatorname{Sym}_i \left[\frac{z_1}{z_1 - z_2} \frac{\mathcal{F}(\sigma(z_1))}{1 - s' z_1 P'(z_1)}\right]_{a_i} \\
&= - \operatorname{Sym}_i \left[\frac{\mathrm{d}z_1\, \mathrm{d}z_2}{(z_1 - z_2)^2} \frac{x_1^2}{(\mathrm{d}x_1)^2} \Omega(\sigma(z_1))\right]_{a_i} \\
&= - \operatorname{Sym}_i \left[\omega_{0,2}(z_1, z_2) \frac{x_1^2}{(\mathrm{d}x_1)^2} \Omega(\sigma(z_1))\right]_{a_i}
\end{aligned}$$

The first equality is a straightforward algebraic manipulation. The second equality uses the fact that the second term is annihilated by d_2 as well as the fact that the third and fourth terms have no pole at $z_1 = a_i$. To obtain the third equality, write $\mathcal{G}(z) = \frac{z}{z-z_2}\frac{1}{1-s'zP'(z)}$ and observe that we have the relation $[(\mathcal{F}(z_1) + \mathcal{F}(\sigma(z_1))(\mathcal{G}(z_1) + \mathcal{G}(\sigma(z_1))]_{a_i} = 0$, since both parentheses are analytic at $z_1 = a_i$. The fourth equality evaluates the derivative and uses $\frac{z_1}{1-s'z_1P'(z_1)} = \frac{\mathrm{d}z_1}{\mathrm{d}x_1}$. Finally, the fifth inequality simply substitutes the definition of $\omega_{0,2}$. □

LEMMA 4.10. *Assuming the linear loop equations of Conjecture 4.1, the following is true for $2g-2+n > 0$.*

$$\Omega_{g,n}(z_1,\ldots,z_n) = -\mathrm{d}_2\cdots\mathrm{d}_n\sum_{i=1}^{d}\left[\frac{\mathrm{d}x_1}{x_1}\frac{1}{(y_1-\sigma_i(y_1))}\mathrm{Sym}_i\left[\frac{\partial}{\partial s'}F_{g,n}(z_1,\ldots,z_n)\right]\right]_{a_i}$$

PROOF. Given that $F_{g,n}(z_1,\ldots,z_n) \in V(z_1)\otimes V(z_2)\otimes\cdots\otimes V(z_n)$, we deduce that

$$\begin{aligned}\mathrm{Sym}_i\left[\frac{\partial}{\partial s'}F_{g,n}(z_1,\ldots,z_n)\right]_{a_i} &= \mathrm{Sym}_i\left[-y_1x_1\frac{\partial}{\partial x_1}F_{g,n}(z_1,\ldots,z_n)\right]_{a_i}\\ &= \left[-(y_1-\sigma_i(y_1))x_1\frac{\partial}{\partial x_1}F_{g,n}(z_1,\ldots,z_n)\right]_{a_i},\end{aligned}$$

for all $i = 1, 2, \ldots, d$.

The first equality holds due to the inclusion $\frac{\partial}{\partial s}V(z_1) \subset V(z_1)$ discussed in Remark 4.5 and the second since $x_1\frac{\partial}{\partial x_1}F_{g,n}(z_1,\ldots,z_n) \in V(z_1)\otimes V(z_2)\otimes\cdots\otimes V(z_n)$. Multiply both sides by $-\frac{\mathrm{d}x_1}{x_1}\frac{1}{y_1-\sigma_i(y_1)}$, which is analytic at $z_1 = a_i$, and take the principal part of both sides again to obtain

$$[\mathrm{d}_1F_{g,n}(z_1,\ldots,z_n)]_{a_i} = \left[-\frac{\mathrm{d}x_1}{x_1}\frac{1}{(y_1-\sigma_i(y_1))}\,\mathrm{Sym}_i\left[\frac{\partial}{\partial s'}F_{g,n}(z_1,\ldots,z_n)\right]\right]_{a_i},$$

for all $i = 1, 2, \ldots, d$.

Since a rational meromorphic form is the sum of its principal parts, summing over $i = 1, 2, \ldots, d$ yields

$$\mathrm{d}_1F_{g,n}(z_1,\ldots,z_n) = -\sum_{i=1}^{d}\left[\frac{\mathrm{d}x_1}{x_1}\frac{1}{(y_1-\sigma_i(y_1))}\,\mathrm{Sym}_i\left[\frac{\partial}{\partial s'}F_{g,n}(z_1,\ldots,z_n)\right]\right]_{a_i}.$$

The desired result then follows by applying $\mathrm{d}_2\cdots\mathrm{d}_n$ to both sides. □

Appendix A. Table of double Hurwitz numbers

g	$\boldsymbol{\mu}$	$DH_{g,n}(\mu_1, \dots, \mu_n)$ evaluated at $s = 1$
0	(1)	q_1
0	(2)	$\frac{1}{2}q_2 + \frac{1}{2}q_1^2$
0	(3)	$\frac{1}{3}q_3 + q_2q_1 + \frac{1}{2}q_1^3$
0	(4)	$\frac{1}{4}q_4 + q_3q_1 + \frac{1}{2}q_2^2 + 2q_2q_1^2 + \frac{2}{3}q_1^4$
0	(5)	$\frac{1}{5}q_5 + q_4q_1 + q_3q_2 + \frac{5}{2}q_3q_1^2 + \frac{5}{2}q_2^2q_1 + \frac{25}{6}q_2q_1^3 + \frac{25}{24}q_1^5$
0	(11)	$q_2 + \frac{1}{2}q_1^2$
0	(21)	$q_3 + 2q_2q_1 + \frac{2}{3}q_1^3$
0	(31)	$q_4 + 3q_3q_1 + \frac{3}{2}q_2^2 + \frac{9}{2}q_2q_1^2 + \frac{9}{8}q_1^4$
0	(22)	$q_4 + 3q_3q_1 + q_2^2 + 4q_2q_1^2 + q_1^4$
0	(41)	$q_5 + 4q_4q_1 + 4q_3q_2 + 8q_3q_1^2 + 8q_2^2q_1 + \frac{32}{3}q_2q_1^3 + \frac{32}{15}q_1^5$
0	(32)	$q_5 + 4q_4q_1 + 3q_3q_2 + \frac{15}{2}q_3q_1^2 + 6q_2^2q_1 + 9q_2q_1^3 + \frac{9}{5}q_1^5$
0	(111)	$3q_3 + 4q_2q_1 + q_1^3$
0	(211)	$4q_4 + 9q_3q_1 + 4q_2^2 + 10q_2q_1^2 + 2q_1^4$
0	(311)	$5q_5 + 16q_4q_1 + 15q_3q_2 + \frac{51}{2}q_3q_1^2 + 24q_2^2q_1 + 27q_2q_1^3 + \frac{9}{2}q_1^5$
0	(221)	$5q_5 + 16q_4q_1 + 12q_3q_2 + 24q_3q_1^2 + 20q_2^2q_1 + 24q_2q_1^3 + 4q_1^5$
0	(1111)	$16q_4 + 27q_3q_1 + 12q_2^2 + 24q_2q_1^2 + 4q_1^4$
0	(2111)	$25q_5 + 64q_4q_1 + 54q_3q_2 + 81q_3q_1^2 + 72q_2^2q_1 + 70q_2q_1^3 + 10q_1^5$
1	(2)	$\frac{1}{4}q_2 + \frac{1}{12}q_1^2$
1	(3)	$q_3 + \frac{3}{2}q_2q_1 + \frac{3}{8}q_1^3$
1	(4)	$\frac{5}{2}q_4 + 6q_3q_1 + \frac{7}{3}q_2^2 + \frac{20}{3}q_2q_1^2 + \frac{4}{3}q_1^4$
1	(5)	$5q_5 + \frac{50}{3}q_4q_1 + \frac{25}{2}q_3q_2 + \frac{625}{24}q_3q_1^2 + \frac{125}{6}q_2^2q_1 + \frac{625}{24}q_2q_1^3 + \frac{625}{144}q_1^5$
1	(11)	$\frac{1}{6}q_2 + \frac{1}{24}q_1^2$
1	(21)	$\frac{3}{2}q_3 + \frac{5}{3}q_2q_1 + \frac{1}{3}q_1^3$
1	(31)	$6q_4 + \frac{45}{4}q_3q_1 + \frac{9}{2}q_2^2 + \frac{81}{8}q_2q_1^2 + \frac{27}{16}q_1^4$
1	(22)	$\frac{14}{3}q_4 + 9q_3q_1 + \frac{10}{3}q_2^2 + 8q_2q_1^2 + \frac{4}{3}q_1^4$
1	(41)	$\frac{50}{3}q_5 + \frac{136}{3}q_4q_1 + \frac{104}{3}q_3q_2 + \frac{176}{3}q_3q_1^2 + 48q_2^2q_1 + \frac{448}{9}q_2q_1^3 + \frac{64}{9}q_1^5$
1	(32)	$\frac{25}{2}q_5 + \frac{104}{3}q_4q_1 + 24q_3q_2 + \frac{355}{8}q_3q_1^2 + 34q_2^2q_1 + \frac{147}{4}q_2q_1^3 + \frac{21}{4}q_1^5$
1	(111)	$\frac{9}{4}q_3 + 2q_2q_1 + \frac{1}{3}q_1^3$
1	(211)	$\frac{40}{3}q_4 + \frac{81}{4}q_3q_1 + 8q_2^2 + \frac{91}{6}q_2q_1^2 + \frac{13}{6}q_1^4$
1	(311)	$\frac{625}{12}q_5 + \frac{352}{3}q_4q_1 + \frac{355}{4}q_3q_2 + \frac{511}{4}q_3q_1^2 + 104q_2^2q_1 + 93q_2q_1^3 + \frac{93}{8}q_1^5$
1	(221)	$\frac{125}{3}q_5 + 96q_4q_1 + 68q_3q_2 + 104q_3q_1^2 + \frac{244}{3}q_2^2q_1 + \frac{224}{3}q_2q_1^3 + \frac{28}{3}q_1^5$
2	(2)	$\frac{1}{48}q_2 + \frac{1}{240}q_1^2$
2	(3)	$\frac{3}{4}q_3 + \frac{27}{40}q_2q_1 + \frac{9}{80}q_1^3$
2	(4)	$\frac{41}{6}q_4 + \frac{54}{5}q_3q_1 + \frac{61}{15}q_2^2 + \frac{364}{45}q_2q_1^2 + \frac{52}{45}q_1^4$
2	(11)	$\frac{1}{120}q_2 + \frac{1}{720}q_1^2$
2	(21)	$\frac{27}{40}q_3 + \frac{91}{180}q_2q_1 + \frac{13}{180}q_1^3$
2	(31)	$\frac{54}{5}q_4 + \frac{567}{40}q_3q_1 + \frac{27}{5}q_2^2 + \frac{729}{80}q_2q_1^2 + \frac{729}{640}q_1^4$
2	(22)	$\frac{122}{15}q_4 + \frac{54}{5}q_3q_1 + \frac{182}{45}q_2^2 + \frac{104}{15}q_2q_1^2 + \frac{13}{15}q_1^4$
3	(2)	$\frac{1}{1440}q_2 + \frac{1}{10080}q_1^2$
3	(3)	$\frac{9}{40}q_3 + \frac{81}{560}q_2q_1 + \frac{81}{4480}q_1^3$
3	(4)	$\frac{73}{9}q_4 + \frac{324}{35}q_3q_1 + \frac{1094}{315}q_2^2 + \frac{328}{63}q_2q_1^2 + \frac{328}{567}q_1^4$

Appendix B. Table of pruned double Hurwitz numbers

g	$\boldsymbol{\mu}$	$PH_{g,n}(\mu_1, \ldots, \mu_n)$ evaluated at $s = 1$
0	(11)	$q_2 + \frac{1}{2}q_1^2$
0	(21)	$q_3 + q_2q_1 + \frac{1}{6}q_1^3$
0	(31)	$q_4 + q_3q_1 + \frac{1}{2}q_2^2 + \frac{1}{2}q_2q_1^2 + \frac{1}{24}q_1^4$
0	(22)	$q_4 + q_3q_1 + q_2^2 + q_2q_1^2 + \frac{1}{6}q_1^4$
0	(41)	$q_5 + q_4q_1 + q_3q_2 + \frac{1}{2}q_3q_1^2 + \frac{1}{2}q_2^2q_1 + \frac{1}{6}q_2q_1^3 + \frac{1}{120}q_1^5$
0	(32)	$q_5 + q_4q_1 + 2q_3q_2 + q_3q_1^2 + \frac{3}{2}q_2^2q_1 + \frac{5}{6}q_2q_1^3 + \frac{11}{120}q_1^5$
0	(111)	$3q_3 + 4q_2q_1 + q_1^3$
0	(211)	$4q_4 + 6q_3q_1 + 4q_2^2 + 6q_2q_1^2 + q_1^4$
0	(311)	$5q_5 + 8q_4q_1 + 12q_3q_2 + 9q_3q_1^2 + 12q_2^2q_1 + 8q_2q_1^3 + q_1^5$
0	(221)	$5q_5 + 8q_4q_1 + 12q_3q_2 + 9q_3q_1^2 + 12q_2^2q_1 + 8q_2q_1^3 + q_1^5$
0	(1111)	$16q_4 + 27q_3q_1 + 12q_2^2 + 24q_2q_1^2 + 4q_1^4$
0	(2111)	$25q_5 + 48q_4q_1 + 54q_3q_2 + 54q_3q_1^2 + 60q_2^2q_1 + 46q_2q_1^3 + 6q_1^5$
1	(2)	$\frac{1}{4}q_2 + \frac{1}{12}q_1^2$
1	(3)	$q_3 + q_2q_1 + \frac{5}{24}q_1^3$
1	(4)	$\frac{5}{2}q_4 + 3q_3q_1 + \frac{11}{6}q_2^2 + \frac{5}{2}q_2q_1^2 + \frac{3}{8}q_1^4$
1	(5)	$5q_5 + \frac{20}{3}q_4q_1 + 9q_3q_2 + \frac{51}{8}q_3q_1^2 + 8q_2^2q_1 + 5q_2q_1^3 + \frac{7}{12}q_1^5$
1	(11)	$\frac{1}{6}q_2 + \frac{1}{24}q_1^2$
1	(21)	$\frac{3}{2}q_3 + \frac{3}{2}q_2q_1 + \frac{7}{24}q_1^3$
1	(31)	$6q_4 + \frac{33}{4}q_3q_1 + \frac{13}{3}q_2^2 + \frac{41}{6}q_2q_1^2 + \frac{25}{24}q_1^4$
1	(22)	$\frac{14}{3}q_4 + 6q_3q_1 + \frac{10}{3}q_2^2 + \frac{29}{6}q_2q_1^2 + \frac{17}{24}q_1^4$
1	(41)	$\frac{50}{3}q_5 + \frac{82}{3}q_4q_1 + \frac{63}{2}q_3q_2 + \frac{223}{8}q_3q_1^2 + \frac{94}{3}q_2^2q_1 + \frac{265}{12}q_2q_1^3 + \frac{65}{24}q_1^5$
1	(32)	$\frac{25}{2}q_5 + \frac{58}{3}q_4q_1 + \frac{45}{2}q_3q_2 + \frac{151}{8}q_3q_1^2 + \frac{64}{3}q_2^2q_1 + \frac{173}{12}q_2q_1^3 + \frac{41}{24}q_1^5$
1	(111)	$\frac{9}{4}q_3 + 2q_2q_1 + \frac{1}{3}q_1^3$
1	(211)	$\frac{40}{3}q_4 + 18q_3q_1 + 8q_2^2 + \frac{79}{6}q_2q_1^2 + \frac{11}{6}q_1^4$
1	(311)	$\frac{625}{12}q_5 + \frac{272}{3}q_4q_1 + \frac{173}{2}q_3q_2 + \frac{707}{8}q_3q_1^2 + 86q_2^2q_1 + \frac{190}{3}q_2q_1^3 + \frac{179}{24}q_1^5$
1	(221)	$\frac{125}{3}q_5 + \frac{208}{3}q_4q_1 + 68q_3q_2 + \frac{263}{4}q_3q_1^2 + \frac{196}{3}q_2^2q_1 + \frac{139}{3}q_2q_1^3 + \frac{16}{3}q_1^5$
2	(2)	$\frac{1}{48}q_2 + \frac{1}{240}q_1^2$
2	(3)	$\frac{3}{4}q_3 + \frac{19}{30}q_2q_1 + \frac{5}{48}q_1^3$
2	(4)	$\frac{41}{6}q_4 + \frac{171}{20}q_3q_1 + \frac{161}{40}q_2^2 + \frac{439}{72}q_2q_1^2 + \frac{119}{144}q_1^4$
2	(11)	$\frac{1}{120}q_2 + \frac{1}{720}q_1^2$
2	(21)	$\frac{27}{40}q_3 + \frac{179}{360}q_2q_1 + \frac{17}{240}q_1^3$
2	(31)	$\frac{54}{5}q_4 + \frac{513}{40}q_3q_1 + \frac{647}{120}q_2^2 + \frac{389}{48}q_2q_1^2 + \frac{637}{640}q_1^4$
2	(22)	$\frac{122}{15}q_4 + \frac{189}{20}q_3q_1 + \frac{182}{45}q_2^2 + \frac{427}{72}q_2q_1^2 + \frac{521}{720}q_1^4$
3	(2)	$\frac{1}{1440}q_2 + \frac{1}{10080}q_1^2$
3	(3)	$\frac{9}{40}q_3 + \frac{361}{2520}q_2q_1 + \frac{103}{5760}q_1^3$
3	(4)	$\frac{73}{9}q_4 + \frac{2403}{280}q_3q_1 + \frac{17497}{5040}q_2^2 + \frac{3437}{720}q_2q_1^2 + \frac{27187}{51840}q_1^4$

References

[1] A. Alexandrov, D. Lewanski, and S. Shadrin, *Ramifications of Hurwitz theory, KP integrability and quantum curves*, J. High Energy Phys. **5** (2016), 124, front matter+30. MR3521843

[2] Gaëtan Borot and Bertrand Eynard, *All order asymptotics of hyperbolic knot invariants from non-perturbative topological recursion of A-polynomials*, Quantum Topol. **6** (2015), no. 1, 39–138, DOI 10.4171/QT/60. MR3335006

[3] Gaëtan Borot, Bertrand Eynard, and Nicolas Orantin, *Abstract loop equations, topological recursion and new applications*, Commun. Number Theory Phys. **9** (2015), no. 1, 51–187, DOI 10.4310/CNTP.2015.v9.n1.a2. MR3339853

[4] Vincent Bouchard and Bertrand Eynard, *Think globally, compute locally*, J. High Energy Phys. **2** (2013), 143, front matter + 34. MR3046532

[5] Vincent Bouchard, Daniel Hernández Serrano, Xiaojun Liu, and Motohico Mulase, *Mirror symmetry for orbifold Hurwitz numbers*, J. Differential Geom. **98** (2014), no. 3, 375–423. MR3263522

[6] Vincent Bouchard, Albrecht Klemm, Marcos Mariño, and Sara Pasquetti, *Remodeling the B-model*, Comm. Math. Phys. **287** (2009), no. 1, 117–178, DOI 10.1007/s00220-008-0620-4. MR2480744

[7] Vincent Bouchard and Marcos Mariño, *Hurwitz numbers, matrix models and enumerative geometry*, From Hodge theory to integrability and TQFT tt*-geometry, Proc. Sympos. Pure Math., vol. 78, Amer. Math. Soc., Providence, RI, 2008, pp. 263–283, DOI 10.1090/pspum/078/2483754. MR2483754

[8] Leonid Chekhov and Bertrand Eynard, *Hermitian matrix model free energy: Feynman graph technique for all genera*, J. High Energy Phys. **3** (2006), 014, 18, DOI 10.1088/1126-6708/2006/03/014. MR2222762

[9] Robbert Dijkgraaf, Hiroyuki Fuji, and Masahide Manabe, *The volume conjecture, perturbative knot invariants, and recursion relations for topological strings*, Nuclear Phys. B **849** (2011), no. 1, 166–211, DOI 10.1016/j.nuclphysb.2011.03.014. MR2795276

[10] N. Do and M. Karev, *Monotone orbifold Hurwitz numbers*, Zap. Nauchn. Sem. S.-Peterburg. Otdel. Mat. Inst. Steklov. (POMI) **446** (2016), no. Kombinatorika i Teoriya Grafov. V, 40–69; English transl., J. Math. Sci. (N.Y.) **226** (2017), no. 5, 568–587. MR3520422

[11] Norman Do, Alastair Dyer, and Daniel V. Mathews, *Topological recursion and a quantum curve for monotone Hurwitz numbers*, J. Geom. Phys. **120** (2017), 19–36, DOI 10.1016/j.geomphys.2017.05.014. MR3712146

[12] Norman Do, Oliver Leigh, and Paul Norbury, *Orbifold Hurwitz numbers and Eynard-Orantin invariants*, Math. Res. Lett. **23** (2016), no. 5, 1281–1327, DOI 10.4310/MRL.2016.v23.n5.a3. MR3601067

[13] Norman Do and David Manescu, *Quantum curves for the enumeration of ribbon graphs and hypermaps*, Commun. Number Theory Phys. **8** (2014), no. 4, 677–701, DOI 10.4310/CNTP.2014.v8.n4.a2. MR3318387

[14] Norman Do and Paul Norbury, *Pruned Hurwitz numbers*, Trans. Amer. Math. Soc. **370** (2018), no. 5, 3053–3084, DOI 10.1090/tran/7021. MR3766841

[15] Olivia Dumitrescu, Motohico Mulase, Brad Safnuk, and Adam Sorkin, *The spectral curve of the Eynard-Orantin recursion via the Laplace transform*, Algebraic and geometric aspects of integrable systems and random matrices, Contemp. Math., vol. 593, Amer. Math. Soc., Providence, RI, 2013, pp. 263–315, DOI 10.1090/conm/593/11867. MR3087960

[16] P. Dunin-Barkowski, M. Kazarian, N. Orantin, S. Shadrin, and L. Spitz, *Polynomiality of Hurwitz numbers, Bouchard-Mariño conjecture, and a new proof of the ELSV formula*, Adv. Math. **279** (2015), 67–103, DOI 10.1016/j.aim.2015.03.016. MR3345179

[17] P. Dunin-Barkowski, N. Orantin, S. Shadrin, and L. Spitz, *Identification of the Givental formula with the spectral curve topological recursion procedure*, Comm. Math. Phys. **328** (2014), no. 2, 669–700, DOI 10.1007/s00220-014-1887-2. MR3199996

[18] Petr Dunin-Barkowski, Nicolas Orantin, Aleksandr Popolitov, and Sergey Shadrin, *Combinatorics of loop equations for branched covers of sphere*, International Mathematics Research Notices (2017), rnx047.

[19] Torsten Ekedahl, Sergei Lando, Michael Shapiro, and Alek Vainshtein, *Hurwitz numbers and intersections on moduli spaces of curves*, Invent. Math. **146** (2001), no. 2, 297–327, DOI 10.1007/s002220100164. MR1864018

[20] B. Eynard, *Invariants of spectral curves and intersection theory of moduli spaces of complex curves*, Commun. Number Theory Phys. **8** (2014), no. 3, 541–588, DOI 10.4310/CNTP.2014.v8.n3.a4. MR3282995

[21] B. Eynard and N. Orantin, *Invariants of algebraic curves and topological expansion*, Commun. Number Theory Phys. **1** (2007), no. 2, 347–452, DOI 10.4310/CNTP.2007.v1.n2.a4. MR2346575

[22] B. Eynard and N. Orantin, *Computation of open Gromov-Witten invariants for toric Calabi-Yau 3-folds by topological recursion, a proof of the BKMP conjecture*, Comm. Math. Phys. **337** (2015), no. 2, 483–567, DOI 10.1007/s00220-015-2361-5. MR3339157

[23] Bertrand Eynard, Motohico Mulase, and Bradley Safnuk, *The Laplace transform of the cut-and-join equation and the Bouchard-Mariño conjecture on Hurwitz numbers*, Publ. Res. Inst. Math. Sci. **47** (2011), no. 2, 629–670, DOI 10.2977/PRIMS/47. MR2849645

[24] Bohan Fang, Chiu-Chu Melissa Liu, and Zhengyu Zong, *All genus mirror symmetry for toric Calabi-Yau 3-orbifolds*, String-Math 2014, Proc. Sympos. Pure Math., vol. 93, Amer. Math. Soc., Providence, RI, 2016, pp. 1–19. MR3524231

[25] I. P. Goulden and D. M. Jackson, *Transitive factorisations into transpositions and holomorphic mappings on the sphere*, Proc. Amer. Math. Soc. **125** (1997), no. 1, 51–60, DOI 10.1090/S0002-9939-97-03880-X. MR1396978

[26] I. P. Goulden, D. M. Jackson, and A. Vainshtein, *The number of ramified coverings of the sphere by the torus and surfaces of higher genera*, Ann. Comb. **4** (2000), no. 1, 27–46, DOI 10.1007/PL00001274. MR1763948

[27] I. P. Goulden, D. M. Jackson, and R. Vakil, *Towards the geometry of double Hurwitz numbers*, Adv. Math. **198** (2005), no. 1, 43–92, DOI 10.1016/j.aim.2005.01.008. MR2183250

[28] Marvin Anas Hahn, *Pruned double Hurwitz numbers*, Electron. J. Combin. **24** (2017), no. 3, Paper 3.66, 32. MR3711108

[29] A. Hurwitz, *Ueber Riemann'sche Flächen mit gegebenen Verzweigungspunkten* (German), Math. Ann. **39** (1891), no. 1, 1–60, DOI 10.1007/BF01199469. MR1510692

[30] Xiaojun Liu, Motohico Mulase, and Adam Sorkin, *Quantum curves for simple Hurwitz numbers of an arbitrary base curve*, (2013).

[31] Paul Norbury, *Quantum curves and topological recursion*, String-Math 2014, Proc. Sympos. Pure Math., vol. 93, Amer. Math. Soc., Providence, RI, 2016, pp. 41–65. MR3524233

[32] Paul Norbury and Nick Scott, *Gromov-Witten invariants of $\mathbb{P}^1$ and Eynard-Orantin invariants*, Geom. Topol. **18** (2014), no. 4, 1865–1910, DOI 10.2140/gt.2014.18.1865. MR3268770

[33] A. Okounkov and R. Pandharipande, *Gromov-Witten theory, Hurwitz numbers, and matrix models*, Algebraic geometry—Seattle 2005. Part 1, Proc. Sympos. Pure Math., vol. 80, Amer. Math. Soc., Providence, RI, 2009, pp. 325–414, DOI 10.1090/pspum/080.1/2483941. MR2483941

[34] Andrei Okounkov, *Toda equations for Hurwitz numbers*, Math. Res. Lett. **7** (2000), no. 4, 447–453, DOI 10.4310/MRL.2000.v7.n4.a10. MR1783622

[35] Shengmao Zhu, *On the recursion formula for double Hurwitz numbers*, Proc. Amer. Math. Soc. **140** (2012), no. 11, 3749–3760, DOI 10.1090/S0002-9939-2012-11235-3. MR2944715

School of Mathematical Sciences, Monash University, Melbourne, VIC 3800, Australia

Email address: `norm.do@monash.edu`

St. Petersburg Department of the Steklov Mathematical Institute, Fontanka 27, St. Petersburg 191023, Russia

Email address: `max.karev@gmail.com`

Proceedings of Symposia in Pure Mathematics
Volume **100**, 2018
https://doi.org/10.1090/pspum/100/01780

Quantization of spectral curves for meromorphic Higgs bundles through topological recursion

Olivia Dumitrescu and Motohico Mulase

ABSTRACT. A geometric quantization using a partial differential equation version of topological recursion is established for the compactified cotangent bundle of a smooth projective curve of an arbitrary genus. In this quantization, the Hitchin spectral curve of a rank 2 meromorphic Higgs bundle on the base curve corresponds to a quantum curve, which is a Rees D-module on the base. The topological recursion then gives an all-order asymptotic expansion of its solution, thus determining a state vector corresponding to the spectral curve as a meromorphic Lagrangian. We propose a generalization of the topological recursion for a singular spectral curve. We show that the PDE version of the topological recursion automatically selects the normal ordering of the canonical coordinates, and determines the unique quantization of the spectral curve. The quantum curve thus constructed has the semi-classical limit that agrees with the original spectral curve. Typical examples of our construction includes classical differential equations, such as Airy, Hermite, and Gauß hypergeometric equations. The topological recursion gives an asymptotic expansion of solutions to these equations at their singular points, relating Higgs bundles and various quantum invariants.

Contents

2010 *Mathematics Subject Classification.* Primary 14H15, 14N35, 81T45; Secondary 14F10, 14J26, 33C05, 33C10, 33C15, 34M60, 53D37.

Key words and phrases. Topological recursion, quantum curve, Hitchin spectral curve, Higgs field, Rees D-module, mirror symmetry, Airy function, hypergeometric functions, quantum invariants, WKB approximation.

The research of the first author has been supported by GRK 1463 *Analysis, Geometry, and String Theory* at the Leibniz Universität Hannover, and and a grant from MPIM, Bonn.

The research of the second author has been supported by MPIM in Bonn, and NSF grants DMS-1104734, DMS-1309298, DMS-1619760, DMS-1642515, and NSF-RNMS: Geometric Structures And Representation Varieties (GEAR Network, DMS-1107452, 1107263, 1107367).

1. Introduction

1.1. Overview. The integral transformation formula known as **topological recursion** of [**25**] was originally conceived as a computational mechanism to find the multi-resolvent correlation functions of random matrices [**11**, **22**]. It has been proposed that the topological recursion is an effective tool for defining a genus g B-model topological string theory on a holomorphic curve (known as an **Eynard-Orantin spectral curve**), that should be the mirror symmetric dual to the genus g Gromov-Witten theory on the A-model side [**9**, **10**, **45**]. This correspondence has been rigorously established for several examples, most notably for an arbitrary toric Calabi-Yau orbifold of 3 dimensions [**27**], and many other enumerative geometry problems [**8**, **16**, **20**, **24**, **26**, **49**].

Quantum curves are introduced in the physics literature (see for example, [**1**, **13**, **14**, **31**, **36**]) as a device to compactly encode the information of quantum invariants arising in Gromov-Witten theory, Seiberg-Witten theory, and knot theory. The *semi-classical limit* of a quantum curve is a holomorphic curve defining a B-model that is mirror dual to the A-model for these quantum invariants. Geometrically, a quantum curve also appears as an $\hbar$-deformation of a generalized Gauß-Manin connection (or Picard-Fuchs differential equation) on a curve, with regular and *irregular* singularities.

Since both quantum curves and the topological recursion produce B-models on a holomorphic curve, it is natural to ask if they are related. Indeed, it was proposed by physicists [**12**, **31**] for the context of knot theory that the topological recursion would give a perturbative construction of quantum curves. So far such a relation is not fully understood in the mathematical examples of quantum curves constructed in [**8**, **21**, **47**, **48**].

The purpose of this paper is to establish a clear geometric relation between quantum curves and a *partial differential equation version* of topological recursion for the **Hitchin spectral curves** associated with Higgs bundles on a base curve C, with arbitrary *meromorphic* Higgs fields. Although the language of **geometric quantization** (as defined, for example, in [**42**]) does not work in this algebraic geometry context, let us use it for a moment as an analogy. Then the main result of this paper could be understood as follows: *the PDE topological recursion is a geometric quantization of* T^*C. A Hitchin spectral curve is a (*meromorphic*) Lagrangian in the holomorphic symplectic manifold T^*C. Using the topological recursion, we construct a *state vector*, which is a solution to the Schrödinger equation on C that is determined by the spectral curve. The state vector is equivalent to a quantum curve in our setting, as a Rees D-module on C.

Due to the impossibility of performing the residue calculations appearing in the integral topological recursion of [**18**, **25**] for spectral curves of positive genera, our construction of quantization in this paper is restricted to be local on an open subset of C. More precisely, we prove the following.

THEOREM 1.1 (Main results). *Let C be a smooth projective curve of an arbitrary genus, and (E,ϕ) a Higgs bundle of rank 2 on C with a meromorphic Higgs field ϕ. Denote by*

$$\overline{T^*C} := \mathbb{P}(K_C \oplus \mathcal{O}_C) \xrightarrow{\pi} C \tag{1.1}$$

*the compactified cotangent bundle of C (see [**40**]), which is a ruled surface on the base C. Here, K_C is the canonical sheaf. The Hitchin spectral curve*

$$\begin{array}{ccc} \Sigma & \xrightarrow{\ i\ } & \overline{T^*C} \\ & \searrow^{\pi} & \downarrow{\scriptstyle \pi} \\ & & C \end{array} \tag{1.2}$$

for a meromorphic Higgs bundle is defined as the divisor of zeros on $\overline{T^*C}$ of the characteristic polynomial of ϕ:

$$\Sigma = (\det(\eta - \pi^*\phi))_0\,, \tag{1.3}$$

where $\eta \in H^0(T^*C, \pi^*K_C)$ is the tautological 1-form on T^*C extended as a meromorphic 1-form on the compactification $\overline{T^*C}$.

- *The integral transformation formula of [**18, 25**], which we call **integral topological recursion** in this paper, is extended to the singular spectral curve Σ, as (6.12). For this purpose, we blow up $\overline{T^*C}$ several times as in (1.6) to construct the normalization $\widetilde{\Sigma}$. The construction of $Bl(\overline{T^*C})$ is given in Definition 4.7. It is the minimal resolution of the support $\Sigma \cup C_\infty$ of the* total *divisor*

$$\Sigma - 2C_\infty = (\det(\eta - \pi^*\phi))_0 - (\det(\eta - \pi^*\phi))_\infty \tag{1.4}$$

of the characteristic polynomial, where

$$C_\infty := \mathbb{P}(K_C \oplus \{0\}) = \overline{T^*C} \setminus T^*C \tag{1.5}$$

is the divisor at infinity. Therefore, in $Bl(\overline{T^*C})$, the proper transform $\widetilde{\Sigma}$ of Σ is smooth and does not intersect with the proper transform of C_∞.

$$\begin{array}{ccccc} \widetilde{\Sigma} & & \xrightarrow{\quad \tilde{i} \quad} & & Bl(\overline{T^*C}) \\ & \searrow^{\nu} & & & \searrow^{\nu} \\ {\scriptstyle \tilde{\pi}}\downarrow & & \Sigma & \xrightarrow{\quad i \quad} & \overline{T^*C} \\ & \swarrow_{\pi} & & \swarrow_{\pi} & \\ C & & & & \end{array} \tag{1.6}$$

- *The genus of the normalization $\widetilde{\Sigma}$ is given by*

$$g(\widetilde{\Sigma}) = 2g(C) - 1 + \frac{1}{2}\delta,$$

where δ is the sum of the number of cusp singularities of Σ and the ramification points of $\pi : \Sigma \longrightarrow C$ (Theorem 4.2).

- *The integral topological recursion thus generalized requires a globally defined meromorphic 1-form $W_{0,1}$ on $\widetilde{\Sigma}$ and a symmetric meromorphic 2-form $W_{0,2}$ on the product $\widetilde{\Sigma} \times \widetilde{\Sigma}$ as the initial data. We choose*

$$\begin{cases} W_{0,1} = \tilde{i}^*\nu^*\eta \\ W_{0,2} = d_1 d_2 \log E_{\widetilde{\Sigma}}, \end{cases} \tag{1.7}$$

*where $E_{\widetilde{\Sigma}}$ is a normalized Riemann prime form on $\widetilde{\Sigma}$ (see [**18**, Section 2]). The form $W_{0,2}$ depends only on the intrinsic geometry of the smooth* *curve* $\widetilde{\Sigma}$. *The geometry of* (1.6) *is encoded in* $W_{0,1}$. *The integral topological recursion produces a symmetric meromorphic* n*-linear differential form* $W_{g,n}(z_1,\dots,z_n)$ *on* $\widetilde{\Sigma}^n$ *for every* (g,n) *subject to* $2g-2+n>0$ *from the initial data* (1.7).

- *Evaluation of residues in the integral topological recursion* (6.12) *is impossible, due to the fact that the integrand of the right-hand side of* (6.12) *is not a globally defined meromorphic* 1*-form on* $\widetilde{\Sigma}$ *in the* z*-component. Therefore, as in* [**18**, (4.7)], *we propose a* ***partial differential equation version*** *of topological recursion* (6.6), *using the same geometric data as above and* (1.7). *It determines the free energy* $F_{g,n}(z_1,\dots,z_n)$, *a symmetric meromorphic function on* $\mathcal{U}^n$ *for* $2g-2+n>0$, *up to a constant. Here,* $\varpi:\mathcal{U}\longrightarrow\widetilde{\Sigma}$ *is the universal covering of* $\widetilde{\Sigma}$.
- *The quantum curve associated with the Hitchin spectral curve* Σ *is defined as a Rees D-module (Definition* 3.1*) on* C. *On a coordinate neighborhood* $U\subset C$ *with coordinate* x, *a generator of the quantum curve is given by*

$$P(x,\hbar)=\left(\hbar\frac{d}{dx}\right)^2-\operatorname{tr}\phi(x)\left(\hbar\frac{d}{dx}\right)+\det\phi(x).$$

In particular, the ***semi-classical limit*** *of the quantum curve recovers the singular spectral curve* $\Sigma|_U$, *not its normalization* $\widetilde{\Sigma}|_U$.

- *We construct the* ***all-order WKB expansion***

$$\Psi(x,\hbar)=\exp\left(\sum_{m=0}^{\infty}\hbar^{m-1}S_m(x)\right) \tag{1.8}$$

of a solution to the Schrödinger equation

$$\left(\left(\hbar\frac{d}{dx}\right)^2-\operatorname{tr}\phi(x)\left(\hbar\frac{d}{dx}\right)+\det\phi(x)\right)\right)\Psi(x,\hbar)=0, \tag{1.9}$$

on an open neighborhood $U\subset C$ *near a critical value of* $\pi:\Sigma\longrightarrow C$, *in terms of the free energies. The formal expression* (1.8) *is a local section of* $K_C^{-\frac{1}{2}}$ *in* x. *We find that* (1.9) *is equivalent to the* ***principal specialization*** *of the differential recusion* (6.6). *The equivalence is given by*

$$S_m(x)=\sum_{2g-2+n=m-1}\frac{1}{n!}F_{g,n}\big(z(x)\big), \tag{1.10}$$

where $F_{g,n}\big(z(x)\big)$ *is the principal specialization of* $F_{g,n}(z_1,\dots,z_n)$ *evaluated at a local section* $z=z(x)$ *of* $\tilde{\pi}:\widetilde{\Sigma}\longrightarrow C$ *on* U.

- *The canonical ordering of the quantization of the local functions on* T^*C *is automatically chosen in the process of the integration from* (6.12) *to* (6.6) *and the principal specialization* (1.10). *This selects the canonical ordering in* (1.9).

REMARK 1.2. Although $\overline{T^*C}$ is not a holomorphic symplectic manifold, in the analogy of geometric quantization mentioned above, our quantization is similar to a holomorphic quantization of T^*C, where the fiber coordinate is quantized to $\hbar\frac{d}{dx}$. A Hitchin spectral curve is a *meromorphic* Lagrangian, and corresponds via the PDE topological recursion to a state vector $\Psi(x,\hbar)$ of (1.8).

REMARK 1.3. The constant ambiguity in the symmetric function $F_{g,n}$ is reflected in a factor $\exp(\sum \hbar^{m-1} c_m)$ multiplied to $\Psi(x, \hbar)$ of (1.8), where c_m is an arbitrary constant. Therefore, our method does not determine the $\hbar$ dependence of the solution to (1.9).

REMARK 1.4. The current paper is a generalization of [**18**]. In the process of establishing a geometric theory of topological recursion and quantum curves, we have discovered in [**18**] that the topological recursion of [**25**] can be naturally generalized to the Hitchin spectral curves for holomorphic Higgs bundles defined on a smooth projective curve C of genus $g(C) \geq 2$. We have then showed that the Hitchin spectral curve for an $SL(2, \mathbb{C})$-Higgs bundle is locally quantizable, and that the PDE version of topological recursion gives an asymptotic expansion of a holomorphic solution to the quantum curve (1.9) with $\operatorname{tr} \phi = 0$.

REMARK 1.5. The singularities of the quantum curve, which are regular and irregular singular points of a differential equation (1.9) on the base curve C, are analyzed by the geometry of the Hitchin spectral curve Σ (Theorem 6.11). For example, the number of resolutions required to desingularize $\Sigma \cup C_\infty$ at $P \in \Sigma$ is always $\lceil r \rceil$ if $\pi(P)$ is an irregular singular point of class $r - 1$.

REMARK 1.6. Already several mathematical examples of quantum curves have been rigorously constructed for enumerative geometry problems, such as Catalan numbers and their generalizations, simple and double Hurwitz numbers and their variants, and Gromov-Witten invariants of a point, the projective line, and a few toric Calabi-Yau threefolds [**8, 16, 21, 47, 48, 55, 56**]. In knot theory, a quantum curve is a q-holonomic operator $\widehat{A}$ that quantizes the **A-polynomial** of a knot and characterizes the corresponding colored Jones polynomial [**28, 29**]. The geometric examples we consider in this paper are rather different from the knot theory case, since the A-polynomials are almost never irreducible, while we can consider only irreducible spectral curves.

REMARK 1.7. Another aspect of quantum curves lies in its relation to **non-Abelian Hodge correspondence**. A quantum curve is an $\hbar$-connection on the base curve C, and the Higgs field is recovered as its *classical limit* $\hbar \to 0$. The non-Abelian Hodge correspondence with irregular singular points has been studied extensively both in mathematics and physics, starting from the fundamental papers [**4, 5**] and to more recent ones, including [**6, 52, 53**].

REMARK 1.8. A global theory of quantum curves of an arbitrary order is developed in our recent paper [**19**] in terms of a totally different method, using the notion of *opers*.

Our current paper is motivated by the following simple question: *If quantum curves are truly fundamental objects, then where do we see them most commonly, in particular, in classical mathematics?* The answer we propose in this paper is that the classical differential equations, such as the Airy, Hermite, and Gauß hypergeometric differential equations, are natural examples of our construction of the quantum curves that are associated with stable meromorphic Higgs bundles defined over the projective line $\mathbb{P}^1$. The topological recursion then gives an all-order asymptotic expansion of their solutions, connecting Higgs bundles to the world of quantum invariants.

Once we study these concrete classical examples, it becomes plausible that the base curve C of the Higgs bundle and a spectral curve $\Sigma \subset \overline{T^*C}$ are *moduli spaces* of certain geometries. For example, in a particular case of the Gauß hypergeometric equations considered in Sections 1.2 and 7.2, the base curve is actually $\overline{\mathcal{M}}_{0,4} \cong \mathbb{P}^1$. The spectral curve for this example is the moduli space of elliptic curves, together with the two eigenvalues of the classical limit of the **Gauß-Manin connection** [**44**] that characterizes the periods of elliptic curves.

More precisely, for every $x \in \mathcal{M}_{0,4}$, we consider the elliptic curve $E(x)$ ramified over $\mathbb{P}^1$ at four points $\{0, 1, x, \infty\}$, and its two periods given by the elliptic integrals [**41**]

$$\omega_1(x) = \int_1^\infty \frac{ds}{\sqrt{s(s-1)(s-x)}}, \qquad \omega_2(x) = \int_x^1 \frac{ds}{\sqrt{s(s-1)(s-x)}}. \tag{1.11}$$

The quantum curve in this case is an $\hbar$ **-deformed Gauß-Manin connection**

$$\nabla^\hbar_{GM} = \hbar d - \begin{bmatrix} & \frac{1}{x} \\ -\frac{1}{4(x-1)} & -\frac{2x-1}{x(x-1)} + \frac{\hbar}{x} \end{bmatrix} dx \tag{1.12}$$

in the trivial bundle $\mathcal{O}_{\overline{\mathcal{M}}_{0,4}} \oplus \mathcal{O}_{\overline{\mathcal{M}}_{0,4}}$ of rank 2 over $\overline{\mathcal{M}}_{0,4}$. Here, d denotes the exterior differentiation acting on the local sections of this trivial bundle. The restriction ∇^1_{GM} of the connection at $\hbar = 1$ is equivalent to the Gauß-Manin connection that characterizes the two periods of (1.11), and the Higgs field is the classical limit of the connection matrix at $\hbar \to 0$:

$$\phi = \begin{bmatrix} & \frac{1}{x} \\ -\frac{1}{4(x-1)} & -\frac{2x-1}{x(x-1)} \end{bmatrix} dx. \tag{1.13}$$

The spectral curve $\Sigma \subset \overline{T^*\overline{\mathcal{M}}_{0,4}}$ as a moduli space consists of the data $\big(E(x), \alpha_1(x), \alpha_2(x)\big)$, where $\alpha_1(x)$ and $\alpha_2(x)$ are the two eigenvalues of the Higgs field ϕ. The spectral curve $\Sigma \subset \overline{T^*\overline{\mathcal{M}}_{0,4}} = \mathbb{F}^2$ as a divisor in the Hirzebruch surface is determined by the characteristic equation

$$y^2 + \frac{2x-1}{x(x-1)} y + \frac{1}{4x(x-1)} = 0 \tag{1.14}$$

of the Higgs field. Geometrically, Σ is a singular rational curve with one ordinary double point at $x = \infty$. As we see in the later sections, the quantum curve is a **quantization** of the characteristic equation (1.14) for the eigenvalues $\alpha_1(x)$ and $\alpha_2(x)$ of $\phi(x)$. It is an $\hbar$-deformed **Picard-Fuchs equation**

$$\left(\left(\hbar \frac{d}{dx} \right)^2 + \frac{2x-1}{x(x-1)} \left(\hbar \frac{d}{dx} \right) + \frac{1}{4x(x-1)} \right) \omega_i(x, \hbar) = 0,$$

and its semi-classical limit agrees with the singular spectral curve Σ. As a second order differential equation, the quantum curve has two independent solutions corresponding to the two eigenvalues. At $\hbar = 1$, these solutions are exactly the two periods $\omega_1(x)$ and $\omega_2(x)$ of the Legendre family of elliptic curves $E(x)$. The topological recursion produces asymptotic expansions of these periods as functions in $x \in \overline{\mathcal{M}}_{0,4}$, at which the elliptic curve $E(x)$ degenerates to a nodal rational curve.

REMARK 1.9. Although we do not deal with quantum curves associated with knots (cf. [**31**]) in our current paper, there a spectral curve is the $SL(2,\mathbb{C})$-character variety of the fundamental group of a knot complement in the 3-sphere S^3. Thus the spectral curve is again a moduli space, this time the moduli of flat $SL(2,\mathbb{C})$-connections on the knot complement.

When we deal with a singular spectral curve $\Sigma \subset \overline{T^*C}$, the key question is how to relate the singular curve with smooth ones. In terms of the Hitchin fibration, a singular spectral curve corresponds to a degenerate Abelian variety in the family. There are two different approaches to this question:

(1) Deform Σ locally in the base of the Hitchin fibration to a family of non-singular curves, and study the quantization associated with this deformation family.
(2) Blow up $\overline{T^*C}$ and obtain a resolution of singularities $\widetilde{\Sigma}$ of the singular spectra curve Σ. Then construct the quantum curve for Σ using the geometry of $\widetilde{\Sigma}$.

In this paper we will pursue the second path, and give a construction of a quantum curve using the geometric information of the blow-up (1.6).

In the Higgs bundle context, a quantum curve is a Rees D-module over the Rees ring $\widetilde{\mathcal{D}_C}$ defined by the canonical filtration of $\mathcal{D}_C$ (see for example, [**30**]), such that its semi-classical limit coincides with the Hitchin spectral curve of a meromorphic Higgs bundle on C. Here, $\mathcal{D}_C$ denotes the sheaf of linear ordinary differential operators on C. A $\widetilde{\mathcal{D}_C}$-module is a particular $\hbar$-deformation family of $\mathcal{D}_C$-modules. Suppose a Rees $\widetilde{\mathcal{D}_C}$-module is written locally as

$$\mathcal{M}(U) = \widetilde{\mathcal{D}_C}(U)/\widetilde{\mathcal{D}_C}(U)\cdot P(x,\hbar)$$

on an open disc $U \subset C$ with a local coordinate x, where $P(x,\hbar) \in \widetilde{\mathcal{D}_C}(U)$ is a linear ordinary differential operator depending on the deformation parameter $\hbar$. This operator then characterizes, by an equation

$$P(x,\hbar)\Psi(x,\hbar) = 0, \tag{1.15}$$

the *partition function* $\Psi(x,\hbar)$ of a topological quantum field theory on a 'space' that is considered to be the *mirror dual* to the spectral curve. The physics theories appearing in this way are related to quantum topological invariants and geometric enumeration problems. The variable x of the base curve C is usually the parameter of generating functions of the quantum invariants that are considered in the theory, and the generating functions determine a particular asymptotic expansion of an analytic solution $\Psi(x,\hbar)$ of (1.15) around its singularity.

1.2. Classical examples. Riemann and Poincaré worked on the interplay between algebraic geometry of curves in a ruled surface and the asymptotic expansion of an analytic solution to a differential equation defined on the base curve of the ruled surface. The theme of the current paper lies exactly on this link, looking at the classical subject from a new point of view.

Let us recall the definition of regular and irregular singular points of a second order differential equation.

DEFINITION 1.10. Let

$$\left(\frac{d^2}{dx^2}+a_1(x)\frac{d}{dx}+a_2(x)\right)\Psi(x)=0 \tag{1.16}$$

be a second order differential equation defined around a neighborhood of $x=0$ on a small disc $|x|<\epsilon$ with meromorphic coefficients $a_1(x)$ and $a_2(x)$ with poles at $x=0$. Denote by k (reps. ℓ) the order of the pole of $a_1(x)$ (resp. $a_2(x)$) at $x=0$. If $k\le 1$ and $\ell\le 2$, then (1.16) has a **regular singular point** at $x=0$. Otherwise, consider the *Newton polygon* of the order of poles of the coefficients of (1.16). It is the upper part of the convex hull of three points $(0,0),(1,k),(2,\ell)$. As a convention, if $a_j(x)$ is identically 0, then we assign $-\infty$ as its pole order. Let $(1,r)$ be the intersection point of the Newton polygon and the vertical line through $(1,0)$. Thus

$$r=\begin{cases} k & 2k\ge\ell,\\ \frac{\ell}{2} & 2k\le\ell.\end{cases} \tag{1.17}$$

The differential equation (1.16) has an **irregular singular point of class** $r-1$ at $x=0$ if $r>1$.

To illustrate the scope of interrelations among the geometry of meromorphic Higgs bundles, their spectral curves, the singularities of quantum curves, $\hbar$-connections, and the quantum invariants, let us tabulate five examples here (see Table 1.2). The differential operators of these equations are listed on the third column. In the first three rows, the quantum curves are examples of classical differential equations known as Airy, Hermite, the Gauß hypergeometric equations. The fourth and the fifth rows are added to show that it is *not* the singularity of the spectral curve that determines the singularity of the quantum curve. In each example, the Higgs bundle (E,ϕ) we are considering consists of the base curve $C=\mathbb{P}^1$ and a vector bundle $E=\mathcal{O}_{\mathbb{P}^1}(1)\oplus\mathcal{O}_{\mathbb{P}^1}(-1)=K_{\mathbb{P}^1}^{-\frac{1}{2}}\oplus K_{\mathbb{P}^1}^{\frac{1}{2}}$ of rank 2 on $\mathbb{P}^1$.

The first column of the table shows the Higgs field $\phi: E\longrightarrow E\otimes K_{\mathbb{P}^1}(5)$. Here, x is the affine coordinate of $\mathbb{P}^1\setminus\{\infty\}$. Due to our specific choice of the vector bundle E, the quantization process is simple in each case, and defines an $\hbar$-connection on the trivial bundle of rank 2. Except for the Gauß hypergeometric case, it is given by

$$\nabla^\hbar=\hbar d-\phi, \tag{1.18}$$

where d is the exterior differentiation operator acting on sections of the trivial bundle. The form of (1.18) is valid because of our choice, $(0,dx)$, as the first row of the Higgs field.

For the third example of a Gauß hypergeometric equation, we use a particular choice of parameters so that the $\hbar$-connection becomes an $\hbar$-deformed Gauß-Manin connection of (1.12). This is a singular connection with simple poles at $0,1,\infty$, and has an explicit $\hbar$-dependence in the connection matrix. The Gauß-Manin connection ∇_{GM}^1 at $\hbar=1$ is equivalent to the Picard-Fuchs equation that characterizes the periods (1.11) of the Legendre family of elliptic curves $E(x)$ defined by the cubic equation

$$t^2=s(s-1)(s-x),\qquad x\in\mathcal{M}_{0,4}=\mathbb{P}^1\setminus\{0,1,\infty\}. \tag{1.19}$$

TABLE 1. Examples of quantum curves.

Higgs Field	Spectral Curve	Quantum Curve
$\begin{bmatrix} & 1 \\ x & \end{bmatrix} dx$	$y^2 - x = 0$ $w^2 - u^5 = 0$ $\Sigma = 2C_0 + 5F$ $p_a = 2, p_g = 0$	Airy $\left(\hbar\frac{d}{dx}\right)^2 - x$ Class $\frac{3}{2}$ irregular singularity at ∞
$\begin{bmatrix} & 1 \\ -1 & -x \end{bmatrix} dx$	$y^2 + xy + 1 = 0$ $w^2 - uw + u^4 = 0$ $\Sigma = 2C_0 + 4F$ $p_a = 1, p_g = 0$	Hermite $\left(\hbar\frac{d}{dx}\right)^2 + x\hbar\frac{d}{dx} + 1$ Class 2 irregular singularity at ∞
$\begin{bmatrix} & \frac{1}{x} \\ \frac{1}{4(1-x)} & \frac{2x-1}{x(1-x)} \end{bmatrix} dx$	$y^2 + \frac{2x-1}{x(x-1)}y + \frac{1}{4x(x-1)} = 0$ $w^2 + 4(u-2)uw$ $-4u^2(u-1) = 0$ $\Sigma = 2C_0 + 4F$ $p_a = 1, p_g = 0$	Gauß Hypergeometric $\left(\hbar\frac{d}{dx}\right)^2 + \frac{2x-1}{x(x-1)}\hbar\frac{d}{dx} + \frac{1}{4x(x-1)}$ Regular singular points at $x = 0, 1, \infty$
$\begin{bmatrix} & 1 \\ -\frac{1}{x+1} & -1 \end{bmatrix} dx$	$y^2 + y + \frac{1}{x+1} = 0$ $w^2 - u(u+1)w$ $+u^3(u+1) = 0$ $\Sigma = 2C_0 + 4F$ $p_a = 1, p_g = 0$	$\left(\hbar\frac{d}{dx}\right)^2 + \hbar\frac{d}{dx} + \frac{1}{x+1}$ Regular singular point at $x = -1$ and a class 1 irregular singularity at $x = \infty$
$\begin{bmatrix} & 1 \\ \frac{1}{x^2-1} & -\frac{2x^2}{x^2-1} \end{bmatrix} dx$	$(x^2-1)y^2 + 2x^2y - 1 = 0$ non-singular $\Sigma = 2C_0 + 4F$ $p_a = p_g = 1$	$\left(\hbar\frac{d}{dx}\right)^2 + 2\frac{x^2}{x^2-1}\hbar\frac{d}{dx} - \frac{1}{x^2-1}$ Regular singular points at $x = \pm 1$ and a class 1 irregular singularity at $x = \infty$

The second column gives the spectral curve of the Higgs bundle (E, ϕ). Since the Higgs fields have poles, the spectral curves are no longer contained in the cotangent bundle $T^*\mathbb{P}^1$. We need the compactified cotangent bundle

$$\overline{T^*\mathbb{P}^1} = \mathbb{P}(K_{\mathbb{P}^1} \oplus \mathcal{O}_{\mathbb{P}^1}) = \mathbb{F}_2,$$

which is a Hirzebruch surface. The parameter y is the fiber coordinate of the cotangent line $T^*_x\mathbb{P}^1$. The first line of the second column is the equation of the spectral curve in the (x, y) affine coordinate of $\mathbb{F}_2$. All but the last example produce a singular spectral curve. Let (u, w) be a coordinate system on another affine chart of $\mathbb{F}_2$ defined by

$$\begin{cases} x = 1/u \\ ydx = vdu, \qquad w = 1/v. \end{cases} \tag{1.20}$$

The singularity of Σ in the (u, w)-plane is given by the second line of the second column. The third line of the second column gives $\Sigma \in \mathrm{NS}(\mathbb{F}_2)$ as an element of the Néron-Severy group of $\mathbb{F}_2$. Here, C_0 is the class of the zero-section of $T^*\mathbb{P}^1$, and F represents the fiber class of $\pi : \mathbb{F}_2 \longrightarrow \mathbb{P}^1$. We also give the arithmetic and geometric genera of the spectral curve.

A solution $\Psi(x, \hbar)$ of (1.15) for the first example is given by the **Airy function**

$$Ai(x, \hbar) = \frac{1}{2\pi}\hbar^{-\frac{1}{6}} \int_{-\infty}^{\infty} \exp\left(\frac{ipx}{\hbar^{2/3}} + i\frac{p^3}{3}\right) dp, \tag{1.21}$$

which is an entire function in x for $\hbar \neq 0$. We will perform the all-order WKB analysis in this paper, and give a closed formula for each term of the WKB expansion. The topological recursion produces the asymptotic expansion

$$Ai(x,\hbar) = \exp\left(\sum_{g=0}^{\infty}\sum_{n=1}^{\infty}\frac{1}{n!}\hbar^{2g-2+n}F_{g,n}^{\text{Airy}}(x)\right) \tag{1.22}$$

at $x = \infty$, where

$$F_{g,n}^{\text{Airy}}(x) := \frac{(-1)^n}{2^{2g-2+n}}\cdot x^{-\frac{(6g-6+3n)}{2}}\sum_{\substack{d_1+\cdots+d_n\\=3g-3+n}}\langle\tau_{d_1}\cdots\tau_{d_n}\rangle_{g,n}\prod_{i=1}^{n}(2d_i-1)!!, \tag{1.23}$$

and the coefficients

$$\langle\tau_{d_1}\cdots\tau_{d_n}\rangle_{g,n} = \int_{\overline{\mathcal{M}}_{g,n}}\psi_1^{d_1}\cdots\psi_n^{d_n}$$

are the cotangent class intersection numbers on the moduli space $\overline{\mathcal{M}}_{g,n}$ of stable curves of genus g with n non-singular marked points. The cases for $(g,n) = (0,1)$ and $(0,2)$ require a subtle care, which will be explained in Section 2. The expansion coordinate $x^{\frac{3}{2}}$ of (1.23) indicates the class of the irregular singularity of the Airy differential equation.

The solutions to the second example are given by confluent hypergeometric functions, such as ${}_1F_1\left(\frac{1}{2\hbar};\frac{1}{2};-\frac{x^2}{2\hbar}\right)$, where

$${}_1F_1(a;c;z) := \sum_{n=0}^{\infty}\frac{(a)_n}{(c)_n}\frac{z^n}{n!} \tag{1.24}$$

is the **Kummer confluent hypergeomtric function**, and the **Pochhammer symbol** $(a)_n$ is defined by

$$(a)_n := a(a+1)(a+2)\cdots(a+n-1). \tag{1.25}$$

For $\hbar > 0$, the topological recursion determines the asymptotic expansion of a particular entire solution known as a **Tricomi confluent hypergeomtric function**

$$\begin{aligned}\Psi^{\text{Catalan}}(x,\hbar) = \left(-\frac{1}{2\hbar}\right)^{\frac{1}{2\hbar}}&\left(\frac{\Gamma[\frac{1}{2}]}{\Gamma[\frac{1}{2\hbar}+\frac{1}{2}]}\,{}_1F_1\left(\frac{1}{2\hbar};\frac{1}{2};-\frac{x^2}{2\hbar}\right)\right.\\ &\left.+\frac{\Gamma[-\frac{1}{2}]}{\Gamma[\frac{1}{2\hbar}]}\sqrt{-\frac{x^2}{2\hbar}}\,{}_1F_1\left(\frac{1}{2\hbar}+\frac{1}{2};\frac{3}{2};-\frac{x^2}{2\hbar}\right)\right).\end{aligned}$$

The expansion is given in the form

$$\begin{aligned}\Psi^{\text{Catalan}}(x,\hbar) &= \left(\frac{1}{x}\right)^{\frac{1}{\hbar}}\sum_{n=0}^{\infty}\frac{\hbar^n\left(\frac{1}{\hbar}\right)_{2n}}{(2n)!!}\cdot\frac{1}{x^{2n}}\\ &= \exp\left(\sum_{g=0}^{\infty}\sum_{n=1}^{\infty}\frac{1}{n!}\hbar^{2g-2+n}F_{g,n}^{\text{Catalan}}(x,\ldots,x)\right).\end{aligned} \tag{1.26}$$

Here,

$$F_{g,n}^{\text{Catalan}}(x_1,\ldots,x_n) = \sum_{\mu_1,\ldots,\mu_n>0}\frac{C_{g,n}(\mu_1,\ldots,\mu_n)}{\mu_1\cdots\mu_n}\prod_{i=1}^{n}x_i^{-\mu_i}$$

is the generating function of the *generalized* Catalan numbers $C_{g,n}(\mu_1, \ldots, \mu_n)$ of [**20**, **50**], which count the number of connected cellular graphs (i.e., the 1-skeletons of cell decompositions) of a compact surface of genus g with n labeled vertices of degrees $(\mu_1, \ldots, \mu_n)$, together with an arrow attached to one of the incident half-edges at each vertex. For more detail of cellular graphs, we refer to [**20**, **48**, **50**]. The expansion variable x^2 in (1.26) indicates the class of irregularity of the Hermite differential equation at $x = \infty$. The cases for $(g, n) = (0, 1)$ and $(0, 2)$ require again a special treatment, as we will see later.

REMARK 1.11. The authors are grateful to Peter Zograf for bringing [**50**] to their attention. The recursion for $C_{g,n}(\mu_1, \ldots, \mu_n)$ ([**48**, Theorem 3.1], which is also equivalent to [**20**, Theorem 1.1]), is exactly the same as [**50**, Equation 6]. The topological recursion (6.12) for the generalized Catalan numbers derived in [**20**, Theorem 1.2] is the Laplace transform of [**50**, Equation 6].

REMARK 1.12. Leonid Chekhov has shown that the asymptotic expansion (1.26) can also be derived from the matrix model of [**2**], by simply setting the matrix size equal to 1. The principal specialization often takes this effect in 1-Hermitian matrix models.

The Hermite differential equation becomes simple for $\hbar = 1$, and we have the asymptotic expansion

$$
\begin{aligned}
(1.27) \qquad i\sqrt{\frac{\pi}{2}}e^{-\frac{1}{2}x^2}\left[1 - \operatorname{erf}\left(\frac{ix}{\sqrt{2}}\right)\right] &= \sum_{n=0}^{\infty} \frac{(2n-1)!!}{x^{2n+1}} \\
&= \exp\left(\sum_{2g-2+n\geq -1} \frac{1}{n!} \sum_{\mu_1,\ldots,\mu_n>0} \frac{C_{g,n}(\mu_1,\ldots,\mu_n)}{\mu_1\cdots\mu_n} \prod_{i=1}^{n} x^{-(\mu_1+\cdots+\mu_n)}\right).
\end{aligned}
$$

Here, $\operatorname{erf}(x) := \frac{2}{\sqrt{\pi}}\int_0^x e^{-z^2}dz$ is the Gauß error function.

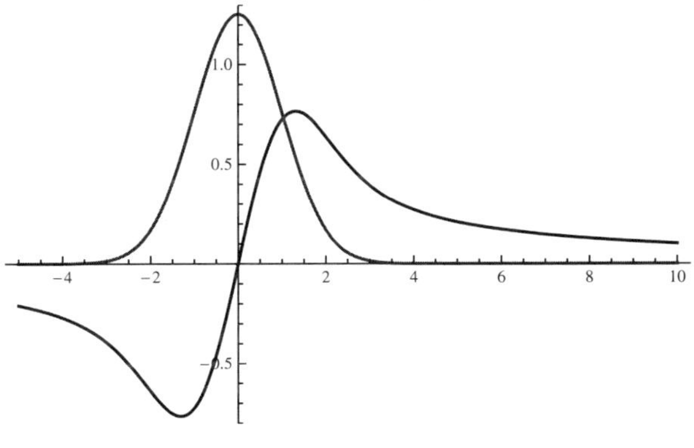

FIGURE 1.1. The imaginary part and the real part of $\Psi^{\text{Catalan}}(x, 1)$. For $x >> 0$, the imaginary part dies down, and only the real part has a non-trivial asymptotic expansion. Thus protect(1.27) is a series with real coefficients.

One of the two independent solutions to the third example, the Gauß hypergeometric equation, that is holomorphic around $x = 0$, is given by

(1.28)
$$\Psi^{\text{Gauß}}(x,\hbar) = {}_2F_1\left(-\frac{\sqrt{(h-1)(h-3)}}{2h}+\frac{1}{h}-\frac{1}{2}, \frac{\sqrt{(h-1)(h-3)}}{2h}+\frac{1}{h}-\frac{1}{2}; \frac{1}{h}; x\right),$$

where

$$(1.29) \qquad {}_2F_1(a,b;c;x) := \sum_{n=0}^{\infty}\frac{(a)_n(b)_n}{(c)_n}\frac{x^n}{n!}$$

is the **Gauß hypergeometric function**. The topological recursion calculates the B-model genus expansion of the periods of the Legendre family of elliptic curves (1.19) at the point where the elliptic curve degenerates to a nodal rational curve. For example, the procedure applied to the spectral curve

$$y^2 + \frac{2x-1}{x(x-1)}y + \frac{1}{4x(x-1)} = 0$$

with a choice of

$$\eta = \frac{-(2x-1)-\sqrt{3x^2-3x+1}}{2x(x-1)}dx,$$

which is an eigenvalue $\alpha_1(x)$ of the Higgs field ϕ, gives a genus expansion at $x = 0$:

$$(1.30) \qquad \Psi^{\text{Gauß}}(x,\hbar) = \exp\left(\sum_{g=0}^{\infty}\sum_{n=1}^{\infty}\frac{1}{n!}\hbar^{2g-2+n}F_{g,n}^{\text{Gauß}}(x)\right).$$

At $\hbar = 1$, we have a topological recursion expansion of the period $\omega_1(x)$ defined in (1.11):

$$(1.31) \qquad \frac{\omega_1(x)}{\pi} = \Psi^{\text{Gauß}}(x,1) = \exp\left(\sum_{g=0}^{\infty}\sum_{n=1}^{\infty}\frac{1}{n!}F_{g,n}^{\text{Gauß}}(x)\right).$$

A subtle point we notice here is that while the Gauß hypergeometric equation has regular singular points at $x = 0, 1, \infty$, the Hermite equation has an irregular singular point of class 2 at ∞. The spectral curve of each case has an ordinary double point at $x = \infty$. But the crucial difference lies in the intersection of the spectral curve Σ with the divisor C_∞. For the Hermite case we have $\Sigma \cdot C_\infty = 4$ and the intersection occurs all at once at $x = \infty$. For the Gauß hypergeometric case, the intersection $\Sigma \cdot C_\infty = 4$ occurs once each at $x = 0, 1$, and twice at $x = \infty$. This *confluence* of regular singular points is the source of the irregular singularity in the Hermite differential equation.

The fourth row indicates an example of a quantum curve that has one regular singular point at $x = -1$ and one irregular singular point of class 1 at $x = \infty$. The spectral curve has an ordinary double point at $x = \infty$, the same as the Hermite case. As Figure 1.2 shows, the class of the irregular singularity at $x = \infty$ is determined by how the spectral curve intersects with C_∞.

The existence of the irregular singularity in the quantum curve associated with a spectral curve has nothing to do with the singularity of the spectral curve. The fifth example shows a non-singular spectral curve of genus 1 (Figure 1.3), for which the quantum curve has a class 1 irregular singularity at $x = \infty$.

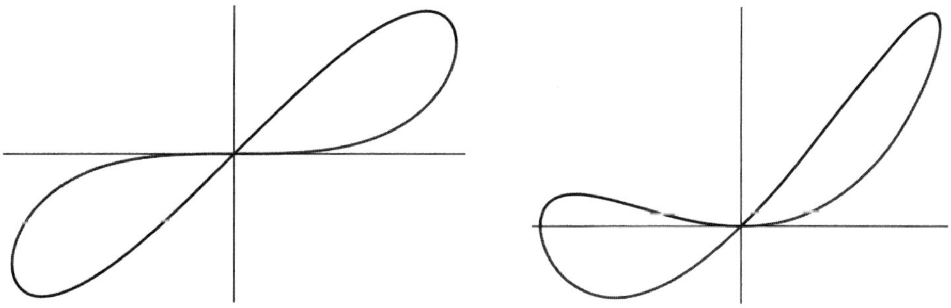

FIGURE 1.2. The spectral curves of the second and the fourth examples. The horizontal line is the divisor C_∞, and the vertical line is the fiber class F at $x = \infty$. The spectral curve intersects with C_∞ a total of four times. The curve on the right has a triple intersection at $x = \infty$, while the one on the left intersects all at once.

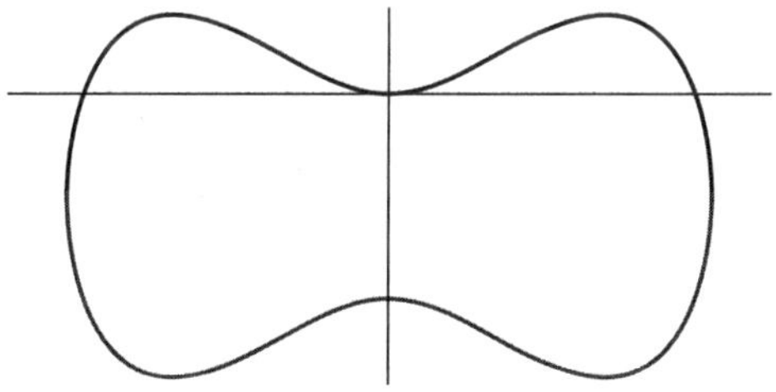

FIGURE 1.3. The spectral curve of the fifth example, which is non-singular. The corresponding quantum curve has two regular singular points at $x = \pm 1$, and a class 1 irregular singular point at $x = \infty$.

The paper is organized as follows. The general structure of the theory is explained using the Airy function as an example in Section 2. The notion of quantum curves as Rees D-modules quantizing Hitchin spectral curves is presented in Section 3. Since our topological recursion depends solely on the geometry of (1.6), the information of Σ and $\widetilde{\Sigma}$, such as their arithmetic genera, becomes important. We will give the genus formulas in Sections 4 and 5. In Section 4 we study the geometry of the Hitchin spectral curves associated with rank 2 meromorphic Higgs bundles. We give the genus formula for the normalization $\widetilde{\Sigma}$ in terms of the characteristic polynomial of the Higgs field ϕ. A more systematic treatment of the spectral curve and its desingularization is given in Section 5. In Section 6, which is the heart of our paper, we prove the main theorem. Two more examples, Hermite differential equations and Gauß hypergeometric differential equations, are studied in Section 7.

2. A walk-through of the simplest example

Before going into the full generality, let us present the simplest example of our construction. With this example we can illustrate the relation between a Higgs bundle, the compactified cotangent bundle of a curve, a quantum curve, a classical differential equation, non-Abelian Hodge correspondence, and the quantum invariants that the quantum curve captures.

As a spectral curve, we take the algebraic curve $\Sigma \subset \mathbb{F}_2 = \mathbb{P}(K_{\mathbb{P}^1} \oplus \mathcal{O}_{\mathbb{P}^1}) = \overline{T^*\mathbb{P}^1}$ embedded in the Hirzebruch surface with the defining equation

$$y^2 - x = 0. \tag{2.1}$$

Here, x is the coordinate of the affine line $\mathbb{A}^1 = \mathbb{P}^1 \setminus \{\infty\}$, and y is the fiber coordinate of the cotangent bundle $T^*\mathbb{P}^1 \subset \mathbb{F}^2$ over $\mathbb{A}^1$. The Hirzebruch surface is the natural compactification of the cotangent bundle $T^*\mathbb{P}^1$, which is the total space of the canonical bundle $K_{\mathbb{P}^1}$. We denote by $\eta \in H^0(T^*\mathbb{P}^1, \pi^*K_{\mathbb{P}^1})$ the tautological 1-form associated with the projection $\pi : T^*\mathbb{P}^1 \longrightarrow \mathbb{P}^1$. It is expressed as $\eta = ydx$ in terms of the affine coordinates. The holomorphic symplectic form on $T^*\mathbb{P}^1$ is given by $-d\eta = dx \wedge dy$. The 1-form η extends to $\mathbb{F}_2$ as a meromorphic differential form and defines a divisor

$$(\eta) = C_0 - C_\infty, \tag{2.2}$$

where C_0 is the zero-section of $T^*\mathbb{P}^1$, and C_∞ the section at infinity of $\overline{T^*\mathbb{P}^1}$. The Picard group $\mathrm{Pic}(\mathbb{F}_2)$ of the Hirzebruch surface is generated by the class C_0 and a fiber class F of π.

Although (2.1) is a perfect parabola in the affine plane, it has a quintic cusp singularity at $x = \infty$. Let (u, w) be a coordinate on another affine piece of $\mathbb{F}_2$ defined by (1.20). Then Σ in the (u, w)-plane is given by

$$w^2 = u^5. \tag{2.3}$$

The expression of Σ as an element of $\mathrm{Pic}(\mathbb{F}_2)$ is thus given by $\Sigma = 2C_0 + 5F$. Define a stable Higgs pair (E, ϕ) with $E = \mathcal{O}_{\mathbb{P}^1}(1) \oplus \mathcal{O}_{\mathbb{P}^1}(-1) = K_{\mathbb{P}^1}^{-\frac{1}{2}} \oplus K_{\mathbb{P}^1}^{\frac{1}{2}}$ and

$$\phi = \begin{bmatrix} & 1 \\ x & \end{bmatrix} dx : E \longrightarrow E \otimes K_{\mathbb{P}^1}(5) = E. \tag{2.4}$$

Here, we choose a meromorphic *quadratic* differential $x(dx)^2 \in H^0\big(\mathbb{P}^1, K_{\mathbb{P}^1}(4)^{\otimes 2}\big)$ that has a simple zero at $0 \in \mathbb{P}^1$ and a pole of order 5 at $\infty \in \mathbb{P}^1$ as the $(2,1)$-component of ϕ. Up to a constant factor, there is only one such differential $x(dx)^2 = (du)^2/u^5$. The spectral curve Σ of (E, ϕ) is given by the characteristic equation

$$\det(\eta - \pi^*\phi) = \eta^2 - \pi^*\mathrm{tr}(\phi) + \pi^*\det(\phi) = 0 \tag{2.5}$$

in $\mathbb{F}_2$. Our *quantization* process applied to ϕ determines a singular $\hbar$-connection [**3**, **43**]

$$\nabla^\hbar = \hbar d - \begin{bmatrix} & 1 \\ x & \end{bmatrix} dx \tag{2.6}$$

on the trivial bundle $\mathcal{O}_{\mathbb{P}^1}^{\oplus 2}$ over $\mathbb{P}^1$.

The **quantization** procedure that we will explain in this paper associates the following differential equation to the spectral curve Σ:

$$\left(\left(\hbar\frac{d}{dx}\right)^2 - x\right)\Psi(x, \hbar) = 0. \tag{2.7}$$

The solution Ψ gives rise to a flat section $\begin{bmatrix}\Psi \\ \hbar\Psi'\end{bmatrix}$ of (2.6) for $\hbar \neq 0$, where $'$ denotes the x differentiation. The differential operator

$$P(x,\hbar) := \left(\hbar\frac{d}{dx}\right)^2 - x \tag{2.8}$$

quantizing (2.1) is an example of what we call a quantum curve. Reflecting the fact (2.3) that Σ has a quintic cusp singularity at $x = \infty$, (2.7) has an *irregular* singular point of *class* $\frac{3}{2}$ at $x = \infty$. This number $\frac{3}{2}$ indicates how the asymptotic expansion of the solution looks like. Indeed, any non-trivial solution has an essential singularity at ∞. We note that every solution of (2.7) is an entire function for any value of $\hbar \neq 0$. Define

$$a_{3n} = a_0 \cdot \frac{\prod_{j=1}^n (3j-2)}{(3n)!}, \qquad a_{3n+1} = a_1 \cdot \frac{\prod_{j=1}^n (3j-1)}{(3n+1)!}, \qquad a_{3n+2} = 0, \tag{2.9}$$

for $n \geq 0$. Then

$$\Psi(x,\hbar) := \sum_{n=0}^{\infty} a_n \left(\frac{x}{\hbar^{2/3}}\right)^n \tag{2.10}$$

gives an arbitrary solution to (2.7), which is entire. The coefficients (2.9) are of no particular interest.

What our quantization procedure tells us is a different, and more interesting, story. Applying our main result of this paper, we construct a particular *all-order* asymptotic expansion of this entire solution

$$\Psi(x,\hbar) = \exp F(x,\hbar), \qquad F(x,\hbar) := \sum_{m=0}^{\infty} \hbar^{m-1} S_m(x), \tag{2.11}$$

valid for $|\mathrm{Arg}(x)| < \pi$, and $\hbar > 0$. Here, the first two terms of the asymptotic expansion are given by

$$S_0(x) = \pm\frac{2}{3}x^{\frac{3}{2}}, \tag{2.12}$$

$$S_1(x) = -\frac{1}{4}\log x. \tag{2.13}$$

Although the *classical limit* $\hbar \to 0$ of (2.7) does not make sense under the expansion (2.11), through the *WKB* analysis

$$\left[e^{-S_1(x)} e^{-\frac{1}{\hbar}S_0(x)} \left(\hbar^2 \frac{d^2}{dx^2} - x\right) e^{\frac{1}{\hbar}S_0(x)} e^{S_1(x)}\right] \exp\left(\sum_{m=2}^{\infty} \hbar^{m-1} S_m(x)\right) = 0, \tag{2.14}$$

the above equation has a well-defined limit $\hbar \to 0$. The result is $S_0'(x)^2 = x$, which gives (2.12), and also (2.1) by defining $dS_0 = \eta$. This process is called the *semi-classical limit*. The vanishing of the $\hbar$-linear terms of (2.14) is $2S_0'(x)S_1'(x) + S_0''(x) = 0$, which gives (2.13) above.

The solution we are talking about is the Airy function (1.21) for the choice of $S_0(x) = -\frac{2}{3}x^{\frac{3}{2}}$. This solution corresponds to (2.10) with the initial condition

$$a_0 = \frac{1}{3^{\frac{2}{3}}\Gamma(\frac{2}{3})}, \qquad a_1 = -\frac{1}{3^{\frac{1}{3}}\Gamma(\frac{1}{3})}.$$

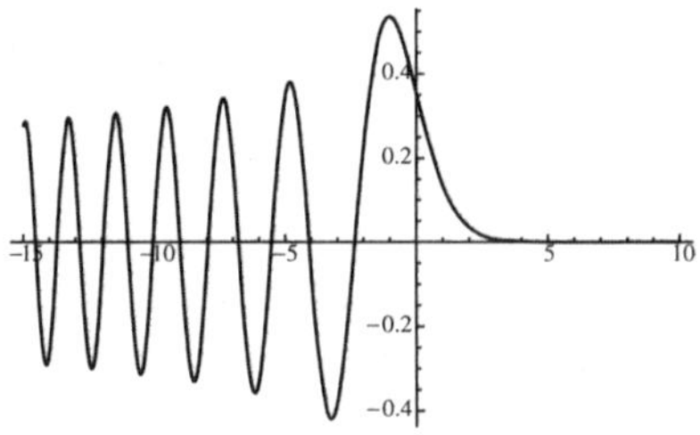

FIGURE 2.1. The Airy function

The surprising discovery of Kontsevich [**39**] is that $S_m(x)$ for $m \geq 2$ has the following *closed* formula:

$$S_m(x) := \sum_{2g-2+n=m-1} \frac{1}{n!} F^{\text{Airy}}_{g,n}(x), \tag{2.15}$$

$$F^{\text{Airy}}_{g,n}(x) := \frac{(-1)^n}{2^{2g-2+n}} \cdot x^{-\frac{(6g-6+3n)}{2}} \sum_{\substack{d_1+\cdots+d_n\\=3g-3+n}} \langle \tau_{d_1} \cdots \tau_{d_n} \rangle_{g,n} \prod_{i=1}^{n} (2d_i - 1)!!. \tag{2.16}$$

Although (2.15) is not a generating function of all intersection numbers, as we will show in the subsequent sections, the quantum curve (2.1) alone actually determines every intersection number $\langle \tau_{d_1} \cdots \tau_{d_n} \rangle_{g,n}$. This mechanism is the *differential recursion equation* of [**18**], based on the theory of integral topological recursion of [**25**], which computes *free energies*

$$F^{\text{Airy}}_{g,n}(x_1, \cdots, x_n) := \frac{(-1)^n}{2^{2g-2+n}} \sum_{\substack{d_1+\cdots+d_n\\=3g-3+n}} \langle \tau_{d_1} \cdots \tau_{d_n} \rangle_{g,n} \prod_{i=1}^{n} \frac{(2d_i - 1)!!}{\sqrt{x_i}^{2d_i+1}} \tag{2.17}$$

as a function in n variables from Σ through the process of blow-ups of $\mathbb{F}_2$.

Let us now give a detailed procedure for this example. We start with the spectral curve Σ of (2.1). Our goal is to come up with (2.7). The first step is to blow up $\mathbb{F}_2$ and to construct (1.6). The discriminant of the defining equation (2.5) of the spectral curve is

$$-\det(\phi) = x(dx)^2 = \frac{1}{u^5}(du)^2.$$

It has a simple zero at $x = 0$ and a pole of order 5 at $x = \infty$. The *Geometric Genus Formula* (4.12) tells us that $\widetilde{\Sigma}$ is a non-singular curve of genus 0, i.e., a $\mathbb{P}^1$, after blowing up $\lfloor \frac{5}{2} \rfloor = 2$ times. The center of blow-up is $(u, w) = (0, 0)$ for the first time. Put $w = w_1 u$, and denote by E_1 the exceptional divisor of the first blow-up. The proper transform of Σ for this blow-up, $w_1^2 = u^3$, has a cubic cusp singularity, so we blow up again at the singular point. Let $w_1 = w_2 u$, and denote by E_2 the exceptional divisor created by the second blow-up. The self-intersection of the proper transform of E_1 is -2. We then obtain the desingularized curve $\Sigma_{\min}$, locally given by $w_2^2 = u$. The proof of Theorem 4.2 also tells us that $\Sigma_{\min} \longrightarrow \mathbb{P}^1$ is ramified at two points. Choose the affine coordinate $t = 2w_2$ of the exceptional divisor added at the second blow-up. Our choice of the constant factor is to make the formula the same as in [**20**]. We have

$$\begin{cases} x = \frac{1}{u} = \frac{1}{w_2^2} = \frac{4}{t^2} \\ y = -\frac{u^2}{w} = -\frac{u^2}{w_2 u^2} = -\frac{2}{t}. \end{cases} \tag{2.18}$$

In the (u, w)-coordinate, we see that the parameter t is a normalization parameter of the quintic cusp singularity:

$$\begin{cases} u - \frac{t^2}{4} \\ w = \frac{t^5}{32}. \end{cases}$$

Note that $\Sigma_{\min}$ intersects transversally with the proper transform of C_∞. We blow up once again at this intersection, and denote by $\widetilde{\Sigma}$ the proper transform of $\Sigma_{\min}$. The blow-up space $Bl(\mathbb{F}^2)$ is the result of $3 = \lceil \frac{5}{2} \rceil$ times blow-ups of the Hirzebruch surface.

Now we apply the differential recursion (6.6) to the geometric data (1.6) and (1.7). We claim that the integral topological recursion of [**25**] for the geometric data we are considering now is exactly the same as the integral topological recursion of [**20**, (6.12)] applied to the curve (2.18) *realized as a plane parabola* in $\mathbb{C}^2$. This is because our integral topological recursion (6.12) has two residue contributions, one each from $t = 0$ and $t = \infty$. As proved in [**20**, Section 6], the integrand on the right-hand side of the integral recursion formula [**20**, (6.12)] does not have any pole at $t = 0$. Therefore, the residue contribution from this point is 0. The differential recursion is obtained by deforming the contour of integration to enclose only poles of the differential forms $W_{g,n}$. Since $t = 0$ is a regular point, the two methods have no difference.

The $W_{0,2}$ of (1.7) is simply $\frac{dt_1 \cdot dt_2}{(t_1 - t_2)^2}$ because $\widetilde{\Sigma} \cong \mathbb{P}^1$. Since t of (2.18) is a normalization coordinate, we have

$$W_{0,1} = \tilde{i}^* \nu^*(\eta) = y(t)dx(t) = \frac{16}{t^4},$$

in agreement of [**20**, (6.8)]. Noting that the solution to the integral topological recursion is unique from the initial data, we conclude that

$$d_1 \cdots d_n F_{g,n}^{\mathrm{Airy}}\big(x(t_1), \ldots, x(t_n)\big) = W_{g,n}.$$

By setting the constants of integration by integrating from $t = 0$ for the differential recursion equation, we obtain the expression (2.17). Then its principal specialization gives (2.16). The equivalence of the differential recursion and the quantum curve equation Theorem 6.1 then proves (2.7) with the expression of (2.11) and (2.15).

In this process, what is truly amazing is that the single differential equation (2.7), which is our quantum curve in this case, knows everything about the free energies (2.17). This is because we can recover the spectral curve Σ from the quantum curve. Then the procedures we need to apply, the blow-ups and the differential recursion equation, are canonical. Therefore, we do recover (2.17) as explained above.

It is surprising to see that a simple entire function (2.10) contains so much geometric information. Our expansion (2.11) is an expression of an entire function viewed from its essential singularity. We can extract rich information of the solution by restricting the region where the asymptotic expansion is valid. If we consider (2.11) only as a formal expression in x and $\hbar$, then we cannot see how the coefficients are related to quantum invariants. The topological recursion [**25**] is a key to connect the two worlds: the world of quantum invariants, and the world of holomorphic functions and differentials. This relation is also knows as a **mirror symmetry**, or in analysis, simply as the *Laplace transform*. The intersection numbers $\langle \tau_{d_1} \cdots \tau_{d_n} \rangle_{g,n}$ belong to the A-model, while the spectral curve Σ of (2.1) and free energies belong to the B-model. We consider (2.17) as an example of the Laplace transform, playing the role of mirror symmetry [**20**].

3. Quantum curves for Higgs bundles

In this section, we give the definition of quantum curves in the context of Hitchin spectral curves. Let C be a non-singular projective algebraic curve defined over $\mathbb{C}$. The sheaf $\mathcal{D}_C$ of differential operators on C is the subalgebra generated by the anti-canonical sheaf K_C^{-1} and the structure sheaf $\mathcal{O}_C$ in the $\mathbb{C}$-linear endomorphism algebra $\mathcal{E}nd_{\mathbb{C}}(\mathcal{O}_C)$. Here, K_C^{-1} acts on $\mathcal{O}_C$ as holomorphic vector fields, and $\mathcal{O}_C$ acts on itself by multiplication. Locally every element of $\mathcal{D}_C$ is written as

$$\mathcal{D}_C \ni P(x) = \sum_{\ell=0}^{r} a_\ell(x) \left(\frac{d}{dx}\right)^{r-\ell}, \qquad a_\ell(x) \in \mathcal{O}_C$$

for some $r \geq 0$. For a fixed r, we introduce the filtration by order of differential operators into $\mathcal{D}_C$ as follows:

$$F_r\mathcal{D}_C = \left\{ P(x) = \sum_{\ell=0}^{r} a_\ell(x) \left(\frac{d}{dx}\right)^{r-\ell} \middle| a_\ell(x) \in \mathcal{O}_C \right\}.$$

The *Rees* ring $\widetilde{\mathcal{D}_C}$ is defined by

$$\widetilde{\mathcal{D}_C} = \bigoplus_{r=0}^{\infty} \hbar^r F_r\mathcal{D}_C \subset \mathbb{C}[\![\hbar]\!] \otimes_{\mathbb{C}} \mathcal{D}_C. \tag{3.1}$$

An element of $\widetilde{\mathcal{D}_C}$ on a coordinate neighborhood $U \subset C$ can be written as

$$P(x,\hbar) = \sum_{\ell=0}^{r} a_\ell(x,\hbar) \left(\hbar \frac{d}{dx}\right)^{r-\ell}. \tag{3.2}$$

DEFINITION 3.1 (Quantum curve). A **quantum curve** is the Rees $\widetilde{\mathcal{D}_C}$-module

$$\widetilde{\mathcal{M}} = \bigoplus_{r=0}^{\infty} \hbar^r F_r\mathcal{M} \tag{3.3}$$

associated with a filtered $\mathcal{D}_C$-module $(\mathcal{M}, F_r)$ defined on C, with the compatibility

$$F_a\mathcal{D}_C \cdot F_b\mathcal{M} \subset F_{a+b}\mathcal{M}.$$

REMARK 3.2. We remark that several examples of quantum curves that are constructed in [**8, 21, 47**], for various Hurwitz numbers and Gromov-Witten theory of $\mathbb{P}^1$, do not fall into our definition in terms of Rees $\mathcal{D}$-modules. This is because in the above mentioned examples, quantum curves involve *infinite*-oder differential operators, or *difference* operators, while we consider only differential operators of finite order in this paper.

Let

$$D = \sum_{j=1}^{n} m_j p_j, \qquad m_j > 0$$

be an effective divisor on C. The point set $\{p_1, \dots, p_n\} \subset C$ is the support of D. A *meromorphic Higgs bundle* with poles at D is a pair (E, ϕ) consisting of an algebraic vector bundle E on C and a Higgs field

$$\phi : E \longrightarrow K_C(D) \otimes_{\mathcal{O}_C} E. \tag{3.4}$$

Since the cotangent bundle

$$T^*C = \mathrm{Spec}\left(\mathrm{Sym}\left(K_C^{-1}\right)\right)$$

is the total space of K_C, we have the tautological 1-form $\eta \in H^0(T^*C, \pi^* K_C)$ on T^*C coming from the projection

$$\begin{array}{ccc} T^*C & \longleftarrow & \pi^* K_C \\ {\scriptstyle \pi}\big\downarrow & & \\ C & \longleftarrow & K_C. \end{array}$$

The natural holomorphic symplectic form of T^*C is given by $-d\eta$. The **compactified cotangent bundle** of C is a ruled surface defined by

$$\overline{T^*C} := \mathbb{P}\left(K_C \oplus \mathcal{O}_C\right) = \mathrm{Proj}\left(\bigoplus_{n=0}^{\infty} \left(K_C^{-n} \cdot I^0 \oplus K_C^{-n+1} \cdot I \oplus \cdots \oplus K_C^0 \cdot I^n\right)\right), \tag{3.5}$$

where I represents $1 \in \mathcal{O}_C$ being considered as a degree 1 element. The divisor at infinity C_∞ of (1.5) is reduced in the ruled surface and supported on the subset $\mathbb{P}\left(K_C \oplus \mathcal{O}_C\right) \setminus T^*C$. The tautological form η extends on $\overline{T^*C}$ as a meromorphic 1-form with simple poles along C_∞. Thus the divisor of η in $\overline{T^*C}$ is given by

$$(\eta) = C_0 - C_\infty, \tag{3.6}$$

where C_0 is the zero section of T^*C.

The relation between the sheaf $\mathcal{D}_C$ and the geometry of the compactified cotangent bundle $\overline{T^*C}$ is the following. First we have

$$\mathrm{Spec}\left(\bigoplus_{m=0}^{\infty} F_m \mathcal{D}_C / F_{m-1} \mathcal{D}_C\right) = \mathrm{Spec}\left(\bigoplus_{m=0}^{\infty} K_C^{-m}\right) = T^*C. \tag{3.7}$$

Let us denote by $gr_m \mathcal{D}_C = F_m \mathcal{D}_C / F_{m-1} \mathcal{D}_C$. By writing $I = 1 \in H^0(C, \mathcal{D}_C)$, we then have

$$\mathrm{Proj}\left(\bigoplus_{m=0}^{\infty} \left(gr_m \mathcal{D}_C \cdot I^0 \oplus gr_{m-1} \mathcal{D}_C \cdot I \oplus gr_{m-2} \mathcal{D}_C \cdot I^{\otimes 2} \oplus \cdots \oplus gr_0 \mathcal{D}_C \cdot I^{\otimes m}\right)\right) = \overline{T^*C}. \tag{3.8}$$

DEFINITION 3.3 (Spectral curve). A **spectral curve** of degree r is a divisor Σ in $\overline{T^*C}$ such that the projection $\pi : \Sigma \longrightarrow C$ defined by the restriction

$$\begin{array}{ccc} \Sigma & \xrightarrow{i} & \overline{T^*C} \\ & \searrow^{\pi} & \downarrow \pi \\ & & C \end{array}$$

is a finite morphism of degree r. The **spectral curve of a Higgs bundle** (E, ϕ) is the divisor of zeros

$$\Sigma = (\det(\eta - \pi^*\phi))_0 \tag{3.9}$$

on $\overline{T^*C}$ of the characteristic polynomial $\det(\eta - \pi^*\phi)$. Here,

$$\pi^*\phi : \pi^*E \longrightarrow \pi^*\left(K_C(D)\right) \otimes_{\mathcal{O}_{\mathbb{P}(K_C \oplus \mathcal{O}_C)}} \pi^*E.$$

REMARK 3.4. The Higgs field ϕ is holomorphic on $C \setminus \operatorname{supp}(D)$. Thus we can define the divisor of zeros

$$\Sigma^\circ = \left(\det\left(\eta - \pi^*\left(\phi|_{C\setminus \operatorname{supp}(D)}\right)\right)\right)_0$$

of the characteristic polynomial on $T^*(C \setminus \operatorname{supp}(D))$. The spectral curve Σ is the complex topology closure of Σ° with respect to the compactification

$$T^*(C \setminus \operatorname{supp}(D)) \subset \overline{T^*C}. \tag{3.10}$$

A left $\mathcal{D}_C$-module $\mathcal{E}$ on C is naturally an $\mathcal{O}_C$-module with a $\mathbb{C}$-linear integrable connection $\nabla : \mathcal{E} \longrightarrow K_C \otimes_{\mathcal{O}_C} \mathcal{E}$. The construction goes as follows:
(3.11)

$$\nabla : \mathcal{E} \xrightarrow{\ \alpha\ } \mathcal{D}_C \otimes_{\mathcal{O}_C} \mathcal{E} \xrightarrow{\nabla_{\mathcal{D}} \otimes id} (K_C \otimes_{\mathcal{O}_C} \mathcal{D}_C) \otimes_{\mathcal{O}_C} \mathcal{E} \xrightarrow{\beta \otimes id} K_C \otimes_{\mathcal{O}_C} \mathcal{E},$$

where

- α is the natural inclusion $\mathcal{E} \ni v \longmapsto 1 \otimes v \in \mathcal{D}_C \otimes_{\mathcal{O}_C} \mathcal{E}$;
- $\nabla_{\mathcal{D}} : \mathcal{D}_C \longrightarrow K_C \otimes_{\mathcal{O}_C} \mathcal{D}_C$ is the connection defined by the $\mathbb{C}$-linear left-multiplication operation of K_C^{-1} on $\mathcal{D}_C$, which satisfies the derivation property

$$\nabla_{\mathcal{D}}(f \cdot P) = f \cdot \nabla_{\mathcal{D}}(P) + df \cdot P \in K_C \otimes_{\mathcal{O}_C} \mathcal{D}_C \tag{3.12}$$

  for $f \in \mathcal{O}_C$ and $P \in \mathcal{D}_C$; and
- β is the canonical right $\mathcal{D}_C$-module structure in K_C.

If we choose a local coordinate neighborhood $U \subset C$ with a coordinate x, then (3.12) takes the following form. Let us denote by $P' = [d/dx, P] = d/dx \cdot P - P \cdot d/dx$, and define

$$\nabla_{\frac{d}{dx}}(P) := P \cdot \frac{d}{dx} + P'.$$

Then we have

$$\nabla_{\frac{d}{dx}}(f \cdot P) = f \cdot \nabla_{\frac{d}{dx}}(P) + \frac{df}{dx} \cdot P.$$

The connection ∇ of (3.11) is integrable because $d^2 = 1$. Actually, the statement is true for any dimensions. We note that there is no reason for $\mathcal{E}$ to be coherent as an $\mathcal{O}_C$-module.

Conversely, if an algebraic vector bundle E on C of rank r admits a holomorphic connection $\nabla : E \longrightarrow K_C \otimes E$, then E acquires the structure of a $\mathcal{D}_C$-module. This is because ∇ is automatically flat, and the covariant derivative ∇_X for $X \in K_C^{-1}$ satisfies

$$\nabla_X(fv) = f\nabla_X(v) + X(f)v \tag{3.13}$$

for $f \in \mathcal{O}_C$ and $v \in E$. A repeated application of (3.13) makes E a $\mathcal{D}_C$-module. The fact that every $\mathcal{D}_C$-module on a curve is principal implies that for every point $p \in C$, there is an open neighborhood $p \in U \subset C$ and a linear differential operator P of odcr r on U, called a generator, such that $E|_U \cong \mathcal{D}_U/\mathcal{D}_U P$. Thus on an open curve U, a holomorphic connection in a vector bundle of rank r gives rise to a differential operator of order r. The converse is true if $\mathcal{D}_U/\mathcal{D}_U P$ is $\mathcal{O}_U$-coherent.

DEFINITION 3.5 ($\hbar$-connection). A formal $\hbar$-connection on a vector bundle $E \longrightarrow C$ is a $\mathbb{C}[\![\hbar]\!]$-linear homomorphism

$$\nabla^\hbar : \mathbb{C}[\![\hbar]\!] \otimes E \longrightarrow \mathbb{C}[\![\hbar]\!] \otimes K_C \otimes_{\mathcal{O}_C} E$$

subject to the derivation condition

$$\nabla^\hbar(f \cdot v) = f\nabla^\hbar(v) + \hbar df \otimes v, \tag{3.14}$$

where $f \in \mathcal{O}_C \otimes \mathbb{C}[\![\hbar]\!]$ and $v \in \mathbb{C}[\![\hbar]\!] \otimes E$.

REMARK 3.6. The **classical limit** of a holomorphic $\hbar$-connection is the evaluation $\hbar = 0$ of $\nabla^\hbar$, which is simply an $\mathcal{O}_C$-module homomorphism

$$\nabla^0 : E \longrightarrow K_C \otimes_{\mathcal{O}_C} E,$$

i.e., a holomorphic Higgs field in the vector bundle E.

REMARK 3.7. An $\mathcal{O}_C \otimes \mathbb{C}[\![\hbar]\!]$-coherent $\widetilde{\mathcal{D}_C}$-module is equivalent to a vector bundle on C equipped with an $\hbar$-connection.

In analysis, the **semi-classical limit** of a differential operator $P(x, \hbar)$ of (3.2) is defined by

$$\lim_{\hbar \to 0} \left(e^{-\frac{1}{\hbar}S_0(x)} P(x,\hbar) e^{\frac{1}{\hbar}S_0(x)} \right) = \sum_{\ell=0}^{r} a_\ell(x,0) \left(S_0'(x)\right)^{r-\ell}, \tag{3.15}$$

where $S_0(x) \in \mathcal{O}_C(U)$. The equation

$$\lim_{\hbar \to 0} \left(e^{-\frac{1}{\hbar}S_0(x)} P(x,\hbar) e^{\frac{1}{\hbar}S_0(x)} \right) = 0 \tag{3.16}$$

then determines the first term of the **singular perturbation expansion**

$$\Psi(x,\hbar) = \exp\left(\sum_{m=0}^{\infty} \hbar^{m-1} S_m(x) \right) \tag{3.17}$$

of the solution $\Psi(x, \hbar)$ of the differential equation

$$P(x,\hbar)\Psi(x,\hbar) = 0$$

on U. Since $dS_0(x)$ is a local section of T^*C on $U \subset C$, $y = S_0'(x)$ gives the local trivialization of $T^*C|_U$, with $y \in T_x^*C$ a fiber coordinate. Equations (3.15) and (3.16) then give an equation

$$\sum_{\ell=0}^{r} a_\ell(x,0) y^{r-\ell} = 0 \tag{3.18}$$

of a curve in $T^*C|_U$. This motivates us to give the following definition.

DEFINITION 3.8 (Semi-classical limit of a Rees differential operator). Let $U \subset C$ be an open subset of C with a local coordinate x such that T^*C is trivial over U with a fiber coordinate y. The semi-classical limit of a local section

$$P(x,\hbar) = \sum_{\ell=0}^{r} a_\ell(x,\hbar)\left(\hbar\frac{d}{dx}\right)^{r-\ell}$$

of the Rees ring $\widetilde{\mathcal{D}_C}$ of the sheaf of differential operators $\mathcal{D}_C$ on U is the holomorphic function

$$\sum_{\ell=0}^{r} a_\ell(x,0)y^{r-\ell}$$

defined on $T^*C|_U$.

DEFINITION 3.9 (Semi-classical limit). Suppose a Rees $\widetilde{\mathcal{D}_C}$-module $\widetilde{\mathcal{M}}$ is written as

$$\widetilde{\mathcal{M}}(U) = \widetilde{\mathcal{D}_C}(U)/\widetilde{\mathcal{D}_C}(U)P_U \tag{3.19}$$

on every coordinate neighborhood $U \subset C$ with a differential operator P_U of the form (3.2). Using this expression (3.2) for P_U, we construct a meromorphic function

$$p_U(x,y) = \sum_{\ell=0}^{r} a_\ell(x,0)y^{r-\ell} \tag{3.20}$$

on $\overline{T^*C}|_U$, where y is the fiber coordinate of T^*C, which is trivialized on U. Define

$$\Sigma_U = (p_U(x,y))_0 \tag{3.21}$$

as the divisor of zero of the function $p_U(x,y)$. If Σ_U's glue together to a spectral curve $\Sigma \subset \overline{T^*C}$, then we call Σ the **semi-classlical limit** of the Rees $\widetilde{\mathcal{D}_C}$-module $\widetilde{\mathcal{M}}$.

REMARK 3.10. For the local equation (3.20) to be consistent globally on C, the coefficients of (3.2) have to satisfy

$$a_\ell(x,0) \in \Gamma\big(U, K_C^{\otimes \ell}\big)\,. \tag{3.22}$$

DEFINITION 3.11 (Quantum curve for holomorphic Higgs bundle). A **quantum curve** associated with the spectral curve $\Sigma \subset T^*C$ of a holomorphic Higgs bundle on a projective algebraic curve C is a Rees $\widetilde{\mathcal{D}_C}$-module $\mathcal{E}$ whose semi-classical limit is Σ.

The main reason we need to extend our framework to meromorphic connections is that there are no non-trivial holomorphic connections on $\mathbb{P}^1$, whereas many important classical examples of differential equations are naturally defined over $\mathbb{P}^1$ with regular and irregular singularities. A $\mathbb{C}$-linear homomorphism

$$\nabla : E \longrightarrow K_C(D) \otimes_{\mathcal{O}_C} E$$

is said to be a *meromorphic connection* with poles along an effective divisor D if

$$\nabla(f\cdot v) = f\nabla(v) + df \otimes v$$

for every $f \in \mathcal{O}_C$ and $v \in E$. Let us denote by

$$\mathcal{O}_C(*D) := \varinjlim \mathcal{O}_C(mD), \qquad E(*D) := E \otimes_{\mathcal{O}_C} \mathcal{O}_C(*D).$$

Then ∇ extends to

$$\nabla : E(*D) \longrightarrow K_C(*D) \otimes_{\mathcal{O}_C(*D)} E(*D).$$

Since ∇ is holomorphic on $C \setminus \mathrm{supp}(D)$, it induces a $\mathcal{D}_{C\setminus \mathrm{supp}(D)}$-module structure in $E|_{C\setminus \mathrm{supp}(D)}$. The $\mathcal{D}_C$-module direct image $\widetilde{E} = j_*\left(E|_{C\setminus \mathrm{supp}(D)}\right)$ associated with the open inclusion map $j : C \setminus \mathrm{supp}(D) \longrightarrow C$ is then naturally isomorphic to

$$\widetilde{E} = j_*\left(E|_{C\setminus \mathrm{supp}(D)}\right) \cong E(*D) \tag{3.23}$$

as a $\mathcal{D}_C$-module. (3.23) is called the *meromorphic extension* of the $\mathcal{D}_{C\setminus \mathrm{supp}(D)}$-module $E|_{C\setminus \mathrm{supp}(D)}$.

Let us take a local coordinate x of C, this time around a pole $p_j \in \mathrm{supp}(D)$. If a generator $\widetilde{P}$ of $\widetilde{E}$ near $x = 0$ has a local expression

$$\widetilde{P}(x, d/dx) = x^k \sum_{\ell=0}^{r} b_\ell(x)\left(x\frac{d}{dx}\right)^{r-\ell} \tag{3.24}$$

around p_j with locally defined holomorphic functions $b_\ell(x)$, $b_0(0) \neq 0$, and an integer $k \in \mathbb{Z}$, then $\widetilde{P}$ has a *regular* singular point at p_j. Otherwise, p_j is an *irregular* singular point of $\widetilde{P}$.

DEFINITION 3.12 (Quantum curve for a meromorphic Higgs bundle). Let (E, ϕ) be a meromorphic Higgs bundle defined over a projective algebraic curve C of any genus with poles along an effective divisor D, and $\Sigma \subset \overline{T^*C}$ its spectral curve. A **quantum curve** associated with Σ is the meromorphic extension of a Rees $\widetilde{\mathcal{D}_C}$-module $\mathcal{E}$ on $C \setminus \mathrm{supp}(D)$ such that the closure of its semi-classical limit $\Sigma^\circ \subset T^*C|_{C\setminus \mathrm{supp}(D)}$ in the compactified cotangent bundle $\overline{T^*C}$ agrees with Σ.

4. Geometry of spectral curves in the compactified cotangent bundle

To construct quantum curves using the topological recursion, we need a smooth Eynard-Orantin spectral curve [**25**] for which we can apply the recursion mechanism. When the given Hitchin spectral curve Σ is singular, we have to find a non-singular model. In this paper we use the normalization $\widetilde{\Sigma}$ of the singular spectral curve. Since the quantum curve reflects the geometry of $\Sigma \subset \overline{T^*C}$, it is important to identify the choice of the blow-up space $Bl(\overline{T^*C})$ of (1.6) in which $\widetilde{\Sigma}$ is realized as a smooth divisor. We then determine the initial value $W_{0,1}$ for the topological recursion.

The geometry of a spectral curve also gives us the information of the singularity of the quantum curve. For example, when we have a component of a spectral curve tangent to the divisor C_∞, the quantum curve has an irregular singular point, and the class of the irregularity is determined by the degree of tangency. We will give a classification of the singularity of the quantum curves in terms of the geometry of spectral curves in Section 6.3.

In this section, we give the construction of the canonical blow-up space $Bl(\overline{T^*C})$, and determine the genus of the normalization $\widetilde{\Sigma}$. This genus is necessary to identify the Riemann prime form on it, which determines another input datum $W_{0,2}$ for the topological recursion.

There are two different ways of defining the spectral curve for Higgs bundles with meromorphic Higgs field. Our definition of the previous section uses the compactified cotangent bundle. This idea also appears in [**40**]. The traditional definition, which assumes the pole divisor D of the Higgs field to be reduced, is suitable for the study of moduli spaces of parabolic Higgs bundles. When we deal with non-reduced effective divisors, parabolic structures do not play any role. Non-reduced divisors appear naturally when we deal with classical equations such as the Airy differential equation, which has an irregular singular point of class $\frac{3}{2}$ at $\infty \in \mathbb{P}^1$.

Although we do not use the terminology, our point of view of spectral curves is also closely related to considering the *stable pairs* of pure dimension 1 on $\overline{T^*C}$. Through Hitchin's abelianization idea, the moduli space of stable pairs and the moduli space of Higgs bundles are identified in [**37**].

Let E be an algebraic vector bundle of rank 2 on a non-singular projective algebraic curve C of genus g, and

$$\phi : E \longrightarrow K_C(D) \otimes_{\mathcal{O}_C} E$$

a meromorphic Higgs field with poles along an effective divisor D. The trace and the determinant of ϕ,

$$a_1 := -\mathrm{tr}(\phi) \in H^0\left(C, K_C(D)\right), \tag{4.1}$$

$$a_2 := \det(\phi) \in H^0\left(C, K_C^{\otimes 2}(2D)\right), \tag{4.2}$$

are well defined and determine the spectral curve Σ of (3.9). For the purpose of investigating the geometry of Σ, we do not need the information of the Higgs bundle (E, ϕ), or even the pole divisor D. Thus in what follows, we only assume that a_1 is a meromorphic section of K_C, and that a_2 a meromorphic section of $K_C^{\otimes 2}$. Then the spectral curve is re-defined as the zero-locus in $\overline{T^*C}$ of a quadratic equation with a_1 and a_2 its coefficients:

$$\Sigma := \left(\eta^2 + \pi^*(a_1)\eta + \pi^*(a_2)\right)_0. \tag{4.3}$$

The only condition we impose here is that *the spectral curve is irreducible.* In the language of Higgs bundles, this condition corresponds to the stability of (E, ϕ).

Recall that $\mathrm{Pic}(\overline{T^*C})$ is generated by the zero section C_0 of T^*C and fibers of the projection map $\pi : \overline{T^*C} \longrightarrow C$. Since the spectral curve Σ is a double covering of C, as a divisor it is expressed as

$$\Sigma = 2C_0 + \sum_{j=1}^{a} \pi^*(p_j) \in \mathrm{Pic}(\overline{T^*C}), \tag{4.4}$$

where $\alpha = \sum_{j=1}^{a} p_j \in \mathrm{Pic}^a(C)$ is a divisor on C of degree a. As an element of the Néron-Severi group

$$\mathrm{NS}(\overline{T^*C}) = \mathrm{Pic}(\overline{T^*C})/\mathrm{Pic}^0(\overline{T^*C}),$$

it is simply

$$\Sigma = 2C_0 + aF \in \mathrm{NS}(\overline{T^*C})$$

for a typical fiber class F. Since the intersection $FC_\infty = 1$, we have $a = \Sigma C_\infty$ in $\mathrm{NS}(\overline{T^*C})$. From the genus formula

$$p_a(\Sigma) = \frac{1}{2}\Sigma \cdot (\Sigma + K_{\overline{T^*C}}) + 1$$

and

$$K_{\overline{T^*C}} = -2C_0 + (4g-4)F \in \mathrm{NS}(\overline{T^*C}),$$

we find that the arithmetic genus of the spectral curve Σ is

$$p_a(\Sigma) = 4g - 3 + a, \tag{4.5}$$

where a is the number of intersections of Σ and C_∞. Now we wish to find the geometric genus of Σ.

Motivated by the completion of square expression of the defining equation (4.3),

$$\eta^2 + \pi^*(a_1)\eta + \pi^*(a_2) - \left(\eta + \frac{1}{2}\pi^*(a_1)\right)^2 - \left(\frac{1}{4}\pi^*(a_1)^2 - \pi^*(a_2)\right) \tag{4.6}$$

as a meromorphic section of $\pi^* K_C^{\otimes 2}$, we give the following definition.

DEFINITION 4.1 (Discriminant divisor). The **discriminant divisor** of the spectral curve (4.3) is a divisor on C defined by

$$\Delta := \left(\frac{1}{4}a_1^2 - a_2\right) = \Delta_0 - \Delta_\infty, \tag{4.7}$$

where

$$\Delta_0 = \sum_i m_i q_i, \quad m_i > 0, \quad q_i \in C, \tag{4.8}$$

$$\Delta_\infty = \sum_j n_j p_j, \quad n_j > 0, \quad p_j \in C. \tag{4.9}$$

Since $\frac{1}{4}a_1^2 - a_2$ is a meromorphic section of $K_C^{\otimes 2}$, we have

$$\deg \Delta = \deg \Delta_0 - \deg \Delta_\infty = \sum_i m_i - \sum_j n_j = 4g - 4. \tag{4.10}$$

THEOREM 4.2 (Geometric genus formula). *Let us define an invariant of the discriminant divisor by*

$$\delta = |\{i \mid m_i \equiv 1 \mod 2\}| + |\{j \mid n_j \equiv 1 \mod 2\}|. \tag{4.11}$$

Then the geometric genus of the spectral curve Σ of (4.3) is given by

$$\tilde{g} := p_g(\Sigma) = 2g - 1 + \frac{1}{2}\delta. \tag{4.12}$$

We note that (4.10) implies $\delta \equiv 0 \mod 2$.

REMARK 4.3. If ϕ is a holomorphic Higgs field, then $a_1 = -\mathrm{tr}(\phi)$, $a_2 = \det(\phi)$, and

$$\deg \Delta_0 = \delta = 4g - 4, \quad \deg \Delta_\infty = 0.$$

Therefore we have $\tilde{g} = 4g - 3$, which agrees with the genus formula of [**18**, Eq.(2.5)].

Before giving the proof of the formula, first we wish to identify the geometric meaning of the invariant δ. Since $\Sigma \subset \overline{T^*C}$ is a double covering of C in a ruled surface, locally at every singular point p, Σ is either irreducible, or reducible and consisting of two components. When irreducible, it is locally isomorphic to

$$t^2 - s^{2m+1} = 0, \qquad m \geq 1. \tag{4.13}$$

If it has two components, then it is locally isomorphic to

$$t^2 - s^{2m} = (t - s^m)(t + s^m) = 0. \tag{4.14}$$

Since the local form of Σ at a ramification point of $\pi : \Sigma \longrightarrow C$ is written as (4.13) with $m = 0$, by extending the terminology "singularity" to "critical points" of the morphism π, we include a ramification point as a cusp with $m = 0$.

PROPOSITION 4.4. *The invariant* δ *of* (4.11) *counts the number of cusps of the spectral curve* Σ.

Thus we have

$$p_g(\Sigma) = 2g(C) - 1 + \frac{1}{2} \text{ (the number of cusps).} \tag{4.15}$$

PROOF OF (4.15), ASSUMING PROPOSITION 4.4. Let $\nu_{\min} : \Sigma_{\min} \longrightarrow \Sigma$ be the minimal resolution of singularities of Σ. Then $\pi_{\min} = \pi \circ \nu_{\min} : \Sigma_{\min} \longrightarrow C$ is a double sheeted covering of C by a smooth curve $\Sigma_{\min}$. If Σ has two components at a singularity P as in (4.14), then $\pi_{\min}^{-1}(P)$ consists of two points and $\pi_{\min}$ is not ramified there. If P is a cusp (4.13), then $\pi_{\min}^{-1}(P)$ is a ramification point of the covering $\pi_{\min}$. If δ counts the total number of cusp singularities and the ramification points of $\pi : \Sigma \longrightarrow C$, then the Riemann-Hurwitz formula gives us

$$2 - 2g(\Sigma_{\min}) - \delta = 2(2 - 2g(C) - \delta),$$

which yields the genus formula (4.15). □

Since we wish to give all information of (1.6) from the defining equation (4.3), we proceed to derive a local structure of Σ at each singularity from the global equation in what follows.

PROOF OF THEOREM 4.2 AND PROPOSITION 4.4. The proof is broken into four cases. Cases 3 and 4 are related to the Newton Polygon we mentioned in Introduction, (1.17).

CASE 1. First we consider the case with a *holomorphic* $a_1 \in H^0(C, K_C)$, and both Δ_0 and Δ_∞ are reduced. As we see below, in this case Σ is non-singular, and the two genera (4.5) and (4.12) agree.

Let us consider the *graph* $\Gamma_{-\frac{1}{2}a_1}$ of the holomorphic 1-form $-\frac{1}{2}a_1$ in T^*C. Since T^*C is the total space of the canonical bundle K_C, the graph is a cross-section of T^*C. We define an involution $\sigma : \overline{T^*C} \longrightarrow \overline{T^*C}$ as a reflection about $\Gamma_{-\frac{1}{2}a_1}$ along each fiber of π. In terms of the fiber coordinate $y \in T_x^*C$, it is written as

$$ydx \longmapsto -ydx - a_1, \qquad ydx \in \overline{T_x^*C}. \tag{4.16}$$

The spectral curve is invariant under the involution, $\sigma : \Sigma \longrightarrow \Sigma$, because of (4.6). By definition, $\Gamma_{-\frac{1}{2}a_1} \subset T^*C$ is a fixed-point set of the involution σ. The divisor C_∞ is also fixed by σ. Note that we have in this case

$$\mathrm{Gal}(\Sigma/C) = \langle \sigma \rangle. \tag{4.17}$$

Thus for a holomorphic a_1, the Galois action of $\pi : \Sigma \longrightarrow C$ extends to the whole ruled surface $\overline{T^*C}$. This does not hold for a meromorphic a_1.

As remarked above, if $a_2 \in H^0(C, K_C^{\otimes 2})$ is also holomorphic, then $\pi : \Sigma \longrightarrow C$ is simply branched over $\Delta = \Delta_0$, and Σ is a smooth curve of genus $4g - 3$. This is in agreement of (4.5) because $n = 0$ in this case.

If a_2 is meromorphic, then the pole divisor of a_2 is given by $(a_2)_\infty = \Delta_\infty$ of degree n. Since Δ_∞ is reduced, from (4.6) we see that $\pi : \Sigma \longrightarrow C$ is ramified at the intersection of C_∞ and $\pi^{-1}(\Delta_\infty)$. The spectral curve is also ramified at its intersection with $\Gamma_{-\frac{1}{2}a_1}$, which occurs along the fibers $\pi^{-1}(\Delta_0)$. Note that $\deg \Delta_0 = 4g - 4 + n$ because of (4.10). Thus $\pi : \Sigma \longrightarrow C$ is simply ramified at a total of $4g-4+2n$ points. Therefore, Σ is non-singular, and we deduce that its genus is given by

$$p_g(\Sigma) = p_a(\Sigma) = 4g-3+n$$

from the Riemann-Hurwitz formula. As a divisor class, we have

$$\Sigma = 2C_0 + \pi^*(\Delta_\infty) \in \mathrm{Pic}(\overline{T^*C}),$$

in agreement of (4.4).

CASE 2. Still $a_1 \in H^0(C, K_C)$ is holomorphic, but Δ is non-reduced. The first example of Table 1.2, the Airy differential equation, at $x=\infty$ falls into this category.

The involution (4.16) is well defined. Let $q_i \in \mathrm{supp}(\Delta_0)$ be a point at which $m_i > 1$. From the global equation (4.6), we see that the curve germ of Σ at its intersection Q with the fiber $\pi^{-1}(q_i)$ is given by a formula

$$y^2 = x^{m_i},$$

where x is the pull-back of the base coordinate on C and y a fiber coordinate, possibly tilted by a holomorphic function in x. We blow up once at $(x,y)=(0,0)$, using a local parameter $y_1 = y/x$ on the exceptional divisor. The proper transform of the curve germ becomes

$$y_1^2 = x^{m_i-2}.$$

Repeat this process at $(x,y_1)=(0,0)$, until we reach the equation

$$y_\ell^2 = x^\epsilon,$$

where $\epsilon = 0$ or 1. The proper transform of the curve germ is now non-singular. We see that after a sequence of $\lfloor \frac{m_i}{2} \rfloor$ blow-ups starting at the point $Q \in \Sigma \cap \pi^{-1}(q_i)$, the proper transform of Σ is simply ramified over $q_i \in C$ if m_i is odd, and unramified if m_i is even. We apply the same sequence of blow-ups at each q_i with higher multiplicity.

Now let $p_j \in \mathrm{supp}(\Delta_\infty)$ with $n_j > 1$. The intersection $P = \Sigma \cap \pi^{-1}(p_j)$ lies on C_∞, and Σ has a singularity at P. Let $z = 1/y$ be a fiber coordinate of $\pi^{-1}(p_j)$ at the infinity. Then the curve germ of Σ at the point P is given by

$$z^2 = x^{n_j}.$$

The involution σ in this coordinate is $z \longmapsto -z$. The blow-up process we apply at P is the same as before. After $\lfloor \frac{n_j}{2} \rfloor$ blow-ups starting at the point $P \in \Sigma \cap \pi^{-1}(p_j)$, the proper transform of Σ is simply ramified over $p_j \in C$ if n_j is odd, and unramified if n_j is even. Again we do this process for all p_j with a higher multiplicity.

Let us define $Bl_{\min}(\overline{T^*C})$ as the application of a total of

$$\sum_{i=1}^{m} \left\lfloor \frac{m_i}{2} \right\rfloor + \sum_{j=1}^{n} \left\lfloor \frac{n_j}{2} \right\rfloor \tag{4.18}$$

times blow-ups on $\overline{T^*C}$ as described above.

$$
(4.19)\qquad
\begin{array}{ccccc}
\Sigma_{\min} & & \xrightarrow{\;\bar{i}\;} & & Bl_{\min}(\overline{T^*C}) \\
\Big\downarrow \bar{\pi} & \searrow^{\nu_{\min}} & & & \searrow^{\nu_{\min}} \\
 & & \Sigma & \xrightarrow{\;i\;} & \overline{T^*C}. \\
 & \swarrow_{\pi} & & \swarrow_{\pi} & \\
C & & & &
\end{array}
$$

The proper transform $\Sigma_{\min}$ is the minimal desingularization of Σ. Note that the morphism

$$
\bar{\pi} = \pi \circ \nu : \Sigma_{\min} \longrightarrow C \tag{4.20}
$$

is a double covering, ramified exactly at δ points. Since $p_a(\Sigma_{\min}) = p_g(\Sigma)$, (4.12) follows from the Riemann-Hurwitz formula applied to $\bar{\pi} : \Sigma_{\min} \longrightarrow C$. It is also obvious that δ counts the number of cusp points of Σ, including smooth ramification points of π.

CASE 3. We are now led to considering a meromorphic a_1. Let $p \in C$ be a pole of a_1 of order $k \geq 1$. Assume that a_2 also has a pole of oder ℓ at p, and that $\ell > k$. The second, the third, and the fourth examples of Table 1.2, all at $x = \infty$, fall into this category.

Choose a local coordinate x of C around p, and express

$$
a_1 = \frac{c_1}{x^k}, \qquad a_2 = \frac{c_2}{x^\ell},
$$

where c_1 and $c_2 \in \mathcal{O}_{C,p}$ are unit elements. Since both of a_1 and a_2 have poles at p, the spectral curve intersects with C_∞ along the fiber $\pi^{-1}(p)$. The curve germ at this intersection point is given by the equation

$$
y^2 + \frac{c_1}{x^k} y + \frac{c_2}{x^\ell} = 0, \tag{4.21}
$$

or equivalently,

$$
z^2 + \frac{c_1}{c_2} x^{\ell-k} z + \frac{1}{c_2} x^\ell = \left(z + \frac{1}{2}\frac{c_1}{c_2} x^{\ell-k}\right)^2 - \frac{1}{4}\left(\frac{c_1}{c_2} x^{\ell-k}\right)^2 + \frac{1}{c_2} x^\ell = 0, \tag{4.22}
$$

where $z = 1/y$ is a fiber coordinate of $\pi^{-1}(p)$ at infinity. Note that the coefficients of (4.22) are all in $\mathcal{O}_{C,p}$. The discriminants of (4.21) and (4.22) are given by

$$
\Delta_y := \frac{1}{4}\frac{c_1^2}{x^{2k}} - \frac{c_2}{x^\ell}, \qquad \Delta_z := \frac{1}{4}\left(\frac{c_1}{c_2} x^{\ell-k}\right)^2 - \frac{1}{c_2} x^\ell. \tag{4.23}
$$

If $2k > \ell$, then the contribution from p in Δ_∞ is $-2kp$, which does not count in δ. Since this inequality is equivalent to $2(\ell - k) < \ell$, the contribution from p in the discriminant Δ_z is $2(\ell - k)p$. Locally around the singularity, the spectral curve is thus reducible with two components. We can apply the blow-up process of Case 2 to (4.22) and obtain a resolved curve germ unramified over $p \in C$. For the case of the Hermite differential equation given as the second example of Table 1.2, we have $k = 3$ and $\ell = 4$.

If $2k \leq \ell$, then the contribution from p in Δ_∞ is $-\ell p$. Therefore, depending on the parity of ℓ, it has a contribution to δ. The infinity point $x = \infty$ of the Gauß hypergeometric equation, the third example of Table 1.2, falls into this case, where we have $k = 1$ and $\ell = 2$. The inequality $2k \leq \ell$ is the same as $2(\ell - k) \geq \ell$, hence the contribution of p in Δ_z is ℓp. Therefore, whether the resolved curve germ is ramified or unramified depends on the parity of ℓ, which is exactly recorded in δ. If it is odd, then the singularity is a cusp, contributing 1 to δ.

The above consideration shows that we need to perform $\ell - k$ times blow-ups if $2k > \ell$, and $\lfloor \frac{\ell}{2} \rfloor$ times blow-ups if $2k \leq \ell$, to construct $Bl_{\text{min}}(\overline{T^*C})$ and Σ_{min}.

CASE 4. Finally, we assume that a_1 has a pole of oder $k \geq 1$ at $p \in C$, and a_2 has a pole of order ℓ at p, with $k \geq \ell$. We allow a_2 to be holomorphic at p. The third example of Table 1.2, the Gauß hypergeomtric equation at $x = 0, 1$, and the final example, at $x = \pm 1, \infty$, fall into this case.

The equation of the spectral curve is the same as (4.21), and its discriminant is given by Δ_y of (4.23). Since $k \geq \ell$, the contribution of p in Δ_∞ is $-2kp$, which is not counted in δ. Let us re-write (4.22) as

$$x^{k-\ell} z^2 + \frac{c_1}{c_2} z + \frac{1}{c_2} x^k = 0. \tag{4.24}$$

Since $c_1/c_2 \in \mathcal{O}_{C,p}$ is a unit, we can see from this equation that the curve germ passes through $(x, z) = (0, 0)$ only once as a regular point. Indeed the discriminant of (4.24) does not vanish at $x = 0$. In particular, Σ is non-singular at its intersection of $\pi^{-1}(p)$. Therefore, $p \in C$ does not contribute into the Rimann-Hurwitz formula, which agrees with the fact that δ does not record p.

This completes the proof of Theorem 4.2, and the fact that δ counts the total number of odd cusps on Σ. □

The proof of the above theorem give us the way to construct the blow-up space $Bl(\overline{T^*C})$ of (1.6). The data we need is not only the discriminant divisor (4.7), but also the pole divisors of the coefficients of the defining equation (4.3) of the spectral curve. Let us write

$$(a_1)_\infty = \sum_{j-1}^{n} k_j p_j, \qquad (a_2)_\infty = \sum_{j-1}^{n} \ell_j p_j, \tag{4.25}$$

where $\{p_1, \ldots, p_j\} = \text{supp}(\Delta_\infty)$. At each p_j, a Newton polygon is defined as the upper part of the convex hull of three points $(0,0), (1, k_j), (2, \ell_j) \in \mathbb{R}^2$, as in Definition 1.10. We also define the invariant

$$r_j = \begin{cases} k_j & 2k_j \geq \ell_j, \\ \frac{\ell_j}{2} & 2k_j \leq \ell_j. \end{cases} \tag{4.26}$$

If our mission is only to resolve the singularities of Σ, then we can use the following blow-up method.

DEFINITION 4.5 (Construction of the minimal blow-up space). The minimal blow-up space $Bl_{\text{min}}(\overline{T^*C})$ of (4.19) is defined by blowing up $\overline{T^*C}$ in the following way, as analyzed in the proof of Theorem 4.2.

- At each q_i of (4.8), blow up at the intersection $\Sigma \cap \pi^{-1}(q_i)$ a total of $\lfloor \frac{m_i}{2} \rfloor$ times.
- At each p_j of (4.9), blow up at the intersection $\Sigma \cap \pi^{-1}(p_j)$ a total of
  (1) $\lfloor \frac{n_j}{2} \rfloor$ times, if $k_j = 0$, or $k_j > 0$ and $\ell_j \geq 2k_j$, and
  (2) $\ell_j - k_j$ times, if $k_j > 0$ and $2\ell_j > 2k_j > \ell_j$.
  Here, k_j (ℓ_j, resp.) is the order of pole of a_1 (a_2, resp.) at p_j.

REMARK 4.6. The last case, $k_j > 0$ and $2\ell_j > 2k_j > \ell_j$, is counter intuitive and does not follow the rest of the pattern. The singularity of the spectral curve of the Hermite differential equation at $x = \infty$ (the second row of Table 1.2) gives a good example. While the pole divisor of the discriminant has order 6, and the intersection of the spectral curve Σ and C_∞ has degree 4, we only need one time blow-up.

The cumbersome definition of $Bl_{\min}(\overline{T^*C})$ becomes simple if we appeal to the Newton polygon.

DEFINITION 4.7 (Construction of the blow-up space). The blow-up space $Bl(\overline{T^*C})$ of (1.6) is defined by blowing up $\overline{T^*C}$ in the following way.

- At each q_i of (4.8), blow up at the intersection $\Sigma \cap \pi^{-1}(q_i)$ a total of $\lfloor \frac{m_i}{2} \rfloor$ times.
- At each p_j of (4.9), blow up at the intersection $\Sigma \cap \pi^{-1}(p_j)$ a total of $\lceil r_j \rceil$ times.

THEOREM 4.8. *In the blow-up space $Bl(\overline{T^*C})$, we have the following.*

- *The proper transform $\widetilde{\Sigma}$ of the spectral curve $\Sigma \subset \overline{T^*C}$ by the birational morphism $\nu : Bl(\overline{T^*C}) \longrightarrow \overline{T^*C}$ is a smooth curve with a holomorphic map $\tilde{\pi} = \pi \circ \nu : \widetilde{\Sigma} \longrightarrow C$.*
- *The proper transform of C_∞ and $\widetilde{\Sigma}$ do not intersect in $Bl(\overline{T^*C})$.*
- *The Galois action $\sigma : \Sigma \longrightarrow \Sigma$ lifts to an involution of $\widetilde{\Sigma}$, and the morphism*

$$\tilde{\pi} : \widetilde{\Sigma} \longrightarrow C \tag{4.27}$$

is a Galois covering with the Galois group $\mathrm{Gal}(\widetilde{\Sigma}/C) = \langle \tilde{\sigma} \rangle \cong \mathbb{Z}/2\mathbb{Z}$.

$$\begin{array}{ccccc} \widetilde{\Sigma} & \xrightarrow{\ \nu\ } & \Sigma & \xrightarrow{\ \pi\ } & C \\ {\scriptstyle \tilde{\sigma}}\downarrow & & \downarrow{\scriptstyle \sigma} & & \| \\ \widetilde{\Sigma} & \xrightarrow[\ \nu\ ]{} & \Sigma & \xrightarrow[\ \pi\ ]{} & C \end{array} \tag{4.28}$$

PROOF. The curve we are trying to desingularize is the support $\Sigma \cup C_\infty$ of the total divisor (1.4) of the characteristic polynomial. Even if Σ is smooth at its intersection with C_∞, the support $\Sigma \cup C_\infty$ is always singular. The key point is that the $\lceil r_j \rceil$ times blow-up at the intersection is exactly what we need to desingularize $\Sigma \cup C_\infty$.

Let $P \in \Sigma \cap C_\infty$ so that $\pi(P) = p_j$. We drop the index j in the rest of this proof.

The only case P is a smooth point of Σ is Case 4. From (4.24), we see that the spectral curve near P is given by $z = x^k$. Thus it is tangent to C_∞ with the multiplicity k. Therefore, we need k times blow-ups to separate the proper transforms of Σ and C_∞. By (4.26), we have $r = k$.

The point P is a cusp singularity only when $0 \le 2k \le \ell$ and ℓ is odd. We have $r = \ell/2$, and we need $\lfloor \ell/2 \rfloor$-times blow-ups to desingularize Σ at P. To separate the proper transforms of Σ and C_∞ in the end, we need one more blow-up. Therefore, we need a total of $\lceil r \rceil$ blow-ups.

If $0 \leq 2k \leq \ell$ and ℓ is even, then still we have $r = \ell/2$. In this case the spectral curve is locally reducible at P, and requires r-times blow-ups for desingularization. Since the proper transform of P consists of two distinct points, the proper transforms of Σ and C_∞ are separated after the r-th blow-up.

The remaining case is $0 < k < \ell < 2k$. We have $r = k$. We need only $(\ell - k)$ times blow-ups to desingularize the spectral curve at P. Let us take a close look at (4.22). We take $z^2 = 0$ to see the infinitesimal relation between Σ and C_∞. Then the equation becomes

$$c_1 z + x^k = 0, \tag{4.29}$$

which represents an irreducible component of the spectral curve near P that is tangent to C_∞. The degree of tangency is k, hence it requires k-times blowing up to separate the proper transforms of Σ and C_∞.

Since the spectral curve Σ is a double covering of C, $\text{Gal}(\Sigma/C) \cong \mathbb{Z}/2\mathbb{Z}$. The involution σ is its generator, which may or may not extend to the whole $\overline{T^*C}$. Since we construct $\Sigma_{\min} \subset Bl(\overline{T^*C})$ as a simply ramified double covering over C in Theorem 4.2, it is non-singular and there is a natural involution on it. The additional blow-ups

$$\bar{\nu} : Bl(\overline{T^*C}) \longrightarrow Bl_{\min}(\overline{T^*C})$$

do not affect the proper transform $\widetilde{\Sigma}$ of $\Sigma_{\min}$, which also has an involution $\tilde{\sigma}$. The involution $\tilde{\sigma}$ agrees with σ on the complement of the singular locus of Σ, thus satisfying $\nu \circ \sigma = \tilde{\sigma} \circ \nu$.

This completes the proof. □

5. The spectral curve as a divisor and its minimal resolution

In this section we give a formula for the minimal resolution $\Sigma_{\min}$ of the spectral curve Σ as an element of the Picard group $\text{Pic}\big(Bl_{\min}(\overline{T^*C})\big)$. We also give a genus formula for $\Sigma_{\min}$ in terms of its geometry in $Bl_{\min}(\overline{T^*C})$. This gives another interpretation of the invariant δ of the genus formula (4.12).

The Picard group of $\overline{T^*C}$ is generated by the pull-back of $\text{Pic}(C)$ and the zero-section C_0 of the cotangent bundle T^*C. Denote by F the fiber over a point of the morphism $\pi : \overline{T^*C} \longrightarrow C$. We have the following intersection table:

$$\begin{aligned} F^2 &= 0 \\ C_0^2 &= 2g(C) - 2 \\ FC_0 &= 1. \end{aligned} \tag{5.1}$$

For the sake of simplicity, in what follows we denote simply

$$\alpha F := \pi^*(\alpha) \tag{5.2}$$

for any divisor $\alpha \in \text{Pic}(C)$ (see [**32**, Chapter V.2]). In this notation, the canonical divisor of $\overline{T^*C}$ is given by

$$K_{\overline{T^*C}} = -2C_0 + \beta F \tag{5.3}$$

with a divisor $\beta \in \text{Pic}^{4g-4}(C)$ of degree $4g - 4$.

In Section 4 we identified a spectral curve Σ as a divisor in the ruled surface $\overline{T^*C}$. With the convention of (5.2), (4.4) reads

$$\Sigma = \alpha F + 2C_0, \tag{5.4}$$

where

$$\alpha = \Sigma C_\infty \in \mathrm{Pic}(C_\infty) \cong \mathrm{Pic}(C)$$

is a divisor of degree a on C.

Since we deal only with the spectral curve of a rank 2 Higgs bundle, its singularities are mild, and we can describe their resolution in detail. A singular spectral curve has *infinitely near* singularities, which require iterative sequence of blow-ups on the ruled surface to be resolved. For every singular point, say $P \in \Sigma$, we introduce a sequence of blow-ups

$$Bl_i^P(\overline{T^*C}) \xrightarrow{\nu_i^P} \cdots \xrightarrow{\nu_2^P} Bl_1^P(\overline{T^*C}) \xrightarrow{\nu_1^P} Bl_0^P(\overline{T^*C}) = \overline{T^*C}. \tag{5.5}$$

Here, $\nu_{j+1}^P : Bl_{j+1}^P(\overline{T^*C}) \longrightarrow Bl_j^P(\overline{T^*C})$, $j = 1, \cdots, i-1$, is a blow-up at the singular point of $\Sigma_j \subset Bl_j^P(\overline{T^*C})$, and Σ_j is the proper transform of $\Sigma_{j-1} \subset Bl_{j-1}^P(\overline{T^*C})$ under the blow-up ν_j^P. Each ν_j^P introduces an exceptional divisor E_j^P with self-intersection -1 on $Bl_j^P(\overline{T^*C})$. By abuse of notation, we also write

$$E_j^P := \left(\nu_{j+1}^P\right)^*(E_j^P) - E_{j+1}^P, \tag{5.6}$$

which is the proper transform of the divisor E_j^P on $Bl_j^P(\overline{T^*C})$ by ν_{j+1}^P. On $Bl_{i+1}^P(\overline{T^*C})$, we have a chain of self-intersection -2 curves with the following intersections properties:

$$\begin{aligned} &\left(E_1^P\right)^2 = \ldots = \left(E_i^P\right)^2 = -2 \\ &\left(E_{i+1}^P\right)^2 = -1 \\ &E_{j-1}^P E_j^P = 1, \text{ for all } 2 \le j \le i+1 \\ &E_j^P E_k^P = 0, \text{ for all } 1 \le j,k \le i+1 \text{ with } |j-k| > 1. \end{aligned} \tag{5.7}$$

From (5.4), we see that Σ has only infinitely near double singularities. We denote by

$$\Sigma_i \xrightarrow{\nu_i^P} \cdots \xrightarrow{\nu_2^P} \Sigma_1 \xrightarrow{\nu_1^P} \Sigma_0 = \Sigma \tag{5.8}$$

the sequence of proper transforms of Σ under (5.5). The **multiplicity** of the singularity $P \in \Sigma$ is defined to be the least i of (5.8) such that Σ_i is non-singular.

THEOREM 5.1. *Let $Q_1, Q_2, \ldots, Q_c$ be the singularities of Σ not on the zero section C_0 or on the divisor C_∞, and $Q_{c+1}, \ldots, Q_s$ the singularities on C_∞. We denote by n_k the multiplicity of Q_k, $k = 1, 2, \ldots, s$. Let*

$$\Sigma_{\min} \subset Bl_{\min}(\overline{T^*C}) \xrightarrow{\nu_{\min}} T^*C \tag{5.9}$$

be the minimal resolution of Σ after performing the blow-ups at each singularity exactly as required, which includes blow-ups on the singularities on C_0. Then the genus of the smooth curve $\Sigma_{\min}$ is given by

$$p_a(\Sigma_{\min}) = 2g - 1 + \frac{N_0 + N_\infty}{2} - \sum_{k=1}^{c} n_k, \tag{5.10}$$

where N_0 (resp. N_∞) is the number of intersection points of $\Sigma_{\min}$ with the proper transform of C_0 (resp. C_∞) in $Bl_{\min}(\overline{T^*C})$.

Denote by E_j^k the exceptional divisor of (5.6) for $P = Q_k$, $k = 1, 2, \ldots, s$. Then as an element of the Picard group, we have

$$\Sigma_{\min} = 2C_0 + \alpha F - 2\sum_{k=1}^{s}\sum_{j=1}^{n_k} jE_j^k \in \operatorname{Pic}\left(Bl_{\min}(\overline{T^*C})\right), \tag{5.11}$$

where $\alpha \in \operatorname{Pic}(C)$ is a divisor on C of degree

$$a = N_\infty + 2\sum_{k=c+1}^{s} n_k. \tag{5.12}$$

REMARK 5.2. If Σ is smooth, then $N_0 - N_\infty = 4g-4$ and $a = N_\infty$. Therefore, (5.10) agrees with (4.5).

PROOF. Let us denote by $P_1, \ldots, P_t$ the singular points of Σ on the zero section C_0 with multiplicities $m_1, \ldots, m_t$. To avoid confusion, we denote by $G_j^k := E_j^{P_k}$ the exceptional divisor of (5.6) at $P = P_k$. The construction of $\Sigma_{\min}$ and $Bl_{\min}(\overline{T^*C})$ requires also sequences of blow-ups at these points. We use the same notation C_0 for the proper transform of the zero section via any of the blow-up appearing in this construction of the minimal resolution.

The Picard group $\operatorname{Pic}\left(Bl_{\min}(\overline{T^*C})\right)$ is generated by $\operatorname{Pic}(C)$, the divisors E_j^k's and G_j^k's, and C_0. These generators satisfy, in addition to (5.7), the following:

$$\begin{aligned} C_0^2 &= 2g - 2 - \sum_{k=1}^{t} m_k \\ C_0 G_{m_k} &= 1, \text{ for every } 1 \le k \le t. \end{aligned} \tag{5.13}$$

Since the singular points of the spectral curve are not in general position, to give an explicit expression for $\Sigma_{\min}$ as a divisor, we consider two separate cases.

(1) Resolving singularities of Σ on the zero section C_0.

For a singular point P_k, the resolution Σ_{m_k} is of the form

$$\Sigma_{m_k} = \alpha_k F + 2C_0, \tag{5.14}$$

where $\alpha_k \in \operatorname{Pic}(C)$. At each step of the blow-ups, the canonical divisor of $Bl_j^{P_k}(\overline{T^*C})$, for $j \le m_k$, is $K_j = (\nu_j)^*(K_{j-1}) + G_j$. Therefore,

$$\begin{aligned} K_j &= (\nu_j)^*(K_{j-1}) + G_j \\ &= -2(C_0 + G_j) - (\nu_j)^*\left(\sum_{\ell=1}^{j-1} \ell G_\ell\right) + \beta F + G_j \\ &= -2C_0 - \sum_{\ell=1}^{j-2} \ell G_\ell - (j-1)(G_{j-1} + G_j) - G_j + \beta F \\ &= -2C_0 + \beta F - \sum_{\ell=1}^{j} \ell G_\ell, \end{aligned}$$

where β is the divisor of (5.3).

(2) Resolving singularities of Σ not on the zero section.

We now consider the singular point Q_k. The proper transform of Σ under n_k iterated blow-ups is

$$\Sigma_{n_k} = \Sigma - 2\sum_{i=1}^{n_k}\left(\sum_{j=1}^{i} E_{n_k-j+1}\right) = \Sigma - 2\sum_{i=1}^{n_k} iE_i. \tag{5.15}$$

The canonical divisor on the blown up ruled surface at the point $Bl_i^{Q_k}(\overline{T^*C})$ is

$$\begin{aligned} K_i &= (\nu_i)^*(K_{i-1}) + E_i \\ &= -2C_0 + \beta F + (\nu_i)^*\left(\sum_{j=1}^{i-1} jE_j\right) + E_i \\ &= -2C_0 + \beta F + \sum_{j=1}^{i-2} jE_j + (i-1)(E_{i-1}+E_i) + E_i \\ &= -2C_0 + \beta F + \sum_{j=1}^{i} jE_j. \end{aligned}$$

Alter all these blowups, we obtain the expression of the canonical divisor on $Bl_{\min}(\overline{T^*C})$:

$$K_{Bl_{\min}(\overline{T^*C})} = -2C_0 + \beta F + \sum_{k=1}^{s}\sum_{i=1}^{n_k} iE_i^k - \sum_{k=1}^{t}\sum_{i=1}^{m_k} iG_i^k. \tag{5.16}$$

From (5.14) and (5.15) we obtain

$$\Sigma_{\min} = (2C_0 + \alpha F) - 2\sum_{k=1}^{s}\sum_{i=1}^{n_k} iE_i^k, \tag{5.17}$$

where α is the sum of all α_k of (5.14).

Let us now turn our attention to determining the degree of α. We recall the genus formula

$$p_a(\Sigma_{\min}) = \frac{\Sigma_{\min}\left(\Sigma_{\min} + K_{Bl_{\min}(\overline{T^*C})}\right)}{2} + 1.$$

Equations (5.16) and (5.17) yield

$$\Sigma_{\min} + K_{Bl_{\min}(\overline{T^*C})} = (\alpha+\beta)F - \sum_{k=1}^{s}\sum_{i=1}^{n_k} iE_i^k - \sum_{k=1}^{t}\sum_{i=1}^{m_k} iG_i^k.$$

Denoting $\deg \alpha = a$, the above genus formula yields

$$\begin{aligned}
&2p_a(\Sigma_{\min}) - 2 \\
&= 2(a+4g-4)C_0F - 2\sum_{k=1}^{t}\sum_{i=1}^{m_k} iC_0G_i^k + 2\left(\sum_{k=1}^{s}\sum_{i=1}^{n_k} iE_i^k\right)^2 \\
&= 2(a+4g-4) - 2\sum_{k=1}^{t} m_kC_0G_{m_k}^k + 2\sum_{k=1}^{s}\left(\sum_{i=1}^{n_k-1}(-2)i^2 - n_k^2 + 2\sum_{i=1}^{n_k-1} i(i+1)\right) \\
&= 2(a+4g-4) - 2\sum_{k=1}^{t} m_k + 2\sum_{k=1}^{s}\left(-n_k^2 + 2\sum_{k=1}^{n_k-1} i\right) \\
&= 2(a+4g-4) - 2\sum_{k=1}^{t} m_k - 2\sum_{k=1}^{s} n_k.
\end{aligned}$$

We therefore conclude that

$$p_a(\Sigma_{\min}) = a + 4g - 3 - \sum_{k=1}^{t} m_k - \sum_{k=1}^{s} n_k. \tag{5.18}$$

Since the proper transform of C_0 on $Bl_{\min}(\overline{T^*C})$ does not intersect with the exceptional divisors E_j^k's, from (5.17) we have

$$\Sigma_{\min} \cdot C_0 = (\alpha F + 2C_0)\, C_0 = N_0.$$

This yields

$$4g - 4 + a = \sum_{i=k}^{t} 2m_k + N_0. \tag{5.19}$$

The proper transform of C_∞ on $Bl_{\min}(\overline{T^*C})$ is given by

$$C_\infty - \sum_{k=1}^{t}\sum_{j=1}^{m_k} jE_j^k, \tag{5.20}$$

which we also denote simply by C_∞ if there is no confusion. We recall that on $Bl_{\min}(\overline{T^*C})$ we have

$$C_\infty F = 1, \qquad C_\infty C_0 = 0,$$

$$C_\infty E_j^k = \begin{cases} n_k & j = n_k, k = c+1, \dots, s \\ 0 & \text{otherwise.} \end{cases}$$

Thus from the intersection of (5.17) and C_∞, we obtain

$$a = N_\infty + 2\sum_{k=c+1}^{s} n_k. \tag{5.21}$$

This proves (5.12).

Substituting (5.21) in (5.19), we obtain

$$4g - 4 = 2\sum_{k=1}^{t} m_k - 2\sum_{k=c+1}^{s} n_k + N_0 - N_\infty,$$

which shows that $N_0 \pm N_\infty$ is even. From (5.18) we have

$$\begin{aligned} p_a(\Sigma_{\min}) &= (a+4g-3) - \sum_{k=1}^{t} m_k - \sum_{k=1}^{c} n_k - \sum_{k=c+1}^{s} n_k \\ &= 1 + N_0 + \sum_{k=1}^{t} m_k - \sum_{k=1}^{c} n_k - \sum_{k=c+1}^{s} n_k. \end{aligned}$$

From the above two equations, we obtain

$$4g-4 = 2p_a(\Sigma_{\min}) - 2 + 2\sum_{k=1}^{c} n_k - N_0 - N_\infty,$$

which yields

$$p_a(\Sigma_{\min}) = 2g - 1 + \frac{N_0+N_\infty}{2} - \sum_{k=1}^{c} n_k.$$

This completes the proof of Theorem 5.1. □

6. Construction of the quantum curve

When we say that the quantization of a characteristic equation

$$\eta^2 - \pi^* \mathrm{tr}(\phi)\eta + \pi^* \det(\phi) = 0 \tag{6.1}$$

of a Higgs field ϕ is a differential equation

$$\left(\left(\hbar\frac{d}{dx}\right)^2 - \mathrm{tr}(\phi(x))\left(\hbar\frac{d}{dx}\right) + \det(\phi(x))\right)\Psi(x,\hbar) = 0, \tag{6.2}$$

it may sound obvious. The point is that since x and the differential operator d/dx do not commute, there are many different differential equations other than (6.2) that correspond to the starting algebraic equation (6.1). The mechanism we use to identify the correct formula for the quantization is the topological recursion. In this section, first we formulate our main theorem of quantization. Then we give the definition of the topological recursion, using the desingularization of the spectral curve (1.6), constructed in Theorem 4.8 and Definition 4.7. The rest of the section is devoted to proving the main theorem.

6.1. The main theorem. The main theorem of this paper is a local construction of the quantum curve guided by the asymptotic expansion of its solutions, which is obtained by the topological recursion.

THEOREM 6.1 (Main Theorem). *Let C be a smooth projective curve of an arbitrary genus, and (E,ϕ) a rank 2 Higgs bundle consisting of a topologically trivial vector bundle E and an arbitrary meromorphic Higgs field ϕ. We denote by $\Sigma \subset \overline{T^*C}$ the spectral curve defined by (3.9). On every coordinate chart $U \subset C$ with a local coordinate x, there exists a Rees D-module $\widetilde{\mathcal{M}}$ whose semi-classical limit agrees with the spectral curve $\Sigma|_U$. A generator of $\widetilde{\mathcal{M}}$ is given by a differential operator*

$$P(x,\hbar) = \left(\hbar\frac{d}{dx}\right)^2 - \mathrm{tr}\phi(x)\left(\hbar\frac{d}{dx}\right) + \det\phi(x) \in \widetilde{D_C}(U), \tag{6.3}$$

so that we have

$$\widetilde{\mathcal{M}} = \widetilde{D_C}(U)/\widetilde{D_C}(U) \cdot P(x,\hbar). \tag{6.4}$$

Let $q \in C$ be one of the critical values of the projection $\pi : \Sigma \longrightarrow C$ that corresponds to a branch point of the desingularized covering $\tilde{\pi} : \widetilde{\Sigma} \longrightarrow C$. Then there exists a coordinate neighborhood $U_q \subset C$ of q with a coordinate x centered at q such that the following holds.

1. *For an arbitrary point $p \in U_q$, there is a contractible open neighborhood $V_p \subset U_q$ of p that does not contain q.*
2. *Choose an eigenvalue α of ϕ on V_p. Then there is an all-order asymptotic solution to the differential equation*

$$P(x,\hbar)\Psi^\alpha(x,\hbar) = 0 \tag{6.5}$$

   that is defined on V_p.

3. *The asymptotic expansion is given by*

$$\Psi^\alpha(x,\hbar) = \exp\left(\sum_{m=0}^{\infty} \hbar^{m-1} S_m^\alpha(x)\right). \tag{6.6}$$

- *The 0-th term $S_0^\alpha(x)$ is determined by solving (6.15).*
- *The first term $S_1^\alpha(x)$ is determined by solving (6.16).*
- *The second term $S_2^\alpha(x)$ is determined by solving (6.17).*
- *Then find $F_{1,1}(z)$ and $F_{0,3}(z_1,z_2,z_3)$ so that*

$$S_2^\alpha(x) = F_{1,1}\big(z(x)\big) + \frac{1}{6}F_{0,3}\big(z(x),z(x),z(x)\big).$$

  This can be achieved as follows. First integrate $W_{1,1}(z)$ of (6.12) to construct $F_{1,1}(z)$. We do the same for the solution $W_{0,3}(z_1,z_2,z_3)$ of (6.12). We now define

$$F_{0,3}(z_1,z_2,z_3) = \int\!\!\int\!\!\int W_{0,3}(z_1,z_2,z_3) + 2\left(f(z_1)+f(z_2)+f(z_3)\right), \tag{6.7}$$

  where

$$f(z) := \widetilde{S}_2(z) - \left(F_{1,1}(z) + \frac{1}{6}\int^z\!\!\int^z\!\!\int^z W_{0,3}(z_1,z_2,z_3)\right), \tag{6.8}$$

  and $\widetilde{S}_2(z)$ is the lift of $S_2^\alpha(x)$ to $\widetilde{\Sigma}$.
- *$S_m^\alpha(x)$ for $m \geq 2$ is given by*

$$S_m^\alpha(x) = \sum_{2g-2+n=m-1} \frac{1}{n!} F_{g,n}^\alpha(x). \tag{6.9}$$

- *The free energies $F_{g,n}(z_1,\cdots,z_n)$ for $2g-2+n>1$ are determined by the differential recursion (6.6) from the initial data $F_{1,1}$ and $F_{0,3}$ as constructed above.*
- *Each $F_{g,n}^\alpha(x)$ is the* ***principal specialization*** *of the restriction of the free energy to the open subset $V_\alpha \subset \Sigma$ of Σ that corresponds to the eigenvalue α on V_p, which we identify with V_p by $\pi : V_\alpha \xrightarrow{\sim} V_p$.*

REMARK 6.2. Since $\operatorname{tr}\phi$ and $\det\phi$ are globally defined meromorphic sections of K_C and $K_C^{\otimes 2}$, respectively, which behave differently from global Rees differential operators, the Rees D-module $\widetilde{\mathcal{M}}$ is only locally defined. Construction of global Rees D-modules as quantum curves is achieved in a totally different method in [**19**].

6.2. The topological recursion and the WKB method. Let us start with defining each terminology in the main theorem.

Although the topological recursion should be a general procedure applicable to an arbitrary ramified covering of a base curve C of any degree, for the purpose of quantization in this paper, we need a Galois covering, and we also need to calculate the residues in the formula. Therefore, we deal with the topological recursion only for a covering of degree 2 in this paper.

DEFINITION 6.3 (Integral topological recursion for a degree 2 covering). Let C be a non-singular projective algebraic curve, and $\tilde{\pi} : \widetilde{\Sigma} \longrightarrow C$ a degree 2 covering by another non-singular curve $\widetilde{\Sigma}$. We denote by R the ramification divisor of $\tilde{\pi}$. In this case the covering $\tilde{\pi}$ is a Galois covering with the Galois group $\mathbb{Z}/2\mathbb{Z} = \langle\tilde{\sigma}\rangle$, and R is the fixed-point divisor of the involution $\tilde{\sigma}$. The **integral topological recursion** is an inductive mechanism of constructing meromorphic differential forms $W_{g,n}$ on the Hilbert scheme $\widetilde{\Sigma}^{[n]}$ of n-points on $\widetilde{\Sigma}$ (which is the n-symmetric product of $\widetilde{\Sigma}$) for all $g \geq 0$ and $n \geq 1$ in the *stable range* $2g - 2 + n > 0$, from given initial data $W_{0,1}$ and $W_{0,2}$.

- $W_{0,1}$ is a meromorphic 1-form on $\widetilde{\Sigma}$.
- $W_{0,2}$ is defined to be

$$W_{0,2}(z_1, z_2) = d_1 d_2 \log E_{\widetilde{\Sigma}}(z_1, z_2), \tag{6.10}$$

  where $E_{\widetilde{\Sigma}}(z_1, z_2)$ is the normalized Riemann prime form on $\widetilde{\Sigma} \times \widetilde{\Sigma}$ (see [**18**, Section 2]).

Let $\omega^{a-b}(z)$ be a normalized Cauchy kernel on $\widetilde{\Sigma}$, which has simple poles at $z = a$ of residue 1 and at $z = b$ of residue -1. Then (see [**18**, Section 2])

$$d_1 \omega^{z_1 - b}(z_2) = W_{0,2}(z_1, z_2).$$

Define

$$\Omega := \tilde{\sigma}^* W_{0,1} - W_{0,1}. \tag{6.11}$$

Then $\tilde{\sigma}^*\Omega = -\Omega$, hence $\operatorname{supp}(R) \subset \operatorname{supp}(\Omega)$, where $\operatorname{supp}(\Omega)$ denotes the support of both zero and pole divisors of Ω. The inductive formula of the topological recursion is then given by the following:

$$W_{g,n}(z_1, \dots, z_n) = \frac{1}{2}\frac{1}{2\pi\sqrt{-1}} \sum_{p \in \operatorname{supp}(\Omega)} \oint_{\gamma_p} \frac{\omega^{\tilde{z}-z}(z_1)}{\Omega(z)} \times \left[W_{g-1,n+1}(z, \tilde{z}, z_2, \dots, z_n) + \sum_{\substack{g_1+g_2=g \\ I \sqcup J = \{2,\dots,n\}}}^{\text{No }(0,1)} W_{g_1,|I|+1}(z, z_I) W_{g_2,|J|+1}(\tilde{z}, z_J) \right]. \tag{6.12}$$

Here,

- γ_p is a positively oriented small loop around a point $p \in \operatorname{supp}(\Omega)$;
- the integration is taken with respect to $z \in \gamma_p$ for each $p \in \operatorname{supp}(\Omega)$;

- $\tilde{z} = \tilde{\sigma}(z)$ is the Galois conjugate of $z \in \widetilde{\Sigma}$;
- the ratio of two global meromorphic 1-forms on the same curve makes sense as a global meromorphic function. The operation $1/\Omega$ applied to a meromorphic 1-form produces this ratio;
- "No $(0,1)$" means that $g_1 = 0$ and $I = \emptyset$, or $g_2 = 0$ and $J = \emptyset$, are excluded in the summation;
- the sum runs over all partitions of g and set partitions of $\{2, \ldots, n\}$, other than those containing the $(0,1)$ geometry;
- $|I|$ is the cardinality of the subset $I \subset \{2, \ldots, n\}$; and
- $z_I = (z_i)_{i \in I}$.

The passage from the topological recursion (6.12) to the quantum curve (1.9) is the evaluation of the residues in the formula.

DEFINITION 6.4 (Free energies). A **free energy** of type (g,n) is a function $F_{g,n}(z_1, \ldots, z_n)$ defined on the universal covering $\mathcal{U}^n$ of $\widetilde{\Sigma}^n$ such that

$$d_1 \cdots d_n F_{g,n} = W_{g,n}.$$

REMARK 6.5. Free energies as defined may contain logarithmic singularities, since it is an integral of a meromorphic function. For example, $F_{0,2}$ is the Riemann prime form itself considered as a function on $\mathcal{U}^2$, which has logarithmic singularities along the diagonal [**18**, Section 2].

DEFINITION 6.6 (Differential recursion for a degree 2 covering). The **differential recursion** is the following partial differential equation for all (g,n) subject to $2g - 2 + n \geq 2$:

$$\begin{aligned} d_1 F_{g,n}(z_1, \ldots, z_n) &= \sum_{j=2}^{n} \left[\frac{\omega^{z_j - \sigma(z_j)}(z_1)}{\Omega(z_1)} \cdot d_1 F_{g,n-1}\big(z_{[\hat{j}]}\big) - \frac{\omega^{z_j - \sigma(z_j)}(z_1)}{\Omega(z_j)} \cdot d_j F_{g,n-1}\big(z_{[\hat{1}]}\big) \right] \\ &+ \frac{1}{\Omega(z_1)} d_{u_1} d_{u_2} \Bigg[F_{g-1,n+1}\big(u_1, u_2, z_{[\hat{1}]}\big) \\ &+ \sum_{\substack{g_1+g_2=g \\ I \sqcup J = [\hat{1}]}}^{\text{stable}} F_{g_1,|I|+1}(u_1, z_I) F_{g_2,|J|+1}(u_2, z_J) \Bigg] \Bigg|_{\substack{u_1=z_1 \\ u_2=z_1}} . \end{aligned}$$

Here, $1/\Omega$ is again the ratio operation, and the index subset $[\hat{j}]$ denotes the exclusion of $j \in \{1, 2, \ldots, n\}$.

REMARK 6.7. As pointed out in [**18**, Remark 4.8], (6.6) is a coordinate-free equation, written in terms of exterior differentiations and the ratio operation on the universal covering of $\widetilde{\Sigma}$.

THEOREM 6.8 (The relation between the differential recursion and the integral recursion). *Suppose that $F_{g,n}$ for $2g-2+n > 0$ are globally meromorphic on $\widetilde{\Sigma}^n$ with poles located only along the divisor of $\widetilde{\Sigma}^n$ when one of the factors lies in the zeros of Ω. Define $W_{g,n} := d_1 \cdots d_n F_{g,n}$ for $2g - 2 + n > 0$, and use* (6.10) *and* (6.11) *for (g,n) in the unstable range. If $F_{g,n}$'s are symmetric and satisfy the differential recursion* (6.6), *and if $W_{1,1}$ and $W_{0,3}$ satisfy the initial equations of the integral* *topological recursion* (6.12), *then* $W_{g,n}$*'s for all values of* (g,n) *satisfy the integral topological recursion.*

REMARK 6.9. The assumption of the theorem holds for $g(C)=0$, and therefore, for all the examples we discuss in this paper. But we are not establishing the general **equivalence** of the integral topological recursion (6.12) and the PDE recursion (6.6). They are never equivalent when $g(\widetilde{\Sigma})>0$. Actually, what is assumed in the above theorem is that $W_{g,n}$ is exact, i.e., integrable by definition. In particular, this implies that $W_{g,n}$ has 0-period for every topological cycle. This does not happen if we start with (6.12) in general. Therefore, the above theorem serves only as a heuristic motivation for our discovery of (6.6) in [**18**].

Although the context of the statement is slightly different, the proof is essentially the same as that of [**18**, Theorem 4.7].

Now let us consider a spectral curve $\Sigma \subset \overline{T^*C}$ of (3.9) defined by a pair of meromorphic sections $a_1 = -\mathrm{tr}\phi$ of K_C and $a_2 = \det\phi$ of $K_C^{\otimes 2}$. Let $\widetilde{\Sigma}$ be the desingularization of Σ in (1.6). We apply the topological recursion (6.12) to the covering $\tilde{\pi}:\widetilde{\Sigma}\longrightarrow C$ of (4.27). The geometry of the spectral curve Σ provides us with a canonical choice of the initial differential forms (1.7). At this point we pay a special attention that the topological recursions (6.12) and (6.6) are both defined on the spectral curve $\widetilde{\Sigma}$, while we wish to construct a Rees D-module on C. Since the free energies are defined on the universal covering of $\widetilde{\Sigma}$, we need to have a mechanism to relate a coordinate on the desingularized spectral curve and that of the base curve C.

Take an arbitrary point $p\in C\setminus \mathrm{supp}(\Delta)$, and a local coordinate x around p. Here, Δ is the discriminant divisor (4.7). By choosing a small disc V around p, we can make the inverse image of $\tilde{\pi}:\widetilde{\Sigma}\longrightarrow C$ consisting of two isomorphic discs. Since V is away from the critical values of $\tilde{\pi}$, the inverse image consists of two discs in the original spectral curve Σ. Note that we choose an eigenvalue α of ϕ on V in Theorem 6.1. We are thus specifying one of the inverse image discs here. Let us name the disc V_α that corresponds to α.

At this point apply the WKB analysis to the differential equation (6.5) with the WKB expansion of the solution

$$\Psi^\alpha(x,\hbar)=\exp\left(\sum_{m=0}^{\infty}\hbar^{m-1}S_m\big(x(z)\big)\right)=\exp F^\alpha(x,\hbar), \tag{6.13}$$

where we choose a coordinate z of V_α so that the function $x=x(z)$ represents the projection $\pi:V_\alpha\longrightarrow V$. The equation $P\Psi^\alpha = Pe^{F^\alpha}=0$ reads

$$\hbar^2\frac{d^2}{dx^2}F^\alpha+\hbar^2\frac{dF^\alpha}{dx}\frac{dF^\alpha}{dx}+a_1\hbar\frac{dF^\alpha}{dx}+a_2=0. \tag{6.14}$$

The $\hbar$-expansion of (6.14) gives

$$\hbar^0\text{-terms}:\quad (S_0'(x))^2+a_1S_0'(x)+a_2=0, \tag{6.15}$$

$$\hbar^1\text{-terms}:\quad 2S_0'(x)S_1'(x)+S_0''(x)+a_1S_1'(x)=0, \tag{6.16}$$

$$\hbar^2\text{-terms}:\quad S_1''(x)+\sum_{a+b=2}S_a'(x)S_b'(x)+a_1S_2'(x)=0, \tag{6.17}$$

$$\hbar^{m+1}\text{-terms}:\quad S_m''(x)+\sum_{a+b=m+1}S_a'(x)S_b'(x)+a_1S_{m+1}'(x)=0,\quad m\ge 2, \tag{6.18}$$

where $'$ denotes the x-derivative. The WKB method is to solve these equations iteratively and find $S_m(x)$ for all $m \geq 0$. Here, (6.15) is the **semi-classical limit** of (6.5), and (6.16) is the *consistency condition* we need to solve the WKB expansion. Since the 1-form $dS_0(x)$ is a local section of T^*C, we identify $y = S_0'(x)$. Then (6.15) is the local expression of the spectral curve equation (4.3). This expression is the same everywhere for $p \in C \setminus \text{supp}(\Delta)$. We note a_1 and a_2 are globally defined. Therefore, we recover the spectral curve Σ from the differential operator (6.3).

The topological recursion provides a closed formula for each $S_m(x)$.

THEOREM 6.10 (Topological recursion and WKB). *Let us determine $S_0(x)$ and $S_1(x)$ from the semi-classical limit and the consistency condition. Then the* ***principal specialization*** *of* (6.6) *is equivalent to* (6.18).

PROOF. First let us take $q \in C$ one of the q_i's of (4.8), above which $\tilde{\pi} : \widetilde{\Sigma} \longrightarrow C$ is simply ramified at $Q := \tilde{\pi}^{-1}(q) \in \widetilde{\Sigma}$. We choose a local coordinate x on C centered at q. The Galois action of $\tilde{\sigma}$ on $\widetilde{\Sigma}$ fixes Q. Let $\widetilde{U} \subset \widetilde{\Sigma}$ be a neighborhood of Q such that $p \in \tilde{\pi}(\widetilde{U})$ and $a_1 \in H^0(\tilde{\pi}(\widetilde{U}), K_C)$, i.e., holomorphic on $\tilde{\pi}(\widetilde{U})$. The defining equation of the spectral curve Σ on $\tilde{\pi}(\widetilde{U})$ is

$$\left(ydx + \frac{1}{2}a_1\right)^2 - \left(\frac{1}{4}a_1^2 - a_2\right) = 0.$$

Since a_1 is holomorphic at $x = q$, the Galois action of σ on the spectral curve Σ extends to $\overline{T^*C}|_{\tilde{\pi}(\widetilde{U})}$ by the formula given in (4.16). As we have shown in Case 1 of the proof of Theorem 4.2, the degree m_i of zero of the discriminant $\frac{1}{4}a_1^2 - a_2$ at $q = q_i$ is odd, say $m_i = 2\mu + 1$, and the construction of $Bl_{\min}(\overline{T^*C})$ contains blow-ups of $\lfloor \frac{2\mu+1}{2} \rfloor = \mu$ times at the singular point above q. In terms of the coordinate x, we can write

$$a_1 = a_1(x)dx, \quad a_2 = a_2(x)(dx)^2, \quad \frac{1}{4}a_1(x)^2 - a_2(x) = cx^{2\mu+1}$$

with a unit $c \in \mathcal{O}_{C,q}$. Define $y_0 := y + \frac{1}{2}a_1(x)$. Then the first blow-up at the singular point above q is done by replacing $y_0 = y_1 x$ so that the proper transform is locally defined by

$$y_1^2 = cx^{2\mu-1}.$$

The coordinate y_1 is the affine coordinate of the exceptional divisor. Repeating this process μ-times, we end up with a coordinate $y_{\mu-1} = y_\mu x$ and an equation

$$y_\mu^2 = cx.$$

Here again, y_μ is the affine coordinate of the last exceptional divisor resulted from the μ-th blow-up. We now write $z = y_\mu$ so that the proper transform of the μ-times blow-ups is given by

$$z^2 = cx. \tag{6.19}$$

Note that the Galois action of $\tilde{\sigma}$ at Q is simply $z \longmapsto -z$. Solving (6.19) as a functional equation, we obtain a Galois invariant local expression

$$x = x(z) = c_Q(z^2)z^2, \tag{6.20}$$

where $c_Q \in \mathcal{O}_{\widetilde{\Sigma},Q}$ is a unit element. This formula (6.20) is precisely the local expression of the morphism $\tilde{\pi} : \widetilde{\Sigma} \longrightarrow C$ at $Q \in \widetilde{\Sigma}$. On the other hand, from the construction we also have

$$y_0 = zx^\mu = y + \frac{1}{2}a_1(x),$$

or equivalently,

$$\eta = y(z)dx = zx^\mu dx - \frac{1}{2}a_1, \qquad y(z) = zx^\mu - \frac{1}{2}a_1(x). \tag{6.21}$$

We have thus obtained the normalization coordinate z on the desingularized curve $\overline{\Sigma}$ near Q:

$$\begin{cases} x = x(z) = c_Q(z^2)z^2, \\ y = y(z) = zx^\mu - \frac{1}{2}a_1(x). \end{cases} \tag{6.22}$$

Notice that we now have a parametric equation for the singular spectral curve Σ:

$$\left(y(z) + \frac{1}{2}a_1(x(z))\right)^2 = z^2x(z)^{2\mu} = cx(z)^{2\mu+1} = \frac{1}{4}a_1(x(z))^2 - a_2(x(z)).$$

The differential form η of (6.21) is the local expression of the form Ω in the differential topological recursion (6.6).

We have now established the local expression of all functions and forms involved in the topological recursion. From here the rest of the proof is parallel to [**18**].

As we have shown in the process of the proof of Theorem 4.2, the situation is the same if $q \in C$ corresponds to a branch point of $\tilde{\pi} : \widetilde{\Sigma} \longrightarrow C$ that comes from an odd cusp of Σ on the divisor C_∞. A similar argument of the above proof works for this case.

This completes the proof. □

We have thus completed the proof of Theorem 6.1.

6.3. Singularity of quantum curves. Let p be a pole of the discriminant divisor Δ of (4.9). The local equation for the spectral curve around p is

$$y^2 + a_1(x)y + a_2(x) = 0.$$

As we have shown above, the local generator of the quantum curve as a Rees D-module is given by a differential operator

$$\left(\hbar\frac{d}{dx}\right)^2 + a_1(x)\left(\hbar\frac{d}{dx}\right) + a_2(x).$$

Therefore, the type of the singularity of the quantum curve is determined by the local geometry of the spectral curve. We have the following.

THEOREM 6.11 (Regular and irregular singular points of the quantum curve). *Let $P \in \Sigma \cap C_\infty$ be a point at the intersection of the spectral curve Σ and the divisor C_∞ at infinity of the ruled surface $\overline{T^*C}$. Suppose it requires ρ times blow up at P to construct $Bl(\overline{T^*C})$. Then the quantum curve of Theorem 6.1 has*

- *a regular singular point at $p = \pi(P)$ if $\rho = 1$.*
- *If $\rho > 1$, then the quantum curve has an irregular singular point at $p = \pi(P)$ of class either $\rho - 1$ or $\rho - \frac{3}{2}$, the latter occurring only when P is a cusp singularity of Σ.*

PROOF. As in the proof of Theorem 4.2, we denote by k (reps. ℓ) the pole order of $a_1(x)$ (reps. $a_2(x)$) at $x = p$. Let r be the invariant defined in (4.26). Then by Definition 1.10, p is a regular singular point of the quantum curve if $0 < r \leq 1$. In this case we need to blow-up once at P for construction of $Bl(\overline{T^*C})$, because $\lceil r \rceil = 1$. If $r > 1$, then the singularity is irregular with class $r-1$, and we need $\lceil r \rceil$ times blow-ups. As we see from the proof of Theorem 4.8, a non-integer r occurs only when P is a cusp. This completes the proof. $\sqcap$

7. The classical differential equations as quantum curves

The key examples of the theory of quantum curves as presented in this paper are the classical differential equations. In this section, we present the Hermite and Gauss hypergeometric differential equations.

7.1. Hermite differential equation. The base curve is $C = \mathbb{P}^1$, as in the Airy case. The stable Higgs bundle (E, ϕ) consists of a vector bundle $E = \mathcal{O}_{\mathbb{P}^1}(1) \oplus \mathcal{O}_{\mathbb{P}^1}(-1)$ and a Higgs field

$$\phi = \begin{bmatrix} & 1 \\ -1 & -x \end{bmatrix} dx : E \longrightarrow E \otimes K_{\mathbb{P}^1}(4) = E. \tag{7.1}$$

In the affine coordinate (x, y) of the Hirzebruch surface $\mathbb{F}^2$, the spectral curve Σ is given by

$$\det\big(\eta - \pi^*(\phi)\big) = (y^2 + xy + 1)(dx)^2 = 0, \tag{7.2}$$

where $\pi : \mathbb{F}_2 \longrightarrow \mathbb{P}^1$ is the projection. In the other affine coordinate (u, w) of (1.20), the spectral curve is singular at $(u, w) = (0, 0)$:

$$u^4 - uw + w^2 = 0. \tag{7.3}$$

These equations tell us that $\Sigma \cdot C_0 = 0$ and $\Sigma \cdot C_\infty = 4$. Therefore,

$$\Sigma = 2C_0 + 4F \in \mathrm{NS}(\mathbb{F}_2).$$

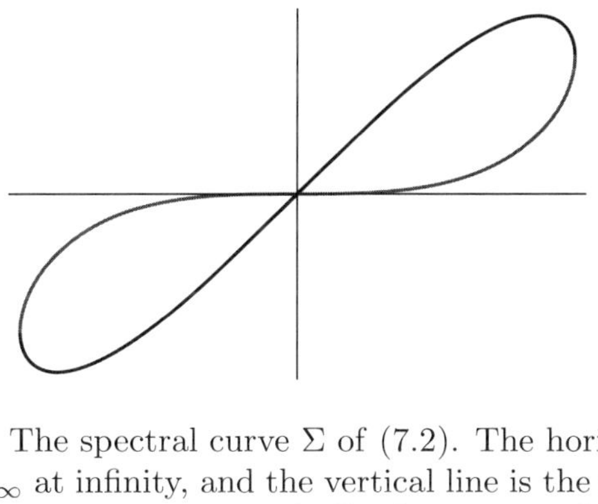

FIGURE 7.1. The spectral curve Σ of (7.2). The horizontal line is the divisor C_∞ at infinity, and the vertical line is the fiber class F. The spectral curve intersects with C_∞ four times. One of the two curve germ components is given by $w = u$, and the other by $w = u^3$.

The discriminant of the defining equation (7.2) is

$$\left(-\frac{1}{4}x^2 + 1\right)(dx)^2 = -\frac{1}{4}(x-2)(x+2)(dx)^2 = \frac{u^2 - \frac{1}{4}}{u^6}(du)^2.$$

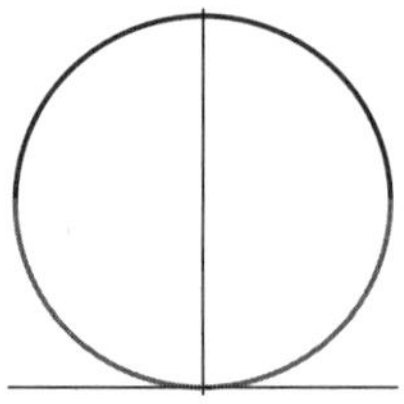

FIGURE 7.2. The desingularization $\widetilde{\Sigma}$ of the spectral curve (7.2).

It has two simple zeros at $x = \pm 2$ and a pole of order 6 at $x = \infty$. We note that

$$\mathrm{tr}(\phi) = -x dx = \frac{du}{u^3}$$

has a cubic pole at $u = 0$. As explained in Case 3 of the proof of Theorem 4.2, we need to compare the poles of $\mathrm{tr}(\phi)$ and

$$\det(\phi) = (dx)^2 = \frac{(du)^2}{u^4}.$$

Since $4 - 3 = 1$, we blow up Σ once at its nodal singularity $(u, w) = (0, 0)$. We introduce $w = w_1 u$. Then (7.3) becomes

$$u^2 + \left(w_1 - \frac{1}{2}\right)^2 = \frac{1}{4}. \tag{7.4}$$

The geometric genus formula (4.12) tells us that $\Sigma_{\min}$ has genus 0, and $\Sigma_{\min} \longrightarrow C$ is ramified at two points, corresponding to the original ramification points $(x, y) = (\pm 2, \mp 1)$ of Σ. The rational parametrization of (7.4) is given by

$$\begin{cases} u = \frac{1}{2} \cdot \frac{t^2-1}{t^2+1} \\ w_1 = \frac{1}{2} - \frac{t}{t^2+1}, \end{cases}$$

where t is an affine coordinate of $\widetilde{\Sigma}$ such that $t = \pm 1$ gives $(u, w) = (0, 0)$. The parameter t is a normalization coordinate of the spectral curve Σ:

$$\begin{cases} x = 2 + \frac{4}{t^2+1} \\ y = -\frac{t+1}{t-1}, \end{cases} \qquad \begin{cases} u = \frac{1}{2} \cdot \frac{t^2-1}{t^2+1} \\ w = \frac{1}{4} \cdot \frac{(t-1)^3(t+1)}{(t^2+1)^2}. \end{cases} \tag{7.5}$$

We notice that the expression of (7.5) is exactly the same as [**20**, (3.13), (3.14)]. The integral topological recursion applied to $\widetilde{\Sigma}$ again agrees with that of [**20**].

The quantum curve construction of [**48**] is thus consistent with our new definition. The result is

$$\left(\left(\hbar\frac{d}{dx}\right)^2 + x\left(\hbar\frac{d}{dx}\right) + 1\right)\Psi(x, \hbar) = 0. \tag{7.6}$$

Since x and d/dx do not commute, the passage from (7.2) to (7.6) is non-trivial, in the sense that the constant term could have contained a term $c \cdot \hbar$. On the affine open subset $U_1 = \mathbb{P}^1 \setminus \{0\}$, the operator of (7.6) has an expression

$$u^4\left(\hbar\frac{d}{du}\right)^2 + (2u^3\hbar - u)\left(\hbar\frac{d}{du}\right) + 1 \in \widetilde{\mathcal{D}_{U_1}}.$$

Thus the point $\infty \in \mathbb{P}^1$ is an irregular singular point of class 2 of (7.6) for $\hbar \neq 0$.

The semi-classical limit (6.15) of (7.6) using the WKB formula (1.8) is

$$S_0'(x)^2 + xS_0'(x) + 1 = 0. \tag{7.7}$$

Following [**20**], define

$$z = \frac{t+1}{t-1} = \sum_{m=0}^{\infty} \frac{C_m}{x^{2m+1}}, \tag{7.8}$$

where $C_m = \frac{1}{m+1}\binom{2m}{m}$ is the m-th Catalan number. The inverse function of (7.8) for $x = \infty \Rightarrow z = 0$ is given by $x = x(z) = z + \frac{1}{z}$. In terms of z, the two solutions of (7.7) are given by

$$S_0\big(x(z)\big) = \begin{cases} -\frac{1}{2}z^2 + \log z + \text{const} \\ -\frac{1}{2}\frac{1}{z^2} - \log z + \text{const}. \end{cases} \tag{7.9}$$

Corresponding to these choices, the solutions to the consistency condition (6.16) are given by

$$S_1\big(x(z)\big) = \begin{cases} -\frac{1}{2}\log(1-z^2) + \text{const} \\ -\frac{1}{2}\log(1-z^2) + \log z + \text{const}. \end{cases} \tag{7.10}$$

Every solution of (7.6) is a linear combination of two solutions, with coefficients given by arbitrary functions in $\hbar$. One is given by the **Kummer confluent hypergeomtric function** of (1.24):

$$\Psi_1(x,\hbar) = {}_1F_1\left(\frac{1}{2\hbar}; \frac{1}{2}; -\frac{x^2}{2\hbar}\right).$$

The other is a bit more complicated function known as the **Tricomi confluent hypergeomtric function**

$$\Psi_2(x,\hbar) = \frac{\Gamma[\frac{1}{2}]}{\Gamma[\frac{1}{2\hbar}+\frac{1}{2}]}\,{}_1F_1\left(\frac{1}{2\hbar}; \frac{1}{2}; -\frac{x^2}{2\hbar}\right) + \frac{\Gamma[-\frac{1}{2}]}{\Gamma[\frac{1}{2\hbar}]}\sqrt{\frac{x^2}{-2\hbar}}\,{}_1F_1\left(\frac{1}{2\hbar}+\frac{1}{2}; \frac{3}{2}; -\frac{x^2}{2\hbar}\right). \tag{7.11}$$

For a positive real $\hbar > 0$, let us consider a special solution

$$\Psi^{\text{Catalan}}(x,\hbar) := \left(-\frac{1}{2\hbar}\right)^{\frac{1}{2\hbar}} \Psi_2(x,\hbar). \tag{7.12}$$

This solution corresponds to the WKB solution (1.8) for the first choices of (7.9) and (7.10), with both constants of integration to be set 0. Then we have a closed formula for the all-order asymptotics of this particular confluent hypergeometric function:

$$\begin{aligned} \Psi^{\text{Catalan}}(x,\hbar) &= \left(\frac{1}{x}\right)^{\frac{1}{\hbar}} \sum_{n=0}^{\infty} \frac{\hbar^n \left(\frac{1}{\hbar}\right)_{2n}}{(2n)!!} \cdot \frac{1}{x^{2n}} \\ &= \exp\left(\sum_{2g-2+n \geq -1} \frac{1}{n!}\hbar^{2g-2+n} F_{g,n}^{\text{Catalan}}(x,\dots,x)\right), \end{aligned} \tag{7.13}$$

where $(1/\hbar)_{2n}$ is the Pochhammer symbol (1.25). The free energies are defined by

$$F_{g,n}^{\text{Catalan}}(x_1,\dots,x_n) = \sum_{\mu_1,\dots,\mu_n>0} \frac{C_{g,n}(\mu_1,\dots,\mu_n)}{\mu_1\cdots\mu_n} \prod_{i=1}^{n} x_i^{-\mu_i} \tag{7.14}$$

for $2g-2+n>0$, and $F_{0,1}^{\text{Catalan}}(x)=S_0(x)$ and $\frac{1}{2}F_{0,2}^{\text{Catalan}}(x)=S_1(x)$. Here, $C_{g,n}(\mu_1,\dots,\mu_n)$ is the generalized Catalan number of genus g and n labeled vertices of degrees $(\mu_1,\dots,\mu_n)$ that counts the number of **cellular graphs** [**20**, **50**]. In [**20**, Theorem 4.3, Proposition A.1], we show that $F_{g,n}^C$ satisfies the differential recursion equation (6.6). (We note that the differential recursion of [**20**] is derived by taking the Laplace transform of [**50**, Equation 6]). The initial geometric data (1.7) are also the same as [**20**, (3.12), (4.3)]. Therefore, the application of the topological recursion to the desingularized spectral curve $\Sigma_{\min}$ produces (7.14), and the quantum curve (7.6).

At the special value $\hbar=1$, the expansion (7.13) has the following simple form

$$\sum_{n=0}^{\infty}\frac{(2n-1)!!}{x^{2n+1}}=\exp\left(\sum_{2g-2+n\geq -1}\frac{1}{n!}\sum_{\mu_1,\dots,\mu_n>0}\frac{C_{g,n}(\mu_1,\dots,\mu_n)}{\mu_1\cdots\mu_n}\prod_{i=1}^{n}x^{-(\mu_1+\cdots+\mu_n)}\right).$$

Note that the sum of the degrees of the vertices $\mu_1+\cdots+\mu_n$ is always even. Therefore, except for the unstable geometries $(g,n)=(0,1)$ and $(0,2)$, the above expansion is in x^{-2}. This indicates that the Hermite equation has an irregular singular point of class 2 at $x=\infty$.

7.2. Gauß hypergeometric differential equation. The Higgs bundle (E,ϕ) on $\mathbb{P}^1$ is given by the same vector bundle $E=\mathcal{O}_{\mathbb{P}^1}(1)\oplus\mathcal{O}_{\mathbb{P}^1}(-1)$, and a Higgs field

$$\phi=\begin{bmatrix} & \frac{1}{x}\\ -\frac{ab}{x-1} & -\frac{(a+b+1)x-c}{x(x-1)}\end{bmatrix}dx, \tag{7.15}$$

where a,b,c are constant parameters. The spectral curve $\Sigma\in\mathbb{F}_2$ is defined by

$$x(x-1)y^2+((a+b+1)x-c)y+ab=0. \tag{7.16}$$

In terms of the (u,w) coordinate of (1.20), the spectral curve is given by

$$abw^2+(cu-a-b)uw-u^2(u-1)=0. \tag{7.17}$$

It has an ordinary double point at $(u,w)=(0,0)$. The discriminant divisor of (4.7) is

$$\Delta^{\text{Gauß}}=\left(\frac{\left(\frac{1}{4}((a+b+1)x-1)^2-abx(x-1)\right)(dx)^2}{x^2(x-1)^2}\right), \tag{7.18}$$

which consists of two simple zeros and 3 double poles at $x=0,1,\infty$. Following Definition 4.7, we blow up $\mathbb{F}_2$ once at the point at infinity of Σ to construct the normalization $\Sigma_{\min}$. The invariant β of (4.11) is equal to 2, and hence $\Sigma_{\min}$ is isomorphic to $\mathbb{P}^1$.

The quantum curve we obtain is a Gauß hypergeometric differential equation

$$\left(x(x-1)\left(\hbar\frac{d}{dx}\right)^2+((a+b+1)x-c)\,\hbar\frac{d}{dx}+ab\right)\Psi^{\text{Gauß}}(x,\hbar)=0. \tag{7.19}$$

One of the two independent solutions that is holomorphic at $x = 0$ is given in terms of a **Gauß hypergeometric function**

$$\begin{aligned}(7.20)\quad \Psi^{\text{Gauß}}(x,\hbar) = {}_2F_1\Bigg(&-\frac{\sqrt{(a+b+1-\hbar)^2-4ab}}{2\hbar}+\frac{a+b+1}{2\hbar}-\frac{1}{2},\\ &\frac{\sqrt{(a+b+1-\hbar)^2-4ab}}{2\hbar}+\frac{a+b+1}{2\hbar}-\frac{1}{2};\frac{c}{\hbar};x\Bigg).\end{aligned}$$

If we choose $\sqrt{(a+b+1-\hbar)^2-4ab} = b-a$ when $\hbar = 1$, then

$$(7.21)\qquad \Psi^{\text{Gauß}}(x,1) = {}_2F_1(a,b;c,x) = \sum_{n=0}^{\infty}\frac{(a)_n(b)_n}{(c)_n}\frac{x^n}{n!}$$

solves the standard form of the Gauß hypergeometric equation

$$(7.22)\qquad \left(x(x-1)\left(\frac{d}{dx}\right)^2+((a+b+1)x-c)\frac{d}{dx}+ab\right)\Psi^{\text{Gauß}}(x,1)=0.$$

Now let us specialize $a = b = \frac{1}{2}, c = 1$. We have the relation between the hypergeometric function and the period function of (1.11):

$$(7.23)\qquad \frac{\omega_1(x)}{\pi} = {}_2F_1\left(\frac{1}{2},\frac{1}{2};1,x\right).$$

The spectral curve (7.16) becomes

$$(7.24)\qquad x(x-1)y^2+(2x-1)y+\frac{1}{4}=0.$$

On the normalization $\widetilde{\Sigma}$, we have $W_{0,1}(x) = ydx$, which actually depends of the sheet of the covering $\tilde{\pi}:\widetilde{\Sigma}\longrightarrow\mathbb{P}^1$. For our purpose, we choose

$$(7.25)\qquad y = y(x) = \frac{-(2x-1)-\sqrt{3x^2-3x+1}}{2x\,(x-1)}.$$

Then

$$\begin{aligned}(7.26)\quad S_0(x) = F_{0,1}(x) &= \int y(x)dx\\ &= \frac{x}{4}-\frac{21\,(4\sqrt{3}-7)}{32\,(2\sqrt{3}-3)^2}x^2+\frac{23\,(26\sqrt{3}-45)}{32\,(2\sqrt{3}-3)^3}x^3-\frac{2547\,(56\sqrt{3}-97)}{1024\,(2\sqrt{3}-3)^4}x^4\\ &+\frac{7281\,(362\sqrt{3}-627)}{2560\,(2\sqrt{3}-3)^5}x^5-\frac{38115\,(780\sqrt{3}-1351)}{4096\,(2\sqrt{3}-3)^6}x^6+\frac{265869\,(5042\sqrt{3}-8733)}{28672\,(2\sqrt{3}-3)^7}x^7+\cdots\end{aligned}$$

solves the semi-classical limit equation (6.15). The solution of the consistency condition (6.16) is given by

$$\begin{aligned}(7.27)\quad S_1(x) &= -\int\frac{y'(x)}{2y(x)+\frac{2x-1}{x(x-1)}}dx\\ &= -\frac{7}{32}x^2-\frac{53}{96}x^3-\frac{1075}{1024}x^4-\frac{4319}{2560}x^5-\frac{28319}{12288}x^6-\frac{72109}{28672}x^7+\cdots.\end{aligned}$$

The solution of (6.18) for $m = 2$ is

$$(7.28)\quad S_2(x) = \frac{7x^2}{32} + \frac{113x^3}{96} + \frac{1821x^4}{512} + \frac{1269x^5}{160} + \frac{56151x^6}{4096} + \frac{487323x^7}{28672} + \cdots .$$

From (7.26), (7.27) and (7.28), we have an expression

$$(7.29)\quad \exp\left(\frac{1}{\hbar}S_0(x) + S_1(x) + \hbar S_2(x)\right)$$
$$= 1 + \frac{x}{4\hbar} + \frac{1 + 7\hbar - 7\hbar^2 + 7\hbar^3}{32\hbar^2}x^2 + \frac{1 + 21\hbar + 71\hbar^2 - 191\hbar^3 + \cdots}{384\hbar^3}x^3$$
$$+ \frac{1 + 42\hbar + 473\hbar^2 + 598\hbar^3 + \cdots}{6144\hbar^4}x^4 + \frac{1 + 70\hbar + 1585\hbar^2 + 1141 - \hbar^3 + \cdots}{122880\hbar^5}x^5 + \cdots .$$

This is in good agreement of the hypergeometric function of (7.20), which can be expanded as

$$\Psi^{\text{Gauß}}(x,\hbar) = 1 + \sum_{n=1}^{\infty} \frac{1}{4^n n!} \frac{\prod_{m=1}^{n}\left(1 + 8(m-1)\hbar + 4(m-1)(m-2)\hbar^2\right)}{\prod_{m=1}^{n}(1 + (m-1)\hbar)} \left(\frac{x}{\hbar}\right)^n$$
$$= 1 + \frac{x}{4\hbar} + \frac{1 + 8\hbar}{32\hbar(1+\hbar)}x^2 + \frac{(1+8\hbar)(1+16\hbar+8\hbar^2)}{384\hbar^3(1+\hbar)(1+2\hbar)}x^3$$
$$+ \frac{(1+8\hbar)(1+16\hbar+8\hbar^2)(1+24\hbar+24\hbar^2)}{6144\hbar^4(1+\hbar)(1+2\hbar)(1+3\hbar)}x^4 + \cdots ,$$

up to order of $\hbar^3$ of the numerator of every coefficient of x^n, $n \geq 0$. The topological recursion applied to the spectral curve (7.16) with $F_{0,1} = S_0(x)$ and the standard Riemann prime form on $\widetilde{\Sigma} = \mathbb{P}^1$ for $F_{0,2}$ then gives a genus expansion of $\Psi^{\text{Gauß}}(x,\hbar)$, constructing a genus g B-model on the curve (7.16).

Acknowledgments

The authors are grateful to the American Institute of Mathematics in Palo Alto, the Banff International Research Station, the Institute for Mathematical Sciences at the National University of Singapore, Kobe University, and Max-Planck-Institut für Mathematik in Bonn, for their hospitality and financial support. A large portion of this work is carried out during the authors' stay in these institutions. They thank Jørgen Andersen, Philip Boalch, Leonid Chekhov, Bertrand Eynard, Tamás Hausel, Kohei Iwaki, Maxim Kontsevich, Alexei Oblomkov, Albert Schwarz, Yan Soibelman, Ruifang Song, and Peter Zograf for useful comments, suggestions, and discussions. The authors also thank the anonymous referee for comments to improve the paper. M.M. thanks the Euler International Mathematical Institute in St. Petersburg for hospitality.

References

[1] Mina Aganagic, Robbert Dijkgraaf, Albrecht Klemm, Marcos Mariño, and Cumrun Vafa, *Topological strings and integrable hierarchies*, Comm. Math. Phys. **261** (2006), no. 2, 451–516, DOI 10.1007/s00220-005-1448-9. MR2191887

[2] J.E. Andersen, L.O Chekhov, P. Norbury, and R.C. Penner, *Cohomological field theories, discretisation of moduli spaces, and Gaussian means*, Private communication.

[3] D. Arinkin, *On λ-connections on a curve where λ is a formal parameter*, Math. Res. Lett. **12** (2005), no. 4, 551–565, DOI 10.4310/MRL.2005.v12.n4.a9. MR2155231

[4] Olivier Biquard and Philip Boalch, *Wild non-abelian Hodge theory on curves*, Compos. Math. **140** (2004), no. 1, 179–204, DOI 10.1112/S0010437X03000010. MR2004129

[5] Philip Boalch, *Symplectic manifolds and isomonodromic deformations*, Adv. Math. **163** (2001), no. 2, 137–205, DOI 10.1006/aima.2001.1998. MR1864833

[6] P. Boalch, *Hyperkähler manifolds and nonabelian Hodge theory of (irregular) curves*, arXiv:1203.6607v1 (2012).

[7] Vincent Bouchard and Bertrand Eynard, *Think globally, compute locally*, J. High Energy Phys. **2** (2013), 143, front matter + 34, DOI 10.1007/JHEP02(2013)143. MR3046532

[8] Vincent Bouchard, Daniel Hernández Serrano, Xiaojun Liu, and Motohico Mulase, *Mirror symmetry for orbifold Hurwitz numbers*, J. Differential Geom. **98** (2014), no. 3, 375–423. MR3263522

[9] Vincent Bouchard, Albrecht Klemm, Marcos Mariño, and Sara Pasquetti, *Remodeling the B-model*, Comm. Math. Phys. **287** (2009), no. 1, 117–178, DOI 10.1007/s00220-008-0620-4. MR2480744

[10] Vincent Bouchard, Albrecht Klemm, Marcos Mariño, and Sara Pasquetti, *Topological open strings on orbifolds*, Comm. Math. Phys. **296** (2010), no. 3, 589–623, DOI 10.1007/s00220-010-1020-0. MR2628817

[11] Leonid Chekhov, Bertrand Eynard, and Nicolas Orantin, *Free energy topological expansion for the 2-matrix model*, J. High Energy Phys. **12** (2006), 053, 31, DOI 10.1088/1126-6708/2006/12/053. MR2276699

[12] Robbert Dijkgraaf, Hiroyuki Fuji, and Masahide Manabe, *The volume conjecture, perturbative knot invariants, and recursion relations for topological strings*, Nuclear Phys. B **849** (2011), no. 1, 166–211, DOI 10.1016/j.nuclphysb.2011.03.014. MR2795276

[13] Robbert Dijkgraaf, Lotte Hollands, and Piotr Sułkowski, *Quantum curves and $\mathcal{D}$-modules*, J. High Energy Phys. **11** (2009), 047, 59, DOI 10.1088/1126-6708/2009/11/047. MR2628886

[14] Robbert Dijkgraaf, Lotte Hollands, Piotr Sułkowski, and Cumrun Vafa, *Supersymmetric gauge theories, intersecting branes and free fermions*, J. High Energy Phys. **2** (2008), 106, 57, DOI 10.1088/1126-6708/2008/02/106. MR2385939

[15] Robbert Dijkgraaf, Herman Verlinde, and Erik Verlinde, *Loop equations and Virasoro constraints in nonperturbative two-dimensional quantum gravity*, Nuclear Phys. B **348** (1991), no. 3, 435–456, DOI 10.1016/0550-3213(91)90199-8. MR1083914

[16] Norman Do and David Manescu, *Quantum curves for the enumeration of ribbon graphs and hypermaps*, Commun. Number Theory Phys. **8** (2014), no. 4, 677–701, DOI 10.4310/CNTP.2014.v8.n4.a2. MR3318387

[17] Boris Dubrovin, *Geometry of 2D topological field theories*, Integrable systems and quantum groups (Montecatini Terme, 1993), Lecture Notes in Math., vol. 1620, Springer, Berlin, 1996, pp. 120–348, DOI 10.1007/BFb0094793. MR1397274

[18] Olivia Dumitrescu and Motohico Mulase, *Quantum curves for Hitchin fibrations and the Eynard-Orantin theory*, Lett. Math. Phys. **104** (2014), no. 6, 635–671, DOI 10.1007/s11005-014-0679-0. MR3200933

[19] O. Dumitrescu and M. Mulase, *Interplay between opers, quantum curves, WKB analysis, and Higgs bundles*, arXiv:1702.00511 (2017).

[20] Olivia Dumitrescu, Motohico Mulase, Brad Safnuk, and Adam Sorkin, *The spectral curve of the Eynard-Orantin recursion via the Laplace transform*, Algebraic and geometric aspects of integrable systems and random matrices, Contemp. Math., vol. 593, Amer. Math. Soc., Providence, RI, 2013, pp. 263–315, DOI 10.1090/conm/593/11867. MR3087960

[21] Petr Dunin-Barkowski, Motohico Mulase, Paul Norbury, Alexander Popolitov, and Sergey Shadrin, *Quantum spectral curve for the Gromov-Witten theory of the complex projective line*, J. Reine Angew. Math. **726** (2017), 267–289, DOI 10.1515/crelle-2014-0097. MR3641659

[22] Bertrand Eynard, *Topological expansion for the 1-Hermitian matrix model correlation functions*, J. High Energy Phys. **11** (2004), 031, 35 pp. (2005), DOI 10.1088/1126-6708/2004/11/031. MR2118807

[23] B. Eynard, *Intersection numbers of spectral curves*, arXiv:1104.0176 (2011).

[24] Bertrand Eynard, Motohico Mulase, and Bradley Safnuk, *The Laplace transform of the cut-and-join equation and the Bouchard-Mariño conjecture on Hurwitz numbers*, Publ. Res. Inst. Math. Sci. **47** (2011), no. 2, 629–670, DOI 10.2977/PRIMS/47. MR2849645

[25] B. Eynard and N. Orantin, *Invariants of algebraic curves and topological expansion*, Commun. Number Theory Phys. **1** (2007), no. 2, 347–452, DOI 10.4310/CNTP.2007.v1.n2.a4. MR2346575

[26] Xiao Bin Li, *Computations of 2-point orbifold Gromov-Witten invariants for local Calabi-Yau 3-folds* (Chinese, with English and Chinese summaries), Sichuan Daxue Xuebao **49** (2012), no. 3, 531–534. MR2985537

[27] Bohan Fang, Chiu-Chu Melissa Liu, and Zhengyu Zong, *All genus mirror symmetry for toric Calabi-Yau 3-orbifolds*, String-Math 2014, Proc. Sympos. Pure Math., vol. 93, Amer. Math. Soc., Providence, RI, 2016, pp. 1–19. MR3524231

[28] Stavros Garoufalidis, *On the characteristic and deformation varieties of a knot*, Proceedings of the Casson Fest, Geom. Topol. Monogr., vol. 7, Geom. Topol. Publ., Coventry, 2004, pp. 291–309, DOI 10.2140/gtm.2004.7.291. MR2172488

[29] Stavros Garoufalidis and Thang T. Q. Lê, *The colored Jones function is q-holonomic*, Geom. Topol. **9** (2005), 1253–1293, DOI 10.2140/gt.2005.9.1253. MR2174266

[30] V. Ginzburg, *Lectures on $\mathcal{D}$-modules*, Lectures delivered at the University of Chicago, Winter 1998, with collaboration of V. Baranovsky and S. Evens (1998).

[31] Sergei Gukov and Piotr Sułkowski, *A-polynomial, B-model, and quantization*, J. High Energy Phys. **2** (2012), 070, front matter+56, DOI 10.1007/JHEP02(2012)070. MR2996110

[32] Robin Hartshorne, *Algebraic geometry*, Springer-Verlag, New York-Heidelberg, 1977. Graduate Texts in Mathematics, No. 52. MR0463157

[33] N. J. Hitchin, *The self-duality equations on a Riemann surface*, Proc. London Math. Soc. (3) **55** (1987), no. 1, 59–126, DOI 10.1112/plms/s3-55.1.59. MR887284

[34] Nigel Hitchin, *Stable bundles and integrable systems*, Duke Math. J. **54** (1987), no. 1, 91–114, DOI 10.1215/S0012-7094-87-05408-1. MR885778

[35] Andrew R. Hodge and Motohico Mulase, *Hitchin integrable systems, deformations of spectral curves, and KP-type equations*, New developments in algebraic geometry, integrable systems and mirror symmetry (RIMS, Kyoto, 2008), Adv. Stud. Pure Math., vol. 59, Math. Soc. Japan, Tokyo, 2010, pp. 31–77. MR2683206

[36] L. Hollands, *Topological strings and quantum curves*, Ph.D. Thesis, University of Amsterdam arXiv:0911.3413 [hep-th] (2009).

[37] D. Huybrechts and M. Lehn, *Stable pairs on curves and surfaces*, J. Algebraic Geom. **4** (1995), no. 1, 67–104. MR1299005

[38] Anton Kapustin and Edward Witten, *Electric-magnetic duality and the geometric Langlands program*, Commun. Number Theory Phys. **1** (2007), no. 1, 1–236, DOI 10.4310/CNTP.2007.v1.n1.a1. MR2306566

[39] Maxim Kontsevich, *Intersection theory on the moduli space of curves and the matrix Airy function*, Comm. Math. Phys. **147** (1992), no. 1, 1–23. MR1171758

[40] Maxim Kontsevich and Yan Soibelman, *Wall-crossing structures in Donaldson-Thomas invariants, integrable systems and mirror symmetry*, Homological mirror symmetry and tropical geometry, Lect. Notes Unione Mat. Ital., vol. 15, Springer, Cham, 2014, pp. 197–308, DOI 10.1007/978-3-319-06514-4_6. MR3330788

[41] Maxim Kontsevich and Don Zagier, *Periods*, Mathematics unlimited—2001 and beyond, Springer, Berlin, 2001, pp. 771–808. MR1852188

[42] Bertram Kostant, *Quantization and unitary representations. I. Prequantization*, Lectures in modern analysis and applications, III, Springer, Berlin, 1970, pp. 87–208. Lecture Notes in Math., Vol. 170. MR0294568

[43] Ph. Maisonobe and C. Sabbah, *Aspects of the theory of $\mathcal{D}$-modules*, Kaiserslautern Lecture Notes (2002).

[44] Ju. I. Manin, *Algebraic curves over fields with differentiation* (Russian), Izv. Akad. Nauk SSSR. Ser. Mat. **22** (1958), 737–756. MR0103889

[45] Marcos Mariño, *Open string amplitudes and large order behavior in topological string theory*, J. High Energy Phys. **3** (2008), 060, 34, DOI 10.1088/1126-6708/2008/03/060. MR2391060

[46] Motohico Mulase and Michael Penkava, *Topological recursion for the Poincaré polynomial of the combinatorial moduli space of curves*, Adv. Math. **230** (2012), no. 3, 1322–1339, DOI 10.1016/j.aim.2012.03.027. MR2921181

[47] M. Mulase, S. Shadrin, and L. Spitz, *The spectral curve and the Schrödinger equation of double Hurwitz numbers and higher spin structures*, Commun. Number Theory Phys. **7** (2013), no. 1, 125–143, DOI 10.4310/CNTP.2013.v7.n1.a4. MR3108774

[48] Motohico Mulase and Piotr Sułkowski, *Spectral curves and the Schrödinger equations for the Eynard-Orantin recursion*, Adv. Theor. Math. Phys. **19** (2015), no. 5, 955–1015, DOI 10.4310/ATMP.2015.v19.n5.a2. MR3487649

[49] Motohico Mulase and Naizhen Zhang, *Polynomial recursion formula for linear Hodge integrals*, Commun. Number Theory Phys. **4** (2010), no. 2, 267–293, DOI 10.4310/CNTP.2010.v4.n2.a1. MR2725053

[50] T. R. S. Walsh and A. B. Lehman, *Counting rooted maps by genus. I*, J. Combinatorial Theory Ser. B **13** (1972), 192–218. MR0314686

[51] Edward Witten, *Two-dimensional gravity and intersection theory on moduli space*, Surveys in differential geometry (Cambridge, MA, 1990), Lehigh Univ., Bethlehem, PA, 1991, pp. 243–310. MR1144529

[52] Edward Witten, *Gauge theory and wild ramification*, Anal. Appl. (Singap.) **6** (2008), no. 4, 429–501, DOI 10.1142/S0219530508001195. MR2459120

[53] Edward Witten, *Mirror symmetry, Hitchin's equations, and Langlands duality*, The many facets of geometry, Oxford Univ. Press, Oxford, 2010, pp. 113–128, DOI 10.1093/acprof:oso/9780199534920.003.0007. MR2681689

[54] Edward Witten, *Analytic continuation of Chern-Simons theory*, Chern-Simons gauge theory: 20 years after, AMS/IP Stud. Adv. Math., vol. 50, Amer. Math. Soc., Providence, RI, 2011, pp. 347–446. MR2809462

[55] J. Zhou, *Intersection numbers on Deligne-Mumford moduli spaces and quantum Airy curve*, arXiv:1206.5896 [math.AG] (2012).

[56] J. Zhou, *Quantum Mirror Curves for* $\mathbb{C}^3$ *and the Resolved Confiold*, arXiv:1207.0598v1 [math.AG] (2012).

Department of Mathematics, Pierce Hall 209, Central Michigan University, Mount Pleasant 48859, Michigan; and Simion Stoilow Institute of Mathematics, Romanian Academy, 21 Calea Grivitei Street, 010702 Bucharest, Romania

Email address: dumit1om@cmich.edu

Department of Mathematics, University of California, Davis, CA 95616–8633; and Kavli Institute for Physics and Mathematics of the Universe, The University of Tokyo, Kashiwa, Japan

Email address: mulase@math.ucdavis.edu

Proceedings of Symposia in Pure Mathematics
Volume **100**, 2018
https://doi.org/10.1090/pspum/100/01773

Topological recursion and Givental's formalism: Spectral curves for Gromov-Witten theories

P. Dunin-Barkowski

Abstract. We describe a way of producing local spectral curves for arbitrary semisimple cohomological field theories (and Gromov-Witten theories in particular) and global spectral curves for semisimple cohomological field theories satisfying certain conditions. By this we mean that applying the topological recursion procedure on the spectral curve reproduces the total potential of the corresponding cohomological field theory.

Present text is based on papers [*Identification of the Givental formula with the spectral curve topological recursion procedure*, Comm. Math. Phys. 328 (2014), pp. 669–700] and [*Dubrovin's superpotential as a global spectral curve*, arXiv:1509.06954] joint work with Nicolas Orantin, Sergey Shadrin and Loek Spitz and with Paul Norbury, Nicolas Orantin, Alexandr Popolitov and Sergey Shadrin, respectively.

Contents

1. Introduction

The contents of the present text were presented as a mini-course at the 2016 AMS von Neumann Symposium on Topological Recursion and its Influence in Analysis, Geometry, and Topology. It is based on papers [**17**] and [**18**], written in collaboration with Nicolas Orantin, Sergey Shadrin and Loek Spitz and with Nicolas Orantin, Paul Norbury, Alexandr Popolitov and Sergey Shadrin, respectively.

In the present text we describe the way of reproducing potentials of semisimple cohomological field theories (of which semisimple Gromov-Witten theories are a subset) with the help of spectral curve topological recursion.

Gromov-Witten theories originated from string theory, where they appeared as part of the "toy model" of topological string theory. It turned out that they

2010 *Mathematics Subject Classification.* Primary 81T45; Secondary 14N35, 53D45.

The author was partially supported by RFBR grant 16-31-60044-mol-a-dk.

are of particular interest in mathematics, as they provided an interesting way to build invariants of complex manifolds. Kontsevich and Manin generalized Gromov-Witten theories to the so-called *cohomological field theories*. This allowed to better incorporate this area of research into the general study of moduli spaces of algebraic curves and their cohomologies. A seminal result was first suggested by Witten and later proved by Kontsevich regarding the fact that the Gromov-Witten theory of a point corresponds to a KdV tau-function.

Spéctral curve topological recursion is a quite general technique which has applications in many different branches of mathematics and physics. This technique, given certain initial data, namely, a *spectral curve* (which is an algebraic curve equipped with some additional structure), produces so-called *n-point functions* on this spectral curve. It turns out that for a vast array of problems in such areas of science as algebraic geometry, mathematical physics, topology and combinatorics, these n-point functions appear to be generating functions for numbers arising in these problems, if one takes appropriate initial data. Spectral curve topological recursion is very interesting since it is one and the same procedure that for appropriate choices of initial data produces answers to many seemingly unrelated problems.

The first aim of the present text is to describe a way [**17**] of reproducing any semisimple cohomological field theory (and any semisimple Gromov-Witten theory in particular) with the help of the so-called *local* version of spectral curve topological recursion. This local version is a generalization of the spectral curve topological recursion where instead of the whole spectral curve we consider just the germs of all objects at branching points. In particular any ordinary spectral curve topological recursion gives rise to a local version, but not necessarily the other way around. This is a quite strong result, but in itself it might be too general, as local topological recursion (to the moment) lacks interesting applications or profound meaning separate from what originates from ordinary (global) spectral curve topological recursion. Thus, at least for now, it seems that the main applications of this result are related to proving spectral curve topological recursion for those CohFTs which do correspond to global spectral curves.

This is precisely the subject of the second part of the present text (which follows [**18**]). In that part we describe certain conditions under which a cohomological field theory has a corresponding spectral curve. Under these conditions we describe (with the help of a construction due to Dubrovin) an explicit way of constructing the global spectral curve using just the information of the Frobenius manifold which underlies the CohFT, and we prove that indeed the topological recursion procedure on the said curve reproduces the total potential of the CohFT. We provide several examples of this. For all these examples the corresponding Frobenius manifold belongs to the wide class of the so-called Hurwitz-Frobenius manifolds. While this is outside of the scope of the present text, we note that in [**19**] it is proved that this technique works for *all* semisimple Hurwitz-Frobenius manifolds; moreover, it turns out that this holds in the other direction as well: all global spectral curves with a mild restriction correspond via topological recursion to CohFTs based on this large class of Hurwitz-Frobenius manifolds.

1.1. Structure of the text. We start by introducing all relevant objects in Section 2.

In Section 3 (following [**17**]), we describe how to construct a local spectral curve for an arbitrary semisimple CohFT. This culminates with Theorem 3.16 which proves, with the help of celebrated Givental formula, that the potential of the CohFT is reproduced by the topological recursion procedure on this curve. We follow it up by describing how to apply this theorem to the particular example of the Gromov-Witten theory of $\mathbb{CP}^1$.

Then, in Section 4 (following [**18**]), we describe a way of constructing global spectral curves for semisimple CohFTs, with the help of Dubrovin's superpotential. Theorems 4.10 and 4.13, prove that, under certain conditions on the CohFT, Dubrovin's superpotential obtained from the corresponding Frobenius manifold can serve as a global spectral curve. We follow these theorems up with examples.

2. Preliminaries

2.1. Frobenius manifolds. A Frobenius manifold is a differential-geometric structure that was introduced by Dubrovin in the early 1990's as a mathematical framework for the study of two-dimensional topological field theory in genus zero [**10, 12**]. It has appeared to be a quite universal structure that has many naturally arising examples. In particular, Frobenius manifolds can serve as a classification tool for (dispersionless) bi-Hamiltonian hierarchies of hydrodynamic type [**14, 15**]. Nowadays there is a number of standard textbooks on Frobenius manifolds, see [**12, 39, 48**].

Here we introduce Frobenius manifolds following [**12**].

DEFINITION 2.1. An algebra A over $\mathbb{C}$ is called (commutative) *Frobenius algebra* if:

(1) It is a commutative associative $\mathbb{C}$-algebra with a unity e.
(2) It is supplied with a $\mathbb{C}$-bilinear symmetric nondegenerate inner product

$$A \times A \to \mathbb{C}, \ a, \, b \mapsto (a, b)$$

being invariant in the following sense:

$$(a \cdot b, c) = (a, b \cdot c)$$

DEFINITION 2.2. M is *Frobenius manifold* if a structure of Frobenius algebra is specified on any tangent plane T_tM at any point $t \in M$ smoothly depending on the point such that

(1) The invariant inner product $(\, , \,)$ is a flat metric on M.
(2) The unity vector field e is covariantly constant w.r.t. the Levi-Civitá connection ∇ for the metric $(\, , \,)$

$$\nabla e = 0$$

(3) Let

$$c(u, v, w) := (u \cdot v, w)$$

(a symmetric 3-tensor). We require the 4-tensor

$$(\nabla_z c)(u, v, w)$$

to be symmetric in the four vector fields u, v, w, z.

(4) A vector field E must be determined on M such that

$$\nabla(\nabla E) = 0$$

and that the correspondent one-parameter group of diffeomorphisms acts by conformal transformations of the metric $(\ ,\ )$ and by rescalings on the Frobenius algebras $T_t M$.

Locally Frobenius manifold can be defined in terms of a *Frobenius prepotential* $F(t)$ as follows.

DEFINITION 2.3. Function $F = F(t)$, $t = (t^1, \dots, t^n)$ is a *Frobenius prepotential* if its third derivatives

$$c_{\alpha\beta\gamma}(t) := \frac{\partial^3 F(t)}{\partial t^\alpha \partial t^\beta \partial t^\gamma}$$

obey the following equations

(1) Normalization:

$$\eta_{\alpha\beta} := c_{1\alpha\beta}(t)$$

is a constant nondegenerate matrix. Let

$$(\eta^{\alpha\beta}) := (\eta_{\alpha\beta})^{-1}.$$

We will use the matrices $(\eta_{\alpha\beta})$ and $(\eta^{\alpha\beta})$ for raising and lowering indices.

(2) Associativity: the functions

$$c_{\alpha\beta}^{\gamma}(t) := \eta^{\gamma\epsilon} c_{\epsilon\alpha\beta}(t)$$

(summation over repeated indices is assumed here) for any t must define in the n-dimensional space with a basis e_1, ..., e_n a structure of an associative algebra A_t

$$e_\alpha \cdot e_\beta = c_{\alpha\beta}^{\gamma}(t) e_\gamma.$$

Note that the vector e_1 will be the unity for all the algebras A_t:

$$c_{1\alpha}^{\beta}(t) = \delta_\alpha^\beta.$$

(3) $F(t)$ must be quasihomogeneous function of its variables:

$$F(c^{d_1} t^1, \dots, c^{d_n} t^n) = c^{d_F} F(t^1, \dots, t^n) \tag{1}$$

for any nonzero c and for some numbers d_1, ..., d_n, d_F.

Coordinates t such as above on the Frobenius manifold are called the *flat coordinates.*

It will be convenient to rewrite the quasihomogeneity condition in the infinitesimal form introducing the *Euler vector field*

$$E = E^\alpha(t) \partial_\alpha$$

as

$$\mathcal{L}_E F(t) := E^\alpha(t) \partial_\alpha F(t) = d_F \cdot F(t)$$

$E(t)$ is a linear vector field

$$E = \sum_\alpha d_\alpha t^\alpha \partial_\alpha$$

generating the scaling transformations (1).

Let us denote $3 - d_F$ by d.

DEFINITION 2.4. Point of a Frobenius manifold is called *semisimple* if at this point the Frobenius algebra structure is nondegenerate, i.e. there is no tangent vector that squares to zero.

Near a semisimple point of a Frobenius manifold one can introduce the so-called *canonical coordinates*, which are defined as coordinates u^i that have the property

$$\frac{\partial}{\partial u^i}\cdot\frac{\partial}{\partial u^j}=\delta_{ij}\frac{\partial}{\partial u^j},\tag{2}$$

where δ_{ij} is the Kronecker delta.

Define $\Delta_i := 1/(\partial_i, \partial_i)$ to be the inverse of the square of the length of the ith canonical basis element, and call $\{\partial/\partial v^i := \Delta_i^{1/2}\partial/\partial u^i\}$ the normalized canonical basis in the tangent space.

Let U be the matrix of canonical coordinates $U = \mathrm{diag}(u^1, \ldots, u^n)$ and denote by Ψ the transition matrix from the flat to the normalized canonical bases. That is, denoting $\mathrm{d}t = (\mathrm{d}t^1, \ldots, \mathrm{d}t^n)^{\mathrm{T}}$ and $\mathrm{d}u = (\mathrm{d}u^1, \ldots, \mathrm{d}u^n)^{\mathrm{T}}$, one has

$$\Delta^{-1/2}\mathrm{d}u = \Psi\mathrm{d}t,\tag{3}$$

where $\Delta = \mathrm{diag}(\Delta_1, \ldots, \Delta_n)$.

Define

$$V_i := (\partial_{u_i}\Psi)\cdot\Psi^{-1}$$

and

$$V := \sum_i u_i V_i.$$

The second metric η' is defined in the following way: for any two vector fields ∂' and ∂'' we define

$$\eta'(\partial', \partial'') := \eta(\partial', E\cdot\partial'').$$

2.2. Superpotential. A convenient way to describe a Frobenius structure is in terms of a so-called Landau-Ginzburg superpotential. We recall the definition from [**12, 13**]. A *superpotential* is a holomorphic function $\lambda(p, u_1, \ldots, u_n)$ of a variable $p \in \mathcal{D}$ in some domain $\mathcal{D}$ (which is an open domain on a Riemann surface realized as a branched covering over complex plane with finite number of sheets) that depends on points $(u_1, \ldots, u_n) \in B_0 \subset B^{ss}$ in a ball in the semisimple part of the Frobenius manifold (Riemann surface which contains domain $\mathcal{D}$ may depend on u, but the projection to the complex plane should remain fixed), and satisfies the following properties:

(1) The critical values of λ as a function on $\mathcal{D}$ are $u_1, \ldots, u_n$.
(2) The critical points are non-degenerate.
(3) If there are several critical points in the inverse image $\lambda^{-1}(u_i)$, then the Hessians of λ at these points must coincide.
(4) For any choice $p_1, \ldots, p_n \in \mathcal{D}$ of the critical preimages of $u_1, \ldots, u_n$ (that is, $\lambda(p_i, u_1, \ldots, u_n) = u_i$) and for any choice of the vector fields ∂', ∂'',

and ∂''' on B_0 we have:

(4) $$\eta(\partial',\partial'') = -\sum_{i=1}^{n} \operatorname*{Res}_{p\to p_i} \frac{\partial'(\lambda dp)\partial''(\lambda dp)}{d_p\lambda};$$

(5) $$\eta'(\partial',\partial'') = -\sum_{i=1}^{n} \operatorname*{Res}_{p\to p_i} \frac{\partial'(\log\lambda dp)\partial''(\log\lambda dp)}{d_p\log\lambda};$$

(6) $$\eta(\partial'\cdot\partial'',\partial''') = -\sum_{i=1}^{n} \operatorname*{Res}_{p\to p_i} \frac{\partial'(\lambda dp)\partial''(\lambda dp)\partial'''(\lambda dp)}{dp\, d_p\lambda}$$

where $\partial'(\lambda dp)$ gives the action of the vector field by derivation in the parameters u_i. In particular, the map $\partial' \mapsto \partial'(\lambda dp)$ from vector fields on B_0 to meromorphic differentials on $\mathcal{D}$ quotiented out by $d_p\lambda$ is injective.

(5) There exist some cycles $Z_1,\dots,Z_n$ in $\mathcal{D}$ such that the integrals

(7) $$\frac{1}{\sqrt{z}}\int_{Z_\alpha} e^{z\lambda}dp, \qquad \alpha = 1,\dots,n$$

converge and give a non-degenerate system of flat coordinates for $\tilde{\nabla}$.

In these terms, the identity vector field ∂_0 of the Frobenius manifold is represented by dp, i.e. $\partial_0(\lambda dp) = dp$. Indeed, since $\eta(\partial_0\cdot\partial,\partial') = \eta(\partial,\partial')$ for all vector fields ∂,∂', then non-degeneracy of η implies that $\partial_0\cdot\partial = \partial$ for all ∂. The Euler vector field is represented in these terms by λdp, i.e. $E(\lambda dp) = \lambda dp$.

2.3. Dubrovin's superpotential. In this Section we recall a construction of a particular Landau-Ginzburg superpotential due to Dubrovin [**13**].

Given a manifold M equipped with a flat metric, a locally defined function t is a *flat coordinate* at $p\in M$, if

(i) $dt(p)\neq 0$ and
(ii) dt is covariantly constant with respect to the Levi-Civita connection.

Condition (i) guarantees that t is a local coordinate, i.e. we can find a coordinate system $(t^1,\dots,t^n)$ with $t^1 = t$ and an open neighbourhood $B\subset M$ of p such that $(t^1,\dots,t^n): B\to B_0\subset\mathbb{R}^n$ is a homeomorphism onto an open set B_0 of $\mathbb{R}^n$. Condition (ii), which uses the induced connection on the cotangent bundle, guarantees that $(t^1,\dots,t^n)$ can be chosen so that the metric is represented by a constant matrix with respect to $(t^1,\dots,t^n)$.

We now consider a flat coordinate $\rho(\lambda,u)$ with respect to the pencil of metrics $\eta'-\lambda\eta$. We study covariant constancy of $d\rho$ via its gradient vector field $\phi(\lambda,u) = \nabla\rho(\lambda,u)$ defined by

$$(\eta'-\lambda\eta)(\phi,\cdot) = d\rho.$$

The Levi-Civita connection of η' with respect to flat coordinates (for η) is given in [**13**, Equation (5.5)]). This leads to the following system of equations for vector fields ϕ expressed in canonical coordinates on a Frobenius manifold (the extended Gauss-Manin system [**13**, Equations (5.31) and (5.32)]):

(8) $$\mathrm{d}\phi = -(U-\lambda)^{-1}\mathrm{d}(U-\lambda)\left(\frac{1}{2}+V\right)\phi + \mathrm{d}\Psi\cdot\Psi^{-1}\phi.$$

Here $\mathrm{d} = \mathrm{d}_\lambda + \mathrm{d}_u$ is the total de Rham differential; by writing ϕ we understand the column of components of vector field ϕ in the normalized canonical basis; $U = \operatorname{diag}(u_1,\dots,u_n)$ and V and Ψ are naturally associated to a Frobenius manifold as defined in Section 2.1. Abusing notation, we use λ for the matrix of multiplication by λ. So (8) encodes the system of PDEs giving covariant constancy of $\phi(\lambda,u) = \nabla\rho(\lambda,u)$ in directions $\partial/\partial\lambda, \partial/\partial u_i$.

One can retrieve ρ from its gradient vector field via

$$\rho(\lambda,u) = \frac{\sqrt{2}}{1-d}\phi^T(U-\lambda)\Psi\mathbb{1}. \tag{9}$$

This is proved in [**11**, Section 2].

This equation has poles at $\lambda = u_1,\dots,u_n$ on the λ-plane, so we choose parallel straight cuts $L_1,\dots,L_n$ from the points u_i to infinity (we assume that $u_j \notin L_i$ for $i \neq j$). On $\mathbb{C}\setminus\cup_{i=1}^n L_i$ we choose branches of functions $\sqrt{u_i-\lambda}$, $i=1,\dots,n$. We denote by $\mathcal{R}_i$ the monodromy of the space of solutions of Equation (8) corresponding to following a small loop around u_1.

Dubrovin proves that there exist a unique system of solutions $\phi^{(1)},\dots,\phi^{(n)}$ to equation (8) satisfying the following properties:

$$\mathcal{R}_j\phi^{(j)} = -\phi^{(j)}, \qquad j=1,\dots,n; \tag{10}$$

$$\phi_j^{(j)} = \frac{1}{\sqrt{u_j-\lambda}} + O(\sqrt{u_j-\lambda}) \text{ for } \lambda\to u_j, \qquad j=1,\dots,n; \tag{11}$$

$$\phi_a^{(j)} = \sqrt{u_j-\lambda}\cdot O(1) \text{ for } \lambda\to u_j, \qquad a\neq j;\ a,j=1,\dots,n; \tag{12}$$

$$\mathcal{R}_j\phi^{(i)} = \phi^{(i)} - 2G^{ij}\phi^{(j)}, \qquad i,j=1,\dots,n; \tag{13}$$

where $G^{ij} := (\phi^{(i)})^T(U-\lambda)\phi^{(j)}$ is a bilinear form that doesn't depend on λ and $u_1,\dots,u_n$.

Assume that G^{ij} is non-degenerate and denote by G_{ij} the inverse matrix. Note that non-degeneracy of G^{ij} is a property of the Frobenius manifold M which holds generically. In fact the proof of Theorem 4.1 does not require the non-degeneracy of G^{ij}—see Remark 4.2. Consider a special solution of Equation (8) given by $\phi := \sum_{i,j=1}^n G_{ij}\phi^{(j)}$. The main property of this solution is that ϕ has the local behavior

$$\phi_j = \frac{1}{\sqrt{u_j-\lambda}} + O(1) \text{ for } \lambda\to u_j, \qquad j=1,\dots,n; \tag{14}$$

$$\phi_a = \sqrt{u_j-\lambda}\cdot O(1) \text{ for } \lambda\to u_j, \qquad a\neq j; a,j=1,\dots,n. \tag{15}$$

We consider the function $p = p(\lambda,u)$ given by the formula

$$p(\lambda,u) := \frac{\sqrt{2}}{1-d}\phi^T(U-\lambda)\Psi\mathbb{1}. \tag{16}$$

This function is analytic in $\mathbb{C}\setminus\cup_{i=1}^n L_i$, with a regular singularity at infinity, and its local behavior for $\lambda\to u_i$ is given by

$$p(\lambda,u) = p(u_i,u) + \Psi_{i,\mathbb{1}}\sqrt{2(u_i-\lambda)} + O(u_i-\lambda), \qquad i=1,\dots,n. \tag{17}$$

The 1-form $d_\lambda p$ has at most a finite number of zeros. We denote them by $r_1, \dots, r_N$ and we assume that they do not belong to the cuts L_i, $i = 1, \dots, n$. Let D be the image of $\mathbb{C} \setminus \cup_{i=1}^n L_i$ under the map $p(\lambda, u)$. This domain has a boundary given by the unfolding of the cuts L_i, $i = 1, \dots, n$. The inverse function $\lambda = \lambda(p, u)$ is a multivalued function on D. Consider the points $p(r_c, u)$, $c = 1, \dots, N$. We glue a finite number of copies of D along the cuts from the points $p(r_c, u)$ to infinity, $c = 1, \dots, N$. In this way we obtain a domain $\hat{D}$, where the function λ is single-valued. This is possible since $p(\lambda, u)$ is analytic as a function of $\lambda \in \mathbb{C} \setminus \cup_{i=1}^n L_i$ with a regular singularity at infinity. In other words, the resulting surface is essentially the surface $\{(\lambda, p) \in (\mathbb{C} \setminus \cup_{i=1}^n L_i) \times \mathbb{C} \mid p = p(\lambda, u)\}$.

We analytically continue the function λ on $\hat{D}$ beyond the boundary. This essentially corresponds to gluing several copies of $\hat{D}$ along the boundaries that are the images of the same cuts on the λ-plane. This is always a correctly defined procedure as this corresponds to simply crossing the cut on the λ-plane. This procedure is not unique as this gluing can happen in various different ways. In any case, we can perform this construction uniformly over a small ball in the space of parameters $u_1, \dots, u_n$. This way we obtain a (not necessarily compact) Riemann surface $\mathcal{D}$, with a function $\lambda = \lambda(\tilde{p}, u) \colon \mathcal{D} \to \mathbb{C}$ (by $\tilde{p}$ we denote some local coordinate on $\mathcal{D}$).

Dubrovin proves in [**13**] that the family of functions $\lambda(\tilde{p}, u)$ defined this way is a superpotential of the Frobenius manifold which was the input of this construction.

2.4. Gromov-Witten theory. Let us introduce Gromov–Witten theory, following, e.g. [**57**].

First, let us introduce moduli spaces of algebraic curves. The moduli space of curves $\mathcal{M}_{g,n}$, $g \geq 0$, $n \geq 0$, $2g - 2 + n > 0$, parametrizes smooth complex curves of genus g with n ordered marked points. It is a smooth complex orbifold of dimension $3g - 3 + n$.

The space $\overline{\mathcal{M}}_{g,n}$ is a compactification of $\mathcal{M}_{g,n}$. It parametrizes stable curves of genus g with n ordered marked points. A stable curve is a possibly reducible curve with possible nodes, such that the order of its automorphism group is finite. Genus of a stable curve is the arithmetic genus, namely, the genus of the smooth curve that we get if the replace each node (given locally by the equation $xy = 0$) with a cylinder (given locally by the equation $xy = \epsilon$). The space $\overline{\mathcal{M}}_{g,n}$ is a smooth compact complex orbifold.

There is a number of natural mappings between the moduli spaces of curves.

First, there are projections $\pi \colon \overline{\mathcal{M}}_{g,n+1} \to \overline{\mathcal{M}}_{g,n}$ that forget the last marked point. Note that there is a subtlety related to the fact that when we forget a marked point a stable curve can become unstable.

Second, there is a 2-to-1 mapping $\sigma \colon \overline{\mathcal{M}}_{g-1,n+2} \to \overline{\mathcal{M}}_{g,n}$ whose image is the boundary divisor of irreducible curves with one node.

Third, there are mappings $\rho \colon \overline{\mathcal{M}}_{g_1,n_1+1} \times \overline{\mathcal{M}}_{g_2,n_2+1} \to \overline{\mathcal{M}}_{g,n}$, $g_1 + g_2 = g$, $n_1 + n_2 = n$, whose images are the other irreducible boundary divisors of the compactification of $\overline{\mathcal{M}}_{g,n}$.

Let L_i the line bundle over $\overline{\mathcal{M}}_{g,n}$, whose fiber over a point $x \in \overline{\mathcal{M}}_{g,n}$ represented by a curve C_g with marked points $x_1, \dots, x_n$ is equal to $T^*_{x_i} C_g$. Denote by $\psi_i \in H^2(\overline{\mathcal{M}}_{g,n})$ the first Chern class of L_i. ψ_i are referred to as *psi-classes*.

Given a complex manifold X, one can consider $[X_{g,k,\mathrm{deg}}]$, the moduli space of degree deg stable maps to X of genus-g curves with k marked points. There is a natural projection from $[X_{g,k,\deg}]$ to $\overline{\mathcal{M}}_{g,k}$ which consists of forgetting the map to X. Note that there is a subtlety related to the fact that an unstable curve can be mapped to X in a stable way. One can consider pullback of psi-classes to $[X_{g,k,\deg}]$ with respect to this projection. We will denote them as ψ_i too.

Gromov–Witten theory studies the so-called Gromov–Witten invariants (also called Gromov–Witten correlators), defined as follows (left hand side is just a notation):

$$(18)\quad \langle\tau_{d_1}(e_{i_1})\tau_{d_2}(e_{i_2})\cdots\tau_{d_k}(e_{i_k})\rangle_g := \sum_{\deg}\int_{[X_{g,k,\deg}]^{\mathrm{vir}}} ev_1^*(e_{i_1})\psi_1^{d_1}\, ev_2^*(e_{i_2})\psi_2^{d_2}\cdots ev_k^*(e_{i_k})\psi_k^{d_k},$$

where , ev_i is the evaluation map at the ith.

Gromov–Witten *potential* $\mathcal{F}_g$ of genus g is the following generating function for the correlators ($v^{d,i}$ are formal variables):

$$(19)\qquad \mathcal{F}_g=\sum\frac{\langle\tau_{d_1}(e_{i_1})\tau_{d_2}(e_{i_2})\cdots\tau_{d_k}(e_{i_k})\rangle_g}{|\operatorname{Aut}((i_m,d_m)_{m=1}^k)|}v^{d_1,i_1}\cdots v^{d_k,i_k},$$

where $|\operatorname{Aut}((i_m,d_m)_{m=1}^k)|$ denotes the number of automorphisms of the collection of multi-indices (i_m,d_m) and where the sum is such that it includes each monomial $v^{d_1,i_1}\cdots v^{d_k,i_k}$ exactly once.

It turns out [**48**] that *the genus zero potential without descendants*, i.e.

$$(20)\qquad F=\mathcal{F}_0\big|_{v^{d,i}=0,\ d>0}$$

is a Frobenius prepotential.

Partition function of the Gromov–Witten theory is defined as follows:

$$(21)\qquad Z=\exp\left(\sum_{g\geq 0}\hbar^{g-1}\mathcal{F}_g\right),$$

where $\hbar$ is a formal parameter.

Celebrated result of Kontsevich–Witten states that the partition function for the Gromov-Witten theory of a point (i.e. for $X=\{\mathrm{pt}\}$) is the tau-function of the KdV integrable hierarchy. We will denote this partition function as Z_{KdV}.

2.5. Cohomological field theory. Here we introduce cohomological field theory, again following [**57**].

Cohomological field theory is a generalization of Gromov-Witten theory where one drops the target complex manifold and takes just some vector space V in place of its cohomology. Roughly speaking, a CohFT is a system of cohomology classes on the moduli spaces of curves with the values in the tensor powers of V, compatible with all natural mappings between the moduli spaces.

The formal definition is the following. We fix a vector space $V=\langle e_1,\dots,e_s\rangle$ (e_1 will play a special role) with a non-degenerate scalar product η. A cohomological field theory is a system of cohomology classes $\alpha_{g,n}\in H^*(\overline{\mathcal{M}}_{g,n},V^{\otimes n})$ satisfying the properties:

(1) $\alpha_{g,n}$ is equivariant with respect to the action of S_n on the labels of marked points and components of $V^{\otimes n}$.

(2) $\sigma^*\alpha_{g,n} = (\alpha_{g-1,n+2}, \eta^{-1})$; $\rho^*\alpha_{g,n} = (\alpha_{g_1,n_1+1} \cdot \alpha_{g_2,n_2+1}, \eta^{-1})$ (in both cases we contract with the scalar product the two components of V corresponding to the two points in the preimage of the node under normalization).
(3) $(\alpha_{0,3}, e_1 \otimes e_i \otimes e_j) = \eta_{ij}$, $\pi^*\alpha_{g,n} = (\alpha_{g,n+1}, e_1)$ (again, we contract the component of V corresponding to the last marked point with e_1).

Then *correlators* in cohomological field theory are defined as follows:

$$\langle \tau_{d_1}(e_{i_1}) \cdots \tau_{d_n}(e_{i_n}) \rangle_g := \int_{\overline{\mathcal{M}}_{g,n}} \prod_{j=1}^n \psi_j^{d_j} \cdot (\alpha_{g,n}, \otimes_{j=1}^n e_{i_j}) \tag{22}$$

Genus-g potentials and partition function are defined for cohomological field theories in exactly the same way as for Gromov–Witten theories above.

The statement about the genus zero potential without the descendants being a Frobenius prepotential is also true for any cohomological field theory.

2.6. Givental theory. Givental theory [**35, 36, 38**] is one of the most important tools in the study of Gromov-Witten invariants of target varieties and general cohomological field theories. It allows, in particular, to obtain explicit relations between the partition functions of different theories, reconstruct higher genera correlators from the genus 0 data, and establish general properties of semi-simple theories.

The core of the theory is Givental's formula that gives a formal Gromov-Witten potential associated to a semi-simple Frobenius structure. Teleman proves [**60**] that the formal Gromov-Witten potential associated to the Frobenius structure of a target variety with semi-simple quantum cohomology coincides with the actual Gromov-Witten potential in all genera.

In order to write Givental formula, let us construct an operator series $R(z) = \sum_{k\geq 0} R_k z^k$ in the following way.

Recursively define the off-diagonal entries of R_k in normalized canonical coordinates by solving the equation

$$\Psi^{-1}\mathrm{d}(\Psi R_{k-1}) = [\mathrm{d}U, R_k]. \tag{23}$$

using $R_0 = \mathbf{I}$ as a base case. Construct the diagonal entries of R_k by integrating the next equation

$$\Psi^{-1}\mathrm{d}(\Psi R_k) = [\mathrm{d}U, R_{k+1}] \tag{24}$$

using the fact that the diagonal entries of $[\mathrm{d}U, R_{k+1}]$ are equal to zero. To fix the integration constant, use the Euler equation

$$R_k = -(i_E \mathrm{d}R_k)/k, \tag{25}$$

where $E = \sum u^i \partial_i$ is the Euler field.

This procedure recursively defines R_k for all k.

Let us consider the following reexpansion:

$$R(z) = \sum_{l=0}^{\infty} R_l z^l = \exp\left(\sum_{l=1}^{\infty} r_l z^l\right). \tag{26}$$

Then we denote by $(r_l z^l)\hat{}$ the following differential operator:

$$(27)\qquad (r_l z^l)\hat{} := -(r_l)^i_{\mathbf{1}} \frac{\partial}{\partial v^{l+1,i}} + \sum_{d=0}^{\infty} v^{d,i} (r_l)^j_i \frac{\partial}{\partial v^{d+l,j}} + \frac{\hbar}{2} \sum_{m=0}^{l-1} (-1)^{m+1} (r_l)^{i,j} \frac{\partial^2}{\partial v^{m,i} \partial v^{l-1-m,j}}.$$

Here the indices $i, j \in \{1, \ldots, N\}$ on r_l correspond to the basis $\{e_1, \ldots, e_N\}$ of V, and the index $\mathbf{1}$ corresponds to the unit vector $e_{\mathbf{1}}$. When we write r_l with two upper-indices we mean as usual that we raise one of the indices using the scalar product η.

The quantization $\hat{R}$ of series $R(z)$ is then defined by

$$(28)\qquad \hat{R} = \exp\left(\sum_{l=1}^{\infty} \left((-1)^l r_l z^l\right)\hat{} \right).$$

Givental formula then allows to reproduce all ancestor correlators of the given (homogenous) cohomological field theory in the following way:

$$(29)\qquad Z = \hat{\Psi} \hat{R} \hat{\Delta} \mathcal{T},$$

where

$$\mathcal{T} := Z_{\mathrm{KdV}}(\{u^{d,1}\}) \cdots Z_{\mathrm{KdV}}(\{u^{d,N}\});$$

$\hat{\Delta}$ replaces the variables of ith KdV τ-function according to $u^{d,i} = \Delta_i^{1/2} v^{d,i}$ and replaces $\hbar$ with $\Delta_i \hbar$, while $\hat{\Psi}$ is the change of variables $v^{d,i} = \Psi^i_\nu t^{d,\nu}$.

2.7. Spectral curve topological recursion. The theory developed by Eynard and Orantin (see [**8**, **27**, **29**]), is a procedure, called spectral curve topological recursion, that takes the following objects as input. First, a particular Riemann surface, which is called the *spectral curve*. Second, two functions x and y on this surface, and, third, a choice of a bi-differential on this surface, which we will call the *two-point function* (which also is often called Bergman kernel). And, occasionally, a particular extra choice of a coordinate on an open part of the Riemann surface. The output of the topological recursion is a set of n-forms $\omega_{g,n}$, whose expansion in this additional coordinate generates interesting numbers.

In some cases these numbers are correlators of a matrix model (that was the original motivation for introducing the topological recursion; it is a natural generalization of the reconstruction procedure for the correlators of a certain class of matrix models, see, e.g. [**3**]), in some other cases they appear to be related to Gromov-Witten theory and to various intersection numbers on the moduli space of curves.

Note that this spectral curve topological recursion is unrelated to the topological recursion occurring in the theory of moduli spaces of curves.

One of the ways to think about the input data of the topological recursion theory is to say that the $(g, n) = (0, 1)$ part of a partition function in some geometrically motivated theory determines the spectral curve; the $(g, n) = (0, 2)$ part of a partition function determines the two-point function, and the rest of the correlators can be reconstructed from these two via topological recursion, in terms of a proper expansion of $\omega_{g,n}$ (see [**16**]).

The topological recursion theory is often used to reproduce known partition functions, extracts from them some higher genus correlators which were up to now unreachable and gives new non-trivial relations for the correlators, see e. g. [**31**].

Local version of the spectral curve topological recursion is defined as follows.

For $N \in \mathbb{N}^*$, we call *times* a set of N families of complex numbers $\left\{h_k^i\right\}_{k\in\mathbb{N}}$ for $i = 1, \ldots, N$ and *jumps* another set of $N \times N$ infinite families of complex numbers $\left\{B_{k,l}^{i,j}\right\}_{(k,l)\in\mathbb{N}^2}$ for $i, j = 1, \ldots, N$. We finally define a set of canonical coordinates $\{a_i\}_{i=1}^N \in \mathbb{C}^N$ subject to $a_i \neq a_j$ for $i \neq j$.

DEFINITION 2.5. For all $i, j \in \{1, \ldots, N\}$, we define the following set of analytic functions and differential forms in a neighborhood of $0 \in \mathbb{C}$:

$$x^i(z) := z^2 + a_i\,, \quad y^i(z) := \sum_{k=0}^{\infty} h_k^i z^k \tag{30}$$

and

$$B^{i,j}(z, z') = \delta_{i,j}\frac{dz \otimes dz'}{(z - z')^2} + \sum_{k,l=0}^{\infty} B_{k,l}^{i,j} z^k z'^l dz \otimes dz'. \tag{31}$$

For $2g - 2 + n > 0$, we define the genus g, n-point correlation functions $\omega_{g,n}^{i_1,\ldots,i_n}(z_1, \ldots z_n)$ recursively by

$$\omega_{g,n+1}^{i_0,i_1,\ldots,i_n}(z_0, z_1, \ldots, z_n) := \sum_{j=1}^{N} \operatorname*{Res}_{z\to 0} \frac{\int_{-z}^{z} B^{i_0,j}(z_0,\cdot)}{2\left(y^j(z) - y^j(-z)\right) \mathrm{d}x^j(z)} \times \tag{32}$$

$$\left(\omega_{g-1,n+2}^{j,j,i_1,\ldots,i_n}(z, -z, z_1, \ldots, z_n) + \sum_{A\cup B=\{1,\ldots,n\}} \sum_{h=0}^{g} \omega_{h,|A|+1}^{j,\mathbf{i}_A}(z, \mathbf{z}_A)\omega_{g-h,|B|+1}^{j,\mathbf{i}_B}(-z, \mathbf{z}_B)\right),$$

where for any set A, we denote by $\mathbf{z}_A$ (resp., $\mathbf{i}_A$) the set $\{z_k\}_{k\in A}$ (resp., $\{i_k\}_{k\in A}$), and where the base of the recursion is given by

$$\omega_{0,1}^i(z) := 0; \qquad \omega_{0,2}^{i,j}(z, z') := B^{i,j}(z, z'). \tag{33}$$

For the global version we have the following.

Consider a Riemann surface Σ with meromorphic functions $x, y \colon \Sigma \to \mathbb{C}$ such that x has a finite number of critical points, $c_1, \ldots, c_n$, and y is holomorphic near these points with a non-vanishing derivative. Let B be a symmetric bi-differential on $\Sigma \times \Sigma$, with a double pole on the diagonal, the double residue equal to 1, and no further singularities.

We define a sequence of symmetric n-forms $\omega_{g,k}(z_1, \ldots, z_k)$ on $\Sigma^{\times k}$, known as *correlation differentials* for the spectral curve, by the following recursion:

$$\omega_{0,1}(z) := y(z)dx(z); \tag{34}$$

$$\omega_{0,2}(z_1, z_2) := B(z_1, z_2); \tag{35}$$

$$\omega_{g,k+1}(z_0, z_1, \ldots, z_k) := \tag{36}$$

$$\sum_{i=1}^{n} \operatorname*{Res}_{z\to c_i} \frac{\int_z^{\sigma_i(z)} \omega_{0,2}(\bullet, z_0)}{2(\omega_{0,1}(\sigma_i(z)) - \omega_{0,1}(z))} \tilde{\omega}_{g,2|k}(z, \sigma_i(z)|z_1, \ldots, z_k),$$

where σ_i is the deck transformation for the function x near the point c_i, $i = 1, \ldots, n$, and $\tilde{\omega}_{g,2|k}$ is defined by the following formula:

$$\tilde{\omega}_{g,2|k}(z', z''|z_1, \ldots, z_k) := \omega_{g-1,k+2}(z', z'', z_1, \ldots, z_k) + \sum_{\substack{g_1+g_2=g \\ I_1 \sqcup I_2 = \{1,\ldots,k\} \\ 2g_1-1+|I_1|\geq 0 \\ 2g_2-1+|I_2|\geq 0}} \omega_{g_1,|I_1|+1}(z', z_{I_1}) \omega_{g_2,|I_2|+1}(z'', z_{I_2}). \tag{37}$$

Here we denote by z_I the sequence $z_{i_1}, \ldots, z_{i_{|I|}}$ for $I = \{i_1, \ldots, i_{|I|}\}$.

REMARK 2.6. In the global recursion we also allow y to be the (multivalued) primitive of a differential ω on Σ. The ambiguity in y consists of periods and residues of ω and hence the ambiguity is locally constant. Since y appears in the recursion only via $y(\sigma_i(z)) - y(z)$ (and there are no poles of ω at the zeros of dx) the locally constant ambiguity disappears and the recursion is well-defined.

3. Local spectral curves for cohomological field theories

In this section we establish the identification of Givental formula and the local spectral curve topological recursion procedure with an appropriate choice of initial data. Then, as a corollary, we prove the Norbury–Scott conjecture on the spectral curve topological recursion for the Gromov–Witten theory of $\mathbb{C}\mathrm{P}^1$.

This section deals with the Givental theory and the spectral curve topological recursion theory; for more information on both theories we refer to [**29, 46, 57**] as possible sources. In Section 3.1 we recall the Givental theory, and present the Givental formula as a sum over graphs. In Section 3.3 we do the same for the topological recursion. In Section 3.4 we prove the theorem on identification of the two theories and provide a corresponding dictionary. In Section 3.5 we provide the computations showing that this identification works for the spectral curve proposed by Norbury and Scott for the Gromov-Witten theory of $\mathbb{C}\mathrm{P}^1$.

3.1. Givental group action as a sum over graphs. In this section we review the Givental group action and we remind the reader how it can be used to write the partition function of an N-dimensional semi-simple cohomological field theory as an operator acting on the product of N KdV τ-functions. Using this, we write the partition function for such a cohomological field theory as a sum over decorated graphs.

3.1.1. *Givental group action.* We remind the reader of the original formulation, due to Y.-P. Lee, of the infinitesimal Givental group action in terms of differential operators [**43–45**].

Consider the space of partition functions for N-dimensional cohomological field theories

$$Z = \exp\left(\sum_{g\geq 0} \hbar^{g-1} \mathcal{F}_g\right) \tag{38}$$

in variables $v^{d,i}$, $d \geq 0$, $i = 1, \ldots, N$. There is a fixed scalar product $\eta_{ij} = \delta_{ij}$ on the vector space $V := \langle e_1, \ldots, e_N \rangle$ of primary fields corresponding to the indices $i = 1, \ldots, N$. Furthermore, we will denote by $\mathbf{e_1}$ the vector in V that plays the role of the unit.

Later on we will also use the so-called *correlators*

$$\langle \tau_{d_1}(e_{i_1})\tau_{d_2}(e_{i_2})\cdots\tau_{d_k}(e_{i_k})\rangle_g$$

which correspond to the coefficients of formal power series $\mathcal{F}_g$ in the following way:

$$\mathcal{F}_g = \sum \frac{\langle \tau_{d_1}(e_{i_1})\tau_{d_2}(e_{i_2})\cdots\tau_{d_k}(e_{i_k})\rangle_g}{|\operatorname{Aut}((i_m,d_m)_{m=1}^k)|} v^{d_1,i_1}\cdots v^{d_k,i_k}, \tag{39}$$

where $|\operatorname{Aut}((i_m,d_m)_{m=1}^k)|$ denotes the number of automorphisms of the collection of multi-indices (i_m,d_m) and where the sum is such that it includes each monomial $v^{d_1,i_1}\cdots v^{d_k,i_k}$ exactly once. Note that in the special case of a Gromov-Witten theory for some manifold X, these correlators carry the following meaning:

$$\langle \tau_{d_1}(e_{i_1})\tau_{d_2}(e_{i_2})\cdots\tau_{d_k}(e_{i_k})\rangle_g = \sum_{\deg}\int_{[X_{g,k,\deg}]} ev_1^*(e_{i_1})\psi_1^{d_1}\, ev_2^*(e_{i_2})\psi_2^{d_2}\cdots ev_k^*(e_{i_k})\psi_k^{d_k}, \tag{40}$$

where $[X_{g,k,\deg}]$ is the moduli space of degree deg stable maps to X of genus-g curves with k marked points, ev_i is the evaluation map at the ith point and ψ correspond to ψ-classes.

Consider a sequence of operators $r_l \in \operatorname{Hom}(V,V)$ for $l \geq 1$, such that the operators with odd (resp., even) indices are symmetric (resp., skew-symmetric). Then we denote by $(r_l z^l)\hat{}$ the following differential operator:

$$(r_l z^l)\hat{} := -(r_l)_{\mathbf{1}}^i \frac{\partial}{\partial v^{l+1,i}} + \sum_{d=0}^{\infty} v^{d,i}(r_l)_i^j \frac{\partial}{\partial v^{d+l,j}} + \frac{\hbar}{2}\sum_{m=0}^{l-1}(-1)^{m+1}(r_l)^{i,j}\frac{\partial^2}{\partial v^{m,i}\partial v^{l-1-m,j}}.$$

Here the indices $i,j \in \{1,\ldots,N\}$ on r_l correspond to the basis $\{e_1,\ldots,e_N\}$ of V, and the index $\mathbf{1}$ corresponds to the unit vector $e_{\mathbf{1}}$. When we write r_l with two upper-indices we mean as usual that we raise one of the indices using the scalar product η.

Given such a sequence of operators r_l, we define an operator series $R(z)$ in the following way

$$R(z) = \sum_{l=0}^{\infty} R_l z^l := \exp\left(\sum_{l=1}^{\infty} r_l z^l\right). \tag{41}$$

The quantization $\hat{R}$ of this series is defined by

$$\hat{R} = \exp\left(\sum_{l=1}^{\infty}((-1)^l r_l z^l)\hat{}\right). \tag{42}$$

Givental observed that the action of such operators $\hat{R}$ on formal power series Z for which the number of ψ-classes (given by the first index of $v^{d,\mu}$) at any monomial of degree n is no more than $3g-3+n$, is well-defined. The main theorem of [**32**] states that this action preserves the property that Z is a generating function of the correlators of a cohomological field theory with target space (V,η) (see also [**40, 60**]).

REMARK 3.1. Note that in order to agree with the conventions of Givental, we label in a matrix by the upper index the column and by the lower index the row.

3.1.2. *Givental operator for a Frobenius manifold.* Let $Z(\{t^{d,\mu}\})$ be the partition function of some N-dimensional semi-simple conformal cohomological field theory. We recall the construction (due to Givental [**35,36,38**], see also Dubrovin [**13**]) of an operator series $R(z)$ as in the previous section whose quantization takes the product of N KdV τ-functions to Z.

Let F be the restriction of $\log(Z)$ to the genus zero part without descendants. Denote $t^\mu := t^{0,\mu}$. Then F can be interpreted as a formal Frobenius manifold with metric

$$\eta_{\alpha\beta} = \frac{\partial^3 F}{\partial t^{\mathbf{1}} \partial t^\alpha \partial t^\beta} \tag{43}$$

and Frobenius algebra structure $c^\gamma_{\alpha\beta}$

$$c_{\alpha\beta\gamma} = \frac{\partial^3 F}{\partial t^\alpha \partial t^\beta \partial t^\gamma}. \tag{44}$$

We can assume that $\eta_{\alpha\beta} = \delta_{\alpha+\beta,n+1}$ and $e_{\mathbf{1}} = e_1$. According to [**12**] it is always possible by an appropriate choice of these flat coordinates t^μ.

3.1.3. *Canonical coordinates.* Another set of coordinates is given by the *canonical coordinates* $\{u^i\}$ which can be found as solutions to Equation (3.54) from [**12**], and have the property that $\{\partial_i := \partial/\partial u^i\}$ forms a basis of canonical idempotents of the Frobenius algebra product. In these coordinates the metric is diagonal and the unit vector field is given by $e_{\mathbf{1}} = \partial_1 + \cdots + \partial_N$.

Define $\Delta_i := 1/(\partial_i, \partial_i)$ to be the inverse of the square of the length of the ith canonical basis element, and call $\{\partial/\partial v^i := \Delta_i^{1/2} \partial/\partial u^i\}$ the normalized canonical basis in the tangent space. We denote the coordinates corresponding to this basis by v^i, and the formal variables corresponding to these coordinates by $v^{d,i}$. They are precisely the formal variables $v^{d,i}$ appearing in the previous section.

Let U be the matrix of canonical coordinates $U = \operatorname{diag}(u^1, \ldots, u^N)$ and denote by Ψ the transition matrix from the flat to the normalized canonical bases. That is, denoting $\mathrm{d}t = (\mathrm{d}t^1, \ldots, \mathrm{d}t^N)^{\mathrm{T}}$ and $\mathrm{d}u = (\mathrm{d}u^1, \ldots, \mathrm{d}u^N)^{\mathrm{T}}$, one has

$$\Delta^{-1/2} \mathrm{d}u = \Psi \mathrm{d}t, \tag{45}$$

where $\Delta = \operatorname{diag}(\Delta_1, \ldots, \Delta_N)$.

REMARK 3.2. Note that Ψ obtained with the help of the definition above depends on the point p of the Frobenius manifold.

3.1.4. *Recursion.* Construct an operator series $R(z) = \sum_{k\geq 0} R_k z^k$ as in the previous section in the following way.

Recursively define the off-diagonal entries of R_k in normalized canonical coordinates by solving the equation

$$\Psi^{-1} \mathrm{d}(\Psi R_{k-1}) = [\mathrm{d}U, R_k]. \tag{46}$$

using $R_0 = \mathbf{I}$ as a base case. Construct the diagonal entries of R_k by integrating the next equation

$$\Psi^{-1} \mathrm{d}(\Psi R_k) = [\mathrm{d}U, R_{k+1}] \tag{47}$$

using the fact that the diagonal entries of $[\mathrm{d}U, R_{k+1}]$ are equal to zero. To fix the integration constant, use the Euler equation

$$R_k = -(i_E \mathrm{d}R_k)/k, \tag{48}$$

where $E = \sum u^i \partial_i$ is the Euler field (here we use the fact that we started with a conformal cohomological field theory).

This procedure recursively defines R_k for all k. The following proposition is essentially proved in Givental's papers [**35**, **36**].

PROPOSITION 3.3. *Let F be a local N-dimensional Frobenius manifold structure, semisimple at the origin, and let (R_k) be the series of operators constructed from this F by the recursive procedure described above, at the origin. Let Ψ and Δ be as above, taken at the origin as well. Then we have the following formula:*

$$\mathcal{F}_0 = \operatorname*{Res}_{\hbar=0} d\hbar \cdot \log \hat{\Psi}\hat{R}\hat{\Delta}\mathcal{T}. \tag{49}$$

Here $\mathcal{F}_0 = \mathcal{F}_0(\{t^{d,\mu}\})$ is the genus 0 descendant potential of cohomological field theory associated to F; $\mathcal{T}$ is the product of N KdV tau-functions,

$$\mathcal{T} := Z_{\mathrm{KdV}}(\{u^{d,1}\}) \cdots Z_{\mathrm{KdV}}(\{u^{d,N}\});$$

$\hat{\Delta}$ replaces the variables of ith KdV τ-function according to $u^{d,i} = \Delta_i^{1/2} v^{d,i}$ and replaces $\hbar$ with $\Delta_i \hbar$, while $\hat{\Psi}$ is the change of variables $v^{d,i} = \Psi^i_\nu t^{d,\nu}$. The unit for the R-action is given by $(\Psi^1_1, \ldots, \Psi^N_1)$.

REMARK 3.4. In fact, using Teleman's result in [**60**], one has a refined version of Equation (49):

$$Z = \hat{\Psi}\hat{R}\hat{\Delta}\mathcal{T}. \tag{50}$$

Note that it holds for cohomological field theories. In the Gromov-Witten case, when quadratic terms in the potential cannot be neglected, there appears an additional complication, see the next remark below.

REMARK 3.5. Givental's formula [**36**] for a Gromov-Witten total descendant potential (without the $(g = 1, n = 0)$-term),

$$Z = \hat{S}^{-1}\hat{\Psi}\hat{R}\hat{\Delta}\mathcal{T}, \tag{51}$$

also includes the operator $\hat{S}$, given by

$$\hat{S} = \exp\left(\sum_{l=1}^{\infty} (s_l z^{-l})\hat{}\right), \tag{52}$$

where the operators $(s_l z^{-l})\hat{}$ are defined in the following way (see, e. g., [**32**, Section 4.2]):

$$\sum_{l=1}^{\infty} (s_l z^{-l})\hat{} = -(s_1)^\mu_1 \frac{\partial}{\partial t^{0,\mu}} + \frac{1}{\hbar}\sum_{d=0}^{\infty} (s_{d+2})_{1,\mu} t^{d,\mu} \tag{53}$$
$$+ \sum_{\substack{d=0\\l=1}}^{\infty} (s_l)^\mu_\nu t^{d+l,\nu} \frac{\partial}{\partial t^{d,\mu}} + \frac{1}{2\hbar} \sum_{\substack{d_1,d_2\\ \mu_2,\mu_2}} (-1)^{d_1} (s_{d_1+d_2+1})_{\mu_1,\mu_2} t^{d_1,\mu_1} t^{d_2,\mu_2}.$$

Note that formula (52) for the quantization of S differs from the analogous formula (42) for R by a factor of $(-1)^l$ in the exponent, which agrees with the definition in Givental's papers [**35**, **36**].

The matrices s_k are defined through the following relation:

$$S(z) = \sum_{k=0}^{\infty} S_k z^{-k} = \exp\left(\sum_{l=0}^{\infty} s_l z^{-l}\right), \tag{54}$$

where for $S(z)$, taken at a point p of the Frobenius manifold, we have (see [**36**]), for any points a and b of the Frobenius manifold,

$$(a, b\, S_p) := (a, b) + \sum_{k=0}^{\infty} \langle \tau_0(a) \exp(\tau_0(p))\, \tau_k(b) \rangle_0\, z^{-1-k}. \tag{55}$$

Here on the left hand side the brackets stand for the scalar product on the tangent space to the Frobenius manifold at p, and we used an identification of the tangent space with the whole Frobenius manifold, since in this case the Frobenius manifold is itself a vector space. If p is the origin, we have just

$$(a, b\, S) := (a, b) + \sum_{k=0}^{\infty} \langle \tau_0(a) \tau_k(b) \rangle\ z^{-1-k}. \tag{56}$$

Note that this S action is defined in the general case when the total descendant genus 0 potential is known. For the case when only a Frobenius potential is specified, the choice of S is then called a *calibration* of the Frobenius manifold, see [**13, 35**] for related details. In the case of cohomological field theory when we disregard quadratic terms, the S action is trivial if p is taken to be the origin.

It turns out that in most of the relevant cases, e.g. for the Gromov-Witten theory of $\mathbb{C}\mathrm{P}^1$ (see section 3.5.1 below), the only relevant term in equation (53) is

$$\sum_{\substack{d=0\\ l=1}}^{\infty} (s_l)^\mu_\nu\, t^{d+l,\nu} \frac{\partial}{\partial t^{d,\mu}},$$

since $(s_1)^\mu_1$ vanishes and all other terms just change the unstable terms in the potential.

This means, that in these cases $\hat{S}^{-1}$ just performs a linear change of formal variables $t^{d,\mu}$ in the following way:

$$t^{d,\mu} \mapsto \sum_{m=d}^{\infty} (S_{m-d})^\mu_\nu t^{m,\nu}. \tag{57}$$

3.2. Expressions in terms of graphs.

NOTATION 3.6. Let γ be any graph. By a half-edge we mean either a leaf or an edge together with a choice of one of the two vertices it is attached to. By $V(\gamma)$, $E(\gamma)$, $H(\gamma)$ and $L(\gamma)$ we denote the sets of vertices, edges, half-edges and leaves of γ. For any vertex v of γ, denote by $H(v)$ the set of half-edges connected to v.

Let $\widetilde{\Gamma}$ be the set of all connected graphs γ together with a choice of disjoint splitting $L(\gamma) = L^*(\gamma) \coprod L^\bullet(\gamma)$, a labelling of the vertices by pairs $(g,i) \in \mathbb{Z}_{\geq 0} \times \{1,\dots,N\}$ and a labelling of the elements of $H(\gamma)$ by non-negative integers, such that the label of a leaf in $L^\bullet$ is always greater than one. The elements of $L^*(\gamma)$ are called *ordinary leaves*, the elements of $L^\bullet$ are called *dilaton leaves*. We denote by Γ the subset of all graphs in $\widetilde{\Gamma}$ that are *stable*; that is, any vertex labelled $(0,i)$ for some i is of valence at least three.

For any graph γ denote by $\mathfrak{g}\colon V(\gamma) \to \mathbb{Z}_{\geq 0}$ and $\mathfrak{i}\colon V(\gamma) \to \{1,\dots,N\}$ the maps that associate to any vertex its first and second label respectively, and by $\mathfrak{k}\colon H(\gamma) \to \mathbb{Z}_{\geq 0}$ the map that associates to any half-edge its label. Denote by $\mathfrak{v}\colon L(\gamma) \to V(\gamma)$ the map that associates to each leaf the corresponding vertex, and by $\mathfrak{v}_1, \mathfrak{v}_2\colon E(\gamma) \to V(\gamma)$ and by $\mathfrak{h}_1, \mathfrak{h}_2\colon E(\gamma) \to H(\gamma)$ the maps that associate to an edge the first and second vertex, and the corresponding half-edges respectively.

REMARK 3.7. The labels introduced above are used to keep track of different data for the trivial cohomological field theory; g is for the genus, i for the primary field in canonical coordinates and the labelling of the marked half-edges is for the power of ψ-class.

REMARK 3.8. Edges of a graph in Γ are considered to be oriented (this allows to define the maps $\mathfrak{v}_1$ and $\mathfrak{v}_2$ unambiguously); the final result does not depend on the orientation.

Let $R(z)^i_j$ be the components of the operator series $R(z)$ in normalized canonical basis as computed in Section 3.1.4. To each part of a graph $\gamma \in \Gamma$ we assign some polynomial in formal variables $\hbar$ and $v^{d,i}$. Here $\hbar$ is used to keep track of the genus, while the first index of $v^{d,i}$ keeps track of the number of ψ-classes and the second index keeps track of the normalized canonical coordinate.

3.2.1. *Leaves.* To each ordinary leaf $l \in L^*$ marked by k attached to a vertex marked by the pair (g, i), we assign

$$(\mathcal{L}^*)^i_k(l) := [z^k]\left(\sum_{d\geq 0}\left((R(-z))^i_j v^{d,j} z^d\right)\right), \tag{58}$$

which corresponds to the second term in (41).

To a dilaton leaf $\lambda \in L^\bullet(\gamma)$ marked by k attached to a vertex marked by (g, i) we assign

$$(\mathcal{L}^\bullet)^i_k(\lambda) := [z^{k-1}]\left(-(R(-z))^i_{\mathbb{1}}\right), \tag{59}$$

which corresponds to the first term in (41), which is called the *dilaton shift.*

3.2.2. *Edges.* To an edge e connecting a vertex v_1 marked by (g_1, i_1) to a vertex v_2 marked by (g_2, i_2) and with markings k_1 and k_2 at the corresponding half-edges, we assign

$$\mathcal{E}^{i_1,i_2}_{k_1,k_2}(e) := [z^{k_1} w^{k_2}]\left(\hbar \cdot \frac{\delta^{i_1 i_2} - \sum_s (R(-z))^{i_1}_s (R(-w))^{i_2}_s}{z+w}\right). \tag{60}$$

Note that this does not depend on the choice of ordering of the vertices and that it follows from the fact that $R(z)$ can be written as $R(z) = \exp(\sum r_l z^l)$ that the numerator on the right-hand side is equal to the product of $(z+w)$ with some power series in z and w, so this definition makes sense.

3.2.3. *Vertices.* Let v be a vertex marked by (g, i) with n half-edges attached to it (this includes all ordinary and dilaton leaves and also half-edges that are parts of internal edges) labelled by $k_1, \dots, k_n$. Then we assign to v the following expression:

$$\mathcal{V}^{(g,i)}_{\{k_1,\dots,k_n\}}(v) := \hbar^{g-1}(\Delta_i)^{\frac{1}{2}(2g-2+n)} \int_{\bar{\mathcal{M}}_{g,n}} \psi_1^{k_1} \cdots \psi_n^{k_n}. \tag{61}$$

3.2.4. *Z as a sum over graphs.* We recover the partition function Z of the cohomological field theory we started with as a sum over Γ:

$$(62)\quad (\hat{R}\hat{\Delta}\mathcal{T})(\{v^{d,j}\}) = \sum_{\gamma\in\Gamma} \frac{1}{|\operatorname{Aut}(\gamma)|}$$

$$\prod_{v\in V(\Gamma)} \hbar^{\mathfrak{g}(v)-1}(\Delta_{\mathfrak{i}(v)})^{\frac{1}{2}(2\mathfrak{g}(v)-2+\mathrm{val}(v))} \left\langle \prod_{h\in H(v)} \tau_{\mathfrak{k}(h)} \right\rangle_{\mathfrak{g}}$$

$$\prod_{e\in E(\gamma)} \mathcal{E}^{\mathfrak{i}(\mathfrak{v}_1(e)),\mathfrak{i}(\mathfrak{v}_2(e))}_{\mathfrak{k}(\mathfrak{h}_1(e)),\mathfrak{k}(\mathfrak{h}_2(e))}(e) \prod_{l\in L^*(\gamma)} (\mathcal{L}^*)^{\mathfrak{i}(\mathfrak{v}(l))}_{\mathfrak{k}(l)}(l) \prod_{\lambda\in L^\bullet(\gamma)} (\mathcal{L}^\bullet)^{\mathfrak{i}(\mathfrak{v}(l))}_{\mathfrak{k}(l)}(\lambda).$$

3.3. Topological recursion. In this section, we define a local version of the topological recursion and write the corresponding invariants as a sum over graphs, which allows us to compare it to the Givental action in the next section.

3.3.1. *Local topological recursion.* We define a local version of the topological recursion in the following way. The term local refers to the fact that the data are all defined locally around the canonical coordinates without any reference to the possible existence of a global manifold where these functions can be defined.

DEFINITION 3.9. For $N \in \mathbb{N}^*$, we call times a set of N families of complex numbers $\{h^i_k\}_{k\in\mathbb{N}}$ for $i = 1,\dots,N$ and jumps another set of $N\times N$ infinite families of complex numbers $\left\{B^{i,j}_{k,l}\right\}_{(k,l)\in\mathbb{N}^2}$ for $i,j = 1,\dots,N$. We finally define a set of canonical coordinates $\{a_i\}_{i=1}^N \in \mathbb{C}^N$ subject to $a_i \neq a_j$ for $i \neq j$.

For all $i,j \in \{1,\dots,N\}$, we define the following set of analytic functions and differential forms in a neighborhood of $0 \in \mathbb{C}$:

$$(63)\qquad x^i(z) := z^2 + a_i\,, \quad y^i(z) := \sum_{k=0}^{\infty} h^i_k z^k$$

and

$$(64)\qquad B^{i,j}(z,z') = \delta_{i,j}\frac{dz\otimes dz'}{(z-z')^2} + \sum_{k,l=0}^{\infty} B^{i,j}_{k,l} z^k z'^l dz\otimes dz'.$$

For $2g-2+n>0$, we define the genus g, n-point correlation functions $\omega^{i_1,\dots,i_n}_{g,n}(z_1,\dots z_n)$ recursively by

$$(65)\quad \omega^{i_0,i_1,\dots,i_n}_{g,n+1}(z_0,z_1,\dots,z_n) := \sum_{j=1}^{N} \operatorname*{Res}_{z\to 0} \frac{\int_{-z}^{z} B^{i_0,j}(z_0,\cdot)}{2\left(y^j(z)-y^j(-z)\right)\mathrm{d}x^j(z)} \times$$

$$\left(\omega^{j,j,i_1,\dots,i_n}_{g-1,n+2}(z,-z,z_1,\dots,z_n) + \sum_{A\cup B=\{1,\dots,n\}} \sum_{h=0}^{g} \omega^{j,\mathbf{i}_A}_{h,|A|+1}(z,\mathbf{z}_A)\omega^{j,\mathbf{i}_B}_{g-h,|B|+1}(-z,\mathbf{z}_B)\right),$$

where for any set A, we denote by $\mathbf{z}_A$ (resp., $\mathbf{i}_A$) the set $\{z_k\}_{k\in A}$ (resp., $\{i_k\}_{k\in A}$), and where the base of the recursion is given by

$$(66)\qquad \omega^i_{0,1}(z) := 0; \qquad \omega^{i,j}_{0,2}(z,z') := B^{i,j}(z,z').$$

For convenience, in the sequel we denote

$$(67)\qquad K^{i,j}(z,z') = \frac{\int_{-z}^{z} B^{i,j}(z',\cdot)}{2(y^j(z)-y^j(-z))dx^j(z)}$$

and

$$\omega_{g,n}(\vec{z}) = \sum_{\vec{i}} \omega^{\vec{i}}_{g,n}(\vec{z}) \,, \tag{68}$$

where the length of $\vec{z}$ and $\vec{i}$ is n.

3.3.2. *Correlation functions and intersection numbers.* The correlation functions built by this topological recursion can actually be written in terms of intersection of ψ classes on the moduli space of Riemann surfaces. This result is a slight generalization of [**24**, **25**] to the local topological recursion.

3.3.3. *One-branch point case.* The link between the topological recursion formalism and intersection numbers on the moduli space of Riemann surfaces comes from the application of this formalism to the Airy curve. This case corresponds to $N = 1$ and:

$$x(z) = z^2 + a\,, \quad y(z) = z \quad \text{and} \quad B(z,z') = \frac{dz \otimes dz'}{(z-z')^2}. \tag{69}$$

REMARK 3.10. Since there is only one branch point in this case, i.e. $N = 1$, we omit the superscript indicating which branch point we consider in the notations of this section.

For further convenience, we introduce two additional parameters by considering the curve

$$x(z) = z^2 + a\,, \quad y(z) = \alpha z \quad \text{and} \quad B(z,z') = \beta\frac{dz \otimes dz'}{(z-z')^2}, \tag{70}$$

the usual Airy curve being $\alpha = \beta = 1$. In this case, the topological recursion reads

$$\omega_{g,n+1}(z_0, z_1, \ldots, z_n) := \operatorname*{Res}_{z\to 0} \frac{\beta}{2\alpha}\frac{dz_0}{2z\,dz}\frac{1}{(z_0^2 - z^2)} \times \tag{71}$$

$$\left(\omega_{g-1,n+2}(z,-z,z_1,\ldots,z_n) + \sum_{A\cup B=\{1,\ldots,n\}} \sum_{h=0}^{g} \omega_{h,|A|+1}(z,\mathbf{z}_A)\omega_{g-h,|B|+1}(-z,\mathbf{z}_B) \right)$$

and one has

LEMMA 3.11. *The correlation functions of the Airy curve can be expressed in terms of intersection numbers:*

$$\omega_{g,n}(z_1,\ldots,z_n) = \tag{72}$$

$$\left(-\frac{\beta}{2\alpha}\right)^{2g+n-2} \beta^{g+n-1} \sum_{\alpha_1,\ldots,\alpha_n \geq 0} \langle \tau_{a_1} \cdots \tau_{a_n} \rangle_{g,n} \prod_{i=1}^{n} \frac{(2\alpha_i+1)!!\, dz_i}{z_i^{2\alpha_i+2}}.$$

This lemma was proved many times by direct computation [**23**, **24**, **29**, **62**], matching the topological recursion with the recursive definition of the intersection numbers.

As a side note, the first few correlation functions are

$$\omega_{0.3}(z_1,z_2,z_3) = -\frac{\beta^3}{2\alpha}\prod_{i=1}^{3}\frac{dz_i}{z_i^2}, \tag{73}$$

$$\omega_{0,4}(z_1,z_2,z_3,z_4) = \frac{\beta^5}{4\alpha^2}\prod_{i=1}^{4}\frac{dz_i}{z_i^2}\sum_{i=1}^{4}\frac{3}{z_i^2}, \tag{74}$$

$$\omega_{1,1}(z) = \frac{-\beta^2}{2\alpha}\frac{dz}{8z^4} \tag{75}$$

and

$$\omega_{1,2}(z_1,z_2) = \frac{\beta^4}{4\alpha^2}\frac{dz_1\,dz_2}{8z_1^2z_2^2}\left(\frac{5}{z_1^4}+\frac{5}{z_2^4}+\frac{3}{z_1^2z_2^2}\right). \tag{76}$$

REMARK 3.12. It is important to remark that there exist different conventions in the literature for defining the topological recursion, mainly differing by a change of sign of the recursion kernel. The latter can be recovered by a change of sign $\alpha \to -\alpha$.

Let us now consider a deformation of the Airy curve which we will refer to as the KdV curve in the following. It has only one branch point, $N = 1$, and reads

$$\begin{cases} x(z) = z^2 + a_i \\ y(z) = \alpha\sum_{k=1}^{\infty} h_k z^k \\ B(z,z') = \beta B_{\mathrm{KdV}}(z,z') = \beta dz\otimes dz'(z-z')^2 \end{cases}. \tag{77}$$

The corresponding correlation functions can also be expressed in terms intersection numbers as follows:

LEMMA 3.13. *The correlation functions of the KdV curve read:*

$$\omega_{g,n}(z_1,\dots,z_n) = \left(-\frac{\beta}{2\alpha h_1}\right)^{2g+n-2}\beta^{g+n-1}\sum_{m=0}^{\infty}\frac{(-1)^m}{m!} \tag{78}$$

$$\sum_{\vec{\alpha}\in\mathbb{N}^{*m}}\prod_{k=1}^{m}(2\alpha_k+1)!!\frac{h_{2\alpha_k+1}}{h_1}\prod_{i=1}^{n}\frac{(2d_i+1)!!\,dz_i}{z_i^{2d_i+2}}\left\langle\prod_{j=1}^{n}\tau_{d_j}\prod_{k=1}^{m}\tau_{\alpha_k+1}\right\rangle_{g,n+m}.$$

PROOF. Once again the proof can be found in the literature [**23, 24, 28**]. However, let us study a graphical interpretation of this result when considering an arbitrary convention for the topological recursion. For $f(z)$ an analytic function around $z \to 0$ and $\{T_k\}_{k\in\mathbb{Z}}$ a set of parameters, one can compute

$$\operatorname*{Res}_{Z_1\to 0}\operatorname*{Res}_{Z_2\to 0} K(Z_1,z)\left\{\left(\sum_{k\geq 1}T_kZ_1^k\right)dZ_1\,K(Z_2,-Z_1)f(Z_2)\,[dZ_2]^2\right. \tag{79}$$
$$\left. -\left(\sum_{k\geq 1}T_k(-Z_1)^k\right)dZ_1\,K(Z_2,Z_1)f(Z_2)\,[dZ_2]^2\right\}$$

where the recursion kernel is the one of the Airy curve, i.e. the one for which $h_k = 0$ for $k \geq 2$:

$$K(z,z_0) = \frac{\beta}{2\alpha h_1}\frac{dz_0}{2z\,dz}\frac{1}{(z_0^2-z^2)}. \tag{80}$$

One can move the integration contours to get

$$\operatorname*{Res}_{Z_1\to 0}\operatorname*{Res}_{Z_2\to 0} = \operatorname*{Res}_{Z_2\to 0}\operatorname*{Res}_{Z_1\to 0} + \operatorname*{Res}_{Z_2\to 0}\operatorname*{Res}_{Z_1\to Z_2} + \operatorname*{Res}_{Z_2\to 0}\operatorname*{Res}_{Z_1\to -Z_2}. \tag{81}$$

The first term of the right hand side vanishes since the integrand does not have any pole at $Z_1 \to 0$. Let us now compute one of the other two terms:

$$(82)\quad \operatorname*{Res}_{Z_2\to 0}\operatorname*{Res}_{Z_1\to Z_2} K(Z_1,z)\left(\sum_{k\geq 1} T_k Z_1^k dZ_1\right) K(Z_2,-Z_1) f(Z_2)\,[dZ_2]^2$$

$$= -\operatorname*{Res}_{Z_2\to 0}\frac{\beta}{2\alpha h_1}\frac{dZ_2}{2Z_2} f(Z_2)\operatorname*{Res}_{Z_1\to Z_2}\frac{dz}{2Z_1}\frac{1}{(z^2-Z_1^2)}\frac{\beta}{2\alpha h_1}\frac{1}{(Z_1^2-Z_2^2)}\left(\sum_{k\geq 1} T_k Z_1^k dZ_1\right)$$

$$= -\operatorname*{Res}_{Z_2\to 0}\frac{\beta}{2\alpha h_1}\frac{dZ_2 dz}{2Z_2} f(Z_2)\frac{1}{(z^2-Z_2^2)}\frac{\beta}{2\alpha h_1}\sum_{k\geq 1}\frac{T_k}{4} Z_2^{k-2}.$$

In the same way,

$$(83)\quad \operatorname*{Res}_{Z_2\to 0}\operatorname*{Res}_{Z_1\to -Z_2} K(Z_1,z)\left(\sum_{k\geq 1} T_k Z_1^k dZ_1\right) K(Z_2,-Z_1) f(Z_2)\,[dZ_2]^2$$

$$= -\operatorname*{Res}_{Z_2\to 0}\frac{\beta}{2\alpha h_1}\frac{dZ_2 dz}{2Z_2} f(Z_2)\frac{1}{(z^2-Z_2^2)}\frac{\beta}{2\alpha h_1}\sum_{k\geq 1}\frac{T_k}{4}(-Z_2)^{k-2}.$$

The sum of these two terms reads

$$(84)\quad \operatorname*{Res}_{Z_2\to 0}\operatorname*{Res}_{Z_1\to \pm Z_2} K(Z_1,z)\left(\sum_{k\geq 1} T_k Z_1^k dZ_1\right) K(Z_2,-Z_1) f(Z_2)\,[dZ_2]^2$$

$$= -\operatorname*{Res}_{Z_2\to 0}\frac{\beta}{2\alpha h_1}\frac{dZ_2 dz}{2Z_2} f(Z_2)\frac{1}{(z^2-Z_2^2)}\frac{\beta}{2\alpha h_1}\sum_{k\geq 1}\frac{T_{2k}}{2}(Z_2)^{2k-2}$$

and finally:

$$(85)\quad \operatorname*{Res}_{Z_1\to 0}\operatorname*{Res}_{Z_2\to 0} K(Z_1,z)\left\{\left(\sum_{k\geq 1} T_k Z_1^k\right) dZ_1\, K(Z_2,-Z_1) f(Z_2)\,[dZ_2]^2\right.$$

$$\left. - \left(\sum_{k\geq 1} T_k(-Z_1)^k\right) dZ_1\, K(Z_2,Z_1) f(Z_2)\,[dZ_2]^2\right\} =$$

$$= \operatorname*{Res}_{Z_2\to 0}\frac{\beta}{2\alpha h_1}\frac{dZ_2 dz}{2Z_2} f(Z_2)\frac{1}{(z^2-Z_2^2)}\left(-\frac{\beta}{2\alpha h_1}\right)\sum_{k\geq 1} T_{2k}(Z_2)^{2k-2}.$$

On the other hand, plugging in the times h_k amounts to computing similar quantities:

$$(86)\quad \operatorname*{Res}_{z\to 0}\frac{\beta}{2\alpha h_1}\frac{dz_0}{2z\,dz}\frac{1}{(z_0^2-z^2)}\frac{1}{\left(1+\sum_{k=1}^{\infty}\frac{h_{2k+1}}{h_1}z^{2k}\right)} f(z)\,[dz]^2$$

$$= \operatorname*{Res}_{z\to 0}\frac{\beta}{2\alpha h_1}\frac{dz_0}{2z\,dz}\frac{1}{(z_0^2-z^2)} f(z)\,[dz]^2\,(1-\sum_{k=1}^{\infty}\frac{h_{2k+1}}{h_1}z^{2k}+\left[\sum_{k=1}^{\infty}\frac{h_{2k+1}}{h_1}z^{2k}\right]^2+\dots)$$

The first term of this sum is the Airy recursion kernel. The second one is of the shape of the preceding one with $T_{2k+2} = \dfrac{2\alpha h_{2k+1}}{h_1}$ for $k \geq 1$ so that:

$$\begin{aligned}(87)\qquad & -\operatorname*{Res}_{z\to 0} \frac{\beta}{2\alpha}\frac{dz_0}{2z\,dz}\frac{1}{(z_0^2 - z^2)} f(z)\,[dz]^2 \sum_{k=1}^{\infty} \frac{h_{2k+1}}{h_1} z^{2k} \\ & - \operatorname*{Res}_{Z_1\to 0}\operatorname*{Res}_{Z_2\to 0} K(Z_1, z_0)\Big\{ g(Z_1)dZ_1\, K(Z_2, -Z_1) f(Z_2)\,[dZ_2]^2 \\ & \qquad\qquad - g(-Z_1)dZ_1\, K(Z_2, Z_1) f(Z_2)\,[dZ_2]^2 \Big\}\end{aligned}$$

where

$$(88)\qquad g(z) := \sum_{k\geq 1} \frac{2\alpha h_{2k+1}}{\beta h_1} z^{2k+2}.$$

This same procedure can be applied to the other terms of the sum. The kth order term can be written as a sequence of $k+1$ residues computed with the Airy recursion kernel with $g(z)dz$ on one of the outgoing legs. This computation shows that introducing non-vanishing times amounts to introducing a non-vanishing $\omega_{0,1}(z) := g(z)dz$ in the topological recursion.

It is often useful to represent the topological recursion in a graphical form by representing the interaction kernel $K(z, z_0)$ by an edge oriented from z_0 towards a trivalent vertex labelled by z and the function $\omega_{0,2}(z_1, z_2)$ by a non-oriented edge (see **[27]** for more details about this set of graphs). In this form, $\omega_{g,n}(z_1, \dots, z_n)$ is a sum over trivalent graphs of genus g with n leaves labelled by the arguments $z_1, \dots, z_n$. The preceding computation shows that the correlation functions of the KdV curve can be obtained from the correlation functions of the Airy curve by introducing a set of new leaves, called dilation leaves, in the definition of the graphs used. A dilation leave decorated by a label k is weighted by

$$(89)\qquad (2d-1)!!\ \operatorname*{Res}_{z\to 0} g(z)\frac{dz}{z^{2d+1}} = (2d-1)!!\ \frac{2\alpha h_{2d-1}}{\beta}.$$

Plugging this expression into the formula for the Airy correlation functions proves the result. □

3.3.4. *General case.* In this section we give a formula for the correlation function of the local topological recursion.

DEFINITION 3.14. Let $\Gamma_{g,n}$ be the subset of Γ (see Notation 3.6) consisting of graphs of genus g' such that $g' + \sum_{v\in V(\Gamma)} \mathfrak{g}(v) = g$ and with n ordinary leaves. Let us also introduce orderings on the ordinary leaves and denote by $\check{\Gamma}_{g,n}$ the set of all graphs from $\Gamma_{g,n}$ with all possible orderings on the ordinary leaves. For a given graph with a fixed ordering $\check{\gamma} \in \check{\Gamma}_{g,n}$ for an ordinary leaf of that graph $l \in L^*(\check{\gamma})$ we denote by $\mathfrak{m}(l)$ the index of this particular leaf (then $\mathfrak{m}(l)$ is an integer from 1 to n such that different leaves have different values $\mathfrak{m}(l)$ assigned to them).

THEOREM 3.15. *The correlation functions can be written as a sum over decorated graphs whose vertices are weighted by intersection of ψ-classes on $\overline{\mathcal{M}}_{g,n}$, edges by the jumps, ordinary leaves by primitives of B and dilaton leaves by the times.*

For $2-2g-n<0$, one has

$$(90)\quad \omega_{g,n}(\vec{z}) = \frac{1}{n!}\sum_{\check{\gamma}\in\check{\Gamma}_{g,n}}\prod_{v\in V(\check{\gamma})}\left(-2h_1^{\mathfrak{i}(v)}\right)^{2-2\mathfrak{g}(v)-\mathrm{val}(v)}\left\langle\prod_{h\in H(v)}\tau_{\mathfrak{k}(h)}\right\rangle_{\mathfrak{g}(v),\mathrm{val}(v)}$$

$$\prod_{e\in E(\check{\gamma})}\check{B}^{\mathfrak{i}(\mathfrak{v}_1(e)),\mathfrak{i}(\mathfrak{v}_2(e))}_{\mathfrak{k}(\mathfrak{h}_1(e)),\mathfrak{k}(\mathfrak{h}_2(e))}\prod_{l\in L^*(\check{\gamma})}\sum_{j=1}^{N}\mathrm{d}\xi^{\mathfrak{i}(\mathfrak{v}(l))}_{\mathfrak{k}(l)}(z_{\mathfrak{m}(l)},j)\prod_{\lambda\in L^\bullet(\check{\gamma})}\check{h}^{\mathfrak{i}(\mathfrak{v}(\lambda))}_{\mathfrak{k}(\lambda)}$$

with

$$(91)\qquad \check{h}^i_k := 2(2k-1)!!h^i_{2k-1},$$

$$(92)\qquad d\xi^i_d(z_\alpha,j) := \operatorname*{Res}_{z\to 0}\frac{(2d+1)!!dz}{z^{2d+2}}\int^z B^{i,j}(z,z_\alpha),$$

$$(93)\qquad \check{B}^{i,j}_{d_1,d_2} := B^{i,j}_{2d_1,2d_2}\,(2d_1-1)!!\,(2d_2-1)!!$$

and

$$(94)\qquad \left\langle\prod_{i=1}^{n}\tau_{k_i}\right\rangle_{g,n} := \int_{\overline{\mathcal{M}}_{g,n}}\psi_1^{k_1}\psi_2^{k_2}\dots\psi_n^{k_n}.$$

PROOF. The proof is very similar to the one presented in [**25, 42**]. However, we prefer to present a completely graphical proof so that the link with the next sections becomes clear.

We follow the proof of [**25**]. From the definition, one can write the correlation functions as a sum over graphs with oriented and non-oriented arrows linking trivalent vertices resulting in the following expression:

$$(95)\qquad \omega^{\vec{i}}_{g,n}(\vec{z}) = \sum_{G\in\widehat{G}_{g,n}}\omega(G)$$

with $\widehat{G}_{g,n}$ the set of genus g trivalent graphs with one root and $n-1$ leaves labelled by the arguments z_i and a skeleton tree of oriented edges pointing from the root towards the leaves weighted by

$$\begin{aligned}\omega(G) \;=\; & \vec{\prod_{v\in V(G)}}\operatorname*{Res}_{Z_v\to 0}\prod_{e\in E_{\mathrm{oriented}}(G)}K^{i(v_1(e)),i(v_2(e))}(Z_{v_1(e)},Z_{v_2(e)})\\ & \prod_{e\in E_{\mathrm{unoriented}}(G)}B^{i(v_1(e)),i(v_2(e))}(Z_{v_1(e)},Z_{v_2(e)})\end{aligned}$$

where each leaf is considered as a one-valent vertex v and one denotes Z_v the variable z_i associated to this leaf in the correlation function, $E_{\mathrm{oriented}}(G)$ is the set of oriented leaves of G and $E_{\mathrm{unoriented}}(G)$ is the set of unoriented leaves of G (see [**27**] for further details). The product of residues $\vec{\prod}_{v\in V(G)}\operatorname{Res}_{Z_v\to 0}$ is oriented following the arrows, i.e. one first computes the residue corresponding to the end of an arrow before the one associated to its root.

It is useful to remark, that, for any edge, oriented or not, one has two types of contributions. Indeed, the functions $B^{i,j}(z,z')$ have a singular part

(97) $$B^{i,j}_{\mathrm{KdV}}(z,z') := \delta_{i,j}\frac{dz\otimes dz'}{(z-z')^2}$$

and a regular part

(98) $$B^{i,j}_{\mathrm{reg}}(z,z') := \sum_{k,l=0}^{\infty} B^{i,j}_{k,l} z^k z'^l dz\otimes dz'$$

when $z\to z'$:

(99) $$B^{i,j}(z,z') = B^{i,j}_{\mathrm{KdV}}(z,z') + B^{i,j}_{\mathrm{reg}}(z,z').$$

In the same way, one has

(100) $$K^{i,j}(z,z') = K^{i,j}_{\mathrm{KdV}}(z,z') + K^{i,j}_{\mathrm{reg}}(z,z').$$

One can translate this by representing $B^{i,j}_{\mathrm{KdV}}(z,z')$ (resp. $B^{i,j}_{\mathrm{reg}}(z,z')$) by dashed (resp. dotted) unoriented edges and $K^{i,j}_{\mathrm{KdV}}(z,z')$ (resp. $K^{i,j}_{\mathrm{reg}}(z,z')$) by dashed (resp. dotted) oriented edges form z to z'. The preceding sum is thus transformed into a sum over graphs where the edges are dotted or dashed and weighted accordingly.

The dashed edges can be expressed in a slightly different way. Indeed, one has

(101) $$B^{i,j}_{\mathrm{reg}}(z,z') = \operatorname*{Res}_{z_1\to z}\operatorname*{Res}_{z_2\to z'} B^{i,i}_{\mathrm{KdV}}(z,z_1)\left[\int^{z_1}\int^{z_2} B^{i,j}_{\mathrm{reg}}(z_1,z_2)\right] B^{j,j}_{\mathrm{KdV}}(z_2,z')$$

and

(102) $$K^{i,j}_{\mathrm{reg}}(z,z') = \operatorname*{Res}_{z_1\to z}\operatorname*{Res}_{z_2\to \pm z'} B^{i,i}_{\mathrm{KdV}}(z,z_1)\left[\int^{z_1}\int^{z_2} B^{i,j}_{\mathrm{reg}}(z_1,z_2)\right] K^{j,j}_{\mathrm{KdV}}(z_2,z')$$

by a simple application of the Cauchy formula.

Remember that such an edge comes with integration of its boundary variables, thus, one typically has to compute

(103) $$\operatorname*{Res}_{z\to 0}\operatorname*{Res}_{z'\to 0} g(z)K^{i,j}_{\mathrm{reg}}(z,z')f(z')$$

which reads

(104) $$\operatorname*{Res}_{z\to 0}\operatorname*{Res}_{z_1\to z}\operatorname*{Res}_{z'\to 0}\operatorname*{Res}_{z_2\to\pm z'} g(z)B^{i,i}_{\mathrm{KdV}}(z,z_1)\left[\int^{z_1}\int^{z_2} B^{i,j}_{\mathrm{reg}}(z_1,z_2)\right] K^{j,j}_{\mathrm{KdV}}(z_2,z')f(z').$$

One can move the integration contours around 0 thanks to:

(105) $$\operatorname*{Res}_{z\to 0}\operatorname*{Res}_{z_1\to z} = \operatorname*{Res}_{z_1\to 0}\operatorname*{Res}_{z\to 0} - \operatorname*{Res}_{z\to 0}\operatorname*{Res}_{z_1\to 0}$$

and

(106) $$\operatorname*{Res}_{z'\to 0}\operatorname*{Res}_{z_2\to\pm z'} = \operatorname*{Res}_{z_2\to 0}\operatorname*{Res}_{z'\to 0} - \operatorname*{Res}_{z'\to 0}\operatorname*{Res}_{z_2\to 0}.$$

Since, the integrand does not have any pole as $z_1\to 0$ nor $z_2\to 0$, this shows that 103 is equal to

(107) $$\operatorname*{Res}_{z_1\to 0}\operatorname*{Res}_{z\to 0}\operatorname*{Res}_{z_2\to 0} g(z)\, B^{i,i}_{\mathrm{KdV}}(z,z_1)\left[\int^{z_1}\int^{z_2} B^{i,j}_{\mathrm{reg}}(z_1,z_2)\right]\operatorname*{Res}_{z'\to 0} K^{j,j}_{\mathrm{KdV}}(z_2,z')\, f(z').$$

In the same way, one gets that

$$\operatorname*{Res}_{z\to 0}\operatorname*{Res}_{z'\to 0} g(z)B^{i,j}_{\rm reg}(z,z')f(z') \tag{108}$$

is equal to

(109)
$$\operatorname*{Res}_{z_1\to 0}\operatorname*{Res}_{z\to 0}\operatorname*{Res}_{z_2\to 0} g(z)\,B^{i,i}_{\rm KdV}(z,z_1)\left[\int^{z_1}\int^{z_2} B^{i,j}_{\rm reg}(z_1,z_2)\right]\operatorname*{Res}_{z'\to 0} B^{j,j}_{\rm KdV}(z_2,z')\,f(z').$$

One can finally proceed in a similar way for re-expressing the weights of the root and the leaves by writing[1]

$$\operatorname*{Res}_{z'\to 0} K^{i,j}(z,z')f(z') = \operatorname*{Res}_{z_2\to 0}\left[\int^{z_2} B^{i,j}(z,z_2)\right]\operatorname*{Res}_{z'\to 0} K^{j,j}_{\rm KdV}(z_2,z')\,f(z') \tag{110}$$

and

$$\operatorname*{Res}_{z\to 0} g(z)B^{i,j}(z,z') = \operatorname*{Res}_{z_1\to 0}\operatorname*{Res}_{z\to 0} g(z)\,B^{i,i}_{\rm KdV}(z,z_1)\left[\int^{z_1} B^{i,j}(z_1,z')\right]. \tag{111}$$

As a result, by applying this transformation to each dotted line, any graph is composed of a set of dotted subgraphs whose vertices have the same label separated by dashed lines. Since each subgraph with label i also includes a root and leaves, it is a contribution to the correlation functions obtained for the case $N=1$, times h^i_k and vanishing jumps $B^{i,i}_{k,l}=0$. In the sum over graphs, one can thus replace every sum over such sub-graphs by vertices of corresponding genus weighted by the correlation function for $N=1$, which reads

$$\omega^{\vec{i}}_{g,n}(\vec{z}) = \sum_{\gamma\in\Gamma_{g,n}} \Omega(\gamma) \tag{112}$$

where

$$\begin{aligned}\Omega(\gamma) \quad = \quad & \prod_{v\in V(\gamma)}\prod_{h\in H(v)} \operatorname*{Res}_{Z_h\to 0}\omega^{{\rm KdV},i(v)}_{g(v),{\rm val}(v)}\left(\{Z_h\}_{h\in H(v)}\right)\\ & \prod_{e\in E(\gamma)}\int^{Z_{h_1(e)}}\int^{Z_{h_2(e)}} B^{i(v_1(h)),i(v_2(h))}_{\rm reg}(Z_{h_1(e)},Z_{h_2(e)})\\ & \prod_{h\in L^*(\gamma)}\int^{Z_h} B^{i,j}(Z_h,z_h)\end{aligned} \tag{113}$$

where $\omega^{{\rm KdV},i}_{g,n}(z_1,\dots,z_n)$ is the genus g, n-pointed correlation function obtained from the topological recursion in the case $N=1$ and the initial data:

$$\left\{\begin{aligned} & x(z)=z^2+a_i\\ & y(z)=\sum_{k=1}^{\infty} h^i_k z^k\\ & B(z,z')=B_{\rm KdV}(z,z')=\frac{dz\otimes dz'}{(z-z')^2}\end{aligned}\right. . \tag{114}$$

[1]Remark that, for the roots and leaves, in opposition to the inner edges, the functions are the full ones, not just the regular part.

As explained in the preceding section, it can be expressed in terms of intersection numbers:

$$\omega_{g,n}^{\mathrm{KdV},i}(z_1,\dots,z_n) = \left(-2h_1^i\right)^{2-2g-n} \sum_{m=0}^{\infty} \frac{(-1)^m}{m!} \sum_{\vec{\alpha}\in\mathbb{N}^{*m}} \prod_{k=1}^{m} (2\alpha_k+1)!! \frac{h^i_{2\alpha_k+1}}{h^i_1} \prod_{i=1}^{n} \frac{(2d_i+1)!!\, dz_i}{z_i^{2d_i+2}} \left\langle \prod_{j=1}^{n} \tau_{d_j} \prod_{k=1}^{m} \tau_{\alpha_k+1} \right\rangle_{g,n+m} \tag{115}$$

which can be made more symmetric under the exchange of the ordinary and dilation leaves by writing

$$\omega_{g,n}^{\mathrm{KdV},i}(z_1,\dots,z_n) = \sum_{m=0}^{\infty} (-2h_1)^{2-2g-n-m} \frac{1}{m!} \sum_{\vec{\alpha}\in\mathbb{N}^{*m}} \prod_{k=1}^{m} (2\alpha_k-1)!! 2h^i_{2\alpha_k-1} \prod_{i=1}^{n} \frac{(2d_i+1)!!\, dz_i}{z_i^{2d_i+2}} \left\langle \prod_{j=1}^{n} \tau_{d_j} \prod_{k=1}^{m} \tau_{\alpha_k} \right\rangle_{g,n+m} \tag{116}$$

Absorbing the factors of the form

$$\frac{(2d+1)!!\mathrm{d}z}{z^{2d+2}} \tag{117}$$

into the corresponding half-edge contribution, the weight of an inner edge becomes

$$\operatorname*{Res}_{z_1\to 0}\operatorname*{Res}_{z_2\to 0} \int^{z_1}\int^{z_2} B^{i,j}_{\mathrm{reg}}(z_1,z_2)\, \frac{(2d_1+1)!!\mathrm{d}z_1}{z_1^{2d_1+2}} \frac{(2d_2+1)!!\mathrm{d}z_1}{z_2^{2d_2+2}} \tag{118}$$

which is equal to

$$\check{B}^{i,j}_{d_1,d_2} := B^{i,j}_{2d_1,2d_2}\, (2d_1-1)!!\, (2d_2-1)!! \tag{119}$$

while the weight of the ordinary leaves becomes

$$\mathrm{d}\xi^i_d(z_\alpha,j) := \operatorname*{Res}_{z\to 0} \frac{(2d+1)!!\mathrm{d}z}{z^{2d+2}} \int^z B^{i,j}(z,z_\alpha), \tag{120}$$

where one considers both the singular and non-singular part of $B^{i,j}(z,z_\alpha)$. Collecting these contributions together proves the theorem. □

3.3.5. *Change of scales.* An important property of the correlation functions built in this way is their homogeneity property which reads

$$\forall\lambda\in\mathbb{C}\,,\ \omega_{g,n}(\vec{z}_N|x,\lambda y,B) = \lambda^{2-2g-n}\omega_{g,n}(\vec{z}_N|x,y,B) \tag{121}$$

One can thus get an additional factor λ^i by replacing $h^i_k \to \lambda^i h^i_k$ resulting in a rescaling of the weight of the vertices by $\left(\lambda^{\mathrm{i}(v)}\right)^{2-2g(v)-\mathrm{val}(v)}$.

3.3.6. *Weights, Laplace transform and recursive definition.* It is interesting to note that the weights of the edges are the coefficient of the Laplace transform of B:

$$\check{B}^{i,j}(u,v) := \sum_{(k,l)\in\mathbb{N}^2} \check{B}^{i,j}_{k,l} u^{-k} v^{-l} \tag{122}$$

is equal to

$$\begin{aligned}\check{B}^{i,j}(u,v) = \delta_{i,j}\frac{uv}{u+v} + \\ \frac{\sqrt{uv}e^{ua_i+va_j}}{2\pi}\int_{x(z)-a_i\in\mathbb{R}^+}\int_{x(z')-a_j\in\mathbb{R}^+} B^{i,j}(z,z')e^{-ux(z)-vx(z')}.\end{aligned} \tag{123}$$

In [**25**], it was proved that, if dx is a meromorphic form defined on a Riemann surface, $\check{B}^{i,j}(u,v)$ can be factorized and expressed in terms of some basic functions. Here, we will consider the converse and build $B^{i,j}_{k,l}$ by induction in such a way that there exist a set of functions $\{f_{i,j}(u)\}_{i,j=1}^N$ such that

$$\check{B}^{i,j}(u,v) = \frac{uv}{u+v}\left(\delta_{i,j} - \sum_{k=1}^N f_{i,k}(u)f_{k,j}(v)\right). \tag{124}$$

Let us define the coefficients $B^{i,j}_{k,l}$ recursively in terms of the initial data $B^{i,j}_{k,0}$ by imposing that

$$\xi^i_{d+1}(z,j) := -2\frac{d\xi^i_d(z,j)}{dx^{[j]}(z)} - \sum_{k=1}^{\mathfrak{b}} \check{B}^{i,k}_{d,0}\,\xi^k_0(z,j), \tag{125}$$

or, in terms of the Laplace transform

$$\begin{aligned} f^i_d(u,j) &:= \frac{\sqrt{u}}{2\sqrt{\pi}}\int_{x(z)-a_j\in\mathbb{R}^+} e^{-u(x(z)-a_j)}dx^j(z)\,\xi^i_d(z) \\ &= \delta_{i,j}(-1)^d u^d - \sum_{d'} \check{B}^{i,j}_{d,d'}u^{-d'-1}, \end{aligned} \tag{126}$$

it reads

$$f^i_{d+1}(u,j) := -2uf^i_d(u,j) - \sum_{k=1}^N \check{B}^{i,k}_{d,0}\, f^k_0(z,j). \tag{127}$$

With this definition, one has

$$\check{B}^{i,j}(u,v) = \frac{uv}{u+v}\left(\delta_{i,j} - \sum_{k=1}^{\mathfrak{b}} f^k_0(u,i)f^k_0(v,j)\right). \tag{128}$$

3.4. Identification of the two theories. In this section we show how to find a local spectral curve corresponding to any semi-simple conformal Frobenius manifold.

Suppose some local spectral curve is given. For any $i\in\{1,\dots,N\}$ and $k\in\mathbb{Z}_{\geq 0}$ define

$$W^i_k(z) := \sum_{j=1}^N d\left(\left(-\frac{1}{z}\frac{\mathrm{d}}{\mathrm{d}z}\right)^k \xi^i_0(z,j)\right).$$

THEOREM 3.16. *Let R be some series of operators on an N-dimensional vector space V as in Section 3.1. Let $Z = \hat{R}\hat{\Delta}\mathcal{T}$, where $\mathcal{T}$ is a product of N KdV τ-functions, be the partition function of the corresponding semi-simple cohomological field theory.*

Define a local spectral curve by the following data

$$\check{B}^{i,j}_{p,q} := [z^p w^q] \frac{\delta^{ij} - \sum_{s=1}^N R^i_s(-z)R(-w)^j_s}{z+w} \tag{129}$$

and

$$\check{h}^i_k := [z^{k-1}] \left(-R(-z))^i_{\mathbb{1}}\right) \tag{130}$$

$$h^i_1 := -\frac{1}{2\sqrt{\Delta^i}}. \tag{131}$$

Let $\omega_{g,n}$ be the genus g, n-pointed topological recursion invariant of this spectral curve and denote by

$$\Omega(\{v^{d,i}\}) = \left(\sum_{g,d} \omega_{g,d}(z_1,\ldots,z_d)\Big|_{W^i_d(z_m)=v^{d,i}} \hbar^{g-1}\right)$$

their sum after a change of variables $W^i_k(z_m) \leftrightarrow v^{d,i}$ for all m. Then the partition function of the cohomological field theory and the topological recursion invariants agree in the following sense:

$$Z(\{v^{d,i}\}) = \exp\left(\Omega(\{v^{d,i}\})\right). \tag{132}$$

PROOF. In Sections 3.1 and 3.3 we have given representations of Z and $\omega_{g,n}$ as sums over the set Γ (in fact, in the case of $\omega_{g,n}$ this set is $\check{\Gamma}$ rather than Γ, but after changing the variables $W^i_k(z_m) \leftrightarrow v^{d,i}$ we can take the sum over orderings and arrive at the sum over Γ acquiring an additional factor of $n!$, which cancels with the corresponding factor in (90)). We prove the theorem by showing that the contribution of each individual graph to Z is equal to the contribution to Ω.

Let $\gamma \in \Gamma$ be some graph. Note that on both sides we assign the same weight to the vertices of γ, namely to a vertex labelled (g,i) with n half-edges attached to it labelled $d_1,\ldots,d_n$ we associate

$$(-2h^i_1)^{2-2g-n}\langle\tau_{d_1}\cdots\tau_{d_n}\rangle_{g,n}. \tag{133}$$

Furthermore, by equation (129), any edge in γ contributes the same to Z and Ω.

Let l be an ordinary leaf of γ labelled by k attached to a vertex labelled by (g,i). We use induction on k to show that the contribution to Z is the same as the contribution to Ω.

The contribution of l to Z is given by

$$\mathcal{L}^i_k(l) = [z^k]\left(\sum_d (R(-z))^i_j v^{d,j} z^d\right) = \sum_{d=0}^k (-1)^{k-d}(R_{k-d})^i_j v^{d,j}. \tag{134}$$

When $k=0$, the contribution of l to Ω is given by

$$\sum_j d\xi^i_0(z_j, j) = W^i_0. \tag{135}$$

Since $(R_0)^i_j = \delta^i_j$, the contributions to Z and Ω agree when $k=0$.

Now suppose that they agree for some $k \in \mathbb{Z}_{\geq 0}$. That is, suppose that

$$\sum_j d\xi_k^i(z^j, j) = \sum_{l=0}^{k} (-1)^{k-l} (R_{k-l})_s^i W_l^s. \tag{136}$$

Then, using Equation (125), the contribution of the leaf to Ω for the index $k+1$ is given by

$$\begin{aligned} \sum_j d\xi_{k+1}^i(z^j, j) &= \sum_j d\left(-2\frac{\partial \xi_k^i(z^j, j)}{\partial x^j} - \sum_{t=1}^{N} \breve{B}_{k,0}^{i,t} \xi_0^t(z^j, j) \right) \\ &= \sum_j d\left(-\frac{1}{z^j}\frac{\partial}{\partial z^j} \xi_k^i(z^j, j) - \sum_{t=1}^{n} -(-1)^{k+1}(R_{k+1})_t^i \xi_0^t(z^j, j) \right) \\ &= \sum_{l=0}^{k} (-1)^l (R_l)_t^i W_{k+1-l}^t + (-1)^{k+1}(R_{k+1})_t^i W_0^t = \sum_{l=0}^{k+1} (-1)^l (R_l)_t^i W_{k+1-l}^t, \end{aligned} \tag{137}$$

where we used equation (129) to write

$$\breve{B}_{k,0}^{i,t} = -(-1)^{k+1}(R_{k+1})_t^i. \tag{138}$$

This completes the induction, and since it is clear that the dilaton leaves contribute the same in both cases, it also completes the proof of the theorem. □

REMARK 3.17. The theorem above deals with the potential of a cohomological field theory written in terms of formal variables $v^{d,i}$ corresponding to normalized canonical basis. In order to pass to flat coordinates one can change the variables in the following way:

$$v^{d,i} = \Psi_\mu^i t^{d,\mu}. \tag{139}$$

On the spectral curve side it will correspond to changing the variables W_k^i in the following way:

$$W_k^i = \Psi_\mu^i V_k^\mu. \tag{140}$$

Thus, the theorem holds in the same form for the potential of cohomological field theory written in terms of formal variables $t^{d,\mu}$, only one should identify $t^{d,\mu}$ with V_d^μ.

REMARK 3.18. Above we established the correspondence between cohomological field theories and symplectic invariants of spectral curves. However, as noted in Remark 3.5, in the case of Gromov-Witten theories we cannot disregard quadratic terms. So, in the formula for the total descendent potential an additional operator $\hat{S}$ appears. In some cases, again see Remark 3.5, it performs only a linear change of formal variables $t^{d,\mu}$ on which the potential depends. Thus, to establish the correspondence in this case, one has to change the variables W_k^i in precisely the same way, and then identify the resulting variables with $t^{d,\mu}$, similar to the case of previous remark. Occasionally, the changes of variables preformed by $\hat{\Psi}$ and $\hat{S}^{-1}$ can be a re-expansion of $\omega_{g,n}$ in a new coordinate on the spectral curve. We explain this procedure in detail for the case of $\mathbb{C}\mathrm{P}^1$ below in section 3.5.

REMARK 3.19. The system of equations obtained via a Laplace transform from the equations of Givental for the R-matrix (that is, the so-called equations of deformed flat connection) is studied in detail in [**13**, Section 5]. This gives, in particular, a recipe to reconstruct the two-point function directly from the Frobenius structure bypassing the reconstruction of the R-matrix. This also explains why we call the critical values $a_1, \dots, a_N$ of x the canonical coordinates.

3.5. The Norbury-Scott conjecture. In this section we recall and prove the Norbury-Scott conjecture on the stationary sector of the Gromov-Witten theory of $\mathbb{C}\mathrm{P}^1$.

3.5.1. *Gromov-Witten theory of* $\mathbb{C}\mathrm{P}^1$. The Gromov-Witten theory of $\mathbb{C}\mathrm{P}^1$ is discussed from the geometric point of view in many sources, see e. g. [**53**]. Givental proved in [**36**] that his formula for the formal Gromov-Witten potential coincides with the geometric Gromov-Witten potential of $\mathbb{C}\mathrm{P}^1$, so we discuss it here only from the Givental point of view, ignoring the geometric background. The same computations one can find in [**58, 59**].

The underlying structure of Frobenius manifold is determined by the following solution of the WDVV equation

$$\frac{1}{2}(t^1)^2t^2 + e^{t^2}, \tag{141}$$

and the scalar product given by

$$\begin{pmatrix} 0 & 1 \\ 1 & 0 \end{pmatrix}. \tag{142}$$

All ingredients of the Givental formula depend on a particular choice of the point on the Frobenius manifold, and in this case we choose the point $(0, 0)$ in the coordinates (t_1, t_2).

We perform a direct computation following the recipe of Givental in [**35**], see also Section 3.1.2. As a possible choice of the canonical coordinates, we use

$$u^1 = t^1 + 2\exp(t^2/2); \tag{143}$$

$$u^2 = t^1 - 2\exp(t^2/2). \tag{144}$$

In particular, for $t^1 = t^2 = 0$ we have $u^1 = -u^2 = 2$. Then,

$$\Delta_1^{-1} = \left\langle \frac{\partial}{\partial u^1}, \frac{\partial}{\partial u^1} \right\rangle = \frac{\exp(-t^2/2)}{2}; \tag{145}$$

$$\Delta_2^{-1} = \left\langle \frac{\partial}{\partial u^2}, \frac{\partial}{\partial u^2} \right\rangle = \frac{-\exp(-t^2/2)}{2}, \tag{146}$$

so we can choose the square roots as

$$\Delta_1^{-1/2} = \frac{\exp(-t^2/4)}{\sqrt{2}}; \tag{147}$$

$$\Delta_2^{-1/2} = \frac{-i\exp(-t^2/4)}{\sqrt{2}}, \tag{148}$$

and for this choice we have the following matrix of transition from the basis given by $(\partial/\partial t^1, \partial/\partial t^2)$ to the normalized canonical basis:

$$\Psi = \begin{pmatrix} \frac{\exp(-t^2/4)}{\sqrt{2}} & \frac{-i\exp(-t^2/4)}{\sqrt{2}} \\ \frac{\exp(t^2/4)}{\sqrt{2}} & \frac{i\exp(t^2/4)}{\sqrt{2}} \end{pmatrix}. \tag{149}$$

It is the matrix $\Psi = \Psi^i_\alpha$, where α labels the rows and corresponds to the flat basis, while i labels the columns and corresponds to the normalized canonical basis.

The recipe of reconstruction of the matrix R from [**35**] gives at the origin the matrix $R(\zeta) = \sum_{k=0}^{\infty} R_k\zeta^k$, where

$$R_k = \frac{(2k-1)!!(2k-3)!!}{2^{4k}k!} \cdot \begin{pmatrix} -1 & (-1)^{k+1}2ki \\ 2ki & (-1)^{k+1} \end{pmatrix} \tag{150}$$

The S matrix is given by the derivatives of the deformed flat coordinates, computed in [**15**, Example 3.7.9] At the origin we have:

$$\begin{aligned} S(\zeta^{-1}) =& \mathbf{I} + \zeta^{-1} \cdot \begin{pmatrix} 0 & 0 \\ 1 & 0 \end{pmatrix} \\ &+ \sum_{k=1}^{\infty} \frac{\zeta^{-2k}}{(k!)^2} \begin{pmatrix} 1 - 2k\left(\frac{1}{1} + \cdots \frac{1}{k}\right) & 0 \\ 0 & 1 \end{pmatrix} \\ &+ \sum_{k=1}^{\infty} \frac{\zeta^{-2k-1}}{(k!)^2} \begin{pmatrix} 0 & -2\left(\frac{1}{1} + \cdots \frac{1}{k}\right) \\ \frac{1}{k+1} & 0 \end{pmatrix}. \end{aligned} \tag{151}$$

(Note once again that we are using the convention that the matrices are acting on vector rows, opposite to the standard one).

The unit vector at the origin in the normalized canonical basis is equal to

$$e = (1,0) \cdot \begin{pmatrix} \frac{1}{\sqrt{2}} & \frac{-i}{\sqrt{2}} \\ \frac{1}{\sqrt{2}} & \frac{i}{\sqrt{2}} \end{pmatrix} = \left(\frac{1}{\sqrt{2}}, \frac{-i}{\sqrt{2}} \right). \tag{152}$$

Therefore, the dilaton leaves (cf. Equation (59)) in the Givental formula for $\mathbb{C}\mathrm{P}^1$ at the origin are

$$(\mathcal{L}^\bullet)^1_{k+1} = \frac{1}{\sqrt{2}} \cdot \frac{(-1)^{k+1}\left((2k-1)!!\right)^2}{k!2^{4k}}; \tag{153}$$

$$(\mathcal{L}^\bullet)^2_{k+1} = \frac{i}{\sqrt{2}} \cdot \frac{\left((2k-1)!!\right)^2}{k!2^{4k}} \tag{154}$$

for $k \geq 0$.

PROPOSITION 3.20. *The Gromov-Witten potential of* $\mathbb{C}\mathrm{P}^1$,

$$Z_{\mathbb{C}\mathrm{P}^1}(\hbar, \{t^{\ell,1}, t^{\ell,2}\}_{\ell=0}^{\infty}), \tag{155}$$

is obtained from $\hat{R}\hat{\Delta}Z_{\mathrm{KdV}}^{\otimes 2}$ *(understood as a sum over graphs in the sense of Section 3.2 and written down in the normalized canonical basis, that is, in the variables* $v^{d,i}$, $d \geq 0$, $i = 1,2$*) via a linear change of variables given by*

$$\sum_{m \geq k}^{\infty} \left(t^{m,1}, t^{m,2}\right) S_k \zeta^{m-k} = \sum_{\ell=0}^{\infty} \left(v^{\ell,1}, v^{\ell,2}\right) \zeta^\ell \cdot \Psi^{-1}, \tag{156}$$

and a correction of the unstable terms (that is, (g,n)*-correlators with* $2g-2+n \leq 0$*).*

PROOF. In order to get the Gromov-Witten potential of $\mathbb{C}\mathrm{P}^1$ as given by the Givental formula, we have to apply the $\hat{\Psi}$- and $\hat{S}^{-1}$-action to the expression in terms of graphs discussed in Section 3.2 that corresponds to $\hat{R}\hat{\Delta}Z_{\mathrm{KdV}}^{\otimes 2}$. The $\hat{\Psi}$-action is just a linear change of variable by definition. The general S-action is discussed in [**32**, Section 4.2]. It is a combination of a shift of variables that vanishes in our case (indeed, $(1,0)S_1 = (0,0)$), the linear change of variables that we have in the statement of Proposition, and a correction of unstable terms that is not essential for us. □

3.5.2. *The Norbury-Scott conjecture.* Norbury and Scott [**52**] propose the following construction. They consider a spectral curve given by

$$\begin{cases} x &= z + \frac{1}{z}; \\ y &= \log z, \end{cases} \tag{157}$$

and the standard two-point function

$$B(z, z') = \frac{dz \otimes dz'}{(z - z')^2}. \tag{158}$$

Via topological recursion they obtain the n-forms $\omega_{g,n}$ that they consider in the global variable x, and they conjecture the following theorem (they prove it for $g = 0, 1$):

THEOREM 3.21. *For $2g - 2 + n > 0$, we have:*

$$\prod_{j=1}^{n} \left(- \operatorname*{Res}_{x_j=\infty} \frac{1}{(a_j+1)!} x_1^{a_j+1} \right) \omega_{g,n}(x_1, \ldots, x_n) = \langle \prod_{j=1}^{n} \tau_{2,a_j} \rangle_g, \tag{159}$$

where $\langle \prod_{j=1}^{n} \tau_{2,a_j} \rangle_g$ is the corresponding correlator in $Z_{\mathbb{C}\mathrm{P}^1}$, that is, the coefficient of $\hbar^{g-1} \prod_{j=1}^{n} t_{2,a_j} / |Aut(a_1, \ldots, a_n)|$ in $\log Z_{\mathbb{C}\mathrm{P}^1}$.

In the rest of this section we prove this theorem, identifying all ingredients of the topological recursion with the corresponding parts of the Givental formula.

3.5.3. *Proof of the Norbury-Scott conjecture.*

3.5.4. *Local coordinates near the branch points.* We denote the local coordinates by $z_1 = \sqrt{x-2}$ and $z_2 = \sqrt{x+2}$. Then we have:

$$x = z_1^2 + 2 \text{ near } x = 2,\ z = 1,\ z_1 = 0; \tag{160}$$

$$x = z_2^2 - 2 \text{ near } x = -2,\ z = -1,\ z_2 = 0. \tag{161}$$

Therefore,

$$z = 1 + \frac{z_1^2}{2} \pm z_1 \sqrt{1 + \frac{z_1^2}{4}}; \tag{162}$$

$$z = -1 + \frac{z_2^2}{2} \pm i z_2 \sqrt{1 - \frac{z_2^2}{4}}. \tag{163}$$

In both cases we choose $+$ for $\pm$.

3.5.5. *Expansion of y.* Recall that $y = \log z$. The direct computation shows:

$$y = \int \frac{dz_1}{\sqrt{1 + \frac{z_1^2}{4}}}; \tag{164}$$

$$y = \int \frac{-i\, dz_2}{\sqrt{1 - \frac{z_2^2}{4}}}; \tag{165}$$

Note that

$$\frac{1}{\sqrt{1+\frac{z_1^2}{4}}} = 1 + \sum_{k=1}^{\infty} z_1^{2k} \cdot \frac{(-1)^k(2k-1)!!}{k!2^{3k}}; \tag{166}$$

$$\frac{-i}{\sqrt{1-\frac{z_2^2}{4}}} = -i + \sum_{k=1}^{\infty} z_2^{2k} \cdot \frac{(-i)\cdot(2k-1)!!}{k!2^{3k}}. \tag{167}$$

Therefore

$$y = z_1 + \sum_{k=1}^{\infty} z_1^{2k+1} \cdot \frac{(-1)^k(2k-1)!!}{k!2^{3k}(2k+1)}; \tag{168}$$

$$y = -iz_2 + \sum_{k=1}^{\infty} z_2^{2k+1} \cdot \frac{(-i)\cdot(2k-1)!!}{k!2^{3k}(2k+1)}. \tag{169}$$

Thus the coefficients $\check{h}^i_{k+1}$, $k \geq 0$, are given by the following formulas:

$$\check{h}^1_{k+1} = 2 \cdot \frac{(-1)^k\left((2k-1)!!\right)^2}{k!2^{3k}}; \tag{170}$$

$$\check{h}^2_{k+1} = 2 \cdot \frac{(-i)\cdot\left((2k-1)!!\right)^2}{k!2^{3k}}. \tag{171}$$

3.5.6. *Matrix $f_{i,j}(w)$.* We use the following definition of the matrix $f_{ij}(w)$ (cf. Equation (124)):

$$f_{ij}(w) = \delta_{ij} - w\check{B}^{[ij]}(0, w^{-1}), \tag{172}$$

where $w = v^{-1}$. We use $\tilde{B}^{ij}_{0,l} = (B^{ij}_{\mathrm{reg}})_{0,2l}(2l-1)!!$, and the following expressions:

$$B^{11}_{\mathrm{reg}}(0, z_1) = \left[\frac{dz(z_1') \otimes dz(z_1)}{(z(z_1') - z(z_1))^2} - \frac{dz_1' \otimes dz_1}{(z_1' - z_1)^2}\right]_{z_1'=0} \tag{173}$$

$$B^{12}_{\mathrm{reg}}(0, z_2) = \left[\frac{dz(z_1') \otimes dz(z_2)}{(z(z_1') - z(z_2))^2}\right]_{z_1'=0} \tag{174}$$

$$B^{21}_{\mathrm{reg}}(0, z_1) = \left[\frac{dz(z_2') \otimes dz(z_1)}{(z(z_2') - z(z_1))^2}\right]_{z_2'=0} \tag{175}$$

$$B^{22}_{\mathrm{reg}}(0, z_2) = \left[\frac{dz(z_2') \otimes dz(z_2)}{(z(z_2') - z(z_2))^2} - \frac{dz_2' \otimes dz_2}{(z_2' - z_2)^2}\right]_{z_2'=0} \tag{176}$$

Therefore,

$$B^{11}_{\rm reg}(0,z_1) = \frac{1}{z_1^2}\left(\frac{1}{\sqrt{1+\frac{z_1^2}{4}}}-1\right) \tag{177}$$

$$B^{12}_{\rm reg}(0,z_2) = \frac{i}{4(1-\frac{z_2^2}{4})^{3/2}} \tag{178}$$

$$B^{21}_{\rm reg}(0,z_1) = \frac{i}{4(1+\frac{z_1^2}{4})^{3/2}} \tag{179}$$

$$B^{22}_{\rm reg}(0,z_2) = \frac{1}{z_2^2}\left(\frac{1}{\sqrt{1-\frac{z_2^2}{4}}}-1\right) \tag{180}$$

So, we have the following expansions:

$$B^{11}_{\rm reg}(0,z_1) = \sum_{k=0}^{\infty} z_1^{2k}\cdot\frac{(-1)^{k+1}(2k+1)!!}{(k+1)!2^{3(k+1)}} \tag{181}$$

$$B^{12}_{\rm reg}(0,z_2) = \sum_{k=0}^{\infty} z_2^{2k}\cdot\frac{i(2k+1)!!}{(k)!2^{3k+2}} \tag{182}$$

$$B^{21}_{\rm reg}(0,z_1) = \sum_{k=0}^{\infty} z_2^{2k}\cdot\frac{i(-1)^k(2k+1)!!}{(k)!2^{3k+2}} \tag{183}$$

$$B^{22}_{\rm reg}(0,z_2) = \sum_{k=0}^{\infty} z_2^{2k}\cdot\frac{(2k+1)!!}{(k+1)!2^{3(k+1)}}. \tag{184}$$

The formulas for $f_{ij}(w)$ are then

$$f_{11}(w) = 1+\sum_{k=1}^{\infty} w^k\cdot\frac{(-1)^{k+1}(2k-1)!!(2k-3)!!}{k!2^{3k}} \tag{185}$$

$$f_{12}(w) = \sum_{k=1}^{\infty} w^k\cdot\frac{-i(2k-1)!!(2k-3)!!}{(k-1)!2^{3k-1}} \tag{186}$$

$$f_{21}(w) = \sum_{k=1}^{\infty} w^k\cdot\frac{(-1)^k i(2k-1)!!(2k-3)!!}{(k-1)!2^{3k-1}} \tag{187}$$

$$f_{22}(w) = 1+\sum_{k=1}^{\infty} w^k\cdot\frac{-(2k-1)!!(2k-3)!!}{k!2^{3k}} \tag{188}$$

This coincides with the formula for the $\sum_{k=0}^{\infty} R_k 2^k(-w)^k$ at the point $(0,0)$.

3.5.7. *Comparison of the coefficient of (g,n,m)-vertex.* In this section we consider a vertex of genus g with n attached half-edges or ordinary leaves, and m dilaton leaves, with an associated intersection number $\langle\prod_{i=1}^n \tau_{d_i}\prod_{i=1}^m \tau_{a_i+1}\rangle_{g,n+m}$. There are vertices of type 1 and type 2, depending on the canonical coordinate that we associate to the vertex. We compare the coefficients that we associate to these vertices in the Givental case, using the data from Section 3.5.1 in Formula (62), and in the case of local topological recursion, using the data from Sections 3.5.4-3.5.6 in Formula (90).

The coefficients that we have in Formula (90) (at the vertex of the type 1 and 2 resp.):

$$(189)\qquad (-2)^{2-2g-n-m} \quad \text{and} \quad (2i)^{2-2g-n-m}.$$

Let us compute how these coefficients change if we take into account all the differences between R-matrix and the dilaton leaves. First, observe that the extra factor of 2^k in R_k and, in addition, an extra factor of $\sqrt{2}$ that we have to put by hand on each ordinary leave give us together the extra factors of

$$(190)\qquad 2^{\sum_{i=1}^{n} d_i} 2^{n/2} \quad \text{and} \quad 2^{\sum_{i=1}^{n} d_i} 2^{n/2}.$$

Then the quotient of the contributions of the dilaton leaves gives us the extra factors of

$$(191)\qquad 2^{\sum_{i=1}^{m}(a_i+1)} 2^{m/2}(-1)^m \quad \text{and} \quad 2^{\sum_{i=1}^{m}(a_i+1)} 2^{m/2}(-1)^m.$$

Let us assign by hand an extra factor of $(-1)^{2g-2+n}$ to each (g,n,m)-vertex. This way we get the following coefficients:

$$(192)\qquad 2^{g-1+n/2+m/2} \quad \text{and} \quad 2^{g-1+n/2+m/2} i^{2g-2+n+m}.$$

These coefficients are precisely

$$(193)\qquad \left((\Delta_1)^{1/2}\right)^{2g-2+n+m} \quad \text{and} \quad \left((\Delta_2)^{1/2}\right)^{2g-2+n+m}.$$

Therefore, the coefficient of $\prod_{k=1}^{\hat{n}} W_{d_k}^{i_k}$ in a graph of global genus $\hat{g}$ with $\hat{n}$ marked leaves in the Formula (90) for the set up of Norbury-Scott, multiplied by

$$(194)\qquad 2^{\hat{n}/2}(-1)^{2\hat{g}-2+\hat{n}} = (-\sqrt{2})^{\hat{n}},$$

is equal to the coefficient of $\prod_{k=1}^{\hat{n}} t^{d_k,i_k}$ in the same graph in Formula (62). This extra factor will be taken into account via rescaling of the variables by $-\sqrt{2}$.

3.5.8. *The Ψ-action.* Let us apply the Ψ-operator to the leaves. After comparing the R-action with graph expansion given formulas (62) and (90), and taking into account the extra factor of $-\sqrt{2}$, we have the following identification of the marking on the leaves:

$$(195)\qquad \sum_{a-b=c} (t^{a,1}, t^{a,2}) S_b = \left(W_c^1, W_c^2\right) \Psi^{-1}/(-\sqrt{2}).$$

Here

$$(196)\qquad W_0^1 = \left.\frac{dz}{(1-z)^2}\right|_{z=z(z_1)} + \left.\frac{dz}{(1-z)^2}\right|_{z=z(z_2)}$$

$$(197)\qquad W_0^2 = \left.\frac{i\,dz}{(1+z)^2}\right|_{z=z(z_1)} + \left.\frac{i\,dz}{(1+z)^2}\right|_{z=z(z_2)},$$

and

$$(198)\qquad W_c^i = d\left(\left(-2\frac{d}{dx}\right)^c \int W_0^i\right),$$

so we can work in the global coordinate z rather than in the local coordinates z_1, z_2.

Since

$$(199)\qquad \Psi^{-1}/(-\sqrt{2}) = \begin{pmatrix} \frac{-1}{2} & \frac{-1}{2} \\ \frac{-i}{2} & \frac{i}{2} \end{pmatrix},$$

we have:

$$\sum_{a-b=c}(t^{a,1},t^{a,2})S_b = \left(U_c^1, U_c^2\right), \tag{200}$$

where

$$U_0^1 = \frac{1}{2}\left(-\frac{dz}{(1-z)^2}+\frac{dz}{(1+z)^2}\right) \tag{201}$$

$$U_0^2 = \frac{-1}{2}\left(\frac{dz}{(1-z)^2}+\frac{dz}{(1+z)^2}\right) \tag{202}$$

and

$$U_c^i = d\left(\left(-2\frac{d}{dx}\right)^c \int U_0^i\right), \qquad i=1,2; c=0,1,2,\dots. \tag{203}$$

3.5.9. *The S-action.* The S-action is just a linear change of variables prescribed by Equation (200). This means that we replace each U_c^i with a linear combination of times $t^{a,j}$, $a \geq c$, where the coefficient of $t^{a,2}$ (this is the series of variables corresponding to the stationary sector) is equal to

$$\begin{cases} 0, & \text{if } a-c \text{ is even;} \\ \frac{1}{(k+1)\cdot(k!)^2}, & \text{if } a-c=2k+1. \end{cases} \tag{204}$$

for $i=1$, and

$$\begin{cases} \frac{1}{(k!)^2}, & \text{if } a-c=2k; \\ 0, & \text{if } a-c \text{ is odd.} \end{cases} \tag{205}$$

for $i=2$.

Norbury and Scott make the same kind of a linear change of variables, with the coefficient of $t^{a,2}$ in U_c^j, $j=1,2$, given by

$$-\operatorname*{Res}_{x=\infty}\frac{1}{(a+1)!}x^{a+1}U_c^j = \frac{1}{(a+1)!}\operatorname*{Res}_{z=0}\left(z+\frac{1}{z}\right)^{a+1}U_c^j. \tag{206}$$

In order to complete the proof of Theorem 3.21, we have to check two things: (1) that the Norbury-Scott formula for the contribution depends only on the difference $a-c$; (2) that for $c=0$ Equation (206) gives exactly the same coefficients as we have in Equations (204) and (205).

The first thing follows directly from the formula. Indeed,

$$\begin{aligned} -\oint \frac{x^{a+1}}{(a+1)!}d\left(\left(-2\frac{d}{dx}\right)^c\int U_0^j\right) &= \oint \frac{x^a}{(a)!}\left(\left(-2\frac{d}{dx}\right)^c\int U_0^j\right)dx \\ &= 2^c\oint\frac{x^{a-c}}{(a-c)!}\left(\int U_0^j\right)dx \\ &= -2^c\oint\frac{x^{a+1-c}}{(a+1-c)!}d\left(\int U_0^j\right). \end{aligned} \tag{207}$$

In particular, we see that the coefficient is equal to 0 if $a<c$.

Then, a direct computation shows that

$$
\begin{aligned}
(208)\quad & \frac{1}{(a+1)!}\operatorname*{Res}_{z=0}\left(z+\frac{1}{z}\right)^{a+1} U_0^1 \\
&= \frac{1}{(a+1)!}\operatorname*{Res}_{z=0}\left(z+\frac{1}{z}\right)^{a+1}\frac{1}{2}\left(-\frac{dz}{(1-z)^2}+\frac{dz}{(1+z)^2}\right) \\
&= \frac{1}{(a+1)!}\operatorname*{Res}_{z=0}\left(z+\frac{1}{z}\right)^{a+1}\frac{-2z\,dz}{(1-z^2)^2} \\
&= \begin{cases} 0, & \text{if } a \text{ is even;} \\ \frac{-2}{(2k+2)!}\left(\binom{2k+2}{0}\cdot(k+1)+\binom{2k+2}{1}\cdot k+\cdots\binom{2k+2}{k}\cdot 1\right) & \text{if } a=2k+1. \end{cases} \\
&= \begin{cases} 0, & \text{if } a \text{ is even;} \\ \frac{-1}{(k+1)(k!)^2} & \text{if } a=2k+1. \end{cases}
\end{aligned}
$$

and

$$
\begin{aligned}
(209)\quad & \frac{1}{(a+1)!}\operatorname*{Res}_{z=0}\left(z+\frac{1}{z}\right)^{a+1} U_0^2 \\
&= \frac{1}{(a+1)!}\operatorname*{Res}_{z=0}\left(z+\frac{1}{z}\right)^{a+1}\frac{-1}{2}\left(\frac{dz}{(1-z)^2}+\frac{dz}{(1+z)^2}\right) \\
&= \frac{-1}{(a+1)!}\operatorname*{Res}_{z=0}\left(z+\frac{1}{z}\right)^{a+2}\frac{z\,dz}{(1-z^2)^2} \\
&= \begin{cases} \frac{-1}{(2k+1)!}\left(\binom{2k+2}{0}\cdot(k+1)+\binom{2k+2}{1}\cdot k+\cdots\binom{2k+2}{k}\cdot 1\right) & \text{if } a=2k; \\ 0, & \text{if } a \text{ is odd} \end{cases} \\
&= \begin{cases} \frac{-1}{(k!)^2} & \text{if } a=2k; \\ 0, & \text{if } a \text{ is odd.} \end{cases}
\end{aligned}
$$

We see that there is an extra factor of (-1) in all coefficients. This means that the (g,n)-correlation functions of Norbury-Scott differ from the stationary Gromov-Witten invariants of $\mathbb{C}\mathrm{P}^1$ by the factor of $(-1)^n$. But this factor is exactly the difference we must have because Norbury and Scott using a different convention on the sign in the topological recursion, cf. Remark 3.12. This completes the proof of Theorem 3.21.

4. Dubrovin's superpotential as a global spectral curve

In this section we describe a way of constructing global spectral curves for cohomological field theories when certain conditions are satisfied.

4.1. Spectral curve topological recursion via CohFTs. Here we reformulate the results of Section 3.4 in a different way which would suit us better for what follows.

Namely, we now consider a global spectral curve. Note that it gives rise to a local spectral curve when one takes the germs of all the functions at the branching points of x. Suppose that this local spectral curve corresponds to a cohomological field theory according to Section 3.4. Then let us formulate what that would mean in terms of the global spectral curve information.

We choose the local coordinates w_i in the domains U_i such that $x|_{U_i} = -w_i^2/2+x(c_i)$, $i=1,\dots,n$. The identification with the data of a CohFT then goes as follows:

$$\Delta_i^{-\frac{1}{2}} = \frac{dy}{dw_i}(0); \tag{210}$$

$$R^{-1}(\zeta^{-1})_i^j = -\frac{1}{\sqrt{2\pi\zeta}}\int_{-\infty}^{\infty} \left.\frac{B(w_i,w_j)}{dw_i}\right|_{w_i=0} \cdot e^{(x(w_j)-x(c_j))\zeta}; \tag{211}$$

$$\sum_{k=1}^{n}(R^{-1}(\zeta^{-1}))_k^i \Delta_k^{-\frac{1}{2}} = \frac{\sqrt{\zeta}}{\sqrt{2\pi}}\int_{-\infty}^{\infty} dy(w_i)\cdot e^{(x(w_i)-x(c_i))\zeta}. \tag{212}$$

Note that Equation (210) is in fact a consequence of Equation (212).

There is an extra condition on the bi-differential B that can be formulated as a requirement on decomposition of its Laplace transform as

$$\begin{aligned} &\frac{\sqrt{\zeta_1\zeta_2}}{2\pi}\int_{-\infty}^{\infty}\int_{-\infty}^{\infty} B(w_i,w_j)e^{(x(w_i)-x(c_i))\zeta_1+(x(w_j)-x(c_j))\zeta_2} \\ &= \frac{\sum_{k=1}^n R^{-1}(\zeta_1^{-1})_k^i R^{-1}(\zeta_2^{-1})_k^j}{\zeta_1^{-1}+\zeta_2^{-1}}. \end{aligned} \tag{213}$$

This assumption is always satisfied if the curve is compact and the differential dx is meromorphic. This uses a general finite decomposition for $B(p,q)$ proven by Eynard in Appendix B of [**25**] together with (211).

This data (the constants $\Delta_i^{-\frac{1}{2}}$ and the matrix $R^{-1}(\zeta^{-1})_i^j$) determine for us a semi-simple CohFT $\{\alpha_{g,k}\}$ with an n-dimensional space of primary fields $V := \langle e_1,\dots,e_n\rangle$. The differentials $\omega_{g,k}$ can be written in terms of the auxiliary functions

$$\xi^i(z) := \int^z \left.\frac{B(w_i,\bullet)}{dw_i}\right|_{w_i=0} \tag{214}$$

as

$$\omega_{g,k} = \sum_{\substack{i_1,\dots,i_k\\ d_1,\dots,d_k}} \int_{\overline{\mathcal{M}}_{g,k}} \alpha_{g,k}(e_{i_1},\dots,e_{i_k}) \prod_{j=1}^{k} \psi_j^{d_j} d\left(\left(\frac{d}{dx}\right)^{d_j}\xi^{i_j}\right). \tag{215}$$

(These kind of formulas are typically of ELSV-type, see [**21**] for explanation.) In terms of the underlying Frobenius manifold structure, the basis $e_1,\dots,e_n$ corresponds to the normalized canonical basis.

4.2. Superpotential and function y**.** The goal of this Section is to prove that Dubrovin's superpotential provides us with a Riemann surface with two functions, $x:=\lambda$ and $y:=p$, such that the local expansion of y near the critical points of x reproduces the unit vector at the point $(u_1,\dots,u_n)$ of the underlying Frobenius manifold as well as the value of the matrix R^{-1} on the unit vector. These two local properties of y are precisely equivalent to the equations (210) and (212).

Consider Dubrovin's construction of a superpotential on the Riemann surface $\mathcal{D}$ described in Section 2.3. It is associated to a Frobenius manifold with given constants $\Delta_i^{-\frac{1}{2}}$ and the matrix $R^{-1}(\zeta^{-1})_i^j$ at the point with canonical coordinates $u_1,\dots,u_n$. Consider the points $c_i = p(u_i,u)\in\mathcal{D}$. These points are the critical points of the function $x:=\lambda$.

THEOREM 4.1. *Given a semi-simple Frobenius manifold M, and Dubrovin's construction of a superpotential $\mathcal{D}$ for M, define spectral curve data by $\Sigma = \mathcal{D}$, $x := \lambda$, $y := p$ (with B yet to be defined). Then equations* (210) *and* (212) *are satisfied for the constants $\Delta_i^{-\frac{1}{2}}$ and the matrix $R^{-1}(\zeta^{-1})_i^j$ associated to M.*

PROOF. Let us prove the first statement, namely, Equation (210) (though it is a corollary of Equation (212), it is convenient to check it directly). Indeed, Equation (17) states that near the points c_i the function p looks like

$$p = c_i + \Psi_{i,\mathbb{1}}(u)\sqrt{2(u_j - \lambda)} + O(u_j - \lambda).$$

Therefore, the derivative of p with respect to the local coordinate $w_i = \sqrt{2(u_i - \lambda)}$ at the point c_i is equal to $\Psi_{i,\mathbb{1}}(u) = \Delta_i^{-\frac{1}{2}}$.

Now we prove Equation (212). We can assume that the contour of integration on the right hand side in Equation (212) is the image of L_i under the map p. Then,

$$\frac{\sqrt{\zeta}}{\sqrt{2\pi}}\int_{p(L_i)} dp\cdot e^{(\lambda-u_i)\zeta} = \frac{\sqrt{\zeta}}{\sqrt{2\pi}}\int_{p(L_i)} \frac{dp}{d\lambda}\cdot e^{(\lambda-u_i)\zeta}d\lambda. \tag{216}$$

Here we treat dp and $d\lambda$ as 1-forms defined on the surface $\mathcal{D}$.

Observe that from equation (8) we have

$$\frac{d\phi^T}{d\lambda} = \phi^T\left(\frac{1}{2} - V\right)(U-\lambda)^{-1}. \tag{217}$$

Therefore, using definition (16), we get

$$\begin{aligned}\frac{dp}{d\lambda} &= \frac{d}{d\lambda}\frac{\sqrt{2}}{1-d}\phi^T(U-\lambda)\Psi\mathbb{1} = \frac{\sqrt{2}}{1-d}\phi^T\left(\frac{1}{2}-V\right)\Psi\mathbb{1} - \frac{\sqrt{2}}{1-d}\phi^T\Psi\mathbb{1}\\ &= \frac{\sqrt{2}}{1-d}\phi^T\Psi\Psi^{-1}\left(-\frac{1}{2}-V\right)\Psi\mathbb{1} = \frac{\sqrt{2}}{1-d}\phi^T\Psi\left(-\frac{1}{2}-\mu\right)\mathbb{1}\\ &= \frac{\sqrt{2}}{1-d}\phi^T\Psi\left(-\frac{1}{2}+\frac{d}{2}\right)\mathbb{1} = -\frac{1}{\sqrt{2}}\phi^T\Psi\mathbb{1}.\end{aligned} \tag{218}$$

(In this computation we used the fact that $\mu\mathbb{1} = (-d/2)\mathbb{1}$.)

Equations (14) and (15) imply that on the contour $p(L_i)$ the vector ϕ is equal to $\phi^{(i)} + E_i$, where E_i is some holomorphic function of $(u_i - \lambda)$. Recall also that $(\Psi\mathbb{1})_k = \Delta_k^{-\frac{1}{2}}$. Therefore,

$$\frac{\sqrt{\zeta}}{\sqrt{2\pi}}\int_{p(L_i)} \frac{dp}{d\lambda}\cdot e^{(\lambda-u_i)\zeta}d\lambda = \sum_{k=1}^{n}\Delta_k^{-\frac{1}{2}}\cdot\frac{-\sqrt{\zeta}}{2\sqrt{\pi}}\int_{p(L_i)}\phi_k^{(i)}\cdot e^{(\lambda-u_i)\zeta}d\lambda. \tag{219}$$

Dubrovin shows in [**13**, Proof of Lemma 5.4] that the second factor in this expression is $(R^{-1}(\zeta^{-1}))_k^i$. Thus the right hand side of Equation (219) coincides with the left hand side of Equation (212). This completes the proof of the Theorem. □

REMARK 4.2. Note that we have not used the specific formula for ϕ in the proof. We used only Equation (8) and the fact that the local expansion of ϕ for $\lambda \to u_i$ coincides with the local expansion of $\phi^{(i)}$ up to some holomorphic non-branching term. Thus, if we have a solution for (8) satisfying this property, we can use it directly in the formula for the superpotential (16), bypassing the requirement for G^{ij} to be non-degenerate. This will be important below in certain applications.

REMARK 4.3. *Flat identity.* Topological recursion satisfies the string equation.

$$\sum_{i=1}^{n} \operatorname*{res}_{p=c_i} y(p)\omega_{g,k+1}(p,p_1,...,p_k) = -\sum_{j=1}^{k} d_{p_j}\partial z_j \left(\frac{\omega_{g,k}(p_1,...,p_k)}{dx(p_j)}\right) \tag{220}$$

where the sum is over the zeros $dx(c_i) = 0$ and d_{p_j} is exterior derivative in the variables p_j. The operator $\omega \mapsto \sum_i \mathrm{Res}_{p=c_i} y(p)\omega(p)$ acts on differentials ω. It is non-zero (and evaluates to 1) on the auxiliary differential $\sum_j a_j d\xi^j$ corresponding to the flat identity and annihilates all others. In particular

$$\sum_i \operatorname*{res}_{p=c_i} d\left(\left(\frac{d}{dx}\right)^{d_j} \xi^{i_j}\right) = 0, \quad d_j > 0.$$

This corresponds to insertion/removal of the identity vector in ancestor invariants.

4.3. Compatibility between B and y. In this section we discuss a necessary condition on a spectral curve to be able to apply the inverse construction of Section 3, i.e. so that a CohFT can be reconstructed from this spectral curve.

More precisely, for a given data of a spectral curve (Σ, x, y, B) (maybe, local) Equations (211) and (212), (210) imply some relation for x, y, and B, and we want to state this relation in a direct geometric way rather than in terms of the Laplace transform.

The compatibility condition below is equivalent to differentiation of the potential of a CohFT by t_1 producing the string equation. In the language of [**27**], $\delta(ydx) = \int (dy/dx)(p')\, B(p,p') = d(dy/dx)$ gives rise to variations of $\omega_{g,k}$ corresponding to the string equation (220).

Recall that x defines a local involution σ_i near each zero c_i of dx, $i = 1,\dots,n$.

THEOREM 4.4. *A CohFT can be reconstructed from a spectral curve (Σ, x, y, B) via the inverse construction of Section 3 reformulated in Section 4.1 if and only if the 1-form on Σ*

$$\eta(z) = d\left(\frac{dy}{dx}(z)\right) + \sum_{i=1}^{n} \operatorname*{res}_{z'=c_i} \frac{dy}{dx}(z')B(z,z'). \tag{221}$$

is invariant under each local involution σ_i, $i = 1,\dots,n$.

PROOF. The construction of Section 3 requires equations (210), (211), and (212) to hold. We will prove that the 1-form (221) is invariant under each local involution σ_i, $i = 1,\dots,n$ if and only if equations (210), (211), and (212) are compatible (as equations for the unknown variables R^{-1} and $\Delta_i^{-\frac{1}{2}}$, $i = 1,\dots,n$).

Recall that $x = x(c_i) - w_i^2/2$ in a neighborhood of c_i. Note that

$$\begin{aligned}\operatorname*{res}_{w_i=0} \frac{dy}{dx}(w_i)B(z,w_i) &= \operatorname*{res}_{w_i=0} \frac{dy}{dw_i}(w_i)\cdot\frac{dw_i}{dx}\cdot B(z,w_i) \\ &= -\operatorname*{res}_{w_i=0} \frac{dy}{dw_i}(w_i)\cdot\frac{dw_i}{w_i}\cdot\frac{B(z,w_i)}{dw_i} = \frac{dy}{dw_i}(0)\cdot\left.\frac{B(z,w_i)}{dw_i}\right|_{w_i=0}.\end{aligned} \tag{222}$$

An equivalent way to say that η is σ_i-invariant is to say that the following Laplace transform of η is equal to zero:

$$\int_{-\infty}^{\infty} \eta(w_i)e^{(x(w_i)-x(c_i))\zeta} = 0. \tag{223}$$

On the other hand,

(224)

$$\int_{-\infty}^{\infty} \eta(w_i) e^{(x(w_i)-x(c_i))\zeta} = -\zeta \int_{-\infty}^{\infty} \frac{dy}{dx}(w_i) e^{(x(w_i)-x(c_i))\zeta} dx \\ - \sum_{j=1}^{n} \frac{dy}{dw_j}(0) \int_{-\infty}^{\infty} \left.\frac{B(w_i,w_j)}{dw_j}\right|_{w_j=0} e^{(x(w_i)-x(c_i))\zeta}.$$

Thus, Equation (223) is satisfied if and only if

$$\frac{\sqrt{\zeta}}{\sqrt{2\pi}} \int_{-\infty}^{\infty} dy(w_i) e^{(x(w_i)-x(c_i))\zeta} \\ = \sum_{j=1}^{n} \frac{dy}{dw_j}(0) \cdot \frac{-1}{\sqrt{2\pi\zeta}} \int_{-\infty}^{\infty} \left.\frac{B(w_i,w_j)}{dw_j}\right|_{w_j=0} e^{(x(w_i)-x(c_i))\zeta}, \tag{225}$$

which is precisely the compatibility condition for Equations (210), (211), and (212). □

We can state (221) in simpler terms when the spectral curve is connected.

COROLLARY 4.5. *For a connected spectral curve, equations* (210), (211), *and* (212) *are compatible if and only if the 1-form defined in* (221) *is a pull-back of a 1-form downstairs, i.e.* $\eta(z) = x^*\omega$.

PROOF. If $\eta(z) = x^*\omega$ for ω a differential downstairs then it is invariant under local involutions hence Theorem 4.4 applies. On a comnnected spectral curve Σ the converse is also true. This follows from the more general fact that any $\eta(z)$ which is invariant under local involutions defined around simple ramification points of $x : \Sigma \to \mathbb{C}$ is the pull-back of a differential downstairs. Take any regular (w.r.t. function x) point $p \in \Sigma$ and a path γ from p to a zero b of dx. Then $x(\gamma)$ is covered by a path $\tilde{\gamma} \subset \Sigma$ that contains p and p' where $x(p) = x(p')$. The local involution defined by x in a neighbourhood of b can be analytically continued along $\tilde{\gamma}$. Since $\eta(z)$ is invariant under the local involution at b, it is invariant under the continued involution above a neighbourhood of $x(\gamma)$. So $\eta(z)$ agrees (via identification of cotangent bundles using x) around p and p'. Connectedness of Σ guarantees that the monodromy of the cover defined by x is transitive and generated by local involutions. Hence we can find paths γ_i that can be used to show that $\eta(z)$ agrees around p and any point in the fibre over $x(p)$. Hence $\eta(z) = x^*\omega$ locally and this pieces together to give the global result. The result isn't true on disconnected curves, in particular local curves, because monodromy is not transitive. □

Let us show how this compatibility test can be used.

PROPOSITION 4.6. *The differential* $\eta \equiv 0$, *hence Equation* (223) *is satisfied, when* Σ *is a global curve equipped with a canonical bidifferential* B *normalized so that* $\int_{p'\in\alpha_i} B(p,p') = 0$ *for a choice of A-cycles* α_i, *and one of the following holds:*

1. Σ *is rational with global coordinate* z *chosen so that* $x(z=\infty) = \infty$;
2. dy *is a meromorphic differential such that* $\frac{dy}{dx}$ *has poles only at the zeros of* dx, *for example* dy *is a holomorphic differential.*

Note that in case (2) above, we take y to be the (multiply-defined) primitive of a differential which is sufficient for the purposes of topological recursion—see Remark (2.6).

PROOF. Recall the property that for any function f on Σ, $\mathrm{Res}_{p'=p} f(p')B(p,p') = df(p)$ (independent of the choice of A-cycles along which B is normalized). For example, in the rational case $B = \frac{dzdz'}{(z-z')^2}$ and this property is the Cauchy integral formula. Since $\frac{dy}{dx}$ has poles only at the zeros of dx

$$\sum_{i=1}^{n} \operatorname*{res}_{p'=c_i} \frac{dy}{dx}(p')B(p,p') = -\operatorname*{res}_{p'=p} \frac{dy}{dx}(p')B(p,p') = -d\left(\frac{dy}{dx}(p)\right)$$

hence $\eta \equiv 0$. □

EXAMPLE 4.7. Consider $x = z+1/z$, $y = p(z)$ a polynomial. Then $\frac{dy}{dx} = \frac{z^2 p'(z)}{z^2-1}$ has poles at $z = \pm 1$ and possibly $z = \infty$. Hence $\eta(z) = dq(z)$ where $q(z)$ is a polynomial given by the principal part of dy/dx at $z = \infty$. A non-trivial polynomial has poles only at $z = \infty$ so if $\eta \neq 0$ it cannot be the pull-back of a differential form downstairs since it would necessarily require poles at $x^{-1}(\infty) = \{0, \infty\}$. Hence this fails the compatibility test, unless $\eta(z) \equiv 0$ i.e. $\deg p(z) \leq 1$. If $\deg p(z) = 1$ then Equation (223) is satisfied.

EXAMPLE 4.8. Consider $x = z + 1/z$, $y = \ln z$. Then $\frac{dy}{dx} = \frac{z}{z^2-1}$ has poles only at $z = \pm 1$ so Equation (223) is satisfied.

EXAMPLE 4.9. Since the compatibility test is a linear condition in y, $x = z+1/z$, $y = \ln z + cz$ also satisfies the compatibility test and leads to a CohFT with a flat unit. This was also observed in [**33**].

4.4. Superpotential as a global spectral curve in genus 0 case. In this Section we discuss a special case of Dubrovin's superpotential defined in Section 2.3 and show that it indeed gives a proper spectral curve for the correponding cohomological field theory.

More precisely, we start with a homogeneous cohomological field theory. Its genus zero part without descendants defines a Frobenius manifold that we assume to be semi-simple. Consider Dubrovin's construction in Section 2.3. Assume that this construction goes through in such a way that

(1) The form $d_\lambda p$ has no zeros in $\mathbb{C} \setminus \cup_{i=1}^n L_i$;
(2) $\lambda(p = \infty) = \infty$;
(3) The resulting curve $\mathcal{D}$ is a compact curve of genus 0 and p is a global coordinate on it;
(4) There is exactly one critical point in each singular fiber of function λ.

THEOREM 4.10. *Under the conditions* (1)-(4) *above, the correlators of the CohFT are related by Equation* (215) *to the correlator differentials obtained through spectral curve topological recursion on a curve $\mathcal{D}$ with $x = \lambda$, $y = p$ and $B(p_1,p_2) = dp_1dp_2/(p_1-p_2)^2$.*

In other words, in this case the ancestor potential of CohFT is reproduced by global topological recursion related to Dubrovin's superpotential. Note that this identification happens over an open ball in the underlying Frobenius manifold.

PROOF. First of all, note that since p is a global coordinate and $\lambda(p=\infty)=\infty$, this spectral curve satisfies the compatibility condition of Theorem 4.4, which means that one can reconstruct a CohFT such that Equation (215) is satisfied. We only need to prove that this CohFT is the same as the original one.

Theorem 4.1 implies that we have the right function y, so, in particular, the functions $\Delta_i^{-\frac{1}{2}}(u)$ are correctly reproduced on an open ball in the space of parameters $u_1,\dots,u_n$. Note that these functions determine completely the structure of Frobenius multiplication, so we can conclude that the CohFT reconstructed from the spectral curve data coincides with the original one in genus zero.

Higher genera correlators of a semi-simple CohFT are determined uniquely by genus 0 data in homogeneous cases [**60**]. Therefore, it is sufficient to prove that the CohFT reconstructed from the spectral curve data is homogeneous. We do this by proving the Euler equation for the corresponding R-matrix. Namely, a CohFT with an R-matrix $R(\xi)$ is homogeneous if and only if the R-matrix satisfies the Euler equation [**36**]:

$$\left(\xi\frac{d}{d\xi}+\sum_{i=1}^{n}u_i\frac{\partial}{\partial u_i}\right)R(\xi,u)=0 \tag{226}$$

(or, equivalently, we can consider the same equation for $R^{-1}(\xi,u)=R(-\xi,u)^T$). Using Equation (213), the Euler equation for the R-matrix can be rewritten as

$$\left(1+\xi_1\frac{\partial}{\partial\xi_1}+\xi_2\frac{\partial}{\partial\xi_2}+\sum_{i=1}^{n}u_i\frac{\partial}{\partial u_i}\right)\check{B}=0 \tag{227}$$

for $\check{B}=\check{B}^{ij}(\xi_1,\xi_2)$ given by

$$\frac{e^{-\frac{u_i}{\xi_1}-\frac{u_j}{\xi_2}}}{2\pi\sqrt{\xi_1\xi_2}}\int_{p(L_i)}\int_{p(L_j)}B\cdot e^{\frac{\lambda_1}{\xi_1}+\frac{\lambda_2}{\xi_2}}. \tag{228}$$

Recall that we consider the case when $d_\lambda p$ does not have zeros in $\mathbb{C}\setminus\cup_{i=1}^n L_i$, and the Riemann surface $\mathcal{D}$ that we get through Dubrovin's construction has genus 0. The Bergman kernel $B(p_1,p_2)$ has the form $dp_1dp_2/(p_1-p_2)^2$.

PROPOSITION 4.11. *Under these conditions Equation* (227) *is satisfied.*

We prove this proposition below. It implies that the R-matrix associated to the Bergman kernel in this case satsifies the Euler equation, and, therefore, the corresponding CohFT is homogeneous. This proposition completes the proof of Theorem 4.10. □

For the proof of Proposition 4.11 we need the following technical lemma:

LEMMA 4.12. *We have:*

$$\left(\lambda\frac{d}{d\lambda}+\sum_{i=1}^{n}u_i\frac{\partial}{\partial u_i}\right)p(\lambda,u)=\frac{1-d}{2}p(\lambda,u). \tag{229}$$

PROOF. Recall Equation (218):

$$d_\lambda p(\lambda,u)=-\frac{1}{\sqrt{2}}\phi^T d\lambda\Psi\mathbb{1}. \tag{230}$$

In the same way we prove that

$$d_u p(\lambda, u) = \frac{1}{\sqrt{2}} \phi^T dU \Psi \mathbb{1} \tag{231}$$

(this is [**13**, equation (5.66)]; note that there is a misprint in this equation in [**13**]). Combining these equations, we get

$$\left(\lambda \frac{d}{d\lambda} + \sum_{i=1}^n u_i \frac{\partial}{\partial u_i}\right) p(\lambda, u) = \frac{1}{\sqrt{2}} \phi^T (U - \lambda) \Psi \mathbb{1} = \frac{1-d}{2} p(\lambda, u). \tag{232}$$

□

PROOF OF PROPOSITION 4.11. We have:

$$\begin{aligned}
&\left(1 + \xi_1 \frac{\partial}{\partial \xi_1} + \xi_2 \frac{\partial}{\partial \xi_2} + \sum_{i=1}^n u_i \frac{\partial}{\partial u_i}\right) \check{B} \\
&= \frac{e^{-\frac{u_i}{\xi_1} - \frac{u_j}{\xi_2}}}{2\pi\sqrt{\xi_1 \xi_2}} \iint \frac{d\lambda_1 d\lambda_2}{(p(\lambda_1) - p(\lambda_2))^2} \frac{dp}{d\lambda}(\lambda_1) \frac{dp}{d\lambda}(\lambda_2)\, e^{\frac{\lambda_1}{\xi_1} + \frac{\lambda_2}{\xi_2}} X,
\end{aligned} \tag{233}$$

where

$$\begin{aligned}
X = &- \frac{\lambda_1}{\xi_1} - \frac{\lambda_2}{\xi_2} - 2 \cdot \frac{\left(\sum_{i=1}^n u_i \frac{\partial}{\partial u_i}\right)(p(\lambda_1) - p(\lambda_2))}{p(\lambda_1) - p(\lambda_2)} \\
&+ \frac{\left(\sum_{i=1}^n u_i \frac{\partial}{\partial u_i}\right) \frac{dp}{d\lambda}(\lambda_1)}{\frac{dp}{d\lambda}(\lambda_1)} + \frac{\left(\sum_{i=1}^n u_i \frac{\partial}{\partial u_i}\right) \frac{dp}{d\lambda}(\lambda_2)}{\frac{dp}{d\lambda}(\lambda_2)}.
\end{aligned}$$

Applying the integration by parts to the terms $-\lambda_1/\xi_1$ and $-\lambda_2/\xi_2$, we can rewrite the right hand side of Equation (233) as

$$\frac{e^{-\frac{u_i}{\xi_1} - \frac{u_j}{\xi_2}}}{2\pi\sqrt{\xi_1 \xi_2}} \iint \frac{d\lambda_1 d\lambda_2}{(p(\lambda_1) - p(\lambda_2))^2} \frac{dp}{d\lambda}(\lambda_1) \frac{dp}{d\lambda}(\lambda_2)\, e^{\frac{\lambda_1}{\xi_1} + \frac{\lambda_2}{\xi_2}} Y, \tag{234}$$

where

$$\begin{aligned}
Y = 2 &+ \frac{\left(\lambda_1 \frac{d}{d\lambda_1} + \sum_{i=1}^n u_i \frac{\partial}{\partial u_i}\right) \frac{dp}{d\lambda}(\lambda_1)}{\frac{dp}{d\lambda}(\lambda_1)} + \frac{\left(\lambda_2 \frac{d}{d\lambda_2} + \sum_{i=1}^n u_i \frac{\partial}{\partial u_i}\right) \frac{dp}{d\lambda}(\lambda_2)}{\frac{dp}{d\lambda}(\lambda_2)} \\
&- 2 \cdot \frac{\left(\lambda_1 \frac{d}{d\lambda_1} + \sum_{i=1}^n u_i \frac{\partial}{\partial u_i}\right) p(\lambda_1) - \left(\lambda_2 \frac{d}{d\lambda_2} + \sum_{i=1}^n u_i \frac{\partial}{\partial u_i}\right) p(\lambda_2)}{p(\lambda_1) - p(\lambda_2)}.
\end{aligned}$$

Using Equation (229), we rewrite Y as

$$Y = 2 + \frac{\left(-1+\frac{1-d}{2}\right)\frac{dp}{d\lambda}(\lambda_1)}{\frac{dp}{d\lambda}(\lambda_1)} + \frac{\left(-1+\frac{1-d}{2}\right)\frac{dp}{d\lambda}(\lambda_2)}{\frac{dp}{d\lambda}(\lambda_2)}$$
$$- 2\cdot\frac{\frac{1-d}{2}p(\lambda_1) - \frac{1-d}{2}p(\lambda_2)}{p(\lambda_1)-p(\lambda_2)}$$
$$= \&2 + \left(-1+\frac{1-d}{2}\right) + \left(-1+\frac{1-d}{2}\right) - 2\cdot\frac{1-d}{2} = 0,$$

which proves the proposition. □

4.5. Superpotential as a global spectral curve for arbitrary genus. In this Section we extend the result of the previous section to the case of a compact global curve of arbitrary genus.

THEOREM 4.13. *Given a conformal Frobenius manifold, construct a superpotential $p(\lambda; u)$ which defines the Riemann surface $\mathcal{D}$ according to Dubrovin's construction of Section 2.3. Assume the following:*

- *$\mathcal{D}$ is a compact curve of genus g;*
- *there is exactly one critical point in each singular fiber of $\lambda : \mathcal{D} \to \mathbb{C}$.*

Fix a symplectic basis $(\mathcal{A}_i, \mathcal{B}_i)_{i=1}^g$ of $H_1(\mathcal{D}, \mathbb{Z})$ and define $B(p_1, p_2)$ as the only Bergman kernel on $\mathcal{D}$ normalized by

$$\forall i = 1, \ldots, g\,, \quad \oint_{p_1\in\mathcal{A}_i} B(p_1, p_2) = 0. \tag{235}$$

Further assume that:

- *the pair $(p, B(p_1, p_2))$ passes the compatibility test of Section 4.3 in any of its possible forms (given by Theorem 4.4, Corollary 4.5, or Proposition 4.6).*

Then the correlators of the CohFT associated to the Frobenius manifold are related by Equation (215) to the correlator differentials obtained through spectral curve topological recursion on the Riemann surface $\mathcal{D}$ with $x = \lambda$, $y = p$ and $B(p_1, p_2)$.

REMARK 4.14. This result extends Theorem 4.10 to an arbitrary compact curve. The new feature is that one needs to normalize the Bergman kernel on an arbitrary basis of cycles. In particular, for each basis, we recover a total ancestor potential for the same CohFT.

PROOF. The proof is very similar to the proof of the genus 0 case presented in the preceding section. However, it is important to remark that this proof only relies on Rauch's variational formula, i.e. it is valid for any compact curve presented as a ramified cover of the Riemann sphere with simple branch points. It does not require any knowledge about an auxiliary meromorphic form such as the super-potential.

Let us first show that the $(0, 3)$ correlators are independent of choice of normalisation cycles for B. $\omega_{0,3}$ depends on these choices, but when decomposed into linear combinations of auxiliary differentials $d\xi^j = B/ds_j$ (for s_j defined by $x = (1/2)s_j^2 + a_j$) the coefficients are independent of A-cycles. By reconstruction, as in the proof of Theorem 4.10, this means that all correlators are the same. The formula

$$\begin{aligned}\omega_{0,3}(z_1,z_2,z_3) &= \sum_{i=1}^{n} \operatorname*{Res}_{p=c_i} B(p,z_1)B(p,z_2)B(p,z_3)/dx(p)dy(p)\\ &= \sum_{i=1}^{n} B(a_i,z_1)B(a_i,z_2)B(a_i,z_3)/x''(a_i)y'(a_i)\\ &= \sum_{i=1}^{n} \langle \dots\rangle d\xi^i(z_1)d\xi^i(z_2)d\xi^i(z_3)\end{aligned}$$

shows the independence of the coefficients $\langle\dots\rangle$ on the choice of B.

For the rest of the proof, the only part differing from the genus 0 case is the proof of the homogeneity of the CohFT, i.e. the fact that the R-matrix satisfies the Euler equation.

The first step consists in proving that there exist a R-matrix. This is due to a lemma of Eynard [**25**]:

LEMMA 4.15. *If $d\lambda$ is a meromorphic form on $\mathcal{D}$ and B the Bergman kernel normalized on a basis of $\mathcal{A}$-cycles as above, then the Laplace transform of the Bergman kernel satisfies Equation* (213) .

The Euler equation for the R-matrix is then equivalent to the following equation for the Laplace transform of B:

$$\left(1+\xi_1\frac{\partial}{\partial\xi_1}+\xi_2\frac{\partial}{\partial\xi_2}+\sum_{i=1}^{n}u_i\frac{\partial}{\partial u_i}\right)\check{B}=0 \tag{236}$$

for $\check{B}=\check{B}^{ij}(\xi_1,\xi_2)$ given by

$$\frac{e^{-\frac{u_i}{\xi_1}-\frac{u_j}{\xi_2}}}{2\pi\sqrt{\xi_1\xi_2}}\int_{p(L_i)}\int_{p(L_j)} B\cdot e^{\frac{\lambda_1}{\xi_1}+\frac{\lambda_2}{\xi_2}}. \tag{237}$$

By inverting the Laplace transform and integration by part, this is equivalent to

$$d_1\left(\frac{\lambda_1\,B(p_1,p_2)}{d\lambda_1}\right)+d_2\left(\frac{\lambda_2\,B(p_1,p_2)}{d\lambda_2}\right)+\sum_{i=1}^{n}u_i\frac{\partial}{\partial u_i}B(p_1,p_2). \tag{238}$$

In order to prove this equation, we remind Rauch's variational formula which expresses the variations of the Bergman kernel under deformation of the spectral curve. In particular

$$\frac{\partial B(p_1,p_2)}{\partial u_i}=\operatorname*{Res}_{r\to a_i}\frac{B(p_1,r)\,B(p_2,r)}{d\lambda(r)} \tag{239}$$

which implies that

$$\sum_{i=1}^{n}u_i\frac{\partial}{\partial u_i}B(p_1,p_2)=\sum_{i=1}^{n}\operatorname*{Res}_{r\to a_i}\frac{\lambda(r)B(p_1,r)\,B(p_2,r)}{d\lambda(r)}. \tag{240}$$

Moving the integration contours around the other poles of the integrands and reminding that the $\mathcal{A}$-periods of $B(p,r)$ are vanishing, this reads

(241)

$$\sum_{i=1}^{n} u_i \frac{\partial}{\partial u_i} B(p_1,p_2) = -\operatorname*{Res}_{r\to p_1,p_2} \frac{\lambda(r)B(p_1,r)\,B(p_2,r)}{d\lambda(r)} = -\left(\frac{d}{d\lambda_1}+\frac{d}{d\lambda_2}\right) B(p_1,p_2)$$

proving Equation 238. □

4.6. Global curves for A_n singularities. In this Section we apply the results of Sections 4.2, 4.3, and 4.4 in order to construct the spectral curve for the ancestor potential of A_n-singularities, $n=1,2,\dots$. The structure of this Frobenius manifolds is described in terms of Saito's theory on the space of polynomials

$$f(p,\tau) = p^{n+1} + \tau_1 p^{n-1} + \cdots \tau_n. \tag{242}$$

We refer to [**12**, **37**] for the detailed description of the structure of this Frobenius manifold. In particular, it is enough to say that $\lambda = f(p,\tau)$ is a superpotential of this Frobenius manifold.

The corresponding CohFT is well-studied. It was a subject of Witten's conjecture [**61**] proved in [**32**]. We refer to [**56**] for an exposition of this CohFT that includes an overview of its constructions; the CohFT whose correlators give the ancestor potential at the point τ of this Frobenius manifold is called there the shifted Witten class of A_n singularity.

THEOREM 4.16. *The correlation differentials of the global spectral curve data $\Sigma := \mathbb{CP}^1$, $y := p$ (the global coordinate), $x := f(p,\tau)$, $B := dp_1dp_2/(p_1-p_2)^2$ are expressed via Equation* (215) *in terms of the shifted Witten class of A_n singularity.*

PROOF. As we have already mentioned, the function $\lambda = f(p,\tau)$ is known to be a superpotential of the corresponding Frobenius manifold. We have to show that this superpotential can be obtained by Dubrovin's construction in Section 2.3. Then it is easy to see that all conditions of Theorem 4.10 are satisfied, which implies this theorem.

We construct solutions of Equation (8) in terms of the integrals over the vanishing cycles. Namely, consider the tangent bundle over the space of polynomials parametrized by $\tau\in\mathcal{T}$. It is identified with the space $\mathbb{C}[p]/(d_pf(p,\tau)/dp)$ by the map $v\mapsto (d_\tau f(p,\tau))(v)$ and equipped with a flat metric given by

$$(v_1,v_2) := \operatorname*{res}_{p=\infty} \frac{d_\tau f(v_1) d_\tau f(v_2)}{d_p f(p,\tau)/dp} dp. \tag{243}$$

For a cycle $\beta\in H_0(f^{-1}(\lambda),\mathbb{C})$ we denote by $I_\beta(\lambda,\tau)$ the section of the tangent bundle specified by the following formula:

$$(I_\beta(\lambda,\tau),v) := \int_\beta d_\tau f(v)\cdot\frac{dp}{d_pf}. \tag{244}$$

In normalized canonical coordinates $I_\beta(\lambda,\tau)$ is represented by the vector $\phi^\beta(\lambda,\tau)$ with components given by

$$\phi_i^\beta(\lambda,\tau) := \left(I_\beta(\lambda,\tau), \frac{\Delta_i^{\frac12}}{\sqrt{2}}\frac{\partial}{\partial u_i}\right) \tag{245}$$

is a solution of Equation (8) (see [**37**]). Let us discuss the singularities of this solution, depending on β.

Consider the λ-plane as the image of the map $\lambda = f(p,\tau)$. Let $u_1,\dots,u_n$ be the critical values of $f(p,\tau)$. We can alway choose a system of cuts L_i, $i=1,\dots,n$, from u_i to infinity such that the preimage $f^{-1}(\mathbb{C}\setminus\cup_{i=1}^n L_i)$ is a union of $n+1$ disks, $D_0,D_1,\dots,D_n$, glued along the boundary cuts in the following way:

- D_0 is glued to D_i along the boundary that is a double cover of L_i; in particular, their common boundary contains the critical preimage of u_i; All other lifts of the cut L_i are just cuts inside D_j, $j\neq 0,i$; the endpoints of these cuts are non-critical preimages of u_i.

For $n=1,2,3$ we give the corresponding pictures for a real orientable blow-up at infinity (that is, the boundary circle on the picture corresponds to the infinity point of the source sphere). The domain D_0 is shadowed there.

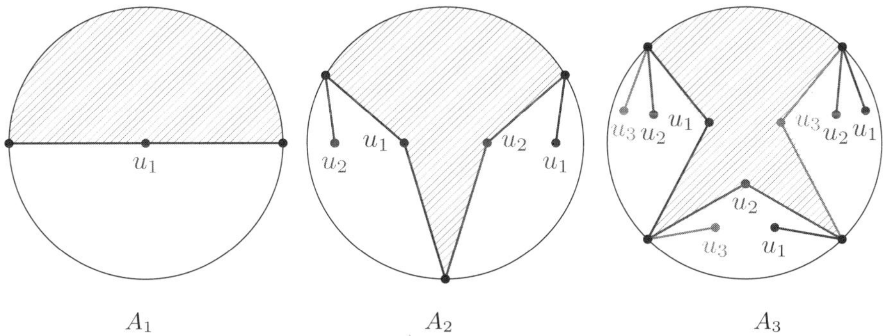

FIGURE 1. A_n-singularities, $1\le n\le 3$

Consider the vanishing cycles $\beta_i\in H_0(f^{-1}(\lambda),\mathbb{C})$ given by $\beta_i := p_0 - p_i$, where $\lambda = f(p_0,\tau) = f(p_i,\tau)$, and $p_0\in D_0$, $p_i\in D_i$. Then the system of solutions of Equation (8) given by $\phi^{(i)}(\lambda,\tau) := \phi^{\beta_i}(\lambda,\tau)$ satisfies the properties given by Equations (10)–(13). In particular, $G^{ij}=1/2$ for $i\neq j$ and $G^{ii}=1$. The inverse matrix is given by $G_{ii} = 2n/(n+1)$ and $G_{ij} = -2/(n+1)$ for $i\neq j$. Therefore, Dubrovin's solution $\phi = \sum_{i,j=1}^n G_{ij}\phi^{(j)}$ is equal to ϕ^{β_0} for

$$\beta_0 = \sum_{i=1}^n \left(\frac{2n}{n+1} - (n-1)\frac{2}{n+1}\right)\beta_i = 2p_0 - \frac{2}{n+1}\sum_{i=0}^n p_i. \tag{246}$$

Recall that for the Frobenius structure A_n, $d = (n-1)/(n+1)$. Also we recall that for the Euler vector field $E = \sum_{i=1}^n u_i\frac{\partial}{\partial u_i}$ and the unit vector field $e = \sum_{i=1}^n \frac{\partial}{\partial u_i}$ so that we have:

$$Ef(p,\tau) = f(p,\tau) - \frac{p}{n+1}\frac{d_p f(p,\tau)}{dp}, \tag{247}$$

$$ef(p,\tau) = 1. \tag{248}$$

The formula $\phi^T(U-\lambda)\Psi\mathbb{1}$ can be written as $\big(I_{\beta_0}(\lambda,\tau),(E-\lambda e)/\sqrt{2}\big)$. Therefore,

$$\begin{aligned}\frac{\sqrt{2}}{1-d}\phi^T(U-\lambda)\Psi\mathbb{1} &= \frac{n+1}{2}\int_{\beta_0}(E-\lambda e)f(p,\tau)\cdot\frac{dp}{d_p f(p,\tau)}\\ &= \frac{n+1}{2}\int_{\beta_0}\left(f(p,\tau) - \frac{p}{n+1}\frac{d_p f(p,\tau)}{dp} - \lambda\right)\cdot\frac{dp}{d_p f(p,\tau)}.\end{aligned} \tag{249}$$

Since the cycle β_0 lies in $f^{-1}(\lambda)$, then $(f(p,\tau)-\lambda)|_{\beta_0}=0$. Therefore, the last integral can be rewritten as

$$\frac{n+1}{2}\int_{\beta_0}\frac{p}{n+1}=p_0(\lambda,\tau)-\frac{1}{n+1}\sum_{i=0}^{n}p_i(\lambda,\tau). \tag{250}$$

Since $\sum_{i=0}^{n}p_i(\lambda,\tau)=0$ (recall the form of the polynomial $f(p,\lambda)$), we conclude that the function $\frac{\sqrt{2}}{1-d}\phi^T(U-\lambda)\Psi\mathbb{1}$ is equal to the branch D_0 of $p=f^{-1}(\lambda,\tau)$.

So, $p(\mathbb{C}\backslash\cup_{i=1}^{n}L_i)=D_0$. It is obvious that $d_\lambda p$ has no zeros in $\mathbb{C}\backslash\cup_{i=1}^{n}L_i$, so $\hat{D}=D$, and one of the possible analytic continuation of the function $\lambda=f(p,\tau)|_{D_0}$ is its extension to the polynomial $f(p,\tau)$ defined on $\mathbb{CP}^1$. All condition of Theorem 4.10 are satisfied, so we apply it here to complete the proof. □

REMARK 4.17 (Relation to Milanov's spectral curve). The global spectral curve that we constructed differs from the one constructed by Milanov in [**50**]. Milanov gets a spectral curve with the same local behavior as $x=f(y,\tau)$ near the critical points, but, in our terms, he chooses a different analytic continuation of $\lambda|_D$. He constructs an analytic continuation using the action of the Weyl group , and obtains a curve where all preimages of the critical points in the x-plane are critical. In our terms, this can be achieved by gluing $n!$ copies of the curve $x=f(y,\tau)$ along the cuts connecting the non-critical preimages of the points u_i, $i=1,\dots,n$ such that each point belongs to exactly one cut. This makes all preimages of $u_1,\dots,u_n$ critical and will produce a curve of genus $1+\frac{n!}{2}(\frac{n^2}{2}-\frac{n}{2}-2)$ where the function x has degree $(n+1)!$, and it has $n!$ poles of degree $(n+1)$ each .

4.6.1. *Bouchard-Eynard recursion.* In this Section we discuss an application of Theorem 4.16. There is a more general formulation of topological recursion that works for functions x with higher order singular points [**6**]. Locally, a higher order singularity is given by $x=y^{n+1}$, $B=dy_1dy_2/(y_1-y_2)^2$. Bouchard and Eynard announced a theorem [**7**] that identifies the coefficients of the local expansion in y at $y=0$ of the correlation differentials of this spectral curve with the coefficients of the string solution of the $(r+1)$-Gelfand-Dickey hierarchy, also known as the total descendant potential of the A_r singularity. The proof of Bouchard and Eynard goes through analysis of matrix models. Here we give a new proof of their theorem, namely, we derive it directly from Theorem 4.16.

THEOREM 4.18. [**7**] *The Bouchard-Eynard recursion applied to* $x=p^{n+1}$, $y=p$, $B=dp_1dp_2/(p_1-p_2)^2$ *produces differentials* $\omega_{g,k}$, *whose expansions near infinity are given by*

$$\omega_{g,k}(p_1,\dots,p_k)=\sum_{\substack{0\le a_1,\dots,a_k\le n-1\\ d_1,\dots,d_k}}\langle\tau_{d_1a_1}\cdots\tau_{d_ka_k}\rangle_{g,k}\times \tag{251}$$
$$\prod_{j=1}^{k}\left(\frac{(a_j+1)(a_j+1+(n+1))\cdots(a_j+1+d_j(n+1))}{(-1)^{d_j}(n+1)^{d_j+1}}\frac{dp_j}{p_j^{(n+1)d_j+a_j+2}}\right),$$

where $\langle\tau_{d_1a_1}\cdots\tau_{d_ka_k}\rangle_{g,k}$ *are the coefficients of the string solution of the* $(n+1)$*-Gelfand-Dickii hierarchy* [**32, 37, 61**].

Note that we don't recall and don't use the definition of the Bouchard-Eynard recursion. The only property that we are using here is that it is compatible with the usual recursion on the curves with simple singularities and the limits [**6**]. In this case, we know that in the neighborhood of infinity the correlation differentials of the Bouchard-Eynard recursion are the limits for $\epsilon \to 0$ of the correlation differentials of the usual recursion applied to $x = y^{n+1} + \epsilon y$.

Let us now prove theorem 4.18.

PROOF. The flat coordinates $t_0 = t_{\mathbb{1}}, t_1, \dots, t_{n-1}$ are given on the space of polynomials $f(p,\tau)$ defined in Equation (242) by the following formula:

$$f(p,\tau)^{\frac{1}{n+1}} = p + \frac{1}{n+1}\left(\frac{t_{n-1}}{p} + \frac{t_{n-2}}{p^2} + \dots + \frac{t_0}{p^n}\right) + O\left(\frac{1}{p^{n+1}}\right). \tag{252}$$

Recall that the canonical coordinates are the critical values $u_1, \dots, u_n$ of $f(p,\tau)$ and it is obvious that $\partial u_i/\partial t_{\mathbb{1}} = 1$. We denote by $c_1, \dots, c_n$ the positions of the critical points of function $f(p,\tau)$; so $u_i = f(c_i, \tau)$, $i = 1, \dots, n$.

We perform all computations only on a special curve in the space of polynomials, namely, $f(p,\tau) = p^{r+1} + \epsilon p$, and we are interested in all results only up $O(\epsilon)$ for $\epsilon \to 0$. In particular, we note that $t_a = O(\epsilon)$, $a = 0, \dots, n-1$.

The full Jacobian of the change from the canonical to flat coordinates is then given by the following computation:

$$\begin{aligned}\frac{\partial u_i}{\partial t_a} &= \frac{\partial f(c_i,\tau)}{\partial t_a} = f(c_i,\tau)^{\frac{n}{n+1}}\left(\frac{1}{c_i^{n-a}} + O(\epsilon)\right) = c_i^a \frac{\partial u_i}{\partial t_0} + O(\epsilon) \\ &= c_i^a + O(\epsilon).\end{aligned} \tag{253}$$

The correlation differentials, written in terms of a CohFT considered in normalized canonical frame in Equation (215), can be rewritten in terms of the correlators of the ancestor potential of Givental [**37**] $\mathcal{A}_t(\{t_{d,\alpha}\})$ considered at the point t in flat coordinates $t_{d,\alpha}$, $d = 0, 1, 2, \dots,$, $\alpha = 0, \dots, n-1$ in the following way:

(254)

$$\begin{aligned}\omega_{g,k} &= \sum_{\substack{i_1,\dots,i_k \\ d_1,\dots,d_k}} \int_{\overline{\mathcal{M}}_{g,k}} \alpha_{g,k}\left(\Delta_{i_1}^{\frac{1}{2}}\frac{\partial}{\partial u_{i_1}}, \dots, \Delta_{i_k}^{\frac{1}{2}}\frac{\partial}{\partial u_{i_k}}\right)\prod_{j=1}^k \psi_j^{d_j} d\left(\left(\frac{d}{dx}\right)^{d_j}\xi_{i_j}\right). \\ &= \sum_{\substack{\alpha_1,\dots,\alpha_k \\ d_1,\dots,d_k}} \langle \tau_{d_1\alpha_1}\cdots\tau_{d_k\alpha_k}\rangle_{g,k}(t)\prod_{j=1}^k d\left(\left(\frac{d}{dx}\right)^{d_j}\frac{-1}{p^{\alpha_j+1}}\right) + O(\epsilon).\end{aligned}$$

(by $\langle \tau_{d_1\alpha_1}\cdots\tau_{d_k\alpha_k}\rangle_{g,k}(t)$ we denote the coefficients of the expansion of $\log \mathcal{A}_t$). Indeed, let us expand the vector $\sum_{i=1}^n \xi_i(p)\Delta_i^{\frac{1}{2}}\frac{\partial}{\partial u_i}$ near $p = \infty$. Recall that we denote by $c_1, \dots, c_n$ the positions of the critical points of function $f(p,\tau)$. We have:

$$\begin{aligned}\sum_{i=1}^n \xi_i(p)\Delta_i^{\frac{1}{2}}\frac{\partial}{\partial u_i} &= \sum_{i=1}^n \left(\frac{1}{z-p}\frac{dz}{d\sqrt{f(p,\tau)-u_i}}\right)\Bigg|_{z=c_i} \Delta_i^{\frac{1}{2}}\frac{\partial}{\partial u_i} \\ &= -\sum_{k=0}^\infty \frac{1}{p^{k+1}}\sum_{i=1}^n c_i^k\frac{\partial}{\partial u_i} = -\sum_{k=0}^{n-1}\frac{1}{p^{k+1}}\frac{\partial}{\partial t_k} + O(\epsilon).\end{aligned} \tag{255}$$

(we use Equation (253) and the fact that $c_i^n = O(\epsilon)$ for the last equality).

Recall [**56**] that the correlators of the ancestor potential $\mathcal{A}_t$ are represented in terms of the correlators of the descendant potential (which is exactly the string solution of the $(n+1)$-Gelfand-Dickey hierarchy) as

$$\langle \tau_{d_1\alpha_1}\cdots\tau_{d_k\alpha_k}\rangle_{g,k}(t) = \langle \tau_{d_1\alpha_1}\cdots\tau_{d_k\alpha_k}\rangle_{g,k} + O(\epsilon). \tag{256}$$

Thus we see that

$$\omega_{g,k} = \sum_{\substack{\alpha_1,\dots,\alpha_k\\ d_1,\dots,d_k}} \langle \tau_{d_1\alpha_1}\cdots\tau_{d_k\alpha_k}\rangle_{g,k} \prod_{j=1}^{k} d\left(\left(\frac{d}{dx}\right)^{d_j}\frac{-1}{p^{\alpha_j+1}}\right) + O(\epsilon). \tag{257}$$

In the limit $\epsilon\to 0$ we get exactly Equation (251). □

4.7. Frobenius manifolds for hypermaps. In this Section we construct a global spectral curve for the Frobenius manifold given by the superpotential $\lambda = f(p,a)$, where $a=(a_0,\dots,a_{n+1})$, $n\geq 1$, and

$$f(p,a) = p^n + a_2p^{n-2} + a_3p^{n-3} + \cdots + a_n + \frac{a_{n+1}}{p-a_1}. \tag{258}$$

This superpotential defines a semi-simple Frobenius manifold (this Frobenius manifold is studied in [**12**, Section 5]). Furthermore the spectral curve

$$(\Sigma,x,y,B) = \left(\mathbb{CP}^1, f(p,a), p, \frac{dp_1dp_2}{(p_1-p_2)^2}\right)$$

satisfies equations (210)-(213) hence it stores the correlators of a CohFT via Equation (215). The following theorem answers the question of whether these two CohFTs coincide.

THEOREM 4.19. *The CohFT associated to the Frobenius manifold given by the superpotential $\lambda = f(p,a)$ coincides with the one reconstructed from the spectral curve $(\mathbb{CP}^1, f(p,a), p, dp_1dp_2/(p_1-p_2)^2)$.*

REMARK 4.20. The correlation differentials for this spectral curve considered for the particular values of the parameters a enumerate hypermaps on the curves. This is proved in [**20**], see also [**9**], where some special case of that was conjectured.

So, Theorem 4.19 is to be used in the converse way: We start with a combinatorial problem that is known to be solved by global topological recursion. It appears that the correlators of this global topological recursion are expressed in terms of a CohFT. This CohFT appears to be homogeneous, so it is associated to a Frobenius manifold, and this Theorem describes precisely the underlying Frobenius manifold.

PROOF. The proof is completely parallel to the proof of Theorem 4.16. Note that as in the case of A_n-singularity, we claim that the spectral curve is the superpotential itself. We use Theorem 4.10, so it is enough to show that we can reproduce the superpotential $\lambda = f(p,a)$ via Dubrovin's construction from Section 2.3.

The canonical coordinates $u_1,\dots,u_{n+1}$ of this Frobenius manifold are the critical values of $f(p,a)$; the Euler vector field is given by

$$E = \sum_{i=1}^{n} u_i\frac{\partial}{\partial u_i} = \sum_{i=1}^{n+1}\frac{i}{n}a_i\frac{\partial}{\partial a_i}; \tag{259}$$

the unit vector field is equal to

$$e = \sum_{i=1}^{n}\frac{\partial}{\partial u_i} = \frac{\partial}{\partial a_n}; \tag{260}$$

and the constant d is equal to $(n-2)/n$. Note that

$$Ef(p,a) = f(p,a) - \frac{p}{n}\frac{df(p,a)}{dp}. \tag{261}$$

As in the case of A_n singularity, the solutions to the Equation (8) are given by the integrals over the cycles $\beta \in H_0(f^{-1}(\lambda),\mathbb{C})$, where the components of the solutions are given by

$$\phi_i^\beta := \int_\beta \frac{\Delta_i^{\frac{1}{2}}}{\sqrt{2}} \frac{\partial f(p,a)}{\partial u_i} \cdot \left(\frac{df(p,a)}{dp}\right)^{-1}. \tag{262}$$

Consider the λ-plane as the image of the map $\lambda = f(p,\tau)$. Recall that $u_1,\dots,$ u_{n+1} are the critical values of $f(p,\tau)$. We can alway choose a system of cuts L_i, $i=1,\dots,n+1$, from u_i to infinity such that the preimage $f^{-1}(\mathbb{C}\setminus\cup_{i=1}^n L_i)$ is a union of $n+1$ disks, $D_0, D_1,\dots,D_n$, glued along the boundary cuts in the following way:

- D_0 is glued to D_i, $i=2,\dots,n$, along the boundary that is a double cover of L_i; in particular, their common boundary contains the critical preimage of u_i;
- D_0 is glued to D_1 along two components of the boundary that are double covers of L_1 and L_{n+1} and these boundary components have common point $p=a_1$. In particular, these boundary components contain the critical preimages of u_1 and u_{n+1};
- All other lifts of the cut L_i are just cuts inside D_j, $j\neq 0,i$ for $i=2,\dots,n$ and $j\neq 0,1$ for $i=1,n+1$; the endpoints of these cuts are non-critical preimages of u_i.

For $n=1,2,3$ we give the corresponding pictures for a real orientable blow-up at infinity (that is, the external boundary circle on the picture corresponds to the infinity point of the source sphere, and the internal circle corresponds to $p=a_1$). The domain D_0 is shadowed.

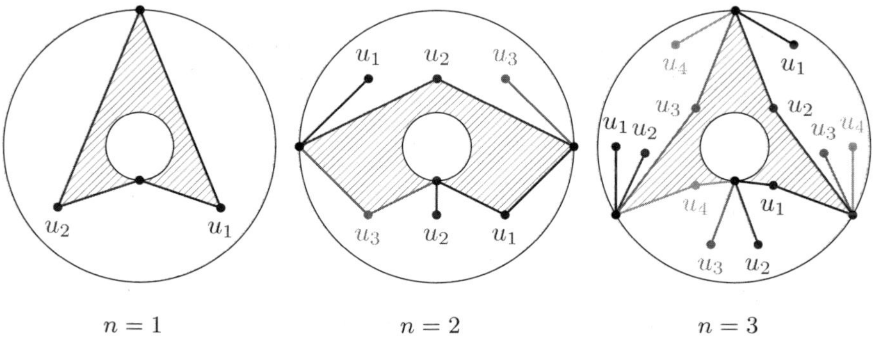

FIGURE 2. Hypermaps $z^n + \frac{1}{z}$, $1\le n\le 3$

Consider the vanishing cycles $\beta_i \in H_0(f^{-1}(\lambda),\mathbb{C})$ given by $\beta_i := p_i - p_0$, where $\lambda = f(p_0,\tau) = f(p_i,\tau)$, and $p_0\in D_0$, $p_i \in D_i$. Then the system of solutions of Equation (8) given by $\phi^{(i)}(\lambda,\tau) := \phi^{\beta_i}(\lambda,\tau)$, $i=1,\dots,n$, $\phi^{(n+1)} = \phi^{(1)}$, satisfies the properties given by Equations (10)–(13). In particular, $G^{ii}=1$ for $i=1,\dots,$

$n+1$, $G^{1,n+1} = G^{n+1,1} = 1$, and for all other $i \neq j$ $G^{ij} = 1/2$. So, this matrix is degenerate.

However, Remark 4.2 specifies the properties of ϕ that are sufficient for Theorems 4.1 and 4.10. Note that

$$\phi := \frac{2}{n+1} \sum_{i=1}^{n} \phi^{(i)} = \phi^{\beta_0} \tag{263}$$

for $\beta_0 := \frac{2}{n+1} \sum_{i=0}^{n} p_i - 2p_0$ satisfies all condition of Remark 4.2. With this choice of ϕ and, therefore, β_0, Dubrovin's superpotential can be presented as

$$\frac{\sqrt{2}}{1-d} \phi^T (U - \lambda) \Psi \mathbb{1} = \frac{n}{2} \int_{\beta_0} (E - \lambda e) f(p,a) \cdot \left(\frac{df(p,a)}{dp} \right)^{-1}. \tag{264}$$

Using Equation (261), we have:

$$\begin{aligned} \frac{\sqrt{2}}{1-d} \phi^T (U - \lambda) \Psi \mathbb{1} &= \frac{n}{2} \int_{\beta_0} \left(f(p,a) - \frac{p}{n} \frac{df(p,a)}{dp} - \lambda \right) \cdot \left(\frac{df(p,a)}{dp} \right)^{-1} \\ &= \int_{\beta_0} -\frac{p}{2} = p_0 - \frac{1}{n+1} \sum_{i=0}^{n} p_i = \begin{cases} p_0 - \frac{\lambda + a_1}{2} & n = 1; \\ p_0 - \frac{a_1}{n+1} & n > 1. \end{cases} \end{aligned} \tag{265}$$

(in the last equality we used that we know the sum of all roots of the equation $f(p,a) = \lambda$).

Let us now discuss the cases that we get. For $n > 1$ Dubrovin's function $p_{\mathrm{Dub}} = p_{\mathrm{Dub}}(\lambda, a)$ is the branch D_0 of the inverse function of $\lambda = f(p,a)$ shifted by a constant. Obviously, $d_\lambda p_{\mathrm{Dub}}$ has no zeros in $\mathbb{C} \setminus \cup_{i=1}^{n+1} L_i$, and we can choose as the analytic extension of $\lambda|_{D_0}$ the function $\lambda = f(p,a)$ defined on $\mathbb{CP}^1$. Then Theorem 4.10 is applied. We get, therefore, not precisely the statement that we want to prove, but we have instead $y = p_{\mathrm{Dub}} = p - a_1/(n+1)$ (the Bergman kernel is still the same). However, it doesn't change anything in topological recursion if we shift y by a constant.

The case $n = 1$ is even more interesting. One can easily check by direct computation that Dubrovin's function $p_{\mathrm{Dub}} = p_{\mathrm{Dub}}(\lambda, a)$ is equal to $\sqrt{(\lambda - u_1)(\lambda - u_2)}/2$. Further construction of the curve gives the following equation:

$$p_{\mathrm{Dub}}^2 - \frac{1}{4} (\lambda - a_1)^2 + a_2 = 0$$

It is a rational curve, and it has a global coordinate $p = p_{\mathrm{Dub}} + (\lambda + a_1)/2$, which is our original coordinate p, that is, $\lambda = p + a_2/(p - a_1)$. Theorem 4.10 can not be applied directly, but in this case we can just check by hand that we get the statement that we want to prove.

Note that Theorem 4.1 suggests that the right choice of function y is $y = p_{\mathrm{Dub}} = p - (\lambda + a_1)/2$ rather than $y = p$. However, it doesn't change anything in topological recursion if we shift y by a function of $x = \lambda$, so there is no contradiction. □

Since in this example we rather start from a combinatorial problem of enumeration of hypermaps and use Theorem 4.19 in order to clarify the structure of the ELSV-type formula (215) for this combinatorial problem, it is interesting to have a description of the underlying Frobenius manifolds (given by superpotentials) in terms of their prepotentials. We know an algorithm which can produce this prepotential for any given n (this algorithm follows from Dubrovin's construction found in [**12**]), but we do not know a general formula which would describe these prepotentials for all $n \geq 1$. Here we list the formulas for cases $n = 1, 2, 3$:

$$\begin{aligned}
n = 1: &\quad \frac{a_1^2 a_2}{2} + \frac{a_2^2}{2}\log a_2; \\
n = 2: &\quad \frac{a_1^3}{6} + a_1 a_2 a_3 + \frac{a_3^2}{2}\log a_3 + \frac{a_1^3 a_3}{6} - \frac{3a_3^2}{4}; \\
n = 3: &\quad \frac{a_1^2 a_4}{2} + a_1 a_2 a_3 - \frac{3a_2^2}{4} + \frac{a_2^2}{2}\log(a_2) + \frac{a_2 a_3^4}{4} + \frac{3a_2 a_3^2 a_4}{2} + \frac{3a_2 a_4^2}{2} - \frac{3a_4^4}{8}
\end{aligned}$$

Note that in the case $n = 1$ the corresponding combinatorial problem has also interpretation in terms of the discrete volumes of the moduli space of curves [**51**] and discrete surfaces/generalized Catalan numbers [**1, 16, 30**]. The relation of these combinatorial problems to a CohFT is also discussed in [**5, 33**], though it is not mentioned there that the underlying Frobenius manifold is given by the prepotential $a_1^2 a_2/2 + a_2^2/2 \cdot \log a_2$.

4.8. Elliptic example. In this section we give an example of a superpotential that satisfies the conditions of Theorem 4.13.

Consider the spectral curve defined by the Weierstrass $\wp$-function

$$\lambda = \wp(z), \quad p = z, \quad B(z, z') = (\wp(z - z') + b)\, dz dz' \tag{266}$$

where $b \in \mathbb{C}$ and p is only defined locally—it is the primitive of a holomorphic differential on the curve—which is sufficient for topological recursion. The compatibility condition (221) is satisfied by Proposition 4.6. It is equivalent to the elliptic identity:

$$\frac{\wp''(z)}{\wp'(z)^2} = \sum_{i=1}^{3} \frac{\wp(z - \omega_i)}{\wp''(\omega_i)} \tag{267}$$

where the sum is over the zeros ω_i of $\wp'(z)$. Hence the spectral curve defines a CohFT.

Introduce three parameters ω, ω' and c into the spectral curve to define the following superpotential taken from [**12**]:

$$\lambda = \wp(z; \omega, \omega') + c, \quad p = \frac{z}{\omega} \tag{268}$$

where

$$\wp(z; \omega, \omega') = \frac{1}{z^2} + \sum_{(m,n) \neq (0,0)} \frac{1}{(z - 2m\omega - 2n\omega')^2} - \frac{1}{(2m\omega + 2n\omega')^2}. \tag{269}$$

The Frobenius manifold structure on $M = \{(\omega, \omega', c)\}$ is given by the formulae (4), (5), (6) where the vector fields ∂ on M are given by, for example ∂_ω, $\partial_{\omega'}$, ∂_c. Note that in (267), $\omega_1 = \omega$, $\omega_2 = \omega'$, $\omega_3 = \omega + \omega'$.

REMARK 4.21. Note that we know that the superpotential (268) defines a Frobenius manifold due to the existence of flat coordinates, proven in [**12**], and also given below. The CohFT produced by topological recursion applied to (266) is homogeneous if we choose b in (266) so that $\int_A B(z,z') = 0$, i.e. $b = \eta/\omega$ where the A and B periods are $2\omega = \oint_A dz$, $2\omega' = \oint_B dz$ and

$$\eta = -\frac{1}{2}\oint_A \wp(z)dz, \quad \eta' = -\frac{1}{2}\oint_B \wp(z)dz.$$

The periods satisfy Legendre's relation

$$\eta\omega' - \eta'\omega = \frac{i\pi}{2}.$$

The homogeneous CohFT corresponds to a conformal Frobenius manifold which gives rise to a superpotential via Dubrovin's construction (actually, since $d = 1$ it is a variant of the construction). What needs to be proven is that the two superpotentials agree.

THEOREM 4.22. *The superpotential* (268) *can be obtained via (a variant of) Dubrovin's construction described in Section* 2.3 *applied to the Frobenius manifold* M. *The conditions of Theorem* 4.13 *are satisfied for this superpotential. Hence the two cohomological field theories—obtained from the superpotential* (268) *and topological recursion applied to the spectral curve* (266) *with* $b = \eta/\omega$*—agree.*

PROOF. To apply Dubrovin's construction to M we construct a solution of the Gauss-Manin system as in the proof of Theorem 4.16.

The flat metric (4) for the superpotential (268) is given by

$$(\partial, \partial') := \sum_{i=1}^{3} \operatorname*{res}_{z=\omega_i} \frac{\partial\lambda \cdot \partial'\lambda}{\wp'} dp. \tag{270}$$

We use this to construct a vector field $I_\beta(\lambda; u)$ on M for any cycle $\beta \in H_0(\lambda^{-1}(\mathrm{pt}), \mathbb{C})$ specified by:

$$(I_\beta(\lambda; u), \partial) := \int_\beta \frac{\partial(\lambda)}{d_p\lambda/dp}. \tag{271}$$

The elliptic curve (268) is built by gluing two copies of the disk $D = \mathbb{C} \setminus \cup_{i=1}^3 L_i$ in the λ plane along L_i. Choose β to be the cycle given by $p_0 - p_1$, for p_0 and p_1 the pre-images of λ in each of the two disks. In normalized canonical coordinates $I_\beta(\lambda; u)$ is represented by a solution $\phi^\beta(\lambda; u) = \sum_i \phi_i^\beta(\lambda; u)\partial_{v_i}$ of the Gauss-Manin system (8) which has components given by

$$\begin{aligned}\phi_i^\beta(\lambda; u) &= \left(I_\beta(\lambda; u), \frac{1}{\sqrt{2}}\Delta_i^{\frac{1}{2}}\partial_{u_i}\right)\\ &= \int_\beta \frac{\Delta_i^{\frac{1}{2}}\partial_{u_i}\lambda}{\sqrt{2}\cdot d_p\lambda/dp}\\ &= \frac{\sqrt{2}\cdot\Delta_i^{\frac{1}{2}}\partial_{u_i}\lambda}{d_p\lambda/dp}\end{aligned}$$

where the integral over β simply doubles the integrand since the integrand is skew symmetric. Since $d = 1$, we cannot use the inversion formula (9) so we directly check that $\frac{1}{\sqrt{2}}\phi^\beta = \nabla_u p$ as follows.

$$\nabla_u p = \eta^{-1} d_u p = \frac{1}{\omega}\sum_{i=1}^{3} \Delta_i \cdot \frac{\partial_{u_i}\lambda}{\wp'(z)}\partial_{u_i} = \frac{1}{\omega}\sum_{i=1}^{3} \frac{\Delta_i^{\frac{1}{2}}\partial_{u_i}\lambda}{\wp'(z)}\partial_{v_i} = \frac{1}{\sqrt{2}}\phi^\beta.$$

Using

$$\partial_{u_i}\lambda = \frac{1}{2\wp''(\omega_i)}\frac{\wp'(z)^2}{\wp(z)-\wp(\omega_i)} + \frac{z\wp(\omega_i)+\zeta(z)}{\wp''(\omega_i)}\wp'(z)$$

which can be proven from the known variations $\sum_i u_i^k\partial_{u_i}$, $k=0,1,2$, we see that the solution $\phi^\beta(\lambda;u)$ satisfies

$$\begin{aligned}\frac{1}{\sqrt{2}}\phi^T\Psi\mathbb{1} &= \frac{1}{\omega}\sum_{i=1}^{3}\frac{\partial_{u_i}\lambda}{\wp'(z)}\\ &= \frac{1}{\omega}\sum_{i=1}^{3}\frac{1}{2\wp''(\omega_i)}\frac{\wp'(z)}{\wp(z)-\wp(\omega_i)} + \frac{z\wp(\omega_i)+\zeta(z)}{\wp''(\omega_i)}\\ &= \frac{1}{\omega\wp'(z)} = \frac{dp}{d\lambda}\end{aligned}$$

which is (218) and hence via Remark 4.2 we see that the properties of ϕ are sufficient for Theorems 4.1 and 4.10. Hence the theorem follows. □

Theorem 4.22 states that we can study the CohFT obtained from the superpotential (268) via topological recursion applied to the spectral curve (266). We will need the three-point function of this CohFT in calculations below. We calculate it in two ways to demonstrate the proof, although we know from the theorem that they coincide.

4.8.1. *Three-point function. Superpotential.* Introduce the canonical coordinates

$$u_i = \wp(\omega_i)+c,\quad i=1,2,3$$

where, as usual, $\omega_1=\omega$, $\omega_2=\omega'$, $\omega_3=\omega+\omega'$. The three-point calculations take place in the ring $\mathbb{C}[E]/\wp' = \mathbb{C}[\wp]/\wp'$ and we have

$$\frac{\partial\lambda}{\partial u_1} \equiv \frac{(\lambda-u_2)(\lambda-u_3)}{(u_1-u_2)(u_1-u_3)},\quad \frac{\partial\lambda}{\partial u_1}\frac{\partial\lambda}{\partial u_j} \equiv \delta_{1j}\frac{\partial\lambda}{\partial u_1},\ j=2,3.$$

and cyclic permutation of the above. This is quite general and also can be proven via elliptic identities. Hence the three-point function for the superpotential is

$$\left\langle\frac{\partial}{\partial u_i},\frac{\partial}{\partial u_i},\frac{\partial}{\partial u_i}\right\rangle = \left\langle\frac{\partial}{\partial u_i},\frac{\partial}{\partial u_i}\right\rangle = \left\langle\frac{\partial}{\partial u_i}\right\rangle = \sum_{j=1}^{3}\operatorname*{res}_{z=\omega_j}\frac{\frac{\partial\lambda}{\partial u_i}}{\wp'(z)}\frac{dz}{\omega^2} = \frac{1}{\omega^2\wp''(\omega_i)}.$$

Thus

$$\left\langle\frac{\partial}{\partial v_i},\frac{\partial}{\partial v_i},\frac{\partial}{\partial v_i}\right\rangle = \omega\sqrt{\wp''(\omega_i)} \tag{272}$$

where $\frac{\partial}{\partial v_i} = \Delta_i^{\frac12}\frac{\partial}{\partial u_i}$ for $\Delta_i^{-1} = \left\langle\frac{\partial}{\partial u_i},\frac{\partial}{\partial u_i}\right\rangle = \omega^2\cdot\wp''(\omega_i)$ give the normalized canonical coordinates.

Topological recursion. The three-point function obtained via topological recursion is

$$\begin{aligned}\omega_{0,3}(z_1,z_2,z_3) &= \sum_{j=1}^{3} \operatorname*{res}_{z=\omega_j} \frac{\omega dz}{\wp'(z)} \prod_{i=1}^{3}(\wp(z_i-z)+b)dz_i \\ &= \sum_{j=1}^{3} \frac{\omega}{\wp''(\omega_j)} \prod_{i=1}^{3}(\wp(z_i-\omega_j)+b)dz_i \\ &= \sum_{j=1}^{3} \omega\sqrt{\wp''(\omega_j)}V_0^j(z_1)V_0^j(z_2)V_0^j(z_3)\end{aligned}$$

for

$$V_0^j(z) = \left.\frac{(\wp(z-\omega_j)+b)dzdz_j}{ds_j}\right|_{s_j=0} = \frac{\wp(z-\omega_j)+b}{\sqrt{\wp''(\omega_i)}}dz$$

where $\lambda = \wp(z)+c = \frac{1}{2}s_j^2+\wp(\omega_j)+c$ defines the local coordinate s_j. The coefficients of $V_0^j(z_i)$ define the three-point function of the cohomological field theory which agree with (272).

4.8.2. *Flat coordinates.* The cohomological field theory is defined on the three-dimensional vector space $\mathbb{C}[\wp]/\wp'$ equipped with its natural ring structure and gives rise to a Frobenius manifold structure on the family M of such rings parametrized by $\{\omega,\omega',c\}$. It will be convenient to express the metric on M with respect to a natural basis of vector fields on M corresponding to the basis $\{1,\wp,\wp^2\}$ of $\mathbb{C}[\wp]/\wp'$ since the metric requires knowledge of the variation of $\wp$ under the action of vector fields on the Frobenius manifold. We will see that $\{\omega,\omega',c\}$ are not flat coordinates and find in Lemma 4.24 flat coordinates $\{t_1,t_2,t_3\}$ on M, i.e. so that the metric on M is constant with respect to them.

Recall that correlation functions of the cohomological field theory arising from topological recursion applied to a spectral curve appear as coefficents of auxiliary differentials on the spectral curve. Proposition 4.26 gives the auxiliary differentials that correspond to the flat basis for the metric.

In the following lemma we calculate the vector fields on M that correspond to the basis elements $1,\wp,\wp^2$ of $\mathbb{C}[\wp]/\wp'$. This uses $g_2 = g_2(\omega,\omega')$ defined by $\wp'(z)^2 = 4\wp(z) - g_2\wp - g_3$.

LEMMA 4.23. *Under the map $TM \to \mathbb{C}[\wp]/\wp'$ defined by $\partial \mapsto \partial\lambda(mod\ \wp')$ for $\lambda = \wp(z;\omega,\omega')+c$*

$$\partial_c \mapsto 1, \quad -\tfrac{1}{2}\left(\omega\partial_\omega + \omega'\partial_{\omega'}\right) \mapsto \wp, \quad -\tfrac{1}{2}\left(\eta\partial_\omega + \eta'\partial_{\omega'}\right) + \tfrac{1}{6}g_2\partial_c \mapsto \wp^2. \tag{273}$$

PROOF. The variation $\partial_c\lambda = 1$ is obvious. The identity

$$\omega\partial_\omega\wp(z) + \omega'\partial_{\omega'}\wp(z) + z\wp'(z) = -2\wp(z) \tag{274}$$

follows immediately from the expansion (269) of $\wp$ and yields $-\frac{1}{2}\left(\omega\partial_\omega + \omega'\partial_{\omega'}\right) \mapsto \wp$. The final identification uses the identity proven in [**34**]

$$\eta\partial_\omega\wp(z) + \eta'\partial_{\omega'}\wp(z) + \zeta(z)\wp'(z) = -2\wp(z)^2 + \tfrac{1}{3}g_2 \tag{275}$$

where $\zeta(z)$ is the Weierstrass ζ-function

$$\zeta(z;\omega,\omega') = \frac{1}{z} + \sum_{(m,n)\neq(0,0)} \frac{1}{z-2m\omega-2n\omega'} + \frac{1}{2m\omega+2n\omega'} + \frac{z}{(2m\omega+2n\omega')^2}$$

which is not an elliptic function (C.63 in [**12**]). Note that $\eta = \zeta(\omega)$, $\eta' = \zeta(\omega')$. □

The metric

$$\langle \wp^j, \wp^k \rangle = \sum_{i=1}^{3} \operatorname*{res}_{z=\omega_i} \frac{\wp^{j+k}}{\wp'(z)} \frac{dz}{\omega^2} = - \operatorname*{res}_{z=0} \frac{\wp^{j+k}}{\wp'(z)} \frac{dz}{\omega^2}$$

is given by

	1	$\wp$	$\wp^2$
1	0	0	$1/2\omega^2$
$\wp$	0	$1/2\omega^2$	0
$\wp^2$	$1/2\omega^2$	0	$g_2/8\omega^2$

LEMMA 4.24 (Dubrovin [**12**]). *Flat coordinates for the metric are given by*

$$t_1 = c - \frac{\eta}{\omega}, \quad t_2 = \frac{1}{\omega}, \quad t_3 = \frac{\omega'}{\omega}.$$

PROOF. This is (5.95) in [**12**]. We simply use change of coordinates given by (273) and the metric calculated above. We have $\partial_c = \partial_{t_1}$. From the identity

$$(\omega\partial_\omega + \omega'\partial_{\omega'})\frac{\eta}{\omega} = -2\frac{\eta}{\omega}$$

which uses the fact that $\omega\partial_\omega + \omega'\partial_{\omega'}$ is the degree operator, $\frac{\eta}{\omega}$ is homogeneous of degree -2 we have $\omega\partial_\omega + \omega'\partial_{\omega'} = \frac{2\eta}{\omega}\partial_{t_1} - \frac{1}{\omega}\partial_{t_2}$. The identity

$$(\eta\partial_\omega + \eta'\partial_{\omega'})\frac{\eta}{\omega} = -\frac{1}{12}g_2 - \frac{\eta^2}{\omega^2}$$

appearing as (C.69) in [**12**] gives $\eta\partial_\omega + \eta'\partial_{\omega'} = \left(\frac{1}{12}g_2 + \frac{\eta^2}{\omega^2}\right)\partial_{t_1} - \frac{\eta}{\omega^2}\partial_{t_2} - \frac{i\pi}{2\omega^2}\partial_{t_3}$. Hence we have

(276)

$$\partial_{t_1} \mapsto 1, \quad -\frac{\eta}{\omega}\partial_{t_1} + \frac{1}{2\omega}\partial_{t_2} \mapsto \wp, \quad \left(\frac{1}{8}g_2 - \frac{\eta^2}{2\omega^2}\right)\partial_{t_1} + \frac{\eta}{2\omega^2}\partial_{t_2} + \frac{i\pi}{4\omega^2}\partial_{t_3} \mapsto \wp^2.$$

Hence the metric is given by:

	∂_{t_1}	∂_{t_2}	∂_{t_3}
∂_{t_1}	0	0	$2/i\pi$
∂_{t_2}	0	2	0
∂_{t_3}	$2/i\pi$	0	0

which is constant so that $\{t_1, t_2, t_3\}$ are flat coordinates. □

REMARK 4.25. We can choose a different $(0,2)$ term $B(z,z')$ on the spectral curve (266) which still satisfies the compatibility condition (221) by varying $b \in \mathbb{C}$. For each b it gives rise to a CohFT with the same genus 0 three-point function since ancestor invariants are coefficients of B-dependent differentials. When b is chosen so that $B(z,z')$ is normalised along a choice of cycle, e.g. $b = \eta'/\omega'$ so $\int_B B(z,z') = 0$, then the CohFT is homogeneous and hence the same CohFT as for $b = \eta/\omega$. Other choices of b gives rise to non-homogeneous CohFTs.

PROPOSITION 4.26. *The flat coordinates correspond to the following auxiliary differentials:*

$$dt_1 \longleftrightarrow T_0^1 = (\omega\wp + b)dz - 2\omega d\left(\frac{\wp^2}{\wp'}\right) + 2\eta d\left(\frac{\wp}{\wp'}\right) + \left(\frac{\omega g_2}{4} + \frac{\eta^2}{\omega}\right) d\left(\frac{1}{\wp'}\right)$$

$$dt_2 \longleftrightarrow T_0^2 = -d\left(\frac{\wp}{\wp'}\right) - \frac{\eta}{\omega} d\left(\frac{1}{\wp'}\right)$$

$$dt_3 \longleftrightarrow T_0^3 = -\frac{i\pi}{2\omega} d\left(\frac{1}{\wp'}\right).$$

PROOF. The auxiliary differentials on the spectral curve corresponding to the normalized canonical basis are straighforward. They are given by $V_k^i dz$ where

$$V_0^i = \frac{\wp(z - \omega_i)}{\sqrt{\wp''(\omega_i)}}$$

and for $k > 0$, V_k^i is the principal part of the kth derivative of V_0^i with respect to $\wp(z)$. We also have the canonical basis $U_0^i = \omega\wp(z - \omega_i)$. The auxiliary differentials $T_k^i dz$ corresponding to flat coordinates are linear combinations of $V_k^i dz$

$$V_k^i = \Psi_\mu^i \cdot T_k^\mu$$

where we recall that Ψ_μ^i is the transition matrix from flat coordinates labeled by μ to normalized canonical coordinates labeled by i. We can calculate Ψ via

$$\begin{pmatrix} 1 & 1 & 1 \\ \wp(\omega_1) & \wp(\omega_2) & \wp(\omega_3) \\ \wp(\omega_1)^2 & \wp(\omega_2)^2 & \wp(\omega_3)^2 \end{pmatrix} \begin{pmatrix} \partial_{u_1} \\ \partial_{u_2} \\ \partial_{u_3} \end{pmatrix} = \begin{pmatrix} 1 & 0 & 0 \\ -\frac{\eta}{\omega} & \frac{1}{2\omega} & 0 \\ \frac{1}{8}g_2 - \frac{\eta^2}{2\omega^2} & \frac{\eta}{2\omega^2} & \frac{i\pi}{4\omega^2} \end{pmatrix} \begin{pmatrix} \partial_{t_1} \\ \partial_{t_2} \\ \partial_{t_3} \end{pmatrix}$$

which we write as $M\frac{\partial}{\partial u} = T\frac{\partial}{\partial t}$ hence $M^{-1}T = \Delta^{1/2}\Psi^T$. The auxiliary differentials corresponding to $1, \wp, \wp^2$ in the Landau-Ginzburg model are:

$$\begin{aligned}
&[U^1, U^2, U^3] \cdot M^{-1} = \\
&2\omega dz \left[\frac{\wp(z-\omega_1)+b}{\wp''(\omega_1)}, \frac{\wp(z-\omega_2)+b}{\wp''(\omega_2)}, \frac{\wp(z-\omega_3)+b}{\wp''(\omega_3)}\right] \begin{pmatrix} \wp(\omega_1)^2 - \frac{1}{4}g_2 & \wp(\omega_1) & 1 \\ \wp(\omega_2)^2 - \frac{1}{4}g_2 & \wp(\omega_2) & 1 \\ \wp(\omega_3)^2 - \frac{1}{4}g_2 & \wp(\omega_3) & 1 \end{pmatrix} \\
&\quad = \left[-2\omega d\left(\frac{\wp^2}{\wp'}\right) + (\omega\wp + b)dz + \frac{\omega g_2}{2} d\left(\frac{1}{\wp'}\right), -2\omega d\left(\frac{\wp}{\wp'}\right), -2\omega d\left(\frac{1}{\wp'}\right)\right]
\end{aligned}$$

which is proven using the elliptic identities

$$\frac{\wp(z)^k}{\wp'(z)^2} = \sum_{i=1}^3 \frac{\wp(\omega_i)^k \wp(z-\omega_i)}{\wp''(\omega_i)^2}, \qquad k = 0, 1, 2$$

and slight generalisations for $k > 2$. Hence

$$\begin{aligned}
[T^1, T^2, T^3] &= [U^1, U^2, U^3] \cdot M^{-1} \cdot T \\
&= \Big[-2\omega d\left(\frac{\wp^2}{\wp'}\right) + (\omega\wp + b)dz + \left(\frac{\omega g_2}{4} + \frac{\eta^2}{\omega}\right) d\left(\frac{1}{\wp'}\right) + 2\eta d\left(\frac{\wp}{\wp'}\right), \\
&\qquad -d\left(\frac{\wp + \eta/\omega}{\wp'}\right), -\frac{i\pi}{2\omega} d\left(\frac{1}{\wp'}\right)\Big]
\end{aligned}$$

□

The following lemma allows us to apply equation (215) to obtain ancestor invariants for the CohFT.

LEMMA 4.27. *The following kernels* K_0^i

$$K_0^1 = y(z), \quad K_0^2 = -2\omega\zeta(z)+2\eta, \quad K_0^3 = \frac{4}{i\pi}\left(\omega z\wp(z)^2 - \left(\frac{\eta^2}{2\omega} + \frac{\omega}{8}g_2\right)z + \eta\zeta(z)\right)$$

are dual (as linear functionals) to T_0^i *for* $i = 1, 2, 3$*, i.e.*

$$\sum_{j=1}^{3} \operatorname*{res}_{z=\omega_j} K_0^j(z)T_k^i(z) = \delta_{ij}\delta_{k0}.$$

PROOF. Each kernel is analytic at $z = \omega_i$, $i = 1, 2, 3$ and hence annihilates differentials analytic at $z = \omega_i$. Consider the action of each kernel on $d(\wp^k/\wp')$ for $k = 0, 1, 2$.

$$\sum_{j=1}^{3} \operatorname*{res}_{z=\omega_j} K_0^i(z)d\left(\frac{\wp^k}{\wp'}\right) = -\sum_{j=1}^{3} \operatorname*{res}_{z=\omega_j} dK_0^i(z)\frac{\wp^k}{\wp'} = \operatorname*{res}_{z=0} dK_0^i(z)\frac{\wp^k}{\wp'}$$

so $K_0^1 = y(z) = z/\omega$ annihilates $d(\wp^k/\wp')$ for $k = 0, 1$ and sends $d(\wp^2/\wp')$ to $-1/2\omega$. Similarly $K_0^2 = \zeta(z)$ annihilates $d(\wp^k/\wp')$ for $k = 0, 2$ and sends $d(\wp/\wp')$ to $1/2$. Apply the kernels to T_0^i given in Proposition 4.26 as linear combinations of $d(\wp^k/\wp')$ (and terms analytic at $z = \omega_i$) to achieve the result.

The kernels K_0^1 and K_0^2 annihilate exact differentials that vanish to order 2 at $z = 0$, in particular T_k^i for $k > 0$ by integration by parts. One can also check that K_0^3 annihilates T_k^i for $k > 0$. □

Remark. One can also produce kernels K_j^i dual to each T_j^i.

The 3-point function in flat coordinates leads to the prepotential given in [**12**] (C.87):

$$F_0 = \frac{1}{i\pi}t_1^2 t_3 + t_1 t_2^2 - \frac{i\pi}{2}t_2^4\left(\frac{1}{24} - \sum_{n=1}^{\infty}\frac{nq^n}{1-q^n}\right), \quad q = e^{2\pi i t_3}.$$

PROPOSITION 4.28.

$$\exp F_1 = t_2^{1/8}\eta(q)^{1/4}, \quad \eta(q) = q^{1/24}\prod_{n=1}^{\infty}(1-q^n).$$

PROOF. Topological recursion—defined in Section 2.7—applied to the spectral curve (266) uses the kernel

$$\begin{aligned} K(z_1, z) &= \frac{\omega}{2}\frac{\int_{\sigma_i(z)}^{z}(\wp(z_1 - w) + b)dw dz_1}{(z - \sigma_i(z))\wp'(z)dz} \\ &= \frac{\omega}{4}\frac{(\zeta(z_1 + z) - \zeta(z_1 - z) + 2\eta_i + 2b(z - \omega_i))dz_1}{(z - \omega_i)\wp'(z)dz} \end{aligned}$$

where $\sigma_i(z) = 2\omega_i - z$. Hence

$$\omega_{1,1}(z_1) = \sum_{j=1}^{3} \operatorname*{res}_{z=\omega_j} K(z_1,z)\wp(2z) = \sum_{j=1}^{3} \operatorname*{res}_{z=\omega_j} \frac{\omega}{4}\frac{(\zeta(z_1+z)-\zeta(z_1-z))}{(z-\omega_i)\wp'(z)}\wp(2z)dzdz_1$$
$$= \frac{\omega}{8}\left(2\sum_{j=1}^{3}\frac{\wp(\omega_i)\wp(z_0-\omega_i)}{\wp''(\omega_i)} - \sum_{j=1}^{3}\frac{\wp(z_0-\omega_i)^2}{\wp''(\omega_i)}\right)dz_1$$
$$= \frac{\omega}{8}\left(\frac{2\wp\wp''}{(\wp')^2} - \frac{(\wp'')^3}{(\wp')^4} + 10g_2\frac{\wp}{(\wp')^2} + 15g_3\frac{1}{(\wp')^2} + 11\right)dz_1$$

where $\eta_i \in \mathbb{C}$ and b are annihilated by the residues. Integrate the kernels K^i_j against $\omega_{1,1}$ to get

$$\omega_{1,1} = 0\cdot T^1_0 + \frac{\omega}{8}T^2_0 + \frac{i\eta\omega}{4\pi}T^3_0 + \frac{1}{8}T^1_1 + \frac{\eta}{4}T^2_1 + \frac{g_2\omega^2 - 12\eta^2}{48i\pi}T^3_1.$$

The primary part uses only T^k_0 and yields

$$F_1 = \frac{1}{8}\log t_2 + f(t_3), \quad f'(t_3) = \frac{i\pi}{2}\left(\frac{1}{24} - \sum_{n=1}^{\infty}\frac{nq^n}{1-q^n}\right)$$

which is obtained from $\omega_{1,1}$ since $\frac{\partial F_1}{\partial t_2} = \frac{1}{8t_2} = \frac{\omega}{8}$ agrees with the coefficient of T^2_0 and $\frac{\partial F_1}{\partial t_3} = \frac{i\pi}{2}\left(\frac{1}{24} - \sum_{n=1}^{\infty}\frac{nq^n}{1-q^n}\right) = \frac{i\eta\omega}{4\pi}$ agrees with the coefficient of T^2_0. □

Acknowledgments

The author is extremely grateful to Paul Norbury, Nicolas Orantin, Alexandr Popolitov, Sergey Shadrin and Loek Spitz, who coauthored the papers which served as the basis of the present text.

The author is also extremely grateful to Motohico Mulase, Bertrand Eynard and Chiu-Chu Melissa Liu, and to AMS for organizing the wonderful symposium where this work was presented, and to NSF for supporting traveling and accommodation costs.

The author would also like to express thanks to the anonymous referee whose remarks allowed to improve the text.

References

[1] A. Alexandrov, A. Morozov, and A. Mironov, *Partition functions of matrix models: first special functions of string theory*, Internat. J. Modern Phys. A **19** (2004), no. 24, 4127–4163, DOI 10.1142/S0217751X04018245. MR2104531

[2] A. Alexandrov, A. Mironov, and A. Morozov, *Solving Virasoro constraints in matrix models*, Fortschr. Phys. **53** (2005), no. 5-6, 512–521, DOI 10.1002/prop.200410212. MR2145334

[3] A. S. Aleksandrov, A. D. Mironov, and A. Yu. Morozov, *M-theory of matrix models* (Russian, with Russian summary), Teoret. Mat. Fiz. **150** (2007), no. 2, 179–192, DOI 10.1007/s11232-007-0011-6; English transl., Theoret. and Math. Phys. **150** (2007), no. 2, 153–164. MR2325922

[4] A. Alexandrov, A. Mironov, and A. Morozov, *Instantons and merons in matrix models*, Phys. D **235** (2007), no. 1-2, 126–167, DOI 10.1016/j.physd.2007.04.018. MR2385123

[5] Jørgen Ellegaard Andersen, Leonid O. Chekhov, Paul Norbury, and Robert C. Penner, *Models of discretized moduli spaces, cohomological field theories, and Gaussian means*, J. Geom. Phys. **98** (2015), 312–339, DOI 10.1016/j.geomphys.2015.08.018. MR3414961

[6] Vincent Bouchard and Bertrand Eynard, *Think globally, compute locally*, J. High Energy Phys. **2** (2013), 143, front matter + 34. MR3046532

[7] V. Bouchard, B. Eynard, Private communication.

[8] Leonid Chekhov, Bertrand Eynard, and Nicolas Orantin, *Free energy topological expansion for the 2-matrix model*, J. High Energy Phys. **12** (2006), 053, 31, DOI 10.1088/1126-6708/2006/12/053. MR2276699

[9] Norman Do and David Manescu, *Quantum curves for the enumeration of ribbon graphs and hypermaps*, Commun. Number Theory Phys. **8** (2014), no. 4, 677–701, DOI 10.4310/CNTP.2014.v8.n4.a2. MR3318387

[10] B. Dubrovin, *Integrable systems and classification of 2-dimensional topological field theories*, Integrable systems (Luminy, 1991), Progr. Math., vol. 115, Birkhäuser Boston, Boston, MA, 1993, pp. 313–359. MR1279829

[11] Boris Dubrovin, *On almost duality for Frobenius manifolds*, Geometry, topology, and mathematical physics, Amer. Math. Soc. Transl. Ser. 2, vol. 212, Amer. Math. Soc., Providence, RI, 2004, pp. 75–132, DOI 10.1090/trans2/212/05. MR2070050

[12] Boris Dubrovin, *Geometry of 2D topological field theories*, Integrable systems and quantum groups (Montecatini Terme, 1993), Lecture Notes in Math., vol. 1620, Springer, Berlin, 1996, pp. 120–348, DOI 10.1007/BFb0094793. MR1397274

[13] Boris Dubrovin, *Painlevé transcendents in two-dimensional topological field theory*, The Painlevé property, CRM Ser. Math. Phys., Springer, New York, 1999, pp. 287–412. MR1713580

[14] Boris Dubrovin and Youjin Zhang, *Bi-Hamiltonian hierarchies in 2D topological field theory at one-loop approximation*, Comm. Math. Phys. **198** (1998), no. 2, 311–361, DOI 10.1007/s002200050480. MR1672512

[15] B. Dubrovin, Y. Zhang, *Normal forms of hierarchies of integrable PDEs, Frobenius manifolds and Gromov-Witten invariants*, arXiv:math/0108160v1.

[16] Olivia Dumitrescu, Motohico Mulase, Brad Safnuk, and Adam Sorkin, *The spectral curve of the Eynard-Orantin recursion via the Laplace transform*, Algebraic and geometric aspects of integrable systems and random matrices, Contemp. Math., vol. 593, Amer. Math. Soc., Providence, RI, 2013, pp. 263–315, DOI 10.1090/conm/593/11867. MR3087960

[17] P. Dunin-Barkowski, N. Orantin, S. Shadrin, and L. Spitz, *Identification of the Givental formula with the spectral curve topological recursion procedure*, Comm. Math. Phys. **328** (2014), no. 2, 669–700, DOI 10.1007/s00220-014-1887-2. MR3199996

[18] P. Dunin-Barkowski, N. Orantin, P. Norbury, A. Popolitov, S. Shadrin, *Dubrovin's superpotential as a global spectral curve*, arXiv:1509.06954.

[19] P. Dunin-Barkowski, N. Orantin, P. Norbury, A. Popolitov, S. Shadrin, *Primary invariants of Hurwitz Frobenius manifolds*, arXiv:1605.07644.

[20] P. Dunin-Barkowski, N. Orantin, A. Popolitov, S. Shadrin, *Combinatorics of loop equations for branched covers of sphere*, arXiv:1412.1698.

[21] P. Dunin-Barkowski, D. Lewanski, A. Popolitov, and S. Shadrin, *Polynomiality of orbifold Hurwitz numbers, spectral curve, and a new proof of the Johnson-Pandharipande-Tseng formula*, J. Lond. Math. Soc. (2) **92** (2015), no. 3, 547–565, DOI 10.1112/jlms/jdv047. MR3431649

[22] Petr Dunin-Barkowski, Sergey Shadrin, and Loek Spitz, *Givental graphs and inversion symmetry*, Lett. Math. Phys. **103** (2013), no. 5, 533–557, DOI 10.1007/s11005-013-0606-9. MR3041758

[23] Bertrand Eynard, *Recursion between Mumford volumes of moduli spaces*, Ann. Henri Poincaré **12** (2011), no. 8, 1431–1447, DOI 10.1007/s00023-011-0113-4. MR2855174

[24] B. Eynard, *Intersection numbers of spectral curves*, arXiv:1104.0176.

[25] B. Eynard, *Invariants of spectral curves and intersection theory of moduli spaces of complex curves*, Commun. Number Theory Phys. **8** (2014), no. 3, 541–588, DOI 10.4310/CNTP.2014.v8.n3.a4. MR3282995

[26] Bertrand Eynard, *An overview of the topological recursion*, Proceedings of the International Congress of Mathematicians—Seoul 2014. Vol. III, Kyung Moon Sa, Seoul, 2014, pp. 1063–1085. MR3729064

[27] B. Eynard and N. Orantin, *Invariants of algebraic curves and topological expansion*, Commun. Number Theory Phys. **1** (2007), no. 2, 347–452, DOI 10.4310/CNTP.2007.v1.n2.a4. MR2346575

[28] B. Eynard, N. Orantin, *Weil-Petersson volume of moduli spaces, Mirzakhani's recursion and matrix models*, arXiv:0705.3600

[29] B. Eynard, N. Orantin, *Algebraic methods in random matrices and enumerative geometry*, arXiv:0811.3531

[30] Bertrand Eynard and Nicolas Orantin, *Topological recursion in enumerative geometry and random matrices*, J. Phys. A **42** (2009), no. 29, 293001, 117, DOI 10.1088/1751-8113/42/29/293001. MR2519749

[31] B. Eynard and N. Orantin, *Computation of open Gromov-Witten invariants for toric Calabi-Yau 3-folds by topological recursion, a proof of the BKMP conjecture*, Comm. Math. Phys. **337** (2015), no. 2, 483–567, DOI 10.1007/s00220-015-2361-5. MR3339157

[32] Carel Faber, Sergey Shadrin, and Dimitri Zvonkine, *Tautological relations and the r-spin Witten conjecture* (English, with English and French summaries), Ann. Sci. Éc. Norm. Supér. (4) **43** (2010), no. 4, 621–658, DOI 10.24033/asens.2130. MR2722511

[33] Bohan Fang, Chiu-Chu Melissa Liu, and Zhengyu Zong, *The Eynard-Orantin recursion and equivariant mirror symmetry for the projective line*, Geom. Topol. **21** (2017), no. 4, 2049–2092, DOI 10.2140/gt.2017.21.2049. MR3654104

[34] L. Stickelberger and G. F. Frobenius, *Ueber die Differentiation der elliptischen Functionen nach den Perioden und Invarianten* (German), J. Reine Angew. Math. **92** (1882), 311–327, DOI 10.1515/crll.1882.92.311. MR1579906

[35] Alexander B. Givental, *Semisimple Frobenius structures at higher genus*, Internat. Math. Res. Notices **23** (2001), 1265–1286, DOI 10.1155/S1073792801000605. MR1866444

[36] Alexander B. Givental, *Gromov-Witten invariants and quantization of quadratic Hamiltonians* (English, with English and Russian summaries), Mosc. Math. J. **1** (2001), no. 4, 551–568, 645. Dedicated to the memory of I. G. Petrovskii on the occasion of his 100th anniversary. MR1901075

[37] Alexander Givental, *A_{n-1} singularities and $nKdV$ hierarchies* (English, with English and Russian summaries), Mosc. Math. J. **3** (2003), no. 2, 475–505, 743. Dedicated to Vladimir I. Arnold on the occasion of his 65th birthday. MR2025270

[38] Alexander B. Givental, *Symplectic geometry of Frobenius structures*, Frobenius manifolds, Aspects Math., E36, Friedr. Vieweg, Wiesbaden, 2004, pp. 91–112. MR2115767

[39] Claus Hertling, *Frobenius manifolds and moduli spaces for singularities*, Cambridge Tracts in Mathematics, vol. 151, Cambridge University Press, Cambridge, 2002. MR1924259

[40] M. Kazarian, *Deformations of cohomological field theories*, preprint 2007.

[41] M. Kontsevich and Yu. Manin, *Gromov-Witten classes, quantum cohomology, and enumerative geometry*, Comm. Math. Phys. **164** (1994), no. 3, 525–562. MR1291244

[42] Ivan Kostov and Nicolas Orantin, *CFT and topological recursion*, J. High Energy Phys. **11** (2010), 056, 30, DOI 10.1007/JHEP11(2010)056. MR2795103

[43] Y.-P. Lee, *Witten's conjecture, Virasoro conjecture, and invariance of tautological equations*, arXiv:math/0311100.

[44] Y.-P. Lee, *Invariance of tautological equations. I. Conjectures and applications*, J. Eur. Math. Soc. (JEMS) **10** (2008), no. 2, 399–413, DOI 10.4171/JEMS/115. MR2390329

[45] Y.-P. Lee, *Invariance of tautological equations. II. Gromov-Witten theory*, J. Amer. Math. Soc. **22** (2009), no. 2, 331–352, DOI 10.1090/S0894-0347-08-00616-4. With an appendix by Y. Iwao and the author. MR2476776

[46] Y.-P. Lee, *Notes on axiomatic Gromov-Witten theory and applications*, Algebraic geometry—Seattle 2005. Part 1, Proc. Sympos. Pure Math., vol. 80, Amer. Math. Soc., Providence, RI, 2009, pp. 309–323, DOI 10.1090/pspum/080.1/2483940. MR2483940

[47] Danilo Lewanski, Alexandr Popolitov, Sergey Shadrin, and Dimitri Zvonkine, *Chiodo formulas for the r-th roots and topological recursion*, Lett. Math. Phys. **107** (2017), no. 5, 901–919, DOI 10.1007/s11005-016-0928-5. MR3633029

[48] Yuri I. Manin, *Frobenius manifolds, quantum cohomology, and moduli spaces*, American Mathematical Society Colloquium Publications, vol. 47, American Mathematical Society, Providence, RI, 1999. MR1702284

[49] Todor Milanov, *The Eynard-Orantin recursion for the total ancestor potential*, Duke Math. J. **163** (2014), no. 9, 1795–1824, DOI 10.1215/00127094-2690805. MR3217767

[50] Todor Milanov, *The Eynard-Orantin recursion for simple singularities*, Commun. Number Theory Phys. **9** (2015), no. 4, 707–739, DOI 10.4310/CNTP.2015.v9.n4.a3. MR3506789

[51] Paul Norbury, *Counting lattice points in the moduli space of curves*, Math. Res. Lett. **17** (2010), no. 3, 467–481, DOI 10.4310/MRL.2010.v17.n3.a7. MR2653682

[52] Paul Norbury and Nick Scott, *Gromov-Witten invariants of* $\mathbb{P}^1$ *and Eynard-Orantin invariants*, Geom. Topol. **18** (2014), no. 4, 1865–1910, DOI 10.2140/gt.2014.18.1865. MR3268770

[53] A. Okounkov and R. Pandharipande, *Gromov-Witten theory, Hurwitz numbers, and matrix models*, Algebraic geometry—Seattle 2005. Part 1, Proc. Sympos. Pure Math., vol. 80, Amer. Math. Soc., Providence, RI, 2009, pp. 325–414, DOI 10.1090/pspum/080.1/2483941. MR2483941

[54] N. Orantin, *From matrix models' topological expansion to topological string theories: counting surfaces with algebraic geometry, Ph.D. thesis Université Paris 6 – Pierre et Marie Curie*, arXiv:0709.2992.

[55] N. Orantin, *Symplectic invariants, Virasoro constraints and Givental decomposition*, arXiv:0808.0635.

[56] Rahul Pandharipande, Aaron Pixton, and Dimitri Zvonkine, *Relations on* $\overline{\mathcal{M}}_{g,n}$ *via 3-spin structures*, J. Amer. Math. Soc. **28** (2015), no. 1, 279–309, DOI 10.1090/S0894-0347-2014-00808-0. MR3264769

[57] Sergey Shadrin, *BCOV theory via Givental group action on cohomological fields theories* (English, with English and Russian summaries), Mosc. Math. J. **9** (2009), no. 2, 411–429, back matter. MR2568443

[58] Jun S. Song and Yun S. Song, *On a conjecture of Givental*, J. Math. Phys. **45** (2004), no. 12, 4539–4550, DOI 10.1063/1.1808486. MR2105206

[59] Jun S. Song, *Descendant Gromov-Witten invariants, simple Hurwitz numbers, and the Virasoro conjecture for* $\mathbb{P}^1$, Adv. Theor. Math. Phys. **3** (1999), no. 6, 1721–1768 (2000), DOI 10.4310/ATMP.1999.v3.n6.a3. MR1812135

[60] Constantin Teleman, *The structure of 2D semi-simple field theories*, Invent. Math. **188** (2012), no. 3, 525–588, DOI 10.1007/s00222-011-0352-5. MR2917177

[61] Edward Witten, *Algebraic geometry associated with matrix models of two-dimensional gravity*, Topological methods in modern mathematics (Stony Brook, NY, 1991), Publish or Perish, Houston, TX, 1993, pp. 235–269. MR1215968

[62] J. Zhou, *Intersection numbers on Deligne-Mumford moduli spaces and quantum Airy curve*, arXiv:1206.5896.

Faculty of Mathematics, National Research University Higher School of Economics, Usacheva 6, 119048 Moscow, Russia

Email address: ptdbar@hse.ru

Proceedings of Symposia in Pure Mathematics
Volume **100**, 2018
https://doi.org/10.1090/pspum/100/01768

Primary invariants of Hurwitz Frobenius manifolds

P. Dunin-Barkowski, P. Norbury, N. Orantin, A. Popolitov, and S. Shadrin

ABSTRACT. Hurwitz spaces parameterizing covers of the Riemann sphere can be equipped with a Frobenius structure. In this review, we recall the construction of such Hurwitz Frobenius manifolds as well as the correspondence between semisimple Frobenius manifolds and the topological recursion formalism. We then apply this correspondence to Hurwitz Frobenius manifolds by explaining that the corresponding primary invariants can be obtained as periods of multidifferentials globally defined on a compact Riemann surface by topological recursion. Finally, we use this construction to reply to the following question in a large class of cases: given a compact Riemann surface, what does the topological recursion compute?

CONTENTS

1. Introduction

Consider a diagonal flat metric on a complex manifold M with local coordinates $u = (u_1, ..., u_n)$

$$ds^2 = \sum_{i=1}^{N} \eta_i(u) du_i^2 \quad \text{is flat,} \tag{1}$$

2010 *Mathematics Subject Classification.* Primary 53D45; Secondary 32G15, 81T45.
Key words and phrases. Frobenius manifolds, spectral curve topological recursion, Hurwitz spaces.
The first author was partially supported by RFBR grants 16-31-60044-mol-a-dk and 15-01-05990, and by joint RFBR-India grant 16-51-45029-Ind.
The fourth author was partially supported by RFBR grant 16-01-00291.
The first and fourth authors were also partially supported by RFBR grant 15-31-20832-mol-a-ved and by joint RFBR-Japan grant 15-52-50041-YaF.
The third and fourth authors were supported by the Netherlands Organisation for Scientific Research.

generated by a potential $H : M \to \mathbb{C}$

$$\eta_i(u) = \partial_{u_i} H, \quad i = 1, ..., N. \tag{2}$$

Metrics satisfying (1) and (2) are known as Darboux-Egoroff metrics [**6**]. Condition (1) is equivalent to a nonlinear PDE in $\eta_i(u)$ which gives vanishing of the Riemann curvature tensor $R_{ijkl} = 0$. The PDE becomes integrable when condition (2) is added. By a metric we mean a smooth family of complex non-degenerate symmetric bilinear forms on the tangent space $T_m M$, so in particular it is not Riemannian.

Analogous to K. Saito's construction [**24**] of flat coordinates on unfolding spaces of singularities, Dubrovin [**6**] and Krichever [**19**] produced beautiful families of Darboux-Egoroff metrics on moduli spaces of pairs (Σ, x) consisting of an algebraic curve Σ equipped with a meromorphic function $x : \Sigma \to \mathbb{C}$. Such a pair (Σ, x) is a point in a Hurwitz space $H_{g,\mu}$ which parametrises covers $x : \Sigma \to \mathbb{P}^1$ of genus g with points above infinity marked and with fixed ramification profile $(\mu_1, \ldots, \mu_d)$. Further, choose a symplectic basis of cycles $(\mathcal{A}_i, \mathcal{B}_i)_{i=1,\ldots,g}$ on Σ to define a point in a cover $\widetilde{H}_{g,\mu}$ of a Hurwitz space. Namely, $\widetilde{H}_{g,\mu}$ consists of the data of a point in a Hurwitz space together with the data of a Torelli marking. One goes from one sheet of the cover $\widetilde{H}_{g,\mu}$ to another one through the action of modular transformations $Sp(2g, \mathbb{Z})$.

DEFINITION 1.1. Given $(\Sigma, x, \{\mathcal{A}_i, \mathcal{B}_i\}_{i=1,\ldots,g})$ define a set of generalised contours $\mathcal{D}$ on Σ as follows. Choose representatives for $\{\mathcal{A}_i, \mathcal{B}_i\}_{i=1,\ldots,g}$ in $H_1(\Sigma \setminus x^{-1}(\infty))$ and choose a set of relative homology classes $\gamma_i \in H_1(\Sigma, x^{-1}(\infty))$, $i = 2, ..., d$ such that $\gamma_i \subset \Sigma \setminus \{\mathcal{A}_1, ...\mathcal{A}_g, \mathcal{B}_1, ..., \mathcal{B}_g\}$ runs from ∞_i to ∞_1 where the poles of x are given by $x^{-1}(\infty) = \{\infty_1, ..., \infty_d\}$ with respective orders $\{\mu_1, ..., \mu_d\}$. Let $\mathcal{C}_{\infty_i}$, $i = 1, ..., d$ be small circles around each pole ∞_i of x. Then define

$$\mathcal{D} = \{x\mathcal{A}_1, ..., x\mathcal{A}_g, \mathcal{B}_1, ..., \mathcal{B}_g\} \bigcup_{i=2,\ldots,d} \{\gamma_i, x\mathcal{C}_{\infty_i}\} \bigcup_{\substack{j = 1, ..., d, \\ k = 1, ..., \mu_j - 1}} \{x^{k/\mu_j}\mathcal{C}_{\infty_j}\}. \tag{3}$$

If x has only simple poles, so each $\mu_i = 1$, then the contours are built out of classes in $H_1(\Sigma \setminus x^{-1}(\infty))$ and $H_1(\Sigma, x^{-1}(\infty))$. A contour $\mathcal{C}$ acts on a differential ω by $\omega \mapsto \int_{\mathcal{C}} \omega$, and by $x\mathcal{C}$ we mean $\omega \mapsto \int_{x\mathcal{C}} \omega := \int_{\mathcal{C}} x\omega$. We often enumerate the elements of $\mathcal{D}$ by $\mathcal{C}_\alpha \in \mathcal{D}$ for $\alpha = 1, ..., N = |\mathcal{D}|$. Note that $N = \dim H_{g,\mu}$—see (39).

DEFINITION 1.2. On any compact Riemann surface $(\Sigma, \{\mathcal{A}_i\}_{i=1,\ldots,g})$ with a given set of $\mathcal{A}$-cycles, define a *Bergman kernel* $B(p, p')$ to be a symmetric bidifferential with a double pole on the diagonal $\Delta \in \Sigma \times \Sigma$, i.e. $B(p, p') \in H^0(\Sigma \times \Sigma, \pi_1^* K_\Sigma \otimes \pi_2^* K_\Sigma \otimes O(2\Delta))$, uniquely defined by the properties that at the pole it has zero residue in each variable and double residue equal to 1, no further singularities and normalised by $\int_{p \in \mathcal{A}_i} B(p, p') = 0$, $i = 1, ..., g$. It satisfies the Cauchy property for any meromorphic function f on Σ

$$df(p) = \operatorname*{Res}_{p'=p} f(p')B(p, p'). \tag{4}$$

On Σ, choose a Bergman kernel $B(p, p')$ normalised to have zero periods over the $\mathcal{A}$-cycles in the Torelli marking $\{\mathcal{A}_i, \mathcal{B}_i\}$. For any $\mathcal{C}_\alpha \in \mathcal{D}$ define a *primary differential* by

$$\phi_\alpha(p) = \oint_{p' \in \mathcal{C}^*_\alpha} B(p, p') \tag{5}$$

where $\mathcal{C}^*_\alpha$ is a cycle dual to $\mathcal{C}_\alpha$ defined in section 4. Each primary differential is locally holomorphic on $\Sigma \setminus x^{-1}(\infty)$.

Denote by $\mathcal{P}_i \in \Sigma$ the finite critical points of x, i.e. $dx(\mathcal{P}_i) = 0$. For a generic point in $\widetilde{H}_{g,\mu}$ the critical points of x are simple and the critical values $u_i = x(\mathcal{P}_i)$, $i = 1, ..., N$ of x are local coordinates in the open dense domain of $\widetilde{H}^s_{g,\mu} \subset \widetilde{H}_{g,\mu}$ defined by $u_i \neq u_j$ for $i \neq j$ and $\{\mathcal{A}_1, ...\mathcal{A}_g, \mathcal{B}_1, ..., \mathcal{B}_g\}$ avoid $x^{-1}(\infty)$.

Define a metric on $\widetilde{H}^s_{g,\mu}$ by

$$\eta = \sum_{i=1}^{N} du_i^2 \cdot \operatorname*{Res}_{\mathcal{P}_i} \frac{\phi \cdot \phi}{dx} \tag{6}$$

for any choice of primary differential $\phi = \phi_\alpha$ on Σ obtained from $\mathcal{C}_\alpha \in \mathcal{D}$ via (5).

THEOREM 1 (Dubrovin [**6**]). *(i) The metric η defined in* (6) *is flat with local flat coordinates given by*

$$t_\beta = \int_{\mathcal{C}_\beta} \phi, \quad \mathcal{C}_\beta \in \mathcal{D}, \quad \beta = 1, ..., N \tag{7}$$

i.e. the metric is constant with respect to the coordinates t_β.

(ii) The flat metric η forms part of a Frobenius manifold structure on $\widetilde{H}^s_{g,\mu}$ with multiplication on the tangent space of $\widetilde{H}^s_{g,\mu}$ defined using the local basis of vector fields ∂_{u_i} by

$$\partial_{u_i} \cdot \partial_{u_j} = \delta_{ij} \partial_{u_i}. \tag{8}$$

The theorem was proven by Dubrovin in [**6**] using a definition of primary differential via deformations of $y_\alpha dx$—see Lemma 4.4. As stated here we use an equivalent definition of primary differential (5) proven by Shramchenko [**27**].

Recall that a Frobenius manifold M comes equipped with a flat metric η together with a commutative, associative product $\cdot$ on its tangent space satisfying the compatibility condition $\eta(u \cdot v, w) = \eta(u, v \cdot w)$ for all $u, v, w \in T_pM$. Associated to each semisimple point p of a Frobenius manifold M is a cohomological field theory [**14, 16, 20, 29**] defined on $(H, \eta) = (T_pM, \eta|_{T_pM})$, which is a sequence of S_n-equivariant maps

$$I_{g,n} : H^{\otimes n} \to H^*(\overline{\mathcal{M}}_{g,n})$$

that satisfy gluing conditions on boundary divisors in $\overline{\mathcal{M}}_{g,n}$ given explicitly in Section 2.2. For any collection of vectors $v_1, ..., v_n \in T_pM$, the integral $\int_{\overline{\mathcal{M}}_{g,n}} I_{g,n}(v_1 \otimes ... \otimes v_n) \in \mathbb{C}$ (which is a function of $p \in M$) is known as a primary invariant of M.

Recently, [**9**] explained that one can compute the primary (and ancestor) invariants of a semisimple Frobenius manifold efficiently using the topological recursion procedure of [**13**]. The present review explains how this result can be applied to Hurwitz Frobenius manifolds.

The main observation of this text is that just as the flat coordinates can be obtained as periods of a primary differential along cycles taken from $\mathcal{D}$ via (7), the primary invariants of the Hurwitz Frobenius manifolds can also be obtained as periods along cycles taken from $\mathcal{D}$. Since we need multiple insertions of vectors into the primary invariants, we need to take periods of symmetric *multidifferentials* on Σ which are meromorphic sections of $\bigotimes_{i=1}^n \pi_i^* K_\Sigma$ defined over $\Sigma^n = \Sigma \times ... \times \Sigma$.

THEOREM 2. *Given a point $p = (\Sigma, x, \{\mathcal{A}_i, \mathcal{B}_i\}) \in \widetilde{H}^s_{g,\mu}$ and a choice of primary differential dy_α that determines a Frobenius manifold structure on $\widetilde{H}^s_{g,\mu}$, there exist multidifferentials $\omega_{g,n}$ defined on Σ whose periods along contours in $\mathcal{D}$ give the primary invariants of the Frobenius manifold at p. More precisely, for flat coordinates $\{t_1, ..., t_N\}$, put $e_\alpha = \partial_{t_\alpha}$ and define the dual vector with respect to the metric* (6) *by $e^\alpha = \sum_\beta \eta^{\alpha\beta} e_\beta$. Then*

$$\int_{\mathcal{C}_{\alpha_1}} \cdots \int_{\mathcal{C}_{\alpha_n}} \omega_{g,n} = \int_{\overline{\mathcal{M}}_{g,n}} I_{g,n}\Big(e^{\alpha_1} \otimes ... \otimes e^{\alpha_n}\Big)$$

where $\mathcal{C}_{\alpha_i}$ and $e_{\alpha_i} = \partial_{t_{\alpha_i}}$ are related by (7).

Theorem 2 is a consequence of the more general Theorem 3 that proves that the $\omega_{g,n}$ store all ancestor invariants of the Frobenius manifold, using a larger class of cycles than those in $\mathcal{D}$.

REMARK 1.3. The statement and conclusion of Theorem 2 can be made for any point in $\widetilde{H}_{g,\mu}$ not just the semisimple points $\widetilde{H}^s_{g,\mu} \subset \widetilde{H}_{g,\mu}$. It would be interesting to prove the theorem with these weaker hypotheses. There are candidate multidifferentials, such as those defined in [**3**] where the zeros of dx are not required to be simple, or in the case of Dubrovin's superpotential, studied from the perspective of topological recursion in [**8**], which applies to *any* semisimple Frobenius manifold, and where there may be multiple zeros of dx above a critical value.

The multidifferentials $\omega_{g,n}$ in Theorem 2 are obtained from the topological recursion procedure associated to the spectral curve (Σ, x, y_α, B) where $B = B(p, p')$ is the Bergman kernel defined in Definition 1.2 using the Torelli marking and y_α is a function defined on $\Sigma \setminus \{\mathcal{A}_i, \mathcal{B}_i\}$ such that locally $\phi_\alpha = dy_\alpha$. In general [**13**], the $\omega_{g,n}$ are a family of symmetric multidifferentials on the spectral curve that encode solutions of a wide array of problems from mathematical physics, geometry and combinatorics. By a spectral curve[1] we mean the data of (Σ, x, y, B) given by a Riemann surface Σ equipped with a meromorphic function x and a locally defined meromorphic function $y\colon \Sigma \to \mathbb{C}$ such that the zeros of dx given by $\{\mathcal{P}_1, ..., \mathcal{P}_N\}$ are simple and dy is analytic and non-vanishing on $\{\mathcal{P}_1, ..., \mathcal{P}_N\}$, and equipped with a symmetric bidifferential B on $\Sigma \times \Sigma$, with a double pole on the diagonal of zero residue, double residue equal to 1, and no further singularities. The spectral curve may be a collection of N open disks, known as a *local spectral curve*, because $\omega_{g,n}$ are defined using only local information about x, y and B around zeros of dx—see Section 3. On a compact spectral curve we relax the condition on y being globally defined, and instead require that dy is a locally defined meromorphic differential (a connection) ambiguous up to $dy + df(x)$ for any rational function f. This gives rise to a locally defined function y on Σ which is sufficient to apply topological recursion. Note that one can relax the condition that dy is analytic at $\{\mathcal{P}_1, ..., \mathcal{P}_N\}$

[1]The term spectral curve is inherited from the matrix model origin of this formalism. In the general formalism, this term is expected to make sense due to the probable existence of an associated integrable system.

and instead allow dy to have poles there, giving rise to an *irregular* spectral curve [**5**]. We consider only regular spectral curves here.

In [**9, 21**], it was proven that, starting from a semisimple CohFT, or equivalently a semisimple Frobenius manifold M, it is possible to compute its correlation functions by the topological recursion procedure applied to a specific local spectral curve:

(9)

$$\{\text{semisimple CohFT}\} \longrightarrow \{\text{topological recursion applied to a local spectral curve}\}$$

Under this correspondence, whose details are reviewed in Section 3.1, the number of zeros of dx on the local spectral curve (Σ, x, y, B) is equal to the dimension N of the Frobenius manifold M. It was then proven in [**8**] that, under some additional assumptions on the Frobenius manifold M, it is possible to arrange that the image of (9) is a *compact* spectral curve producing the same correlation functions. This compact Riemann surface is given by Dubrovin's superpotential [**6, 7**] which is a family of compact Riemann surfaces parametrised by the semisimple points of M and constructed out of flat coordinates of a pencil of metrics on M.

We can now try to reverse the direction of the arrow in (9). Given a *compact* spectral curve, when does it lie in the image of (9), and can we reconstruct the corresponding CohFT (or, equivalently, the Frobenius manifold)? The following theorem answers this question. It begins with the observation that a compact spectral curve gives a point (Σ, x) in a Hurwitz space $H_{g,\mu}$.

THEOREM 3. *Given a generic point $(\Sigma, x) \in H_{g,\mu}$, equip it with a bidifferential B normalised over a given set of $\mathcal{A}$-cycles, and choose $\mathcal{C} \in \mathcal{D}$ to define a locally defined function y on Σ by $dy(p) := \oint_{\mathcal{C}} B(p, p')$. The topological recursion procedure applied to the spectral curve (Σ, x, y, B) computes the ancestor invariants of the CohFT associated to Dubrovin's Frobenius manifold structure on the cover $\widetilde{H}^s_{g,\mu}$ via:*

$$\omega_{g,n}(p_1, ..., p_n) = \sum_{\substack{i_1,...,i_n \\ d_1,...,d_n}} \int_{\overline{\mathcal{M}}_{g,n}} I_{g,n}(e_{i_1}, \ldots, e_{i_n}) \prod_{j=1}^{n} \psi_j^{d_j} \cdot \bigotimes_{j=1}^{n} V_{d_j}^{i_j}(p_j) \tag{10}$$

where $V_k^i(p)$, $i = 1, ..., N$, $k = 0, 1, ...,$ are canonical differentials on Σ defined by (36) *in Section* 3.1.

The proof of Theorem 3—contained in Section 4.3—is a simple combination of results from [**7, 9, 15, 28**] which are reviewed in this paper. The main tool in the proof is the map (9) from [**9**] which shows how topological recursion relates to Givental's construction [**15**] of the total ancestor potential associated to each semisimple point of a Frobenius manifold. To apply the reverse construction of (9) one needs a specific relationship between the Bergman kernel B on the spectral curve and the R-matrix of the Frobenius manifold which is proven in [**28**].

REMARK 1.4. Theorem 3 also answers the following question. Given a compact spectral curve, what does the topological recursion procedure compute? For a large class of spectral curves—where B and y are determined almost canonically by Σ and x—the answer is that it produces generating functions for ancestor invariants of a Hurwitz space to which the branched cover underlying the spectral curve belongs. In particular, it completes the picture drawn by Zhou in [**30**] for relating Frobenius manifolds and spectral curves. Even when the spectral curve does not belong to this class, theorem 9 shows that the topological recursion is equivalent to the Givental reconstruction formula for an R-matrix similar to the one coming from a Frobenius manifold.

REMARK 1.5. Theorem 2 concerns only the primary invariants in (10), corresponding to $d_j = 0$, $j = 1, ..., n$. One can also construct generalised contours $\mathcal{C}_{\alpha,k} = p_k(x)\mathcal{C}_\alpha$, for $\mathcal{C}_\alpha \in \mathcal{D}$ and $p_k(x) = x^k + ...$ a monic polynomial of degree k in x, so that the ancestor invariants, corresponding to $d_j \geq 0$, appear as periods thus generalising Theorem 2:

$$\int_{\mathcal{C}_{\alpha_1,k_1}} \cdots \int_{\mathcal{C}_{\alpha_n,k_n}} \omega_{g,n} = \int_{\overline{\mathcal{M}}_{g,n}} I_{g,n}\left(e^{\alpha_1} \otimes ... \otimes e^{\alpha_n}\right) \cdot \prod_{j=1}^{n} \psi_j^{k_j}. \tag{11}$$

Theorems 2 and 3 enable one to generate primary invariants and all ancestor invariants of $\widetilde{H}^s_{g,\mu}$ of all genera. Previously only genus 0 and genus 1 primary invariants were known. Theorems 2 and 3 also have applications to the topological recursion procedure. Using the generalised contours in $\mathcal{D}$ one gets a direct map from $\omega_{g,n}$ to primary invariants via integration over the cycles.

The paper is organised as follows. In Sections 2 and 3, we remind the reader of the general theory of Frobenius manifolds and topological recursion, as well as the correspondence between the two, following [**9**]. In Section 4, we describe the construction of Dubrovin of Frobenius manifold structures on covers of Hurwitz spaces and prove Theorems 2 and 3. In Section 5, we discuss an extension of the results to non-semisimple points and spectral curves which do not come from a Frobenius manifold. In Section 6, we revisit some of the variational formulae satisfied by the differentials $\omega_{g,n}$ from the point of view of Hurwitz Frobenius manifolds.

Acknowledgments. The authors would like to thank Gaëtan Borot for fruitful discussions on the subject of this paper and the referees for useful comments. N.O. would like to thank the KdV-Institute where part of this work was done. P.N. would like to thank Ludwig-Maximillians-Universitát, Munich and the Max-Planck-Insititut für Mathematik, Bonn for hosting him while part of this work was carried out.

2. Frobenius manifolds

In this section we give a short introduction to Frobenius manifolds. An important construction for this paper is Givental's R-matrix defined in Section 2.3.

2.1. Frobenius manifold.

DEFINITION 2.1. A Frobenius algebra $(H, \eta, \cdot)$ is a finite-dimensional vector space H equipped with a metric $\eta = \langle\ ,\ \rangle$ and a commutative, associative product $\cdot$ satisfying $\langle u \cdot v, w\rangle = \langle u, v \cdot w\rangle$.

EXAMPLE 2.2.

$$H \cong \mathbb{C} \oplus \mathbb{C} \oplus ... \oplus \mathbb{C}, \quad \langle e_i, e_j\rangle = \delta_{ij}\eta_i, \quad e_i \cdot e_j = \delta_{ij} e_i$$

for any $\eta_i \in \mathbb{C}\backslash\{0\}$, $i = 1, ..., N$ and where $\{e_i\}$ is the standard basis. This Frobenius algebra is *semisimple*—as a module over itself it decomposes into simple modules. Conversely, any *semisimple* Frobenius algebra is isomorphic to this example, and is determined uniquely by N non-zero complex numbers $\{\eta_i\}$.

A Frobenius manifold is defined by the data of a Frobenius algebra on the tangent space at each point of the manifold and such that the metric is flat. In terms of flat coordinates $\{t^\alpha\}$ a Frobenius manifold can be defined locally as follows. Consider a function $F(t^1,\dots,t^N)$ defined on a ball $B\subset\mathbb{C}^N$ and a constant inner product $\eta^{\alpha\beta}$ such that the triple derivatives of F with one raised index,

$$C^\gamma_{\alpha\beta} := \frac{\partial^3 F}{\partial t^\alpha \partial t^\beta \partial t^\lambda}\eta^{\lambda\gamma}, \tag{12}$$

are the structure constants of a commutative associative Frobenius algebra with the scalar product given by $\eta_{\alpha\beta}$. We can think about this structure as defined on the tangent bundle of $B\subset\mathbb{C}^N$ (and we denote the corresponding multiplication of vector field by $\cdot$), and we require that ∂_{t^1} is the unit of the algebra in each fibre.

REMARK 2.3. We use the following summation convention standard in the theory of Frobenius manifolds: we always take the sum over the repeated Greek indices, as, for instance, in Equation (12), and we never assume summation over the repeated Latin indices, as, for instance, in Equation (15).

We further consider structures (almost) homogeneous under a vector field $E := \sum_{\alpha=1}^N((1-q_\alpha)t^\alpha+r_\alpha)\partial_{t^\alpha}$, where q_α and r_α are constants for $\alpha=1,\dots,N$, satisfying $q_1=0$ and $r_\alpha\neq 0$ only in the case $1-q_\alpha=0$. We require that there exists a constant d such that $E.F-(3-d)F$ is a polynomial of order at most 2 in $t^1,\dots,t^N$.

The triple (F,η,E) that satisfies all conditions above gives us the structure of a (conformal) Frobenius manifold of rank N and conformal dimension d with flat unit. The function F is called the prepotential; the vector field E is called the Euler vector field. The coordinate-free description of this structure requires a flat metric with associated Levi-Civita connection, unit and Euler vector fields satisfying compatibility conditions—see [**6**] for details.

In this paper we only consider semisimple Frobenius manifolds, that is, we require that the algebra structure at each point on an open subset $B^{ss}\subset B$ is semisimple hence isomorphic to Example 2.2. In a neighborhood of a semisimple point we have a system of canonical coordinates $u_1,\dots,u_N$, defined up to permutations, such that the vector fields ∂_{u_i}, $i=1,\dots,N$, are the idempotents of the algebra product, and the Euler vector field has the form $E=\sum_{i=1}^N u_i\partial_{u_i}$. This gives rise to two important systems of coordinates: flat coordinates, leading to a fixed metric and varying product, and canonical coordinates, leading to a fixed product and varying metric. With respect to the canonical coordinates, the flat metric on M is diagonal with diagonal terms generated by a potential $H: M\to\mathbb{C}$ via (2) which satisfies (1).

Define the rotation coefficients

$$\beta_{ij} = \frac{\partial_{u_j}\eta_i}{\sqrt{\eta_i\eta_j}}. \tag{13}$$

Then (1) and (2) imply that $\beta_{ij}(u)$ satisfy the Darboux-Egoroff system

$$\beta_{ij} = \beta_{ji}, \tag{14}$$

$$\partial_{u_k}\beta_{ij} = \beta_{ik}\beta_{jk}. \tag{15}$$

Flatness of the identity and conformality imply

$$\sum_k \partial_{u_k}\beta_{ij} = 0, \tag{16}$$

$$\sum_k u_k\partial_{u_k}\beta_{ij} = -\beta_{ij}. \tag{17}$$

Assemble the rotation coefficients into a symmetric $N \times N$ matrix $\Gamma = \Gamma(u)$—whose diagonal is not a part of the structure—by $\Gamma_{ij} = \beta_{ij}$. Then equations (15-17) are equivalent to the Darboux-Egoroff equation

$$d[\Gamma, U] = [[\Gamma, U], [\Gamma, dU]] \tag{18}$$

where $U = \mathrm{diag}(u_1, \ldots, u_N)$.

The rotation coefficients give less information than the metric, i.e. there are different solutions of (1) and (2) that give rise to the same rotation coefficients. The system

$$\partial_{u_j}\psi_i = \beta_{ij}\psi_j, \quad i \neq j$$

$$\sum_{j=1}^{N} \partial_{u_j}\psi_i = 0, \quad i = 1, \ldots, N$$

has an N-dimensional space of solutions $\psi = (\psi_1(u), \ldots, \psi_N(u))$ which enables one to retrieve a metric for each solution from the rotation coefficients. Put N independent solutions of this system into the columns of a matrix Ψ, so the system becomes

$$d\Psi = [\Gamma, dU]\Psi. \tag{19}$$

The canonical coordinate vector fields ∂_{u_i} are orthogonal but not orthonormal. We can normalise them to produce a so-called normalised canonical frame in each tangent space, that is, if $\eta_i = \eta(\partial_{u_i}, \partial_{u_i})$, then the orthonormal basis is given by $\partial_{v_i} := \eta_i^{-1/2}\partial_{u_i}$, $i = 1, \ldots, N$. The matrix Ψ in (19) is the transition matrix from the flat basis to the normalised canonical basis.

2.2. Cohomological field theory. A *cohomological field theory* is a pair (H, η) composed of a finite-dimensional complex vector space H equipped with a metric η and a sequence of S_n-equivariant maps.

$$I_{g,n} : H^{\otimes n} \to H^*(\overline{\mathcal{M}}_{g,n})$$

that satisfy compatibility conditions from inclusion of strata:

$$\psi : \overline{\mathcal{M}}_{g-1,n+2} \to \overline{\mathcal{M}}_{g,n}, \quad \phi_I : \overline{\mathcal{M}}_{g_1,|I|+1} \times \overline{\mathcal{M}}_{g_2,|J|+1} \to \overline{\mathcal{M}}_{g,n}, \quad I \sqcup J = \{1, \ldots, n\}$$

given by

$$\phi_I^* I_{g,n}(v_1 \otimes \ldots \otimes v_n) = I_{g_1,|I|+1} \otimes I_{g_2,|J|+1}\Big(\bigotimes_{i\in I} v_i \otimes \Delta \otimes \bigotimes_{j\in J} v_j\Big) \tag{20}$$

$$\psi^* I_{g,n}(v_1 \otimes \ldots \otimes v_n) = I_{g-1,n+2}(v_1 \otimes \ldots \otimes v_n \otimes \Delta) \tag{21}$$

where $\Delta \in H \otimes H$ is dual to the metric. In local coordinates it is given by $\Delta = \eta^{\alpha\beta} e_\alpha \otimes e_\beta$.

The metric $\eta = \langle \cdot, \cdot \rangle$ and the 3-point function $I_{0,3}$ induce a product $\cdot$ on H via

$$\langle u \cdot v, w \rangle = I_{0,3}(u, v, w) \in H^*(\overline{\mathcal{M}}_{0,3}) \cong \mathbb{C}.$$

The metric makes H into a Frobenius algebra. We define a CohFT to be semisimple if its associated Frobenius algebra is semisimple.

Correlators, or ancestor invariants, of the CohFT make use of the Chern classes $\psi_j = c_1(L_j)$ of the tautological line bundles $L_j, j = 1, ..., n$ over $\overline{\mathcal{M}}_{g,n}$. The correlators are defined by:

$$\langle\tau_{k_1}(e_{\nu_1})...\tau_{k_n}(e_{\nu_n})\rangle_g := \int_{\overline{\mathcal{M}}_{g,n}} I_{g,n}(e_{\nu_1}, ..., e_{\nu_n}) \cdot \prod_{j=1}^{n} \psi_j^{k_j} \tag{22}$$

for $k_i \in \mathbb{N}$, $\{e_\nu, {}_{\nu=1,...,N}\} \subset H$. When $k_i = 0$, $i = 1, ..., n$ the ancestor invariants are also known as *primary invariants* of the CohFT.

Givental [**15**] introduced a group action on genus 0 potentials of a CohFT, and used it to propose a formula for higher genera. Faber, Zvonkine and the last author [**14**] proved that the higher genera formula satisfies all properties that might be imposed on correlators of CohFT, hence the Givental group acts on partition functions of CohFTs in all genera. The interpretation of the action on correlators as an action on cohomology classes was constructed by several people independently, namely, by Teleman [**29**], Katzarkov-Kontsevich-Pantev (unpublished), and Kazarian (unpublished)—see [**25**]. There is a good account of this action on cohomology classes by Pandharipande-Pixton-Zvonkine [**22**]. Hence we can associate a CohFT to a semisimple point of a Frobenius manifold. Conversely a CohFT gives rise to a Frobenius manifold structure on (a neighborhood inside) H using the constant metric η as the flat metric and a varying family of products using $I_{0,n}$ in place of $I_{0,3}$. See [**20**] for details.

If the Frobenius manifold has flat identity—meaning that the identity vector field for the product on the tangent bundle is parallel with respect to the Levi-Civita connection of the flat metric η—then this is realised on the CohFT level by an extra relation involving the forgetful map

$$\pi : \overline{\mathcal{M}}_{g,n+1} \to \overline{\mathcal{M}}_{g,n}$$

given by

$$I_{g,n+1}(v_1 \otimes \cdots \otimes v_n \otimes \mathbb{1}) = \pi^* I_{g,n}(v_1 \otimes \cdots \otimes v_n), \quad I_{0,3}(v_1 \otimes v_2 \otimes \mathbb{1}) = \eta(v_1 \otimes v_2) \tag{23}$$

where $\mathbb{1}$ is the unit vector for the product.

2.3. Classification of semisimple cohomological field theories. The Givental-Teleman theorem [**15, 29**] states that a semisimple CohFT is equivalent to the pair (H, η) together with a so-called R-matrix. An R-matrix

$$R(z) = \sum_{k=0}^{\infty} R_k z^k$$

is a formal series whose coefficients are $N \times N$ matrices where $N = \dim H$ is the rank of the Frobenius manifold. Givental used $R(z)$ to produce a differential operator, a so-called quantisation of $R(z)$, which acts on a known tau-function to produce a generating series for the correlators of the CohFT.

The coefficients R_k are defined using Ψ, the transition matrix from flat coordinates to normalised canonical coordinates determined by (19), via $R_0 = I$ and the inductive equation

$$d\left(\Psi^{-1} R(z)\right) = \Psi^{-1} \frac{[dU, R(z)]}{z} \tag{24}$$

which uniquely determines $R(z)$ up to right multiplication by a diagonal matrix $D(z)$ independent of u with $D(0) = I$. We recall that this equation is a consequence of the fact that $R(z)$ is the regular part of the expansion of the solution of a linear system associated by Dubrovin to any semisimple Frobenius manifold around its essential singularity (see for example [**6**, Lecture 3] or [**7**, Section 2 and 3] for more details).

Using $d\Psi = [\Gamma, dU]\,\Psi$ one can write

$$d\left(\Psi^{-1}R(z)\right) = \Psi^{-1}dR(z) + d[\Psi^{-1}]R(z) = \Psi^{-1}dR(z) - \Psi^{-1}\left[\Gamma, dU\right]R(z).$$

Together with equation (24) and the invertibility of Ψ, this gives

$$dR(z) = \frac{[dU, R(z)]}{z} + [\Gamma, dU]\,R(z) \tag{25}$$

This re-expresses the equation for $R(z)$ in terms of the rotation coefficients, which uses less information than the full metric, encoded in Ψ. Since $dU(\mathbb{1}) = I$, an immediate consequence of (25) is

$$\mathbb{1} \cdot R(z) = 0. \tag{26}$$

If the theory is homogenous, then invariance under the action of the Euler field

$$(z\partial_z + E) \cdot R(z) = 0 \tag{27}$$

fixes the diagonal ambiguity in $R(z)$.

The first non-trivial term R_1 of $R(z)$ is given by the rotation coefficients

$$R_1 = \Gamma. \tag{28}$$

This follows from comparing the constant (in z) term in (24) which is

$$d\Psi = [R_1, dU]\Psi$$

to equation (19) given by $d\Psi = [\Gamma, dU]\,\Psi$. Since dU is diagonal with distinct diagonal terms, we see that (28) holds for off-diagonal terms, and the ambiguity in the diagonal term for both is unimportant—it can be fixed in R_1 by (24) together with $E \cdot R_1 = -R_1$.

3. Topological recursion and cohomological field theory

In this section, we give a brief overview of topological recursion defined in [**13**]. Consider a Riemann surface Σ equipped with meromorphic functions $x, y\colon \Sigma \to \mathbb{C}$ such that the zeros of dx, given by $\{\mathcal{P}_1, ..., \mathcal{P}_N\}$ are simple and dy is analytic and non-vanishing on $\{\mathcal{P}_1, ..., \mathcal{P}_N\}$. Let B be a Bergman kernel on $\Sigma \times \Sigma$ as in definition 1.2.

Define a sequence of symmetric multidifferentials $\omega_{g,n}(p_1, \ldots, p_n)$ on $\Sigma^{\times n}$ by the following recursion:

$$\omega_{0,1}(p) := y(p)dx(p); \tag{29}$$

$$\omega_{0,2}(p_1, p_2) := B(p_1, p_2); \tag{30}$$

$$\begin{aligned}\omega_{g,n+1}(p_0, p_1, \ldots, p_n) :=& \\ &\sum_{i=1}^{N} \operatorname*{Res}_{p=\mathcal{P}_i} \frac{\int_p^{\sigma_i(p)} \omega_{0,2}(\bullet, p_0)}{2(\omega_{0,1}(\sigma_i(p)) - \omega_{0,1}(p))} \tilde{\omega}_{g,2|n}(p, \sigma_i(p)|p_1, \ldots, p_n),\end{aligned} \tag{31}$$

where σ_i is the local involution defined by x near the point $\mathcal{P}_i$, $i = 1, \dots, N$, and $\tilde{\omega}_{g,2|n}$ is defined by the following formula:

$$\begin{aligned}\tilde{\omega}_{g,2|n}(p', p''|p_1, \dots, p_n) &:= \omega_{g-1,n+2}(p', p'', p_1, \dots, p_n) \\ &+ \sum_{\substack{g_1+g_2=g \\ I_1 \sqcup I_2=\{1,\dots,n\} \\ 2g_1-1+|I_1|\geq 0 \\ 2g_2-1+|I_2|\geq 0}} \omega_{g_1,|I_1|+1}(p', p_{I_1})\omega_{g_2,|I_2|+1}(p'', p_{I_2}).\end{aligned} \tag{32}$$

Here we denote by p_I the sequence $p_{i_1}, \dots, p_{i_{|I|}}$ for $I = \{i_1, \dots, i_{|I|}\}$.

REMARK 3.1. The recursion was defined on so-called *local* spectral curves in [**10**] as follows. Consider small neighborhoods $U_i \subset \Sigma$ of the points $\mathcal{P}_i$. If we look at just the restrictions of $\omega_{g,n}$ to the products of these disks, $U_{i_1} \times \cdots \times U_{i_n}$, we can still proceed by topological recursion, using as an input the restrictions of $\omega_{0,1}$ to U_i, $i = 1, \dots, N$, and $\omega_{0,2}$ to $U_i \times U_j$, $i, j = 1, \dots, N$. Indeed, Equation (31) uses only local expansion data around the points $\mathcal{P}_i$. Hence, the word local refers to the unique knowledge of these local data.

REMARK 3.2. In the topological recursion on a compact spectral curve we also allow y to be the (multivalued) primitive of a differential ω on Σ. The ambiguity in y consists of periods and residues of ω and hence the ambiguity is locally constant. Since y appears in the recursion formula (32) only via $y(\sigma_i(p)) - y(p)$ (and there are no poles of ω at the zeros of dx) the locally constant ambiguity disappears and the recursion is well-defined. We go even further and allow ω to be a locally defined meromorphic differential (a connection) ambiguous up to $dy + df(x)$ for any rational function f. In this case the ambiguity $y \mapsto y + f(x)$ is no longer constant, but again $y(\sigma_i(p)) - y(p)$ is unchanged.

3.1. Topological recursion from CohFTs. We recall the relation (9) of topological recursion on a local spectral curve to the Givental formulae for cohomological field theories obtained in [**9**].

DEFINITION 3.3. For a Riemann surface equipped with a meromorphic function (Σ, x) we define evaluation of any meromorphic differential ω at a simple zero $\mathcal{P}$ of dx by

$$\omega(\mathcal{P}) := \operatorname*{Res}_{p=\mathcal{P}} \frac{\omega(p)}{\sqrt{2(x(p) - x(\mathcal{P}))}}$$

where we choose a branch of $\sqrt{x(p) - x(\mathcal{P})}$ once and for all at each $\mathcal{P}$ to remove the ± 1 ambiguity.

THEOREM 4. [**9**] *Given a semisimple CohFT presented via the R-matrix* $R(z) = \sum_{k=0}^{\infty} R_k z^k$ *and constants* $\eta_1, \dots, \eta_N$ *define a local spectral curve* (Σ, x, y, B)*, presented as (the Laplace transform of) local series for* $dy(p)$ *and* $B(p, p')$ *around each zero* $p = \mathcal{P}_i$, $p' = \mathcal{P}_j$ *of* dx *(which is locally canonical) as follows:*

$$\left[R^{-1}(z)\right]_i^j = -\frac{\sqrt{z}}{\sqrt{2\pi}} \int_{\Gamma_j} B(\mathcal{P}_i, p) \cdot e^{\frac{(u_j - x(p))}{z}} \tag{33}$$

$$\sum_{k=1}^{N} \left[R^{-1}(z)\right]_k^i \cdot \eta_k^{1/2} = \frac{1}{\sqrt{2\pi z}} \int_{\Gamma_i} dy(p) \cdot e^{\frac{(u_i - x(p))}{z}} \tag{34}$$

(35)

$$\frac{1}{2\pi\sqrt{z_1z_2}}\int_{\Gamma_i}\int_{\Gamma_j} B(p_1,p_2)e^{\frac{(u_i-x(p_1))}{z_1}+\frac{(u_j-x(p_2))}{z_2}} = -\frac{\sum_{k=1}^N\left[R^{-1}(z_1)\right]_k^i\left[R^{-1}(z_2)\right]_k^j}{z_1+z_2}$$

where Γ_i is a path containing $u_i = x(\mathcal{P}_i)$. Here the row and column indices are given by the superscript, respectively subscript, indices. Then the multidifferentials $\omega_{g,n}(p_1,...,p_n)$ obtained via topological recursion applied to the local spectral curve (Σ, x, y, B) are polynomials in differentials $V_k^i(p_j)$ defined by

$$V_0^i(p) = B(\mathcal{P}_i, p), \quad V_{k+1}^i(p) = d\left(\frac{V_k^i(p)}{dx(p)}\right), \ k = 0, 1, 2, ... \tag{36}$$

with coefficients given by ancestor invariants of the CohFT:

$$\omega_{g,n}(p_1,...,p_n) = \sum_{\substack{i_1,...,i_n \\ d_1,...,d_n}} \int_{\overline{\mathcal{M}}_{g,n}} I_{g,n}(e_{i_1},\ldots,e_{i_n}) \prod_{j=1}^n \psi_j^{d_j} \cdot \bigotimes_{j=1}^n V_{d_j}^{i_j}(p_j).$$

REMARK 3.4. The spectral curve thus obtained is local, i.e. a collection of open sets U_i each containing a unique zero $\mathcal{P}_i$ of dx. Thus Γ_i is defined only locally, which is fine since we are interested only in the asymptotic expansion for R around $z = 0$. Let us also remind the reader that this result is valid for any semisimple Frobenius manifold. We shall see in the next section that, in the case of Hurwitz Frobenius manifolds, one can make these Laplace transform globally well-defined by choosing carefully the integration cycles to consider.

REMARK 3.5. This data (the constants η_i and the matrix $R(z)_i^j$) determine for us a semisimple CohFT $\{I_{g,n}\}$ with an N-dimensional space of primary fields $V := \langle e_1, \ldots, e_N\rangle$ corresponding to a chosen point $(u_1, ..., u_N)$ on a Frobenius manifold—see Section 2.3. In terms of the underlying Frobenius manifold structure, the basis $e_1, \ldots, e_N$ corresponds to the normalised canonical basis.

REMARK 3.6. Note that the limit of (34) at $z = 0$ yields:

$$\eta_i^{1/2} = dy(\mathcal{P}_i) \tag{37}$$

which tells us that dy encodes the metric.

REMARK 3.7. Compatibility of (33) and (35) is a condition on the bidifferential B, not satisfied in general, nevertheless always satisfied if the spectral curve is compact and the differential dx is meromorphic. Compatibility for compact spectral curves uses a general finite decomposition for $B(p_1, p_2)$ proven by Eynard in Appendix B of [**11**] together with (33). This is recalled in section 5.1.

Theorem 4 produces a map

$$\{\text{semisimple CohFT}\} \longrightarrow \{\text{topological recursion applied to a local spectral curve}\}$$

with image consisting of spectral curves with B and y necessarily satisfying compatibility conditions—compatibility of (33), (34) and (35). A general spectral curve will not satisfy such compatibility conditions, i.e. in general one can choose B and y independently. For example, the rational spectral curve $(\mathbb{P}^1, x, y, B)$ for $x = z+1/z$, $B = dzdz'/(z-z')^2$, $dy = z^m dz$, $m \in \{-1, 0, 1, 2, ...\}$, lies in the image of the map only for $m = -1$ or 0.

Compatibility of (33) and (35) is discussed in Remark 3.7 and compatibility of (33) and (34) is characterised by the following theorem.

THEOREM 5 ([**8**]). *Equations* (33) *and* (34) *are compatible (as equations for the unknown variables* R^{-1} *and* η_i*),* $i = 1, \ldots, N$*) if and only if the 1-form*

$$\omega(p) = d\left(\frac{dy}{dx}(p)\right) + \sum_{i=1}^{N} \operatorname*{Res}_{p'=\mathcal{P}_i} \frac{dy}{dx}(p')B(p,p'). \tag{38}$$

is invariant under each local involution σ_i*,* $i = 1, \ldots, N$*.*

The characterisation in Remark 3.7 and Theorem 5 allows a *converse* construction of semisimple CohFTs from compact spectral curves. The following is a sufficient condition for compatibility of (33) and (34).

DEFINITION 3.8. We say that a compact spectral curve (Σ, x, y, B) is *dominant* if x and dy are meromorphic and the poles of dx dominate the poles of dy.

COROLLARY 3.9. *A dominant compact spectral curve* (Σ, x, y, B) *lies in the image of* (9) *and hence gives rise to a semisimple CohFT.*

PROOF. Any Bergman kernel satisfies the Cauchy property (4). If the poles of dx dominate the poles of dy then dy/dx has poles only at the zeros $\mathcal{P}_i$ of dx. Then $\omega(p) \equiv 0$ since

$$\sum_{i=1}^{N} \operatorname*{Res}_{p'=\mathcal{P}_i} \frac{dy}{dx}(p')B(p,p') = -\operatorname*{Res}_{p'=p} \frac{dy}{dx}(p')B(p,p') = -d\left(\frac{dy}{dx}(p)\right)$$

and hence it is invariant under each local involution σ_i. Since the Riemann surface Σ is compact it automatically satisfies (35) hence the claim is proven. $\square$

REMARK 3.10. In fact Corollary 3.9 allows a weaker hypothesis which we will need. We can instead allow dy to be a locally defined meromorphic differential, essentially a connection, which is ambiguous up to $dy \mapsto dy + \lambda dx$. The conclusion of Corollary 3.9 still holds since $d\left(\frac{dy}{dx}\right)$ is globally defined.

4. Hurwitz Frobenius manifolds

In this section we first remind the reader of Dubrovin's construction of a Frobenius manifold on a cover of Hurwitz space and then prove a number of deformation lemmas, which will be useful in the following sections.

4.1. Dubrovin's construction. As defined in the introduction, denote by $\widetilde{H}_{g,\mu}$ the moduli space of tuples $(\Sigma, x, \{\mathcal{A}_i, \mathcal{B}_i\}_{i=1,\ldots,g})$ consisting of covers $x : \Sigma \to \mathbb{P}^1$ of genus g with fixed ramification profile above infinity $\mu = (\mu_1, \ldots, \mu_d)$ together with a choice of a symplectic basis of cycles $(\mathcal{A}_i, \mathcal{B}_i)_{i=1,\ldots,g}$ and marked branches of x at each point above ∞.

Given such a generic cover x, we denote its simple branch points by

$$u_i = x(\mathcal{P}_i) \quad \text{where} \quad dx|_{\mathcal{P}_i} = 0, \quad i = 1, \ldots, N.$$

By the Riemann-Hurwitz formula:

$$N = 2g - 2 + d + |\mu|$$

and since an element of the Hurwitz space is defined up to a finite information by its critical values, this gives the dimension

$$\dim_{\mathbb{C}} \widetilde{H}_{g,\mu} = 2g - 2 + d + |\mu|.$$

In the introduction we claimed that

$$\#\mathcal{D} = N = \#\{p \mid dx(p) = 0\} \tag{39}$$

i.e. the number of generalised contours, defined by $\mathcal{D}$ in (3), coincides with $\dim_{\mathbb{C}} \widetilde{H}_{g,\mu}$. This follows from the fact that dx is a meromorphic differential so its divisor $(dx) = Z - P$ has degree $2g-2$, where Z and P are the zeros and poles of dx. Hence

$$\dim_{\mathbb{C}} \widetilde{H}_{g,\mu} = |Z| = 2g - 2 + |P| = \dim H_1(\Sigma, x^{-1}(\infty)) - 1 + \sum_{i=1}^{d} \mu_i = \#\mathcal{D}.$$

The last equality is clear since the elements of $\mathcal{D}$ consist of firstly $\{x\mathcal{A}_1, \dots, x\mathcal{A}_g, \mathcal{B}_1, \dots, \mathcal{B}_g, \gamma_2, \dots, \gamma_d\}$ which has cardinality equal to $\dim H_1(\Sigma, x^{-1}(\infty)) = 2g - 1 + d$, together with $-1 + \sum_{i=1}^{d} \mu_i = |P| - d - 1$ extra elements that are given by $x^{k/\mu_i}\mathcal{C}_{\infty_i}$ for $i = 1, \dots, d$ and $k = 1, \dots, \mu_i$, except for $i = 1$, in which case $k = \mu_1$ is not included. This makes in total $\#\mathcal{D} = 2g - 2 + d + |\mu|$ as required by the Riemann-Hurwitz formula.

We use the critical values u_i as local coordinates in an open dense domain of $\widetilde{H}^s_{g,\mu} \subset \widetilde{H}_{g,\mu}$ where $u_i \neq u_j$ for $i \neq j$. The vector fields ∂_{u_i} give a basis of $T\widetilde{H}_{g,\mu}$ and define a multiplication $\cdot$ given by:

$$\partial_{u_i} \cdot \partial_{u_j} = \delta_{ij}\partial_{u_i}. \tag{40}$$

We denote the unity and the Euler vector fields:

$$e = \sum_{i=1}^{N} \partial_{u_i}, \quad E = \sum_{i=1}^{N} u_i \partial_{u_i}. \tag{41}$$

Let us now define one-forms on $\widetilde{H}_{g,\mu}$. For any quadratic differential Q on Σ, define the one-form

$$\Omega_Q = \sum_{i=1}^{N} du_i \operatorname*{Res}_{p=\mathcal{P}_i} \frac{Q(p)}{dx}.$$

Dubrovin defines a set of differentials ϕ on Σ, defined in (5) and described in more detail below, which have poles dominated by the poles of dx. They are known as *primary differentials* and used to produce a quadratic differential $Q = \phi^2$.

THEOREM 6 (Dubrovin [**6**]). *For any primary differential ϕ, $\widetilde{H}^s_{g,\mu} \bigcap \{u \mid \phi(\mathcal{P}_i) \neq 0\}$ is equipped with a structure of a Frobenius manifold with multiplication* (40), *unity and Euler vector fields* (41) *and metric*

$$\eta := \sum_{i=1}^{N} du_i^2 \cdot \operatorname*{Res}_{p=\mathcal{P}_i} \frac{\phi^2(p)}{dx(p)} = \sum_{i=1}^{N} du_i^2 \cdot \phi(\mathcal{P}_i)^2 \tag{42}$$

In addition, the corresponding flat coordinates $(t_\alpha)_{\alpha=1,\dots,N}$ can be explicitly written in terms of periods of ϕ via

$$t_\alpha = \int_{\mathcal{C}_\alpha} \phi$$

for any $\mathcal{C}_\alpha \in \mathcal{D}$.

REMARK 4.1. In the Equation (42) in the statement of Theorem 6 we used the notation of Definition 3.3 for the evaluation of a one-form at a point. In particular, it is easy to check the second equality in (42) by a computation of the residues in local coordinates near the points $\mathcal{P}_i$, $i = 1, \ldots, N$.

This means that the data of such a Frobenius manifold structure on $\widetilde{H}^s_{g,\mu}$ is given by the choice of a primary differential ϕ. The definition of a primary differential uses the Torelli marking of Σ as follows. Fix a point in $\widetilde{H}_{g,\mu}$, i.e. a pair (Σ, x) (a point in a Hurwitz space) together with a basis $(\mathcal{A}_i, \mathcal{B}_i)_{i=1,\ldots,g}$ (a sheet of $\widetilde{H}_{g,\mu}$ seen as a cover). Recall from the introduction that there is a unique Bergman Kernel $B(p,p')$ which is a bidifferential of the second kind normalised to have zero periods over the $\mathcal{A}$-cycles in the Torelli marking $\{\mathcal{A}_i, \mathcal{B}_i\}$. For any generalised contour $\mathcal{C}_\alpha \in \mathcal{D}$ we define a primary differential by

$$\phi_\alpha(p) = dy_\alpha(p) = \oint_{\mathcal{C}^*_\alpha} B(p,p')$$

which is locally holomorphic on $\Sigma \setminus x^{-1}(\infty)$. Here, the dual $\mathcal{C}^*_\alpha = \eta_{\alpha\beta}\mathcal{C}_\beta$ with respect to the metric η.

Following Dubrovin, let us classify into 5 types the cycles C_α on which we integrate the differential $f(p) = \phi_\alpha(p)$ or $f(p) = \omega_{g,n+1}(p, p_1, ..., p_n)$.

- Type (1): for $i = 1, \ldots, d$ and $k = 1, \ldots, \mu_i - 1$:

$$\int_{p\in\mathcal{C}_{t_{i,k}}} f(p) = \frac{1}{\mu_i - 1} \operatorname*{Res}_{p\to\infty_i} x(p)^{\frac{k}{\mu_i}} f(p);$$

- Type (2) : for $i = 2, \ldots, d$:

$$\int_{p\in\mathcal{C}_{v_i}} f(p) = \operatorname*{Res}_{p\to\infty_i} x(p)\, f(p);$$

- Type (3): for $i = 2, \ldots, d$:

$$\int_{p\in\mathcal{C}_{w_i}} f(p) = \text{v.p.} \int_{\infty_1}^{\infty_i} f(p),$$

  where v.p. in front of the integral means the principal value with respect to $\zeta_i = x(p)^{-\frac{1}{\mu_i}}$, which is calculated by taking an asymptotic expansion of a primitive of the integrand $f(p)$ with respect to $\zeta_i \to 0$ and removing the non-analytic parts;
- Type (4): for $i = 1, \ldots, g$:

$$\int_{p\in\mathcal{C}_{r_i}} f(p) = -\oint_{\mathcal{A}_i} x(p)\, f(p);$$

- Type (5): for $i = 1, \ldots, g$:

$$\int_{p\in\mathcal{C}_{s_i}} f(p) = \frac{1}{2i\pi} \oint_{\mathcal{B}_i} f(p).$$

Note that the principal value in Type (3) is not required for $f(p) = \omega_{g,n+1}(p, p_1, ..., p_n)$ which is regular at each ∞_i. We see that the two important systems of coordinates—flat coordinates and canonical coordinates—correspond to cycles in $\mathcal{D}$, respectively zeros of dx. These sets have the same cardinality by (39).

4.2. Vector fields, cycles and meromorphic differentials. Let us now introduce a correspondence between vector fields and meromorphic forms using the Bergman kernel B which allows us to express all the quantities defining the Hurwitz Frobenius manifold in terms of meromorphic forms. For flat coordinates

$$\partial_{t_\alpha} \mapsto \phi_\alpha(p) = \int_{p' \in \mathcal{C}^*_\alpha} B(p,p').$$

By linearity, for any vector field v, we can define a cycle $\mathcal{C}_v$ by

$$\mathcal{C}_v = \sum_\alpha \langle v, \partial_{t_\alpha} \rangle_\eta \mathcal{C}_\alpha \tag{43}$$

a meromorphic differential

$$\phi_v(p) = \int_{p' \in \mathcal{C}_v} B(p,p')$$

and the metric η by

$$\langle v_1, v_2 \rangle_\phi = \sum_i \operatorname*{Res}_{p=\mathcal{P}_i} \frac{\phi_{v_1}\phi_{v_2}}{dx(p)}. \tag{44}$$

Note that (43) and (44) are proven by verifying them on a basis. We choose the flat basis, to prove (43). Substitute $v = \partial_{t_\alpha}$ into (43) to get $\mathcal{C}_{\partial_{t_\alpha}} = \sum_\beta \left\langle \partial_{t_\alpha}, \partial_{t_\beta} \right\rangle_\eta \mathcal{C}_\beta = \sum_\beta \eta_{\alpha\beta} \mathcal{C}_\beta = \mathcal{C}^*_\alpha$ as required. We choose the canonical basis to prove (44) as follows.

Apply (43) to the canonical vector fields to get

$$\mathcal{C}_{\partial_{u_i}} = \sum_\alpha \langle \partial_{u_i}, \partial_{t_\alpha} \rangle_\eta \mathcal{C}_\alpha = \sum_\alpha \phi(\mathcal{P}_i) \Psi^i_\alpha \mathcal{C}_\alpha$$

and hence

$$\begin{aligned}
\phi_{\partial_{u_i}}(p) = \int_{\mathcal{C}_{\partial_{u_i}}} B(p,p') &= \sum_\alpha \phi(\mathcal{P}_i)\Psi^i_\alpha \int_{\mathcal{C}_\alpha} B(p,p') = \sum_{\alpha,\beta} \phi(\mathcal{P}_i)\Psi^i_\alpha \eta^{\alpha\beta} \int_{\mathcal{C}^*_\beta} B(p,p') \\
&= \sum_{\alpha,\beta} \phi(\mathcal{P}_i)\Psi^i_\alpha \eta^{\alpha\beta} \phi_\beta(p)
\end{aligned}$$

We will study $\phi_{\partial_{u_i}}$ via evaluation at $\mathcal{P}_j$.

$$\phi_{\partial_{u_i}}(\mathcal{P}_j) = \sum_{\alpha,\beta} \phi(\mathcal{P}_i)\Psi^i_\alpha \eta^{\alpha\beta} \phi_\beta(\mathcal{P}_j) = \sum_{\alpha,\beta} \phi(\mathcal{P}_i)\Psi^i_\alpha \eta^{\alpha\beta} \Psi^j_\beta = \delta_{ij}\phi(\mathcal{P}_i)$$

which uses the relation $\phi_\beta(\mathcal{P}_j) = \Psi^j_\beta$ proven in Proposition 4.5. Since $\phi_{\partial_{u_i}}(p)$ vanishes at $\mathcal{P}_j$ for $j \neq i$, (44) becomes rather simple:

$$\left\langle \partial_{u_i}, \partial_{u_j} \right\rangle_\phi = \sum_k \operatorname*{Res}_{p=\mathcal{P}_k} \frac{\phi_{\partial_{u_i}}\phi_{\partial_{u_j}}}{dx(p)} = \delta_{i,j} \operatorname*{Res}_{p=\mathcal{P}_i} \frac{\phi^2(p)}{dx(p)}$$

in agreement with (42) and hence proving (44) for all vector fields.

The product in terms of the canonical basis gives us a formula in terms of the matrix Ψ of change of basis from flat to canonical which takes the form of the Verlinde formula, or Krichever formula depending on the context. (The Verlinde formula is actually for the degree 0 part of the theory.) This can be written for example following [**6**] equation (5.61)

$$C_{\alpha\beta\gamma} = \sum_i \operatorname*{Res}_{p=\mathcal{P}_i} \frac{\phi_\alpha(p)\phi_\beta(p)\phi_\gamma(p)}{dx(p)\phi(p)}. \tag{45}$$

This depends on the choice of Frobenius structure through ϕ which appears in the denominator and a point in the Frobenius manifold through the dependence on x.

Let us finally identify the identity and the Euler field. The consistency condition for the identity vector field $\mathbb{1} = \partial_{t_{\alpha_0}}$

$$\left\langle \partial_{t_\alpha}, \partial_{t_\beta} \right\rangle_\phi = C_{\alpha\beta\alpha_0}$$

imposes

$$\phi_{\mathbb{1}} = \phi_{\alpha_0} = \phi$$

and the Euler vector field

$$\phi_E = -E \cdot ydx|_{x \text{ fixed}} = E \cdot xdy|_{y \text{ fixed}} = xdy = x\phi$$

uses variations of the structures which are described below.

4.3. Rauch variational formula. An important tool used in this paper is Rauch variational formula expressing the variation of the Bergman kernel with respect to the position of the critical values.

$$\frac{\partial}{\partial u_i} B(p_1, p_2) = \operatorname*{Res}_{p=\mathcal{P}_i} \frac{B(p,p_1)B(p,p_2)}{dx(p)}. \tag{46}$$

Rauch originally derived the dependence of the Riemann matrix of periods of a Hurwitz cover on the critical values of the covering map in [**23**]. It later led to the expression of the variation of the Bergman kernel in [**18**].

In the present context, the meaning of the variation is as follows. Over the Frobenius manifold $M = \widetilde{H}^s_{g,\mu}$ we have a universal curve $\pi : \tilde{C} \to M$ and a function $x : \tilde{C} \to M \times \overline{\mathbb{C}}$ satisfying:

- (i) Each fibre $C = C_u = \pi^{-1}(u)$ is a Riemann surface.
- (ii) x is meromorphic on each fibre C.
- (iii) The critical values $\{u_1, ..., u_n\}$ of x on each fibre above a semisimple point are canonical coordinates for M.

For any vector field $\partial \in \Gamma(TM)$, we choose a lift $\tilde{\partial} \in \Gamma(TC)$ so that $\tilde{\partial} x = 0$. We abuse terminology and write $\tilde{\partial} = \partial$. Hence we make sense of variations of a function $f(p_1, p_2)$ on $C \times C$ by identifying $p_i \in C_u$ with $p_i' \in C_{u'}$ when $x(p_1) = x(p_1')$.

Rauch's variational formula for the Bergman kernel leads to variational formulae for other quantities, in particular primary differentials.

$$\begin{aligned} \partial_{u_i} dy(p) &= \partial_{u_i} \int_{\mathcal{C}_\alpha} B(p,p') = \int_{\mathcal{C}_\alpha} \partial_{u_i} B(p,p') = \int_{\mathcal{C}_\alpha} B(p,\mathcal{P}_i)B(p',\mathcal{P}_i) \\ &= dy(\mathcal{P}_i)B(p,\mathcal{P}_i). \end{aligned} \tag{47}$$

We apply this to give a short proof of flatness of the metric (42) and refer to [**6**] for the full proof of Theorem 6 which gives a different proof of flatness. The tangent space to $\widetilde{H}_{g,\mu}$ is spanned by primary differentials constructed from contours in $\mathcal{D}$. Hence the following lemma proves flatness of the metric.

LEMMA 4.2. *When $\mathcal{C}, \mathcal{C}' \in \mathcal{D}$ then $\langle \phi_{\mathcal{C}}, \phi_{\mathcal{C}'} \rangle_\phi$ is constant in $\{u_1, ..., u_N\}$.*

PROOF. From the Rauch's variational formula (46), we have $\partial_{u_i} \phi_{\mathcal{C}}(p) = \phi_{\mathcal{C}}(\mathcal{P}_i)B(p,\mathcal{P}_i)$. This uses the fact that the contour $\mathcal{C}$ depends only on a geometric contour independent of the choice of u_i, and possibly a function of x which is constant, i.e. $\partial_{u_i} x = 0$ by assumption. Hence

$$\begin{aligned}\partial_{u_j}\langle\phi_{\mathcal{C}},\phi_{\mathcal{C}'}\rangle &= \sum_i \operatorname*{Res}_{p=\mathcal{P}_i} \frac{\partial_{u_j}(\phi_{\mathcal{C}}(p)\phi_{\mathcal{C}'}(p))}{dx(p)} \\ &= \sum_i \operatorname*{Res}_{p=\mathcal{P}_i} \frac{B(p,\mathcal{P}_j)(\phi_{\mathcal{C}}(\mathcal{P}_j)\phi_{\mathcal{C}'}(p)+\phi_{\mathcal{C}}(p)\phi_{\mathcal{C}'}(\mathcal{P}_j))}{dx(p)} = 0.\end{aligned}$$

Note that the integrand potentially has poles at $\mathcal{P}_j$ and ∞_k but since each $\phi_{\mathcal{C}}(p)$ is dominated by $dx(p)$ at each $p = \infty_k$ the poles at ∞_k are removable. Hence the last equality uses the fact that the integrand has poles only at $\mathcal{P}_i$, $i = 1, ..., N$ so that the sum of its residues at $\mathcal{P}_i$ is 0. □

The following theorem proven by Shramchenko identifies the $R(z)$ matrix of the Hurwitz Frobenius manifold with the Laplace transform of the Bergmann kernel. It uses the Rauch's variational formula.

THEOREM 7 (Shramchenko, [**28**]). *Given a point $(\Sigma, x, (\mathcal{A}_i, \mathcal{B}_i)_{i=1,\dots,g})$ in the cover of a Hurwitz space with $B(p,p')$ normalised on the $\mathcal{A}$-cycles together with a choice of admissible differential ϕ the $R(z)$ matrix of the Hurwitz Frobenius manifold is given by:*

$$[R^{-1}(z)]_i^j := -\frac{\sqrt{z}}{\sqrt{2\pi}} \int_{\Gamma_j} e^{-\frac{(x(p)-u_j)}{z}} B(p,\mathcal{P}_i). \tag{48}$$

REMARK 4.3. Note that the way we state Theorem 7 differs in some details from [**28**, Theorem 1]. Let us explain the differences. Shramchenko proves that the matrix

$$[S(w)]_j^i := \frac{1}{\mathrm{i}\sqrt{2\pi w}} \int_{\Gamma_j} e^{wx(p)} B(p,\mathcal{P}_i) \tag{49}$$

is an asymptotic solution of the equations

$$\begin{aligned}\frac{dS(w)}{dw} &= \left(U + \frac{[\Gamma,U]}{w}\right) S(w); \\ dS(w) &= (wdU + [\Gamma,dU])\, S(w),\end{aligned} \tag{50}$$

of the shape $S(w) = (1 + \mathcal{O}(w^{-1}))e^{wU}$. Consider $S(z^{-1})$. It satisfies the equations

$$\begin{aligned}\frac{dS(z)}{dz} &= \left(-\frac{U}{z^2} - \frac{[\Gamma,U]}{z}\right) S(z); \\ dS(z) &= \left(\frac{dU}{z} + [\Gamma,dU]\right) S(z),\end{aligned} \tag{51}$$

which are equivalent to the Equations (25) and (27) determining $R(z)$ if $S(z) = R(z)e^{U/z}$. Thus, Shramchenko's formula gives us the following expression for the matrix $R(z)$ (in addition to the substitution $w = z^{-1}$ we multiply both sides of Equation (49) by $e^{-U/z}$):

$$[R(z)]_j^i := \frac{\sqrt{z}}{\mathrm{i}\sqrt{2\pi}} \int_{\Gamma_j} e^{(x(p)-u_j)/z} B(p,\mathcal{P}_i) \tag{52}$$

If we furthermore change z to $-z$ and transpose the matrix, we obtain an expression for $R^t(-z) = R^{-1}(z)$, which is exactly the formula (48) above.

The resemblance of (48) and (33) means we are now in a position to prove Theorem 3. Let us also remark that Shramchenko's result goes further than a formal series in z. Indeed, [**28**] defines integration cycles Γ_i such that $R(z)$ is the regular part of the expansion of a solution to Dubrovin's linear system which is well defined in a half plane.

PROOF OF THEOREM 3. The proof combines Theorem 4, Theorem 5 and Theorem 7.

Define the spectral curve (Σ, x, y, B) by a generic point $(\Sigma, x) \in H_{g,\mu}$ equipped with a bidifferential B normalised over a given set of $\mathcal{A}$-cycles, and a primary differential by $dy(p) := \oint_{\mathcal{C}} B(p,p')$ for some $\mathcal{C} \in \mathcal{D}$, defined in (3). If the spectral curve satisfies the conditions (33)-(35) of Theorem 4 for the $R(z)$ matrix of the Hurwitz Frobenius manifold then topological recursion applied to the spectral curve produces the ancestor invariants of the Frobenius manifold via the decomposition of $\omega_{g,n}$ given by (10) and the theorem is proven.

By Theorem 7 the $R(z)$ matrix of the Hurwitz Frobenius manifold is given by (48) hence condition (33) is satisfied. Next we need to show that the choice of y is the correct one. But since $dy(p) := \oint_{\mathcal{C}} B(p,p')$ the poles of dy are dominated by the poles of dx—the pole behaviour of the integrals over generalised cycles described in Section 4.1 is given in [**6**]—hence the spectral curve is dominant and Corollary 3.9 applies, proving that condition (34) is satisfied. Finally condition (35) is satisfied by Lemma 5.4 since Σ is compact and x is meromorphic. □

4.4. Shramchenko's deformation. Following methods of Kokotov-Korotkin [**17**], Shramchenko [**26**] defined deformations of Dubrovin's Frobenius manifold structures on $\widetilde{H}_{g,\mu}$. See also Buryak-Shadrin [**4**]. Recall that once we are given $(\Sigma, x, \{\mathcal{A}_i, \mathcal{B}_i\}_{i=1,\dots,g})$ and $\mathcal{D}$, we define a Bergman kernel and use that to define primary differentials ϕ_α for $\alpha \in \mathcal{D}$. Instead of the Bergman kernel $B(p,p')$ Shramchenko considered arbitrary Bergman kernels $\omega_{0,2}^{[\kappa]}(p,p')$ on Σ which are symmetric bidifferentials on $\Sigma \times \Sigma$, with a double pole on the diagonal of zero residue, double residue equal to 1, and no further singularities. The set of such kernels is parameterised by symmetric matrices κ of size $g \times g$. We denote by $\omega_{0,2}^{[0]} = B$ the Bergman kernel normalised in the basis of cycles chosen, i.e.

$$\forall i = 1, \dots, g\,, \oint_{\mathcal{A}_i} \omega_{0,2}^{[0]} = 0.$$

The key ingredients in the proofs of Theorems 2 and 3 are Rauch's variational principle for $B(p,p')$ which holds more generally for Bergman kernels normalised on geometric cycles and Eynard's formula (35) which is valid for any $B = \omega_{0,2}^{[\kappa]}$.

THEOREM 8. *The conclusion of Theorems 2 and 3 holds for Frobenius manifold structures on $\widetilde{H}_{g,\mu}$ defined by $\omega_{0,2}^{[\kappa]}(p,p')$ when κ is such that there exist a basis of geometric cycles $\left(\mathcal{A}_i^{[\kappa]}, \mathcal{B}_i^{[\kappa]}\right)_i$ satisfying*

$$\forall i = 1, \dots, g\,, \oint_{p' \in \mathcal{A}_i^{[\kappa]}} \omega_{0,2}^{[\kappa]}(p,p') = 0.$$

For arbitrary κ, a generalised version of Rauch's variational formula can be written by considering covariant derivatives (see [**13**]) but Theorems 2 and 3 do not hold as such. It would be interesting to understand how the conclusion of these theorems gets deformed when considering arbitrary κ. Indeed, such a general Bergman kernel proved to be important in bringing the topological recursion into the setup of Gromov-Witten invariants of Toric Calabi-Yau threefolds where a good choice of κ defines correlation functions which satisfy a set of equations known as holomorphic anomaly equations [**12**]. These equations, which allow one to fix some generating functions of the Gromov-Witten invariants up to some holomorphic ambiguity, can be understood as stating that the generating functions can be obtained as wave functions of some quantum system obtained by geometric quantisation. More precisely, they state that the partition function $Z = \exp\left(\sum_{g=0}^{\infty} \hbar^{g-1}\omega_{g,0}\right)$ is a flat section of some Hitchin connection. From this perspective, the choice of κ is identified with a choice of polarisation for this quantisation. The κ corresponding to geometric cycles give rise to real polarisations, whereas arbitrary κ's browse all possible choices of complex polarisations. While the possibility of considering arbitrary κ is natural from this perspective it is still not clear how to interpret it from the Hurwitz Frobenius manifold perspective.

4.5. Landau-Ginzburg model. In Section 4.2 we described a map from the tangent space at $p \in \widetilde{H}_{g,\mu}$ to the vector space spanned by primary differentials, denoted by V_p^{prim}. It was defined via a map to contours which are linear combinations of contours in $\mathcal{D}$. For $v \in T_p\widetilde{H}_{g,\mu}$ we defined

$$v \mapsto \mathcal{C}_v \mapsto \phi_v(p)$$

A more direct path uses variations. It is known as a Landau-Ginzburg model for (Σ, x, dy) and defined by:

$$\begin{array}{ccc} T_p\widetilde{H}_{g,\mu} & \to & V_p^{\text{prim}} \\ v & \mapsto & v \cdot (-ydx) \end{array}.$$

So the claim is that the variation gives the composition of the two maps above, i.e. $v \cdot (-ydx(p)) = \phi_v(p)$ for any $v \in T_p\widetilde{H}_{g,\mu}$. We prove this relation in terms of flat coordinates.

LEMMA 4.4. [**6**, Lemma 5.2] *For $\mathcal{C}_\alpha \in \mathcal{D}$, the coordinate $t_\alpha = \int_{\mathcal{C}_\alpha} dy$ is associated to the differential $\phi_\alpha(p)$ via*

$$\partial_{t_\alpha}[-y(p)dx(p)] = \phi_\alpha(p) = \int_{\mathcal{C}_\alpha^*} B(p,p').$$

PROOF. The main idea of the proof is to consider evaluation of $\partial_{t_\alpha} ydx(p)$ at $p = \mathcal{P}_i$ in order to be able to integrate by parts. From the variation of dy with respect to canonical coordinates given in (47) we have

(53)

$$\begin{aligned}\partial_{t_\alpha} dy(p) &= \sum_i \Psi^i_\alpha \partial_{v_i} dy(p) = \sum_i \frac{\Psi^i_\alpha}{dy(\mathcal{P}_i)} \partial_{u_i} dy(p) = \sum_i \frac{\Psi^i_\alpha}{dy(\mathcal{P}_i)} dy(\mathcal{P}_i) B(p,\mathcal{P}_i) \\ &= \sum_i \Psi^i_\alpha B(p,\mathcal{P}_i).\end{aligned}$$

Then

$$\begin{aligned}\partial_{t_\alpha}[-ydx](\mathcal{P}_i) &= -\operatorname*{Res}_{p=\mathcal{P}_i}\frac{1}{\sqrt{2(x(p)-u_i)}}\partial_{t_\alpha}[ydx]\\ &= \operatorname*{Res}_{p=\mathcal{P}_i}\sqrt{2(x(p)-u_i)}\partial_{t_\alpha}dy\\ &= \operatorname*{Res}_{p=\mathcal{P}_i}\sqrt{2(x(p)-u_i)}\sum_j \Psi^j_\alpha B(p,\mathcal{P}_j)\\ &= \operatorname*{Res}_{p=\mathcal{P}_i}\sqrt{2(x(p)-u_i)}B(p,\mathcal{P}_i)\Psi^i_\alpha\\ &= \Psi^i_\alpha = \phi_\alpha(\mathcal{P}_i)\end{aligned}$$

where the second line uses (47), the third line uses the fact that $B(p,\mathcal{P}_j)$ has no pole at $\mathcal{P}_i$ for $j\neq i$, the third line uses $\operatorname{Res}_{p=\mathcal{P}_i}\sqrt{2(x(p)-u_i)}B(p,\mathcal{P}_i)=1$, and the final equality uses Proposition 4.5.

Hence

$$\partial_{t_\alpha}[ydx](\mathcal{P}_i)=\phi_\alpha(\mathcal{P}_i),\quad i=1,\dots,N$$

which is nearly enough to guarantee that the differentials $\partial_{t_\alpha}ydx$ and ϕ_α agree. Define the function on Σ by

$$f(p)=\frac{\partial_{t_\alpha}ydx(p)-\phi_\alpha(p)}{dx(p)}.$$

Then $f(p)$ has no poles since the numerator of $f(p)$ vanishes at $p=\mathcal{P}_i$ and $dx(p)$ has simple zeros at $p=\mathcal{P}_i$. Also, from (53) we see that $\partial_{t_\alpha}ydx$ has no poles at $x=\infty$ hence $\partial_{t_\alpha}ydx-\phi_\alpha$ has poles only at $x=\infty$, dominated by poles of dx, since this is true of ϕ_α. In particular $f(p)$ has no poles at $x=\infty$.

Thus $f(p)=c$ constant and $\partial_{t_\alpha}ydx(p)=\phi_\alpha(p)+cdx(p)$. In [**6**] Dubrovin proves that the differential $\phi_\alpha(p)$ is either strictly dominated by dx at at least one point $\infty_i\in x^{-1}(\infty)$, in which case $f(\infty_i)=0$, or $\phi_\alpha(p)$ is a connection with ambiguity given by cdx for any constant c. Hence we may assume $c=0$ and the lemma is proven. □

We can now identify the transition matrix Ψ between flat and normalised canonical vector fields in an elegant way. Flat coordinates correspond to periods along generalised contours while canonical coordinates correspond to (finite) critical points of x. The Bergman kernel allows a natural marriage of the two.

Proposition 4.5 ([**26**]). *The transition matrix Ψ between flat and normalised canonical vector fields, defined in* (19) *is given by*

$$\Psi^i_\alpha=\int_{p\in\mathcal{C}^*_\alpha}B(p,\mathcal{P}_i)=\phi_\alpha(\mathcal{P}_i).$$

As usual the indices $i=1,\dots,N$ are associated to the canonical coordinates and $\alpha=1,\dots,N$ are associated to the flat coordinates.

Proof. We have

$$\partial_{u_i}t_\alpha=\partial_{u_i}\int_{\mathcal{C}_\alpha}dy=\int_{\mathcal{C}_\alpha}\partial_{u_i}dy=dy(\mathcal{P}_i)\int_{\mathcal{C}_\alpha}B(p,\mathcal{P}_i)$$

where the last equality uses (47), hence

$$\partial_{v_i} = \sum_\alpha \int_{\mathcal{C}_\alpha} B(p, \mathcal{P}_i)\, \partial_{t_\alpha}. \tag{54}$$

Now

$$[\partial_{v_1}, ..., \partial_{v_N}]\Psi = [\partial_{t_1}, ..., \partial_{t_N}]$$

and since $\Psi^T\Psi = \eta$, or $\Psi\eta^{-1}\Psi^T = I$ we have

$$[\partial_{v_1}, ..., \partial_{v_N}] = [\partial_{t_1}, ..., \partial_{t_N}]\eta^{-1}\Psi^T$$

hence

$$\partial_{v_i} = \sum_{\alpha,\beta} \eta^{\alpha\beta}\Psi^i_\beta \cdot \partial_{t_\alpha}$$

and comparing this with (54) we see that

$$\sum_\beta \eta^{\alpha\beta}\Psi^i_\beta = \int_{\mathcal{C}_\alpha} B(p, \mathcal{P}_i)$$

so

$$\Psi^i_\gamma = \sum_{\alpha,\beta} \eta_{\gamma\alpha}\eta^{\alpha\beta}\Psi^i_\beta = \sum_\alpha \eta_{\gamma\alpha} \int_{\mathcal{C}_\alpha} B(p, \mathcal{P}_i) = \int_{\sum_\alpha \eta_{\gamma\alpha}\mathcal{C}_\alpha} B(p, \mathcal{P}_i) = \int_{\mathcal{C}^*_\gamma} B(p, \mathcal{P}_i)$$

as required. The second equality in the statement of the proposition simply uses the definition $\phi_\alpha(p) := \int_{p\in\mathcal{C}^*_\alpha} B(p, p')$. □

REMARK 4.6. The column of Ψ corresponding to the vector field gives the square root of the diagonal coefficients $\eta_i^{1/2}$ of the metric η in canonical coordinates. From Proposition 4.5, we have $\eta_i = \phi(\mathcal{P}_i)^2$ which agrees with (42).

The transition matrix Ψ gives rise to the R matrix of the Frobenius manifold built from a choice of point $(\Sigma, x, (\mathcal{A}_i, \mathcal{B}_i)_{i=1,...,g})$ in $\widetilde{H}_{g,\mu}$ given in Theorem 7 together with a choice of admissible differential η.

4.6. Flat coordinates. Let us now explain how to recover the expression of the correlators of the CohFT in flat coordinates out of integration along contours in $\mathcal{D}$.

LEMMA 4.7. *For any generalised contour $\mathcal{C} \in \mathcal{D}$ and any $(g, n) \in \mathbb{N} \times \mathbb{N}^*$, the map*

$$\omega_{g,n} \mapsto \int_{\mathcal{C}} \omega_{g,n}$$

defining the action of integration of the correlation functions is well defined.

PROOF. Since $\mathcal{C}$ is only an isotopy class of contours (with coefficients that are functions of x) in $\Sigma \setminus x^{-1}(\infty)$ and $\omega_{g,n}$ has poles in $\Sigma \setminus x^{-1}(\infty)$ we need to prove that the integral is independent of the choice of contour. This is a consequence of the fact that $\omega_{g,n}$ and $x\omega_{g,n}$ have zero residues at $\mathcal{P}_i$ proven in [**13**, Corollary 4.1]. Note that the residues at ∞_j might not be zero, but the contours are not allowed to deform through ∞_j. □

PROPOSITION 4.8. *For any $\mathcal{C} \in \mathcal{D}$ define $\phi_{\mathcal{C}}(p) = \int_{\mathcal{C}} B(p, p') = df_{\mathcal{C}}$ where $f_{\mathcal{C}}$ is locally valued. As operators acting on $\omega_{g,n}$ for $2g - 2 + n > 0$,*

$$\sum_i \operatorname*{Res}_{\mathcal{P}_i} f_{\mathcal{C}} \cdot = \int_{\mathcal{C}} \cdot$$

in other words,

$$\sum_i \operatorname*{Res}_{p=\mathcal{P}_i} f_{\mathcal{C}}(p)\omega_{g,n}(p, p_2, ..., p_n) = \int_{p\in\mathcal{C}} \omega_{g,n}(p, p_2, ..., p_n).$$

PROOF. Recall Riemann's bilinear relation. For meromorphic differentials ϕ and ω such that ϕ is residueless

$$\sum_P \operatorname*{Res}_P f \cdot \omega = \frac{1}{2\pi i} \sum_{j=1}^{g} \left[\oint_{\mathcal{A}_i} \phi \oint_{\mathcal{B}_i} \omega - \oint_{\mathcal{B}_i} \phi \oint_{\mathcal{A}_i} \omega \right] \tag{55}$$

where $df = \phi$ for a locally defined primitive f and the sum is over all poles P of ϕ and ω.

Primary differentials $\phi_{\mathcal{C}}$ of types 1, 2 and 5, with respect to the classification given in Section 4.1, are residueless so apply (55) to $\phi = \phi_{\mathcal{C}}$ and $\omega = \omega_{g,n}$.

For $\mathcal{C} = \mathcal{B}_j$, $i = j, ..., g$, $\phi_{\mathcal{C}}(p) = \int_{\mathcal{B}_j} B(p, p') = \theta_j = df_{\mathcal{C}}$ (f defined locally) is a holomorphic differential satisfying $\int_{\mathcal{A}_k} \theta_j = 2\pi i \cdot \delta_{jk}$. Then (55) becomes:

$$\sum_i \operatorname*{Res}_{\mathcal{P}_i} f_{\mathcal{C}} \cdot \omega_{g,n} = \frac{1}{2\pi i} \sum_{k=1}^{g} \left[\oint_{\mathcal{A}_k} \phi_{\mathcal{C}} \oint_{\mathcal{B}_k} \omega_{g,n} - \oint_{\mathcal{B}_k} \phi_{\mathcal{C}} \oint_{\mathcal{A}_k} \omega_{g,n} \right] = \oint_{\mathcal{B}_j} \omega_{g,n} = \oint_{\mathcal{C}} \omega_{g,n}$$

since $\oint_{\mathcal{A}_k} \omega_{g,n} = 0$.

For $\mathcal{C} = x^{k/(n_i+1)}\mathcal{C}_i$, $k = 1, ..., n_i + 1$, $i = 1, ..., d$, (which this includes both types 1 and 2) then $\phi_{\mathcal{C}} = \operatorname{Res}_{\infty_i} x^{k/(n_i+1)} B = df_{\mathcal{C}}$ is residueless and normalised so that $\int_{\mathcal{A}_i} \phi_{\mathcal{C}} = 0$. Then (55) becomes

$$\sum_{P=\mathcal{P}_k, \infty_\ell} \operatorname*{Res}_P f_{\mathcal{C}} \cdot \omega_{g,n} = \frac{1}{2\pi i} \sum_{j=1}^{g} \left[\oint_{\mathcal{A}_j} \phi_{\mathcal{C}} \oint_{\mathcal{B}_j} \omega_{g,n} - \oint_{\mathcal{B}_j} \phi_{\mathcal{C}} \oint_{\mathcal{A}_j} \omega_{g,n} \right] = 0$$

$$\Rightarrow \sum_k \operatorname*{Res}_{\mathcal{P}_k} f_{\mathcal{C}} \cdot \omega_{g,n} = - \operatorname*{Res}_{\infty_i} f_{\mathcal{C}} \cdot \omega_{g,n} = \oint_{\mathcal{C}} \omega_{g,n}$$

where the last equality uses the fact that $f_{\mathcal{C}} \sim -x^{k/(n_i+1)}$ near ∞_i.

For $\mathcal{C} = x\mathcal{C}_i$, $i = 1, ..., d$, $\phi_{\mathcal{C}} = \int_{\Gamma_i} B$ is a differential of the 3rd kind with simple poles at ∞_1 and ∞_i normalised so that $\int_{\mathcal{A}_k} \phi_{\mathcal{C}} = 0$. Since $\omega_{g,n}$ is residueless we switch the roles of ϕ and ω in (55). Choose $F_{g,n}$ such that $dF_{g,n} = \omega_{g,n}$, i.e. a primitive with respect to one variable. Then

$$\begin{aligned}
\sum_{p=\mathcal{P}_k,\infty_\ell} \operatorname*{Res}_p F_{g,n}(p,p_2,...,p_n)\phi_{\mathcal{C}}(p) &= \frac{1}{2\pi i}\sum_{j=1}^{g}\left[\oint_{\mathcal{A}_j}\phi_{\mathcal{C}}\oint_{\mathcal{B}_j}\omega_{g,n} - \oint_{\mathcal{B}_j}\phi_{\mathcal{C}}\oint_{\mathcal{A}_j}\omega_{g,n}\right]\\
&= 0\\
\Rightarrow \sum_k \operatorname*{Res}_{\mathcal{P}_k} f_{\mathcal{C}}(p)\omega_{g,n}(p,p_2,...,p_n) &= -\sum_k \operatorname*{Res}_{\mathcal{P}_k} F_{g,n}(p,p_2,...,p_n)\phi_{\mathcal{C}}(p)\\
&= \operatorname*{Res}_{p=\infty_i} F_{g,n}(p,p_2,...,p_n)\phi_{\mathcal{C}}(p)\\
&\quad + \operatorname*{Res}_{p=\infty_1} F_{g,n}(p,p_2,...,p_n)\phi_{\mathcal{C}}(p)\\
&= F_{g,n}(\infty_i,p_2,...,p_n) - F_{g,n}(\infty_1,p_2,...,p_n)\\
&= \int_{\infty_1}^{\infty_i}\omega_{g,n} = \int_{\mathcal{C}}\omega_{g,n}
\end{aligned}$$

For $\mathcal{C} = x\mathcal{A}_i$, $i = 1,...,g$, we cannot apply (55) directly since $\phi_{\mathcal{C}}$ is not a globally defined differential. Instead we need to apply the proof of (55) as follows. Cut Σ along $\mathcal{A}$ and $\mathcal{B}$ cycles meeting at a common point P_0 to leave a simply-connected region $R \subset \Sigma$ on which $\phi_{\mathcal{C}}$ and a primitive (with respect to one variable) $F_{g,n}(p)$ of $\omega_{g,n}(p)$ (suppress variables $p_2,...,p_n$) are well-defined. As in the proof of (55) integrate $\frac{1}{2\pi i}\phi_{\mathcal{C}}F_{g,n}$ along the boundary of R given by the $\mathcal{A}$ and $\mathcal{B}$ cycles to get

$$\begin{aligned}
\sum_{p=\mathcal{P}_k} \operatorname*{Res}_p F_{g,n}(p)\phi_{\mathcal{C}}(p) &= \sum_{j=1}^{g}\left[\oint_{\mathcal{A}_j}\phi_{\mathcal{C}}\oint_{\mathcal{B}_j}\omega_{g,n} - \oint_{\mathcal{B}_j}\phi_{\mathcal{C}}\oint_{\mathcal{A}_j}\omega_{g,n}\right]\\
&\quad - \int_{P_0+\mathcal{B}_i}^{P_0+\mathcal{B}_i+\mathcal{A}_i} F_{g,n}(p)dx(p)\\
&= -\int_{P_0+\mathcal{B}_i}^{P_0+\mathcal{B}_i+\mathcal{A}_i} F_{g,n}(p)dx(p)\\
&= -\int_{\mathcal{A}_i} x(p)\omega_{g,n}(p) = -\int_{\mathcal{C}}\omega_{g,n}(p)\\
\Rightarrow \sum_{p=\mathcal{P}_k} \operatorname*{Res}_p f_{\mathcal{C}}(p)\omega_{g,n}(p) &= \int_{\mathcal{C}}\omega_{g,n}(p).
\end{aligned}$$

□

Theorem 3 proved that topological recursion applied to the spectral curve $(\Sigma, x, (\mathcal{A}_i,\mathcal{B}_i)_{i=1,\ldots,g})$ with a choice of admissible differential $\phi = dy$ and $\omega_{0,2} = B$, stores the ancestor invariants of the Hurwitz Frobenius manifold and hence proves Theorem 2. We now prove the remainder of the statement of Theorem 2 by showing how to extract the ancestor invariants via integration over generalised contours.

PROPOSITION 4.9. *Integration over flat contours $\mathcal{C}_\alpha \in \mathcal{D}$ produces primary invariants:*

$$\int_{\mathcal{C}_{\alpha_1}}\cdots\int_{\mathcal{C}_{\alpha_n}}\omega_{g,n} = \int_{\overline{\mathcal{M}}_{g,n}} I_{g,n}\Big(e^{\alpha_1}\otimes\ldots\otimes e^{\alpha_n}\Big)$$

PROOF. We will prove the dual statement

$$\int_{\mathcal{C}^*_{\alpha_1}}\cdots\int_{\mathcal{C}^*_{\alpha_n}}\omega_{g,n} = \int_{\overline{\mathcal{M}}_{g,n}} I_{g,n}\Big(e_{\alpha_1}\otimes\ldots\otimes e_{\alpha_n}\Big). \tag{56}$$

For $k > 0$,

$$\sum_i \operatorname*{Res}_{\mathcal{P}_i} y_\alpha \cdot V_k^i = 0, \quad k > 0$$

where V_k^i are defined in (36) and $dy_\alpha = \phi_\alpha$. Hence the operator $\sum_i \operatorname{Res}_{\mathcal{P}_i} y_\alpha \cdot$ only detects coefficients of $V_0^i(p) = B(\mathcal{P}_i, p)$ in $\omega_{g,n}$ which stores the primary invariants by (10). Now

$$\operatorname*{Res}_{\mathcal{P}_j} y_\alpha(p) \cdot V_0^i(p) = \operatorname*{Res}_{\mathcal{P}_j} y_\alpha(p) \cdot B(\mathcal{P}_i, p) = \delta_{ij}\phi_\alpha(\mathcal{P}_j) = \Psi_\alpha^i$$

since B acts as a Cauchy kernel which sends y_α to evaluation of dy_α. Hence $\sum_i \operatorname*{Res}_{\mathcal{P}_i} y_\alpha \cdot$ acts as insertion of the vector

$$\Psi_\alpha^i \cdot \partial_{v_i} = \partial_{t_\alpha} = e_\alpha$$

into the ancestor invariant. Thus, using Proposition 4.8 we see that as an operator on $\omega_{g,n}$

$$\int_{\mathcal{C}_\alpha^*} \cdot = \sum_i \operatorname*{Res}_{\mathcal{P}_i} y_\alpha \cdot$$

acts as insertion of the vector e^α into the ancestor invariant and in particular (56) holds. Note that since $\mathcal{C}_\alpha^* = \sum_\beta \eta_{\alpha\beta}\mathcal{C}_\beta$ is a constant linear combination of contours in $\mathcal{D}$, then Proposition 4.8 applies also to $\mathcal{C}_\alpha^*$. □

REMARK 4.10. Let us apply Proposition 4.9, or more precisely (56), to the simplest case of $\omega_{0,3}$ to get the following.

$$\begin{aligned}
C_{\alpha\beta\gamma} &= \int_{\overline{\mathcal{M}}_{0,3}} I_{0,3}\Big(e_\alpha \otimes e_\beta \otimes e_\gamma\Big) \\
&= \int_{\mathcal{C}_\alpha^*}\int_{\mathcal{C}_\beta^*}\int_{\mathcal{C}_\gamma^*} \omega_{0,3} \\
&= \int_{\mathcal{C}_\alpha^*}\int_{\mathcal{C}_\beta^*}\int_{\mathcal{C}_\gamma^*} \sum_i \operatorname*{Res}_{p=\mathcal{P}_i} \frac{B(p_1,p)B(p_2,p)B(p_3,p)}{dx(p)dy(p)} \\
&= \sum_i \operatorname*{Res}_{p=\mathcal{P}_i} \int_{\mathcal{C}_\alpha^*}\int_{\mathcal{C}_\beta^*}\int_{\mathcal{C}_\gamma^*} \frac{B(p_1,p)B(p_2,p)B(p_3,p)}{dx(p)dy(p)} \\
&= \sum_i \operatorname*{Res}_{p=\mathcal{P}_i} \frac{\phi_\alpha(p)\phi_\beta(p)\phi_\gamma(p)}{dx(p)dy(p)}
\end{aligned}$$

which agrees with (45) as expected. Here we have used the formula

$$\omega_{0,3}(p_1,p_2,p_3) = \sum_i \operatorname*{Res}_{p=\mathcal{P}_i} \frac{B(p_1,p)B(p_2,p)B(p_3,p)}{dx(p)dy(p)}$$

proven in [**13**].

The following proposition generalises Theorem 2.

PROPOSITION 4.11. *There exist generalised contours $\mathcal{C}_{\alpha,k} = p_k(x)\mathcal{C}_\alpha$, for $\mathcal{C}_\alpha \in \mathcal{D}$ and $p_k(x) = x^k + ...$ a monic polynomial of degree k in x, so that the ancestor* *invariants, corresponding to $d_j \geq 0$, appear as periods.*

$$\int_{\mathcal{C}_{\alpha_1,k_1}} \dots \int_{\mathcal{C}_{\alpha_n,k_n}} \omega_{g,n} = \int_{\overline{\mathcal{M}}_{g,n}} I_{g,n}\left(e^{\alpha_1} \otimes \dots \otimes e^{\alpha_n}\right) \cdot \prod_{j=1}^{n} \psi_j^{k_j}. \tag{57}$$

PROOF. Using integration by parts, we see that the contour $x^k \mathcal{C}_i$ acts on the differential $V_k^i(p)$ by

$$\int_{x^k \mathcal{C}_i} V_k^i(p) = \int_{\mathcal{C}_i} V_0^i(p).$$

Hence there exists a monic polynomial $p_k(x) = x^k + \dots$ of degree k in x such that

$$\int_{p_k(x)\mathcal{C}_i} V_m^j(p) = \delta_{ij}\delta_{km}.$$

Define $\mathcal{C}^*_{\alpha,k} = p_k(x)\mathcal{C}^*_\alpha$, for $\mathcal{C}_\alpha \in \mathcal{D}$ then we have

$$\int_{\mathcal{C}^*_{\alpha_1,k_1}} \dots \int_{\mathcal{C}^*_{\alpha_n,k_n}} \omega_{g,n} = \langle \prod_{j=1}^{n} \tau_{k_j}(e_{\alpha_j}) \rangle. \tag{58}$$

□

5. Topological recursion for compact spectral curves

In this section we associate to a spectral curve a so-called $\hat{R}(z)$ matrix, which in the case of spectral curves lying in the image of the map (9) from CohFTs to spectral curves, coincides with the $R(z)$ matrix of the Frobenius manifold.

DEFINITION 5.1. Given a spectral curve (Σ, x, y, B) define a formal series

$$\hat{R}(z) = \sum_{k=0}^{\infty} \hat{R}_k z^k$$

with coefficients $N \times N$ matrices where $N =$ number of zeros of dx by

$$\left[\hat{R}^{-1}(z)\right]_i^j := -\frac{\sqrt{z}}{\sqrt{2\pi}} \int_{\Gamma_j} e^{-\frac{(x(p)-u_j)}{z}} B(p, \mathcal{P}_i). \tag{59}$$

In fact $\hat{R}(z)$ depends only on (Σ, x, B). This definition begins with the spectral curve and produces $\hat{R}(z)$ which reverses the direction of (33) where one begins with a Frobenius manifold and its associated $R(z)$ and produces a spectral curve. In general $\hat{R}(z)$ will not arise out of a Frobenius manifold.

REMARK 5.2. Note that $\left[\hat{R}^{-1}(z)\right]_i^j$ is well-defined for $i = j$ because the integrand has a pole of residue zero at $\mathcal{P}_i$, so Γ_i can be deformed to avoid $\mathcal{P}_i$ in a well-defined manner.

REMARK 5.3. The paths Γ_i were defined only locally in a neighbourhood of $\mathcal{P}_i$ in Section 3.1. That is also sufficient here, because again we are only concerned with the asymptotic expansion of $\hat{R}(z)$ at $z = 0$. Nevertheless, we can choose paths along which $x/z \to \infty$ in both directions, such as a path of steepest descent of $-x/z$ so that the series $\hat{R}(z)$ converges.

Let us denote $\left[\hat{R}(z)\right]_i^j = \sum_k \left[\hat{R}_k\right]_i^j z^{-k}$. In particular, one has

$$\left[\hat{R}_1\right]_i^j = B_{0,0}^{i,j} = B(\mathcal{P}_i, \mathcal{P}_j). \tag{60}$$

5.1. Factorisation property. On a compact spectral curve $\hat{R}(z)$ shares the symplectic property of any $R(z)$ associated to a Frobenius manifold. This is proven below as a consequence of a factorisation formula for the (Laplace transform of the) bidifferential B in terms of $\hat{R}(z)$. The factorisation formula is also required in the proof of Corollary 3.9.

LEMMA 5.4 (Eynard, [**11**]). *Whenever the spectral curve is a Hurwitz cover of* $\mathbb{P}^1$ *with* dx *a meromorphic form with simple zeroes,* $\hat{R}(z)$*—defined in* (59)*—satisfies the symplectic condition*

$$\hat{R}(z)\hat{R}^T(-z) = Id. \tag{61}$$

Furthermore, the Laplace transform of a Bergman kernel

$$\breve{B}^{i,j}(z_1, z_2) = \frac{e^{\frac{u_i}{z_1}+\frac{u_j}{z_2}}}{2\pi\sqrt{z_1 z_2}} \int_{\Gamma_i}\int_{\Gamma_j} B(p,p')e^{-\frac{x(p)}{z_1}-\frac{x(p')}{z_2}}.$$

satisfies

$$\breve{B}^{i,j}(z_1, z_2) = -\frac{\sum_{k=1}^N \left[\hat{R}^{-1}(z_1)\right]_k^i \left[\hat{R}^{-1}(z_2)\right]_k^j}{z_1 + z_2}. \tag{62}$$

This means that the coefficients $B_{k,l}^{i,j}$ of the expansion of the Bergman kernel around the branch points $\mathcal{P}_i$ and $\mathcal{P}_j$ can be defined recursively in terms of the initial data $B_{k,0}^{i,j}$. We give a proof here that differs from the proof in [**11**].

PROOF. We have

$$\begin{aligned} \sum_{i=1}^N \operatorname*{Res}_{q=\mathcal{P}_i} \frac{B(p,q)B(p',q)}{dx(q)} &= -\operatorname*{Res}_{q=p} \frac{B(p,q)B(p',q)}{dx(q)} - \operatorname*{Res}_{q=p'} \frac{B(p,q)B(p',q)}{dx(q)} \\ &= -d_p\left(\frac{B(p,p')}{dx(p)}\right) - d_{p'}\left(\frac{B(p,p')}{dx(p')}\right) \end{aligned} \tag{63}$$

where the first equality uses the fact that the only poles of the integrand are $\{p, p', \mathcal{P}_i, i = 1, ..., N\}$, and the second equality uses the Cauchy formula (4) satisfied by the Bergman kernel. The Laplace transform of the LHS of (63) is

$$\begin{aligned} &\frac{e^{\frac{u_i}{z_1}+\frac{u_j}{z_2}}}{2\pi\sqrt{z_1 z_2}} \int_{\Gamma_i}\int_{\Gamma_j} e^{-\frac{x(p)}{z_1}-\frac{x(p')}{z_2}} \sum_{k=1}^N \operatorname*{Res}_{q=\mathcal{P}_k} \frac{B(p,q)B(p',q)}{dx(q)} \\ &= \sum_{k=1}^N \frac{e^{\frac{u_i}{z_1}+\frac{u_j}{z_2}}}{2\pi\sqrt{z_1 z_2}} \int_{\Gamma_i} e^{-\frac{x(p)}{z_1}} B(p,\mathcal{P}_k) \int_{\Gamma_j} e^{-\frac{x(p')}{z_2}} B(p',\mathcal{P}_k) \\ &= \sum_{k=1}^N \frac{\left[\hat{R}^{-1}(z_1)\right]_k^i \left[\hat{R}^{-1}(z_2)\right]_k^j}{z_1 z_2} \end{aligned}$$

and the Laplace transform of the RHS of (63) is

$$-\frac{e^{\frac{u_i}{z_1}+\frac{u_j}{z_2}}}{2\pi\sqrt{z_1z_2}}\int_{\Gamma_i}\int_{\Gamma_j}e^{-\frac{x(p)}{z_1}-\frac{x(p')}{z_2}}\left\{d_p\left(\frac{B(p,p')}{dx(p)}\right)+d_{p'}\left(\frac{B(p,p')}{dx(p')}\right)\right\}$$
$$=-\left(\frac{1}{z_1}+\frac{1}{z_2}\right)\check{B}^{i,j}(z_1,z_2)$$

since the Laplace transform satisfies

$$\int_{\Gamma_i}d\left(\frac{\omega(p)}{dx(p)}\right)e^{-\frac{x(p)}{z}}=\frac{1}{z}\int_{\Gamma_i}\omega(p)e^{-\frac{x(p)}{z}}.$$

for any differential $\omega(p)$, by integration by parts. Hence we see that the Laplace transform of (63) gives (62) as required. Then (61) is a consequence of (62) and the finiteness of $\check{B}^{i,j}(z_1,z_2)$ at $z_2=-z_1$. □

5.2. Defining equation for $\hat{R}_{ij}(z)$. In the preceding section we defined an $\hat{R}$-matrix in terms of the Bergman kernel. When the Bergman kernel is normalised on a basis of geometric cycles, we can go further and compute all the terms $\left[\hat{R}_k\right]_{ij}$ in terms of some minimal quantities. This uses the Rauch variational formula (46) which allows us to derive an equation for the $\hat{R}$ matrix.

When a spectral curve lies in the image of the map (9) from CohFTs to spectral curves, the following theorem is a consequence of the properties (25), (26), (27) of the R matrix of a CohFT. For compact spectral curves, by Theorem 7 generically $\hat{R}=R$, but $\hat{R}$ is a little more general, and for example exists when critical values u_i coincide and R is problematic, since it is defined over the semisimple part of the Frobenius manifold. The outcome of the following theorem is that $\hat{R}$ resembles R and it can be used to give an alternative proof of Theorem 3.

THEOREM 9. *Given a triple* (Σ,x,B) *consisting of a compact Riemann surface* Σ*, a meromorphic function* $x:\Sigma\to\mathbb{C}$ *with zeros of* dx *simple, and a Bergman kernel* B*, then* $\hat{R}(z)$ *satisfies* (25), (26), (27), *i.e.*

$$d\hat{R}(z)=\frac{\left[dU,\hat{R}(z)\right]}{z}+[\Gamma,dU]\,\hat{R}(z),\tag{64}$$

$$\mathbb{1}\cdot\hat{R}(z)=0,\tag{65}$$

$$(z\partial_z+E)\cdot\hat{R}(z)=0.\tag{66}$$

PROOF. Although (65) is a consequence of (64) we first prove (65) and use this to prove (64).

Proof of (65): Differentiate Eynard's formula (62)

$$\check{B}^{i,j}(z_1,z_2)=\frac{e^{\frac{u_i}{z_1}+\frac{u_j}{z_2}}}{2\pi\sqrt{z_1z_2}}\int_{\Gamma_i}\int_{\Gamma_j}B(p,p')e^{-\frac{x(p)}{z_1}-\frac{x(p')}{z_2}}=-\sum_{k=1}^{N}\frac{\left[\hat{R}^{-1}(z_1)\right]_k^i\left[\hat{R}^{-1}(z_2)\right]_k^j}{z_1+z_2}.$$

to get

$$\sum_{k=1}^{N}\frac{\partial}{\partial u_k}\check{B}^{i,j}(z_1,z_2)$$
$$=\sum_{k=1}^{N}\frac{e^{\frac{u_i}{z_1}+\frac{u_j}{z_2}}}{2\pi\sqrt{z_1z_2}}\int_{\Gamma_i}B(p,\mathcal{P}_k)e^{-\frac{x(p)}{z_1}}\int_{\Gamma_j}B(p',\mathcal{P}_k)e^{-\frac{x(p')}{z_2}}+\left(\frac{1}{z_1}+\frac{1}{z_2}\right)\check{B}^{i,j}(z_1,z_2)$$
$$=\sum_{k=1}^{N}\frac{\left[\hat{R}^{-1}(z_1)\right]_k^i\left[\hat{R}^{-1}(z_2)\right]_k^j}{z_1z_2}-\sum_{k=1}^{N}\frac{\left[\hat{R}^{-1}(z_1)\right]_k^i\left[\hat{R}^{-1}(z_2)\right]_k^j}{z_1z_2}=0.$$

Since $\left[\hat{R}^{-1}(z_1)\right]_j^i=-z\check{B}^{i,j}(z,0)$, we have

$$\mathbb{1}\cdot\left[\hat{R}^{-1}(z_1)\right]_j^i=\sum_{k=1}^{N}\frac{\partial}{\partial u_k}\left[\hat{R}^{-1}(z_1)\right]_j^i=-z\sum_{k=1}^{N}\frac{\partial}{\partial u_k}\check{B}^{i,j}(z,0)=0,\quad\forall i,j$$

and since $\mathbb{1}\cdot\hat{R}(z)=0\Leftrightarrow\mathbb{1}\cdot\hat{R}^{-1}(z)=0$ this proves (65).
Proof of (64): For $k\neq i$,

$$\begin{aligned}\frac{\partial\left[\hat{R}^{-1}(z)\right]_i^j}{\partial u_k}&=-\frac{\partial}{\partial u_k}\frac{\sqrt{z}}{\sqrt{2\pi}}\int_{\Gamma_j}e^{-\frac{(x(p)-u_j)}{z}}B(p,\mathcal{P}_i)\\&=\delta_{k,j}\frac{1}{z}\left[\hat{R}^{-1}(z)\right]_i^j-\frac{\sqrt{z}}{\sqrt{2\pi}}\int_{\Gamma_j}e^{-\frac{(x(p)-u_j)}{z}}B(p,\mathcal{P}_k)B(\mathcal{P}_k,\mathcal{P}_i)\\&=\delta_{k,j}\frac{\left[\hat{R}^{-1}(z)\right]_i^j}{z}+\left[\hat{R}^{-1}(z)\right]_k^j\beta_{ki}.\end{aligned}\tag{67}$$

For $k=i$, by (65),

$$\frac{\partial\left[\hat{R}^{-1}(z)\right]_i^j}{\partial u_i}=-\sum_{m\neq i}\frac{\partial\left[\hat{R}^{-1}(z)\right]_i^j}{\partial u_m}=(\delta_{i,j}-1)\frac{\left[\hat{R}^{-1}(z)\right]_i^j}{z}-\sum_{m\neq i}\left[\hat{R}^{-1}(z)\right]_m^j\beta_{mi}$$

which gives the (j,i) component of the equation

$$d\hat{R}^{-1}(z)=\frac{\left[dU,\hat{R}^{-1}(z)\right]}{z}-\hat{R}^{-1}(z)\left[\Gamma,dU\right].$$

Hence

$$\begin{aligned}d\hat{R}(z)=-\hat{R}(z)d\hat{R}^{-1}(z)\hat{R}(z)&=-\hat{R}(z)\left(\frac{\left[dU,\hat{R}^{-1}(z)\right]}{z}-\hat{R}^{-1}(z)\left[\Gamma,dU\right]\right)\hat{R}(z)\\&=\frac{\left[dU,\hat{R}(z)\right]}{z}+\left[\Gamma,dU\right]\hat{R}(z)\end{aligned}$$

and (64) holds.

Proof of (66): We begin with a variation of the proof of (62), replacing the identity vector field with the Euler vector field. We have

$$\begin{aligned}
\sum_i u_i \frac{\partial}{\partial u_i} B(p,p') &= \sum_{i=1}^N u_i \operatorname*{Res}_{q=\mathcal{P}_i} \frac{B(p,q)B(p',q)}{dx(q)} \\
&= \sum_{i=1}^N \operatorname*{Res}_{q=\mathcal{P}_i} \frac{x(q)B(p,q)B(p',q)}{dx(q)} \\
&= -\operatorname*{Res}_{q=p} \frac{x(q)B(p,q)B(p',q)}{dx(q)} - \operatorname*{Res}_{q=p'} \frac{x(q)B(p,q)B(p',q)}{dx(q)} \\
&= -d_p\left(\frac{x(p)B(p,p')}{dx(p)}\right) - d_{p'}\left(\frac{x(p')B(p,p')}{dx(p')}\right).
\end{aligned} \tag{68}$$

Then

$$\begin{aligned}
&\left(1 + z_1\frac{\partial}{\partial z_1} + z_2\frac{\partial}{\partial z_2} + \frac{u_i}{z_1} + \frac{u_j}{z_2}\right)\check{B}^{i,j}(z_1,z_2) \\
&= \frac{e^{\frac{u_i}{z_1}+\frac{u_j}{z_2}}}{2\pi\sqrt{z_1z_2}}\left(z_1\frac{\partial}{\partial z_1} + z_2\frac{\partial}{\partial z_2}\right)\int_{\Gamma_i}\int_{\Gamma_j} e^{-\frac{x(p)}{z_1}-\frac{x(p')}{z_2}} B(p,p') \\
&= -\frac{e^{\frac{u_i}{z_1}+\frac{u_j}{z_2}}}{2\pi\sqrt{z_1z_2}}\int_{\Gamma_i}\int_{\Gamma_j}\sum_k u_k\frac{\partial}{\partial u_k}B(p,p') \\
&= \left(\frac{u_i}{z_1} + \frac{u_j}{z_2} - \sum_k u_k\frac{\partial}{\partial u_k}\right)\check{B}^{i,j}(z_1,z_2)
\end{aligned}$$

where the second equality uses (68) and integration by parts to show that for any differential $\omega(p)$ the Laplace transform satisfies

$$\int_{\Gamma_i} e^{-\frac{x(p)}{z}} d\left(\frac{x(p)\omega(p)}{dx(p)}\right) = z\frac{d}{dz}\int_{\Gamma_i} e^{-\frac{x(p)}{z}}\omega(p).$$

Hence we are left with the following equation which is essentially the Laplace transform of (68):

$$\left[1 + z_1\frac{\partial}{\partial z_1} + z_2\frac{\partial}{\partial z_2} + \sum_i u_i\frac{\partial}{\partial u_i}\right]\check{B}(z_1,z_2) = 0. \tag{69}$$

We will now take the $z_2 \to 0$ limit of (69). From Eynard's formula (62) we see that $\frac{\partial}{\partial z_2}\check{B}(z_1,z_2)$ is well-defined at $z_2 = 0$, hence $\lim_{z_2\to 0} z_2\frac{\partial}{\partial z_2}\check{B}(z_1,z_2) = 0$. We also have $\check{B}^{j,k}(z_1,0) = -\frac{1}{z_1}\left[\hat{R}^{-1}(z_1)\right]^j_k$. Thus the $z_2 \to 0$ limit of (69) becomes

$$0 = \left[1 + z_1\frac{\partial}{\partial z_1} + \sum_i u_i\frac{\partial}{\partial u_i}\right]\frac{1}{z_1}\left[\hat{R}^{-1}(z_1)\right]^j_k = \frac{1}{z_1}\left[z_1\frac{\partial}{\partial z_1} + \sum_i u_i\frac{\partial}{\partial u_i}\right]\left[\hat{R}^{-1}(z_1)\right]^j_k$$

which gives (66). □

REMARK 5.5. A spectral curve (Σ, x, y, B) with dy a primary differential is dominant—see Definition 3.8—hence by Corollary 3.9 it corresponds to a CohFT, which we have identified with the Hurwitz Frobenius manifold corresponding to the primary differential. More generally we can take dy to be any linear combination of primary differentials, which is no longer a primary differential hence Theorem 3 does not apply, but the spectral curve is still dominant and hence corresponds to a CohFT.

6. Topological recursion for families of spectral curves

Vector fields on the Frobenius manifold $\widetilde{H}^s_{g,\mu}$ can give rise to recursion relations between ancestor invariants. In this section we show how the vector fields act on the multidifferentials $\omega_{g,n}$ arising out of topological recursion and give rise to the recursion relations between ancestor invariants.

Over the Frobenius manifold $\widetilde{H}^s_{g,\mu}$ is a *universal curve* which is a family of spectral curves constructed via the underlying Hurwitz map (Σ, x) together with natural cycles on Σ used to define the full spectral curve (Σ, x, y, B). Note that topological recursion applied to a single spectral curve produces a CohFT which extends uniquely to a family of CohFTs, nicely encoded in a Frobenius manifold, and each giving rise to a corresponding spectral curve. Hence in this way the family of spectral curves is reconstructed from any single spectral curve in the family.

Consider the family of multidifferentials $\omega_{g,n}$ obtained by applying topological recursion to the universal curve. We can differentiate the multidifferentials $\omega_{g,n}$ with respect to vector fields on the Frobenius manifold $\widetilde{H}_{g,\mu}$. As usual, for any vector field $v \in \Gamma(T\widetilde{H}_{g,\mu})$ we choose a lift $\tilde{v} \in \Gamma(TC)$ where C is the universal curve over $\widetilde{H}_{g,\mu}$, so that $\tilde{v} \cdot x = 0$. We abuse terminology and write $\tilde{v} = v$.

First order deformations of topological recursion are described in [**13**]. There it is shown that deformations of $\omega_{0,1}$ propagate via the recursion to determine deformations of $\omega_{g,n}$. Specifically, for v a vector field on $\widetilde{H}_{g,\mu}$, if we can express the variation of ydx as an integral of B over a generalised contour $\mathcal{C}$, then the variation of $\omega_{g,n}$ uses the same contour as follows.

$$(70)\quad v \cdot ydx(p) = \int_{\mathcal{C}} B(p', p) \quad \Rightarrow \quad v \cdot \omega_{g,n}(p_1, ..., p_n) = \int_{\mathcal{C}} \omega_{g,n+1}(p', p_1, ..., p_n).$$

Deformations with respect to natural vector fields on the Frobenius manifold correspond to relations between correlators in the CohFT. In the remainder of this section we describe the dictionary between deformations by the unit and Euler vector fields and their realisations via topological recursion.

6.1. Identity vector field. When $v = \mathbb{1}$ is the identity vector, we have

$$\mathbb{1} \cdot ydx|_{x \text{ fixed}} = -\mathbb{1} \cdot xdy|_{y \text{ fixed}} = -dy = -\operatorname*{Res}_{p'=p} y(p')B(p,p') = \sum_P \operatorname*{Res}_{p'=P} y(p')B(p,p')$$

where the sum is over the poles P of y. Hence by (70)

$$(71)\qquad \mathbb{1} \cdot \omega_{g,n} = \sum_P \operatorname*{Res}_{p'=P} y(p')\omega_{g,n+1} = -\sum_i \operatorname*{Res}_{p'=\mathcal{P}_i} y(p')\omega_{g,n+1}.$$

We can calculate the action of $\mathbb{1}$ on $\omega_{g,n}$ in a different way via the lift of $\mathbb{1}$ to the universal curve. Note that there are flat coordinates $t_1, ., ., t_N$ such that $\mathbb{1} = \partial/\partial t_1$ where t_1 appears in x as $x = x_0 + t_1$ for x_0 independent of t_1. The lift $\mathbb{1}$ necessarily annihilates x so with respect to a local parameter z on Σ

$$0 = \mathbb{1} \cdot x = x'(z)\mathbb{1} \cdot z + 1 \quad \Rightarrow \quad \mathbb{1} \cdot z = -1/x'(z)$$

where we used the explicit partial derivative $\partial_{t_1} x = 1$. Hence for any differential ξ with no explicit t_1 dependence, locally $\xi = df$ so we have

$$\mathbb{1} \cdot \xi(z) = d\mathbb{1} \cdot f(z) = d(f'(z)\mathbb{1} \cdot z) = -d(f'(z)/x'(z)) = -d(df/dx) = -d(\xi/dx).$$

In other words the lift of $\mathbb{1}$ coincides on fibres with the operator

$$\mathbb{1} = -\frac{d}{dx}$$

which acts on functions or differentials. It is proven in [**11**] that the multidifferentials $\omega_{g,n}$ are polynomials in the differentials

$$\xi_k^i(p) := \operatorname*{Res}_{p'=\mathcal{P}_i} \frac{\omega_{0,2}(p',p)}{(x(p')-u_i)^k \sqrt{2(x(p')-u_i)}}.$$

In particular, ξ_k^α have no explicit t_1 dependence so

$$\mathbb{1} \cdot \xi_k^\alpha = -d\left(\frac{\xi_k^\alpha}{dx}\right) = -\xi_{k+1}^\alpha.$$

Furthermore $\omega_{g,n}$ has no explicit t_1 dependence since topological recursion is unchanged under $x \mapsto x + t_1$. Hence

$$\mathbb{1} \cdot \omega_{g,n} = -\sum_{j=1}^{n} d\left(\frac{\omega_{g,n}(p_1,\dots,p_n)}{dx(p_j)}\right).$$

The relation

$$\sum_i \operatorname*{Res}_{p'=\mathcal{P}_i} y(p')\omega_{g,n+1} = -\sum_{j=1}^{n} d\left(\frac{\omega_{g,n}(p_1,\dots,p_n)}{dx(p_j)}\right)$$

is proven in a different way in [**13**] as a direct consequence of topological recursion for arbitrary spectral curves. Here we have shown it to be a consequence of the action of the lift of the identity vector field on the universal curve.

The Hurwitz Frobenius manifolds from Section 4 have flat identity hence the CohFT satisfies the pull-back relation (23). A consequence of (23) on correlators is known as the string equation which expressed in tau notation:

$$\left\langle \prod_{i=1}^{n} \tau_{k_i}(v_i) \right\rangle_g := \int_{\overline{\mathcal{M}}_{g,n}} I_{g,n}(v_1 \otimes \cdots \otimes v_n) \prod_{i=1}^{n} \psi_i^{k_i}$$

is given by

$$\langle \tau_0(1)\tau_{k_1}(v_1)\cdots\tau_{k_n}(v_n)\rangle_g = \sum_{i=1}^{n} \langle \tau_{k_1}(v_1)\cdots\tau_{k_i-1}(v_i)\cdots\tau_{k_n}(v_n)\rangle_g.$$

But this is precisely equivalent to the relation (71) since

$$d\left(\frac{\xi_k^\alpha}{dx(p_j)}\right) = \xi_{k+1}^\alpha$$

where coefficients of ξ_k^α correspond to insertions of the vector field $\partial/\partial t_\alpha$.

6.2. Euler vector field. When $v = E$ is the Euler vector field, we have

$$E \cdot ydx|_{x \text{ fixed}} = -E \cdot xdy|_{y \text{ fixed}} = -xdy$$

(72)
$$= \operatorname*{Res}_{p'=p}(\Phi - xy)(p')B(p,p') = -\sum_P \operatorname*{Res}_{p'=P}(\Phi - xy)B(p,p')$$

where $d\Phi = ydx$ and the sum is over the poles P of $\Phi - xy$. Hence

(73)
$$E \cdot \omega_{g,n} = -\sum_P \operatorname*{Res}_{p'=P}(\Phi - xy)\omega_{g,n+1} - \sum_i \operatorname*{Res}_{p'=\mathcal{P}_i}(\Phi - xy)\omega_{g,n+1}$$
$$= (2g-2+n)\omega_{g,n}(p_1,...,p_n) - \sum_{j=1}^n d\left(\frac{x(p_j)\omega_{g,n}(p_1,...,p_n)}{dx(p_j)}\right)$$

where the last equality uses the dilaton and second string equation satisfied quite generally by the $\omega_{g,n}$, proven in [**13**]. Analogous to the string equation above, which enables one to remove or insert the identity vector field, this last expression enables one to remove or insert the Euler vector field in correlators. For example, in the Gromov-Witten case, it is given by the divisor equation.

A conformal Frobenius manifold corresponds to a homogeneous CohFT. A CohFT is *homogeneous* of weight d if

(74)
$$((g-1)d+n)I_{g,n} = \deg I_{g,n}(v_1 \otimes ... \otimes v_n) - \sum_{j=1}^n I_{g,n}(v_1 \otimes ... \otimes [E, v_j] \otimes ... \otimes v_n)$$

(75)
$$+ \pi_* I_{g,n+1}(v_1 \otimes ... \otimes v_s \otimes E)$$

where E is the Euler vector field and $\pi : \overline{\mathcal{M}}_{g,n+1} \to \overline{\mathcal{M}}_{g,n}$ is the forgetful map. Equation (74) allows one to remove or insert the Euler vector field in correlators and (72) is equivalent to this relation.

References

[1] J. E. Andersen, G. Borot and N. Orantin. *Modular functors, cohomological field theories and topological recursion.* arXiv:1509.01387

[2] Michael Atiyah, *Topological quantum field theories*, Inst. Hautes Études Sci. Publ. Math. **68** (1988), 175–186 (1989). MR1001453

[3] Vincent Bouchard and Bertrand Eynard, *Think globally, compute locally*, J. High Energy Phys. **2** (2013), 143, front matter + 34. MR3046532

[4] Alexandr Buryak and Sergey Shadrin, *A remark on deformations of Hurwitz Frobenius manifolds*, Lett. Math. Phys. **93** (2010), no. 3, 243–252, DOI 10.1007/s11005-010-0410-8. MR2679973

[5] Norman Do and Paul Norbury, *Topological recursion for irregular spectral curves*, J. Lond. Math. Soc. (2) **97** (2018), no. 3, 398–426, DOI 10.1112/jlms.12112. MR3816393

[6] Boris Dubrovin, *Geometry of 2D topological field theories*, Integrable systems and quantum groups (Montecatini Terme, 1993), Lecture Notes in Math., vol. 1620, Springer, Berlin, 1996, pp. 120–348, DOI 10.1007/BFb0094793. MR1397274

[7] Boris Dubrovin, *Painlevé transcendents in two-dimensional topological field theory*, The Painlevé property, CRM Ser. Math. Phys., Springer, New York, 1999, pp. 287–412. MR1713580

[8] P. Dunin-Barkowski, P. Norbury, N. Orantin, A. Popolitov, and S. Shadrin, *Dubrovin's superpotential as a global spectral curve.* To appear in J. Inst. Math. Jussieu. arXiv:1509.06954

[9] P. Dunin-Barkowski, N. Orantin, S. Shadrin, and L. Spitz, *Identification of the Givental formula with the spectral curve topological recursion procedure*, Comm. Math. Phys. **328** (2014), no. 2, 669–700, DOI 10.1007/s00220-014-1887-2. MR3199996

[10] B. Eynard, *Intersection numbers of spectral curves.* arXiv:1104.0176.

[11] B. Eynard, *Invariants of spectral curves and intersection theory of moduli spaces of complex curves*, Commun. Number Theory Phys. **8** (2014), no. 3, 541–588, DOI 10.4310/CNTP.2014.v8.n3.a4. MR3282995

[12] Bertrand Eynard, Nicolas Orantin, and Marcos Mariño, *Holomorphic anomaly and matrix models*, J. High Energy Phys. **6** (2007), 058, 20, DOI 10.1088/1126-6708/2007/06/058. MR2326593

[13] B. Eynard and N. Orantin, *Invariants of algebraic curves and topological expansion*, Commun. Number Theory Phys. **1** (2007), no. 2, 347–452, DOI 10.4310/CNTP.2007.v1.n2.a4. MR2346575

[14] Carel Faber, Sergey Shadrin, and Dimitri Zvonkine, *Tautological relations and the r-spin Witten conjecture* (English, with English and French summaries), Ann. Sci. Éc. Norm. Supér. (4) **43** (2010), no. 4, 621–658, DOI 10.24033/asens.2130. MR2722511

[15] Alexander B. Givental, *Gromov-Witten invariants and quantization of quadratic Hamiltonians* (English, with English and Russian summaries), Mosc. Math. J. **1** (2001), no. 4, 551–568, 645. Dedicated to the memory of I. G. Petrovskii on the occasion of his 100th anniversary. MR1901075

[16] Alexander B. Givental, *Semisimple Frobenius structures at higher genus*, Internat. Math. Res. Notices **23** (2001), 1265–1286, DOI 10.1155/S1073792801000605. MR1866444

[17] A. Kokotov and D. Korotkin, *A new hierarchy of integrable systems associated to Hurwitz spaces*, Philos. Trans. R. Soc. Lond. Ser. A Math. Phys. Eng. Sci. **366** (2008), no. 1867, 1055–1088, DOI 10.1098/rsta.2007.2061. MR2377685

[18] A. Kokotov and D. Korotkin, *Tau-functions on Hurwitz spaces*, Math. Phys. Anal. Geom. **7** (2004), no. 1, 47–96, DOI 10.1023/B:MPAG.0000022835.68838.56. MR2053591

[19] I. M. Krichever, *Algebraic-geometric n-orthogonal curvilinear coordinate systems and the solution of associativity equations* (Russian, with Russian summary), Funktsional. Anal. i Prilozhen. **31** (1997), no. 1, 32–50, 96, DOI 10.1007/BF02466001; English transl., Funct. Anal. Appl. **31** (1997), no. 1, 25–39. MR1459831

[20] Yuri I. Manin, *Frobenius manifolds, quantum cohomology, and moduli spaces*, American Mathematical Society Colloquium Publications, vol. 47, American Mathematical Society, Providence, RI, 1999. MR1702284

[21] Todor Milanov, *The Eynard-Orantin recursion for the total ancestor potential*, Duke Math. J. **163** (2014), no. 9, 1795–1824, DOI 10.1215/00127094-2690805. MR3217767

[22] Rahul Pandharipande, Aaron Pixton, and Dimitri Zvonkine, *Relations on $\overline{\mathcal{M}}_{g,n}$ via 3-spin structures*, J. Amer. Math. Soc. **28** (2015), no. 1, 279–309, DOI 10.1090/S0894-0347-2014-00808-0. MR3264769

[23] H. E. Rauch, *Weierstrass points, branch points, and moduli of Riemann surfaces*, Comm. Pure Appl. Math. **12** (1959), 543–560, DOI 10.1002/cpa.3160120310. MR0110798

[24] Kyoji Saito, *Period mapping associated to a primitive form*, Publ. Res. Inst. Math. Sci. **19** (1983), no. 3, 1231–1264, DOI 10.2977/prims/1195182028. MR723468

[25] Sergey Shadrin, *BCOV theory via Givental group action on cohomological fields theories* (English, with English and Russian summaries), Mosc. Math. J. **9** (2009), no. 2, 411–429, back matter. MR2568443

[26] Vasilisa Shramchenko, *Deformations of Frobenius structures on Hurwitz spaces*, Int. Math. Res. Not. **6** (2005), 339–387, DOI 10.1155/IMRN.2005.339. MR2131384

[27] Vasilisa Shramchenko, *"Real doubles" of Hurwitz Frobenius manifolds*, Comm. Math. Phys. **256** (2005), no. 3, 635–680, DOI 10.1007/s00220-005-1321-x. MR2161275

[28] Vasilisa Shramchenko, *Riemann-Hilbert problem associated to Frobenius manifold structures on Hurwitz spaces: irregular singularity*, Duke Math. J. **144** (2008), no. 1, 1–52, DOI 10.1215/00127094-2008-030. MR2429320

[29] Constantin Teleman, *The structure of 2D semi-simple field theories*, Invent. Math. **188** (2012), no. 3, 525–588, DOI 10.1007/s00222-011-0352-5. MR2917177

[30] J. Zhou, *Frobenius manifolds, spectral curves and integrable hierarchies*. arXiv:1512.05466

Faculty of Mathematics, National Research University Higher School of Economics, Usacheva 6, 119048 Moscow, Russia
Email address: ptdbar@hse.ru

School of Mathematics and Statistics, University of Melbourne, Australia 3010
Email address: norbury@unimelb.edu.au

Département de mathématiques, Ecole Polytechnique Fédérale de Lausanne, CH-1015 Lausanne, Switzerland
Email address: nicolas.orantin@epfl.ch

Department of Physics and Astronomy, Uppsala University, Box 516, 751 20 Uppsala, Sweden; Institute for Information Transmission Problems, Moscow 127994, Russia; and ITEP, Moscow 117218, Russia
Email address: popolit@gmail.com

Korteweg-de Vries Institute for Mathematics, University of Amsterdam, Postbus 94248, 1090 GE Amsterdam, The Netherlands
Email address: s.shadrin@uva.nl

Proceedings of Symposia in Pure Mathematics
Volume **100**, 2018
https://doi.org/10.1090/pspum/100/01770

Hopf algebras and topological recursion

João N. Esteves

ABSTRACT. We first review our previous work where we considered a model for topological recursion based on the Hopf algebra of planar binary trees of Loday and Ronco and showed that extending this Hopf algebra by identifying pairs of nearest neighbour leaves and thus producing graphs with loops we obtain the full recursion formula of Eynard and Orantin. Then we discuss the algebraic structure of the spaces of correlation functions in $g = 0$ and in $g > 0$. By taking a classical and a quantum product respectively we endow both spaces with a ring structure. This is an extended version of the contributed talk given at the 2016 von Neumann Symposium on Topological Recursion and its Influence in Analysis, Geometry and Topology, from 4 to 8 July 2016 at Hilton Charlotte University Place, USA.

CONTENTS

1. Introduction

In the paper [**10**] we showed how the Hopf algebra of planar binary trees of Loday and Ronco [**14**] can be seen as a sort of representation of the space generated by correlation functions that obey the Eynard-Orantin recursion formula. These correlation functions are graded by the Euler characteristic and we can consider for each degree the vector space over a field k of characteristic 0 generated by them and then take the direct sum of these vector spaces for all degrees. Here we will discuss, among other things and after reviewing the work in [**10**] if this vector space can be given additional algebraic structures.

2010 *Mathematics Subject Classification.* Primary 81Q30, 05C30, 16T05, 16T30, 15B52.
Key words and phrases. Hopf algebras, topological recursion, matrix models.
The author was supported by Fundação para a Ciência e a Tecnologia through the grant SFRH/BPD/77123/2011. He greatly thanks the organizers of the 2016 von Neumann Symposium for the opportunity of giving this talk.

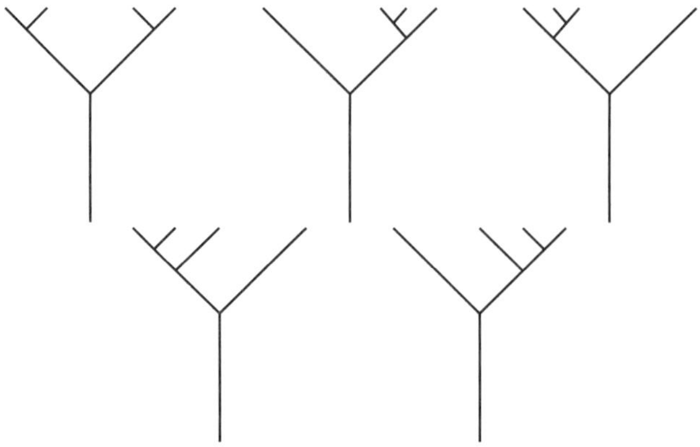

FIGURE 1.1. Planar binary trees of order 3 as generators of the correlation functions W_5^0, W_3^1 and W_1^2 with $\chi = -3$. Reprinted with permission from [**10**].

First we consider planar binary trees of order n, that is with n vertices's and $n+1$ leaves, as a representation of genus $g = 0$ correlation function $W_k^0(p, p_1, \dots, p_{k-1})$ of Euler characteristic $\chi = 2-2g-k$ equal to $-n$. Here the Euler characteristic is the one of Riemann or topological surfaces of genus g with k punctures or boundaries to which the correlation functions $W_k^g(p, p_1, \dots, p_{k-1})$ are usually related in some concrete problems. For $g = 0$ we label the root with p and the $n+1$ leaves with the $p_1, \dots, p_{n+1}$ variables. Then by connecting the nearest neighbours leaves with a single edge and reducing the number of pairs of labels in the same way as increasing the genus we obtain graphs with loops that we see as a representation of higher genus correlation functions with the same Euler characteristic.

As an example take $W_1^2(p)$ which has $\chi = -3$. Its underlying generating trees are planar binary trees of order 3 which are also models for $W_5^0(p, p_1, p_2, p_3, p_4)$: Identifying pairs of nearest neighbour leaves in the left and right branches independently we get the second term of topological recursion. Identifying pairs of leaves each taken from the left and the right branches gives the first term. Note that in this last case not every planar binary tree of order 3 contributes to W_1^2. In fact the first tree of fig. 1.1 does not give a genus 2 correlation function by identifying the nearest neighbour leaves in opposite branches.

2. The topological recursion of Eynard and Orantin

The topological recursion formula of Eynard and Orantin has its origin in Matrix Models, for general reviews see for instance [**8, 11**]. In the hermitian 1-matrix form of the theory the purpose is to compute connected correlation functions W_{k+1} depending on a set of variables $p, p_1, \dots, p_k$

$$W_{k+1}(p, p_1, \dots p_k) = \left\langle \mathrm{Tr}\frac{1}{p-M}\mathrm{Tr}\frac{1}{p_1-M}\dots\mathrm{Tr}\frac{1}{p_k-M}\right\rangle_c \tag{1}$$

starting with $W_1(p)$ and $W_2(p, p_1)$. These functions which are solutions of the so-called loop equations are only well defined over Riemann surfaces because in $\mathbb{C}$ they are multi-valued. They admit an expansion on the order N of the random matrix M, with components $W_{k+1}^g(p, p_1, \dots p_k)$ related to a definite genus. We will not be concerned here with the actual computation of correlation functions in specific models.

Let $K = (p_1, \ldots, p_k)$ be a vector of variables. For instance in concrete cases these can be coordinates of punctures on Riemann surfaces, labels of borders on topological surfaces or variables in Matrix Models, but we just leave them as labels of leaves of planar binary trees or of graphs obtained from planar binary trees. We assign the label p to the root of a tree or of a graph with loops obtained from a tree. The topological recursion formula is

$$W^g_{k+1}(p,K) = \sum_{\text{branch points } \alpha} \mathrm{Res}_{p\to\alpha} K_p(q,\bar{q})$$

$$\left(W^{g-1}_{k+2}(q,\bar{q},K) + \sum_{L\cup M=K, h=0}^{g} W^h_{|L|+1}(q,L) W^{g-h}_{|M|+1}(\bar{q},M) \right) \tag{2}$$

where the sum is restricted to terms with Euler characteristic equal or smaller than 0. For instance if $h = 0$ then $|L| \geq 1$. For a very clear exposition about this setup from the point of view of Algebraic Geometry see for instance [**9**] but some comments are in order. The branch points are the ones from a meromorphic function x defined on a so called spectral curve $\mathcal{E}(x,y) = 0$. The recursion kernel $K_p(q,\bar{q})$ is, roughly speaking, a meromorphic (1,1) tensor that depends on a regular point p in the neighbourhood of a branch point and on q and its conjugated point $\bar{q}$ for which $x(q) = x(\bar{q})$ and $y(q) = -y(\bar{q})$. In fact it can be computed from $W^0_1(p)$ and $W^0_2(p,p)$ which are symmetric differentials of order one and two respectively. Actually, all W^g_k are meromorphic symmetric differentials but we will continue to refer to them as correlation functions. Since our approach will be purely algebraic and in order to soften the notation we will not explicitly mention the sum of the residues over the branch points when referring to this formula.

3. The Loday-Ronco Hopf algebra of planar binary trees

We collect here some important facts of the Loday-Ronco Hopf algebra. Details and proofs can be found in [**1, 14**]. Let S_n be the symmetric group of order n with the usual product $\rho \cdot \sigma$ given by the composition of permutations. When necessary we denote a permutation ρ by its image $(\rho(1)\rho(2)\ldots\rho(n))$. Recall that a shuffle $\rho(p,q)$ of type (p,q) in S_n is a permutation such that $\rho(1) < \rho(2) < \cdots < \rho(p)$ and $\rho(p+1) < \rho(p+2) < \cdots < \rho(p+q)$. For instance the shuffles of type $(1,2)$ in S_3 are $(123), (213)$ and (312). We denote the set of (p,q) shuffles by $S(p,q)$. Take

$$k[S^\infty] = \oplus_{n=0}^{\infty} k[S_n] \tag{3}$$

with S_0 identified with the empty permutation. $k[S^\infty]$ is a vector space over a field k of characteristic 0 generated by linear combinations of permutations. It is graded by the order of permutations and $k[S_0]$ which contains the empty permutation is identified with the field k. For two permutations $\rho \in S_p$ and $\sigma \in S_q$ there is a natural product on S^∞ denoted by $\rho \times \sigma$ which is a permutation on S_{p+q} given by letting ρ acting on the first p variables and σ acting on the last q variables.

There is a unique decomposition of any permutation $\sigma \in S_n$ in two permutations $\sigma_i \in S_i$ and $\sigma'_{n-i} \in S_{n-i}$ for each i such that

$$\sigma = (\sigma_i \times \sigma'_{n-i}) \cdot w^{-1} \tag{4}$$

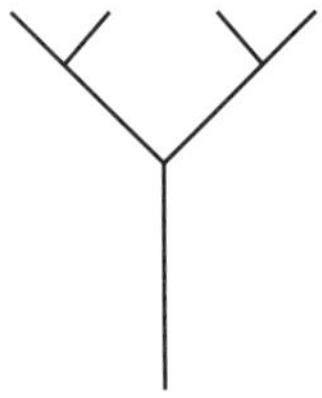

FIGURE 3.1. A planar binary tree of order 3. Reprinted with permission from [**10**].

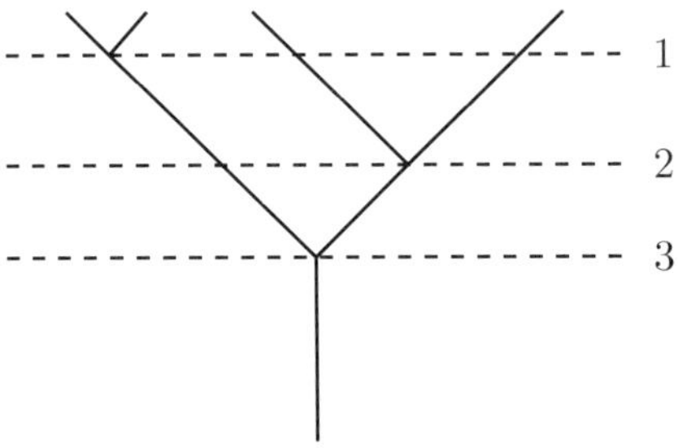

FIGURE 3.2. Planar binary tree with levels that is the image of (**132**). Reprinted with permission from [**10**].

where w is a shuffle of type $(i, n-i)$. With the $*$ product

$$\rho * \sigma = \sum_{\alpha_{n,m} \in S_{(n,m)}} \alpha_{n,m} \cdot (\rho \times \sigma) \tag{5}$$

and the co-product

$$\Delta\sigma = \sum \sigma_i \otimes \sigma'_{n-i} \tag{6}$$

$k[S^\infty]$ becomes a bi-algebra and since it is graded and connected it is automatically a Hopf algebra.

A planar binary tree is a graph with no loops embedded in the plane with only trivalent vertices. In every planar binary tree there are paths that start on a special edge called the root and end on the terminal edges called leaves. The leaves can be left or right oriented. The order $|t|$ of a planar binary tree t is the number of its vertices and on each planar binary tree of order n there are $n+1$ leaves that usually are numbered from 0 to n from left to right. It is frequent to visualize planar binary trees from the bottom to the top, with the root as its lowest vertical edge and the leaves as the highest edges, oriented SW-NE or SE-NW. We will denote the set of planar binary trees of order n by Y^n and by $k[Y^\infty]$ the vector space over k generated by planar binary trees of all orders. Additionally a planar binary tree with levels is a planar binary tree such that on each horizontal line there is at most one vertex. It is clear that reading the vertices from left to right and from top to bottom it is possible to assign a permutation of order n to a planar binary tree with levels and that this assignment is unique. For example in fig. 3.2 the tree corresponds to the permutation (132). In this way it is completely equivalent to consider the Hopf algebra $k[S^\infty]$ or the Hopf algebra of planar binary trees with levels because they are isomorphic. However Loday and Ronco show in [**14**] that

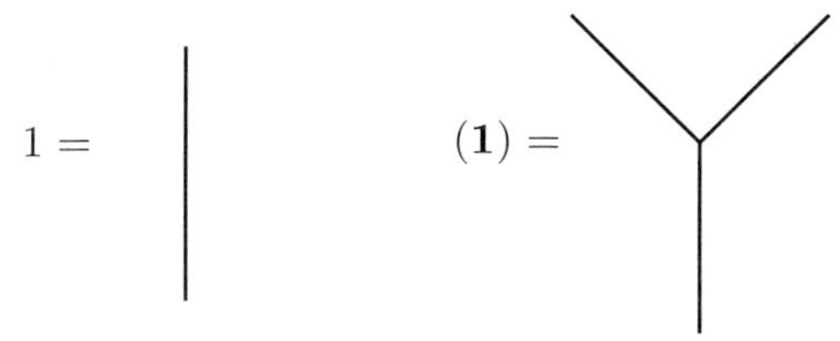

FIGURE 3.3. The identity and the generator in $k[Y^\infty]$. Reprinted with permission from [**10**].

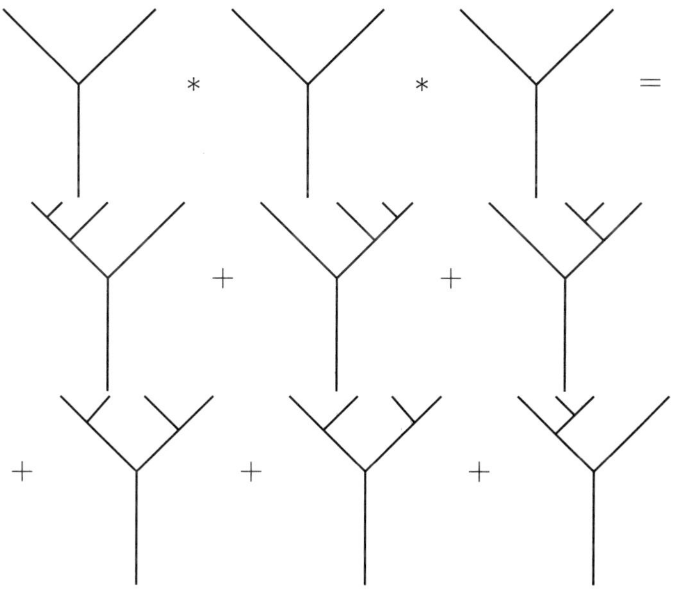

FIGURE 3.4. $(\mathbf{1}) * (\mathbf{1}) * (\mathbf{1}) = (\mathbf{123}) + (\mathbf{321}) + (\mathbf{312}) + (\mathbf{132}) + (\mathbf{231}) + (\mathbf{213})$ computed in $k[S^\infty]$. Note that in $k[Y^\infty]$ the fourth and the fifth trees are the same. Reprinted with permission from [**10**].

the $*$ product and the co-product are internal on the algebra of planar binary trees which is then isomorphic to a sub-Hopf algebra of $k[S^\infty]$ with the same product and co-product. The identity of the Hopf algebra $k[Y^\infty]$ is the tree with a single edge and no vertices, following the convention of considering only internal vertices, which represents the empty permutation, and the trivial permutation of S_1 is represented by the tree with one vertex and two leaves, see fig 3.3. In fact this element is the generator of the augmented algebra by the $*$ product. See fig. (3.4) for an example of an order 3 product.

The grafting $t_1 \vee t_2$ of two trees t_1 and t_2 is the operation of producing a new tree t by inserting t_1 on the left and t_2 on the right leaves of $(\mathbf{1})$. It is clear that any tree of order n can be written as $t_1 \vee t_2$ with t_1 of order p, t_2 of order q and $n = p+q+1$. If a tree has only leaves on the right branch besides the first leaf then it can be written as $| \vee t_2$ and reciprocally if it has only leaves on the left branch besides the last leaf. Note that in particular $(\mathbf{1}) = | \vee |$.

We can give an explicit expression for the correspondence between a permutation $\sigma \in S^n$ and a planar binary tree $t \in Y^n$. A known notion in combinatorics is the standard permutation associated with a set of different positive integers: say that we have a sequence of n positive integers $a = (a_1, a_2 \dots a_n)$ all distinct. The standard permutation $\mathrm{std}(a)$ is the $n-$permutation that has the same ordering of the integers as the sequence a. For instance, for $a = (10, 5, 6)$, we have that $\mathrm{std}(a) = (312)$. It is also possible to generalize it to sequences of repeated positive integers, but we will not need this. To compute the tree t start by identifying the position of n on a permutation $\sigma \in S^n$, $i = \sigma^{-1}(n)$. Then do the standard permutations $\mathrm{std}(\sigma(1)\sigma(2)\dots\sigma(i-1))$ and $\mathrm{std}(\sigma(i+1)\sigma(i+2)\dots\sigma(n))$. Finally the tree t will be the grafting of the two trees that correspond to the two standard permutations. Obviously this is a recursive procedure, starting with $|$ for the empty permutation and $\curlyvee$ for the single permutation $(\mathbf{1})$.

To give a concrete example, take the permutation $(\mathbf{312})$. The left of 3 is the empty permutation which corresponds to $|$. The right of 3 is the permutation $(\mathbf{12})$ which by this iteration procedure can easily be seen to correspond to $\curlyvee \vee |$. In this way we get the third tree on the right of the equation of fig. 3.4.

In [**14**] Loday and Ronco show that the $*$ product restricted to planar binary trees satisfies the identity

$$t * t' = t_1 \vee (t_2 * t') + (t * t_1^{'}) \vee t_2^{'} \tag{7}$$

and

$$t * | = | * t = t \tag{8}$$

with $t = t_1 \vee t_2$ and $t' = t_1^{'} \vee t_2^{'}$.

If t_1 is a tree of order p and a representative element of $W^0_{p+2}(p, L)$ and t_2 is a tree of order q and a representative element of $W^0_{q+2}(p, M)$ then $t = t_1 \vee t_2$ is a tree of order $n = p + q + 1$ and a representative element of

$$W^0_{n+2}(p, K) = K_p(q, \bar{q}) W^0_{p+2}(q, L) W^0_{q+2}(\bar{q}, M).$$

with $K = \{p_1, \dots, p_{n+1}\} = L \cup M$. We will clarify this in what follows.

4. The solution of topological recursion[1]

4.1. Genus 0. The existence of a sort of representation map ψ from the vector space of correlation functions of genus g to the vector space of graphs with loops was suggested in [**10**] such that in particular to a correlation function $W^0_{n+2}(p, p_1, \dots, p_{n+1})$ of Euler characteristic $\chi = -n$ and genus 0 would correspond the trees of order n. In fact the way that this map was defined implied that the representation of $W^0_{n+2}(p, p_1, \dots, p_{n+1})$ is the sum of all trees of order n. In what follows, in particular in section 6, we will see that the map ψ allows to give a ring structure to the space of correlation functions that obey the Eynard-Orantin formula, with an identity that will be given by the cylinder, $W^0_2(p, p_1)$. In fact it is a consequence of the axioms of topological quantum field theory as stated by Atiyah for example in [**2**] that the cylinder $\Sigma \times I$, where Σ is a topological surface without border and I is a interval of real numbers, may be identified with the identity map in a vector space.

[1]In this section we will omit most proofs that can be found in [**10**].

DEFINITION 1. Consider the planar binary tree with one vertex (**1**). The 3-point correlation function $W_3^0(p, p_1, p_2)$ is represented by the sum of two planar binary trees with one vertex, obtained by the permutation of the leaf labels p_1 and p_2:

$$\psi\left(W_3^0(p, p_1, p_2)\right) = \sum_{\text{perm. of leaf labels } \{p_1, p_2\}} (\mathbf{1}) \tag{9}$$

The trees that represent $W_3^0(p, p_1, p_2)$ are given by the permutations of the leaf labels of $| \vee |$. Then it is natural to represent the operation of grafting two trees by the insertion of the recursion kernel $K_p(q, \bar{q})$ on its roots. Therefore the symbol $\curlyvee$ has two meanings. When isolated it represents $W_3^0(p, p_1, p_2)$ because the two cylinders $W_2^0(q, p_1)$ and $W_2^0(\bar{q}, p_2)$ are implicitly identified with its leaves. When it is an internal vertex of a more complex tree it is the recursion kernel $K_p(q, \bar{q})$ with suitable labels of its variables.

DEFINITION 1′. The propagator or cylinder (also named Bergman kernel in the literature) $W_2^0(q, \bar{q})$ is represented through ψ by the empty permutation $|$ and the recursion kernel is represented through ψ by $\curlyvee$ when in an internal vertex of some tree. Then each planar binary tree of order n is a representation of an instance of some correlation function in genus 0 with each vertex identified with a recursion kernel and each left leaf identified with the cylinder $W_2^0(q_i, p_j)$ or each right leaf identified with the cylinder $W_2^0(\bar{q}_i, p_k)$. Finally the image under ψ of a correlation function $W_{n+2}^0(p, p_1, \ldots, p_{n+1})$ with $\chi = -n$ is the sum of all planar binary trees of order n considering all permutations of their leaf labels and with the identifications mentioned above,

$$\psi\left(W_{n+2}^0(p, p_1, \ldots, p_{n+1})\right) = \sum_{\substack{t_i \in Y^n \\ \text{perm. of leaf labels } \{p_1, \ldots, p_{n+1}\}}} t_i. \tag{10}$$

Hence Definition 1 becomes the following example:

EXAMPLE 1. Consider the planar binary tree with one vertex. The 3-point correlation function $W_3^0(p, p_1, p_2)$ is represented by the sum of two planar binary trees with one vertex, obtained by the permutation of the leaf labels p_1 and p_2.

$$\begin{aligned}
\psi\left(W_3^0(p, p_1, p_2)\right) &= \psi\left(K_p(q, \bar{q}) W_2^0(q, p_1) W_2^0(\bar{q}, p_2)\right) + \text{ perm. of } \{p_1, p_2\} \\
&= \sum_{\text{perm. of } \{p_1, p_2\}} | \vee | \\
&= \sum_{\text{perm. of } \{p_1, p_2\}} (\mathbf{1})
\end{aligned} \tag{11}$$

PROPOSITION 1. *If $W_{n+2}^0(p, p_1, \ldots, p_{n+1})$ is a correlation function with Euler characteristic $\chi = -n$ that is a solution of (2) then we have*

$$\begin{aligned}
\psi\left(W_{n+2}^0(p, p_1, \ldots, p_{n+1})\right) = &\sum_{\substack{p+q+1=n \\ |t_1|=p, |t_2|=q}} t_1 \vee t_2 \\
&+ \textit{perm. of leaf labels } \{p_1, \ldots, p_{n+1}\}
\end{aligned} \tag{12}$$

PROOF. This is the topological recursion in genus 0 written with planar binary trees. For $n = 1$ this is the example 1. For n arbitrary by Definition 1′

$$\psi\left(W^0_{n+2}(p, p_1, \ldots, p_{n+1})\right) = \sum_{\substack{t \in Y^n \\ \text{perm. of leaf labels } \{p_1, \ldots, p_{n+1}\}}} t \tag{13}$$

Decompose uniquely any t of order n into $t = t_1 \vee t_2$ of orders $|t_1| = p$ and $|t_2| = q$ with $p + q + 1 = n$ to get

$$\psi\left(W^0_{n+2}(p, p_1, \ldots, p_{n+1})\right) = \sum_{\substack{t_1 \in Y^p, t_2 \in Y^q, p+q+1=n \\ \text{perm. of leaf labels } \{p_1, \ldots, p_{n+1}\}}} t_1 \vee t_2. \tag{14}$$

Then t_1 and t_2 are in the image under ψ of W_{p+2} and W_{q+2} for p and q varying from 0 to $n-1$ and constrained by $p+q+1 = n$. Since the operation of grafting two trees is represented by attaching the recursion kernel to its roots then, summing for all $t_1 \in Y^p, t_2 \in Y^q$ and for $p + q + 1 = n$, we get the topological recursion formula for $g = 0$ after taking the preimage of (14) by ψ:

$$\begin{aligned} &W^0_{n+2}(p, p_1, \ldots, p_{n+1}) \\ &\quad = K_p(q, \bar{q}) \sum_{\substack{L \cup M = \{p_1, \ldots, p_{n+1}\}, \\ |L| = p+1, |M| = q+1}} W^0_{|L|+1}(q, L) W^0_{|M|+1}(\bar{q}, M). \end{aligned} \tag{15}$$

□

REMARK 1. Note that by $W^0_{n+2}(p, K)$ with $|K| = n + 1$ we understand all instances of the correlation function with $g = 0$ and $n + 2$ labels. This is similar to the situation in High Energy Physics where for the same physical process described by a scattering amplitude there are several Feynman diagrams that contribute.

It is well known that the dimension of the vector space generated by planar binary trees of order n is given by the Catalan number (see for instance [**13**])

$$c_n = \frac{2n!}{n!(n+1)n!}.$$

It is also known that correlation functions in Matrix Models have a large N or planar expansion that is given in terms of Catalan numbers. Therefore it is of no surprise that there exists a correspondence between planar binary trees and correlation functions of genus 0.

THEOREM 1. *The n-order solution $W^0_{n+2}(p_1, \ldots, p_{n+1})$ of the topological recursion in genus 0 is represented by the linear combination*

$$\sum t = (\mathbf{1}) * (\mathbf{1}) * \cdots * (\mathbf{1})$$

with n factors of $(\mathbf{1})$ followed by the sum over all permutations of its labels.

In this way $W^0_{n+2}(p, p_1, \ldots, p_{n+1})$ is represented by $\sum t$ followed by the identification of cylinders $W^0_2(q_i, p_j)$ with the left leaves or $W^0_2(\bar{q}_i, p_k)$ with the right leaves and finally by summing over all permutations of the labels $p_1, p_2, \ldots, p_{n+1}$. In other words, the $*$ product $(\mathbf{1}) * (\mathbf{1}) * \cdots * (\mathbf{1})$ gives all possible insertions of recursion kernels of W^0_{n+2}.

4.2. Genus higher than 0. The procedure of attaching an edge to two consecutive leaves and producing a graph with loops allows to represent correlation functions with genus $g > 0$. This is equivalent to extracting the outermost cylinders $W_2^0(x, p_j), W_2^0(y, p_{j+1}), x = q_i$ or $\bar{q}_i$, $y = \bar{q}_j$ or q_j and to couple a cylinder $W_2^0(x, y)$ to two recursion kernels $K_{q_l}(q_i, \bar{q}_i)$ and $K_{q_m}(q_j, \bar{q}_j)$, for some convenient choice of indices, that are identified with two internal vertices. This procedure does not change the Euler characteristic of the associated correlation functions because the number of pairs of leaf labels is reduced exactly as the genus is increased. For instance with this procedure we can make the sequence

$$W_5^0(p, p_1, \ldots p_4) \longrightarrow W_3^1(p, p_1, p_2) \longrightarrow W_1^2(p) \tag{16}$$

and remain in the same graded vector space that contains $k[Y^3]$. How this changes the Hopf algebra structure is not yet clear. For now, we define the operation ${}_i \leftrightarrow_{i+1}$ on a planar binary tree.

DEFINITION 2. Starting with a planar binary tree of order n and $n + 2$ labels (including the root label p) the operation ${}_i \leftrightarrow_{i+1}$ consists in erasing the labels of the leaves i and $i + 1$ then connecting them by an edge and finally relabeling the remaining leaves, now numbered j with $j = 0, \ldots, n - 2$, with the p_{j+1} labels, producing in this way a graph with one loop.

Therefore we represent a correlation function $W_k^g(p, p_1, \ldots, p_{k-1})$ of genus g by graphs with loops t^g that are obtained by successive applications of the ${}_i \leftrightarrow_{i+1}$ operation. We denote by $(Y^n)^g$ the set of different graphs with g loops that are obtained from trees $t \in Y^n$.

DEFINITION 3. A correlation function $W_k^g(p, p_1, \ldots, p_{k-1})$ of genus g and Euler characteristic $\chi = 2 - 2g - k$ is represented by a sum of all different graphs with loops $t^g \in (Y^n)^g$ for $n = -\chi$:

$$\psi\left(W_k^g(p, p_1, \ldots, p_{k-1})\right) = \sum_{t^g \in (Y^n)^g} t^g \tag{17}$$

REMARK 2. Two graphs $(t)^g, (t')^g \in (Y^n)^g$ are considered different in the obvious way. Either the underlying binary trees $t, t' \in Y^n$ are distinct as base elements of $k[Y^\infty]$ or the tree t has a pair of leaves, say $(i, i+1)$ that are identified with an edge in $(t)^g$ producing a loop and are free in $(t')^g$ (and reciprocally because the two graphs have the same genus).

In particular $W_1^1(p)$ is represented by a single graph with one loop denoted $(\mathbf{1})^1$ whose underlying planar binary tree is $(\mathbf{1})$. More generally we have

PROPOSITION 2. *The second summand of the topological recursion formula for the correlation function $W_n^1(p, p_1, \ldots, p_{n-1})$ with $\chi = -n$ is represented by the sum*

$$\sum_{\substack{(t_1)^1 \in (Y^p)^1, t_2 \in Y^q \\ p+q+1=n}} (t_1)^1 \vee t_2 + \sum_{\substack{t_1 \in Y^p, (t_2)^1 \in (Y^q)^1 \\ p+q+1=n}} t_1 \vee (t_2)^1 \tag{18}$$

where each underlying planar binary tree t of order n is decomposed as $t = t_1 \vee t_2$, with $|t_1| = p$, $|t_2| = q$, $p + q + 1 = n$.

Now we consider the first term in topological recursion which for $W_n^1(p, p_1, \ldots, p_{n-1})$ is

$$K_p(q, \bar{q}) W_{n+1}^0(q, \bar{q}, p_1, \ldots, p_{n-1}). \tag{19}$$

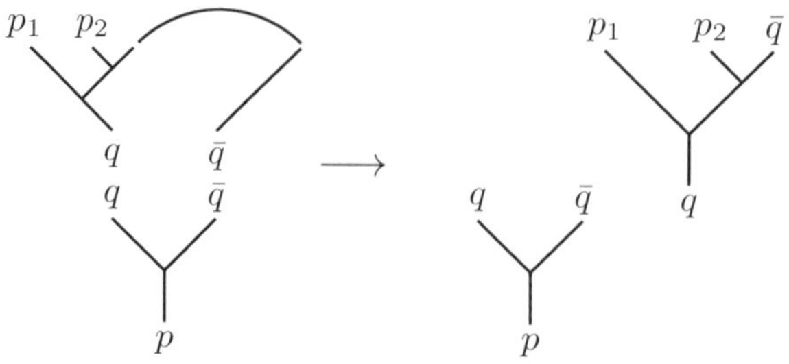

FIGURE 4.1. Ungrafting a 1-loop graph of $W_3^1(p, p_1, p_2)$. The resulting tree is (**21**). Reprinted with permission from [**10**].

We start by a definition:

DEFINITION 4. The ungrafting operation $\overline{\vee}$ is defined by removing from t the tree (**1**) that contains the root producing a forest with two trees t_1 and t_2. When t represents an instance of a correlation function then the roots of t_1 and t_2 are labeled by q and $\bar{q}$ and as before the tree (**1**) represents $K_p(q, \bar{q})$.

REMARK 3. The operations $\vee$ and $\overline{\vee}$ are similar to the operations B^+ and B^- of the Connes-Kreimer Hopf algebra described, for instance, in [**7**].

If we start with the planar binary tree $t \in Y^n$ with leaves labels $p_1, \ldots, p_{n+1}$ and root label p and identify two nearest neighbour leaf in opposite branches then we get a 1-loop graph $t^1 \in (Y^n)^1$ with a relabeling $p_1, \ldots, p_{n-1}$. Then, by applying $\overline{\vee}$, we get another tree t' with two more edges with labels q and $\bar{q}$ besides the leaves labelled by $p_1, \ldots, p_{n-1}$. This tree is isomorphic as a graph to a planar binary tree in Y^{n-1} that we denote also t' by promoting the edge with the label q to the root and the other edge to the rightmost leaf, see fig 4.1 for an example with $W_3^1(p, p_1, p_2)$. In this way we get a representation of (19) by summing over all planar binary trees $t' \in Y^{n-1}$.

In [**10**] we prove the following proposition

PROPOSITION 3. *The representation of the first term of the topological recursion formula for $W_n^1(p, p_1, \ldots, p_{n-1})$ is given by the identification of leaves on opposite branches of the decomposition $t = t_1 \vee t_2$ with $t \in Y^n, t_1 \in Y^p, t_2 \in Y^q, p+q+1 = n$:*

$$\psi\left(K_p(q, \bar{q}) W_{n+1}^0(q, \bar{q}, p_1, \ldots, p_{n-1})\right) = \sum_{\substack{t_1 \in Y^p, t_2 \in Y^q \\ p+q+1=n}} t_1 \widehat{\vee} t_2 \tag{20}$$

The obvious notation $\widehat{\vee}$ means that two consecutive leaves in opposite branches are identified.

Therefore we have exhausted all possibilities of obtaining 1-loop graphs from planar binary trees of order n and the two previous propositions imply the following theorem.

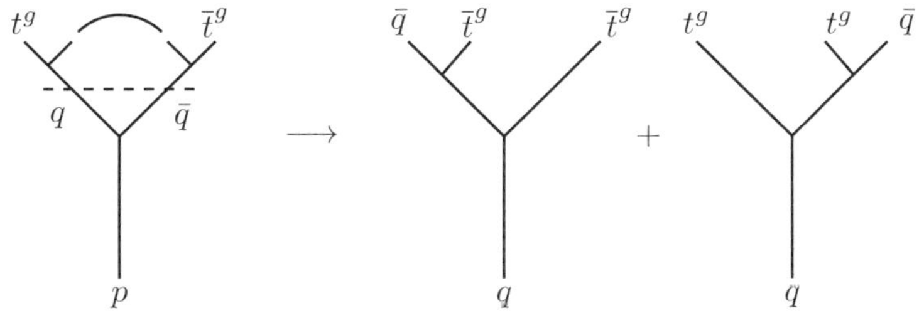

FIGURE 4.2. A symmetric $2g+1$-loop graph with t^g a $g-$loop graph. The graph $\bar{t}^g$ is the reflection of t^g in the vertical axis that passes through the root. After the ungrafting operation the first graph is obtained by exchanging q and $\bar{q}$ on the original graph. Reprinted with permission from [**10**].

THEOREM 2. *The n order solution $W_n^1(p, p_1, \ldots, p_{n-1})$ of the topological recursion in genus 1 is given by $(\mathbf{1}) * (\mathbf{1}) * \cdots * (\mathbf{1})$, with n factors, followed by the identification of pairs of nearest neighbour leaves producing 1-loop graphs and finally by summing over all permutations of leaves labels $p_1, p_2, \ldots, p_{n-1}$:*

$$\psi\left(W_n^1(p, p_1, \ldots, p_{n-1})\right)$$

$$\sum_{perm.\ of\ \{p_1,\ldots,p_{n-1}\}} \sum_{i=0}^{n-1} =_{i \leftrightarrow i+1} \left((\mathbf{1}) * (\mathbf{1}) * \cdots * (\mathbf{1})\right) \tag{21}$$

Next we proved in [**10**] a simple lemma regarding symmetric graphs as in fig. 4.2. Note that the resulting ungrafted graphs have the left-right order of the right branch of the original graph exchanged:

LEMMA 1. *If a graph that enters in the representation of the correlation function $W_{k+1}^{2g+1}(p, K)$ has nearest neighbour leaves identified in different branches and is symmetric with respect to the vertical axis that passes through the root then it has a weight factor of 2, that is, it appears two times in the complete graph representation of $W_{k+1}^{2g+1}(p, K)$.*

EXAMPLE 2. See fig. 4.3 for the graph representation of $W_3^1(p, p_1, p_2)$ with a graph of weight 2.

A simple but important fact is that ${}_i \leftrightarrow_{i+1}$ acts as a derivation when applied independently to the left and right branches of a tree $t = t_1 \vee t_2$. This is apparent in Proposition 2. However when summing over all graphs we must take care with overcounting. If we start with $t^1 = (t_1)^1 \vee t_2 + t_1 \vee (t_2)^1$ with $|t| = n, |t_1| = p, |t_2| = q, p + q + 1 = n$ and apply ${}_i \leftrightarrow_{i+1}$ to the two branches independently and sum over

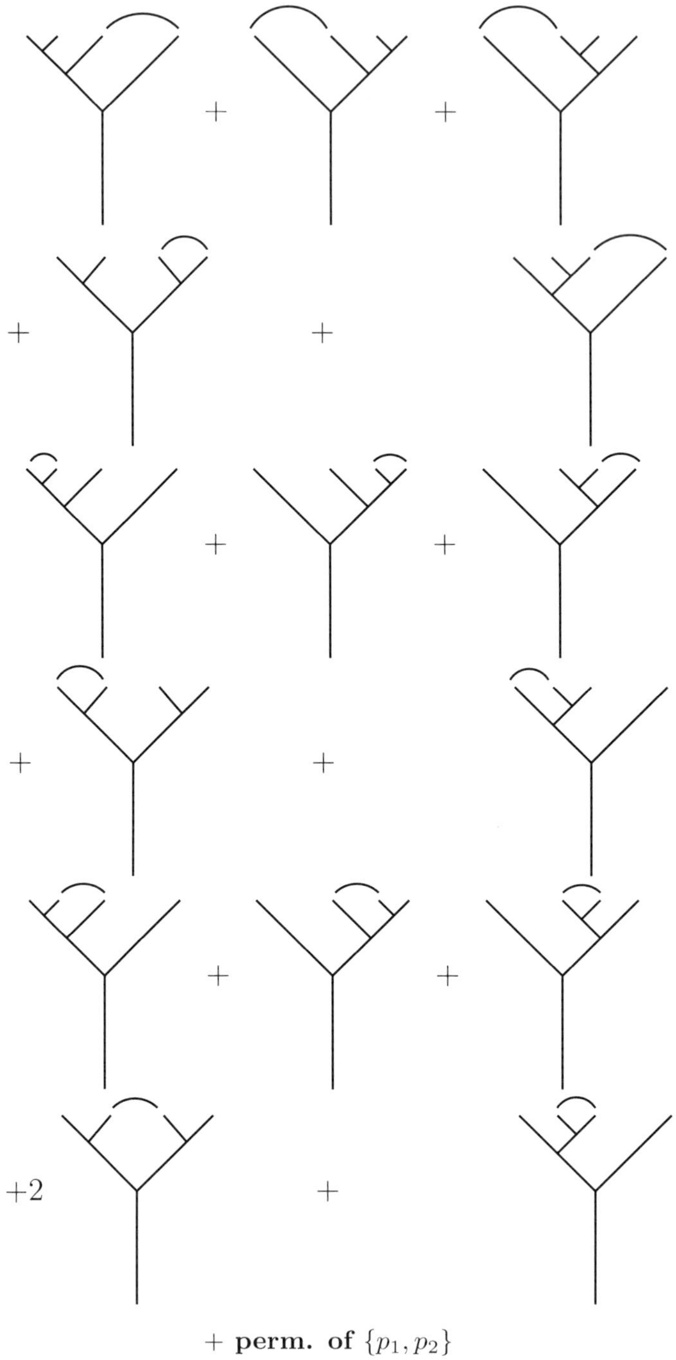

FIGURE 4.3. $W_3^1(p, p_1, p_2)$. The root label p and the leaf labels $\{p_1, p_2\}$ are omitted. Reprinted with permission from [**10**].

all different graphs we get

$$\sum_{i=0}^{n-1} ({}_i \leftrightarrow_{i+1})_{\text{same branches}} \left(\sum_{\substack{(t_1)^1 \in (Y^p)^1, t_2 \in Y^q \\ p+q+1=n}} (t_1)^1 \vee t_2 + \sum_{\substack{t_1 \in Y^p, (t_2)^1 \in (Y^q)^1 \\ p+q+1=n}} t_1 \vee (t_2)^1 \right)$$

$$= \sum_{\substack{(t_1)^2 \in (Y^p)^2, t_2 \in Y^q \\ p+q+1=n}} 2(t_1)^2 \vee t_2 + \sum_{\substack{(t_1)^1 \in (Y^p)^1, (t_2)^1 \in (Y^q)^1 \\ p+q+1=n}} 2(t_1)^1 \vee (t_2)^1$$

$$+ \sum_{\substack{t_1 \in Y^p, t_2 \in Y^q, (t_2)^2 \in (Y^q)^2 \\ p+q+1=n}} 2t_1 \vee (t_2)^2, \tag{22}$$

whenever the operation is well defined[2]. The reasoning for the factors of 2 is the double counting of identical graphs. For instance if $(t_1)^1$ has a loop starting at leaf 2 and $(t_1)^2$ was obtained by producing a second loop starting at leaf 0, then this $(t_1)^2$ is identical to the 2-loop graph that was obtained by producing a second loop at leaf 2 in $(t_1)^1$ that had already a loop starting at leaf 0.

Therefore the sum of different 2-loop graphs obtained from all planar binary trees $t = t_1 \vee t_2$ is

$$\sum_{(t)^2 \in (Y^n)^2} (t)^2 = \sum_{\substack{(t_1)^2 \in (Y^p)^2, t_2 \in Y^q \\ p+q+1=n}} (t_1)^2 \vee t_2 + \sum_{\substack{(t_1)^1 \in (Y^p)^1, (t_2)^1 \in (Y^q)^1 \\ p+q+1=n}} (t_1)^1 \vee (t_2)^1 + \sum_{\substack{t_1 \in Y^p, (t_2)^2 \in (Y^q)^2 \\ p+q+1=n}} t_1 \vee (t_2)^2. \tag{23}$$

As for the identification of two consecutive leafs in separate branches, we start again from $t^1 = (t_1)^1 \vee t_2 + t_1 \vee (t_2)^1$ to get

$$t^2 = (t_1)^1 \widehat{\vee} t_2 + t_1 \widehat{\vee} (t_2)^1 \tag{24}$$

The fact that it may not be always possible to contract two leaves in opposite branches is important to count the dimensions of the vector spaces $k[(Y^n)^g]$ generated by graphs with g loops but we will not consider this.

It is clear that we can continue this procedure and generate graphs $t^g \in (Y^n)^g$ with an increasing number of g loops and k labels (including the root) as far as the consistence of the relation $-n = 2 - 2g - k$ permits. Hence we have the following proposition.

[2]The operation is not well defined if there is only one leaf available before and/or after a certain loop or if there are no more leaves to contract. In this case we set to 0 the result of acting with ${}_i \leftrightarrow_{i+1}$.

PROPOSITION 4. *The graphs with loops $t^g \in (Y^n)^g$ and k labels, including the root, that are compatible with $-n = 2 - 2g - k$ are obtained from planar binary trees $t = t_1 \vee t_2$ by successive applications of $({}_i \leftrightarrow_{i+1})_{same\ branches}$ and $({}_i \leftrightarrow_{i+1})_{opposite\ branches}$ to all $t \in Y^n$:*

$$
\begin{aligned}
\sum_{(t)^g \in (Y^n)^g} (t)^g = &\sum_{k=0}^{g} \sum_{\substack{(t_1)^k \in (Y^p)^k, (t_2)^{g-k} \in (Y^q)^{g-k} \\ p+q+1=n}} \left((t_1)^k \vee (t_2)^{g-k}\right) \\
&+ \sum_{k=0}^{g-1} \sum_{\substack{(t_1)^k \in (Y^p)^k, (t_2)^{g-k} \in (Y^q)^{g-k} \\ p+q+1=n}} \left((t_1)^{g-1-k} \widehat{\vee} (t_2)^k\right)
\end{aligned} \tag{25}
$$

THEOREM 3. *The n order solution W^g_{2-2g+n} of the topological recursion in genus $g > 0$ and $k = 2 - 2g + n > 0$ variables is given by $(\mathbf{1}) * (\mathbf{1}) * \cdots * (\mathbf{1})$, with n factors, followed by the identification of pairs of nearest neighbour leaves producing graphs with loops as in Proposition 4 and finally by summing over all permutations of $p_1, p_2, \ldots, p_{1-2g+n}$.*

REMARK 4. After being ungrafted the second term in (25) can still have leaves in opposite branches identified. In general this happens if $(t)^g = (t_1)^{g_1} \widehat{\vee} (t_2)^{g_2}$ with $g_1 + g_2 = g - 1$ and say $(t_1)^{g_1}$ has a decomposition $(t_1)^{g_1'} \widehat{\vee} (t_2)^{g_2'}$ with $g'_1 + g'_2 = g_1 - 1$.

EXAMPLE 3. See fig. 4.4 for the graph representation of $W_1^2(p)$.

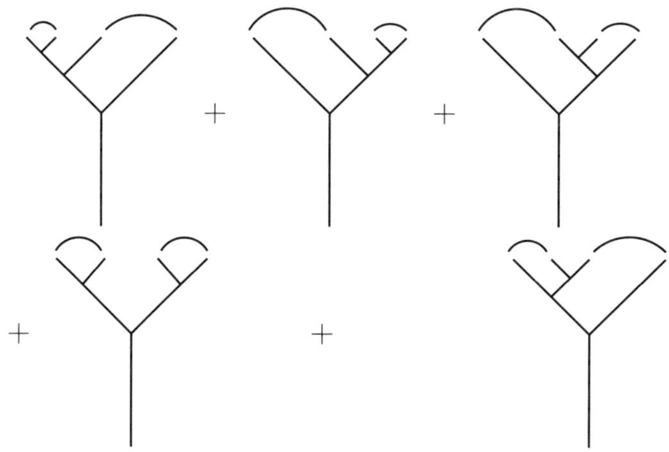

FIGURE 4.4. $W_1^2(p)$ The root label p is omitted. Reprinted with permission from [**10**].

5. The antipode

In a graded connected Hopf algebra there is a canonical antipode S whose expression is given by the convolution inverse of the identity:

$$
m\,(S \otimes \mathrm{Id})\,\Delta = m\,(I \otimes \mathrm{S})\,\Delta = \eta \cdot \epsilon \tag{26}
$$

with m the product, Δ the co-product, η the unit and ϵ the co-unit. Explicitly, in the Loday-Ronco Hopf algebra, we have

$$S(t) = -t - S(t_1) * t_2 \tag{27}$$

where in Sweedler notation

$$\Delta t = \sum t_1 \otimes t_2 \tag{28}$$

is the co-product in $k[Y^\infty]$ induced by (6). For instance, $S(\mathbf{1}) = -\mathbf{1}$ because $\mathbf{1}$ is primitive and $S(\mathbf{12}) = (\mathbf{21})$ and also $S(\mathbf{21}) = (\mathbf{12})$. This suggests that a map $\psi^* S$ induced by the antipode on the vector space of correlations functions should give

$$(\psi^* S)(W_4^0) = W_4^0.$$

More generally, from $S((\mathbf{1})) = -(\mathbf{1})$ we see that

$$S((\mathbf{1}) * (\mathbf{1}) * \cdots * (\mathbf{1})) = (-1)^n (\mathbf{1}) * (\mathbf{1}) * \cdots * (\mathbf{1})$$

with n factors in the $*$ product and then

$$(\psi^* S)(W_{n+2}^0) = (-1)^n W_{n+2}^0.$$

Since n is identified with the Euler characteristic we see that the induced map respects the grading of $k[Y^\infty]$ for $g = 0$.

6. Discussion and other topics

This section contains some new material which may not be in a complete final form.

We would like to give a more direct approach to Topological Recursion in genus 0 in terms of the co-product of the Loday-Ronco Hopf algebra and analyze the algebraic structure of the space of correlation functions. We start by recalling that the elementary tree $\curlyvee$ is the same as the elementary permutation $(\mathbf{1})$ and that this tree is primitive in the Loday-Ronco Hopf algebra:

$$\Delta \curlyvee = 1 \otimes \curlyvee + \curlyvee \otimes 1. \tag{29}$$

Now by Theorem 1 a solution of Topological Recursion of Euler characteristic $-n$ in genus 0 is given by the product of n terms of $(\mathbf{1})$, modulo the permutations of leaves labels, and since the co-product is a homomorphism we have that

$$\begin{aligned} \Delta\left((\mathbf{1})^n\right) &= \left(\Delta(\mathbf{1})\right)^n \\ &= \left(1 \otimes (\mathbf{1}) + (\mathbf{1}) \otimes 1\right)^n \\ &= \sum_{k=0}^{n} C_k^n (\mathbf{1})^k \otimes (\mathbf{1})^{n-k} \end{aligned} \tag{30}$$

where in the powers the Loday-Ronco $*$ product is understood. If we absorb the factor $n!$ in a normalization of n factors of $(\mathbf{1})$ then the co-product is

$$\Delta\left((\mathbf{1})^n\right) = \sum_{k=0}^{n} (\mathbf{1})^k \otimes (\mathbf{1})^{n-k} \tag{31}$$

This is exactly the same situation that appears in the elementary Hopf algebra $k[X]$ of polynomials in one variable X over some field k of characteristic 0. By declaring

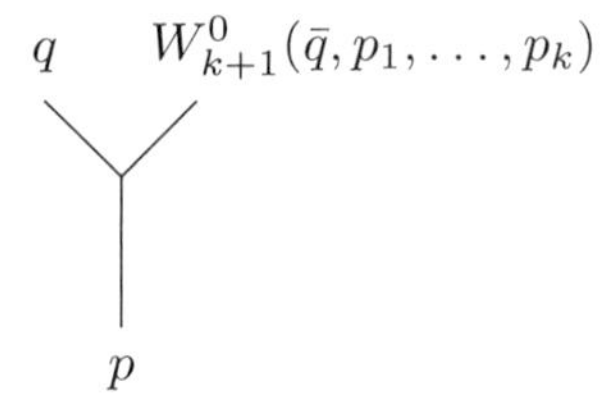

FIGURE 6.1. $W^0_{k+2}(p, q, p_1, \ldots, p_k)$

the variable X to be primitive and normalizing X^n with $n!$ we get

$$\Delta\left(\frac{X^n}{n!}\right) = \sum_{k=0}^{n} \frac{X^k}{k!} \otimes \frac{X^{n-k}}{(n-k)!} \tag{32}$$

With the map ψ we can translate this to the language of correlation functions in genus zero $W^0_{n+2}(p, p_1, \ldots, p_{n+1})$ with Euler characteristic $\chi = -n$:

$$\begin{aligned} &\Delta\left(W^0_{n+2}(p, p_1, \ldots, p_{n+1})\right) \\ &= \sum_{k=0}^{n} W^0_{k+2}(p, p_1, \ldots, p_k, \bar{q}) \otimes W^0_{n-k+2}(\bar{q}, p_{k+1}, \ldots, p_{n+1}) \\ &\qquad + \sum_{k=0}^{n} W^0_{k+2}(q, p_1, \ldots, p_{k+1}) \otimes W^0_{n-k+2}(p, q, p_{k+2}, \ldots, p_{n+1}) \end{aligned} \tag{33}$$

where a normalization of $n!$ for W^0_{n+2} is implicit. An explanation for the extra labels $q, \bar{q}$ and also for the two sums is in order. Note that the co-product in (31) just amounts to cut the product of n (**1**)'s into two consecutive factors. At the level of trees this amounts to cut a complex tree into two more elementary trees, where the cut is done along an internal vertex, giving a tree that contains the original root and a new leaf just below the cut and another tree that contains a new root just above the cut and some leaves of the original tree. In this way, every time a tree is cut along an internal vertex we assign the label q if the new leaf is left oriented and $\bar{q}$ if it is right oriented. Moreover the new root of the second tree is also labeled q and $\bar{q}$ respectively. That is why it is necessary to take into account the two decompositions, say a left and a right one. Hence in our approach the object $W^0_{k+2}(p, q, p_1, \ldots, p_k)$ that contains simultaneously the root p and an internal leaf label q_i (or $\bar{q}_i$) is an intermediate object between a correlation function and a recursion kernel in the sense that the internal label does not couple to a cylinder, see fig. 6.1 As an example take for instance $W^0_4(p, p_1, p_2, p_3)$. We get

$$\begin{aligned} \Delta W^0_4(p, p_1, p_2, p_3) &= 1 \otimes W^0_4(p, p_1, p_2, p_3) + W^0_4(p, p_1, p_2, p_3) \otimes 1 \\ &+ W^0_3(p, p_1, \bar{q}) \otimes W^0_3(\bar{q}, p_2, p_3) + W^0_3(q, p_1, p_2) \otimes W^0_3(p, q, p_3). \end{aligned} \tag{34}$$

If we look at the two trees in fig. 6.2 we see that only the tree (**21**) admits a decomposition like the third term and only the tree (**12**) admits a decomposition like the fourth term. Next translating $W^0_3(p, p_1, q)$ into the expression

$$W^0_3(p, p_1, q) = K_p(q, \bar{q}) W^0_2(\bar{q}, p_1) \tag{35}$$

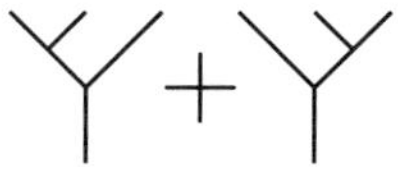

FIGURE 6.2. Representation of $W_4^0(p, p_1, p_2, p_3)$. Root and leaves labels are not shown.

we get the terms of Topological Recursion for $W_4^0(p, p_1, p_2, p_3)$. Note that $W_3^0(p, p_1, q)$ carries an internal index that results from an internal cut, so this index does not couple to a cylinder.

Now is easy to see that on the one hand the expression (33) gives all the terms of the first iteration of Topological Recursion in genus 0 for W_{n+2}^0 and on the other hand that by successive iterations of this co-product we end up with the complete decomposition of the solution in terms of recursion kernels. More specifically what we need is the reduced co-product Δ' that is equal to the co-product minus the primitive part:

$$\Delta' a = \Delta a - 1 \otimes a - a \otimes 1 \tag{36}$$

for a generic element a in some Hopf algebra. Note that $\Delta'(a) = 0$ for a primitive element a, and also by definition $\Delta' 1 = 0$. In particular,

$$\Delta' W_3^0(p, p_1, p_2) = 0. \tag{37}$$

Then we take the iteration $\Delta'^{(n)}$:

$$\begin{aligned} \Delta'^{(1)} &= \Delta' \\ \Delta'^{(2)} &= (\Delta' \otimes \mathrm{Id}) \otimes \Delta'^{(1)} \\ \Delta'^{(n)} &= (\Delta' \otimes \mathrm{Id}) \Delta'^{(n-1)}. \end{aligned} \tag{38}$$

Let us take a product of n terms of $(\mathbf{1})$ and apply $\Delta'^{(n-1)}$:

$$\begin{aligned} \Delta'\left((\mathbf{1})^n\right) &= (1 \otimes (\mathbf{1}) + (\mathbf{1}) \otimes 1)^n - (\mathbf{1})^n \otimes 1 - 1 \otimes (\mathbf{1})^n \\ &= \sum_{k=1}^{n-1} (\mathbf{1})^k \otimes (\mathbf{1})^{n-k}; \\ \Delta'^{(2)}\left((\mathbf{1})^n\right) &= (\Delta' \otimes \mathrm{Id}) \left(\sum_{k_1=1}^{n-1} (\mathbf{1})^{k_1} \otimes (\mathbf{1})^{n-k_1} \right) \\ &= \left(\sum_{k_1=2}^{n-1} \sum_{k_2=1}^{k_1-1} (\mathbf{1})^{k_2} \otimes (\mathbf{1})^{k_1-k_2} \otimes (\mathbf{1})^{n-k_1} \right) \end{aligned}$$

$$\Delta'^{(n-1)}\left((\mathbf{1})^n\right) = (\mathbf{1}) \otimes (\mathbf{1}) \otimes \cdots \otimes (\mathbf{1}), n \text{ factors}. \tag{39}$$

where we have absorbed the combinatorial factors on a normalization of $(\mathbf{1})$ factors. However the simplicity of this expression is misleading because we are ignoring the root and leaves labels. In fact all $(\mathbf{1})$'s represent recursion kernels that carry coupled internal indexes.

Now that we have discussed the co-product at the level of correlation functions of genus 0 and not at the level of planar binary trees and have given the decomposition of Topological Recursion in terms of recursion kernels using this co-product, we should see if there is a product in the first place. But this is immediately induced by our solution stated by Theorem 1. Given two correlation functions W^0_{k+2} and W^0_{l+2} (here we let drop the map ψ which we implicitly assume)

$$W^0_{k+2}(p_1,\ldots,p_{k+1}) = (\mathbf{1}) * (\mathbf{1}) * \cdots * (\mathbf{1}), k \text{ factors},$$

$$W^0_{l+2}(p_1,\ldots,p_{l+1}) = (\mathbf{1}) * (\mathbf{1}) * \cdots * (\mathbf{1}), l \text{ factors}, \tag{40}$$

then their product is

$$\begin{aligned}&W^0_{k+2}(p,p_1,\ldots,p_{k+1}) * W^0_{l+2}(p',p'_1,\ldots,p'_{l+1})\\ &= W^0_{m+2}(p,p_1,\ldots,p_{m+1}) = (\mathbf{1}) * (\mathbf{1}) * \cdots * (\mathbf{1}), m = k+l \text{ factors}\end{aligned} \tag{41}$$

which is clearly associative, commutative and with the unit $W^0_2(p,p_1)$. However we are ignoring the role of the root and leaves labels. It is possible to generalize this product in order to take into account the labels. First we consider the product of a cylinder $W^0_2(p,p_1)$ and a generic correlation function $W^0_{n+2}(p',p'_1,\ldots,p'_{n+1})$:

$$\begin{aligned}W^0_2(p,p_1) * W^0_{n+2}(p',p'_1,\ldots,p'_{n+1}) =& W^0_{n+2}(p',p'_1,\ldots,p'_{n+1}) * W^0_2(p,p_1)\\ =& W^0_{n+2}(p,p'_1,\ldots,p'_{n+1})\end{aligned} \tag{42}$$

This amounts to setting $p' = p_1$ and replacing the root label of W^0_{n+2} by p. Next we use the expression (7) for the product of two planar binary trees in order to get the product of two W^0_3:

$$\begin{aligned}&W^0_3(p,p_1,p_2) * W^0_3(p',p'_1,p'_2)\\ &= W^0_2(q,p_1) \vee (W^0_2(\bar{q},p_2) * W^0_3(p',p'_1,p'_2)\\ &\quad + (W^0_3(p,p_1,p_2) * W^0_2(q',p'_1)) \vee W^0_2(\bar{q}',p'_2)\\ &= W^0_2(q,p_1) \vee W^0_3(\bar{q},p'_1,p'_2) + W^0_3(q',p_1,p_2) \vee W^0_2(\bar{q}',p'_2)\\ &= K_p(q,\bar{q}) W^0_2(q,p_1) W^0_3(\bar{q},p'_1,p'_2) + K_{p'}(q',\bar{q}') W^0_3(q',p_1,p_2) W^0_2(\bar{q}',p'_2),\end{aligned} \tag{43}$$

where we have used the decomposition

$$W^0_3(p,p_1,p_2) = K_p(q,\bar{q}) W^0_2(q,p_1) W^0_2(\bar{q},p_2) \tag{44}$$

In the particular case of equal root labels and $p'_1 = p_2, p'_2 = p_3$, we get

$$W^0_3(p,p_1,p_2) * W^0_3(p,p_2,p_3) = W^0_4(p,p_1,p_2,p_3). \tag{45}$$

Note that this product is no longer commutative and is additive on the Euler characteristic. Finally we define in a similar way the product for two generic correlation functions of $g = 0$:

$$\begin{aligned}&W^0_{l+2}(p,p_1,\ldots,p_{l+1}) * W^0_{m+2}(p',p'_1\ldots,p'_{m+1})\\ &= \delta_{p,p'}\delta_{p_{l+1},p'_1} W^0_{k+2}(p,p_1,\ldots,p_{k+1}), k = l+m.\end{aligned} \tag{46}$$

with the redefinition $p_{l+2} = p'_2,\ldots,p_{l+m+1} = p'_{m+1}$. We show that this product is compatible with Topological Recursion. Start with

$$\begin{aligned}&W^0_{l+2}(p,p_1,\ldots,p_{l+1})\\ &= K_p(q,\bar{q}) \sum_{k=0}^{l-1} W^0_{k+2}(q,\ldots,p_{k+1}) W^0_{l-k+1}(\bar{q},p_{k+2},\ldots,p_{l+1})\\ &= \sum_{t_1\in Y^p, t_2\in Y^q, p+q+1=l} t_1 \vee t_2\end{aligned} \tag{47}$$

and similarly for $W^0_{m+2}(p,p_1,\dots,p_{m+1})$. In terms of planar trees and using (7) we get

$$
\begin{aligned}
&W^0_{l+2}(p,p_1,\dots,p_{l+1}) * W^0_{m+2}(p',p_1'\dots,p_{m+1}') \\
(48)\qquad &= \sum_{t_1\in Y^p, t_2\in Y^q, p+q+1=l} \; \sum_{t_1'\in Y^{p'}, t_2'\in Y^{q'}, p'+q'+1=m} t_1\vee(t_2 * t') + (t*t_1')\vee t_2'.
\end{aligned}
$$

Translating this into correlation functions,

$$
\begin{aligned}
&W^0_{l+2}(p,p_1,\dots,p_{l+1}) * W^0_{m+2}(p',p_1'\dots,p_{m+1}') = K_p(q,\bar q) \\
&\quad \times\Big(\sum_{k=0}^{l-1} W^0_{k+2}(q,p_1,\dots,p_{k+1})\vee \\
&\Big(W^0_{l-k+1}(\bar q,p_{k+2},\dots,p_{l+1}) * W^0_{m+2}(p',p_1'\dots,p_{m+1}')\Big)\Big) \\
&\quad + K_{p'}(q',\bar q')\times \\
&\left(\sum_{n=0}^{m-1}\Big(W^0_{l+2}(p,p_1,\dots,p_{l+1}) * W^0_{n+2}(q',p_1',\dots,p_{n+1}')\Big)\right. \\
&\quad \left.\vee W^0_{m-n+1}(\bar q',p_{n+2}',\dots,p_{m+1}')\right) \\
&= K_p(q,\bar q)\times \\
&\left(\sum_{k=0}^{l-1} W^0_{k+2}(q,p_1,\dots,p_{k+1}) W^0_{l+m-k+1}(\bar q,p_{k+2},\dots,p_{l+m+1})\right. \\
&\quad \left. + \sum_{n=0}^{m-1} W^0_{l+n+2}(q,p_1,\dots,p_{l+n+1}) W^0_{m-n+1}(\bar q,p_{n+2}',\dots,p_{m+1}')\right) \\
&= K_p(q,\bar q)\times \\
&\left(\sum_{k=0}^{l-1} W^0_{k+2}(q,p_1\dots,p_{k+1}) W^0_{l+m-k+1}(\bar q,p_{k+2},\dots,p_{l+m+1})\right. \\
&\quad \left. + \sum_{k=l}^{m+l-1} W^0_{k+2}(q,p_1,\dots,p_{k+1}) W^0_{l+m-k+1}(\bar q,p_{k-l+2}',\dots,p_{m+1}')\right) \\
(49)\qquad &= W^0_{l+m+2}(p,p_1,\dots,p_{l+m+1})
\end{aligned}
$$

for $p=p'$ and with obvious redefinitions of the variables. Therefore we recover (41) and confirm the fact that the space of correlation functions has an algebra structure. It is also a bi-algebra as the product and the co-product are induced by these same structures in the Loday-Ronco Hopf algebra. The antipode has been discussed in the previous section so this algebra is indeed a Hopf algebra.

6.1. The Connes-Kreimer Hopf algebra. Here we won't have the opportunity to give a detailed exposition of the intricacies of the Renormalization procedure in Quantum Field Theory [**6**] and how the combinatorics of this procedure are well described by Connes-Kreimer Hopf algebra. A good review can be found in [**7**]. We just mention that the space of this Hopf algebra is the vector space generated by rooted trees over $\mathbb{Q}$, that are graphs with no loops, with a preferred vertex called the root that has only child edges, and with vertices's of any order. These trees are not planar and the product in the algebra is just disjoint union. For instance, the elementary tree $\bullet$ describes any Feynman diagram that is divergent but that has no subdivergent diagrams, that are subdiagrams of a more complex diagram that are by themselves divergent and whose divergent integral is a function of the integration variable of the diagram to which they are a part of[3]. This situation goes on order-by-order in perturbation theory, or equivalently in loop order, giving more and more nested subdiagrams and subdivergences. The first example of a nested divergence is the tree $\begin{smallmatrix}\bullet\\ |\\ \bullet\end{smallmatrix}$ that describes any divergent Feynman diagram with a subdivergent diagram that doesn't have by itself any subdiagram. Any rooted tree of any order with an arbitrary number of branches can be used to describe some complicated divergent integrals associated to some Feynman diagrams. See for instance [**3**–**5**]. Note that a particularly interesting aspect is the fact that rooted trees in the Connes-Kreimer algebra describe what are known as One Particle Irreducble (OPI) Feynman diagrams. These are connected diagrams that represent the Legendre transform of connected correlation functions. Its is well known that in Quantum Field Theory connected correlation functions are generated by the free energy F while products of connected correlation functions are generated by the partition function Z:

$$Z = \exp F. \tag{50}$$

We will come back to this later.

There is a generalization due to Foissy [**12**] of the Connes-Kreimer Hopf algebra called the noncommutative Connes-Kreimer Hopf algebra that considers planar rooted trees. The product is ordered disjoint union and is no longer commutative as for instance one distinguishes $\bullet\begin{smallmatrix}\bullet\\ |\\ \bullet\end{smallmatrix}$ from $\begin{smallmatrix}\bullet\\ |\\ \bullet\end{smallmatrix}\bullet$. The article [**1**] describes a nice inductive map ϕ to compute the planar binary tree that corresponds to a planar rooted tree or a planar forest in the noncommutative Connes-Kreimer Hopf algebra. We need to introduce two operations $\backslash$ and $/$ on trees t_1 and t_2, $t_1\backslash t_2$ and t_1/t_2. The first amounts to graft t_2 on the right most leaf of t_1 and the second is the grafting of t_1 on the left most leaf of t_2, see fig. 6.3 for a simple example (there are corresponding operations on permutations that can also be found in [**1**]). First, $\phi(\emptyset) = |$. Next, removing the root from a planar rooted tree t gives a planar forest $f = (t_1, t_2, \ldots, t_n)$ with each t_i a planar rooted tree and we set $\phi(t) = \phi(f)/Y$. Finally for a forest f we do $\phi(f) = \phi(t_1)\backslash\phi(t_2)\backslash\ldots\backslash\phi(t_n)$. For instance, $\phi(\bullet) = Y$, $\phi(\begin{smallmatrix}\bullet\\ |\\ \bullet\end{smallmatrix}) = \phi(\bullet)/\phi(\bullet) = (\mathbf{12})$, $\phi(\bullet\bullet) = \phi(\bullet)\backslash\phi(\bullet) = (\mathbf{21})$.

[3]We should say in passing that a regularization procedure is necessary to give a meaning to the manipulation of these integrals and this is tacitly assumed.

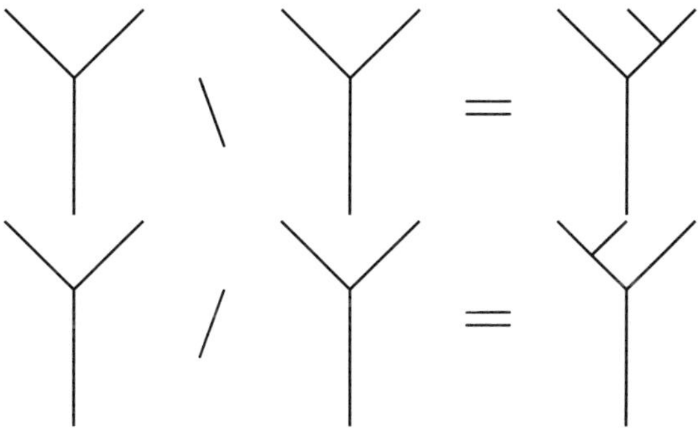

FIGURE 6.3. The \ and / operation on trees.

Now exponentiate the elementary tree (**1**) multiplied by a constant g that we would like to see as a coupling constant in some physical theory:

$$\exp(g\Upsilon) = 1 + g\Upsilon + \frac{g^2}{2}\Upsilon * \Upsilon + \frac{g^3}{3!}\Upsilon * \Upsilon * \Upsilon + \dots$$

(51)

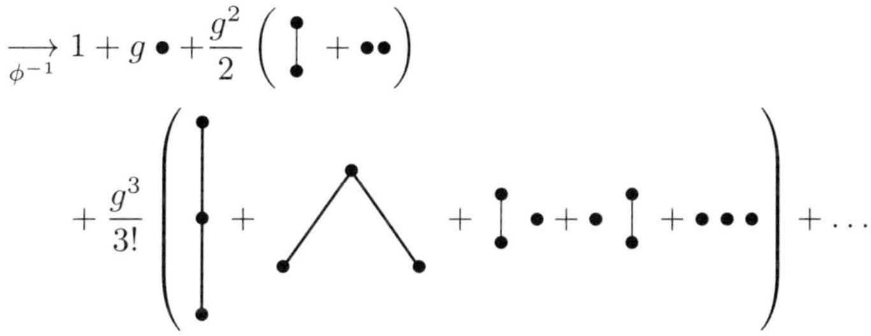

We see that the exponential series of Υ generates at some specific order in g all possible divergences in loops of the Connes-Kreimer Hopf algebra. Remember that this algebra is in correspondence with nested divergences of the OPI diagrams, that are connected and have a definite loop order, and so is directly related with the free energy F. Also the elementary tree $\bullet$ may represent any OPI diagram of any loop order that does not have subdivergences. On the other hand the Loday-Ronco algebra seems to generate all divergences in all possible loop orders compatible with a specific order in the coupling constant g. Hence it appears to be related with the expansion of the partition function Z in terms of (not connected) correlation functions of some physical theory. We would like to clarify this in future work.

6.2. Genus higher than 0. We start with the space of correlation functions of genus 0 that are solutions of Topological Recursion and which we denote by **Corr**. We consider it as the direct sum of spaces of correlation functions of specific Euler characteristic $\chi = -n$, which we denote by $\mathbf{Corr^{(n)}}$:

$$\mathbf{Corr} = \bigoplus_{n=0}^{\infty} \mathbf{Corr^{(n)}} \tag{52}$$

We have seen already that **Corr** carries an algebra structure with the $*$ product given by (41). We see it as a classical space and consider its quantization $\mathbf{Corr_h}$

to be the space of correlation functions with $g > 0$. In the same way we have

$$\mathbf{Corr_h} = \overset{\infty}{\underset{n=0}{\oplus}} \mathbf{Corr_h^{(n)}} \tag{53}$$

$\mathbf{Corr_h^{(n)}}$ is by definition the free module generated by correlation functions $W_k^g(p, p_1, \ldots, p_{k-1})$ with a fixed Euler characteristic $\chi = -n$ and varying genus $g > 0$, up to the consistency of the formula $\chi = 2 - 2g - k$, over the ring $k[[]h]]$ of formal power series in a parameter h with coefficients in a field k of characteristic 0. In this way every correlation function $W^{(n)} \in \mathbf{Corr_h^{(n)}}$ admits an expansion in terms of h. For $\chi = -n$ even,

$$\begin{aligned} W^{(n)} =& W_{n+2}^0(p, p_1, \ldots, p_{n+1}) + hW_n^1(p, p_1, \ldots, p_{n-1}) + \\ & \cdots + h^{n/2} W_2^{n/2}(p, p_1). \end{aligned} \tag{54}$$

and for $\chi = -n$ odd,

$$\begin{aligned} W^{(n)} =& W_{n+2}^0(p, p_1, \ldots, p_{n+1}) + hW_n^1(p, p_1, \ldots, p_{n-1}) + \\ & \cdots + h^{(n+1)/2} W_1^{(n+1)/2}(p). \end{aligned} \tag{55}$$

We would like to describe a procedure of passing from **Corr** to $\mathbf{Corr_h}$ that encodes Topological Recursion for arbitrary genus. Hopefully this will give some insight to what would be the good direction towards the quantization of the Loday-Ronco Hopf algebra.

First we obtain the one-loop graph of order one, $t^1 \in (Y^1)^1$ as a result of applying a quantization map Q to the tree $(\mathbf{1}) \in Y^1$:

(56) 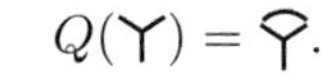

We also set $Q(|) = 0$. As a concrete model we can take the Teichmuller space of Riemann surfaces, take $(\mathbf{1}) =$  as a pair of pants and  as the result of attaching a cylinder to two of the borders of . Now we obtain by induction the first order quantization of a tree $t \in Y^n, t = t_1 \vee t_2$:

$$Q(t) = Q(t_1) \vee t_2 + t_1 \vee Q(t_2) + t_1 \widehat{\vee} t_2. \tag{57}$$

By successive applications of Q we get graphs with more and more loops starting from a planar binary tree, that correspond to higher order terms in quantization, up to exhausting the available pairs of leaves. Hence the operator Q is nilpotent. For instance, the quantization of  $=$  $\vee \,|$ is

(58) 

It is clear that Q^2  $= 0$. We define three more operators Q_L, Q_M and Q_R:

$$Q_L(t) = Q(t_1) \vee t_2, \tag{59}$$

$$Q_M(t) = t_1 \widehat{\vee} t_2, \tag{60}$$

$$Q_R(t) = t_1 \vee Q(t_2). \tag{61}$$

We see that

$$Q(t) = Q_L(t) + Q_M(t) + Q_R(t). \tag{62}$$

We would like in the future to study more deeply the quantization operator Q and its derivate operators Q_L, Q_M and Q_R.

Next we define a quantum product of two Υ as a first order expansion in h:

$$\Upsilon *_h \Upsilon = \Upsilon * \Upsilon + hQ(\Upsilon * \Upsilon)$$

(63) 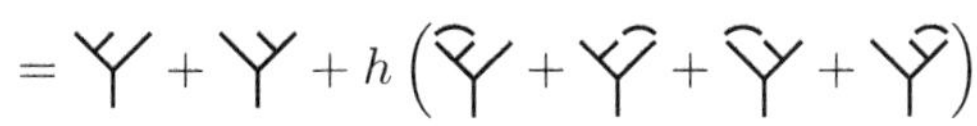

Then we see that

$$\Upsilon *_h \Upsilon = \Upsilon * \Upsilon + h\left(\sum_{i=1}^{2} {}_{i\leftrightarrow i+1}((\mathbf{1}) * (\mathbf{1}))\right). \tag{64}$$

The next step would be to define the product between Υ and . We do it in a way such that

(65) 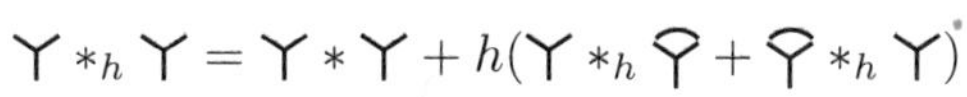

We use the formula (7) to set

(66) 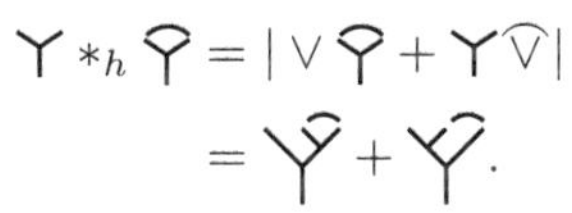

The formula for  $*_h \Upsilon$ should now be obvious.

Having settled the low order cases we now move to the general case. We take the product of a graph $t^{g_1} \in (Y^p)^{g_1}$ with a graph $t^{g_2} \in (Y^q)^{g_2}$ in such a way as to give the $h^{g_1+g_2}$ contribution to the quantum product $t_1 *_h t_2$, for $t_1 \in Y^p$ and $t_2 \in Y^q$ the underlying trees of $t_1^{g_1}$ and $t_2^{g_2}$:

$$t_1 *_h t_2 = t_1 * t_2 + \cdots + h^{g_1+g_2}(t_1^{g_1} *_h t_2^{g_2} + t_1^{g_2} *_h t_2^{g_1}) + \ldots \tag{67}$$

Reversing the argument we see that the quantum product becomes a sum over the genus of graphs:

$$t_1 *_h t_2 = \sum_{k=0}^{(p+q-k_{\min})/2+1} \sum_{l=0}^{k} h^k t_1^l *_h t_2^{k-l}, t_1^l \in (Y^p)^l, t_2^{k-l} \in (Y^q)^{k-l} \tag{68}$$

where $k_{\min}$ is 1 or 2. By summing over all planar binary trees and the corresponding graphs with loops we get the corresponding formula for the quantum product of correlation functions:

$$\begin{aligned} &W^0_{m+2} *_h W^0_{n+2} = \\ &\sum_{k=0}^{(m+n-k_{\min})/2+1} \sum_{l=0}^{k} h^k W^l_{m+2-2l} *_h W^{k-l}_{n+2-2(k-l)} \end{aligned} \tag{69}$$

In terms of correlation functions $W^{(n)} \in \mathbf{Corr}_{\mathbf{h}}^{(\mathbf{n})}$ the above examples give

$$W^{(1)} = W_3^0(p, p_1, p_2) + hW_1^1(p)$$

(70) 

$$W^{(2)} = W_4^0(p, p_1, p_2, p_3) + hW_2^1(p, p_1)$$

(71) 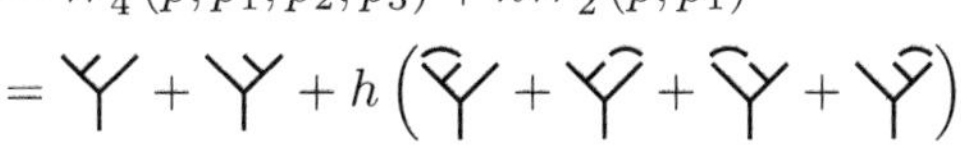

Now giving the module $\mathbf{Corr}_{\mathbf{h}}^{(\mathbf{n})}$ the obvious ring structure that consists in multiplying term by term the expansion in h we have

$$
\begin{aligned}
W^{(2)} &= W^{(1)} \cdot W^{(1)} \\
&= W_3^0(p, p_1, p_2) * W_3^0(p', p_1', p_2') \\
&\quad + h\left(W_1^1(p) *_h W_3^0(p', p_1', p_2') + W_3^0(p, p_1, p_2) *_h W_1^1(p')\right). \qquad (72)
\end{aligned}
$$

This allows to identify the terms in the expansion of $W^{(2)}$:

$$
\begin{aligned}
W_4^0(p, p_1, p_2, p_3) &= W_3^0(p, p_1, p_2) * W_3^0(p, p_2, p_3) \\
&= K_p(q, \bar{q}) \left(W_3^0(q, p_1, p_2) W_2^0(\bar{q}, p_3) + W_2^0(q, p_1) W_3^0(\bar{q}, p_2, p_3)\right); \qquad (73) \\
W_2^1(p, p_1) &\dot{=} W_1^1(p) *_h W_3^0(p, p_1, p_2) + W_3^0(p, p_1, p_2) *_h W_1^1(p) \\
&= K_p(q, \bar{q}) \left(W_3^0(q, \bar{q}, p_1) + W_1^1(q) W_2^0(\bar{q}, p_1) + W_2^0(q, p_1) W_1^1(\bar{q})\right). \qquad (74)
\end{aligned}
$$

where we have used (66) and the corresponding formula for $\curlyvee *_h \curlyvee$. Hence the term-by-term product of $W^{(1)}$ by itself gives the corresponding terms in the Topological Recursion formula for the components of $W^{(2)}$.

References

[1] Marcelo Aguiar and Frank Sottile, *Structure of the Loday-Ronco Hopf algebra of trees*, J. Algebra **295** (2006), no. 2, 473–511, DOI 10.1016/j.jalgebra.2005.06.021. MR2194965

[2] Michael Atiyah, *Topological quantum field theories*, Inst. Hautes Études Sci. Publ. Math. **68** (1988), 175–186 (1989). MR1001453

[3] C. Brouder, *Runge-Kutta methods and renormalization*, Eur.Phys.J. **C12** (2000), 521–534.

[4] Christian Brouder, *On the trees of quantum fields*, Eur.Phys.J. **C12** (2000), 535–549.

[5] Christian Brouder and Alessandra Frabetti, *QED Hopf algebras on planar binary trees*, J. Algebra **267** (2003), no. 1, 298–322, DOI 10.1016/S0021-8693(03)00331-4. MR1993478

[6] John C. Collins, *Renormalization*, Cambridge Monographs on Mathematical Physics, Cambridge University Press, Cambridge, 1984. An introduction to renormalization, the renormalization group, and the operator-product expansion. MR778558

[7] A. Connes and D. Kreimer, *Hopf algebras, renormalization and noncommutative geometry*, Quantum field theory: perspective and prospective (Les Houches, 1998), NATO Sci. Ser. C Math. Phys. Sci., vol. 530, Kluwer Acad. Publ., Dordrecht, 1999, pp. 59–108. MR1725011

[8] P. Di Francesco, P. Ginsparg, and J. Zinn-Justin, *2D gravity and random matrices*, Phys. Rep. **254** (1995), no. 1-2, 133, DOI 10.1016/0370-1573(94)00084-G. MR1320471

[9] Olivia Dumitrescu, Motohico Mulase, Brad Safnuk, and Adam Sorkin, *The spectral curve of the Eynard-Orantin recursion via the Laplace transform*, Algebraic and geometric aspects of integrable systems and random matrices, Contemp. Math., vol. 593, Amer. Math. Soc., Providence, RI, 2013, pp. 263–315, DOI 10.1090/conm/593/11867. MR3087960

[10] João N. Esteves, *Hopf algebras and topological recursion*, J. Phys. A **48** (2015), no. 44, 445205, 19, DOI 10.1088/1751-8113/48/44/445205. MR3417992

[11] B. Eynard and N. Orantin, *Invariants of algebraic curves and topological expansion*, Commun. Number Theory Phys. **1** (2007), no. 2, 347–452, DOI 10.4310/CNTP.2007.v1.n2.a4. MR2346575

[12] L. Foissy, *Les algèbres de Hopf des arbres enracinés décorés. II* (French, with English and French summaries), Bull. Sci. Math. **126** (2002), no. 4, 249–288, DOI 10.1016/S0007-4497(02)01113-2. MR1909461

[13] Alessandra Frabetti, *Simplicial properties of the set of planar binary trees*, J. Algebraic Combin. **13** (2001), no. 1, 41–65, DOI 10.1023/A:1008723801201. MR1817703

[14] Jean-Louis Loday and María O. Ronco, *Hopf algebra of the planar binary trees*, Adv. Math. **139** (1998), no. 2, 293–309, DOI 10.1006/aima.1998.1759. MR1654173

CAMGSD, Departamento de Matemática, Instituto Superior Técnico, Av. Rovisco Pais 1, 1049-001 Lisboa, Portugal

Email address: joao.n.esteves@tecnico.ulisboa.pt

Proceedings of Symposia in Pure Mathematics
Volume **100**, 2018
https://doi.org/10.1090/pspum/100/01767

Graph sums in the remodeling conjecture

Bohan Fang and Zhengyu Zong

ABSTRACT. The BKMP Remodeling Conjecture (see [V. Bouchard et al., *Remodeling the B-model*, Comm. Math. Phys. 287 (2009), pp. 117–178], [V. Bouchard et al. *Topological open strings on orbifolds*, Comm. Math. Phys. 296 (2010), pp. 589–623] and [M. Mariño, *Open string amplitudes and large order behavior in topological string theore*, J. High Energy Phys. 3 (2008)]) preditcs all genus open-closed Gromov-Witten invariants for a toric Calabi-Yau 3-orbifold by Eynard-Orantin's topological recursion [*Invariants of algebraic curves and topological expansion*, Commun. Number Theory Phys. 1 (2007), pp. 347–452] on its mirror curve. The proof of the Remodeling Conjecture by the authors ([*Equivariant Gromov-Witten theory of affine smooth toric Deligne-Mumford stacks*, Int. Math. Res. Not. IMRN 7 (2016), pp. 2127–2144] and [*On the remodeling conjecture for toric Calabi-Yau 3-orbifolds*, arXiv:1604.07123]) relies on comparing two Feynman-type graph sums in both A and B-models. In this paper, we will survey these graph sum formulae and discuss their roles in the proof of the conjecture.

CONTENTS

1. Introduction

1.1. Outline of the proof. The Remodeling Conjecture can be viewed as an all genus open-closed mirror symmetry for toric Calabi-Yau 3-orbifolds. On A-model side, one has the higher genus open-closed Gromov-Witten potential for a toric Calabi-Yau 3-orbifold. On B-model side, the higher genus B-model potential comes from the Eynard-Orantin topological recursion on the mirror curve. The Remodeling Conjecture identifies the A-model and B-model higher genus potentials

2010 *Mathematics Subject Classification.* Primary 14J33, 14H70.

under the mirror map. At first glance, the Remodeling Conjecture may seem to be mysterious. However, we will see that the proof of the Remodeling Conjecture is quite natural under the point of view of quantization of semisimple Frobenius structures. In this subsection, we give the outline of the proof of the Remodeling Conjecture.

Recall that the genus zero mirror theorem [**27**] works for general semi-projective toric orbifolds. On A-model side, one considers an n–dimensional semi-projective toric orbifold $\mathcal{X}$ over $\mathbb{C}$. There is an n-dimensional torus $\mathbb{T} \cong (\mathbb{C}^*)^n$ in $\mathcal{X}$ as a dense open subset. The natural $\mathbb{T}$-action on itself extends to a $\mathbb{T}$-action on $\mathcal{X}$. Under this action, there are finitely many fixed points and finitely many 1-dimensional $\mathbb{T}$–invariant orbits. In practice, one may choose a smaller torus $\mathbb{T}' \subset \mathbb{T}$ acting on $\mathcal{X}$ such that fixed points and the 1-dimensional $\mathbb{T}'$–invariant orbits are the same as those of the $\mathbb{T}$-action.

One can consider the $\mathbb{T}'$–equivariant orbifold quantum cohomology $QH^*_{\mathrm{CR},\mathbb{T}'}(\mathcal{X})$ of $\mathcal{X}$. The quantum product $\star_{\boldsymbol{\tau}}$ on $QH^*_{\mathrm{CR},\mathbb{T}'}(\mathcal{X})$ is defined by the genus 0 Gromov-Witten invariants of $\mathcal{X}$. There is also a $\mathbb{T}'$– equivariant Poincare pairing $(\cdot,\cdot)_{\mathcal{X},\mathbb{T}'}$ on $\mathcal{X}$. Then one obtains a Frobenius algebra

$$(QH^*_{\mathrm{CR},\mathbb{T}'}(\mathcal{X}), \star_{\boldsymbol{\tau}}, (\cdot,\cdot)_{\mathcal{X},\mathbb{T}'}).$$

On B-model side, one has the Landau-Ginzburg mirror of $\mathcal{X}$. More concretely, there is a $\mathbb{T}'$-equivariant superpotential

$$W^{\mathbb{T}'}: (\mathbb{C}^*)^n \to \mathbb{C},$$

which is determined by the toric data of $\mathcal{X}$ and the $\mathbb{T}'$-action on $\mathcal{X}$. The $\mathbb{T}'$-equivariant super potential $W^{\mathbb{T}'}$ defines the Jacobi ring $\mathrm{Jac}(W^{\mathbb{T}'})$:

$$\mathrm{Jac}(W^{\mathbb{T}'}) = \mathbb{C}[X_1^{\pm}, \cdots, X_n^{\pm}]/\langle X_1 \frac{\partial W^{\mathbb{T}'}}{\partial X_1}, \cdots, X_n \frac{\partial W^{\mathbb{T}'}}{\partial X_n}\rangle.$$

There is a natural product $\cdot$ on $\mathrm{Jac}(W^{\mathbb{T}'})$ induced from the product on $\mathbb{C}[X_1^{\pm}, \cdots, X_n^{\pm}]$. One can also consider the residue pairing on the Jacobi ring $\mathrm{Jac}(W^{\mathbb{T}'})$:

$$(f,g) := \frac{1}{(2\pi\sqrt{-1})^n} \int_{|dW^{\mathbb{T}'}|=\epsilon} \frac{fg \frac{dX_1 \cdots dX_n}{X_1 \cdots X_n}}{X_1 \frac{\partial W^{\mathbb{T}'}}{\partial X_1} \cdots X_n \frac{\partial W^{\mathbb{T}'}}{\partial X_n}},$$

where $f, g \in \mathrm{Jac}(W^{\mathbb{T}'})$. These data form a Frobenius algebra on B-model:

$$(\mathrm{Jac}(W^{\mathbb{T}'}), \cdot, (\cdot,\cdot)).$$

The genus 0 mirror theorem can be viewed as an identification of Frobenius structures on A-model and B-model: under the mirror map, we have an isomorphism of Frobenius algebras[1]

$$(QH^*_{\mathrm{CR},\mathbb{T}'}(\mathcal{X}), \star_{\boldsymbol{\tau}}, (\cdot,\cdot)_{\mathcal{X},\mathbb{T}'}) \cong (\mathrm{Jac}(W^{\mathbb{T}'}), \cdot, (\cdot,\cdot)).$$

Both A-model and B-model Frobenius algebras are semisimple.

When we restrict the above mirror theorem to the case of toric Calabi-Yau 3-orbifolds, we obtain an identification of the Frobenius algebra on A-model and the Frobenius algebra on its Landau-Ginzburg mirror. On the other hand, the Remodeling Conjecture concerns the higher genus (open-closed) Gromov-Witten

[1] We require $\boldsymbol{\tau} \in H^{\leq 2}(\mathcal{X})$, i.e. we only consider *small* quantum cohomology. The mirror symmetry is then established after the mirror map tranformation.

invariants instead of the quantum cohomology which is about the genus 0 data. So a natural question is: what is the higher genus B-model? In [**35**], Eynard and Orantin introduce the topological recursion on spectral curves. The output of the topological recursion is a symmetric n-form $\omega_{g,n}$ on the curve, where $g, n \geq 0$. In [**13**] and [**14**], Bouchard, Klemm, Mariño, and Pasquetti conjecture that when we apply the topological recursion to the mirror curve, we obtain the higher genus B-model potential $\check{F}_{g,n}$ is defined from $\omega_{g,n}$. The Remodeling Conjecture identifies the higher genus open-closed Gromov-Witten potential $F_{g,n}^{\mathcal{X},(\mathcal{L},f)}$ ($\mathcal{L}$ is an Aganagic-Vafa brane and $f \in \mathbb{Z}$ is the framing) of a toric Calabi-Yau 3-orbifold $\mathcal{X}$ and the higher genus B-model potential $\check{F}_{g,n}$ under the open-closed mirror map. Thus the Remodeling Conjecture is an all genus open-closed mirror symmetry for toric Calabi-Yau 3-orbifolds.

The mirror curve as the B-model in the Remodeling Conjecture is related to the Landau-Ginzburg model by a dimensional reduction [**56, 57**]. More concretely, for any toric *Calabi-Yau* 3-fold $\mathcal{X}$, the mirror curve C is defined by the following equation:

$$C = \{(X,Y)|H(X,Y) = 0\} \subset (\mathbb{C}^*)^2,$$

while the $\mathbb{T}'$-equivariant super potential $W^{\mathbb{T}'} : (\mathbb{C}^*)^3 \to \mathbb{C}$ is given by

$$W^{\mathbb{T}'} = H(X,Y)Z - \mathsf{u}_1 \log X - \mathsf{u}_2 \log Y.$$

Here $(\mathbb{C}^*)^2 \cong \mathbb{T}' \subset \mathbb{T} \cong (\mathbb{C}^*)^3$ is the Calabi-Yau sub-torus which acts trivially on the canonical bundle of $\mathcal{X}$ and $\mathsf{u}_1, \mathsf{u}_2$ are the equivariant parameters i.e. $H^*(\mathcal{B}\mathbb{T}') = \mathbb{C}[\mathsf{u}_1, \mathsf{u}_2]$.

The key idea in the proof of the Remodeling Conjecture is that one can realize both A-model and B-model higher genus potentials as quantizations on the same semisimple Frobenius structure. On A-model side, this is given by a generalization of Givental formula to the orbifold case [**102**]. In [**102**], the Givental formula is generalized to any GKM orbifolds (there are finitely many torus fixed points and finitely many 1-dimensional invariant orbits) and one can apply this formula to the case of toric Calabi-Yau 3-orbifolds which are a special kind of GKM orbifolds. Givental formula is expressed in terms of quantization of quadratic Hamiltonians which involves the exponential of quadratic differential operators.

On B-model side, the quantization procedure is given by the topological recursion on the mirror curve. The higher genus data $\omega_{g,n}$ is obtained recursively from the initial data $\omega_{0,1}, \omega_{0,2}$. By the dimensional reduction, the data $\omega_{0,1}, \omega_{0,2}$ is equivalent to the data of the Frobenius structure $(\mathrm{Jac}(W^{\mathbb{T}'}), \cdot, (\cdot, \cdot))$ of the Landau-Ginzburg B-model.

The bridge connecting the A-model quantization (orbifold Givental formula) and the B-model quantization (topological recursion) is the graph sum formula. On A-model side, one can apply Wick's formula to the Givental formula to rewrite this formula (involving differential operators) in terms of Feynman graphs. On B-model side, by [**36**] [**34**], the topological recursion is equivalent to a graph sum formula. The Remodeling Conjecture is proved by identifying each factor in the graph sum formulas on A-model and B-model.

It turns out that all the factors in the graph sum formulas are determined by the R-matrix, which appears in the fundamental solutions of the quantum differential equation, and the disk potential $F_{0,1}^{\mathcal{X},(\mathcal{L},f)}$ ($\check{F}_{0,1}$ on B-model). The genus 0 mirror theorem identifies the Frobenius structures on A-model and B-model and hence identifies the quantum differential equations. By the uniqueness of the fundamental solution, the A-model and B-model R-matrices can be identified up to a constant matrix. On A-model side, this constant matrix is fixed by the orbifold quantum Riemann-Roch theorem [**91**] while on B-model side, this constant matrix is obtained by direct computation. It turns out that the A-model and B-model R-matrices are indeed equal. In the end, the identification of the disk potentials $F_{0,1}^{\mathcal{X},(\mathcal{L},f)}$ and $\check{F}_{0,1}$ is given in [**41**]. Therefore, the Remodeling Conjecture follows immediately.

1.2. Overview of the paper. We fix the notation on toric orbifolds and Gromov-Witten invariants in Section 2. In Section 3 we introduce 3 related and equivalent B-models to a toric Calab-Yau 3-orbifold $\mathcal{X}$, with an emphasis on mirror curves. In Section 4 we will give a quick review on the genus 0 mirror theorem, and on the identification of the Frobenius structures. In Section 5, we study the A-model quantization. The orbifold Givental formula expresses the higher genus Gromov-Witten potentials of a toric Calabi-Yau 3-orbifold in terms of its Frobenius structure, which is the genus zero data. We also discuss the graph sum version of the orbifold Givental formula. In Section 6, we move on to the B-model quantization, which is the topological recursion on a spectral curve. We also discuss the graph sum formula of the topological recursion. Then we specialize to the case of the mirror curve of a toric Calabi-Yau 3-orbifold. In Section 7 we discuss the dimensional reduction from the Landau-Ginzburg B-model. Then we expresses the factors in the graph sum formula of B-model in terms of the B-model Frobenius structure. Then comparing the graph sum components leads to a complete proof of the Remodeling Conjecture. We will illustrate many facts by a running example.

2. Geometry and the A-model of a toric Calabi-Yau 3-orbifold

The Remodeling Conjecture [**13, 14, 81**] concerns the open-closed all genus Gromov-Witten invariants of a semi-projective toric Calabi-Yau 3-orbifold. We fix the notations in this section.

2.1. Toric Calabi-Yau 3-orbifolds. A Calabi-Yau 3-fold X is toric if it contains the algebraic torus $\mathbb{T} = (\mathbb{C}^*)^3$ as a Zariski dense open subset, and the action of $\mathbb{T}$ on itself extends to X. All Calabi-Yau 3-folds are non-compact. There is a rank 2 subtorus $\mathbb{T}' \subset \mathbb{T}$ which acts trivially on the canonical line bundle of X. We call $\mathbb{T}'$ the Calabi-Yau torus. Then $\mathbb{T} \cong \mathbb{T}' \times \mathbb{C}^*$. Let $\mathbb{T}'_{\mathbb{R}} \cong U(1)^2$ be the maximal compact subgroup of $\mathbb{T}'$.

Let $M' = \mathrm{Hom}(\mathbb{T}', \mathbb{C}^*) \cong \mathbb{Z}^2$ and $N' = \mathrm{Hom}(\mathbb{C}^*, \mathbb{T}')$ be the character lattice and the cocharacter lattice of $\mathbb{T}'$, respectively. Then M' and N' are dual lattices. Let X_Σ be a toric Calabi-Yau 3-fold defined by a simplicial fan $\Sigma \subset N'_{\mathbb{R}} \times \mathbb{R}$, where $N'_{\mathbb{R}} := N' \otimes_{\mathbb{Z}} \mathbb{R} \cong \mathbb{R}^2$ can be identified with the Lie algebra of $\mathbb{T}'_{\mathbb{R}}$. Then X_Σ has at most quotient singularities. We assume that X_Σ is semi-projective, i.e., X_Σ contains at least one $\mathbb{T}$-fixed point, and X_Σ is projective over its affinization $X_0 := \mathrm{Spec} H^0(X_\Sigma, \mathcal{O}_{X_\Sigma})$. Then the support of the fan Σ is a strongly convex rational polyhedral cone $\Sigma_0 \subset N'_{\mathbb{R}} \times \mathbb{R} \cong \mathbb{R}^3$, and X_0 is the affine toric variety defined by the 3-dimensional cone Σ_0. There exists a convex polytope $P \subset N'_{\mathbb{R}} \cong \mathbb{R}^2$ with vertices in the lattice $N' \cong \mathbb{Z}^2$, such that Σ_0 is the cone over $P \times \{1\} \subset N'_{\mathbb{R}} \times \mathbb{R}$, i.e. $\Sigma_0 = \{(tx, ty, t) : (x, y) \in P, t \in [0, \infty)\}$. The fan Σ determines a triangulation of P: the 1-dimensional, 2-dimensional, and 3-dimensional cones in Σ are in one-to-one correspondence with the vertices, edges, and faces of the triangulation of P,

TABLE 1. Table of notations

Symbols	Explanation	Remark
$\mathcal{X}$	a toric CY 3-(orbi)fold, defined by a defining polytope P and its triangulation	fixed throughout the paper; fan Σ is a cone over P
$\mathbb{T}$	torus acting on the toric CY 3-fold $\mathcal{X}$	$\mathbb{T} \cong (\mathbb{C}^*)^3$
$\mathbb{T}'$	Calabi-Yau torus $\subset \mathbb{T}$ preserving the CY form	$\mathbb{T}' \cong (\mathbb{C}^*)^2$
$\mathcal{L}$	a fixed outer Aganagic-Vafa brane	location (phase) given by $\sigma_0 \in \Sigma(3), \tau_0 \in \Sigma(2)$
I_Σ	set of canonical basis for $QH^*_{\mathbb{T}}(\mathcal{X}), H^*_{\mathrm{CR},\mathbb{T}}(\mathcal{X})$, etc.	$I_\Sigma = \{\boldsymbol{\sigma} = (\sigma,\gamma)\}, \sigma \in \Sigma(3), \gamma \in G^*_\sigma$.
I_C	set of critical points on a spectral curve $\mathcal{C}$	$I_C = I_\Sigma$ by the mirror thm if $\mathcal{C}$ = mirror curve
$P_{\boldsymbol{\alpha}}$	critical points of $W^{\mathbb{T}'}$, in $(\mathbb{C}^*)^3$	
$p_{\boldsymbol{\alpha}}$	critical points of the mirror curve $C_q \subset (\mathbb{C}^*)^2$	$P_{\boldsymbol{\alpha}} = (p_{\boldsymbol{\alpha}}, Z_{\boldsymbol{\alpha}})$
$\mathrm{Jac}(W^{\mathbb{T}'})$	Jacobian ring of $W^{\mathbb{T}'}$	$\cong QH^*_{\mathbb{T}'}(\mathcal{X})$ under the mirror map, as a Frobenius alg.
$V_{\boldsymbol{\alpha}}$	canonical basis of $\mathrm{Jac}(W^{\mathbb{T}'})$	$V_{\boldsymbol{\alpha}}(P_{\boldsymbol{\beta}}) = \delta_{\boldsymbol{\alpha}\boldsymbol{\beta}}$
$\mathfrak{p}$	$h^2_{\mathrm{CR}}(\mathcal{X})$, number of twisted Kähler parameters	also #(number of integral points in Δ) -3
$\mathfrak{g}$	$h^4_{\mathrm{CR}}(\mathcal{X})$, also the genus of the compactified mirror curve $\overline{C}_q$	also number of integral points in $\mathrm{Int}(\Delta)$
$q_a,\ a = 1\dots\mathfrak{p}$	complex parameters mirror to extended Kähler parameters	depending on a choice of extended Kähler basis $H_1,\dots,H_{\mathfrak{p}}$
C_q	Affine mirror curve at parameter q	
$\overline{C}_q$	Compactified mirror curve at parameter q	
$F^{\mathcal{X},(\mathcal{L},f)}_{g,n}$	A-model open GW potential	depends on $\mathcal{X}$, $\mathcal{L}$ and the framing f
$\omega_{g,n}$	B-model higher genus invariants from the EO recursion	symmetric meromorphic form on $(\overline{C}_q)^n$.
$\check{F}_{g,n}$	B-model open potential	defined as the indefinite integral of $\omega_{g,n}$

respectively. This triangulation of P is known as the toric diagram or the dual graph of the simplicial toric Calabi-Yau 3-fold X_Σ.

Let $\Sigma(d)$ be the set of d-dimensional cones in Σ, and let $\mathfrak{p}' = |\Sigma(1)| - 3$. We label $\rho_1,\dots,\rho_{\mathfrak{p}'+3} \in \Sigma(1)$ (and usually denote the their generators in $P \times \{1\}$ by $b_1,\dots,b_{\mathfrak{p}'+3}$). Then X_Σ is a GIT quotient

$$X_\Sigma = \mathbb{C}^{3+\mathfrak{p}'} /\!\!/ G_\Sigma = (\mathbb{C}^{3+\mathfrak{p}'} - Z_\Sigma)/G_\Sigma$$

where G_Σ is a $\mathfrak{p}'$-dimensional subgroup of $(\mathbb{C}^*)^{3+\mathfrak{p}'}$ and Z_Σ is a Zariski closed subset of $\mathbb{C}^{3+\mathfrak{p}'}$ determined by the fan Σ. If X_Σ is a smooth toric Calabi-Yau 3-fold then $G_\Sigma \cong (\mathbb{C}^*)^{\mathfrak{p}'}$ and G_Σ acts freely on $\mathbb{C}^{3+\mathfrak{p}'} - Z_\Sigma$. In general we have $(G_\Sigma)_0 \cong (\mathbb{C}^*)^{\mathfrak{p}'}$, where $(G_\Sigma)_0$ is the connected component of the identity, and the stabilizers of the G_Σ-action on $\mathbb{C}^{3+\mathfrak{p}'} - Z_\Sigma$ are at most finite and generically trivial. The stacky quotient

$$\mathcal{X} = [(\mathbb{C}^{3+\mathfrak{p}'} - Z_\Sigma)/G_\Sigma]$$

is a toric Calabi-Yau 3-orbifold; it is a toric Deligne-Mumford stack in the sense of Borisov-Chen-Smith [**11**]. For any $\sigma \in \Sigma(d)$ there is a codimensional d closed substack $\mathfrak{l}_\sigma$ associated to σ. We denote its generic stabilizer to be G_σ. When $\sigma \in \Sigma(3)$, then $\mathfrak{l}_\sigma$ is a $\mathbb{T}$-fixed (probably stacky) point.

We denote the $\bar{D}_i$ to be the first Chern class of divisors $\widetilde{D}_i = \{Z_i = 0\}$, where Z_i are homogeneous coordinates for $i = 1, \dots, \mathfrak{p}'$.

2.2. Toric crepant resolution and extended Kähler classes. Given a semi-projective simplicial toric Calabi-Yau 3-fold X_Σ which is not smooth, there exists a subdivision Σ' of Σ, such that

$$X_{\Sigma'} = (\mathbb{C}^{3+\mathfrak{p}} - Z_{\Sigma'})/G_{\Sigma'} \longrightarrow X_\Sigma = \big((\mathbb{C}^{3+\mathfrak{p}'} - Z_\Sigma) \times (\mathbb{C}^*)^{\mathfrak{s}}\big)/G_{\Sigma'}$$

is a crepant toric resolution, where $X_{\Sigma'}$ is a smooth toric Calabi-Yau 3-fold, $\mathfrak{p} + 3 = |\Sigma'(1)|$, and $G_{\Sigma'} \cong (\mathbb{C}^*)^{\mathfrak{p}}$. We denote $\rho_{\mathfrak{p}'+4}, \dots, \rho_{\mathfrak{p}+3} \in \Sigma'(1) \setminus \Sigma(1)$ as new 1-cones in Σ', and their generators in $P \times \{1\}$ by $b_{\mathfrak{p}'+4}, \dots, b_{\mathfrak{p}+3}$. $X_{\Sigma'}$ and X_Σ are GIT quotients of the same $G_{\Sigma'}$-action on $\mathbb{C}^{3+\mathfrak{p}}$ with respect to different stability conditions. We set $b_i = (m_i, n_i, 1) \in P \times \{1\}$.

Let $K_{\Sigma'} \cong U(1)^{\mathfrak{p}}$ be the maximal compact subgroup of $G_{\Sigma'} \cong (\mathbb{C}^*)^{\mathfrak{p}}$. The $G_{\Sigma'}$-action on $\mathbb{C}^{3+\mathfrak{p}}$ restricts to a Hamiltonian $K_{\Sigma'}$-action on the Kähler manifold $(\mathbb{C}^{3+\mathfrak{p}}, \omega_0 = \sqrt{-1}\sum_{i=1}^{3+\mathfrak{p}} dz_i \wedge d\bar{z}_i)$, with moment map $\widetilde{\mu} : \mathbb{C}^{3+\mathfrak{p}} \to \mathrm{Lie}(K_{\Sigma'})^\vee = H^2_{G_{\Sigma'}}(\mathbb{C}^{3+\mathfrak{p}}; \mathbb{R}) = \mathbb{R}^{\mathfrak{p}}$. There exist two (open) cones C and C' in $\mathbb{R}^{\mathfrak{p}}$ such that

$$\begin{aligned} &\widetilde{\mu}^{-1}(\vec{r})/K_{\Sigma'} \\ &= \begin{cases} (\mathbb{C}^{3+\mathfrak{p}} - Z_{\Sigma'})/G_{\Sigma'} = X_{\Sigma'}, & \vec{r} \in \mathsf{C}', \\ \big((\mathbb{C}^{3+\mathfrak{p}'} - Z_\Sigma) \times (\mathbb{C}^*)^{\mathfrak{p}-\mathfrak{p}'}\big)/G_{\Sigma'} = (\mathbb{C}^{3+\mathfrak{p}'} - Z_\Sigma)/G_\Sigma = X_\Sigma, & \vec{r} \in \mathsf{C} \end{cases}\end{aligned}$$

Here $\mathsf{C}' \subset \mathbb{R}^{\mathfrak{p}} = H^2(X_{\Sigma'}; \mathbb{R})$ is the Kähler cone of $X_{\Sigma'}$ and $\mathsf{C} \subset \mathbb{R}^{\mathfrak{p}}$ is the extended Kähler cone of X_Σ. Let $D_i \in H^2_{G_{\Sigma'}}(\mathbb{C}^{3+\mathfrak{p}}; \mathbb{R}) = \mathrm{Lie}(K_{\Sigma'})^\vee$ for $i = 1, \dots, \mathfrak{p}$ be the equivariant Poincaré dual to $\{Z_i = 0\}$ in $\mathbb{C}^{3+\mathfrak{p}}$ (Z_i are coordinates).

The parameter $\vec{r} \in \mathsf{C}$ determines a Kähler form $\omega(\vec{r})$ on the toric Calabi-Yau 3-orbifold $\mathcal{X} = [(\mathbb{C}^{3+\mathfrak{p}'} - Z_\Sigma)/G_\Sigma]$. As shown in [**60**], there is a canonical decomposition

$$\mathrm{Lie}(G_{\Sigma'})^\vee \cong H^2(X_\Sigma; \mathbb{C}) \oplus \bigoplus_{j=\mathfrak{p}'+4}^{\mathfrak{p}+3} \mathbb{C} D_j. \tag{1}$$

In particular, we write $\bar{H}$ for the projection of $H \in \mathrm{Lie}(G_{\Sigma'})^\vee$ to $H^2(X_\Sigma; \mathbb{C})$. Our notation $\bar{D}_i \in H^2(X_\Sigma; \mathbb{C})$ indeed satisfies this convention, and in particular $\bar{D}_j = 0$ for $j = \mathfrak{p}' + 4, \dots, \mathfrak{p} + 3$. This splitting also applies to the Käher cone

$$\mathsf{C} = (\text{Kähler cone of } X_\Sigma) \bigoplus \Big(\sum_{i=\mathfrak{p}'+4}^{\mathfrak{p}+3} \mathbb{R}_{>0} D_i \Big).$$

EXAMPLE 2.1. The polytope P and the triangulation is the following. The vertices are $(0,0), (0,2), (1,0), (2,-1)$ (Figure 1). The fan Σ is a cone over the triangulated P. If one adds the dashed line to the triangulation, we get the fan Σ'.

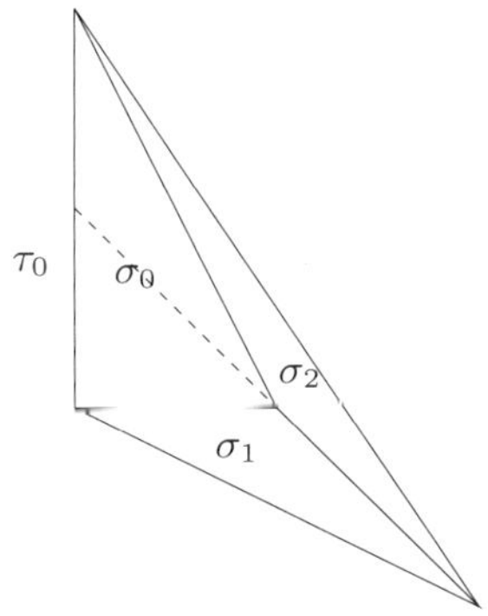

FIGURE 1. The defining polytope of our main example $\mathcal{X}$.

2.3. Chen-Ruan orbifold cohomology. Let $U = \mathbb{C}^{3+\mathfrak{p}'} - Z_\Sigma$, so that $\mathcal{X} = [(\mathbb{C}^{3+\mathfrak{p}'} - Z_\Sigma)/G_\Sigma]$. Given $v \in G_\Sigma$, let $U^v = \{z \in U : v \cdot z = z\}$. The inertia stack of $\mathcal{X}$ is

$$\mathcal{I}\mathcal{X} = \bigcup_{v \in \mathrm{Box}(\Sigma)} \mathcal{X}_v$$

where $\mathrm{Box}(\Sigma) = \{v \in G_\Sigma : U^v \neq \varnothing\}$ and $\mathcal{X}_v = [U^v/G_\Sigma]$.

We consider cohomology with $\mathbb{C}$-coefficient. As a graded $\mathbb{C}$-vector space, the Chen-Ruan orbifold cohomology [**24**] of $\mathcal{X}$ is

$$H^*_{\mathrm{CR}}(\mathcal{X};\mathbb{C}) = \bigoplus_{v \in \mathrm{Box}(\Sigma)} H^*(\mathcal{X}_v;\mathbb{C})[2\mathrm{age}(v)], \quad \mathrm{age}(v) \in \{0,1,2\}.$$

So

$$H^2_{\mathrm{CR}}(\mathcal{X};\mathbb{C}) = H^2(X_\Sigma;\mathbb{C}) \bigoplus \Big(\bigoplus_{\mathrm{age}(v)=1} \mathbb{C}\mathbf{1}_v \Big).$$

One can show $H^2_{\mathrm{CR}}(\mathcal{X};\mathbb{C}) \cong \mathrm{Lie}(G_{\Sigma'})^\vee \cong \mathbb{C}^{\mathfrak{p}}$ as a vector space. Under this isomorphism $\mathbf{1}_v$ corresponds to a D_i for $i = \mathfrak{p}'+4, \ldots, \mathfrak{p}+3$ (c.f. Equation (1)).

Let $\mathfrak{g} := |\mathrm{Int}(P) \cap N'|$ be the number of lattice points in $\mathrm{Int}(P)$, the interior of the polytope P, and let $\mathfrak{n} := |\partial P \cap N'|$ be the number of lattice points on ∂P, the boundary of the polytope P. Then

$$\begin{aligned}
\mathfrak{p}' &= |\Sigma(1)| - 3 = \dim_{\mathbb{C}} H^2(X_\Sigma;\mathbb{C}),\\
\mathfrak{p} &= |\Sigma'(1)| - 3 = |P \cap N'| - 3 = \dim_{\mathbb{C}} \mathrm{Lie}(G_{\Sigma'})^\vee = \dim_{\mathbb{C}} H^2_{\mathrm{CR}}(\mathcal{X};\mathbb{C})\\
&= \mathfrak{g} + \mathfrak{n} - 3,\\
\mathfrak{g} &= |\mathrm{Int}(P) \cap N'| = \dim_{\mathbb{C}} H^4_{\mathrm{CR}}(\mathcal{X};\mathbb{C}),\\
N &:= \chi(\mathcal{X}) = |\Sigma'(3)| = 2\mathrm{Area}(P) = \dim_{\mathbb{C}} H^*_{\mathrm{CR}}(\mathcal{X};\mathbb{C})\\
&= 1 + \mathfrak{p} + \mathfrak{g} = 2\mathfrak{g} - 2 + \mathfrak{n}.
\end{aligned}$$

We choose $H_1, \ldots, H_{\mathfrak{p}}$ in the closure of the extended Kähler cone $\bar{\mathsf{C}} \subset \mathrm{Lie}(K_{\Sigma'})^\vee$ such that the following is true.

- $\{H_1, \ldots, H_{\mathfrak{p}}\}$ is a basis of $\mathrm{Lie}(K_{\Sigma'})^\vee = \mathbb{R}^{\mathfrak{p}}$.
- $\{\bar{H}_1, \ldots, \bar{H}_{\mathfrak{p}'}\}$ is a basis of $H^2(X_\Sigma;\mathbb{R})$. We require $H_i = \bar{H}_i$ under the identification (1) for $i = 1, \ldots, \mathfrak{p}'$, and H_i is in the Kähler cone of X_Σ.
- $H_a = D_{3+a}$ for $a = \mathfrak{p}'+1, \ldots, \mathfrak{p}$.
- We choose H_a in the lattice generated by all D_i.

The $\mathfrak{p}$ parameters $\vec{r} = (r_1, \ldots, r_{\mathfrak{p}})$ are extended Kähler parameters of $\mathcal{X}$, where $r_1, \ldots, r_{\mathfrak{p}'}$ are Kähler parameters of $\mathcal{X}$. The A-model closed string flat coordinates are complexified extended Kähler parameters

$$\tau_a = -r_a + \sqrt{-1}\theta_a, \quad a = 1, \ldots, \mathfrak{p}.$$

EXAMPLE 2.2 (Example 2.1,continued).

$$\begin{aligned}
&\mathcal{X} = (\mathbb{C}^4 - Z_{\Sigma'})/G_{\Sigma'} = (\mathbb{C}^5 - Z_{\Sigma})/G_{\Sigma},\\
&\mathfrak{p} = 2,\ \mathfrak{p}' = 1,\ \mathfrak{g} = 1,\ \mathfrak{n} = 4,\ N = 4,\\
&\text{Kähler cone} = \mathbb{R}_{>0}H_1,\\
&\text{extended Kähler cone} \mathsf{C} = \mathbb{R}_{>0}H_1 \oplus \mathbb{R}_{>0}H_2.
\end{aligned}$$

The moment map of K_Σ is given by

$$|Z_1| - 4|Z_2|^2 + |Z_3|^2 + 2|Z_4|^2,$$

while the moment map $\widetilde{\mu}$ of $K_{\Sigma'}$ is given by

$$(|Z_1| - 4|Z_2|^2 + |Z_3|^2 + 2|Z_4|^2, |Z_1|^2 + |Z_3|^2 - 2|Z_5|^2).$$

H_1, H_2 are $(1,0)$ and $(0,-1)$ respectively in the image of the moment map of K_Σ.

2.4. Equivariant cohomology and its canonical basis. We can work with the equivariant version of the Chen-Ruan (or ordinary) cohomology. We set the notions here for $\mathbb{T}$-equivariant cohomology, while the notions for $\mathbb{T}'$-equivariant cohomology are self-evident if we change $\mathbb{T}$ to $\mathbb{T}'$.

The torus $\mathbb{T}$ fits into the following exact sequence

$$1 \to G_\Sigma \to \widetilde{\mathbb{T}} \cong (\mathbb{C}^*)^{\mathfrak{p}+3} \to \mathbb{T} \to 1,$$

where $\widetilde{\mathbb{T}}$ acts on $\mathbb{C}^{\mathfrak{p}+3}$ in the standard way.

Let $R_\mathbb{T} = H^*_\mathbb{T}(\mathrm{pt})$. Then

$$R_\mathbb{T} = \mathbb{C}[\mathsf{u}_1, \mathsf{u}_2, \mathsf{u}_3],\ S_\mathbb{T} = \mathbb{C}(\mathsf{u}_1, \mathsf{u}_2, \mathsf{u}_3).$$

These $\mathsf{u}_1, \mathsf{u}_2, \mathsf{u}_3$ are basis of M, and characters of $\mathbb{T}$. Choose u_3 such that $\mathbb{T}' = \ker(\mathsf{u}_3)$. Setting $\mathsf{u}_3 = 0$ passes into the equivariant setting for $\mathbb{T}'$. We have $R_{\mathbb{T}'} = H^*_{\mathbb{T}'}(\mathrm{pt})$. Then

$$R_{\mathbb{T}'} = \mathbb{C}[\mathsf{u}_1, \mathsf{u}_2],\ S_{\mathbb{T}'} = \mathbb{C}(\mathsf{u}_1, \mathsf{u}_2).$$

One defines $D_i^\mathbb{T} \in H^2_{\widetilde{\mathbb{T}}}(\mathbb{C}^{\mathfrak{p}+3};\mathbb{C}) \cong H^2_{\mathrm{CR},\mathbb{T}}(\mathcal{X};\mathbb{C})$ as the $\widetilde{\mathbb{T}}$-equivariant Poincaré dual of $\{Z_i = 0\} \subset \mathbb{C}^{\mathfrak{p}+3}$ – it is a lift of D_i into the equivariant cohomology. Similarly, we denote $\bar{D}_i^\mathbb{T}$ as the equivariant first Chern class of $\{Z_i = 0\} \subset X_\Sigma$. This is an equivariant lift of $\bar{D}_i$. In particular, $\bar{D}_i^\mathbb{T} = 0$ for $i = \mathfrak{p}' + 4, \ldots, \mathfrak{p} + 3$.

We have

$$H^2_{\mathrm{CR},\mathbb{T}}(\mathcal{X};\mathbb{C}) = H^2_\mathbb{T}(X_\Sigma;\mathbb{C}) \bigoplus \Big(\bigoplus_{\mathrm{age}(v)=1} \mathbb{C}\mathbf{1}_v \Big).$$

We still denote the projection of H to $H^2_\mathbb{T}(X_\Sigma;\mathbb{C})$ by $\bar{H}$. We choose an equivariant lift $\bar{H}_a^\mathbb{T}$ of $\bar{H}_a$ for $a = 1, \ldots, \mathfrak{p}'$. Then

$$H^2_{\mathrm{CR},\mathbb{T}}(\mathcal{X};\mathbb{C}) = \bigoplus_{i=1}^{3} \mathbb{C}\mathsf{u}_i \bigoplus_{a=1}^{\mathfrak{p}'} \mathbb{C}\bar{H}_a^\mathbb{T} \bigoplus \Big(\bigoplus_{\mathrm{age}(v)=1} \mathbb{C}\mathbf{1}_v \Big).$$

In the rest of this paper, whenever an equivariant (quantum, Chen-Ruan or classical) cohomology $QH^*_\mathbb{T}, H^*_\mathbb{T}, H^*_{\mathrm{CR},\mathbb{T}}$ is omitting the coefficient, we always regard it as over $\mathbb{C}$.

We define the following "extended Mori cone".

$$\mathbb{K}_{\text{eff}} = \bigcup_{\sigma\in\Sigma(3)} \{\beta \in \text{Lie}(K_{\Sigma'})|\langle\beta, D_i\rangle \in \mathbb{Z}_{\geq 0}, \Sigma(1) \ni \rho_i \notin \sigma\}.$$

For any $\beta \in \mathbb{K}_{\text{eff}}$, $v(\beta) = \sum_{i=1}^{p+3}\lceil\langle D_i, \beta\rangle\rceil b_i$. We choose a finite extension field $\bar{S}_{\mathbb{T}}$ of $S_{\mathbb{T}}$ such that $H^*_{\text{CR},\mathbb{T}}(\mathcal{X};\mathbb{C}) \otimes_{R_{\mathbb{T}}} S_{\mathbb{T}}$ is a semisimple algebra. We describe the canonical basis here.

For any $\sigma \in \Sigma(3)$, $\mathcal{X}_\sigma = [\mathbb{C}^3/G_\sigma] \subset \mathcal{X}$ is an affine toric CY 3-orbifold. An $\bar{S}_{\mathbb{T}}$-basis of $H^*_{\text{CR},\mathbb{T}}(\mathcal{X}_\sigma;\mathbb{C}) \otimes_{R_{\mathbb{T}}} S_{\mathbb{T}}$ are $\mathbf{1}_h$, $h \in G_\sigma$, each corresponding to a connected component of $\mathcal{I}\mathcal{X}_\sigma$. Their products are

$$\mathbf{1}_h \cup_{\mathcal{B}G_\sigma} \mathbf{1}_{h'} = \mathbf{1}_{hh'}.$$

For any $\gamma \in G^*_\sigma$, introduce

$$\phi_\gamma = \sum_{h\in G^*_\sigma} \chi_\gamma(h^{-1})\frac{\mathbf{1}_h}{\prod_{i=1}^3 \mathsf{w}_i(\sigma)^{c_i^\sigma(h)}}.$$

Here $\mathsf{w}_i(\sigma)$ is the weight of the $\mathbb{T}$ action on $T_{\mathfrak{l}_\sigma}\mathcal{X}_\sigma$, while $h \in G_\sigma$ acts on the i-th factor of $(\mathbb{C}^*)^3$ by multiplying $e^{2\pi\sqrt{-1}c_i^\sigma(h)}$. They are canonical basis of $H^*_{\text{CR},\mathbb{T}}(\mathcal{X}_\sigma;\mathbb{C})\otimes_{R_{\mathbb{T}}} S_{\mathbb{T}}$

$$\phi_\gamma \cup_{\mathcal{X}_\sigma} \phi_{\gamma'} = \delta_{\gamma,\gamma'}\phi_\gamma.$$

The sum of the pullbacks of the inclusion maps $\mathcal{X}_\sigma \hookrightarrow \mathcal{X}$ identifies

$$H^*_{\text{CR},\mathbb{T}}(\mathcal{X};\mathbb{C}) \otimes_{R_{\mathbb{T}}} S_{\mathbb{T}} \cong \bigoplus_{\sigma\in\Sigma(3)} H^*_{\text{CR},\mathbb{T}}(\mathcal{X}_\sigma;\mathbb{C}) \otimes_{R_{\mathbb{T}}} S_{\mathbb{T}}.$$

The basis $\{\phi_{\boldsymbol{\sigma}}|\boldsymbol{\sigma} = (\sigma,\gamma),\ \sigma \in \Sigma(3), \gamma \in G^*_\sigma\}$ is a canonical basis. We denote the set of A-model canonical basis by $I_\Sigma = \{\boldsymbol{\sigma} = (\sigma,\gamma)|\sigma \in \Sigma(3), \gamma \in G^*_\sigma\}$, and $N = \#I_\Sigma = \dim_{\mathbb{C}} H^*_{\text{CR}}(\mathcal{X};\mathbb{C})$. Let $\phi^{\boldsymbol{\sigma}}$ be the dual basis to $\phi_{\boldsymbol{\sigma}}$ under the equvariant Poincaré pairing.

2.5. Toric graphs. The action of the Calabi-Yau torus $\mathbb{T}'$ on $\mathcal{X}$ restricts to a Hamiltonian $\mathbb{T}'_{\mathbb{R}}$-action on the Kähler orbifold $(\mathcal{X},\omega(\vec{r}))$, with moment map $\mu' : \mathcal{X} \to M'_{\mathbb{R}} = \mathbb{R}^2$. The 1-skeleton $\mathcal{X}^1$ of the toric Calabi-Yau 3-fold $\mathcal{X}$ is the union of 0-dimensional and 1-dimensional orbits of the $\mathbb{T}$-action on $\mathcal{X}$. The image $\mu'(\mathcal{X}^1) \subset \mathbb{R}^2$ is a planar trivalent graph, which is known as the toric graph of the symplectic toric Calabi-Yau 3-orbifold $(\mathcal{X},\omega(\vec{r}))$. The toric graph depends also on the symplectic structure of $\mathcal{X}$.

2.6. Aganagic-Vafa Lagrangian branes. An Aganagic-Vafa Lagrangian brane in a toric Calabi-Yau 3-orbifold $\mathcal{X}$ is a Lagrangian sub-orbifold of the form

$$\mathcal{L} = [\widetilde{L}/K_{\Sigma'}] \subset \mathcal{X} = [\widetilde{\mu}^{-1}(\vec{r})/K_{\Sigma'}]$$

where

$$\widetilde{L} = \Bigg\{(z_1,\ldots,z_{3+p+s}) \in \widetilde{\mu}^{-1}(\vec{r}) : \sum_{i=1}^{3+p+s} \hat{l}_i^1|z_i|^2 = c_1,\ \sum_{i=1}^{3+p+s} \hat{l}_i^2|z_i|^2 = c_2,\ \arg(z_1\cdots z_{3+p+s}) = c_3\Bigg\},$$

c_1, c_2, c_3 are constants, and

$$\sum_{i=1}^{3+p+s} \hat{l}_i^\alpha = 0, \quad \alpha = 1, 2.$$

The compact 2-torus $\mathbb{T}'_{\mathbb{R}} \cong U(1)^2$ acts on $\mathcal{L}$, and under its moment map μ', the image $\mu'(\mathcal{L})$ is a point on the toric graph $\Gamma = \mu'(\mathcal{X}^1)$ and it is not a vertex. The Lagrangian $\mathcal{L}$ intersects a unique 1-dimensional $\mathbb{T}$ orbit $\mathfrak{l} \subset \mathcal{X}$ such that $\bar{\mathfrak{l}} = \mathfrak{l}_{\tau_0}$ where $\tau_0 \in \Sigma(2)$. We have $\mathfrak{l} \cong \mathbb{C}^* \times \mathcal{B}\boldsymbol{\mu}_{\mathfrak{m}}$ for some positive integer $\mathfrak{m}$ where $G_{\tau_0} \cong \boldsymbol{\mu}_{\mathfrak{m}}$. Here $\boldsymbol{\mu}_{\mathfrak{m}} \cong \mathbb{Z}_{\mathfrak{m}}$ is a multiplicative subgroup of $U(1)$. When $\mathfrak{m} = 1$, $\mathcal{L} \cong S^1 \times \mathbb{C}$ is smooth; when $\mathfrak{m} > 1$, $\mathcal{L}$ is smooth away from $\mathcal{L} \cap \mathfrak{l} \cong S^1 \times \mathcal{B}\boldsymbol{\mu}_m$. We require our Lagrangian $\mathcal{L}$ is *outer*, i.e. the closure of $\mathfrak{l}$ in $\mathcal{X}$ is not compact ($\bar{\mathfrak{l}} = \mathbb{C} \times \mathcal{B}\boldsymbol{\mu}_{\mathfrak{m}}$). Then τ_0 lies in a unique $\sigma_0 \in \Sigma(3)$. By a rearrangement of order, we require b_1, b_2, b_3 span σ_0, while b_2, b_3 span τ_0 and b_1, b_2, b_3 are labeled counterclockwisely as the vertices of σ_0. We have a short exact sequence of finite abelian groups

$$1 \to G_{\tau_0} \cong \boldsymbol{\mu}_{\mathfrak{m}} \to G_{\sigma_0} \to \boldsymbol{\mu}_{\mathfrak{r}} \to 1.$$

The stabilizer at the $\mathbb{T}$-fixed point $\mathfrak{l}_{\sigma_0}$ is G_{σ_0} while the generic stabilizer on $\mathfrak{l}_{\tau_0} = \bar{\mathfrak{l}}$ is G_{τ_0}.

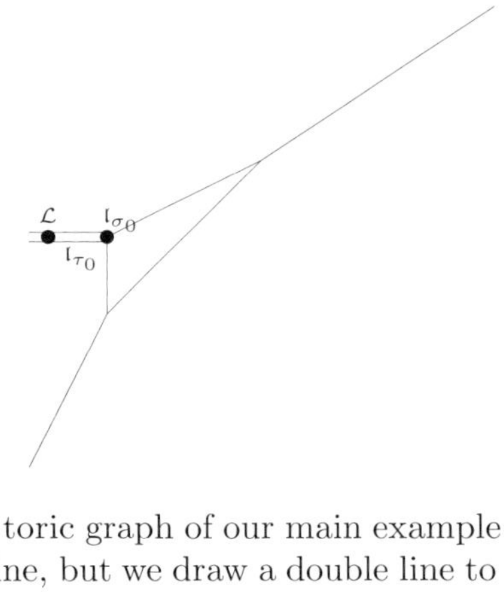

FIGURE 2. The toric graph of our main example $\mathcal{X}$. The gerby leg $\mathfrak{l}_{\tau_0}$'s image is a line, but we draw a double line to denote it is gerby.

EXAMPLE 2.3 (Example 2.2, continued). Given the choice of σ_0 and τ_0 as in Figure 1, the toric graph for $\mathcal{X}$ and the phase of the Aganagic-Vafa brane $\mathcal{L}$ is in Figure 2. We have

$$\begin{aligned} &G_{\tau_0} \cong \boldsymbol{\mu}_2, \ \mathfrak{m} = 2, \ G_{\sigma_0} \cong \boldsymbol{\mu}_2, \ \mathfrak{r} = 1, \\ &\mathfrak{l}_{\tau_0} \cong \mathbb{C} \times \mathcal{B}\boldsymbol{\mu}_2, \ \mathfrak{l}_{\sigma_0} = [\mathrm{pt}/\boldsymbol{\mu}_2], \\ &G_{\sigma_1} = G_{\sigma_2} = \{1\}, \ \mathfrak{l}_{\tau_1} = \mathfrak{l}_{\tau_2} = \mathrm{pt}. \end{aligned}$$

There is also a *framing* datum $f \in \mathbb{Z}$. It prescribes a subtorus $\mathbb{T}'_f \subset \mathbb{T}'$ by $\mathbb{T}'_f = \ker(\mathsf{u}_2 - f\mathsf{u}_1)$. We denote $H^*_{\mathbb{T}'_f}(\mathrm{pt}) = \mathbb{C}[\mathsf{v}]$ such that v is the image of u_1 under the restriction, while the image of u_2 is $f\mathsf{v}$.

2.7. Primary closed Gromov-Witten invariants and A-model free energies. We define genus g, degree β primary closed Gromov-Witten invariants:

$$\langle\gamma_1,\ldots,\gamma_\ell\rangle^{\mathcal{X}}_{g,\ell,\beta} = \int_{[\overline{\mathcal{M}}_{g,\ell}(\mathcal{X},\beta)^{\mathbb{T}'}]^{\mathrm{vir}}} \frac{\prod_{i=1}^{\ell}\mathrm{ev}_i^*(\gamma_i)}{e_{\mathbb{T}'}(N^{\mathrm{vir}})} \in R_{\mathbb{T}'}.$$

The A-model genus g free energy $\widetilde{F}^{\mathcal{X}}_g$ is a generating function of primary genus g closed Gromov-Witten invariants, as a function of $\boldsymbol{\tau}\in H^2_{\mathrm{CR},\mathbb{T}'}(\mathcal{X})$.

$$\widetilde{F}^{\mathcal{X}}_g(\boldsymbol{\tau}) = \sum_{\beta,\ell\geq 0}\frac{\langle\boldsymbol{\tau},\ldots,\boldsymbol{\tau}\rangle^{\mathcal{X}}_{g,\ell,\beta}}{\ell!}.$$

2.8. Open Gromov-Witten invariants and A-model open potentials. The BKMP Remodeling Conjecture builds the mirror symmetry for the open Gromov-Witten potentials $\widetilde{F}^{\mathcal{X},(\mathcal{L},f)}_{g,n}(\widetilde{X}_1,\ldots,\widetilde{X}_n,\boldsymbol{\tau})$ as well as free energies $\widetilde{F}^{\mathcal{X}}_g(\boldsymbol{\tau})$.

The Aganagic-Vafa Lagrangian brane $\mathcal{L}$ is homotopic to $S^1\times\mathcal{B}\boldsymbol{\mu}_{\mathfrak{m}}$, so

$$H_1(\mathcal{L};\mathbb{Z}) = \mathbb{Z}\times\boldsymbol{\mu}_{\mathfrak{m}}.$$

Open GW invariants of $(\mathcal{X},\mathcal{L})$ count holomorphic maps

$$u:(\Sigma,x_1,\ldots,x_\ell,\partial\Sigma=\coprod_{j=1}^{n}R_j)\to(\mathcal{X},\mathcal{L})$$

where Σ is a bordered Riemann surface with stacky points $x_i=\mathcal{B}\mathbb{Z}_{r_i}$ and $R_j\cong S^1$ are connected components of $\partial\Sigma$. These invariants depend on the following data:

(1) the topological type (g,n) of the coarse moduli of the domain, where g is the genus of Σ and n is the number of connected components of $\partial\Sigma$,
(2) the degree $\beta' = u_*[\Sigma]\in H_2(\mathcal{X},\mathcal{L};\mathbb{Z})$,
(3) the winding numbers $\mu_1,\ldots,\mu_n\in\mathbb{Z}$ and the monodromies $k_1,\ldots,k_n\in\boldsymbol{\mu}_{\mathfrak{m}}$, where $(\mu_j,k_j)=u_*[R_j]\in H_1(\mathcal{L};\mathbb{Z})=\mathbb{Z}\times\boldsymbol{\mu}_{\mathfrak{m}}$,
(4) the framing $f\in\mathbb{Z}$ of $\mathcal{L}$.

We call the pair $(\mathcal{L},f)$ a framed Aganagic-Vafa Lagrangian brane. We write $\vec{\mu} = ((\mu_1,k_1),$
$\ldots,(\mu_n,k_n))$. Here $k_i\in\{0,\ldots,\mathfrak{m}-1\}$ is regarded as an element $e^{2\pi\sqrt{-1}k_i/\mathfrak{m}}\in\boldsymbol{\mu}_{\mathfrak{m}}$. Let $\mathcal{M}_{(g,n),\ell}(\mathcal{X},\mathcal{L}\mid\beta',\vec{\mu})$ be the moduli space parametrizing maps described above, and let $\overline{\mathcal{M}}_{(g,n),\ell}(\mathcal{X},\mathcal{L}\mid\beta',\vec{\mu})$ be the partial compactification: we allow the domain Σ to have nodal singularities, and an orbifold/stacky point on Σ is either a marked point x_j or a node; we require the map u to be stable in the sense that its automorphism group is finite. Evaluation at the i-th marked point x_i gives a map $\mathrm{ev}_i:\overline{\mathcal{M}}_{(g,n),\ell}(\mathcal{X},\mathcal{L}\mid\beta',\vec{\mu})\to\mathcal{I}\mathcal{X}$.

Given $\gamma_1,\ldots,\gamma_\ell\in H^*_{\mathrm{CR},\mathbb{T}'}(\mathcal{X};\mathbb{C})$, we define

$$\langle\gamma_1,\ldots,\gamma_\ell\rangle^{\mathcal{X},(\mathcal{L},f)}_{g,\beta,\vec{\mu}} := \int_{[\overline{\mathcal{M}}_{(g,n),\ell}(\mathcal{X},\mathcal{L}\mid\beta',\vec{\mu})^{\mathbb{T}'_{\mathbb{R}}}]^{\mathrm{vir}}}\frac{\prod_{i=1}^{\ell}\mathrm{ev}_i^*\gamma_i}{e_{\mathbb{T}'_{\mathbb{R}}}(N^{\mathrm{vir}})}\Bigg|_{(\mathbb{T}_f)_{\mathbb{R}}} \in \mathbb{C}\mathsf{v}^{\sum_{i=1}^{\ell}\frac{\deg\gamma_i}{2}-1} \tag{2}$$

where $\mathbb{T}'_{\mathbb{R}}$ and $(\mathbb{T}_f)_{\mathbb{R}}$ are the corresponding real sub-torus of $\mathbb{T}'$ and $\mathbb{T}_f$, which preserves the Lagrangian $\mathcal{L}$, $H^*_{\mathbb{T}'_f}(\mathrm{pt})=\mathbb{C}[\mathsf{v}]$, $\beta\in H_2(\mathcal{X};\mathbb{Z})$ and $\beta'=\beta+\sum\mu_i\in H_2(\mathcal{X},\mathcal{L};\mathbb{Z})$.

For $\boldsymbol{\tau} \in H^2_{\mathrm{CR},\mathbb{T}'}(\mathcal{X};\mathbb{C})$, we define generating functions $\widetilde{F}^{\mathcal{X},(\mathcal{L},f)}_{g,n}$ of open Gromov-Witten invariants as follows.

$$\begin{aligned}
(3)\qquad & \widetilde{F}^{\mathcal{X},(\mathcal{L},f)}_{g,n}(\widetilde{X}_1,\dots,\widetilde{X}_n,\boldsymbol{\tau}) \\
&= \sum_{\beta,\ell\geq 0}\sum_{(\mu_j,k_j)\in\mathbb{Z}\times\boldsymbol{\mu}_{\mathfrak{m}}} \frac{\langle \boldsymbol{\tau}^{\ell}\rangle^{\mathcal{X},(\mathcal{L},f)}_{g,\beta,(\mu_1,k_1)\cdots,(\mu_n,k_n)} Q^{\beta}}{\ell!} \\
&\cdot \otimes_{j=1}^{n}\Big(\widetilde{X}_j^{\mu_j}(-(-1)^{\frac{-k_j}{\mathfrak{m}}})\mathbf{1}'_{\frac{-k_j}{\mathfrak{m}}}\Big) \in H^*_{\mathrm{CR}}(\mathcal{B}\boldsymbol{\mu}_{\mathfrak{m}};\mathbb{C})^{\otimes n}
\end{aligned}$$

where $H^*_{\mathrm{CR}}(\mathcal{B}\boldsymbol{\mu}_{\mathfrak{m}};\mathbb{C}) = \oplus_{k=0}^{m-1}\mathbb{C}\mathbf{1}'_{\frac{k}{\mathfrak{m}}}$, and $\mathbf{1}'_{-\frac{k}{\mathfrak{m}}} = \mathbf{1}'_{\frac{\mathfrak{m}-k}{\mathfrak{m}}}$ for $k \in \{1,\dots,\mathfrak{m}-1\}$. The closed Gromov-Witten invariants and potential can be viewed as a special case for $n = 0$ i.e. there is no boundary on the domain curve. The variable Q is the Novikov variable

$$Q^{\beta} = Q_1^{\langle H_1,\beta\rangle}\cdots Q_{\mathfrak{p}'}^{\langle H_{\mathfrak{p}'},\beta\rangle}.$$

2.9. Descendant closed Gromov-Witten invariants. Given $\gamma_1,\dots,\gamma_n$, we define a generating function of genus g, n-point descendant closed Gromov-Witten invariants:

$$\langle\!\langle \gamma_1\psi_1^{k_1},\dots,\gamma_n\psi_n^{k_n}\rangle\!\rangle^{\mathcal{X}}_{g,n} = \sum_{\beta,\ell\geq 0}\frac{Q^{\beta}}{\ell!}\langle \gamma_1\psi_1^{k_1},\dots,\gamma_n\psi^{k_n},\boldsymbol{\tau}^{\ell}\rangle^{\mathcal{X}}_{g,n+\ell,\beta},$$

where $\psi_i = c_1(\mathbb{L}_i)$ and $\mathbb{L}_i \to \overline{\mathcal{M}}_{g,n+\ell}(\mathcal{X},\beta)$ is line bundle whose fiber at moduli point $[u : (C,x_1,\dots,x_{n+\ell}) \to \mathcal{X}]$ is the cotangent line $T^*_{x_i}C$ at the i-th marked point to (the coarse moduli space of) the domain curve. In this notation $\widetilde{F}_g(\boldsymbol{\tau}) = \langle\!\langle\rangle\!\rangle^{\mathcal{X}}_{g,0}$.

3. Mirror curves and the Landau-Ginzburg mirror

3.1. Three B-models. The Remodeling Conjecture by Mariño and Bouchard-Klemm-Mariño-Pasquetti [**13, 14, 81**] expresses all genus open-closed Gromov-Witten invariants in terms of the Eynard-Orantin recursion on its mirror curve. The mirror curve plays the central role in the B-model, and is one of three related and equivalent B-models.

3.1.1. *Landau-Ginzburg model.* Recall we have a choice of Kähler basis $\{H_1,\dots,H_{\mathfrak{p}}\}$ as given in section 2.3. The non-equivariant superpotential is

$$W = \sum_{i=1}^{\mathfrak{p}+3} X_i,$$

where

$$(4)\qquad \prod_{i=1}^{\mathfrak{p}+3} X_i^{\langle D_i,\beta\rangle} = q^{\beta},\ \ \forall\beta\in\mathbb{K}_{\mathrm{eff}}.$$

The Equation (4) prescribe a 3-dimenionsal algebraic torus $\mathcal{Y}\subset(\mathbb{C}^*)^{\mathfrak{p}+3}$. We regard this W as the superpotential in the LG-model on this algebraic torus $\mathcal{Y}$. The parameters $q_1,\dots,q_{\mathfrak{p}}$ are complex parameters of the B-model. We use $q_K = (q_1,\dots,q_{\mathfrak{p}'})$ to denote the parameters corresponding to the *Kähler* part, while $q_{\mathrm{orb}} = (q_{\mathfrak{p}'+1},\dots,q_{\mathfrak{p}})$ denotes the *twisted* part. Under mirror symmetry, heuristically $\log q_K$ measures the mutual distances between vertices (corresponding to 3-cones, or torus fixed points in $\mathcal{X}$) in the toric graph.

We define

$$a_{m_i,n_i} = 1,\ i = 1,2,3,$$
$$\prod_{a=1}^{\mathfrak{p}} (a_{m_{a+3},n_{a+3}}(q))^{\langle D_{a+3},\beta\rangle} = q^{\beta}, \forall \beta \in \mathbb{K}.$$

Under the large radius limit $q_K \to 0$, $a_{m_{a+3},n_{a+3}}(q) \to 0$ for $a = 1,\ldots,\mathfrak{p}'$. The coordinates X, Y and $a_{m_{a+3},n_{a+3}}(q)$ are specific to the choice of σ_0 and τ_0 (coming from the position (phase) of the Aganagic-Vafa brane). The parameters $\log a_{m_{a+3},n_{a+3}}(q) \to \infty$ for $a = 1,\ldots,\mathfrak{p}'$ – heuristically, in the toric graph they measure "distances" from other 3-cones to σ_0, while $q_{\mathfrak{p}'+1},\ldots,q_{\mathfrak{p}}$ are parameters for the orbifolds twisted sectors.

One can write the non-equivariant superpotential as the following form

$$W = H_q(X,Y)Z,$$
$$H_q(X,Y) = X^{\mathfrak{r}}Y^{-\mathfrak{s}} + Y^{\mathfrak{m}} + 1 + \sum_{a=1}^{\mathfrak{p}} a_{m_{3+a},n_{3+a}}(q)X^{m_{3+a}}Y^{n_{3+a}}.$$

The equivariantly-perturbed superpotential is

$$W^{\mathbb{T}'} = W + \hat{x},$$

where $\hat{x} = \mathsf{u}_1 x + \mathsf{u}_2 y$. It is a holomorphic function defined on the universal cover $\widetilde{\mathcal{Y}} = \mathbb{C}^3$ of $\mathcal{Y}$.[2]

3.1.2. *Mirror curve.* The mirror curve $C_q \subset (\mathbb{C}^*)^2$ is defined by the equation $H_q(X,Y)$
$= 0$. The defining polytope P defines a polarized 2-dimensional toric surface $\mathbb{S}_P$ with an ample line bundle L_P, and $H_q(X,Y)$ extends to a section in $H^0(\mathbb{S}_P; L_P)$. The zero section is the compactified mirror curve $\overline{C}_q \in \mathbb{S}_P$. It is of genus $\mathfrak{g}$, and intersects transversally with $\partial\mathbb{S}_P$ at $\mathfrak{n}$ points (see Section 2.3 for the definition of $\mathfrak{g}, \mathfrak{n}$).

In fact, there is an explicit construction of a flat family of toric surfaces over a neighborhood of $q = 0$ in [**45**]. Each generic fiber is a toric surface isomorphic to $\mathbb{S}_P$, while the central fiber is $\cup_{\sigma\in\Sigma(3)}\mathbb{S}_{P_\sigma}$, a normal crossing union of several toric surfaces – each corresponds to the polytope P_σ in the triangulation of the defining polytope P. The toric surface $\mathbb{S}_P$ degenerates into $\bigcup_{\sigma\in\Sigma(3)}\mathbb{S}_{P_\sigma}$ at $q = 0$. The family $\mathfrak{C}$ of mirror curves is the zero section of a fiberwise ample line bundle – on each non-degenerate fiber this line bundle restricts to L_P. At a generic point q, the fiber $\mathfrak{C}_q$ is just $\overline{C}_q$. At $q = 0$, $\mathfrak{C}_q$ degenerates into a nodal curve while each piece lives inside $\mathbb{S}_{P_\sigma}$. We denote this neighborhood of $q = 0$ by $\mathfrak{B}$, and denote

$$\mathfrak{B}^\circ = \mathfrak{B} \cap \left(\bigcap_{a=1}^{\mathfrak{p}} \{q_a \neq 0\} \right).$$

Denote $\mathbb{D}^\infty = \mathfrak{C} \cap (\partial\mathbb{S}_P)$. This is the family of punctures $\overline{C}_q \setminus C_q$

[2]On the B-model side, we regard u_1 and u_2 as complex numbers.

EXAMPLE 3.1 (Example 2.3, continued).

$$\begin{aligned}
&W^{\mathbb{T}'} = H_q(X,Y) + \mathsf{u}_1 x + \mathsf{u}_2 y,\\
&\mathfrak{r} = 1,\ \mathfrak{s} = 0,\ \mathfrak{m} = 2,\\
&H_q(X,Y) = X + Y^2 + 1 + q_1 X^2 Y^{-1} + q_2 Y,\\
&q_K = q_1,\ q_{\mathrm{orb}} = q_2.
\end{aligned}$$

The mirror curve is illustrated in Figure 3. It is a fattened tube around the toric graph Figure 2. Notice that the gerby leg contributes to two punctures p_0, p_1. The degenerated mirror curve $\mathfrak{C}_0$ is illustrated in Figure 4.

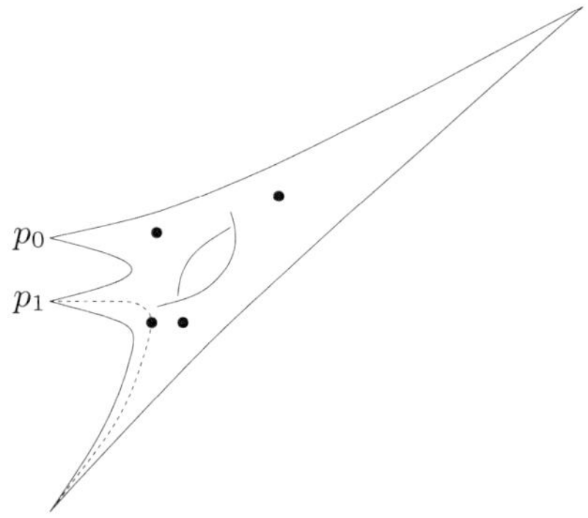

FIGURE 3. The mirror curve of $\mathcal{X}$. It can be regarded as a curve in the toric surface $\mathbb{S}_P$. Black dots are the ramification points, while the dashed curve is the Lefschetz thimble passing through one ramification point with $\hat{x} \to \infty$.

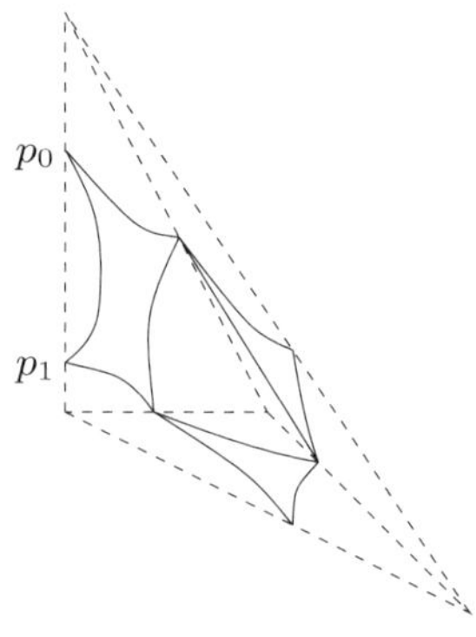

FIGURE 4. The degenerated mirror curve of $\mathcal{X}$ at $q = 0$. It can be regarded as a curve in the degenerated toric surface - a normal crossing of two $\mathbb{P}^2$ and a $\mathbb{P}(1,1,2)$, whose moment polytopes are dashed lines.

3.1.3. *Hori-Vafa mirror.* The Hori-Vafa mirror $(\check{\mathcal{X}}_q, \Omega_q)$ is a non-compact Calabi-Yau 3-fold (Ω_q is the Calabi-Yau form):

$$\check{\mathcal{X}}_q = \{(u,v,X,Y) \in \mathbb{C}^2 \times (\mathbb{C}^*)^2 : uv = H_q(X,Y)\},$$
$$\Omega_q = \mathrm{Res}_{\check{X}_q}\left(\frac{1}{H(X,Y,q) - uv} du \wedge dv \wedge \frac{dX}{X} \wedge \frac{dY}{Y}\right).$$

3.2. Open and closed mirror maps from the B-model. Throughout this section, we choose *framing* $f \in \mathbb{Z}$ the same as in Section 2.6 and let $u_1 = 1$ and $u_2 = f$ in the B-model setting. So

$$\hat{x} = x + fy,\ \hat{y} = y.$$

The mirror curve equation for C_q becomes

$$H(\hat{X},\hat{Y}) = \hat{X}^{\mathfrak{r}}\hat{Y}^{-\mathfrak{s}-\mathfrak{r}f} + \hat{Y}^{\mathfrak{m}} + 1 + \text{other terms}.$$

The revised coordinates $\hat{X}, \hat{Y}$ are specific to the position ("phase") of the Aganagic-Vafa brane *and* its framing. We denote the Seiberg-Witten form

$$\Phi = \hat{y}d\hat{x}.$$

3.2.1. *Closed mirror maps as periods.* Let us recall the geometry of the mirror curve C_q and its compactification $\overline{C}_q$ (c.f. Section 2.3).

- The genus of $\overline{C}_q$ is $\mathfrak{g} = h^4_{\mathrm{CR}}(\mathcal{X})$.
- The number of punctures of C_q is $\mathfrak{n} = \mathfrak{p} - \mathfrak{g} + 3$ ($\mathfrak{p} = h^2_{\mathrm{CR}}(\mathcal{X})$, i.e. the number of extended Kähler parameters.)
- $h_1(\overline{C}_q) = 2\mathfrak{g}$, $h_1(C_q) = h_1(\overline{C}_q, \mathbb{D}^\infty) = \mathfrak{g} + \mathfrak{n} - 1 = \mathfrak{p} + \mathfrak{g} + 2$.

The lore for toric Calabi-Yau 3-folds is that the mirror maps are obtained by integrations

$$\tau_i = \frac{1}{2\pi\sqrt{-1}} \int_{A_i} \Phi.$$

Here A_i are 1-cycles in $H_1(C_q)$, probably with non-integral coefficients in the presence of a toric orbifold. Since $\Phi = \hat{y}d\hat{x}$, this map is not well-defined. For practical computation in explicit examples, one can easily make sense of the result by ignoring the constants due to the non-trivial monodromy of $\hat{y}$. We make this statement a little more precise here.

Consider the inclusion $I : C_q \to (\mathbb{C}^*)^2$ where $q \in \mathfrak{B}^\circ$. Denote the kernel of the map

$$K_1(C_q;\mathbb{Z}) = \ker I_* \subset H_1(C_q;\mathbb{Z}).$$

This sublattice $K_1(C_q;\mathbb{Z}) \cong \mathbb{Z}^{2\mathfrak{g}+\mathfrak{n}-3}$ consists of cycles with trivial x and y monodromy. One can lift any element $\gamma \in K_1(C_q;\mathbb{Z})$ to a loop $\widetilde{\gamma} \in \pi_1(\widetilde{C}_q)$. Here $\widetilde{C}_q = \pi^{-1}(C_q)$ under the universal covering map $\pi : \mathbb{C}^2 \to (\mathbb{C}^*)^2$. The following integral

$$\frac{1}{2\pi\sqrt{-1}} \int_{\widetilde{\gamma}} \Phi \in \mathbb{C}$$

does depend on the choice of $\widetilde{\gamma}$, but only up to an integral multiple of $2\pi\sqrt{-1}$. If the toric Calabi-Yau 3-fold $\mathcal{X}$ is smooth, then one can choose $A_a \in K_1^{\mathrm{inv}}(C_q;\mathbb{Z})$ for

$a = 1, \dots, \mathfrak{p}$, the *monodromy invariant* cycles,[3] such that

$$\frac{1}{2\pi\sqrt{-1}}\int_{A_a}\Phi \in \mathbb{C} = \log q_a + O(q) \mod 2\pi\sqrt{-1}.$$

Since for each $A \in K_1^{\rm inv}(C_q;\mathbb{Z})$ can be extended a family of *flat* cycles over $\mathfrak{B}^\circ$, we have a map

$$A \mapsto \frac{1}{2\pi\sqrt{-1}}\int_A \Phi \in \mathcal{H}ol(\mathfrak{B}^\circ, \mathbb{C}/(2\pi\sqrt{-1}\mathbb{Z})).$$

Here $\mathcal{H}ol$ is the set of holomorphic maps (in paricular $\log q_a$ is in this set). More subtleties arise when $\mathcal{X}$ is an orbifold. To acquire the desired leading order behavior, we have to use cycles in $K_1^{\rm inv}(C_q;\mathbb{C})$. The explicit construction of cycles $A_1, \dots, A_{\mathfrak{p}}$ is in [**45**]. We list the result here

PROPOSITION 3.2. *There exists cycles $A_1, \dots, A_{\mathfrak{p}'}, \dots, A_{\mathfrak{p}} \in K_1^{\rm inv}(C_q;\mathbb{C})$ such that*

$$\frac{1}{2\pi\sqrt{-1}}\int_{A_a}\Phi = \begin{cases} \log q_a + O(q_K) + O(|q_{\rm orb}|^2), & a = 1, \dots, \mathfrak{p}' \\ q_a + O(q_K) + O(|q_{\rm orb}|^2), & a = \mathfrak{p}', \dots, \mathfrak{p}. \end{cases}$$

One has to understand the right hand side of this proposition as holomorphic functions in $q_1, \dots, q_{\mathfrak{p}}$ up to constants.

3.2.2. *Large radius limit for the open parameter and open mirror map.* A point on C_q heuristically correponds to a *B-brane*. The large radius limit of the open parameter should correspond to moving the Aganagic-Vafa brane $\mathcal{L}$ to infinity along the leg. We define the *large radius points* $\bar{p}_0(q), \dots, \bar{p}_{\mathfrak{m}-1}(q)$ on $\mathfrak{C}_q$ by requiring $\hat{X} = 0$ and

$$\hat{Y}^{\mathfrak{m}} = -1, \text{ when } q = 0.$$

We also denote D_q^ℓ to be a small neighborhood of $\bar{p}_\ell(q)$ on $\overline{C}_q = \mathfrak{C}_q$.

There is a $\boldsymbol{\mu}_{\mathfrak{m}}^*$-action permuting these points $\bar{p}_\ell$, given by mapping the $\hat{Y}$-coordinate $(-1)^{\frac{k}{\mathfrak{m}}}$ to $(-1)^{\frac{1}{\mathfrak{m}}}\chi(1^{\frac{k}{\mathfrak{m}}})$ for $\chi \in \boldsymbol{\mu}_{\mathfrak{m}}^*$. We can identify these points $\bar{p}_\ell$, $\ell = 1, \dots, \mathfrak{m}$ with elements in $\boldsymbol{\mu}_{\mathfrak{m}}^*$ in a non-canonical way. Define

$$\psi_\ell := \frac{1}{\mathfrak{m}}\sum_{k=0}^{\mathfrak{m}-1}\omega_{\mathfrak{m}}^{-k\ell}\mathbf{1}'_{\frac{k}{\mathfrak{m}}}, \quad \ell = 0, 1, \dots, \mathfrak{m}-1,$$

where $\omega_{\mathfrak{m}} = e^{2\pi\sqrt{-1}/\mathfrak{m}}$. Then ψ_ℓ for $\ell \in \boldsymbol{\mu}_{\mathfrak{m}}^*$ form a dual basis to $\mathbf{1}'_{\frac{k}{m}}$ for $k \in \boldsymbol{\mu}_{\mathfrak{m}}$.

As indicated in [**6**], there is an explicit constuction of a (linear combination of) path $\widetilde{\gamma}_0$ in $\widetilde{C}_q$. Each component in $\widetilde{\gamma}_0$ is a path starting at the point where $\hat{x}$-coordinate is $\hat{x}_0$ near $\bar{p}_\ell$, such that

- $\widetilde{\gamma}_0$ descends to a cycle A_0 in $H_1(C_q;\mathbb{C})$ – it has trivial monodromy in $\hat{x}$ while the monodromy in $\hat{y}$ is precisely $\hat{y} \mapsto \hat{y} - 2\pi\sqrt{-1}$;
- The integration depends on the starting point:

$$\frac{1}{2\pi\sqrt{-1}}\int_{\widetilde{\gamma}_0}\hat{y}d\hat{x} = \hat{x}_0 + O(q_K) + O(|q_{\rm orb}|^2). \tag{5}$$

This will be the open mirror (24).

[3]Monodromy under the Gauss-Manin connection around $\mathfrak{B}^\circ$, i.e. around hyperplanes $\{q_a = 0\}$.

4. A quick review of the genus zero mirror theorem for toric orbifolds

4.1. Frobenius structures for toric Calabi-Yau 3-orbifolds: Quantum cohomology. Let

$$\chi = \dim_{\mathbb{C}} H^*_{\mathrm{CR}}(\mathcal{X}) = \dim_{\bar{S}_{\mathbb{T}}} H^*_{\mathrm{CR},\mathbb{T}}(\mathcal{X}; \bar{S}_{\mathbb{T}}).$$

We choose a $\bar{S}_{\mathbb{T}}$-basis of $H^*_{\mathrm{CR},\mathbb{T}}(\mathcal{X};\bar{S}_{\mathbb{T}})$ $\{T_i : i = 0, 1, \dots, \chi - 1\}$ such that

$$T_0 = 1, \quad T_a = \bar{H}^{\mathbb{T}}_{3+a} \text{ for } a = 1, \dots, \mathfrak{p}', \quad T_a = \mathbf{1}_{b_{3+a}} \text{ for } a = \mathfrak{p}' + 1, \dots, \mathfrak{p},$$

and for $i = \mathfrak{p} + 1, \dots, \chi - 1$, T_i is of the form $T_a T_b$ for some $a, b \in \{1, \dots, \mathfrak{p}\}$. Write $t = \sum_{a=0}^{\chi-1} \tau^a T_a$, and let $\tau' = (\tau_1, \dots, \tau_{\mathfrak{p}'})$, $\tau'' = (\tau_0, \tau_{\mathfrak{p}'+1}, \dots, \tau_{\chi-1})$. By the divisor equation,

$$\langle\!\langle T_i, T_j, T_k \rangle\!\rangle_{0,3}^{\mathcal{X},\mathbb{T}} \in S_{\mathbb{T}}[[\widetilde{Q}, \tau'']], \quad \langle\!\langle \hat{\phi}_{\boldsymbol{\sigma}}, \hat{\phi}_{\boldsymbol{\sigma}'}, \hat{\phi}_{\boldsymbol{\sigma}''} \rangle\!\rangle_{0,3}^{\mathcal{X},\mathbb{T}} \in \bar{S}_{\mathbb{T}}[\![\widetilde{Q}, \tau'']\!],$$

where $\widetilde{Q}^d = Q^d \exp(\sum_{a=1}^{\mathfrak{p}'} \tau_a \langle T_a, d\rangle)$. Let $S := \bar{S}_{\mathbb{T}}[\![\widetilde{Q}, \tau'']\!]$. Given $a, b \in H^*_{\mathrm{CR},\mathbb{T}}(\mathcal{X}; \bar{S}_{\mathbb{T}})$, define the *quantum product*

$$a \star_t b := \sum_{\boldsymbol{\sigma} \in I_\Sigma} \langle\!\langle a, b, \hat{\phi}_{\boldsymbol{\sigma}} \rangle\!\rangle \hat{\phi}_{\boldsymbol{\sigma}} \in H^*_{\mathrm{CR},\mathbb{T}}(\mathcal{X}; \bar{S}_{\mathbb{T}}) \otimes_{\bar{S}_{\mathbb{T}}} S.$$

Then $A := H^*_{\mathrm{CR},\mathbb{T}}(\mathcal{X}; \bar{S}_{\mathbb{T}}) \otimes_{\bar{S}_{\mathbb{T}}} S$ is a free S-module of rank χ, and $(A, \star_t)$ is a commutative, associative algebra over S. Let $I \subset S$ be the ideal generated by $\widetilde{Q}, \tau''$, and define

$$S_n := S/I^n, \quad A_n := A \otimes_S S_n$$

for $n \in \mathbb{Z}_{\geq 0}$. Then A_n is a free S_n-module of rank χ, and the ring structure $\star_t$ on A induces a ring structure $\star_{\underline{n}}$ on A_n. In particular,

$$S_1 = \bar{S}_{\mathbb{T}}, \quad A_1 = H^*_{\mathrm{CR},\mathbb{T}}(\mathcal{X}; \bar{S}_{\mathbb{T}}),$$

and $\star_{\underline{1}} = \star_{\mathcal{X}}$ is the orbifold cup product. So

$$\{\phi^{(1)}_{\boldsymbol{\sigma}} := \phi_{\boldsymbol{\sigma}} : \boldsymbol{\sigma} \in I_\Sigma\}$$

is an idempotent basis of $(A_1, \star_{\underline{1}})$. For $n \geq 1$, let $\{\phi^{(n+1)}_{\boldsymbol{\sigma}} : \boldsymbol{\sigma} \in I_\Sigma\}$ be the unique idempotent basis of $(A_{n+1}, \star_{\underline{n+1}})$ which is the lift of the idempotent basis $\{\phi^{(n)}_{\boldsymbol{\sigma}} : \boldsymbol{\sigma} \in I_\Sigma\}$ of $(A_n, \star_{\underline{n}})$ [**67**, Lemma 16]. Then

$$\{\phi_{\boldsymbol{\sigma}}(t) := \lim \phi^{(n)}_{\boldsymbol{\sigma}} : \boldsymbol{\sigma} \in I_\Sigma\}$$

is an idempotent basis of $(A, \star_t)$. The ring $(A, \star_t)$ is called the *equivariant big quantum cohomology ring*, which is also denoted by $QH^*_{\mathbb{T}}(\mathcal{X})$.

Set

$$\bar{\Lambda}^{\mathbb{T}}_{\mathrm{nov}} := \bar{S}_{\mathbb{T}} \otimes_{\mathbb{C}} \Lambda_{\mathrm{nov}} = \bar{S}_{\mathbb{T}}[[E(\mathcal{X})]].$$

Then $H := H^*_{\mathrm{CR},\mathbb{T}}(\mathcal{X}; \bar{\Lambda}^{\mathbb{T}}_{\mathrm{nov}})$ is a free $\bar{\Lambda}^{\mathbb{T}}_{\mathrm{nov}}$-module of rank χ. Any point $t \in H$ can be written as $t = \sum_{\boldsymbol{\sigma} \in I_\Sigma} t^{\boldsymbol{\sigma}} \hat{\phi}_{\boldsymbol{\sigma}}$. We have

$$H = \mathrm{Spec}(\bar{\Lambda}^{\mathbb{T}}_{\mathrm{nov}}[t^{\boldsymbol{\sigma}} : \boldsymbol{\sigma} \in I_\Sigma]).$$

Let $\hat{H}$ be the formal completion of H along the origin:

$$\hat{H} := \mathrm{Spec}(\bar{\Lambda}^{\mathbb{T}}_{\mathrm{nov}}[[t^{\boldsymbol{\sigma}} : \boldsymbol{\sigma} \in I_\Sigma]]).$$

Let $\mathcal{O}_{\hat{H}}$ be the structure sheaf on $\hat{H}$, and let $\mathcal{T}_{\hat{H}}$ be the tangent sheaf on $\hat{H}$. Then $\mathcal{T}_{\hat{H}}$ is a sheaf of free $\mathcal{O}_{\hat{H}}$-modules of rank χ. Given an open set in $\hat{H}$,

$$\mathcal{T}_{\hat{H}}(U) \cong \bigoplus_{\sigma \in I_\Sigma} \mathcal{O}_{\hat{H}}(U) \frac{\partial}{\partial t^\sigma}.$$

The big quantum product and the $\mathbb{T}$-equivariant Poincaré pairing defines the structure of a formal Frobenius manifold on $\hat{H}$:

$$\frac{\partial}{\partial t^\sigma} \star_t \frac{\partial}{\partial t^{\sigma'}} = \sum_{\rho \in I_\Sigma} \langle\!\langle \hat{\phi}_\sigma, \hat{\phi}_{\sigma'}, \hat{\phi}_\rho \rangle\!\rangle_{0,3}^{\mathcal{X},\mathbb{T}} \frac{\partial}{\partial t^\rho} \in \Gamma(\hat{H}, \mathcal{T}_{\hat{H}}).$$

$$\left(\frac{\partial}{\partial t^\sigma}, \frac{\partial}{\partial t^{\sigma'}}\right)_{\mathcal{X},\mathbb{T}} = \delta_{\sigma,\sigma'}.$$

The length of the canonical basis in the equivariant Chen-Ruan cohomology and the equivariant quantum cohomology are denoted as

$$(\phi_\sigma, \phi_\sigma)_{\mathcal{X},\mathbb{T}} = \frac{1}{\Delta^\sigma}, \quad (\phi_\sigma(t), \phi_\sigma(t))_{\mathcal{X},\mathbb{T}} = \frac{1}{\Delta^\sigma(t)}.$$

By replacing $\mathbb{T}$ by $\mathbb{T}'$ in the above discussion, we obtain the equivariant big quantum cohomology ring $QH^*_{\mathbb{T}'}(\mathcal{X})$.

4.2. The B-model Frobenius structure: The Jacobian ring. We can define a Frobenius algebra for each q:

$$\mathrm{Jac}(W^{\mathbb{T}'}) = \frac{\mathbb{C}[X^\pm, Y^\pm, Z^\pm]}{\langle \frac{\partial W^{\mathbb{T}'}}{\partial X}, \frac{\partial W^{\mathbb{T}'}}{\partial Y}, \frac{\partial W^{\mathbb{T}'}}{\partial Z} \rangle}.$$

The ring structure is self-evident in the definition, while the metric is

$$(f,g) = \frac{1}{(2\pi\sqrt{-1})^3} \int_{|dW^{\mathbb{T}'} = \epsilon|} \frac{f g dx \wedge dy \wedge dz}{\frac{\partial W^{\mathbb{T}'}}{\partial x} \frac{\partial W^{\mathbb{T}'}}{\partial y} \frac{\partial W^{\mathbb{T}'}}{\partial z}} = \sum_{\alpha} \frac{f(P_\alpha) g(P_\alpha)}{\det \mathrm{Hess}_{P_\alpha}(W^{\mathbb{T}'})}.$$

In this expression P_α runs through all critical points of $W^{\mathbb{T}'}$. An element in $\mathrm{Jac}(W^{\mathbb{T}'})$ is a Laurent polynomial in $\mathbb{C}[X^\pm, Y^\pm, Z^\pm]$ while two Laurent polynomials are identified if their values on critical points of $W^{\mathbb{T}'}$ are the same. The canonical basis of $\mathrm{Jac}(W^{\mathbb{T}'})$ is a function taking value 1 on one critical point of $W^{\mathbb{T}'}$ while vanishing at all other critical points.

Notice the current set-up does not constitute a Frobenius *manifold* – we just have a family of Frobenius algebras when varying q.

4.3. Genus 0 mirror symmetry for toric Calabi-Yau 3-orbifolds: Identification of I–function and J-function. Given the choice of $H_1, \ldots, H_{\mathfrak{p}}$ (and the lift $\bar{H}_1^{\mathbb{T}}, \ldots, \bar{H}_{\mathfrak{p}'}^{\mathbb{T}} \in H^2_{\mathbb{T}}(\mathcal{X}) \subset H^2_{\mathrm{CR},\mathbb{T}}(\mathcal{X})$), for any $\beta \in \mathbb{K}_{\mathrm{eff}}$, we define

$$q^\beta = \prod_{a=1}^{\mathfrak{p}} q_a^{\langle H_a, \beta \rangle}.$$

It is a monomial of $q_1, \ldots, q_{\mathfrak{p}}$. We define $\mathbb{T}$-equivariant small I-function as follows. Recall that $q_K = (q_1, \ldots, q_{\mathfrak{p}'})$ for the Kähler part, while $q_{\mathrm{orb}} = (q_{\mathfrak{p}'+1}, \ldots, q_{\mathfrak{p}})$ for the twisted sector.

DEFINITION 4.1.

$$\begin{aligned} I_{\mathbb{T}}(t_0,q,z) &= e^{(\sum_{a=1}^{\mathfrak{p}'}\bar{H}_a^{\mathbb{T}}\log q_a)/z}\sum_{\beta\in\mathbb{K}_{\rm eff}} q^\beta\prod_{i=1}^{3+\mathfrak{p}'}\frac{\prod_{m=\lceil\langle D_i,\beta\rangle\rceil}^{\infty}(\bar{D}_i^{\mathbb{T}}+(\langle D_i,\beta\rangle-m)z)}{\prod_{m=0}^{\infty}(\bar{D}_i^{\mathbb{T}}+(\langle D_i,\beta\rangle-m)z)} \\ &\cdot\prod_{i=4+\mathfrak{p}'}^{3+\mathfrak{p}}\frac{\prod_{m=\lceil\langle D_i,\beta\rangle\rceil}^{\infty}(\langle D_i,\beta\rangle-m)z}{\prod_{m=0}^{\infty}(\langle D_i,\beta\rangle-m)z}\mathbf{1}_{v(\beta)}. \end{aligned}$$

DEFINITION 4.2. The equivariant J-function is

$$J_{\mathbb{T}}(\boldsymbol{\tau},z)=\sum_{\boldsymbol{\sigma}\in I_\Sigma}\langle\!\langle\frac{\phi_{\boldsymbol{\sigma}}}{z-\psi_1},1\rangle\!\rangle_{0,2}^{\mathcal{X}}\phi^{\boldsymbol{\sigma}}.$$

The main result in [**27**] implies the following $\mathbb{T}$-equivariant mirror theorem:

THEOREM 4.3 (Coates-Corti-Iritani-Tseng).

$$J_{\mathbb{T}}(\boldsymbol{\tau},z)|_{Q=1}=I_{\mathbb{T}}(t_0,q,z),$$

where the equivariant closed mirror map $q\mapsto\boldsymbol{\tau}(q)\in H^2_{\mathrm{CR},\mathbb{T}}(\mathcal{X})$ is determined by the first-order term in the asymptotic expansion of the I-function

$$I(t_0,q,z)=1+\frac{\boldsymbol{\tau}(q)}{z}+o(z^{-1}).$$

More explicitly, the equivariant closed mirror map is given by

$$\boldsymbol{\tau}=\tau_0(q)+\sum_{a=1}^{\mathfrak{p}'}\tau_a(q)\bar{H}_a^{\mathbb{T}}+\sum_{a=\mathfrak{p}'+1}^{\mathfrak{p}}\tau_a(q)\mathbf{1}_{b_{a+3}},$$

where each $\tau_a(q)$ can be obtained readily from the expansion of the I-function. They have the following asymptotic behavior:

$$\begin{aligned} \tau_0(q) &= O(q)\in\sum_{i=1}^{3}\mathbb{C}\mathsf{u}_i, \\ \tau_a(q) &= \begin{cases}\log(q_a)+O(q), & 1\le a\le\mathfrak{p}',\\ q_a+\textit{higher order terms}, & \mathfrak{p}'+1\le a\le\mathfrak{p}.\end{cases} \end{aligned}\tag{6}$$

It does make sense to set $Q=1$ – the Nokikov variable could eventually be dropped, since in principle by the divisor equation e^τ carries the same information as Q. See Remark 5.2 for a precise argument.

Under this mirror map, the B-model large radius/orbifold mixed-type limit $q\to 0$ corresponds to the A-model large radius/orbifold mixed type limit $\tau'\to-\infty,\tau''\to 0$.

The mirror maps in Equation (6) and the period integrals in Proposition 3.2 are solutions to certain GKZ system with prescribed asymptotic behavior. They have same leading order, and this ensures that they are equal. The closed mirror map τ_i is the period integrals along A_i cycle.

PROPOSITION 4.4. *In the mirror map Equation* (6)

$$\tau_a=\frac{1}{2\pi\sqrt{-1}}\int_{A_a}\Phi \quad \text{mod const.}$$

4.4. Genus 0 mirror symmetry for toric Calabi-Yau 3-orbifolds: Identification of Frobenius algebras. The genus 0 mirror theorem of [**27**] implies the following under the closed mirror map $\boldsymbol{\tau} = \boldsymbol{\tau}(q)$

$$\mathrm{Jac}(W^{\mathbb{T}'}) \cong QH^*_{\mathbb{T}'}(\mathcal{X})|_{Q=1}. \tag{7}$$

This statement should be understood a pointwise isomorphism of Frobenius algebras.[4] This statement identifies $\mathrm{Jac}(W^{\mathbb{T}'})$ with a *slice* of $QH^*_{\mathbb{T}'}(\mathcal{X})$. From this fact, there is an identification of canonical basis for $\mathrm{Jac}(W^{\mathbb{T}'})$ and $QH^*_{\mathbb{T}'}(\mathcal{X})$ when $\boldsymbol{\tau} = \boldsymbol{\tau}(q)$. So the index set of the canonical basis I_Σ is identified with the index set of critical points of $W^{\mathbb{T}'}$, which we also denote by I_Σ. We use bold greek letters like $\boldsymbol{\alpha}$ and $\boldsymbol{\sigma}$ to denote such an index. The dimension $N = \dim QH^*_{\mathbb{T}'}(\mathcal{X})$ and *it is also the number of critical points of* $W^{\mathbb{T}'}$.

Each critical point $P_{\boldsymbol{\alpha}}$ of $W^{\mathbb{T}'}$ in $(\mathbb{C}^*)^3$ (although $W^{\mathbb{T}'}$ is defined on the universal cover of $(\mathbb{C}^*)^3$, $dW^{\mathbb{T}'}$ is well-defined on $(\mathbb{C}^*)^3$), by direct calculation, is characterized by the following

$$\begin{aligned}&P_{\boldsymbol{\alpha}} = (X_{\boldsymbol{\alpha}}, Y_{\boldsymbol{\alpha}}, Z_{\boldsymbol{\alpha}}(X_{\boldsymbol{\alpha}}; Y_{\boldsymbol{\alpha}})) \text{ is a critical point of } W^{\mathbb{T}'} \iff \\ &p_{\boldsymbol{\alpha}} = (X_{\boldsymbol{\alpha}}, Y_{\boldsymbol{\alpha}}) \text{ is a critical point of } d\hat{x} \text{ on the mirror curve } C_q.\end{aligned}$$

Let's denote the canonical basis $V_{\boldsymbol{\alpha}} \in \mathrm{Jac}(W^{\mathbb{T}'})$ to be the function taking value 1 on $P_{\boldsymbol{\alpha}}$ and 0 on $P_{\boldsymbol{\beta}}, \boldsymbol{\beta} \neq \boldsymbol{\alpha}$. In particular, $W^{\mathbb{T}'}(P_{\boldsymbol{\alpha}}) = \hat{x}(p_{\boldsymbol{\alpha}})$, i.e. the *canonical coordinate* for each $V_{\boldsymbol{\alpha}}$ is the *critical value* of $W^{\mathbb{T}'}$, which is also the branch value of $\hat{x}_{0,\boldsymbol{\alpha}} := \hat{x}(p_{\boldsymbol{\alpha}})$. A straightforward calculation shows

- One sets $\hat{x} = \hat{x}_{0,\boldsymbol{\alpha}} + \zeta_{\boldsymbol{\alpha}}^2$. In the expansion $\hat{y} = \hat{y}_{0,\boldsymbol{\alpha}} + \sum_{d=1}^\infty h_d^{\boldsymbol{\alpha}} \zeta_{\boldsymbol{\alpha}}^d$,
$$h_1^{\boldsymbol{\alpha}} = \sqrt{\frac{2}{\frac{d^2\hat{x}}{d\hat{y}^2}}};$$
- $\det \mathrm{Hess}_{P_{\boldsymbol{\alpha}}}(W^{\mathbb{T}'}) = -\frac{d^2\hat{x}}{d\hat{y}^2}(p_{\boldsymbol{\alpha}})$;
- The squared norm of the canonical basis
$$(V_{\boldsymbol{\alpha}}, V_{\boldsymbol{\alpha}}) = \frac{1}{\det \mathrm{Hess}_{P_{\boldsymbol{\alpha}}}(W^{\mathbb{T}'})}.$$

Since we denote $(\hat{\phi}_{\boldsymbol{\alpha}}(\boldsymbol{\tau}), \hat{\phi}_{\boldsymbol{\alpha}}(\boldsymbol{\tau})) = \frac{1}{\Delta^{\boldsymbol{\alpha}}(\boldsymbol{\tau})}$, by the isomorphism of Frobenius algebra under the closed mirror map $\boldsymbol{\tau} = \boldsymbol{\tau}(q)$, we conclude that

$$\frac{h_1^{\boldsymbol{\alpha}}}{\sqrt{-2}} = \frac{1}{\sqrt{\Delta^{\boldsymbol{\alpha}}(\boldsymbol{\tau})}}. \tag{8}$$

We will see in Section 7.2 this identifies the vertex terms of A and B-model graph sum formulae (Theorem 5.7 and 6.7).

[4] A reasonably enlarged B-model with more parameters than q should produce a Frobenius manifold isomorphic to $QH^*_{\mathbb{T}'}$. However we only need pointwise isomorphism of Frobenius algebras for the purpose of proving the Remodeling Conjecture.

5. A-model quantization: The orbifold Givental formula

5.1. The equivariant big quantum differential equation. We consider the Dubrovin connection ∇^z, which is a family of connections parametrized by $z \in \mathbb{C} \cup \{\infty\}$, on the tangent bundle $T_{\hat{H}}$ of the formal Frobenius manifold $\hat{H}$:

$$\nabla^z_{\sigma} = \frac{\partial}{\partial t^{\sigma}} - \frac{1}{z}\hat{\phi}_{\sigma} \star_t$$

The commutativity (resp. associativity) of $\star_t$ implies that ∇^z is a torsion free (resp. flat) connection on $T_{\hat{H}}$ for all z. The equation

$$\nabla^z \mu = 0 \tag{9}$$

for a section $\mu \in \Gamma(\hat{H}, \mathcal{T}_{\hat{H}})$ is called the $\mathbb{T}$-*equivariant big quantum differential equation* ($\mathbb{T}$-equivariant big QDE). Let

$$\mathcal{T}^{f,z}_{\hat{H}} \subset \mathcal{T}_{\hat{H}}$$

be the subsheaf of flat sections with respect to the connection ∇^z. For each z, $\mathcal{T}^{f,z}_{\hat{H}}$ is a sheaf of $\bar{\Lambda}^{\mathbb{T}}_{\mathrm{nov}}$-modules of rank χ.

A section $L \in \mathrm{End}(T_{\hat{H}}) = \Gamma(\hat{H}, \mathcal{T}^*_{\hat{H}} \otimes \mathcal{T}_{\hat{H}})$ defines an $\mathcal{O}_{\hat{H}}(\hat{H})$-linear map

$$L : \Gamma(\hat{H}, \mathcal{T}_{\hat{H}}) = \bigoplus_{\sigma \in I_{\Sigma}} \mathcal{O}_{\hat{H}}(\hat{H}) \frac{\partial}{\partial t^{\sigma}} \to \Gamma(\hat{H}, \mathcal{T}_{\hat{H}})$$

from the free $\mathcal{O}_{\hat{H}}(\hat{H})$-module $\Gamma(\hat{H}, \mathcal{T}_{\hat{H}})$ to itself. Let $L(z) \in \mathrm{End}(T_{\hat{H}})$ be a family of endomorphisms of the tangent bundle $T_{\hat{H}}$ parametrized by z. $L(z)$ is called a *fundamental solution* to the $\mathbb{T}$-equivariant QDE if the $\mathcal{O}_{\hat{H}}(\hat{H})$-linear map

$$L(z) : \Gamma(\hat{H}, \mathcal{T}_{\hat{H}}) \to \Gamma(\hat{H}, \mathcal{T}_{\hat{H}})$$

restricts to a $\bar{\Lambda}^{\mathbb{T}}_{\mathrm{nov}}$-linear isomorphism

$$L(z) : \Gamma(\hat{H}, \mathcal{T}^{f,\infty}_{H}) = \bigoplus_{\sigma \in I_{\Sigma}} \bar{\Lambda}^{\mathbb{T}}_{\mathrm{nov}} \frac{\partial}{\partial t^{\sigma}} \to \Gamma(\hat{H}, \mathcal{T}^{f,z}_{H}).$$

between rank χ free $\bar{\Lambda}^{\mathbb{T}}_{\mathrm{nov}}$-modules.

5.2. The $\mathcal{S}$-operator. The $\mathcal{S}$-operator is defined as follows. For any cohomology classes $a, b \in H^*_{\mathrm{CR},\mathbb{T}}(\mathcal{X}; \bar{S}_{\mathbb{T}})$,

$$(a, \mathcal{S}(b))_{\mathcal{X},\mathbb{T}} = (a, b)_{\mathcal{X},\mathbb{T}} + \langle\!\langle a, \frac{b}{z - \psi} \rangle\!\rangle^{\mathcal{X},\mathbb{T}}_{0,2}$$

where

$$\frac{b}{z - \psi} = \sum_{i=0}^{\infty} b\psi^i z^{-i-1}.$$

The $\mathcal{S}$-operator can be viewed as an element in $\mathrm{End}(T_{\hat{H}})$ and is a fundamental solution to the $\mathbb{T}$-equivariant big QDE (9). The proof for $\mathcal{S}$ being a fundamental solution can be found in [**28**] for the smooth case and in [**60**] for the orbifold case.

REMARK 5.1. One may notice that since there is a formal variable z in the definition of the $\mathbb{T}$-equivariant big QDE (9), one can consider its solution space over different rings. Here the operator $\mathcal{S} = \mathbf{1} + \mathcal{S}_1/z + \mathcal{S}_2/z^2 + \cdots$ is viewed as a formal power series in $1/z$ with operator-valued coefficients.

REMARK 5.2. Given $t \in H^*_{\mathrm{CR},\mathbb{T}}(\mathcal{X}) \otimes_{R_\mathbb{T}} \bar{S}_\mathbb{T}$, let $t = t' + t''$ where $t' \in H^2_\mathbb{T}(\mathcal{X}) \otimes_{R_\mathbb{T}} \bar{S}_\mathbb{T}$ and t'' is a linear combination of elements in $H^{\neq 2}_{\mathrm{CR},\mathbb{T}}(\mathcal{X}) \otimes_{R_\mathbb{T}} \bar{S}_\mathbb{T}$ and elements in degree 2 twisted sectors. Then by divisor equation, we have

$$\begin{aligned}
&(a,b)_{\mathcal{X},\mathbb{T}} + \langle\!\langle a, \frac{b}{z-\psi} \rangle\!\rangle^{\mathcal{X},\mathbb{T}}_{0,2} \\
&= (a, be^{t'/z})_{\mathcal{X},\mathbb{T}} + \sum_{m=0}^{\infty} \sum_{d \in E(\mathcal{X})(d,m)\neq(0,0)} \frac{Q^d e^{\int_d t'}}{m!} \langle a, \frac{be^{t'/z}}{z-\psi}, (t'')^m \rangle^{\mathcal{X},\mathbb{T}}_{0,2+m,d}.
\end{aligned}$$

In the above expression, if we fix the power of z^{-1}, then only finitely many terms in the expansion of $e^{t'/z}$ contribute. Therefore, the factor $e^{\int_d t'}$ can play the role of Q^d and hence the restriction $\langle\!\langle a, \frac{b}{z-\psi} \rangle\!\rangle^{\mathcal{X},\mathbb{T}}_{0,2}|_{Q=1}$ is well-defined. So the operator $\mathcal{S}|_{Q=1}$ is well-defined.

We consider several different (flat) bases for $H^*_{\mathrm{CR},\mathbb{T}}(\mathcal{X}; \bar{S}_\mathbb{T})$:

(1) The classical canonical basis $\{\phi_{\boldsymbol{\sigma}} : \boldsymbol{\sigma} \in I_\Sigma\}$ defined in Section 2.4.
(2) The basis dual to the classical canonical basis with respect to the $\mathbb{T}$-equivariant Poincare pairing: $\{\phi^{\boldsymbol{\sigma}} = \Delta^{\boldsymbol{\sigma}} \phi_{\boldsymbol{\sigma}} : \boldsymbol{\sigma} \in I_\Sigma\}$, where $1/\Delta^{\boldsymbol{\sigma}} = (\phi_{\boldsymbol{\sigma}}, \phi_{\boldsymbol{\sigma}})$.
(3) The classical normalized canonical basis $\{\hat{\phi}_{\boldsymbol{\sigma}} = \sqrt{\Delta^{\boldsymbol{\sigma}}}\phi_{\boldsymbol{\sigma}} : \boldsymbol{\sigma} \in I_\Sigma\}$ which is self-dual: $\{\hat{\phi}^{\boldsymbol{\sigma}} = \hat{\phi}_{\boldsymbol{\sigma}} : \boldsymbol{\sigma} \in I_\Sigma\}$.

We also consider several different non-flat basis:

(1) The quantum canonical basis $\{\phi_{\boldsymbol{\sigma}}(t) : \boldsymbol{\sigma} \in I_\Sigma\}$, such that at LRL
$$\lim_{\widetilde{Q}\to 0, t''\to 0} \phi_{\boldsymbol{\sigma}}(t) = \phi_{\boldsymbol{\sigma}}.$$
(2) The basis dual to the qunatum canonical basis with respect to the $\mathbb{T}$-equivariant Poincare pairing: $\{\phi^{\boldsymbol{\sigma}}(t) = \Delta^{\boldsymbol{\sigma}}(t)\phi_{\boldsymbol{\sigma}}(t) : \boldsymbol{\sigma} \in I_\Sigma\}$, where $(\phi_{\boldsymbol{\sigma}}(t), \phi_{\boldsymbol{\sigma}}(t)) = 1/\Delta^{\boldsymbol{\sigma}}(t)$.
(3) The quantum normalized canonical basis $\{\hat{\phi}_{\boldsymbol{\sigma}}(t) = \sqrt{\Delta^{\boldsymbol{\sigma}}(t)}\phi_{\boldsymbol{\sigma}}(t) : \boldsymbol{\sigma} \in I_\Sigma\}$ which is self-dual: $\hat{\phi}^{\boldsymbol{\sigma}}(t) = \hat{\phi}_{\boldsymbol{\sigma}}(t) : \boldsymbol{\sigma} \in I_\Sigma$.

The "classical" basis are flat while the "quantum" ones are not. For most of our application, we will set $Q = 1$, and $t = \boldsymbol{\tau}$ for an $H^{\leq 2}$ element.

For $\boldsymbol{\sigma}, \boldsymbol{\sigma}' \in I_\Sigma$, define

$$S^{\boldsymbol{\sigma}'}{}_{\boldsymbol{\sigma}}(z) := (\phi^{\boldsymbol{\sigma}}, \mathcal{S}(\phi_{\boldsymbol{\sigma}})).$$

Then $(S^{\boldsymbol{\sigma}'}{}_{\boldsymbol{\sigma}}(z))$ is the matrix of the $\mathcal{S}$-operator with respect to the canonical basis $\{\phi_{\boldsymbol{\sigma}} : \boldsymbol{\sigma} \in I_\Sigma\}$:

$$\mathcal{S}(\phi_{\boldsymbol{\sigma}}) = \sum_{\boldsymbol{\sigma}' \in I_\Sigma} \phi_{\boldsymbol{\sigma}'} S^{\boldsymbol{\sigma}'}{}_{\boldsymbol{\sigma}}(z). \tag{10}$$

For $\boldsymbol{\sigma}, \boldsymbol{\sigma}' \in I_\Sigma$, define

$$S_{\boldsymbol{\sigma}'}{}^{\widehat{\boldsymbol{\sigma}}}(z) := (\phi_{\boldsymbol{\sigma}'}, \mathcal{S}(\hat{\phi}^{\boldsymbol{\sigma}})).$$

Then $(S_{\boldsymbol{\sigma}'}{}^{\widehat{\boldsymbol{\sigma}}})$ is the matrix of the $\mathcal{S}$-operator with respect to the bases $\{\hat{\phi}^{\boldsymbol{\sigma}} : \boldsymbol{\sigma} \in I_\Sigma\}$ and $\{\phi^{\boldsymbol{\sigma}} : \boldsymbol{\sigma} \in I_\Sigma\}$:

$$\mathcal{S}(\hat{\phi}^{\boldsymbol{\sigma}}) = \sum_{\boldsymbol{\sigma}' \in I_\Sigma} \phi^{\boldsymbol{\sigma}'} S_{\boldsymbol{\sigma}'}{}^{\widehat{\boldsymbol{\sigma}}}(z). \tag{11}$$

Introduce

$$S_z(a,b) = (a, \mathcal{S}(b))_{\mathcal{X},\mathbb{T}},$$

$$V_{z_1,z_2}(a,b) = \frac{(a,b)_{\mathcal{X},\mathbb{T}}}{z_1+z_2} + \langle\!\langle \frac{a}{z_1-\psi_1}, \frac{b}{z_2-\psi_2} \rangle\!\rangle_{0,2}^{\mathcal{X},\mathbb{T}}.$$

A well-known WDVV-like argument says

$$V_{z_1,z_2}(a,b) = \frac{1}{z_1+z_2} \sum_i S_{z_1}(T_i,a) S_{z_2}(T^i,b), \tag{12}$$

where T_i is any basis of $H^*_{\mathrm{CR},\mathbb{T}}(\mathcal{X};\bar{S}_{\mathbb{T}})$ and T^i is its dual basis. In particular,

$$V_{z_1,z_2}(a,b) = \frac{1}{z_1+z_2} \sum_{\boldsymbol{\sigma}\in I_\Sigma} S_{z_1}(\hat{\phi}_{\boldsymbol{\sigma}},a) S_{z_2}(\hat{\phi}_{\boldsymbol{\sigma}},b).$$

5.3. Quantization of quadratic Hamiltonians. In this section, we review the basic concepts of the quantization of quadratic Hamiltonians (see [**51**] for more details). The quantization procedure provides a way to recover the higher genus theory from the genus zero data which we will use in the next section.

5.3.1. *Symplectic space formalism.* So far, we have been working on (a formal neighborhood of) the space $H = \mathrm{Spec}(\bar{\Lambda}^{\mathbb{T}}_{\mathrm{nov}}[t^{\boldsymbol{\sigma}} : \boldsymbol{\sigma} \in I_\Sigma])$ which provides us the Frobenius structure and state space of the corresponding Gromov-Witten theory. When we consider the descendent theory of $\mathcal{X}$, however, additional parameters are needed. Let $\mathbf{t}(\psi) = t_0 + t_1\psi + t_2\psi^2 + \cdots$ be a formal power series in ψ with an integer index that keeps track in the power of ψ. Here each t_a lies in $H^*_{\mathrm{CR},\mathbb{T}'}(\mathcal{X})$. We define

$$\langle \mathbf{t}(\psi_1), \cdots, \mathbf{t}(\psi_k)\rangle_{g,k,\beta}^{\mathcal{X},\mathbb{T}} = \int_{[\overline{\mathcal{M}}_{g,k}(\mathcal{X},\beta)^{\mathbb{T}}]^{\mathrm{vir}}} \frac{\prod_{j=1}^k (\sum_{a=0}^\infty (\mathrm{ev}_j^* t_a)\psi_j^a)}{e_{\mathbb{T}}(N^{\mathrm{vir}})}.$$

The additional index a leads to the study of the symplectic space formalism.

Let z be a formal variable. We consider the space $\mathbb{H}$ which is the space of Laurent polynomials in one variable z with coefficients in H. We define the symplectic form Ω on $\mathbb{H}$ by

$$\Omega(f,g) = \mathrm{Res}_{z=0}(f(-z), g(z))_{\mathcal{X},\mathbb{T}} dz$$

for any $f,g \in \mathbb{H}$. Note that we have $\Omega(f,g) = -\Omega(g,f)$. There is a natural polarization $\mathbb{H} = \mathbb{H}_+ \oplus \mathbb{H}_-$ corresponding to the decomposition $f(z,z^{-1}) = f_+(z) + f_-(z^{-1})z^{-1}$ of laurent polynomials into polynomial and polar parts. It is easy to see that $\mathbb{H}_+$ and $\mathbb{H}_-$ are both Lagrangian subspaces of $\mathbb{H}$ with respect to Ω.

Introduce a Darboux coordinate system $\{p_a^{\boldsymbol{\sigma}}, q_b^{\boldsymbol{\rho}}\}$ on $\mathbb{H}$ with respect to the above polarization. This means that we write a general element $f \in \mathbb{H}$ in the form

$$\sum_{a\geq 0,\boldsymbol{\sigma}\in I_\Sigma} p_a^{\boldsymbol{\sigma}} \hat{\phi}^{\boldsymbol{\sigma}}(-z)^{-a-1} + \sum_{b\geq 0,\boldsymbol{\rho}\in I_\Sigma} q_b^{\boldsymbol{\rho}} \hat{\phi}_{\boldsymbol{\rho}} z^b.$$

Denote

$$\begin{aligned} \mathbf{p}(z): &= p_0(-z)^{-1} + p_1(-z)^{-2} + \cdots \\ \mathbf{q}(z): &= q_0 z + q_1 z^2 + \cdots, \end{aligned}$$

where $p_a = \sum_{\boldsymbol{\sigma}} p_a^{\boldsymbol{\sigma}} \hat{\phi}^{\boldsymbol{\sigma}}$ and $q_b = \sum_{\boldsymbol{\rho}} q_b^{\boldsymbol{\rho}} \hat{\phi}_{\boldsymbol{\rho}}$.

Recall that when we discussed the Gromov-Witten theory of $\mathcal{X}$, we introduced the formal power series $\mathbf{t}(z) = t_0 + t_1 z + t_2 z^2 + \cdots$. With z replaced by ψ, $\mathbf{t}$ appears as the insertion in the genus g correlator. We relate $\mathbf{t}(z)$ to the Darboux coordinates by introducing the *dilaton shift*: $\mathbf{q}(z) = \mathbf{t}(z) - \mathbf{1}z$. The dilaton shift appears naturally in the quantization procedure. We will explain this phenomenon as a group action on Cohomological field theories in the next section.

5.3.2. *Quantization of quadratic Hamiltonians.* Let $A : \mathbb{H} \to \mathbb{H}$ be a linear infinitesimally symplectic transformation, i.e. $\Omega(Af, g) + \Omega(f, Ag) = 0$ for any $f, g \in \mathbb{H}$. Under the Darboux coordinates, the quadratic Hamiltonian

$$f \to \frac{1}{2}\Omega(Af, f)$$

is a series of homogeneous degree two monomials in $\{p_a^\sigma, q_b^\rho\}$. Let $\hbar$ be a formal variable and define the quantization of quadratic monomials as

$$\widehat{q_a^\sigma q_b^\rho} = \frac{q_a^\sigma q_b^\rho}{\hbar}, \widehat{q_a^\sigma p_b^\rho} = q_a^\sigma \frac{\partial}{\partial q_b^\rho}, \widehat{p_a^\sigma p_b^\rho} = \hbar \frac{\partial}{\partial q_a^\sigma}\frac{\partial}{\partial q_b^\rho}.$$

We define the quantization $\widehat{A}$ by extending the above equalities linearly. The differential operators $\widehat{q_a^\sigma q_b^\rho}, \widehat{q_a^\sigma p_b^\rho}, \widehat{p_a^\sigma p_b^\rho}$ act on the so called Fock space *Fock* which is the space of formal functions in $\mathbf{t}(z) \in \mathbb{H}_+$. For example, the descendent potential and ancestor potential are regarded as elements in *Fock*. The quantization operator $\widehat{A}$ does not act on *Fock* in general since it may contain infinitely many monomials. However, the actions of quantization operators in our paper are well-defined. The quantization of a symplectic transform of the form $\exp(A)$, with A infinitesimally symplectic, is defined to be $\exp(\widehat{A}) = \sum_{n\geq 0} \frac{\widehat{A}^n}{n!}$.

REMARK 5.3. Let $A : \mathbb{H} \to \mathbb{H}$ be a linear infinitesimally symplectic transformation. The quantization of $Az^m, m \geq 0$ is studied in [**26**, equation (1.3)] and in [**91**, Appendix C]. There is a sign error in the second term of [**26**, equation (1.3)] and the corresponding identity in [**91**, Appendix C]. The sign $(-1)^k$ should be replaced by $(-1)^{k+m-1}$.

5.4. Givental's formula. Let U denote the diagonal matrix whose diagonal entries are the canonical coordinates. So $u^{\boldsymbol{\sigma}}$ are canonical coordinates for each $\boldsymbol{\sigma} \in I_\Sigma$. The results in [**50**] and [**102**] imply the following statement.

THEOREM 5.4. *There exists a unique matrix power series* $R(z) = \mathbf{1} + R_1 z + R_2 z^2 + \cdots$ *satisfying the following properties.*

(1) *The entries of* R_d *lie in* $\bar{S}_{\mathbb{T}}[\![\widetilde{Q}, \tau'']\!]$.
(2) $\widetilde{S} = \Psi R(z) e^{U/z}$ *is a fundamental solution to the* $\mathbb{T}$*-equivariant big QDE* (9).
(3) R *satisfies the unitary condition* $R^T(-z)R(z) = \mathbf{1}$.
(4)

$$\begin{aligned}(13)\qquad &\lim_{\widetilde{Q},\tau''\to 0} R_{\rho,\delta}{}^{\sigma,\gamma}(z) \\ &= \frac{\delta_{\rho,\sigma}}{|G_\sigma|}\sum_{h\in G_\sigma}\chi_\rho(h)\chi_\gamma(h^{-1})\prod_{i=1}^{3}\exp\Big(\sum_{m=1}^{\infty}\frac{(-1)^m}{m(m+1)}B_{m+1}(c_i^\sigma(h))(\frac{z}{\mathsf{w}_i(\sigma)})^m\Big)\end{aligned}$$

Each matrix in (2) of Theorem 5.4 represents an operator with respect to the classical canonical basis $\{\hat{\phi}_{\boldsymbol{\sigma}} : \boldsymbol{\sigma} \in I_\Sigma\}$. So R^T is the adjoint of R with respect to the $\mathbb{T}$-equivariant Poincaré pairing $(\ ,\ )_{\mathcal{X},\mathbb{T}}$. The matrix $(\widetilde{S}_{\boldsymbol{\sigma}'}{}^{\widehat{\boldsymbol{\sigma}}})(z)$ is of the form

$$(14)\qquad \widetilde{S}_{\boldsymbol{\sigma}'}{}^{\widehat{\boldsymbol{\sigma}}}(z) = \sum_{\boldsymbol{\rho}\in I_\Sigma}\Psi_{\boldsymbol{\sigma}'}{}^{\boldsymbol{\rho}}R_{\boldsymbol{\rho}}{}^{\boldsymbol{\sigma}}(z)e^{u^{\boldsymbol{\sigma}}/z} = (\Psi R(z))_{\boldsymbol{\sigma}'}{}^{\boldsymbol{\sigma}}e^{u^{\boldsymbol{\sigma}}/z}$$

where $R(z) = (R_{\boldsymbol{\rho}}{}^{\boldsymbol{\sigma}}(z)) = \mathbf{1} + \sum_{k=1}^{\infty} R_k z^k$.

We call the unique $R(z)$ in Theorem 5.4 the *A-model R-matrix*. The A-model R-matrix plays a central role in the quantization formula of the descendent potential of $\mathbb{T}$-equivariant Gromov-Witten theory of $\mathcal{X}$.

Let $\bar{\psi}_i$ be the pullback of the i-ith ψ-class on $\overline{\mathcal{M}}_{g,k}$ to $\overline{\mathcal{M}}_{g,k}(\mathcal{X};\beta)$. We define the ancestor potential to be

$$\mathcal{A}_{\mathcal{X}}(\boldsymbol{\tau}) = \sum_{g\geq 0}\sum_{k\geq 0}\frac{\hbar^{g-1}}{k!}\langle\!\langle \mathbf{t}(\bar{\psi}_1),\ldots,\mathbf{t}(\bar{\psi}_k)\rangle\!\rangle^{\mathcal{X}}_{g,k}$$

Before we move on to the quantization process, let us consider the potential functions of the trivial cohomological field theory I. Define the correlator $\langle\rangle^I_{g,k}$ to be

$$\langle \tau_{a_1}(\hat{\phi}_{\boldsymbol{\sigma}_1}),\cdots,\tau_{a_k}(\hat{\phi}_{\boldsymbol{\sigma}_k})\rangle^I_{g,k} = \begin{cases} \Delta_{\boldsymbol{\sigma}}^{g-1+k/2}(\boldsymbol{\tau})\int_{\overline{\mathcal{M}}_{g,k}}\psi_1^{a_1}\cdots\psi_k^{a_k}, & \text{if } \boldsymbol{\sigma}_1=\boldsymbol{\sigma}_2=\cdots=\boldsymbol{\sigma}_k=\boldsymbol{\sigma},\\ 0, & \text{otherwise}\end{cases}$$

where $a_1,\cdots,a_k$ are nonnegative integers. Let

$$\mathcal{D}_I = \exp\Big(\sum_{g\geq 0}\sum_{k\geq 0}\sum_{a_1,\cdots,a_k\geq 0}\sum_{\boldsymbol{\sigma}_1,\cdots,\boldsymbol{\sigma}_k}\frac{\hbar^{g-1}t^{\boldsymbol{\sigma}_1}_{a_1}\cdots t^{\boldsymbol{\sigma}_k}_{a_k}}{a_1!\cdots a_k!}\langle \tau_{a_1}(\hat{\phi}_{\boldsymbol{\sigma}_1}),\cdots,\tau_{a_k}(\hat{\phi}_{\boldsymbol{\sigma}_k})\rangle^I_{g,k}\Big).$$

In [**102**], the third author generalizes Givental's formula for the total descendant potential of equivariant Gromov-Witten theory of GKM manifolds to GKM orbifolds. When we apply this formula to the case of a toric Calabi-Yau 3-orbifold, we obtain the following theorem:

THEOREM 5.5 (Zong [**102**]). *Let $\mathcal{A}_{\mathcal{X}}(\boldsymbol{\tau})$ be the ancestor potential of $\mathcal{X}$. Then*

$$\mathcal{A}_{\mathcal{X}}(\boldsymbol{\tau}) = \widehat{\Psi}\widehat{R}\mathcal{D}_I.$$

Here $\widehat{\Psi}$ is the operator $\mathcal{G}(\Psi^{-1}\mathbf{q})\mapsto\mathcal{G}(\mathbf{q})$ for any element $\mathcal{G}$ in the Fock space.

Similarly, there is a Givental formula for the descendent potential of $\mathcal{X}$:

THEOREM 5.6. *For $2g-2+n>0$, we have the following relation*

$$\langle\!\langle \mathbf{t}(\psi_1),\ldots,\mathbf{t}(\psi_n)\rangle\!\rangle^{\mathcal{X}}_{g,n} = \langle\!\langle [\mathcal{S}\mathbf{t}]_+(\bar{\psi}_1),\ldots,[\mathcal{S}\mathbf{t}]_+(\bar{\psi}_n)\rangle\!\rangle^{\mathcal{X}}_{g,n}.$$

Here we consider $\mathbf{t}=\mathbf{t}(z)$ as element in $\mathbb{H}$ and $[\mathcal{S}\mathbf{t}]_+(z)$ is the part of $\mathcal{S}\mathbf{t}$ containing nonnegative powers of z.

5.5. The graph sum formula. In order to state the graph sum formula, we need to introduce some definitions.

- We define
$$S^{\widehat{\underline{\boldsymbol{\sigma}}}}{}_{\widehat{\underline{\boldsymbol{\sigma}}'}}(z) := (\hat{\phi}_{\boldsymbol{\sigma}}(t),\mathcal{S}(\hat{\phi}_{\boldsymbol{\sigma}'}(t))).$$
Then $(S^{\widehat{\underline{\boldsymbol{\sigma}}}}{}_{\widehat{\underline{\boldsymbol{\sigma}}'}}(z))$ is the matrix of the $\mathcal{S}$-operator with respect to the normalized canonical basis $\{\hat{\phi}_{\boldsymbol{\sigma}}(t):\boldsymbol{\sigma}\in I_\Sigma\}$:
$$\mathcal{S}(\hat{\phi}_{\boldsymbol{\sigma}'}(t)) = \sum_{\boldsymbol{\sigma}\in I_\Sigma}\hat{\phi}_{\boldsymbol{\sigma}}(t)S^{\widehat{\underline{\boldsymbol{\sigma}}}}{}_{\widehat{\underline{\boldsymbol{\sigma}}'}}(z). \tag{15}$$
- We define
$$S^{\widehat{\underline{\boldsymbol{\sigma}}}}{}_{\boldsymbol{\sigma}'}(z) := (\hat{\phi}_{\boldsymbol{\sigma}}(t),\mathcal{S}(\phi_{\boldsymbol{\sigma}'})).$$
Then $(S^{\widehat{\underline{\boldsymbol{\sigma}}}}{}_{\boldsymbol{\sigma}'}(z))$ is the matrix of the $\mathcal{S}$-operator with respect to the basis $\{\phi_{\boldsymbol{\sigma}}:\boldsymbol{\sigma}\in I_\Sigma\}$ and $\{\hat{\phi}_{\boldsymbol{\sigma}}(t):\boldsymbol{\sigma}\in I_\Sigma\}$:
$$\mathcal{S}(\phi_{\boldsymbol{\sigma}'}) = \sum_{\boldsymbol{\sigma}\in I_\Sigma}\hat{\phi}_{\boldsymbol{\sigma}}(t)S^{\widehat{\underline{\boldsymbol{\sigma}}}}{}_{\boldsymbol{\sigma}'}(z). \tag{16}$$

Given a connected graph Γ, we introduce the following notation.

(1) $V(\Gamma)$ is the set of vertices in Γ.
(2) $E(\Gamma)$ is the set of edges in Γ.
(3) $H(\Gamma)$ is the set of half edges in Γ.
(4) $L^o(\Gamma)$ is the set of ordinary leaves in Γ. The ordinary leaves are ordered: $L^o(\Gamma) = \{l_1, \dots, l_n\}$ where n is the number of ordinary leaves.
(5) $L^1(\Gamma)$ is the set of dilaton leaves in Γ. The dilaton leaves are unordered.

With the above notation, we introduce the following labels:

(1) (genus) $g : V(\Gamma) \to \mathbb{Z}_{\geq 0}$.
(2) (marking) $\boldsymbol{\sigma} : V(\Gamma) \to I_\Sigma$. This induces $\boldsymbol{\sigma} : L(\Gamma) = L^o(\Gamma) \cup L^1(\Gamma) \to I_\Sigma$, as follows: if $l \in L(\Gamma)$ is a leaf attached to a vertex $v \in V(\Gamma)$, define $\boldsymbol{\sigma}(l) = \boldsymbol{\sigma}(v)$.
(3) (height) $k : H(\Gamma) \to \mathbb{Z}_{\geq 0}$.

Given an edge e, let $h_1(e), h_2(e)$ be the two half edges associated to e. The order of the two half edges does not affect the graph sum formula in this paper. Given a vertex $v \in V(\Gamma)$, let $H(v)$ denote the set of half edges emanating from v. The valency of the vertex v is equal to the cardinality of the set $H(v)$: $\mathrm{val}(v) = |H(v)|$. A labeled graph $\vec{\Gamma} = (\Gamma, g, \boldsymbol{\sigma}, k)$ is *stable* if

$$2g(v) - 2 + \mathrm{val}(v) > 0$$

for all $v \in V(\Gamma)$.

Let $\mathbf{\Gamma}(\mathcal{X})$ denote the set of all stable labeled graphs $\vec{\Gamma} = (\Gamma, g, \boldsymbol{\sigma}, k)$. The genus of a stable labeled graph $\vec{\Gamma}$ is defined to be

$$g(\vec{\Gamma}) := \sum_{v \in V(\Gamma)} g(v) + |E(\Gamma)| - |V(\Gamma)| + 1 = \sum_{v \in V(\Gamma)} (g(v) - 1) + (\sum_{e \in E(\Gamma)} 1) + 1.$$

Define

$$\mathbf{\Gamma}_{g,n}(\mathcal{X}) = \{\vec{\Gamma} = (\Gamma, g, \boldsymbol{\sigma}, k) \in \mathbf{\Gamma}(\mathcal{X}) : g(\vec{\Gamma}) = g, |L^o(\Gamma)| = n\}.$$

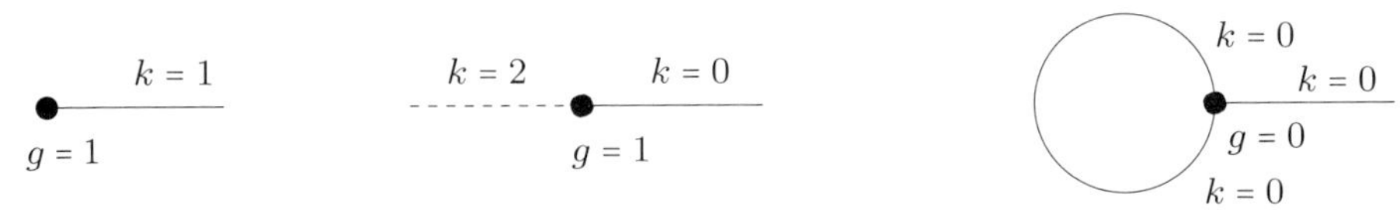

FIGURE 5. A set $\mathbf{\Gamma}_{1,1}(\mathcal{X})$. Each graph here corresponds to several graphs in $\mathbf{\Gamma}_{1,1}(\mathcal{X})$: one can label each vertex by any element in I_Σ. Dashed lines are dilaton leafs, whose lowest possible height is 2.

We assign weights to leaves, edges, and vertices of a labeled graph $\vec{\Gamma} \in \mathbf{\Gamma}(\mathcal{X})$ as follows.

(1) *Ordinary leaves.* To each ordinary leaf $l_j \in L^o(\Gamma)$ with $\boldsymbol{\sigma}(l_j) = \boldsymbol{\sigma} \in I_\Sigma$ and $k(l) = k \in \mathbb{Z}_{\geq 0}$, we assign the following descendant weight:

$$(\mathcal{L}^{\mathbf{u}})^{\boldsymbol{\sigma}}_k(l_j) = [z^k]\Big(\sum_{\boldsymbol{\sigma}', \rho \in I_\Sigma} \left(\frac{\mathbf{u}_j^{\boldsymbol{\sigma}'}(z)}{\sqrt{\Delta^{\boldsymbol{\sigma}'}(t)}} S^{\underline{\hat{\rho}}}{}_{\underline{\hat{\boldsymbol{\sigma}}'}}(z)\right)_+ R(-z)_\rho{}^{\boldsymbol{\sigma}}\Big), \tag{17}$$

where $(\cdot)_+$ means taking the nonnegative powers of z.

(2) *Dilaton leaves.* To each dilaton leaf $l \in L^1(\Gamma)$ with $\boldsymbol{\sigma}(l) = \boldsymbol{\sigma} \in I_\Sigma$ and $2 \le k(l) = k \in \mathbb{Z}_{\ge 0}$, we assign

$$(\mathcal{L}^1)^{\boldsymbol{\sigma}}_k := [z^{k-1}](-\sum_{\boldsymbol{\sigma}' \in I_\Sigma} \frac{1}{\sqrt{\Delta^{\boldsymbol{\sigma}'}(t)}} R_{\boldsymbol{\sigma}'}{}^{\boldsymbol{\sigma}}(-z)).$$

(3) *Edges.* To an edge connected a vertex marked by $\boldsymbol{\sigma} \in I_\Sigma$ to a vertex marked by $\boldsymbol{\sigma}' \in I_\Sigma$ and with heights k and l at the corresponding half-edges, we assign

$$\mathcal{E}^{\boldsymbol{\sigma},\boldsymbol{\sigma}'}_{k,l} := [z^k w^l]\Big(\frac{1}{z+w}(\delta_{\boldsymbol{\sigma}\boldsymbol{\sigma}'} - \sum_{\boldsymbol{\rho}\in I_\Sigma} R_{\boldsymbol{\rho}}{}^{\boldsymbol{\sigma}}(-z) R_{\boldsymbol{\rho}}{}^{\boldsymbol{\sigma}'}(-w)\Big).$$

(4) *Vertices.* To a vertex v with genus $g(v) = g \in \mathbb{Z}_{\ge 0}$ and with marking $\boldsymbol{\sigma}(v) = \boldsymbol{\sigma}$, with n ordinary leaves and half-edges attached to it with heights $k_1, ..., k_n \in \mathbb{Z}_{\ge 0}$ and m more dilaton leaves with heights $k_{n+1}, \dots, k_{n+m} \in \mathbb{Z}_{\ge 0}$, we assign

$$\Big(\sqrt{\Delta^{\boldsymbol{\sigma}}(t)}\Big)^{2g(v)-2+\mathrm{val}(v)} \int_{\overline{\mathcal{M}}_{g,n+m}} \psi_1^{k_1} \cdots \psi_{n+m}^{k_{n+m}}.$$

We define the weight of a labeled graph $\vec{\Gamma} \in \mathbf{\Gamma}(\mathcal{X})$ to be

$$\begin{aligned} w^{\mathbf{u}}_A(\vec{\Gamma}) \quad = \quad & \prod_{v\in V(\Gamma)} (\sqrt{\Delta^{\boldsymbol{\sigma}(v)}(t)})^{2g(v)-2+\mathrm{val}(v)} \langle \prod_{h\in H(v)} \tau_{k(h)} \rangle_{g(v)} \prod_{e\in E(\Gamma)} \mathcal{E}^{\boldsymbol{\sigma}(v_1(e)),\boldsymbol{\sigma}(v_2(e))}_{k(h_1(e)),k(h_2(e))} \\ & \cdot \prod_{l\in L^1(\Gamma)} (\mathcal{L}^1)^{\boldsymbol{\sigma}(l)}_{k(l)} \prod_{j=1}^{n} (\mathcal{L}^{\mathbf{u}})^{\boldsymbol{\sigma}(l_j)}_{k(l_j)}(l_j). \end{aligned}$$

With the above definition of the weight of a labeled graph, we have the following theorem which expresses the $\mathbb{T}$–equivariant descendent Gromov-Witten potential of $\mathcal{X}$ in terms of graph sum.

THEOREM 5.7 (Zong [**102**]). *Suppose that* $2g - 2 + n > 0$. *Then*

$$\langle\!\langle \mathbf{u}_1, \dots, \mathbf{u}_n \rangle\!\rangle^{\mathcal{X},\mathbb{T}}_{g,n} = \sum_{\vec{\Gamma}\in\mathbf{\Gamma}_{g,n}(\mathcal{X})} \frac{w^{\mathbf{u}}_A(\vec{\Gamma})}{|\mathrm{Aut}(\vec{\Gamma})|}.$$

REMARK 5.8. In the above graph sum formula, we know that the restriction $S^{\widehat{\boldsymbol{\rho}}}{}_{\widehat{\boldsymbol{\sigma}'}}(z)|_{Q=1}$ is well-defined by Remark 5.2. Meanwhile by (1) in Theorem 5.4, we know that the restriction $R(z)|_{Q=1}$ is also well-defined. Therefore by Theorem 5.7, we have $\langle\!\langle \mathbf{u}_1, \dots, \mathbf{u}_n \rangle\!\rangle^{\mathcal{X},\mathbb{T}}_{g,n}|_{Q=1}$ is well-defined.

5.6. Open-closed Gromov-Witten invariants. The BKMP Remodeling Conjecture predicts open-closed Gromov-Witten potentials (3) from the B-model.

Let $\mathcal{L} \subset \mathcal{X}$ be an outer Aganagic-Vafa Lagrangian brane. Let $G_0 := G_{\sigma_0}$ be the stabilizer of the stacky point $\mathfrak{l}_{\sigma_0}$. Our notation is similar to that in [**43**, Section 5]. In particular, the interger f is a framing, and $\mathbb{T}'_f := \mathrm{Ker}(\mathsf{u}_2 - f\mathsf{u}_1)$. The morphism $H^*(B\mathbb{T}';\mathbb{Q}) = \mathbb{Q}[\mathsf{u}_1, \mathsf{u}_2] \to H^*(B\mathbb{T}'_f;\mathbb{Q}) = \mathbb{Q}[\mathsf{v}]$ is given by $\mathsf{u}_1 \mapsto \mathsf{v}$, $\mathsf{u}_2 \mapsto f\mathsf{v}$. The weights of $\mathbb{T}'$-action on $T_{\mathfrak{l}_{\sigma_0}}\mathcal{X}$ are

$$\mathsf{w}'_1 = \frac{1}{\mathfrak{r}}\mathsf{u}_1, \quad \mathsf{w}'_2 = \frac{\mathfrak{s}}{\mathfrak{r}\mathfrak{m}}\mathsf{u}_1 + \frac{1}{\mathfrak{m}}\mathsf{u}_2, \quad \mathsf{w}'_3 = -\frac{\mathfrak{s}+\mathfrak{m}}{\mathfrak{r}\mathfrak{m}}\mathsf{u}_1 - \frac{1}{\mathfrak{m}}\mathsf{u}_2,$$

so the weights of $\mathbb{T}'_f$-action on $T_{\mathfrak{p}_{\sigma_0}}\mathcal{X}$ are $w_1\mathsf{v}$, $w_2\mathsf{v}$, $w_3\mathsf{v}$, where

$$w_1 = \frac{1}{\mathfrak{r}}, \quad w_2 = \frac{\mathfrak{s}+\mathfrak{r}f}{\mathfrak{r}\mathfrak{m}}, \quad w_3 = \frac{-\mathfrak{m}-\mathfrak{s}-\mathfrak{r}f}{\mathfrak{r}\mathfrak{m}}.$$

Recall that the correlator $\langle\tau^\ell\rangle^{\mathcal{X},(\mathcal{L},f)}_{g,d,(\mu_1,k_1),\dots,(\mu_n,k_n)}$ is the the equivariant open-closed Gromov-Witten invariant defined in Equation (2) (see [**41**, Section 3] for more details). The open Gromov-Witten potential is defined in (3).

We introduce some notation.

(1) Given $d_0 \in \mathbb{Z}$ and $k \in \boldsymbol{\mu}_{\mathfrak{m}}$ let $D'(d_0,k)$ be the disk factor defined by Equation (13) in [**43**], and define

$$h(d_0,k) := (e^{2\pi\sqrt{-1}d_0w_1}, e^{2\pi\sqrt{-1}(d_0w_2-\frac{k}{\mathfrak{m}})}, e^{2\pi\sqrt{-1}(d_0w_3+\frac{k}{\mathfrak{m}})}) \in G_0 \subset \mathbb{T} = (\mathbb{C}^*)^3.$$

(2) Given $h\in G_0$, define

$$\begin{aligned}\Phi^h_0(\widetilde{X}) &:= \frac{1}{|G_0|}\sum_{\substack{(d_0,k)\in\mathbb{Z}_{\geq 0}\times\mathbb{Z}_{\mathfrak{m}}\\ h(d_0,k)=h}} D'(d_0,k)\widetilde{X}^{d_0}(-(-1)^{-k/\mathfrak{m}})\mathbf{1}'_{\frac{-k}{\mathfrak{m}}}\\ &= \frac{1}{|G_0|}\sum_{\substack{(d_0,k)\in\mathbb{Z}_{\geq 0}\times\mathbb{Z}_{\mathfrak{m}}\\ h(d_0,k)=h}} \frac{1}{\mathfrak{m}\mathsf{v}}e^{\sqrt{-1}\pi(d_0w_3-c_3(h))}\left(\frac{\mathsf{v}}{d_0}\right)^{\mathrm{age}(h)-1}\\ &\quad\cdot\frac{\Gamma(d_0(w_1+w_2)+c_3(h))}{\Gamma(d_0w_1-c_1(h)+1)\Gamma(d_0w_2-c_2(h)+1)}\widetilde{X}^{d_0}\mathbf{1}'_{\frac{-k}{\mathfrak{m}}}.\end{aligned}$$

Then $\Phi^h_0(\widetilde{X})$ takes values in $\bigoplus_{k=0}^{\mathfrak{m}-1}\mathbb{C}\mathsf{v}^{\mathrm{age}(v)-2}\mathbf{1}'_{\frac{k}{\mathfrak{m}}}$.

For $a\in\mathbb{Z}$ and $h\in G_0$, we define

$$\Phi^h_a(\widetilde{X}) := \frac{1}{|G_0|}\sum_{\substack{d_0>0\\ h(d_0,k)=h}} D'(d_0,k)\left(\frac{d_0}{\mathsf{v}}\right)^a\widetilde{X}^{d_0}(-(-1)^{-k/\mathfrak{m}})\mathbf{1}'_{\frac{-k}{\mathfrak{m}}}.$$

Then $\Phi^h_a(\widetilde{X})$ takes values in $\bigoplus_{k=0}^{\mathfrak{m}-1}\mathbb{C}\mathsf{v}^{\mathrm{age}(v)-2-a}\mathbf{1}'_{\frac{k}{\mathfrak{m}}}$, and

$$\Phi^h_{a+1}(\widetilde{X}) = \left(\frac{1}{\mathsf{v}}\widetilde{X}\frac{d}{d\widetilde{X}}\right)\Phi^h_a(\widetilde{X}).$$

(3) For $a\in\mathbb{Z}$ and $\alpha\in G_0^*$, we define

$$\widetilde{\xi}^\alpha_a(\widetilde{X}) := |G_0|\sum_{h\in G}\chi_\alpha(h^{-1})\left(\prod_{i=1}^3(w_i\mathsf{v})^{1-c_i(h)}\right)\Phi^h_a(\widetilde{X}). \tag{18}$$

Then $\widetilde{\xi}^\alpha_a(\widetilde{X})$ takes values in $\bigoplus_{k=0}^{\mathfrak{m}-1}\mathbb{C}\mathsf{v}^{1-a}\mathbf{1}'_{\frac{k}{\mathfrak{m}}}$. Denote $\widetilde{\xi}^\alpha(z,\widetilde{X}) = \sum_{a\in\mathbb{Z}_{\geq -2}} z^a\widetilde{\xi}^\alpha_a(\widetilde{X})$.

(4) Given $h\in G_0 = G_{\sigma_0}$, recall that $\mathbf{1}_{\sigma_0,h}$ be characterized by $\mathbf{1}_{\sigma_0,h}|_{\mathfrak{l}_{\sigma_0}} = \delta_{\sigma,\sigma_0}\mathbf{1}_h$. We define $\mathbf{1}^*_{\sigma_0,h} = |G|\mathbf{e}_h\mathbf{1}_{\sigma_0,h^{-1}}$, where $\mathbf{e}_h = \prod_{i=1}^3(w_i\mathsf{v})^{\delta_{0,c_i(h)}}$.

With the above notation, the following proposition comes from localization computation (see [**41**]).

PROPOSITION 5.9.

(1) Disk invariants

$$\begin{aligned}
&\widetilde{F}_{0,1}^{\mathcal{X},(\mathcal{L},f)}(\boldsymbol{\tau},Q;\widetilde{X})\\
=&\ \Phi_{-2}^{1}(\widetilde{X})+\sum_{a=1}^{p}\tau_a\Phi_{-1}^{h_a}(\widetilde{X})+\sum_{a\in\mathbb{Z}_{\geq 0}}\sum_{h\in G_0}\big(\langle\!\langle \mathbf{1}^*_{\sigma_0,h}\psi^a\rangle\!\rangle_{0,1}^{\mathcal{X},\mathbb{T}_f}\big|_{t=\boldsymbol{\tau}}\big)\Phi_a^h(\widetilde{X})\\
=&\ \frac{1}{|G_0|^2w_1w_2w_3}\Big(\sum_{\alpha\in G_0^*}\widetilde{\xi}_{-2}^{\alpha}(\widetilde{X})+\sum_{a=1}^{p}\tau_a\prod_{i=1}^{3}w_i^{c_i(h_a)}\sum_{\alpha\in G_0^*}\chi_\alpha(h_a)\widetilde{\xi}_{-1}^{\alpha}(\widetilde{X})\Big)\Big|_{\mathsf{v}=1}\\
&+\sum_{a\in\mathbb{Z}_{\geq 0}}\sum_{\alpha\in G_0^*}\big(\langle\!\langle\phi_{\sigma_0,\alpha}\psi^a\rangle\!\rangle_{0,1}^{\mathcal{X},\mathbb{T}_f}\big|_{t=\boldsymbol{\tau}}\big)\widetilde{\xi}_a^{\alpha}(\widetilde{X})\\
=&\ [z^{-2}]\sum_{\alpha\in G_0^*}S_z(1,\phi_{\sigma_0,\alpha})\widetilde{\xi}^{\alpha}(z,\widetilde{X}).
\end{aligned}$$

(2) (annulus invariants)

$$\begin{aligned}
&\widetilde{F}_{0,2}^{\mathcal{X},(\mathcal{L},f)}(\boldsymbol{\tau};\widetilde{X}_1,\widetilde{X}_2)-F_{0,2}^{\mathcal{X},(\mathcal{L},f)}(0;\widetilde{X}_1,\widetilde{X}_2)\\
=&\sum_{a_1,a_2\in\mathbb{Z}_{\geq 0}}\sum_{h_1,h_2\in G_0}\big(\langle\!\langle\mathbf{1}^*_{\sigma_0,h_1}\psi^{a_1},\mathbf{1}^*_{\sigma_0,h_2}\psi^{a_1}\rangle\!\rangle_{0,2}^{\mathcal{X},\mathbb{T}_f}\big|_{t=\boldsymbol{\tau}}\big)\Phi_{a_1}^{h_1}(\widetilde{X}_1)\Phi_{a_2}^{h_2}(\widetilde{X}_2)\\
=&\sum_{a_1,a_2\in\mathbb{Z}_{\geq 0}}\sum_{\alpha_1,\alpha_2\in G_0^*}\big(\langle\!\langle\phi_{\sigma_0,\alpha_1}\psi^{a_1},\phi_{\sigma_0,\alpha_2}\psi^{a_1}\rangle\!\rangle_{0,2}^{\mathcal{X},\mathbb{T}_f}\big|_{t=\boldsymbol{\tau}}\big)\widetilde{\xi}_{a_1}^{\alpha_1}(\widetilde{X}_1)\widetilde{\xi}_{a_2}^{\alpha_2}(\widetilde{X}_2)
\end{aligned}$$

where

$$\begin{aligned}
&(\widetilde{X}_1\frac{\partial}{\partial\widetilde{X}_1}+\widetilde{X}_2\frac{\partial}{\partial\widetilde{X}_2})\widetilde{F}_{0,2}^{\mathcal{X},(\mathcal{L},f)}(0;\widetilde{X}_1,\widetilde{X}_2)\\
=&|G_0|\Big(\sum_{h\in G_0}\mathbf{e}_h\Phi_0^h(\widetilde{X}_1)\Phi_0^{h^{-1}}(\widetilde{X}_2)\Big)\Big|_{\mathsf{v}=1}\\
=&\frac{1}{|G_0|^2w_1w_2w_3}\Big(\sum_{\gamma\in G_0^*}\big(\widetilde{\xi}_0^{\gamma}(\widetilde{X}_1)\widetilde{\xi}_0^{\gamma}(\widetilde{X}_2)\big)\Big|_{\mathsf{v}=1}.
\end{aligned}\tag{19}$$

So we have

$$\begin{aligned}
&\widetilde{F}_{0,2}^{\mathcal{X},(\mathcal{L},f)}(\boldsymbol{\tau};\widetilde{X}_1,\widetilde{X}_2)\\
=&[z_1^{-1}z_2^{-1}]\sum_{\alpha_1,\alpha_2\in G_0^*}V_{z_1,z_2}(\phi_{\sigma_0,\alpha_1},\phi_{\sigma_0,\alpha_2})\widetilde{\xi}^{\alpha_1}(z_1,\widetilde{X}_1)\widetilde{\xi}^{\alpha_2}(z_2,\widetilde{X}_2).
\end{aligned}$$

(3) *For* $2g-2+n>0$,

$$\begin{aligned}
&\widetilde{F}_{g,n}^{\mathcal{X},(\mathcal{L},f)}(\boldsymbol{\tau};\widetilde{X}_1,\ldots,\widetilde{X}_n)\\
=&\sum_{a_1,\ldots,a_n\in\mathbb{Z}_{\geq 0}}\sum_{h_1,\ldots,h_n\in G_0}\Big(\langle\!\langle\mathbf{1}^*_{h_1}\psi^{a_1},\ldots,\mathbf{1}^*_{h_n}\psi^{a_n}\rangle\!\rangle_{g,n}^{\mathcal{X},\mathbb{T}_f}\Big|_{t=\boldsymbol{\tau}}\Big)\prod_{j=1}^{n}\Phi_{a_j}^{h_j}(\widetilde{X}_j)\\
=&\sum_{a_1,\ldots,a_n\in\mathbb{Z}_{\geq 0}}\sum_{\alpha_1,\ldots,\alpha_n\in G_0^*}\Big(\langle\!\langle\phi_{\sigma_0,\alpha_1}\psi^{a_1},\ldots\phi_{\sigma_0,\alpha_n}\psi^{a_n}\rangle\!\rangle_{g,n}^{\mathcal{X},\mathbb{T}_f}\Big|_{t=\boldsymbol{\tau}}\Big)\prod_{j=1}^{n}\widetilde{\xi}_{a_j}^{\alpha_j}(\widetilde{X}_j).\\
=&\ [z_1^{-1}\ldots z_n^{-1}]\sum_{\alpha_1,\ldots,\alpha_n\in G_0^*}\Big(\langle\!\langle\frac{\phi_{\sigma_0,\alpha_1}}{z_1-\psi_1},\frac{\phi_{\sigma_0,\alpha_2}}{z_2-\psi_2},\ldots,\frac{\phi_{\sigma_0,\alpha_n}}{z_n-\psi_n}\rangle\!\rangle_{g,n}^{\mathcal{X},\mathbb{T}_f}\Big|_{t=\boldsymbol{\tau}}\Big)\prod_{j=1}^{n}\widetilde{\xi}^{\alpha_j}(z_j,\widetilde{X}_j).
\end{aligned}$$

REMARK 5.10. $\widetilde{F}_{0,2}^{\mathcal{X},(\mathcal{L},f)}(0;\widetilde{X}_1,\widetilde{X}_2)$ is an $H^*(\mathcal{B}\boldsymbol{\mu}_m;\mathbb{C})^{\otimes 2}$-valued power series in $\widetilde{X}_1,\widetilde{X}_2$ which vanishes at $(\widetilde{X}_1,\widetilde{X}_2)=(0,0)$, so it is determined by (19).

We now combine Section 5.5 and the above Proposition 5.9 to obtain a graph sum formula for $\widetilde{F}^{\mathcal{X},(\mathcal{L},f)}_{g,n}$. We use the notation in Section 5.5, and introduce the notation

$$\widetilde{\xi}^{\boldsymbol{\sigma}}(z,\widetilde{X}) = \begin{cases} \widetilde{\xi}^{\alpha}(z,\widetilde{X}), & \text{if } \boldsymbol{\sigma} = (\sigma_0,\alpha), \\ 0, & \text{if } \boldsymbol{\sigma} = (\sigma,\alpha) \text{ and } \sigma \neq \sigma_0. \end{cases}$$

- Given a labelled graph $\vec{\Gamma} \in \mathbf{\Gamma}_{g,n}(\mathcal{X})$, to each ordinary leaf $l_j \in L^o(\gamma)$ with $\boldsymbol{\sigma}(l_j) = \boldsymbol{\sigma} \in I_\Sigma$ and $k(l_j) \in \mathbb{Z}_{\geq 0}$ we assign the following weight (open leaf)

$$(\widetilde{\mathcal{L}}^{\widetilde{X}_j})^{\boldsymbol{\sigma}}_k(l_j) = [z^k]\Big(\sum_{\boldsymbol{\rho},\boldsymbol{\sigma}\in I_\Sigma}\Big(\widetilde{\xi}^{\boldsymbol{\sigma}}(z,\widetilde{X}_j)S^{\widehat{\boldsymbol{\rho}}}{}_{\boldsymbol{\sigma}}\Big|_{\substack{t=\boldsymbol{\tau}\\ \mathsf{w}_i=w_i\mathsf{v}}}\Big)_+ R(-z)_{\boldsymbol{\rho}}{}^{\boldsymbol{\sigma}}\Big|_{\substack{t=\boldsymbol{\tau}\\ \mathsf{w}_i=w_i\mathsf{v}}}\Big). \tag{20}$$

- Given a labelled graph $\mathbf{\Gamma}_{g,n}(\mathcal{X})$, we define a weight

$$\widetilde{w}^{\widetilde{X}}_A(\vec{\Gamma}) = \prod_{v\in V(\Gamma)}\Big(\sqrt{\Delta^{\boldsymbol{\sigma}(v)}(t)}\Big|_{\substack{t=\boldsymbol{\tau}\\ \mathsf{w}_i=w_i\mathsf{v}}}\Big)^{2g(v)-2+\mathrm{val}(v)}\Big\langle\prod_{h\in H(v)}\tau_{k(h)}\Big\rangle_{g(v)}$$
$$\cdot\Big(\prod_{e\in E(\Gamma)}\mathcal{E}^{\boldsymbol{\sigma}(v_1(e)),\boldsymbol{\sigma}(v_2(e))}_{k(h_1(e)),k(h_2(e))}\cdot\prod_{l\in L^1(\Gamma)}(\mathcal{L}^1)^{\boldsymbol{\sigma}(l)}_{k(l)}\Big)\Big|_{\substack{t=\boldsymbol{\tau}\\ \mathsf{w}_i=w_i\mathsf{v}}}\prod_{j=1}^n(\widetilde{\mathcal{L}}^{\widetilde{X}_j})^{\boldsymbol{\sigma}(l_j)}_{k(l_j)}(l_j)$$

Then we have the following graph sum formula for $F^{\mathcal{X},(\mathcal{L},f)}_{g,n}$.

THEOREM 5.11.

$$\widetilde{F}^{\mathcal{X},(\mathcal{L},f)}_{g,n} = \sum_{\vec{\Gamma}\in\mathbf{\Gamma}_{g,n}(\mathcal{X})}\frac{\widetilde{w}^{\widetilde{X}}_A(\vec{\Gamma})}{|\mathrm{Aut}(\vec{\Gamma})|}.$$

PROOF. This follows from Theorem 5.7 and Proposition 5.9. □

DEFINITION 5.12 (Restriction to $Q=1$). We define

$$F^{\mathcal{X}}_g(\boldsymbol{\tau}) := \widetilde{F}^{\mathcal{X}}_g(\boldsymbol{\tau})|_{Q=1}$$
$$F^{\mathcal{X},(\mathcal{L},f)}_{g,n}(\boldsymbol{\tau};\widetilde{X}_1,\ldots,\widetilde{X}_n) := \widetilde{F}^{\mathcal{X},(\mathcal{L},f)}_{g,n}(\boldsymbol{\tau};\widetilde{X}_1,\ldots,\widetilde{X}_n)|_{Q=1}.$$

By Remark 5.8 and Propsition 5.9, $F^{\mathcal{X},(\mathcal{L},f)}_{g,n}$ is well-defined. Theorem 5.11 implies

COROLLARY 5.13.

$$F^{\mathcal{X},(\mathcal{L},f)}_{g,n} = \sum_{\vec{\Gamma}\in\mathbf{\Gamma}_{g,n}(\mathcal{X})}\frac{w^{\widetilde{X}}_A(\vec{\Gamma})}{|\mathrm{Aut}(\vec{\Gamma})|}.$$

where $w^{\widetilde{X}}_A(\vec{\Gamma}) = \widetilde{w}^{\widetilde{X}}_A(\vec{\Gamma})\big|_{Q=1}$.

6. B-model quantization: The topological recursion

In this section we review the definition of a general spectral curve and the topological recursion on it [**35**]. The variables $\hat{x},\hat{y},\hat{X},\hat{Y}$, when restricted to a mirror curve, are indeed the same variables in the previous sections.

6.1. Spectral curves. Let C be a smooth affine algebraic curve in $(\mathbb{C}^*)^2$. The coordinates $\hat{X}, \hat{Y}$ map Σ to the first and the second component of $(\mathbb{C}^*)^2$ respectively. They are holomorphic functions on Σ and we require them to be Morse. Let $\hat{X} = e^{-x}, \hat{Y} = e^{-y}$, where $\hat{x}, \hat{y}$ are well-defined on the universal cover $\mathbb{C}^2$ of $(\mathbb{C}^*)^2$. We denote the covering map π

$$p : \mathbb{C}^2 \to (\mathbb{C}^*)^2,$$
$$(\hat{x}, \hat{y}) \mapsto (e^{-\hat{x}}, e^{-\hat{y}}).$$

Let $\widetilde{C}$ be the lift of C under this map, and let $\overline{C}$ be a choice of smooth compactification of C, which is a compact Riemann surface. We denote the genus of $\overline{C}$ by $\mathfrak{g}$.

The intersection pairing $H_1(\overline{C}; \mathbb{C}) \times H_1(\overline{C}; \mathbb{C}) \to \mathbb{C}$ is a symplectic pairing. We choose a *Lagrangian* subspace $\mathcal{A}$ of $H_1(\overline{C}; \mathbb{C})$. A *Torelli marking* on $\overline{C}$ is a choice of symplectic basis $A_1, \ldots, A_{\mathfrak{g}}, B_1, \ldots, B_{\mathfrak{g}}$ in $H_1(\overline{C}; \mathbb{C})$, such that $A_i \cap B_j = \delta_{i,j}$ and $A_i \cap A_j = B_i \cap B_j = 0$.[5] Following the notions from [**46**], we define the fundamental normalized differential of the second kind (a.k.a. Bergman kernel in Eynard-Orantin [**35**]).

DEFINITION 6.1. The *fundamental normalized differential of the second kind* (abbreviated as *fundamental differential* in this paper) associated to a Lagrangian subspace $\mathcal{A} \subset H_1(\overline{C}; \mathbb{C})$ is the symmetric meromorphic form on $(\overline{C})^2$ satisfying the following conditions.

- The only pole is the double pole along the diagonal, i.e. given any local coordinate ζ near a point $p \in \overline{C}$, the differential $\omega_{0,2}$ has the following form near $(p, p) \in (\overline{C})^2$
$$\omega_{0,2} = \frac{d\zeta_1 d\zeta_2}{(\zeta_1 - \zeta_2)^2} + \text{holomorphic part}.$$
- It is normalized by the choice of A-cycles in $\mathcal{A}$
$$\int_{q \in A} \omega_{0,2}(p, q) = 0, \ \forall A \in \mathcal{A}.$$

DEFINITION 6.2. A spectral curve $\mathcal{C} = (C, \mathcal{A})$ consists of the following data:

- a smooth affine algebraic curve C in $(\mathbb{C}^*)^2$ where the coordinate functions $\hat{X}, \hat{Y}$ are holomorphic Morse;
- the compactification $\overline{C}$ of C as a smooth projective curve;
- a choice of Lagrangian subspace $\mathcal{A} \subset H_1(\overline{C}; \mathbb{C})$;
- a fundamental normalized differential of the second kind B on $\overline{C}$ with respect to such choice of $\mathcal{A}$.

Fix a spectral curve $\mathcal{C}$. Since $\hat{X}$ (and then $\hat{x}$) is Morse, the critical points ($d\hat{x} = 0$) form a finite set $\{p_\sigma : \sigma \in I_{\mathcal{C}}\}$ – here $I_{\mathcal{C}}$ is the index set for the ramification points on C. Define the Liouville form $\Phi = \hat{y} d\hat{x} = -\log Y d\hat{x}$. It is a well-defined holomorphic form on $\widetilde{C}$.

At each critical point p_σ, we define the local coordinate ζ_σ by

$$\hat{x} = \zeta_\sigma^2 + \hat{x}_{0,\sigma},$$

[5] We allow such cycles to be non-geometric, i.e. elements in $H_1(\overline{C}; \mathbb{C})$, and not necessarily in $H_1(\overline{C}; \mathbb{Z})$.

where $\hat{x}_{0,\sigma}$ is the value of $\hat{x}$ at p_σ (well-defined up to an integral multiple of $2\pi\sqrt{-1}$). For any p near p_σ, let $\bar{p}$ be the point on C such that $\zeta_\sigma(\bar{p}) = -\zeta_\sigma(p)$. We also denote $\hat{y}(p_\sigma) = \hat{y}_{0,\sigma}$, and

$$\hat{y} = \hat{y}_{0,\sigma} + \sum_{i=d}^{\infty} h_d^\sigma \zeta_\sigma^d.$$

By the smoothness of the curve C, $h_1^\sigma \neq 0$ for all $\boldsymbol{\sigma}$. Here $\hat{y}_{0,\sigma}$ is also well-defined up to an integral multiple of $2\pi\sqrt{-1}$.

6.2. Eynard-Orantin's topological recursion.

DEFINITION 6.3. Given a spectral curve $\mathcal{C}$, the Eynard-Orantin recursive algorithm defines a sequence of symmetric meromorphic forms $\omega_{g,n}$ on $(\overline{C})^n$ for $g \in \mathbb{Z}_{\geq 0}, n \in \mathbb{Z}_{>0}$ as follows.

- Initial conditions:

$$\omega_{0,1} = \Phi = \hat{y}d\hat{x}, \quad \omega_{0,2} \text{ given by the choice of } \mathcal{A}.$$

- Recursive algorithm:

$$\omega_{g,n}(p_1,\dots,p_n) = \sum_{p'\in I_C} \mathrm{Res}_{p=p'} \frac{\int_{\xi=p}^{\bar{p}} B(p_n,\xi)}{2(\Phi(p)-\Phi(\bar{p}))}\Big(\omega_{g-1,n+1}(p,\bar{p},p_1,\dots,p_{n-1})$$
$$+ \sum_{g_1+g_2=g,\ I\sqcup J=\{1,\dots,n-1\}}{}' \omega_{g_1,|I|+1}(p,p_I)\omega_{g_2,|J|+1}(\bar{p},p_J)\Big).$$

Here the sum symbol $\sum'$ excludes the case $(g_1,|I|) = (0,1),(0,n-1),(g,1)$ or $(g,n-1)$.

The resulting $\omega_{g,n}$ are well-behaved.

PROPOSITION 6.4 (Appendix A of [**35**]). *When $2g-2+n>0$, the poles of $\omega_{g,n}$ are at $d\hat{x}_i = 0$ (critical points), where $\hat{x}_i = -\log \hat{X}_i$ is the $\hat{x}$-coordinate on i-th copy of C^n.*

REMARK 6.5. The original recursion algorithm in [**35**] sets $\omega_{0,1} = 0$ while it does not exclude these four special cases by a special sum symbol $\sum'$ – it is the same as the recursion here. Another different convention is to introduce a minus sign in the recursion kernel, i.e. using $2(\Phi(\bar{p})-\Phi(p))$ in the denominator. Adopting this convention is equivalent to changing all $\omega_{g,n}$ to $(-1)^{g-1}\omega_{g,n}$. We stick to the convention in Definition 6.3 throughout this paper.

6.3. Differential forms on the spectral curve. Given a spectral curve $\mathcal{C} = (C,\mathcal{A})$, for each ramification point p_σ, $\sigma \in I_C$, we associate a path γ_σ as the Lefschetz thimble

$$\hat{x}(\gamma_\sigma) = [\hat{x}_{0,\sigma}, +\infty)$$

Following [**36, 38**], given any $\boldsymbol{\sigma} \in I_C$ and $d \in \mathbb{Z}_{\geq 0}$, define

$$\theta_{\boldsymbol{\sigma}}^d(p) := (2d-1)!!2^{-d}\mathrm{Res}_{p'\to p_\sigma}\omega_{0,2}(p,p')\zeta_\sigma^{-2d-1}.$$

Then $\theta_{\boldsymbol{\sigma}}^d$ satisfies the following properties.

(1) $\theta_{\boldsymbol{\sigma}}^d$ is a meromorphic 1-form on $\overline{C}$ with a single pole of order $2d+2$ at p_σ.

(2) In local coordinate $\zeta_{\sigma} = \sqrt{\hat{x} - \hat{x}_{0,\sigma}}$ near p_{σ},

$$\theta^d_{\sigma} = \Big(\frac{-(2d+1)!!}{2^d \zeta_{\sigma}^{2d+2}} + f(\zeta_{\sigma})\Big) d\zeta_{\sigma},$$

where $f(\zeta_{\sigma})$ is analytic around p_{σ}. The residue of θ^0_{σ} at p_{σ} is zero, so θ^d_{σ} is a differential of the second kind.

(3)

$$\int_A \theta^d_{\sigma} = 0, \ \forall A \in \mathcal{A}$$

The meromorphic 1-form θ^d_{σ} is characterized by the above properties; θ^d_{σ} can be viewed as a section in $H^0(\overline{C}, \omega_{\overline{C}}((2d+2)p_{\sigma}))$.

REMARK 6.6. The meromorphic 1-form θ^d_{σ} corresponds to $d\xi^d_{\sigma}$ in [**36**] and [**37**]. However there is a sign error in [**37**, equation (4.7)]: there is no minus sign on the right hand side. The expansion in [**37**, equation (4.8)] is consistent with (2) in the above properties of θ^d_{σ}. The expansion in [**37**, equation (4.8)] will be consistent with [**37**, equation (4.7)] after correcting this sign error.

Besides, we use the notation θ^d_{σ} instead of $d\xi^d_{\sigma}$ since in general θ^d_{σ} is not an exact form.

6.4. Graph sum formula from Dunin-Barkowski–Orantin–Shardin–Spitz [34]. Following [**34**], the B-model invariants $\omega_{g,n}$ are expressed in terms of graph sums. We first introduce some notation.

- For any $\boldsymbol{\sigma} \in I_{\Sigma}$, we define

$$\check{h}^{\sigma}_k := \frac{(2k-1)!!}{2^{k-1}} h^{\sigma}_{2k-1}. \tag{21}$$

Then by expanding $\hat{x}$ and compute the integral term-by-term (for $u > 0$)

$$\check{h}^{\sigma}_k = [u^{1-k}] \frac{u^{3/2}}{\sqrt{\pi}} e^{u\hat{x}_{0,\sigma}} \int_{p \in \gamma_{\sigma}} e^{-u\hat{x}(p)} \Phi(p).$$

- For any $\boldsymbol{\sigma}, \boldsymbol{\sigma}' \in I_C$, we expand

$$\omega_{0,2}(p_1, p_2) = \Big(\frac{\delta_{\sigma,\sigma'}}{(\zeta_{\sigma} - \zeta_{\sigma'})^2} + \sum_{k,l \in \mathbb{Z}_{\geq 0}} B^{\sigma,\sigma'}_{k,l} \zeta^k_{\sigma} \zeta^l_{\sigma'}\Big) d\zeta_{\sigma} d\zeta_{\sigma'},$$

near $p_1 = p_{\sigma}$ and $p_2 = p_{\sigma'}$, and define

$$\check{B}^{\sigma,\sigma'}_{k,l} := \frac{(2k-1)!!(2l-1)!!}{2^{k+l+1}} B^{\sigma,\sigma'}_{2k,2l}. \tag{22}$$

Then

$$\begin{aligned} \check{B}^{\sigma,\sigma'}_{k,l} &= [u^{-k} v^{-l}] \left(\frac{uv}{u+v} (\delta_{\sigma,\sigma'} - \sum_{\gamma \in I_C} f_{\gamma}{}^{\sigma}(u) f_{\gamma}{}^{\sigma'}(v)) \right) \\ &= [z^k w^l] \left(\frac{1}{z+w} (\delta_{\sigma,\sigma'} - \sum_{\gamma \in I_C} f_{\gamma}{}^{\sigma}(\frac{1}{z}) f_{\gamma}{}^{\sigma'}(\frac{1}{w})) \right). \end{aligned}$$

Given a labeled graph $\vec{\Gamma} \in \mathbf{\Gamma}_{g,n}(I_{\mathcal{C}})$ with $L^o(\Gamma) = \{l_1, \dots, l_n\}$, and $\vec{s} = (s_1, \dots, s_n) \in \overline{C}^n$, we define its weight to be

$$w_B^{\vec{s}}(\vec{\Gamma}) = (-1)^{g(\vec{\Gamma})-1} \prod_{v \in V(\Gamma)} \Big(\frac{h_1^{\boldsymbol{\alpha}(v)}}{\sqrt{-2}}\Big)^{2-2g(v)-\mathrm{val}(v)} \langle \prod_{h \in H(v)} \tau_{k(h)} \rangle_{g(v)} \prod_{e \in E(\Gamma)} \check{B}^{\boldsymbol{\alpha}(v_1(e)),\boldsymbol{\alpha}(v_2(e))}_{k(e),l(e)} \cdot \prod_{l \in \mathcal{L}^1(\Gamma)} (\check{\mathcal{L}}^1)^{\boldsymbol{\alpha}(l)}_{k(l)} \prod_{j=1}^{n} (\check{\mathcal{L}}^{\vec{s}}_B)^{\boldsymbol{\alpha}(l_j)}_{k(l_j)}(l_j)$$

where

- (dilaton leaf)
$$(\check{\mathcal{L}}^1)^{\boldsymbol{\sigma}}_k = -\frac{1}{\sqrt{-2}} \check{h}^{\boldsymbol{\sigma}}_k.$$
- (descendant leaf)
$$(\check{\mathcal{L}}^{\vec{s}}_B)^{\boldsymbol{\sigma}}_k(l_j) = \frac{-1}{\sqrt{-2}} \theta^k_{\boldsymbol{\sigma}}(p_j)$$

Notice that $\check{B}^{\boldsymbol{\sigma},\boldsymbol{\sigma}'}_{k,l}$ plays the role of the edge contribution, while the vertex contribution is $(-1)^{g(\vec{\Gamma})-1} \prod_{v \in V(\Gamma)} \Big(\frac{h_1^{\boldsymbol{\alpha}(v)}}{\sqrt{-2}}\Big)^{2-2g(v)-\mathrm{val}(v)} \langle \prod_{h \in H(v)} \tau_{k(h)} \rangle_{g(v)}$.

In our notation [**34**, Theorem 3.7] is equivalent to:

THEOREM 6.7 (Dunin-Barkowski–Orantin–Shadrin–Spitz [**34**]). *For* $2g - 2 + n > 0$,

$$\omega_{g,n}(s_1, \dots, s_n) = \sum_{\Gamma \in \mathbf{\Gamma}_{g,n}(I_{\mathcal{C}})} \frac{w_B^{\vec{s}}(\vec{\Gamma})}{|\mathrm{Aut}(\vec{\Gamma})|}.$$

REMARK 6.8. Our convention for the factors in the above graph sum formula is different from that in [**34**, Theorem 3.7]. We summarize the following convention differences.

(1) Our $\check{h}^{\boldsymbol{\sigma}}_k$ is $\frac{1}{2^k}$ times the $\check{h}^{\boldsymbol{\sigma}}_k$ in [**34**, Theorem 3.7].
(2) Our $\check{B}^{\boldsymbol{\sigma},\boldsymbol{\sigma}'}_{k,l}$ is $\frac{1}{2^{k+l+1}}$ times the $\check{B}^{\boldsymbol{\sigma},\boldsymbol{\sigma}'}_{k,l}$ in [**34**, Theorem 3.7].
(3) Our $\theta^k_{\boldsymbol{\sigma}}$ is $\frac{1}{2^k}$ times the $d\xi^{\boldsymbol{\sigma}}_k$.

6.5. An equivalent graph sum formula from Eynard. In [**36**, Theorem 5.1], Eynard obtains a graph sum formula for $\omega_{g,n}$ on a general spectral curve. In this subsection, we show that this graph sum formula is equivalent to the graph sum formula in [**34**, Theorem 3.7] by direct computation.

The formula in [**36**, Theorem 5.1] sums over all the stable degeneracies of the moduli space $\mathcal{M}_{g,n}$. So by the dual graph of a stable curve, this is equivalent to summing over the graphs in $\mathbf{\Gamma}_{g,n}(I_{\mathcal{C}})$. The ordinary leaf term in our graph sum formula matches the factor $d\xi$ in [**36**, Theorem 5.1] up to the factor $\frac{-1}{\sqrt{-2}}$ (But we should be careful with the sign problem in $d\xi$ see Remark 6.6). The $\hat{B}$ factor in [**36**, Theorem 5.1] corresponds to our $\check{B}^{\boldsymbol{\sigma},\boldsymbol{\sigma}'}_{k,l}$ which appears in the edges factor. The $\hat{B}$ factor which appears directly in the graph sum formula in [**36**, Theorem 5.1] corresponds to the loop factor i.e. an edge connecting one vertex. The correlator in [**36**, equation (5.1)] involves the κ classes by [**36**, equation (5.4)]. The second factor in [**36**, equation (5.4)] gives us the factors for an edge which connects two different vertices.

The only nontrivial factor is the first factor in [**36**, equation (5.4)] which involves the κ classes and $\tilde{t}_{\boldsymbol{\sigma},k}$ which is called the dual time in [**36**]. By [**36**, equation (6.7)], we have

$$e^{-\sum_{k=0}^{\infty}\tilde{t}_{\boldsymbol{\sigma},k}u^{-k}} = 2\sum_{k=0}^{\infty}\check{h}_{k+1}^{\boldsymbol{\sigma}}u^{-k}. \tag{23}$$

In particular, $e^{-\tilde{t}_{\boldsymbol{\sigma},0}} = 2\check{h}_1^{\boldsymbol{\sigma}}$. Notice that $\kappa_0 = 2g(v) - 2 + \mathrm{val}(v)$ on $\overline{\mathcal{M}}_{g(v),\mathrm{val}(v)}$ and so we have

$$e^{\sum_{k=0}^{\infty}\tilde{t}_{\boldsymbol{\sigma},k}\kappa_k} = (2\check{h}_1^{\boldsymbol{\sigma}})^{(2-2g(v)-\mathrm{val}(v))}e^{\sum_{k=1}^{\infty}\tilde{t}_{\boldsymbol{\sigma},k}\kappa_k}.$$

The first factor on the right hand side is consistent with our vertex factor up to the power of 2 and roots of unity. For the second factor, we apply (23) and [**80**, Lemma 2.2] and it is easy to see that this will give us the dilaton leaf factor and the correlator in our vertex factor. In the end, one only needs to notice the identity

$$\sum_{v\in\Gamma}(2g(v) - 2 + \mathrm{val}(v)) = 2g - 2 + n$$

to match the factors involving powers of 2 and the roots of unity in [**36**, Theorem 5.1] and in Theorem 6.7.

6.6. B-model open potentials from the Eynard-Orantin's recursion. For any $q \in \mathfrak{B}^\circ$, the mirror curve C_q comes with a compactification $\overline{C}_q \subset \mathbb{S}_P$. The images of $A_1, \ldots, A_{\mathfrak{p}} \in H_1(C_q;\mathbb{C})$ in $H_1(\overline{C}_q;\mathbb{C})$ under the map $H_1(C_q;\mathbb{C}) \to H_1(\overline{C}_q;\mathbb{C})$ span a Lagrangian subspace in $H_1(\overline{C}_q;\mathbb{C})$. We also have $\hat{X}$ and $\hat{Y}$ as two holomorphic Morse functions on C_q and they are meromorphic on $\overline{C}_q$. Thus this is a spectral curve for $q \in \mathcal{B}^\circ$, denoted by $\mathcal{C}_q$.

Let $\rho_q^\ell : D_\delta \hookrightarrow D_q^\ell$ be an embedding of a small disk $\{|\underline{X}| < \delta\}$ into D_q^ℓ by mapping $\underline{X}$ to the point whose $\hat{X}$-coordinate is $\underline{X}$, while $\rho_q^{\ell_1,\ldots,\ell_n} = \rho_q^{\ell_1} \times \cdots \times \rho_q^{\ell_n} : (D_\delta)^n \to D_q^{\ell_1} \times \cdots \times D_q^{\ell_n} \subset (\overline{C}_q)^n$. The Eynard-Orantin's topological recursion (Definition 6.3) produces symmetric meromorphic forms $\omega_{g,n}$ on $(\overline{C}_q)^n$. From them we define the B-model open potentials as below.

(1) (disk invariants) When $q = 0$, $\hat{Y}(\bar{p}_\ell)^{\mathfrak{m}} = -1$ for all ℓ. On each D_q^ℓ, $\hat{Y}$ is a holomorphic function. Since D_q^ℓ is very small, the real part of $\hat{Y}$ on it is still negative – so we can choose a branch of logarithm $\log : \mathbb{C} \setminus [0,\infty) \to \mathbb{C}$, and define a function $\hat{y}^\ell$ on D_q^ℓ

$$\hat{y}^\ell = -\log\hat{Y}.$$

The function $\hat{y}^\ell - \hat{y}^\ell(\bar{p}^\ell)$ does not depend on the choice of logarithm. So define

$$\check{F}_{0,1}(q;\hat{X}) = -\sum_{\ell\in\boldsymbol{\mu}_{\mathfrak{m}}^*}\int_0^{\hat{X}}(\rho_q^\ell)^*(\hat{y}^\ell - \hat{y}^\ell(\bar{p}_\ell))\frac{dX'}{X'}\psi_\ell.$$

(2) (annulus invariants) The meromorphic form $\omega_{0,2}$ is not holomorphic on $D_q \times D_q \subset \overline{C}_q \times \overline{C}_q$. One removes the singular part, and defines the following

$$\check{F}_{0,2}(q;\hat{X}_1,\hat{X}_2) := \sum_{\ell_1,\ell_2 \in \boldsymbol{\mu}_{\mathfrak{m}}^*} \int_0^{\hat{X}_1}\int_0^{\hat{X}_2} \left((\rho_q^{\ell_1,\ell_2})^* \omega_{0,2} - \frac{dX_1' dX_2'}{(X_1'-X_2')^2} \right) \psi_{\ell_1} \otimes \psi_{\ell_2}.$$

(3) (stable cases: $2g-2+n>0$) $(\rho_q \ell_1,\dots,\ell_n)^*\omega_{g,n}$ is holomorphic on $(D_q)^n$. We define

$$\check{F}_{g,n}(q;\hat{X}_1,\dots,\hat{X}_n) := \sum_{\ell_1,\dots,\ell_n \in \boldsymbol{\mu}_{\mathfrak{m}}^*} \int_0^{\hat{X}_1}\cdots\int_0^{\hat{X}_n} (\rho_q^{\ell_1,\dots,\ell_n})^* \omega_{g,n} \psi_{\ell_1} \otimes \cdots \otimes \psi_{\ell_n}.$$

7. Comparing the graph sums: Proving the Remodeling Conjecture

In this section we survey the proof from [**45**] on how to match the graph sums. The key idea is that *graph sum ingredients are genus 0 information, and the genus 0 open-closed mirror theorem can be used to match them.*

Throughout this section we use the 1-dimensional torus $\mathbb{T}'_f$ for all equivariant cohomology.

7.1. The statement of open mirror theorems (BKMP Remodeling Conjecture).

There is an open mirror map

$$\widetilde{X} = \hat{X}(1+O(q)). \tag{24}$$

We do not give its explicit formula here, which can be directly written down since it is a solution to certain GKZ system with a prescribed asymptotic behavior [**68**] (see also [**41**] for orbifolds). We call Equation (6) and (24) the *open-closed mirror map.*

REMARK 7.1. This open mirror map is the same as Equation (5), which is the geometric origin of (24). One does not need this fact to prove the related mirror symmetry statements, such as for disk invariants and higher-genus (BKMP).

The full genus mirror symmetry statements are the following

THEOREM 7.2. *Under the open-closed mirror map,*

- *(Aganagic-Klemm-Vafa's conjecture on disk invariants, the general case proved in* [**41**]*)*

$$\check{F}_{0,1}(q;\hat{X}) = |G_0| F_{0,1}^{\mathcal{X},(\mathcal{L},f)}(\boldsymbol{\tau},\widetilde{X}); \tag{25}$$

- *(BKMP's Remodeling Conjecture)*
  - *(Annulus invariants)*

$$\check{F}_{0,2}(q;\hat{X}_1,\hat{X}_2) = -|G_0|^2 F_{0,2}^{\mathcal{X},(\mathcal{L},f)}(\boldsymbol{\tau};\widetilde{X}_1,\widetilde{X}_2);$$

  - *(Stable cases) For $2g-2+n>0$,*

$$\check{F}_{g,n}(q;\hat{X}_1,\dots,\hat{X}_n) = |G_0|^n (-1)^{g-1+n} F_{g,n}^{\mathcal{X},(\mathcal{L},f)}(\boldsymbol{\tau};\widetilde{X}_1,\dots,\widetilde{X}_n).$$

7.2. Graph sum components: Vertices. As discussed in Section 4.4, the identification of Frobenius algebras implies that the length of the canonical basis matches, as in Equation (8)

$$\frac{h_1^{\boldsymbol{\alpha}}}{\sqrt{-2}} = \frac{1}{\sqrt{\Delta^{\boldsymbol{\alpha}}(\boldsymbol{\tau})}}.$$

This equates the A-model vertex contribution

$$\left(\frac{1}{\sqrt{\Delta^{\boldsymbol{\alpha}(v)}(\boldsymbol{\tau})})}\right)^{2-2g(v)-\mathrm{val}(v)} |_{Q=1}\langle \prod_{h\in H(v)} \tau_{k(h)}\rangle_{g(v)}$$

with the B-model vertex contribution

$$\left(\frac{h_1^{\boldsymbol{\alpha}(v)}}{\sqrt{-2}}\right)^{2-2g(v)-\mathrm{val}(v)} \langle \prod_{h\in H(v)} \tau_{k(h)}\rangle_{g(v)}$$

7.3. Oscillatory integrals and $\check{R}$-matrices. The functions $V_{\boldsymbol{\alpha}}$ are canonical basis for the Landau-Ginzburg B-model $((\mathbb{C}^*)^3, W^{\mathbb{T}'})$.

The roles of the meromorphic 1-forms $\theta_{\boldsymbol{\alpha}}^0$ are similar to the canonical basis $V_{\boldsymbol{\alpha}}$, in the following sense.

PROPOSITION 7.3 (Dimensional reduction). *For any $\boldsymbol{\alpha} \in I_\Sigma$ (i.e. index sets of the canonical basis are the same for both A and B-models since the Frobenius algebra are isomorphic), we have a 3-cycle (non-compact) $\Gamma_{\boldsymbol{\alpha}} \ni P_{\boldsymbol{\alpha}}$ in $(\mathbb{C}^*)^3$, and a non-compact 1-cycle $\gamma_{\boldsymbol{\alpha}} \ni p_{\boldsymbol{\alpha}}$ in C_q such that*

$$\int_{\Gamma_{\boldsymbol{\alpha}}} e^{-\frac{W^{\mathbb{T}'}}{z}}\Omega = 2\pi\sqrt{-1}\int_{\gamma_{\boldsymbol{\alpha}}} e^{-\frac{\hat{x}}{z}}\Phi;$$

$$\int_{\Gamma_{\boldsymbol{\alpha}}} e^{-\frac{W^{\mathbb{T}'}}{z}}\overline{V}_{\boldsymbol{\beta}}\Omega = 2\pi\sqrt{-1}z^2\int_{\gamma_{\boldsymbol{\alpha}}} e^{-\frac{\hat{x}}{z}}\frac{h_1^{\boldsymbol{\beta}}\theta_{\boldsymbol{\beta}}^0}{2}.$$

Here $z > 0$, and $\gamma_{\boldsymbol{\alpha}} = \{\hat{x} \in [\hat{x}_{0,\boldsymbol{\alpha}}, \infty)\}$.

REMARK 7.4. The modified canonical basis $\overline{V}_{\boldsymbol{\alpha}}$ is a quadratic polynomial in z – the degree 0 term is $V_{\boldsymbol{\alpha}}$. There is a unique way to construct $\overline{V}_{\boldsymbol{\alpha}}$ from $V_{\boldsymbol{\alpha}}$. See [**45**, Section 6.1]

REMARK 7.5. It is expected that these oscillatory integrals [**39**, **44**], under the open-closed mirror map $\boldsymbol{\tau} = \boldsymbol{\tau}(q)$, should be equal to

$$\langle\!\langle \phi_{\boldsymbol{\beta}}(\boldsymbol{\tau}), \frac{\kappa(\mathcal{E}_{\boldsymbol{\alpha}})}{z-\psi}\rangle\!\rangle_{0,2},$$

where $\kappa(\mathcal{E}_{\boldsymbol{\alpha}})$ is certain characteristic classes invovling Gamma functions of a *mirror* coherent sheaf $\mathcal{E}_{\boldsymbol{\alpha}}$ (mirror to $\gamma_{\boldsymbol{\alpha}}$ or $\Gamma_{\boldsymbol{\alpha}}$) on $\mathcal{X}$.

There are related definition (following [**38**])

$$\int_{\Gamma_{\boldsymbol{\sigma}}} e^{-\frac{W^{\mathbb{T}'}}{z}}\frac{\sqrt{-2V_{\boldsymbol{\sigma}'}}}{h_1^{\boldsymbol{\sigma}'}} \sim (-2\pi z)^{\frac{3}{2}}\check{R}_{\boldsymbol{\sigma}'}^{\;\boldsymbol{\sigma}}(z)e^{-\frac{\hat{x}_{0,\boldsymbol{\sigma}}}{z}} \tag{26}$$

$$f_{\boldsymbol{\sigma}'}{}^{\boldsymbol{\sigma}}(u) \sim \frac{e^{\hat{x}_{\boldsymbol{\sigma},0}}}{2\sqrt{\pi u}}\int_{\gamma_{\boldsymbol{\sigma}}} e^{-u\hat{x}}\theta_{\boldsymbol{\sigma}'}^0.$$

The dimensional reduction Proposition 7.3 implies

$$\check{R}_{\boldsymbol{\alpha}}^{\boldsymbol{\beta}}(z) = f_{\boldsymbol{\alpha}}^{\boldsymbol{\beta}}(\frac{1}{z}).$$

The A-side QDE is (9), while the B-side equation comes from the following simple calculus

$$-z\frac{\partial}{\partial \tau_i}\int e^{-\frac{W^{\mathbb{T}'}}{z}}\omega = (\frac{\partial W^{\mathbb{T}'}}{\partial \tau_i})\int e^{-\frac{W^{\mathbb{T}'}}{z}}\omega.$$

Some remarks on this simple fact:

- The integral is over any flat half-dimensional cycle on which the integral is converging.
- ω needs to be *flat*, i.e. it does not depend on the parameter q (or $\boldsymbol{\tau}$, differing with q by a mirror map), or invariant under the Gauss-Manin connection. Notice $V_{\boldsymbol{\sigma}}$ or $\overline{V}_{\boldsymbol{\sigma}}$ is not flat – they are canonical basis and vary with the parameters q. So the integral in Equation (26) does not satisfy this equation *per se*. A linear combination of $V_{\boldsymbol{\sigma}}$ (with coefficients dependent on q or $\boldsymbol{\tau}$) which produces a flat form does satisfy this equation – for example, using the canonical to flat change of basis matrices $\Psi_{\boldsymbol{\sigma}'}^{\ \boldsymbol{\sigma}}$ as in Section 5.2.
- The identification of the genus 0 mirror symmetry identifies $\frac{\partial W^{\mathbb{T}'}}{\partial \tau_i}$ with H_i – so the above differential equation is the same as the A-side QDE (9).

The A-model S-matrix (14) and the B-model oscillatory integral (26) satisfy the QDE (9). By a theorem in [**33**, **49**], we know R and $\check{R}$ are uniquely determined up to constants. We can fix these constants at the large radius limit point $q = 0$. The value of $\check{R}$ is explicitly computed in [**43**] (here $\boldsymbol{\alpha} = (\alpha, \gamma), \boldsymbol{\beta} = (\beta, \delta)$)

$$\begin{aligned}&\check{R}_{\boldsymbol{\alpha}}^{\boldsymbol{\beta}}(-z)|_{q=0}\\ =&\frac{\delta_{\alpha,\beta}}{|G_\alpha|}\sum_{h\in G_\alpha}\chi_\gamma(h)\chi_\delta(h^{-1})\prod_{i=1}^{3}\exp\Big(\sum_{m=1}^{\infty}\frac{(-1)^m}{m(m+1)}B_{m+1}(c_i^\alpha(h))(\frac{z}{\mathsf{w}_i(\alpha)})^m\Big)\end{aligned}$$

which is precisely $R_{\boldsymbol{\alpha}}^{\boldsymbol{\beta}}(z)|_{q=0}$ given in Equation (13). We have matched $R_{\boldsymbol{\alpha}}^{\boldsymbol{\beta}}(z) = \check{R}_{\boldsymbol{\alpha}}^{\boldsymbol{\beta}}(-z)$.

7.4. Open leafs. The matching of A and B-model R-matrices $R_{\boldsymbol{\alpha}}^{\boldsymbol{\beta}}(z)$ with $\check{R}_{\boldsymbol{\alpha}}^{\boldsymbol{\beta}}(-z)$ identifies all graph components other than open leafs. Open leaf requires more than R-matrices. Identifying R-matrices we rely on that the Frobenius structures on both sides are equal, while identifying open leafs we rely on genus 0 *open* mirror theorem, namely identifying the disk Gromov-Witten potential with $\Phi = \hat{y}d\hat{x}$ (Φ corresponds to disk invariants with *no inseration.*)

Recall the A-model open leaf at vertex $\boldsymbol{\sigma}$ with height k ($k \geq 0$) is (Equation (20))

$$(\mathcal{L}^{\widetilde{X}})_k^{\boldsymbol{\sigma}}(l_j) = [z^k]\left(\sum_{\boldsymbol{\sigma}',\boldsymbol{\rho}\in I_\Sigma}(\widetilde{\xi}^{\boldsymbol{\sigma}'}(z,\widetilde{X})S(\hat{\phi}_{\boldsymbol{\rho}}(\boldsymbol{\tau}),\phi_{\boldsymbol{\sigma}'}))_+R(-z)_{\boldsymbol{\rho}}^{\ \boldsymbol{\sigma}}\right). \tag{27}$$

The open leaf weight, as a power series in $\widetilde{X}$, determines each other for different height k – they are related by (c.f. Equation (18) for the definition of $\widetilde{\xi}$)

$$(\mathcal{L}^{\widetilde{X}})_{k+1}^{\boldsymbol{\sigma}} = (\frac{1}{\mathsf{v}}\widetilde{X}\frac{d}{d\widetilde{X}})(\mathcal{L}^{\widetilde{X}})_k^{\boldsymbol{\sigma}}.$$

The localization computation says

$$[z^0]\sum_{\sigma'\in I_\Sigma}\widetilde{\xi}^{\sigma'}(z,\widetilde{X})S(1,\phi_{\sigma'})=(\widetilde{X}\frac{d}{d\widetilde{X}})^2F_{0,1}^{\mathcal{X},(\mathcal{L},f)}(\boldsymbol{\tau},\widetilde{X}),$$

while the open mirror theorem of [**41**] further relates $F_{0,1}^{\mathcal{X},(\mathcal{L},f)}(\boldsymbol{\tau},\widetilde{X})$ to

$$|G_0|F_{0,1}^{\mathcal{X},(\mathcal{L},f)}(\boldsymbol{\tau},\widetilde{X})=\check{F}_{0,1}(q;\hat{X}).$$

So immediately one obtains

$$[z^0]\sum_{\sigma'\in I_\Sigma}\widetilde{\xi}^{\sigma'}(z,\widetilde{X})S(1,\phi_{\sigma'})=-\frac{1}{|G_0|}\sum_{\ell\in\boldsymbol{\mu}_{\mathfrak{m}}^*}(\rho_q^\ell)^*(\frac{d\hat{y}}{d\hat{x}})\psi_\ell.$$

The part in $()_+$ of the open leaf (27) involves the insertion of $\hat{\phi}_\rho(\boldsymbol{\tau})$, while the genus zero mirror theorem only deals with the insertion of 1 in the S-function. However, since $S(1,\phi_{\sigma'})$ is a solution to QDE (Equation (9)), taking derivatives with respect to $\boldsymbol{\tau}$ we obtain the following

$$-z\frac{\partial}{\partial\tau_i}=S(H_i^{\mathbb{T}},\phi_{\sigma'}).$$

Since $\{H_i^{\mathbb{T}}\}_{i=1}^{\mathfrak{p}}$ multiplicatively generate the cohomology, we choose a_i,b_i ($i=1,\dots,\mathfrak{g}$) such that $H_{a_i}\star H_{b_i}$ form a basis of $H^4(\mathcal{X})$. This choice is not unique. In principle one can express in terms of

$$\hat{\phi}_{\boldsymbol{\sigma}}(\boldsymbol{\tau})=\sum_{i=1}^{\mathfrak{g}}\hat{A}_{\boldsymbol{\sigma}}^i(\boldsymbol{\tau})H_{a_i}^{\mathbb{T}}\star H_{b_i}^{\mathbb{T}}+\sum_{a=1}^{\mathfrak{p}}\hat{B}_{\boldsymbol{\sigma}}^a(\boldsymbol{\tau})H_a^{\mathbb{T}}+\hat{C}_{\boldsymbol{\sigma}}(\boldsymbol{\tau})\mathbf{1}.$$

The coefficients are complicated. An very important observation is that we have the *same* expression of $\theta_{\boldsymbol{\sigma}}$ with *same* coefficients

$$\frac{\theta_{\boldsymbol{\sigma}}}{\sqrt{-2}}=\sum_{i=1}^{\mathfrak{g}}\hat{A}_{\boldsymbol{\sigma}}^i(\boldsymbol{\tau}(q))\frac{\partial^2\Phi}{\partial\tau_{a_i}\partial\tau_{b_i}}+\sum_{a=1}^{\mathfrak{p}}\hat{B}_{\boldsymbol{\sigma}}^a(\boldsymbol{\tau}(q))d(\frac{\frac{\partial\Phi}{\partial\tau_a}}{d\hat{x}})+\hat{C}_{\boldsymbol{\sigma}}(\boldsymbol{\tau}(q))d(\frac{d\hat{y}}{d\hat{x}}).$$

This allows us to compute the $()_+$-part in (27). We end up with

$$|G_0|\left(\sum_{\sigma'\in I_\Sigma}\widetilde{\xi}^{\sigma'}(z,\widetilde{X})S(\hat{\phi}_{\boldsymbol{\sigma}}(\boldsymbol{\tau}),\phi_{\sigma'})\right)_+=\sum_{\ell\in\boldsymbol{\mu}_{\mathfrak{m}}^*}\int_0^{\hat{X}}\frac{(\rho_q^\ell)^*\hat{\theta}_{\boldsymbol{\sigma}}(z)}{\sqrt{-2}}.$$

We see that $\theta_{\boldsymbol{\sigma}}^0$ plays the role of disk invariants where one has *a closed insertion* $\hat{\phi}_{\boldsymbol{\sigma}}(\boldsymbol{\tau})$, after some constant factor. Therefore, comparing the leaf terms of A-model and B-model graph sums

$$|G_0|(\mathcal{L}_O^{\widetilde{X}})_k^{\boldsymbol{\sigma}}(l_j)=-(\check{\mathcal{L}}_O^{\hat{X}})_k^{\boldsymbol{\sigma}}(l_j).$$

Thus all graph components are matched – the factor $|G_0|$ here results in the factor $|G_0|^n$ in the conjecture, and the sign contributes to $(-1)^n$.

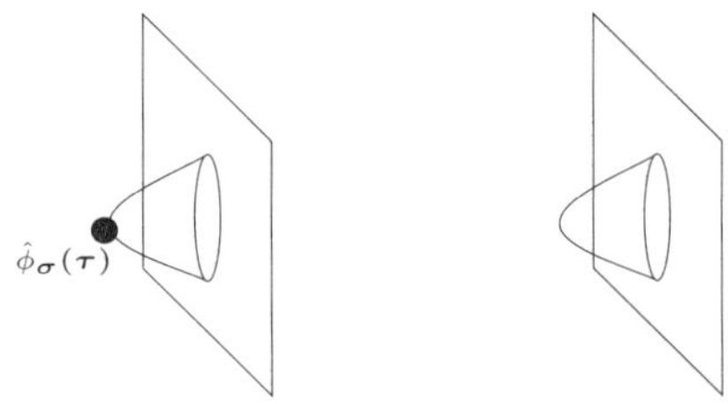

FIGURE 6. The disk invariants with $\hat{\phi}_{\boldsymbol{\sigma}}(\boldsymbol{\tau})$ inserted, versus with no insertion. The former corresponds to $\theta_{\boldsymbol{\sigma}}$ while the later corresponds to Φ on the B-model.

We summarize the comparison of graph components in the following table.

A-model GW	B-model spectral curve	Remark
# of toric fixed pts	# of ramification points	dimension of Frobenius algebra
$\sqrt{\frac{1}{\Delta^{\alpha}(\tau)}}$	$\frac{h_1^{\alpha}}{\sqrt{-2}}$	length of canonical basis, matching vertices
R-matrix $R_{\alpha}^{\ \beta}(z)$	$\check{R}_{\alpha}^{\ \beta}(-z)$	matching edges, dilaton leafs
canonical coordinate u^{α}	critical value $\check{x}_{0,\alpha}$	
meromorphic form $\frac{\theta_{\alpha}^0}{\sqrt{-2}}$	disk potential inserted by $\hat{\phi}^{\alpha}(\tau)$	matching open leafs, see Figure 6
$\hat{y}d\hat{x}$	disk potential (no insertion)	see Figure 6

Acknowledgments

We would like to thank Chiu-Chu Melissa Liu for the wonderful collaboration towards a proof of the BKMP Remodeling Conjecture, and her constant encouragement during the writing of this survey. We would also like to thank Bertrand Eynard for pointing out the relation between the graph sums in [**36**] and [**34**].

References

[1] Mina Aganagic, Vincent Bouchard, and Albrecht Klemm, *Topological strings and (almost) modular forms*, Comm. Math. Phys. **277** (2008), no. 3, 771–819, DOI 10.1007/s00220-007-0383-3. MR2365453

[2] Dan Abramovich and Barbara Fantechi, *Orbifold techniques in degeneration formulas*, Ann. Sc. Norm. Super. Pisa Cl. Sci. (5) **16** (2016), no. 2, 519–579. MR3559610

[3] Dan Abramovich, Tom Graber, and Angelo Vistoli, *Algebraic orbifold quantum products*, Orbifolds in mathematics and physics (Madison, WI, 2001), Contemp. Math., vol. 310, Amer. Math. Soc., Providence, RI, 2002, pp. 1–24, DOI 10.1090/conm/310/05397. MR1950940

[4] Dan Abramovich, Tom Graber, and Angelo Vistoli, *Gromov-Witten theory of Deligne-Mumford stacks*, Amer. J. Math. **130** (2008), no. 5, 1337–1398, DOI 10.1353/ajm.0.0017. MR2450211

[5] Mina Aganagic, Albrecht Klemm, Marcos Mariño, and Cumrun Vafa, *The topological vertex*, Comm. Math. Phys. **254** (2005), no. 2, 425–478, DOI 10.1007/s00220-004-1162-z. MR2117633

[6] Mina Aganagic, Albrecht Klemm, and Cumrun Vafa, *Disk instantons, mirror symmetry and the duality web*, Z. Naturforsch. A **57** (2002), no. 1-2, 1–28, DOI 10.1515/zna-2002-9-1001. MR1906661

[7] M. Aganagic, C. Vafa, "Mirror symmetry, D-branes and counting holomorphic discs," `arXiv:hep-th/0012041`.

[8] Victor V. Batyrev and David A. Cox, *On the Hodge structure of projective hypersurfaces in toric varieties*, Duke Math. J. **75** (1994), no. 2, 293–338, DOI 10.1215/S0012-7094-94-07509-1. MR1290195

[9] M. Bershadsky, S. Cecotti, H. Ooguri, and C. Vafa, *Holomorphic anomalies in topological field theories*, Nuclear Phys. B **405** (1993), no. 2-3, 279–304, DOI 10.1016/0550-3213(93)90548-4. MR1240687

[10] M. Bershadsky and A. Radul, *Conformal field theories with additional Z_N symmetry*, Internat. J. Modern Phys. A **2** (1987), no. 1, 165–178, DOI 10.1142/S0217751X87000053. MR880789

[11] Lev A. Borisov, Linda Chen, and Gregory G. Smith, *The orbifold Chow ring of toric Deligne-Mumford stacks*, J. Amer. Math. Soc. **18** (2005), no. 1, 193–215, DOI 10.1090/S0894-0347-04-00471-0. MR2114820

[12] Vincent Bouchard, Andrei Catuneanu, Olivier Marchal, and Piotr Sułkowski, *The remodeling conjecture and the Faber-Pandharipande formula*, Lett. Math. Phys. **103** (2013), no. 1, 59–77, DOI 10.1007/s11005-012-0588-z. MR3004817

[13] Vincent Bouchard, Albrecht Klemm, Marcos Mariño, and Sara Pasquetti, *Remodeling the B-model*, Comm. Math. Phys. **287** (2009), no. 1, 117–178, DOI 10.1007/s00220-008-0620-4. MR2480744

[14] Vincent Bouchard, Albrecht Klemm, Marcos Mariño, and Sara Pasquetti, *Topological open strings on orbifolds*, Comm. Math. Phys. **296** (2010), no. 3, 589–623, DOI 10.1007/s00220-010-1020-0. MR2628817

[15] Vincent Bouchard, Daniel Hernández Serrano, Xiaojun Liu, and Motohico Mulase, *Mirror symmetry for orbifold Hurwitz numbers*, J. Differential Geom. **98** (2014), no. 3, 375–423. MR3263522

[16] Vincent Bouchard and Piotr Sułkowski, *Topological recursion and mirror curves*, Adv. Theor. Math. Phys. **16** (2012), no. 5, 1443–1483. MR3056954

[17] Andrea Brini, *Open topological strings and integrable hierarchies: remodeling the A-model*, Comm. Math. Phys. **312** (2012), no. 3, 735–780, DOI 10.1007/s00220-012-1489-9. MR2925133

[18] Jim Bryan, Charles Cadman, and Ben Young, *The orbifold topological vertex*, Adv. Math. **229** (2012), no. 1, 531–595, DOI 10.1016/j.aim.2011.09.008. MR2854183

[19] Philip Candelas, Xenia C. de la Ossa, Paul S. Green, and Linda Parkes, *A pair of Calabi-Yau manifolds as an exactly soluble superconformal theory*, Nuclear Phys. B **359** (1991), no. 1, 21–74, DOI 10.1016/0550-3213(91)90292-6. MR1115626

[20] Sergio Cecotti and Cumrun Vafa, *Massive orbifolds*, Modern Phys. Lett. A **7** (1992), no. 19, 1715–1723, DOI 10.1142/S0217732392001415. MR1168619

[21] Weimin Chen and Yongbin Ruan, *Orbifold Gromov-Witten theory*, Orbifolds in mathematics and physics (Madison, WI, 2001), Contemp. Math., vol. 310, Amer. Math. Soc., Providence, RI, 2002, pp. 25–85, DOI 10.1090/conm/310/05398. MR1950941

[22] L. Chen, "Bouchard-Klemm-Marino-Pasquetti Conjecture for $\mathbb{C}^3$," `arXiv:0910.3739`.

[23] Weimin Chen and Yongbin Ruan, *Orbifold Gromov-Witten theory*, Orbifolds in mathematics and physics (Madison, WI, 2001), Contemp. Math., vol. 310, Amer. Math. Soc., Providence, RI, 2002, pp. 25–85, DOI 10.1090/conm/310/05398. MR1950941

[24] Weimin Chen and Yongbin Ruan, *A new cohomology theory of orbifold*, Comm. Math. Phys. **248** (2004), no. 1, 1–31, DOI 10.1007/s00220-004-1089-4. MR2104605

[25] T.-M. Chiang, A. Klemm, S.-T. Yau, and E. Zaslow, *Local mirror symmetry: calculations and interpretations*, Adv. Theor. Math. Phys. **3** (1999), no. 3, 495–565, DOI 10.4310/ATMP.1999.v3.n3.a3. MR1797015

[26] Thomas Henry Coates, *Riemann-Roch theorems in Gromov-Witten theory*, ProQuest LLC, Ann Arbor, MI, 2003. Thesis (Ph.D.)–University of California, Berkeley. MR2705177

[27] Tom Coates, Alessio Corti, Hiroshi Iritani, and Hsian-Hua Tseng, *A mirror theorem for toric stacks*, Compos. Math. **151** (2015), no. 10, 1878–1912, DOI 10.1112/S0010437X15007356. MR3414388

[28] David A. Cox and Sheldon Katz, *Mirror symmetry and algebraic geometry*, Mathematical Surveys and Monographs, vol. 68, American Mathematical Society, Providence, RI, 1999. MR1677117

[29] Duiliu-Emanuel Diaconescu and Bogdan Florea, *Localization and gluing of topological amplitudes*, Comm. Math. Phys. **257** (2005), no. 1, 119–149, DOI 10.1007/s00220-005-1323-8. MR2163571

[30] Lance Dixon, Daniel Friedan, Emil Martinec, and Stephen Shenker, *The conformal field theory of orbifolds*, Nuclear Phys. B **282** (1987), no. 1, 13–73, DOI 10.1016/0550-3213(87)90676-6. MR869937

[31] L. Dixon, J. A. Harvey, C. Vafa, and E. Witten, *Strings on orbifolds*, Nuclear Phys. B **261** (1985), no. 4, 678–686, DOI 10.1016/0550-3213(85)90593-0. MR818423

[32] L. Dixon, J. Harvey, C. Vafa, and E. Witten, *Strings on orbifolds. II*, Nuclear Phys. B **274** (1986), no. 2, 285–314, DOI 10.1016/0550-3213(86)90287-7. MR851703

[33] Boris Dubrovin, *Geometry of 2D topological field theories*, Integrable systems and quantum groups (Montecatini Terme, 1993), Lecture Notes in Math., vol. 1620, Springer, Berlin, 1996, pp. 120–348, DOI 10.1007/BFb0094793. MR1397274

[34] P. Dunin-Barkowski, N. Orantin, S. Shadrin, and L. Spitz, *Identification of the Givental formula with the spectral curve topological recursion procedure*, Comm. Math. Phys. **328** (2014), no. 2, 669–700, DOI 10.1007/s00220-014-1887-2. MR3199996

[35] B. Eynard and N. Orantin, *Invariants of algebraic curves and topological expansion*, Commun. Number Theory Phys. **1** (2007), no. 2, 347–452, DOI 10.4310/CNTP.2007.v1.n2.a4. MR2346575

[36] B. Eynard, "Intersection number of spectral curves," arXiv:1104.0176.

[37] B. Eynard, *Invariants of spectral curves and intersection theory of moduli spaces of complex curves*, Commun. Number Theory Phys. **8** (2014), no. 3, 541–588, DOI 10.4310/CNTP.2014.v8.n3.a4. MR3282995

[38] B. Eynard and N. Orantin, *Computation of open Gromov-Witten invariants for toric Calabi-Yau 3-folds by topological recursion, a proof of the BKMP conjecture*, Comm. Math. Phys. **337** (2015), no. 2, 483–567, DOI 10.1007/s00220-015-2361-5. MR3339157

[39] B. Fang, "Central charges of T-dual branes for toric varieties", `arXiv:1611.05153`.

[40] Bohan Fang and Chiu-Chu Melissa Liu, *Open Gromov-Witten invariants of toric Calabi-Yau 3-folds*, Comm. Math. Phys. **323** (2013), no. 1, 285–328, DOI 10.1007/s00220-013-1771-5. MR3085667

[41] B. Fang, C.-C. M. Liu, H.-H. Tseng, "Open-closed Gromov-Witten invariants of 3-dimensional Calabi-Yau smooth toric DM stacks," `arXiv:1212.6073`.

[42] Bohan Fang, Chiu-Chu Melissa Liu, and Zhengyu Zong, *Equivariant Gromov-Witten theory of affine smooth toric Deligne-Mumford stacks*, Int. Math. Res. Not. IMRN **7** (2016), 2127–2144, DOI 10.1093/imrn/rnv201. MR3509949

[43] Bohan Fang, Chiu-Chu Melissa Liu, and Zhengyu Zong, *All genus mirror symmetry for toric Calabi-Yau 3-orbifolds*, String-Math 2014, Proc. Sympos. Pure Math., vol. 93, Amer. Math. Soc., Providence, RI, 2016, pp. 1–19. MR3524231

[44] Bohan Fang, Chiu-Chu Melissa Liu, and Zhengyu Zong, *The Eynard-Orantin recursion and equivariant mirror symmetry for the projective line*, Geom. Topol. **21** (2017), no. 4, 2049–2092, DOI 10.2140/gt.2017.21.2049. MR3654104

[45] B. Fang, C.-C. M. Liu, Z. Zong, "On the remodeling conjecture for toric Calabi-Yau 3-orbifolds," arXiv:1604.07123.

[46] John D. Fay, *Theta functions on Riemann surfaces*, Lecture Notes in Mathematics, Vol. 352, Springer-Verlag, Berlin-New York, 1973. MR0335789

[47] Alexander B. Givental, *Equivariant Gromov-Witten invariants*, Internat. Math. Res. Notices **13** (1996), 613–663, DOI 10.1155/S1073792896000414. MR1408320

[48] Alexander Givental, *A mirror theorem for toric complete intersections*, Topological field theory, primitive forms and related topics (Kyoto, 1996), Progr. Math., vol. 160, Birkhäuser Boston, Boston, MA, 1998, pp. 141–175. MR1653024

[49] Alexander Givental, *Elliptic Gromov-Witten invariants and the generalized mirror conjecture*, Integrable systems and algebraic geometry (Kobe/Kyoto, 1997), World Sci. Publ., River Edge, NJ, 1998, pp. 107–155. MR1672116

[50] Alexander B. Givental, *Semisimple Frobenius structures at higher genus*, Internat. Math. Res. Notices **23** (2001), 1265–1286, DOI 10.1155/S1073792801000605. MR1866444

[51] Alexander B. Givental, *Gromov-Witten invariants and quantization of quadratic Hamiltonians* (English, with English and Russian summaries), Mosc. Math. J. **1** (2001), no. 4,

551–568, 645. Dedicated to the memory of I. G. Petrovskii on the occasion of his 100th anniversary. MR1901075

[52] T. Graber and R. Pandharipande, *Localization of virtual classes*, Invent. Math. **135** (1999), no. 2, 487–518, DOI 10.1007/s002220050293. MR1666787

[53] Tom Graber and Eric Zaslow, *Open-string Gromov-Witten invariants: calculations and a mirror "theorem"*, Orbifolds in mathematics and physics (Madison, WI, 2001), Contemp. Math., vol. 310, Amer. Math. Soc., Providence, RI, 2002, pp. 107–121, DOI 10.1090/conm/310/05400. MR1950943

[54] Mark Gross, Ludmil Katzarkov, and Helge Ruddat, *Towards mirror symmetry for varieties of general type*, Adv. Math. **308** (2017), 208–275, DOI 10.1016/j.aim.2016.03.035. MR3600059

[55] Reese Harvey and H. Blaine Lawson Jr., *Calibrated geometries*, Acta Math. **148** (1982), 47–157, DOI 10.1007/BF02392726. MR666108

[56] K. Hori, C. Vafa, "Mirror symmetry," `arXiv:hep-th/0002222`.

[57] K. Hori, A. Iqbal, C. Vafa, "D-branes and mirror symmetry," `arXiv:hep-th/0005247`.

[58] Shahram Hamidi and Cumrun Vafa, *Interactions on orbifolds*, Nuclear Phys. B **279** (1987), no. 3-4, 465–513, DOI 10.1016/0550-3213(87)90006-X. MR867241

[59] M.-x. Huang, A. Klemm, and S. Quackenbush, *Topological string theory on compact Calabi-Yau: modularity and boundary conditions*, Homological mirror symmetry, Lecture Notes in Phys., vol. 757, Springer, Berlin, 2009, pp. 45–102. MR2596635

[60] Hiroshi Iritani, *An integral structure in quantum cohomology and mirror symmetry for toric orbifolds*, Adv. Math. **222** (2009), no. 3, 1016–1079, DOI 10.1016/j.aim.2009.05.016. MR2553377

[61] Tyler J. Jarvis and Takashi Kimura, *Orbifold quantum cohomology of the classifying space of a finite group*, Orbifolds in mathematics and physics (Madison, WI, 2001), Contemp. Math., vol. 310, Amer. Math. Soc., Providence, RI, 2002, pp. 123–134, DOI 10.1090/conm/310/05401. MR1950944

[62] Jerzy Kaczorowski and Alberto Perelli, *A uniform version of Stirling's formula*. part 1, Funct. Approx. Comment. Math. **45** (2011), no. part 1, 89–96, DOI 10.7169/facm/1317045234. MR2865415

[63] Sheldon Katz and Chiu-Chu Melissa Liu, *Enumerative geometry of stable maps with Lagrangian boundary conditions and multiple covers of the disc*, Adv. Theor. Math. Phys. **5** (2001), no. 1, 1–49, DOI 10.4310/ATMP.2001.v5.n1.a1. MR1894336

[64] Albrecht Klemm and Eric Zaslow, *Local mirror symmetry at higher genus*, Winter School on Mirror Symmetry, Vector Bundles and Lagrangian Submanifolds (Cambridge, MA, 1999), AMS/IP Stud. Adv. Math., vol. 23, Amer. Math. Soc., Providence, RI, 2001, pp. 183–207. MR1876069

[65] V. G. Knizhnik, *Analytic fields on Riemann surfaces. II*, Comm. Math. Phys. **112** (1987), no. 4, 567–590. MR910579

[66] Ivan Kostov and Nicolas Orantin, *CFT and topological recursion*, J. High Energy Phys. **11** (2010), 056, 30, DOI 10.1007/JHEP11(2010)056. MR2795103

[67] Y.-P. Lee, R. Pandharipande, *Frobenius manifolds, Gromov-Witten theory, and Virasoro constraints*, `http://www.math.utah.edu/~yplee/research/`

[68] P. Mayr, *Mirror symmetry, $N = 1$ superpotentials and tensionless strings on Calabi-Yau four-folds*, Nuclear Phys. B **494** (1997), no. 3, 489–545, DOI 10.1016/S0550-3213(97)00196-X. MR1453771

[69] An-Min Li and Yongbin Ruan, *Symplectic surgery and Gromov-Witten invariants of Calabi-Yau 3-folds*, Invent. Math. **145** (2001), no. 1, 151–218, DOI 10.1007/s002220100146. MR1839289

[70] Jun Li, *Stable morphisms to singular schemes and relative stable morphisms*, J. Differential Geom. **57** (2001), no. 3, 509–578. MR1882667

[71] J. Li, "A degeneration formula of Gromov-Witten invariants," J. Diff. Geom. **60** (2002), 199–293.

[72] Chiu-Chu Melissa Liu, Kefeng Liu, and Jian Zhou, *A proof of a conjecture of Mariño-Vafa on Hodge integrals*, J. Differential Geom. **65** (2003), no. 2, 289–340. MR2058264

[73] Chiu-Chu Melissa Liu, Kefeng Liu, and Jian Zhou, *A formula of two-partition Hodge integrals*, J. Amer. Math. Soc. **20** (2007), no. 1, 149–184, DOI 10.1090/S0894-0347-06-00541-8. MR2257399

[74] Jun Li, Chiu-Chu Melissa Liu, Kefeng Liu, and Jian Zhou, *A mathematical theory of the topological vertex*, Geom. Topol. **13** (2009), no. 1, 527–621, DOI 10.2140/gt.2009.13.527. MR2469524

[75] Bong H. Lian, Kefeng Liu, and Shing-Tung Yau, *Mirror principle. I*, Asian J. Math. **1** (1997), no. 4, 729–763, DOI 10.4310/AJM.1997.v1.n4.a5. MR1621573

[76] Bong H. Lian, Kefeng Liu, and Shing-Tung Yau, *Mirror principle. II*, Asian J. Math. **3** (1999), no. 1, 109–146, DOI 10.4310/AJM.1999.v3.n1.a6. Sir Michael Atiyah: a great mathematician of the twentieth century. MR1701925

[77] Bong H. Lian, Kefeng Liu, and Shing-Tung Yau, *Mirror principle. III*, Asian J. Math. **3** (1999), no. 4, 771–800, DOI 10.4310/AJM.1999.v3.n4.a4. MR1797578

[78] Junho Lee, *Family Gromov-Witten invariants for Kaehler surfaces*, ProQuest LLC, Ann Arbor, MI, 2001. Thesis (Ph.D.)–Michigan State University. MR2702374

[79] Chiu-Chu Melissa Liu, *Localization in Gromov-Witten theory and orbifold Gromov-Witten theory*, Handbook of moduli. Vol. II, Adv. Lect. Math. (ALM), vol. 25, Int. Press, Somerville, MA, 2013, pp. 353–425. MR3184181

[80] Kefeng Liu and Hao Xu, *Recursion formulae of higher Weil-Petersson volumes*, Int. Math. Res. Not. IMRN **5** (2009), 835–859, DOI 10.1093/imrn/rnn148. MR2482127

[81] Marcos Mariño, *Open string amplitudes and large order behavior in topological string theory*, J. High Energy Phys. **3** (2008), 060, 34, DOI 10.1088/1126-6708/2008/03/060. MR2391060

[82] D. Maulik, N. Nekrasov, A. Okounkov, and R. Pandharipande, *Gromov-Witten theory and Donaldson-Thomas theory. I*, Compos. Math. **142** (2006), no. 5, 1263–1285, DOI 10.1112/S0010437X06002302. MR2264664

[83] D. Maulik, A. Oblomkov, A. Okounkov, and R. Pandharipande, *Gromov-Witten/Donaldson-Thomas correspondence for toric 3-folds*, Invent. Math. **186** (2011), no. 2, 435–479, DOI 10.1007/s00222-011-0322-y. MR2845622

[84] D. Maulik and R. Pandharipande, *A topological view of Gromov-Witten theory*, Topology **45** (2006), no. 5, 887–918, DOI 10.1016/j.top.2006.06.002. MR2248516

[85] A. Okounkov and R. Pandharipande, *Hodge integrals and invariants of the unknot*, Geom. Topol. **8** (2004), 675–699, DOI 10.2140/gt.2004.8.675. MR2057777

[86] Shi-Shyr Roan, *On Calabi-Yau orbifolds in weighted projective spaces*, Internat. J. Math. **1** (1990), no. 2, 211–232, DOI 10.1142/S0129167X90000137. MR1060636

[87] Dustin Ross, *Localization and gluing of orbifold amplitudes: the Gromov-Witten orbifold vertex*, Trans. Amer. Math. Soc. **366** (2014), no. 3, 1587–1620, DOI 10.1090/S0002-9947-2013-05835-7. MR3145743

[88] Dustin Ross and Zhengyu Zong, *The gerby Gopakumar-Mariño-Vafa formula*, Geom. Topol. **17** (2013), no. 5, 2935–2976, DOI 10.2140/gt.2013.17.2935. MR3190303

[89] Dustin Ross and Zhengyu Zong, *Cyclic Hodge integrals and loop Schur functions*, Adv. Math. **285** (2015), 1448–1486, DOI 10.1016/j.aim.2015.08.023. MR3406532

[90] Dustin Ross, *On the Gromov-Witten/Donaldson-Thomas correspondence and Ruan's conjecture for Calabi-Yau 3-orbifolds*, Comm. Math. Phys. **340** (2015), no. 2, 851–864, DOI 10.1007/s00220-015-2438-1. MR3397033

[91] Hsian-Hua Tseng, *Orbifold quantum Riemann-Roch, Lefschetz and Serre*, Geom. Topol. **14** (2010), no. 1, 1–81, DOI 10.2140/gt.2010.14.1. MR2578300

[92] Johannes Walcher, *Opening mirror symmetry on the quintic*, Comm. Math. Phys. **276** (2007), no. 3, 671–689, DOI 10.1007/s00220-007-0354-8. MR2350434

[93] Johannes Walcher, *Extended holomorphic anomaly and loop amplitudes in open topological string*, Nuclear Phys. B **817** (2009), no. 3, 167–207, DOI 10.1016/j.nuclphysb.2009.02.006. MR2522663

[94] Johannes Walcher, *Evidence for tadpole cancellation in the topological string*, Commun. Number Theory Phys. **3** (2009), no. 1, 111–172, DOI 10.4310/CNTP.2009.v3.n1.a3. MR2504755

[95] Satoshi Yamaguchi and Shing-Tung Yau, *Topological string partition functions as polynomials*, J. High Energy Phys. **7** (2004), 047, 20, DOI 10.1088/1126-6708/2004/07/047. MR2095047

[96] Eric Zaslow, *Topological orbifold models and quantum cohomology rings*, Comm. Math. Phys. **156** (1993), no. 2, 301–331. MR1233848

[97] J. Zhou, "Local Mirror Symmetry for One-Legged Topological Vertex," `arXiv:0910.4320`; "Local Mirror Symmetry for the Topological Vertex," `arXiv:0911.2343`.

[98] J. Zhou, "Open string invariants and mirror curve of the resolved conifold," arXiv:1001.0447.

[99] Shengmao Zhu, *On a proof of the Bouchard-Sulkowski conjecture*, Math. Res. Lett. **22** (2015), no. 2, 633–643, DOI 10.4310/MRL.2015.v22.n2.a14. MR3342249

[100] Aleksey Zinger, *The reduced genus 1 Gromov-Witten invariants of Calabi-Yau hypersurfaces*, J. Amer. Math. Soc. **22** (2009), no. 3, 691–737, DOI 10.1090/S0894-0347-08-00625-5. MR2505298

[101] Zhengyu Zong, *Generalized Mariño-Vafa formula and local Gromov-Witten theory of orbicurves*, J. Differential Geom. **100** (2015), no. 1, 161–190. MR3326577

[102] Zhengyu Zong, *Equivariant Gromov-Witten Theory of GKM Orbifolds*, ProQuest LLC, Ann Arbor, MI, 2015. Thesis (Ph.D.)–Columbia University. MR3312922

Beijing International Center for Mathematical Research, Peking University, 5 Yiheyuan Road, Beijing 100871, China
Email address: bohanfang@gmail.com

Yau Mathematical Sciences Center, Tsinghua University, Jin Chun Yuan West Building, Tsinghua University, Haidian District, Beijing 100084, China
Email address: zyzong@math.tsinghua.edu.cn

Proceedings of Symposia in Pure Mathematics
Volume **100**, 2018
https://doi.org/10.1090/pspum/100/01762

Double quantization of Seiberg–Witten geometry and W-algebras

Taro Kimura

ABSTRACT. We show that the double quantization of Seiberg–Witten spectral curve for Γ-quiver gauge theory defines the generating current of W(Γ)-algebra in the free field realization. We also show that the partition function is given as a correlator of the corresponding W(Γ)-algebra, which is equivalent to the AGT relation under the gauge/quiver (spectral) duality.

CONTENTS

1. Introduction and summary

Recent progress on non-perturbative aspects of supersymmetric gauge theory exhibits interesting connections with various concepts of mathematical physics. In particular, Seiberg–Witten theory for four-dimensional $\mathcal{N}=2$ gauge theory [**1**, **2**] provides a geometrical interpretation for the Coulomb branch moduli space as a base of the algebraic integrable system [**3**–**6**]. As a result, the Seiberg–Witten curve can be identified as a classical spectral curve of the affine Toda chain for the pure SYM theory, the $SU(2)$ rational spin chain for $\mathcal{N}=2$ SQCD with the fundamental hypermultiplets [**7**].

2010 *Mathematics Subject Classification.* Primary 81R10; Secondary 81T60, 14D21, 81R50.

This work was supported in part by Keio Gijuku Academic Development Funds, JSPS Grant-in-Aid for Scientific Research (No. JP17K18090), the MEXT-Supported Program for the Strategic Research Foundation at Private Universities "Topological Science" (No. S1511006), JSPS Grant-in-Aid for Scientific Research on Innovative Areas "Topological Materials Science" (No. JP15H05855), and "Discrete Geometric Analysis for Materials Design" (No. JP17H06462).

TABLE 1. Gauge theory geometry and IIA/M/F theory realization associated with rational/trigonometric/elliptic integrable systems. See, for example, [**16**].

Gauge theory geometry	Geometric realization	Integrable system
$\mathbb{R}^4$	IIA string on CY3	rational
$\mathbb{R}^4 \times S^1$	M theory on CY3 $\times S^1$	trigonometric
$\mathbb{R}^4 \times T^2$	F theory on CY3 $\times T^2$	elliptic

The relation between gauge theory and integrable system has various generalizations:

Dimensional hierarchy

Replacing the spacetime on which the gauge theory is defined with five and six dimensional manifolds compactified on S^1 and T^2, it corresponds to trigonometric and elliptic deformations of rational integrable system [**8, 9**], as shown in Table 1. The Seiberg–Witten curve for 4d/5d/6d theory coincides with the spectral curve of rational/trigonometric/elliptic integrable system.

Quiver gauge theory

In the relation to integrable system, change of the gauge symmetry doesn't lead to change of underlying symmetry on the integrable system side. For this purpose, we need quiver gauge theory labeled by a quiver graph Γ. The corresponding symmetry group for integrable system turns out to be Lie group G_Γ whose Dynkin diagram is given by Γ, and the Seiberg–Witten geometry is described by the fundamental characters of G_Γ group [**9**].

Ω-background and quantization

The four-dimensional spacetime for gauge theory allows an equivariant deformation with Ω-background [**10, 11**], and its specific limit $(\epsilon_1, \epsilon_2) \to (\hbar, 0)$, called the Nekrasov–Shatashvili (NS) limit, corresponds to quantization of integrable system [**12, 13**]. Then the Seiberg–Witten curve characterized by the algebraic relation is promoted to difference equation, and this quantum spectral curve is identified as Baxter's TQ-relation for G_Γ-spin chain, which is also equivalent to the corresponding Bethe ansatz equation [**14**] described by the polynomial equation called the q-character [**15**] (Table 2).

From this point of view, we still have another deformation parameter: We have used only one of equivariant parameters to quantize the integrable system. Thus it is natural to ask *what is the underlying algebraic structure for generic Ω-background parametrization.* It has been recently shown that the polynomial equation still holds for generic (ϵ_1, ϵ_2) just by replacing the q-character with another algebraic structure, called the qq-character [**17–19**]. See also [**20–28**] for its realization in gauge theory and topological string theory.

TABLE 2. Ω-background and (double) quantization of Seiberg–Witten geometry described by G_Γ-character/q-character/qq-character for Γ-quiver gauge theory.

Ω-background	SW geometry	Integrable system
$(\epsilon_1, \epsilon_2) = (0, 0)$	character	classical
$(\epsilon_1, \epsilon_2) = (\hbar, 0)$	q-character	quantum
$(\epsilon_1, \epsilon_2) \neq (0, 0)$ (w/ t-extention)	qq-character	doubly quantum (W-algebra)

We proceed with this program to construct W-algebras using quiver gauge theory [**29**–**31**]. In our construction, in addition to consider generic equivariant parameters (ϵ_1, ϵ_2), we incorporate the higher time variables, which deform UV prepotential by the local operators [**32**]. Our claim is as follows:

The qq-character together with double quantization of Seiberg–Witten spectral curve for the t-extended Γ-quiver gauge theory defines the generating current of W(Γ)-algebra.

The doubly quantum spectral curve is an operator relation, in contrast to the classical and quantum curves which are relations between correlation functions. This double quantization scheme will be discussed in Sec. 3 in more detail.

Another claim about the partition function is:

The t-extended Γ-quiver partition function is a state (Z-*state*) in the Hilbert space, and the non-t-extended partition function is given by a correlation function of W(Γ)-algebra.

This statement is analogous, but slightly different from the AGT relation, which claims that the partition function of gauge theory enjoying gauge symmetry G is given by a correlation function of W(G)-algebra [**33**, **34**].[1] This reflects that our construction of W(Γ) is equivalent to AGT for W(G) by the gauge/quiver duality $G \leftrightarrow \Gamma$ [**38**–**40**], which is confirmed for A-type quiver (Fig. 1). In the classical limit, this duality becomes the Nahm–Fourier–Mukai duality between the phase space of Hitchin system on a cylinder and the phase space of periodic monopoles [**9**]. We remark a similar realization of W-algebra correlators associated with the quiver structure [**41**–**43**].

Let us summarize our construction of quiver W-algebras [**29**, **30**]. We start with the classical Seiberg–Witten curve given in the limit $(\epsilon_1, \epsilon_2) \to (0, 0)$. For Γ-quiver gauge theory, this spectral curve is given by the fundamental characters of G_Γ group, if the quiver diagram Γ coincides with a Dynkin diagram for finite simple Lie algebras. This construction is also applicable to generic quiver which

[1] Our construction naturally gives rise to the q-deformation of W(Γ)-algebra from 5d quiver gauge theory, while the ordinary AGT relation claims a connection between 4d gauge theory and W(G)-algebra without q-deformation. The counterpart of our construction is thus the q-AGT relation for 5d gauge theory and the q-deformed W(G)-algebra [**35**]. A similar idea to realize the Virasoro/W-algebra through quantization of geometry is found in [**36**, **37**] for non-q-deformed case.

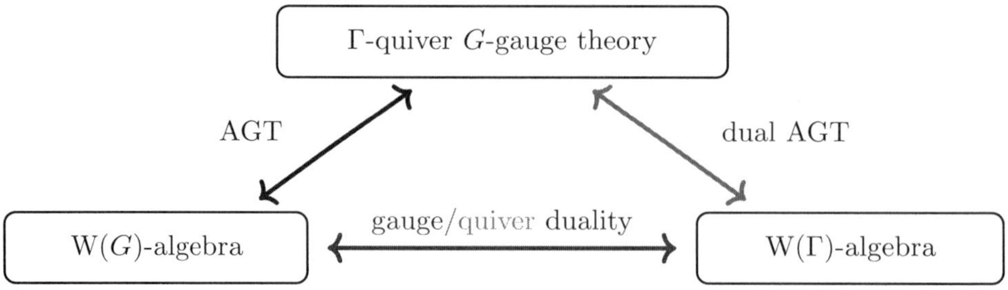

FIGURE 1. AGT and dual AGT

doesn't correspond to any finite Lie algebra. Tuning on the equivariant parameter as $(\epsilon_1, \epsilon_2) \to (\hbar, 0)$, we obtain the quantum Seiberg–Witten curve, which is identified as the TQ-relation of the integrable G_Γ-spin chain. We then consider the double quantization of Seiberg–Witten curve by taking into account the higher time variables with the fully generic equivariant parameters (ϵ_1, ϵ_2). We obtain an operator relation, which is identified as the quantum Sugawara construction of the generating current of W(Γ)-algebra.

This article is a self-contained review based on [**29**, **30**]. Our construction exhibits several new features of W-algebras:

Edge (bifundamental) mass-deformation of W-algebra

In general, quiver gauge theory consists of the vector multiplet assigned to a node of quiver i, and the hypermultiplet in bifundamental representation for an edge e with multiplicative mass parameter μ_e. (We will deal with (anti)fundamental hypermultiplets separately.) We will show that the bifundamental mass gives rise to new mass-deformation of W-algebra, which is reduced to Frenkel–Reshetikhin's construction in the limit $\mu_e \to 1$ [**44**, **45**].

Affine quiver W-algebra

Our construction of W-algebra is applicable to generic quiver gauge theory. Applying this construction to affine quiver theory, e.g., $\mathcal{N} = 2^*$ theory ($\widehat{A}_0$-quiver), we obtain new W-algebra associated to affine Lie algebra. The Seiberg–Witten curve (and its quantization) is described by the affine character, which is given as an infinite series.

Quiver elliptic W-algebra

We can construct elliptically deformed W-algebras using six-dimensional quiver gauge theory on $\mathbb{R}^4 \times T^2$, which corresponds to the elliptic integrable system.

This article contains several new features compared with the original papers [**29**, **30**]: We introduce the A-operator to discuss the iWeyl reflection in Sec. 6. We show the explicit OPE relations for A_2 quiver in Sec. 6.2. We also provide the fundamental qq-characters for D_4 quiver in Sec. 6.4. Furthermore, generalization to fractional quiver, which gives rise to non-simply-laced algebra, is found in [**31**].

The rest of this article is organized as follows. We start in Sec. 2 to discuss the classical Seiberg–Witten geometry and its 1st quantization. The resultant quantum spectral curve obtained in the NS limit is a difference operator, which turns out to be Baxter's TQ-relation, and the Q-operator plays a role of the wavefunction. In Sec. 3 we discuss 2nd quantization of the spectral curve. To obtain an operator relation, we introduce the higher time variable as a Heisenberg oscillator, leading to modification of the UV prepotential. We then show in Sec. 4 how the generating current of q-deformed Virasoro algebra arises in 5d gauge theory. The construction using the qq-character yields the free field realization of the current. In Sec. 5 we show that the t-extended gauge theory partition function is given by a state, which we call the Z-state, in the Hilbert space through the operator/state correspondence. In particular, the Z-state is constructed with the screening charge of the corresponding conformal algebra. We also show that the (anti)fundamental matter contribution is realized by a vertex operator applied to the state. In Sec. 6 we show how to construct W(Γ)-algebra associated with generic quiver Γ from the mass-deformed q-Cartan matrix characterizing the quiver gauge theory. We discuss the regularity of the qq-character which provides the generating current, and show it commutes with the screening charge of W(Γ)-algebra. We demonstrate our construction with several examples reproducing known results in the limit. In Sec. 7 our construction is applied to affine quiver theory, involving the adjoint matter, corresponding to 4d $\mathcal{N} = 2^*$ (5d $\mathcal{N} = 1^*$) theory. This quiver doesn't correspond to finite Lie algebra, and thus provides a new W-algebra such that the generating current is given by an infinite sum of the fundamental current. In Sec. 8 we discuss elliptic generalization of quiver W-algebra using 6d gauge theory defined on $\mathbb{R}^4 \times T^2$. The whole construction discussed above is straightforwardly lifted, and thus we obtain elliptically deformed W-algebras associated with generic quiver.

2. Seiberg–Witten spectral curve and 1st quantization

The low-energy effective dynamics of 4d $\mathcal{N} = 2$ gauge theory is completely determined by the algebraic curve called the Seiberg–Witten curve [**1, 2**]. For $U(n)$ SYM theory it is characterized by the algebraic relation

$$y(x) + \frac{1}{y(x)} = T_n(x) \tag{2.1}$$

where $T_n(x)$ is a degree n monic polynomial in x^{-1} variable,[2] and the vev of Coulomb moduli parameter is encoded into its coefficients.[3] This algebraic relation can be obtained from the saddle point analysis of gauge theory partition function [**9, 11, 46**], and its derivation is quite analogous to the large N analysis of matrix models: y-variable is realized as a one-point function of a generating function of gauge invariant observables (the Y-operator) [**14**]

$$y(x) = \Big\langle \mathsf{Y}(x) \Big\rangle , \tag{2.3}$$

[2]While we usually use the convention such that $T_n(x)$ is a polynomial in x, we apply the opposite convention in this article for latter convenience.

[3]In this article we basically use the 5d notation:

$$q_1 = e^{\epsilon_1}, \qquad q_2 = e^{\epsilon_2}, \qquad q = q_1 q_2, \qquad \{\nu_{i,n}\}_{i\in\Gamma_0,\alpha\in[1,\dots n_i]} = \{e^{a_{i,n}}\}_{i\in\Gamma_0,\alpha\in[1,\dots n_i]}. \tag{2.2}$$

These parameters correspond to the standard notation of q-deformed W-algebras: $(q_1, q_2) = (t^{-1}, q)$.

which plays a similar role to the resolvent function having a singularity in the complex plane. Incorporating the one-form differential defined on the curve, called the Seiberg–Witten differential,[4]

$$\lambda = \begin{cases} (4d) & x\, d\log y \qquad (x,y) \in \mathbb{C}\times\mathbb{C}^\times \\ (5d) & \log x\, d\log y \qquad (x,y)\in\mathbb{C}^\times\times\mathbb{C}^\times \end{cases}, \tag{2.4}$$

they define the Seiberg–Witten geometry describing the SUSY vacuum structure. We remark that the curve given by (2.1) coincides with the classical spectral curve of the affine n-site Toda chain, and also the inhomogeneous limit of $SU(2)$ rational/trigonometric/elliptic spin chain for 4d/5d/6d gauge theory.

Although the spectral curve is obtained in the classical limit $\epsilon_1, \epsilon_2 \to 0$ ($q_1, q_2 \to 1$), a similar structure can be found even after turning on the equivariant parameters. Taking into account one of the equivariant parameters $(\epsilon_1, \epsilon_2) \to (\hbar, 0)$, which is called the Nekrasov–Shatashvili limit, we obtain a difference equation instead of the algebraic relation for 5d theory [**13, 14**][5]

$$y(x) + \frac{1}{y(q_1^{-1}x)} = T_n(x)\,. \tag{2.5}$$

This is again obtained by the saddle point analysis with respect to $\epsilon_2 \to 0$, and the other finite parameter $\epsilon_1 = \hbar$ plays a role of the Planck constant. Indeed the difference equation (2.5) is interpreted as a quantization of the spectral curve (2.1) with respect to a canonical pair $(\log x, \log y)$,

$$[\log y, \log x] = \hbar\,, \tag{2.6}$$

which corresponds to the one-form (2.4) or the associated symplectic two-form

$$\Omega = d\log x \wedge d\log y\,. \tag{2.7}$$

This quantization implies the y-variable behaves as a shift operator $y = \exp(\hbar\partial_{\log x})$, and thus the classical curve (2.1) becomes a difference operator

$$\left[e^{\hbar\partial_{\log x}} + e^{-\hbar\partial_{\log x}} - T_n(x)\right] Q(x) = 0\,, \tag{2.8}$$

which is equivalent to (2.5) under identification[6]

$$y(x) = \frac{Q(q_1 x)}{Q(x)}\,. \tag{2.9}$$

In this way, quantization of spectral curve gives a differential/difference equation. In particular, in this case, the quantum curve (2.8) is equivalent to Baxter's TQ-relation for $SU(2)$ spin chain,[7] which is a consequence of the connection between gauge theory and integrable system at the quantum level.

---

[4]The higher-dimensional lift is equivalent to imposing periodicity: The cylindrical coordinate for 5d theory on $\mathbb{R}^4\times S^1$ is given by $\mathbb{C}^\times = \mathbb{C}/\mathbb{Z}$. The geometry for 6d gauge theory compactified on a torus $\mathbb{R}^4 \times T^2$ is parametrized by a coordinate $(x,y)\in E_{T^2}\times\mathbb{C}^\times$ where the elliptic curve E_{T^2} is doubly periodic, $E_{T^2} = \mathbb{C}/(\mathbb{Z}+\tau\mathbb{Z})$ with the torus modulus τ.

[5]This is also called the q-character due to its representation theoretical origin in quantum affine algebra [**15**].

[6]This transformation is analogous to that for the Riccati-type differential equation, widely used to derive quantum spectral curves.

[7]To obtain a precise agreement, we have to take into account the Chern–Simons factor and the (anti)fundamental hypermultiplets. See [**13, 14**] for more arguments.

In addition, it has been recently revealed that a similar polynomial relation holds even away from the saddle point analysis, which is called the qq-character equation [**17, 20, 23, 25, 26**]

$$y(x) + \frac{1}{y(q^{-1}x)} = T_n(x)\,. \tag{2.10}$$

Precisely speaking, the second factor should be $y(q^{-1}x)^{-1} \to \langle \mathsf{Y}(q^{-1}x)^{-1} \rangle$ as discussed below. By the analogy with the matrix model and other related theories, this is quite natural for quantum curve. For example, in the case of the matrix model, the characteristic polynomial average plays a similar role to the wavefunction $Q(x)$ in (2.8), and actually satisfies the quantum curve equation at finite N under identification $\hbar = 1/N$. In general, quantization scheme is a way to go back to the finite coupling regime from the classical limit. We remark that, in the unrefined limit $q_1 q_2 = q = 1$ ($\epsilon_1 + \epsilon_2 = 0$), it is not distinguishable from the classical relation (2.1), since the summation plays a role of the Planck constant in this case, $\epsilon_1 + \epsilon_2 = \hbar$. However, when we consider the higher-weight character, an extra factor has to be taken into account as discussed later.

3. Operator formalism and 2nd quantization

As mentioned above, the y-variable plays a similar role to the resolvent function, and thus the classical spectral curve (2.1) and its quantizations (2.5) and (2.10) yield relations between (one-point) correlation functions in gauge theory. To emphasize it, let us rewrite the qq-character with the Y-operator

$$\Big\langle \mathsf{Y}(x) \Big\rangle + \Big\langle \frac{1}{\mathsf{Y}(q^{-1}x)} \Big\rangle = T_n(x)\,. \tag{3.1}$$

Now the degree of $T_n(x)$ depends on the gauge group rank. However, the meaning of average $\langle\, \mathcal{O}(x) \,\rangle$ actually depends on the theory that we consider (gauge group, matter content, and so on). From this point of view, it is natural to define T-operator whose average with $U(n)$ SYM theory is given by

$$\Big\langle T(x) \Big\rangle = T_n(x)\,. \tag{3.2}$$

Then we obtain a universal operator relation

$$\mathsf{Y}(x) + \frac{1}{\mathsf{Y}(q^{-1}x)} = T(x)\,. \tag{3.3}$$

This is the 2nd quantization of Seiberg–Witten spectral curve in our sense, and this equation defines the T-operator in terms of the Y-operator. We remark this operator relation is independent of gauge group rank, in contrast to the polynomial relation. In this sense, it can be seen as a universal relation for gauge theory.

Precisely speaking, in order to obtain a proper operator relation, we need to take into account the higher time variables, namely the potential term in gauge theory

$$Z_{\text{pot}}(t) = \exp\left(\sum_{n=1}^{\infty} t_n \mathbf{Y}^{[n]} \right), \tag{3.4}$$

so that the t-extended partition function plays a role of generating function of the gauge theory observable $\mathbf{Y}^{[n]}$, which is a trace of the n-th power adjoint scalar $\operatorname{Tr} \Phi^n$ for 4d, or its loop operator along a compactified circle for 5d theory. Inclusion of the potential term corresponds to deformation of the UV prepotential with the holomorphic operators [**32**]

$$\mathcal{F}_{\mathrm{UV}} \longrightarrow \mathcal{F}_{\mathrm{UV}} + \sum_{n=1}^{\infty} t_n \operatorname{Tr} \Phi^n \, . \tag{3.5}$$

Then since the situation is completely parallel with the matrix model, let us explain the reason why we need the potential term (3.4) by the analogy. The resolvent operator of matrix model is a generating function of the gauge invariant observable, which is a single trace operator,

$$\operatorname{Tr} \frac{1}{x - X} = \sum_{n=0}^{\infty} x^{-n-1} \operatorname{Tr} X^n \, . \tag{3.6}$$

This observable is equivalent to derivative with respect to the coupling constant, playing a role of the time variable, in the potential term

$$\frac{\partial}{\partial t_n} e^{\operatorname{Tr} V(X)} = \operatorname{Tr} X^n \, e^{\operatorname{Tr} V(X)} \tag{3.7}$$

where

$$V(x) = \sum_{n=1}^{\infty} t_n \, x^n \, . \tag{3.8}$$

In this sense the single trace observable is replaced with the derivative with t_n under the presence of the potential term.

4. From double quantization to Virasoro/W-algebra

In terms of the gauge theory observable, the Y-operator (in particular, its average) is obtained [**9, 14**]

$$\Big\langle \mathsf{Y}(x) \Big\rangle = \exp\left(- \sum_{n=1}^{\infty} \frac{x^{-n}}{n} \mathbf{Y}^{[n]} \right) . \tag{4.1}$$

Taking into account the higher time variables, the Y-operator shows t-dependence, which is written in terms of (q_1, q_2)-modified free field operators [**29**]

$$\mathsf{Y}(x) = q_1^{\frac{1}{2}} : \exp\left(y_0 + \sum_{n \neq 0} y_n \, x^{-n} \right) : \tag{4.2}$$

where the free fields are defined

(4.3)

$$(n > 0) \qquad y_{-n} = \frac{(1 - q_1^n)(1 - q_2^n)}{1 + q^n} t_n \, , \qquad y_n = - \frac{1}{n} \frac{\partial}{\partial t_n} \, , \qquad y_0 = - \frac{1}{2} t_0 \log q_2 \, .$$

A constant and zero mode factors are added for convenience. According to the operator relation (3.3), it defines the T-operator in the free field realization, and it turns out to coincide with the generating current of q-deformed Virasoro algebra [**44, 47**]. Indeed the generator appearing in the mode expansion

$$T(x) = \sum_{n \in \mathbb{Z}} T_n \, x^{-n} \tag{4.4}$$

satisfies the algebraic relation

(4.5)
$$[T_n, T_m] = -\sum_{k=1}^{\infty} f_k \left(T_{n-k}T_{m+k} - T_{m-k}T_{n+k}\right) - \frac{(1-q_1)(1-q_2)}{1-q}\left(q^n - q^{-n}\right)\delta_{n+m,0}$$

where

$$\sum_{k=0}^{\infty} f_k\, x^k = \exp\left(\sum_{n=1}^{\infty} \frac{1}{n}\frac{(1-q_1^n)(1-q_2^n)}{1+q^n} x^n\right) =: f(x)\,. \tag{4.6}$$

In practice, it is convenient to see the q-analog of OPE from the two-point function of the generating current, which is equivalent to the algebraic relation (4.5),

(4.7)
$$f\left(\frac{y}{x}\right)T(x)T(y) - f\left(\frac{x}{y}\right)T(y)T(x) = -\frac{(1-q_1)(1-q_2)}{1-q}\left(\delta\left(\frac{qy}{x}\right) - \delta\left(\frac{qx}{y}\right)\right)$$

where the delta function is defined

$$\delta(x) = \sum_{n\in\mathbb{Z}} x^n\,. \tag{4.8}$$

This shows that double quantization of Seiberg–Witten spectral curve provides the generating current of q-Virasoro algebra in the free field realization. We remark that such an idea on the realization of generating current through the quantization of Seiberg–Witten curve was already discussed in the context of the (non q-deformed) AGT relation [**33**].

What we should remark is that the conformal algebra obtained here using the gauge theory does not depend on the gauge group rank: we obtain (q-deformed) Virasoro algebra for $U(n)$ gauge theory for arbitrary $n \in \mathbb{Z}_{\geq 1}$. This situation is actually different from the AGT relation where the underlying conformal algebra is associated to the gauge group.

5. Z-state

Let us see how the gauge theory partition function is interpreted in our construction. In particular, the partition function extended with the time variables (3.4) explicitly depends on them. Since the time variable behaves as an operator, the t-extended partition function $Z(t)$ has to be identified as an operator. This interpretation directly leads to the notion of Z-state through the operator/state correspondence of CFT.

We show that the t-extended full partition function (including both one-loop and instanton contributions) for $U(n)$ SYM theory is a state given by [**29**]

$$|Z\rangle = \sum_{\mathcal{X}\in\mathfrak{M}^{\mathsf{T}}} \prod_{x\in\mathcal{X}}^{\succ} S(x)\,|1\rangle \tag{5.1}$$

where the product is a radial-ordered product, and $\mathfrak{M}^{\mathsf{T}}$ is the torus action fixed point in the instanton moduli space, characterized by a set of partitions [**11**] with

$$\mathcal{X} = \{x_{\alpha,k}\}_{\alpha\in[1\ldots n],k\in[1\ldots\infty]}\,, \qquad x_{\alpha,k} = \nu_\alpha q_1^{k-1} q_2^{\lambda_{\alpha,k}}\,. \tag{5.2}$$

The vacuum state is denoted by $|1\rangle$, which is a constant with respect to the time variables, $\partial_{t_n}|1\rangle = 0$, namely a primary state, and $S(x)$ is a screening current operator

$$(5.3)\qquad S(x) = :\exp\left(s_0 \log x + \tilde{s}_0 + \frac{\kappa}{2}\left(\log_{q_2}^2 x - \log_{q_2} x\right) + \sum_{n\neq 0} s_n x^{-n}\right): .$$

As well as the Y-operator, they are (q_1, q_2)-deformed free field operators

$$(5.4)\qquad (n>0) \qquad s_{-n} = (1-q_1^n)t_n\,, \qquad s_n = -\frac{1+q^{-n}}{n(1-q_2^{-n})}\frac{\partial}{\partial t_n}\,, \qquad s_0 = t_0$$

obeying the commutation relation

$$(5.5)\qquad \left[s_n, s_m\right] = -\frac{1}{n}\frac{1-q_1^n}{1-q_2^{-n}}(1+q^{-n})\,\delta_{n+m,0}\,.$$

The zero mode t_0 corresponds to the gauge coupling $t_0 = \log_{q_2}\mathfrak{q}$, but we need some redefinition to obtain precise agreement with the gauge theory result. The additional zero mode satisfies

$$(5.6)\qquad \left[\tilde{s}_0, s_n\right] = -2\beta\delta_{n,0}$$

where $\beta = -\epsilon_1/\epsilon_2$, and the κ-term is added to reproduce the Chern–Simons term in gauge theory. See [**29**] for details.

The Z-state can be also written using the screening charge, which commutes with the generating current $T(x)$. First of all, the summation over the fixed points $\mathfrak{M}^{\mathsf{T}}$ in the partition function can be replaced with $\mathbb{Z}^{\mathcal{X}_0}$, which is a set of arbitrary integer sequences terminating by zeros. $\mathcal{X}_0$ is the ground configuration $\lambda_{\alpha,k} = 0$ for $k \in \mathbb{Z}_{>0}$, and $\mathring{x}_{\alpha,k} = \nu_\alpha q_1^{k-1}$. If the partition $\lambda_{\alpha,k}$ is not a non-increasing sequence, there appears a zero factor in the expression

$$(5.7)\qquad \prod_{x\in\mathcal{X}}^{\succ} S(x)\,|1\rangle = 0$$

for $\mathcal{X} \in \mathbb{Z}^{\mathcal{X}_0}$ but $\mathcal{X} \notin \mathfrak{M}^{\mathsf{T}}$. Therefore the Z-state is given by

$$(5.8)\qquad |Z\rangle = \prod_{\mathring{x}\in\mathcal{X}_0}^{\succ} \mathsf{S}(\mathring{x})\,|1\rangle$$

where we define the screening charge[8]

$$(5.10)\qquad \mathsf{S}(\mathring{x}) = \sum_{k\in\mathbb{Z}} S(q_2^k \mathring{x})\,.$$

For q-deformed Virasoro/W-algebra, the screening charge is a discrete sum of the screening current. Precisely speaking, this sum is given by the Jackson integral to obtain a proper convergence property.

[8]This infinite sum can be written as the Jackson integral with respect to q_2,

$$(5.9)\qquad \mathsf{S}(x) = \int dx_{q_2} S(x)\,,$$

which is more convenient to discuss its convergence property. In the limit $q_2 \to 1$, the discrete sum is replaced with the usual integral.

Let us mention an important property of the screening charge. The OPE between the Y-operator and the screening current is given by

(5.11)
$$\mathsf{Y}(x)S(x') = \frac{1 - x'/x}{1 - q_1 x'/x} : \mathsf{Y}(x)S(x') : , \quad S(x')\mathsf{Y}(x) = q_1^{-1} \frac{1 - x/x'}{1 - q_1^{-1} x/x'} : \mathsf{Y}(x)S(x') :$$

and thus the commutation relation shows

$$[\mathsf{Y}(x), S(x')] = (1 - q_1^{-1})\, \delta\left(q_1 \frac{x'}{x}\right) : \mathsf{Y}(x)S(x') : \tag{5.12}$$

where the delta function defined (4.8). Similarly we obtain

$$\left[\mathsf{Y}(q^{-1}x)^{-1}, S(x')\right] = -(1 - q_1^{-1})\, \delta\left(q \frac{x'}{x}\right) : \mathsf{Y}(x)S(x') : \tag{5.13}$$

where we used the formula

$$q_1^{-1} : \mathsf{Y}(x)\mathsf{Y}(q^{-1}x) : = : S(q^{-1}x)S(q_1^{-1}x)^{-1} : \tag{5.14}$$

obtained by comparing the y and s fields, (4.3) and (5.4). We have almost the same OPEs for Y and Y^{-1} except for the q_2-shift. This discrepancy is canceled in the screening charge because it is defined as a summation over the q_2-shifted screening currents. Therefore the T-operator turns out to commute with the screening charge

$$\left[T(x), \mathsf{S}(x')\right] = 0\,. \tag{5.15}$$

This assures the regularity of the T-operator, and the mode expansion (4.4). This commutativity also corresponds to the pole cancellation mechanism in the qq-character.

In order to obtain the non-extended plain partition function, we apply a dual vacuum state $\langle 1|$ because it plays a role of the projector into the $t = 0$ sector, $\langle 1|\, t_n = 0$. Thus the plain partition function is given by a correlator

$$Z(t=0) = \langle 1|Z\rangle = \langle 1| \prod_{\mathring{x} \in \mathcal{X}_0}^{\succ} \mathsf{S}(\mathring{x})\, |1\rangle\,. \tag{5.16}$$

See also [**41**–**43**] for a similar construction using little string theory.[9] This expression resembles the AGT relation [**33**], which is the coincidence between the gauge theory partition function and the Liouville/Toda CFT correlator. In the case of the AGT relation, we need to subtract the $U(1)$ factor to obtain agreement between the instanton partition function and the conformal block. In this case, on the other hand, we don't need to care about such an extra factor. We can directly see the agreement. In addition, as mentioned before, the underlying conformal algebra does not depend on the gauge group. We will come back to this issue later.

So far we have focused on the vector multiplet contribution. We can also incorporate the fundamental hypermultiplet in the operator formalism. Since the

[9] In these papers, the authors discuss a correlator which consists of finite screening currents by considering the defect operators.

fixed point contribution to the gauge theory observable is given by[10]

$$\mathbf{Y}^{[n]} = (1-q_1^n) \sum_{x\in\mathcal{X}} x^n \,, \tag{5.18}$$

the potential term (3.4) yields the fundamental matter contribution by the shift of time variables

$$t_n \longrightarrow t_n + \frac{1}{n} \frac{1}{(1-q_1^n)(1-q_2^n)} \mu^{-n} \tag{5.19}$$

where $\mu \in \mathbb{C}^\times$ is the multiplicative mass parameter. We introduce an operator which induces the time shift

$$\mathsf{V}(x) = : \exp\left(\sum_{n\neq 0} v_n \, x^{-n} \right) : \tag{5.20}$$

where

$$(n>0) \qquad v_{-n} = -\frac{1}{1+q^n} t_n \,, \qquad v_n = \frac{1}{n} \frac{1}{(1-q_1^n)(1-q_2^n)} \frac{\partial}{\partial t_n} \,. \tag{5.21}$$

We compute the OPE of V and S operators

(5.22)

$$\mathsf{V}(x) S(x') = \left(\frac{x'}{x}; q_2 \right)_\infty^{-1} : \mathsf{V}(x) S(x') : \,, \qquad S(x') \mathsf{V}(x) = \left(\frac{q_2 x}{x'}; q_2 \right)_\infty : \mathsf{V}(x) S(x') : \,.$$

They correspond to the fundamental and antifundamental hypermultiplet contributions, while the OPE of V and V does not yield dynamical contribution. Thus the extended partition function in the presence of (anti)fundamental matters is obtained by inserting the V-operators

$$|Z\rangle = \left(\prod_{x\in\mathcal{X}_\mathrm{f}} \mathsf{V}(x) \right) \left(\prod_{\mathring{x}\in\mathcal{X}_0}^{\succ} \mathsf{S}(\mathring{x}) \right) \left(\prod_{x\in\tilde{\mathcal{X}}_\mathrm{f}} \mathsf{V}(x) \right) |1\rangle \tag{5.23}$$

where $\mathcal{X}_\mathrm{f} = \{\mu_f\}_{f\in[1\ldots n^\mathrm{f}]}$ and $\tilde{\mathcal{X}}_\mathrm{f} = \{\tilde{\mu}_f\}_{f\in[1\ldots \tilde{n}^\mathrm{f}]}$ are sets of fundamental and antifundamental mass parameters. This V-operator puts a singularity on the spectral curve at $x = \mu_f$. Then the plain partition function is given by a correlator with additional vertex operators

$$Z(t=0) = \langle 1| \left(\prod_{x\in\mathcal{X}_\mathrm{f}} \mathsf{V}(x) \right) \left(\prod_{\mathring{x}\in\mathcal{X}_0}^{\succ} \mathsf{S}(\mathring{x}) \right) \left(\prod_{x\in\tilde{\mathcal{X}}_\mathrm{f}} \mathsf{V}(x) \right) |1\rangle \,. \tag{5.24}$$

This correlator is given by an infinite discrete sum, which can be interpreted as a q-analog of Dotsenko–Fateev's integral formula of the conformal block [**48, 49**].

[10]This expression can be interpreted as a q-deformation of n-th Casimir element

$$C_n = \frac{1}{n} \sum_{k=1}^{\infty} \left(\left(\lambda_k - k + \frac{1}{2} \right)^n - \left(-k + \frac{1}{2} \right)^n \right) . \tag{5.17}$$

6. Quiver W-algebra

The construction shown above is applicable to generic quiver gauge theory, and we correspondingly obtain the W-algebra associated to the quiver structure. Let us first fix the notations. Let Γ be a quiver with the set of nodes Γ_0 and the set of edges Γ_1. A quiver Γ defines a $|\Gamma_0| \times |\Gamma_0|$ matrix

$$c_{ij}^{[n]} = (1 + q^{-n})\delta_{ij} - \sum_{e:i\to j} \mu_e^{-n} - \sum_{e:j\to i} \mu_e^{n} q^{-n} \tag{6.1}$$

where $\mu_e \in \mathbb{C}^\times$ is the multiplicative bifundamental mass parameter associated to each edge $e \in \Gamma_1$. This is (q, μ_e)-deformation of Cartan matrix, reduced to the ordinary Cartan matrix in the limit $n \to 0$. In the analogy with matrix model, the quiver theory with gauge group $\times_{i\in\Gamma_0} U(n_i)$ corresponds to the ADE multi-matrix model [**50, 51**], so that we have $|\Gamma_0|$ sets of *eigenvalues*

$$\mathcal{X}_i = \{x_{i,\alpha,k}\}_{\alpha\in[1\ldots \mathsf{n}_i],k\in[1\ldots\infty]}, \qquad \mathcal{X} = \bigsqcup_{i\in\Gamma_0} \mathcal{X}_i\,. \tag{6.2}$$

Let i: $\mathcal{X} \to \Gamma_0$ be the node label so that $\mathsf{i}(x) = i$ for $x \in \mathcal{X}_i$.

Then the construction is quite parallel with the previous situation: The t-extended partition function is given by a state

$$|Z\rangle = \prod_{\mathring{x}\in\mathcal{X}_0}^{\succ} \mathsf{S}_{\mathsf{i}(\mathring{x})}(\mathring{x})\,|1\rangle\,. \tag{6.3}$$

The screening charge is defined as a summation over the screening current defined

$$S_i(x) = :\exp\left(s_{i,0}\log x + \tilde{s}_{i,0} + \frac{\kappa_i}{2}\left(\log_{q_2}^2 x - \log_{q_2} x\right) + \sum_{n\neq 0} s_{i,n}\, x^{-n}\right): . \tag{6.4}$$

where the (q_1, q_2) modified free fields are written using the deformed Cartan matrix

(6.5)

$$(n>0) \qquad s_{i,-n} = (1 - q_1^n)t_{i,n}\,, \qquad s_{i,n} = -\frac{1}{n(1-q_2^{-n})}\, c_{ji}^{[n]} \frac{\partial}{\partial t_{j,n}}\,, \qquad s_{i,0} = t_{i,0}$$

which obey the commutation relation

$$\left[s_{i,n}, s_{j,m}\right] = -\frac{1}{n}\frac{1-q_1^n}{1-q_2^{-n}}\, c_{ji}^{[n]}\, \delta_{n+m,0}\,. \tag{6.6}$$

Again the zero mode is related to the gauge coupling $t_{i,0} = \log_{q_2} \mathfrak{q}_i$, under some redefinition. The additional zero mode similarly satisfies

$$\left[\tilde{s}_{i,0}, s_{j,n}\right] = -\beta\,\delta_{n,0}\, c_{ji}^{[0]}\,. \tag{6.7}$$

We can show that the Z-state (6.3) coincides wit the t-extended gauge theory partition function [**29**].

Let us then discuss how the generating current is constructed using the Y-operator. For quiver gauge theory, the Y-operator is defined for each node $i \in \Gamma_0$, corresponding to the fundamental weight,

$$\mathsf{Y}_i(x) = q_1^{\tilde{\rho}_i} :\exp\left(y_{i,0} + \sum_{n\neq 0} y_{i,n}\, x^{-n}\right): \tag{6.8}$$

where $\tilde{\rho}_i = \sum_{j \in \Gamma_0} \tilde{c}^{[0]}_{ji}$ is the Weyl vector for non-affine quiver, and $\tilde{\rho}_i = 0$ for affine quiver. The modified free fields are given by[11]

$$(6.9) \quad (n > 0) \quad y_{i,-n} = (1 - q_1^n)(1 - q_2^n)\, \tilde{c}^{[-n]}_{ji} t_{j,n}\,, \qquad y_{i,n} = -\frac{1}{n}\frac{\partial}{\partial t_{i,n}}\,, \qquad y_{i,0} = -\tilde{c}^{[0]}_{ji} t_{j,0} \log q_2$$

with the commutation relation

$$\begin{aligned} \Big[y_{i,n}, y_{j,m} \Big] &= -\frac{1}{n}(1 - q_1^n)(1 - q_2^n)\tilde{c}^{[-n]}_{ij} \delta_{n+m,0} \\ &= -\frac{1}{n}(1 - q_1^{-n})(1 - q_2^{-n})\tilde{c}^{[n]}_{ji} \delta_{n+m,0}\,. \end{aligned} \tag{6.10}$$

In this case, we can define the T-operator which starts with each operator $\mathsf{Y}_i(x)$ for $i \in \Gamma_0$. It is generated by the local reflection, which is the mass deformed iWeyl reflection [**17**][12]

$$(6.11) \quad \mathsf{Y}_i(x) \longrightarrow \mathsf{Y}_i(x) \times \left(\frac{1}{\mathsf{Y}_i(x)\mathsf{Y}_i(q^{-1}x)} \prod_{e:i\to j} \mathsf{Y}_j(\mu_e^{-1}x) \prod_{e:j\to i} \mathsf{Y}_j(\mu_e q^{-1}x) \right)$$

where we assume that there is no self-connecting edge in the quiver. See Sec. 7 for $\widehat{A}_0$ quiver. This reflection formula is obtained by comparing configurations at the fixed points which differ by one instanton factor. This means that we shall see the behavior under the partition shift $\lambda_{i,\alpha,k} \to \lambda_{i,\alpha,k} + 1$, which is equivalent to $x \to q_2 x$ for $x \in \mathcal{X}_i$. Although the Y-operator itself has a singularity, it is cancelled by the reflection. Then we obtain

$$(6.12) \quad T_i(x) = \mathsf{Y}_i(x) + :\mathsf{Y}_i(q^{-1}x)^{-1} \prod_{e:i\to j} \mathsf{Y}_j(\mu_e^{-1}x) \prod_{e:j\to i} \mathsf{Y}_j(\mu_e q^{-1}x) : + \cdots\,.$$

If the reflection gives rise to another singularity, we apply another reflection to cancel it. For the finite-type quiver, this procedure closes within a finite time.

Since the Z-state (6.3) is constructed by the screening charges, we consider the shift of screening current,

$$(6.13) \quad : \frac{S_i(x)}{S_i(q_2 x)} : = q_1^{-1} \mathsf{A}_i(x)$$

where the A-operator is defined

$$(6.14) \quad \mathsf{A}_i(x) = q_1 : \exp\left(a_{i,0} + \sum_{n\neq 0} a_{i,n}\, x^{-n} \right) :$$

with the free field operators

$$(6.15) \quad a_{i,n} = (1 - q_2^{-n}) s_{i,n}\,, \qquad a_{i,0} = -t_{i,0} \log q_2\,.$$

[11]For the quiver whose Cartan matrix is not invertible $\det \mathsf{c}^{[0]} = 0$, we cannot incorporate the coupling constant into the zero mode $t_{i,0} = \log_{q_2} \mathfrak{q}_i$. One needs to put the $\mathfrak{q}_i$ factor for every iWeyl reflection by hand.

[12]In the classical limit $q_1, q_2, \mu_e \to 1$, this reflection is reduced to the ordinary Weyl reflection, and thus the classical Seiberg–Witten geometry is expressed using the character of group G_Γ, associated to Dynkin-quiver diagram Γ [**9**].

These A and Y operators correspond to "root" and "fundamental weight" respectively [**45**]. Thus the oscillators are related as follows:

$$a_{i,n} = y_{j,n}\, c_{ji}^{[n]}\,, \tag{6.16}$$

or explicitly,

$$\mathsf{A}_i(x) = :\frac{\mathsf{Y}_i(x)\mathsf{Y}_i(qx)}{\prod_{e:i\to j}\mathsf{Y}_j(\mu_e^{-1}qx)\prod_{e:j\to i}\mathsf{Y}_j(\mu_e x)}: . \tag{6.17}$$

This leads to the pole cancellation in the iWeyl reflection

$$\operatorname*{Res}_{x'\to q^{-1}x}\Big[\mathsf{Y}_i(x)S_i(q_2x')\Big] + \operatorname*{Res}_{x'\to q^{-1}x}\Big[:\mathsf{Y}_i(x)\mathsf{A}_i(q^{-1}x)^{-1}:S_i(x')\Big] = 0\,. \tag{6.18}$$

Similarly we can show that the T-operator commutes with the screening charge defined as a summation over the screening operator (5.10),

$$\Big[T_i(x), \mathsf{S}_j(x')\Big] = 0\,. \tag{6.19}$$

This shows that the T-operator yields the generating current of $\mathrm{W}(\Gamma)$-algebra associated to Γ-quiver in the free field realization.

We remark that when there appears a product of the Y-operators from the same node $i \in \Gamma_0$ in the reflection (6.12), one needs an extra structure

$$\begin{aligned} :\mathsf{Y}_i(x)\mathsf{Y}_i(x'): + \mathcal{S}\left(\frac{x'}{x}\right) :\frac{\mathsf{Y}_i(x)\mathsf{Y}_i(x')}{\mathsf{A}_i(q^{-1}x)}: + \mathcal{S}\left(\frac{x}{x'}\right) :\frac{\mathsf{Y}_i(x)\mathsf{Y}_i(x')}{\mathsf{A}_i(q^{-1}x')}: \\ + :\frac{\mathsf{Y}_i(x)\mathsf{Y}_i(x')}{\mathsf{A}_i(q^{-1}x)\mathsf{A}_i(q^{-1}x')}: \end{aligned} \tag{6.20}$$

where the factor corresponding to the OPE between Y and A operators defined

$$\mathcal{S}(x) = \frac{(1-q_1x)(1-q_2x)}{(1-qx)(1-x)}\,. \tag{6.21}$$

Furthermore, in the collision limit $x' \to x$, it involves a derivative term

$$\begin{aligned} :\mathsf{Y}_i(x)^2: + :\left(\mathfrak{c}(q_1,q_2) - \frac{(1-q_1)(1-q_2)}{1-q}\partial_{\log x}\log \mathsf{A}_i(q^{-1}x)\right)\frac{\mathsf{Y}_i(x)^2}{\mathsf{A}_i(q^{-1}x)}: \\ + :\frac{\mathsf{Y}_i(x)^2}{\mathsf{A}_i(q^{-1}x)^2}: \end{aligned} \tag{6.22}$$

where the constant defined

$$\mathfrak{c}(q_1,q_2) = \lim_{x\to 1}\left(\mathcal{S}(x) + \mathcal{S}(x^{-1})\right) = \frac{1-6q+q^2+(q_1+q_2)(1+q)}{(1-q)^2}\,. \tag{6.23}$$

In the NS limit $q_2 \to 1$, the derivative term vanishes, and the factor becomes $\mathfrak{c}(q_1,1) = 2$ with $\mathcal{S}(x) \to 1$. Thus the reflection structure becomes simple. In general, one can consider the product of more than two Y-operators. In that case, we obtain correspondingly higher derivative terms.

We notice that the free fields (6.9) explicitly depend on the bifundamental mass parameter, which may affect the algebraic relation of W-algebra. Since our construction coincides with the definition of W-algebra given by [**45**] in the mass less limit $\mu_e \to 1$ for $e \in \Gamma_1$, the algebra obtained from quiver gauge theory yields a new mass deformation of W-algebra. Furthermore, the bifundamental mass can be absorbed by redefinition of the Coulomb moduli parameter when there is no loop edge [**9**, **14**].

6.1. A_1 quiver. We show more explicit examples. The simplest example is the A_1 quiver. The T-operator is given by

$$T_1(x) = \mathsf{Y}_1(x) + \frac{1}{\mathsf{Y}_1(q^{-1}x)}, \tag{6.24}$$

which is a double quantum deformation of the fundamental representation character for $G_\Gamma = SU(2)$

$$\chi_{\mathbf{2}}^{SU(2)}(y) = y + y^{-1}. \tag{6.25}$$

In addition, we compute the degree n current, corresponding to $(n/2)$-spin representation of $SU(2)$, with the weight $(w_1, \ldots, w_n)$

$$\begin{aligned} T_1^{[\mathbf{w}]}(x) &= :\mathsf{Y}_1(w_1 x)\mathsf{Y}_1(w_2 x)\cdots\mathsf{Y}_1(w_n x): + \cdots \\ &= \sum_{I \cup J = \{1\ldots n\}} \prod_{i\in I, j\in J} \mathcal{S}\left(\frac{w_i}{w_j}\right) : \prod_{i\in I} \mathsf{Y}_1(w_i x) \prod_{j\in J} \mathsf{Y}_1(w_j q^{-1} x)^{-1} : . \end{aligned} \tag{6.26}$$

The degree two current is especially used to define the algebraic relation for the q-Virasoro algebra (4.7).

6.2. A_2 quiver. The next example is the A_2 quiver, consisting of two nodes and an edge connecting them. We have two T-operators

$$T_1(x) = \mathsf{Y}_1(x) + :\frac{\mathsf{Y}_2(\mu^{-1}x)}{\mathsf{Y}_1(q^{-1}x)}: + \frac{1}{\mathsf{Y}_2(\mu^{-1}q^{-1}x)}, \tag{6.27}$$

$$T_2(x) = \mathsf{Y}_2(x) + :\frac{\mathsf{Y}_1(\mu q^{-1}x)}{\mathsf{Y}_2(q^{-1}x)}: + \frac{1}{\mathsf{Y}_1(\mu q^{-2}x)}, \tag{6.28}$$

where $\mu = \mu_{1\to2} = \mu_{2\to1}^{-1} q$. They correspond to the fundamental and antifundamental representation characters of $G_\Gamma = SU(3)$

(6.29)
$$\chi_{\mathbf{3}}^{SU(3)}(y_1, y_2) = y_1 + y_1^{-1}y_2 + y_2^{-1}, \qquad \chi_{\bar{\mathbf{3}}}^{SU(3)}(y_1, y_2) = y_2 + y_1 y_2^{-1} + y_1^{-1}.$$

The OPEs for these T-operators are given by

$$f_{11}\left(\frac{y}{x}\right) T_1(x)T_1(y) - f_{11}\left(\frac{x}{y}\right) T_1(y)T_1(x)$$

(6.30)
$$= -\frac{(1-q_1)(1-q_2)}{1-q}\left(\delta\left(q\frac{y}{x}\right) T_2(\mu^{-1}x) - \delta\left(q^{-1}\frac{y}{x}\right) T_2(\mu^{-1}qx)\right)$$

$$f_{12}\left(\frac{y}{x}\right) T_1(x)T_2(y) - f_{21}\left(\frac{x}{y}\right) T_2(y)T_1(x)$$

(6.31)
$$= -\frac{(1-q_1)(1-q_2)}{1-q}\left(\delta\left(\mu q\frac{y}{x}\right) - \delta\left(\mu q^{-2}\frac{y}{x}\right)\right)$$

$$f_{22}\left(\frac{y}{x}\right) T_2(x)T_2(y) - f_{22}\left(\frac{x}{y}\right) T_2(y)T_2(x)$$

(6.32)
$$= -\frac{(1-q_1)(1-q_2)}{1-q}\left(\delta\left(q\frac{y}{x}\right) T_1(\mu q^{-1}x) - \delta\left(q^{-1}\frac{y}{x}\right) T_1(\mu x)\right)$$

where

$$f_{ij}(x) = \exp\left(\sum_{n=1}^{\infty} \frac{(1-q_1^n)(1-q_2^n)}{n}\, \tilde{c}_{ij}^{[-n]} x^n\right). \tag{6.33}$$

Now the symmetry under exchange $\mathsf{Y}_1 \leftrightarrow \mathsf{Y}_2$ is obvious. We can read off the algebraic relation for the (q,μ)-deformed $\mathrm{W}(A_2)$ algebra from these OPEs, which explicitly depends on the mass parameter μ. We remark that the diagonal element $f_{ii}(x)$ does not depend on the mass parameter $(\mu_e)_{e\in\Gamma_1}$ in general. It appears only in the off-diagonal element $f_{ij}(x)$ for $i \neq j$.

6.3. A_r quiver. Let us demonstrate our construction with the linear quiver theory $\Gamma = A_r$. The T-operators are computed using the iWeyl reflection

$$T_i(\mu_i^{-1}x) = \sum_{1\le j_1\le\cdots\le j_i\le r+1} : \prod_{k=1}^{i} \Lambda_{j_k}(q^{-i+k}x) : \tag{6.34}$$

where

$$\Lambda_i(x) = \mathsf{Y}_i(\mu_i^{-1}x)\mathsf{Y}_{i-1}(\mu_{i-1}^{-1}q^{-1}x)^{-1} \tag{6.35}$$

with $\mathsf{Y}_0(x) = \mathsf{Y}_{r+1}(x) = 1$, and the mass product defined

$$\mu_i := \mu_{1\to 2}\mu_{2\to 3}\cdots\mu_{i-1\to i}\,. \tag{6.36}$$

We remark that these T-operators are given by (q,μ)-deformation of fundamental representation characters for $G_\Gamma = SU(r+1)$. This construction coincides with the quantum Miura transformation for the q-deformed $\mathrm{W}(A_r)$-algebra in the massless limit $\mu_e \to 1$ [**44, 52**]

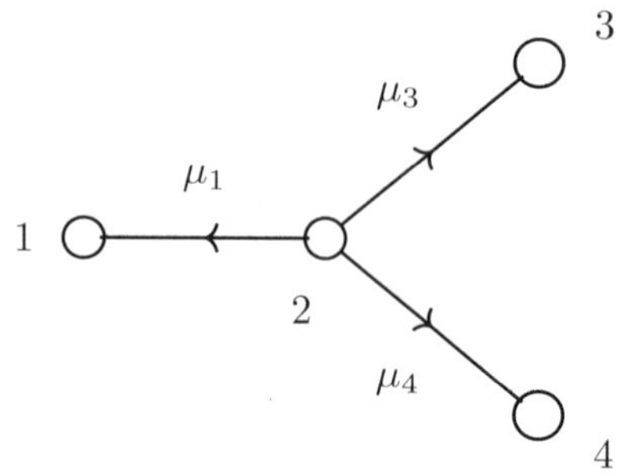

FIGURE 2. D_4 quiver decorated with bifundamental mass $\mu_i = \mu_{2\to i}$.

6.4. D_r quiver. We consider D-type quiver gauge theory. The simplest case is $\Gamma = D_4$, shown in Fig. 2. This quiver shows a symmetry under exchange $1 \leftrightarrow 3 \leftrightarrow 4$, which is known as the $SO(8)$ triality. Putting the bifundamental mass parameters $\mu_i := \mu_{2\to i}$ for $i = 1, 3, 4$, we compute the T-operator

$$\begin{aligned} T_1(x) = \mathsf{Y}_1(x) &+ :\frac{\mathsf{Y}_2(\mu_1 q^{-1}x)}{\mathsf{Y}_1(q^{-1}x)}: + :\frac{\mathsf{Y}_3(\mu_1\mu_3^{-1}q^{-1}x)\mathsf{Y}_4(\mu_1\mu_4^{-1}q^{-1}x)}{\mathsf{Y}_2(\mu_1 q^{-2}x)}: \\ &+ :\frac{\mathsf{Y}_4(\mu_1\mu_4^{-1}q^{-1}x)}{\mathsf{Y}_3(\mu_1\mu_3^{-1}q^{-2}x)}: \\ &+ :\frac{\mathsf{Y}_3(\mu_1\mu_3^{-1}q^{-1}x)}{\mathsf{Y}_4(\mu_1\mu_4^{-1}q^{-2}x)}: + :\frac{\mathsf{Y}_2(\mu_1 q^{-2}x)}{\mathsf{Y}_3(\mu_1\mu_3^{-1}q^{-2}x)\mathsf{Y}_4(\mu_1\mu_4^{-1}q^{-2}x)}: \\ &+ :\frac{\mathsf{Y}_1(q^{-2}x)}{\mathsf{Y}_2(\mu_1 q^{-3}x)}: + \frac{1}{\mathsf{Y}_1(q^{-3}x)}. \end{aligned} \tag{6.37}$$

The operators $T_3(x)$ and $T_4(x)$ are similarly obtained by permutation. These three T-operators correspond to three **8**-representations of $SO(8)$. The remaining $T_2(x)$ operator, corresponding to **28**-representation, involves collision and derivative terms, (6.20) and (6.22),

$$\begin{aligned} &T_2(x) \\ &= T_2^+(x) + T_2^-(x) + \mathcal{S}(q)\left(:\frac{\mathsf{Y}_1(\mu_1^{-1}x)}{\mathsf{Y}_1(\mu_1^{-1}q^{-1}x)}: + :\frac{\mathsf{Y}_3(\mu_3^{-1}x)}{\mathsf{Y}_3(\mu_3^{-1}q^{-1}x)}: + :\frac{\mathsf{Y}_4(\mu_4^{-1}x)}{\mathsf{Y}_4(\mu_4^{-1}q^{-1}x)}: \right) \\ &+ \left(\mathfrak{c}(q_1, q_2) - \frac{(1-q_1)(1-q_2)}{1-q}\partial_{\log x} \log\left(\frac{\mathsf{Y}_2(q^{-1}x)\mathsf{Y}_2(q^{-2}x)}{\mathsf{Y}_1(\mu_1^{-1}q^{-1}x)\mathsf{Y}_3(\mu_3^{-1}q^{-1}x)\mathsf{Y}_4(\mu_4^{-1}q^{-1}x)} \right)\right): \\ &\frac{\mathsf{Y}_2(q^{-1}x)}{\mathsf{Y}_2(q^{-2}x)}: \end{aligned} \tag{6.38}$$

where

(6.39)

$$
\begin{aligned}
T_2^+(x) = \mathsf{Y}_2(x) &+ : \frac{\mathsf{Y}_1(\mu_1^{-1}x)\mathsf{Y}_3(\mu_3^{-1}x)\mathsf{Y}_4(\mu_4^{-1}x)}{\mathsf{Y}_2(q^{-1}x)} : + : \frac{\mathsf{Y}_3(\mu_3^{-1}x)\mathsf{Y}_4(\mu_4^{-1}x)}{\mathsf{Y}_1(\mu_1^{-1}q^{-1}x)} : \\
&+ : \frac{\mathsf{Y}_1(\mu_1^{-1}x)\mathsf{Y}_4(\mu_4^{-1}x)}{\mathsf{Y}_3(\mu_3^{-1}q^{-1}x)} : + : \frac{\mathsf{Y}_1(\mu_1^{-1}x)\mathsf{Y}_3(\mu_3^{-1}x)}{\mathsf{Y}_4(\mu_4^{-1}q^{-1}x)} : \\
&+ : \frac{\mathsf{Y}_1(\mu_1^{-1}x)\mathsf{Y}_2(q^{-1}x)}{\mathsf{Y}_3(\mu_3^{-1}q^{-1}x)\mathsf{Y}_4(\mu_4^{-1}q^{-1}x)} : + : \frac{\mathsf{Y}_3(\mu_3^{-1}x)\mathsf{Y}_2(q^{-1}x)}{\mathsf{Y}_1(\mu_1^{-1}q^{-1}x)\mathsf{Y}_4(\mu_4^{-1}q^{-1}x)} : \\
&+ : \frac{\mathsf{Y}_4(\mu_4^{-1}x)\mathsf{Y}_2(q^{-1}x)}{\mathsf{Y}_1(\mu_1^{-1}q^{-1}x)\mathsf{Y}_3(\mu_3^{-1}q^{-1}x)} : + : \frac{\mathsf{Y}_2(q^{-1}x)^2}{\mathsf{Y}_1(\mu_1^{-1}q^{-1}x)\mathsf{Y}_3(\mu_3^{-1}q^{-1}x)\mathsf{Y}_4(\mu_4^{-1}q^{-1}x)} : \\
&+ : \frac{\mathsf{Y}_1(\mu_1^{-1}x)\mathsf{Y}_1(\mu_1^{-1}q^{-1}x)}{\mathsf{Y}_2(q^{-2}x)} : + : \frac{\mathsf{Y}_3(\mu_3^{-1}x)\mathsf{Y}_3(\mu_3^{-1}q^{-1}x)}{\mathsf{Y}_2(q^{-2}x)} : \\
&+ : \frac{\mathsf{Y}_4(\mu_4^{-1}x)\mathsf{Y}_4(\mu_4^{-1}q^{-1}x)}{\mathsf{Y}_2(q^{-2}x)} : ,
\end{aligned}
$$

(6.40)

$$
\begin{aligned}
T_2^-(x) =& : \frac{\mathsf{Y}_1(\mu_1^{-1}q^{-1}x)\mathsf{Y}_3(\mu_3^{-1}q^{-1}x)\mathsf{Y}_4(\mu_4^{-1}q^{-1}x)}{\mathsf{Y}_2(q^{-2}x)^2} : + : \frac{\mathsf{Y}_2(q^{-1}x)}{\mathsf{Y}_1(\mu_1^{-1}q^{-1}x)\mathsf{Y}_1(\mu_1^{-1}q^{-2}x)} : \\
&+ : \frac{\mathsf{Y}_2(q^{-1}x)}{\mathsf{Y}_3(\mu_3^{-1}q^{-1}x)\mathsf{Y}_3(\mu_3^{-1}q^{-2}x)} : + : \frac{\mathsf{Y}_2(q^{-1}x)}{\mathsf{Y}_4(\mu_4^{-1}q^{-1}x)\mathsf{Y}_4(\mu_4^{-1}q^{-2}x)} : \\
&+ : \frac{\mathsf{Y}_3(\mu_3^{-1}q^{-1}x)\mathsf{Y}_4(\mu_4^{-1}q^{-1}x)}{\mathsf{Y}_1(\mu_1^{-1}q^{-2}x)\mathsf{Y}_2(q^{-2}x)} : + : \frac{\mathsf{Y}_1(\mu_1^{-1}q^{-1}x)\mathsf{Y}_4(\mu_4^{-1}q^{-1}x)}{\mathsf{Y}_3(\mu_3^{-1}q^{-2}x)\mathsf{Y}_2(q^{-2}x)} : \\
&+ : \frac{\mathsf{Y}_1(\mu_1^{-1}q^{-1}x)\mathsf{Y}_3(\mu_3^{-1}q^{-1}x)}{\mathsf{Y}_4(\mu_4^{-1}q^{-2}x)\mathsf{Y}_2(q^{-2}x)} : + : \frac{\mathsf{Y}_1(\mu_1^{-1}q^{-1}x)}{\mathsf{Y}_3(\mu_3^{-1}q^{-2}x)\mathsf{Y}_4(\mu_4^{-1}q^{-2}x)} : \\
&+ : \frac{\mathsf{Y}_3(\mu_1^{-1}q^{-1}x)}{\mathsf{Y}_1(\mu_1^{-1}q^{-2}x)\mathsf{Y}_4(\mu_4^{-1}q^{-2}x)} : + : \frac{\mathsf{Y}_4(\mu_4^{-1}q^{-1}x)}{\mathsf{Y}_1(\mu_1^{-1}q^{-2}x)\mathsf{Y}_3(\mu_3^{-1}q^{-2}x)} : \\
&+ : \frac{\mathsf{Y}_2(q^{-2}x)}{\mathsf{Y}_1(\mu_1^{-1}q^{-2}x)\mathsf{Y}_3(\mu_3^{-1}q^{-2}x)\mathsf{Y}_4(\mu_4^{-1}q^{-2}x)} : + \mathsf{Y}_2(q^{-3}x)^{-1} .
\end{aligned}
$$

In the classical limit, this T_2-operator is reduced to the character of 28 dimensional representation of $SO(8)$. The $\mathcal{S}$-factor appears at the zero weight terms.

7. Affine quiver W-algebra

Applying our construction to affine quiver gauge theory, we can define a new W-algebra, associated to affine Lie algebra. Let us consider the simplest affine quiver $\Gamma = \widehat{A}_0$, corresponding to 4d $\mathcal{N} = 2^*$ (5d $\mathcal{N} = 1^*$) theory, whose deformation of Cartan matrix is given by

$$
c^{[n]} = 1 + q^{-n} - \mu^{-n} - \mu^n q^{-n} \stackrel{n \to 0}{\longrightarrow} 0 . \tag{7.1}
$$

This Cartan matrix can be nontrivial only in the presence of the multiplicative adjoint mass parameter $\mu\,(\neq 1)$. This means that the mass deformation of the Cartan matrix is essential for the affine quiver, and this deformation goes beyond

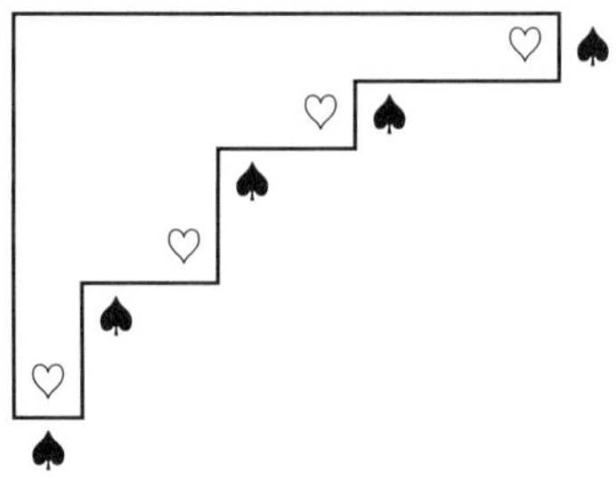

FIGURE 3. The outer and inner boundaries of the partition, $\partial_+\lambda$ and $\partial_-\lambda$, denote the to-be-added and to-be-removed boxes, ♠ and ♡, respectively. The number of ♠ is always bigger than ♡ by one.

Frenkel–Reshetikhin's construction of W-algebra [**44, 45**]. The free field operators for Y and S fields obey the commutation relation

$$\left[y_{1,n}, y_{1,m}\right] = -\frac{1}{n}\frac{(1-q_1^n)(1-q_2^n)}{(1-\mu^n)(1-\mu^{-n}q^n)}\delta_{n+m,0}\,, \tag{7.2}$$

and

$$\left[s_{1,n}, s_{1,m}\right] = -\frac{1}{n}\frac{1-q_1^n}{1-q_2^{-n}}(1-\mu^{-n})(1-\mu^n q^{-n})\,\delta_{n+m,0}\,. \tag{7.3}$$

The s-oscillator becomes trivial in the limit $\mu \to 1$ (and also $\mu \to q$), which means that the Nekrasov factor becomes trivial in $\mathcal{N} = 4$ theory, while the y-oscillator becomes singular in this limit.

The T-operator in this case is given by the iWeyl reflection as well, but it does not terminate within a finite sum: the affine character needs an infinite sum parametrized by the partition

$$\begin{aligned} T_1(x) &= \mathsf{Y}_1(x) + \mathfrak{q}\,\mathsf{S}(\mu^{-1}) : \mathsf{Y}_1(q^{-1}x)^{-1}\mathsf{Y}_1(\mu^{-1}x)\mathsf{Y}_1(\mu q^{-1}x) : + \cdots \\ &= \sum_\lambda \mathfrak{q}^{|\lambda|} Z_\lambda^{\widehat{A}_0}(\tilde{q}_1, \tilde{q}_2, \tilde{\mu}) : \prod_{s\in\partial_+\lambda} \mathsf{Y}_1(qx/\tilde{x}(s)) \prod_{s\in\partial_-\lambda} \mathsf{Y}_1(x/\tilde{x}(s))^{-1} : \end{aligned} \tag{7.4}$$

where the coupling constant $\mathfrak{q}$ is now explicitly appearing because the Cartan matrix $c = (0)$ is not invertible for affine quiver $\Gamma = \widehat{A}_0$. $\partial_+\lambda$ and $\partial_-\lambda$ are the outer and inner boundary of the partition λ, as shown in Fig. 3, and we define

$$\tilde{x}(s) = (\mu^{-1}q)^{s_1-1}\mu^{s_2-1}q\,, \tag{7.5}$$

which reads the dual variables

$$\tilde{q}_1 = \mu^{-1}q\,, \qquad \tilde{q}_2 = \mu\,, \qquad \tilde{\mu} = q_2\,, \qquad \tilde{\nu} = q\,. \tag{7.6}$$

We remark that $\tilde{q} := \tilde{q}_1\tilde{q}_2 = q$. The Nekrasov instanton function $Z_\lambda^{\widehat{A}_0}(\tilde{q}_1, \tilde{q}_2, \tilde{\mu})$ is again for $\widehat{A}_0$ quiver, but $U(1)$ gauge theory

$$Z_\lambda^{\widehat{A}_0}(q_1, q_2, \mu) = \prod_{(s_1,s_2)\in\lambda} \frac{(1-\mu\, q_1^{l(s)+1} q_2^{-a(s)})(1-\mu^{-1}q_1^{l(s)} q_2^{-a(s)-1})}{(1-q_1^{l(s)+1} q_2^{-a(s)})(1-q_1^{l(s)} q_2^{-a(s)-1})} \tag{7.7}$$

where the arm and leg lengths are defined: $a(s) = \lambda_{s_1} - s_2$, $l(s) = \lambda^t_{s_2} - s_1$. The higher degree current is given as a summation over multiple partition. This dual expansion is naturally explained using the 8-dimensional setup [**17**].

8. Quiver elliptic W-algebra

We have constructed the q-deformed W-algebra $\mathrm{W}_{q_1,q_2}(\Gamma)$ associated with Γ-quiver gauge theory in 5d, which corresponds to trigonometric (relativistic) integrable system. According to such a connection with integrable systems, it is natural to consider its elliptic deformation using 6d gauge theory compactified on the elliptic curve $\mathbb{R}^4 \times T^2$ [**30**]. See also [**21, 24, 25, 53–58**] for recent results along this direction.

In this case the spectral parameter is periodic

$$x \simeq px \tag{8.1}$$

where τ is the modulus of T^2 that defines $p = e^{2\pi i \tau} \in \mathbb{C}^\times$. One needs to impose this periodicity on the index functor. Recalling that the 5d gauge theory partition function on $\mathbb{R}^4 \times S^1$ is given by the index along S^1 defined

$$\mathbb{I}\left[\sum_\alpha x_\alpha\right] = \prod_\alpha \left(1 - x_\alpha^{-1}\right) , \tag{8.2}$$

the elliptic index (equivariant elliptic genus on T^2) yields the 6d partition function for $\mathbb{R}^4 \times T^2$

$$\mathbb{I}_p\left[\sum_\alpha x_\alpha\right] = \prod_\alpha \theta(x_\alpha^{-1}; p) \tag{8.3}$$

where the Jacobi theta function is given by[13]

$$\begin{aligned} \theta(x;p) &= (x;p)_\infty (px^{-1};p)_\infty \\ &= \exp\left(- \sum_{n\neq 0} \frac{1}{n} \frac{x^n}{1-p^n} \right) . \end{aligned} \tag{8.7}$$

We assume $|p| < 1$ and this elliptic system is reduced to 5d theory in the limit $p \to 0$.

---

[13]In this article we apply the Dolbeault convention, which obeys

$$\mathbb{I}\left[x^{-1}\right] = -x^{-1}\mathbb{I}[x] , \qquad \mathbb{I}_p\left[x^{-1}\right] = -x^{-1}\mathbb{I}_p[x] . \tag{8.4}$$

This convention can be converted into the Dirac index,

$$\mathbb{I}\left[\sum_\alpha x_\alpha\right] = \prod_\alpha \left(x_\alpha^{1/2} - x_\alpha^{-1/2}\right) , \qquad \mathbb{I}_p\left[\sum_\alpha x_\alpha\right] = \prod_\alpha \theta_1(x_\alpha;p) \tag{8.5}$$

with

$$\mathbb{I}\left[x^{-1}\right] = -\mathbb{I}\left[x\right] , \qquad \mathbb{I}_p\left[x^{-1}\right] = -\mathbb{I}_p\left[x\right] . \tag{8.6}$$

Under the conformal condition (8.14), these two conventions are consistent.

Since the elliptic index (8.3) involves both positive and negative powers due to the expansion (8.7), we need to introduce "positive" and "negative" oscillators [**30**, **59**, **60**]

$$\left[y_{i,n}^{(\pm)}, y_{j,m}^{(\pm)}\right] = \mp\frac{1}{n}\frac{(1-q_1^{\mp n})(1-q_2^{\pm n})}{1-p^{\pm n}}\,\tilde{c}_{ij}^{[\mp n]}\delta_{n+m,0}\,, \tag{8.8}$$

$$\left[s_{i,n}^{(\pm)}, s_{j,m}^{(\pm)}\right] = \mp\frac{1}{n}\frac{1-q_1^{\pm n}}{(1-p^{\pm n})(1-q_2^{\mp n})}\,c_{ij}^{[\pm n]}\delta_{n+m,0}\,, \tag{8.9}$$

and thus

$$\mathsf{Y}_i(x) = q_1^{\tilde{\rho}_i} : \exp\left(y_{i,0} + \sum_{n\neq 0} y_{i,n}^{(+)}x^{-n} + \sum_{n\neq 0} y_{i,n}^{(-)}x^{+n}\right) :, \tag{8.10}$$

$$S_i(x) =: \exp\left(s_{i,0}\log x + \tilde{s}_{i,0} + \sum_{n\neq 0} s_{i,n}^{(+)}x^{-n} + \sum_{n\neq 0} s_{i,n}^{(-)}x^{+n}\right) : . \tag{8.11}$$

Then the remaining construction is completely parallel with 5d theory: The T-operator is generated by the iWeyl reflection with the Y-operator, and the Z-state is constructed using the screening charge, built with the elliptic oscillators (8.8) and (8.9). In this procedure, the whole rational functions are replaced with the elliptic function. For example, the scalar factor (6.21) due to the Y and A OPE becomes

$$\mathcal{S}(x) = \frac{\theta(q_1 x;p)\theta(q_2 x;p)}{\theta(qx;p)\theta(x;p)}\,, \tag{8.12}$$

and the commutation relation between the Y and S operators is given by

$$\left[\mathsf{Y}_i(x), S_j(x')\right] = \delta_{i,j}\frac{\theta(q_1^{-1};p)}{(p;p)_\infty^2}\delta\left(q_1\frac{x'}{x}\right) : \mathsf{Y}_i(x)S_j(x') : \tag{8.13}$$

which becomes zero in the limit $q_1 \to 1$.

We can obtain the non-extended partition function from the Z-state as a correlator of the elliptic algebra as well as (5.16). In this case, the elliptic correlator can be also written in terms of 5d operators[14]

$$Z(t=0) = \langle 1| \prod_{\mathring{x}\in\mathcal{X}_0}^{\succ} \mathsf{S}_{\mathrm{i}(\mathring{x})}^{\mathrm{6d}}(\mathring{x})|1\rangle = \operatorname{Tr}\left[p^{L_0}\prod_{\mathring{x}\in\mathcal{X}_0}^{\succ} \mathsf{S}_{\mathrm{i}(\mathring{x})}^{\mathrm{5d}}(\mathring{x})\right] \tag{8.15}$$

where the energy operator is defined

$$L_0 = \sum_{i\in\Gamma_0}\sum_{m=1}^{\infty} m\, t_{i,m}\frac{\partial}{t_{i,m}}\,. \tag{8.16}$$

The meaning of this trace formula is as follows. One possibility to impose the periodicity (8.1) is doubling the free fields. Another possible way is to change the definition of the correlator to the trace, as shown in the bottom of Fig. 4, as well as the usual CFT correlator on a torus. They are actually two equivalent methods

[14] To assure the modularity of (non-extended) gauge theory partition function, we need to assign the anomaly-free condition for 6d theory

$$\sum_{j\in\Gamma_0} c_{ij}\, n_j = n_i^{\mathrm{flavor}} \quad \text{for} \quad \forall i \in \Gamma_0\,. \tag{8.14}$$

This is equivalent to the conformal condition in 4d quiver gauge theory.

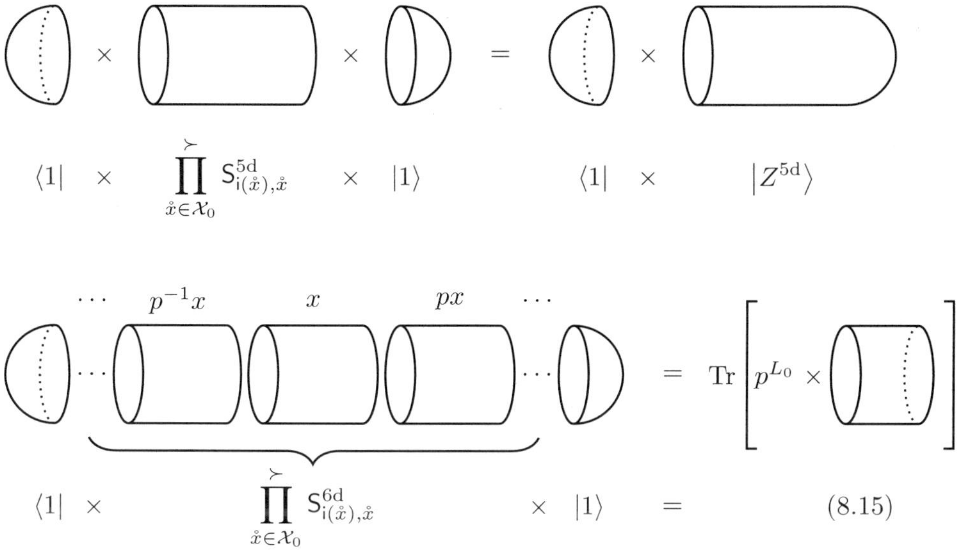

FIGURE 4. Conformal blocks as the partition function of five-dimensional (top) and six-dimensional theory (bottom). The six-dimensional block has two equivalent expressions.

to compute the torus correlation function [**30, 59**]. From this point of view, the Z-state itself is not necessarily modular invariant, while the torus correlator should be.

Let us show the generating current of elliptic W-algebra with A_r linear quiver examples. The generating current itself is constructed by the Y-operator in the same way as before. The elliptic $\mathrm{W}(A_1)$ generating current is given by

$$T_1(x) = \mathsf{Y}_1(x) + \frac{1}{\mathsf{Y}_1(q^{-1}x)}\,, \tag{8.17}$$

as discussed in Sec. 6.1, and $\mathrm{W}(A_2)$ currents are given by

$$T_1(x) = \mathsf{Y}_1(x) + :\frac{\mathsf{Y}_2(\mu^{-1}x)}{\mathsf{Y}_1(q^{-1}x)}: + \frac{1}{\mathsf{Y}_2(\mu^{-1}q^{-1}x)}\,, \tag{8.18}$$

$$T_2(x) = \mathsf{Y}_2(x) + :\frac{\mathsf{Y}_1(\mu q^{-1}x)}{\mathsf{Y}_2(q^{-1}x)}: + \frac{1}{\mathsf{Y}_1(\mu q^{-2}x)}\,, \tag{8.19}$$

where $\mu = \mu_{1\to 2} = \mu_{2\to 1}^{-1} q$, as shown in Sec. 6.2. The construction is also parallel for the linear quiver A_r shown in Sec. 6.3.

These generating currents exhibit a similar OPE relation. For A_1 quiver we have [**56**]

$$\begin{aligned} & f_{11}\left(\frac{y}{x}\right) T_1(x)T_1(y) - f_{11}\left(\frac{x}{y}\right) T_1(y)T_1(x) \\ & \qquad = -\frac{\theta(q_1;p)\theta(q_2;p)}{(p;p)_\infty^2\theta(q;p)} \left(\delta\left(\frac{qy}{x}\right) - \delta\left(\frac{qx}{y}\right)\right) \end{aligned} \tag{8.20}$$

and for A_2 quiver we have

$$f_{11}\left(\frac{y}{x}\right)T_1(x)T_1(y) - f_{11}\left(\frac{x}{y}\right)T_1(y)T_1(x)$$

(8.21)

$$= -\frac{\theta(q_1;p)\theta(q_2;p)}{(p;p)_\infty^2\theta(q;p)}\left(\delta\left(q\frac{y}{x}\right)T_2(\mu^{-1}x) - \delta\left(q^{-1}\frac{y}{x}\right)T_2(\mu^{-1}qx)\right)$$

$$f_{12}\left(\frac{y}{x}\right)T_1(x)T_2(y) - f_{21}\left(\frac{x}{y}\right)T_2(y)T_1(x)$$

(8.22)

$$= -\frac{\theta(q_1;p)\theta(q_2;p)}{(p;p)_\infty^2\theta(q;p)}\left(\delta\left(\mu q\frac{y}{x}\right) - \delta\left(\mu q^{-2}\frac{y}{x}\right)\right)$$

$$f_{22}\left(\frac{y}{x}\right)T_2(x)T_2(y) - f_{22}\left(\frac{x}{y}\right)T_2(y)T_2(x)$$

(8.23)

$$= -\frac{\theta(q_1;p)\theta(q_2;p)}{(p;p)_\infty^2\theta(q;p)}\left(\delta\left(q\frac{y}{x}\right)T_1(\mu q^{-1}x) - \delta\left(q^{-1}\frac{y}{x}\right)T_1(\mu x)\right)$$

where

$$f_{ij}(x) = \exp\left(\sum_{n\neq 0}^{\infty}\frac{(1-q_1^n)(1-q_2^n)}{n(1-p^n)}\tilde{c}_{ij}^{[-n]}x^n\right). \tag{8.24}$$

These OPE relations characterize the algebraic relation for the elliptic W-algebra generators.

Acknowledgments

The author would like to thank V. Pestun for fruitful collaboration which materializes this article and useful comments on the draft. This article is based on a talk presented at 2016 AMS von Neumann Symposium: Topological Recursion and its Influence in Analysis, Geometry, and Topology, July 4–8, 2016, Hilton Charlotte University Place, Charlotte, NC. The author is grateful to the symposium organizers for giving the opportunity to present our work in a stimulating atmosphere.

References

[1] N. Seiberg and E. Witten, *Electric-magnetic duality, monopole condensation, and confinement in $N = 2$ supersymmetric Yang-Mills theory*, Nuclear Phys. B **426** (1994), no. 1, 19–52, DOI 10.1016/0550-3213(94)90124-4. MR1293681

[2] N. Seiberg and E. Witten, *Monopoles, duality and chiral symmetry breaking in $N = 2$ supersymmetric QCD*, Nuclear Phys. B **431** (1994), no. 3, 484–550, DOI 10.1016/0550-3213(94)90214-3. MR1306869

[3] A. Gorsky, I. M. Krichever, A. Marshakov, A. Mironov, and A. Morozov, *Integrability and Seiberg-Witten exact solution*, Phys. Lett. B **355** (1995), no. 3-4, 466–474, DOI 10.1016/0370-2693(95)00723-X. MR1345598

[4] E. J. Martinec and N. P. Warner, *Integrable systems and supersymmetric gauge theory*, Nuclear Phys. B **459** (1996), no. 1-2, 97–112, DOI 10.1016/0550-3213(95)00588-9. MR1372221

[5] R. Donagi and E. Witten, *Supersymmetric Yang-Mills theory and integrable systems*, Nuclear Phys. B **460** (1996), no. 2, 299–334, DOI 10.1016/0550-3213(95)00609-5. MR1377167

[6] N. Seiberg and E. Witten, *Gauge dynamics and compactification to three dimensions*, The mathematical beauty of physics (Saclay, 1996), Adv. Ser. Math. Phys., vol. 24, World Sci. Publ., River Edge, NJ, 1997, pp. 333–366. MR1490862

[7] A. Gorsky, A. Marshakov, A. Mironov, and A. Morozov, *$N = 2$ supersymmetric QCD and integrable spin chains: rational case $N_f < 2N_c$*, Phys. Lett. B **380** (1996), no. 1-2, 75–80, DOI 10.1016/0370-2693(96)00480-7. MR1396968

[8] N. Nekrasov, *Five-dimensional gauge theories and relativistic integrable systems*, Nuclear Phys. B **531** (1998), no. 1-3, 323–344, DOI 10.1016/S0550-3213(98)00436-2. MR1657857

[9] N. Nekrasov and V. Pestun, *Seiberg–Witten geometry of four dimensional $N = 2$ quiver gauge theories* (2012), `arXiv:1211.2240 [hep-th]`.

[10] G. Moore, N. Nekrasov, and S. Shatashvili, *Integrating over Higgs branches*, Comm. Math. Phys. **209** (2000), no. 1, 97–121, DOI 10.1007/s002200050016. MR1736944

[11] N. A. Nekrasov, *Seiberg-Witten prepotential from instanton counting*, Adv. Theor. Math. Phys. **7** (2003), no. 5, 831–864. MR2045303

[12] N. A. Nekrasov and S. L. Shatashvili, *Supersymmetric vacua and Bethe ansatz*, Nuclear Phys. B Proc. Suppl. **192/193** (2009), 91–112, DOI 10.1016/j.nuclphysbps.2009.07.047. MR2570974

[13] N. A. Nekrasov and S. L. Shatashvili, "Quantization of Integrable Systems and Four Dimensional Gauge Theories," in *XVIth International Congress on Mathematical Physics* (2009), pp. 265–289. `arXiv:0908.4052 [hep-th]`.

[14] N. Nekrasov, V. Pestun, and S. Shatashvili, *Quantum geometry and quiver gauge theories*, Comm. Math. Phys. **357** (2018), no. 2, 519–567, DOI 10.1007/s00220-017-3071-y. MR3767702

[15] E. Frenkel and N. Reshetikhin, *The q-characters of representations of quantum affine algebras and deformations of $\mathcal{W}$-algebras*, Recent developments in quantum affine algebras and related topics (Raleigh, NC, 1998), Contemp. Math., vol. 248, Amer. Math. Soc., Providence, RI, 1999, pp. 163–205, DOI 10.1090/conm/248/03823. MR1745260

[16] T. Hollowood, A. Iqbal, and C. Vafa, *Matrix models, geometric engineering and elliptic genera*, J. High Energy Phys. **3** (2008), 069, 81, DOI 10.1088/1126-6708/2008/03/069. MR2391051

[17] N. Nekrasov, *BPS/CFT correspondence: non-perturbative Dyson–Schwinger equations and qq-characters*, J. High Energy Phys. **1603** (2016) 181. `arXiv:1512.05388 [hep-th]`.

[18] N. Nekrasov, *BPS/CFT correspondence II: instantons at crossroads, moduli and compactness theorem*, Adv. Theor. Math. Phys. **21** (2017), no. 2, 503–583, DOI 10.4310/ATMP.2017.v21.n2.a4. MR3678005

[19] N. Nekrasov, *BPS/CFT correspondence III: Gauge origami partition function and qq-characters*, Comm. Math. Phys. **358** (2018), no. 3, 863–894, DOI 10.1007/s00220-017-3057-9. MR3778345

[20] J.-E. Bourgine, Y. Mastuo, and H. Zhang, *Holomorphic field realization of SH^c and quantum geometry of quiver gauge theories*, J. High Energy Phys. **1604** (2016) 167.

[21] A. Mironov, A. Morozov, and Y. Zenkevich, *On elementary proof of AGT relations from six dimensions*, Phys. Lett. **B756** (2016) 208–211.

[22] H.-C. Kim, *Line defects and 5d instanton partition functions*, J. High Energy Phys. **1603** (2016) 199.

[23] J.-E. Bourgine, M. Fukuda, Y. Matsuo, H. Zhang, and R.-D. Zhu, *Coherent states in quantum $\mathcal{W}_{1+\infty}$ algebra and qq-character for 5d super Yang-Mills*, PTEP. Prog. Theor. Exp. Phys. **12** (2016), 123B05, 41. MR3628699

[24] A. Mironov, A. Morozov, and Y. Zenkevich, *Spectral duality in elliptic systems, six-dimensional gauge theories and topological strings*, J. High Energy Phys. **5** (2016), 121, front matter+43. MR3521840

[25] A. Mironov, A. Morozov, and Y. Zenkevich, *Ding-Iohara-Miki symmetry of network matrix models*, Phys. Lett. B **762** (2016), 196–208, DOI 10.1016/j.physletb.2016.09.033. MR3569486

[26] H. Awata, H. Kanno, T. Matsumoto, A. Mironov, A. Morozov, A. Morozov, Y. Ohkubo, and Y. Zenkevich, *Explicit examples of DIM constraints for network matrix models*, J. High Energy Phys. **7** (2016), 103, front matter+66. MR3550198

[27] H. Awata, H. Kanno, A. Mironov, A. Morozov, A. Morozov, Y. Ohkubo, and Y. Zenkevich, *Toric Calabi-Yau threefolds as quantum integrable systems. $\mathcal{R}$-matrix and $\mathcal{RTT}$ relations*, J. High Energy Phys. **10** (2016), 047, front matter+48. MR3577516

[28] T. Kimura, H. Mori, and Y. Sugimoto, *Refined geometric transition and qq-characters*, J. High Energy Phys. **1** (2018), 025, front matter+40, DOI 10.1007/jhep01(2018)025. MR3764250

[29] T. Kimura and V. Pestun, *Quiver W-algebras*, Lett. Math. Phys. **108** (2018), no. 6, 1351–1381, DOI 10.1007/s11005-018-1072-1. MR3797750

[30] T. Kimura and V. Pestun, *Quiver elliptic W-algebras*, Lett. Math. Phys. **108** (2018), no. 6, 1383–1405, DOI 10.1007/s11005-018-1073-0. MR3797751

[31] T. Kimura and V. Pestun, *Fractional quiver W-algebras*, Lett. Math. Phys. (2018), `arXiv:1705.04410 [hep-th]`.

[32] A. Marshakov and N. A. Nekrasov, *Extended Seiberg-Witten theory and integrable hierarchy*, J. High Energy Phys. **1** (2007), 104, 39, DOI 10.1088/1126-6708/2007/01/104. MR2285911

[33] L. F. Alday, D. Gaiotto, and Y. Tachikawa, *Liouville correlation functions from four-dimensional gauge theories*, Lett. Math. Phys. **91** (2010), no. 2, 167–197, DOI 10.1007/s11005-010-0369-5. MR2586871

[34] N. Wyllard, *A_{N-1} conformal Toda field theory correlation functions from conformal $\mathcal{N}=2$ $\mathrm{SU}(N)$ quiver gauge theories*, J. High Energy Phys. **11** (2009), 002, 22, DOI 10.1088/1126-6708/2009/11/002. MR2628905

[35] H. Awata and Y. Yamada, *Five-dimensional AGT conjecture and the deformed Virasoro algebra*, J. High Energy Phys. **1** (2010), 125, 11, DOI 10.1007/JHEP01(2010)125. MR2660773

[36] J. Teschner, *Quantization of the Hitchin moduli spaces, Liouville theory and the geometric Langlands correspondence I*, Adv. Theor. Math. Phys. **15** (2011), no. 2, 471–564. MR2924236

[37] M. Manabe and P. Sułkowski, *Quantum curves and conformal field theory*, Phys. Rev. **D95** (2017) 126003. `arXiv:1512.05785 [hep-th]`.

[38] S. Katz, P. Mayr, and C. Vafa, *Mirror symmetry and exact solution of 4D $N=2$ gauge theories. I*, Adv. Theor. Math. Phys. **1** (1997), no. 1, 53–114, DOI 10.4310/ATMP.1997.v1.n1.a2. MR1489296

[39] O. Aharony, A. Hanany, and B. Kol, *Webs of (p,q) five-branes, five-dimensional field theories and grid diagrams*, J. High Energy Phys. **01** (1998) 002. `arXiv:hep-th/9710116 [hep-th]`.

[40] L. Bao, E. Pomoni, M. Taki, and F. Yagi, *M5-branes, toric diagrams and gauge theory duality*, J. High Energy Phys. **4** (2012), 105, front matter+58, DOI 10.1007/JHEP04(2012)105. MR2967598

[41] M. Aganagic, N. Haouzi, C. Kozçaz, and S. Shakirov, *Gauge/Liouville Triality* (2013), `arXiv:1309.1687 [hep-th]`.

[42] M. Aganagic, N. Haouzi, and S. Shakirov, *A_n-Triality* (2014), `arXiv:1403.3657 [hep-th]`.

[43] M. Aganagic and N. Haouzi, *ADE Little String Theory on a Riemann Surface (and Triality)* (2015), `arXiv:1506.04183 [hep-th]`.

[44] E. Frenkel and N. Reshetikhin, *Quantum affine algebras and deformations of the Virasoro and $\mathcal{W}$-algebras*, Comm. Math. Phys. **178** (1996), no. 1, 237–264. MR1387950

[45] E. Frenkel and N. Reshetikhin, *Deformations of $\mathcal{W}$-algebras associated to simple Lie algebras*, Comm. Math. Phys. **197** (1998), no. 1, 1–32. MR1646483

[46] N. A. Nekrasov and A. Okounkov, *Seiberg-Witten theory and random partitions*, The unity of mathematics, Progr. Math., vol. 244, Birkhäuser Boston, Boston, MA, 2006, pp. 525–596, DOI 10.1007/0-8176-4467-9_15. MR2181816

[47] J. Shiraishi, H. Kubo, H. Awata, and S. Odake, *A quantum deformation of the Virasoro algebra and the Macdonald symmetric functions*, Lett. Math. Phys. **38** (1996), no. 1, 33–51, DOI 10.1007/BF00398297. MR1401054

[48] Vl. S. Dotsenko and V. A. Fateev, *Conformal algebra and multipoint correlation functions in 2D statistical models*, Nuclear Phys. B **240** (1984), no. 3, 312–348, DOI 10.1016/0550-3213(84)90269-4. MR762194

[49] Vl. S. Dotsenko and V. A. Fateev, *Four-point correlation functions and the operator algebra in 2D conformal invariant theories with central charge $C \leq 1$*, Nuclear Phys. B **251** (1985), no. 5-6, 691–734, DOI 10.1016/S0550-3213(85)80004-3. MR789026

[50] S. Kharchev, A. Marshakov, A. Mironov, A. Morozov, and S. Pakuliak, *Conformal matrix models as an alternative to conventional multi-matrix models*, Nuclear Phys. B **404** (1993), no. 3, 717–750, DOI 10.1016/0550-3213(93)90595-G. MR1234891

[51] I. K. Kostov, *Gauge invariant matrix model for the $\hat{A}$-$\hat{D}$-$\hat{E}$ closed strings*, Phys. Lett. B **297** (1992), no. 1-2, 74–81, DOI 10.1016/0370-2693(92)91072-H. MR1199768

[52] H. Awata, H. Kubo, S. Odake, and J. Shiraishi, *Quantum $\mathcal{W}_N$ algebras and Macdonald polynomials*, Comm. Math. Phys. **179** (1996), no. 2, 401–416. MR1400746

[53] M.-C. Tan, *An M-Theoretic Derivation of a 5d and 6d AGT Correspondence, and Relativistic and Elliptized Integrable Systems*, J. High Energy Phys. **1312** (2013) 031. `arXiv:1309.4775 [hep-th]`.

[54] P. Koroteev and A. Sciarappa, *Quantum hydrodynamics from large-n supersymmetric gauge theories*, Lett. Math. Phys. **108** (2018), no. 1, 45–95, DOI 10.1007/s11005-017-0996-1. MR3740429

[55] A. Iqbal, C. Kozçaz, and S.-T. Yau, *Elliptic Virasoro Conformal Blocks* (2015), `arXiv:1511.00458 [hep-th]`.

[56] F. Nieri, *An elliptic Virasoro symmetry in 6d*, Lett. Math. Phys. **107** (2017), no. 11, 2147–2187, DOI 10.1007/s11005-017-0986-3. MR3714778

[57] P. Koroteev and A. Sciarappa, *On elliptic algebras and large-n supersymmetric gauge theories*, J. Math. Phys. **57** (2016), no. 11, 112302, 32, DOI 10.1063/1.4966641. MR3570998

[58] M.-C. Tan, *Higher AGT Correspondences, W-algebras, and Higher Quantum Geometric Langlands Duality from M-Theory* (2016), `arXiv:1607.08330 [hep-th]`.

[59] L. Clavelli and J. A. Shapiro, *Pomeron factorization in general dual models*, Nucl. Phys. **B57** (1973) 490–535.

[60] Y. Saito, *Elliptic Ding-Iohara algebra and the free field realization of the elliptic Macdonald operator*, Publ. Res. Inst. Math. Sci. **50** (2014), no. 3, 411–455, DOI 10.4171/PRIMS/139. MR3262444

KEIO UNIVERSITY, HIYOSHI 4-1-1, YOKOHAMA, 223-8521 JAPAN
Email address: `taro.kimura@keio.jp`

Proceedings of Symposia in Pure Mathematics
Volume **100**, 2018
https://doi.org/10.1090/pspum/100/01765

Airy structures and symplectic geometry of topological recursion

Maxim Kontsevich and Yan Soibelman

Contents

1. Introduction

1.1. Linear algebra data of the topological recursion. Conventional approach to topological recursion (TR for short, see [**EO07**]) produces a collection of meromorphic polydifferentials $\omega_{g,n}$ on a spectral curve Σ starting with the "initial data" $\omega_{0,1}, \omega_{0,2}$. From $\omega_{0,1}, \omega_{0,2}$ one derives a certain integral kernel $K(z,w)$, which behaves like a differential form in one variable and vector field in the other one. Forms $\omega_{g,n}$ except of $\omega_{0,1}, \omega_{0,2}$ belong to the symmetric powers $Sym^n(V)$ of a certain infinite-dimensional vector space V of meromorphic 1-forms. The form $\omega_{0,1}$ is auxiliary and is used only via so-called recursion kernel. The form $\omega_{0,2}$ is not an honest element of $Sym^2(V)$.

In this paper we suggest to change the set of initial data, by replacing the above forms by four tensors assigned to V, which we denote by A, B, C, ϵ. These tensors can be derived from $\omega_{0,1}, \omega_{0,2}$. The tensor A is essentially $\omega_{0,3}$, and the tensor ϵ is $\omega_{1,1}$. There are explicit formulas for the tensors B and C which we will discuss in the paper. The vector space V is the space of such meromorphic 1-forms on the spectral curve which have trivial residues at the ramification points.

2010 *Mathematics Subject Classification.* Primary 51P05, 81Q30.
The second author's work was partially supported by an NSF grant.

From this new point of view the topological recursion becomes a special case of the deformation quantization. More precisely it gives a cyclic vector (wave function) of a deformation quantization module (DQ-module for short) corresponding to a certain quadratic[1] Lagrangian subvariety in the infinite-dimensional symplectic vector space $V \oplus V^*$. The Lagrangian subvariety can be described explicitly in terms of the tensors A, B, C, ϵ. The construction is purely algebraic and it does not depend on the notion of spectral curve. Furthermore, it makes V into an algebra over a certain PROP defined by the above four tensors.

From this perspective spectral curves are not fundamental objects in the formalism of topological recursion, contrary to the conventional point of view.

In particular, the vector space V can be arbitrary (e.g. finite-dimensional or infinite-dimensional).[2]

1.2. Lie algebra of the topological recursion. The notion of *Airy structure* underlying our approach to TR is named after the archetypical example of the "Airy spectral curve" $x = y^2$. In fact there are two notions: *classical Airy structure* and *quantum Airy structure*.

Starting with the tensors A, B, C we define a *classical* Airy structure. It is equivalent to the above-mentioned quadratic Lagrangian subvariety. Skew-symetrization of the tensor C endows V with a Lie coalgebra structure, hence a Lie algebra structure on $\mathfrak{g} := V^*$.

The tensor ϵ is responsible for the quantization of the classical Airy structure. It gives a trivialization (i.e. a way to represent the cocycle as a coboundary) of a certain 2-cocycle on the Lie algebra $\mathfrak{g}$. The cocycle itself is a restriction to the above Lie algebra of the famous "Japanese cocycle" for the Lie algebra $gl(V \oplus V^*)$. In the case when $dim\, V < \infty$ the restricted cocycle is naturally a coboundary. In the infinite-dimensional case (which we are mostly interested in) a choice of trivializing 1-cochain ϵ is an additional piece of information. In the case of the Airy spectral curve one finds that $\mathfrak{g} \simeq \mathbf{C}[[x]]\,\partial_x \ltimes (\mathbf{C}[[x]]/\mathbf{C})$.

1.3. Spectral curves and the corresponding Airy structures. Our notion of spectral curve is more general than the one in the conventional TR. Namely, we consider a smooth complex Poisson surface P (which can be non-compact), such that the divisor D of zeros of the Poisson bivector field is smooth. We assume that the open symplectic leaf $S := P - D$ is endowed with a Lagrangian foliation $\mathcal{F}$, which extends by continuity to P and the extension is tangent to D. Then an $(\mathcal{F}, D)$-transversal spectral curve is a smooth compact curve $\Sigma \subset P$ which intersects transversally D at finitely many points p_β with multiplicities $k_\beta \geq 1$, and which is tangent with order 1 to leaves of the foliation $\mathcal{F}$ at finitely many points r_α.

These data determine a germ of complex analytic space $\mathcal{B}_1$, which is the moduli space of deformations of Σ in the class of spectral curves. It is a subspace of a bigger moduli space $\mathcal{B}_0$ of spectral curves defined in the same way, but without the tangency condition.

[1] Here we mean that the subvariety under consideration is the set of common zeroes of a collection of polynomials of degree at most 2.

[2] In [**ABCO**] the authors study the dynamics of Young diagrams via an "abstract" version of topological recursion which seems to give an Airy structure in our sense.

Consider the family of symplectic vector spaces

$$\mathcal{H}_\Sigma := \mathbb{H}^1(\Sigma, \Omega^0_\Sigma(-\sum_\beta k_\beta p_\beta) \to \Omega^1_\Sigma(\sum_\beta k_\beta p_\beta)) .$$

We prove that the family $\mathcal{H}_\Sigma$ gives rise to a symplectic vector bundle on $\mathcal{B}_0$, endowed with an *affine* flat connection. It follows that (locally) $\mathcal{B}_0$ can be embedded into any affine symplectic vector space $\mathcal{H}_\Sigma$ as a Lagrangian subvariety. In other words we have a (germ of) Lagrangian submanifold in an affine symplectic space.

Then the general story of deformation quantization ensures that there is a cyclic DQ-module which quantizes the Lagrangian image of $\mathcal{B}_1$. This cyclic DQ-module can be realized as a representation of the (completed) Weyl algebra of $\mathcal{H}_\Sigma$. Locally (along the Lagrangian germ) it is unique up to isomorphism. Nevertheless there is no preferable choice of a cyclic vector.

It turns out that our quantum Airy structure gives a canonical cyclic vector $\psi_{\mathcal{B}_1}$ (wave function). In fact it exists even for a bigger space $\mathcal{B}_0$. More precisely, the finite-dimensional simplectic vector space $\mathcal{H}_\Sigma$ can be represented as the quotient $G/G^\perp$, where G is an infinite-dimensional coisotropic subspace of a symplectic vector space W of meromorphic 1-forms with trivial residue on the disjoint union of formal discs about r_α (here $G^\perp$ is the skew-orthogonal to G). Then $\mathcal{B}_1$ can be obtained as a symplectic (a.k.a Hamiltonian) reduction of an infinite-dimensional quadratic Lagrangian submanifold $L \subset W$, i.e. $\mathcal{B}_1$ is isomorphic to the image of $L \cap G$ under the natural projection $G \to G/G^\perp = \mathcal{H}_\Sigma$. The Lagrangian submanifold L gives rise to a classical Airy structure on any Lagrangian complement to a tangent space to L.

Any Lagrangian complement $\mathcal{V}_\Sigma$ to the finite-dimensional Lagrangian subspace $T_\Sigma \mathcal{B}_0 \subset \mathcal{H}_\Sigma$ gives (after the pullback by epimorphism $G \twoheadrightarrow \mathcal{H}_\Sigma$) an infinite-dimensional Lagrangian complement V_Σ to the tangent space to L. This gives us a classical Airy structure on V_Σ, which can be quantized. The corresponding wave function ψ is given by the TR, and it corresponds to $\psi_{\mathcal{B}_1}$. Notice that the choice of V_Σ corresponds to a choice of the Bergman kernel in the language of conventional TR.

1.4. Topological recursions. Another interesting observation is that there are different versions of TR. In particular, we define two sets of tensors: the set $(A_{TR}, B_{TR}, C_{TR}, \epsilon_{TR})$ corresponding to the conventional (Eynard-Orantin) version of the TR, and the set (A, B, C, ϵ) mentioned above, which is based on certain residue constraints arising in the deformation theory of the spectral curve Σ. These two sets of tensors are different. Furthermore, the set $(A_{TR}, B_{TR}, C_{TR}, \epsilon_{TR})$ does not satisfy the axiomatics of Airy structures. Nevertheless the comparison of the two approaches can be made. For that we introduce a certain subspace $V_{odd} \subset V$ respected by both sets of tensors. Furthermore they define on it the same Airy structure (which is a natural Airy substructure of the one given by (A, B, C, ϵ)). This leads to a new recursion formulas, which are different from those of Eynard-Orantin, but produce the same collection of polydifferentials $\omega_{g,n}, n \geq 1, 2-2g-n < 0$.

Moreover, our modified recursion is in a sense simpler than the original Eynard-Orantin TR. Our formulas do not involve local involutions (compare Sections 3.1 and 3.4).

1.5. Contents of the paper. Classical and quantum Airy structures are introduced in Section 2. We also prove there the result that connects TR with deformation quantization. The proof is purely algebraic, so it works for Airy structures over an arbitrary field of characteristic zero. An important corollary of our formalism is the fact that the tensors arising in this "abstract" topological recursion are symmetric.

In Section 3 we consider the local version of TR. First we recall the basics on the conventional TR and derive the corresponding tensors $(A_{TR}, B_{TR}, C_{TR}, \epsilon_{TR})$. As we have already pointed out they do not satisfy the axioms of the Airy structure, contrary to the alternative approach via the residue constraints. We explain in Section 3 that two approaches agree if we restrict both tensors to the smaller subspace V_{odd}.

Finally we compare two approaches in general. The comparison reveals the role of Bergman kernel and the notion of primitive Airy structure which we introduced in Section 2.

In Section 4 we discuss the notion of spectral curve as well the deformation theory of spectral curves. It turns out that the natural framework for that is the one of a complex Poisson surface which has an open symplectic leaf endowed with an analytic foliation. This allows us to consider e.g. non-compact spectral curves, like those which are associated with irregular Hitchin systems (see e.g. [**KS14**]).

In Section 5 we discuss the notion of affine symplectic connection which allows us to embed the germ of the moduli space of spectral curves into an affine symplectic space as a germ of Lagrangian submanifold.

In Section 6 we discuss an infinite-dimensional version of the material of Section 5 in the case of spaces of embedded formal discs. There we use the ideas of formal differential geometry of Gelfand and Kazhdan (see [**GelKaz**]).

In Section 7 we explain the formal version of the embedded Lagrangian germ. It is described in terms of the Hamiltonian reduction of a certain infinite-dimensional Lagrangian submanifold of a Tate vector space with respect to a coisotropic subspace. In that construction only the coistoropic space depends on the global geometry of the spectral curve. Rest of the data are local: they are derived from the formal neighborhoods of points, where the spectral curve is tangent to the foliation. An important corollary is that TR solves the so-called Holomorphic Anomaly Equation, i.e. quantizes the Lagrangian image of $\mathcal{B}_0$ in a covariantly constant way.

In Section 8 we discuss a possible extension of the quantization of the moduli space of smooth spectral curves to the case when there is no foliation at all. In the case when the foliation exists, the moduli space of smooth spectral curves can be embedded locally as a Lagrangian submanifold in an affine symplectic space (e.g. in examples it can be an open dense subvariety of the base of Hitchin integrable system). In the case when there is no foliation, the moduli space still can be embedded locally as a Lagrangian submanifold, but the ambient symplectic manifold does not have a global affine structure. The ambient space has a weaker structure which we call a *semi-affine symplectic structure*.

In Section 9 we propose two speculations. One is about a possible relation of the quantization of the moduli space of smooth spectral curves (e.g. the above-mentioned open subset of the Hitchin base) with the quantization of an individual spectral curve. Another speculation is about differential-graded Airy structures derived from the moduli spaces of *compact* 3-dimensional Calabi-Yau varieties.

Few final remarks.

We warn the reader that this paper is not about applications of the topological recursion. It is about the origin of the recursion relations. For that reason we make only a basic comparison of the formalism of Airy structures with the one of Eynard-Orantin. We leave for the future work many interesting aspects of the story, which (as we hope) admit a natural explanation in the framework of Airy structures. Those include e.g. the relation of the TR to WKB quantization of an *individual* spectral curve (see [**DuMu**]), symplectic invariance in TR (see [**EynOr2**]). We do not discuss the passage from the multi-differentials $\omega_{g,n}$ to the corresponding functions $F_{g,n}$. There are many other questions which arise naturally in the framework of our formalism, but which we prefer to leave for the future work. This includes e.g. the generalization of TR to non-smooth spectral curves and also to so-called quantum curves.

Acknowledgments. The authors thank to J. Andersen, G. Borot, L. Chekhov, B. Eynard, M. Mulase for discussions on various aspects of topological recursion. Y.S. thanks to IHES for excellent research conditions.

2. Airy structures

2.1. Classical Airy structures. Let $\mathbf{k}$ be a field of characteristic zero, V an n-dimensional $\mathbf{k}$-vector space. Let us choose an ordered basis on V and denote by $y_1, ..., y_n$ the corresponding coordinates. Denote by $x^1, ..., x^n$ the dual coordinates on $V^* = Hom_{\mathbf{k}}(V, \mathbf{k})$.

On the vector space $W := T^*(V^*) = V \oplus V^*$ we have the standard symplectic structure. The corresponding Poisson bracket in the coordinates are given by the formulas $\{y_i, x^j\} = \delta_{ij}$, $\{x^i, x^j\} = \{y_i, y_j\} = 0$.

We denote by $Sym_{\leq 2}(W)$ the Lie algebra $Sym^0(W) \oplus Sym^1(W) \oplus Sym^2(W)$ of polynomial functions on W^* of degree less or equal than 2, endowed with the natural Lie algebra structure induced by the above Poisson bracket.

DEFINITION 2.1.1. Classical Airy structure on V is given by a collections of at most quadratic polynomials (i.e. elements of $Sym_{\leq 2}(W)$) of the form

$$H_i := -y_i + \sum_{jk} a_{ijk} x^j x^k + 2 \sum_{jk} b_{ij}^k x^j y_k + \sum_{jk} c_i^{jk} y_j y_k, 1 \leq i \leq n,$$

such that the vector space $\oplus_{1 \leq i \leq n} \mathbf{k} \cdot H_i$ is closed under the Poisson bracket.

To save the language we will often skip the word "at most" and call such polynomials *quadratic*. The same terminology will apply to the variety defined by equations $H_i = 0, 1 \leq i \leq n$.

In a coordinate-free form a classical Airy structure on V is defined by three tensors:

a) tensor $A = (a_{ijk}) \in V^{\otimes 3}$,
b) tensor $B = (b_{ij}^k) : V \to V \otimes V$,
c) tensor $C = (c_i^{jk}) : V \otimes V \to V$,

such that A and C are symmetric with respect to the permutation of the indices j, k.

Let us analyze the constraints on A, B, C.

By definition, any classical Airy structure gives rise to a Lie subalgebra

$$\mathfrak{g} := V^* \simeq \oplus_{1 \leq i \leq n} \mathbf{k} \cdot H_i \subset Sym_{\leq 2}(W.)$$

Let us compute its structure constants. We have

$$\{H_{i_1}, H_{i_2}\} = \{-y_{i_1} + 2\sum_{j,k} b^k_{i_1 j} x^j y_k + ..., -y_{i_2} + 2\sum_{j,k} b^k_{i_2 j} x^j y_k + ...\}$$
$$= 2\sum_k (b^k_{i_1 i_2} - b^k_{i_2 i_1}) y_k + \dots .$$

Since in the RHS we must have a linear combination of $H_1, ..., H_n$ we conclude that

$$\{H_{i_1}, H_{i_2}\} = \sum_k g^k_{i_1 i_2} H_k,$$

where $g^k_{i_1 i_2} = 2(b^k_{i_2 i_1} - b^k_{i_1 i_2})$ are the structure constants of $\mathfrak{g}$.

Since there are no linear terms in $x^i, 1 \leq i \leq n$ in the expression for the generator H_k, we can see (by directly computing the coefficient in front of x^i in the LHS of the above formula) that the tensor a_{ijk} is symmetric with respect to arbitrary permutations of the indices i, j, k.

In a similar way comparing in the above formula the coefficients in front of the quadratic monomials in x^i and y_j we can reformulate the notion of a classical Airy structure in the following way: it is a collection of tensors

$$A \in (V^{\otimes 3})^{S_3} \subset V^{\otimes 3},$$
$$B \in Hom(V, V \otimes V),$$
$$C \in Hom((V \otimes V)_{S_2}, V) = Hom(V \otimes V, V)^{S_2} \subset Hom(V \otimes V, V),$$

where the notation like E^{S_m} (resp E_{S_m}) refers to invariants (resp. coinvariants) of the action of the symmetric group S_m on an S_m-vector space E. The above three tensors are required to satisfy the following three relations: for any $1 \leq i_1, i_2, i_3, i_4 \leq n$

$$\text{Coeff}_{x^{i_3} x^{i_4}} \Delta_{i_1, i_2} = 0, \ \text{Coeff}_{x^{i_3} y_{i_4}} \Delta_{i_1, i_2} = 0, \ \text{Coeff}_{y_{i_3} y_{i_4}} \Delta_{i_1, i_2} = 0$$
$$\text{where } \Delta_{i_1, i_2} := \{H_{i_1}, H_{i_2}\} - 2\textstyle\sum_k (b^k_{i_2 i_1} - b^k_{i_1 i_2}) H_k. \tag{1}$$

These relations can be considered in a coordinate-free way as certain cumbersome purely quadratic in A, B, C expressions with values in respectively

$$\wedge^2 V \otimes Sym^2 V, \ \ Hom(V, \wedge^2 V \otimes V), \ \ Hom(Sym^2 V, \wedge^2 V).$$

Furthermore, we can reformulate the notion of classical Airy structure geometrically. Let M be a $\mathbf{k}$-affine symplectic space of dimension $2n$ and $\mathfrak{g}$ is a Lie algebra over $\mathbf{k}$ of dimension n which acts on M by Hamiltonian transformations. Assume that the corresponding moment map $\mu : M \to \mathfrak{g}^*$ is at most quadratic. This means that the moment map is given explicitly by $\mu(p) = (H_1(p), ..., H_n(p))$, where each H_i is a polynomial of degree less or equal than two on $M \simeq \mathbf{k}^{2n}$.

Let $p \in \mu^{-1}(0)$ be a smooth point. The germ L at p of the algebraic variety $\mu^{-1}(0)$ is a smooth coisotropic subvariety of M of dimension n, hence it is Lagrangian. We will call such Lagrangian submanifolds (or germs of such submanifolds) *quadratic Lagrangian submanifolds*.

The tangent space $T_pL \subset T_pM$ is a Lagrangian subspace. Let us choose a Lagrangian vector subspace $V \subset T_pM$ which is transversal to T_pL (we call it *Lagrangian complement*).

PROPOSITION 2.1.2. *The vector space V carries a natural Airy structure. Conversely any classical Airy structure is isomorphic in the natural sense to the one given by this geometric construction.*

PROOF. Indeed the above choices identify symplectically M with $V \oplus V^* = T^*(V^*)$, as long as we identified V^* with T_pL. The quadratic Hamiltonians H_i has the form

$$H_i := -y_i + \sum_{jk} a_{ijk}x^j x^k + 2\sum_{jk} b_{ij}^k x^j y_k + \sum_{jk} c_i^{jk} y_j y_k, 1 \le i \le n.$$

Hence we have a classical Airy structure on V.

Conversely, if we have a classical Airy structure on V, we observe that $\mathfrak{g} = V^*$ acts in the Hamiltonian way on $M = V \oplus V^*$, the moment map $\mu : V \oplus V^* \to \mathfrak{g}^* \simeq V$ is given by the formula $\mu(p) = (H_1(p), ..., H_n(p))$ with the Hamiltonians as above, and that $p = 0$ is a smooth point of $\mu^{-1}(0)$. This completes the proof. $\square$

REMARK 2.1.3. The space $\mathcal{O}(M)_{\le 2}$ of polynomial functions on M of degree less or equal than 2 is a Lie subalgebra of the algebra $\mathcal{O}(M)$ of polynomial functions on M. Then the above considerations mean that a classical Airy structure is given by a choice of an n-dimensional Lie algebra $\mathfrak{g}$ over $\mathbf{k}$ together with a monomorphism of Lie algebras $\phi : \mathfrak{g} \to \mathcal{O}(M)_{\le 2}$, a choice of a point $p \in \mu^{-1}(0)$ where $\mu^{-1}(0)$ is a non-empty smooth variety of dimension n and a choice of a transversal Lagrangian complement to the tangent space to $\mu^{-1}(0)$ at p .

Suppose that the field $\mathbf{k} = \mathbf{C}$ or $\mathbf{R}$. Then we have the Lie group $G = exp(\mathfrak{g})$ (say, connected and simply-connected). This group acts on $\mu^{-1}(0)$ by affine symplectic isomorphisms. On a small neighborhood of the smooth point p the action is locally transitive with the discrete stabilizer. This means that for any two points p_1, p_2 which are sufficiently close to p there is a unique element $g = g_{p_1,p_2} \in G$ close to id_G which maps p_1 to p_2 and hence inducing an identifications of affine spaces of Lagrangian complements to $T_{p_i}\mu^{-1}(0) \subset T_{p_i}M$, and an equivalence of the corresponding Airy structures. It is not clear a priori (although very likely) that the obtained equivalence depends *algebraically* (and not only analytically) on the Lie subalgebra $\mathfrak{g} \subset \mathcal{O}(M)_{\le 2}$, and a pair of smooth points $p_1, p_2 \in \mu^{-1}(0)$.

Finally, let us write the formula for the change of tensors A, B, C with respect to the change of the transversal Lagrangian subspace at a given smooth point p. Any two Lagrangian complements $V_p^{(1)}, V_p^{(2)}$ to $T_pL \subset T_pM$ are both canonically identified with $T_p^*L = T_pM/T_pL$, hence identified with each other. One of them can be seen as the graph of linear map $V \to V^*$ which comes from a symmetric bilinear pairing $Sym^2(V) \to \mathbf{k}$. In local coordinates the calculation reduces to the application of a linear symplectic change of variables

$$y_i \to y_i,\ \ x^i \to x^i + \sum_j s^{ij} y_j,$$

where $s=(s^{ij})_{1\le i,j\le n}$ is a symmetric matrix. A direct calculation gives us the following **gauge transformaton**:

$$\begin{aligned} a_{ijk} &\to a_{ijk}, \\ b_{ij}^k &\to b_{ij}^k+\textstyle\sum_l a_{ijl}s^{lk}, \\ c_i^{jk} &\to c_i^{jk}+\textstyle\sum_{l_1,l_2} a_{il_1l_2}s^{l_1j}s^{l_2k}+\sum_l b_{il}^k s^{lj}+\sum_l b_{il}^j s^{lk}. \end{aligned} \tag{2}$$

2.2. Quantum Airy structures. We keep the notation of the previous subsection. The vector space of at most quadratic functions $Sym_{\le 2}(W), W=V\oplus V^*$ is spanned by

$$\{1,x^i,y_i,x^ix^j,x^iy_j,y_iy_j\},\quad 1\le i,j\le n\,.$$

We define an associative $\mathbf{k}$-algebra $Sym^{quant}(W)$ generated by $1,\hbar,x^i,\hbar\partial_i$, with 1 and $\hbar$ being central elements, subject to the relations $(\hbar\partial_i)x^j-x^j(\hbar\partial_i)=\hbar\delta_{ij}$ (i.e. $\partial_i=\partial/\partial x^i$ if we allow the division by $\hbar$). We define $Sym_{\le 2}^{quant}(W)$ as the vector subspace of $Sym^{quant}(W)$ spanned by

$$\{1,\hbar,x^i,\hbar\partial_i,x^ix^j,\hbar(x^i\partial_j+\partial_jx^i),\hbar^2\partial_i\partial_j\},\quad 1\le i,j\le n\,.$$

It is a Lie algebra with the bracket equal to $(ab-ba)/\hbar$. Clearly modulo $\hbar$ it is isomorphic to the Lie algebra $Sym_{\le 2}(W)$.

DEFINITION 2.2.1. A **quantization** of a classical Airy structure is given by a monomorphism $\mathfrak{g}\simeq V^*\to Sym_{\le 2}^{quant}(W)$ such that modulo $\hbar$ it coincides with the monomorphism $\mathfrak{g}\to Sym_{\le 2}(W)$. We will call the former monomorphism quantum Airy structure.

We have quantized Hamilitonians

$$\widehat{H}_i=-\hbar\partial_i+\sum_{j,k}a_{ijk}x^jx^k+2\hbar\sum_{j,k}b_{ij}^kx^j\partial_k+\hbar^2\sum_{j,k}c_i^{jk}\partial_j\partial_k+\hbar\varepsilon_i\cdot 1,$$

where $\varepsilon=(\varepsilon_i\in\mathbf{k})_{1\le i\le n}$ are some constants.

In the coordinate-free notation ε is an element of V. The constraint on tensor ε follows from the relation saying that

$$\forall i_1,i_2\ \ \mathrm{Coeff}_{\hbar\cdot 1}\left(\frac{1}{\hbar}\left(\widehat{H}_{i_1}\widehat{H}_{i_2}-\widehat{H}_{i_2}\widehat{H}_{i_1}\right)-2\sum_k(b_{i_2i_1}^k-b_{i_1i_2}^k)\widehat{H}_k\right)=0. \tag{3}$$

In the coordinate-free form, this constraint means vanishing of certain purely quadratic expression in A,B,C,ε with values in $\wedge^2V$. Together with classical constraints (1) we obtain an explicit algebraic description of quantum Airy structures.

The gauge transformation formula (2) extends to the quantum case as

$$\varepsilon_i\ \to\ \varepsilon_i+\sum_{jk}a_{ijk}s^{jk}. \tag{4}$$

(Apply transformation $\hbar\partial_i\to\hbar\partial_i,\ x^i\to x^i+\sum_j s^{ij}\hbar\partial_j$ where $s=(s^{ij})_{1\le i,j\le n}$ is a symmetric matrix).

For any classical Airy structure on a *finite-dimensional* space V there exists a *canonical* choice of its quantization[3]. Namely, we set

$$\widehat{H}_i^{can} := -\hbar\partial_i + \sum_{j,k} a_{ijk}x^jx^k + \hbar\sum_{j,k} b_{ij}^k x^j\partial_k + \hbar\sum_{j,k} b_{ij}^k\partial_k x^j + \hbar^2\sum_{j,k} c_i^{jk}\partial_j\partial_k.$$

In other words, we set $\varepsilon_i^{can} := \sum_j b_{ij}^j$. A general solution to (3) is

(5) $\quad \varepsilon = \varepsilon^{can} + \chi,$ where $\chi \in V = \mathfrak{g}^*$ is any character of $\mathfrak{g} \Leftrightarrow \chi \in H^1(\mathfrak{g}, \mathbf{k})$.

2.3. Airy structures in the infinite-dimensional case. Four tensors A, B, C, ε satisfy four relations, see (1) and (3).

Notice that since all four relations include acyclic tensors, our definitions of Airy structures (both classical and quantum) make sense in the case of any $\mathbf{k}$-linear tensor category. In particular, they make sense in the case when $dim\, V = \infty$. Notice that in the infinite-dimensional case Lie algebra $\mathfrak{g}$ is naturally endowed with linear topology being dual to an ordinary vector space V (endowed with the discrete topology). In other words, V is a Lie coalgebra.

Let us analyze constraint (3). Rewrite it as

(6) $$\forall i_1, i_2 \;\; 2\sum_{j,k}(c_{i_1}^{jk}a_{i_2jk} - c_{i_2}^{jk}a_{i_1jk}) = \sum_k g_{i_1i_2}^k\varepsilon_k.$$

where $g_{i_1i_2}^k := 2(b_{i_2i_1}^k - b_{i_1i_2}^k)$ are structure constants of $\mathfrak{g}$. One sees that the RHS is the (i_1, i_2)-component of the coboundary of the continuous cochain (ε_k) on the Lie algebra $\mathfrak{g} = V^*$.

Let us look more closely on the expression in the LHS in (6). As the indices i_1, i_2 are fixed, we see that $a_{i_1jk}, a_{i_2jk} \in Sym^2(V)$, and $c_{i_1}^{jk}, c_{i_2}^{jk} \in \widehat{Sym^2}(V^*)$, where the hat notation over a vector space indicates the natural topological completion.[4]

Notice that the Lie algebra $gl(W)$ carries a natural 2-cocycle ψ ("Japanese cocycle"). Namely, splitting $W = V \oplus V^*$ allows us to represent any linear map S as a block 2×2 matrix $(s_{ij})_{1\le i,j\le 2}$. Then the cocycle ψ can be written as $\psi(S,T) = Tr(t_{21}s_{12} - s_{21}t_{12})$. Comparing it with the above expression for the LHS of (6), we see that the latter is twice of the pull-back of the Japanese 2-cocycle to the Lie algebra $\mathfrak{g}$ under the composition of natural morphisms

$$\mathfrak{g} \to \widehat{Sym_{\le 2}W} \twoheadrightarrow \widehat{Sym^2}(W) \simeq sp(W) \hookrightarrow gl(W).$$

Therefore existence of a quantization of a classical Airy structure in these terms corresponds to triviality of the pullback of the cohomology class ψ.

Ansatz (5) shows that in the finite-dimensional case one can recover the 1-cochain ε from the other data by taking $\varepsilon_i = \sum_j b_{ij}^j + \chi(H_j)$, where $\chi : \mathfrak{g} \to \mathbf{k}$ is any character. In the infinite-dimensional case this formula does not make sense since it involves taking the trace of the tensor B. In this case the *continuous* 1-cochain ε is a new datum for the quantum Airy structure.

[3]In the next section we will talk about Airy structures in the infinite-dimensional case where the analogous assertion is false in general.

[4]We hope that the reader will not mix it with the hat indicating quantization, like in the previous section.

In general, if a quantization of a classical Airy structure given by a Lie algebra $\mathfrak{g}$ exists, all such quantizations form a torsor over the space of characters $Hom(\mathfrak{g}, \mathbf{k}) = H^1_{cont}(\mathfrak{g}, \mathbf{k})$. In particular, if $H^1_{cont}(\mathfrak{g}, \mathbf{k}) = H^2_{cont}(\mathfrak{g}, \mathbf{k}) = 0$ then there exists a unique quantization of the classical Airy structure.

Finally, we remark that the tensors A, B, C, ε satisfying the above four quadratic relations make V into an algebra over a certain PROP. Obviously, (quantum or classical) Airy structures form a category (there is a straightforward notion of a morphism). Also, for any family of Airy strructures on spaces $(V_i)_{i\in I}$ the direct sum space $\oplus_{i\in I} V_i$ carries a natural direct sum Airy structure.

2.4. Cyclic DQ-modules. We start with considerations which hold in a bigger generality than the framework of Airy structures. The corollary of these considerations will be the (non-trivial) fact that the polydifferential forms $\omega_{g,n}(z_1, ..., z_n)$ defined via the topological recursion are symmetric with respect to the variables $z_1, ..., z_n$.

Let us first assume that V is a finite-dimensional vector space over a field $\mathbf{k}$ of characteristic zero. For convenience, we introduce the coordinates $x^i, 1 \le i \le \dim V$ on V^* as we discussed previously[5]. In order to keep track of the invariant (= coordinate-free) meaning, we use the natural identification of the algebra of formal power series $\mathbf{k}[[x^\bullet]]$ with $\prod_{n\ge 0} Sym^n(V)$, by associating with every series f the collection of tensors (Taylor coefficients)

$$Taylor_n(f) \in \left(V^{\otimes n}\right)^{Sym_n} \subset V^{\otimes n}, \quad Taylor_n(f)_{i_1,\dots,i_n} := \partial_{i_1} \dots \partial_{i_n}(f)(0).$$

Let $\mathbf{D}$ be the completed algebra of differential operators on V^*. More precisely, $\mathbf{D} = \mathbf{k}[[\hbar]]\,[[x^\bullet]]\,[[\hbar\partial_\bullet]]$ endowed with the standard product. Let us introduce a grading on $\mathbf{D}$ such that $deg\, \hbar = 2, deg\, x^i = deg\, \hbar\partial_i = 1$. Denote by $\mathbf{I} \subset \mathbf{D}$ the closed 2-sided ideal spanned by elements of degree ≥ 2.

Let now $\widehat{H}_i, 1 \le i \le \dim V$ be elements of $\mathbf{D}$ which satisfy the following properties:

1) $\widehat{H}_i = -\hbar\partial_i + P_i$, where $P_i \in \mathbf{I}$;

2) for any $i \ne j$ the commutator $[\widehat{H}_i, \widehat{H}_j]$ belongs to the left ideal generated by $\hbar\widehat{H}_\bullet$, i.e. there exists (non-unique) "structure constants" $g^k_{i,j} \in \mathbf{D}$ such that

$$\frac{1}{\hbar}[\widehat{H}_i, \widehat{H}_j] = \sum_k g^k_{i,j} \widehat{H}_k.$$

Notice that the case when $\widehat{H}_i, 1 \le i \le \dim V$ form a basis of a Lie algebra over $\mathbf{k}$, i.e. $g^k_{i,j} \in \mathbf{k} \subset \mathbf{D}$ (and, more generally, over $\mathbf{k}[[\hbar]]$: $g^k_{i,j} \in \mathbf{k}[[\hbar]] \subset \mathbf{D}$) is a particular case of the above condition.

Having a formal expression

$$\psi = \exp(S) = \exp\left(S_0/\hbar + \sum_{g\ge 1} \hbar^{g-1} S_g\right),$$

with

$$S_g = \sum_{n\ge 0} S_{g,n} \in \prod_{n\ge 0} Sym^n(V) = \mathbf{k}[[x^\bullet]], \forall\, g \ge 0$$

[5]In this section we *do not* denote the dimension of V by n, the letter n is reserved for the homogeneity degree of polynomials in $x^\bullet$.

we define a continuous automorphism $T_\psi : \mathbf{D} \to \mathbf{D}$ which corresponds informally to the conjugation $D \mapsto \psi^{-1} D \psi$. Explicitly, the automorphism T_ψ is defined on the generators by the formulas $T_\psi(x^i) = x^i, T_\psi(\hbar\partial_i) = \hbar\partial_i + \partial_i(S_0 + \sum_{g\geq 1} \hbar^g S_g)$. Of course this is motivated by the well-known formula $e^{-f}\partial_i e^f = \partial_i + \partial_i(f)$. By definition we have $T_{\psi\phi} = T_\psi T_\phi$, where the product $\psi\phi$ is defined naturally, as the exponent of the sum of two expressions. The formula $\widehat{H}(e^S f) = e^S(T_{e^S}(\widehat{H})(f))$ which is valid for the differential operators $\widehat{H}$ and functions S, f motivates the following definition.

DEFINITION 2.4.1. We say that the above formal expression ψ satisfies the system of equations $\widehat{H}_i(\psi) = 0$ if $T_\psi(\widehat{H}_i)(1) = 0, 1 \leq i \leq \dim V$.

In the language of quantum mechanics, ψ is a wave function in WKB form.

THEOREM 2.4.2. *Suppose that the Hamiltonians $\widehat{H}_i, 1 \leq i \leq \dim V$ satisfy the above conditions 1) and 2). Then there exists a unique solution to the system of equations $\widehat{H}_i(\psi) = 0, 1 \leq i \leq \dim V$ of the form $\psi = exp(S_0/\hbar + \sum_{g\geq 1} \hbar^{g-1} S_g)$, where $S_0 = \sum_{n\geq 3} S_{0,n}$ and for $g \geq 1$ $S_g = \sum_{n\geq 1} S_{g,n}$ with $S_{g,n} \in Sym^n(V)$ for any $g, n \geq 0$ (in particular $S_{0,\leq 2} = S_{\geq 0,0} = 0$).*

PROOF. We will proceed by induction in g, and for each g we will proceed by induction in n. We start with $g = 0$ case. The equation for $g = 0$ says that for any i

$$H_i(x^\bullet, \partial_\bullet S_0) = 0 \in \mathbf{k}[[x^\bullet]].$$

Here the classical limit $H_i \in \mathbf{k}[[x^\bullet, y_\bullet]]$ is defined as the image of $\widehat{H}_i$ in $\mathbf{D}/\hbar\mathbf{D} \simeq \mathbf{k}[]x^\bullet]] [[\hbar\partial_\bullet]]$, when we identify variables y_i with the images of $\hbar\partial_i$, $1 \leq i \leq \dim V$ in the quotient algebra $\mathbf{D}/\hbar\mathbf{D}$. The condition 1) implies that $H_i = -y_i$ modulo quadratic and higher order terms in $x^\bullet, y_\bullet$. The classical limit of the commutation relation 2) implies that $H_\bullet$ generate a Poisson ideal. Therefore, the formal germ at 0 of a submanifold L in $\mathbf{k}^{2\dim V} = V \oplus V^*$ given by equation $H_i(x^\bullet, y_\bullet) = 0$ is coisotropic. It is of dimension $\dim V$, hence it is Lagrangian. This implies that it is given by the graph of dS_0 for some formal power series $S_0 \in \mathbf{k}[[x_\bullet]]$. We can assume that the constant term is zero. The tangent space to L at 0 is given by equations $y_i = 0$, $1 \leq i \leq \dim V$, hence S_0 starts with a cubic term.

It is useful to formulate the proof purely in terms of linear algebra using the induction on $n \geq 3$ (the order of the term in the Taylor expansion). In what follows we will use a shorthand notation where the upper index describes the constraints on the homogeneity degree of monomials in variables $x^\bullet$ or $y_\bullet$. First, let us prove that there exists a unique cubic polynomial $S_{0,3} = x^{=3} \in \mathbf{k}[x^\bullet]$ such that $H_i(x^\bullet, \partial_\bullet S_{0,3}) = x^{\geq 3}$ $\forall i$. Indeed, we have

$$H_i = -y_i + H_{i,2} + x^{\geq 3} + x^{\geq 1} y^{=1} + x^{\geq 0} y^{\geq 2},$$

where $H_{i,2} = x^{=2}$ is a homogeneous polynomial in x of degree 2. The equation on $S_{0,3}$ says that $-\partial_i S_{0,3} + H_{i,2} = 0$ $\forall i$. The Poisson bracket

$$\{H_{i_1}, H_{i_2}\} = \partial_{i_1} H_{i_2,2} - \partial_{i_2} H_{i_1,2} + x^{\geq 2} + x^{\geq 0} y^{\geq 1}$$

belongs to the ideal generated by $H_\bullet$. Hence $-\partial_{i_1} H_{i_2,2} + \partial_{i_2} H_{i_1,2} = 0$ $\forall i_1, i_2$, i.e. 1-form $\sum_i H_{i,2} dx^i$ is closed, therefore exact and given by $dS_{0,3}$ for a homogeneous cubic polynomial $S_{0,3}$.

Now, let us assume that we have already a solution of $H_i(x^\bullet, \partial_\bullet S_0) = 0$ up to order $n \geq 3$, and want to extend it to a solution up to order $n+1$. Applying the conjugation by $\exp(S_0/\hbar)$ we may assume that the solution up to order n vanishes, i.e.

$$S_0 = x^{\geq(n+1)}, \quad H_i(x^\bullet, \partial_\bullet S_0) = x^{\geq(n+1)} \ \forall i \ .$$

This implies that H_i has the form

$$H_i = -y_i + H_{i,n} + x^{\geq(n+1)} + x^{\geq 1}y^{=1} + x^{\geq 0}y^{\geq 2}.$$

Similarly to the previous arguments, the Poisson bracket has the form

$$\{H_{i_1}, H_{i_2}\} = -\partial_{i_1} H_{i_2,n} + \partial_{i_2} H_{i_1,n} + x^{\geq n} + x^{\geq 0}y^{\geq 1}.$$

Hence homogeneous 1-form $\sum_i H_{i,n}dx^i$ is closed and can be written as $dS_{0,n+1}$ where $S_{0,n+1} = x^{=(n+1)}$. This proves the n-th step of induction for the case $g = 0$.

Finally, we proceed with the induction on g. Assume that for $g \geq 1$ and $n \geq 0$ we have a solution S up to $\hbar^{g-1}x^{\geq(n+1)} + \hbar^{\geq g}x^{\geq 1}$. At beginning of our induction, when $n = 0$, we may assume that $S_{g,0} = 0$. Indeed adding a multiple of $\hbar^{g-1}$ to S does not change the constraint $\widehat{H}_i(\exp(S)) = 0$. Applying the conjugation, we may assume that

$$S = \hbar^{g-1}x^{\geq(n+1)} + \hbar^{\geq g}x^{\geq 1}, \quad \widehat{H}_i(\exp(S)) = 0 \text{ modulo } \hbar^g x^{\geq n} + \hbar^{\geq(g+1)}x^{\geq 0},$$

where $\exp(S) := 1+S+S^2/2!+\dots$ is a well-defined invertible element of $\mathbf{k}[[\hbar]][[x^\bullet]] \subset \mathbf{D}$ (unlike the formal expression $\psi = \exp(S_0/\hbar + \dots)$ when $S_0 \neq 0$). The latter condition on $\widehat{H}_i, S$ implies that

$$\widehat{H}_i = -\hbar\partial_i + \hbar^g x^{\geq n} + \hbar^{\geq(g+1)}x^{\geq 0} + \hbar^{\geq 0}x^{\geq 1}(\hbar\partial)^{\geq 1} + \hbar^{\geq 1}x^{\geq 0}(\hbar\partial)^{\geq 1}.$$

Let us denote by $H_{g;i,n} \in \mathbf{k}[x^\bullet]$ the homogeneous term of the type $\hbar^g x^{=n}$ in $\widehat{H}_i$. The commutator has the form

$$[\widehat{H}_{i_1}, \widehat{H}_{i_2}] = -\hbar\partial_{i_1} H_{g;i_2,n} + \hbar\partial_{i_2} H_{g;i_1,n} + \text{ higher order terms},$$

where the higher order terms are of the form

$$\hbar^{g+1}x^{\geq n} + \hbar^{\geq(g+2)}x^{\geq 0} + \hbar^{\geq 0}x^{\geq 1}(\hbar\partial)^{\geq 1} + \hbar^{\geq 1}x^{\geq 0}(\hbar\partial)^{\geq 1}.$$

The constraint 2)

$$[\widehat{H}_{i_1}, \widehat{H}_{i_2}] \in \hbar \sum_j \mathbf{D} \cdot \widehat{H}_j$$

implies that $\hbar\partial_{i_1} H_{g;i_2,n} = \hbar\partial_{i_2} H_{g;i_1,n}$, hence there exists a unique homegenous $S_{g,n} \in \mathbf{k}[x^\bullet]$ of degree $n+1$ such that $H_{g;i,n} = \hbar^g \partial_i S_{g,n} \ \forall i$. Then $S = \hbar^{g-1}S_{g,n}+\dots$ is a solution up to higher order terms $\hbar^{g-1}x^{\geq(n+2)} + \hbar^{\geq g}x^{\geq 1}$. We proved the induction step $(g,n) \to (g,n+1)$, hence proved the theorem. □

Let us discuss algebraic meaning of the above theorem. First of all, the algebra $\mathbf{D}$ is a topologically free module over the central subalgebra $\mathbf{k}[[\hbar]]$. Furthermore $[\mathbf{D}, \mathbf{D}] \subset \hbar\mathbf{D}$. We claim that the condition 2) on the collection of elements $\widehat{H}_i$ is equivalent to the constraint *only* on the left ideal $J := \sum_i \mathbf{D} \cdot \widehat{H}_i \subset \mathbf{D}$:

$$J = \mathbf{D}J, \quad [J, J] \subset \hbar J.$$

The implication in one direction is obvious. Conversely, assuming condition 2) we have for $\forall a_1, a_2 \in \mathbf{D}$, $\forall i_1, i_2$

$$[a_1 \widehat{H}_{i_1}, a_2 \widehat{H}_{i_2}] = [a_1 \widehat{H}_{i_1}, a_2]\, \widehat{H}_{i_2} + a_2 a_1 [\widehat{H}_{i_1}, \widehat{H}_{i_2}] + a_2\, [a_1, \widehat{H}_{i_2}]\, \widehat{H}_{i_1} \in \hbar J.$$

Theorem means that under conditions 1), 2) the cyclic module $\mathcal{M} := \mathbf{D}/J$ is isomorphic to the canonical holonomic $\mathbf{D}$-module associated with the germ of Lagrangian submanifold $graph(dS_0)$. In particular it is a topologically free $\mathbf{k}[[\hbar]]$-module.

In the case (relevant to Airy structures) when $\mathfrak{g} := \oplus_i \mathbf{k} \cdot \widehat{H}_i$ is closed under Lie bracket, the fact that cyclic module $\mathcal{E} := \mathbf{D}/\sum_i \mathbf{D}\widehat{H}_i$ is topologically free over $\mathbf{k}[[\hbar]]$, has an easy alternative proof. Indeed, let us consider the $\hbar$-enveloping algebra $U_\hbar(\mathfrak{g})$ of $\mathfrak{g}$ (i.e. the commutation relations are $ab - ba = \hbar[a,b], a, b, \in \mathfrak{g}$). It is a deformation quantization of $Sym^*(\mathfrak{g})$ endowed with the Lie-Kirillov-Kostant Poisson bracket. The natural completion $\widehat{U}_\hbar(\mathfrak{g})$ is a topologically free $\mathbf{k}[[\hbar]]$-algebra, which is naturally embedded to the topologically free $\mathbf{k}[[\hbar]]$-algebra $\mathbf{k}[[x_i, \hbar\partial_i, \hbar]]$ of $\hbar$-differential operators (it is an embedding modulo $\hbar$ which implies an embedding over $\mathbf{k}[[\hbar]]$). Moreover, $\mathbf{D}$ is topologically free as right $U_\hbar(\mathfrak{g})$-module. Finally,

$$\mathcal{E} = \mathbf{D}/\sum_i \mathbf{D}\widehat{H}_i = \mathbf{D} \otimes_{U_\hbar(\mathfrak{g})} \mathbf{k}[[\hbar]],$$

where we treat $\mathbf{k}[[\hbar]]$ as the trivial rank one representation of $\widehat{U}_\hbar(\mathfrak{g})$. Hence $\mathcal{E}$ is topologically free over $\mathbf{k}[[\hbar]]$. This argument generalizes immediately to the case of Lie algebra over $\mathbf{k}[[\hbar]]$.

2.5. Application to quantum Airy structures: Abstract topological recursion. Assume that our Hamiltonians $\widehat{H}_i, 1 \le i \le n$ come from a quantum Airy structure. The general result from the previous subsection implies that there exists a unique WKB solution $\exp(S_0/\hbar + \dots)$. We obtain a collection of symmetric tensors

$$S_{g,n} \in Sym^n(V),\ g = 0, n \ge 3 \text{ or } g \ge 1, n \ge 1,$$
$$S_{g,n;i_1,\dots,i_n} := \partial_{i_1} \dots \partial_{i_n} S_g(0).$$

The equation $\widehat{H}_i(\exp(S)) = 0$ for $S = \sum_g \hbar^{g-1} S_g,\ S_g = \sum_n S_{g,n}$ reads as

$$\hbar\partial_i S = \sum_{jk} a_{ijk}\, x^j x^k + 2 \sum_{jk} b_{ij}^k\, x^j\, \hbar\partial_k S +$$
$$+ \sum_{jk} c_i^{jk}\, \hbar\partial_j(S)\hbar\partial_k(S) + \sum_{jk} c_i^{jk}\, \hbar^2 \partial_j \partial_k(S) + \hbar\varepsilon_i.$$

Expanding the RHS of the above equation into series in $\hbar, x^\bullet$ we obtain recursive formulas for $\partial_i S_{g,n}$ in terms of tensors A, B, C, ε and previously known $S_{g',n'}$ with either $g' < g$ or $g' = g, n' < n$. It is *not obvious* a priori that the RHS which should be equal by induction to the 1-form $dS_{g,n} = \sum_i \partial_i S_{g,n} dx^i$ is in fact a closed 1-form (which would imply exactness). Theorem guarantees that this is the case.

The recursion starts with the cases $g = 0, n = 3$ and $g = 1, n = 1$. Then

$$\text{at the level of tensors } S_{0,3} = 2A \iff \text{as a polynomial } S_{0,3} = \frac{1}{3}\sum_{ijk} a_{ijk} x^i x^j x^k;$$

$$\text{at the level of tensors } S_{1,1} = \varepsilon \iff \text{as a polynomial } S_{1,1} = \sum_i \varepsilon_i x^i.$$

In general, for all other possible values of g, n (i.e. for $g = 0, n \geq 4$, or $g = 1, n \geq 2$ or $g \geq 2, n \geq 1$) we have

$$S_{g,n;i,i_1,\ldots,i_{n-1}} = 2\sum_{\alpha=1}^{n-1}\sum_{k} b^k_{ii_\alpha} S_{g,n-1;k,i_{\{1,\ldots,n-1\}\setminus\{\alpha\}}}$$

$$+ \sum_{g_1+g_2=g;J_1\sqcup J_2=\{1,\ldots,n-1\}} \sum_{jk} c_i^{jk} S_{g_1,\#J_1+1;j,i_{J_1}} S_{g_2,\#J_2+1;k,i_{J_2}} \tag{7}$$

$$+\sum_{jk} c_i^{jk} S_{g-1,n+2;j,k,i_1,\ldots,i_{n-1}},$$

where at the RHS we use only terms $S_{g',n'}$ with either $g' = 0, n' \geq 3$ or $g' \geq 1, n' \geq 1$. Also, here for any subset $J \subset \{1, \ldots, n-1\}$ such that $J = \{\alpha_1, \ldots, \alpha_{\#J}\}$ we denote by i_J the sequence $i_{\alpha_1}, \ldots, i_{\alpha_{\#J}}$. This sequence *depends* on the order in which we enumerate elements of J. Nevertheless this dependence is irrelevant for the recursion formula because tensors $S_{g,n}$ are symmetric.

We call the above formula **abstract topological recursion**.

EXAMPLE 2.5.1. Suppose we are given a homogeneous cubic polynomial P_3 as well as the non-degenerate quadratic form P_2, both in variables $y_\bullet$.

Consider the $1/\hbar$-Fourier transform

$$I(\hbar, x) = \int_Z e^{-\frac{1}{\hbar}xy} e^{\frac{1}{\hbar}(P_2(y)+P_3(y))} dy.$$

Here we understand $I(\hbar, x)$ in terms of formal Feynman diagram expansion as $\hbar \to 0$. The "integration cycle" Z is the germ at the origin of a real closed n-dimensional ball in $\mathbf{C}^n$ such that $Re(P_2 + P_3)_{|Z-\{0\}} < 0$.

Then one has

$$\log(I(\hbar, x)) \sim const + \sum_{g\geq 0, n\geq 0, 2-2g-n<0} \hbar^{g-1} x^n S_{g,n}$$

with the coefficients computed by Feynman rules. This means that

$$S_{g,n} = \sum_{G(g,n)} R_{G(g,n)}/|Aut(G(g,n)|,$$

where the sum is taken over all connected 3-valent graphs of genus g with n numbered tails, each tail is assigned the corresponding variable $x^\bullet$. The degree n homogeneous polynomial $R_{G(g,n)}(x)$ is obtained by formal application of the Feynman rules (i.e. every edge is assigned the inverse to the quadratic form P_2, while each 3-valent vertex is assigned the cubic form P_3). Then our theorem claims that $S_{g,n}$ satisfy the abstract topological recursion formulas. The corresponding Lie algebra $\mathfrak{g}$ is abelian, and the quantum Airy structure is obtained by the gauge transformation (see (2) and (4)) from the following Ansatz:

$$\widehat{H}_i := -\hbar\partial_i + \sum_{j,k} a_{ijk}x^j x^k, \ \ \widehat{H}_i \left(\exp\left(\hbar^{-1}S_{0,3}(x)\right)\right) = 0.$$

Here $S_{0,3} = \frac{1}{3}\sum_{ijk} a_{ijk}x^i x^j x^k$ is obtained by a change of variables in P_3 from $y_\bullet$ to $x^\bullet$ via the inverse of the quadratic form P_2.

2.6. Primitive Airy structures. The terminology below will be used both in classical and quantum cases. For that reason sometimes we will skip the words "classical" or "quantum", and will speak about Airy structures, etc. The mathematical content of this section is almost empty, we introduce it only to handle the fact that the topological recursion as originally defined by Eynard and Orantin, does *not* satisfy the axiomatics of Airy structures. Nevertheless, the data of the conventional TR contain an Airy substructure. Hence by our general theory of Airy structures the conventional TR produces a collection of symmetric tensors (usually denoted by $\omega_{g,n}$).

DEFINITION 2.6.1. A **classical pre-Airy structure** on vector space V given by a tuple (A, B, C) such that

$$A \in V^{\otimes 3},\ B(V) \subset V \otimes V,\ C(V \otimes V) \subset V.$$

A **quantum pre-Airy structure** on V is given by a tuple (A, B, C, ε) such that (A, B, C) is a classical pre-Airy structure and $\varepsilon \in V$ is a vector. A **substructure** of a (quantum or classical) pre-Airy structure is given by a vector subspace $U \subset V$ such that

$$A \in U^{\otimes 3},\ B(U) \subset U \otimes U,\ C(U \otimes U) \subset U.$$

and $\varepsilon \in U$ in the quantum case.

Obviously, any quantum pre-Airy structure on V induces a classical pre-Airy structure.

DEFINITION 2.6.2. A (quantum or classical) pre-Airy structure on space V is called **primitive** if it does not contain a nontrivial substructure (i.e. the one with $U \neq V$).

Obviously, if U is an Airy substructure, then it carries the induced classical Airy structure. Equivalently, an Airy substructure is the same as a morphism of Airy structures which is an inclusion at the level of underlying vector spaces. It is easy to see that any (quantum or classical) pre-Airy structure contains a unique primitive substructure.

In both quantum or classical cases, an Airy structure is a particular case of a pre-Airy structure. A substructure (in the sense of pre-structures) of an Airy structure is again an Airy structure.

If a classical Airy structure has an Airy substructure then the generating series S_0 from the Theorem 2.4.2 is constant along fibers of the linear surjection $V^* \twoheadrightarrow U^*$. Equivalently, for a basis in V such that U is spanned by a part of the basis, the series S_0 depends only on the corresponding subset of the coordinates in V^*. Classical Hamiltonians corresponding to elements of $(V/U)^* \subset V^*$ are linear (i.e. they do not contain non-trivial quadratic forms).

Geometrically, the existence of a non-trivial classical Airy substructure implies that the germ L of the corresponding Lagrangian submanifold in an affine symplectic space M lies in a non-trivial affine coisotropic submanifold $N \subset M$. Hence L is the pullback of a Lagrangian germ $L' \subset M'$, where M' is a symplectic affine space which is the quotient of N by the constant foliation given by the skew-orthogonal $T_N^{\perp} \subset T_N$ (equivalently, by the kernel of the symplectic form restricted to N).

Similarly to the above, one can speak about substructures and primitivity of *quantum* Airy structures, by adding the condition $\varepsilon \in U$. Quantum Hamiltionians corresponding to $(V/U)^* \subset V^*$ are just derivations along corresponding directions, and the equation on wave function means that $\exp(S)$ does not depend on a part of coordinates.

It follows from the above considerations that if one uses a quantum Airy structure only as a technical tool for producing the series S_g, $g \geq 0$ then it suffices to work with any quantum Airy substructure, in particular with the primitive one.

REMARK 2.6.3. The modification formulas (2) and (4) make sense for pre-Airy structures as well, where the symmetric marrix $s = (s^{ij})$ controlling the gauge change is understood in coordinate-free manner as a symmetric bilinear form on V. Obviosuly, if $U \subset V$ is a substructure, then the restriction of bilinear form s to U gives a gauge change of the induced structure on U.

3. Comparison with topological recursion

3.1. Reminder: Topological recursion. Here we recall the classical Eynard-Orantin formalism for topological recursion. Let Σ be a smooth complex curve (not necessarily connected, e.g. a disjoint union of discs). We assume that we are given a finite collection of distinct points $(r_\alpha \in \Sigma)_{\alpha \in Ram}$ where *Ram* is a finite set, a collection of germs of involutions $\sigma = \sigma_\alpha$ defined in the vicinity of r_α such that

$$\sigma(r_\alpha) = r_\alpha,\ \sigma_{|T_{r_\alpha}\Sigma} = -id.$$

In what follows we will consider "tensor fields" on the product of copies of Σ. They are sections of tensor products of pull-backs the line bundles K_Σ and K_Σ^{-1}, where $K_\Sigma = T_\Sigma^*$ is the canonical bundle of Σ. For example, the expression $dz_1 \dots dz_n$ on $\mathbf{C}^n$ where $\Sigma = \mathbf{C}$ and $(z_i)_{i=1,\dots,n}$ are coordinates on Σ^n is not the top degree differential form, it is a polydifferential. Almost all our tensor fields will be polydifferentials, i.e. the pull-backs of K_Σ^{-1} will be absent (exception is the recursion kernel below). We denote by $(\pi_i)_{i=1,\dots,n}$ the projections $\Sigma^n \to \Sigma$ to factors.

The input data for the topological recursion consist of:

(1) The *Bergman kernel* $\omega_{0,2} \in \Gamma(\Sigma \times \Sigma \setminus \{diag\}, \pi_1^* K_\Sigma \otimes \pi_2^* K_\Sigma)$ such that $\omega_{0,2} = \omega_{0,2}(z_1, z_2)$ is symmetric under permutation $z_1 \leftrightarrow z_2$ and has second order pole at the diagonal with the leading coefficient of the polar part equal to 1: in local coordinates near every point

$$\omega_{0,2}(z_1, z_2) = \frac{dz_1\, dz_2}{(z_1 - z_2)^2} + \text{ regular terms for } z_1 \text{ close to } z_2.$$

(2) A germ of 1-form $\omega_{0,1} = \omega_{0,1}(z_1)$ defined in a vicinity of each point r_α, satisfying the condition that

$$\omega_{0,1}(z_1) - \omega_{0,1}(\sigma(z_1)) \text{ vanishes with the order 2 at } r_\alpha.$$

In particular the form $\omega_{0,1}$ vanishes at r_α.

In a typical example, Σ is an affine curve in the plane $\mathbf{C} \times \mathbf{C}$ endowed with coordinates (x, y), such that the projection to x-coordinate $\mathbf{C}$ has only double ramification points r_α (hence the notation *Ram* for the set of ramification points). The local involution σ_α interchanges two branches of the projection $x : \Sigma \to \mathbf{C}$ near r_α, and (the germ of) the 1-form $\omega_{0,1}$ is $y\, dx$.

DEFINITION 3.1.1. The **recursion kernel** $K = K(z_1, z_2 = z)$ is a collection labeled by $\alpha \in Ram$ of sections of $\pi_1^* K_\Sigma \otimes \pi_2^* K_\Sigma^{-1}$ defined for arbitrary $z_1 \in \Sigma$ such that $z_1 \neq r_\alpha$ for *all* $\alpha \in Ram$, and for arbitrary z lying in the vicinity of r_α for *some* α, $z \neq z_1, r_\alpha$, given by the formula

$$K(z_1, z) := \frac{-1}{2} \cdot \frac{\int_{z'=\sigma(z)}^{z'=z} \omega_{0,2}(z_1, z')}{\omega_{0,1}(z) - \omega_{0,1}(\sigma(z))},$$

where in the numerator we use arbitrary integration path from $z' = \sigma(z)$ to $z' = z$ in the vicinity of r_α.

EXAMPLE 3.1.2. The most important for us example is the *Airy curve* $\Sigma = \{(x, y) \in \mathbf{C}^2 | x = y^2\}$ parametrized by the coordinate z so that $x = z^2, y = z$. In this case we have only one ramification point $z = 0$. The involution is given by $\sigma(z) = -z$. The most natural choice of the Bergman kernel is $\omega_{0,2}(z_1, z_2) := \frac{dz_1\, dz_2}{(z_2 - z_1)^2}$. Also, the 1-form $\omega_{0,1}$ is $\omega_{0,1}(z) := y(z)dx(z) = 2z^2 dz$.

The integral in the definition of the recursion kernel is

$$\int_{z'=-z}^{z'=z} \frac{dz'\, dz_1}{(z' - z_1)^2} = dz_1 \cdot \int_{z'=-z}^{z'=z} d\frac{1}{z_1 - z'} = dz_1 \cdot \left(\frac{1}{z_1 - z} - \frac{1}{z_1 + z}\right) = \frac{2z\, dz_1}{(z_1^2 - z^2)}$$

$$\Longrightarrow K(z_1, z) := \frac{1}{4(z^2 - z_1^2)z} \frac{dz_1}{dz}.$$

Finally, given the input data $\omega_{0,1}, \omega_{0,2}$ one defines a collection of new tensor fields $\omega_{g,n}$ for $g = 0, n \geq 3$ or $g \geq 1, n \geq 1$ which are sections of $\pi_1^* K_\Sigma \otimes \cdots \otimes \pi_n^* K_\Sigma$ on $(\Sigma - \{r_\alpha | \alpha \in Ram\})^n$ inductively by the formula

$$\text{for } g = 0, n \geq 3 \text{ or } g \geq 1, n \geq 1 \qquad \omega_{g,n}(z_1, z_2, \ldots, z_n)$$

$$(8) \qquad = \sum_\alpha Res_{z=r_\alpha} K(z_1, z) \cdot \omega_{g-1,n+1}(z, \sigma(z), z_2, \ldots, z_n)$$

$$+ \sum_\alpha \sum_{\substack{g_1+g_2=g \\ I_1 \sqcup I_2 = \{2,\ldots,n\}}}^{*} Res_{z=r_\alpha} K(z_1, z) \cdot \omega_{g_1,1+\#I_1}(z, z_{I_1}) \cdot \omega_{g_2,1+\#I_2}(\sigma(z), z_{I_2}).$$

The first term in the RHS appears only if $g \geq 1$. The symbol $\sum^*$ in the second term indicates that we *do not* allow cases $g_1 = 0, \#I_1 = 0$ or $g_2 = 0, \#I_2 = 0$, i.e. only $\omega_{0,2}$, $\omega_{0,\geq 3}$ and $\omega_{\geq 1, \geq 1}$ can appear in the RHS. Notice that $\omega_{0,1}$ does not appear explicitly in the RHS, it participates in the whole formalism only via the denominator $\omega_{0,1}(z) - \omega_{0,1}(\sigma(z))$ for the recursion kernel.

An amazing property of the above formalism is that the tensor fields $\omega_{g,n}$ defined recursively are invariant under the action of the symmetric group S_n. The recursion formulas for cases $g = 0, n = 3$ and $g = 1, n = 1$ are somewhat different from the rest:

$$(9) \quad \textbf{A-type} \ : \omega_{0,3}(z_1, z_2, z_3) = 2 \sum_\alpha Res_{z=r_\alpha} K(z_1, z)\, \omega_{0,2}(z, z_2)\, \omega_{0,2}(\sigma(z), z_3),$$

$$(10) \qquad \boldsymbol{\varepsilon}\textbf{-type} \ : \omega_{1,1}(z_1) = \sum_\alpha Res_{z=r_\alpha} K(z_1, z)\, \omega_{0,2}(z, \sigma(z)).$$

In further cases $g=0, n\geq 4$, or $g=1, n\geq 2$ or $g=2, n\geq 1$ in the RHS of the recursion formula for $\omega_{g,n}$ one has *two* types of summands:

(1) **B-type**: the terms including decomposition $g=g_1+g_2, \{2,\dots,n\} = I_1\sqcup I_2$ with *either* $g_1=0, \#I_1=1$ or $g_2=0, \#I_2=1$. These terms are given by the action of a linear operator whose argument $f(z)\,dz$ is a collection of germs of meromorphic 1-forms at points r_α,

$$f(z)dz\mapsto \sum_\alpha Res_{z=r_\alpha} K(z_1,z)\,\omega_{0,2}(\sigma(z),z_2)\,f(z)dz. \tag{11}$$

(2) **C-type**: all other terms, corresponding to a bilinear operator on pairs $(f(z_1)\,dz_1, g(z_2)\,dz_2)$ of collections of germs of meromorphic forms as above,

$$f(z_1)dz_1\otimes g(z_2)dz_2\mapsto \sum_\alpha Res_{z=r_\alpha} K(z_1,z)\,f(z)dz\,g(\sigma(z))\,d(\sigma(z))\,. \tag{12}$$

We can recast the topological recursion by saying that $\omega_{0,3},\omega_{1,1}$ are the *initial data* (of correspondingly A-type and ε-type), and the recursive steps use only operations of B-type or C-type. Comparing formula (8) with the formula (7) from Section 2.5, we see that the perfect match:

$$\omega_{g,n}\longleftrightarrow S_{g,n},\quad g=0, n\geq 3 \text{ or } g\geq 1, n\geq 0\,.$$

Nevertheless, in order to identify the topological recursion with the formalism of Airy structures, one has to choose an appropriate space V of 1-forms on $\Sigma-\{r_\alpha|\,\alpha\in Ram\}$.

3.2. Topological recursion, the case of Airy curve. Recall from Example 3.1.2 Bergman and recursion kernels for the Airy curve:

$$\omega_{0,2}(z_1,z_2):=\frac{dz_1\,dz_2}{(z_2-z_1)^2},\quad K(z_1,z):=\frac{1}{4(z^2-z_1^2)z}\frac{dz_1}{dz}.$$

Let $V:=z^{-1}\mathbf{C}[z^{-1}]\frac{dz}{z}$. We endow it with the basis

$$e_n:=z^{-n}\frac{dz}{z},\quad n=1,2,\dots$$

Define 4 tensors of topological recursion

$$A_{TR}\in Sym^3(V), B_{TR}\in Hom(V,V\otimes V), C_{TR}\in Hom(V\otimes V,V)^{S_2}, \varepsilon_{TR}\in V$$

via formulas (see (9),(10),(11), (12)):

$$\begin{array}{ll}
A_{TR}:= & Res_z K(z_1,z)\,\omega_{0,2}(z,z_2)\,\omega_{0,2}(-z,z_3),\\
B_{TR}: f(z)dz\mapsto & Res_z K(z_1,z)\,\omega_{0,2}(z,z_2)\,f(-z)d(-z),\\
C_{TR}: f(z)dz\otimes g(z)dz\mapsto & Res_z K(z_1,z)\,f(z)dz\,g(-z)d(-z),\\
\varepsilon_{TR}:= & Res_z K(z_1,z)\,\omega_{0,2}(z,-z),
\end{array}$$

where $Res_z := Res_{z=0}$. Let us calculate the structure constants of these tensors in the basis $(e_n)_{n\geq 1}$. We have:

$$A_{TR} = -\frac{1}{4} Res_z \frac{1}{(z_1^2 - z^2)z} \frac{dz_1}{dz} \frac{dz\, dz_2}{(z_2 - z)^2} \frac{-dz\, dz_3}{(z_3 + z)^2}$$

$$= \frac{dz_1\, dz_2\, dz_3}{4} Res_z \frac{dz}{z} \left(\frac{1}{z_1^2} + \frac{z^2}{z_1^4} + \dots \right) \left(\frac{1}{z_2^2} + \frac{2z}{z_2^3} + \dots \right) \left(\frac{1}{z_3^2} - \frac{2z}{z_3^3} + \dots \right)$$

$$= \frac{1}{4} \frac{dz_1}{z_1^2} \frac{dz_2}{z_2^2} \frac{dz_3}{z_3^2} = \frac{1}{4} e_1 \otimes e_1 \otimes e_1,$$

$$B_{TR} : e_n = z^{-n} \frac{dz}{z} \mapsto -\frac{1}{4} Res_z \frac{1}{(z_1^2 - z^2)z} \frac{dz_1}{dz} \frac{dz\, dz_2}{(z_2 - z)^2} (-z)^{-n} \frac{dz}{z}$$

$$= -\frac{1}{4} dz_1\, dz_2\, Res_z \frac{dz}{z} (-1)^n z^{-n-1} \left(\frac{1}{z_1^2} + \frac{z^2}{z_1^4} + \dots \right) \left(\frac{1}{z_2^2} + \frac{2z}{z_2^3} + \frac{3z^2}{z_2^4} + \dots \right)$$

$$= -\frac{1}{4} dz_1\, dz_2\, (-1)^n \sum_{n_1, n_2 \geq 0;\, 2n_1 + n_2 = n+1} (n_2 + 1)\, z_1^{-2n_1 - 2}\, z_2^{-n_2 - 2}$$

$$= -\frac{1}{4} \sum_{n_1, n_2 \geq 0;\, 2n_1 + n_2 = n+1} (-1)^{n_2 + 1} (n_2 + 1)\, e_{2n_1 + 1} \otimes e_{n_2 + 1},$$

$$C_{TR} : e_{n_1} \otimes e_{n_2} = \frac{dz}{z^{n_1 + 1}} \otimes \frac{dz}{z^{n_2 + 1}} \mapsto -\frac{1}{4} Res_z \frac{1}{(z_1^2 - z^2)z} \frac{dz_1}{dz} \frac{dz\, d(-z)}{z^{n_1 + 1} (-z)^{n_2 + 1}}$$

$$= -\frac{1}{4} (-1)^{n_2}\, dz_1\, Res_z \frac{dz}{z} z^{-(n_1 + n_2 + 2)} \left(\frac{1}{z_1^2} + \frac{z^2}{z_1^4} + \dots \right)$$

$$= \begin{cases} 0 & \text{if } n_1 + n_2 \in 1 + 2\mathbf{Z} \\ -\frac{1}{4}(-1)^{n_2}\, e_{2m+1} & \text{if } n_1 + n_2 + 2 = 2m \geq 2, \end{cases}$$

$$\varepsilon_{TR} = -\frac{1}{4} Res_z \frac{1}{(z_1^2 - z^2)z} \frac{dz_1}{dz} \frac{dz\, d(-z)}{4z^2} = -\frac{1}{4} dz_1\, Res_z \frac{dz}{z} \frac{-1}{4z^2} \left(\frac{1}{z_1^2} + \frac{z^2}{z_1^4} + \dots \right)$$

$$= \frac{dz_1}{16\, z_1^4} = \frac{1}{16} e_3.$$

For the future reference we combine above formulas:

$$A_{TR} = \frac{1}{4} e_1 \otimes e_1 \otimes e_1, \tag{13}$$

$$B_{TR} : e_n \mapsto \frac{1}{4} \sum_{n_1, n_2 \geq 1;\, n_1 + n_2 = n+3;\, n_1\ odd} (-1)^{n_2 + 1} n_2\, e_{n_1} \otimes e_{n_2}, \tag{14}$$

$$C_{TR} : e_{n_1} \otimes e_{n_2} \mapsto \begin{cases} 0 & \text{if } n_1 + n_2 \text{ odd} \\ \frac{1}{4}(-1)^{n_2 + 1}\, e_{n_1 + n_2 + 3} & \text{if } n_1 + n_2 \text{ even}, \end{cases} \tag{15}$$

$$\varepsilon_{TR} = \frac{1}{16} e_3. \tag{16}$$

We will see below that the data $(V, A_{TR}, B_{TR}, C_{TR}, \varepsilon_{TR})$ define a quantum pre-Airy structure on V, which is *not* a quantum Airy structure. In order to fix the problem let us consider $V_{odd} := z^{-1}\mathbf{C}[z^{-2}] \frac{dz}{z} \subset V$, which is the vector subspace spanned by $e_1, e_3, e_5, \dots$. We will see that the restrictions of the tensors $A_{TR}, B_{TR}, C_{TR}, \varepsilon_{TR}$ to V_{odd} give rise to a quantum Airy structure.

3.3. Airy structure from residue constraints. Let us consider the infinite-dimensional symplectic vector space

$$W_{Airy} := \{\eta \in \mathbf{C}((z))\, dz \mid Res_z\,(\eta) = 0\} = \left\{ J \cdot \frac{dz}{z} \;\middle|\; J = \sum_{n\in\mathbf{Z};n\neq 0;n\ll+\infty} J_n z^{-n} \right\}$$

with the symplectic pairing given by

$$(17)\quad \langle df_1, df_2\rangle := Res_z\,(f_1\, df_2) \Longleftrightarrow \left\langle z^{-n}\frac{dz}{z}, z^n\frac{dz}{z}\right\rangle = \frac{1}{n} \Longleftrightarrow \{J_n, J_{-n}\} = n.$$

Let us introduce a family of residue constraints on the elements η which are polynomials of degree ≤ 2 in the coordinates[6] x, y defined below[7]:

$$\forall n \geq 1\ Res_z\,(y\, x^n\, dx) = 0, \qquad \forall m \geq 0\ Res_z\,(y^2\, x^m\, dx) = 0,$$

where

$$x := z^2 \in \mathbf{C}((z)),\ \ y = y(\eta) := z - \frac{J}{2x} = z - \frac{J}{2z^2},\ \ J = \frac{\eta}{z^{-1}dz}.$$

Equivalently, $y(\eta)\, dx = zd(z^2) - \eta$.

The first constraint $Res_z\,(y\, x^n\, dx) = 0$ gives

$$Res_z(\frac{J}{z^2}\, z^{2n}\, z\, dz) = Res_z(\frac{dz}{z}\, J\, z^{2n}) = 0 \Longleftrightarrow J_{2n} = 0.$$

The second constraint $Res_z\,(y^2\, x^m\, dx) = 0$ for $m \geq 0$ gives

$$Res_z\left(z - \frac{J}{2z^2}\right)^2 z^{2m}\, z\, dz = Res_z\,\frac{dz}{z}\left(-J\, z^{2m+1} + \frac{1}{4}J^2 z^{2m-2}\right) = 0 \Longleftrightarrow$$

$$-J_{2m+1} + \frac{1}{4}\sum_{k+l=2m-2;k,l\neq 0} J_k J_l = 0.$$

Therefore, the first and the second residue constraints can be written together as a system of equations $(H_n = 0)_{n\geq 1}$ where:

$$\begin{aligned}
H_1 &:= -J_1 + \tfrac{1}{4}J_{-1}^2 + \tfrac{1}{2}\left(J_{-3}J_1 + J_{-4}J_2 + \dots\right),\\
H_2 &:= -J_2,\\
H_3 &:= -J_3 + \tfrac{1}{2}\left(J_{-1}J_1 + J_{-2}J_2 + \dots\right),\\
H_4 &:= -J_4,\\
H_5 &:= -J_5 + \tfrac{1}{4}J_1^2 + \tfrac{1}{2}\left(J_{-1}J_3 + J_{-2}J_4 + \dots\right),\\
H_6 &:= -J_6,\\
H_7 &:= -J_7 + \tfrac{1}{4}J_2^2 + \tfrac{1}{2}J_1J_3 + \tfrac{1}{2}\left(J_{-1}J_5 + J_{-2}J_6 + \dots\right),\\
&\dots
\end{aligned}$$

REMARK 3.3.1. The infinite-dimensional topological vector or affine spaces considered in this paper are Tate spaces, i. e. they are direct sum of vector or affine spaces endowed with discrete topology and dual to such spaces. One can also say that Tate spaces are vector spaces endowed with the topology of the vector space of Laurent series. Formal germ at $0 \in W$ of the Tate space W (e.g. $W = W_{Airy}$)

---

[6] In formulas below x^n means the power of the variable x. The reader should not confuse it with the notation for the coordinates $(x^1, ..., x^n)$ in Section 2.

[7] The geometric meaning of the coordinates and the constraints will be explained below in Section 6.3. At the moment the reader should consider all that as a "black box" producing an Airy structure.

is understood as a functor on the category of finite-dimensional commutative local Artin algebras:

$$R \mapsto m_R^* \otimes W,$$

where $m_R \subset R$ is the maximal ideal. In this way one can also define the notion of subvariety, intersection of subvarieties, etc.

LEMMA 3.3.2. *The complete topological vector subspace of the space $Sym_{\leq 2}(W)$ spanned by $(H_n)_{n\geq 1}$ is closed under the Poisson bracket given in coordinates $(J_k)_{k\neq 0}$ by $\{J_k, J_{-k}\} = k$. Moreover, there exists a unique lift of $\prod_{n\geq 1} \mathbf{C}\cdot H_n$ to a Lie algebra of $\hbar$-differential operators.*

PROOF. It is convenient to add a central variable J_0 and denote for $\forall n \in \mathbf{Z}$

$$L_n := \frac{1}{4} \sum_{k+l=2n} J_k J_l.$$

In the end we will specialize results to the hypersurface $J_0 = 0$. Functions $L_\bullet$ form a Lie algebra with respect to the Poisson bracket:

$$\begin{aligned}
\{L_{n_1}, L_{n_2}\} &= \frac{1}{16} \sum_{k_1+l_1=2n_1, k_2+l_2=2n_2} \{J_{k_1}J_{l_1}, J_{k_2}J_{l_2}\} \\
&= \frac{1}{4} \sum_{k_1+l_1=2n_1, -k_1+l_2=2n_2} k_1 J_{l_1} J_{l_2} \\
= \frac{1}{4}\sum_k kJ_{2n_1-k}J_{2n_2+k} &= \frac{1}{8} \sum_{k_1+k_2=2(n_1+n_2)} ((2n_1-k_1)-(2n_2-k_2))J_{k_1}J_{k_2} \\
&= \frac{1}{8} \sum_{k_1+k_2=2(n_1+n_2)} 2(n_1-n_2)J_{k_1}J_{k_2} = (n_1-n_2)L_{n_1+n_2}.
\end{aligned}$$

The reader can recognize here the structure constants of the Lie algebra $\mathbf{C}[[x]]\partial_x$ of formal vector fields on the line. Also, let us calculate the bracket

$$\{L_n, J_m\} = \frac{1}{4} \sum_{k+l=2n} \{J_k J_l, J_m\} = \frac{1}{2}(-m)J_{2n+m}.$$

We have

$$H_{2n+3} = -J_{2n+3} + L_n \,\forall n \geq -1, \quad H_{2m} = -J_{2m}\,\forall m \geq 1.$$

This formulas implies

$$\begin{aligned}
&\{H_{2n_1+3}, H_{2n_2+3}\} = (n_1-n_2)H_{2n_1+2n_2+3}, \\
&\{H_{2n+3}, H_{2m}\} = (-m)H_{2n+2m}, \\
&\{H_{2m_1}, H_{2m_2}\} = 0.
\end{aligned} \tag{18}$$

The above calculation shows that the Lie algebra $\mathfrak{g} = \prod_{n\geq 1}\mathbf{C}\cdot H_n$ is isomorphic to the quotient of the algebra of differential operators of order ≤ 1 on $\mathbf{C}[[x]]$ by constants $\mathbf{C}\cdot 1$:

$$H_{2n+3} \leftrightarrow -x^{n+1}\partial_x \,\forall n \geq -1, \qquad H_{2m} \leftrightarrow x^m \quad \forall m \geq 1.$$

Let us introduce coordinates $x_1, x_2, \ldots$ on an infinite-dimensional space, and realize quantum operators $(\widehat{J}_n)_{n\in\mathbf{Z}, n\neq 0}$ as

$$\widehat{J}_n = \hbar\frac{\partial}{\partial x_n}, \quad \widehat{J}_{-n} := nx_n \;\forall n \geq 1.$$

The direct calculation shows that the same commutation relations as in Equation (18) hold for

$$\begin{array}{ll}\widehat{H}_{2n+3} & := -\widehat{J}_{2n+3} + \frac{1}{4}\sum_{k+l=2n;k,l\neq 0}\widehat{J}_k\widehat{J}_l,\ \ n=-1,1,2,3,\ldots,\\ \widehat{H}_3 & := -\widehat{J}_3 + \frac{1}{2}(\widehat{J}_{-1}\widehat{J}_1 + \widehat{J}_{-2}\widehat{J}_2 + \ldots) + \frac{\hbar}{16},\\ \widehat{H}_{2m} & := -\widehat{J}_{2m},\ \ m\geq 1.\end{array}$$

More precisely, instead of the Poisson bracket we use the Lie bracket $(ab-ba)/\hbar$.

The uniqueness of the quantum lift follows from the vanishing $H^1_{cont}(\mathfrak{g},\mathbf{C})$. The existence was proven above via explicit formulas.[8]

Let us split $W = W_{Airy}$ into the direct sum of two complementary Lagrangian subspaces:

$$W = V^*\oplus V,\ \ V^* := \mathbf{C}[[z]]dz,\ \ V := z^{-2}\mathbf{C}[z^{-1}]dz.$$

This splitting gives us a quantum Airy structure on V. More precisely, the germ of the quadratic Lagrangian submanifold $L = L_{Airy}$ defined by the system of equations $H_n = 0, n\geq 1$ has the tangent space at 0 given by the system of linear equations $J_1 = J_2 = \ldots. = 0$, which is isomorphic to V^*.

From the above splitting we derive the tensors A, B, C, ε. E.g. the tensor B is encoded in the bracketed sum in the formulas for $(H_n)_{n\geq 1}$.

In the basis $(e_n)_{n\geq 1}$ we have

$$\begin{aligned} &A = \tfrac{1}{4}e_1\otimes e_1\otimes e_1,\\ &B : e_n \mapsto \tfrac{1}{4}\textstyle\sum_{n_1,n_2\geq 1;\, n_1+n_2=n+3;\, n_1\ odd} n_2\, e_{n_1}\otimes e_{n_2},\\ &C : e_{n_1}\otimes e_{n_2} \mapsto \begin{cases} 0 & \text{if } n_1+n_2 \text{ odd}\\ \frac{1}{4}\, e_{n_1+n_2+3} & \text{if } n_1+n_2 \text{ even},\end{cases}\\ &\varepsilon = \tfrac{1}{16}e_3.\end{aligned}\tag{19}$$

One can write formulas for tensors B, C using residues:

$$\begin{aligned} B : f(z)dz &\mapsto \quad -Res_z\, K(z_1,z)\,\omega_{0,2}(z,z_2)\, f(z)dz,\\ C : f(z)dz\otimes g(z)dz &\mapsto \quad -Res_z\, K(z_1,z)\, f(z)dz\, g(z)dz.\end{aligned}\tag{20}$$

3.4. Common Airy substructure and modified topological recursion. We see (formulas (13), (14), (15), (16) versus (19)) that in the example of Airy curve we have *two* collections of tensors $(A_{TR}, B_{TR}, C_{TR}, \varepsilon_{TR})$ and (A, B, C, ε) which can be used for the recursive definition of $\omega_{g,n}$ with $g = 0, n\geq 3$ or $g\geq 1, n\geq 1$.

It turns out that they produce the *same* tensors $\omega_{g,n}$. It will follow from the fact that both collections of tensors (i.e. the corresponding pre-Airy structures) define quantum Airy structures on the vector subspace

$$V_{odd} := \oplus_{m\geq 0}\mathbf{C}\cdot e_{2m+1}\subset V,$$

and moreover these quantum structures coincide, since the restrictions of the above tensors on V_{odd} coincide.

Furthermore V_{odd} endowed with this quantum Airy structure is a primitive Airy structure (and the corresponding classical Airy structure is also primitive).

[8]Notice that $H^2_{cont}(\mathfrak{g},\mathbf{C})\neq 0$ as $\mathfrak{g}$ has a non-trivial central extension which is the Atiyah Lie algebra $\mathbf{C}[[x]]\partial_x\oplus\mathbf{C}[[x]]$, hence the existence of a lift is not automatic.

For this primitive *classical* structure the existence and uniqueness of the quantum lift is obvious, as $H^1_{cont}(\mathfrak{g}_0, \mathbf{C}) = H^2_{cont}(\mathfrak{g}_0, \mathbf{C}) = 0$ for the Lie algebra $\mathfrak{g}_0 \simeq \mathbf{C}[[x]]\partial_x$ of formal vector fields on the line (sometimes it is called Witt Lie algebra).

REMARK 3.4.1. One can characterize V_{odd} as a subspace of V intrinsically. Namely, it is the space of such meromorphic 1-forms from V that their direct image to the x-line under the projection $z \mapsto x = z^2$ has no poles.

Subspaces $V, V_{odd} \subset \mathbf{C}((z))\,dz$ are naturally associated with kernels $\omega_{0,2}(z_1, z)$ and $K(z_1, z)$ in the following way. Let us expand these kernels in the iterated Laurent series in the domain $0 \ll |z| \ll |z_1| \ll 1$:

$$\omega_{0,2}(z_1, z)_{|z|\ll|z_1|} = \left(\frac{1}{z_1^2} + \frac{2z}{z_1^3} + \frac{3z^2}{z_1^4} + \dots\right) dz_1 dz \in \mathbf{C}((z_1))((z))\,dz_1 dz,$$

$$K(z_1, z)_{|z|\ll|z_1|} = \frac{-1}{4}\cdot\left(\frac{z^{-1}}{z_1^2} + \frac{z}{z_1^4} + \frac{z^3}{z_1^6} + \dots\right)\frac{dz_1}{dz} \in \mathbf{C}((z_1))((z))\,\frac{dz_1}{dz}.$$

We see that $V^{z\to z_1}, V_{odd}^{z\to z_1}$ are the *minimal* subspaces of $\mathbf{C}((z_1))\,dz_1$ such that

$$\omega_{0,2}(z_1, z)_{|z|\ll|z_1|} \in V((z))\,dz, \quad K(z_1, z)_{|z|\ll|z_1|} \in V_{odd}((z))(dz)^{-1}.$$

Here the notation $z \to z_1$ means that we replace variable z by z_1 and realize V, V_{odd} as subspaces of $\mathbf{C}((z_1))\,dz_1$ instead of $\mathbf{C}((z))\,dz$.

Let us define the recursion based on tensors A, B, C, ε and using the residue formulas from equation (20). We call it the **modified topological recursion**. Explicitly, it is given by:

$$\text{initial data: } \omega_{0,3} = 2A = \frac{1}{2}\frac{dz_1\,dz_2\,dz_3}{z_1^2 z_2^2 z_3^2},\ \omega_{1,1} = \epsilon = \frac{1}{16}\frac{dz_1}{z_1^4},$$

for $g = 0, n \geq 4$ or $g = 1, n \geq 2$ or $g \geq 2, n \geq 1$

$$\omega_{g,n}(z_1, z_2, \dots, z_n) =$$
$$-\sum_\alpha Res_{z=r_\alpha} K(z_1, z)\cdot\omega_{g-1,n+1}(z, z, z_2, \dots, z_n)$$
$$-\sum_\alpha \sum^*_{\substack{g_1+g_2=g\\ I_1\sqcup I_2=\{2,\dots,n\}}} Res_{z=r_\alpha} K(z_1, z)\cdot\omega_{g_1,1+\#I_1}(z, z_{I_1})\cdot\omega_{g_2,1+\#I_2}(z, z_{I_2}).$$

The first sum in the RHS appears only if $g \geq 1$. Also, similarly to the Eynard-Orantin topological recursion, the notation $\sum^*$ in the second sum means that we do not allow cases $g_1 = 0, \#I_1 = 0$ or $g_2 = 0, \#I_2 = 0$. In fact, the above formula holds also in the a priori forbidden case $g = 0, n = 3$ (but not for $g = 1, n = 1$ as $\omega_{0,2}(z, z)$ is ill-defined).

Let us summarize differences between the modified topological recursion and the usual one:

a) the overall minus sign;

b) the absence of the local involution in the RHS of the formula for the modified topological recursion;

c) lesser range of validity of the modified topological recursion (i.e. $\omega_{0,3}$ and $\omega_{1,1}$ are treated in the modified recursion as initial data).

REMARK 3.4.2. An interesting (and a priori unexpected) corollary of the above discussion is the fact that the tensors $\omega_{g,n}$ derived from the usual (i.e. Eynard-Orantin) topological recursion satisfy also the modified topological recursion. This observation seems to be new.

3.5. Comparison: TR versus Airy structures in general. In general, in the formalism of topological recursion, one can replace the curve Σ by a disjoint union of small discs containing ramification points $\{r_\alpha\}$. Moreover, one can pick a local coordinate z on each disc (uniquely up to $\mathbf{Z}/3\mathbf{Z}$=action $z \mapsto \xi z$, $\xi^3 = 1$) in such a way that

$$\sigma(z) = -z, \qquad \omega_{1,0}(z) - \omega_{1,0}(-z) = 4z^2\, dz.$$

The remaining input is the Bergman kernel $\omega_{0,2}$ which can be written as the sum of canonical kernels $dz_1dz_2/(z_1 - z_2)^2$ on each disc, plus a regular (= without poles) S_2-symmetric section of $\pi_1^*K_\Sigma \otimes \pi_2^*K_\Sigma$. As a matrix-valued 2-differential

$$\omega_{0,2}(z_1, z) = \left(\frac{\delta_{\alpha_1,\alpha_2}\, dz_1 dz}{(z_1 - z)^2} + \sum_{n_1,n_2 \geq 1} P^{\alpha_1,\alpha_2}_{n_1,n_2}\, z_1^{n_1} z^{n_2} \frac{dz_1 dz}{z_1 z} \right)_{\alpha_1,\alpha_2 \in Ram},$$

where $(P^{\alpha_1,\alpha_2}_{n_1,n_2} \in \mathbf{C})_{\alpha_1,\alpha_2 \in Ram; n_1,n_2 \geq 1}$ obey the symmetry $P^{\alpha_1,\alpha_2}_{n_1,n_2} = P^{\alpha_2,\alpha_1}_{n_2,n_1}$.

Define the Tate topological vector space

$$W := W^{Ram}_{Airy} = \mathbf{C}^{Ram} \otimes \{\gamma \in \mathbf{C}((z))\, dz \mid Res_z(\gamma) = 0\} \subset \mathbf{C}^{Ram}((z))dz.$$

It has topological basis $([\alpha] \otimes z^n dz/z)_{\alpha \in Ram, n \neq 0}$ where $([\alpha])_{\alpha \in Ram}$ is the coordinate basis of $\mathbf{C}^{Ram}$. We endow W with the symplectic pairing

$$\begin{aligned} \langle [\alpha_1] \otimes df_1, [\alpha_2] \otimes df_2 \rangle &:= \delta_{\alpha_1,\alpha_2} Res_z(f_1 df_2) \\ \Longleftrightarrow \langle [\alpha_1] \otimes z^{n_1} \frac{dz}{z}, [\alpha_2] \otimes z^{n_2} \frac{dz}{z} \rangle &:= \frac{1}{n_2} \delta_{\alpha_1,\alpha_2} \delta_{n_1,-n_2}. \end{aligned}$$

Define a Lagrangian subspace $W_+ := \mathbf{C}^{Ram}[[z]]dz \subset W$, which is topologically spanned by the base vectors $[\alpha] \otimes z^n \frac{dz}{z}$ with $n \geq 1$. It is complementary to the Lagrangian subspace $W_- \simeq W_+^*$ spanned by $[\alpha] \otimes z^{-n} \frac{dz}{z}$ with $n \geq 1$.

LEMMA 3.5.1. *Under the above assumptions formal Bergman kernels (formal in the sense that we do not put convergence condition on the regular part of $\omega_{0,2}$) are in one-to-one correspondence with Lagrangian subspaces $V \subset W$ complementary to W_+. The correspondence is defined as follows: expand $\omega_{0,2}(z_1, z)$ into the iterated Laurent series:*

$$\omega_{0,2}(z_1, z)_{|z| \ll |z_1|} \in \mathbf{C}^{Ram} \widehat{\otimes} \mathbf{C}^{Ram}((z_1))((z))\, dz_1 dz.$$

Then one has

$$\omega_{0,2}(z_1, z)_{|z| \ll |z_1|} \in V^{z \to z_1} \widehat{\otimes} \mathbf{C}^{Ram}[[z]]\, dz. \tag{21}$$

Proof. In the above notation one has

$$\omega_{0,2}(z_1, z) = \sum_\alpha \frac{[\alpha] \otimes [\alpha]\, dz_1 dz}{(z_1 - z)^2} + \omega^{reg}_{0,2},$$

where $\omega^{reg}_{0,2} \in \widehat{Sym^2} W_+$ (recall that the hat means topological completion). The polar part is equal to

$$\sum_{n \geq 1, \alpha} \left([\alpha] \otimes n z^{-n} \frac{dz}{z} \right) \otimes \left([\alpha] \otimes z^n \frac{dz}{z} \right) = (\text{canonical pairing})^{-1} \in W_- \widehat{\otimes} W_+.$$

The regular part is given by

$$\omega_{0,2}^{reg} = \sum_{\alpha_1,\alpha_2} \sum_{n_1,n_2\geq 1} P^{\alpha_1\alpha_2}_{n_1 n_2} \left([\alpha_1]\otimes z_1^{n_1}\frac{dz_1}{z_1}\right)\otimes\left([\alpha_2]\otimes z_2^{n_2}\frac{dz_2}{z_2}\right).$$

Any Lagrangian subspace $V \subset W$ complementary to W_+ is the graph in $W \simeq T^*W_-$ of the differential of a quadratic form on W_-. Such a form corresponds to an element of $\widehat{Sym^2} W_+$ □

From now on we denote by V the Lagrangian subspace of W associated to the given Bergman kernel $\omega_{0,2}$ whose regular part could be even formal (non-convergent) matrix-valued power series in two variables.[9] Space V has canonical basis $(e_n^\alpha)_{\alpha\in Ram;n\geq 1}$:

$$e_n^\alpha := [\alpha]\otimes z^{-n}\frac{dz}{z} + \sum_{\alpha';n'\geq 1} \frac{1}{n} P^{\alpha',\alpha}_{n',n}\,[\alpha']\otimes z^{n'}\frac{dz}{z}. \tag{22}$$

Vector e_n^α is the unique vector in V such that

$$e_n^\alpha = [\alpha]\otimes z^{-n}\frac{dz}{z} + \left(\text{regular part } \in \mathbf{C}^{Ram}[[z]]\,dz\right).$$

It will be convenient to write Bergman kernel as

$$\omega_{0,2}(z_1,z) = \sum_{\alpha;n>0} (e_n^\alpha)^{z\to z_1}\otimes\left([\alpha]\otimes n z^{n-1}dz\right) \tag{23}$$

where notation $(e_n^\alpha)^{z\to z_1}$ means the element of $\mathbf{C}^{Ram}((z_1))\,dz_1$ obtained by the replacement of variables $z\to z_1$ from $e_n^\alpha \in V \subset \mathbf{C}^{Ram}((z))\,dz$.

The recursion kernel $K(z_1,z)$ exists in general only as a $(Ram\times Ram)$-matrix with coefficients in $\mathbf{C}((z_1))((z))\,\frac{dz_1}{dz}$. Let us expand the recursion kernel in double series:

$$\begin{aligned} K(z_1,z)_{|z|\ll|z_1|} &= \sum_\alpha [\alpha]\otimes[\alpha]\cdot\frac{-1}{4}\cdot\left(\frac{z^{-1}}{z_1^2}+\frac{z}{z_1^4}+\dots\right)\frac{dz_1}{dz} \\ &+ \sum_{\alpha_1,\alpha_2}[\alpha_1]\otimes[\alpha_2]\sum_{n_1,n_2\geq 1;n_2\ odd}\frac{-1}{4n_2}P^{\alpha_1,\alpha_2}_{n_1,n_2} z_1^{n_1} z^{n_2-2}\frac{dz_1}{dz}. \end{aligned}$$

This formula implies

$$K(z_1,z) = -\frac{1}{4}\sum_{\alpha;n\geq 1\ odd} (e_n^\alpha)^{z\to z_1}\otimes\left([\alpha]\otimes\frac{z^{n-2}}{dz}\right) \tag{24}$$

Denote by $W_{odd}\subset W$ the coisotropic subspace of meromorphic 1-forms on Σ whose pushfoward to the quotient Σ/σ has no poles at the images of ramification points $(r_\alpha)_{\alpha\in Ram}$. Subspace W_{odd} is the topological span of base vectors $[\alpha]\otimes z^n\frac{dz}{z}$ with $n\geq 1$ arbitrary and $n\leq -1$ odd. In particular, W_+ is a subspace of W_{odd}. Let

$$V_{odd} := V\cap W_{odd}$$

be the subspace of V spanned by basis vectors $(e_n^\alpha)_{\alpha\in Ram,n\geq 1\ odd}$.

[9]In fact the whole formalism of the topological recursion can be generalized straightforwardly to the formal case. We are going to use this observation without further comments.

Combining together the above formulas we conclude that the residue formulas from the conventional topological recursion (see formulas (9),(10),(11), (12) from Section 3.1) give tensors (no completions for the tensor products):

$$\begin{aligned} A_{TR}^{univ} &\in V_{odd}\otimes V\otimes V, \\ B_{TR}^{univ} &: W\to V_{odd}\otimes V, \\ C_{TR}^{univ} &: W\otimes W\to V_{odd}, \\ \varepsilon_{TR}^{univ} &\in V_{odd}. \end{aligned}$$

Restricting B_{TR}^{univ} and C_{TR}^{univ} to V we see that $(A_{TR}^{univ}, B_{TR}^{univ}, C_{TR}^{univ}, \varepsilon_{TR}^{univ})$ induce a pre-Airy structure on V. We denote the corresponding tensors by $(A_{TR}^{(1)}, B_{TR}^{(1)}, C_{TR}^{(1)}, \varepsilon_{TR}^{(1)})$. Below we will calculate these tensors explicitly.

LEMMA 3.5.2. *Tensor $A_{TR}^{(1)}$ is equal to $\frac{1}{4}\sum_\alpha e_1^\alpha\otimes e_1^\alpha\otimes e_1^\alpha$.*

PROOF. The definition of tensor A_{TR}^{univ} by the formula (9) gives

$$\begin{aligned} A_{TR}^{univ} &= -\frac{1}{4}\sum_\alpha Res_z\left(e_1^\alpha\frac{1}{z\,dz}+\dots\right)\otimes(e_1^\alpha dz+\dots)\otimes(-e_1^\alpha dz+\dots) \\ &= \frac{1}{4}\sum_\alpha e_1^\alpha\otimes e_1^\alpha\otimes e_1^\alpha. \end{aligned}$$

In particular, we see that A_{TR}^{univ} is a *symmetric* tensor. □

LEMMA 3.5.3. *Vector $\varepsilon_{TR}^{(1)}$ is equal to the sum of two terms*

$$\sum_\alpha\left(\frac{e_3^\alpha}{16}+\frac{P_{1,1}^{\alpha,\alpha}e_1^\alpha}{4}\right).$$

PROOF. By definition epsilon-formula general we have

$$\varepsilon_{TR}^{(1)} = -\frac{1}{4}\sum_{\alpha;n\geq 1\ odd} e_n^\alpha\times Res_z\left(\frac{z^{n-2}}{dz}Coeff_{[\alpha]\otimes[\alpha]}(\omega_{0,2}(z,-z))\right).$$

For each value of $\alpha\in Ram$ the canonical pole part of $\omega_{0,2}$ gives $e_3^\alpha/16$, and the regular part $\omega_{0,2}^{reg}$ gives $P_{1,1}^{\alpha,\alpha}e_1^\alpha/4$. □

LEMMA 3.5.4. *Operator $B_{TR}^{(1)}$ maps e_n^α to the sum of two terms*

$$\frac{1}{4}\sum_{n_1,n_2\geq 1;\, n_1+n_2=n+3;\, n_1\ odd}(-1)^{n_2+1}n_2\, e_{n_1}^\alpha\otimes e_{n_2}^\alpha+\frac{1}{4}\sum_{\alpha'}P_{1,n}^{\alpha',\alpha}\,e_1^{\alpha'}\otimes e_1^{\alpha'}.$$

PROOF. For any $\alpha'\in Ram, n'\in\mathbf{Z}_{\neq 0}$ tensor $B_{TR}^{univ}([\alpha']\otimes z^{n'}\frac{dz}{z})\in V\otimes V$ is given (using (11)) by formula

$$-\frac{1}{4}\sum_{n_1\geq 1\ odd}\ \sum_{n_2\geq 1} e_{n_1}^{\alpha'}\otimes e_{n_2}^{\alpha'}\cdot Res\left(\frac{z^{n_1-2}}{dz}\frac{(-1)^{n_2}}{n_2}z^{n_2-1}dz\cdot z^{n'}\frac{dz}{z}\right). \tag{25}$$

In particular, we see that this expression vanishes if $n'\geq 2$. Hence, in the calculation of $B_{TR}^{(1)}(e_n^\alpha)$ we should use only the polar term $(n'=-n)$

$$e_n^\alpha = [\alpha]\otimes z^{-n}\frac{dz}{z}+\dots$$

(giving the main part of the formula), and constant terms ($n' = 1$)

$$e_n^\alpha = \cdots + \sum_{\alpha'} P_{1,n}^{\alpha',\alpha}[\alpha'] \otimes z^1 \frac{dz}{z} + \ldots$$

contributing in (25) only for $n_1 = n_2 = 1$. □

LEMMA 3.5.5. *Operator $C_{TR}^{(1)}$ maps $e_{n_1}^{\alpha_1} \otimes e_{n_2}^{\alpha_2}$ to the sum of four terms:*

(1) $\frac{1}{4}(-1)^{n_2+1} e_{n_1+n_2+3}^{\alpha_1}$ *if $\alpha_1 = \alpha_2$ and $n_1 + n_2$ even,*

(2) $\frac{1}{4}\sum_{n_2' \geq 1; 2+n_2'-n_1 \in 2\mathbf{Z}_{\geq 0}} \frac{(-1)^{n_2'+1}}{n_2'} P_{n_2',n_2}^{\alpha_1,\alpha_2} e_{3+n_1-n_2'}^{\alpha_1}$,

(3) $\frac{1}{4}\sum_{n_2' \geq 1; 2+n_2'-n_1 \in 2\mathbf{Z}_{\geq 0}} \frac{(-1)^{n_2'+1}}{n_2'} P_{n_2',n_2}^{\alpha_1,\alpha_2} e_{3+n_1-n_2'}^{\alpha_1}$,

(4) $\frac{1}{4}\sum_\alpha P_{1,n_1}^{\alpha,\alpha_1} P_{1,n_2}^{\alpha,\alpha_2} e_1^\alpha$.

PROOF. For any $\alpha_1' \neq \alpha_2'$ and any $n_1', n_2' \in \mathbf{Z}_{\neq 0}$ we have

$$C_{TR}^{univ}(([\alpha_1'] \otimes z^{n_1'}\frac{dz}{z}) \otimes ([\alpha_2'] \otimes z^{n_2'}\frac{dz}{z})) = 0\,,$$

while for coinciding ramification points $r_{\alpha_1'} = r_{\alpha_2'}$, $\alpha_1' = \alpha_2' =: \alpha'$ we have (using (12)):

$$C_{TR}^{univ}(([\alpha'] \otimes z^{n_1'}\frac{dz}{z}) \otimes ([\alpha'] \otimes z^{n_1'}\frac{dz}{z}))$$

$$= -\frac{1}{4}\sum_{n' \geq 1\ odd} e_{n'}^{\alpha'} \cdot Res\left(\frac{z^{n'-2}}{dz} z^{n_1'-1}dz\,(-1)^{n_2'+1}z^{n_2'-1}dz\right).$$

Therefore, we get a general formula

$$C_{TR}^{(1)}(([\alpha_1'] \otimes z^{n_1'}\frac{dz}{z}) \otimes ([\alpha_2'] \otimes z^{n_2'}\frac{dz}{z}))$$

$$= \begin{cases} \frac{(-1)^{n_2'}}{4} e_{3-n_1'-n_2'}^{\alpha_1'} & \text{if } \alpha_1' = \alpha_2',\ 2 - n_1' - n_2' \in 2\mathbf{Z}_{\geq 0} \\ 0 & \text{otherwise} \end{cases} \tag{26}$$

Using (22) write $e_{n_1}^{\alpha_1} \otimes e_{n_2}^{\alpha_2}$, $n_1, n_2 \geq 1$ as

$$+ ([\alpha_1] \otimes z^{-n_1}\frac{dz}{z}) \otimes ([\alpha_2] \otimes z^{-n_2}\frac{dz}{z}) \tag{27}$$

$$+ \sum_{\alpha_2'; n_2' \geq 1} \frac{1}{n_2'} P_{n_2',n_2}^{\alpha_2',\alpha_2} ([\alpha_1] \otimes z^{-n_1}\frac{dz}{z}) \otimes ([\alpha_2'] \otimes z^{n_2'}\frac{dz}{z}) \tag{28}$$

$$+ \sum_{\alpha_1'; n_1' \geq 1} \frac{1}{n_1'} P_{n_1',n_1}^{\alpha_1',\alpha_1} ([\alpha_1'] \otimes z^{n_1'}\frac{dz}{z}) \otimes ([\alpha_2] \otimes z^{-n_2}\frac{dz}{z}) \tag{29}$$

$$+ \sum_{\alpha_1'\alpha_2'; n_1', n_2' \geq 1} \frac{1}{n_1'\, n_2'} P_{n_1',n_1}^{\alpha_1',\alpha_1} P_{n_2',n_2}^{\alpha_2',\alpha_2} ([\alpha_1'] \otimes z^{n_1'}\frac{dz}{z}) \otimes ([\alpha_2'] \otimes z^{n_2'}\frac{dz}{z}). \tag{30}$$

Applying (26) to four lines (27),(28),(29),(30) we obtain the statement of the lemma (notice that for term (30) the only relevant part will be $\alpha_1' = \alpha_2' =: \alpha$ and $n_1' = n_2' = 1$). □

Let us consider the quantum pre-Airy structure $(A^{(0)}_{TR}, B^{(0)}_{TR}, C^{(0)}_{TR}, \varepsilon^{(0)}_{TR})$ on the vector space W_- which is equal to the direct sum of copies labeled by set *Ram* of the canonical quantum pre-Airy structure from Section 3.2. The vector space W_+ is the tangent space at 0 to the germ of the associated Lagrangian subvariety. We have shown in Lemma 3.5.1 that a choice of the formal Bergman kernel corresponds to the choice of a complementary Lagrangian subspace V to W_+. Hence it gives a way to modify the pre-structure $(A^{(0)}_{TR}, B^{(0)}_{TR}, C^{(0)}_{TR}, \varepsilon^{(0)}_{TR})$ by a gauge equivalence (see formulas (2) and (4) from Section 2). The role of symmetric matrix $s = (s^{ij})$ in these formulas is played by $\omega^{reg}_{0,2} \in \widehat{Sym^2 W_+}$ understood as an infinite symmetric matrix:

$$P := \left(P^{\alpha_1,\alpha_2}_{n_1,n_2}\right)_{\alpha_1,\alpha_2\in Ram; n_1,n_2>0}, \quad \text{index set} = Ram \times \mathbf{Z}_{>0}.$$

We identify vector spaces W_- and V via

$$W_- \xleftarrow{\sim} W/W_+ \xrightarrow{\sim} V\,.$$

THEOREM 3.5.6. *The gauge transformation of $(A^{(0)}_{TR}, B^{(0)}_{TR}, C^{(0)}_{TR}, \varepsilon^{(0)}_{TR})$ as described above, coincides with $(A^{(1)}_{TR}, B^{(1)}_{TR}, C^{(1)}_{TR}, \varepsilon^{(1)}_{TR})$.*

PROOF. Just read formulas from Lemmas 3.5.2, 3.5.3, 3.5.4, 3.5.5, as well as formulas (13), (14), (15), (16) for the standard pre-Airy structure, and compare with corresponding expressions (2) and (4). □

COROLLARY 3.5.7. *Pre-Airy structure $(A^{(1)}_{TR}, B^{(1)}_{TR}, C^{(1)}_{TR}, \varepsilon^{(1)}_{TR})$ contains a primitive quantum Airy substructure on the subspace V_{odd}.*

PROOF. Here it is useful to recall (essentially trivial) Remark 2.6.3. The quantum pre-Airy structure $(A^{(0)}_{TR}, B^{(0)}_{TR}, C^{(0)}_{TR}, \varepsilon^{(0)}_{TR})$ on V contains the quantum Airy substructure given by the restriction of all tensors to V_{odd}. The gauge transformation from the above Theorem transforms this Airy substructure to the Airy substructure of the pre-Airy structure $(A^{(1)}_{TR}, B^{(1)}_{TR}, C^{(1)}_{TR}, \varepsilon^{(1)}_{TR})$. This proves the result. □

There is a variation of the above theorem. Namely, consider the quantum Airy structure $(A^{(0)}, B^{(0)}, C^{(0)}, \varepsilon^{(0)})$ on W_- which is the direct sum of copies labeled by set *Ram* of the canonical quantum Airy structure from Section 3.3. Also, denote by $(A^{(1)}, B^{(1)}, C^{(1)}, \varepsilon^{(1)})$ the quantum pre-Airy structure on V given by the modified recursion as in Section 3.4.

THEOREM 3.5.8. *Pre-Airy structure $(A^{(1)}, B^{(1)}, C^{(1)}, \varepsilon^{(1)})$ is in fact an Airy structure, and it is obtained from $(A^{(0)}, B^{(0)}, C^{(0)}, \varepsilon^{(0)})$ by the gauge transformation associated with $\omega^{reg}_{0,2} \in \widehat{Sym^2 W_+}$. Moreover, this structure has the primitive substructure on subspace V_{odd}, and this substructure coincides with the one from Theorem* 3.5.6.

PROOF. The fact that as a pre-structure $(A^{(0)}, B^{(0)}, C^{(0)}, \varepsilon^{(0)})$ transforms to $(A^{(1)}, B^{(1)}, C^{(1)}, \varepsilon^{(1)})$ follows from the calculations almost identical to those in the proof of Theorem 3.5.6, with the only correction that factors $(-1)^{n_2+1}, (-1)^{n_2'+1}$ are removed. The rest of statements of the Theorem follows immediately. □

4. Spectral curves

4.1. Poisson surfaces with foliation. Any complex analytic Poisson surface which does not consists entirely of 0-dimensional symplectic leaves, contains an open dense symplectic leaf. The complement to the symplectic leaf is the set of zeros of the Poisson structure. Hence it is an analytic divisor, which consists of 0-dimensional symplectic leaves.

DEFINITION 4.1.1. A **good foliated Poisson surface** is a complex Poisson surface P endowed with an analytic 1-dimensional foliation $\mathcal{F}$, such that the Poisson structure is non-degenerate on an open dense symplectic leaf $S \subset P$, the effective divisor D of zeros of the Poisson structure is the disjoint union of smooth components with positive multiplicities, and $\mathcal{F}$ is tangent to D.

The inverse to the non-degenerate Poisson structure on S is the symplectic form. Hence we can (and will) denote the data from the above Definition by $(P, \omega, \mathcal{F})$, where ω is the symplectic form on $S = P - D$ understood as a meromorphic 2-form on P with poles at D.

DEFINITION 4.1.2. Let $(P, \omega, \mathcal{F})$ be a good foliated Poisson surface. A smooth complex curve $\Sigma \subset P$ is called a D-**transversal** spectral curve if it is a smooth connected compact complex curve which intersects D at finitely many points $\{p_\beta\}_{\beta \in Pol}$, and the intersection is transversal.

Such curve is called $(\mathcal{F}, D)$-**transversal**[10] if Σ is tangent to leaves of $\mathcal{F}$ at finitely many points $\{r_\alpha\}_{\alpha \in Ram}$, with the tangency order 1 (i.e. locally near the tangency point it looks like the curve $x = y^2$ tangent to the line $x = 0$). Here Pol (resp. Ram) is the notation for the above finite set of "poles" (resp. "ramificiation points"). [11]

We denote by $\mathcal{B}_0$ (resp. $\mathcal{B}_1$) the moduli space of D-transversal (resp. $(\mathcal{F}, D)$-transversal) curves. The moduli spaces are defined in terms of complex analytic families over complex analytic bases in the usual way. Notice that the families of sets $Pol = Pol(\Sigma)$ and $Ram = Ram(\Sigma)$ are locally constant over $\mathcal{B}_0$ or $\mathcal{B}_1$ (here we speak about D-transversal or resp. $(\mathcal{F}, D)$-transversal spectral curves depending on the class of curves Σ).

For any D-transversal spectral curve Σ we denote the corresponding point by $[\Sigma] \in \mathcal{B}_0$ (and, sometimes, simply by Σ in order to alleviate the notation).

Obviously, $\mathcal{B}_1$ is an open subset of $\mathcal{B}_0$ (dense in typical situation). We will see below (see Lemma 4.2.1) that $\mathcal{B}_0$ (and hence $\mathcal{B}_1$) is smooth.

Notice that for a $(\mathcal{F}, D)$-transversal curve, we can choose near each tangent point r_α local analytic coordinates (x, y) on S such that the point r_α has coordinates $x = 0, y = 0$, the symplectic form on S is equal to $\omega = dx \wedge dy$, the leaves of foliation $\mathcal{F}$ are given by the equations $x = const$ and the spectral curve Σ is given by the equation $x = y^2$. Such coordinates are unique modulo analytic automorphism $x \mapsto \xi, y \mapsto \xi^{-1} y$, where $\xi = e^{2\pi i/3}$ is the primitive cubic root of 1. We will call such coordinates *canonical*.

EXAMPLE 4.1.3. Let C be a smooth projective curve of genus $g \geq 2$. Let $\pi : T^*C \to C$ be the standard projection, and $\Sigma \subset T^*C$ be a smooth compact

[10] Probably a better name would be $\mathcal{F}$-Morse.

[11] There are further possibilities for generalizations here. In particular we can consider the case of immersed rather than embedded curves Σ.

connected curve. Take $P = S := T^*C$ and $\mathcal{F}$ to be the foliation by cotangent fibers. Then $Pol = \emptyset$ and Σ is an D-transversal spectral curve. For any integer $d \geq 1$, the connected component of $\mathcal{B}_0$ corresponding to smooth spectral curves Σ such that $\pi_{|\Sigma} : \Sigma \to C$ has degree d, is a Zariski open part of the base of Hitchin fibration for the group $GL(d)$. Recall that this base is

$$\mathcal{B} = \oplus_{i=1}^{d} \Gamma\left(C, (\Omega_c^1)^{\otimes i}\right) \simeq \mathbf{C}^{d^2(g-1)+1}$$

and it can be interpreted as the linear system of compact effective divisors in T^*C of degree d over C. The complement $\mathcal{B} - \mathcal{B}_0$ is the discriminant hypersurface parametrizing singular spectral curves. Zariski open subvariety $\mathcal{B}_1 \subset \mathcal{B}_0$ parametrizes smooth Σ such that $\pi_{|\Sigma}$ has only double ramification points.

This example suggests that in general the moduli space $\mathcal{B}_0$ should be considered as an open subspace of a larger space $\mathcal{B}$ parametrizing compact effective divisors in P. It is not clear whether the unobstructedness result (Lemma 4.2.1 from the next section) extends to the larger context of effective divisors.

EXAMPLE 4.1.4. More generally, let C be a smooth projective curve endowed with irregular data in the sense of [**KS14**]. The irregular data consist of a collection of marked points, and Puiseux polynomials (called singular terms in the loc. cit.) with multiplicities. They encode the irregular singular behavior of Higgs fields or connections at the marked points. In the case of Higgs bundles the irregular data modify the definition of the spectral curve from the previous example.

By [**KS14**] one can construct canonically a Poisson surface P, which is obtained by a sequence of blow-ups of the fiberwise compactification of T^*C.

Then there are two possibilities for the choice of P and the symplectic leaf $S \subset P$. Both possibilities are described in the loc. cit. :

1) If one does not fix eigenvalues of the Higgs field (by definition, the residues of the restriction of the Liouville 1-form to Σ) then there is a choice of P which has an open dense symplectic leaf S isomorphic to T^*C. Then the spectral curve Σ is defined as the compact closure in P of a curve in S. Eigenvalues of the Higgs field correspond to intersections of Σ with $P - S$.

2) If one fixes the eigenvalues of the Higgs field, then one makes additional blow-ups in 1) at the intersection points of Σ with the divisor $P - S$. In this case P contains an open symplectic leaf S which is strictly bigger than T^*C. The spectral curve $\Sigma \subset S$ is compact.

The Poisson surface P endowed with the mermorphic foliation $\mathcal{F}$ induced from the foliation along fibers $T^*C \to C$ is not in general a good foliated Poisson surface in the sense of definition 4.1.1. One has to remove various types of "bad" points: double points of divisor $D = P - S$, undeterminacy points of $\mathcal{F}$, components of D which are generically transversal to $\mathcal{F}$. If we are fortunate, the closure Σ of a smooth spectral curve in T^*C will not meet bad points and will be a smooth compact curve transversal to D. In this case we will obtain a point in $\mathcal{B}_0$.

Notice that in [**KS14**] we considered compactified spectral curves which intersect divisor D only along the components with multiplicity 1, while here we allow transversal intersections of Σ with components D of arbitrary multiplicity. One can make a sequence of blow-ups at the transversal intersection point with a component of multiplicity $k > 1$ and arrive to the situation when the multiplicity of the intersection is equal to one. Nevertheless, this procedure will produce *different* moduli space of spectral curves. The conclusion is that the framework of the present paper is indeed more general than the one in [**KS14**], even for the case of cotangent bundles.

EXAMPLE 4.1.5. One can consider a "multiplicative" version of the previous example. Namely, let us start with $S = \mathbf{C}^* \times \mathbf{C}^*$ endowed with the symplectic form $\frac{dx}{x} \wedge \frac{dy}{y}$ and endow with it with the foliation $\mathcal{F}$ with leaves $x = const$. Let us compactify our symplectic surface to a toric Poisson surface P and make a chain of Poisson blow-ups similar to those mentioned in the previous example. After removing bad points as in the previous example, we obtain a good foliated Poisson surface with all components of D having multiplicity one. The corresponding spectral curves are naturally related to q-difference equations. Similarly to the previous example they could happen to be smooth and D-transversal, hence giving points in $\mathcal{B}_0$.

4.2. Deformation theory of spectral curves. We are interested in the deformation theory of Σ in the class of D-transversal spectral curves. The deformation complex is $\mathbf{R}\Gamma(\Sigma, N_\Sigma)$, where $N_\Sigma = (T_P)_{|\Sigma}/T_\Sigma$ is the normal bundle to Σ in P. Using the symplectic structure on $S \subset P$ we identify

$$N_\Sigma \simeq \Omega^1_\Sigma(D_\Sigma), \quad D_\Sigma := D \cap \Sigma = \sum_{\beta \in Pol} k_\beta p_\beta,$$

where Ω^1_Σ is the sheaf of holomorphic 1-forms on Σ, and $k_\beta \geq 1$ is the local multiplicity of D at p_β, i.e. the order of pole of ω at divisor D near the point p_β.

LEMMA 4.2.1. *The deformation theory of Σ is unobstructed.*

PROOF. In the case when $Pol \neq \emptyset$ we have $H^1(\Sigma, \Omega^1_\Sigma(D_\Sigma)) = 0$ by Serre duality. Hence the deformation theory of Σ is unobstructed.

In the more classical case when $Pol = \emptyset$ the unobstructedness is presumably known. Alternatively, one can adapt the proof of Proposition 8.2.1 in [**KS14**], where a similar statement (unobstructedness) was proven for compact complex Lagrangian submanifolds in quasi-projective varieties. In the present framework we do not require S to be quasi-projective, it is just an analytic surface. Nevertheless, the proof from the loc. cit. still works in our case, because the degeneration of the Hodge-to-de Rham spectral sequence (see Step 2 of the proof in the loc.cit.) holds for a family of compact complex analytic curves over the spectrum of a local Artin algebra. The Lemma is proved. □

By Lemma 4.2.1 the complex analytic moduli space space $\mathcal{B}_0$ of D-transversal curves is a disjoint union of complex analytic manifolds. The dimension of the the germ $\mathcal{B}_{[\Sigma]}$ of $\mathcal{B}_0$ at point $[\Sigma]$ is

$$\dim_{\mathbf{C}} \mathcal{B}_{[\Sigma]} = \operatorname{rk} H^0(\Sigma, \Omega^1_\Sigma(D_\Sigma)) = \begin{cases} g & \text{if } Pol = \emptyset \\ g + \sum_{\beta \in Pol} k_\beta - 1 & \text{if } Pol \neq \emptyset \end{cases}.$$

5. Affine symplectic connection and local embedding of the moduli space of spectral curves

5.1. Generalities on local Lagrangian embeddings. Let us recall the notion of an affine flat connection on a vector bundle.

DEFINITION 5.1.1. For a complex analytic manifold X and an analytic vector bundle $\mathcal{H}$ on X, an **affine flat connection** on $\mathcal{H}$ is a pair (∇, ϕ) where ∇ is an analytic flat linear connection, and $\phi \in \Gamma(X, \Omega^1_X \otimes \mathcal{H})$ is such that $\nabla(\phi) = 0$.

An affine flat connection is the same as a flat connection $\nabla + \phi$ with values in the Lie algebra of affine transformations. Hence it gives an identification of fibers $\mathcal{H}_x$ as *affine spaces* for all points $x \in X$ close to any reference point x_0. Under this identification the zero point $0_x \in \mathcal{H}_x$ is mapped to a certain point in $\mathcal{H}_{x_0}$. Therefore, we obtain a germ of an analytic map

$$\text{germ of } X \text{ at } x_0 \mapsto \text{germ of } \mathcal{H}_{x_0} \text{ at } 0.$$

This map is an immersion if and only if ϕ is a monomorphism (if it is understood as a map of vector bundles $T_X \to \mathcal{H}$). Finally, let us assume that the bundle $\mathcal{H}$ carries a non-degenerate skew-symmetric pairing $\langle \bullet, \bullet \rangle$ which is covariantly constant, and map $\phi : T_X \to \mathcal{H}$ is a linear Lagrangian embedding. Then (∇, ϕ) induces a Lagrangian embedding from the germ of X at x_0 to the symplectic affine space $\mathcal{H}_{x_0}$. Moreover, these embeddings form a formally flat family, as we vary the reference point x_0.

REMARK 5.1.2. Suppose that with any point $x_0 \in X$ and any pair of points x_1, x_2 sufficiently close to x_0, we associate a vector $v(x_1, x_2)_{x_0} \in \mathcal{H}_{x_0}$, which is covariantly constant if we vary x_0 (and keep points x_1, x_2 fixed), and satisfying the cocycle condition

$$v(x_1, x_2)_{x_0} + v(x_2, x_3)_{x_0} = v(x_1, x_3)_{x_0} .$$

It is easy to see that such data are equivalent to a *closed* 1-form ϕ with values in a flat vector bundle $\mathcal{H}$ on X. More precisely, suppose we have a linear flat connection ∇ and ∇-covariant 1-form ϕ. Given a triple of points (x_0, x_1, x_2) which are sufficiently close to each other we can find a contractible open subset $U \subset X$ containing all of them. Then on $U^3 \subset X^3$ we have a trivial bundle $pr^*_{(x_0,x_1,x_2)\mapsto x_0}(\mathcal{H})$, endowed with the pull-backs of ∇ and ϕ. It can be trivialized, such that the fiber of the bundle is identified with $\mathcal{H}_{x_0}$. Then for a short path γ_{x_1,x_2} joining x_1 and x_2 we define $v(x_1, x_2)_{x_0} = \int_{\gamma_{x_1,x_2}} \phi \in \mathcal{H}_{x_0}$. Conversely, one can reconstruct ϕ from v by taking the first derivative of $v(x_1, x_2)_{x_0}$ w.r.t. x_2 at $x_2 = x_1$.

5.2. Local Lagrangian embedding of $\mathcal{B}_0$. Our goal is to apply the general construction from the previous subsection to the case $X = \mathcal{B}_0$ and obtain for every smooth spectral curve Σ the corresponding germ of the Lagrangian embedding. The construction will depend on the foliation $\mathcal{F}$, but only through the collection of its $(2k-1)$-germs near components of D at which ω has pole of order $k \geq 1$.

First, we define a symplectic vector bundle such that its fiber at $[\Sigma] \in \mathcal{B}_0$ is the hypercohomology

$$\mathcal{H}_\Sigma = \mathbb{H}^0(\Sigma, Cone(d_{dR} : \mathcal{O}_\Sigma(-D_\Sigma) \to \Omega^1_\Sigma(D_\Sigma))), \tag{31}$$

where $D_\Sigma := \sum_{\beta \in Pol} k_\beta p_\beta$ is an effective divisor on Σ as in Section 4.2. Symplectic pairing $\langle \bullet, \bullet \rangle$ on $\mathcal{H}_\Sigma$ is given by the Serre duality. It is easy to see that

$$\dim \mathcal{H}_\Sigma = \begin{cases} 2g & \text{if } Pol = \emptyset \\ 2g + 2\sum_{\beta \in Pol} k_\beta - 2 & \text{if } Pol \neq \emptyset \end{cases} .$$

The 1-form ϕ interpreted as a morphism $T_{\mathcal{B}_0} \to \mathcal{H}$ is given by the linear map

$$T_{[\Sigma]}\mathcal{B}_0 = H^0(\Sigma, \Omega^1_\Sigma(D_\Sigma)) \to \mathcal{H}_\Sigma \tag{32}$$

which is the boundary map in the standard long exact sequence. Equivalently, it comes from the exact sequence of morphisms of cones

$$Cone(0 \to \Omega^1_\Sigma(D_\Sigma)) \to Cone(\mathcal{O}_\Sigma(-D_\Sigma) \to \Omega^1_\Sigma(D_\Sigma))\,.$$

It is easy to see that ϕ always gives a Lagrangian embedding.

It is left to construct a linear flat symplectic connection on the bundle $\mathcal{H}$ and prove that ϕ is covariant. In the case $Pol = \emptyset$ there is an obvious candidate, namely, the Gauss-Manin connection. For $Pol \neq \emptyset$ the story is more complicated.

Let us use analytic topology, and consider a disjoint union of open discs

$$\sqcup_{\beta \in Pol}\mathbb{D}_{p_\beta} \subset \Sigma, \quad p_\beta \in \mathbb{D}_{p_\beta}\ \forall \beta$$

containing points p_β. Let us choose a coordinate $x = x_{(\beta)}$ on each disc $\mathbb{D}_{p_\beta}$ for each β, such that $x_{(\beta)}$ vanishes at the point p_β. Using these local coordinates we will construct an isomorphism

$$\psi : \mathcal{H}_\Sigma \xrightarrow{\sim} \Big(H^1(\Sigma - \{p_\beta\}_{\beta\in Pol}, i_\epsilon(Pol); \mathbf{Z}) \otimes \mathbf{C}\Big) \oplus \bigoplus_{\beta \in Pol} \Big(\oplus_{i=-k_\beta}^{k_\beta-2} \mathbf{C}\cdot x^i dx / \mathbf{C}x^{-1}dx\Big)\,, \tag{33}$$

where $i_\epsilon : Pol \hookrightarrow \Sigma - \{p_\beta\}_{\beta \in Pol}$ is the embedding which associates with β the point in $\mathbb{D}_{p_\beta}$ with $x_{(\beta)}$-coordinate equal to ϵ, $0 < \epsilon \ll 1$. The definition makes sense as for small ϵ the relative cohomology groups are canonically identified.

In order to describe the morphism ψ it will be convenient to use the following explicit presentation of $\mathcal{H}_\Sigma$ in terms of global differential forms with arbitrary poles at a finite *non-empty* collection of distinct points $(r_i \in \Sigma)_{i\in I}$, which are also distinct from points $(p_\beta)_{\beta \in Pol}$[12]:

$$\mathcal{H}_\Sigma \simeq \{\eta \in \Gamma(\Sigma, \Omega^1_\Sigma(D_\Sigma + \sum_i \infty r_i) | \forall i\ Res_{r_i}\eta = 0\} / d\Gamma(\Sigma, \mathcal{O}_\Sigma(-D_\Sigma + \sum_i \infty r_i)). \tag{34}$$

For a class $[\eta] \in \mathcal{H}_\Sigma$ represented by meromorphic 1-form η, we define $\psi([\eta])$ as the sum of two terms

$$\psi([\eta]) = \psi_1(\eta) + \psi_2(\eta)$$

described such as follows.

The term $\psi_2(\eta)$ is defined as the sum over $\beta \in Pol$ of the image of the formal power series decomposition of η in the coordinate x at the point $p_\beta \in \mathbb{D}_{p_\beta}$, under the natural map

$$x^{-k_\beta}\mathbf{C}[[x]]dx \to \oplus_{i=-k_\beta}^{k_\beta-2} \mathbf{C}\cdot x^i dx / \mathbf{C}x^{-1}dx, \quad x = x_{(\beta)}\ .$$

The term $\psi_1(\eta) \in H^1(\Sigma - \{p_\beta\}_{\beta\in Pol}, i_\epsilon(Pol); \mathbf{Z}) \otimes \mathbf{C}$ is defined via the pairing with relative homology classes such as follows. Let $\gamma = \gamma(\epsilon)$ be a 1-chain on $\Sigma - \{p_\beta\}_{\beta \in Pol}$ with the boundary lying in the finite set $i_\epsilon(Pol)$ and depending continuously on the parameter ϵ (and hence representing the same relative homology class). Let us

[12]The points r_i at this time have nothing to do with ramification points r_α. But our notation is suggestive: we will later choose the set $(r_i \in \Sigma)_{i\in I}$ in such a way that it will coincide with the set $(r_\alpha)_{\alpha \in Ram}$.

assume that this chain avoids the set $\{r_i\}_{i\in I}$. Then the integral $\int_{\gamma(\epsilon)}\eta$ considered as a function of ϵ has an expansion as $\epsilon\to 0$ of the form:

$$(35)\qquad \int_{\gamma(\epsilon)}\eta=\sum_{i=-N}^{\infty}c_i\epsilon^i+c_{log}\cdot\log(\epsilon),\ \ N=\max_{\beta}(k_\beta)-1\,.$$

Then we define the cohomology class $\psi_1(\eta)$ by the formula

$$(36)\qquad \psi_1(\eta)([\gamma]):=\ \text{coefficient } c_0 \text{ in the above expansion}\,.$$

PROPOSITION 5.2.1. *Cohomology class $\psi_1(\eta)$ does not depend on the choice of representatives $\gamma(\epsilon)$. Furthermore it depends only on the class $[\eta]$ in $\mathcal{H}_\Sigma$. Also, $\psi_1(\eta)$ depends only on the class $[\eta]$ in $\mathcal{H}_\Sigma$.*

The proof is a straightforward check, and is left as an exercise to the reader. The above Proposition implies that the map ψ is well-defined.

THEOREM 5.2.2. *Map ψ defined in* (33) *is an isomorphism.*

PROOF. The dimensions of the RHS and LHS in (33) are equal to each other, hence it is sufficient to check that $Ker\,\psi=0$. Let us assume that $\psi([\eta])=0$ for some meromorphic 1-form η representing the cohomology class $[\eta]$ via (34). Then η should have zero residue at each point p_β (otherwise it will produce a non-zero class in $H^1(\Sigma-\{p_\beta\}_{\beta\in Pol},i_\epsilon(Pol);\mathbf{C})$). Vanishing of other Taylor coefficients (guaranteed by the condition $\psi_2(\eta)=0$) implies that

$$\eta\in\Gamma(\Sigma,\Omega^1_\Sigma(-\sum_{\beta\in Pol}(k_\beta-1)p_\beta+\sum_i\infty r_i)),\ \ Res_{r_i}(\eta)=0\ \forall i\,.$$

We see that η is non-singular at all points $p_\beta,\beta\in Pol$. Therefore in the formula (35) there are no terms $c_i, i<0$ and no term c_{log}. Hence we can define c_0 as the limit as $\varepsilon\to 0$ of $\int_{\gamma(\varepsilon)}\eta$. The condition $\psi_1(\eta)=0$ implies that η represents zero cohomology class in $H^1(\Sigma-\{r_i\}_{i\in I},\{p_\beta\}_{\beta\in Pol};\mathbf{C})$, hence it is equal to df for some function f with poles at $\{r_i\}_{i\in I}$ and vanishing at $\{p_\beta\}_{\beta\in Pol}$. Moreover, vanishing of $\psi_2(\eta)$ implies that for every $\beta\in Pol$ the function f vanishes at p_β with multiplicity $\geq k_\beta$. Therefore $[\eta]=0\in\mathcal{H}_\Sigma$. □

Now we are ready to define the desired flat connection ∇ on the vector bundle $\mathcal{H}$ on the moduli space $\mathcal{B}_0$ of smooth D-transversal spectral curves. For a given curve $[\Sigma]\in\mathcal{B}_0$ and every $\beta\in Pol:=Pol(\Sigma)$ let us choose a local coordinate $x_{(\beta)}$ on a small disc parametrizing leaves of the foliation $\mathcal{F}$ which are close to p_β. We also assume that $x_{(\beta)}$ vanishes on the leaf passing through p_β. This local coordinate defines a local coordinate on the disc $\mathbb{D}'_{p_\beta}\subset\Sigma'$ associated to $\beta\in Pol(\Sigma)=Pol(\Sigma')$ for any spectral curve Σ' which is sufficiently close to Σ. We identify $\psi(\mathcal{H}_\Sigma)$ and $\psi(\mathcal{H}_{\Sigma'})$ using homotopy equivalence or surfaces with marked points (Gauss-Manin connection)

$$H^1(\Sigma-\{p_\beta\}_{\beta\in Pol},i_\epsilon(Pol);\mathbf{Z})\simeq H^1(\Sigma'-\{p'_\beta\}_{\beta\in Pol},i'_\epsilon(Pol);\mathbf{Z}),$$

and using the *identity* isomorphism on spaces $\oplus_{i=-k_\beta}^{k_\beta-2}\mathbf{C}\cdot x^i dx/\mathbf{C}x^{-1}dx$, $x=x_{(\beta)}$ for each β.

THEOREM 5.2.3. *The above construction gives a flat connection ∇ on the bundle $\mathcal{H}$ which does not depend on the choices of local coordinates $x_{(\beta)}$, and also on the choice of auxiliary punctures $(r_i)_{i\in I}$. Moreover, the connection depends only on the* *collection of $(2k-1)$-jets of foliation $\mathcal{F}$ near components of divisor $D = P - S$ where symplectic form ω has pole of order k. Furthermore, the connection ∇ preserves the natural symplectic pairing on $\mathcal{H}$, and the $\mathcal{H}$-valued 1-form ϕ defined in* (32) *is closed.*

PROOF. The independence of the choice of the auxiliary finite nonempty set $\{r_i\}_{i \in I} \subset \Sigma - \{p_\beta\}_{\beta \in Pol}$ is obvious, as one can pass from this subset to arbitrary larger finite subset. The independence of the choice of local coordinates $\{x_{(\beta)}\}_{\beta \in Pol}$ follows from the fact that a change of coordinates $x_{(\beta)} \to \widetilde{x}_{(\beta)}$ modifies the isomorphism $\psi = \psi_1 + \psi_2$ in certain universal way which is independent of a spectral curve Σ' close to Σ.

The covariance of the symplectic pairing follows from the description of connection in terms of Hamiltonian reduction in Section 7. Notice that considerations of Section 7 are given for $\Sigma \in \mathcal{B}_1$. On the other hand, as one can show that $\mathcal{B}_1$ is dense in $\mathcal{B}_0$, it follows that the symplectic pairing is covariant by the argument of continuity. Alternatively, one can describe the symplectic structure on $\mathcal{H}_\Sigma$ in terms of the RHS of the formula (33) and deduce the covariance from that description.

The fact that the 1-form ϕ is closed, follows from Remark 5.1.2 as well as the alternative description of the local embedding of the germ of $\mathcal{B}_0$ at $[\Sigma]$ to the affine space $\mathcal{H}_\Sigma$ given in the next subsection. The symplectic pairing is explained in Section 7, see Remark 7.1.5. Technically considerations there deal with the open dense subspace $\mathcal{B}_1 \subset \mathcal{B}_0$. Continuity arguments show that the result holds for $\mathcal{B}_0$ as well. □

REMARK 5.2.4. In Section 8 we will use a slightly different approach which gives an alternative (and maybe more natural) proof of Theorem 5.2.3.

5.3. Connection and 1-form via regularized integrals. In this subsection we will use Remark 5.1.2 in order to prove that $\nabla(\phi) = 0$.

For that let us take three points $x_0 = \Sigma, x_1 = \Sigma_0, x_2 = \Sigma_1 \in \mathcal{B}_0$ which are sufficiently close to each other and consider an analytic path $\Sigma(\lambda), \lambda \in [0,1]$ such that $\Sigma(0) = \Sigma_0, \Sigma(1) = \Sigma_1$.

Following Remark 5.1.2 we would like to define the corresponding vector $v(\Sigma_0, \Sigma_1)_\Sigma \in \mathcal{H}_\Sigma$ which satisfies the cocycle condition. As we see the curve Σ plays an auxiliary role in the considerations, so we can rename Σ_0 to Σ and consider just a pair of sufficiently close spectral curves. The path $\Sigma(\lambda)$ can be thought of as an analytic map $f : T \to S$ of the total space T of the analytic family of spectral curves $\Sigma(\lambda)$ with the above boundary conditions as $\lambda = 0, 1$. Then $\eta(\lambda) := \frac{f^*(\omega)}{d\lambda}$ is a family of meromorphic 1-forms on $\Sigma(\lambda)$.

Recall that the RHS of the formula (33) consists of two summands. Let us consider the first one, which denote by H_{top}. Relative cohomology classes in that formula are dual to the corresponding relative homology classes.

Let $T' := T - \cup_{\beta \in Pol}\{(\lambda, p_\beta(\lambda))\}, \lambda \in [0,1]$. We denote by Q_ε the subset it T' given by $(0, \Sigma(0)) \cup (1, \Sigma(1)) \cup (\lambda, i_\varepsilon(\beta)(\lambda))$, where $\lambda \in [0,1]$. Here we use the self-explained notation like $(\lambda, p), \lambda \in [0,1], p \in \Sigma(\lambda)$ for points of T, There is a natural isomorphism $g : H_1(\Sigma - \{p_\beta\}_{\beta \in Pol}, i_\epsilon(Pol); \mathbf{Z}) \simeq H_2(T', Q_\varepsilon; \mathbf{Z})$ which follows from the observation that families of points $p_\beta(\lambda), i_\varepsilon(\beta)(\lambda), \lambda \in [0,1]$ is a union of $|Pol|$ closed intervals, and for the interval $[0,1]$ we have the natural isomorphism $H_1([0,1], \{0\} \cup \{1\}; \mathbf{Z}) \simeq \mathbf{Z}$.

Let $[\delta(\varepsilon)] = g([\gamma(\varepsilon)])$ be the class of a relative 2-chain corresponding to the class of the 1-chain $\gamma(\varepsilon)$. Then $\int_{f^*([\delta(\varepsilon)])} f^*(\omega)$ has the asymptotic expansion (as $\varepsilon \to 0$) given by the RHS of the formula (35). Take the constant terms of this expansion (i.e. the coefficient c_0). Then we obtain the functional $c_0 = c_0(\Sigma_0, \Sigma_1)$ on H_{top}.

In order to take care about the second summand in the RHS of the isomorphism (33) let us use as before the coordinate $x(\lambda) = x_{(\beta)}(\lambda)$ on the disc $\mathbb{D}_{p_\beta}(\lambda)$ containing $p_\beta(\lambda)$, and identify all coordinates $x(\lambda)$ for $\lambda \in [0,1]$. Let us denote this coordinate by x. Then we take the image of the Laurent expansion in x of $\int_{[0,1]} \eta(\lambda)d\lambda$ in $\oplus_{i=-k_\beta}^{k_\beta-2} \mathbf{C} \cdot x^i dx/\mathbf{C}x^{-1}dx$ and after that take the direct sum over all $\beta \in Pol$.

Let us denote the sum of the above two summands by $v(\Sigma_0, \Sigma_1)_\Sigma$. It is clear from the construction that for a triple of spectral curves $\Sigma_0, \Sigma_1, \Sigma_2$ we have $v(\Sigma_0, \Sigma_1)_\Sigma + v(\Sigma_1, \Sigma_2)_\Sigma = v(\Sigma_0, \Sigma_2)_\Sigma$. Indeed having 2-chains adjacent one to another one we naturally compose them into a single 2-chain, which gives the additivity of the corresponding integrals over the chains. Similarly the Laurent expansions are added because the corresponding integrals of 1-forms $\eta(\lambda)$ are added.

PROPOSITION 5.3.1. *The map which assigns $\eta(0)$ to the tangent vector at $\lambda = 0$ to the path $\Sigma(\lambda)$ coincides with ψ (after the above identification with the RHS of the formula (33)).*

PROOF. The map ψ assigns to a tangent vector from $T_{[\Sigma]}\mathcal{B}_0$ the corresponding 1-form on Σ. By construction $\eta(0)$ can be identified with the same. □

This concludes the alternative description of the connection and the 1-form as well as the proof of the Theorem 5.2.3.

REMARK 5.3.2. In the simplest case $Pol = \emptyset$ considerations of this subsection have simple (and well-known) meaning. Indeed, the family of the first homology groups $H_1(\Sigma', \mathbf{Z})$ form a local system over a neighborhood of $[\Sigma] \in \mathcal{B}_0$. The map $\kappa : l \mapsto \int_l \omega, l \in H_1(\Sigma', \mathbf{Z}) \simeq H_1(\Sigma, \mathbf{Z})$ defines a local embedding of $\mathcal{B}_0$ into the symplectic vector space $H^1(\Sigma, \mathbf{C})$. The integer homology groups of the spectral curve carry a linear flat connection along $\mathcal{B}_0$, since they do not change when we deform Σ. The closed 1-form κ defines an embedding of the germ $\mathcal{B}_\Sigma$ into $\mathcal{H}_\Sigma \simeq H^1(\Sigma, \mathbf{C})$.

6. Formal discs and universal Airy structure

6.1. Space of formal discs. Let (S, ω) be a complex analytic symplectic surface endowed with an analytic 1-dimensional foliation $\mathcal{F}$ (i.e. we consider only the sympletic leaf in a good foliated Poisson surface, and endow the leaf with the induced foliation).

For every $n \geq 2$ denote by $Discs^{(n)}$ the space of n-jets of embedded formal discs in S which are tangent to a leaf of foliation $\mathcal{F}$ with the tangency order equal to one. Technically speaking, a point of $Discs^{(n)}$ is given by a point $s \in S$, an epimorphism from the local ring $\mathcal{O}_s$ (germs of analytic functions at s) to a local Artin algebra R isomorphic to the algebra of truncated polynomials $\mathbf{C}[z]/(z^{n+1})$, such that the image of the local coordinate on the space of leaves belongs to $\mathfrak{m}_R^2 - \mathfrak{m}_R^3$, where $\mathfrak{m}_R \subset R$ is the maximal ideal.

The space $Discs^{(n)}$ is a smooth variety fibered over S with fibers isomorphic to $\mathbf{C}^* \times \mathbf{C}^{n-2}$. Also, we have an obvious projective system of fibrations

$$Discs^{(2)} \leftarrow Discs^{(3)} \leftarrow \dots$$

We define $Discs$ to be the projective limit, understood e.g. as a functor on finite-dimensional analytic spaces. In the algebro-geometric framework we assume that S is an algebraic surface and line subbundle $T_{\mathcal{F}} \subset T_S$ is algebraic as well. Then $Discs$ is a scheme of infinite type.

Our goal in this section will be to construct germs of Lagrangian embeddings of $Discs$ into infinite-dimensional affine symplectic spaces, in terms of the general formalism from Section 5.1.

For a point $t \in Discs$ we denote by $\widehat{\mathbb{D}}_t$ the corresponding formal disc embedded in S. Algebra of function $\mathcal{O}(\widehat{\mathbb{D}}_t)$ is isomorphic *non-canonically* to $\mathbf{C}[[z]]$. Denote by $UnivDiscs$ the total space of the fibration over $Discs$ whose fiber at t is fomal scheme $\mathbb{D}_t$. We have following universal maps:

(i) fibration $UnivDiscs \to Discs$;
(ii) section $Discs \to UnivDiscs$ of the above fibration, associating to the disc its unique tangency point to $\mathcal{F}$;
(iii) $UnivDiscs \to S$, the universal embedding $\widehat{\mathbb{D}}_t \hookrightarrow S$ for $t \in Discs$.

As in the case of D-transversal spectral curves, we have a natural isomorphism

$$T_t Discs = \Gamma\big(\widehat{\mathbb{D}}_t, (T_S)_{|\widehat{\mathbb{D}}_t}/T_{\widehat{\mathbb{D}}_t}\big) \simeq \Omega^1(\widehat{\mathbb{D}}_t). \tag{37}$$

Denote by $UnivDiscs^*$ the fibration over $Discs$ with fiber $\widehat{\mathbb{D}}_t^*$ at $t \in Discs$ given by the *punctured* formal disc, i.e. $\widehat{\mathbb{D}}_t$ with the removed tangency point. Algebra of functions on $\widehat{\mathbb{D}}_t^*$ is isomorphic to $\mathbf{C}((z))$.

We will define a canonical non-linear flat connection on the fibration $UnivDiscs^* \to Discs$, using the foliation $\mathcal{F}$ (and *not* using the symplectic form ω). In general, in the case of *non-punctured* discs, if one wants to make a non-linear connection on $UnivDiscs \to Discs$ along a tangent vector $v \in T_t Discs$, one has to make a lift

$$v \in \Gamma\big(\widehat{\mathbb{D}}_t, (T_S)_{|\widehat{\mathbb{D}}_t}/T_{\widehat{\mathbb{D}}_t}\big) \rightsquigarrow \widetilde{v} \in \Gamma\big(\widehat{\mathbb{D}}_t, (T_S)_{|\widehat{\mathbb{D}}_t}\big)\,.$$

In the case of punctured discs, one should find a lift in $\Gamma\big(\widehat{\mathbb{D}}_t^*, (T_S)_{|\widehat{\mathbb{D}}_t^*}\big)$. We define the **canonical** lift to be a unique lift which is the section of the subbundle $T_{\mathcal{F}} \subset T_S$. Such a lift exists and unique because $\mathcal{F}$ is transversal to $\widehat{\mathbb{D}}_t^*$. As a result we obtain a universal non-linear connection ∇_{univ} on the fibration $UnivDiscs^* \to Discs$.

EXAMPLE 6.1.1. Let us consider in local coordinates (x, y) on S such that $T_{\mathcal{F}} = \mathbf{C}\partial_y$ (leaves of $\mathcal{F}$ are given by $x = const$). Consider 1-parameter family of embeddings $(f_\lambda)_{\lambda \in \mathbf{C}}$ of the standard coordinate formal disc into S with the first order tangency to $\mathcal{F}$:

$$\text{parameter } \lambda \in \mathbf{C} \mapsto \Big(f_\lambda : Spf(\mathbf{C}[[z]]) \to Spec(\mathbf{C}[x, y]),\ \ z \mapsto (x = z^2 + \lambda, y = z)\Big)\,.$$

In this way we obtain an analytic path $\mathbf{C} \to Discs$, $\lambda \mapsto t(\lambda)$ is in $Discs$. The tangent vector ∂_λ to the path can be lifted naively using identification of standard coordinates z on discs $\widehat{\mathbb{D}}_{t(\lambda)}$. The corresponding section of $(T_S)_{\widehat{\mathbb{D}}_{t(\lambda)}}$ is given by ∂_x. The *canonical* lift of ∂_λ is different, it is equal to $-\frac{1}{2z}\partial_y$. The difference between the canonical and the naive lifts is a meromorphic section of $(T_S)_{\widehat{\mathbb{D}}_{t(\lambda)}}$ which is tangent to $\widehat{\mathbb{D}}_{t(\lambda)}$, and corresponds to the vector field $-\frac{1}{2z}\partial_z$ on $\widehat{\mathbb{D}}^*_{t(\lambda)}$:

$$T_{f_\lambda} : -\frac{1}{2z}\partial_z \mapsto -\frac{1}{2z}\partial_y - \partial_x \,.$$

PROPOSITION 6.1.2. *The non-linear connection ∇_{univ} defined above is flat.*

This statement can be proven by direct calculation. If one replaces formal punctured discs by actual embedded annuli transversal to the foliation, an analogous statement will be almost a tautology.

Let us return to the formal case. Lie algebra of vector fields on a punctured formal disc $\widehat{\mathbb{D}}_t$ naturally acts on the symplectic vector space

$$\mathcal{H}_t := \{\gamma \in \Omega^1(\widehat{\mathbb{D}}^*_t) \,|\, Res(\gamma) = 0\},$$

where the symplectic form is given by (17) (and is coordinate-independent). Therefore, the non-linear connection ∇_{univ} gives rise to a flat linear symplectic connection on the infinite-dimensional[13] bundle $\mathcal{H}$ with fibers defined above.

Define a $\mathcal{H}$-valued 1-form ϕ on *Discs* as the natural embedding (here we use only the symplectic form ω, and not the foliation):

$$T_t Discs = \Omega^1(\widehat{\mathbb{D}}_t) \hookrightarrow Ker\left(Res : \Omega^1(\widehat{\mathbb{D}}^*_t) \to \mathbf{C}\right) = \mathcal{H}_t \,.$$

THEOREM 6.1.3. *Form ϕ defined above is closed under the covariant derivative, and defines a Lagrangian embedding of the tangent bundle T_{Discs} into the symplectic bundle $\mathcal{H}$.*

PROOF. The fact that ϕ gives a Lagrangian embedding is obvious, as in local coordinates $\mathbf{C}[[z]]dz$ is a Lagrangian subspace in $Ker\,(Res : \mathbf{C}((z))dz \to \mathbf{C})$. The fact that ϕ is closed can be checked by direct calculation.

Alternatively, it can be explained by direct construction of a germ of Lagrangian embedding in the non-formal case and the Remark 5.1.2. Namely, consider a germ of a finite-dimensional family of embedded analytic discs $\mathbb{D}_{t(\lambda)}$ where $\lambda \in \Lambda$ (parameter space) is close to λ_0. then one can pick an open domain $\mathcal{U} \subset S$ close to the tangency point of $\mathbb{D}_{t(\lambda_0)}$ such that $\mathcal{U} \cap \mathbb{D}_{t(\lambda_0)}$ is an annulus $A \simeq \{z \in \mathbf{C} \,|\, \epsilon_1 < |z| < |\epsilon_2\}$ for $0 < \epsilon_1 < \epsilon_2 \ll 1$, and $\mathcal{U}$ is *fibered* over A with fibers being leaves of $\mathcal{F}$. Symplectic structure ω gives an *affine* structure on fibers $\mathcal{U} \to A$, and moreover makes fibers to be parallel to the cotangent bundle T^*A[14]. Then for any λ, λ' close to λ_0 we have two sections $\mathcal{U} \cap \mathbb{D}_{t(\lambda)}$ and $\mathcal{U} \cap \mathbb{D}_{t(\lambda')}$ of the fibration $\mathcal{U} \to A$. Their difference is a 1-form $\eta_{\lambda_1,\lambda_2}$ on A (here we use affine structure along leaves of $\mathcal{F}$).

LEMMA 6.1.4. *The contour integral of this 1-form over a non-trivial loop in A is equal to zero.*

Proof. Indeed, one can choose locally near the tangency point of $\mathbb{D}_{t(\lambda_0)}$ a local system of coordinates (x, y) (canonical coordinates) such that $\omega = dx \wedge dy$, and the foliation is given by $x = const$. The contour integral of ydx over a loop in $\mathbb{D}_{t(\lambda)}$ vanishes for any λ close to λ_0, because this loop bounds a disc. Therefore, the countour integral over a loop in A of the pullback ξ_λ of ydx via the map

[13]Technically, one should consider $\mathcal{H}$ as an inductive limit of projective limits of finite-dimensional vector bundles.

[14]This is a general feature of Lagrangian foliations in arbitrary dimensions.

$A \to \mathcal{U} \cap \mathbb{D}_{t(\lambda)}$ (inverse to the projection) also vanishes. Finally, the 1-form $\eta_{\lambda_1,\lambda_2}$ is equal to the difference $\xi_{\lambda_1} - \xi_{\lambda_2}$. Therefore its contour integral vanishes as well. The Lemma is proved. □

We see that in the analytic situation we obtain a flat familiy of local embeddings of the space of discs into infinite-dimensional affine space. The cocycle condition 5.1.2 for the closedness of our vector-valued 1-form follows from the trivial identity

$$\eta_{\lambda_1,\lambda_2} + \eta_{\lambda_2,\lambda_3} = \eta_{\lambda_1,\lambda_3}\,.$$

This explains that $\nabla\phi = 0$. □

6.2. Relation to the universal classical Airy structure. For a given symplectic surface (S,ω) endowed with (automatically Lagrangain) one-dimensional foliation $\mathcal{F}$, define an infinite-dimensional space *Coord* to be the space of tuples (s,x,y) where $s \in S$ is a point, (x,y) is a formal coordinate system at s such that

(i) symplectic form ω is equal to $dx \wedge dy$;
(ii) tangent bundle $T_{\mathcal{F}}$ to the foliation $\mathcal{F}$ is spanned by ∂_y.

This space is projective limit of a tower of fibrations of finite-dimensional smooth varieties (like the space *Discs* in the previous subsection).

The tangent space to *Coord* at any point is canonically isomorphic to the Lie algebra $\mathfrak{g}$ where:

$$\mathfrak{g} := \{\xi = a(x,y)\,\partial_x + b(x,y)\,\partial_y \mid a,b \in \mathbf{C}[[x,y]],\ \xi \text{ preserves } \omega \text{ and } T_{\mathcal{F}}\}\ .$$

As in the formal differential geometry of Gelfand and Kazhdan (see [**GelKaz**]), we have a *free transitive action* of $\mathfrak{g}$ on *Coord*. An easy calculation shows that $\mathfrak{g}$ is canonically isomorphic to the quotient of the Lie algebra of differential operators of order ≤ 1 on $\mathbf{C}[[x]]$ by the one-dimensional center consisting of constants, i.e. $\mathfrak{g} \simeq \mathbf{C}[[x]]\partial_x \ltimes (\mathbf{C}[[x]]/\mathbf{C})$, where [15]

$$f(x)\,\partial_x + g(x)/const \mapsto f(x)\,\partial_x - yf'(x)\,\partial_y - g'(x)\,\partial_y = \{f(x)y + g(x), \cdot\}\,.$$

Consider the map $\pi : Coord \to Discs$ given by

$$(s,x,y) \mapsto \text{ formal disc with coordinate } z \text{ embedded via } \left(z \mapsto (x = z^2, y = z)\right).$$

Map π is a $3:1$ cover. Indeed $(s, x' = \xi^2 x, y' = \xi)$ gives the same embedded formal disc as (s,x,y) for any $\xi \in \mathbf{C}, \xi^3 = 1$.

For any point $(s,x,y) \in Coord$ and the corresponding point $t = \pi(s,x,y) \in Discs$, the local transitive free action of Lie algebra $\mathfrak{g}$ on *Coord* gives an isomorphism

$$\mathfrak{g} \simeq T_{(s,x,y)}Coord \simeq T_t Discs \simeq \Omega^1(\widehat{\mathbb{D}}_t) \simeq \mathbf{C}[[z]]dz\,.$$

PROPOSITION 6.2.1. *The above isomorphism does not depend on $(s,x,y) \in$ Coord, and is given by the formula*

$$f(x)\,\partial_x + g(x)/const \mapsto d(f(z^2)z + g(z^2)) \in \mathbf{C}[[z]]dz\,.$$

The proof is given by direct calculation.

The pullback $\pi^*\mathcal{H}$ is a canonically trivialized symplectic vector bundle, each fiber is identified with the space $W := W_{Airy} := Ker(Res : \mathbf{C}((z))dz \to \mathbf{C})$ from Section 3.3.

[15] Here $g(x)/const$ denotes the image of the element $g = g(x)$ of $\mathbf{C}[[x]]$ in the quotient space $\mathbf{C}[[x]]/\mathbf{C}$.

THEOREM 6.2.2. *Consider the affine flat connection on the trivial bundle with fiber W over $Coord$ given by the homomorphism*

$$\mathfrak{g} \to spAff(W) = sp(W) \ltimes W$$

which is associated with the standard classical Airy structure from Section 3.3. Then, after the above identifications, it coincides with the affine flat connection constructed in the Section 6.1.

PROOF. We have to check first that the linear connections match. Let us calculate the linear part of the flat affine connection from Section 6.1 following the definitions. We obtain it as the composition of two homomorphisms

$$\mathfrak{g} \to \mathbf{C}((z))\,\partial_z \to sp(W)$$

where the second maps is the natural action by coordinate changes. The first map comes from the geometric description of the non-linear flat connection ∇_{univ} via foliation. Namely, with any element $f(x)\partial_x + g(x)/const$ we associate the corresponding Hamiltonian vector field v in $Spf(\mathbf{C}[[x,y]])$. Let us consider restriction of v to the parabola $x = y^2$, and find another meromorphic vector field v' on parabola which is vertical and such that $u = v - v'$ is tangent to the parabola. Then u expanded in coordinate z gives an element in $\mathbf{C}((z))\,\partial_z$ corresponding to $v \in \mathfrak{g}$. The direct calculation gives

$$v_{\widehat{D}_t} = f(z^2)\,\partial_x - zf'(z^2)\,\partial_y - g'(z^2)\,\partial_y\,.$$

The map from vector fields tangent to parabola in coordinate z, to section of the pullback T_S to the parabola is given by

$$h(z)\,\partial_z \mapsto 2zh(z)\,\partial_x + h(z)\,\partial_y\,.$$

Therefore, the vector field tangent to parabola which differ from v by a purely vertical field is

$$u = \frac{f(z^2)}{2z}\,\partial_z\,.$$

This gives a homomorphism $\mathfrak{g} \to sp(W)$. Direct calculation shows that it coincides with the linear part of the homomorphism from Section 3.3. Similarly, the shift part (i.e. 1-form ϕ with values in the trivial bundle with the fiber W) matches, as we identify the tangent space to $Coord$ with the space of 1-forms $\mathbf{C}[[z]]\,dz$ as in Proposition 6.2.1. □

6.3. Geometric meaning of residue constraints. Let $t \in Discs$ and $\widehat{\mathbb{D}}_t$ the corresponding formal disc. We endow it with canonical coordinates (x, y), so that the foliation $\mathcal{F}$ is given by the equation $x = const$ and the symplectic form $\omega = dx \wedge dy$. Disc $\widehat{\mathbb{D}}_t$ is endowed with coordinate z such that the embedding into S is $z \mapsto (x = z^2, y = z)$.

Tate vector space $W = W_{Airy} = Ker(Res_z : \mathbf{C}((z))dz \to \mathbf{C})$ is a hyperplane in $W' := \mathbf{C}((z))dz$. The formal germ $\widehat{W'}$ of W' at 0 controls deformations of the punctured formal disc $\widehat{\mathbb{D}}_t^*$. Its formal subgerm $\widehat{W}$ is determined by the condition that the residue of the pull-back of the 1-form ydx to the deformed punctured formal disc is trivial. The formal germ $\widehat{W}$ contains a formal germ of $t \in Discs$ corresponding to deformations of the non-punctured formal disc $\mathbb{D}_t$.[16] The latter

[16] We understand all submanifolds as functors on commutative local Artin algebras, see Remark 3.3.1.

germ is *contained* in the formal germ of the Lagrangian submanifold L which by definition is singled out in $\widehat{W}$ by the system of equations

$$Res_z(x^k y\,dx) = 0, k \geq 1, Res_z(x^k y^2 dx) = 0, k \geq 0\,.$$

These are exactly the residue constraints from Section 3.3. They hold on the formal completion of *Discs* since the residue of a holomorphic form is trivial.

Notice that the tangent space at zero $T_0 L \simeq \mathbf{C}[[z]]dz$ and hence it coincides with the tangent space to *Discs*. Hence L coincides the formal neighborhood of $t \in Discs$ embedded in $\widehat{W}$. This explains the residue constraints from Section 3.3.

We will also explain below the invariant meaning the residue constraints.

Let U_1 be the vector space of 1-forms on the formal completion of of S at $(x, y) = (0, 0) \in \mathbb{D}_t$ which are equal to zero on the leaves of the foliation $\mathcal{F}$ with the order ≤ 2 (the latter notion makes sense even globally because of the intrinsically defined affine structure on the leaves).

In the canonical coordinates (x, y) an element of U_1 can be written as $F(x, y)dx$, where $F(x, y)$ is a polynomial of degree ≤ 2 in y for a fixed x. Let $U_2 \subset U_1$ be a vector subspace consisting of such 1-forms μ that $d\mu = c \cdot dx \wedge dy, c \in \mathbf{C}$.

Spaces U_1, U_2 have topological bases

$$\{x^{\geq 0}dx, x^{\geq 0}ydx, x^{\geq 0}y^2dx\}, \quad \{x^{\geq 0}dx, ydx\}$$

respectively. Therefore , the quotient U_1/U_2 has the topological basis

$$\{x^{\geq 1}ydx, x^{\geq 0}y^2dx\}\,.$$

The pull-back of $\mu \in U_2$ to a deformed $\mathbb{D}_t^*$ has trivial residue (we deform the disc inside of $\widehat{W} \subset \widehat{W}'$. Hence U_1/U_2 is naturally mapped to the space of formal functions on the formal neighborhood of $0 \in \widehat{W}$. Vanishing of these functions is the same as the residue constraints from Section 3.3.

7. Hamiltonian reduction and Holomorphic Anomaly Equation

7.1. Global Hamiltonian reduction. In this subsection we explain how to treat the germ $\mathcal{B}_\Sigma$ as a Hamiltonian reduction of a germ of an infinite-dimensional Lagrangian submanifold of a Tate affine symplectic space (see Remark 3.3.1) that we understand Tate spaces as functors on Artin algebras.

We start with a finite-dimensional model. Let M be a finite-dimensional symplectic manifold, $G \subset M$ a coisotropic submanifold and $L \subset M$ a Lagrangian submanifold. Let $x \in L \cap G$ be a point satisfying the following equivalent properties:

1) $T_x L + T_x G = T_x M$,
2) $T_x L \cap (T_x G)^{\perp} = 0$,

where $\perp$ means the skew-orthogonal with respect to the symplectic form.

Notice that the family of vector spaces $(T_x G)^{\perp}$ defines an integrable distribution of planes. Hence we have the corresponding foliation of G with (locally) symplectic quotient. The following proposition is well-known and is easy to prove.

PROPOSITION 7.1.1. *The image $\mathcal{L}_x$ of the germ of $L_x \cap G_x$ at x is a germ of Lagrangian submanifold in the (germ of the) above symplectic quotient.*

This Proposition can be generalized to the case of Tate spaces and formal germs.

Let now Σ be a $(\mathcal{F}, D)$-transversal curve Σ in a good foliated Poisson surface. Let us choose local coordinates (x, y) on S and z on Σ near each ramification point so we get locally a standard parabola, as in the previous section. The germ $\mathcal{B}_\Sigma$ of the moduli space $\mathcal{B}_1$ at $[\Sigma]$[17] maps to the product $Discs^{Ram}$ of copies of the space $Discs$ labeled by ramification points. We denote by $\widehat{\mathcal{B}}_\Sigma$ the formal germ of $\mathcal{B}_\Sigma$ at the point $[\Sigma] \in \mathcal{B}_1$. Thus, we obtain a local immersion of the finite-dimensional formal germ $\widehat{\mathcal{B}}_\Sigma$ into an infinite-dimensional affine vector space $W := W^{Ram}_{Airy}$. On the other side, we have a natural germ L of infinite-dimensional Lagrangian subvariety in W^{Ram}_{Airy}, given by the formal neighborhood of a point in $Discs^{Ram}$. Clearly $L = L^{Ram}_{Airy}$ is the product of copies of formal germs of infinite-dimensional Lagrangian submanifolds L_{Airy} from Section 3.3.

In order to simplify the notation we will denote by $\Delta = \sum_{\alpha \in Ram} r_\alpha$ the ramification divisor, and as before we will denote by $D_\Sigma = \sum_{\beta \in Pol} k_\beta p_\beta$ the pole divisor.

PROPOSITION 7.1.2. *The finite-dimensional germ $\widehat{\mathcal{B}}_\Sigma$ is equal to the intersection of the infinite-dimensional germ of $Discs^{Ram}$ with the coisotropic affine subspace $G_\Sigma \subset W^{Ram}_{Airy}$ defined as the space of 1-forms*

$$G_\Sigma := \{\eta \in \Gamma(\Sigma, \Omega^1_\Sigma(D_\Sigma + \sum_{\alpha \in Ram} \infty r_\alpha)) | \ Res_{r_\alpha}(\eta) = 0\} \tag{38}$$

expanded in canonical local coordinates z near each ramification poijnt r_α (compare with (34)). Moreover, the subbundle with the fiber G_Σ of the trivial bundle with the fiber W^{Ram}_{Airy}, is covariantly constant as an affine subbundle.

PROOF. We will explain below both statements in the complex analytic setting. There is a similar but less transparent proof in the framework of Tate spaces.

Notice that by our assumptions, the foliation $\mathcal{F}$ is transversal to Σ everywhere except of ramification points. Therefore, there exists an open domain $\mathcal{U} \subset P$ such that

$$\mathcal{U} \cap \Sigma = \Sigma^0 := \Sigma - \sqcup_{\alpha \in Ram} \mathbb{D}_{\alpha, \epsilon_1}$$

where $\mathbb{D}_{\alpha,\epsilon_1}$ is a closed disc of radius ϵ_1 in the standard coordinate $z = z_{(\alpha)}$ near r_α, $\forall \alpha \in Ram$, and ϵ_1, $0 < \epsilon_1 \ll 1$ is sufficiently small. Moreover, we may assume that the projection to the space of leaves of $\mathcal{F}$ gives a fibration of $\mathcal{U}$ over Σ^0. Any spectral curve Σ' which is sufficiently close to Σ gives a section $\Sigma' \cap \mathcal{U}$ of the fibration $\mathcal{U} \to \Sigma^0$, which differs form the section given by Σ by a 1-form on Σ^0 with poles of order $\leq k_\beta$ at p_β, $\beta \in Pol$. This follows from the fact that fibers of the fibration $\mathcal{U} \to \Sigma^0$ are naturally affine lines parallel to the cotangent bundle to Σ^0, twisted by divisor D_Σ (here we use the symplectic structure to ensure that fibers of a Lagrangian foliation carry affine structure). This 1-form has contour integrals over all annuli $\epsilon_1 < |z^{(\alpha)}| < \epsilon_2 \ll 1$ equal to zero (because these annuli are contained in discs). This gives an element in the (analytic version of) G_Σ .

We see that spectral curves Σ' can be described as deformations of annuli $\epsilon_1 < |z^{(\alpha)}| < \epsilon_2$ along the foliation $\mathcal{F}$ satisfying the vanishing of residue condition as well as two other constraints:

- each annulus is contained in a holomorphic disc (geometrically it means that we land into the product of infinite-dimensional Lagrangian subvarieties given by quadratic equations of the standard classical Airy structure)

[17] We will use the notation $\mathcal{B}_\Sigma$ and $\mathcal{B}_{[\Sigma]}$ for this germ on the equal footing.

- the collection of 1-forms on annuli extends to a global 1-form from the analytic version of G_Σ, equivalently the deformed annuli lie in a shifted section of the fibration $\mathcal{U} \to \Sigma^0$.

This description identifies the formal germ $\widehat{\mathcal{B}}_\Sigma$ with the intersection of a Lagrangian subvariety coming from the classical Airy structure and an affine coisotrpic subvariety. □

Next, observe that the 2-term complex of sheaves $\mathcal{O}_\Sigma(-D_\Sigma) \to \Omega^1_\Sigma(D_\Sigma)$ endowed with de Rham differential is quasi-isomorphic to the following complex of sheaves:

$$\mathcal{O}_\Sigma(-D_\Sigma + *\Delta) \to \Omega^1_\Sigma(D_\Sigma + *\Delta) \to \oplus_{r_\alpha \in Ram}\mathbf{C} \to 0,$$

where $\Delta = \sum_{\alpha \in Ram} r_\alpha$, first two terms are sheaves of functions/ forms having poles/zeros at the points p_β bounded by k_β as well as and poles of arbitrary order at the points r_α (the map is the de Rham differential), and second (surjective) arrow is the map which assigns to a differential form the sum of residues at all points r_α. Each sheaf in the above complex has trivial cohomology, except of (possibly) in degree zero. It follows that $\mathcal{H}_\Sigma \simeq G_\Sigma/d(\mathcal{O}_\Sigma(-D_\Sigma + *\Delta))$. Thus we have the natural map $G_\Sigma \to \mathcal{H}_\Sigma$.

PROPOSITION 7.1.3. *We have the following isomorphism of Tate spaces:*

$$d(H^0(\mathcal{O}_\Sigma(-D_\Sigma + *\Delta))) = G_\Sigma^\perp,$$

and hence $G_\Sigma/G_\Sigma^\perp \simeq \mathcal{H}_\Sigma$.

PROOF. It suffices to check that $d(H^0(\mathcal{O}_\Sigma(-D_\Sigma + *\Delta)))^\perp = G_\Sigma$.

By definition the LHS consists of collections of forms $(\mu_\alpha)_{\alpha \in Ram}$ such that $\mu_\alpha \in \mathbf{C}((z_{(\alpha)}))dz_{(\alpha)}$ (here $z_{(\alpha)}$ is the canonical local parameter at r_α), satisfying the conditions

- $Res(\mu_\alpha) = 0 \ \forall\alpha$,
- $\sum_\alpha Res(f\mu_\alpha) = 0$ for any $f \in H^0(\mathcal{O}_\Sigma(-D_\Sigma + *\Delta))$.

Let us introduce the Tate vector space G'_Σ as the space of collections of 1-forms $\mu_\alpha, \alpha \in Ram$ as above, but without the condition $Res(\mu_\alpha) = 0 \ \forall\alpha$.

Before proceeding further, let us recall the following well-known result about Tate spaces associated with vector bundles on curves (Serre duality for curves formulated in terms of local fields, see [**Tat68**], Corollary to Theorem 5).

Let Σ be a smooth projective curve over a field $\mathbf{k}$, and $\mathcal{E}$ be a vector bundle over Σ. Let $\{r_i\}_{i\in I}$ be a finite subset of $\Sigma(\mathbf{k})$. Let us denote by $\mathcal{E}'$ the vector bundle $\mathcal{E}^\vee \otimes \Omega^1_\Sigma$. For each r_i denote by K_{r_i} the completed local field of r_i. Each K_{r_i} is isomorphic to the field $\mathbf{k}((z_{(i)}))$, where $z_{(i)}$ is a local coordinate at r_i. For each $i \in I$ we have the finite-dimensional K_{r_i}-vector spaces $\widehat{\mathcal{E}}_{r_i}$ and $\widehat{\mathcal{E}}_{r'_i}$ of sections over the formal punctured discs at r_i. They are Tate $\mathbf{k}$-vector spaces endowed with the topological perfect duality $(v,w) \mapsto Res_{r_i}(v,w)$. Direct sum $\oplus_{i\in I}\widehat{\mathcal{E}}_{r_i}$ contains the subspace $\Gamma(\Sigma - \{r_i\}, \mathcal{E})$, and similarly $\oplus_{i\in I}\widehat{\mathcal{E}}_{r_i}$ contains the subspace $\Gamma(\Sigma - \{r_i\}, \mathcal{E}')$. The direct sums themselves are naturally dual to each other.

LEMMA 7.1.4. *The duality between above Tate spaces makes the subspaces orthogonal to each other.*

We will apply the Lemma to $\mathcal{E} = \mathcal{O}_\Sigma(-D_\Sigma), \mathcal{E}' = \Omega^1_\Sigma(D_\Sigma), \{r_i\}_{i\in I} = \{r_\alpha\}_{\alpha\in Ram}$.

Then the lemma implies that

$$G'_\Sigma \simeq \{\eta \in \Gamma(\Sigma, \Omega^1_\Sigma(D_\Sigma + \sum_{\alpha\in Ram} \infty\, r_\alpha))\}.$$

Imposing the condition $Res_{r_\alpha}(\eta) = 0$ on elements of G'_Σ we obtain G_Σ. This proves the Proposition. □

REMARK 7.1.5. The Proposition gives also an alternative explanation to the fact that the hypercohomology is a symplectic vector space.

The Proposition 7.1.3 gives rise to a formal symplectic vector space and formal connection only, since in the framework of Tate spaces we understand varieties and bundles only as functors on Artin algebras.

Recall that we have an embedding of the germ $\mathcal{B}_1$ (in fact even a bigger space $\mathcal{B}_0$) into $\mathcal{H}_\Sigma$ derived from the Theorem 5.2.2. Let us compare these two embeddings.

PROPOSITION 7.1.6. *The formal Lagrangian submanifold of $\mathcal{H}_\Sigma$ defined in this subsection is a formal completion at zero of the germ of the Lagrangian submanifold defined in Section 5.2.*

PROOF. Recall the RHS of the formula (33) which consists of two summands. Recall also the global description of $\mathcal{H}_\Sigma$ given in the formula (34). Consider a complex manifold with boundary obtained from Σ by removing small open discs $\mathbb{D}_{r_\alpha}, \alpha \in Ram$ which contain ramification points r_α. We have the corresponding vector space of holomorphic 1-forms, which are analytic on the boundary, and which have trivial integrals over each boundary component. We can use this space of 1-forms in order to obtain an presentation for $\mathcal{H}_\Sigma$ analogous to (34):

(39)
$$\mathcal{H}_\Sigma = \{\eta \in \Gamma(\Sigma - \cup_\alpha \mathbb{D}_{r_\alpha}, \Omega^1_\Sigma(D_\Sigma))| \int_{\partial\overline{\mathbb{D}}_{r_\alpha}} \eta = 0\}/d\Gamma(\Sigma - \cup_\alpha \mathbb{D}_{r_\alpha}, \mathcal{O}_\Sigma(-D_\Sigma)).$$

Using the foliation $\mathcal{F}$ we can identify as abstract complex manifolds with boundary

$$\Sigma - \cup_\alpha \mathbb{D}_{r_\alpha} \simeq \Sigma' - \cup_\alpha \mathbb{D}'_{r_\alpha},$$

for any Σ' which is sufficiently close to Σ and the corresponding $\mathbb{D}'_{r_\alpha}$. Hence the RHS of the formula (39) is invariant with respect to the deformations of Σ. As a result we obtain a linear flat symplectic connection described in terms of spectral curves with small discs removed.

Notice the RHS of the formulas (39) and (34) are the only cohomologies in degree 1 of a 2-term complexes. There is a natural quasi-isomorphism of these complexes induced by the embedding $\Sigma - \cup_\alpha \mathbb{D}_{r_\alpha} \to \Sigma - \cup_\alpha\{r_\alpha\}$. Moreover the isomorphsim ψ from formula (33) can be generalized to the description of $\mathcal{H}_\Sigma$ given in the formula (39).

This ensures equivalence of both descriptions of $\mathcal{H}_\Sigma$, which in turn ensures equivalence of two descriptions of the affine symplectic connection on $\mathcal{H}_\Sigma$.

Notice that the Hamiltonian reduction in the framework of Tate spaces should be understood in terms of functors on Artin algebras (see Remark 3.3.1). Then we see that the embedding of the moduli space $\mathcal{B}_1$ obtained from the Proposition 7.1.3 coincides (as a functor on Artin algebras) with the description given in Section 5. This proves the statement. □

7.2. Quantum Airy structures and quantum Hamiltonian reduction of the wave function. In the case $Pol = \emptyset$ let us choose a Lagrangian complement $\mathcal{V}_\Sigma \subset H^1(\Sigma, \mathbf{C})$ to the Lagrangian subspace given by the first term of the Hodge filtration $F^1 H^1_{dR}(H^1(\Sigma, \mathbf{C})) := H^{1,0}(\Sigma) = T_\Sigma \mathcal{B} \subset H^1(\Sigma, \mathbf{C}) = \mathcal{H}_\Sigma$ (a splitting of the Hodge filtration is equivalent to a choice of Bergman kernel in the traditional language of topological recursion).

Let us consider the infinite-dimensional vector space $V_\Sigma \subset G_\Sigma$ consisting of holomorphic 1-forms $\gamma \in \Omega^1(\Sigma - \cup_{\alpha \in Ram}\{r_\alpha\})$, which are meromorphic on Σ, have trivial residue at all r_α, and such that $[\gamma] \in \mathcal{V}_\Sigma$. It is easy to see that V_Σ is a Lagrangian complement to the tangent space to the Lagrangian submanifold $L = L^{Ram}_{Airy} \subset W = W^{Ram}_{Airy}$.

The story is similar in the case $Pol \neq \emptyset$. In this case we choose a transversal Lagrangian complement $\mathcal{V}_\Sigma$ in $\mathcal{H}_\Sigma$ to the Lagrangian subspace $H^0(\Sigma, \Omega^1_\Sigma(D_\Sigma))$. Its preimage $V_\Sigma := (G_\Sigma \twoheadrightarrow \mathcal{H}_\Sigma)^{-1}\mathcal{V}_\Sigma$ is a Lagrangian complement to $T_0 L$. Thus, we have in general the following result.

COROLLARY 7.2.1. *The Lagrangian complement V_Σ carries both classical and quantum Airy structures.*

It follows that we have the wave function $\psi = \exp(\sum_{g \geq 0} S_g \hbar^{g-1})$ associated with the cyclic DQ-module corresponding to L and the Lagrangian complement V_Σ. If we don't choose the Lagrangian complement then we still have a cyclic DQ-module with a cyclic vector, i.e. a left ideal in the completed Weyl algebra. The choice of Lagrangian complement allows us to encode the left ideal in terms of the wave function.

REMARK 7.2.2. Notice that in the global version of the story the space $V_{\Sigma,odd}$ is the subspace of the previously defined "local" space V_{odd} which consists of such meromorphic forms γ on $\Sigma - \cup_\alpha\{r_\alpha\}$ that $\gamma + \sigma^*(\gamma)$ has no poles at each r_α. Here σ is the local involution from Section 3.1. The subspace $V_{\Sigma,odd} \subset V_\Sigma$ carries a quantum Airy substructure.

Let us explain how the Hamiltonian reduction gives us the cyclic DQ-module $\mathcal{E}_\Sigma$ with a cyclic vector quantizing the Lagrangian germ $\mathcal{B}_\Sigma \subset \mathcal{H}_\Sigma$. This is a part of the following more general story.

Suppose we are given an affine symplectic vector space M an affine coisotropic subspace $G \subset M$, a point $x \in G$ and a formal germ of Lagrangian submanifold L_x at x satisfying the conditions 1) or 2) from Section 7.1. Let us work for simplicity over the field of complex numbers $\mathbf{C}$, although it is not necessary.

Notice that with the point x we can associate the completed Weyl algebra $Weyl_x$ of the tangent space $T_x M$. It is a topologically free algebra over $\mathbf{C}[[\hbar]]$. Suppose we are given a cyclic module $\mathcal{E}$ over $Weyl_x$ with the cyclic vector $e \in \mathcal{E}$. Assume that modulo $\hbar$ we have $\mathcal{E} \simeq \mathcal{O}(L_x)$ and $e = 1 \in \mathcal{O}(L_x)$. By Proposition 7.1.1 we have a symplectic affine space $M' = G/G^\perp$ and the Hamiltonian reduction $L'_x \subset M'$ of the germ L_x. We also have the completed Weyl algebra $Weyl'_x$ associated with M'. Then G is naturally embedded an affine Lagrangian subspace in the symplectic affine space $M' \times \overline{M}$, where $\overline{M}$ denote M with the opposite symplectic form. By general theory we have a **canonical** cyclic DQ modules $\mathcal{E}^{can}_G$ with the canonical cyclic vector e^{can}_G quantizing G.

We treat $\mathcal{E}^{can}_G$ as a $Weyl'_x - Weyl_x$-bimodule.

DEFINITION 7.2.3. Quantum Hamiltonian reduction of $(\mathcal{E}, e)$ is the $Weyl'_x$-module $\mathcal{E}' = \mathcal{E}_G^{can} \otimes \mathcal{E}$ endowed with the cyclic vector $e' = e_G^{can} \otimes e$.

Clearly modulo $\hbar$ we obtain the Hamiltonian reduction L'_x of L_x.

The above construction works for our infinite-dimensional affine spaces and germs of formal Lagrangian manifolds. As a result we obtain a cyclic DQ-module $\mathcal{E}_\Sigma$ associated with the spectral curve Σ. In general a choice of Lagrangian complement to $T_x(L_x)$ allows us to translate the story from the language o cyclic DQ-modules to the language of wave functions. In the case of spectral curves this translation looks as follows.

Recall that we have a "local" symplectic vector space W^{Ram} together with a germ of formal Lagrangian submanifold $L \subset W^{Ram}$, both depending only[18] on the set *Ram* of ramification points. We have also defined a coisotropic vector space $G_\Sigma \subset W^{Ram}$ depending on Σ.

Then the quantization procedure give us a cyclic DQ-module $\mathcal{E}_\Sigma$ with the cyclic vector ψ_L. Let us split coordinates in each of the symplectic spaces, so that the dual coordinates in W^{Ram} will be denoted by (q,p) and in $\mathcal{H}_\Sigma$ by (q',p') in such a way that $\psi_L = \psi_L(q)$ is the wave function corresponding to the quantized Airy structure.

We have the wave function $\psi_{G_\Sigma} = \psi_{G_\Sigma}(q,q') = e^{Q_2(q,q')/\hbar}$ corresponding to the coistoropic subspace G_Σ, where Q_2 is the quadratic form such that $graph(dQ_2) = G_\Sigma$. Then the Hamiltonian reduction of $L \subset W^{Ram}$ to $\widehat{\mathcal{B}}_\Sigma \subset \mathcal{H}_\Sigma$ at the level of wave functions becomes the integral operator

$$\psi_L \mapsto \psi_{\widehat{\mathcal{B}}} := \int e^{Q_2(q,q')/\hbar}\psi_L(q)dq$$

understood as a formal integration with respect to the Gaussian measure.

7.3. Holomorphic Anomaly Equation. Let us recall the framework and notation from Section 5.1. Thus we have a tuple $(X, \mathcal{H}, \langle \bullet, \bullet \rangle, \nabla, \phi)$ consisting of an analytic manifold X, analytic vector bundle $\mathcal{H}$ over X, linear flat connection ∇ on $\mathcal{H}$, $\mathcal{H}$-valued ∇-flat 1-form ϕ, and a ∇-covariant symplectic structure $\langle \bullet, \bullet \rangle$ on $\mathcal{H}$.

Assume that for any point $x \in X$ and assume that the embedding of the germ of X_x of X at x into $\mathcal{H}_x$ defined in Section 5.1 is Lagrangian (i.e. the image is a germ L_x of Lagrangian submanifold in $\mathcal{H}_x$). Notice that the affine flat connection $\nabla + \phi$ induces a flat connection $\widehat{\nabla}$ on the vector bundle over X with the fiber $Weyl_x$ (completed at zero Weyl algebra of $\mathcal{H}_x$). This comes from the fact that the group of affine symplectic transformations acts on the non-completed Weyl algebra.

DEFINITION 7.3.1. A **quantization** of the tuple $(X, \mathcal{H}, \langle \bullet, \bullet \rangle, \nabla, \phi)$ is given by a $\widehat{\nabla}$-covariantly constant family of closed left ideals $J_x \subset Weyl_x$ such that the cyclic module $Weyl_x/J_x$ is a quantization the Lagrangian germ $L_x := \phi(X_x)$ (i.e. modulo $\hbar$ it gives the algebra of function $\mathcal{O}(L_x)$, and the cyclic vector $1 \in Weyl_x/J_x$ corresponds to $1 \in \mathcal{O}(L_x)$).

Let us choose a ∇-covariantly constant subspace V_x which is transversal to L_x for all $x \in X$. Then the condition that J_x form a covariantly constant family can be translated into a system of differential equations on the components $(S_{g,n})_x \in$

[18]Here we choose a canonical coordinate near each ramification point, determined a priori only up to multiplication by a third root of 1.

$Sym^n(V_x)$ of the cyclic vector $\psi_x = \exp(\sum_{g,n} \hbar^{g-1}(S_{g,n})_x)$. The summation is over pairs (g, n) with either $g = 0, n \geq 3$ or $g \geq 1, n \geq 1$. This system of equations is called (quantum) *Holomorphic anomaly equation* (HAE for short).

REMARK 7.3.2. Originally HAE was formulated in the context of *compact* Calabi-Yau 3-folds. In this case X (probably should be better denoted by $\mathcal{L}$, as it is locally a Lagrangian cone) is a $\mathbf{C}^*$-bundle over a connected component of the moduli space $\mathcal{M}$ of projective algebraic 3-dimensional varieties Y with $H^1(Y, \mathbf{C}) = 0$ and trivial canonical class. Space $X = \mathcal{L}$ parametrizes pairs $(Y, \Omega_Y^{3,0})$ where $\Omega_Y^{3,0} \in \Gamma(Y, \Omega_Y^3) - \{0\}$ a non-zero holomorphic volume form, and it is embedded locally as holomorphic Lagrangian cone into sympletic vector space $H^3(Y, \mathbf{C})$ via

$$(Y, \Omega_Y^{3,0}) \mapsto [\Omega_Y^{3,0}] \in H^3(Y, \mathbf{C}) .$$

It was predicted in [**BCOV94**][19] that this Lagrangian cone has a **canonical** quantization, equivariant with respect to $\mathbf{C}^*$-action acting by rescaling

$$[\Omega_Y^{3,0}] \mapsto \lambda[\Omega_Y^{3,0}], \quad \hbar \mapsto \lambda^2\hbar, \quad \lambda \in \mathbf{C}^* .$$

The rigorous construction of this quantization was proposed by K. Costello in [**Cos**]. Our framework here is more general, and is applicable e.g. to the moduli spaces of non-compact Calabi-Yau 3-folds, as in [**KS14**].

Let us now set $X = \mathcal{B}_1$. We have constructed the symplectic vector bundle $\mathcal{H}$ with the fiber $\mathcal{H}_\Sigma$ previously, same for the 1-form ϕ. It follows from considerations of the Sections 7.1, 7.2 that we have a tuple $(X, \mathcal{H}, \langle \bullet, \bullet \rangle, \nabla, \phi)$ and its quantization.

Notice that the larger space $\mathcal{B}_0 \supset \mathcal{B}_1$ does not depend on the choice of foliation $\mathcal{F}$. Its local Lagrangian embeddings into symplectic affine spaces depend only on finite jets of $\mathcal{F}$ at D_Σ.

QUESTION 7.3.3. Can this quantization be extended to $\mathcal{B}_0$?

We expect that the answer is positive. Notice that since $\mathcal{B}_0$ does not depend on the foliation then the corollary would be symplectic invariance in topological recursion which does not have so far a conceptual explanation (see [**EynOr2**]).

QUESTION 7.3.4. If the answer to the previous question is positive, does the quantization for $\mathcal{B}_0$ depend only on the collection of $(2 \cdot k_\beta - 1)$-jets of $\mathcal{F}$ at D near points $p_\beta, \beta \in Pol$ (see Section 5.2)?

In the next section we will ask a stronger question 8.2.4, positive answer to which implies independence of the quantization of the choice of finite jets of $\mathcal{F}$.

Notice that although consideration of the Sections 7.1, 7.2 ensures that we have formal germs and quantizations only, the condition to be covariantly constant guarantees that everything is analytic (e.g. $S_{g,n}$ are analytic germs at each point $x \in X$).

REMARK 7.3.5. Our definition 7.3.1 of quantization can be improved further. Namely, for any covariantly constant family of left ideals $J_x \subset Weyl_x$ one can look for cyclic vectors (after choosing a Lagrangian complement to $T_xX \subset \mathcal{H}_x$) of the form $\exp(\sum_{g,n} \hbar^{g-1}(S_{g,n})_x)$ where the constraint $n > 0$ is omitted. The

[19] In [**BCOV94**] at each point $x = [(Y, \Omega_Y^{3,0})] \in \mathcal{L}$ the Lagrangian complement to the tangent space $T_x\mathcal{L} = H^{3,0} \oplus H^{2,1} \subset H^3(Y, \mathbf{C})$ was secretly chosen as the complex conjugate $H^{1,2} \oplus H^{0,3}$. E. Witten in [**Wit**] explained the invariant meaning of HAE as a quantization.

ambiguity for the lift is a torsor over the abelian group $\hbar^{-1}\mathbf{C}[[\hbar]]$ (or, equivalently, the multiplicative group of formal expressions $\exp(\sum_{g\geq 0}(S_{g,0})_x\hbar^{g-1})$ where $(S_{0,0})_x, (S_{1,0})_x, \dots \in \mathbf{C}$ is a sequence of constants). If we ignore terms $(S_{0,0})_x$ and $(S_{1,1})_x$, the rest does not depend on the choice of Lagrangian splitting, and one can speak about covariantly constant formal wave functions modulo the factor $\exp(c_0\hbar^{-1} + c_1)$. Therefore, for each $g \geq 2$ we obtain a canonical $\mathbf{C}$-torsor with flat connection over X. The monodromy gives for every $g \geq 2$ a homomorphism $\pi_1(X) \to \mathbf{C}$. Combining them together we obtain an "obstruction" class in $H^1(X, \hbar\,\mathbf{C}[[\hbar]])$.

DEFINITION 7.3.6. We call an **enhanced** quantization a choice of flat sections of the above torsors for all $g \geq 2$

The enhanced quantization exists if and only if the obstruction class vanishes. In the case of a conical quantization, like for compact Calabi-Yau 3-folds, there is a canonical conical flat trivialization of all above torsors for $g \geq 2$, hence a canonical enhancement.

QUESTION 7.3.7. Does our quantization of the moduli space of spectral curves $\mathcal{B}_1$ extend (in some canonical way) to an enhanced quantization?

REMARK 7.3.8. The datum of a (non-enhanced) quantization corresponding to $g = 0$ is canonically determined by the tuple $(X, \mathcal{H}, \langle\bullet,\bullet\rangle, \nabla, \phi)$, hence carry no more additional information. It is well-known that $g = 1$ part of the quantization is equivalent to a choice of flat connection on virtual line bundle $(\Omega_X^{\dim X})^{\otimes(1/2)}$ (well-defined locally up to involution $\times(-1)$). Equivalently, by taking the square, one gets a flat connection on the honest line bundle $\Omega_X^{\dim X}$, but it is still preferable to speak about its square root as the wave function is invariantly a formal section of $(\Omega_X^{\dim X})^{\otimes(1/2)}$. In the example of compact Calabi-Yau 3-folds the monodromy of the induced flat connection on $\Omega_X^{\dim X}$ is non-trivial and is equal to multiplication by $\exp(2\pi i\chi(Y)/24)$ along the loop in $\mathbf{C}^*$ in the conical direction. One may ask the question whether there exists a canonical flat section well-defined up to $\sqrt[24]{1}$ (as an analog of enhanced quantization for $g = 1$).

8. Semi-affine Lagrangian embeddings

The goal of this section is to define a generalization of local Lagrangian embedding

$$\text{germ at } [\Sigma] \text{ of } \mathcal{B}_0 \hookrightarrow \text{ germ at } 0 \text{ of } \mathcal{H}_\Sigma$$

which *does not* use foliation $\mathcal{F}$, and formulate questions about the irrelevance of the choice of $\mathcal{F}$ for the canonical quantization of locally embedded $\mathcal{B}_1$ discussed in the previous section.

8.1. Semi-affine symplectic structures.

DEFINITION 8.1.1. A **semi-affine structure** on a complex analytic symplectic manifold $(M, \omega = \omega_M)$ is given by a coisotropic foliation $\mathcal{F}$ and an affine structure on leaves of $\mathcal{F}$ such that near each point there exists a system of local coordinates

$$(40) \qquad x_1, \dots, x_n, y^1, \dots, y^n; z_1, \dots, z_{2m}, \ \omega = \sum_{i=1}^{n} dx_i \wedge dy^i + \sum_{j=1}^{m} dz_j \wedge dz_{j+m},$$

the leaves of $\mathcal{F}$ are given by $x_\bullet = const$ and the affine structure on leaves is given by coordinates $y^\bullet, z_\bullet$.

There are two particular cases:

(1) $n = 0$, then M is the only leaf of the foliation and carries an affine symplectic structure,
(2) $m = 0$, then M is endowed with a Lagrangian foliation, and the affine structure on leaves is automatic (e.g. $M = T^*X$ is a cotangent bundle endowed with the cotangent foliation).

In general, local coordinates in the above definition are determined up to the action of the infinitesimal group of transformations whose Lie algebra $\mathfrak{g}_{m,n}$ is given by Poisson brackets with analytic functions of the form

$$F^{(1)}(x_\bullet) + \sum_{i=1}^{n} F_i^{(2)}(x_\bullet)y^i + \sum_{j=1}^{m} F_j^{(3)}(x_\bullet)z_j + \sum_{j_1=1}^{m}\sum_{j_2=1}^{m} F_{j_1 j_2}^{(4)}(x_\bullet)z_{j_1}z_{j_2}\,.$$

In what follows we will use an alternative description of germs of semi-affine symplectic structures:

LEMMA 8.1.2. *A semi-affine structure on a* germ *at $m \in M$ of symplectic manifold M is equivalent to a Poisson fibration*

$$\pi : \textit{germ of } M \textit{ at } m \to \textit{ germ of Possion manifold } P \textit{ at } p = \pi(m)$$

such that P is regular *(i.e. the rank of Poisson structure is constant), and symplectic leaves of P near p are endowed with affine structures compatible with the natural symplectic structures (i.e. the symplectic structure is constant in affine coordinates).*

PROOF. Our assumptions (*without* the use of affine structures) imply that one can choose local coordinates on M at m as in (40) such that $\pi : M \to P$ is the coordinate projection

$$\pi : (x_\bullet, y^\bullet, z_\bullet) \mapsto (x_\bullet, z_\bullet)\,.$$

The natural map π_1 from the germ of P at p to the "center" of P (the germ of the space of symplectic leaves in P) is given by the coordinate projection

$$\pi_1 : (x_\bullet, z_\bullet) \mapsto (z_\bullet)\,.$$

The Lie algebra of germs of vector field at m preserving ω_M and acting on $\pi : M \to P$ is given by Poisson brackets with analytic functions of the form

$$\sum_i f_i(x_\bullet)y^i + G(x_\bullet, z_\bullet)\,. \tag{41}$$

In particular, it maps *epimorphically* to the Lie algebra of germs of Poisson vector fields on P at p. The latter contains as a Lie subalgebra germs of symplectic vector fields in coordinates $z_\bullet$ depending arbitrarily on center coordinates $x_\bullet$. Therefore, we can assume that the affine symplectic structures on symplectic leaves of P near p (see the Lemma) are the *standard* ones in coordinates $z_\bullet$. It is straightforward to see that Poisson brackets with functions from (41) act as affine transformations in $z_\bullet$ if and only if $G = G(x_\bullet, z_\bullet)$ is at most quadratic in $z_\bullet$. Hence we obtain exactly the Lie algebra $\mathfrak{g}_{m,n}$ of local symmetries of a semi-affine symplectic structure, where n is number of $x_\bullet$-coordinates, and $2m$ is number of $z_\bullet$-coordinates. □

DEFINITION 8.1.3. A **flat family of semi-affine symplectic germs** on a complex manifold X is a holomorphically depending on $x \in X$ family of germs of semi-affine symplectic manifolds (M_x, m_x) (with symplectic forms ω_x, and germs of Poisson quotients (P_x, p_x) as in Lemma 8.1.2), endowed with a non-linear flat connection preserving semi-affine symplectic structures, but not necessary based points $(m_x)_{x\in X}$.

Such a family gives for any $x_0 \in X$ a map from the germ of X at x_0 to the germ M_{x_0} at m_{x_0} given by $x \mapsto m_x$ for x sufficiently close to x_0, where we idenitify locally fibers with M_{x_0} via the flat connection.

DEFINITION 8.1.4. A flat family of semi-affine symplectic germs is said to give a **flat family of local Lagrangian embeddings** if the above-defined maps of germs gives a Lagrangian embedding

$$\text{germ of } X \text{ at } x_0 \hookrightarrow M_{x_0}$$

for any $x_0 \in X$.

This definition generalizes tuples $(X, \mathcal{H}, \langle\bullet,\bullet\rangle, \nabla, \phi)$ from Sections 5.1, 7.3 which give local Lagrangian embeddings into affine symplectic spaces. One can also generalize straightforwardly the previous definitions to the case when germs of semi-affine manifolds are replaced by **formal** germs.

REMARK 8.1.5. Omitting semi-affine symplectic structures in the definition 8.1.3 we obtain a more basic notion of a **flat family of germs** (or formal germs) on a manifold X. For example, if X is endowed with a foliation $\mathcal{F}$, then on X we have a flat family of germs of the "space of leaves" of $\mathcal{F}$.

There is an embedding of the Lie algebra $\mathfrak{g}_{m,n}$ into the Lie algebra of $\hbar$-differential operators in $(x_i)_{1\le i\le n}, (z_j)_{1\le j\le m}$ under which $y^j \mapsto \hbar\partial/\partial x_j, z_{l+m} \mapsto \hbar\partial/\partial z_l, 1 \le j \le n, 1 \le l \le m$, and the rest of variables are mapped to themselves. This give rise to a *canonical* completed Weyl algebra associated with a formal germ of a semi-affine symplectic manifold.

For any flat family of formal semi-affine symplectic germs we obtain a bundle of completed Weyl algebras endowed with a flat connection. Similarly to Section 7.3 we define **quantization** and **enhanced quantization** of a flat family of formal local Lagrangian embeddings.

A particular example is the one of a global Lagrangian submanifold $X = L$ in a cotangent bundle $M = T^*Y$ endowed with the natural semi-affine structure given by the foliation by cotangent fibers. Taking formal germs of M at points of L we obtain a flat family of local Lagrangian embeddings. The quantization of this structure in our sense is the same as a cyclic DQ-module on T^*Y with the classical limit $\mathcal{O}_L$ endowed with cyclic vector 1, i.e. a WKB-module.

DEFINITION 8.1.6. For a given symplectic manifold (M, ω) and two semi-affine structures given by coisotropic foliations $\mathcal{F}, \mathcal{F}'$ with affine structures on leaves, we say that $\mathcal{F}$ is **weaker** than $\mathcal{F}'$ if leaves of $\mathcal{F}$ are contained in leaves of $\mathcal{F}'$ and the inclusion of any leaf is an affine map.

It is easy to see that in such a case the flat bundles of completed Weyl algebras associated with two semi-affine structures are *canonically* isomorphic. For example, the Weyl algebra associated with the cotangent space T^*V to an affine space V with the natural semi-affine structure of the cotangent bundle is canonically isomorphic to the Weyl algebra associated with T^*V considered as an affine symplectic space. Same is true for completed Weyl algebras.

8.2. Canonical semi-affine structure on $\mathcal{B}_0$. The setup in this section is the following: P is a complex analytic Poisson surface, with open dense symplectic leaf $(S, \omega = \omega_S)$ and such that the effective divisor D of zeroes of Poisson structure is smooth. We denote as usual by $\mathcal{B}_0$ the moduli space of D-transversal smooth connected compact curves $\Sigma \subset P$. Our goal here is to construct a flat family of Lagrangian embeddings into formal semi-affine symplectic germs

$$\text{germ } \mathcal{B}_{[\Sigma]} \text{ of } \mathcal{B}_0 \text{ at } [\Sigma] \hookrightarrow \text{ germ } M_{[\Sigma]}$$

as in the previous section.

For a given D-transversal spectral curve Σ we denote as before by $\{p_\beta\}_{\beta \in Pol}$ the set $D \cap \Sigma$, by $(k_\beta \geq 1)_{\beta \in Pol}$ the corresponding multiplicities, and by D_Σ the effective divisor $\sum_\beta k_\beta p_\beta$ on Σ. **We assume that** $Pol \neq \emptyset$.

In order to define a formal germ of symplectic manifold $M_{[\Sigma]}$ and a Lagrangian embedding of germ $B_{[\Sigma]}$, we proceed as follows.

DEFINITION 8.2.1. For $[\Sigma] \in \mathcal{B}_0$ define a **collection of movable loops** as a tuple consisting of

(1) a *non-empty* finite collection of disjoint closed discs $(\mathbb{D}_i)_{i \in I}$ in $\Sigma - \{p_\beta\}_{\beta \in Pol}$ with real-analytic boundaries,
(2) a collection of germs of holomorphic 1-forms ν_i on S defined in a neighborhood of of $\mathbb{D}_i \subset S$ such that $d\nu_i = \omega$ and ν_i is everywhere non-vanishing 1-form near $\partial \mathbb{D}_i$ with the kernel *transversal* to $\partial \mathbb{D}_i$.

We denote by $ML(\Sigma)$ the set of such collections. We endow $ML(\Sigma)$ with the partial order:

$$\sigma_1 \prec \sigma_2 \,,$$

if σ_2 is obtained from σ_1 by adding new discs and new germs of 1-forms satisfying the above conditions.

Obviously, the nerve of this poset is *contractible*. The reason for that is the simple fact that for any finite collection of closed discs (maybe intersecting each other) in the non-compact (since $Pol \neq \emptyset$) surface $\Sigma - \{p_\beta\}_{\beta \in Pol}$, there exists another disc in $\Sigma - \{p_\beta\}_{\beta \in Pol}$ disjoint from our collection.

For every collection of movable loops $\sigma \in ML(\Sigma)$ we define *a formal germ of infinite-dimensional manifold* G_σ by considering formal deformations Σ_0' of the complex curve

$$\Sigma_0 := \Sigma - \sqcup_i int(\mathbb{D}_i) \subset P$$

with real-analytic boundaries, such that the boundary is locally a shift of $\partial \mathbb{D}_i$ along the foliation $\mathrm{Ker}(\nu_i)$, and the integral of ν_i over the corresponding boundary component vanishes[20].

The germ G_σ is naturally embedded into the infinite-dimensional vector space

$$W_\sigma = \left\{ (\eta_i)_{i \in I} \,|\eta_i \text{ is } \mathbf{C}\text{-valued real analytic 1-form on } \partial \mathbb{D}_i, \text{ and } \forall i \int_{\partial \mathbb{D}_i} \eta_i = 0 \right\}$$

[20]The germ G_σ can be thought of as a non-linear analog of the coisotropic subspace G_Σ defined in (38).

endowed with the usual skew-symmetric pairing:

$$\langle (\eta_i^{(1)})_{i\in I}, (\eta_i^{(2)})_{i\in I}\rangle := \sum_i \int_{\partial \mathbb{D}_i} f_i^{(1)} \eta_i^{(2)}, \quad df_i^{(1)} = \eta_i^{(1)}.$$

Therefore, G_σ carries a closed 2-form ω_σ, which is the pullback of the translation-invariant closed 2-form on W_σ associated with the above pairing.

Let us describe the tangent space to G_σ and its skew-orthogonal. The tangent space to G_σ at any point Σ_0' close to Σ_0 [21] is given by formula (compare with the proof of Proposition 7.1.2)

$$T_{\Sigma_0'} G_\sigma = \left\{ \eta \in \Gamma(\Sigma_0', \Omega^1_{\Sigma_0'}(D_{\Sigma_0'})) \mid \int_{\partial_i \Sigma_0'} \eta = 0, \ \forall i \in I \right\}. \tag{42}$$

It is naturally embedded into W_σ, using foliations $Ker(\nu_i)$. The skew orthogonal to $T_{\Sigma_0'} G_\sigma$ is the subspace

$$(T_{\Sigma_0'} G_\sigma)^\perp = d\Gamma(\Sigma_0', \mathcal{O}(-D_{\Sigma_0'})) \subset T_{\Sigma_0'} G_\sigma. \tag{43}$$

The quotient space $T_{\Sigma_0'} G_\sigma/(T_{\Sigma_0'} G_\sigma)^\perp$ is a finite-dimensional symplectic vector space which we denote by $\mathcal{H}_{\Sigma_0', D_{\Sigma_0'}}$. This space depends only on Σ_0' (which is considered as an abstract complex curve with real-analytic boundary) and on an effective divisor $D_{\Sigma_0'}$ with the support in the interior of Σ_0'.

Suppose we have a germ of a complex manifold endowed with a closed 2-form of constant rank. It is well-known that the subbundle of the tangent bundle given by the kernel of the form defines a foliation. The germ of the space of leaves of this foliation is naturally a germ of symplectic manifold. We apply these general considerations to the infinite-dimensional manifold $(G_\sigma, \omega_\sigma)$ and obtain a well-defined finite-dimensional formal symplectic germ, which we denote by M_σ (abusing the notation we will denote the symplectic form on M_σ also by ω_σ as the 2-form on G_σ is its pullback).

Let us fix Σ. Now we are going to identify *canonically* all symplectic germs $(M_\sigma, \omega_\sigma)$ for $\sigma \in ML(\Sigma)$. Recall the partial order $\prec$ on $ML(\Sigma)$. Obviously, if $\sigma_1 \prec \sigma_2$, we have a natural embedding

$$i_{\sigma_1,\sigma_2} : G_{\sigma_1} \hookrightarrow G_{\sigma_2}$$

and in this way we get a functor from the poset $ML(\Sigma)$ to the category of infinite-dimensional formal germs and embeddings. One can see immediately that $i^*\omega_{\sigma_2} = \omega_{\sigma_1}$, and moreover i_{σ_1,σ_2} induces a symplectomorphism of formal germs $M_{\sigma_1} \overset{\sim}{\to} M_{\sigma_2}$. The connectedness and 1-connectedness of the nerve of poset $ML(\Sigma)$ implies that we identify all formal germs M_σ with each other.

DEFINITION 8.2.2. For any point $[\Sigma] \in \mathcal{B}_0$ define the associated symplectic formal germ $M_{[\Sigma]}$ as the result of the identifications of formal germs M_σ for $\sigma \in ML(\Sigma)$ described above.

It is obvious from our description that formal germs $(M_{[\Sigma]})_{[\Sigma]\in\mathcal{B}_0}$ form a *flat family* of formal germs, and symplectic structures on fibers are covariantly constant. Also it is straightforward to see that the flat family of germs $([\mathcal{B}_\Sigma])_{[\Sigma]\in\mathcal{B}_0}$ has natural covariantly constant Lagrangian embeddings into $(M_{[\Sigma]})_{[\Sigma]\in\mathcal{B}_0}$.

[21] Since we speak about formal germs, we mean that Σ_0' is a deformation of Σ_0 over the spectrum of a local Artin algebra over $\mathbf{C}$.

Finally let us describe a natural covariantly constant family of semi-affine structures on $M_{[\Sigma]}$. For given curve Σ and $\sigma \in ML(\Sigma)$, let us consider a subbundle of finite codimension of the tangent bundle to G_σ. The subbundle is given by first order deformations such that the corresponding 1-form η with poles at $D_{\Sigma_0'}$ on Σ_0' (which is a small deformation of Σ_0 as a curve with boundary) is *holomorphic*, i.e. has no poles. It is easy to see that this distribution is *integrable*, i.e. it defines a foliation $\mathcal{F}_\sigma$ on G_σ. This follows immediately from the geometric interpretation of the above constraint on first order deformations. More precisely, it means that intersection points p_β of the moving curve with the divisor D stay constant, and moreover $(k_\beta - 1)$-jet of Σ_0 at p_β stays constant. Also it is clear from this description that leaves of $\mathcal{F}_\sigma$ contain leaves of the foliation $Ker(\omega_\sigma)$. Hence $\mathcal{F}_\sigma$ induces a foliation on $M_{[\Sigma]}$, and it is easy to see that it does not depend on σ (since the maps i_{σ_1,σ_2} identify induced foliations).

Let us denote by $\mathcal{F}^{co}_{M_{[\Sigma]}}$ the resulting foliation on $M_{[\Sigma]}$. Pick any $\sigma \in ML(\Sigma)$. Since the restriction of ω_σ to $T_{\mathcal{F}_\sigma}$ vanishes, it follows that $\mathcal{F}^{co}_{M_{[\Sigma]}}$ is coisotropic. The codimension of leaves of $\mathcal{F}^{co}_{M_{[\Sigma]}}$ coincides with the codimension of leaves of $\mathcal{F}_\sigma$, and is equal to $\sum_\beta k_\beta - 1$ (the term -1 comes from the condition that the sum of residues of 1-forms η from (42) at points $p'_\beta \in \Sigma_0'$ vanish).

Notice that there is a 1-form ξ on $\mathcal{F}^{co}_{M_{[\Sigma]}}$. In order to see this, let us replace in (43) the space of functions vanishing at $p_\beta, \beta \in Pol$ by the space of all functions. Then the quotient of the tangent space to $\mathcal{F}^{co}_{M_{[\Sigma]}}$ by this bigger subspace is isomorphic to $H^1(\Sigma, \mathbf{C})$. Natural (surjective) projection to the quotient gives rise to an $H^1(\Sigma, \mathbf{C})$-valued 1-form ξ along leaves of $\mathcal{F}^{co}_{M_{[\Sigma]}}$.

LEMMA 8.2.3. *The form ξ is closed.*

PROOF. We will check vanishing of $d\xi$ using the lift to G_σ for any given $\sigma \in ML(\Sigma)$. In order to do this, we will construct an $H^1(\Sigma, \mathbf{C})$-valued functional on pairs of points of G_σ lying in the same leaf of $\mathcal{F}_\sigma$, and satisfying the cocycle property. The construction is similar to what we had before. We interpret $H^1(\Sigma, \mathbf{C})$ as the space of $\mathbf{C}$-valued linear functionals on $H_1(\Sigma_0', \mathbf{Z})$ vanishing on classes of boundary loops $\partial_i \Sigma_0'$ (here as before, Σ_0' is a small deformation of Σ_0). With any two deformed surfaces Σ_0', Σ_0'' and closed 1-chains ρ', ρ'' on these surfaces representing homology classes identified by the Gauss-Manin connection, we associate a complex number by integrating $\omega = \omega_S$ along 2-chain in S with the boundary equal to $\rho'' - \rho'$. This assignment depends only on homology classes, vanishes on boundary loops, satisfies the cocycle property from Remark 5.3.2, and gives the pullback of ξ as the first derivative at the diagonal. $\square$

Thus we have a foliation on each $\mathcal{F}^{co}_\sigma$ with fibers which are fibers of the map $\xi : T^{co}_{\mathcal{F}_\sigma} \to H^1(\Sigma, \mathbf{C})$. Tangent space to a fiber is isomorphic to $(T^{co}_{\mathcal{F}_\sigma})^\perp$, which is the skew-orthogonal with respect to the symplectic form ω_σ. Thus the skew-orthogonal subbundle $(T_{\mathcal{F}^{co}_{M_{[\Sigma]}}})^\perp \subset TM_{[\Sigma]}$ gives an isotropic subfoliation. The quotient bundle

$$T_{\mathcal{F}^{co}_{M_{[\Sigma]}}}/(T_{\mathcal{F}^{co}_{M_{[\Sigma]}}})^\perp$$

is naturally isomorphic to a flat symplectic bundle of rank $2g$ (where g is the genus of Σ) with the fiber at the base point equal to $H^1(\Sigma, \mathbf{C})$.

This finishes the construction of the canonical flat family of Lagrangian embeddings of $\mathcal{B}_0$ into a family of semi-affine symplectic germs.

Also, we see immediately that in the case when the foliation $\mathcal{F}$ on P exists, the semi-affine structure coming from general considerations is automatically weaker (see Definition 8.1.6) than the semi-affine structure coming from $\mathcal{F}$ (which is in fact a symplectic affine structure on M_Σ). According to the discussion after the Definition 8.1.6 in that case we obtain isomorphic flat bundles of completed Weyl algebras. It is naturally to inquire whether the corresponding quantizations depend on chosen foliations. Hence we ask the following question.

QUESTION 8.2.4. Is it true that the flat family of quantizations of locally Lagrangian embeddings $\mathcal{B}_{[\Sigma]} \hookrightarrow M_\Sigma$ does not depend on the choice of foliation $\mathcal{F}$?

Presumably the answer is positive. We expect that it is related to the deep result about symplectic invariance in TR proven by Eynard and Orantin in [**EynOr2**]. Also, the positive answer will imply (via finite jet approximations of the foliation) that the canonical quantization can be constructed even in the case when there is no foliation $\mathcal{F}$ at all.

9. Two more speculations

9.1. Wave function for an individual spectral curve. One can generalize considerations of this paper to the case of immersed (rather than embedded) spectral curves which can have singularities. The moduli space $\mathcal{B}$ of immersed singular spectral curves contains an open subspace $\mathcal{B}_0$, the moduli space of immersed smooth spectral curves.

Having in mind the above remark let us make a comment about a quantization of an individual spectral curve Σ, assuming that the open symplectic leaf S is isomorphic to T^*C, where C is a compact curve. If the answer to the Question 8.2.4 is positive, then one can study the behavior of the corresponding wave function $\psi_{\mathcal{B}_{[\Sigma]}}$ near the complement $\mathcal{B}-\mathcal{B}_0$ (discriminant locus) in the bigger space $\mathcal{B}$.

Let $T^*_x, x \in C$ denote the cotangent fiber at $x \in C$. For generic collection of points $x_1,\dots,x_n \in C$ let us pick some pre-images $y_1,\dots,y_n \in \Sigma$. Then the abstract curve $\Sigma_{y_1,\dots,y_n}$ is defined as the disjoint union $T^*_{x_1} \sqcup \ldots \sqcup T^*_{x_n} \sqcup \Sigma$ with identified pairs of points of Σ and $T_{x_i}, 1 \le i \le n$ corresponding to the same point from the collection $y_1,\dots,y_n \in \Sigma$. This singular curve is naturally immersed into the Poisson surface P, and hence gives a point in the discriminant $\mathcal{B}-\mathcal{B}_0$. W can try to deform it to an immersed smooth curve.

REMARK 9.1.1. The reason for picking *more* than 1 point in the collection $(x_i)_{1\le i\le n}$ is the following. If we want to deform an immersion of the singular curve $\Sigma_{y_1,\dots,y_n}$, we need (at first order deformation) a meromorphic 1-form on Σ with first order poles at $y_1,\dots,y_n \in \Sigma$. In the case $n=1$ such a form is automatically holomorphic, because the sum of residues vanishes. Hence there is no smoothening of the singular curve in the case $n=1$.

It is plausible that with the Lagrangian subvarieties $\Sigma, \Sigma_{y_1,\dots,y_n} \subset S = T^*C$ one can associate cyclic DQ-modules with cyclic vectors ψ_Σ and $\psi_{\Sigma_{y_1,\dots,y_n}}$ respectively.

QUESTION 9.1.2. How to relate ψ_Σ and $\psi_{\Sigma_{y_1,\dots,y_n}}$ to $\psi_{\mathcal{B}_{[\Sigma]}}$?

Probably the possibility to extract ψ_Σ from $\psi_\mathcal{B}$ is encoded into some constraints which allow us to extend differential forms from $\Omega^1(\Sigma - \cup_\alpha\{r_\alpha\})$ to the whole curve Σ. Relations of ψ_Σ to the topological recursion were discussed in [**DuMu**].

9.2. dg-Airy structure associated with compact Calabi-Yau 3-fold. In this subsection we discuss a differential-graded version of an Airy structure associated with the moduli space of complex Calabi-Yau 3-folds. It is well-known that the analog of the above-discussed moduli space $\mathcal{B}$ is a complex Lagrangian cone in the middle cohomology. The corresponding wave function contains all the information about Gromov-Witten invariants. Moreover, it satisfies the HAE.

Let Y be a compact complex 3-dimensional Calabi-Yau manifold (CY 3-fold for short). Deformation theory of complex structures on Y is controlled by the differential-graded Lie algebra (DGLA for short)

$$\mathfrak{g}_Y = \Gamma(Y, \oplus_{i\geq 0} \wedge^i T_Y^{1,0} \otimes \Omega_Y^{0,*})[1] .$$

We split $\mathfrak{g}_Y$ into the direct sum of $\mathbf{Z}$-graded subspaces:

$$\mathfrak{g}_Y = \oplus_{i\geq 0}\mathfrak{g}^i , \quad \mathfrak{g}_i = \mathfrak{g}_Y^i = \Gamma(Y, \wedge^i T_Y^{1,0} \otimes \Omega_Y^{0,*})[1] .$$

The differential is $d_{\mathfrak{g}} = \overline{\partial} + div$, where div is the divergence operator corresponding to the holomorphic volume for $\Omega^{3,0}$ (see e.g. [**BK98**]). The corresponding homological vector field on $\mathfrak{g}_Y[1]$ is given

$$\dot{\gamma} = (\overline{\partial} + div)(\gamma) + \frac{[\gamma, \gamma]}{2}$$

and the Maurer-Cartan equation (MC equation in short) is the equation for zeroes of this field. The space of solutions of the MC equation carries a foliation with the tangent space at γ consisting of

$$\delta(\gamma) = (\overline{\partial} + div)(\epsilon) + [\epsilon, \gamma], \qquad \epsilon \in \mathfrak{g}_Y .$$

The *extended moduli space of complex structures* is the quotient of the space of solutions to the MC-equation by this foliation. The extended moduli space is a smooth finite-dimensional $\mathbf{Z}$-graded analytic supermanifold by a generalization of Tian-Todorov theorem (see loc.cit.).

Let us make the change of variables $\mu = e^{\gamma}$. Then the above homological vector field becomes

$$\dot{\mu} = (\overline{\partial} + div)(\mu) \tag{44}$$

and the foliation becomes also linear:

$$\delta(\mu) = (\overline{\partial} + div)(\epsilon), \qquad \epsilon \in \mathfrak{g}_Y .$$

In the new variable μ the image of an element $\gamma_0 + \gamma_1$ belonging to the graded Lie sublagebra $\mathfrak{g}^0 \oplus \mathfrak{g}^1 \subset \mathfrak{g}_Y$ has the form $\exp(\gamma_0 + \gamma_1) = \mu_0 + \mu_1 + \mu_2 + \mu_3$ where:

$$\mu_0 = \exp(\gamma_0),\ \mu_1 = \exp(\gamma_0) \cdot \gamma_1,\ \mu_2 = \exp(\gamma_0) \cdot (\gamma_1^2/2),\ \mu_3 = \exp(\gamma_0)(\gamma_1^3/6).$$

Using contraction with volume element $\Omega^{3,0}$ we identify (as a $\mathbf{Z}/2$-graded supermanifold) the vector space $\mathfrak{g}_Y[1]$ endowed with the linear homological vector field (44) with the space of forms

$$W := \Gamma(Y, \oplus_{p.q\geq 0}\Omega_Y^{p,q})[3]$$

endowed with the linear homological vector field $\dot{\alpha} = d\alpha$. The space W is an infinite-dimensional symplectic vector space endowed with the symplectic form $\langle \omega_1, \omega_2 \rangle = \int_X \omega_1 \wedge \omega_2$.

PROPOSITION 9.2.1. *Let us define* $\mathcal{L} = \exp(\mathfrak{g}^0 \oplus \mathfrak{g}^1)$ *considered as a subvariety of* W. *Then:*

(1) $\mathcal{L}$ *is a quadratic Lagrangian subvariety in* W.
(2) *The vector space* $Ker(d) \subset W$ *is a coisotropic vector subspace and* $Im(d)$ *is symplectic orthogonal to* $Ker(d)$.
(3) *The natural projection (symplectic reduction)* $pr : \mathcal{L} \cap Im(d) \to H^\bullet(Y) \to H^3(Y, \mathbf{C})$ *maps* $\mathcal{L}$ *into the Lagrangian cone* $L \subset H^3(Y, \mathbf{C})$ *which is the moduli space of pairs consisting of a complex structure on* X *and a choice of holomorphic volume form* vol_Y.

PROOF. Parts 2) and 3) follow from the deformation theory of the Calabi-Yau structure on X, and can be derived in many ways (see e.g. [**BK98**]).

The fact that W is symplectic is general. It holds for any real compact manifold Y such that $dim_{\mathbf{R}} Y \in 2 + 4\mathbf{Z}_{\geq 0}$ (e.g. for any odd-dimensional compact complex manifold).

Notice that the Calabi-Yau pairing implies the following isomorphisms (properly understood in the infinite-dimensional context): $\mathfrak{g}^0 \simeq (\mathfrak{g}^3)^*, \mathfrak{g}^1 \simeq (\mathfrak{g}^2)^*$. Quadratic equations for $\mathcal{L}$ require passing to the coordinates $\mu_i, i = 0, 1, 2, 3$, in which $\mathcal{L}$ can be described by the equations

$$\mu_0\mu_2 = \frac{\mu_1^2}{2},$$

$$\mu_0\mu_3 = \frac{\mu_1\mu_2}{3}.$$

The fact that $\mathcal{L}$ is Lagrangian can be checked directly. Namely, one checks first that the dimension $\mathcal{L}$ understood in the sense of infinite-dimensional smooth topology on $\mathfrak{g}$ is "one half" of the dimension of $\mathfrak{g}$. Then one shows by a direct computation, that the algebra of functions vanishing on $\mathcal{L}$ is closed under the Poisson bracket. Such functions have form

$$f_{a_0,a_1}(\mu_0, ..., \mu_3) = Tr\left(a_0(\mu_0\mu_2 - \mu_1^2/2) + a_1(\mu_0\mu_3 - \mu_1\mu_2/3)\right),$$

where $a_i \in \mathfrak{g}^i, i = 0, 1$ and $Tr : \mathfrak{g}^3 \to \mathbf{C}$ is the functional given by the holomorphic volume form vol_X. Then one computes $\{f_{a_0,a_1}, f_{b_0,b_1}\}$ and sees that it gives a solvable Lie algebra structure on functions vanishing on $\mathcal{L}$.

This completes the proof. □

We should warn the reader that despite the existence of quadratic Lagrangian, there is no Airy structure in the compact case. The corresponding tensors A, B, C belong to the *completed* tensor products. As a result the series which should give the wave function are divergent and do not give a meaningful answer.

References

[ABCO] J. Andersen, G. Borot, L. Chekhov, N. Orantin, The ABCD of topological recursion, in progress.

[ACNP15] Ĭ. É. Andersen, L. O. Chekhov, P. Norberi, and R. K. Penner, *Topological recursion for Gaussian means and cohomological field theories* (Russian, with Russian summary), Teoret. Mat. Fiz. **185** (2015), no. 3, 371–409, DOI 10.4213/tmf8951; English transl., Theoret. and Math. Phys. **185** (2015), no. 3, 1685–1717. MR3438626

[DK11] Andrea D'Agnolo and Masaki Kashiwara, *On quantization of complex symplectic manifolds*, Comm. Math. Phys. **308** (2011), no. 1, 81–113, DOI 10.1007/s00220-011-1325-7. MR2842971

[BK98] Sergey Barannikov and Maxim Kontsevich, *Frobenius manifolds and formality of Lie algebras of polyvector fields*, Internat. Math. Res. Notices **4** (1998), 201–215, DOI 10.1155/S1073792898000166. MR1609624

[BCOV94] M. Bershadsky, S. Cecotti, H. Ooguri, and C. Vafa, *Kodaira-Spencer theory of gravity and exact results for quantum string amplitudes*, Comm. Math. Phys. **165** (1994), no. 2, 311–427. MR1301851

[Cos] K. Costello, The Gromov-Witten potential associated to a TCFT, arXiv:0509264.

[DuMu] O. Dumitrescu, M.Mulase, Quantization of spectral curves for meromorphic Higgs bundles through topological recursion, arXiv:1411.1023.

[EO07] B. Eynard and N. Orantin, *Invariants of algebraic curves and topological expansion*, Commun. Number Theory Phys. **1** (2007), no. 2, 347–452, DOI 10.4310/CNTP.2007.v1.n2.a4. MR2346575

[Eyn0r2] B. Eynard, N. Orantin, About the $x - y$ symmetry of the F_g algebraic invariants, arXiv:1311.4993.

[GelKaz] I. Gelfand, D. Kazhdan, Some problems of differential geometry and the calculation of the cohomologies of Lie algebras of vector fields, Soviet Math. Doklady 12, 1367-1370, 1971.

[KS14] Maxim Kontsevich and Yan Soibelman, *Wall-crossing structures in Donaldson-Thomas invariants, integrable systems and mirror symmetry*, Homological mirror symmetry and tropical geometry, Lect. Notes Unione Mat. Ital., vol. 15, Springer, Cham, 2014, pp. 197–308, DOI 10.1007/978-3-319-06514-4_6. MR3330788

[Tat68] John Tate, *Residues of differentials on curves*, Ann. Sci. École Norm. Sup. (4) **1** (1968), 149–159. MR0227171

[Wit] E. Witten, Quantum background independence in String Theory, hep-th 9306122.

IHES, 35 ROUTE DE CHARTRES, BURES-SUR-YVETTE, F-91440, FRANCE
Email address: maxim@ihes.fr

DEPARTMENT OF MATHEMATICS, KANSAS STATE UNIVERSITY, MANHATTAN, KS 66506
Email address: soibel@math.ksu.edu

Proceedings of Symposia in Pure Mathematics
Volume **100**, 2018
https://doi.org/10.1090/pspum/100/01763

Periods of meromorphic quadratic differentials and Goldman bracket

D. Korotkin

ABSTRACT. We study symplectic properties for monodromy map of second order linear equation with meromorphic potential having only simple poles on a Riemann surface. We show that the canonical symplectic structure on the cotangent bundle $T^*\mathcal{M}_{g,n}$ implies the Goldman bracket on the corresponding character variety under the monodromy map, thereby extending the recent results of the paper of M.Bertola, C.Norton and the author from the case of holomorphic to meromorphic potentials with simple poles.

1. Introduction

The goal of this paper is to study symplectic aspects of monodromy map for second order linear equation on a Riemann surface of genus g with meromorphic potential having n simple poles. We generalize results of the recent paper by M.Bertola, C.Norton and the author [**4**] where the case of holomorphic potential was treated in detail (see also an earlier paper by S.Kawai [**21**] for another approach to this problem).

The condition of coordinate invariance of the equation $\varphi'' - U\varphi = 0$ on a Riemann surface $\mathcal{C}$ of genus g implies that the function $-2U$ transforms as a projective connection under a coordinate change while the solution φ locally transforms as $-1/2$-differential [**16**]. Thus any meromorphic potential U with simple poles can be represented as $-S_0/2 + Q$ where S_0 is a fixed holomorphic projective connection on $\mathcal{C}$ and Q is a meromorphic quadratic differential with n simple poles. This parameterization of the space of meromorphic potentials on a given Riemann surface is natural since meromorphic quadratic differentials with n simple poles form the cotangent space $T^*\mathcal{M}_{g,n}$ to moduli space of Riemann surfaces of genus g with n marked points. However, there remains a freedom in the choice of the "base" holomorphic projective connection S_0 for a given Riemann surface. In this paper, following [**4, 21**] we assume that S_0 holomorphically depends on moduli of $\mathcal{C}$ (in particular, this requirements rules out the use of Fuchsian projective connection as S_0). Furthermore, following [**4**] we choose S_0 to be the Bergman projective connection S_B (up to the factor $1/6$ this projective connection is the zeroth order term in the expansion of the canonical normalized meromorphic bidifferential $B(x,y) = d_x d_y \log E(x,y)$ near the diagonal $x = y$ [**6**]; $E(x,y)$ is the prime-form).

2010 *Mathematics Subject Classification.* Primary 53D30, Secondary 34M45.
The research of the author was supported by NSERC and FQRNT.

Therefore, we are going to study the equation written in the form

$$\varphi'' + \left(\frac{1}{2}S_B + Q\right)\varphi = 0\,. \tag{1.1}$$

The ratio $f = \varphi_1/\varphi_2$ of two linearly independent solutions of (1.1) solves the Schwarzian equation

$$\mathcal{S}(f,\xi) = S_B(\xi) + Q(\xi)\,,$$

where ξ is an arbitrary local parameter on $\mathcal{C}$ and $\mathcal{S}$ denotes Schwarzian derivative. The Schwarzian equation determines a $PSL(2,\mathbb{C})$ monodromy representation of the fundamental group $\pi_1(\mathbb{C}\setminus\{y_i\}_{i=1}^n)$ which turns out to be liftable to an $SL(2,\mathbb{C})$ representation [**10**] (the lift to an $SL(2,C)$ representation is a non-trivial fact due to spinorial nature of solutions $\varphi_{1,2}$). Following [**4**], we define an $SL(2,\mathbb{C})$ monodromy representation of equation (1.1) directly, by an appropriate change of dependent variable φ.

Denote the standard generators of the fundamental group by $\alpha_1,\beta_1,\dots,\alpha_g,\beta_g$, $\kappa_1,\dots,\kappa_n$; these generators obey the relation $(\kappa_1\dots\kappa_n)\prod_{i=1}^g \alpha_i\beta_i\alpha_i^{-1}b_i^{-1} = id$. Since all poles y_i of the potential of equation (1.1) are simple then both eigenvalues of monodromy matrices M_{κ_i} equal to ± 1 for each i; all other monodromies are $SL(2)$ matrices satisfying certain genericity conditions [**10**].

Our goal is to study symplectic properties of the monodromy map for equation (1.1). Since monodromy matrices depend on the choice of normalization point of solutions of (1.1) it is natural to work with the corresponding character variety which we denote by $\mathbb{V}^0_{g,n}$; a point of $\mathbb{V}^0_{g,n}$ is an equivalence class of monodromy representations which differ by a simultaneous conjugation with the same matrix. The index 0 indicates that the monodromies around poles $y_1,\dots,y_n$ are not generic: all of their eigenvalues are equal to ± 1.

The main result of this paper states that the canonical symplectic structure on $T^*\mathcal{M}_{g,n}$ implies the Goldman bracket on the character variety $V^0_{g,n}$ under the monodromy map of equation (1.1), therefore generalizing the result of [**4**] to potentials with simple poles. Moreover, similar to [**4**] and [**21**], we prove that the same statement holds for equation (1.1) where the Bergman base projective connection is replaced either by Schottky or quasi-fuchsian (Bers) projective connection.

We follow the same strategy as in [**4**]; it is based on the use of the "Abelian periods" of the quadratic differential Q as coordinates on the underlying moduli space $T^*\mathcal{M}_{g,n}$. These coordinates are periods of the Abelian differential $v = \sqrt{Q}$ on the canonical covering $\widehat{\mathcal{C}}$ of $\mathcal{C}$ (the genus of $\widehat{\mathcal{C}}$ equals $4g-3+n$). We call them "Abelian periods" or simply "periods" to distinguish from monodromy matrices of equation (1.1) which are also called sometimes "periods of quadratic differentials" [**32**]; following the terminology of [**32**] the monodromy matrices should probably be rather called "non-Abelian periods". In the theory of dynamical systems the abelian periods of the quadratic differential Q are known under the name of "homological coordinates" (see [**5**] and numerous recent papers; a substantial reference list can be found in [**8**]).

The phase space associated to equation (1.1) is the space of pairs (Riemann surface $\mathcal{C}$ of genus g, meromorphic quadratic differential Q on $\mathcal{C}$ with n simple poles). This phase space is nothing but the cotangent bundle $T^*\mathcal{M}_{g,n}$ (up to subspaces of codimension one and higher; these subspaces contain differentials with multiple zeros). The natural symplectic structure on $T^*\mathcal{M}_{g,n}$ is defined by $\omega_{can} =$

$\sum_{i=1}^{3g-3+n} \mathrm{d}p_i \wedge \mathrm{d}q_i$, where q_i are local coordinates on the moduli space $\mathcal{M}_{g,n}$ while p_i are corresponding momenta (coefficients in the decomposition of a cotangent vector in the basis $\{\mathrm{d}q_i\}$).

The set of holomorphic local coordinates $\{q_i\}$ on $\mathcal{M}_{g,n}$ can be chosen as follows. To determine locally the conformal structure of $\mathcal{C}$ we pick (outside of hyperelliptic locus and for $g > 1$) a set of $3g-3$ entries of the period matrix Ω of $\mathcal{C}$; in different neighbourhoods of the moduli space these entries might have to be chosen differently. The quadratic differentials corresponding to cotangent vectors $d\Omega_{jk}$ are products $v_j v_k$ of normalized holomorphic differentials. An additional set of n coordinates which determine the positions of punctures $\{y_i\}$ on $\mathcal{C}$ we choose to be $q_k = (v_i/v_j)(y_k)$ where v_i and v_j form a pair of normalized holomorphic 1-forms on $\mathcal{C}$ corresponding to some Torelli marking (these coordinates are also local: in different coordinate charts on $\mathcal{M}_{g,n}$ one might need to choose another pair of normalized holomorphic differentials and/or different Torelli markings). The quadratic differential corresponding to cotangent vector dq_k is the meromorphic quadratic differential (given by the formula (5.7) below) whose only simple pole is at y_k. The momenta p_i are then defined to be coefficients of decomposition of an arbitrary meromorphic quadratic differential with simple poles in the basis described above.

An alternative set of Darboux coordinates on $T^*\mathcal{M}_{g,n}$ (more precisely, on the subset $\mathcal{Q}_{g,n}^0 \subset T^*\mathcal{M}_{g,n}$ which contains quadratic differentials with all simple zeros) is given by abelian periods of the quadratic differential Q which are defined as integrals of $v = \sqrt{Q}$ over odd part of homology group $H_1(\widehat{\mathcal{C}}, \mathbb{R})$ of canonical covering $\widehat{\mathcal{C}}$. The canonical two-sheeted covering $\widehat{\mathcal{C}}$ is defined by equation $v^2 = Q$ in $T^*\mathcal{C}$; the branch points of $\widehat{\mathcal{C}}$ lie at zeros and poles of Q (in modern literature $\widehat{\mathcal{C}}$ is sometimes attributed to Seiberg-Witten [**9**] or Hitchin [**27**], although this two-sheeted covering was extensively used starting from early days of Teichmüller theory [**1**, **5**]). The genus of $\widehat{\mathcal{C}}$ equals $4g-3+n$; it admits a natural holomorphic involution which we denote by μ. The homology group $H_1(\widehat{\mathcal{C}}, \mathbb{R})$ can be decomposed into direct sum $H_+ \oplus H_-$ of even and odd subspaces under the action of μ; the dimension of H_+ equals $2g$ while $\dim H_- = 6g-6+2n$. Choosing a symplectic basis $\{a_i^-, b_i^-\}$ in H_- with the intersection matrix $a_i^- \circ b_j^- = \delta_{ij}/2$ we define periods of Q by

$$A_i = \int_{a_i^-} v\,, \qquad B_i = \int_{b_i^-} v\,. \tag{1.2}$$

The intersection pairing in H_- defines the natural symplectic form

$$\omega = 2 \sum_{i=1}^{3g-3+n} \mathrm{d}A_i \wedge \mathrm{d}B_i$$

which turns out to coincide with the canonical symplectic form ω_{can} on $T^*\mathcal{M}_{g,n}$ restricted to the space of quadratic differentials with simple zeros. Moreover, we show that the function generating the transformation from canonical Darboux coordinates (p_i, q_i) to Darboux coordinates given by periods $(\sqrt{2}A_i, \sqrt{2}B_i)$ is given by

$$G = \sum_{i=1}^{3g-3+n} A_i B_i$$

which generalizes the formula obtained in [**4**] to the case $n \neq 0$.

The symplectic form $\omega = \omega_{can}$ induces a symplectic structure on the character variety $\mathbb{V}^0_{g,n}$ via the monodromy map of equation (1.1). We emphasize that this monodromy map essentially depends on the choice of the base projective connection. The choice of Bergman projective connection S_B as the base is not unique since S_B transforms non-trivially under the change of Torelli marking of $\mathcal{C}$. The Bers projective connection chosen as the base in [**21**] carries even more freedom, since it depends on a choice of a point in the Teichmüller space as a parameter. Nevertheless, as it was shown in [**4**], the Poisson structure induced on the character variety is the same for the Bergman projective connection (independently of Torelli marking used) and Bers projective connection used in [**21**] (independently of the choice of the initial point in the Teichmüller space).

Technically, it is convenient to work with the matrix first order equation constructed by introducing functions $\psi_{1,2} = \varphi_{1,2}\sqrt{v}$ where $\varphi_{1,2}$ are two linearly independent solutions of (1.1). Denote by Ψ the Wronskian matrix of ψ_1 and ψ_2. The matrix Ψ satisfies the first order matrix equation

$$\mathrm{d}\Psi = \begin{pmatrix} 0 & v \\ uv & 0 \end{pmatrix} \Psi \,, \tag{1.3}$$

where the meromorphic function u on $\mathcal{C}$ is given by

$$u = -\frac{S_B - S_v}{2Q} - 1 \tag{1.4}$$

and $S_v(\xi) = \mathcal{S}(z(\xi), \xi)$ is the Schwarzian derivative of the coordinate $z(x) = \int_{x_1}^{x} v$ with respect to a local coordinate ξ. Notice that the coefficients v and uv in (1.3) are (holomorphic and meromorphic respectively) differentials on $\widehat{\mathcal{C}}$, not on $\mathcal{C}$ itself.

The Poisson bracket between $u(z)$ and $u(\zeta)$ (assuming that the coordinates z and ζ are independent of moduli) is given by the following expression:

$$\frac{4\pi i}{3}\{u(z), u(\zeta)\} = \mathcal{L}_z \left[\int^z h(z,\zeta)dz\right] - \mathcal{L}_\zeta \left[\int^\zeta h(z,\zeta)d\zeta\right] \,, \tag{1.5}$$

where the bimeromorphic function $h(z,\zeta)$ on $\mathcal{C} \times \mathcal{C}$ is given by $h(z,\zeta) = \frac{B^2(z,\zeta)}{Q(z)Q(\zeta)}$; the differential operator $\mathcal{L}_z = \frac{1}{2}\partial_z^3 - 2u(z)\partial_z - u_z(z)$ is known as the "Lenard's operator" in the theory of integrable systems.

The computation of the Poisson bracket (1.5) from the fundamental Poisson bracket $\{A_i, B_j\} = \delta_{ij}/2$ (and, therefore, also from fundamental Poisson bracket on $T^*\mathcal{M}_{g,n}$) is based on variational formulas for the canonical bidifferential B proved in [**23, 28**]. The monodromy map for equation (1.3) gives an $SL(2,\mathbb{C})$ representation of $\pi_1(\mathcal{C} \setminus \{y_i\}_{i=1}^n, x_0)$. A technical computation originally performed in [**4**] allows to find the Poisson bracket between traces of monodromy matrices of equation (1.3) along two arbitrary loops γ and $\tilde{\gamma}$. The result is the Goldman's bracket [**11**]:

$$\{\mathrm{tr}M_\gamma,\ \mathrm{tr}M_{\tilde{\gamma}}\} = \frac{1}{2}\sum_{p\in\gamma\circ\tilde{\gamma}} \nu(p)\left(\mathrm{tr}M_{\gamma_p\tilde{\gamma}} - \mathrm{tr}M_{\gamma_p\tilde{\gamma}^{-1}}\right) \,, \tag{1.6}$$

where the monodromy matrices $M_\gamma, M_{\tilde{\gamma}} \in SL(2,\mathbb{C})$; $\gamma_p\tilde{\gamma}$ and $\gamma_p\tilde{\gamma}^{-1}$ are paths obtained by resolving the intersection point p in two different ways (see [**11**]); $\nu(p) = \pm 1$ is the contribution of the point p to the intersection index of γ and $\tilde{\gamma}$.

The paper is organized as follows. In Section 2 we describe main objects associated to the canonical two-sheeted covering. In Section 3 we define the $SL(2,\mathbb{C})$

monodromy representation for equation (1.1) via an appropriate matrix reformulation. In section 4 we derive variational formulas for the coefficients of the matrix equation and for its monodromy matrices. In Section 5 we show that the period coordinates are Darboux coordinates for the canonical symplectic structure on $T^*\mathcal{M}_{g,n}$. In Section 6 we outline the modifications which have to be made in the scheme of [**4**] to cover the case of meromorphic differentials with first order poles.

2. Canonical covering of a Riemann surface

2.1. Setup. Denote the moduli space of meromorphic quadratic differentials on Riemann surface of genus g with n simple poles and $4g-4+n$ simple zeros by $\mathcal{Q}^0_{g,n}$. Here we list a few basic facts about the canonical covering of a Riemann surface determined by any quadratic differential $Q \in \mathcal{Q}^0_{g,n}$. The presentation is parallel to the case of canonical covering defined by a holomorphic quadratic differential which was considered in detail in [**4, 24, 28**].

Let $\mathcal{C}$ be a Riemann surface of genus g. The Torelli marking is a choice of canonical basis $\{a_i, b_i\}_{i=1}^g$ in $H_1(\mathcal{C}, \mathbb{Z})$ with the intersection index $a_i \circ b_j = \delta_{ij}$. Let $\{v_i\}_{i=1}^g$ be the dual basis in $H^{(1,0)}(\mathcal{C})$ normalized by $\oint_{a_i} v_j = \delta_{ij}$; integrals of v_i over b-cycles give the period matrix $\Omega_{ij} = \oint_{b_i} v_j$. Introduce the canonical meromorphic bidifferential $B(x,y) = d_x d_y \log E(x,y)$, where E is the prime-form (sometimes B is called the "Bergman kernel" following the paper by Hawley-Schiffer [**16**] of 1966, although this bidifferential was already used by Klein [**22**]). The bidifferential $B(x,y)$ is symmetric, $B(x,y) = B(y,x)$ and its only singularity is on the diagonal: as $y \to x$ in a local coordinate ξ one has

$$B(x,y) = \left(\frac{1}{(\xi(x) - \xi(y))^2} + \frac{1}{6} S_B(\xi(x)) + \dots \right) d\xi(x) d\xi(y)\,; \tag{2.1}$$

moreover, $B(x,y)$ is normalized by the requirement that all of its a-periods vanish with respect to each variable (its b_i-period with respect, say, to y-variable, equals $2\pi i v_i(x)$). Therefore, $B(x,y)$ depends on Torelli marking of $\mathcal{C}$; under the change of Torelli marking it transforms according to formula given at page 21 of [**6**]. The term S_B in (2.1) transforms as projective connection under a change of coordinate ξ; it is called the "Bergman projective connection" (this name is inherited from the "Bergman kenel" terminology adopted in [**16, 32**] although, probably, it would be historically more appropriate to call it "Klein's projective connection"). The projective connection S_B is important both in physics (it equals to the (zz)-component of the energy-momentum tensor of free bosons on a Riemann surface [**31**]) and in mathematics (it describes variations of a holomorphic section of determinant of Hodge vector bundle over various moduli spaces [**25, 26, 28**], and also variations of the determinant of Laplacian on a Riemann surface [**23, 33**]).

2.2. Canonical covering. Let $Q \in \mathcal{Q}^0_{g,n}$ be a meromorphic quadratic differential on $\mathcal{C}$ with n simple poles (denoted by $y_1, \dots, y_n$) and $4g-4+n$ simple zeros (denoted by $x_1, \dots, x_{4g-4+n}$). The canonical covering $\widehat{\mathcal{C}}$ is defined by equation $v^2 = Q$ in $T^*\mathcal{M}_g$. The two-sheeted covering $\pi : \widehat{\mathcal{C}} \to \mathcal{C}$ is branched at all poles and zeros of Q; thus the total number of branch points is $4g-4+2n$ and the genus of $\widehat{\mathcal{C}}$ equals $\widehat{g} = 4g-3+n$. The Abelian differential v is holomorphic on $\widehat{\mathcal{C}}$; it has zeros of order 2 (on $\widehat{\mathcal{C}}$!) at the branch points $\{x_i\}_{i=1}^{4g-4+n}$ and no other zeros. Denote the natural involution on $\widehat{\mathcal{C}}$ by μ and decompose the homology group $H_1(\widehat{\mathcal{C}}, \mathbb{R})$

into direct sum of the even and odd subspaces $H_+ \oplus H_-$, where $\dim H_+ = 2g$ and $\dim H_- = 6g - 6 + 2n$. The holomorphic part of cohomology group $H^{(1,0)}(\widehat{\mathcal{C}}, \mathbb{R})$ is similarly decomposed as $H^+ \oplus H^-$ where $\dim H^+ = g$ and $\dim H^- = 3g - 3 + n$. The differential v belongs to H^-.

Introduce a set of generators of H_- denoted by $\{a_i^-, b_i^-\}$ with the intersection index $a_i^- \circ b_j^- = \delta_{ij}/2$. The integrals of v over $\{a_i^-, b_i^-\}$,

$$A_i = \int_{a_i^-} v \qquad B_i = \int_{b_i^-} v \ , \qquad i = 1, \dots 3g - 3 + n \ , \tag{2.2}$$

are called the (abelian) periods of the quadratic differential Q. The periods can be used as local coordinates on the space $Q_{g,n}^0$ which are called "homological coordinates" in the theory of dynamical systems [**5, 8**] [1].

Another way of looking at coordinates (A_i, B_i) is to consider them as combinations of integrals of the differential v between different branch points of $\widehat{\mathcal{C}}$.

The differential v can be used to introduce a special system of local coordinates on $\widehat{\mathcal{C}}$ and $\mathcal{C}$. Namely, if x is a point of $\widehat{\mathcal{C}}$ which does not coincide with branch points $\{x_i\}$ then the local parameter (the "flat coordinate") on $\widehat{\mathcal{C}}$ in a neighbourhood of x is given by

$$z(x) = \int_{x_1}^{x} v \ , \tag{2.3}$$

where x_1 is a chosen "first" zero of v.

Near branch points $\{x_i\}$ of $\widehat{\mathcal{C}}$ the local coordinates on $\widehat{\mathcal{C}}$ are given by (these coordinates are called "distinguished"):

$$\hat{\zeta}_i(x) = \left[\int_{x_i}^{x} v\right]^{1/3} \ , \qquad i = 1, \dots, 3g - 3 + n \ ; \tag{2.4}$$

near y_k the distinguished local coordinate is

$$\hat{\xi}_k(x) = z(x) - z(y_k) = \int_{y_k}^{x} v, \qquad i = 1, \dots, n \ .$$

The "flat" local coordinates on $\mathcal{C}$ near every point except zeros or poles of Q are the same as on $\widehat{\mathcal{C}}$ (also defined up to a sign).

On the base curve $\mathcal{C}$ the distinguished local coordinate near point x_i is given by

$$\zeta_i(x) = \left[\int_{x_i}^{x} v\right]^{2/3} \ , \tag{2.5}$$

and near y_k:

$$\xi_k(x) = \left[\int_{y_k}^{x} v\right]^{2} \ . \tag{2.6}$$

To work with the canonical covering $\widehat{\mathcal{C}}$ it is convenient (following Lemma 3.1 of [**4**]) to choose a special set of generators $(\{\alpha_i, \beta_i\}_{i=1}^g, \gamma_{x_1}, \dots, \gamma_{x_{4g-4+n}}, \gamma_{y_1}, \dots \gamma_{y_n})$

[1] In recent physics literature these periods are denoted by a_i and a_i^D and are attributed to Seiberg and Witten, see for example [**9**] and references therein.

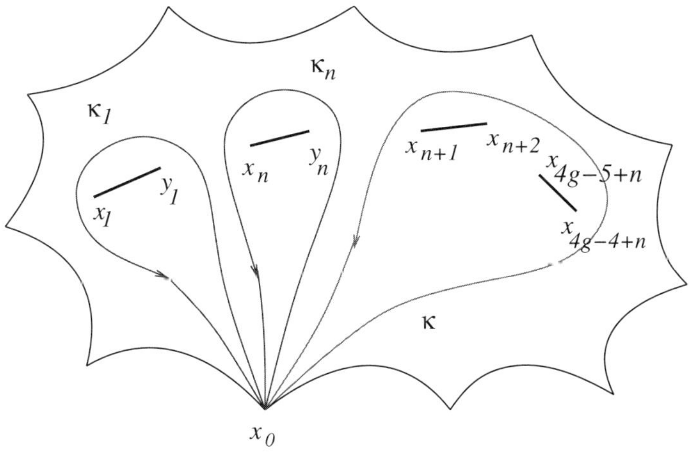

FIGURE 1. Fundamental polygon $\mathcal{C}_0$ with loops κ_i and κ.

of the fundamental group $\pi_1(\mathcal{C}\setminus\{x_i,y_i\})$ which satisfy the relation

$$\left(\prod_{i=1}^{4g-4+n}\gamma_{x_i}\right)\left(\prod_{i=1}^{n}\gamma_{y_i}\right)\prod_{i=1}^{g}\alpha_i\beta_i\alpha_i^{-1}\beta_i^{-1}=id\,. \tag{2.7}$$

The topology of the covering $\widehat{\mathcal{C}}$ determines a homomorphism h of $\pi_1(\mathbb{C}\setminus\{x_i,y_i\})$ to the symmetric group S_2. In parallel to Lemma 3.1 of [**4**] one can prove that the generators α,β,γ can be chosen in such a way that

$$h(\alpha_i)=h(\beta_i)=(21)\qquad i=1,\dots,g\,, \tag{2.8}$$

while

$$h(\gamma_{x_i})=h(\gamma_{y_j})=id\,,\qquad i=1,\dots,4g-4+n\,,\qquad j=1,\dots,n\,. \tag{2.9}$$

Then one can cut $\widehat{\mathcal{C}}$ along loops representing generators α_i and β_i on both sheets of the covering; in this way one gets the two-sheeted covering $\widehat{\mathcal{C}}_0$ of the fundamental polygon $\mathcal{C}_0$ of $\mathcal{C}$ branched at $4g-4+2n$ points $\{x_i\}$ and $\{y_i\}$.

It is convenient to connect the branch points by the following system of branch cuts:

$$[x_i,y_i]\,,\qquad i=1,\dots,n\,, \tag{2.10}$$

$$[x_{n+2i-1},x_{n+2i}]\,,\qquad i=1,\dots,2g-2\,. \tag{2.11}$$

Under this choice of branch cuts $\widehat{\mathcal{C}}_0$ becomes a two-sheeted branched covering of the fundamental polygon $\mathcal{C}_0$ with branch cuts chosen as shown in Fig.1.

Denote by x_0 the corner of $\mathcal{C}_0$. To study the monodromy group of the equation (1.1) we are going to introduce a system of n loops $\kappa_1,\dots,\kappa_n$ on $\widehat{\mathcal{C}}_0$ (Fig.1) such that the loop κ_i goes around the branch cut $[x_i,\,y_i]$. Projections of these loops to $\mathcal{C}_0$ will be denoted by the same letters.

3. Second order equation with meromorphic potential on a Riemann surface

We are going to write the second order linear equation on a Riemann surface (the "Schrödinger equation") in the form

$$\varphi'' + \left(\frac{1}{2}S_B + Q\right)\varphi = 0\,, \tag{3.1}$$

where the derivative is taken with respect to some local coordinate ξ on $\mathcal{C}$. The solution φ of (3.1) is locally a $-1/2$-differential which we write as $\varphi = \varphi(\xi)(d\xi)^{-1/2}$. Choosing two linearly independent solutions of (3.1) construct the Wronskian matrix

$$\Phi(\xi) = \begin{pmatrix} \varphi_1 & \varphi_2 \\ \varphi_{1\xi} & \varphi_{2\xi} \end{pmatrix}. \tag{3.2}$$

This matrix satisfies the equation

$$\frac{d\Phi}{d\xi} = \begin{pmatrix} 0 & 1 \\ -\frac{1}{2}S_B(\xi) - Q(\xi) & 0 \end{pmatrix}\Phi\,, \tag{3.3}$$

where we use the notation $S_B = S_B(\xi)(d\xi)^2$ and $Q = Q(\xi)(d\xi)^2$.

Let the quadratic differential Q have n simple poles $\{y_i\}_{i=1}^n$ and $4g-4+n$ simple zeros $\{x_i\}_{i=1}^{4g-4+n}$. Introduce the canonical cover $\widehat{\mathcal{C}}$ by the equation $v^2 = Q$. The zeros of the Abelian differential v on $\widehat{\mathcal{C}}$ all have multiplicity 2, and therefore $v^{1/2}$ is a section (holomorphic on $\widehat{\mathcal{C}}$ and unique up to a sign) of a spin line bundle over $\widehat{\mathcal{C}}$. Following [**4**] we define

$$\psi(\xi) := \varphi(\xi)v^{1/2}(\xi)\,. \tag{3.4}$$

Then the scalar equation (3.1) takes the following form in terms of ψ:

$$d\left(\frac{d\psi}{v}\right) - uv\psi = 0\,, \tag{3.5}$$

where u is the meromorphic function on $\mathcal{C}$ given by

$$u = -\frac{S_B - S_v}{2v^2} - 1\,. \tag{3.6}$$

and S_v is the meromorphic projective connection on $\mathcal{C}$ given by the Schwarzian derivative of the flat coordinate z:

$$S_v(\xi) = \mathcal{S}\left(\int_{x_1}^{x} v, \xi\right) = \left(\frac{v'}{v}\right)' - \frac{1}{2}\left(\frac{v'}{v}\right)^2. \tag{3.7}$$

Choosing two linearly independent solutions $\psi_{1,2}$ of (3.5) we construct the Wronskian matrix Ψ which is related to the matrix Φ (3.2) as follows

$$\Psi = \begin{pmatrix} \psi_1 & \psi_2 \\ d\psi_1/v & d\psi_2/v \end{pmatrix} \equiv v^{1/2}(\xi)\left[I + \frac{v_\xi}{2v}\begin{pmatrix} 0 & 1 \\ 0 & 0 \end{pmatrix}\right]\Phi\,. \tag{3.8}$$

The matrix Ψ satisfies the equation

$$d\Psi = \begin{pmatrix} 0 & v \\ uv & 0 \end{pmatrix}\Psi\,. \tag{3.9}$$

Matrix entries v and uv of the coefficient matrix of (3.9) are differentials on $\widehat{\mathcal{C}}$ (v is holomorphic while uv is meromorphic) which are anti-symmetric under the involution μ.

The advantage of the matrix Ψ over the original Wronskian matrix Φ is that the matrix entries of the former are functions i.e. they are independent of the choice of the local parameter ξ while the matrix entries of the latter non-trivially depend on the choice of ξ. The price which is paid for this independence is the appearance of the canonical covering $\widehat{\mathcal{C}}$ in the equation satisfied by Ψ. Therefore, Ψ gets additional monodromies $\pm i$ around zeros x_i (Φ is of course monodromy-free around x_i). The same factors $\pm i$ appear when comparing monodromies of (3.3) and (3.9) around poles y_i.

Another way to rewrite equation (3.9) is to use the flat coordinate $z = \int_{x_1}^{x} v$ on $\widehat{\mathcal{C}}$ (z can be used as a local coordinate on $\mathcal{C}$ and $\widehat{\mathcal{C}}$ outside of neighbourhoods of points x_i and y_k) . Then $v = dz$ and the scalar equation (3.5) can be written as

$$\psi_{zz} + u\psi = 0\,. \tag{3.10}$$

The matrix equation (3.9) takes the form

$$\frac{d\Psi}{dz} = \begin{pmatrix} 0 & 1 \\ u(z) & 0 \end{pmatrix} \Psi\,. \tag{3.11}$$

3.1. Monodromy representation. There are various ways to associate a monodromy representation of the fundamental group $\pi_1(\mathcal{C}\backslash\{y_i\}_{i=1}^n, x_0)$ to equations (3.1), (3.3) and (3.9). Due to the sign ambiguity of the spinors $\varphi_{1,2}$ the equation for Φ (3.3) determines only a $PSL(2,\mathbb{C})$ monodromy representation which coincides with the monodromy representation of the Schwarzian equation for function $f = \varphi_1/\varphi_2$:

$$\mathcal{S}(f,\xi) = S_B + 2Q\,. \tag{3.12}$$

It is a non-trivial problem whether or not this representation can be lifted to an $SL(2,\mathbb{C})$ representation (see [**10**]).

On the other hand, if we write down the equation in terms of matrix Ψ (3.9) an $SL(2,\mathbb{C})$ representation can be naturally constructed (as shown in [**4**] for $n = 0$).

Let us show how an $SL(2,\mathbb{C})$ monodromy representation of $\pi_1(\mathcal{C} \setminus \{y_i\}_{i=1}^n, x_0)$ can be associated to equation (3.9) for $n > 0$. The subtlety is in the fact that both differentials, v and uv, forming the matrix of coefficients of the system, are differentials on $\widehat{\mathcal{C}}$, not on $\mathcal{C}$. However a natural $SL(2,\mathbb{C})$ representation of the fundamental group of $\mathcal{C} \setminus \{y_k\}_{k=1}^n$ can be defined as follows.

Assume that the generators of the fundamental group $\pi_1(\mathcal{C}\backslash\{x_i, y_i\})$ are chosen according to (2.8), (2.9) and consider the first sheet $\widehat{\mathcal{C}}_0^{(1)}$ of the two-sheeted covering $\widehat{\mathcal{C}}_0$ of the fundamental polygon $\mathcal{C}_0$ (Fig.1). As before, the loops κ_i, $i = 1, \dots, n$ are chosen to encircle the branch cuts $[x_i, y_i]$; introduce also the loop κ excircling all the remaining branch cuts $[x_i, x_{i+1}]$ for $i = n+1, \dots, 4g-4+n-1$.

Since the coefficients of equation (3.9) are single-valued in the multiply-connected domain obtained by deleting all branch cuts from $\widehat{\mathcal{C}}_0^{(1)}$, one can define monodromy matrices along the loops α_i, β_i, κ_i and κ corresonding to initial point x_0.

The monodromy matrix along κ equals I since (see Lemma 6.2 of [**4**]) the monodromy around each x_i arises only due to the factor of $v^{1/2}$ relating matrices Φ (which is monodromy-free around x_i) and Ψ; this monodromy equals i. Since the number of zeros encircled by κ is a multiple of 4, the total monodromy of Ψ along κ equals I.

Therefore the monodromies along remaining loops κ_i, $\{\alpha_i\}$, $\{\beta_i\}$ satisfy the required relation

$$M_{\kappa_n} \dots M_{\kappa_1} \prod_{i=g}^{1} M_{a_i} M_{\beta_i} M_{\alpha_i}^{-1} M_{\beta_i}^{-1} = I \,, \tag{3.13}$$

i.e. these monodromies form an anti-representation of the fundamental group of $\mathcal{C} \setminus \{[x_i, y_i]\}_{i=1}^n$. Since the monodromies of Ψ around x_i arising from elementary transformation between Φ and Ψ are equal to iI we shall *define* the monodromies around poles y_i to be matrices M_{κ_i}. These matrices obviously correspond to $PSL(2,\mathbb{C})$ monodromies of the associate Schwarzian equation

$$\mathcal{S}(f,\xi) = S_B(\xi) + Q(\xi) \,. \tag{3.14}$$

Therefore, we get the following

PROPOSITION 3.1. *The constructed $SL(2,\mathbb{C})$ monodromy representation of matrix equation (3.9) is a lift of the $PSL(2,\mathbb{C})$ monodromy representation of the Schwarzian equation (3.14).*

The equation (1.1) can be written in terms of a local coordinate ξ near a pole y_i in the form

$$\varphi_{\xi\xi} + \left(\frac{C}{\xi} + O(1)\right)\varphi = 0 \,.$$

The local analysis shows that the monodromy of Φ around y_i has two coinciding eigenvalues equal to 1. Due to an additional factor of $-i$ and i arising from monodromies of $v^{1/2}$ around y_i and x_i, respectively, the eigenvalues of monodromy M_{κ_i} of Ψ along κ_i are also coinciding and equal to $+1$.

4. Variational formulas

4.1. Variational formulas on $Q^0_{g,n}$. The variational formulas on the moduli space $Q^0_{g,n}$ describe the dependence of the period matrix Ω, normalized differentials v_i and the canonical bidifferential B on moduli, which in present setting are given by periods (A_i, B_i). To write down these formulas we introduce a set of generators $s_1 \dots, s_{6g-6+2n}$ in H_- (for example one can choose this set to coincide with the set $\{a_i^-, b_i^-\}$) and introduce the periods $\mathcal{P}_i = \int_{s_i} v$. Denote by $s_1^*, \dots, s_{6g-6+2n}^*$ the set of generators dual to $\{s_i\}$ with intersection index $s_i^* \circ s_j = \delta_{ij}$; the set of dual periods corresponding to $\{a_i^-, b_i^-\}$ is given by $\{-2b_i^-, 2a_i^-\}$.

The variational formulas on the space $Q^0_{g,n}$ given below can be obtained by reduction of the variational formulas on spaces of holomorphic Abelian differentials [**23**] to a subspace consisting of Riemann surfaces admitting a holomorphic involution. These formulas are only a slight modification of variational formulas on spaces of holomorphic quadratic differentials [**4, 28**]. Let us introduce g functions on $\widehat{\mathcal{C}}$:

$$f_i(x) = \frac{v_i(x)}{v(x)} \tag{4.1}$$

and the function of two variables on $\widehat{\mathcal{C}}$

$$b(x,y) = \frac{B(x,y)}{v(x)v(y)} \,. \tag{4.2}$$

The variational formula for the period matrix Ω looks as follows:

$$\frac{\partial \Omega_{jk}}{\partial \mathcal{P}_{s_i}} = \frac{1}{2} \int_{s_i^*} f_j f_k v \,. \tag{4.3}$$

For the holomorphic normalized differentials and for the canonical bidifferential one has:

$$\frac{\partial f_j(x)}{\partial \mathcal{P}_{s_i}} = \frac{1}{2} \int_{s_i^*} f_j(t)\, b(t,y) v(t) \tag{4.4}$$

and

$$\frac{\partial b(x,y)}{\partial \mathcal{P}_{s_i}} = \frac{1}{4\pi i} \int_{s_i^*} b(x,t)\, b(t,y) v(t) \,, \tag{4.5}$$

where $\mathcal{P}_{s_i} := \oint_{s_i} v$ and all derivatives are computed keeping $z(x) = \int_{x_1}^{x} v$ and $z(y) = \int_{x_1}^{y} v$ constant. The fundamental polygon of $\widehat{\mathcal{C}}$ used to define $z(x)$ and $z(y)$ must be invariant under the involtion μ, such that the coordinate $z(x)$ satisfies the relation $z(x^\mu) = -z(x)$ for x lying inside of the fundamental polygon.

Notice that, although the integrand in the right-hand side of (4.5) has poles of second order at the branch points x_i, its residues at these points vanish since this integrand is anti-symmetric under μ.

Integrating (4.5) with respect to variable y between any two points p_1 and p_2 along a contour l one gets a variational formula for the normalized (integrals over all a-periods not intersecting contour l vanish) differential $W_{p_1 p_2}$ of third kind with poles at $p_{1,2}$ and residues ± 1:

$$\frac{\partial W_{p_1,p_2}(x)}{\partial \mathcal{P}_{s_i}} = \frac{1}{4\pi i} \int_{s_i^*} W_{p_1,p_2}(t)\, b(t,x) v(t) \,. \tag{4.6}$$

Taking the limit $y \to x$ in the formula (4.5) one gets the variational formula for the potential $u = -\frac{1}{2}\frac{S_B - S_v}{Q} - 1$ (also at $z(x)$ constant):

$$\frac{\partial u(x)}{\partial \mathcal{P}_{s_i}} = -\frac{3}{8\pi i} \int_{s_i^*} h(x,t) v(t) \,, \tag{4.7}$$

where

$$h(x,t) = \frac{B^2(x,t)}{Q(x)Q(y)} \,. \tag{4.8}$$

The differential $h(x,t)v(t)$ has poles of order 4 on $\widehat{\mathcal{C}}$ at $t = x$ and $t = x^\mu$ with non-trivial residues. Therefore, the choice of the class of the integration path in $H_1(\widehat{\mathcal{C}} \setminus \{x, x^\mu\})$ in the right-hand side of (4.7) is important; this choice must be made in such a way that the integration goes along parts of the boundary of the fundamental polygon of $\widehat{\mathcal{C}}$; the fundamental polygon should coincide with the one used to define the coordinate $z(x)$ which is kept fixed in the left-hand side of (4.7).

4.2. Variational formulas for solution and monodromies of the second order equation. Let $\Psi(z)$ be solution of the matrix equation (3.11) normalized at the point $z_0 = z(x_0)$ via $\Psi(z_0) = I$.

Let us introduce the following auxiliary matrix:

$$\Lambda(x) = \Psi^{-1}(x) \sigma_- \Psi(x) \,, \tag{4.9}$$

where $\sigma_- = \begin{pmatrix} 0 & 0 \\ 1 & 0 \end{pmatrix}$. The matrix Λ satisfies the third order equation [**4**]

$$\Lambda_{zzz} - 4u(z)\Lambda_z - 2u_z\Lambda = 0\ . \tag{4.10}$$

The components of the matrix Λ are given by ψ_1^2, ψ_2^2 and $\psi_1\psi_2$; therefore (4.10) is nothing but a well-known third order equation for products of two solutions of second order equation (3.10).

Dependence of $\Psi(x)$ (for fixed $z(x)$) on moduli is given by the following proposition:

PROPOSITION 4.1. *The following variational formula holds for solution of (3.9) normalized by* $\Psi(x_0) = I$*:*

$$\Psi^{-1}\frac{\partial\Psi(x)}{\partial\mathcal{P}_{s_i}}\Big|_{z(x)=const} = -\frac{3}{8\pi i}\int_{x_0}^{x}\Lambda(x)\left[\int_{t\in s_i^*} h(x,t)v(t)\right] \tag{4.11}$$

for any $s_i \in H_-$*. The integration path* s_i^* *in the r.h.s. of (4.11) goes along the boundary of the fundamental polygon of* $\widehat{\mathcal{C}}$ *used to define the Abelian integral* $z(x)$*.*

The proof of this proposition is a direct application of the variation of parameters formula for the non-homogeneus linear equation using the variational formula (4.7) for the coefficient of the equation.

An immediate corollary of the formula (4.11) is the following variational formula for monodromy matrices.

COROLLARY 4.1. *Let* γ *be any element of* $\pi_1(\mathcal{C}\setminus\{y_i\}_{i=1}^n, x_0)$*. Denote by* M_γ *the monodromy matrix of equation (3.9) along* γ *(as defined in Section 3.1). Then the following variational formulas hold:*

$$M_\gamma^{-1}\frac{\partial M_\gamma}{\partial\mathcal{P}_{s_i}} = -\frac{3}{8\pi i}\int_\gamma\Lambda(x)\left[\int_{t\in s_i^*} h(x,t)v(t)\right]\ . \tag{4.12}$$

5. Canonical symplectic structure on $T^*\mathcal{M}_{g,n}$ via periods of Q

Let $\{q_i\}_{i=1}^{3g-3+n}$ be a set of holomorphic local coordinates on $\mathcal{M}_{g,n}$. Then for any cotangent vector p we can define its coordinates p_i via $p = \sum_{i=1}^{3g-3+n} p_i \mathrm{d}q_i$. Introduce the canonical symplectic structure ω_{can} on $T^*\mathcal{M}_{g,n}$:

$$\omega_{can} = \sum_{i=1}^{3g-3+n} \mathrm{d}p_i \wedge \mathrm{d}q_i \tag{5.1}$$

and the corresponding symplectic potential (the Liouville 1-form)

$$\theta_{can} = \sum_{i=1}^{3g-3+n} p_i \mathrm{d}q_i\ . \tag{5.2}$$

Define the "homological" symplectic structure on $\mathcal{Q}_{g,n}^0$ in terms of periods (A_i, B_i):

$$\omega_{hom} = 2\sum_{i=1}^{3g-3+n} \mathrm{d}A_i \wedge \mathrm{d}B_i \tag{5.3}$$

and its symplectic potential

$$\theta_{hom} = 2 \sum_{i=1}^{3g-3+n} A_i dB_i \ . \tag{5.4}$$

Notice that the form ω_{hom} is independent of the choice of basis (a_i^-, b_i^-) in H_- but θ_{hom} depends on this choice. On the other hand, the 1-form $\theta = \sum_{i=1}^{3g-3+n} A_i dB_i - B_i dA_i$ (such that $d\theta = d\theta_{hom} = \omega_{hom}$) is also independent of a choice of symplectic basis in H_-.

5.1. Local coordinate systems on $\mathcal{M}_{g,n}$ and dual quadratic differentials. Consider first the case $n = 0$ and exclude the hyperelliptic locus from consideration. The natural local coordinates on the moduli space $\mathcal{M}_g$ of unpunctured Riemann surfaces of genus g can be obtained using the Torelli theorem stating that the complex structure of a Riemann surface $\mathcal{C}$ is uniquely determined by its period matrix Ω. Therefore, in a neighbourhood of any point of $\mathcal{M}_g$ there exists a subset D of $3g - 3$ matrix entries of Ω which can be chosen as local coordinates $\{q_{jk} = \Omega_{jk}\}$, $(jk) \in D$ on $\mathcal{M}_g$. The cotangent vectors $d\Omega_{jk}$ to $\mathcal{M}_g$ at a given point can be identified with holomorphic quadratic differentials $v_j v_k$:

$$d\Omega_{jk} \quad \sim \quad v_j v_k \tag{5.5}$$

since variation of Ω_{jk} under an infinitesimal deformation δ_μ of a Riemann surface defined by a Beltrami differential μ (δ_μ can be viewed as an element of the tangent space to $\mathcal{M}_g$ at a given point) is given by the Rauch variational formula $\delta_\mu \Omega_{jk} = \int_{\mathcal{C}} v_j v_k \mu$. The analog of Rauch's formulas on the space $\mathcal{Q}_g$ is given by (4.3) which also confirms the identification of the tangent vector $d\Omega_{jk}$ with holomorphic quadratic differential $v_j v_k$.

For an arbitrary n we are going to consider separately the case $g \geq 2$ and the low genus cases $g = 1$ and $g = 0$.

5.1.1. *Coordinates on $\mathcal{M}_{g,n}$ for $g \geq 2$.* Local holomorphic coordinates on $\mathcal{M}_{g,n}$ can be introduced in various ways. We are going to use the following set of coordinates:

- A set of $3g - 3$ entries Ω_{jk} $(jk) \in D$ of the period matrix. These coordinates determine the complex structure of $\mathcal{C}$. The cotangent vector $\mathrm{d}\Omega_{jk}$ is represented by the holomorphic quadratic differential $v_j v_k$. These holomorphic quadratic differentials span a $3g-3$-dimensional subspace in $3g - 3 + n$ - dimensional cotangent space $T^*_{\mathcal{C}}\mathcal{M}_{g,n}$.
- The additional n coordinates (outside, possibly, a a subset of $\mathcal{M}_{g,n}$ of codimension 1) we are going to choose as follows:

$$w_k = \frac{v_i}{v_j}(y_k) \ , \tag{5.6}$$

where v_i and v_j is an arbitrary pair of normalized holomorphic differentials on $\mathcal{C}$ such that $v_j(y_k) \neq 0$. The differential v_k has to be chosen differently in different neighbourhoods on the moduli space. [The existence of a holomorphic differential non-vanishing at any given point y_k can be easily proved. Namely, assuming that all holomorphic differentials vanish at y_k we conclude that all periods of the second kind differential $B(x, y_k)$ vanish; thus the antiderivative of $B(x, y_k)$ is a meromorphic function whose only simple pole is at y_k which is a contradiction].

The cotangent vector dw_k can be represented by the following (generically meromorphic) differential Q_k whose only pole of first order is at the marked point y_k:

$$Q_k(t) = \frac{1}{4\pi i}\frac{v_i(t)v_j(y_k) - v_i(y_k)v_j(t)}{v_j^2(y_k)}B(t, y_k)\,, \tag{5.7}$$

where B is the canonical bimeromorphic differential. If the numerator of (5.7) vanishes at y_k then the quadratic differential Q_k is holomorphic; thus the function w_k can not be used as local coordinate on the subspace (of codimension 1) of $\mathcal{M}_{g,n}$ defined by equation $(v_i'v_j - v_iv_j')(y_k) = 0$. For our purposes, however, it is sufficient to cover $\mathcal{M}_{g,n}$ by coordinate charts outside of a subspace of codimension one since all equalities of forms we derive can be extended to the whole space by analyticity.

To verify that (5.7) indeed corresponds to cotangent vector dw_k we use variational formulas (4.4) to get

$$\frac{\partial w_k}{\partial \mathcal{P}_i} = \int_{s_i^*} \frac{Q_k(t)}{v(t)}\,. \tag{5.8}$$

5.1.2. *Coordinates on $\mathcal{M}_{1,n}$.* Let $n \geq 2$. The dimension of $\mathcal{M}_{1,n}$ equals n. The first coordinate q_1 on $\mathcal{M}_{1,n}$ is the period Ω. The holomorphic quadratic differential corresponding to the cotangent vector $d\Omega$ is given by v_1^2, in analogy to the higher genus case.

The remaining $n-1$ coordinates $q_2, \dots, q_n$ are defined by integrating the normalized differential v_1 between poles of Q:

$$q_k = \int_{y_1}^{y_k} v_1 \tag{5.9}$$

for $k = 2, \dots, n$.

The meromorphic quadratic differential with simple poles at y_1 and y_k which corresponds to the cotangent vector dq_k equals

$$Q_k(t) = \frac{1}{4\pi i} v_1(t) W_{y_1 y_k}(t)\,, \tag{5.10}$$

where $W_{xy}(t)$ is the normalized (its a-period vanishes) differential of third kind on $\mathcal{C}$ with simple poles at x and y and residues $+1$ and -1, respectively.

The correspondence between the cotangent vector dq_k and the meromorphic quadratic differential Q_k can be verified by integrating variational formulas (4.4) for the differential v_1 between y_1 and y_k which gives

$$\frac{\partial q_k}{\partial \mathcal{P}_i} = \int_{s_i^*} \frac{Q_k}{v}\,, \qquad k = 2, \dots, n\,. \tag{5.11}$$

5.1.3. *Coordinates on $\mathcal{M}_{0,n}$.* Consider the non-trivial case $n \geq 4$; then $\dim M_{0,n} = n-3$. Define coordinates q_k on $M_{0,n}$ as follows:

$$q_k = \int_{y_3}^{y_{k+3}} W_{y_1 y_2} \tag{5.12}$$

for $k = 1, \dots, n-3$, where as before

$$W_{y_1 y_2} = \left(\frac{1}{x - y_1} - \frac{1}{x - y_{k+1}}\right) dx \tag{5.13}$$

is the meromorphic differential of third kind on the Riemann sphere.

According to general variational formulas for the third kind differential $W_{p_1p_2}(x)$ (4.6) the quadratic differential corresponding to cotangent vector dq_k is given by

$$Q_k = \frac{1}{4\pi i} W_{y_1y_2} W_{y_3y_{k+3}} . \tag{5.14}$$

Computing the coordinates q_k (5.12) and the corresponding meromorphic differentials Q_k (5.14) explicitly, we get

$$q_k - \log \frac{(y_{k+3} - y_1)(y_3 - y_2)}{(y_3 - y_1)(y_{k+3} - y_2)} \tag{5.15}$$

and

$$Q_k = \frac{1}{4\pi i} \frac{(y_1 - y_2)(y_3 - y_{k+3})}{(x - y_1)(x - y_2)(x - y_3)(x - y_{k+3})} (dx)^2 . \tag{5.16}$$

5.2. Equivalence of canonical and homological symplectic structures on $T^*\mathcal{M}_{g,n}$. The following theorem generalizes Theorem 4.1 of [**4**] to the case $n > 0$.

THEOREM 5.1. *Being restricted to the moduli space $\mathcal{Q}^0_{g,n}$, the canonical symplectic form (5.1) coincides with the homological symplectic form (5.3):*

$$\omega_{can} = \omega_{hom} . \tag{5.17}$$

The canonical symplectic potential (5.2) is expressed in terms of homological coordinates as follows:

$$\theta_{can} = \sum_{i=1}^{3g-3+n} (A_i dB_i - B_i dA_i) , \tag{5.18}$$

and, therefore, the function generating the transition between canonical and homological Darboux coordinates is given by

$$G = \sum_{i=1}^{3g-3+n} A_i B_i . \tag{5.19}$$

PROOF. For $n = 0$ the proof is given in [**4**], Th.4.1. Let $n \neq 0$. Consider the case $g \geq 2$. At generic point of the moduli space $\mathcal{Q}^0_{g,n}$ we can represent the quadratic differential $Q \in \mathcal{Q}^0_{g,n}$ as a linear combination of differentials Q_k (5.7) and holomorphic differentials $v_j v_k$ where $(jk) \in D$ for some subset D of matrix entries of the period matrix Ω:

$$Q = \sum_{(jk)\in D} p_{jk} v_j v_k + \sum_{l=1}^{n} p_l Q_l , \tag{5.20}$$

or, since $Q = v^2$,

$$v = \sum_{(jk)\in D} p_{jk} \frac{v_j v_k}{v} + \sum_{l=1}^{n} p_l \frac{Q_l}{v} . \tag{5.21}$$

Integrating this relation over cycles a_i^- and b_i^- we get

$$A_i = \sum_{(jk)\in D} p_{jk} \int_{a_i^-} \frac{v_j v_k}{v} + \sum_{l=1}^{n} p_l \int_{a_i^-} \frac{Q_l}{v} \tag{5.22}$$

and

$$B_i = \sum_{(jk)\in D} p_{jk} \int_{b_i^-} \frac{v_j v_k}{v} + \sum_{l=1}^{n} p_l \int_{b_i^-} \frac{Q_l}{v} . \tag{5.23}$$

Taking into account variational formulas (5.8) and (4.3) we rewrite (5.22) and (5.23) as follows:

$$A_i = \sum_{(jk)\in D} p_{jk} \frac{\partial q_{jk}}{\partial B_i} + \sum_{l=1}^{n} p_l \frac{\partial q_l}{\partial B_i} \tag{5.24}$$

and

$$B_i = -\sum_{(jk)\in D} p_{jk} \frac{\partial q_{jk}}{\partial A_i} - \sum_{l=1}^{n} p_l \frac{\partial q_l}{\partial A_i} . \tag{5.25}$$

Therefore,

$$\begin{aligned} \sum_{i=1}^{3g-3+n} (A_i \mathrm{d}B_i - B_i \mathrm{d}A_i) &= \sum_{(jk)\in D} \sum_{i=1}^{3g-3+n} p_{jk} \left(\frac{\partial q_{jk}}{\partial A_i} \mathrm{d}A_i + \frac{\partial q_{jk}}{\partial B_i} \mathrm{d}B_i \right) \\ &\quad + \sum_{l=1}^{n} \sum_{i=1}^{3g-3+n} p_{jk} \left(\frac{\partial q_l}{\partial A_i} \mathrm{d}A_i + \frac{\partial q_l}{\partial B_i} \mathrm{d}B_i \right) \\ &= \sum_{(jk)\in D} p_{jk} \mathrm{d}q_{jk} + \sum_{l=1}^{n} p_l \mathrm{d}q_l = \theta_{can} . \end{aligned} \tag{5.26}$$

This completes the proof of (5.18) for $g \geq 2$; applying d-operator to this relation we get (5.17).

For $g = 0, 1$ the proof is analogous. Say, in $g = 1$ case the sum of residues of Abelian differential Q/v_1 equals 0; therefore, Q can be represented as a linear combination of v_1^2 and quadratic differentials (5.10):

$$Q = p_1 v_1^2 + \sum_{k=2}^{n} p_k Q_k . \tag{5.27}$$

Dividing (5.27) by v, integrating the result over cycles a_i^- and b_i^- and using variational formulas for Ω and Abelian differentials $W_{y_1 y_k}$ we get

$$A_i = p_1 \frac{\partial q_1}{\partial B_i} + \sum_{k=2}^{n} p_k \frac{\partial q_k}{\partial B_i} \tag{5.28}$$

and

$$B_i = -p_1 \frac{\partial q_1}{\partial A_i} - \sum_{k=2}^{n} p_k \frac{\partial q_k}{\partial A_i} , \tag{5.29}$$

which again implies (5.18).

Consider now the case $g = 0$. In genus 0 any quadratic differential with at most simple poles at $y_1, \ldots, y_n$ can be represented as a linear combination of $n - 3$

quadratic differentials Q_k (5.14); thus

$$v = \sum_{k=1}^{g-3} p_k \frac{Q_k}{v} \tag{5.30}$$

for some coefficients p_k. Furthermore, variational formulas (4.6) for $W_{y_1 y_2}$ imply

$$A_i = \sum_{k-1}^{n-3} p_k \frac{\partial q_k}{\partial B_i}\,, \qquad B_i = -\sum_{k=1}^{n-3} p_k \frac{\partial q_k}{\partial A_i}$$

leading to (5.18). □

REMARK 5.1. The statement (5.17) about coincidence of the homological and canonical symplectic structures was contained in the paper **[9]** written in 2012 (Section 7.3.2) in the context of meromorphic quadratic differentials with second or higher order poles.[2] However, the supporting argument in **[9]** was given on the "physics" level of rigour. On the other hand, Theorem 4.1 of **[4]** for $n = 0$ and Theorem 5.1 of this paper (for any n) are proved rigorously. Moreover, in our Theorem 5.1 and Corollary 4.1 of **[4]** we get a stronger result: we don't only prove the coincidence of the symplectic forms but also explicitly compute the generating function (5.19) between two systems of Darboux coordinates. The correspondence of notations used here (as well as in **[4]**) and notations used in **[9]** is as follows: the homological coordinates (the "abelian periods" $\{A_i, B_i\}$) of Q are called in **[9]** the "Seibeg-Witten central charge functions" and denoted by (a_i, a_i^D). The generating function (5.19) of this paper coincides with function $\mathcal{F}$ of **[9]** called there the "Seiberg-Witten prepotential". The terminology of this paper follows classical works on Teichmüller theory; in particular, periods of $\sqrt{Q}$ were already used in 1975 paper **[5]** devoted to the theory of Strebel differentials.

6. From canonical symplectic structure on $T^*\mathcal{M}_{g,n}$ to Goldman bracket

The canonical symplectic structure on $T^*\mathcal{M}_{g,n}$ (or, equivalently, the symplectic structure (5.3) on the space $\mathcal{Q}^0_{g,n}$) induces the Poisson structure on the space of coefficients u of the equation (3.9). The Poisson bracket between $u(z)$ and $u(\zeta)$ (for constant z and ζ) can be computed using the variational formula (4.7) and Theorem 5.1:

$$\{u(z), u(\zeta)\} = \frac{1}{2} \sum_{i=1}^{3g-3+n} \frac{\partial u(z)}{\partial A_i}\frac{\partial u(\zeta)}{\partial B_i} - \frac{\partial u(z)}{\partial B_i}\frac{\partial u(\zeta)}{\partial A_i}\,,$$

which gives, in analogy to the proof of Proposition 4.4 of **[4]**:

$$\frac{4\pi i}{3}\{u(z), u(\zeta)\} = \mathcal{L}_z\left[\int^z h(z,\zeta)dz\right] - \mathcal{L}_\zeta\left[\int^\zeta h(z,\zeta)d\zeta\right]\,, \tag{6.1}$$

where

$$\mathcal{L}_z = \frac{1}{2}\partial_z^3 - 2u(z)\partial_z - u_z(z)$$

is called the "Lenard's operator" in the theory of KdV equation **[2]**. In the computation of the bracket (6.1) it is assumed that the arguments z and ζ of u are independent of moduli.

[2] The author thanks Joerg Teschner for pointing out this reference.

The bracket (6.1) does not imply Poisson brackets between monodromy matrices of the equation (3.9) themselves since these monodromies depend also on the choice of the basepoint x_0. However, the bracket (6.1) defines a Poisson structure on the character variety $\mathbb{V}^0_{g,n}$ which consists of equivalence classes of monodromy representations (moduli simultaneous conjugation by some matrix). We add the index 0 to the notation $\mathbb{V}^0_{g,n}$ to indicate that each of the monodromy matrices around poles y_i have coinciding eigenvalues (1 or -1).

The coordinates on $\mathbb{V}^0_{g,n}$ can be chosen to be traces of monodromy matrices M_γ for a sufficiently large set of loops γ.

The following proposition is rather technically tedious; it is proved in complete analogy to the case $n = 0$ given in Theorems 7.2 and 7.3 of [**4**]:

PROPOSITION 6.1. *Let γ and $\tilde\gamma$ be two closed contours on $\mathcal{C}$. If γ and $\tilde\gamma$ do not intersect then the traces of the corresponding monodromy matrices of equation (3.9) Poisson-commute,*

$$\{\operatorname{tr} M_\gamma,\, \operatorname{tr} M_{\tilde\gamma}\} = 0\,. \tag{6.2}$$

If γ and $\tilde\gamma$ intersect transversally at one point x_0 with $\tilde\gamma \circ \gamma = 1$ then

$$\{\operatorname{tr} M_\gamma,\, \operatorname{tr} M_{\tilde\gamma}\} = \frac{1}{2}(\operatorname{tr} M_\gamma M_{\tilde\gamma} - \operatorname{tr} M_\gamma M_{\tilde\gamma}^{-1})\,. \tag{6.3}$$

From this proposition one can deduce the validity of Goldman's bracket for any two contours γ and $\tilde\gamma$ by finding a sufficiently large supply of loops on $\mathcal{C}$ which intersect at no more than one point.

THEOREM 6.1. *The canonical Poisson structure on $T^*\mathcal{M}_{g,n}$ implies the Goldman's bracket for traces of monodromies of equation (3.9), i.e. for any two loops $\gamma, \tilde\gamma \in \pi_1(\mathcal{C}, x_0)$ one has*

$$\{\operatorname{tr} M_\gamma,\, \operatorname{tr} M_{\tilde\gamma}\} = \frac{1}{2}\sum_{p\in\gamma\cap\tilde\gamma} \nu(p)(\operatorname{tr} M_{\gamma_p\tilde\gamma} - \operatorname{tr} M_{\gamma_p\tilde\gamma^{-1}})\,, \tag{6.4}$$

where $\gamma_p\tilde\gamma$ and $\gamma_p\tilde\gamma^{-1}$ are two ways to resolve the intersection point p to get two new contours $\gamma_p\tilde\gamma$ and $\gamma_p\tilde\gamma^{-1}$ for each p; $\nu(p) = \pm 1$ is the contribution of the point p to the intersection index of γ and $\tilde\gamma$.

PROOF. The proof is parallel to the proof of Theorem 7.4 of [**4**]. Namely, any two loops from the set consisting of the following $g^2 + 2g + 2gn + n$ elements of $\pi_1(\mathcal{C} \setminus \{y_i\}_{i=1}^n, x_0)$:

$$\mathfrak{S} = \{\alpha_i,\ \beta_i,\ \ 1 \le i \le g\,;\ \ \alpha_i\alpha_j,\ i<j\,;\ \ \alpha_i\beta_j,\ i\le j\,;\\ \kappa_k,\ \kappa_k\alpha_i,\ \ \kappa_k\beta_i,\ \ k=1,\dots,n\,,\ \ i,j=1,\dots,g\} \tag{6.5}$$

either don't intersect or intersect at only one point; thus (6.4) holds for any pair of loops $\gamma, \tilde\gamma \in \mathfrak{S}$. Notice that the number of functions $\operatorname{tr} M_\gamma$ for $\gamma \in \mathfrak{S}$ is always greater than the dimension $6g - 6 + 2n$ of the character variety $\mathbb{V}^0_{g,n}$; however, it remains to show that at generic point of $\mathbb{V}^0_{g,n}$ the differentials of $\operatorname{tr} M_\gamma$ generate the whole cotangent space to $\mathbb{V}^0_{g,n}$.

By simultaneous conjugation of all monodromies we can transform the matrix M_{α_g} to a diagonal form (generically), and the matrix M_{β_g} to a matrix whose fixed point (of associated Möbius transformation) equals 1. Then from the triple of numbers $(\operatorname{tr} M_{\alpha_g}, \operatorname{tr} M_{\beta_g}, \operatorname{tr} M_{\alpha_g} M_{\beta_g})$ we find matrices M_{α_g} and M_{β_g}. Then to find any other matrix M_{α_i} one can use the triple of loops $(\alpha_i, \alpha_i\alpha_g, \alpha_i\beta_g)$ also contained in in the set $\mathfrak{S}$.

In turn, knowing all matrices M_{α_i} as well as M_{β_g} one can reproduce the matrices $M_{\beta_2}, \ldots, M_{\beta_{g-1}}$ since for each j the set $\mathfrak{S}$ contains β_j and $\alpha_i\beta_j$ for $i = 1, \ldots, j$, which is sufficient to determine M_{β_j} for $j \geq 2$ (up to a binary choice).

Now we have enough information to determine M_{κ_i} from knowing $\operatorname{tr} M_{\kappa_i \alpha_j}$ for all α_j (actually, each of M_{κ_i} has only one unknown entry. The only remaining unknown matrix is M_{β_1}; this matrix can be determined knowing $\operatorname{tr} M_{\beta_1}$, $\operatorname{tr} M_{\alpha_1} M_{\beta_1}$ and using the relation (3.13) as discussed in [**4**]. □

6.1. Admissible holomorphic sections of $\mathbb{S}_g$. The map from the space of meromorphic quadratic differentials to the character variety $\mathbb{V}^0_{g,n}$ via the monodromy map of the Schrödinger equation is highly non-canonical: it essentially depends on the choice of the base holomorphic section of the affine bundle of projective connections over an appropriate covering of the moduli space. For example, the Bergman projective connection used in [**4**] (as well as in this paper) depends on the choice of canonical basis in homology group of $\mathcal{C}$, and therefore is a holomorphic section of the affine bundle of projective connections over the Torelli space. In general, to be able to compare different holomorphic sections of the affine bundle of projective connections it's convenient to consider the affine bundle $\mathbb{S}_g$ of holomorphic projective connections over the Teichmüller space $\mathcal{T}_g$.

Choosing any other holomorphic section S_0 of $\mathbb{S}_g$ we could also write the Schrödinger equation in the form

$$\varphi'' + \left(\frac{1}{2}S_0 + Q\right)\varphi = 0\,. \tag{6.6}$$

In [**4**] we have discussed alternative ways of fixing S_0: the Schottky or quasi-fuchsian (Bers) projective connections (the most common Fuchsian projective connection is not suitable in this framework since it depends non-holomorphically on moduli of $\mathcal{C}$). In particular, we have shown that if S_0 is the Schottky projective connection then the canonical symplectic structure on T^*M_g also implies the Goldman Poisson structure on the character variety $\mathbb{V}_g$. The same statement for the case when S_0 is given by Bers projective connection follows from the paper by S.Kawai [**21**]. [We recall that the Schottky projective connection is given by Schwarzian derivative $\mathcal{S}(z_S, t)$ with respect to any local coordinate t, where z_S is the Schottky uniformization coordinate. The Bers projective is given by the Schwarzian derivative $\mathcal{S}(z_{\mathcal{C}_0}, t)$ where $z_{\mathcal{C}_o}$ is the coordinate in the fundamental domain of the Kleinian group defining simultaneous uniformization of $\mathcal{C}$ and some "fixed" Riemann surface $\mathcal{C}_0$ with anti-holomorphic complex structure, i.e. there exists in fact infinitely many Bers projective connections; these projective connections are labeled by points of Teichmüller space].

Therefore, all of the choices of the base projective connection S_0 listed above are equivalent from symplectic point of view, which inspires the following definition:

DEFINITION 6.1. A holomorphic section S_0 of the affine line bundle $\mathbb{S}_g$ over the Teichmüller space $\mathcal{T}_g$ is called *admissible* if the canonical symplectic structure on $T^*\mathcal{M}_{g,n}$ implies Goldman's bracket on the character variety $\mathbb{V}_{g,n}$ under the monodromy map of equation (6.6).

According to discussion of [**4**], two holomorphic sections, S_0 and S_1, of $\mathbb{S}_g$ are equivalent iff there exists a holomorphic function G_{01} on $\mathcal{T}_g$ such that

$$\delta_\mu G_{01} = \int_C \mu(S_0 - S_1) \,, \tag{6.7}$$

where μ is an arbitrary Beltrami differential. The function G_{01} is the generating function of the corresponding symplectomorphizm $Q \to Q+\frac{1}{2}(S_0-S_1)$ of $T^*\mathcal{M}_g$ [**4**]. Clearly, the equivalence of S_0 and S_1 does not depend on the number of punctures n. Since the equivalence of Bergman, Schottky and Bers projective connections follows from the analysis of $n = 0$ case contained in [**4, 21**], we can formulate the following corollary of Theorem 6.1.

COROLLARY 6.1. *If the holomorphic section of $\mathbb{S}_g$ in equation (6.6) is chosen to be either Bergman (corresponding to any Torelli marking), Schottky (corresponding to any choice of generators) or Bers (corresponding to any "base" Riemann surface $\mathcal{C}_0$) projective connections then the canonical symplectic structure on $T^*\mathcal{M}_{g,n}$ induces the Goldman bracket on the character variety $\mathbb{V}^0_{g,n}$ under the monodromy map of equation (6.6).*

The Definition 6.1 allows to formulate our Theorem 5.1 simply as follows:

"*The Bergman projective connection is admissible for any n and g*".

Actually, the set of admissible holomorphic sections of $\mathbb{S}_g$ is rather tiny; one of examples of non-admissible holomorphic section of $\mathbb{S}_g$ is given in Remark 5.1 of [**4**].

Suppose now that S_0 and S_1 are two admissible holomorphic sections of $\mathbb{S}_g$ and write down the same equation in two ways:

$$\varphi'' - \left(\frac{1}{2}S_0 + Q_0\right)\varphi = 0 \tag{6.8}$$

and

$$\varphi'' - \left(\frac{1}{2}S_1 + Q_1\right)\varphi = 0 \,, \tag{6.9}$$

where Q_0 and Q_1 are two meromorphic quadratic differentails with simple poles related by

$$Q_0 - Q_1 = \frac{1}{2}(S_1 - S_0) \,. \tag{6.10}$$

Then the canonical coverings $\widehat{\mathcal{C}}_0$ defined by $v_0^2 = Q_0$ and $\widehat{\mathcal{C}}_1$ defined by $v_1^2 = Q_1$ have different conformal structure. Moreover, the periods (A_i^0, B_i^0) of v_0 and (A_i^1, B_i^1) of v_1 are not related to each other in any simple way (even if S_0 and S_1 are Bergman projective connections corresponding to two different Torelli markings). However, since both S_0 and S_1 are admissible, the Goldman bracket on the character variety of equation (6.8), (6.9) implies the coincidence of the homological symplectic forms defined by Q_0 and Q_1:

$$\sum_{i=1}^{3g-3+n} dA_i^0 \wedge dB_i^0 = \sum_{i=1}^{3g-3+n} dA_i^1 \wedge dB_i^1 \tag{6.11}$$

i.e. each admissible projective connection defines its own set of Darboux coordinates for Goldman's bracket. We don't know how the relation (6.11) can be verified directly, without using the link with Goldman's brackets.

6.2. An analog of Jimbo-Miwa tau-function for the system (1.3). In the theory of isomonodromic deformations of a linear system

$$\frac{d\Psi}{dx} = \sum_{i=1}^{N} \frac{A_i}{x - x_i}\Psi$$

the main role is played by the isomonodromic Jimbo-Miwa tau-function defined by equations

$$\frac{\partial \log \tau_{JM}}{\partial x_i} = \frac{1}{2}\mathrm{res}|_{x_i} \frac{\mathrm{tr}\,(d\Psi\Psi^{-1})^2}{dx} . \tag{6.12}$$

A straightforward analog of the definition (6.12) in the context of equation (1.3):

$$d\Psi = \begin{pmatrix} 0 & v \\ uv & 0 \end{pmatrix} \Psi , \tag{6.13}$$

where

$$uv = -\frac{S_B - S_v}{2v} - v , \tag{6.14}$$

looks as follows:

$$\frac{\partial \log \tau}{\partial \mathcal{P}_{s_i}} := \frac{1}{4\pi i}\int_{s_i^*} \left(\frac{\mathrm{tr}\,(d\Psi\Psi^{-1})^2}{v} + 2v \right) = \frac{1}{2\pi i}\int_{s_i^*} (u+1)v , \tag{6.15}$$

where the cycles $\{s_i, s_i^*\}$ form a symplectic basis in the odd part H_- of $H_1(\widehat{\mathcal{C}})$; the term $2v$ is added to provide compatibility of equations (6.15). We notice that the addition of the analogous term proportional to dx to the formula (6.12) does not change the right-hand side since all residues of dx vanish. In the case of (6.15) the addition of this counter-term is crucial.

Using the form (6.14) of the potential u the equations (6.15) can be equivalently rewritten as follows in terms of the Bergman projective connection:

$$\frac{\partial \log \tau}{\partial \mathcal{P}_{s_i}} = -\frac{1}{4\pi i}\int_{s_i^*} \frac{S_B - S_v}{v} . \tag{6.16}$$

The tau-function defined by (6.16) equals the 1/8th power of the Bergman tau-function τ_B on the moduli space of quadratic differentials studied in [**28**]:

$$\tau = \tau_B^{1/8} . \tag{6.17}$$

Alternatively the definition (6.16) of the function τ can be written as follows:

$$d \log \tau_B = \frac{1}{6\pi i}\sum_{i=1}^{3g-3+n} (\tilde{A}_i dB_i - \tilde{B}_i dA_i) , \tag{6.18}$$

where

$$\tilde{A}_i = \int_{a_i^-} (u+1)v \qquad \tilde{B}_i = \int_{b_i^-} (u+1)v \tag{6.19}$$

and (A_i, B_i) are, as before, the periods of the differential v over cycles a_i^- and b_i^-.

The function τ is holomorphic and non-singular if zeros and poles of Q don't merge. It satisfies the following two main properties:

- Transformation under change of Torelli marking (used to define the reference projective connection S_B) by an $Sp(2g,\mathbb{Z})$ matrix $\sigma = \begin{pmatrix} A & B \\ C & D \end{pmatrix}$:

$$\tau^\sigma = \kappa \det{}^6(C\Omega + D)\,\tau \,, \tag{6.20}$$

where κ is an 8th root of unity
- Homogeneity property:

$$\tau(\epsilon Q, \mathcal{C}) = \epsilon^{\frac{2}{9}(5g-5+n)}\tau(Q,\mathcal{C}) \,, \tag{6.21}$$

which implies in particular the following relation between periods (A_i, B_i) of the differential v and the periods $(\tilde{A}_i, \tilde{B}_i)$ of the differential $(u+1)v$:

$$\sum_{i=1}^{3g-3+n} (\tilde{A}_i B_i - \tilde{B}_i A_i) = \frac{2\pi i}{3}(5g-5+n) \,. \tag{6.22}$$

Variational formulas on spaces of quadratic differentials which are the main analytical tool for computation of Poisson brackets in this paper and in [**4**] were developed in [**23**, **28**] with the goal of studying the properties of the Bergman tau-function τ_B; the function τ_B plays an important role in various areas - from the theory of isomonodromic deformations, random matrices and the theory of Frobenius manifolds to spectral theory.

We expect the function τ defined in present context to be closely related to the so-called "Yang-Yang" function F of [**29**] which is the generating function between two systems of Darboux coordinates - the first is the natural system of Darboux coordinates on $T^*\mathcal{M}_{g,n}$, and the second is the system of complex Fenchel-Nielsen coordinates on the character variety. In particular, as it was noticed in [**4**], the function F transforms under a change of Torelli marking of $\mathcal{C}$ in the same was as $\log\tau$. A complete elucidation of the link between τ and F remains a challenging problem.

7. Riemann sphere with four marked points

Assume that the poles of Q are placed to 0, 1, ∞ and t in coordinate x. Then the differential Q can be written as

$$Q = \frac{\mu}{x(x-1)(x-t)}(dx)^2 \,. \tag{7.1}$$

As we see, the moduli space $\mathcal{Q}^0_{0,4}$ has dimension 2; $\log t$ can be chosen as coordinate on $\mathcal{M}_{0,4}$. To find the corresponding momentum we choose in (5.12) $y_1 = 0$, $y_2 = \infty$, $y_3 = 1$ and $y_4 = t$; then the corresponding quadratic differential Q_1 (5.14) equals

$$Q_1 = \frac{1}{4\pi i}\frac{1-t}{x(x-1)(x-t)}(dt)^2 \,.$$

The comparison with (7.1) shows that the momentum p_1 equals to

$$p_1 = 4\pi i \frac{\mu}{1-t} \,, \tag{7.2}$$

and, therefore, the canonical Poisson bracket $\{p_1, q_1\} = 1$ implies

$$\{\mu, t\} = \frac{t(1-t)}{4\pi i} \,.$$

The canonical bidifferential in terms of coordinate x is written as

$$B(x,y) = \frac{dxdy}{(x-y)^2} \ . \tag{7.3}$$

Therefore, in the x-coordinate the Bergman projective connection S_B is identically vanishing and the equation (1.1) takes the form

$$\varphi'' + \frac{\mu}{x(x-1)(x-t)}\varphi = 0 \ , \tag{7.4}$$

which is a special case of Heun's equation [**18**].

The canonical covering $\widehat{\mathcal{C}}$ is the elliptic curve

$$w^2 = x(x-1)(x-t) \tag{7.5}$$

and the differential v on $\widehat{\mathcal{C}}$ is given by

$$v = \frac{\sqrt{\mu}}{\sqrt{x(x-1)(x-t)}}dx \ . \tag{7.6}$$

Choosing two basic cycles a and b on $\widehat{\mathcal{C}}$ (say, a goes around 0 and 1 and b goes around 1 and t) we get homological coordinates as elliptic integrals of first kind:

$$A = \sqrt{\mu}\int_a \frac{dx}{\sqrt{x(x-1)(x-t)}} \ , \qquad B = \sqrt{\mu}\int_b \frac{dx}{\sqrt{x(x-1)(x-t)}} \ . \tag{7.7}$$

The equation (7.4) has 4 monodromy matrices M_0, M_1, M_∞ and M_t; all of their eigenvalues equal to 1. We can choose the generators $\kappa_0, \kappa_1, \kappa_\infty$ and κ_t of the fundamental group of the 4-punctured sphere such that the matrices M_i satisfy the relation

$$M_0 M_1 M_\infty M_t = I \ .$$

The equations (4.12) give derivatives of monodromy matrices with respect to A and B; in present case $h(x,t) = (x-t)^{-4}$.

It is also easy to derive directly the formulas for derivatives of any monodromy matrix M_γ corresponding to a basepoint x_0 with respect to the natural coordinates (t,μ) on the space of differentials (7.1). Namely, introduce a Wronskian matrix Φ of two linearly independent solutions $\varphi_{1,2}$ of (7.4) satisfying the initial condition $\Phi(x_0) = I$. This matrix solves the equation

$$\frac{\mathrm{d}\Phi}{\mathrm{d}x} = \begin{pmatrix} 0 & 1 \\ -Q & 0 \end{pmatrix} \Phi \ . \tag{7.8}$$

Introduce cuts connecting x_0 with the singular points $0, 1, \infty, t$. Differentiating (7.8) with respect to t and μ and using the variation of parameters formula for non-homogeneous equations for Φ_t and Φ_μ we get the following result:

$$\frac{\mathrm{d}\Phi(x)}{\mathrm{d}t} = -\mu\Phi(x)\int_{x_0}^{x} \frac{\Lambda(y)dy}{y(y-1)(y-t)^2} \tag{7.9}$$

and

$$\frac{\mathrm{d}\Phi(x)}{\mathrm{d}\mu} = -\Phi(x)\int_{x_0}^{x} \frac{\Lambda(y)dy}{y(y-1)(y-t)} \ , \tag{7.10}$$

where the integration contour is not supposed to intersect the branch cuts $\Lambda_(y) = \Phi^{-1}(y)\sigma_-\Phi(y)$.

Closing the integration contour along some loop γ we get derivatives of any monodromy matrix M_γ with respect to t and μ:

$$\frac{\mathrm{d}M_\gamma}{\mathrm{d}t} = -\mu M_\gamma \int_\gamma \frac{\Lambda(y)dy}{y(y-1)(y-t)^2} \tag{7.11}$$

and

$$\frac{\mathrm{d}M_\gamma}{\mathrm{d}\mu} = -M_\gamma \int_\gamma \frac{\Lambda(y)dy}{y(y-1)(y-t)}. \tag{7.12}$$

Acknowledgments

The author thanks Marco Bertola and Chaya Norton for numerous useful conversations.

References

[1] William Abikoff, *The real analytic theory of Teichmüller space*, Lecture Notes in Mathematics, vol. 820, Springer, Berlin, 1980. MR590044

[2] Olivier Babelon, Denis Bernard, and Michel Talon, *Introduction to classical integrable systems*, Cambridge Monographs on Mathematical Physics, Cambridge University Press, Cambridge, 2003. MR1995460

[3] Lipman Bers, *Simultaneous uniformization*, Bull. Amer. Math. Soc. **66** (1960), 94–97, DOI 10.1090/S0002-9904-1960-10413-2. MR0111834

[4] Marco Bertola, Dmitry Korotkin, and Chaya Norton, *Symplectic geometry of the moduli space of projective structures in homological coordinates*, Invent. Math. **210** (2017), no. 3, 759–814, DOI 10.1007/s00222-017-0739-z. MR3735629

[5] A. Douady and J. Hubbard, *On the density of Strebel differentials*, Invent. Math. **30** (1975), no. 2, 175–179, DOI 10.1007/BF01425507. MR0396936

[6] John D. Fay, *Theta functions on Riemann surfaces*, Lecture Notes in Mathematics, Vol. 352, Springer-Verlag, Berlin-New York, 1973. MR0335789

[7] Clifford J. Earle, *On variation of projective structures*, Riemann surfaces and related topics: Proceedings of the 1978 Stony Brook Conference (State Univ. New York, Stony Brook, N.Y., 1978), Ann. of Math. Stud., vol. 97, Princeton Univ. Press, Princeton, N.J., 1981, pp. 87–99. MR624807

[8] Alex Eskin, Maxim Kontsevich, and Anton Zorich, *Sum of Lyapunov exponents of the Hodge bundle with respect to the Teichmüller geodesic flow*, Publ. Math. Inst. Hautes Études Sci. **120** (2014), 207–333, DOI 10.1007/s10240-013-0060-3. MR3270590

[9] D. Gaiotto and J. Teschner, *Irregular singularities in Liouville theory and Argyres-Douglas type gauge theories*, J. High Energy Phys. **12** (2012), 050, front matter + 76. MR3045283

[10] Daniel Gallo, Michael Kapovich, and Albert Marden, *The monodromy groups of Schwarzian equations on closed Riemann surfaces*, Ann. of Math. (2) **151** (2000), no. 2, 625–704, DOI 10.2307/121044. MR1765706

[11] William M. Goldman, *The symplectic nature of fundamental groups of surfaces*, Adv. in Math. **54** (1984), no. 2, 200–225, DOI 10.1016/0001-8708(84)90040-9. MR762512

[12] William M. Goldman, *Invariant functions on Lie groups and Hamiltonian flows of surface group representations*, Invent. Math. **85** (1986), no. 2, 263–302, DOI 10.1007/BF01389091. MR846929

[13] William M. Goldman, *Trace coordinates on Fricke spaces of some simple hyperbolic surfaces*, Handbook of Teichmüller theory. Vol. II, IRMA Lect. Math. Theor. Phys., vol. 13, Eur. Math. Soc., Zürich, 2009, pp. 611–684, DOI 10.4171/055-1/16. MR2497777

[14] R. C. Gunning, *Special coordinate coverings of Riemann surfaces*, Math. Ann. **170** (1967), 67–86, DOI 10.1007/BF01362287. MR0207978

[15] R. C. Gunning, *Affine and projective structures on Riemann surfaces*, Riemann surfaces and related topics: Proceedings of the 1978 Stony Brook Conference (State Univ. New York, Stony Brook, N.Y., 1978), Ann. of Math. Stud., vol. 97, Princeton Univ. Press, Princeton, N.J., 1981, pp. 225–244. MR624816

[16] N. S. Hawley and M. Schiffer, *Half-order differentials on Riemann surfaces*, Acta Math. **115** (1966), 199–236, DOI 10.1007/BF02392208. MR0190326

[17] Dennis A. Hejhal, *Monodromy groups and linearly polymorphic functions*, Acta Math. **135** (1975), no. 1, 1–55, DOI 10.1007/BF02392015. MR0463429

[18] Karl Heun, *Zur Theorie der Riemann'schen Functionen zweiter Ordnung mit vier Verzweigungspunkten* (German), Math. Ann. **33** (1888), no. 2, 161–179, DOI 10.1007/BF01443849. MR1510535

[19] John H. Hubbard, *The monodromy of projective structures*, Riemann surfaces and related topics: Proceedings of the 1978 Stony Brook Conference (State Univ. New York, Stony Brook, N.Y., 1978), Ann. of Math. Stud., vol. 97, Princeton Univ. Press, Princeton, N.J., 1981, pp. 257–275. MR624819

[20] Michael Kapovich, *On monodromy of complex projective structures*, Invent. Math. **119** (1995), no. 2, 243–265, DOI 10.1007/BF01245182. MR1312500

[21] Shingo Kawai, *The symplectic nature of the space of projective connections on Riemann surfaces*, Math. Ann. **305** (1996), no. 1, 161–182, DOI 10.1007/BF01444216. MR1386110

[22] Felix Klein, *Zur Theorie der Abel'schen Functionen* (German), Math. Ann. **36** (1890), no. 1, 1–83, DOI 10.1007/BF01199432. MR1510611

[23] Aleksey Kokotov and Dmitry Korotkin, *Tau-functions on spaces of abelian differentials and higher genus generalizations of Ray-Singer formula*, J. Differential Geom. **82** (2009), no. 1, 35–100. MR2504770

[24] Kokotov, A., Korotkin, D., "Tau-functions on spaces of Abelian and quadratic differentials and determinants of Laplacians in Strebel metrics of finite volume" math.SP/0405042, preprint No. 46 of Max-Planck Institut for Mathematics in Science, Leipzig (2004)

[25] A. Kokotov, D. Korotkin, and P. Zograf, *Isomonodromic tau function on the space of admissible covers*, Adv. Math. **227** (2011), no. 1, 586–600, DOI 10.1016/j.aim.2011.02.005. MR2782203

[26] D. Korotkin and P. Zograf, *Tau function and moduli of differentials*, Math. Res. Lett. **18** (2011), no. 3, 447–458, DOI 10.4310/MRL.2011.v18.n3.a6. MR2802579

[27] Olivia Dumitrescu and Motohico Mulase, *Quantum curves for Hitchin fibrations and the Eynard-Orantin theory*, Lett. Math. Phys. **104** (2014), no. 6, 635–671, DOI 10.1007/s11005-014-0679-0. MR3200933

[28] D. Korotkin and P. Zograf, *Tau function and the Prym class*, Algebraic and geometric aspects of integrable systems and random matrices, Contemp. Math., vol. 593, Amer. Math. Soc., Providence, RI, 2013, pp. 241–261, DOI 10.1090/conm/593/11874. MR3087959

[29] Nekrasov, N., Rosly, A., Shatashvili, S., *Darboux coordinates, Yang-Yang functional, and gauge theory*, arXiv: 1103.3919 [hep-th]

[30] Seiberg, N., Witten, E., *Monopole Condensation and Confinement In N=2 Supersymmetric Yang-Mills Theory*, Nucl. Phys. **B426** 19-52 (1994), hep-th/9407087

[31] Hidenori Sonoda, *Functional determinants on punctured Riemann surfaces and their application to string theory*, Nuclear Phys. B **294** (1987), no. 1, 157–192, DOI 10.1016/0550-3213(87)90578-5. MR909430

[32] A. N. Tjurin, *Periods of quadratic differentials* (Russian), Uspekhi Mat. Nauk **33** (1978), no. 6(204), 149–195, 272. MR526014

[33] P. G. Zograf and L. A. Takhtadzhyan, *On the uniformization of Riemann surfaces and on the Weil-Petersson metric on the Teichmüller and Schottky spaces* (Russian), Mat. Sb. (N.S.) **132(174)** (1987), no. 3, 304–321, 444, DOI 10.1070/SM1988v060n02ABEH003170; English transl., Math. USSR-Sb. **60** (1988), no. 2, 297–313. MR889594

Department of Mathematics and Statistics, Concordia University, 1455 de Maisonneuve W., Montréal, Québec, Canada H3G 1M8

Email address: dmitry.korotkin@concordia.ca

Proceedings of Symposia in Pure Mathematics
Volume **100**, 2018
https://doi.org/10.1090/pspum/100/01764

On ELSV-type formulae, Hurwitz numbers and topological recursion

D. Lewanski

ABSTRACT. We present several recent developments on ELSV-type formulae and topological recursion concerning Chiodo classes and several kind of Hurwitz numbers.

1. Introduction

ELSV-type formulae relate connected Hurwitz numbers to the intersection theory of certain classes on the moduli spaces of curves. Both Hurwitz theory and the theory of moduli spaces of curves benefit from them, since ELSV formulae provide a bridge through which calculations and results can be transferred from one to the other. The original ELSV formula [**19**] relates simple connected Hurwitz numbers and Hodge integrals. It plays a central role in many of the alternative proofs of Witten's conjecture that appeared after the first proof by Kontsevich (for more details see [**31**]).

1.0.1. *Examples of ELSV-type formulae:* The simple connected Hurwitz numbers $h^{\circ}_{g;\vec{\mu}}$ enumerate connected Hurwitz coverings of the 2-sphere of degree $|\vec{\mu}|$ and genus g, where the partition $\vec{\mu}$ determines the ramification profile over zero, and all other ramifications are simple, i.e. specified by a transposition of two sheets of the covering $(a_i\, b_i) \in \mathfrak{S}_{|\vec{\mu}|}$. By the Riemann Hurwitz formula, the number of these simple ramifications is $b = 2g-2+l(\vec{\mu})+|\vec{\mu}|$ (for an introduction on Hurwitz theory see, e.g., [**5**]).

The celebrated ELSV formula [**19**] expresses these numbers in terms of the intersection theory of moduli spaces of curves:

$$\frac{h^{\circ}_{g;\vec{\mu}}}{b!} = \prod_{i=1}^{\ell(\vec{\mu})} \frac{\mu_i^{\mu_i}}{\mu_i!} \int_{\overline{\mathcal{M}}_{g,\ell(\vec{\mu})}} \left(\sum_{l=0}^{g} (-1)^l \lambda_l \right) \prod_{j=1}^{\ell(\vec{\mu})} \sum_{d_j=0} \mu_j^{d_j} \psi_j^{d_j}$$

A different Hurwitz problem arises from Harish-Chandra-Itzykson-Zuber matrix model and leads to the study of the simple monotone connected Hurwitz numbers $h^{\circ;\leq}_{g,\vec{\mu}}$ (see [**25**]), in which an extra *monotone* condition is imposed on the coverings — if $(a_i\, b_i)_{i=1,\ldots,b}$ are written such that $a_i < b_i$, the condition requires that $b_i \geq b_{i+1}$

2010 *Mathematics Subject Classification.* Primary 14N10, 14N35; Secondary 30F30, 14E20.

for all $i = 1, \ldots, b-1$. For these Hurwitz numbers an ELSV-type formula is also known [**1**, **12**]:

$$h^{\circ,\leq}_{g,\vec{\mu}} = \prod_{i=1}^{\ell(\vec{\mu})} \binom{2\mu_i}{\mu_i} \int_{\overline{\mathcal{M}}_{g,\ell(\vec{\mu})}} \exp\left(\sum_{l=1} A_l \kappa_l\right) \prod_{j=1}^{\ell(\vec{\mu})} \sum_{d_j=0} \psi_j^{d_j} \frac{(2(\mu_j + d_j) - 1)!!}{(2\mu_j - 1)!!},$$

where the generating series for the coefficients A_i of the kappa classes reads

$$\exp\left(-\sum_{l=1}^{\infty} A_l U^l\right) = \sum_{k=0}^{\infty} (2k+1)!! U^k.$$

More examples exist in the literature, and two of them will be subject of this paper: the ELSV formula due to Johnson, Pandharipande and Tseng (JPT) [**28**] and a conjectural formula due to Zvonkine [**38**], see also [**35**] (for the precise formulae see the table at the end of this paper).

1.0.2. *Structure of ELSV-type formulae:* All four examples above express numbers enumerating connected Hurwitz covers of a certain kind, depending on a genus parameter and a partition, in terms of some *non - polynomial* factor in the entries of a partition $\vec{\mu}$ (in the formulae above $\prod \frac{\mu_i^{\mu_i}}{\mu_i!}$ and $\binom{2\mu_i}{\mu_i}$) and an integral over moduli spaces of curves of a certain class intersected with ψ class. This integral is clearly a *polynomial* of degree $3g - 3 + \ell(\vec{\mu})$ in the μ_i. Conceptually:

$$h^{\circ,condition}_{g,\vec{\mu}} = \mathrm{NonPoly}(\vec{\mu}) \int_{\overline{\mathcal{M}}_{g,\ell(\vec{\mu})}} (\mathrm{Class}) \prod_{j=1}^{\ell(\vec{\mu})} \sum_{d_j=0} c_{d_j}(\mu_j)\psi_j^{d_j} \tag{1.1}$$

where $c_{d_j}(\mu_j)$ is a polynomial of degree d_j in μ_j.

1.0.3. *Topological recursion and ELSV formulae.* The Chekhov, Eynard and Orantin (CEO) topological recursion procedure associates to a spectral curve $\mathcal{S} = (\Sigma, x(z), y(z), B(z_1, z_2))$ (where Σ is a complex curve, for a precise definition see, e.g., [**6**, **23**]) a collection of symmetric correlation differentials $\omega_{g,n}$ defined on the product of the curve $\Sigma^{\times n}$ through a universal recursion on $2g-2+n$. The expansion of these differentials near particular points can unveil interesting invariants, or solutions to enumerative geometric problems. More explicitly, the expansion of the correlation differentials can always be written in the form

$$\omega^{\mathcal{S}}_{g,n}(x_1, \ldots, x_n) = \mathrm{d}_1 \otimes \cdots \otimes \mathrm{d}_n \sum_{\mu_1, \ldots, \mu_n} N^{\mathcal{S}}_{g,\vec{\mu}} \prod_{i=1}^{n} \tilde{x}(x_i)^{\mu_i} \tag{1.2}$$

for some coefficients $N^{\mathcal{S}}_{g,\vec{\mu}}$ and some function $\tilde{x}$ of x. We say that certain numbers $N_{g,\vec{\mu}}$ satisfy the topological recursion if there exists a spectral curve $\mathcal{S}$ and a suitable $\tilde{x}$ such that each coefficient $N^{\mathcal{S}}_{g,\vec{\mu}}$ computed via that particular expansion is equal to $N_{g,\vec{\mu}}$ for each $g \geq 0$ and each partition $\vec{\mu}$. For instance, both the simple Hurwitz and the monotone Hurwitz numbers satisfy the topological recursion (see [**2**, **3**, **11**, **22**, **32**]), and their spectral curves are given respectively by

$$\left(\mathbb{CP}^1, x = -z + \log(z), z, \frac{\mathrm{d}z_1 \mathrm{d}z_2}{(z_1 - z_2)^2}\right), \left(\mathbb{CP}^1, x = \frac{z-1}{z^2}, -z, \frac{\mathrm{d}z_1 \mathrm{d}z_2}{(z_1 - z_2)^2}\right). \tag{1.3}$$

In the simple Hurwitz case $\tilde{x} = e^x$, whereas in the monotone case $\tilde{x} = x$. The expansion of the correlation differentials is hence performed near $e^x = 0$ and near $x = 0$, respectively. On the other hand it was proved that (see Theorem 3.1), roughly speaking, the expansion of the correlation differentials *has the same structure as the right-hand side of ELSV-type formulae* described above, depending on same ingredients that are functions of the spectral curve. More precisely, Theorem 3.1 implies that the coefficients $N^{\mathcal{S}}_{g,\vec{\mu}}$ can be written in the form

$$N^{\mathcal{S}}_{g,\vec{\mu}} = \mathrm{NonPoly}^{\mathcal{S}}(\vec{\mu}) \int_{\overline{\mathcal{M}}_{g,\ell(\vec{\mu})}} (\mathrm{Class}^{\mathcal{S}}) \prod_{j=1}^{\ell(\vec{\mu})} \sum_{d_j=0} c^{\mathcal{S}}_{d_j}(\mu_j)\psi_j^{d_j}, \tag{1.4}$$

for some cohomological class $Class^{\mathcal{S}} \in H^*(\overline{\mathcal{M}}_{g,\ell(\vec{\mu})})$, some function $\mathrm{NonPoly}^{\mathcal{S}}$ (what we call the non-polynomial part), and some polynomials $c^{\mathcal{S}}_{d_j}(v) \in \mathbb{C}[v]$. At this point comes the key observation: if one can compute these ingredients explicitly for a given spectral curve $\mathcal{S}$ and a given $\tilde{x}$, one can formulate equivalence statements in the following sense.

We say that a *TR-ELSV equivalence statement* for a certain Hurwitz problem $h^{\circ,condition}_{g,\vec{\mu}}$ and a spectral curve $\mathcal{S}$ with a chosen expansion function $\tilde{x}$, is a statement that asserts the equivalence between the following two propositions:

i) The numbers $h^{\circ,condition}_{g,\vec{\mu}}$ satisfy the topological recursion with input spectral curve $\mathcal{S}$ and expansion near $\tilde{x} = 0$ (i.e. $h^{\circ,condition}_{g,\vec{\mu}} = N^{\mathcal{S}}_{g,\vec{\mu}}$)

ii) The Hurwitz numbers $h^{\circ,condition}_{g,\vec{\mu}}$ are equal to

$$\mathrm{NonPoly}^{\mathcal{S}}(\vec{\mu}) \int_{\overline{\mathcal{M}}_{g,\ell(\vec{\mu})}} (\mathrm{Class}^{\mathcal{S}}) \prod_{j=1}^{\ell(\vec{\mu})} \sum_{d_j=0} c^{\mathcal{S}}_{d_j}(\mu_j)\psi_j^{d_j},$$

where all the ingredients appearing in the formula are explicit.

Clearly, it makes sense to formulate an equivalence statement for certain Hurwitz numbers $h^{\circ,condition}_{g,\vec{\mu}}$ and a certain spectral curve $\mathcal{S}$ if for at least one of the two propositions there exists some evidence or a proof. Thus, if one establishes *i)* independently of *ii)*, then *ii)* follows immediately (and vice versa), and hence this equivalence relationship has received much attention in the literature. For example, for the case of simple Hurwitz numbers and the first curve in Equation (1.3), proposition *i)* was conjectured to hold by Bouchard and Mariño [**3**], while *ii)* is the original ELSV formula. Proposition *i)* was proved in [**2, 22, 32**], the equivalence statement was proved in [**21**], see also [**35**]. The equivalence statement immediately provides a new proof of *i)* from *ii)*. The proofs [**22, 32**] of *i)* though, make use of a polynomiality property that is extracted from ELSV, hence the equivalence cannot be used in the other direction without falling into a circular argument, unless this polynomiality property can be proved without using ELSV formula. This was done in [**14**], see also [**15, 29**], and thus *ii)* follows from *i)* by the equivalence statement.

In the case of r-spin Hurwitz numbers, proposition *i)* is known as r-Bouchard-Mariño conjecture [**3**], proposition *ii)* is the r-ELSV formula conjectured by Zvonkine [**38**], see also [**35**]. The equivalence of *i)* and *ii)* was established in [**35**], but since neither of the two have been proven, both remain conjectural.

1.0.4. *Givental theory.* The class $\mathrm{Class}^{\mathcal{S}}$ describes a semi-simple cohomological field theory (CohFT), possibly with non-flat unit. Semi-simple CohFTs with unit are classified by Givental - Teleman [**24, 37**] by the action of a Givental R-matrix on a topological field theory. On the other hand, Givental theory has been identified with the CEO topological recursion, in the sense of [**17**]. This identification makes

Givental theory a powerful and explicit tool to compute the ingredients above, and hence to prove equivalence statements.

1.0.5. *Main result.* In this section we consider a particular case of what has been explained above. The main result in [**30**] is the computation of the ingredients $\mathrm{NonPoly}^{\mathcal{S}_{r,s}}$, $\mathrm{Class}^{\mathcal{S}_{r,s}}$, and $c_{d_j}^{\mathcal{S}_{r,s}}(\mu_j)$ for the specific spectral curve

$$\mathcal{S}_{r,s} := \left(\mathbb{C}\mathrm{P}^1, x(z) = -z^r + \log z, y(z) = z^s, B(z,z') = \frac{\mathrm{d}z\,\mathrm{d}z'}{(z-z')^2}\right).$$

The class $\mathrm{Class}^{\mathcal{S}_{r,s}}$ turns out to coincide with Chiodo classes [**8**]. Its specialisation for $r = s = 1$ is used in the ELSV-type formulae for simple Hurwitz numbers, the case $s = 1$ involves the r-spin Hurwitz numbers, while the case $r = s$ is used in the ELSV formula for r-orbifold Hurwitz numbers, derived by Johnson, Pandharipande and Tseng (JPT) [**28**]. The equivalence statement for general r and s is derived, and it specialises to the equivalence statements already known for simple and r-spin Hurwitz numbers. For r-orbifold Hurwitz instead, the statement is new (see Corollary 4.5): this gives a new proof of the topological recursion for r-orbifold Hurwitz numbers from JPT. On the other hand, the topological recursion was already proved in [**4, 13**], but extracting some polynomiality property from JPT formula itself. This polynomiality property was then proved in [**15**], see also [**29**], providing together with the equivalence statement a new proof of JPT formula.

1.0.6. *Plan and organisation of the paper.* In this note we give a short exposition of the results obtained in [**30**]. On the one hand, the technical proofs are omitted and we refer to the original paper for them; conversely, some of the details that are omitted in the paper are here worked out.

In Section 2 we recall the definition of Chiodo classes, we express them in terms of stable graphs and prove that they are given by the action of a particular Givental R-matrix.

In Section 3 we review the key steps of DOSS Identification (here DOSS is the acronym of the authors of [**17**]) and we show the result of the computation for the spectral curve $\mathcal{S}_{r,s}$. This allows to state the main result (see Theorem 4.1).

In Section 4 we treat the equivalence statement for r-orbifold Hurwitz numbers and Johnson-Pandharipande-Tseng formula.

Acknowledgments. This note is based on a talk given by the author at the 2016 AMS von Neumann Symposium *Topological Recursion and its Influence in Analysis, Geometry, and Topology*, July 4 - 8, 2016, in Charlotte, NC. I would like to thank the organisers of the Symposium and the AMS for this opportunity.

I would moreover like to thank N. Do, P. Dunin-Barkowski, M. Karev, R. Kramer, A. Popolitov, S. Shadrin and D. Zvonkine for interesting discussions, special thanks to S. Shadrin for having introduced me to the topic and for very useful remarks. The author is supported by the Netherlands Organization for Scientific Research.

2. Chiodo classes

In this section we recall Chiodo classes and we show their Givental decomposition, for more details we refer the reader to [**8**–**10, 26, 35**]. For $2g - 2 + n > 0$, consider a nonsingular curve with distinct markings $[C, p_1, \dots, p_n] \in \mathcal{M}_{g,n}$, and let

$\omega_{\log} = \omega_C(\sum p_i)$ be its log canonical bundle. Let $r \geq 1$, $1 \leq a_1, \ldots, a_n \leq r$ and $0 \leq s \leq r$ be integers satisfying the condition

$$(2g-2+n)s - \sum_{i=1}^{n} a_i = 0 \mod r \tag{2.1}$$

This condition guarantees the existence of rth tensor roots L of the line bundle $\omega_{\log}^{\otimes s}(-\sum a_i p_i)$ on C. For the moduli space of such rth tensor roots, a natural compactification $\overline{\mathcal{M}}_{g;a_1,\ldots,a_n}^{r,s}$ was constructed in [**7**, **27**]. Let $\pi : \mathcal{C}_{g;a_1,\ldots,a_n}^{r,s} \to \overline{\mathcal{M}}_{g;a_1,\ldots,a_n}^{r,s}$ be the universal curve, let $\mathcal{L} \to \mathcal{C}_{g;a_1,\ldots,a_n}^{r,s}$ be the universal rth root, and let $\epsilon : \overline{\mathcal{M}}_{g;a_1,\ldots,a_n}^{r,s} \to \overline{\mathcal{M}}_{g,n}$ be the forgetful map (in order for ϵ to be unramified in the orbifold sense, the target $\overline{\mathcal{M}}_{g,n}$ is changed into the moduli space r-stable curves, meaning that for each stable curve there is an extra $\mathbb{Z}_r$ stabilizer at each node, see [**7**]). Recall the generating series for the Bernoulli polynomials

$$\sum_{l=0} B_l(x)\frac{t^l}{l!} = \frac{te^{xt}}{e^t-1}$$

where the usual Bernoulli numbers are $B_l(0) = B_l$. We are interested in the Chiodo classes [**8**]

$$\begin{aligned} \mathrm{C}_{g,n}(r,s;a_1,\ldots,a_n) &:= \epsilon_* c\big(-R^*\pi_*\mathcal{L}\big) \\ &= \epsilon_* \exp\left(\sum_{l=1}^{\infty}(-1)^l(l-1)!\mathrm{ch}_l(r,s;a_1,\ldots,a_n)\right) \in H^{even}(\overline{\mathcal{M}}_{g,n}), \end{aligned}$$

where Chiodo formula for the Chern characters reads

$$\begin{aligned} \mathrm{ch}_l(r,s;a_1,\ldots,a_n) &= \frac{B_{l+1}(\frac{s}{r})}{(l+1)!}\kappa_l - \sum_{i=1}^{n}\frac{B_{l+1}(\frac{a_i}{r})}{(l+1)!}\psi_i^l \\ &+\frac{r}{2}\sum_{a=1}^{r}\frac{B_{l+1}(\frac{a}{r})}{(l+1)!}(j_a)_*\frac{(\psi')^l+(-1)^{l-1}(\psi'')^l}{\psi'+\psi''}. \end{aligned} \tag{2.2}$$

Here j_a is the boundary map that represents the boundary divisor with remainder a at one of the two half edges, and ψ', ψ'' are the ψ-classes at the two branches of the node.

For the specialisation $r = s = 1$, and moreover $a_i = 1$, for $i = 1, \ldots, n$, the map ϵ is the identity map and we recover Mumford formula [**33**] for the total Chern class of the dual of the Hodge bundle $c(\Lambda_g^\vee)$:

$$\begin{aligned} C_{g,n}(1,1;1,\ldots,1) &= \exp\Bigg(-\Bigg[\sum_{l=1}^{\infty}\frac{B_{l+1}}{l(l+1)}\kappa_l - \sum_{i=1}^{n}\frac{B_{l+1}}{l(l+1)}\psi_i^l \\ &\quad + \frac{1}{2}\frac{B_{l+1}}{l(l+1)}j_*\frac{(\psi')^l+(-1)^{l-1}(\psi'')^l}{\psi'+\psi''}\Bigg]\Bigg) \\ &= c(\Lambda_g^\vee) = 1 - \lambda_1 + \lambda_2 - \cdots + (-1)^g\lambda_g \end{aligned} \tag{2.3}$$

where the identity $B_l(1) = (-1)^l B_l$ is used. The formula in [**33**] is slightly different due to a different Bernoulli number convention and a missprint in the κ term.

2.1. Expression in terms of stable graphs. Let us recall the expression of the Chiodo class in terms of the sum of products of contributions decorating stable graphs, in order to compare it with the Givental action, for more details see [**26**]. The strata of the moduli space of curves correspond to stable graphs

$$\Gamma = (\mathrm{V}, \mathrm{E}, \mathrm{H}, \mathrm{L}, g, n : \mathrm{V} \to \mathbb{Z}_{\geq 0}, v : \mathrm{H} \to \mathrm{V}, \iota : \mathrm{H} \to \mathrm{H})$$

where $\mathrm{V}(\Gamma)$, $\mathrm{E}(\Gamma)$, $\mathrm{H}(\Gamma)$ and $\mathrm{L}(\Gamma)$ respectively denote the sets of vertices, edges, half-edges and leaves of Γ; self-edges are permitted. A half-edge indicates either a leaf or an edge together with a choice of one of the two vertices it is attached to. The function v associates to each half-edge its vertex assignment, while ι is the involution that swaps the two half-edges of the same edge, or leaves the half-edge invariant if it is a leaf. The function $n(v)$ denotes the valence of Γ at v, including both half-edges and legs, and $g(v)$ denotes the genus function. Every vertex v is required to satisfy the stability condition $2g(v) - 2 + n(v) > 0$, and the genus of a stable graph Γ is defined by $g(\Gamma) := \sum_{v \in V} g(v) + h^1(\Gamma)$. Let $\mathrm{Aut}(\Gamma)$ denote the group of automorphisms of the sets V and H which leave the structures L, g, v, and ι invariant. Let $\mathsf{G}_{g,n}$ be the finite set of isomorphism classes of stable graphs of genus g with n legs. Let moreover $\mathsf{W}_{\Gamma,r,s,\vec{a}}$ be the set of *weightings* $w : \mathrm{H}(\Gamma) \to \{0, \dots, r-1\}$ satisfying the following three properties:

(i) The i-th leaf l_i has weight $w(l_i) = a_i \mod r$, for $i \in \{1, \dots, n\}$.
(ii) For any two half-edges h' and h'' corresponding to the same edge, we have $w(h') + w(h'') = 0 \mod r$.
(iii) The condition in Equation (2.1) is satysfied locally on each component: for any vertex v the sum of the weights associated to the half-edges incident to v is $\sum_{v(h)=v} w(h) = s\big(2g(v) - 2 + n(v)\big) \mod r$.

PROPOSITION 2.1. [**26**]. The Chiodo class $\mathrm{C}_{g,n}(r, s; a_1, \dots, a_n) \in R^*(\overline{\mathcal{M}}_{g,n})$ is equal to

$$\sum_{\Gamma \in \mathsf{G}_{g,n}} \sum_{w \in \mathsf{W}_{\Gamma,r,s,\vec{a}}} \frac{r^{|E(\Gamma)| + \sum_{v \in V(\Gamma)} 2g(v) - 1}}{|\operatorname{Aut}(\Gamma)|} \xi_{\Gamma *} \Bigg[\prod_{v \in \mathrm{V}(\Gamma)} e^{-\sum\limits_{l \geq 1} (-1)^{l-1} \frac{B_{l+1}(s/r)}{l(l+1)} \kappa_l(v)} \cdot \tag{2.4}$$

$$\prod_{i=1}^{n} e^{\sum\limits_{l \geq 1} (-1)^{l-1} \frac{B_{l+1}(a_i/r)}{l(l+1)} \psi_{h_i}^l} \cdot \prod_{\substack{e \in \mathrm{E}(\Gamma) \\ e = (h', h'')}} \frac{1 - e^{\sum\limits_{l \geq 1} (-1)^{l-1} \frac{B_{l+1}(w(h)/r)}{l(l+1)} [(\psi_{h'})^l - (-\psi_{h''})^l]}}{\psi_{h'} + \psi_{h''}} \Bigg].$$

where ξ_Γ is the canonical morphism $\xi_\Gamma : \prod_{v \in \mathrm{V}(\Gamma)} \overline{\mathcal{M}}_{g(v),n(v)} \to \overline{\mathcal{M}}_{g,n}$ of the boundary stratum corresponding to Γ.

2.2. Expression in terms of Givental action. In this section we express Chiodo classes in terms of Givental theory.

Fix a vector space V and a symmetric bilinear form η on V. A Givental R-matrix is a $\mathrm{End}(V)$-valued power series

$$R(\zeta) = 1 + \sum_{l=1} R_l \zeta^l = \exp\left(\sum_{l=1} r_l \zeta^l \right), \qquad R_l, r_l \in \mathrm{End}(V) \tag{2.5}$$

satisfying the symplectic condition

$$R(\zeta)R^*(-\zeta) = 1 \in \mathrm{End}(V)$$

where R^* is the adjoint of R with respect to η.

By Givental - Teleman classification [**24, 37**], every semi-simple cohomological field theory (CohFT) with unit is obtained by the action of a Givental R-matrix on a topological field theory. We will show that the action of the R-matrix

$$R^{-1}(\zeta) := \exp\left(-\sum_{l=1}^{\infty} \frac{\mathrm{diag}_{a=1}^{r} B_{l+1}\left(\frac{a}{r}\right)}{l(l+1)}(-\zeta)^l\right) \tag{2.6}$$

defined as power series valued in the endomorphisms for the vector space

$$V = \langle v_1, \ldots, v_r \rangle$$

with

$$\eta(v_a, v_b) = \frac{1}{r}\delta_{a+b \mod r},$$

acting on the topological field theory

$$\alpha_{g,n}^{top}(v_{a_1} \otimes \cdots \otimes v_{a_n}) = r^{2g-1}\delta_{a_1+\cdots+a_n-s(2g-2+n) \mod r}, \tag{2.7}$$

produces the Chiodo classes. Therefore Chiodo classes determine a CohFT with a known Givental decomposition. The action of the Givental R-matrix is defined as sum over stable graphs Γ weighted by $|\,\mathrm{Aut}(\Gamma)|^{-1}$, with contributions on the leaves, on the edges, on special leaves called *dilaton* leaves, and the topological field theory contributes on the verteces. Chiodo classes are already expressed as a sum over stable graphs in Equation (2.4) with a very similar structure. Let us match the Givental contributions one by one:

2.2.1. *Ordinary leaf contributions.* The contribution of the i-th leaf reads

$$\exp\left(-\sum_{l=1} \frac{B_{l+1}(a_i/r)}{l(l+1)}(-\psi_{h_i})^l\right) = \sum_{j=1}^{r} (R^{-1})_{a_i}^{j}(\psi_{h_i})$$

2.2.2. *Dilaton leaf contributions.* Recall that the kappa classes are defined as $\kappa_l = \pi_*(\psi_{n+1}^{l+1})$ under the map that forgets the last marked point $\pi : \overline{\mathcal{M}}_{g,n+1} \to \overline{\mathcal{M}}_{g,n}$. The contributions on the dilaton leaves correspond to the contributions on the vertices in Equation (2.4) before forgetting the corresponding marked point. For the dilaton leaf marked with label $n+i$, for some positive integer i, the contribution reads:

$$\exp\left(-\sum_{l=1} \frac{B_{l+1}(s/r)}{l(l+1)}(-\psi_{n+i})^l(-\psi_{n+i})\right)$$

We check that v_s is the neutral element $\mathbb{1}$ for the quantum product $\bullet$ in flat basis:

$$\begin{aligned} \eta(v_s \bullet v_a, v_b) &= \alpha_{0,3}^{top}(v_s \otimes v_a \otimes v_b) \\ &= r^{-1}\delta_{s+a+b-s \mod r} = r^{-1}\delta_{a+b \mod r} \\ &= \eta(v_a, v_b) \end{aligned} \tag{2.8}$$

Hence the contribution of the dilaton leaf $n+i$ is

$$\psi_{n+i}\Big[\mathrm{Id} - \sum_{j=1}^{r}(R^{-1})^j_{\mathbb{1}}(\psi_{n+i})\Big]$$

2.2.3. *Edge contributions.* The edge contribution in Equation (2.4), multiplied by the factor $(\psi_{h'}+\psi_{h''})$ and after applying the property of Bernoulli numbers $(-1)^{p+1}B_{p+1}\left(\frac{w(h')}{r}\right) = B_{p+1}\left(\frac{r-w(h')}{r}\right)$, reads

$$1 - \exp\left(-\sum_{l=1} \frac{B_{l+1}(\frac{w(h')}{r})}{l(l+1)}(-\psi_{h'})^l\right)\exp\left(-\sum_{p=1}\frac{B_{p+1}(\frac{r-w(h'')}{r})}{p(p+1)}(-\psi_{h'})^p\right).$$

Note that the condition on the weightings $w(h')+w(h'') = 0 \mod r$ can be taken care by the scalar product η. Hence we can write the Givental contribution on the edges as

$$\sum_{j_1,j_2} \frac{\eta^{j_1,j_2} - (R^{-1})^{j_1}_{w(h')}(\psi_{h'})\eta^{w(h'),w(h'')}(R^{-1})^{j_2}_{w(h'')}(\psi_{h''})}{\psi_{h'}+\psi_{h''}}$$

2.2.4. *Weightings.* Out of the three conditions on the weightings, condition (i) becomes $w(l_i) = a_i$, condition (ii) on the edges is taken care by the bilinear form η, condition (iii) can be substituted by the topological field theory condition.

2.2.5. *Powers of* r. Every stable graph contributes with r to the power

$$|E(\Gamma)| + \sum_{v\in V(\Gamma)} 2g(v) - 1.$$

Indeed, the topological field theory in the vertex v contributes with r to the power $2g(v)-1$, and the inverse of η contributes with r^1 for each edge of Γ.

2.2.6. *The expression of Givental action.* Let us indicate with

$$\{l_1,\dots,l_n,l_{n+1},\dots,l_{n+k}\} = L(\Gamma)$$

the set of legs, corresponding to marked points of the curves in $\overline{\mathcal{M}}_{g,n+k}$, and let

$$\xi^{(k)}_\Gamma : \prod_{v\in V(\Gamma)} \overline{\mathcal{M}}_{g(v),n(v)} \to \overline{\mathcal{M}}_{g,n}$$

be the canonical morphism of the boundary stratum corresponding to Γ that forgets the last k marked points. Let us consider functions $w^\vee : H(\Gamma) \to \mathbb{Z}_{\geq 0}$ without *any* further condition. We use here the notation $w^\vee$, instead of w, to remark that the weighting $w^\vee$ decorates the half-edges *after* the application of the endomorphisms

R_l^{-1}. Collecting the contributions and the considerations above, we have:

$$C_{g,n}(r,s;a_1,\dots,a_n) = \sum_{k=0} \sum_{\substack{\Gamma\in G_{g,n+k}\\ w^\vee : H(\Gamma)\to\mathbb{Z}_{\geq 0}}} \frac{1}{|\operatorname{Aut}(\Gamma)|}\left(\xi_\Gamma^{(k)}\right)_* \Bigg[$$

$$\prod_{v\in V(\Gamma)} \alpha^{top}_{g(v),n(v)}\Bigg(\bigotimes_{\substack{h\in H(\Gamma):\\ v(h)=v}} v_{w^\vee(h)}\Bigg)$$

$$\prod_{i=1}^{n}(R^{-1})_{a_i}^{w^\vee(l_i)}(\psi_i)$$

$$\prod_{i=1}^{k}\psi_{n+i}\Big[\operatorname{Id}-(R^{-1})_{1\!\!1}^{w^\vee(l_{n+i})}(\psi_{n+i})\Big]$$

$$\prod_{e=(h',h'')\in V(\Gamma)} \frac{\eta^{w^\vee(h'),w^\vee(h'')} - \sum_{k_1,k_2}(R^{-1})_{k_1}^{w^\vee(h')}(\psi_{h'})\eta^{k_1,k_2}(R^{-1})_{k_2}^{w^\vee(h'')}(\psi_{h''})}{\psi_{h'}+\psi_{h''}}\Bigg]$$

The expression above is equivalent to $\left(R.\alpha^{top}\right)_{g,n}(v_{a_1}\otimes\cdots\otimes v_{a_n})$, i.e. the Givental action of the matrix R on α^{top} correspondent of genus g and n marked points, evaluated in the element $v_{a_1}\otimes\cdots\otimes v_{a_n}$, (see [**17, 24, 34**]).

Consider then $\mathrm{C}_{g,n}(r,s;a_1,\dots,a_n)$ as the evaluation of a map

$$\mathrm{C}_{g,n}(r,s)\colon V^{\otimes n}\to H^{even}(\overline{\mathcal{M}}_{g,n}), \tag{2.9}$$

where $V=\langle v_1,\dots,v_r\rangle$, and

$$\mathrm{C}_{g,n}(r,s)\colon v_{a_1}\otimes\cdots\otimes v_{a_n}\mapsto \mathrm{C}_{g,n}(r,s;a_1,\dots,a_n). \tag{2.10}$$

The previous calculation shows

PROPOSITION 2.2 ([**30**]). *For $0\leq s\leq r$ the collection of maps $\{\mathrm{C}_{g,n}(r,s)\}$ defined by the Chiodo classes form a semi-simple cohomological field theory with flat unit, obtained by the action of the Givental matrix R on the topological field theory $\alpha^{top}_{g,n}$:*

$$\left(R.\alpha^{top}\right)_{g,n} = C_{g,n}(r,s).$$

3. From the spectral curve to the Givental R-matrix

In this section we recall the main result of [**17, 20**], which expresses the correlation differentials $\omega_{g,n}$ of the CEO topological recursion in terms of integral over moduli spaces of curves (Theorem 3.1). We then recall the DOSS identification [**17**] between topological recursion and Givental theory and apply it to a particular spectral curve (Definition 3.4). The result is an explicit expression for the coefficients of $\omega_{g,n}$ (Theorem 3.6).

3.1. Local topological recursion. The local version of the CEO topological recursion takes as input the following set of data $\mathcal{S}=(\Sigma,x,y,B)$:

I). A local spectral curve $\Sigma=\sqcup_{i=1}^r U_i$, given by the disjoint union of open disks with the center points p_i, $i=1,\dots,r$.

II). A holomorphic function $x\colon \Sigma \to \mathbb{C}$ such that the zeros of its differential dx are $p_1, \dots, p_r$. We will assume the zeroes of dx to be simple.

III). A holomorphic function $y\colon \Sigma \to \mathbb{C}$ which does not vanish on the zeroes of dx.

IV). A symmetric bidifferential B defined on $\Sigma \times \Sigma$ with a double pole on the diagonal with residue 1.

The output of the topological recursion procedure consists of a collection of symmetric differentials $\omega_{g,n}^{\mathcal{S}}$ defined on the topological product of the curve $\Sigma^{\times n}$. The data $I-IV$ determine the first differentials $\omega_{0,1}$ and $\omega_{0,2}$. The other correlators are obtained recursively through a universal recursion on $2g-2+n$, where, roughly speaking, $\omega_{g,n}$ is obtained by summing over all possible pair of pants decompositions of a surface of genus g and n marked points. These correlation differentials take the following form.

THEOREM 3.1. [**17**, **20**] *The correlation differentials $\omega_{g,n}^{\mathcal{S}}$ produced via the topological recursion procedure from the spectral curve $\mathcal{S} = (\Sigma, x, y, B)$ can be written as*

$$C^{2g-2+n} \sum_{\substack{i_1,\dots,i_n \\ d_1,\dots,d_n}} \int_{\overline{\mathcal{M}}_{g,n}} \left(\mathrm{Class}^{\mathcal{S}}\right)_{g,n} (e_{i_1} \otimes \cdots \otimes e_{i_n}) \prod_{j=1}^{n} \psi_j^{d_j} \mathrm{d}\left(\left(-\frac{1}{w_j}\frac{\mathrm{d}}{\mathrm{d}w_j}\right)^{d_j} \xi_{i_j}^{\mathcal{S}}\right),$$

where the ingredients of the formula are explained below.

REMARK 3.2. The $\mathrm{Class}^{\mathcal{S}}$ obtained in this way always defines a semi-simple CohFT, possibly with a non-flat unit. In this paper we will restrict the attention to CohFT with flat unit and we will write

$$\left(\mathrm{Class}^{\mathcal{S}}\right)_{g,n} = \left(R^{\mathcal{S}}.\alpha^{\mathcal{S},top}\right)_{g,n} \tag{3.1}$$

to indicate its Givental decomposition (see Section 2.2; for CohFT with non-flat unit see [**34**] or [**30**], Section 2.3).

Let us describe the ingredients in the formula above in terms of the data of the spectral curve, following [**17**]. The only difference with the usual representation is that we incorporate a torus action on cohomological field theories, fixing a point $(C, C_1, \dots, C_r) \in (\mathbb{C}^*)^{r+1}$. It is important to remark that this formula does not depend on these parameters, though all its ingredients do (although we omit their dependence to keep the notation clean). The quantity in equation (3.1) is therefore well-defined for each choice of $(C, C_1, \dots, C_r)$.

i). The local coordinates w_i on U_i, $i = 1, \dots, r$, are chosen such that $w_i(p_i) = 0$ and $x = (C_i w_i)^2 + x_i$

ii). The underlying topological field theory is given in idempotent basis by

$$\eta(e_i, e_j) = \delta_{ij},$$

$$\alpha_{g,n}^{\mathcal{S},top}(e_{i_1} \otimes \cdots \otimes e_{i_n}) = \delta_{i_1 \dots i_n} \left(-2C_i^2 C \frac{\mathrm{d}y}{\mathrm{d}w_i}(0)\right)^{-2g+2-n}.$$

iii). The Givental matrix $R^{\mathcal{S}}(\zeta)$ is given by

$$-\frac{1}{\zeta}(R^{\mathcal{S}})^{-1}(\zeta)_i^j = \frac{1}{\sqrt{2\pi\zeta}} \int_{-\infty}^{\infty} \left.\frac{B(w_i, w_j)}{\mathrm{d}w_i}\right|_{w_i=0} \cdot e^{-\frac{w_j^2}{2\zeta}}.$$

iv). The auxiliary functions $\xi_i^{\mathcal{S}} : \Sigma \to \mathbb{C}$ are given by

$$\xi_i^{\mathcal{S}}(x) := \int^x \frac{B(w_i, w)}{dw_i}\bigg|_{w_i=0}$$

v). *DOSS Test* (see [**16**], Section 4): The following condition for the function y is necessary and sufficient in order for the unit of the cohomological field theory $R^{\mathcal{S}}.\alpha^{\mathcal{S}}$ to be flat.

$$\frac{2C_i^2 C}{\sqrt{2\pi\zeta}} \int_{-\infty}^{\infty} dy \cdot e^{-\frac{w_i^2}{2\zeta}} = \sum_{k=1}^{r} (R^{-1})_k^i \left(2C_k^2 C \frac{dy}{dw_k}(0) \right)$$

NOTATION 3.3. For clarity, we remark that in this section the notation w is reserved to local coordinates, and is not used for weightings. Moreover C and C_i indicate scalars, whereas Chiodo classes $C_{g,n}$ are always written with two indices.

3.2. The spectral curve $\mathcal{S}_{r,s}$ and its Givental R-matrix. Let us apply the formula in Theorem 3.1 to the specral curve introduced in [**30**].

DEFINITION 3.4. For $0 \le s \le r$, with r and s integer parameters, let the spectral curve $\mathcal{S}_{r,s}$ be defined by the following initial data in terms of a global coordinate z on the Riemann sphere:

$$\mathcal{S}_{r,s} := \left(\mathbb{C}\mathrm{P}^1, x(z) = -z^r + \log z, y(z) = z^s, B(z,z') = \frac{dz\, dz'}{(z-z')^2} \right) \tag{3.2}$$

PROPOSITION 3.5 ([**30**]). The ingredients described above for the spectral curve $\mathcal{S}_{r,s}$ are given by:

i). Choose the constants $C_i = 1/\sqrt{-2r}$ for $i = 1, \dots, r$, and $C = r^{1+s/r}/s$. With this choice the local coordinates w_i on U_i, $i = 1, \dots, r$ satisfy $x = -\frac{w_i^2}{2r} + x(p_i)$.

ii). The underlying topological field theory is given by

$$\eta(v_a, v_b) = \frac{1}{r}\delta_{a+b \mod r}$$

$$\alpha_{g,n}^{\mathcal{S}_{r,s},top}(v_{a_1} \otimes \cdots \otimes v_{a_n}) = r^{2g-1}\delta_{a_1+\cdots+a_n - s(2g-2+n) \mod r}.$$

where the flat coordinates v_a are defined in terms of the idempotents by

$$v_a := \sum_{i=0}^{r-1} \frac{J^{ai}}{r} e_i, \qquad a = 1, \dots, r.$$

iii). The Givental matrix $R^{\mathcal{S}_{r,s}}(\zeta)$ is given by

$$R^{\mathcal{S}_{r,s}}(\zeta) = \exp\left(-\sum_{k=1}^{\infty} \frac{\mathrm{diag}_{a=1}^r B_{k+1}\left(\frac{a}{r}\right)}{k(k+1)} \zeta^k \right).$$

iv). The auxiliary functions $\xi_a^{\mathcal{S}_{r,s}} : \Sigma \to \mathbb{C}$ are given by

$$\xi_a^{\mathcal{S}_{r,s}} = r^{\frac{r-a}{r}} \sum_{n=0}^{\infty} \frac{(nr+r-a)^n}{n!} e^{(nr+r-a)x}$$

v). DOSS Test is satisfied.

Let us now put together the ingredients for the correlation differentials $\omega_{g,n}^{\mathcal{S}_{r,s}}$, as in Theorem 3.1. First of all, in the coordinates v_i, the TFT and the Givental R-matrix for $\mathcal{S}_{r,s}$ coincide with the ones resulting in the Chiodo CohFT for the same parameters r and s by Proposition 2.2. Secondly, since $\frac{1}{r}\frac{d}{dx} = -\frac{1}{w_i}\frac{d}{dw_i}$, the computation of the derivatives of the auxiliary functions, divided by the powers of r, read

$$\begin{aligned} \mathrm{d}\left(\left(-\frac{\psi_j}{w_j}\frac{\mathrm{d}}{\mathrm{d}w_j}\right)^{d_j}\xi_{i_j}^{\mathcal{S}_{r,s}}\right) &= \mathrm{d}\left[\left(\frac{\psi_j}{r}\frac{\mathrm{d}}{\mathrm{d}x_j}\right)^{d_j}\sum_{l_j=0}^{\infty}\frac{(l_jr+r-a_j)^n}{l_j!}e^{(l_jr+r-a_j)x_j}\right] \\ &= \mathrm{d}_j\sum_{l_j=0}^{\infty}\frac{(l_jr+r-a_j)^{l_j+d_j}}{l_j!}\left(\frac{\psi_j}{r}\right)^{d_j}e^{(l_jr+r-a_j)x_j} \\ &= \mathrm{d}_j\sum_{\mu_j=1}^{\infty}\frac{\mu_j^{[\mu_j]}}{[\mu_j]!}\left(\frac{\mu_j}{r}\psi_j\right)^{d_j}e^{\mu_jx_j}, \end{aligned}$$

where we write the euclidean division by r as $\mu = [\mu]r + \langle\mu\rangle$, with $\langle\mu\rangle < r$. The powers of s only come from C^{2g-2+n}, and hence they are equal to $-(2g-2+n)$. The remaining powers of r to compute amount to

$$2g-2+n+\frac{(2g-2+n)s+\sum_i(r-a_i)}{r} = 2g-2+n+\frac{(2g-2+n)s+\sum_i\langle\mu_i\rangle}{r},$$

though it is handy to collect an extra $r^{\sum_i[\mu_i]}$ outside the product. This proves:

THEOREM 3.6 ([**30**]). *The correlation differentials* $\omega_{g,n}^{\mathcal{S}_{r,s}}$ *of the spectral curve* (3.2) *are equal to*

$$\mathrm{d}_1\otimes\cdots\otimes\mathrm{d}_n\frac{r^{2g-2+n+b}}{s^{2g-2+n}}\sum_{\mu_1,\ldots,\mu_n=1}^{\infty}\prod_{j=1}^{n}\frac{\left(\frac{\mu_j}{r}\right)^{[\mu_j]}}{[\mu_j]!}\int_{\overline{\mathcal{M}}_{g,n}}\frac{\mathrm{C}_{g,n}\left(r,s;r-\langle\vec{\mu}\rangle\right)}{\prod_{j=1}^{n}(1-\frac{\mu_i}{r}\psi_i)}e^{\sum\mu_jx_j}$$

where $b(r,s) = \left((2g-2+n)s+\sum_{j=1}^{n}\mu_j\right)/r$ *and* $C_{g,n}$, *are Chiodo classes defined in section* 2.

REMARK 3.7. Note that the case $s = 1$ reproduces Theorem 1.7 in [**35**].

Expanding the correlation differentials as

$$\omega_{g,n}^{\mathcal{S}_{r,s}} = \mathrm{d}_1\otimes\cdots\otimes\mathrm{d}_n\sum_{\mu_1,\ldots,\mu_n=1}^{\infty}\frac{N_{g,\vec{\mu}}^{\mathcal{S}_{r,s}}}{b(r,s)!}e^{\sum_{j=1}^{n}\mu_jx_j}, \tag{3.3}$$

we obtain the following corollary.

COROLLARY 3.8 ([**30**]).

$$N_{g,\vec{\mu}}^{\mathcal{S}_{r,s}} = b(r,s)!\frac{r^{b(r,s)+2g-2+n}}{s^{2g-2+n}}\prod_{i=1}^{n}\frac{\left(\frac{\mu_i}{r}\right)^{[\mu_i]}}{[\mu_i]!}\int_{\overline{\mathcal{M}}_{g,n}}\frac{C_{g,n}(r,s;r-\langle\vec{\mu}\rangle)}{\prod_{j=1}^{n}(1-\frac{\mu_j}{r}\psi_j)}. \tag{3.4}$$

4. Equivalence statements: A new proof of the Johnson-Pandharipande-Tseng formula

In this section we consider the case $s = r$ of Theorem 3.6. In this case, the correlation differentials of this spectral curve are known to give the r-orbifold Hurwitz numbers $h^{\circ,[r]}_{g;\vec{\mu}}$ [**4, 13, 15**], which enumerate connected Hurwitz coverings of the 2-sphere of degree $|\vec{\mu}|$ and genus g, where the partition $\vec{\mu}$ determines the ramification profile over zero, the ramification over infinity is of cycle type $(r, r, \ldots, r)$ and all other ramifications are simple.

Corollary 3.8 for $s = r$ specialises to

COROLLARY 4.1 ([**30**]).

$$N^{\mathcal{S}_{r,r}}_{g,\vec{\mu}} = b(r,r)!r^{b(r,r)} \prod_{i=1}^{n} \frac{\left(\frac{\mu_i}{r}\right)^{[\mu_i]}}{[\mu_i]!} \int_{\overline{\mathcal{M}}_{g,n}} \frac{C_{g,n}(r,r;r-\langle\vec{\mu}\rangle)}{\prod_{j=1}^{n}(1-\frac{\mu_j}{r}\psi_j)}. \tag{4.1}$$

Plugging the $r-$orbifold Hurwitz numbers and the curve $\mathcal{S}_{r,r}$ into the TR-ELSV equivalence statement of section 1.0.3, we get:

COROLLARY 4.2. The two propositions are equivalent:

i) $h^{\circ,[r]}_{g,\vec{\mu}} = N^{\mathcal{S}_{r,r}}_{g,\vec{\mu}}$

ii) $$h^{\circ,[r]}_{g,\vec{\mu}} = b(r,r)!r^{b(r,r)} \prod_{i=1}^{n} \frac{\left(\frac{\mu_i}{r}\right)^{[\mu_i]}}{[\mu_i]!} \int_{\overline{\mathcal{M}}_{g,n}} \frac{C_{g,n}(r,r;r-\langle\vec{\mu}\rangle)}{\prod_{j=1}^{n}(1-\frac{\mu_j}{r}\psi_j)}.$$

On the other hand, r-orbifold Hurwitz numbers are also known to satisfy the Johnson-Pandharipande-Tseng (JPT) ELSV-type formula [**28**] (specialised here to the case $G = \mathbb{Z}/r\mathbb{Z}$, U equal to the representation that sends 1 to $e^{\frac{2\pi i}{r}}$, and empty γ):

$$h^{\circ,[r]}_{g;\vec{\mu}} = b(r,r)!r^{b(r,r)} \prod_{i=1}^{n} \frac{\left(\frac{\mu_i}{r}\right)^{[\mu_i]}}{[\mu_i]!} \int_{\overline{\mathcal{M}}_{g,n}} \frac{p_* \sum_{i\geq 0}(-1)^i\lambda_i}{\prod_{j=1}^{n}(1-\frac{\mu_j}{r}\psi_j)}, \tag{4.2}$$

REMARK 4.3. Note that the powers of r are here slightly rearranged to easily match the equation for the correlation differentials.

The class $p_* \sum_{i\geq 0}(-r)^i\lambda_i$ is described in [**28**] via admissible covers, while Chiodo's classes rely on the moduli space of r-th tensor roots. These two approaches are in fact equivalent:

PROPOSITION 4.4 ([**30**]). $p_* \sum_{i\geq 0}(-r)^i\lambda_i = \mathrm{C}_{g,n}(r,r;r-\langle\vec{\mu}\rangle)$.

By Proposition 4.4 and Formula (4.2), we can re-state Corollary 4.2 as:

COROLLARY 4.5 ([**30**]). The two statements are equivalent:

i) The r-orbifold Hurwitz numbers satisfy the topological recursion from the spectral curve $\mathcal{S}_{r,r}$.
ii) The JPT formula holds.

Since the JPT formula is proved independently from the topological recursion, Corollary 4.5 provides a new proof of the topological recursion for the numbers $h^{\circ,[r]}_{g;\vec{\mu}}$.

On the other hand, both [**4**, **13**] derive the topological recursion for $h^{\circ,[r]}_{g;\vec{\mu}}$ combining the cut-and-join equation with a polynomiality property which is extracted from the JPT formula itself. Hence one cannot conclude that Corollary 4.5 provides a new proof of the the JPT formula, unless this polynomiality property is derived independently from JPT. This polynomiality property is proved in [**15**] with no use of JPT, see also [**29**]. Therefore these results together, and via Corollary 4.5, provide a new proof of Johnson-Pandharipande-Tseng formula.

Case	Topological recursion		ELSV-type formula
$s=1$ $r=1$	*The standard Hurwitz numbers* $h^{\circ}_{g,\vec{\mu}}$ *are generated by* $\left.\begin{array}{r}-z+\log z = x(z)\\ z = y(z)\end{array}\right\}$	$\Longleftrightarrow$	*The ELSV formula* $\frac{h^{\circ}_{g;\vec{\mu}}}{b!} = \prod_{i=1}^{n} \frac{\mu_i^{\mu_i}}{\mu_i!} \int_{\overline{\mathcal{M}}_{g,n}} \frac{\sum_{i=0}^{g}(-1)^i\lambda_i}{\prod_{j=1}^{n}(1-\mu_j\psi_j)}$, *holds, where:* $b = 2g-2+n+\vert\vec{\mu}\vert$.
$s=r$	*The r-orbifold Hurwitz numbers* $h^{\circ,r}_{g,\vec{\mu}}$ *are generated by* $\left.\begin{array}{r}-z^r+\log z = x(z)\\ z^r = y(z)\end{array}\right\}$	$\Longleftrightarrow$	*The Johnson-Pandharipande-Tseng formula* $\frac{h^{\circ,[r]}_{g;\vec{\mu}}}{b!} = r^b \prod_i \frac{\left(\frac{\mu_i}{r}\right)^{[\mu_i]}}{[\mu_i]!} \int_{\overline{\mathcal{M}}_{g,n}} \frac{p_* \sum_{i\geq 0}^{g}(-1)^i\lambda_i}{\prod_{j=1}^{n}(1-\frac{\mu_j}{r}\psi_j)}$, *holds, where:* $b = 2g-2+n+\vert\vec{\mu}\vert/r$.
$s=1$	*The r-spin Hurwitz numbers* $h^{\circ,r-spin}_{g,\vec{\mu}}$ *are generated by* $\left.\begin{array}{r}-z^r+\log z = x(z)\\ z = y(z)\end{array}\right\}$	$\Longleftrightarrow$	*The r-spin ELSV formula* $\frac{h^{\circ,r-spin}_{g;\vec{\mu}}}{b!} = r^{b-\chi} \prod_i \frac{\left(\frac{\mu_i}{r}\right)^{[\mu_i]}}{[\mu_i]!} \int_{\overline{\mathcal{M}}_{g,n}} \frac{C_{g,n}(r,1,r-\langle\vec{\mu}\rangle)}{\prod_{j=1}^{n}(1-\frac{\mu_j}{r}\psi_j)}$, *holds, where:* $b = (2g-2+n+\vert\vec{\mu}\vert)/r$.

REMARK 4.6. In the table all the spectral curves are of genus zero, i.e. $\Sigma = \mathbb{CP}^1$, and hence with the standard kernel $B(z,z') = \frac{\mathrm{d}z\,\mathrm{d}z'}{(z-z')^2}$.

References

[1] A. Alexandrov, D. Lewanski, and S. Shadrin, *Ramifications of Hurwitz theory, KP integrability and quantum curves*, J. High Energy Phys. **5** (2016), 124, front matter+30. MR3521843

[2] Gaëtan Borot, Bertrand Eynard, Motohico Mulase, and Brad Safnuk, *A matrix model for simple Hurwitz numbers, and topological recursion*, J. Geom. Phys. **61** (2011), no. 2, 522–540, DOI 10.1016/j.geomphys.2010.10.017. MR2746135

[3] Vincent Bouchard and Marcos Mariño, *Hurwitz numbers, matrix models and enumerative geometry*, From Hodge theory to integrability and TQFT tt*-geometry, Proc. Sympos. Pure Math., vol. 78, Amer. Math. Soc., Providence, RI, 2008, pp. 263–283, DOI 10.1090/pspum/078/2483754. MR2483754

[4] Vincent Bouchard, Daniel Hernández Serrano, Xiaojun Liu, and Motohico Mulase, *Mirror symmetry for orbifold Hurwitz numbers*, J. Differential Geom. **98** (2014), no. 3, 375–423. MR3263522

[5] Renzo Cavalieri and Eric Miles, *Riemann surfaces and algebraic curves*, London Mathematical Society Student Texts, vol. 87, Cambridge University Press, Cambridge, 2016. A first course in Hurwitz theory. MR3585685

[6] Leonid Chekhov and Bertrand Eynard, *Matrix eigenvalue model: Feynman graph technique for all genera*, J. High Energy Phys. **12** (2006), 026, 29, DOI 10.1088/1126-6708/2006/12/026. MR2276715

[7] Alessandro Chiodo, *Stable twisted curves and their r-spin structures* (English, with English and French summaries), Ann. Inst. Fourier (Grenoble) **58** (2008), no. 5, 1635–1689. MR2445829

[8] Alessandro Chiodo, *Towards an enumerative geometry of the moduli space of twisted curves and rth roots*, Compos. Math. **144** (2008), no. 6, 1461–1496, DOI 10.1112/S0010437X08003709. MR2474317

[9] Alessandro Chiodo and Yongbin Ruan, *Landau-Ginzburg/Calabi-Yau correspondence for quintic three-folds via symplectic transformations*, Invent. Math. **182** (2010), no. 1, 117–165, DOI 10.1007/s00222-010-0260-0. MR2672282

[10] A. Chiodo and D. Zvonkine, *Twisted r-spin potential and Givental's quantization*, Adv. Theor. Math. Phys. **13** (2009), no. 5, 1335–1369. MR2672465

[11] Norman Do, Alastair Dyer, and Daniel V. Mathews, *Topological recursion and a quantum curve for monotone Hurwitz numbers*, J. Geom. Phys. **120** (2017), 19–36, DOI 10.1016/j.geomphys.2017.05.014. MR3712146

[12] N. Do and M. Karev, *Monotone orbifold Hurwitz numbers*, Zap. Nauchn. Sem. S.-Peterburg. Otdel. Mat. Inst. Steklov. (POMI) **446** (2016), no. Kombinatorika i Teoriya Grafov. V, 40–69; English transl., J. Math. Sci. (N.Y.) **226** (2017), no. 5, 568–587. MR3520422

[13] Norman Do, Oliver Leigh, and Paul Norbury, *Orbifold Hurwitz numbers and Eynard-Orantin invariants*, Math. Res. Lett. **23** (2016), no. 5, 1281–1327, DOI 10.4310/MRL.2016.v23.n5.a3. MR3601067

[14] P. Dunin-Barkowski, M. Kazarian, N. Orantin, S. Shadrin, and L. Spitz, *Polynomiality of Hurwitz numbers, Bouchard-Mariño conjecture, and a new proof of the ELSV formula*, Adv. Math. **279** (2015), 67–103, DOI 10.1016/j.aim.2015.03.016. MR3345179

[15] P. Dunin-Barkowski, D. Lewanski, A. Popolitov, and S. Shadrin, *Polynomiality of orbifold Hurwitz numbers, spectral curve, and a new proof of the Johnson-Pandharipande-Tseng formula*, J. Lond. Math. Soc. (2) **92** (2015), no. 3, 547–565, DOI 10.1112/jlms/jdv047. MR3431649

[16] P. Dunin-Barkowski, P. Norbury, N. Orantin, A. Popolitov, S. Shadrin, Dubrovin's superpotential as a global spectral curve, arXiv:1509.06954.

[17] P. Dunin-Barkowski, N. Orantin, S. Shadrin, and L. Spitz, *Identification of the Givental formula with the spectral curve topological recursion procedure*, Comm. Math. Phys. **328** (2014), no. 2, 669–700, DOI 10.1007/s00220-014-1887-2. MR3199996

[18] Petr Dunin-Barkowski, Sergey Shadrin, and Loek Spitz, *Givental graphs and inversion symmetry*, Lett. Math. Phys. **103** (2013), no. 5, 533–557, DOI 10.1007/s11005-013-0606-9. MR3041758

[19] Torsten Ekedahl, Sergei Lando, Michael Shapiro, and Alek Vainshtein, *Hurwitz numbers and intersections on moduli spaces of curves*, Invent. Math. **146** (2001), no. 2, 297–327, DOI 10.1007/s002220100164. MR1864018

[20] B. Eynard, *Invariants of spectral curves and intersection theory of moduli spaces of complex curves*, Commun. Number Theory Phys. **8** (2014), no. 3, 541–588, DOI 10.4310/CNTP.2014.v8.n3.a4. MR3282995

[21] B. Eynard, Intersection numbers of spectral curves, arXiv:1104.0176.

[22] Bertrand Eynard, Motohico Mulase, and Bradley Safnuk, *The Laplace transform of the cut-and-join equation and the Bouchard-Mariño conjecture on Hurwitz numbers*, Publ. Res. Inst. Math. Sci. **47** (2011), no. 2, 629–670, DOI 10.2977/PRIMS/47. MR2849645

[23] B. Eynard and N. Orantin, *Invariants of algebraic curves and topological expansion*, Commun. Number Theory Phys. **1** (2007), no. 2, 347–452, DOI 10.4310/CNTP.2007.v1.n2.a4. MR2346575

[24] Alexander B. Givental, *Gromov-Witten invariants and quantization of quadratic Hamiltonians* (English, with English and Russian summaries), Mosc. Math. J. **1** (2001), no. 4, 551–568, 645. Dedicated to the memory of I. G. Petrovskii on the occasion of his 100th anniversary. MR1901075

[25] I. P. Goulden, Mathieu Guay-Paquet, and Jonathan Novak, *Monotone Hurwitz numbers and the HCIZ integral* (English, with English and French summaries), Ann. Math. Blaise Pascal **21** (2014), no. 1, 71–89. MR3248222

[26] F. Janda, R. Pandharipande, A. Pixton, and D. Zvonkine, *Double ramification cycles on the moduli spaces of curves*, Publ. Math. Inst. Hautes Études Sci. **125** (2017), 221–266, DOI 10.1007/s10240-017-0088-x. MR3668650

[27] Tyler J. Jarvis, *Geometry of the moduli of higher spin curves*, Internat. J. Math. **11** (2000), no. 5, 637–663, DOI 10.1142/S0129167X00000325. MR1780734

[28] P. Johnson, R. Pandharipande, and H.-H. Tseng, *Abelian Hurwitz-Hodge integrals*, Michigan Math. J. **60** (2011), no. 1, 171–198, DOI 10.1307/mmj/1301586310. MR2785870

[29] R. Kramer, D. Lewanski, S. Shadrin, Quasi-polynomiality of monotone orbifold Hurwitz numbers and Grothendieck's dessins d'enfants, arXiv:1610.08376.

[30] Danilo Lewanski, Alexandr Popolitov, Sergey Shadrin, and Dimitri Zvonkine, *Chiodo formulas for the r-th roots and topological recursion*, Lett. Math. Phys. **107** (2017), no. 5, 901–919, DOI 10.1007/s11005-016-0928-5. MR3633029

[31] Chiu-Chu Melissa Liu, *Lectures on the ELSV formula*, Transformation groups and moduli spaces of curves, Adv. Lect. Math. (ALM), vol. 16, Int. Press, Somerville, MA, 2011, pp. 195–216. MR2883688

[32] Motohico Mulase and Naizhen Zhang, *Polynomial recursion formula for linear Hodge integrals*, Commun. Number Theory Phys. **4** (2010), no. 2, 267–293, DOI 10.4310/CNTP.2010.v4.n2.a1. MR2725053

[33] David Mumford, *Towards an enumerative geometry of the moduli space of curves*, Arithmetic and geometry, Vol. II, Progr. Math., vol. 36, Birkhäuser Boston, Boston, MA, 1983, pp. 271–328. MR717614

[34] Rahul Pandharipande, Aaron Pixton, and Dimitri Zvonkine, *Relations on $\overline{\mathcal{M}}_{g,n}$ via 3-spin structures*, J. Amer. Math. Soc. **28** (2015), no. 1, 279–309, DOI 10.1090/S0894-0347-2014-00808-0. MR3264769

[35] S. Shadrin, L. Spitz, and D. Zvonkine, *Equivalence of ELSV and Bouchard-Mariño conjectures for r-spin Hurwitz numbers*, Math. Ann. **361** (2015), no. 3-4, 611–645, DOI 10.1007/s00208-014-1082-y. MR3319543

[36] S. Shadrin, L. Spitz, and D. Zvonkine, *On double Hurwitz numbers with completed cycles*, J. Lond. Math. Soc. (2) **86** (2012), no. 2, 407–432, DOI 10.1112/jlms/jds010. MR2980918

[37] Constantin Teleman, *The structure of 2D semi-simple field theories*, Invent. Math. **188** (2012), no. 3, 525–588, DOI 10.1007/s00222-011-0352-5. MR2917177

[38] D. Zvonkine, A preliminary text on the r-ELSV formula, Preprint 2006.

Korteweg-de Vries Institute for Mathematics, University of Amsterdam, Postbus 94248, 1090 GE Amsterdam, The Netherlands

Email address: ilgrillodani@mpim-bonn.mpg.de

Proceedings of Symposia in Pure Mathematics
Volume **100**, 2018
https://doi.org/10.1090/pspum/100/01769

Quantum curves for simple Hurwitz numbers of an arbitrary base curve

Xiaojun Liu, Motohico Mulase, and Adam Sorkin

Abstract. Various generating functions of simple Hurwitz numbers of the projective line are known to satisfy many properties. They include a heat equation, a cut-and-join recursion, an infinite-order differential equation called a quantum curve equation, and a Schrödinger like partial differential equation. In this paper we generalize these properties to simple Hurwitz numbers with an arbitrary base curve. For projective line case, the equivalency between the cut-and-join recursion and the Chekhov-Eynard-Orantin topological recursion has been proved. However, the relation between these two recursions are not discussed for the simple Hurwitz number we considered here.

1. Introduction and the main results

The purpose of this paper is to determine functional properties of various generating functions of simple Hurwitz numbers with an arbitrary fixed base curve B. We derive a partial differential equation for the Laplace transform of these Hurwitz numbers. The equation is completely analogous to the result of [**28**] for the usual simple Hurwitz numbers based on $\mathbb{P}^1$. We also obtain an infinite-order differential equation, or a *quantum curve*, for the case of an arbitrary base curve, that is parallel to the various Hurwitz problems studied in [**3**, **25**, **26**] with the base curve $\mathbb{P}^1$ and its orbifold version $\mathbb{P}^1[a]$.

Our main motivation is to examine whether the topological recursion of Chekhov-Eynard-Orantin [**14**, **15**] and the existence of a quantum curve of [**1**, **7**–**9**, **19**] hold for the enumeration problem of simple Hurwitz numbers over an arbitrary curve B. Consider a holomorphic map $f : C \longrightarrow B$ of a non-singular algebraic curve C onto a fixed base curve B. We choose a general point $0 \in B$ and fix it once for all. The quantity we are interested in this paper is the base B Hurwitz number $H^B_{g,n}(\mu_1, \ldots, \mu_n)$, which counts the automorphism weighted number of the topological types of holomorphic maps f of a genus g domain curve with n labeled

2010 *Mathematics Subject Classification.* Primary 14H15, 14N35, 05C30, 11P21; Secondary 81T30.

The first author received the China Scholarship Council grant CSC-2010811063, which allowed him to conduct research at the Department of Mathematics, University of California, Davis. He is also supported by the National Science Foundation of China grants No.11201477, 11171175, 11471182 and the Chinese Universities Scientific Fund No.2011JS041.

The research of the second author is supported by NSF grant DMS-1104734.

The third author received support from the U.S. Government.

preimages of a fixed point of B with multiplicity $(\mu_1, \dots, \mu_n) \in \mathbb{Z}_+^n$, such that f is simply ramified apart from at these n points. Define its discrete Laplace transform

$$
\begin{aligned}
F_{g,n}^B(x_1, \dots, x_n) &= \sum_{\vec{\mu} \in \mathbb{Z}_+^n} H_{g,n}^B(\mu_1, \dots, \mu_n) \prod_{i=1}^n e^{-w_i \mu_i} \\
&= \sum_{\vec{\mu} \in \mathbb{Z}_+^n} H_{g,n}^B(\mu_1, \dots, \mu_n) \prod_{i=1}^n x_i^{\mu_i}, \quad x_i = e^{-w_i},
\end{aligned}
\tag{1.1}
$$

which we call the *free energy* of type (g, n). Since $\mathbb{P}^1$ does not cover the base curve B of genus $g(B) > 0$,

$$F_{0,n}^B(\mu_1, \dots, \mu_n) = 0$$

for any $n \geq 1$ and any value of $(\mu_1, \dots, \mu_n) \in \mathbb{Z}_+^n$ in this case. Our main result is the following.

THEOREM 1.1. *For $2g - 2 + n > 0$, the free energies $F_{g,n}^B(x_1, \dots, x_n)$ satisfy the following partial differential equation:*

$$
\begin{aligned}
&\left(2g - 2 + n - (1 - \chi(B)) \sum_{i=1}^n x_i \frac{\partial}{\partial x_i}\right) F_{g,n}^B(x_{[n]}) \\
&= \frac{1}{2} \sum_{i \neq j} \frac{x_i x_j}{x_i - x_j} \left(\frac{\partial}{\partial x_i} F_{g,n-1}^B(x_{[\hat{j}]}) - \frac{\partial}{\partial x_j} F_{g,n-1}^B(x_{[\hat{i}]}) \right) \\
&\quad + \frac{1}{2} \sum_{i=1}^n u_1 \frac{\partial}{\partial u_1} u_2 \frac{\partial}{\partial u_2} \Bigg|_{u_1 = u_2 = x_i} \\
&\left[F_{g-1,n+1}^B(u_1, u_2, x_{[\hat{i}]}) + \sum_{\substack{g_1 + g_2 = g \\ I \sqcup J = [\hat{i}]}} F_{g_1,|I|+1}^B(u_1, x_I) F_{g_2,|J|+1}^B(u_2, x_J) \right].
\end{aligned}
\tag{1.2}
$$

Here $[n] = \{1, \dots, n\}$ is an index set, $[\hat{i}] = [n] \setminus \{i\}$, and for any subset $I \subset [n]$, we denote $x_I = (x_i)_{i \in I}$. We denote by $\chi(B) = 2 - 2g(B)$ the Euler characteristic of the base curve B. Note that the complexity $2g - 2 + n$ is reduced by 1 on the right-hand side of the recursion when $g(B) \geq 1$, similar to the Chekhov-Eynard-Orantin integral recursion of [**14, 15**].

Let us define the ***wave function*** *of the base B Hurwitz numbers by*

$$Z^B(x, \hbar) = \exp\left(\sum_{g=1}^\infty \sum_{n=1}^\infty \frac{1}{n!} \hbar^{2g-2+n} F_{g,n}^B(x, x, \dots, x) \right). \tag{1.3}$$

Then it satisfies a ***quantum curve*** *of infinite-order differential equation*

$$\hbar x \frac{d}{dx} \left[1 - \hbar^{1-\chi(B)} x \, e^{\hbar x \frac{d}{dx}} \left(\frac{d}{dx} x \right)^{1-\chi(B)} \right] Z^B(x, \hbar) = 0. \tag{1.4}$$

If we introduce

$$y = \hbar x \frac{d}{dx}$$

and regard it as a commuting variable, then the total symbol of the above operator produces

$$x = y^{\chi(B)-1}e^{-y}. \tag{1.5}$$

We call (1.5) *a* ***Lambert curve****, due to its similarity to the usual Lambert curve* $x = ye^{-y}$*. The wave function also satisfies a* ***Schrödinger*** *like partial differential equation*

$$\left[\frac{\partial}{\partial \hbar} - \frac{1}{2}\left(x\frac{\partial}{\partial x}\right)^2 + \left(\frac{1}{2} - \frac{1-\chi(B)}{\hbar}\right)x\frac{\partial}{\partial x}\right] Z^B(x,\hbar) = 0. \tag{1.6}$$

If we denote the above operators by P *and* Q*, respectively, then they satisfy a commutation relation:*

$$[P,Q] = -\frac{1}{\hbar}\,P. \tag{1.7}$$

This shows that the system of equations (1.4) *and* (1.6) *are compatible. Moreover, the partition function has the following simple expression*

$$Z^B(x,\hbar) = \sum_{m=0}^{\infty}(m!)^{1-\chi(B)}e^{\frac{1}{2}m(m-1)\hbar}\left(\hbar^{1-\chi(B)}x\right)^m. \tag{1.8}$$

REMARK 1.2. It is interesting to note here that the exact same formulas (1.2), (1.5), (1.6), and (1.8) hold for the case of an orbifold base $B = \mathbb{P}^1[a]$, $a > 0$, if we evaluate $\chi(B) = 1 + \frac{1}{a}$, interpret the sum in (1.8) as running over non-negative multiples am of a, and replace $(am)!$ by $(am)! \longmapsto m!a^m$. Since the orbifold case requires a different set of preparations (see [**3, 10, 25**]), we will report the generalization to the case of higher-genus orbifold curve as a base elsewhere.

When the base curve B is an elliptic curve E, a special case of simple Hurwitz numbers exhibits a *quasi-modular* properties. Dijkgraaf [**6**] considered a generating function

$$F_g(q) = \sum_{n=1}^{\infty}\frac{1}{n!}H^E_{g,n}(1,1,\ldots,1)q^n \tag{1.9}$$

for $g > 1$, and

$$F_1(q) = -\frac{1}{24}\log q + \sum_{n=1}^{\infty}\frac{1}{n!}H^E_{1,n}(1,1,\ldots,1)q^n. \tag{1.10}$$

The significant fact here is that $F_g(q)$ for $g > 1$ is a quasi-modular form [**6, 21**]. The relation between the free energies of type (g,n) for all n and $F_g(q)$ is given by

$$F_g(q) = \lim_{\lambda\to 0}\sum_{n=1}^{\infty}\frac{1}{n!}\frac{1}{\lambda^n}F^E_{g,n}(\lambda q,\lambda q,\ldots,\lambda q). \tag{1.11}$$

To motivate the present work, let us recall the corresponding counting problem of simple Hurwitz number $H^{\mathbb{P}^1}_{g,n}(\mu_1,\ldots,\mu_n)$ for a pointed projective line $(\mathbb{P}^1,\infty)$, which has a long history since Hurwitz [**20**]. Its modern interest is due, among other things, to its rich functional properties [**16, 17, 22, 32, 33**], its relation with linear Hodge integrals on $\overline{\mathcal{M}}_{g,n}$ [**12, 18, 30**], and the Bouchard-Mariño conjecture [**5**] and its solutions [**2**, **13**, **28**]. The theorem established in [**13**, **28**] shows that the discrete Laplace transform

$$F^{\mathbb{P}^1}_{g,n}(x_1,\dots,x_n) = \sum_{\vec{\mu}\in\mathbb{Z}^n_+} H^{\mathbb{P}^1}_{g,n}(\mu_1,\dots,\mu_n)\prod_{i=1}^n x_i^{\mu_i} \tag{1.12}$$

satisfies an integral recursion formula that was originally proposed by Chekhov, Eynard and Orantin [**14**, **15**], as conjectured in [**5**]. The input curve for the recursion, the *spectral curve*, is shown to be the original Lambert curve

$$x = ye^{-y}. \tag{1.13}$$

Then in [**26**, **34**] it is discovered that there is a quantum curve of [**1**, **7**–**9**, **19**], which is a generator of the holonomic system that characterizes the wave function of the simple Hurwitz numbers of $\mathbb{P}^1$. The wave function for $B = \mathbb{P}^1$ is

$$Z^{\mathbb{P}^1}(x,\hbar) = \exp\left(\sum_{g=0}^{\infty}\sum_{n=1}^{\infty}\frac{1}{n!}\hbar^{2g-2+n}F^{\mathbb{P}^1}_{g,n}(x,x,\dots,x)\right).$$

It is established in [**26**] that the $\mathbb{P}^1$ wave function has an expression

$$Z^{\mathbb{P}^1}(x,\hbar) = \sum_{m=0}^{\infty}\frac{1}{m!}e^{\frac{1}{2}m(m-1)\hbar}\left(\frac{x}{\hbar}\right)^m, \tag{1.14}$$

and that it satisfies two equations:

$$\left[\hbar x\frac{\partial}{\partial x} - xe^{\hbar x\frac{\partial}{\partial x}}\right]Z^{\mathbb{P}^1}(x,\hbar) = 0, \tag{1.15}$$

$$\left[\frac{\partial}{\partial\hbar} - \frac{1}{2}\left(x\frac{\partial}{\partial x}\right)^2 + \left(\frac{1}{2}+\frac{1}{\hbar}\right)x\frac{\partial}{\partial x}\right]Z^{\mathbb{P}^1}(x,\hbar) = 0. \tag{1.16}$$

If we denote the above operators as P_1 and Q_1, respectively, then they satisfy a commutation relation

$$[P_1, Q_1] = -\frac{1}{\hbar}P_1. \tag{1.17}$$

The semi-classical limit of each of the equations (1.15) and (1.16) recovers the Lambert curve (1.13), as shown in [**26**]. Then from this curve as the spectral curve, and using the functions x and y on it, the Chekhov-Eynard-Orantin integral recursion [**14**] determines the differential forms

$$d_1 d_2\cdots d_n F^{\mathbb{P}^1}_{g,n} \tag{1.18}$$

defined on Σ^n for all (g,n), where Σ is the Lambert curve given by (1.13). It is a simple consequence of the results in [**13**, **26**, **28**] that the differential form (1.18) uniquely recovers the primitive $F^{\mathbb{P}^1}_{g,n}$ as a function in $(y_1,\dots,y_n)\in\Sigma^n$. From this point of view, the generating function $F^{\mathbb{P}^1}_{g,n}(x(y_1),\dots,x(y_n))$ is completely determined by the equation (1.15) or (1.16), where x as a function in y is also given by (1.13).

Our motivating question is, *do the similar properties hold when we consider the simple Hurwitz numbers with an arbitrary curve as the base?* Since the genus 0 base B Hurwitz numbers $H^E_{0,n}(\mu_1,\dots,\mu_n)$ do not exist for a base curve B with $g(B) > 0$, the philosophy of [**11**] does not produce any spectral curve of the Chekhov-Eynard-Orantin integral recursion for this counting problem. Yet we have a topological recursion (1.2) in the form of a partial differential equation. We note the straightforward generalizations (1.4), (1.6), (1.8), and (1.5), that reduce to the $\mathbb{P}^1$ case (1.15), (1.16), (1.14), and (1.13), respectively, when $\chi(B) = 2$.

The paper is organized as follows. In Section 2, we derive the cut-and-join equation of simple Hurwitz numbers with an arbitrary base curve. Then in Section 3, we consider the Laplace transform of the cut-and-join equation, which is exactly (1.2). Using this result we derive the Schrödinger equation (1.6) in Section 4. Section 5 is devoted to explaining the heat equation expression of the cut-and-join equation [**16**, **22**, **33**] and deriving its consequences. In Section 6 we first prove the expansion (1.8), derive the quantum curve (1.4), and then establish the commutator relation (1.7). In the final section we deduce the Lambert curve (1.5) as the semi-classical limit.

2. A cut-and-join equation for simple Hurwitz numbers

The enumeration problem we consider in this paper is the number of topological types of holomorphic maps $f : C \longrightarrow B$ from a nonsingular curve C of genus g to a fixed base curve B, with an arbitrary ramification over one point on B and simple ramification at all other critical points. Let us denote by $0 \in B$ a *general* point arbitrarily chosen and fixed. Such a holomorphic map f is referred to as a *simple Hurwitz cover* of B. We label each inverse image of $0 \in B$ via f, and denote by $(\mu_1, \dots, \mu_n) \in \mathbb{Z}_+^n$ the orders of f at each inverse image of 0. The base *B Hurwitz number* $H_{g,n}^B(\mu_1, \dots .\mu_n)$ counts the automorphism weighted number of the topological types of simple Hurwitz covers.

The degree of the map f is given by

$$d = |\mu| = \sum_{i=1}^{n} \mu_i. \tag{2.1}$$

Here use the notation $|\mu|$, borrowing from the convention in the theory of symmetric functions. The Riemann-Hurwitz formula tells us that there are

$$r = r(g, \mu) = 2g - 2 + n - d\big(2g(B) - 1\big) \tag{2.2}$$

simple ramification points. Instead of counting the *topological types* of f, we can fix r simple branch points on B other than $0 \in B$ and count the automorphism weighted holomorphic maps f. We note that (2.1) and (2.2) imply a condition for the degree and the genus for a base curve with $g(B) \geq 1$:

$$n\big(2g(B) - 1\big) \leq d\big(2g(B) - 1\big) \leq 2g - 2 + n, \tag{2.3}$$

$$n\big(g(B) - 1\big) \leq g - 1. \tag{2.4}$$

In particular, $H_{g,n}(\vec{\mu}) = 0$ for any $n \geq 1$ if $g < g(B)$. This poses a sharp contrast to the Gromov-Witten theory of curves [**31**].

The map $f : C \longrightarrow B$ of degree d is determined by a monodromy representation

$$\rho \in \mathrm{Hom}\big(\pi_i(B \setminus \{0, 1, \dots, r\}), S_d\big)$$

of the fundamental group of an $(r + 1)$-punctured curve into the symmetric group of d letters. Here $\{1, \dots, r\} \subset B$ is the label of r simple branched points on B. Let γ_j, $j = 0, \dots, r$,be based simple loops around each point $j \in B$. Then a simple Hurwitz cover is constructed by assigning each γ_j for $j \geq 1$ to a transposition and

γ_0 to a product of n disjoint cycles of length $\mu_1, \cdots \mu_n$, subject to the commutator relation

$$\rho(\gamma_1)\cdots\rho(\gamma_r)\rho(\gamma_0) = [\alpha_1, \beta_1]\cdots[\alpha_{g(B)}, \beta_{g(B)}],$$

where $\alpha_i, \beta_i \in S_d$ are some elements in the permutation group.

The *cut-and-join equation* [**16**, **17**, **20**, **32**] is the result of an analysis of what happens when we multiply a transposition $\rho(\gamma_r)$ to a product of disjoint cycles $\rho(\gamma_0)$. Since the fundamental group of the base curve does not make any effect on this multiplication, the exact same formula for $H^B_{g,n}(\vec{\mu})$ holds, as for the simple Hurwitz numbers for $\mathbb{P}^1$. The only difference is the number r of simple ramification points.

PROPOSITION 2.1 (The cut-and-join equation). *For all $g \geq 0$, $n \geq 1$, and $(\mu_1, \ldots, \mu_n) \in \mathbb{Z}^n_+$ subject to*

$$2g - 2 + n - \big(2g(B) - 1\big)\sum_{i=1}^{n} \mu_i > 0, \tag{2.5}$$

the simple Hurwitz numbers $H^B_{g,n}(\mu_1, \ldots, \mu_n)$ of degree d satisfy the following equation:

(2.6)

$$\big(2g - 2 + n - \big(2g(B) - 1\big)d\big)\, H^B_{g,n}(\mu_{[n]}) = \frac{1}{2}\sum_{i\neq j}(\mu_i + \mu_j)\, H^B_{g,n-1}\Big(\mu_i + \mu_j, \mu_{[\hat{i},\hat{j}]}\Big)$$

$$+\frac{1}{2}\sum_{i=1}^{n}\sum_{\alpha+\beta=\mu_i}\alpha\beta\left(H^B_{g-1,n+1}(\alpha, \beta, \mu_{[\hat{i}]}) + \sum_{\substack{g_1+g_2=g\\ I\sqcup J=[\hat{i}]}} H^B_{g_1,|I|+1}(\alpha, \mu_I)\, H^B_{g_2,|J|+1}(\beta, \mu_J)\right).$$

Here and throughout the paper we use the following notational convention. $[n] = \{1, \ldots, n\}$ is an index set, and $[\hat{i}] = [n]\backslash\{i\}$, etc. For a subset $I \subset [n]$, $\mu_I = (\mu_i)_{i\in I}$.

REMARK 2.2. Unlike the case of $\mathbb{P}^1$, the cut-and-join equation does not determine all values of simple Hurwitz numbers. When the degree d takes its maximum value $\frac{2g-2+n}{2g(B)-1}$, the equation gives a trivial equality $0 = 0$.

The genus 1 simple Hurwitz numbers based on an elliptic curve $B = E$ are easy to calculate. From (2.3) we have $n = d$ and $r = 0$ when $g = g(B) = 1$. Therefore, the covering is unramified. If we allow disconnected domain, then the total number is equal to the number of partitions

$$p(d) = \big|\mathrm{Hom}\big(\pi_1(E), S_d\big)/\!\!/ S_d\big|$$

of degree d (see for example, [**6**]). Let

$$\phi(q) = \prod_{m=1}^{\infty}(1 - q^m)$$

be the Euler function. Then we have

$$\sum_{n=1}^{\infty}\frac{1}{n!}H^E_{1,n}(1, \ldots, 1)q^n = -\log\phi(q) = \sum_{n=1}^{\infty}\left(\sum_{m|n}\frac{1}{m}\right)q^n,$$

hence

$$H^E_{1,n}(1,\dots,1) = n!\sum_{m|n}\frac{1}{m} = (n-1)!\sigma(n), \tag{2.7}$$

where σ is the sum of divisors function. This is an integer sequence, and its first ten terms are $1, 3, 8, 42, 144, 1440, 5760, 75600, 524160, 6531840$. This sequence has many interesting properties (see OEIS, A038048).

More generally, let us consider the case when the equality holds in (2.4) We need this analysis in Section 4. Because of (2.3), $g-1 = n\big(g(B)-1\big)$ implies $d = n$ and $r = 0$, hence the covering $f : C \longrightarrow B$ that is counted is totally unramified. Here again the number of disconnected unramified coverings is given by the classical dimension formula (see for example [**27**] for elementary derivations):

$$\big|\mathrm{Hom}\big(\pi_1(B), S_d\big)/\!\!/S_d\big| = \sum_{\lambda\vdash d}\left(\frac{\dim\lambda}{d!}\right)^{\chi(B)}.$$

Here $\lambda \vdash d$ parametrizes irreducible representations of S_d, and $\dim\lambda$ is its dimension.

3. The discrete Laplace transform

The discrete Laplace transform $F^B_{g,n}(x_1,\dots,x_n)$ of (1.1) is a polynomial of degree $\frac{2g-2+n}{2g(B)-1}$ with the lowest degree term $H^E_{g,n}(1,1,\dots,1)x_1\cdots x_n$. The following proposition is an analogue of the case of simple Hurwitz numbers of $\mathbb{P}^1$, and the proof is exactly the same as that of [**3**, Lemma 4.1].

PROPOSITION 3.1. *The discrete Laplace transform of the cut-and-join equation* (2.6) *is precisely the partial differential equation* (1.2).

Reflecting Remark 2.2, the differential equation (1.2) alone does not determine free energies $F^B_{g,n}(x_1,\dots,x_n)$. This is because every homogeneous function of degree $\frac{2g-2+n}{2g(B)-1}$ is in the kernel of the Euler differential operator

$$2g-2+n-\big(2g(B)-1\big)\sum_{i=1}^n x_i\frac{\partial}{\partial x_i}$$

on the left-hand side of (1.2). Thus the highest degree terms of free energies are not determined by this differential equation. We will determine the homogeneous highest degree terms of the free energies in a different method in Section 5.

The genus 1 elliptic Hurwitz numbers (2.7) yields

$$F^E_{1,n}(x_1,\dots,x_n) = \left(n!\sum_{m|n}\frac{1}{m}\right)x_1\cdots x_n. \tag{3.1}$$

4. A Schrödinger equation

In this section we prove (1.6), provided that $g(B) > 0$. We remark here that for $g(B) = 0$, the proof is rather different [**26**], yet the same formula holds.

THEOREM 4.1. *The wave function* (1.3) *satisfies the Schrödinger-type equation* (1.6):

$$\left[\frac{\partial}{\partial\hbar}-\frac{1}{2}\left(x\frac{\partial}{\partial x}\right)^2+\left(\frac{1}{2}-\frac{2g(B)-1}{\hbar}\right)x\frac{\partial}{\partial x}\right]Z^B(x,\hbar)=0.$$

PROOF. We use the same method of the proof of [**26**, Theorem 5.1, Theorem 5.3, Appendix A]. When (2.3) holds, the diagonal evaluation of (1.2) yields

(4.1)

$$\begin{aligned}
&\left(2g-2+n-(2g(B)-1)x\frac{d}{dx}\right)F^B_{g,n}(x,\dots,x)\\
&=\frac{n(n-1)}{2}x^2\left.\frac{\partial^2}{\partial u^2}\right|_{u=x}F^B_{g,n-1}(u,x,\dots,x)\\
&+\frac{n}{2}x^2\left.\frac{\partial^2}{\partial u_1\partial u_2}\right|_{u_1=u_2=x}F^B_{g-1,n+1}(u_1,u_2,x,\dots,x)\\
&+\frac{n!}{2}\sum_{\substack{g_1+g_2=g\\ n_1+n_2=n-1}}\frac{x^2}{(n_1+1)!(n_2+1)!}\frac{d}{dx}F^B_{g_1,n_1+1}(x,\dots,x)\cdot\frac{d}{dx}F^B_{g_2,n_2+1}(x,\dots,x).
\end{aligned}$$

For $m\ge 2g(B)$, let us define

$$S_m(x):=\sum_{2g-2+n=m-1}\frac{1}{n!}F^B_{g,n}(x,\dots,x).$$

Note that S_m contains a contribution from domain curves with $g-1=n(g(B)-1)$, or equivalently, $d=n$ and $r=0$, for which the cut-and-join equation is not valid. Therefore, when we deduce an equation for S_m's, we need to remove these boundary terms from the equation. Now from (4.1) we obtain

(4.2)

$$\begin{aligned}
&\left(m-(2g(B)-1)x\frac{d}{dx}\right)S_{m+1}(x)\\
&-\frac{1}{2}x^2\frac{d^2}{dx^2}S_m(x)-\frac{1}{2}\sum_{m_1+m_2=m+1}x\frac{d}{dx}S_{m_1}(x)\cdot x\frac{d}{dx}S_{m_2}(x)\\
&=\sum_{\substack{2g-2+n=m\\ g-1=n(g(B)-1)}}\frac{1}{n!}\left(m-(2g(B)-1)x\frac{d}{dx}\right)F^B_{g,n}(x,\dots,x)\\
&-\frac{1}{2}\sum_{\substack{2g-2+n=m-1\\ g-1=(n+1)(g(B)-1)}}\frac{1}{(n-1)!}x^2\left.\frac{\partial^2}{\partial u^2}\right|_{u=x}F^B_{g,n}(u,x,\dots,x)\\
&-\frac{1}{2}\sum_{\substack{2g-2+n=m-1\\ g=(n-1)(g(B)-1)}}\frac{1}{(n-2)!}x^2\left.\frac{\partial^2}{\partial u_1\partial u_2}\right|_{u_1=u_2=x}F^B_{g,n}(u_1,u_2,x,\dots,x)\\
&-\frac{1}{2}\sum_{\substack{2g-2+n=m\\ g-1=n(g(B)-1)}}\sum_{\substack{g_1+g_2=g\\ n_1+n_2=n+1}}\frac{x^2}{n_1!\,n_2!}\left(\frac{d}{dx}F^B_{g_1,n_1}(x,\dots,x)\right)\left(\frac{d}{dx}F^B_{g_2,n_2}(x,\dots,x)\right).
\end{aligned}$$

We note here that in the boundary contribution $g-1=n\big(g(B)-1\big)$, since $n=d$, the free energies are single monomials proportional to $x_1x_2\cdots x_n$. Let us look at the right-hand side of (4.2). The first line of the right-hand side is 0 because $F^B_{g,n}(x,\dots,x) = H^B_{g,n}(1,\dots,1)x^n$, and $m = 2g-2+n = n\big(2g(B)-1\big)$, since $r=0$. The second line is also 0, because $F^B_{g,n}(u,x,\dots,x)$ is linear in u.

The third summation on the right-hand side is empty, because we need $n\big(g(B)-1\big) \le g-1 = (n-1)\big(g(B)-1\big)-1$, which does not happen. Similarly, the fourth line summation is also empty, because we are requiring $g = g_1+g_2, n+1 = n_1+n_2$, and

$$\begin{aligned} n_1\big(g(B)-1\big) &\le g_1 - 1 \\ n_2\big(g(B)-1\big) &\le g_2 - 1 \\ n\big(g(B)-1\big) &= g-1. \end{aligned}$$

We have thus obtained a recursion equation

$$\begin{aligned} &\left(m - (2g(B)-1)x\frac{d}{dx}\right) S_{m+1}(x) \\ &\qquad = \frac{1}{2}x^2\frac{d^2}{dx^2}S_m(x) + \frac{1}{2}\sum_{m_1+m_2=m+1} x\frac{d}{dx}S_{m_1}(x)\cdot x\frac{d}{dx}S_{m_2}(x). \end{aligned} \tag{4.3}$$

In terms of the generating function

$$F(x,\hbar) = \sum_{m=1}^{\infty} \hbar^{m-1} S_m(x),$$

(4.3) becomes

$$\left(\frac{\partial}{\partial\hbar} - \frac{2g(B)-1}{\hbar}x\frac{d}{dx}\right)F(x,\hbar) = \frac{1}{2}x^2\frac{d^2}{dx^2}F(x,\hbar) + \frac{1}{2}\left(x\frac{d}{dx}F(x,\hbar)\right)^2. \tag{4.4}$$

Since $Z^B(x,\hbar) = \exp F(x,\hbar)$, (1.6) follows directly from (4.4). This completes the proof. □

Using the Schrödinger equation (1.6), we can determine the form of the solution $Z^B(x,\hbar)$.

LEMMA 4.2. *The wave function has the following expansion:*

$$Z^B(x,\hbar) = \sum_{m=0}^{\infty} c_m e^{\frac{1}{2}m(m-1)\hbar}\left(\hbar^{1-\chi(B)}x\right)^m, \tag{4.5}$$

where c_m is a constant.

PROOF. The wave function $Z^B(x,\hbar)$ of (1.3) is a formal power series in x and $\hbar$. Thus it has an expansion of the form

$$Z^B(x,\hbar) = \sum_{m=0}^{\infty} c_m f_m(\hbar)x^m$$

with $f_m(\hbar) \in \mathbb{C}[[\hbar]]$. Then the Schrödinger equation (1.6) yields an ordinary differential equation

$$f_m' = \left(\frac{1}{2}m(m-1) + m\frac{1-\chi(B)}{\hbar}\right)f_m,$$

whose solution is

$$f_m(\hbar) = c\, e^{\frac{1}{2}m(m-1)\hbar}\hbar^{m\left(1-\chi(B)\right)}.$$

This completes the proof. □

5. The heat equation and its consequences

As we have seen in Section 4, the Schrödinger equation (1.6) determines the solution $Z^B(x, \hbar)$ only up to the form (4.5). To determine the coefficients, we need another technique.

In this section we use a heat equation technique of [**22**]. First we remark that the cut-and-join equation gives rise to a heat equation for another generating function of base B Hurwitz numbers. To determine a solution of a heat equation, we need to identify the initial value. We will show that the initial condition exactly corresponds to determining the highest degree terms of $F^B_{g,n}(x_1, \dots, x_n)$. The exponential generating function of these highest degree terms can be determined by a character formula of [**31**]. Thus we obtain the unique solution of the heat equation, which in turn gives all base B Hurwitz numbers.

The generating function for base B Hurwitz numbers we consider is

$$\mathbf{H}(t, \mathbf{p}) = \sum_{g=0}^{\infty} \sum_{n=1}^{\infty} \mathbf{H}_{g,n}(t, \mathbf{p}), \tag{5.1}$$

$$\mathbf{H}_{g,n}(t, \mathbf{p}) = \frac{1}{n!} \sum_{\vec{\mu} \in \mathbb{Z}_+^n} H^B_{g,n}(\vec{\mu}) \mathbf{p}_\mu t^{r(g,\mu)}, \tag{5.2}$$

where $r(g, \mu) = 2g - 2 + n - |\mu|(1 - \chi(B))$ is the number of simple ramification points, and $\mathbf{p}_\mu = p_{\mu_1} \cdots p_{\mu_n}$. The same argument of [**16, 22, 23, 29, 33**] shows that $e^{\mathbf{H}(t,\mathbf{p})}$ satisfies a *heat equation*, that is obtained from the cut-and-join equation. For a partition $\mu = (\mu_1 \geq \mu_2 \geq \cdots)$ of a finite length $\ell(\mu)$, we define the **shifted power-sum function** by

$$\mathbf{p}_r[\mu] := \sum_{i=1}^{\infty} \left[\left(\mu_i - i + \frac{1}{2} \right)^r - \left(-i + \frac{1}{2} \right)^r \right]. \tag{5.3}$$

This is a finite sum of $\ell(\mu)$ terms. Then we have [**16, 33**]

$$\sum_{i,j \geq 1} \left((i+j) p_i p_j \frac{\partial}{\partial p_{i+j}} + ij p_{i+j} \frac{\partial^2}{\partial p_i \partial p_j} \right) s_\mu(\mathbf{p}) = \mathbf{p}_2[\mu] \cdot s_\mu(\mathbf{p}), \tag{5.4}$$

where $s_\mu(\mathbf{p})$ is the Schur function defined by

$$s_\mu(\mathbf{p}) = \sum_{|\lambda| = |\mu|} \frac{\chi_\mu(\lambda)}{z_\lambda} \mathbf{p}_\lambda, \quad z_\mu = \prod_{i=1}^{\ell(\mu)} m_i! i^{m_i}, \tag{5.5}$$

$m_i =$ the number of parts in μ of length i, and $\chi_\mu(\lambda)$ is the value of the irreducible character of the representation μ of the symmetric group evaluated on any element of the conjugacy class described by λ. Let us denote the *cut-and-join* operator by

$$\triangle = \frac{1}{2} \sum_{i,j \geq 1} \left((i+j) p_i p_j \frac{\partial}{\partial p_{i+j}} + ij p_{i+j} \frac{\partial^2}{\partial p_i \partial p_j} \right).$$

Then Schur functions are eigenfunctions of the operator:

$$\triangle s_\mu(\mathbf{p}) = \frac{1}{2} \mathbf{p}_2[\mu] \cdot s_\mu(\mathbf{p}),$$

and the cut-and-join equation of simple Hurwitz numbers (2.6) yields a heat equation

$$\frac{\partial}{\partial t} e^{\mathbf{H}(t,\mathbf{p})} = \triangle e^{\mathbf{H}(t,\mathbf{p})}. \tag{5.6}$$

We can solve a heat equation by the method of eigenfunction expansion. Thus we have an expression

$$e^{\mathbf{H}(t,\mathbf{p})} = \sum_{\mu} a_\mu s_\mu(\mathbf{p}) e^{\frac{1}{2}\mathbf{p}_2[\mu]t} \tag{5.7}$$

for a constant a_μ associated with every partition μ. These constants are determined by the initial value $t = 0$. From (5.2) we see that the initial value comes from the cases when $r(g,\mu) = 0$.

First we note that (2.2) implies that simple Hurwitz numbers with $r = 0$ correspond to the case with only 1 branched point on the base curve B. If we allow disconnected domain curves to cover B, then the number of coverings of degree d with a prescribed ramification data given by a partition $\mu \vdash d$ over $0 \in B$ with no other ramification points is easy to calculate. Let us denote by $H_d^{B\bullet}(\mu)$ such a number, where we do not label the inverse images of $0 \in B$ this time. Then from [**31**, Eq.(0.10)] we obtain

$$H_d^{B\bullet}(\mu) = \sum_{\lambda \vdash d} \left(\frac{\dim\lambda}{d!}\right)^{\chi(B)} |C_\mu| \frac{\chi_\lambda(\mu)}{\dim\lambda}. \tag{5.8}$$

Here C_μ is the conjugacy class of the permutation group S_d determined by the cycle type μ, a partition $\lambda \vdash d$ is a label of an irreducible representation of S_d, and $\dim\lambda$ is its dimension. The cardinality of the conjugacy class is given by

$$|C_\mu| = \frac{d!}{\prod_i m_i! i^{m_i}} = \frac{d!}{z_\mu}. \tag{5.9}$$

The generating function of these disconnected simple Hurwitz numbers can be calculated, appealing to (5.5) and (5.9), as follows:

$$\begin{aligned}
\sum_{d=1}^{\infty}\sum_{\mu\vdash d} H_d^{B\bullet}(\mu)\mathbf{p}_\mu &= \sum_{d=1}^{\infty}\sum_{\mu\vdash d}\sum_{\lambda\vdash d} \left(\frac{\dim\lambda}{d!}\right)^{\chi(B)} |C_\mu| \frac{\chi_\lambda(\mu)}{\dim\lambda}\mathbf{p}_\mu \\
&= \sum_{d=1}^{\infty}\sum_{\lambda\vdash d} \frac{d!}{\dim\lambda}\left(\frac{\dim\lambda}{d!}\right)^{\chi(B)} \sum_{\mu\vdash d}\frac{\chi_\lambda(\mu)}{z_\mu}\mathbf{p}_\mu \\
&= \sum_{\lambda}\left(\frac{|\lambda|!}{\dim\lambda}\right)^{1-\chi(B)} s_\lambda(\mathbf{p}),
\end{aligned} \tag{5.10}$$

where the last sum runs over all partitions λ.

Note that $e^{\mathbf{H}(t,\mathbf{p})}$ is the generating function of simple Hurwitz numbers allowing disconnected domain curves. Therefore, the initial value $e^{\mathbf{H}(0,\mathbf{p})}$ counts the disconnected base B Hurwitz numbers with only one branched point at $0 \in B$. In other words, we have determined the initial condition by (5.10). The result is

$$e^{\mathbf{H}(0,\mathbf{p})} = \sum_{\mu}\left(\frac{|\mu|!}{\dim\mu}\right)^{1-\chi(B)} s_\mu(\mathbf{p}) = \sum_{\mu} a_\mu s_\mu(\mathbf{p}). \tag{5.11}$$

Since Schur functions are linear basis for symmetric functions, we establish the following.

THEOREM 5.1. *The exponential generating function of the base B simple Hurwitz numbers is given by*

$$e^{\mathbf{H}(t,\mathbf{p})} = \sum_{\mu} \left(\frac{|\mu|!}{\dim \mu}\right)^{1-\mathcal{X}(B)} s_\mu(\mathbf{p}) e^{\frac{1}{2}\mathbf{p}_2[\mu]t}. \tag{5.12}$$

6. The quantum curve

As noted in [**26**], the diagonal specialization $F^E_{g,n}(x,x,\dots,x)$ corresponds to substituting the power-sum symmetric function

$$p_j = x_1^j + x_2^j + x_3^j + \cdots \tag{6.1}$$

by its principal specialization $p_j = x^j$. More precisely, we have the following.

LEMMA 6.1. *Let us define the* **principal specialization** *of the power-sum symmetric functions by*

$$p_j(s) = \left(s^{1-\mathcal{X}(B)}x\right)^j$$
$$\mathbf{p}_\mu(s) = p_{\mu_1}(s)\cdots p_{\mu_n}(s) = \left(s^{1-\mathcal{X}(B)}x\right)^{|\mu|}.$$

Then we have

$$Z^B(x,\hbar) = e^{\mathbf{H}(\hbar,\mathbf{p}(\hbar))}. \tag{6.2}$$

PROOF. We have

$$\mathbf{H}(\hbar,\mathbf{p}(\hbar)) = \sum_{g,n\geq 1}\sum_{\mu\in\mathbb{Z}_+^n} \frac{1}{n!}\hbar^{2g-2+n-|\mu|\left(1-\mathcal{X}(B)\right)} H^B_{g,n}(\mu_1,\dots,\mu_n)\mathbf{p}_\mu(\hbar)$$
$$= \sum_{g,n\geq 1} \frac{1}{n!}\hbar^{2g-2+n} F^E_{g,n}(x,x,\dots,x),$$

which yields (6.2). □

From the expansion formulas (5.12) and (4.5), together with the equality (6.2), we obtain

$$\sum_{\mu}\left(\frac{|\mu|!}{\dim \mu}\right)^{1-\mathcal{X}(B)} e^{\frac{1}{2}\mathbf{p}_2[\mu]\hbar} s_\mu\big(\mathbf{p}(\hbar)\big) = \sum_{m=0}^{n} c_m e^{\frac{1}{2}m(m-1)\hbar}\left(\hbar^{1-\mathcal{X}(B)}x\right)^m. \tag{6.3}$$

As explained in [**26**, Section 6] and also in [**25**], this equality is exactly the effect of the principal specialization of Lemma 6.1, which turns the summation over all partitions into the sums over just one-row partitions. We have thus determined the coefficients c_m in the expansion (4.5). It is given by

$$c_m = (m!)^{1-\mathcal{X}(B)}. \tag{6.4}$$

This completes the proof of (1.8).

The expansion (1.8) allows us to derive the quantum curve-type equation (1.4). Since

$$\left[\hbar^{1-\chi(B)}xe^{\hbar x\frac{d}{dx}}\left(\frac{d}{dx}x\right)^{1-\chi(B)}\right](m!)^{1-\chi(B)}e^{\frac{1}{2}m(m-1)\hbar}\left(\hbar^{1-\chi(B)}x\right)^{m}$$
$$= \left((m+1)!\right)^{1-\chi(B)}e^{\frac{1}{2}m(m+1)\hbar}\left(\hbar^{1-\chi(B)}x\right)^{m+1}$$

for every $m \geq 0$, we have

$$\left(1-\hbar^{1-\chi(B)}xe^{\hbar x\frac{d}{dx}}\left(\frac{d}{dx}x\right)^{1-\chi(B)}\right)Z^{B}(x,\hbar)=1. \tag{6.5}$$

By differentiating (6.5) we obtain (1.4).

LEMMA 6.2. *Let*

$$P=\hbar x\frac{d}{dx}\left(1-\hbar^{1-\chi(B)}xe^{\hbar x\frac{d}{dx}}\left(\frac{d}{dx}x\right)^{1-\chi(B)}\right), \tag{6.6}$$

$$Q=\frac{\partial}{\partial\hbar}-\frac{1}{2}\left(x\frac{\partial}{\partial x}\right)^{2}+\left(\frac{1}{2}-\frac{1-\chi(B)}{\hbar}\right)x\frac{\partial}{\partial x}. \tag{6.7}$$

Then

$$[P,Q]=-\frac{1}{\hbar}P.$$

PROOF. We first note that Q commutes with $x\frac{d}{dx}$ and $\frac{d}{dx}x$. Since

$$\left[\hbar^{1-\chi(B)}xe^{\hbar x\frac{d}{dx}},Q\right]$$
$$=\left[\hbar^{1-\chi(B)}xe^{\hbar x\frac{d}{dx}},\frac{\partial}{\partial\hbar}\right]+\left[\hbar^{1-\chi(B)}x,-\frac{1}{2}\left(x\frac{\partial}{\partial x}\right)^{2}+\left(\frac{1}{2}-\frac{1-\chi(B)}{\hbar}\right)x\frac{\partial}{\partial x}\right]e^{\hbar x\frac{d}{dx}}$$
$$=-\left(1-\chi(B)\right)\hbar^{-\chi(B)}xe^{\hbar x\frac{d}{dx}}-\hbar^{1-\chi(B)}x^{2}\frac{d}{dx}e^{\hbar x\frac{d}{dx}}$$
$$+\hbar^{1-\chi(B)}x^{2}\frac{d}{dx}e^{\hbar x\frac{d}{dx}}+\frac{1}{2}\hbar^{1-\chi(B)}xe^{\hbar x\frac{d}{dx}}-\hbar^{1-\chi(B)}\left(\frac{1}{2}-\frac{1-\chi(B)}{\hbar}\right)xe^{\hbar x\frac{d}{dx}}$$
$$=0,$$

(1.7) follows from

$$0=\left[\frac{1}{\hbar}P,Q\right]=\frac{1}{\hbar}[P,Q]+\left[\frac{1}{\hbar},Q\right]P=\frac{1}{\hbar}[P,Q]+\frac{1}{\hbar^{2}}P.$$

□

7. Semi-classical limit

The semi-classical analysis of the operators P and Q are performed in the following way. Suppose our counting problem had genus 0 contributions $F^{B}_{0,n}(x_1,\ldots,x_n)$. Then we define

$$S_m(x):=\sum_{2g-2+n=m-1}\frac{1}{n!}F^{B}_{g,n}(x,\ldots,x) \tag{7.1}$$

as before, and consider a formal expression

$$\overline{Z}^B(x,\hbar) = e^{\sum_{m=0}^{\infty} \hbar^{m-1} S_m(x)} = e^{\frac{1}{\hbar}S_0(x)+S_1(x)} Z^B(x,\hbar). \tag{7.2}$$

Let us introduce a variable u such that $x = e^u$, and regard the coefficients S_m as functions in u. Since

$$x\frac{d}{dx} = \frac{d}{du},$$

we have

$$P = \hbar\frac{d}{du}\left(1 - \hbar^{1-\chi(B)} e^u e^{\hbar\frac{d}{du}}\left(1+\frac{d}{du}\right)^{1-\chi(B)}\right).$$

Then

$$\begin{aligned} e^{-S_1}e^{-\frac{1}{\hbar}S_0}Pe^{\frac{1}{\hbar}S_0}e^{S_1} &= S_0' - e^u\left(S_0'\right)^{2-\chi(B)} e^{\frac{1}{\hbar}\left(S_0(u+\hbar)-S_0(u)\right)} + O(\hbar) \\ &= S_0' - e^u\left(S_0'\right)^{2-\chi(B)} e^{S_0'} + O(\hbar), \end{aligned} \tag{7.3}$$

where $' = \frac{d}{du} = x\frac{d}{dx}$, and by $O(\hbar)$ we mean an operator whose application to the wave function $Z^B(x,\hbar)$ produces a function of order 1 or higher in $\hbar$. Now define a new variable

$$y = S_0'. \tag{7.4}$$

Then the semi-classical limit $\hbar \to 0$ of (7.3) yields an equation

$$y - xy^{2-\chi(B)}e^y = y\left(1 - xy^{1-\chi(B)}e^y\right) = 0, \tag{7.5}$$

since $Z^B(x,0) = 1$. Note that this is exactly the *total symbol* of the operator P, where $\hbar x\frac{d}{dx}$ is represented by a commutative variable y. The second factor of (7.5) also recovers the Lambert curve (1.5).

Although the formal manipulation seems to work, however, we have to remember that we have derived the operator P assuming the shape of the solution $Z^B(x,\hbar)$ as in (1.8). Since we are imposing

$$P\overline{Z}^B(x,\hbar) = 0,$$

it forces that $y = 0$, which makes (7.5) trivially correct.

On the other hand, since the kernel of Q assumes only the expansion (4.5), where the summation index m can be negative, the semi-classical analysis of Q does go through.

$$\begin{aligned} &e^{-S_1}e^{-\frac{1}{\hbar}S_0}Qe^{\frac{1}{\hbar}S_0}e^{S_1} \\ &= e^{-S_1}e^{-\frac{1}{\hbar}S_0}\left[\frac{\partial}{\partial\hbar} - \frac{1}{2}\frac{\partial^2}{\partial u^2} + \left(\frac{1}{2} - \frac{1-\chi(B)}{\hbar}\right)\frac{\partial}{\partial u}\right]e^{\frac{1}{\hbar}S_0}e^{S_1} \\ &= -\frac{1}{\hbar^2}\left(S_0 + \frac{1}{2}\left(S_0'\right)^2 + (1-\chi(B))S_0'\right) \\ &\quad - \frac{1}{\hbar}\left(\frac{1}{2}S_0'' - \frac{1}{2}S_0' + S_0'S_1' + (1-\chi(B))S_1'\right) + O(1). \end{aligned} \tag{7.6}$$

For the $\hbar \to 0$ limit to exist, we need

$$S_0 + \frac{1}{2}\left(S_0'\right)^2 + \big(1-\chi(B)\big)S_0' = 0, \tag{7.7}$$

$$\frac{1}{2}S_0'' - \frac{1}{2}S_0' + S_0'S_1' + \big(1-\chi(B)\big)S_1' = 0. \tag{7.8}$$

From (7.7) we obtain

$$S_0 = -\frac{1}{2}y^2 - \big(1-\chi(B)\big)y, \tag{7.9}$$

or equivalently,

$$y = -\big(y+1-\chi(B)\big)\frac{dy}{du}. \tag{7.10}$$

Its solution is

$$y + \log y^{1-\chi(B)} = -u + \text{const}.$$

If we take the constant of integration to be 0, then we obtain

$$e^u = x = y^{\chi(B)-1}e^{-y},$$

recovering (1.5). From (7.8) we have

$$\frac{1}{2}y' - \frac{1}{2}y + \big(y+1-\chi(B)\big)S_1' = 0,$$

or equivalently

$$\frac{dS_1}{du} = -\frac{1}{2}\frac{1}{y+1-\chi(B)}\frac{dy}{du} + \frac{1}{2}\frac{y}{y+1-\chi(B)}\frac{dy}{dy}.$$

Since we know du/dy from (7.10), we can integrate the above equation to obtain

$$S_1 = -\frac{1}{2}y + \frac{1}{2}\log\big(y+1-\chi(B)\big) + \text{const}. \tag{7.11}$$

The above derivation of the semi-classical limit of the operator Q is valid if $Z^B(x,\hbar)$ contains negative powers of $\hbar$. If we assume the expansion (1.8), then such a situation occurs when the base curve B satisfies $\chi(B) > 1$. Indeed, our formulas (7.9) and (7.11) agree with those of [**26**, Section 5] for $B = \mathbb{P}^1$ with $y = z$, and [**3**, Section 6] for $B = \mathbb{P}^1[a]$ with $y = z^a$.

References

[1] Mina Aganagic, Robbert Dijkgraaf, Albrecht Klemm, Marcos Mariño, and Cumrun Vafa, *Topological strings and integrable hierarchies*, Comm. Math. Phys. **261** (2006), no. 2, 451–516, DOI 10.1007/s00220-005-1448-9. MR2191887

[2] Gaëtan Borot, Bertrand Eynard, Motohico Mulase, and Brad Safnuk, *A matrix model for simple Hurwitz numbers, and topological recursion*, J. Geom. Phys. **61** (2011), no. 2, 522–540, DOI 10.1016/j.geomphys.2010.10.017. MR2746135

[3] Vincent Bouchard, Daniel Hernández Serrano, Xiaojun Liu, and Motohico Mulase, *Mirror symmetry for orbifold Hurwitz numbers*, J. Differential Geom. **98** (2014), no. 3, 375–423. MR3263522

[4] Vincent Bouchard, Albrecht Klemm, Marcos Mariño, and Sara Pasquetti, *Remodeling the B-model*, Comm. Math. Phys. **287** (2009), no. 1, 117–178, DOI 10.1007/s00220-008-0620-4. MR2480744

[5] Vincent Bouchard and Marcos Mariño, *Hurwitz numbers, matrix models and enumerative geometry*, From Hodge theory to integrability and TQFT tt*-geometry, Proc. Sympos. Pure Math., vol. 78, Amer. Math. Soc., Providence, RI, 2008, pp. 263–283, DOI 10.1090/pspum/078/2483754. MR2483754

[6] Robbert Dijkgraaf, *Mirror symmetry and elliptic curves*, The moduli space of curves (Texel Island, 1994), Progr. Math., vol. 129, Birkhäuser Boston, Boston, MA, 1995, pp. 149–163, DOI 10.1007/978-1-4612-4264-2_5. MR1363055

[7] Robbert Dijkgraaf, Lotte Hollands, and Piotr Sułkowski, *Quantum curves and $\mathcal{D}$-modules*, J. High Energy Phys. **11** (2009), 047, 59, DOI 10.1088/1126-6708/2009/11/047. MR2628886

[8] Robbert Dijkgraaf, Lotte Hollands, Piotr Sułkowski, and Cumrun Vafa, *Supersymmetric gauge theories, intersecting branes and free fermions*, J. High Energy Phys. **2** (2008), 106, 57, DOI 10.1088/1126-6708/2008/02/106. MR2385939

[9] R. Dijkgraaf and C. Vafa, *Two Dimensional Kodaira-Spencer Theory and Three Dimensional Chern-Simons Gravity*, `arXiv:0711.1932` (2007).

[10] Norman Do, Oliver Leigh, and Paul Norbury, *Orbifold Hurwitz numbers and Eynard-Orantin invariants*, Math. Res. Lett. **23** (2016), no. 5, 1281–1327, DOI 10.4310/MRL.2016.v23.n5.a3. MR3601067

[11] Olivia Dumitrescu, Motohico Mulase, Brad Safnuk, and Adam Sorkin, *The spectral curve of the Eynard-Orantin recursion via the Laplace transform*, Algebraic and geometric aspects of integrable systems and random matrices, Contemp. Math., vol. 593, Amer. Math. Soc., Providence, RI, 2013, pp. 263–315, DOI 10.1090/conm/593/11867. MR3087960

[12] Torsten Ekedahl, Sergei Lando, Michael Shapiro, and Alek Vainshtein, *Hurwitz numbers and intersections on moduli spaces of curves*, Invent. Math. **146** (2001), no. 2, 297–327, DOI 10.1007/s002220100164. MR1864018

[13] Bertrand Eynard, Motohico Mulase, and Bradley Safnuk, *The Laplace transform of the cut-and-join equation and the Bouchard-Mariño conjecture on Hurwitz numbers*, Publ. Res. Inst. Math. Sci. **47** (2011), no. 2, 629–670, DOI 10.2977/PRIMS/47. MR2849645

[14] B. Eynard and N. Orantin, *Invariants of algebraic curves and topological expansion*, Commun. Number Theory Phys. **1** (2007), no. 2, 347–452, DOI 10.4310/CNTP.2007.v1.n2.a4. MR2346575

[15] Leonid Chekhov, Bertrand Eynard, and Nicolas Orantin, *Free energy topological expansion for the 2-matrix model*, J. High Energy Phys. **12** (2006), 053, 31, DOI 10.1088/1126-6708/2006/12/053. MR2276699

[16] I. P. Goulden, *A differential operator for symmetric functions and the combinatorics of multiplying transpositions*, Trans. Amer. Math. Soc. **344** (1994), no. 1, 421–440, DOI 10.2307/2154724. MR1249468

[17] I. P. Goulden and D. M. Jackson, *Transitive factorisations into transpositions and holomorphic mappings on the sphere*, Proc. Amer. Math. Soc. **125** (1997), no. 1, 51–60, DOI 10.1090/S0002-9939-97-03880-X. MR1396978

[18] Tom Graber and Ravi Vakil, *Hodge integrals and Hurwitz numbers via virtual localization*, Compositio Math. **135** (2003), no. 1, 25–36, DOI 10.1023/A:1021791611677. MR1955162

[19] Sergei Gukov and Piotr Sułkowski, *A-polynomial, B-model, and quantization*, Homological mirror symmetry and tropical geometry, Lect. Notes Unione Mat. Ital., vol. 15, Springer, Cham, 2014, pp. 87–151, DOI 10.1007/978-3-319-06514-4_4. MR3330786

[20] A. Hurwitz, *Ueber Riemann'sche Flächen mit gegebenen Verzweigungspunkten* (German), Math. Ann. **39** (1891), no. 1, 1–60, DOI 10.1007/BF01199469. MR1510692

[21] M. Kaneko and D. Zagier, *A generalized Jacobi theta function and quasi-modular forms*, in: R.H. Dijkgraaf et al. (eds) The Moduli Space of Curves. Progress in Mathematics **129**, Birkhäuser Boston, 165–172 (1995).

[22] M. Kazarian, *KP hierarchy for Hodge integrals*, Adv. Math. **221** (2009), no. 1, 1–21, DOI 10.1016/j.aim.2008.10.017. MR2509319

[23] M. E. Kazarian and S. K. Lando, *An algebro-geometric proof of Witten's conjecture*, J. Amer. Math. Soc. **20** (2007), no. 4, 1079–1089, DOI 10.1090/S0894-0347-07-00566-8. MR2328716

[24] Marcos Mariño, *Open string amplitudes and large order behavior in topological string theory*, J. High Energy Phys. **3** (2008), 060, 34, DOI 10.1088/1126-6708/2008/03/060. MR2391060

[25] M. Mulase, S. Shadrin, and L. Spitz, *The spectral curve and the Schrödinger equation of double Hurwitz numbers and higher spin structures*, Commun. Number Theory Phys. **7** (2013), no. 1, 125–143, DOI 10.4310/CNTP.2013.v7.n1.a4. MR3108774

[26] Motohico Mulase and Piotr Sułkowski, *Spectral curves and the Schrödinger equations for the Eynard-Orantin recursion*, Adv. Theor. Math. Phys. **19** (2015), no. 5, 955–1015, DOI 10.4310/ATMP.2015.v19.n5.a2. MR3487649

[27] M. Mulase and J. Yu, *A generating function of the number of homomorphisms from a surface group into a finite group*, arXiv:math.QA/0209008, (2002).

[28] Motohico Mulase and Naizhen Zhang, *Polynomial recursion formula for linear Hodge integrals*, Commun. Number Theory Phys. **4** (2010), no. 2, 267–293, DOI 10.4310/CNTP.2010.v4.n2.a1. MR2725053

[29] Andrei Okounkov, *Toda equations for Hurwitz numbers*, Math. Res. Lett. **7** (2000), no. 4, 447–453, DOI 10.4310/MRL.2000.v7.n4.a10. MR1783622

[30] A. Okounkov and R. Pandharipande, *Gromov-Witten theory, Hurwitz numbers, and matrix models*, Algebraic geometry—Seattle 2005. Part 1, Proc. Sympos. Pure Math., vol. 80, Amer. Math. Soc., Providence, RI, 2009, pp. 325–414, DOI 10.1090/pspum/080.1/2483941. MR2483941

[31] A. Okounkov and R. Pandharipande, *Gromov-Witten theory, Hurwitz theory, and completed cycles*, Ann. of Math. (2) **163** (2006), no. 2, 517–560, DOI 10.4007/annals.2006.163.517. MR2199225

[32] R. Vakil, *Enumerative Geometry of Curves via Degeneration Methods*, Ph.D. thesis, Harvard University (1997).

[33] J. Zhou, *Hodge integrals, Hurwitz numbers, and symmetric groups*, arXiv:math/0308024, (2003).

[34] J. Zhou, *Quantum Mirror Curves for* $\mathbb{C}^3$ *and the Resolved Confiold*, arXiv:1207.0598v1, (2012).

Department of Applied Mathematics, China Agricultural University, Beijing, 100083, China; and Department of Mathematics, University of California, Davis, California 95616–8633

Email address: xjliu@cau.edu.cn

Department of Mathematics, University of California, Davis, California 95616–8633

Email address: mulase@math.ucdavis.edu

Department of Mathematics, University of California, Davis, California 95616–8633

Email address: azsorkin@math.ucdavis.edu

Selected Published Titles in This Series

100 **Chiu-Chu Melissa Liu and Motohico Mulase, Editors,** Topological Recursion and its Influence in Analysis, Geometry, and Topology, 2018

99 **Vicente Muñoz, Ivan Smith, and Richard P. Thomas, Editors,** Modern Geometry, 2018

98 **Amir-Kian Kashani-Poor, Ruben Minasian, Nikita Nekrasov, and Boris Pioline, Editors,** String-Math 2016, 2018

97 **Tommaso de Fernex, Brendan Hassett, Mircea Mustaţă, Martin Olsson, Mihnea Popa, and Richard Thomas, Editors,** Algebraic Geometry: Salt Lake City 2015 (Parts 1 and 2), 2018

96 **Si Li, Bong H. Lian, Wei Song, and Shing-Tung Yau, Editors,** String-Math 2015, 2017

95 **Izzet Coskun, Tommaso de Fernex, and Angela Gibney, Editors,** Surveys on Recent Developments in Algebraic Geometry, 2017

94 **Mahir Bilen Can, Editor,** Algebraic Groups: Structure and Actions, 2017

93 **Vincent Bouchard, Charles Doran, Stefan Méndez-Diez, and Callum Quigley, Editors,** String-Math 2014, 2016

92 **Kailash C. Misra, Daniel K. Nakano, and Brian J. Parshall, Editors,** Lie Algebras, Lie Superalgebras, Vertex Algebras and Related Topics, 2016

91 **V. Sidoravicius and S. Smirnov, Editors,** Probability and Statistical Physics in St Petersburg, 2016

90 **Ron Donagi, Sheldon Katz, Albrecht Klemm, and David R. Morrison, Editors,** String-Math 2012, 2015

89 **D. Dolgopyat, Y. Pesin, M. Pollicott, and L. Stoyanov, Editors,** Hyperbolic Dynamics, Fluctuations and Large Deviations, 2015

88 **Ron Donagi, Michael R. Douglas, Ljudmila Kamenova, and Martin Rocek, Editors,** String-Math 2013, 2014

87 **Helge Holden, Barry Simon, and Gerald Teschl, Editors,** Spectral Analysis, Differential Equations and Mathematical Physics: A Festschrift in Honor of Fritz Gesztesy's 60th Birthday, 2013

86 **Kailash C. Misra, Daniel K. Nakano, and Brian J. Parshall, Editors,** Recent Developments in Lie Algebras, Groups and Representation Theory, 2012

85 **Jonathan Block, Jacques Distler, Ron Donagi, and Eric Sharpe, Editors,** String-Math 2011, 2012

84 **Alex H. Barnett, Carolyn S. Gordon, Peter A. Perry, and Alejandro Uribe, Editors,** Spectral Geometry, 2012

83 **Hisham Sati and Urs Schreiber, Editors,** Mathematical Foundations of Quantum Field Theory and Perturbative String Theory, 2011

82 **Michael Usher, Editor,** Low-dimensional and Symplectic Topology, 2011

81 **Robert S. Doran, Greg Friedman, and Jonathan Rosenberg, Editors,** Superstrings, Geometry, Topology, and C^*-algebras, 2010

80 **D. Abramovich, A. Bertram, L. Katzarkov, R. Pandharipande, and M. Thaddeus, Editors,** Algebraic Geometry, 2009

79 **Dorina Mitrea and Marius Mitrea, Editors,** Perspectives in Partial Differential Equations, Harmonic Analysis and Applications, 2008

78 **Ron Y. Donagi and Katrin Wendland, Editors,** From Hodge Theory to Integrability and TQFT, 2008

77 **Pavel Exner, Jonathan P. Keating, Peter Kuchment, Toshikazu Sunada, and Alexander Teplyaev, Editors,** Analysis on Graphs and Its Applications, 2008

For a complete list of titles in this series, visit the AMS Bookstore at **www.ams.org/bookstore/pspumseries/**.

ISBN 978-1-4704-3541-7

9 781470 435417

PSPUM/100